Freshwater Animal Diversity Assessment

Developments in Hydrobiology 198

Series editor
K. Martens

Freshwater Animal Diversity Assessment

Edited by

E.V. Balian[1], C. Lévêque[2], H. Segers[1] & K. Martens[3]

[1]Belgian Biodiversity Platform, Freshwater Laboratory, Royal Belgian Institute of Natural Sciences, Vautierstraat 29 B-1000, Brussels, Belgium

[2]Antenne IRD, MNHN-DMPA, 43 rue Cuvier, Case Postale 26, Paris cedex 05 75231, France

[3]Freshwater Laboratory, Royal Belgian Institute of Natural Sciences, Vautierstraat 29 B-1000, Brussels, Belgium; Department of Biology, University of Ghent, K.L. Ledeganckstraat 35, Gent 9000, Belgium

Reprinted from Hydrobiologia, Volume 595 (2008)

Biodiversity.be museum BELGIAN SCIENCE POLICY

Springer

Library of Congress Cataloging-in-Publication Data

A C.I.P. Catalogue record for this book is available from the Library of Congress.

ISBN-13: 978-1-4020-8258-0

Published by Springer,
P.O. Box 17, 3300 AA Dordrecht, The Netherlands

Cite this publication as Hydrobiologia vol. 595 (2008).

Cover illustration: A few inhabitants of fresh water. (clockwise from top left): *Simulium arcticum* (larva) - photo by Michael Spironello; *Crangonyx richmondensis* - photo by Jonathan Witt; *Protorthemis coronata* - photo by Vincent J Kalkman; *Altolamprologus calvus* (Chisanse) - photo by Ad Konings

Frontispiece: Diadeco Bild & Produktionsbyrå, Sweden

Printed on acid-free paper

Printed in the Netherlands

TABLE OF CONTENTS

Colour section ix, xiv–xvi

Foreword
R.J. Naiman 1–2

An introduction to the Freshwater Animal Diversity Assessment (FADA) project
E.V. Balian, H. Segers, C. Lévêque, K. Martens 3–8

Global diversity of aquatic macrophytes in freshwater
P.A. Chambers, P. Lacoul, K.J. Murphy, S.M. Thomaz 9–26

Global diversity of sponges (Porifera: Spongillina) in freshwater
R. Manconi, R. Pronzato 27–33

Global diversity of inland water cnidarians
T. Jankowski, A.G. Collins, R. Campbell 35–40

Global diversity of free living flatworms (Platyhelminthes, "Turbellaria") in freshwater
E.R. Schockaert, M. Hooge, R. Sluys, S. Schilling, S. Tyler, T. Artois 41–48

Global diversity of rotifers (Rotifera) in freshwater
H. Segers 49–59

Global diversity of nemerteans (Nemertea) in freshwater
P. Sundberg, R. Gibson 61–66

Global diversity of nematodes (Nematoda) in freshwater
E. Abebe, W. Decraemer, P. De Ley 67–78

Global diversity of hairworms (Nematomorpha: Gordiaceae) in freshwater
G. Poinar Jr. 79–83

Global diversity of gastrotrichs (Gastrotricha) in fresh waters
M. Balsamo, J.-L. d'Hondt, J. Kisielewski, L. Pierboni 85–91

Global diversity of bryozoans (Bryozoa or Ectoprocta) in freshwater
J.A. Massard, G. Geimer 93–99

Global diversity of tardigrades (Tardigrada) in freshwater
J.R. Garey, S.J. McInnes, P.B. Nichols 101–106

Global diversity of polychaetes (Polychaeta; Annelida) in freshwater
C.J. Glasby, T. Timm 107–115

Global diversity of oligochaetous clitellates ("Oligochaeta"; Clitellata) in freshwater
P. Martin, E. Martinez-Ansemil, A. Pinder, T. Timm, M.J. Wetzel 117–127

Global diversity of leeches (Hirudinea) in freshwater
B. Sket, P. Trontelj 129–137

Global diversity of freshwater mussels (Mollusca, Bivalvia) in freshwater
A.E. Bogan 139–147

Global diversity of gastropods (Gastropoda; Mollusca) in freshwater
E.E. Strong, O. Gargominy, W.F. Ponder, P. Bouchet 149–166

Global diversity of large branchiopods (Crustacea: Branchiopoda) in freshwater
L. Brendonck, D.C. Rogers, J. Olesen, S. Weeks, W.R. Hoeh 167–176

Global diversity of cladocerans (Cladocera; Crustacea) in freshwater
L. Forró, N.M. Korovchinsky, A.A. Kotov, A. Petrusek 177–184

Global diversity of ostracods (Ostracoda, Crustacea) in freshwater
K. Martens, I. Schön, C. Meisch, D.J. Horne 185–193

Global diversity of copepods (Crustacea: Copepoda) in freshwater
G.A. Boxshall, D. Defaye 195–207

Global diversity of fishlice (Crustacea: Branchiura: Argulidae) in freshwater
W.J. Poly 209–212

Global diversity of mysids (Crustacea-Mysida) in freshwater
M.L. Porter, K. Meland, W. Price 213–218

Global diversity of spelaeogriphaceans & thermosbaenaceans (Crustacea; Spelaeogriphacea & Thermosbaenacea) in freshwater
D. Jaume 219–224

Global diversity of cumaceans & tanaidaceans (Crustacea: Cumacea & Tanaidacea) in freshwater
D. Jaume, G.A. Boxshall 225–230

Global diversity of Isopod crustaceans (Crustacea; Isopoda) in freshwater
G.D.F. Wilson 231–240

Global diversity of amphipods (Amphipoda; Crustacea) in freshwater
R. Väinölä, J.D.S. Witt, M. Grabowski, J.H. Bradbury, K. Jazdzewski, B. Sket 241–255

Global diversity of syncarids (Syncarida; Crustacea) in freshwater
A.I. Camacho, A.G. Valdecasas 257–266

Global diversity of crabs (Aeglidae: Anomura: Decapoda) in freshwater
G. Bond-Buckup, C.G. Jara, M. Pérez-Losada, L. Buckup, K.A. Crandall 267–273

Global diversity of crabs (Crustacea: Decapoda: Brachyura) in freshwater
D.C.J. Yeo, P.K.L. Ng, N. Cumberlidge, C. Magalhães, S.R. Daniels, M.R. Campos 275–286

Global diversity of shrimps (Crustacea: Decapoda: Caridea) in freshwater
S. De Grave, Y. Cai, A. Anker 287–293

Global diversity of crayfish (Astacidae, Cambaridae, and Parastacidae—Decapoda) in freshwater
K.A. Crandall, J.E. Buhay 295–301

Global diversity of water mites (Acari, Hydrachnidia; Arachnida) in freshwater
A. Di Sabatino, H. Smit, R. Gerecke, T. Goldschmidt, N. Matsumoto, B. Cicolani 303–315

Global diversity of halacarid mites (Halacaridae: Acari: Arachnida) in freshwater
I. Bartsch 317–322

Global diversity of oribatids (Oribatida: Acari: Arachnida)
H. Schatz, V. Behan-Pelletier 323–328

Global diversity of springtails (Collembola; Hexapoda) in freshwater
L. Deharveng, C.A. D'Haese, A. Bedos 329–338

Global diversity of mayflies (Ephemeroptera, Insecta) in freshwater
H.M. Barber-James, J.-L. Gattolliat, M. Sartori, M.D. Hubbard 339–350

Global diversity of dragonflies (Odonata) in freshwater
V.J. Kalkman, V. Clausnitzer, K.-D.B. Dijkstra, A.G. Orr, D.R. Paulson, J. van Tol 351–363

Global diversity of stoneflies (Plecoptera; Insecta) in freshwater
R. Fochetti, J.M. Tierno de Figueroa 365–377

Global diversity of true bugs (Heteroptera; Insecta) in freshwater
J.T. Polhemus, D.A. Polhemus 379–391

Global diversity of caddisflies (Trichoptera: Insecta) in freshwater
F.C. de Moor, V.D. Ivanov 393–407

Global diversity of dobsonflies, fishflies, and alderflies (Megaloptera; Insecta) and spongillaflies, nevrorthids, and osmylids (Neuroptera; Insecta) in freshwater
M.R. Cover, V.H. Resh 409–417

Global diversity of water beetles (Coleoptera) in freshwater
M.A. Jäch, M. Balke 419–442

Global biodiversity of Scorpionflies and Hangingflies (Mecoptera) in freshwater
L.C. Ferrington Jr. 443–445

Global diversity of non-biting midges (Chironomidae; Insecta-Diptera) in freshwater
L.C. Ferrington Jr. 447–455

Global diversity of craneflies (Insecta, Diptera: Tipulidea or Tipulidae *sensu lato*) in freshwater
H. de Jong, P. Oosterbroek, J. Gelhaus, H. Reusch, C. Young 457–467

Global diversity of black flies (Diptera: Simuliidae) in freshwater
D.C. Currie, P.H. Adler 469–475

Global diversity of mosquitoes (Insecta: Diptera: Culicidae) in freshwater
L.M. Rueda 477–487

Global diversity of dipteran families (Insecta Diptera) in freshwater (excluding Simulidae, Culicidae, Chironomidae, Tipulidae and Tabanidae)
R. Wagner, M. Barták, A. Borkent, G. Courtney, B. Goddeeris, J.-P. Haenni, L. Knutson, A. Pont, G.E. Rotheray, R. Rozkošný, B. Sinclair, N. Woodley, T. Zatwarnicki, P. Zwick 489–519

Global diversity of butterflies (Lepidotera) in freshwater
W. Mey, W. Speidel 521–528

Global diversity of hymenopterans (Hymenoptera; Insecta) in freshwater
A.M.R. Bennett 529–534

Global diversity of true and pygmy grasshoppers (Acridomorpha, Orthoptera) in freshwater
C. Amédégnato, H. Devriese 535–543

Global diversity of fish (Pisces) in freshwater
C. Lévêque, T. Oberdorff, D. Paugy, M.L.J. Stiassny, P.A. Tedesco 545–567

Global diversity of amphibians (Amphibia) in freshwater
M. Vences, J. Köhler 569–580

Global diversity of lizards in freshwater (Reptilia: Lacertilia)
A.M. Bauer, T. Jackman 581–586

Global diversity of crocodiles (Crocodilia, Reptilia) in freshwater
S. Martin 587–591

Global diversity of turtles (Chelonii; Reptilia) in freshwater
R. Bour 593–598

Global diversity of snakes (Serpentes; Reptilia) in freshwater
O.S.G. Pauwels, V. Wallach, P. David 599–605

Global diversity of mammals (Mammalia) in freshwater
G. Veron, B.D. Patterson, R. Reeves 607–617

Global diversity of freshwater birds (Aves)
O. Dehorter, M. Guillemain 619–626

The Freshwater Animal Diversity Assessment: an overview of the results
E.V. Balian, H. Segers, C. Lévèque, K. Martens 627–637

LEGENDS TO COLOUR SECTION

Copepoda

1. *Acanthocyclops trajani* (female)	Danielle Defaye
2. *Ergasilus sieboldi* on gills of its fish host	Geoff Boxshall
3. *Hemidiaptomus ingens* (male)	Danielle Defaye
4. *Argulus* on host	Geoff Boxshall

Gastrotricha

5. *Chaetonotus schultzei*	Maria Balsamo
6. *Heterolepidoderma ocellatum*	Maria Balsamo
7. *Chaetonotus zelinkai*	Lara Pierboni

Cladocera

8. *Daphnia similis* (ephippial female)	Adam Petrusek
9. *Daphnia hispanica* (female)	Adam Petrusek
10. *Acroperus harpae*	Jan Fott
11. *Polyphemus pediculus*	Jan Fott

Oligochaeta

12. *Branchiodrilus hortensis*	Jane McRae
13. *Cernosvitoviella atrata*	Enrique Martínez-Ansemil
14. *Spirosperma velutinus*	C. Caramelo & Enrique Martínez-Ansemil
15. *Nais elinguis*	C. Caramelo & Enrique Martínez-Ansemil
16. *Stylaria lacustris*	C. Caramelo & Enrique Martínez-Ansemil

Isopoda

17. *Notamphisopus dunedinensis*	George D.F. Wilson
18. *Eophreatoicus kershawi*	George D.F. Wilson

Platyhelminthes, "Turbellaria"

19. *Gyratrix*	Bart Tessens
20. *Dugesia sp.*	Ronald Sluys

Ostracoda

21. *Lacrimicypris kumbar*	S. Halse & J. McRae
22. *Repandocypris austinensis*	S. Halse & J. McRae

Amphipoda

28. *Brachyuropus reichertii*	Risto Väinölä
29. *Niphargus valachicus*	Boris Sket
30. *Acanthogammarus victorii*	Risto Väinölä
31. *Typhlogammarus mrazeki*	Boris Sket
32. *Macrohectopus branickii*	Boris Sket
33. *Crangonyx richmondensis*	JonathanWitt
34. *Spinacanthus parasiticus*	Boris Sket

2
3
4
7
8
9
10
11
13
14
15
16
21
18
19
20
22
24
25
26
27
28
29
30
31
33
34

35
36
37
38
39
40
41
42
43
44
45
46
47
48
49
50
51
52
53
54
55
56
57
58
59
60
61
62

63

64

65

66

67

68

69

70

71

72

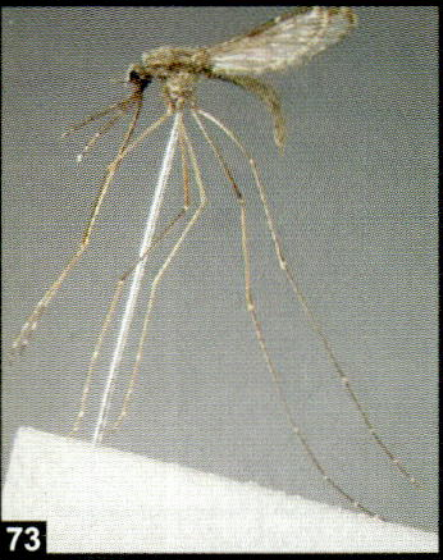
73

74

75

76

77

78

79

80

81

82

83

84

85

86

87

88

89

90

91

92

93

94

95

96

97

98

99

100

101

102

103

104

105

106

107

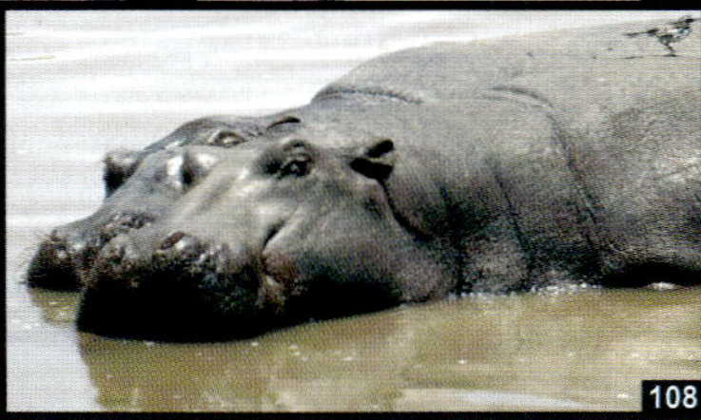
108

109

110

Trichoptera

23. *Barbarochthon brunneum*	Ferdy de Moor
24. *Ceraclea (Pseudoleptocerus) schoutedeni*	Ferdy de Moor
25. *Petrothrincus circularis*	Ferdy de Moor

Megaloptera /Neuroptera

26. *Neohermes filicornis*	Matthew Cover
27. *Sisyra vicaria*	Matthew Cover

Coleoptera (water beetles)

35. *Dytiscus marginalis*	Manfred Jäch
36. *Porrorhynchus latilimbus landaisi*	Harald Schillhammer
37. *Ancyronyx hjarnei*	Harald Schillhammer

Plecoptera

38. *Perla marginata larva*	Romolo Fochetti

Hydrachnidia

39. *Panisopsis curvifrons*	Reinhard Gerecke
40. *Piona sp.*	Reinhard Gerecke

Nematomorpha

41. Hairworms in New Zealand stream	George Poinar
42. Hairworm from orthoptera	George Poinar

Gastropoda

43. *Valvata studeri*	Gerhard Falkner
44. *Bythiospeum*	Annette Schultheiss
45. *Viviparus acerosus*	Gerhard Falkner
46. *Planorbis carinatus*	Gerhard Falkner

Mysidacea

47. *Praunus flexuosus*	Erling Svensen
48. *Americamysis almyra*	Ernst Peebles
49. *Taphromysis bowmani*	Ernst Peebles
50. *Spelaeomysis*	Ernst Peebles

Diptera Simulidae

51. *Simulium arcticum* (larva)	Michael Spironello

Caridea

52 *Xiphocaris elongata*	F. Fasquel
53 *Macrobrachium carcinus*	F. Fasquel
54 *Macrobrachium lar*	A. Anker

Polychaeta

55. *Marifugia vjetrenica*	Boris Sket

Porifera

56. *Spongilla lacustris*	Roberto Pronzato

Brachyura

57. *Johora punicea*	Peter KL Ng
58. *Demanietta khirikhan*	Darren CJ Yeo
59. *Johora tiomanensis*	Peter KL Ng

Ephemeroptera

60. *Compsoneuria njalensis* (adult)	Ferdy de Moor
61. *Tricorythus* (male and female nymphs)	Helen Barber-James
62. *Guloptiloides* (nymph)	Jean-Luc Gattolliat

Odonata

63. *Trithemis annulata* on *Ictinogomphus ferox*	Viola Clausnitzer
64. *Protorthemis coronata*	Vincent J Kalkman
65. *Huonia epinephela*	Vincent J Kalkman

Lepidoptera (Water Moths)

66. *Paracymoriza sp.,* (male. Borneo)	Wolfram Mey
67. *Margarosticha sp.*, (male. Sulawesi)	Wolfram Mey
68. *Eoophyla boernickei*, (male. Borneo)	Wolfram Mey
69. *Parapoynx leucographa*, (male.Borneo)	Wolfram Mey

Heteroptera

70. *Ranatra magna*	J.T. Polhemus
71. *Ptilomera tigrina*	D. A. Polhemus
72. *Laccotrephes pfeifferiae*	J. & D. Polhemus

Diptera Culicidae

73. *Anopheles (Anopheles) sinensis*	Walter Reed Biosystematics Unit
74. *Aedes (Stegomyia) albopictus (Skuse)*	H. J. Harlan, AFPMB

Pisces (Cichlidae)

75. *Mikrogeophagus ramirezi*	Ad Konings
76. *Aethiomastacembelus elipsifer*	Ad Konings
77. *Cyrtocara moorii*	Ad Konings
78. *Eretmodus cyanostictus*	Ad Konings
79. *Ophthalmotilapia ventralis*	Ad Konings
80. *Neolamprologus pulcher*	Ad Konings

81. *Julidochromis regani*	Ad Konings
82. *Cyatopharynx foae*	Ad Konings
83. *Cyphotilapia frontosa*	Ad Konings
84. *Labidochromis caeruleus*	Ad Konings
85. *Aulonocara jacobfreibergi*	Ad Konings
86. *Synodontis multipunctatus*	Ad Konings
87. *Vieja synspila*	Ad Konings
88. *Aequidens rivulatus*	Ad Konings

Reptilia: "Lacertilia (Lizards)

89. Shinisaurus crocodilurus	Le Khac Quyet
90. Varanus salvator	Aaron Bauer

Amphibia

91. *Karsenia koreana*	David R. Vieites
92. *Scaphiophryne madagascariensis*	Miguel Vences and Frank Glaw
93. *Guibemantis_liber_Ranomafana*	Miguel Vences and Frank Glaw
94. *Xenopus muelleri*	Miguel Vences
95. *Boophis occidentalis*	Miguel Vences and Frank Glaw

Reptilia: "Crocodilia," (Crocodiles)

96. *Alligator mississippiensis*	Samuel Martin
97. *Crocodylus niloticus*	Samuel Martin
98. *Gavialis gangeticus*	Michel Gunther

Reptilia "Chelonii" (Turtles)

99. *Macrochelys temminckii*	Roger Bour
100. *Emys orbicularis*	Roger Bour
101. *Geoemyda spengleri*	Roger Bour
102. *Sternotherus odoratus*	Roger Bour
103. *Mauremys rivulata*	Roger Bour
104. *Chelodina canni*	Roger Bour
105. *Mesoclemmys nasuta*	Roger Bour

Reptilia: "Serpentes" (Snakes)

109. *Enhydris jagori*	Olivier Pauwels

Mammalia

106. *Trichechus manatus latirostris* (Florida Manatee)	Todd Pusser
107. *Limnogale mergulus*	S. Zack & B. D. Patterson
108. *Hippopotamus amphibius*	B. D. Patterson
110. *Pusa sibirica* (Baikal seal)	Geoff Boxshall

Hydrobiologia (2008) 595:1–2
DOI 10.1007/s10750-007-9168-0

FRESHWATER ANIMAL DIVERSITY ASSESSMENT

Foreword

Robert J. Naiman

This is a critical time for organisms living in continental waters. Quite literally, the hydrological regime of the Earth is being drastically altered to meet the needs of rapidly expanding societies or in response to alterations of the land and the atmosphere (Vörösmarty et al., 2004). Water regimes that helped shape the evolution of freshwater diversity and the life history adaptations of individual species will be different from now on. These major changes, to one of the Earth's most basic biophysical systems, is taking place with only a rudimentary understanding of the organisms being affected or the large-scale consequences of those changes (Dudgeon et al., 2006). Unfortunately, despite centuries of investigations of the Earth's biota, the taxonomy of freshwater organisms and their distributional patterns are just beginning to become clear—and therein lays the great value of this volume.

One of the most telling graphics about the state of fresh waters is from the recent Millennium Ecosystem Assessment (2005). Between 1970 and 2002—a mere 30 years, freshwater biodiversity declined $\sim$55%, while that of terrestrial systems and marine systems, each declined $\sim$32%. One must suspect that the actual value for continental waters was considerably higher considering the incompleteness of the taxonomic database on freshwater biodiversity. I find this to be a sobering statistic as well as a call to action for freshwater-related sciences and for conservation.

In reading the chapters I was struck by just how many described species were in some phyla—and even more, so by how many new species are described annually, how many are estimated to be awaiting description, and how little is known about distributional patterns. Clearly, the overall task is a daunting challenge for science and for science administration. Is enough emphasis being given to training a new generation of taxonomists? Are the most up-to-date techniques being widely used to assist with timely descriptions? Are existing and emerging data on species and distributions being compiled into databases where the broader research community has reasonable access? These and other key questions underpin deep concerns that freshwater taxonomy needs a 'fresh' start—and better coordination—if it is to fully contribute to global concerns about the condition and the management of continental waters.

Fortunately, there are a number of emerging global initiatives to assist the process of discovering the taxonomic richness of the Earth's fresh waters, and to understand the goods and services they provide to societies. The leadership by the editors in organizing

Guest editors: E. V. Balian, C. Lévêque, H. Segers & K. Martens
Freshwater Animal Diversity Assessment

R. J. Naiman (✉)
School of Aquatic & Fishery Sciences, University of Washington, Seattle, WA 98195, USA
e-mail: naiman@u.washington.edu

the initial workshop and compiling this volume cannot be under-estimated. It not only summarizes a vast array of data on a large number of freshwater phyla but perhaps more importantly, it has also acted as a catalyst to garner the interest and support of international programs focused on understanding and conserving freshwater environments (e.g., UNESCO's International Hydrological Programme, DIVERSITAS International, The Nature Conservancy). The remaining tasks represent a grand scientific challenge but, with this volume as a starting point, the path forward seems much clearer.

References

Dudgeon, D., A. H. Arthington, M. O. Gessner, Z. Kawabata, D. Knowler, C. Lévêque, R. J. Naiman, A.-H. Prieur-Richard, D. Soto, M. L. J. Stiassny & C. A. Sullivan, 2006. Freshwater biodiversity: importance, status, and conservation challenges. Biological Reviews 81: 163–182.

Millennium Ecosystem Assessment, 2005. Island Press, Washington, DC.

Vörösmarty, C., D. Lettenmaier, C. Lévêque, M. Meybeck, C. Pahl-Wostl, J. Alcamo, W. Cosgrove, H. Grassl, H. Hoff, P. Kabat, F. Lansigan, R. Lawford & R.J. Naiman, 2004. Humans transforming the global water system. EOS, American Geophysical Union Transactions 85: 509–514.

Hydrobiologia (2008) 595:3–8
DOI 10.1007/s10750-007-9235-6

FRESHWATER ANIMAL DIVERSITY ASSESSMENT

An introduction to the Freshwater Animal Diversity Assessment (FADA) project

E. V. Balian · H. Segers · C. Lévêque · K. Martens

Abstract The Freshwater Animal Diversity Assessment (FADA) project aims at compiling an overview of genus- and species-level diversity of animals in the continental, aquatic ecosystems of the world. It is a collective effort of 163 experts, and presents 59 articles treating the diversity and endemism of different animal taxa, ranging from microscopic worms to mammals, at global and regional scales. Given their structural importance, an article on macrophytes is also added. Here, we give an overview of the project's history, and outline the common framework of the various articles, as well as the conventions the experts agreed to adhere to in their treatises. Furthermore, we briefly introduce future prospects.

Guest editors: E. V. Balian, C. Lévêque, H. Segers & K. Martens
Freshwater Animal Diversity Assessment

E. V. Balian · H. Segers
Belgian Biodiversity Platform, Brussels, Belgium

E. V. Balian (✉) · H. Segers · K. Martens
Freshwater Biology, Royal Belgian Institute of Natural Sciences, Vautierstraat 29, Brussels 1000, Belgium
e-mail: estelle.balian@naturalsciences.be

C. Lévêque
Antenne IRD, MNHN-DMPA, Case Postale 26, 43 rue Cuvier, Paris Cedex 05 75231, France

Keywords Global biodiversity · Endemism · Metazoa · Aquatic · Non-marine · Freshwater · Review

Introduction

Notwithstanding decades, if not centuries, of taxonomic and faunistic work, it remains difficult to obtain a global overview of biodiversity of freshwater ecosystems. Available knowledge on the matter was never thoroughly compiled and is largely scattered, localised and focuses on a few well-studied groups. Consequently, answering the simple question: "How many species are there in the freshwaters of the world, on continents or in major biogeographic regions?" remained difficult. In addition to constituting relevant basic scientific knowledge on freshwater biodiversity, such an estimate would be a valuable tool for conservation purposes in the face of increasing pressure on freshwater ecosystems. Indeed, more and more evidence documents the major crises faced by biodiversity and biological resources of inland waters, and which are directly correlated to water resource integrity (Postel & Richter, 2003). In addition to their intrinsic value, freshwater ecosystems provide essential goods and services to humankind (Postel & Carpenter, 1997), especially in the third world communities that traditionally depend directly on the availability of natural resources.

Drawing a global picture of freshwater biodiversity has not raised much interest, mainly because of the peculiarities of freshwater habitats. Their island-like nature complicates a global approach, and most taxonomists are overwhelmed by local faunas, especially when studying the highly diverse communities inhabiting ancient lakes or the diversity of groundwater fauna. However, the recognition of changes at a global scale and their impact on freshwater ecosystems (Dudgeon et al., 2006) as well as the need to stop the loss of freshwater biodiversity, motivated the Convention on Biological Diversity (CBD) to support global assessments of status of and trends in freshwater biodiversity, for example Groombridge & Jenkins (1998, 2000) and Revenga & Kura (2003). However, till now, no exhaustive literature review had been performed across all taxonomic animal groups, and a more extensive approach was required to provide information on the diversity and distribution of freshwater species and genera of the world. The Freshwater Animal Diversity Assessment (FADA) project took up the challenge of compiling this information. At the same time, a global assessment was completed on macrophyte diversity, as vascular plants play a key role in structuring the habitat of, and providing food to, many freshwater animals.

In this article, we present a short history of the FADA project, describe its specific objectives, and the common standards and agreements the different FADA experts accepted in order to maintain coherence between the 59 articles of this special issue.

History of the FADA project

Previous assessments

In conjunction with the CBD, some prior attempts to estimate the number of freshwater organisms, and to identify priority areas for conservation, have been made, although these mostly focused on some better-known groups (Groombridge & Jenkins, 1998, 2000; Revenga & Kura, 2003). The latter paper not only compiled a wide range of information on water resources, water system characteristics, threats and conservation aspects, but also included a fairly detailed report of taxonomic diversity for many freshwater taxa. In addition, Revenga & Kura (2003) highlighted the need for additional work on species diversity and distribution in order to better define conservation priorities.

Toward a global assessment of freshwater animal diversity

A preliminary phase of the FADA project lasted from September 2002 to June 2003 and received support from DIVERSITAS and the "Centre National pour la Recherche Scientifique"—French National Research Institute (CNRS). The main objective was to produce a discussion document that identified gaps in our knowledge of freshwater biodiversity, and could be used to triggering experts reactions (Lévêque et al., 2005). This first study led to a gross estimate based on existing databases, published reviews and contacts with taxonomists. The study estimated that known freshwater animal species diversity worldwide was in the order of magnitude of 100,000, half of these being insects. Among other groups, some 20,000 vertebrates; 10,000 crustaceans and 5,000 mollusc species were reported as truly aquatic or water-dependent species.

The preliminary study highlighted gaps in the basic knowledge of species richness at continental and global scales:

1. Some groups such as freshwater nematodes or annelids are understudied and data on their diversity and distribution is scarce. Because current richness estimates for such groups are greatly biased by knowledge availability, we can expect real species numbers to be much higher;
2. Research intensity in the different zoogeographic regions is unbalanced: reliable regional estimates of diversity on the Neotropical and the Oriental regions are lacking for many groups, even for some usually well-known ones such as molluscs or insects.

In addition, the preliminary study of Lévêque et al. (2005) generated numerous comments from the taxonomic community, highlighting that certain key data had not been included. We welcomed these comments by inviting the concerned taxonomic experts to join efforts in the consecutive phase of the project.

Implementation of the FADA project

The Belgian Science Policy (BelSPO), the Belgium Biodiversity Platform and the Royal Belgium Institute of Natural Sciences (RBINS, Brussels, Belgium) provided the necessary support to launch the "Freshwater Animal Diversity Assessment" project in March 2005. Taxonomic experts were invited to join a team of authors to write an article on the diversity of each animal group. These coordinating authors participated in a workshop during which they presented the data on their taxonomic group, and together discussed standards of a common approach (October 13–16, 2005). The resulting reviews are included in the present special issue of *Hydrobiologia.*

As mentioned before, the main goal of FADA is to provide an expert assessment of animal species diversity in the continental (fresh) waters of the world, focusing on taxonomic and biogeographic diversity. The main three objectives for each group are:

1. to give an as accurate as possible estimate of global species and generic diversity;
2. to report on geographic distribution (by zoogeographic region, as described below), and to identify possible gaps;
3. to highlight the main areas of endemicity.

Because extant patterns are the results of historical processes, the project also emphasises phylogenetic aspects and processes of evolution and speciation. In addition, information on human-related issues, such as economical and medical uses, threats, conservation issues, is included when pertinent.

Characteristics of this special issue

Our assessment aims to cover the whole range of freshwater taxa from sponges and nematodes or bryozoans to mammals and birds, including a specific article on macrophytes, but excluding microbes, virus, protists, and algae. In addition, all groups, which are exclusively parasitic and not entirely aquatic are also excluded[1] (i.e., Acanthocephala, Monogenea, Digenea and others); a total of 59 groups/articles are included in this issue. Some articles address a whole Phylum (Rotifera, Porifera...), other papers address a class, an order or even a family, depending on factors like the number of species concerned, level of knowledge on the taxon, available expertise, or historical treatment of the taxon. For instance, an article addressing a relatively species-poor taxon (i.e., Halacaridae), has nevertheless been included, as little comprehensive information had previously been published. On the other hand, the insect order Diptera, is far too diverse, both in number of species and ecology, to be treated in a single article. Consequently, key freshwater families are treated in separate articles (Chironomidae, Culicidae, Simulidae, Tipulidae), and one article addresses the remaining Diptera families. Only the family Tabanidae is not included, as no global expertise appeared to be available.

Article framework

Strict space limits, especially regarding references, were imposed on the authors in order to achieve a single-volume compilation: for each article, space was allocated according to an initial estimate of the diversity of the concerned taxon. A model article framework was imposed to ensure that all standard, required data and information be included, and to maintain coherence amongst reviews, as well as to allow analyses of the data across all taxa.

1. As the main focus of these compilations is not on biology or ecology, only a brief summary of these aspects and some key references are provided in the introduction of each article.
2. The first and main section of each contribution is the "*species and generic diversity section*", which provides information on the known number of species and genera, per relevant higher-level taxon (family, subfamily...). Depending on the group, optional material in this section includes diversity of higher taxa, diversity of groups in selected habitats, data on fossil diversity and estimates of unknown diversity. Only the Gastropoda and the Coleoptera sections do not provide data on generic diversity, but the respective authors provide their arguments for not submitting this information.

[1] Micrognathozoa, a monotypic taxon of moss-dwelling microscopic organisms of which only two disjunct records exist (Disco Island, Greenland and the subantarctic Crozet Islands: De Smet, 2002), is not treated in a full article.

3. The second, optional, section deals with "*phylogeny and historical processes*". Most articles include a brief treatise on evolutionary origin, age, and history of the group. Supplementary information can be added on speciation and diversification processes over time in various areas of the world, and on morphological and molecular phylogenies. Some authors address the main drivers of change: natural and anthropogenic processes of selection and the factors influencing spatial and temporal changes in the genetic stock, in population size, and/or regarding habitat fragmentation.
4. The following, compulsory section on "*Present distribution and endemicity*" provides synthetic maps of species and generic diversity at the level of the main zoogeographical regions (Palaearctic, Nearctic...). The section can include reports on historical patterns and processes, e.g., how the break-up of Gondwana contributed to the present-day distribution. In addition, authors report on endemicity at the species and genus level, and identify hotspots of endemicity.
5. Finally, in a last optional section, "*Human-related issues*" are discussed. This deals with the (potential) economic or medical relevance of the taxon treated, its relevance to fundamental or applied research, or concern for conservation, e.g., IUCN's Red Data Book species, special reserves established or needed, and main threats.

Changes to this framework were allowed for short articles in which it was more logical to address species diversity and distribution together, especially if the optional section on phylogeny was not included.

Terminology

To ensure coherency and homogeneity between articles, the different experts agreed to adhere to common concepts and definitions. An overview of these is as follows.

1. Hotspot: This term is used in relation to richness or endemicity, however, not necessarily with reference to specific threats. In this we deviate from the definition by Myers et al. (2000), in which the term is used in relation with threats and conservation priorities.
2. Endemism/Endemicity: Use of these terms should always include a reference to the relevant geographical unit. In general, endemicity is discussed in relation to the main biogeographic units as defined below. In some cases, endemicity is treated regarding circumscribed local areas, such as Lake Baïkal, Lake Victoria, the Mississippi drainage, or others.
3. Cosmopolitan species: A taxon is considered cosmopolitan if it is present in all zoogeographical regions except Antarctica, unless stated otherwise.
4. Regarding terms related to conservation issues authors refer to the IUCN categories and the IUCN Red list (IUCN, 2006). For example, the term "extinct" is used only in the situation where no more living specimens exist on earth, versus "extirpated" indicating that a taxon or population has disappeared locally.
5. Aquatic and water-dependent species: Defining what exactly constitutes a freshwater species proved to be controversial. For practical reasons, we limited ourselves to non-marine species of inland waters in two categories:

   (1) The 'real aquatic species' accomplish all, or part of their lifecycle in, or on, water.

   (2) "Water-dependent" or "paraquatic" species show close/specific dependency upon aquatic habitats (e.g., for food or habitat).
   Limno-terrestrial species, i.e., species that require an aqueous matrix in strictly terrestrial habitats for active life, like the water film retained by some mosses, are not included in the total numbers. However, they can be discussed in the article when considered pertinent by the expert.
   For some groups, attributing taxa to these ecological categories (water-dependent, limno-terrestrial and terrestrial) turned out to be particularly difficult, mostly owing to a lack of information on life history or ecological requirements of the taxa concerned. Authors were asked to argument their decision on the inclusion or omission of taxa in the total count.

6. Fresh and brackish water species: While the present assessment focuses on diversity of non-marine taxa, a number of thalassic or athalassic

brackish water ecosystems are nevertheless considered. Regarding interface environments (estuaries, anchialine ponds), only the non-marine fauna is included from such habitats. Euryhaline species in estuaries are included in the record, if they show a genuine tolerance to freshwater (<3 g/l). Species that are restricted to such interface environments, and that are therefore absent from both purely marine or fresh waters are not normally included in the total count of freshwater taxa. These cases are specifically addressed in the separate articles, and they can be recorded separately, according to the relevant expert's judgement.

7. Geographical distribution: zoogeographical regions: Regarding the global distribution, reference is made to standard zoogeographic regions as defined in classic textbooks (e.g., Wallace 1876; Cox 2001). We acknowledge that it is impossible to strictly delineate the world's major biogeographic regions. Issues were raised regarding the transitional zone between the Palaearctic and Oriental regions in China and India, the limits between the Oriental and Australasian regions, and the Mexican plateau between the Nearctic and Palaearctic regions. For standardisation purposes, we use the following names and delineations for regions (Fig. 1):

- The *Palaearctic Region (PA)* consists of Europe and Russia, North Africa (not including the Sahara) and Northern and Central Arabian Peninsula, Asia to south edge of Himalayas.
- The *Nearctic Region (NA)* consists of North America, Greenland and the high-altitude regions of Mexico.
- The *Afrotropical Region (AT)* consists of Africa south of the Sahara, the Southern Arabian Peninsula and Madagascar.
- The *Neotropical Region (NT)* consists of Southern and coastal parts of Mexico, Central America, and the Caribbean islands together with South America.
- The *Oriental Region (OL)* consists of India and Southeast Asia south of Himalayas (including lowland southern China) to Indonesia down to the Wallace's Line. It extends

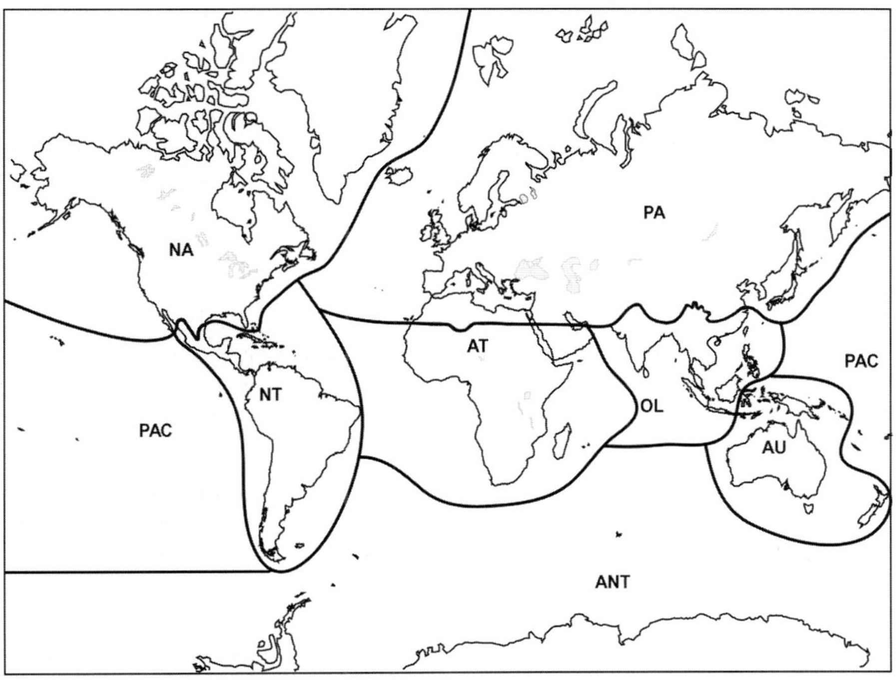

Fig. 1 Standard map of the zoogeographical regions. PA: Palaearctic Region, NA: Nearctic Region, AT: Afrotropical Region, NT: Neotropical Region, OL: Oriental Region, AU: Australasian Region, ANT: Antartic Region, PAC: Pacific Region and Oceanic Islands

through Indonesia as far as Java, Bali, and Borneo to Wallace's line, and includes the Philippines, lowland Taiwan and Japan's Ryukyu Islands.

- The *Australasian Region (AU)* consists of Australia and New Zealand, New Guinea including Papua New Guinea and the Indonesian province of Papua, and Indonesian Islands south and east of Wallace's Line. It includes the island of Sulawesi, the Moluccan islands (the Indonesian provinces of Maluku and North Maluku) and islands of Lombok, Sumbawa, Sumba, Flores, and Timor.
- The *Antarctic Region (ANT)* includes the Antarctic continent and the Antarctic and subantarctic islands south of the Antarctic convergence.
- *The Pacific Region and Oceanic Islands (PAC)*: includes the islands in the North and South Pacific ocean, with the Bismarck Archipelago, Vanuatu, the Solomon Islands, and New Caledonia.

In the few cases where experts were unable to clearly attribute a taxon to a specific region, arguments are listed in support of the final decision on the matter.

Conclusion

This is the first publication of the FADA project, and we are convinced that the information it contains will prove to be useful. In parallel to the production of this work, we are developing a database in which the taxonomic and distributional data on which the treatments presented here are based. This on-going task aims not only to provide access to the raw data the FADA experts have compiled, but we envisage developing a web portal containing additional functionalities like, for example, a repository for local distributional data (see Segers, 2007). These services and any supplementary information resulting from the project will be made accessible through http://fada.biodiversity.be (Balian et al., 2007).

Acknowledgements We are greatly indebted to all experts involved in the project, who contributed their expertise and passion to the daunting task of producing the present volume, and who always showed patience and enthusiasm despite the delays and difficulties encountered during the publishing process. Also, we gratefully acknowledge the numerous reviewers who offered their time and precious advice to improve the contributions.The project was supported by the Belgian Science Policy, the Belgian Biodiversity Platform, and the Royal Belgian Institute of Natural Sciences.

References

Balian, E., H. Segers, C. Lévêque & K. Martens, 2007. Freshwater Animal Diversity Assessment. http://Fada.Biodiversity.be (20 November 2007).

Cox, C. B., 2001. The biogeographic regions reconsidered. Journal of Biogeography 28(4): 511–523.

De Smet, W. H., 2002. A new record of *Limnognathia maerski* Kristensen & Funch, 2000 (Micrognathozoa) from the subantarctic Crozet Islands, with redescription of the trophi. Journal of Zoology, London 258: 381–393.

Dudgeon, D., A. H. Arthington, M. O. Gessner, Z. Kawabata, D. J. Knowler, C. Lévêque, R. J. Naiman, A. Prieur-Richard, D. Soto, M. L. J. Stiassny & C. A. Sullivan, 2006. Freshwater biodiversity: Importance, threats, status and conservation challenges. Biological Reviews 81: 163–182.

Groombridge, B. & M. Jenkins, 1998. Freshwater Biodiversity: A Preliminary Global Assessment. World Conservation Monitoring Centre, Cambridge, U.K.

Groombridge, B. & M. Jenkins, 2000. Global Biodiversity. Earth's Living Resources in the 21st Century. World Conservation Monitoring Centre, Cambridge, U.K.

IUCN, 2006. 2006 IUCN Red List of Threatened Species. http://www.iucnredlist.org (20 November 2007).

Lévêque, C., E. V. Balian & K. Martens, 2005. An assessment of animal species diversity in continental water systems. Hydrobiologia 542: 39–67.

Myers, N., R. A. Mittermeier, C. G. Mittermeier, G. A. B. da Fonseca & J. Kent, 2000. Biodiversity hotspots for conservation priorities. Nature 403: 853–858.

Postel, S., & S. Carpenter, 1997. Freshwater ecosystem services. In: Daily, G. C. (ed.), Nature's Services: Societal Dependence on Ecosystem Services. Island Press, Washington D.C., U.S.A., 195–214.

Postel, S. & B. Richter, 2003. Rivers for Life: Managing Water for People and Nature. Island Press, Washington D.C., U.S.A.

Revenga, C. & Y. Kura, 2003. Status and Trends of Biodiversity of Inland Water Ecosystems. Secretariat of the Convention on Biological Diversity, Montreal, Technical Series 11.

Segers, H., 2007. Annotated checklist of the rotifers (Phylum Rotifera) with notes on nomenclature, taxonomy and distribution. Zootaxa 1564: 1–104.

Wallace, A. R. 1876. The Geographical Distribution of Animals, 2 vol. Harper, New York, xxiii + 503, xi + 553 pp. (reprinted 1962, Hafner, New York).

Hydrobiologia (2008) 595:9–26
DOI 10.1007/s10750-007-9154-6

Global diversity of aquatic macrophytes in freshwater

P. A. Chambers · P. Lacoul · K. J. Murphy · S. M. Thomaz

Abstract Aquatic macrophytes are aquatic photosynthetic organisms, large enough to see with the naked eye, that actively grow permanently or periodically submerged below, floating on, or growing up through the water surface. Aquatic macrophytes are represented in seven plant divisions: Cyanobacteria, Chlorophyta, Rhodophyta, Xanthophyta, Bryophyta, Pteridophyta and Spermatophyta. Species composition and distribution of aquatic macrophytes in the more primitive divisions are less well known than for the vascular macrophytes (Pteridophyta and Spermatophyta), which are represented by 33 orders and 88 families with about 2,614 species in c. 412 genera. These c. 2,614 aquatic species of Pteridophyta and Spermatophyta evolved from land plants and represent only a small fraction ($\sim 1\%$) of the total number of vascular plants. Our analysis of the numbers and distribution of vascular macrophytes showed that whilst many species have broad ranges, species diversity is highest in the Neotropics, intermediate in the Oriental, Nearctic and Afrotropics, lower in the Palearctic and Australasia, lower again in the Pacific Oceanic Islands, and lowest in the Antarctic region. About 39% of the c. 412 genera containing aquatic vascular macrophytes are endemic to a single biogeographic region, with 61–64% of all aquatic vascular plant species found in the Afrotropics and Neotropics being endemic to those regions. Aquatic macrophytes play an important role in the structure and function of aquatic ecosystems and certain macrophyte species (e.g., rice) are cultivated for human consumption, yet several of the worst invasive weeds in the world are aquatic plants. Many of the threats to fresh waters (e.g., climate change, eutrophication) will result in reduced macrophyte diversity and will, in turn, threaten the faunal diversity of aquatic ecosystems and favour the establishment of exotic species, at the expense of native species.

Keywords Aquatic macrophyte · Aquatic weeds · Macroalgae · Diversity · Distribution · Composition · Lakes · Rivers

Guest editors: E. V. Balian, C. Lévêque, H. Segers & K. Martens
Freshwater Animal Diversity Assessment

P. A. Chambers (✉)
Environment Canada, 867 Lakeshore Blvd, P.O. Box 5050, Burlington, ON, Canada L7R 4A6
e-mail: Patricia.Chambers@ec.gc.ca

P. Lacoul
Department of Biology, Dalhousie University, Halifax, NS, Canada B3H 4J1

K. J. Murphy
Division of Environmental and Evolutionary Biology, Institute of Biomedical and Life Sciences, Graham Kerr Building, University of Glasgow, Glasgow G12 8QQ, UK

S. M. Thomaz
Universidade Estadual de Maringá, Nupelia, Maringa 87020-900, Brazil

Introduction

The term 'aquatic macrophytes' refers to a diverse group of aquatic photosynthetic organisms, all large enough to see with the naked eye. It includes macroalgae of the divisions Chlorophyta (green algae), Xanthophyta (yellow-green algae) and Rhodophyta (red algae) and the "blue-green algae" (more correctly known as Cyanobacteria), Bryophyta (mosses and liverworts), Pteridophyta (ferns) and Spermatophyta (seed-bearing plants), the vegetative parts of which actively grow either permanently or periodically (for at least several weeks each year) submerged below, floating on, or growing up through the water surface (Denny, 1985; Pieterse, 1990) (Table 1). Aquatic macrophytes range in size from *Victoria amazonica* with a leaf diameter up to 2.5 m, to the smallest angiosperms, tiny *Wolffia* spp. with a frond diameter less than 0.5 mm. Aquatic macrophytes include emergent macrophytes (plants that are rooted in submersed soils or soils that are periodically inundated, with foliage extending into the air), floating-leaved macrophytes (plants rooted to the lake or stream bottom with leaves that float on the surface of the water), submersed macrophytes (plants that grow completely submerged under the water, with roots or root-analogues in, attached to, or closely associated with the substrate) and free-floating macrophytes (plants that typically float on or under the water surface). Plant species which occur in ephemeral waterbodies (seasonally filled and refilled waters, such as floodplains and temporary ponds) challenge this definition. Our decision has been to include such species as "aquatic macrophytes", only if their environmental survival is clearly dependent upon regular refilling of their aquatic habitat with a source of fresh to brackish water.

The freshwater macroalgae are primarily represented by the green algae, especially the Charales, commonly known as the stoneworts or brittleworts (e.g., *Chara* and *Nitella* spp.). The Charales are often mistaken for higher plants because they have erect central stalks that are divided into short nodes and long internodes of elongated multinucleate cells, with a whorl of "branchlets" at each node (Fig. 1). Individual plants can vary greatly in size, from 5 cm to 1 m in length. This conspicuous stage is the haploid generation. Sexual reproduction commences with production by the haploid plant of complex oogonia and antheridia (often orange in colour and nested in the bases of the branchlets). Flagellated sperm produced in antheridia fertilize egg (oospheres) retained in oogonia, with the result being a diploid oospore. Germination commences with meiosis of the diploid oospore; a haploid protonemal stage develops from one product of meiosis and develops into the haploid plant. Only six genera and a few hundred species of Charales are extant, although a rich fossil record reveals far greater species diversity extending back to the Silurian (Tappan, 1980). The Charales are found in fresh and brackish waters on all continents except Antarctica, generally

Table 1 Freshwater macrophyte divisions and representative genera

Kingdom	Freshwater Macrophyte Divisions	Descriptive Term	Representative Freshwater Macrophyte Genera
Monera	Cyanobacteria	Blue-green algae	*Oscillatoria, Lyngbya*
Protista	Chlorophyta	Green algae	*Chara, Nitella, Cladophora, Enteromorpha*
	Rhodophyta	Red algae	*Lemanea, Batrachospermum*
	Xanthophyta	Yellow-green algae	*Vaucheria*
Plantae	Bryophyta	Mosses and liverworts	*Fontinalis, Riella, Ricciocarpus*
	Pteridophyta	Ferns and allies	*Azolla, Salvinia, Isoetes*
	Spermatophyta	Seed-bearing plants	*Sagittaria, Alisma, Butomus, Brasenia, Cabomba, Callitriche, Ceratophyllum, Scirpus, Carex, Myriophyllum, Elodea, Vallisneria, Juncus, Lemna, Utricularia, Nelumbo, Nymphaea, Nuphar, Spartina, Eichhornia, Potamogeton, Ranunculus, Sparganium, Typha*

in slow-flowing water or in lakes, where they can colonize down to great depths (100 m) in very clear water. In addition to the Charales, freshwater macroalgae include certain other genera of green algae (Chlorophyta: e.g., *Cladophora* and *Enteromorpha* spp.), yellow-green algae (Xanthophyta: e.g., *Vaucheria)* and red algae (Rhodophyta: e.g., *Lemanea* and *Batrachospermum* spp.). Multicellular filamentous "blue-green algae" (Cyanobacteria: e.g., *Oscillatoria* spp.) are also sometimes included in the "macroalgae", particularly species which form large tangled floating mats which can cause a nuisance in freshwater systems (Pieterse & Murphy, 1993). The brown algae (Phaeophyta), so characteristic of marine rocky shore systems, include seven periphytic species that occur in freshwaters but their inclusion in the "macroalgae" is debatable as filamentous forms are typically <10-mm long (Wehr, 2003).

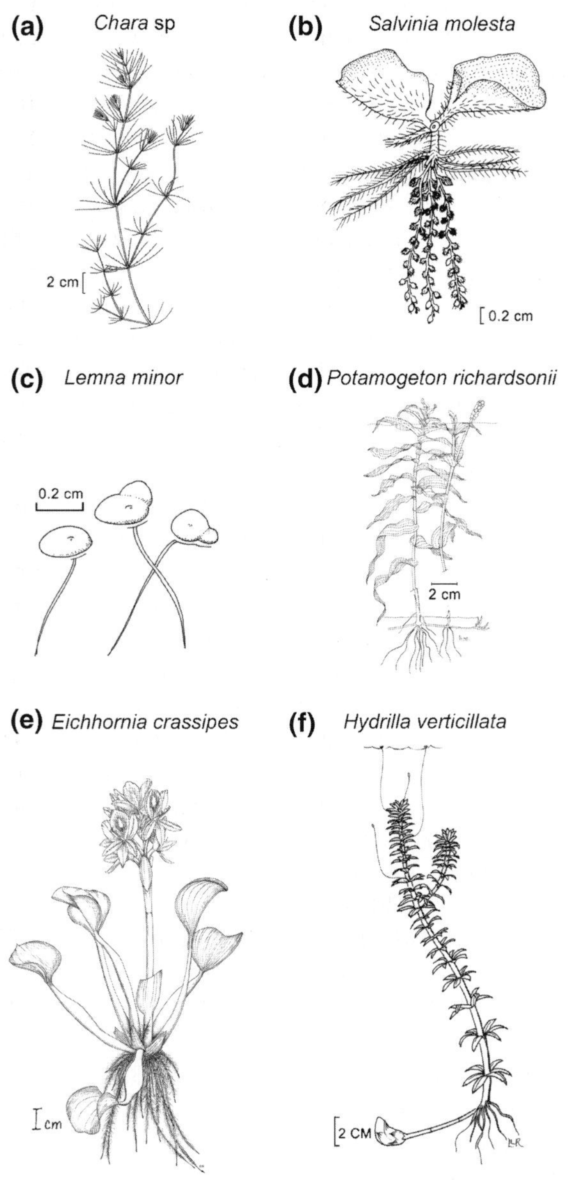

Fig. 1 Examples of aquatic macrophytes: (**a**) macroalgae *Chara* sp. (Order Charales), (**b**) *Salvinia molesta* (Division Pteridophyta), (**c**) *Lemna minor* (angiosperm), (**d**) *Potamogeton richardsonii* (angiosperm), (**e**) *Eichhornia crassipes* (angiosperm) and (f) *Hydrilla verticillata* (angiosperm). Line drawings are from the University of Florida, IFAS Center for Aquatic and Invasive Plants

The mosses, ferns and seed plants are all embryophytes, in that they have a common life cycle involving alternation of sporophyte and gametophyte generations, with the embryo sporophyte retained within the gametophyte at least initially. The bryophytes (mosses and liverworts) differ, however, from ferns and seed plants in that the haploid gametophyte generation, rather than the diploid sporophyte generation, is the most conspicuous. Thus the green moss, with its erect shoot bearing tiny leaf-like structures arranged in spirals, or the thin leathery liverwort are haploid gametophytes. The diploid generation arises after egg and sperm from male and female gametophytes fuse to produce a diploid zygote. The latter grows into a sporophyte, a stalked structure bearing a capsule that produces haploid spores (the future gametophyte generation). The sporophyte is never independent of the gametophyte, remaining attached for provision of water and nutrients. Also unlike ferns and seed plants, bryophytes lack true roots and vascular tissues for uptake and transport of water and organic and inorganic nutrients. About 0.5% of the 20,000 to 25,000 species of bryophytes are truly aquatic macrophytes, in that they require submergence in water to complete their life cycle (Cook, 1999). Other non-aquatic bryophyte species still require water for transfer of spermatozoids, but this can be accomplished simply by raindrops splashing sperm from male to female organs. Aquatic mosses and liverworts are often seen growing attached to rocks in mountain streams, but some (e.g., *Fontinalis antipyretica*) also grow in the shallow to moderately deep water of lakes and in slow-flowing lowland streams and canals. Bryophytes often dominate the macrophyte community found in polar lakes.

The Pteridophyta (ferns and allies) differ from the more primitive bryophytes in that the sporophyte is the dominant and more conspicuous generation, typified by the leafy frond of terrestrial ferns. However, unlike the more advanced seed plants, Pteridophyta lack seeds. Sporophyte plants develop sporangia that contain spores and for most ferns, the spores are identical (i.e., homosporous) and develop into a gametophyte with both antheridia and archegonia. However, some aquatic ferns (e.g., *Isoetes*) are heterosporous, producing separate male spores (microspores) that develop into male gametophytes with antheridia and female spores (megaspores) that develop into female gametophytes with archegonia. Unlike seed plants, the fern gametophyte is a free-living organism typically consisting of a small (<10-mm broad and long) green one-cell thick structure (the prothallus) with single greatly elongated cells (rhizoids) for absorption of water and minerals. The prothallus produces gametes (sperm and egg) that then fuse to form a zygote that grows by mitosis into the sporophyte. Of the 10,500–12,500 species of ferns and fern allies, there are about 171 species (1–2% of all species) that are truly aquatic macrophytes. Aquatic ferns and fern allies include horsetail or scouring rush (*Equisetum* spp.), quillwort (*Isoetes* spp.) and giant salvinia (*Salvinia molesta*), the latter being one of the world's worst aquatic pests (Fig. 1).

The Spermatophyta or seed-bearing plants, consist of two major groups: angiosperms, which have seeds enclosed in an ovary (which matures to become a fruit), and gymnosperms, in which the seeds are not so enclosed. Only the angiosperms, however, have aquatic species. Sporophytes are the dominant generation, and produce haploid microspores and megaspores that divide to form gametophytes. Haploid microspores develop by mitosis into haploid male gametophytes that contain a tube cell and two nonmotile sperm cells. Male gametophytes (pollen grains) are distributed by wind, rain, insects or other organisms. Haploid megaspores develop by mitosis into a haploid female gametophyte, which is composed of seven cells including a large central cell with two polar nuclei and an egg cell with one nucleus. The female gametophyte is retained in the megasporangium in the ovule. During a process that is unique to angiosperms and known as double fertilization, the nucleus of one sperm cell fuses with the nucleus of the haploid egg cell to produce a diploid zygote, and the nucleus of the other sperm cell fuses with the two polar nuclei of the large central cell to produce a triploid endosperm cell. Both the zygote and the endosperm cell divide by mitosis, producing a diploid embryo (the new immature sporophyte) and triploid endosperm (a food reserve for the embryo). Once this embryonic stage is reached, growth is temporarily halted. This stage is known as a seed and consists of the diploid embryo, triploid endosperm and diploid seed coat (from the female gametophyte). Of the 250,000–400,000 angiosperm species, there are only about 2,443 species (<1% of all species) that are aquatic. Aquatic angiosperms include the small free-floating duckweeds (e.g., *Lemna* and *Wolffia* spp.), the cosmopolitan submerged pondweeds (*Potamogeton* spp.) and invasive weeds such as water hyacinth (*Eichhornia crassipes*) and hydrilla (*Hydrilla verticillata*) (Fig. 1).

Species and generic diversity

Aquatic macrophytes are represented in seven plant divisions: Cyanobacteria, Chlorophyta, Rhodophyta, Xanthophyta, Bryophyta, Pteridophyta and Spermatophyta, consisting of at least 41 orders and 103 families. Including the filamentous green algae, the Chlorophyta contribute some 20 genera of aquatic macrophytes, comprising a few hundred species (mostly in the Orders Cladophorales and Charales). There are a few additional freshwater macrophyte species in the Rhodophyta and Xanthophyta, and probably fewer than 20 genera (though the taxonomy is confused) of Cyanobacteria which could be considered as macrophytes. The Bryophyta contribute 22 genera of aquatic macrophytes with about 110 freshwater species (Cook, 1999). Species composition and distribution of aquatic macrophytes in these more primitive divisions are less well known than for the vascular macrophytes (Pteridophyta and Spermatophyta); the remainder of this article focuses on the latter two plant divisions only.

Vascular aquatic macrophytes are represented by 33 orders and 88 families, with about 2,614 species (Table 2) in c. 412 genera (Table 3). Exact numbers are not possible to determine because it is not known whether many so-called 'wetland' species are truly

Table 2 Number of vascular aquatic macrophyte species currently known in the major biogeographic areas

Taxon	PA	NA	AT	NT	OL	AU	PAC	ANT	World
Pteridophyta									
Azollaceae	2	3	2	4	1	1			7
Blechnaceae	1		2		2	3	4		7
Equisetaceae	3	2	1		1		1		3
Isoetaceae	8	27	1	12	18	8			70
Marsileaceae	11	4	24	12	12	11	2		66
Polypodiaceae					1	1			1
Pteridaceae	1	3	2	3	3	2	2		5
Salviniaceae	1	2	2	8	2				10
Thelypteridaceae			2	1	2	2	2		2
Spermatophyta (Angiosperms)									
Acanthaceae	3	3	2	3	10	4	2		18
Acoraceae	1	1			2				2
Alismataceae	19	32	14	39	18	7	1		96
Amaranthaceae	1		1	5	2				7
Amaryllidaceae		1	1	2	1				4
Apiaceae	17	30	3	11	2	1			55
Apocynaceae	1				1				1
Aponogetonaceae			31		10	14			54
Araceae	15	22	19	31	90	19	7		139
Araliaceae		3	2	4		2			5
Asteraceae	1	12	16	29	18	3	1		56
Balsaminaceae	1				1				1
Boraginaceae	2		5	2	6	1			6
Brassicaceae	6	3	2	3	2				12
Burmanniaceae					3	1			3
Butomaceae	1				1		1		1
Cabombaceae	1	3	1	6	1	1			6
Campanulaceae	2	22	8	6	4	7			41
Cannaceae		1		1					1
Ceratophyllales	4	3	3	3	3	2	2		4
Commelinaceae	4	1	5	1	13	3	2		13
Convolvulaceae	2	1		2	1				3
Crassulaceae	2	1	2	3		1			8
Cyperaceae	73	123	78	149	87	67	35	3	276
Droseraceae	1		1		1	1			1
Elatinaceae	10	11	2	6	3	1			25
Eriocaulaceae	6	12	7	45	17	1	1		71
Euphorbiaceae		1		4					4
Fabaceae		6	1	13	2				17
Haloragaceae	10	15	4	11	7	41			65
Hanguanaceae					3	1	1		3

Table 2 continued

Taxon	PA	NA	AT	NT	OL	AU	PAC	ANT	World
Hydatellaceae					1	8			9
Hydrocharitaceae	20	12	43	15	40	23	5		108
Hydroleaceae	1	2		2	1	1			4
Hydrostachyaceae			29						29
Hypericaceae	1								1
Hypoxidaceae			1						1
Iridaceae	1	8	1	1					10
Juncaceae	7	9	4	3	4	2	2	1	14
Juncaginaceae	1	1	1	1		3			5
Lamiaceae	7	8	6	1	9	2	1		23
Lentibulariaceae	11	21	17	26	12	13			70
Limnocharitaceae		2	1	7	1	1			8
Linderniaceae	2	2	1	5	2	1			7
Lythraceae	13	8	13	33	26	6			78
Marantaceae		1	2	1	1	1	1		3
Mayacaceae		1	1	4					5
Melastomataceae				6					6
Menyanthaceae	8	5	16	8	15	36			73
Myrsinaceae	1	3	2						5
Nelumbonaceae	1	1		1	1	1			2
Nymphaeaceae	12	15	15	22	13	14			68
Onagraceae	2	7	4	11	5	4	1		17
Orobanchaceae			1						1
Oxalidaceae			2						2
Pedaliaceae	1								1
Philydraceae	1				1	1	1		1
Phrymaceae		1	1			7			8
Phyllanthaceae			1	1					2
Plantaginaceae	20	28	31	41	16	11	2	2	91
Poaceae	65	78	54	84	64	51	21	1	190
Podostemaceae	7	3	84	188	47	3			330
Polemoniaceae		3		1					4
Polygonaceae	7	9	3	9	3	2			20
Pontederiaceae	2	9	4	23	4	4			33
Portulacaceae	1	2	1	2	1	2		1	3
Potamogetonaceae	46	28	23	31	28	29	9	2	117
Primulaceae	1	1							2
Ranunculaceae	19	13		19	1	1		2	39
Rapateaceae				1					1
Rubiaceae			1	5					6
Saururaceae	1	1			2				3
Sparganiaceae	20	9		1	6	2			22
Sphenocleaceae			2		1				2
Tetrachonraceae				1		1			2

Table 2 continued

Taxon	PA	NA	AT	NT	OL	AU	PAC	ANT	World
Theophrastaceae		2	1	3					3
Thurniaceae			1	2					3
Typhaceae	8	3	3	3	7	2	1		9
Xyridaceae				3	1	1			4
Total	497	644	614	984	664	439	108	12	2614

PA: Palaearctic; NA: Nearctic; AT: Afrotropical ; NT: Neotropical; OL: Oriental; AU: Australasian; PAC: Pacific Oceanic Islands; ANT: Antarctic. Notes: Introduced species not considered. Species were identified as "aquatic" on the basis of published records (in particular Cook, 1996a, 1996b, 2004; Preston & Croft, 1997; Crow & Hellquist, 2000; Ritter, 2000) and the knowledge of the authors. Taxonomy (division, order, family, genera) was updated to APG 2003. Geographic distributions were obtained primarily from the Royal Botanical Gardens, Kew, England checklists for monocots and other selected families (Govaerts et al., 2007a, b) and for grass flora (Clayton et al., 2006), US Department of Agriculture's Germplasm Resources Information Network (GRIN, 2007), the Missouri Botanical Garden's VAST (VAScular Tropicos) nomenclatural database (Missouri Botanical Garden, 2007) and the International Plant Names Index (2004)

aquatic (i.e., their vegetative parts actively grow either permanently or periodically submerged below, floating on, or growing up through the water surface). We have been conservative in our identification of aquatic macrophytes, including only those species that have been determined by the authors or other experts to meet the above definition of 'aquatic'. In addition, previously unknown species continue to be discovered, particularly in tropical areas, thus confounding our estimates of species numbers and geographic distribution. Finally, recent advances in molecular phylogenetics have resulted and will continue to result in revisions of classification at nearly all levels. We based our classification at the ordinal, family and generic levels on the schema of the Angiosperm Phylogeny Group (APG, 2003).

Overall, vascular macrophyte species diversity is highest in the Neotropics (984 spp), intermediate in the Orient, Nearctic and Afrotropics (664, 644 and 614, respectively), lower in the Palearctic and Australasia (497 and 439, respectively), and lower again in the Pacific region and Oceanic islands (108 spp), whilst only very few vascular macrophyte species have been found in the Antarctica bioregion, all confined to sub-Antarctic freshwater habitats (Fig. 2). The higher number of species in the Neotropics is in great part due to the large contribution from the Podostemaceae (188 species) compared to other regions. In terms of both number of genera and species, the Podostemaceae is the largest exclusively aquatic family of angiosperms. Plants in this family are confined to fast-flowing waters, mainly in the tropics, and many species have narrow distributions, such as a single watershed. For all regions (except Antarctica), two of the three most species-rich families were Cyperaceae and Poaceae. The other species-rich family varied amongst regions: Alismataceae for the Nearctic, Araceae for the Orient, Haloragaceae for Australasia, Podostemaceae for the Afrotropics and Neotropics, and Potamogetonaceae for the Pacific and Palaearctic.

Generic diversity of vascular aquatic macrophytes is much less variable compared to species diversity (Table 3). The total number of genera ranged between 152 and 196 for 6 of the 8 bioregions and was highest (192–196) for the Afrotropical, Neotropical and Oriental regions (Fig. 2). As with species diversity, lower generic diversity occurred in the Pacific and Antarctic regions. Within the families, approximately 47% (41 families) have only one genus that includes aquatic plants, although there are often other genera of terrestrial and wetland plants, not meeting the criteria for true aquatic habit, in each of these families. The occurrence of isolated genera that are completely or partially aquatic suggests that the aquatic species in these genera are relatively recent returns to water compared to orders or families that are entirely aquatic and therefore likely returned to water early in the divergence of their lineages.

Twelve genera encompass about 28% of the total vascular macrophyte species richness worldwide (Table 4). With the exception of the genus *Apinagia* that is found only in South America, the other genera have a wide range extension, being present in at least three bioregions. Two of the genera are ferns; the remaining 10 are angiosperms. The 12 species-rich

Table 3 Number of vascular aquatic macrophyte genera currently known in the major biogeographic areas

Taxon	PA	NA	AT	NT	OL	AU	PAC	ANT	World
Pteridophyta									
Azollaceae	1	1	1	1	1	1			1
Blechnaceae	1		1		2	2	3		3
Equisetaceae	1	1	1		1		1		1
Isoetaceae	1	1	1	1	1	1			1
Marsileaceae	2	2	1	3	1	2	1		3
Polypodiaceae					1	1			1
Pteridaceae	1	1	1	1	1	1	1		1
Salviniaceae	1	1	1	1	1				1
Thelypteridaceae			2	1	2	2	2		2
Spermatophyta (Angiosperms)									
Acanthaceae	1	2	1	2	1	1	1		3
Acoraceae	1	1			1				1
Alismataceae	7	4	8	2	7	5	1		12
Amaranthaceae	1		1	1	2				2
Amaryllidaceae		1	1	1	1				1
Apiaceae	6	11	2	3	3	2			14
Apocynaceae	1				1				1
Aponogetonaceae			1		1	1			1
Araceae	8	9	9	12	11	8	6		24
Araliaceae		1	1	1		1			1
Asteraceae	1	9	11	10	9	3	2		24
Balsaminaceae	1				1				1
Boraginaceae	1		2	2	2	1			2
Brassicaceae	3	3	2	2	2				5
Burmanniaceae					1	1			1
Butomaceae	1				1		1		1
Cabombaceae	1	2	1	2	1	1			2
Campanulaceae	1	5	4	3	1	4			9
Cannaceae		1		1					1
Ceratophyllales	1	1	1	1	1	1	1		1
Commelinaceae	3	1	2	1	4	3	2		4
Convolvulaceae	1	1		1	1				1
Crassulaceae	1	1	1	1		1			1
Cyperaceae	14	18	25	24	22	20	14	3	33
Droseraceae	1		1		1	1			1
Elatinaceae	2	2	2	2	2	1			2
Eriocaulaceae	1	2	3	5	1	1	1		6
Euphorbiaceae		1		1					1
Fabaceae		1	1	2	2				2
Haloragaceae	2	2	2	2	4	3			5
Hanguanaceae					1	1	1		1

Table 3 continued

Taxon	PA	NA	AT	NT	OL	AU	PAC	ANT	World
Hydatellaceae					1	2			2
Hydrocharitaceae	7	4	8	6	7	7	4		14
Hydroleaceae	1	1		1	1	1			1
Hydrostachyaceae			1						1
Hypericaceae	1								1
Hypoxidaceae			1						1
Iridaceae	1	1	1	1					2
Juncaceae	1	1	1	1	1	1	1	1	1
Juncaginaceae	1	1	1	1		1			2
Lamiaceae	4	3	3	1	4	2	1		6
Lentibulariaceae	1	2	2	2	1	1			2
Limnocharitaceae		1	1	2	1	1			3
Lindemaceae	1	1	2	1	1	1			2
Lythraceae	4	4	5	4	4	3	1		9
Marantaceae		1	2	1	1	1	1		3
Mayacaceae		1	1	1					1
Melastomataceae				2					2
Menyanthaceae	3	3	2	1	3	3			5
Myrsinaceae	1	1	1						2
Nelumbonaceae	1	1		1	1	1			1
Nymphaeaceae	3	3	1	1	3	3			6
Onagraceae	1	2	1	1	1	1	1		2
Orobanchaceae			1		1				1
Oxalidaceae			1						1
Pedaliaceae	1								1
Philydraceae	1				1	1	1		1
Phrymaceae		1	1			2			3
Phyllanthaceae			1	1					1
Plantaginaceae	7	8	6	6	6	7	1	2	15
Poaceae	31	28	30	25	32	27	10	1	59
Podostemaceae	1	1	16	21	13	2			49
Polemoniaceae		1		1					1
Polygonaceae	2	2	1	1	1	2			2
Pontederiaceae	1	3	3	4	1	1			6
Portulacaeae	1	1	1	1	1	2			2
Potamogetonaceae	4	3	4	3	3	4	2	1	5
Primulaceae	1	1							1
Ranunculaceae	2	2	1	1	1	1		1	2
Rapateaceae				1					1
Rubiaceae			1	2					2
Saururaceae	1	1			2				2
Sparganiaceae	1	1		1	1	1			1
Sphenocleaceae			1		1				1
Tetrachonraceae				1					2

Table 3 continued

Taxon	PA	NA	AT	NT	OL	AU	PAC	ANT	World
Theophrastaceae		1	1	1					1
Thurniaceae			1	1		1			2
Typhaceae	1	1	1	1	1	1	1		1
Xyridaceae				2	1	1			2
Total	154	172	196	192	192	152	62	9	412

PA: Palaearctic; NA: Nearctic; AT: Afrotropical; NT: Neotropical; OL: Oriental; AU: Australasian; PAC: Pacific Oceanic Islands; ANT: Antarctic. Notes are the same as for Table 2

genera also span the full range of plants that are permanently submerged below, floating on, or growing up through the water surface.

Phylogeny and Historical processes

In the early Paleozoic, ancestral marine plants colonized land, giving rise to evolution of vascular plants. Land plant fossils (small, dispersed spores dating from the Ordovician; Wellman et al., 2003) as well as molecular analysis (Sanderson, 2003) place the origin of land plants at 450–475 Mya. Most major land plant lineages (i.e., bryophytes, lycophytes, ferns, gymnosperms) date to the Paleozoic, however the first unequivocal angiosperm fossils appeared ~135 Mya and thereafter radiated into most of the major angiosperm lineages over a period of ~10–15 million years (see review of Feild & Arens, 2007 and references therein). Biologists have long acknowledged a link between green algae and terrestrial plants (Lemieux et al., 2000; Chapman & Waters, 2002; Pombert et al., 2005; Turmel et al., 2006) with some suggesting specifically that the green algae known as stoneworts (Order Charales) are the extant sister group to all land plants (reviewed by McCourt et al., 2004).

Of the many species of terrestrial vascular plants (Pteridophyta and Spermatophyta), only a small fraction of these land plants returned to life in aquatic and marine environments. Since aquatic vascular plants evolved at different times, the return to water was not a single, or even an infrequent, event. Cook (1999), in a survey of the number of plants which

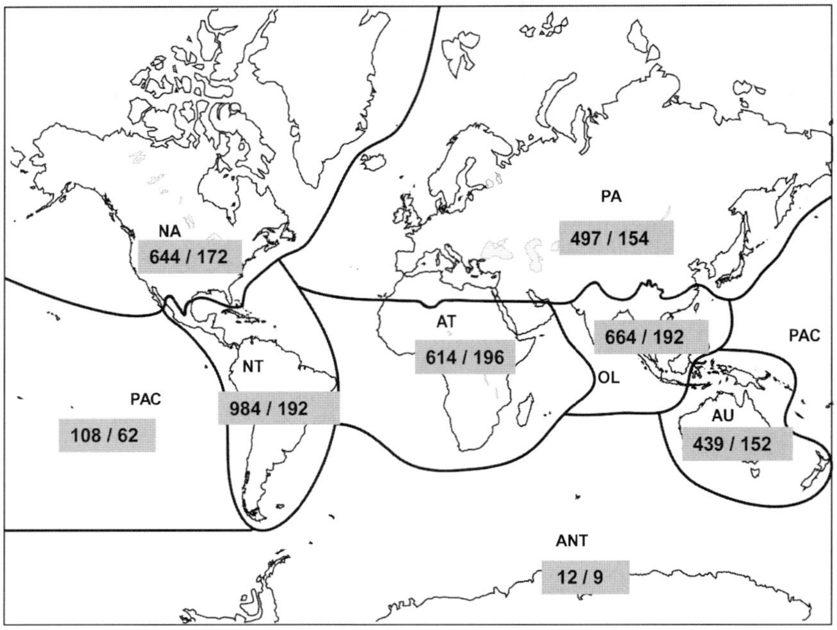

Fig. 2 Diversity of vascular aquatic macrophytes: number of species/number of genera per biogeographic region. PA: Palaearctic, NA: Nearctic, NT: Neotropical, AT: Afrotropical, Au: Australasian, PAC: Pacific Oceanic Islands, ANT: Antarctic

Table 4 Primary distribution and habitat of vascular plant genera with more than 50 aquatic species

Genus	Family	Number of Aquatic Species in Genus	Total Number of Species in Genus	Distribution	Habitat of aquatic species
Potamogeton	Potamogetonaceae	99	99	All regions	Leaves submerged or floating
Isoetes	Isoetaceae	70	~150	All regions except Pacific and Antarctic	Permanently or periodically submerged
Eleocharis	Cyperaceae	70	~200	All regions except Antarctic	Emergent
Marsilea	Marsileaceae	60	60	All regions except Antarctic	Leaves floating on surface or emergent
Apinagia	Podostemaceae	57	57	South America only	Permanently or periodically submerged
Cryptocoryne	Araceae	56	56	Paleoarctic, Orient, Australasia only	Leaves submerged or emergent
Aponogeton	Aponogetonaceae	54	54	Afrotropics, Orient, Australasia only	Leaves submerged or floating
Myriophyllum	Haloragaceae	54	54	All regions except Afrotropics, Pacific and Antarctic	Leaves submerged or emergent
Nymphaea	Nymphaeaceae	53	53	All regions except Pacific and Antarctic	Leaves floating on surface
Cyperus	Cyperaceae	53	~900	All regions except Antarctic	Emergent
Nymphoides	Menyanthaceae	53	53	All regions except Pacific and Antarctic	Leaves floating on surface
Utricularia	Lentibulariaceae	52	216	All regions except Pacific and Antarctic	Leaves submerged or floating

have become secondarily aquatic, estimated that 11 of ~315 genera (or 3%) of ferns and fern allies (i.e., Pteridophyta) and 407 of ~13,200 genera (or 3%) of angiosperms include aquatic species. The evolutionary step of becoming secondarily aquatic probably took place at least 211 times but more likely 252 times (possibly more), with reversion to aquatic life having taken place at least seven times in the Pteridophyta and 204–245 times in the angiosperms (Cook, 1999). In cases where entire orders or families are aquatic, the return to water likely occurred early in the divergence of the lineage. In a review of early angiosperms, Feild & Arens (2007) observed that most molecular analyses place the New Caledonian shrub *Amborella trichopoda* as diverging closest to the root of the angiosperm phylogenetic tree, with the second basal lineage being the entirely aquatic families of Cabombaceae, Nymphaeaceae and Hydatellaceae, the third basal lineage being the Austrobaileyales (lianes occurring in Australia), and the fourth basal lineage being the entirely aquatic family Ceratophyllaceae along with the terrestrial Chloranthaceae. Fossils of water lilies (Nymphaeaceae) have been recorded back to the Early Cretaceous (125–115 Mya) (Friis et al., 2001). The remaining angiosperms form three, well-supported monophyletic lineages (the magnoliids, dicots and monocots), although relations amongst these lineages are still in flux.

As a result of this return to water from the terrestrial environment, aquatic angiosperms have evolved numerous physiological and morphological adaptations to cope with limited carbon dioxide (including the problem of scarcity of CO_2 in solution in many waters, compared to HCO_3^-) and oxygen availability, and reduced light. Aquatic plants operate under dramatically increased diffusion resistance for CO_2 and oxygen as a result of high aqueous resistance to gas diffusion and formation of boundary layers, especially in lentic habitats. To enhance carbon acquisition, submerged leaves are often highly dissected so as to increase surface area (e.g., the thread-like filiform leaves of *Cabomba* and *Ceratophyllum*) and show concentration of the chloroplasts near the leaf surface. Macrophytes in relatively shallow water overcome aqueous inorganic carbon limitations to photosynthesis by drawing on atmospheric CO_2 via aerial or floating leaves. Higher concentrations of CO_2 in bottom sediments (as a result of microbial activity) are also exploited by some macrophytes (e.g., *Isoetes*) whereby CO_2 in the interstitial sediment water diffuses into the roots and then through gas-filled lacunae to the leaves (Raven et al., 1988). In addition to morphological changes, physiological strategies such as utilization of bicarbonate (in addition to CO_2) as an inorganic carbon source and additional biochemical carboxylation pathways (including crassulacean acid metabolism, found, for example, in *Isoetes, Crassula, Littorella, Sagittaria* and *Vallisneria*, and C4—like metabolism found in *Hydrilla verticillata* and *Egeria densa*) have evolved to cope with reduced availability of CO_2 and the prevalence of HCO_3^- as the dominant form of inorganic carbon in higher-pH waters (Maberly & Madsen, 2002). The limited availability of oxygen in aquatic systems has also resulted in development of aerenchyma—tissue containing enlarged gas spaces—for transport of oxygen from shoot to roots and venting of gases (carbon dioxide, ethylene, methane) from the root and soil (Sculthorpe, 1967). Roots are often buried in anoxic sediments and translocated oxygen serves to sustain their aerobic metabolism, at the same time contributing to increased uptake of mineral nutrients as a result of oxygenation of the rhizosphere. To cope with light limitation and changes in spectral quality underwater, many species of submerged plants also evolved strategies such as rapid elongation and physiology, typical of shade plants (Kirk, 1996). In addition, many species considered as nuisance weeds, such as the elodeids *E. densa* and *H. verticillata*, increase their competitive attributes by concentrating their photosynthethic tissues close to the water surface ("canopy forming" strategy). In contrast to adaptations specifically developed by macrophytes for life underwater, many morphological characteristics that evolved to cope with the terrestrial environment have been reduced or eliminated, notably the stomata and cuticles of the leaves, the vascular tissue such as xylem, and structural tissue such as lignin (Sculthorpe, 1967).

Present distribution and main areas of endemicity

Vascular aquatic macrophytes have a world-wide distribution, being found in all biogeographic regions of the world. The broad distributional ranges of vascular macrophytes were noted as early as the mid-1800s by investigators such as de Candolle (1855) and Darwin (1859), and our analyses confirm that many vascular macrophytes are cosmopolitan: 11% of all species occurred in at least three bioregions and 41% of all families spanned ≥6 bioregions (Tables 2 and 3). Species with broad ranges, found in at least seven of the eight bioregions, are *Arundo donax, Brachiaria mutica, Brachiaria subquadripara, Carex echinata, Ceratophyllum demersum, Ceratophyllum muricatum, Ceratopteris thalictroides, Cladium mariscus, Cyperus digitatus, C. odoratus, Echinochloa colona, Echinochloa crus-galli, Echinochloa crus-pavonis, Fimbristylis dichotoma, Fimbristylis littoralis, Ischaemum rugosum, Juncus bufonius, Landoltia punctata, Lemna aequinoctialis, Leptochloa fusca, Montia fontana, Najas marina, Oryza sativa, Panicum repens, Paspalum distichum, Paspalum notatum, Paspalum vaginatum, Pistia stratiotes, Potamogeton nodosus, Ruppia maritima, Schoenoplectus tabernaemontani, Spirodela polyrrhiza* and *Typha domingensis*. Many aquatic vascular plant families can be classed into one of three floristic groups on the basis of species richness: cosmopolitan (e.g., Cyperaceae, Juncaceae, Poaceae), north-temperate (e.g., Potamogetonaceae, Sparganiaceae, Haloragaceae, Elatinaceae and Hippuridaceae) or pan-tropical (e.g., Podostemaceae, Hydrocharitaceae, Limnocharitaceae, Mayacaceae, Pontederiaceae, and Aponogetonaceae) (Crow, 1993). It should be noted that whilst families classed as pan-tropical or north-temperate show much higher species richness in these climatic regions, they may still include species that occur outside their climatic region: a good example being the Haloragaceae, with its numerous Australasian representatives.

The wide distributional ranges of aquatic plants have traditionally been explained by long-distance dispersal by migratory birds (Darwin, 1859; Arber, 1920; Sculthorpe, 1967; Hutchinson, 1975) and human activity (Cook, 1985). However, observations such as the disjunct distributions of aquatic families at the base of the angiosperm phylogenetic tree (i.e., Cabombaceae, Nymphaeaceae and Hydatellaceae) contributed to acceptance of continental drift as a major explanatory factor for modern angiosperm distributions (Raven & Axelrod, 1974). Recently, Les et al. (2003) examined the role of dispersal versus displacement in the distribution of aquatic macrophytes. Using molecular estimates of divergence time involving 71 aquatic angiosperm species from phylogenetically related aquatic taxa that exhibit discontinuous intercontinental distributions, Les et al. (2003) found that for 79 of 87 comparisons, divergence times were far too recent (<30 Mya) to implicate continental drift as a major determinant of these discontinuous distributions. Even *Ceratophyllum demersum*, which is found in all continents except Antarctica, had divergence times of <2.5 Mya for comparisons of specimens from North America, Asia and Australia, indicating recent dispersal rather than a paleodistribution amongst these continents. In an analysis of aquatic macrophyte species and subspecies endemic to Europe and portions of North Africa bordering the Mediterranean, Cook (1983) considered that c. 75% of 61 endemic taxa evolved after the ice age whereas only c. 25% were relicts left by extinction. Long-distance dispersal by birds as well as human activity (both active, through introduction of useful crop plants, and inadvertent) remain viable explanations for widely disjunct aquatic plant distributions although, as Les et al. (2003) note, continental drift may have influenced dispersal patterns by facilitating successful transoceanic dispersal between continents that were previously physically closer in proximity. The successful long-distance dispersal of aquatic plants has been facilitated by the broad ecological tolerances and plastic responses of many aquatic plants, their enhanced survivorship because of clonal growth (very common in macrophytes) and the abundance of easily dislodged propagules (Santamaria, 2002; Les et al., 2003).

Our results showed that vascular macrophyte generic diversity is highest in the tropics (Afrotropics, Neotropics and Orient) and lower in the Nearctic, Palaeoarctic and Australasia (Fig. 2). Species diversity is highest in the Neotropics followed by the Orient, with the Nearctic showing the third highest species diversity (Fig. 2). Previous assessments of macrophyte diversity between temperate and tropical regions indicated that richness (S) was similar, or even richer, in temperate regions (Crow, 1993; Jacobsen & Terneus, 2001). Whilst we have not specifically tallied species numbers in tropical versus temperate latitudes, our comparisons amongst bioregions indicate that vascular macrophyte generic diversity for the tropics is greater than for temperate regions. Species diversity may also be greater for certain tropical compared to temperate regions. Considering the relative lack of data from the tropics compared with temperate regions, this difference may

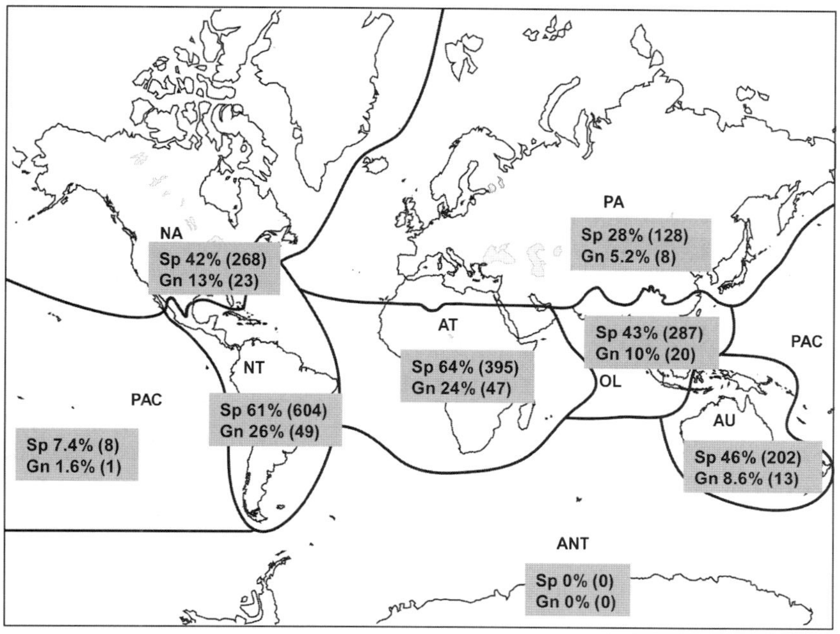

Fig. 3 Vascular aquatic macrophyte endemism, by species (Sp) and genera (Gn) presented as percentage (and number) of endemics per biogeographic region. PA: Palaearctic, NA: Nearctic, NT: Neotropical, AT: Afrotropical, Au: Australasian, PAC: Pacific Oceanic Islands, ANT: Antarctic

increase with time as more investigations are undertaken in the tropics, leading to discoveries of new species or genera. However, even given the high probability of new macrophyte species being found in tropical regions, differences in richness between tropical and temperate regions will likely remain less for aquatic than for terrestrial plants because conditions favouring greater richness in tropical regions (e.g., higher and more uniform temperature) may be offset by increased precipitation in tropical regions (resulting in water level fluctuation and lower underwater light) and greater inorganic carbon availability in temperate regions (Payne, 1986).

Similar to the latitudinal differences in macrophyte distribution, aquatic macrophytes also show decreases in species numbers with altitudinal gain (Jones et al., 2003; Lacoul & Freedman, 2006a). Whereas certain species such as *Callitriche palustris* cover a wide altitudinal range, from sea-level up to 2,500 m in Europe, 3,000 m in Californian mountains and >4,000 m in mountains in the Andes and Himalayas (Beger, 1932; Schotsman, 1954; McLaughlin, 1974; Lacoul, 2004), others such as *Isoetes bolanderi*, *Myriophyllum exalbescens*, *Nuphar lutea* and *Potamogeton alpinus* have restricted distributions in cold high-altitude waters (usually softwater lakes: Murphy, 2002) similar to the restricted distributions observed in the arcto-boreal environment. Some of the highest published altitude records for the aquatic angiosperms include *Zannichellia* sp. at 5,350–5,400 m in Cerro Cóndor, Argentina (Kühn & Rohmeder, 1943; Halloy, 1981, 1983); *Potamogeton* sp., *Myriophyllum* sp., *Isoetes* sp., and *Nitella* at 4,880 m in Peru (Halloy et al., 2005; Seimon et al., 2007); *Myriophyllum* cf. *elatinoides*, *Potamogeton* cf. *pectinatus* and *Isoetes* sp. at 4,400–5,244 m in Peru (Seimon et al., 2007); *Chara* sp. (algae) at 5,030 m in Tibet (Mitamura et al., 2003) and *Ranunculus trichophyllus* at 4,680–4,750 m in Nepal (Lacoul & Freedman, 2006b). Moreover, it is not only the number of macrophyte species that are less at higher altitudes but also the number of endemic species, an example being the fewer endemic species in the northern mountainous regions of Northern India, Nepal and Bhutan compared to peninsular south India and Sri Lanka.

There is strong evidence that within-system diversity (alpha-diversity) of aquatic macrophytes is related not only to geographical factors (e.g., latitude, altitude, as discussed above), and size of waterbody (e.g., Rørslett, 1991), but also to within-system heterogeneity of environmental factors affecting macrophyte growth (e.g., Murphy et al., 2003; Feldmann & Nõges, 2007), and to the intensity of environmental and human-related stress and disturbance pressures acting upon the system. In relation to the last point, data from Swiss lake macrophyte communities (Lachavanne, 1985), for example, show strong evidence that environmental stress associated with nutrient availability (trophic status) of individual lakes is related to macrophyte alpha-diversity, following a classic "hump-back" distribution. Ultra-oligotrophic and oligotrophic lakes at one end of the scale support few species. Mesotrophic lakes, in the middle, tend to support the richest macrophyte diversity, whilst macrophyte richness declines again in eutrophic and hypertrophic lakes.

In contrast to the widely distributed genera (Table 3), it is worth noting that 39% of the genera containing aquatic vascular macrophytes (ignoring any terrestrial species in such genera) are endemic to a single realm. Many of these are genera with single or few aquatic species, but others are multi-species genera, especially in the Podostemaceae. Endemism is rich in two tropical regions (Afrotropical—64% of total species present; Neotropical—61%); intermediate in Australasia (46%), the Oriental region (43%) and the Nearctic (42%); low in the Palaearctic (28%); and negligible or absent in the Pacific (7.4%) and Antarctic (no endemic macrophyte species) (Fig. 3). On a smaller geographic scale, endemism is still rich in some tropical and subtropical regions but also in some temperate systems: 119 endemic species were recorded by Cook (2004) in South Africa; 100 endemic species were recorded in a region including South Brazil, Uruguay and Paraguay and North Argentina (Irgang & Gastal Jr., 2003); 61 endemic species and subspecies were reported for Europe and the portions of North African countries that border the Mediterranean (Cook 1983); 38 endemic aquatic plant species were recorded for New Zealand (Coffey & Clayton, 1988). Surprisingly, ancient large lakes such as Baikal and Biwa are poor in endemic aquatic macrophytes: no endemic aquatic macrophyte has been reported in Lake Baikal, Russia (Kozhova & Izmestéva, 1998) and Lake Biwa, Japan has only two endemics (*Vallisneria biwaensis* and *Potamogeton biwaensis*; Nakajima, 1994).

Human related issues

Aquatic macrophytes play an important role in the structure and function of aquatic ecosystems by altering water movement regimes (flow and wave impact conditions), providing shelter and refuge, serving as a food source, and altering water and sediment quality (e.g., Chambers & Prepas, 1994; Sand-Jensen, 1998; Chambers et al., 1999). They provide a structurally complex environment over spatial scales ranging from millimetres (e.g., foliage structure of macrophytes: Dibble et al., 2006) to hundreds of metres (e.g., distance between weed beds in a lake; Dibble et al., 1996; Rennie & Jackson, 2005). This environmental heterogeneity can increase numbers and types of niches, and can uncouple interacting predators and prey (Harrel & Dibble, 2001). As a result, aquatic macrophyte habitats often represent the most diversified, productive and heterogeneous portions of water bodies. In addition to their important role in maintaining aquatic biodiversity, diverse macrophyte communities also contribute to the maintenance of aquatic ecosystem functioning, for example by sustaining filamentous algal growth (that potentially supports a greater abundance of fish and wildlife) and reducing phosphorus concentrations in the water (Engelhardt & Ritchie, 2001). Eutrophication is one of the greatest environmental problems worldwide and aquatic macrophytes may prove to be "biological engineers" to aid in restoring water quality (Byers et al., 2006).

Perhaps because many vascular macrophyte species exhibit high productivity, broad ecological tolerances and easily dispersed propagules, several of the worst invasive weeds in the world are aquatic macrophytes (Pieterse & Murphy, 1993). Originating in South America, the aquatic fern *Salvinia molesta* and the water hyacinth *Eichhornia crassipes* have become serious aquatic weed problems in the southern USA, Australia, South-East Asia, the Pacific and south, central and eastern Africa. Considered two of the world's worst aquatic pests, these plants are aggressive, competitive species that can cover the surface of lakes and slow-moving rivers, thereby impacting aquatic environments, local economies and human health. Under favourable conditions, plants can double their dry mass in 3–7 days with mats, in some cases, being up to 3-m thick. Another serious aquatic weed is hydrilla (*Hydrilla verticillata*), arguably the most problematic invasive aquatic plant in North America. Native to central and south Asia, it was introduced to Florida in the 1950s or 1960s via the aquarium trade and is now well established in the southern United States and in the west coast states of California and Washington. Hydrilla forms dense submerged mats of vegetation (which may reach to the surface) that interfere with recreation and destroy fish and wildlife habitat. Each year, US agencies spend millions of dollars for hydrilla control involving aquatic herbicides, biological agents, mechanical removal and physical habitat manipulation. Many aquatic weed species are tropical to sub-tropical in origin and global warming will certainly extend the potential range and frequency of occurrence of such species in temperate regions.

In contrast to the threat posed by invasive aquatic macrophytes, a number of macrophyte species are cultivated for human use. Rice (*Oryza* spp.) is the world's most important staple food crop. In 2005, rice production exceeded 6×10^8 Mt (FAO, 2006) with China, India and Indonesia being the top three producers. More than 2.7 billion people rely on rice as their major source of food with this number expected to grow to 3.9 billion by the year 2025. There is increasing concern about current rice production practices being unable to meet future demands as a result of constant or declining yields in many Asian countries, limited possibilities for arable area expansion, and fewer water resources for expanding rice planted areas, as well as concerns related to environmental degradation, genetic erosion and nutritional quality of rice. Whilst rice is probably the most widely used macrophyte by humankind, many other species receive local or widespread use, for example in pulp production (e.g., *Phragmites*), as thatch for houses, mats, etc (e.g., *Cyperus*), in medicine (e.g., *Alternanthera philoxeroides* and *Sagittaria rhombifolia*) and for aesthetic value (e.g., *Nymphaea* spp., *Hydrocleys* spp. and *Victoria amazonica*). The use of several species in phytoremediation has increased recently as an alternative technique for treatment of domestic as well as industrial effluents.

Large gaps still exist in our knowledge of aquatic macrophyte abundance and distribution. Several aquatic vascular macrophytes are recognized as critically endangered (*Isoetes sinensis*, *I. taiwanensis*, *Ledermanniella keayi* and *Saxicolella marginalis*), endangered (*Ledermanniella letouzeyi*, *L. onanae* and

Macropodiella pellucida) or vulnerable (*Ledermanniella thalloidea*), primarily as a result of habitat loss (caused by forestry and agricultural expansion) and water pollution (IUCN, 2004). Many of the threats to fresh waters (e.g., climate change, eutrophication, acidification, alien species introductions) will lead to reduced macrophyte diversity and will, as a result, threaten the faunal diversity of aquatic ecosystems, favour the establishment and expansion of exotic species at the expense of native species, and challenge our abilities to sustain productions of aquatic macrophytes that are needed to meet human consumptive demands.

Acknowledgements We thank the Royal Belgian Institute of Natural Sciences (RBINSc), in particular K. Marten and E. Balian, for the opportunity to participate in a workshop on aquatic biodiversity and the invitation to contribute to this volume. We also thank S. Jubeau, A. Lewin and D. McGoldrick for assistance in assembling data on macrophyte distribution, and Prof. Dr. C. D. K. Cook and Dr. P. Denny for their advice and reviews of an earlier version of this article. S. M. Thomaz is a researcher of the Brazilian Council of Research (CNPq) and acknowledges this agency for long-term funding.

References

APG [Angiosperm Phylogeny Group] II, 2003. An update of the Angiosperm Phylogeny Group classification for the orders and families of flowering plants: APG II. Botanical Journal of the Linnean Society 141: 399–436. More recent updates published on the Internet; http://www.mobot.org/MOBOT/research/APweb/welcome.html. [Accessed April 11 2007].

Arber, A., 1920. Water Plants: A Study of Aquatic Angiosperms. University Press, Cambridge.

Beger, H., 1932. Familie Callitrichaceae. In von Kirchner, O., E. Loew & C. Schroter (eds) Lebensgeschichte der Blutenpflanzen Mitteleuropas III: 3. E. Ulnier, Stuttgart.

Byers, E. J., K. Cuddington, C. G. Jones, T. S. Talley, A. Hastings, J. G. Lambrinos, J. A. Crooks & W. G. Wilson, 2006. Using ecosystem engineers to restore ecological systems. Trends in Ecology and Evolution 21: 493–500.

Chambers, P. A., & E. E. Prepas, 1994. Nutrient dynamics in riverbeds: the impact of sewage effluent and aquatic macrophytes. Water Research 28: 453–464.

Chambers, P. A., R. E. DeWreede, E. A. Irlandi, & H. Vandermeulen, 1999. Management issues in aquatic macrophyte ecology: a Canadian perspective. Canadian Journal of Botany 77: 471–487.

Chapman, R. L., & D. A. Waters, 2002. Green algae and land plants – an answer at last? Journal of Phycology 38: 237–240.

Clayton, W. D., K.T. Harman & H. Williamson, (2006 onwards). GrassBase – The Online World Grass Flora. http://www.kew.org/data/grasses-db.html. [Accessed February–May 2007].

Coffey, B.T., & J. S. Clayton, 1988. New Zealand waterplants: a guide to plants found in New Zealand freshwaters. Raukura Agricultural Centre, Hamilton.

Cook, C. D. K., 1983. Aquatic plants endemic to Europe and the Mediterranean. Botanischer Jahrbücher für Systematik. Pflanzengeschichte und Pflanzengeographie 103: 539–582.

Cook, C. D. K., 1985. Range extensions of aquatic vascular plant species. Journal of Aquatic Plant Management 23: 1–6.

Cook, C. D. K., 1996a. Aquatic and Wetland Plants of India. Oxford University Press, New York, USA.

Cook, C. D. K., 1996b. Aquatic Plant Book. SPB Academic Publishing, The Hague, The Netherlands.

Cook, C. D. K., 1999. The number and kinds of embryo-bearing plants which have become aquatic: a survey. Perspectives in Plant Ecology, Evolution and Systematics 2/1: 79–102.

Cook, C. D. K., 2004. Aquatic and Wetland Plants of Southern Africa. Backhuys Publishers, Leiden, The Netherlands.

Crow, G. E., 1993. Species diversity in aquatic angiosperms: latitudinal patterns. Aquatic Botany 44: 229–258.

Crow, G. E., & C. B. Hellquist, 2000. Aquatic and Wetland Plants of Northeastern North America. Volume 1 Pteridophytes, Gymnosperms and Angiosperms: Dicotyledons. Volume 2 Angiosperms: Monocotyledons. The University of Wisconsin Press, Madison, WI, USA.

Darwin, C. R., 1859. On the Origin of Species by Means of Natural Selection. J. Murray, London.

de Candolle, A., 1855. Géographie botanique raisonnée; ou, exposition des faits principaux et des lois concernant la distribution géographique des plantes de l'époque actuelle. 2 vols. Victor Masson, Paris.

Denny, P., 1985. The Ecology and Management of African Wetland Vegetation. W. Junk, The Hague.

Dibble, E. D., K. J. Killgore & S. L. Harrel, 1996. Assessment of fish-plant interactions. In Miranda L. E. & D. R. DeVries (eds), Multidimensional approaches to reservoir fisheries management. American Fisheries Society, Symposium 16, Bethesda, Maryland: 357–372.

Dibble, E. D., S. M. Thomaz, & A. A. Padial, 2006. Spatial complexity measured at a multi-scale in three aquatic plant species. Journal of Freshwater Ecology 21: 239–247.

Engelhardt, K. A. M., & M. E. Ritchie, 2001. Effects of macrophyte species richness on wetland ecosystem functioning and services. Nature 411: 687–689.

FAO (Food and Agriculture Organization of the United Nations), 2006. FAOSTAT. Published on the Internet; http://faostat.fao.org/ [Accessed June 22 2007].

Feild, T. S., & N. C. Arens, 2007. The ecophysiology of early angiosperms. Plant, Cell and Environment 30: 291–309.

Feldmann, T., & P. Nõges, 2007. Factors controlling macrophyte distribution in large, shallow Lake Võrtsjärv. Aquatic Botany 87: 15–21.

Friis, E. M., K. R. Pedersen, & P. R. Crane, 2001. Fossil evidence of water lilies (Nymphaeales) in the Early Cretaceous. Nature 410: 357–360.

Govaerts, R., M. Ruhsam, L. Andersson, E. Robbrecht, D. Bridson, A. Davis, I. Schanzer & B. Sonké, 2007a. World Checklist of Rubiaceae. The Board of Trustees of the

Royal Botanic Gardens, Kew. Published on the Internet; http://www.kew.org/wcsp/rubiaceae [Accessed February–May 2007].

Govaerts, R., M. Ruhsam, L. Andersson, E. Robbrecht, D. Bridson, A. Davis, I. Schanzer, & B. Sonké, 2007b. World Checklist of Monocotyledons. The Board of Trustees of the Royal Botanic Gardens, Kew. Published on the Internet; http://www.kew.org/wcsp/monocots [Accessed February–May 2007].

GRIN (Germplasm Resources Information Network). 2007. Published on the Internet; http://www.ars-grin.gov/ [Accessed February–May 2007].

Halloy, S. R. P., 1981. La presión de anhidrido carbónico como limitante altitudinal de las plantas. Lilloa 35: 159–167.

Halloy, S. R. P., 1983. El límite superior de aridez, límite de vegetación y el problema de los lagos, nevés y glaciares activos en el ''Núcleo Arido'' de la Cordillera Andina. Actas 1era Reunión Grupo Periglacial Argentino, Mendoza. Anales 83, IANIGLA, 5: 91–108.

Halloy S. R. P., A. Seimon K. Yager & A. Tupayachi Herrera, 2005. Multidimensional (climate, biodiversity, socioeconomics, agriculture) context of changes in land use in the Vilcanota watershed, Peru. In Spehn E. M., M. Liberman Cruz & C. Kőrner (eds), Land Use Changes and Mountain Biodiversity. Boca Raton, Florida: CRC Press: 323–337.

Harrel, S. L., & E. D. Dibble, 2001. Foraging efficiency of juvenile bluegill (*Lepomis macrochirus*) among different vegetated habitats. Environmental Biology of Fishes 62: 441–453.

Hutchinson, G. E., 1975. A Treatise on Limnology, Vol. 3. Limnological botany. Wiley, New York.

IUCN, 2004. IUCN Red List of Threatened Species. Published on the internet www.iucnredlist.org. [Accessed 15 August 2007] .

The International Plant Names Index, 2004. Published on the Internet http://www.ipni.org [Accessed February–May 2007].

Irgang B. E. & C. V. S. Gastal Júnior, 2003. Problemas taxonômicos e distribuição geográfica de macrófitas aquáticas do sul do Brasil. In Thomaz, S. M. & L. M. Bini (eds), Ecologia e manejo de macrófitas. Eduem, Maringá: 163–169.

Jacobsen, D., & E. Terneus, 2001. Aquatic macrophytes in cool aseasonal and seasonal streams: a comparison between Ecuadorian highland and Danish lowland streams. Aquatic Botany 71: 281–295.

Jones, J. I., W. Li, & S. C. Maberly, 2003. Area, altitude and aquatic plant diversity. Ecography 26: 411–420.

Kirk, J. T. O., 1996. Light and Photosynthesis in Aquatic Ecosystems. Cambridge University Press, Cambridge.

Kozhova, O. M., & L. R. Izmestéva, 1998. Lake Baikal: Evolution and Biodiversity. Backhuys Publishers, Leiden, The Netherlands.

Kühn, F. & G. Rohmeder, 1943. Estudio Fisiográfico de las Sierras de Tucumán. Monografías del Instituto de Estudios Geográficos de la Universidad Nacional de Tucumán, 3, Tucumán.

Lachavanne, J. B., 1985. The influence of accelerated eutrophication on the macrophytes of Swiss lakes: abundance and distribution. Verhandlungen Internationale Vereinigung Limnologie 22: 2950–2955.

Lacoul, P., 2004. Aquatic Macrophyte Distribution in Response to Physical and Chemical Environment of the Lakes Along an Altitudinal Gradient in the Himalayas, Nepal. Ph.D. Thesis. Dalhousie University, Halifax, Canada.

Lacoul, P., & B. Freedman, 2006a. Relationships between aquatic plants and environmental factors along a steep Himalayan altitudinal gradient. Aquatic Botany 84: 3–16.

Lacoul, P., B. Freedman, 2006b. Recent observation of a proliferation of *Ranunculus trichophyllus* Chaix. in high-altitude lakes of the Mount Everest region. Arctic, Antarctic, and Alpine Research 38: 394–398.

Lemieux C., C. Otis, & M. Turmel, 2000. Ancestral chloroplast genome in *Mesostigma viride* reveals an early branch of green plant evolution. Nature 403: 649–652.

Les, D. H., D. J. Crawford, R. T. Kimball, M. L. Moody, & E. Landoltk, 2003. Biogeography of discontinuously distributed hydrophytes: a molecular appraisal of intercontinental disjunctions. International Journal of Plant Sciences 164: 917–932.

Maberly, S. C., & T. V. Madsen, 2002. Freshwater angiosperm carbon concentrating mechanisms: processes and patterns. Functional Plant Biology 29: 393–405.

McCourt, R. M., C. F. Delwiche, & K. G. Karol, 2004. Charophyte algae and land plant origins. Trends in Ecology and Evolution 19: 661–666.

McLaughlin, E. G., 1974, Autecological studies of three species of *Callitriche* native in California. Ecological Monographs 44: 1–16.

Missouri Botanical Garden, 2007. VAST (VAScular Tropicos) nomenclatural database. Published on the internet http://www.mobot.mobot.org/W3T/Search/vast.html. [Accessed February–May 2007].

Mitamura, O., Y. Seike, K. Kondo, N. Goto, K. Anbutsu, T. Akatsuka, M. Kihira, T. Qung, & T.M. Nishimura, 2003. First investigation of ultraoligotrophic alpine Lake Puma Yumco in the pre-Himalayas, China. Limnology 4: 167–175.

Murphy, K. J., 2002. Plant communities and plant diversity in softwater lakes of Northern Europe. Aquatic Botany 73: 287–324.

Murphy K. J., G. Dickinson, S. M. Thomaz, L. M. Bini, K. Dick, K. Greaves, M. Kennedy, S. Livingstone, H. McFerran, J. Milne, J. Oldroyd, & R. Wingfield, 2003. Aquatic plant communities and predictors of diversity in a sub-tropical river floodplain: the Upper Rio Paraná, Brazil. Aquatic Botany 77: 257–276.

Nakajima, T., 1994. Lake Biwa. Ergebnisse der Limnologie 44: 43–54.

Payne, A. I., 1986. The Ecology of Tropical Lakes and Rivers. John Wiley, New York.

Pieterse A. H., 1990. Introduction (Chapter 1). In Pieterse A. H., & K. J. Murphy (eds), Aquatic Weeds. Oxford University Press, Oxford, UK: 3–16.

Pieterse, A. H., & K. J. Murphy. 1993. Aquatic weeds, 2nd ed. Oxford University Press, Oxford, UK.

Pombert, J-F, C. Otis, C. Lemieux, & M. Turmel, 2005. The chloroplast genome sequence of the green alga *Pseudendoclonium akinetum* (Ulvophyceae) reveals unusual structural features and new insight into the branching order of Chlorophyte lineages. Molecular Biology and Evolution 22: 1903–1918.

Preston, C. D., & J. M. Croft, 1997. Aquatic Plants in Britain and Ireland. Botanical Society of the British Isles, London, UK.

Raven, J. A., L. L. Handley, J. J. MacFarlane, S. McInroy, L. McKenzie, J.H. Richards, & G. Samuelsson, 1988. The role of CO_2 uptake by roots and CAM in acquisition of inorganic C by plants of the isoetid life-form: a review with new data on *Eriocaulon decangulare* L. New Phytologist 108: 125–148.

Raven, P. H., & D. I. Axelrod, 1974. Angiosperm biogeography and past continental movements. Annals of the Missouri Botanical Garden 61: 539–673.

Rennie, M. D., & L. J. Jackson, 2005. The influence of habitat complexity on littoral invertebrate distributions: patterns differ in shallow prairie lakes with and without fish. Canadian Journal of Fisheries and Aquatic Sciences 62: 2088–2099.

Ritter, N. P., 2000. Biodiversity and phytogeography of Bolivia's wetland flora. Ph.D. thesis, University of New Hampshire, USA (also updates at the Neoaquatica website: http://www.botanize.com/).

Rørslett, B., 1991. Principal determinants of aquatic macrophyte richness in northern European lakes. Aquatic Botany 39, 173–193.

Sand-Jensen, K., 1998. Influence of submerged macrophytes on sediment composition and near-bed flow in lowland streams. Freshwater Biology 39: 663–679.

Sanderson, M. J., 2003. Molecular data from 27 proteins do not support a Precambrian origin of land plants. American Journal of Botany 90: 954–956.

Santamaria, L., 2002. Why are most aquatic plants widely distributed? Dispersal, clonal growth and small-scale heterogeneity in a stressful environment. Acta Oecologica 23: 137–154.

Schotsman, H. D., 1954. A taxonomic spectrum of the section *Eu-Callitriche* in the Netherlands. Acta Botanica Neerlandica 3: 313–384.

Sculthorpe, C. D., 1967. The Biology of Aquatic Vascular Plants. Edward Arnold Publishers, London.

Seimon, A., S. R. P. Halloy, & T. A. Seimon, 2007. Recent observation of a proliferation of *Ranunculus trichophyllus* Chaix. in high-altitude lakes of the Mount Everest region: Comment. Arctic, Antarctic and Alpine Research 39: 340–341.

Tappan, H., 1980. The Paleobiology of Plant Protists. W.H. Freeman and Co., San Francisco, USA.

Turmel, M., C. Otis, & C. Lemieux, 2006. The chloroplast genome sequence of *Chara vulgaris* sheds new light into the closest green algal relatives of land plants. Molecular Biology and Evolution 23: 1324–1338.

Wehr J.D., 2003. Brown algae. In Wehr J.D. & R. G. Sheath (eds), Freshwater Algae of North America. Academic Press, San Diego, USA: 757–773.

Wellman, C. H., P. L. Osterloff, & U. Mohiuddin, 2003. Fragments of the earliest land plants. Nature 425: 282–285.

Hydrobiologia (2008) 595:27–33
DOI 10.1007/s10750-007-9000-x

FRESHWATER ANIMAL DIVERSITY ASSESSMENT

Global diversity of sponges (Porifera: Spongillina) in freshwater

R. Manconi · R. Pronzato

Abstract Porifera is a primarily marine phylum comprising more than 15,000 species. The successful and wide adaptive radiation of freshwater sponges (Haplosclerida: Spongillina) has resulted in the colonization of an extremely wide variety of habitats at all latitudes. Colonization is dated back to the Mesozoic, and the mono- or poly-phyletism of Spongillina, and the number of potential sponge invasions into freshwater is still under debate. Living freshwater sponges belong to 45 genera in six families for a total of 219 species. The highest diversity, at the scale of zoogeographic regions, is recorded from the Neotropical (65 species), Palaearctic (59 species), and Afrotropical regions (49 species). Endemic freshwater sponge species are 103 (47%) out of 219. All species belonging to the families Lubomirskiidae, Metschnikowiidae, and Malawispongiidae are endemic. Endemic species among the other families are 72% for Potamolepidae, 38% for Spongillidae, and 32% for Metaniidae. Data on some wide geographic areas are scattered and fragmentary if not almost completely lacking. Species richness is probably underestimated and doubtless destined to increase with further research.

Keywords Taxonomic richness · Geographic distribution · Endemicity · Habitat

Guest editors: E. V. Balian, C. Lévêque, H. Segers & K. Martens
Freshwater Animal Diversity Assessment

R. Manconi (✉)
Dipartimento di Zoologia e Genetica Evoluzionistica, Università di Sassari, Via Muroni 25, Sassari 07100, Italy
e-mail: rmanconi@uniss.it

R. Pronzato
Dip.Te.Ris. (Dipartimento per lo Studio del Territorio e delle sue Risorse), Università di Genova, Corso Europa 26, Genova 16132, Italy
e-mail: pronzato@dipteris.unige.it

Introduction

The successful and wide adaptive radiation of freshwater sponges has resulted in the colonization of an extremely wide variety of habitats at all latitudes, from cold deserts of the Arctic Circle and Patagonia, to the tropical and equatorial rain forests. Freshwater sponges are dispersed in both lentic and lotic habitats, in continental and insular waters with perennial or temporary regimes, from coast lines to high plain and from high mountains to subterranean environments.

An extremely wide variety of habitats have been colonized by sponges such as springs, streams, rapids, falls, swamps, rivers, estuaries, lakes, thermal vents and springs, caldera lakes, tectonic lakes, alpine lakes, ancient lakes, salt lakes, karstic caves, anchialine caves, ephemeral water bodies in both temperate and strictly arid climates (pools, billabongs, oued in the

Sahara, and pans in the Namibia deserts), man-made basins from pools in zoological–botanical gardens to fonts in archaeological sites, to water tanks, pipelines, reservoirs, and channels.

Recorded bathymetric distribution ranges from habitats at hundreds meters of depth in some lakes (Baikal, Tanganyika, Poso) to the surface of water bodies exposed to direct sun irradiation during low-water levels (e.g. tropical swamps, oueds, and coastal lines of rivers and lakes).

Freshwater sponges are able to tolerate and survive extreme chemico-physical conditions ranging from permafrost, thermal waters, long-lasting dry-up, fluctuating water levels, stagnant to high speed waters, anoxy, oligotrophy to eutrophy, high levels of chemicals, and natural or man-made pollution by hydrocarbons and heavy metals (Harrison, 1974; Pronzato & Manconi, 2002).

Colonised substrata include rocks, boulders, pebbles, shells of bivalves and gastropods, wood debris, roots or branches of riparian trees and bushes, aquatic plants, and various man-made substrata such as glass, cement, plastic, and metallic objects (Pronzato & Manconi, 2002).

Freshwater sponges display highly variable body shape and dimensions, consistency and colour. Specimens range from thin whitish crusts a few mm thick strictly adhering to the substratum, to dark brown massive cushions, to branching or erected growth forms. In most species the body texture is soft and fragile while other species are hard and massive. In unfavourable conditions, such as hard climate, freshwater sponges are represented on the substratum exclusively by carpets of small spherules or resting bodies known as gemmules.

The main diagnostic traits that enable us to identify freshwater sponges are skeletal architecture, range of spicule geometry, size and shape of spicules, and gemmular traits. Skeletal network is a reticulum of siliceous spicules associated to a notably variable amount of spongin. Spicules, megasclere and microscleres, are typically monaxial. Gemmules of freshwater sponges are subspherical, 100–1200 μm in diameter, and bear a structured coat of spongin to protect a mass of totipotent cells contained inside. Morphological characters of this resting stage (such as gemmular cage, gemmular theca, gemmular foramen, arrangement of spicules, architecture of spongin structures, shape, and ornamentations of spicules) are notably diversified and diagnostic at the genus and species level (Manconi & Pronzato, 2002). Identification of gemmule-producing sponges, however, is not possible when gemmules are absent from specimens according to their life cycle phase. Moreover, a number of freshwater sponges, mostly belonging to taxa endemic to ancient lakes, do not produce gemmules.

Diversity

The knowledge on diversity and distribution of sponges from inland waters is reported in a number of historical synopses (Potts, 1887; Weltner, 1895; Annandale, 1911; Arndt, 1926) and in more recent syntheses (Penney & Racek, 1968; Racek, 1969; Volkmer-Ribeiro, 1981; Poirrier, 1982; Frost, 1991; Ricciardi & Reiswig, 1993; Silva & Volkmer-Ribeiro, 2001; Manconi & Pronzato, 1994, 2002, 2004, 2005; Pronzato & Manconi, 2002; Efremova, 2004).

Living freshwater sponges are ascribed, at present, to 219 species belonging to 45 genera in six families, namely Lubomirskiidae Rezvoi, 1936 (4 genera, 10 species), Malawispongiidae Manconi & Pronzato, 2002 (5 genera, 6 species), Metaniidae Volkmer-Ribeiro, 1986 (5 genera, 25 species), Metschnikowiidae Czerniawsky, 1880 (one genus, one species), Potamolepidae Brien, 1967 (6 genera, 29 species), and Spongillidae Gray, 1867 (21 genera, 145 species) (Tables 1, 2). The family Palaeospongillidae Volkmer-Ribeiro & Reitner, 1991 contains exclusively one monotypic genus of fossil sponges. Three monotypic genera of living sponges, namely *Balliviaspongia*, *Makedia*, and *Ohridospongilla* are *incertae sedis*. The most speciose genera contains 15–17 species as in the case of *Corvospongilla*, *Radiospongilla*, and *Eunapius*.

Species richness of freshwater sponges is high when compared to that of the other freshwater sessile invertebrates belonging to Cnidaria and Bryozoa. The total value of 219 species, considered valid at present by the authors, is emended since the Systema Porifera (Manconi & Pronzato, 2002). Species richness is, however, probably underestimated in both temperate and tropical latitudes, where new findings often correspond to the discovery of a new species or genus (Manconi & Pronzato, 2004, 2005). The present diversity values appear doubtless destined to increase with further research on unexplored or

Table 1 Species richness of Spongillina and distribution at the scale of zoogeographic regions

	PA	NA	NT	AT	OL	AU	PAc	ANT	World
Lubomirskiidae	10	0	0	0	0	0	0	0	10
Malawispongiidae	3	0	0	2	0	1	0	0	6
Metaniidae	0	2	17	3	3	3	0	0	25
Metschnikowiidae	1	0	0	0	0	0	0	0	1
Potamolepidae	0	0	11	15	0	0	2	0	29
Spongillidae	44	30	35	28	34	29	3	0	145
Incertae sedis	1	0	1	1	0	0	0	0	3
Total	59	32	65	49	37	33	5	0	219

PA: Palaearctic, NA: Nearctic, NT: Neotropical, AT: Afrotropical, OL: Oriental, AU: Australasian, PAc: Pacific oceanic islands, ANT: Antartic

Table 2 Genera diversity of Spongillina and distribution at the scale of zoogeographic regions

	PA	NA	NT	AT	OL	AU	PAc	ANT	World
Lubomirskiidae	4	0	0	0	0	0	0	0	4
Malawispongiidae	2	0	0	2	0	1	0	0	5
Metaniidae	0	1	5	1	1	1	0	0	5
Metschnikowiidae	1	0	0	0	0	0	0	0	1
Potamolepidae	0	0	3	4	0	0	1	0	6
Spongillidae	13	12	14	9	10	11	3	0	21
Incertae sedis	1	0	1	1	0	0	0	0	3
Total	21	13	23	17	11	13	4	0	45

PA: Palaearctic, NA: Nearctic, NT: Neotropical, AT: Afrotropical, OL: Oriental, AU: Australasian, PAc: Pacific oceanic islands, ANT: Antartic

poorly sampled areas, and on the basis of morphological-molecular analyses focused mainly on cosmopolitan and widespread species presently assumed to consist of complexes of cryptic species.

Phylogeny and historical processes

Porifera is a primarily marine phylum comprising more than 15,000 marine species ascribed to three classes, Hexactinellida, Calcarea and Demospongiae (Hooper & van Soest, 2002). Only the suborder Spongillina of the highly evolved order Haplosclerida of the Demospongiae is represented in freshwater (Manconi & Pronzato, 2002).

The oldest fossil records of freshwater sponges are so far known only from the genera *Palaeospongilla* and *Eospongilla* dating back to the Cretaceous and Jurassic from Patagonia and Colorado (Manconi & Pronzato, in press). *Palaeospongilla chubutensis* shares most gemmular traits with living Spongillidae (Volkmer-Ribeiro & Reitner, 1991) strongly suggesting that gemmular architecture is highly conservative, since the Mesozoic.

The successful colonization of inland waters by these primitive invertebrates seems to be strictly related to cryptobiosis and to the evolutionary novelty represented by gemmules. These peculiar survival devices have a double functional role as resistant resting bodies to persist in situ, and propagules for dispersal in the same or in distant hydrographic basins. Gemmules allow sponges to overcome critical or extreme environmental conditions and to re-establish an active sponge by the rapid proliferation of the totipotent cells contained within them. The pluriannual life cycle of gemmule-producing sponges is characterised by four steps: vegetative growth phase, gemmulation/sexual reproduction, cryptobiosis, hatching of gemmules, and regeneration (Pronzato & Manconi, 1994).

Spongillina share many reproductive characters with marine Haplosclerida. Sexual reproduction occurs by gonochorism, with the development of brooded parenchymula larvae with a short planktonic life. The adaptation to inland fragmented and ephemeral habitats constrained sponges to particular changes resulting in differences between marine and freshwater haplosclerids at the level of sperms, eggs, embryos and larvae supporting an evidence of an early divergence (Ereskovskii, 2004).

Sponge phylogeny at the level of higher taxa (orders) has been outlined based on morphological traits by van Soest (1991). The mono- or polyphyletism of Spongillina, and the number of potential sponge invasions into freshwater is still under debate (Brien, 1969; Manconi & Pronzato, 2002; Pronzato & Manconi, 2002). The absence of gemmules in some families of freshwater sponges and the high possibility of convergence/parallelism at the level of gemmular morpho-traits resulted in inconsistencies with the nearby well-settled systematics (Penney & Racek, 1968; Manconi & Pronzato, 2002), and biased the attempts to match phylogenetic relationships at the genus/species level.

Geographic distribution

At present it is impossible to define the precise geographic ranges for several genera/species without firstly undertaking a systematic revision of materials in both historical and unstudied collections, and a critical analysis of taxonomic data from the literature. Several species and some genera have been recorded only once or exclusively from very restricted geographic areas and the knowledge on some geographic areas is scattered and fragmentary if not almost completely lacking (e.g. Madagascar, Central Asia, Wallacea, West Indies).

Freshwater sponges occur worldwide except so far in the Antarctica region, and their geographical distribution is related to both geological and climatic vicissitudes of the continents, and to the long-term dynamics of hydrographic basins. Different biogeographic patterns are evident, with some species being very common and widespread, as in the case of the Holoarctic *Spongilla lacustris* (Manconi & Pronzato, 2000), while other species are apparently discontinuously distributed, rare or monotopic, as for *Spongilla prespensis* and *Spongilla stankovici* endemic to the Balkanian area (Pronzato & Manconi, 2002).

The highest diversity, at the scale of zoogeographic regions, is recorded from the Neotropical with 65 species (23 genera, 3 families), Palaearctic with 59 species (21 genera, 4 families), and Afrotropical regions with 49 species (17 genera, 4 families). Diversity is lower in the other regions, namely the Oriental with 37 species (11 genera, 2 families), the Australasian with 33 species (13 genera, 3 families), and the Nearctic with 32 species (13 genera, 2 families). The lowest diversity is known from the Oceanic Pacific Islands with five species (4 genera, 2 families) (Tables 1, 2; Fig. 1).

The geographic range of families varies from a cosmopolitan diffusion (Spongillidae), to a condition of an extremely restricted area (Lubomirskiidae in the Baikal, and Metschnikowiidae in the Caspian Sea). A peculiar case is represented by Malawispongiidae recorded only from ancient lakes (e.g. Tanganyika, Malawi, Tiberias) along the African Great Rift Valley in SE-Africa to the Sirian-Palestine Jordan Rift Valley, to the Ohrid lake in the Balkanian area and Poso lake in the Sulawesi microplate (Manconi & Pronzato, 2002).

Potamolepidae are present in the rainforests of the Neotropical, Afrotropical and Pacific Oceanic Islands, whereas Metaniidae shows a true Gondwanian pattern being spread in the circum-tropical rainforests of the Neotropical, Afrotropical, Oriental and Australian regions with an enclave (one genus) in the Nearctic. *Incertae sedis* genera are endemics to old lakes scattered in South America, Africa and Asia (Titicaca, Tana and Ohrid lakes) (Manconi & Pronzato, 2002).

Endemic freshwater sponge species *sensu stricto* (endemic to small areas) are 103 (47%) out of 219. All species belonging to the families Lubomirskiidae (10) from Lake Baikal, Metschnikowiidae (1) from the Caspian Sea, and Malawispongiidae (6) from Tanganyika, Malawi, Tiberias, Ohrid and Poso Lakes are exclusively endemic. Also, the three *incertae sedis* species are endemic each to a single lake. The highest value of endemic species among the most speciose families is 72% for Potamolepidae, 38% for Spongillidae, and 32% for Metaniidae.

The highest values of endemicity *s.s.* are known for the Oceanic Pacific Islands (60%), Palaearctic

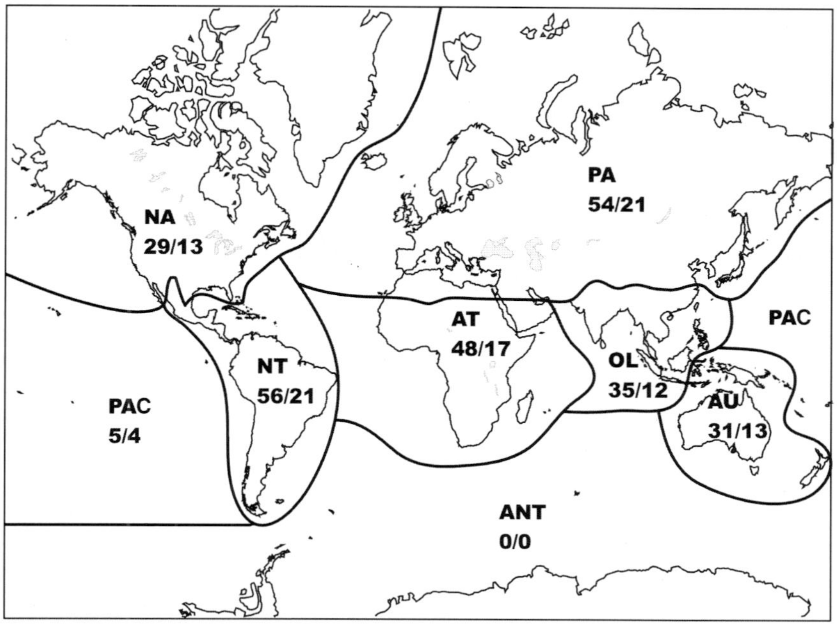

Fig. 1 Distribution of Porifera species and genera per biogeographic region. PA: Palaearctic; NA: Nearctic; NT: Neotropical; AT: Afrotropical; OL: Oriental; AU: Australasian; PAC: Pacific Oceanic Islands, ANT: Antartic

(58%) and Afrotropical (51%) regions. Lower values are shown by the Neotropical (35%), Oriental (24%), Australasian (15%), and Nearctic (16%) regions.

Endemic sponges also occur in ancient, old and/or crateric basins such as Titicaca (South America), Yunnan lakes (W-China), Tana (Ethiopia), Mweru and Luapula (E-Africa), Barombi ma Mbu and Soden (W-Africa).

As far as insular freshwater are concerned, endemics are recorded from Japan, New Zealand, Cuba, Fiji, New Caledonia, Philippines, and Indonesian islands. Endemics are reported also from coastal basins such as those of Louisiana, Florida, Western North America and Brazil. Ancient hydrographic basins such as Amazonian, Orinoco, Paranà-Uruguay-Paraguay, Zaire, and several others host a rich endemic sponge fauna. Main biodiversity hotspots (i.e. highest species richness) are ancient hydrographic basins in tropical latitudes (Manconi & Pronzato, 2002).

Sponges as a natural resource

Sponges represent a natural resource for their functional role involved in natural processes of water purification in freshwater ecosystems. These active filter feeders and their pumping activity play a key-role in the re-cycling of organic matter and contribute to the energetic equilibrium of freshwater ecosystems. The feeding activity is performed by phagocytosis on a high fraction of organic particles and bacteria, and by absorption of dissolved organic matter. Pumping rate is high, *S. lacustris* can filter more than 6 ml/h/mg dry mass; at this rate, a finger-sized sponge could filter more than 125 l in a day (Frost, 1980; Pronzato & Manconi, 2002).

Sponges are centres of biological associations, representing a suitable but selective refuge microhabitat, and are host to a notably diverse assemblage of organisms ranging from other metazoans and protists to bacteria and algae. Representatives of most freshwater invertebrate taxa are recorded in sponges, namely hydrozoans, turbellarians, nematods, oligochaetes, leeches, bivalvs, gastropods, amphipods, copepods, ostracods, hydracarina, and bryozoans to several families of insects (Pronzato & Manconi, 2002). The inter-specific relationships range from endocellular symbiosis, to commensalism or highly specialized predation as in the case of Neuroptera (spongillaflies or Sisyridae), Tricoptera, and Diptera. Some fishes and amphibians were also reported to nest their fertilized eggs in sponges (Pronzato & Manconi, 2002).

Freshwater sponges have been used also by *Homo sapiens*, since ancient times. Some Amazonian tribes perform sponge farming using gemmules as seeds to obtain conspicuous quantities of sponges to strength pottery. An unusual practice is known in cosmetics as in the case of young ladies in Russia that, in the 19th century, used dried spongillids to scrub their skin to have rosy cheeks, but some modern cosmetics are based on the same material. At present bioactive products extracted from sponges are considered one of the most promising source of natural compounds for the pharmacological and biomedical fields. It is well known that *S. lacustris* and *Ephydatia fluviatilis* are homeopathic remedy, since 1700 with the common prelinnean Russian name of Badiaga (Pronzato & Manconi, 2002). No freshwater sponges are currently listed on the official threatened species lists, although in some cases they are indirectly protected being sympatric or syntopic with more "important" taxa such as fishes and amphibians. Successful experimental transplants of freshwater sponges in natural habitats have also been attempted. Conservation of the freshwater sponge fauna and their increase in prevalence and abundance in waters receiving high inputs of organic matter would represent an approach to maintain biodiversity, and to improve the sustainable management of freshwater natural resources (Pronzato & Manconi, 2002). A potential role of sponges to control invasive alloctonous species (e.g. dreissenid bivalves) has been also highlighted (Ricciardi et al., 1995).

Acknowledgements We thank the anonymous referees for helpful comments on the manuscript. R. Manconi is grateful to the organizers of the workshop "A global Assessment of Animal Diversity in Freshwater" for their kind invitation and financial support. Research supported in part by the Italian Ministero dell'Istruzione, dell'Università e della Ricerca Scientifica e Tecnologica (MIUR-PRIN 2004057217 'Zoogeography of Mediterranean-Southern African disjunct distributions by a multimethod approach'), the European program INTERREG Sardinia-Corsica-Tuscany on Biodiversity, Università di Sassari and Università di Genova.

References

Annandale, N., 1911. Freshwater sponges, hydroids and polyzoa. Porifera. In Shipley, A. E. (ed.), Fauna of British India, Including Ceylon and Burma. Taylor & Francis, London, 27–126, 241–245.

Arndt, W., 1926. Die Spongillidenfauna Europas. Archiv für Hydrobiologie 17: 337–365.

Brien, P., 1969. Les Potamolepides africaines. Polyphyletisme des eponges d'eau douce. Archives de Zoologie Experimentale et Generale 110: 527–561.

Efremova, S. M., 2004. New genus and new sponge species of the family Lubomirskiidae Rezvoy, 1936. Index of animal species inhabiting Lake Baikal and its catchment area. Novosibirsk, "Nauka" Vol. 1, Lake Baikal, book 2, 1261–1278 (in Russian).

Ereskovskii, A. V., 2003 (2004). Comparative embriology of sponges and its application for poriferan phylogeny. In Pansini, M., R. Pronzato, G. Bavestrello & R. Manconi (eds), Sponge Science in the New Millennium, Bollettino dei Musei e degli Istituti Biologici dell'Università di Genova 68: 301–318.

Frost, T. M., 1980. Clearance rate determinations for the freshwater sponge *Spongilla lacustris*: effect of temperature, particle type and concentration, and sponge size. Archiv für Hydrobiologie 90(3): 330–356.

Frost, T. M., 1991. Porifera. In Thorp, J. H. & A. P. Covich (eds), Ecology and Classification of North American Freshwater Invertebrates. Academic Press Inc., San Diego, 95–124.

Harrison, F. W., 1974. Sponges (Porifera: Spongillidae). In Hart C. V. Jr., & S. L. H. Fuller (eds), Pollution Ecology of Freshwater Invertebrates. Academic Press, New York, 29–66.

Hooper, J. N. A. & van Soest R. W. M. (eds), 2002. Systema Porifera. A guide to the classification of sponges. I. Kluwer Academic/Plenum Publisher, New York, 1101 pp.

Manconi, R. & R. Pronzato, 1994. Spongillids of Mediterranean Islands. In van Soest, R. W. M., Th. M. G. van Kempen & J. C. Braekman (eds), Sponges in Time and Space. Balkema, Rotterdam, 333–340.

Manconi, R. & R. Pronzato, 2000. Rediscovery of the type material of *Spongilla lacustris* (L., 1759) from the Linnean Herbarium. Italian Journal of Zoology 67(1): 89–92.

Manconi, R. & R. Pronzato, 2002. Spongillina n. subord. Freshwater sponges. Lubomirskiidae, Malawispongiidae n. fam., Metaniidae, Metschnikowiidae, Palaeospongillidae, Potamolepidae, Spongillidae. In Hooper, J. N. A. & R. W. M. van Soest (eds), Systema Porifera. A Guide to the Classification of Sponges, Vol. 1. Kluwer Academic/Plenum Publishers, New York, 921–1019.

Manconi, R. & R. Pronzato, 2004. The genus *Corvospongilla* Annandale (Haplosclerida, Spongillina, Spongillidae) with description of a new species from eastern Mesopotamia, Iraq. Archiv für Hydrobiologie, suppl. Monographic Studies 151(1–2): 161–189.

Manconi, R. & R. Pronzato, 2005. Freshwater sponges of the West Indies: discovery of Spongillidae (Haplosclerida, Spongillina) from Cuba with biogeographic notes and a checklist for the Caribbean area. Journal of Natural History 39(36): 3235–3253.

Manconi, R. & R. Pronzato, in press. Gemmules as a key structure for the adaptive radiation of freshwater sponges: a morpho-functional and biogeographic study. In Custodio, M. R., E. Hajdu, G. Lobo-Hadju & G. Muricy (eds), Porifera Research: Biodiversity, Innovation and Sustainability. Proceedings of the 7th International Symposium 2006.

Penney, J. T. & A. A. Racek, 1968. Comprehensive revision of a world-wide collection of freshwater sponges (Porifera:

Spongillidae). United States National Museum Bulletin 272: 1–184.

Poirrier, M. A., 1982. Porifera. In Hulbert, S. H. & A. Villalobos-Figueroa (eds), Aquatic Biota of Mexico, Central America and West Indies. San Diego State University, San Diego, 59–61.

Potts, E., 1887. Fresh water sponges: a monograph. Proceedings of the Academy of Natural Sciences of Philadelphia 158–279.

Pronzato, R. & R. Manconi, 1994. Adaptive strategies of sponges in inland waters. Bollettino di Zoologia 61: 395–401.

Pronzato, R. & R. Manconi, 2002. Atlas of European Freshwater Sponges. Annali del Museo Civico di Storia naturale di Ferrara 4: 3–64.

Racek, A.A., 1969. The freshwater sponges of Australia (Porifera: Spongillidae). Australian Journal of Marine and Freshwater Research 20: 267–310.

Ricciardi, A. & H. M. Reiswig, 1993. Freshwater sponges (Porifera, Spongillidae) of Eastern Canada: taxonomy, distribution and ecology. Canadian Journal of Zoology 71: 665–682.

Ricciardi, A., F. L. Snyder, D. O. Kelch & H. M. Reiswig, 1995. Lethal and sublethal effects of sponge overgrowth on introduced dreissenid mussels in the Great Lakes, St. Lawrence River System. Canadian Journal of Fisheries and Aquatic Sciences 52: 2695–2703.

de Silva, C. M. M. & C. Volkmer-Ribeiro, 2001. Key to the Ethiopian species of the genus *Metania* Gray, 1867 (Porifera, Metaniidae) with redescription of *Metania rhodesiana* and *Metania godeauxi* comb. n. Bulletin de l'Institut royal des Sciences naturelles de Belgique 71: 127–138.

van Soest, R. W. M., 1991. Demosponge higher taxa classification re-examination. In Reitner, J., & H. Keupp (eds), Fossil and recent sponges. Springer-Verlag, Berlin, 54–71.

Volkmer-Ribeiro, C., 1981. Key to the presently known families and genera of Neotropical freshwater sponges. Revista Brasileira de Biologia 41(4): 803–808.

Volkmer-Ribeiro, C. & J. Reitner, 1991. Renewed study of the type material of *Palaeospongilla chubutensis* Ott & Volkheimer (1972). In Reitner, J., & H. Keupp (eds), Fossil and recent sponges. Springer-Verlag, Berlin, 121–133.

Weltner, W., 1895. Spongillidenstudien III. Katalog und verbreitung der bekannten susswasserschwamme. Archiv für Naturgeschichte 61(1): 114–144.

Hydrobiologia (2008) 595:35–40
DOI 10.1007/s10750-007-9001-9

Global diversity of inland water cnidarians

Thomas Jankowski · Allen G. Collins · Richard Campbell

Abstract Global diversity of inland water cnidarians is low, containing <40 species belonging to phylogenetically distinct groups representing independent invasion events: the common and cosmopolitan hydras (12–15 species); the sporadically occurring freshwater medusae (6–16 sp.); the Cordylophorinae (2 sp.); the parasitic *Polypodium* (1 sp.); the medusae occurring in saline lakes (4 sp.). Freshwater cnidarians inhabit nearly all types of freshwater on all continents (except Antarctica), but only a few species have cosmopolitan distributions. Due to uncertainty in species knowledge, fine scale regions of endemicity are not yet clear.

Guest editors: E. V. Balian, C. Lévêque, H. Segers & K. Martens
Freshwater Animal Diversity Assessment

T. Jankowski (✉)
Water Resources Department, Swiss Federal Institute of Aquatic Science and Technology (Eawag), Dubendorf 8600, Switzerland
e-mail: thomas.jankowski@eawag.ch

A. G. Collins
NMFS, National Systematics Laboratory, National Museum of Natural History, MRC-153, Smithsonian Institution, P.O. Box 37012, Washington, DC 20013-7012, USA

R. Campbell
Department of Developmental and Cell Biology, University of California, Irvine, CA 92697, USA

Keywords Hydra · Polypodium · Cordylophora · Craspedacusta · Distribution · Species diversity · Freshwater cnidarian

Introduction

The Cnidaria is composed of medusae, anemones, corals, and other polyps. Although the phylum is remarkably successful in the marine realm (7000+ species), there are few cnidarian representatives in inland waters. The freshwater species fall into four phylogenetically disparate groups, all save perhaps one belonging to Hydrozoa (Bouillon & Boero, 2000a, b; Collins, 2002): (1): the common *Hydra*, a group of secondarily simple, solitary polyps (Fig. 1A) without medusae; (2) Cordylophorinae, an anthoathecate group that contains freshwater colonial hydroids (*Cordylophora* and *Pachycordyle*) (Fig. 1 C); (3) freshwater medusae, e.g., *Craspedacusta* and *Limnocnida*, which have simple polyp stages that lack tentacles (Fig. 1B); and (4) *Polypodium*, an unusual parasite of fish eggs recently assigned to its own class, Polypodiozoa (Bouillon & Boero, 2000a). Medusae species from saline lakes belong to two distinct groups within Anthoathecata.

Cnidarians are found in nearly all types of freshwater, i.e., streams, rivers, ponds, and lakes, but they mainly occur in mesotrophic to eutrophic habitats. When they are abundant, they can be major predators on small invertebrates (Dumont, 1994;

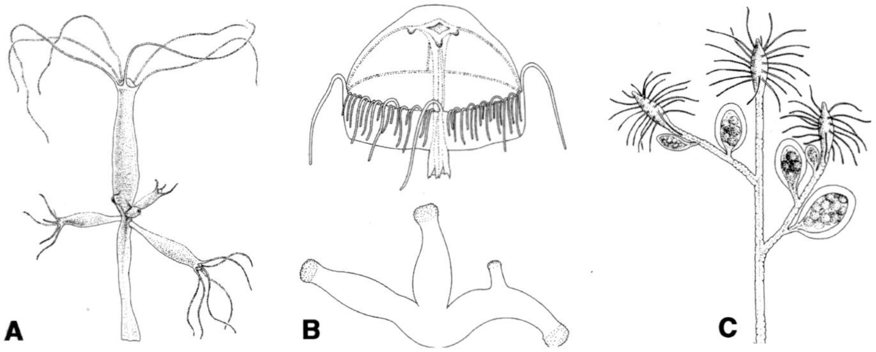

Fig. 1 Habitus of freshwater cnidarians. (**A**) *Hydra* (3–10 mm). (**B**) Medusa (3–20 mm) and Polyp stage (3 polyp colony, 0.5 mm) of *Craspedacusta sowerbii*. (**C**) Part of colony of *Cordylophora* (5 mm). (**A** and **C** from Holstein, 1995 and **B** from Slobodkin & Bossert, 2001)

Jankowski et al., 2005) and occasionally tiny fish, which they catch and immobilize with their characteristic stinging cells, cnidocytes. They are basically planktivorous (Dumont, 1994), though polyps are also benthivorous.

Freshwater cnidarians are of minor economic or medical interest. *Cordylophora* occasional grow such massive colonies that they foul boats and clog waterways, hydras are considered pests in fish hatcheries, and *Polypodium* is a threat to the caviar industry.

Species diversity

Worldwide diversity of inland water cnidarians is low, probably less than 40 species (in <15 genera, Tables 1, 2).

Freshwater medusae—More than 20 species (in 6 genera) have been recorded. However, about half of them may not be valid, because the specific value of many characters is presently uncertain (Bouillon & Boero, 2000b; Jankowski, 2001). Within *Craspedacusta*, *Astrohydra*, and *Limnocnida*, only three to five, one, and six species, respectively, are certain. It is even possible that *Limnocnida* contains just two species, one each in India and Africa (Bouillon & Boero, 2000b). The Indian genera *Mansariella* and *Keralika* are uncertain (Bouillon & Boero, 2000b), as is the holarctic *Calpasoma* (Holstein, 1995). In sum, the number of accepted freshwater medusae species ranges from 6 to 16, though the true diversity may be higher.

Hydras—Of the 80 described species, probably fewer than 15 are distinct. Species are clustered into four groups (Campbell, 1987) that reflect and extend Schulze's (1917) genera, *Hydra*, *Pelmatohydra*, and *Chlorohydra*, which are no longer recognized. These groups are: viridissima group (green, due to intracellular symbiotic algae), probably consisting of a single

Table 1 Species diversity by Family of inland water cnidarian in different biogeographic regions

Biogeographic region	PA	NA	NT	AT	OL	AU	PAC	ANT	World
Olindiidae[a]	4–8	1[c]	1	2–4	2–6	1	1	0	6–16
Australomedusidae[b]	0	0	0	0	0	2	0	0	2
Moerisiidae[b]	1	0	0	1	0	0	0	0	2
Hydridae[a]	4–6	6–7	2–3	2–3	4–5	2–4	0	0	0
Polypodiidae[a]	1	1	0	0	0	0	–	0	1
Cordylophoridae[a]	2	1	1	1	1	1	–	0	2
Total	12–18	9–10	4–5	6–9	7–12	6–8	2	0	13–23

PA: Palaearctic, NA: Nearctic, NT: Neotropical, AT: Afrotropical, OL: Oriental, AU: Australasian, PAC: Pacific Oceanc Islands, ANT: Antarctic

[a] Freshwater species

[b] Salt lake species

[c] *Halmomises lacustris*—found only once in a lagoon in Trinidad—was not considered due to the uncertain status (see Jankowski 2001 for discussion)

Table 2 Genera diversity by Family of inland water cnidarian in different biogeographic regions

Biogeographic region	PA	NA	NT	AT	OL	AU	PAC	ANT	World
Olindiidae[a]	2	1[c]	1	2	2–4	1	1	0	2–4
Australomedusidae[b]	0	0	0	0	0	1	0	0	1
Moerisiidae[b]	1	0	0	1	0	0	0	0	1
Hydridae[a]	1	1	1	1	1	1	0	0	0
Polypodiidae[a]	1	1	0	0	0	0	–	0	1
Cordylophoridae[a]	2	1	1	1	1	1	–	0	2
Total	7	4	3	5	4–6	4	1	0	11

PA: Palaearctic, NA: Nearctic, NT: Neotropical, AT: Afrotropical, OL: Oriental, AU: Australasian, PAC: Pacific Oceanc Islands, ANT: Antarctic

[a] Freshwater species

[b] Salt lake species

[c] *Halmomises lacustris*—found only once in a lagoon in Trinidad—was not considered due to the uncertain status (see Jankowski 2001 for discussion)

species; oligactis group (large stalked hydras), consisting of 3–5 species; braueri group (small hermaphroditic hydras), consisting of 3–5 species, and the remaining vulgaris group (sometimes called common hydra), consisting of 4–6 species.

Polypodium hydriforme is the only described species of *Polypodium*.

Cordylophorinae—Cordylophora and *Pachycordyle* are usually considered to each contain a single species in freshwater.

Saline lake medusae—Australomedusa and *Moerisia* each have two species described from saline lakes.

Phylogeny and historical processes

Not surprisingly, given their small sizes and soft bodies, there is no fossil record for freshwater cnidarians. Nevertheless, their morphologies and

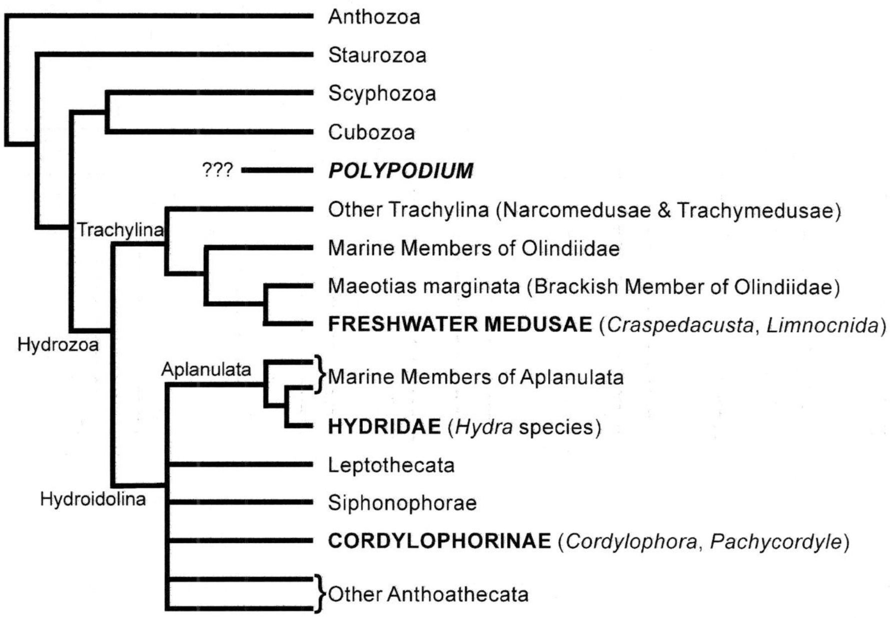

Fig. 2 Hypothesis of cnidarian relationships highlighting independent origins of at least four freshwater groups (bold, all caps), based on Collins (2002), Collins et al. (2005), and Collins et al. (2006)

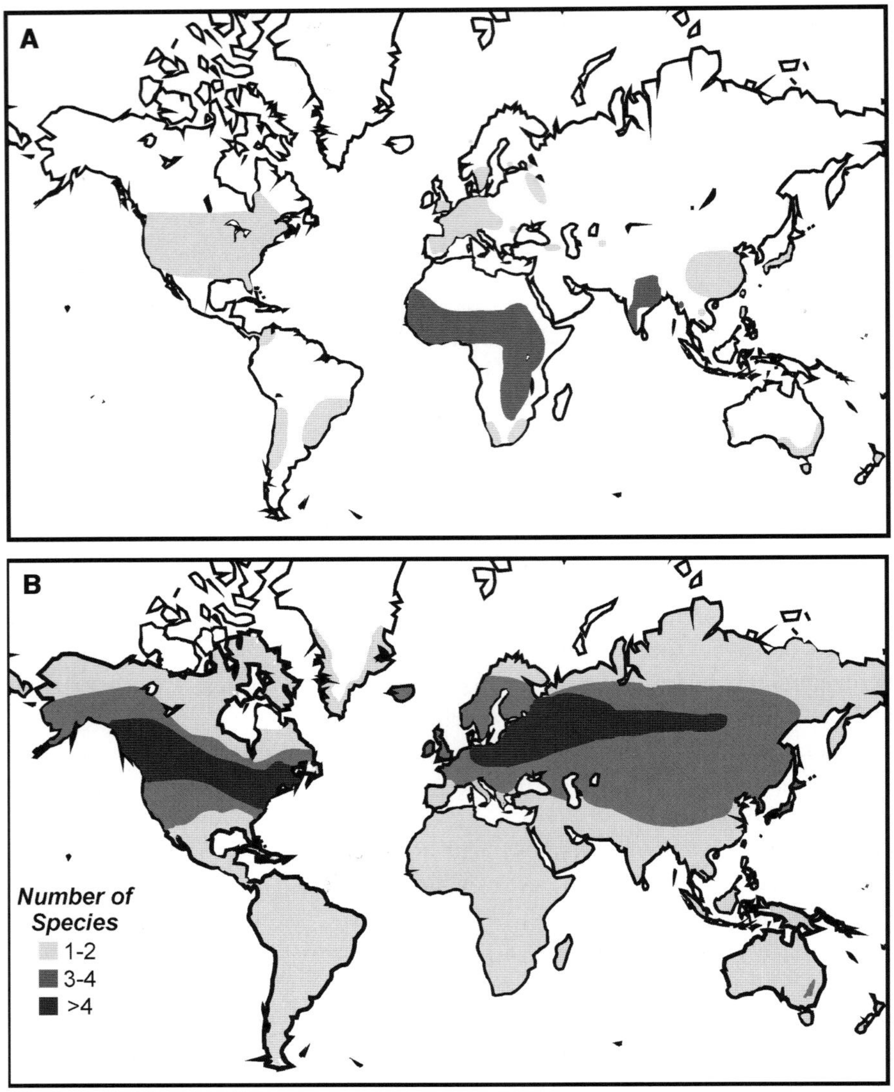

Fig. 3 Distribution of freshwater cnidarians. (**A**) Distribution of the freshwater medusae genera *Craspedacusta* (light gray) and *Limnocnida* (dark gray) (extended after Dumont, 1994). *C. sowerbii* is the only cosmopolitan species. East Asia (China and Japan) is the only area with more than one *Craspedaucsta* species (2–5 species). *Limnocnida* is distributed in Africa (1–3 species) and India (1–3). From India two other species with uncertain status were described. (**B**) Diversity of *Hydra*. There are no distribution data for large dry areas of Africa, Australia and Asia. These areas have been filled in according to the surrounding areas. Hydra are present on continental islands (Japan, Madagascar, New Zealand, New Caledonia, Greenland, Sri Lanka, and British Isles including Orkney and Shetland Islands). They are absent from most oceanic islands. Hydra have been reported from Faroe Islands, Iceland and La Reunion but not from Antarctica

distributions have been used to infer some evolutionary histories. Analysis of molecular sequence data is now putting some of these relationships on a firm basis (Fig. 2). Freshwater medusae originated within Trachylina and form the sister group to the brackish species *Maeotias marginata* (Collins, 2002; Collins et al., 2006). *Hydra* (Hydridae) falls within a clade (Aplanulata) of anthoathecate hydrozoans that develop from egg to polyp via a nonciliated stereogastrula stage, i.e., lacking the characteristic ciliated planula (Collins et al., 2005, 2006). These data show that *Moerisia* is not part of Aplanulata, but they have not provided resolution among the many lineages comprising Anthoathecata. Molecular data have yet to be published for *Cordylophora*, *Pachycordyle*, or *Australomedusa*, but they are classified in the anthoathecate group Filifera. There may have been multiple invasions of freshwater within Cordylophorinae, as most species within the group are adapted to brackish conditions (Stepanjants et al. 2000). Molecular data from the 18S ribosomal gene have been gathered for *Polypodium*, but this gene has undergone such a high rate of divergence in *Polypodium*, that it appears to be an unreliable indicator of its phylogenetic position (Kim et al., 1999). Unfortunately, no molecular clock estimates have been published for the divergences of lineages of freshwater cnidarians.

Although the freshwater cnidarian groups have independent phylogenetic origins, three out of the four have some tie to the Ponto-Caspian basin

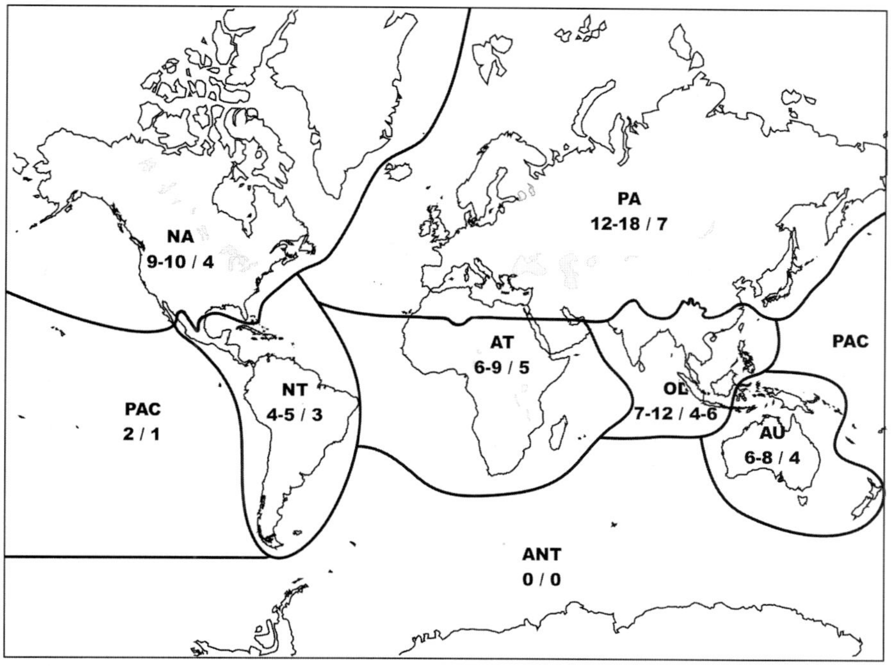

Fig. 4 Distribution of cnidarian species and genera in each zoogeographical region (species number/genus number). PA—Palaearctic, NA—Nearctic, NT—Neotropical, AT—Afrotropical, OL—Oriental, AU—Australasian, PAC—Pacific Oceanc Islands, ANT—Antarctic

encompassing the Black, Azov, Caspian and Aral Sea regions. By providing relatively stable brackish water conditions over many millions of years, this basin may have been critical for the origin of freshwater groups (Croghan, 1983). Three observations fit with such a scenario for three of the freshwater cnidarian groups: (1) the living sister group to the freshwater medusae is a brackish species (*Maeotias marginata*) from the Black Sea (Collins et al., 2006); (2) *Cordylophora caspia* was originally identified from the Caspian Sea; and (3) the Volga River empties into the Caspian Sea and it is in this region that *Polypodium* is most prevalent (Raikova, 2002).

Present distribution and main areas of endemicity

Tables 1 and 2 summarize the species and genera diversity of inland water cnidaria in particular biogeographic regions. Freshwater cnidarians are distributed worldwide (Figs. 3, 4).

Freshwater medusae—Craspedacusta sowerbii is the most widespread freshwater medusa (Fig. 3A), and has successfully colonized all continents, except Antarctica, during the 20th century (Dumont, 1994). This still ongoing expansion is probably related to intercontinental human mediated co-transportation of drought-resistant resting stages with plants and fish (Dumont, 1994) and climate changes. The probable origin and most diverse region of *Craspedacusta* is the Yangtze River basin, in which up to 4 species are endemic (Jankowski, 2001). Whereas *Craspedacusta* seems to have mainly a subtropical to temperate distribution, *Limnocnida* is tropical from West-Africa to India and Myanmar.

Hydras—Hydra are probably unable to disperse across oceans (they are absent from oceanic islands) and this is reflected in their geographical distributions. The viridissima and vulgaris hydras are essentially cosmopolitan, and were probably present before the continents separated. But boreal and austral vulgaris hydra have diverged slightly from each other. The oligactis and braueri hydra are restricted to the northern continents and presumably arose after the separation of northern and southern land masses. In these two groups there has been some divergence between species of N. America and Eurasia. Species diversity is lower at low and very high latitudes and higher in mountainous regions (Fig. 3B). Most species are broadly distributed within one or several continents.

Polypodium—Polypodium is known from water basins of Russia, Romania, Iran, and North America (Raikova, 2002).

Cordylophorinae—Cordylophora is normally found in brackish water, but its unusual tolerance of salinity allows it to span ecosystems from oceans to fresh water. It has been recorded sporadically but widely in freshwater on all continents except Antarctica (Folino, 2000). *Pachycordyle kubotai* is known only from Lake Biwa in Japan (Stepanjants et al., 2000).

Salt lake medusae—Australomedusa (2 sp.) is only known from Australia. *Moerisia* (2 sp.) is known from Lake Qurun (Egypt) and the Caspian Sea (Jankowski, 2001).

Acknowledgements This work was partly funded by the Swiss Federal Office of Education and Science within the framework of the European Union Environment and Climate projects CLIME (EVK1-CT-2002-00121) and Eurolimpacs (GOCE-CT-2003-505540).

References

Bouillon, J. & F. Boero, 2000a. The hydrozoa: a new classification in the light of old knowledge. Thalassia Salentina 24: 3–45.

Bouillon, J. & F. Boero, 2000b. Synopsis of the families and genera of the Hydromedusae of the world, with a list of the worldwide species. Thalassia Salentina 24: 47–296.

Campbell, R. D., 1987. A new species of *Hydra* (Cnidaria: Hydrozoa) from North America with comments on species clusters within the genus. Zoological Journal of the Linnean Society 91: 253–263.

Collins, A. G., 2002. Phylogeny of Medusozoa and the evolution of cnidarian life cycles. Journal of Evolutionary Biology 18: 418–432.

Collins, A. G., P. Schuchert, A. C. Marques, T. Jankowski, M. Medina & B. Schierwater, 2006. Medusozoan phylogeny and character evolution clarified by new large and small subunit rDNA data and an assessment of the utility of phylogenetic mixture models. Systematic Biology 55: 97–115.

Collins, A. G., S. Winkelman, H. Hadrys & B. Schierwater, 2005. Phylogeny of Capitata and Corynidae (Cnidaria, Hydrozoa) in light of mitochondrial 16S rDNA data. Zoological Scripta 34: 91–99.

Croghan, P. C., 1983. Osmotic regulation and the evolution of brackish- and fresh-water faunas. Journal of the Geological Society London 140: 39–46.

Dumont, H. J., 1994. The distribution and ecology of the fresh- and brackish-water medusae of the world. Hydrobiologia 272: 1–12.

Folino, N. C., 2000. The freshwater expansion and classification of the colonial hydroid *Cordylophora* (Phylum Cnidaria, Class Hydrozoa). In Pederson, J. (ed.), Marine Bioinvasions: Proceedings of the First National Conference, January 24–27, 1999. Massachusetts Institute of Technology Sea Grant College Program, Cambridge, MA: 139–144.

Holstein, T., 1995. Cnidaria: Hydrozoa. In Schwoerbel, J. & P. Zwick (eds), Süßwasserfauna von Mitteleuropa, Vol. 1–2. Gustav Fischer, Stuttgart: 67–101.

Jankowski, T., 2001. The freshwater medusae of the world – a taxonomic and systematic literature study with some remarks on other inland water jellyfish. Hydrobiologia 462: 91–113.

Jankowski, T., T. Strauss & H. T. Ratte, 2005. Trophic interactions of the freshwater jellyfish *Craspedacusta sowerbii*. Journal of Plankton Research 27: 811–823.

Kim, J. H., W. Kim & C. W. Cunningham, 1999. A new perspective on lower metazoan relationships from 18S rDNA sequences. Molecular Biology and Evolution 16: 423–427.

Raikova, E. V., 2002. *Polypodium hydriforme* infection in the eggs of acipenseriform fishes. Journal of Applied Ichthyology 18: 405–415.

Schulze, P., 1917. Neue Beiträge zu einer Monographie der Gattung *Hydra*. Archiv für Biontologie 4: 29–119.

Slobodkin, L. B. & P. E. Bossert, 2001. Cnidaria. In Thorp, J. E. & A. P. Covich (eds), Ecology and Classification of North American Freshwater Invertebrates. Academic Press: 135–154.

Stepanjants, S. D., O. A. Timoshkin, B. A. Anokhin & T. A. Napara, 2000. A new species of *Pachycordyle* (Hydrozoa, Clavidae) from Lake Biwa (Japan), with remarks on this and related Clavid genera. Scientia Marina 64(Suppl. 1): 225–236.

Hydrobiologia (2008) 595:41–48
DOI 10.1007/s10750-007-9002-8

Global diversity of free living flatworms (Platyhelminthes, "Turbellaria") in freshwater

Ernest R. Schockaert · Matthew Hooge · Ronald Sluys · Steve Schilling · Seth Tyler · Tom Artois

Abstract This contribution reviews diversity of turbellarian species by biogeographical regions, with comments on species biology. The review draws on the database available at http://www.devbio.umesci.maine.edu/styler/turbellaria. Comparisons between regions suggest that species richness may be at least one order of magnitude higher than the currently reported number of species. In the context of the recent reconstructions of phylogeny of Platyhelminthes based on molecular data, the paper allows inferences as to the history of colonization of freshwaters by turbellarians. Specifically, four, or perhaps six, major invasions of freshwater habitats may have occurred in the Pangean period, each of which gave rise to a monophyletic freshwater taxon. In addition, several occasional invasions by representatives of marine taxa must have taken place.

Keywords Platyhelminthes · Freshwater · Distribution · Phylogeny · History

Guest editors: E. V. Balian, C. Lévêque, H. Segers & K. Martens
Freshwater Animal Diversity Assessment

E. R. Schockaert (✉) · T. Artois
Center for Environmental Sciences, Hasselt University, Agoralaan, Diepenbeek 3590, Belgium
e-mail: ernest.schockaert@uhasselt.be

M. Hooge · S. Schilling · S. Tyler
Department of Biological Sciences, University of Maine, 5751 Murray Hall, Orono, ME 04469-5751, USA

R. Sluys
Institute for Biodiversity and Ecosystem Dynamics & Zoological Museum, University of Amsterdam, P.O. Box 94766, Amsterdam 1090 GT, The Netherlands

Introduction

The taxon Platyhelminthes is traditionally divided into four or five "classes", one of which is the "Turbellaria", characterised by the ciliated epidermis. The other "classes" are all parasites and constitute the monophyletic taxon Neodermata, where, at some stage of their development, the original ciliated epidermis is shed and replaced by a new body lining, the neodermis. The ciliated epidermis is clearly a plesiomorphy, and the "Turbellaria" is thus a paraphyletic assemblage, sometimes referred to as "free-living Platyhelminthes". Since some of them are symbionts, we prefer to use "Turbellaria" (between quotation marks) or the vernacular name turbellarians. The turbellarian database (http://turbellaria.unimaine.edu), compiled and maintained by Tyler and co-workers (2005), lists close to 6,500 species (with a valid name), of which 1/5 have been found in freshwater. Far more turbellarian species are thus known from marine habitats and the marine taxa are more diverse as well.

Platyhelminthes are hermaphrodites, mostly simultaneously male and female, with an internal fertilisation. The reproductive system may be rather complex, especially in the Neoophora (Fig. 1H for an example) where yolk is stored in yolk cells, produced in separate vitellaria, a unique feature in animals. The organisation of the reproductive apparatus and of the digestive system—along with some other morphological characters—have traditionally been the major basis for taxonomy (Fig. 3).

Turbellarians are seldom, if ever, taken into account in biodiversity studies of freshwater habitats, even though they are mostly present in high numbers of species and of individuals. About 1/3 of the freshwater species known are the larger triclads (known as "planarians"). Due to their size (1–5 cm and more) and their "popularity", they have often received more attention than the other taxa. Representatives of the other taxa, only a few millimetres large, must preferably be studied alive for a proper identification. Once fixed, they become opaque and hard, and the internal anatomy, necessary for the identification, can barely be seen under the microscope. Moreover, they contract at fixation and appear as a little sphere that is not even recognised as an animal in a bulk sample. If living material is available, identification is relatively easy. With some training, the major taxa can be recognised and many turbellarians have hard parts in the copulatory organ that provide unambiguous species characters.

Flatworms are bottom dwellers, the triclads often under stones, or live on immersed plants. Only very few species are occasionally found in plankton. Many are heavy predators. Several Dalyelliidae and some Typhloplanidae carry symbiotic algae. The rhabdocoel freshwater flatworms produce dormant and subitaneous eggs (unknown for the other turbellarian taxa), some are viviparous. Several species of temporal waters have been described from individuals that developed in the laboratory from dormant eggs after immersion of sediment (e.g. Artois et al., 2004).

The planarians are known for their tremendous capacity to regenerate, but also other and smaller species of turbellarians are able to regenerate. This regeneration capacity is exclusively due to a reserve of undifferentiated cells, stemcells or neoblasts, which are the only cells able to divide by mitosis, a unique feature in the animal kingdom. Somatic cells do not divide, as in nematodes; they may grow and die and, contary to what happens in nematodes can be replaced by differentiating stemcells. The turbellarians have recently been "discovered" by cell biologists for stemcell research, research on the processes of differentiation and other similar topics.

Other human related issues are accidental invasions, only known for triclads. Invasions of the smaller flatworms must have occurred but are not documented for the reasons explained above. In the first half of the 20th century, *Girardia tigrina* (Girard, 1850) has been introduced in Europe from N. America, while the European *Schmidtea polychroa* (Schmidt, 1861) was introduced in N. America. *Girardia dorotocephala* (Woodworth, 1897) has also undoubtedly been imported in Hawai from the North American continent.

Species diversity and present distribution

Turbellarians can be found in almost all aquatic habitats, marine and freshwater, or in damp terrestrial locations. The Tricladida Terricola (with about 830 species) are exclusively terrestrial. Some 20–25 species of Rhabdocoela have been found in wet terrestrial habitats. They are included in the numbers in Table 1, since some have been found also in fresh water and we suspect that several of the other species may also occur in water bodies.

The number of freshwater species of the various biogeographic regions in fact reflects the scientific activities of the past. In the 19th and 20th century, up to about 1970, the European and Russian continental waters have been investigated rather intensively by e.g. von Graff, Reisinger and Steinböck in Austria, Luther in Finland, Nassonov and Beklemischev in the former USSR, and several other authors. A number of references can be found in Cannon (1986) and in Schockaert (1996). With the on-going research in the Lake Baikal, several species have more recently been added to the list for the Palearctic (see Timoshkin, 2004). Many fewer species have been recorded in North America (see Kenk, 1989; Kolasa, 2000 and the references therein), while the species from South America are mainly known through the activity of Marcus in Brasil in the 1940s and 1950s (see Marcus, 1958 and references in Sluys et al., 2005) and recently of Noreña-Janssen (e.g. Noreña et al., 2005) and Damborenea (for Temnocephalida: Damborenea & Cannon, 2001) in and around Argentina. Records from Africa are all from occasional sampling

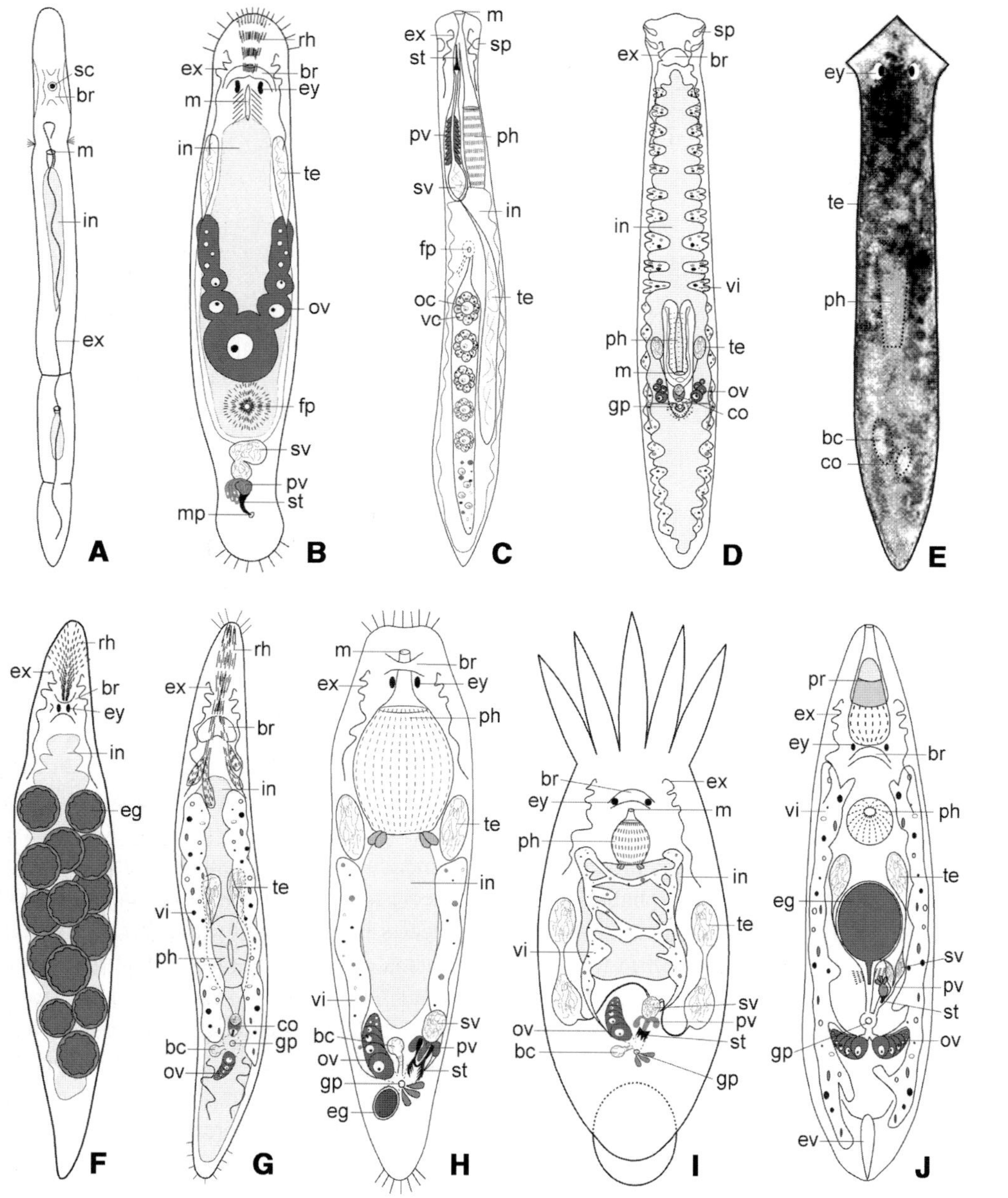

Fig. 1 Some representatives of the major freshwater taxa, as seen alive—(**A**) *Catenula lemnae*, ±1 mm (Catenulida) reproducing asexually—(**B**) *Macrostomum* spec., 1–2 mm (Macrostomida); note the absence of vitellaria ("archoophoran" organisation)—(**C**) *Prorhynchus stagnalis*, ±5 mm (Lecithoepitheliata) the vitellocytes form a follicle around the ovocytes; the male pore is combined with the mouth—(**D**) *Bothrioplana semperi*, ±5 mm (uncertain taxonomic position)—(**E**) *Dugesia* spec., 10–50 mm; position of some structures can be seen—(**F**) *Mesostoma lingua*, ±5 mm (Mesostomidae) with the uterus filled with dormant eggs—(**G**) *Olisthanella* spec., ±1 mm (Mesostomidae)—(**H**) *Microdalyellia* spec. 1–3 mm (Dalyelliidae)—(**I**) *Temnocephala* spec. ±10 mm (Temnocephalida: in the Temnocephalida the number of tentacles ranges from 2 to 10)—(**J**) *Opistocystis goettei* ±2 mm (Eukalyptorhynchia) Abbreviations: bc: bursa copulatrix, br: brain, co: copulatory organ, eg: egg (in uterus), ev: excretory vessel, ex: excretory canal (protonephridium), ey: eye, fp: female pore, gp: common male and female genital pore, in: intestine, m: mouth, mp: male pore, oc: ovocyte, ov: ovary, ph: pharynx, pr: proboscis, pv: prostate vesicle, rh: rhabdite tracks, sc: statocyst, sp: sensory pits, st: stylet, sv: seminal vesicle, te: testis, vc: vitellocyte, vi: vitellarium

Table 1 Number of species recorded in the various biogeographical regions

Taxon	PA	NA	NT	AT	OL	AU	PAc	ANT	TOT	OBS	>1
Acoela	2								2	2	0
Catenulida	36	36	45	10	1	1			90	129	30
Macrostomida	43	26	3	14	2	1			84	89	5
Polycladida					1				1	1	0
Lecithoepitheliata	20	4	4	3		3		1	31	35	2
Proseriata[a]	6	1	3	1				1	11	12	1
Prolecithophora	12	2			1	1			12	12	0
Dalyellioida	98	28	25	13	1	3			159	168	10
Typhloplanoida	233	56	13	19	4	10		1	307	336	26
Temnocephalida	18		20	1	3	56			98	98	0
Kalyptorhynchia	82	2	1	1	1	1			82	88	2
Tricladida	238	66	36	23	23	40	2	3	426	431	3
Total	788	221	150	85	36	116	2	5	1,303	1,404	79
% obs. of total obs.	56.2	15.8	10.7	6.1	2.6	8.3	0.1	0.4	–	–	5.6

TOT: number of species; OBS: total number of observations of those species; >1: number of species observed in more than one region

[a] Including *Bothrioplana semperi* Hofsten, 1907. PA: Palaearctic; NA: Nearctic; NT: Neotropical; AT: Afrotropical; OL: Oriental; AU: Australasian; PAc: Pacific & Oceanic Islands; ANT: Antarctic

campaigns (see Marcus, 1955; Young, 1976); virtually nothing is known of the Oriental region, except some records of triclads, one prolecithophoran and the only known freshwater polyclad, *Limnostylochus borneensis* (Stummer-Traunfels, 1902); of the Australian region only the Temnocephalida and Tricladida are relatively well known (see Sewell & Cannon, 1998; Sluys & Kawakatsu, 2001). In some areas almost only triclads have been studied, as in Japan by Kawakatsu and the Japanese "school" (cf. Kawakatsu, 1991).

The number of species known today in each region is listed in Table 1 and Fig.2 (following the traditional taxonomy: see below). Questionable species, i.e. species we consider insufficiently described or impossible to identify with the existing data, are not included in the counts.

Of the 1,403 records of turbellarian species, 56% were in the Palaearctic, 16% in the Nearctic and 28% in the rest of the world. All together 1,303 different species were recorded. Only 79 species were observed in more than one region, representing 5.6% of the observations and 6.1% of the species. Of those, 16 have been found in three or more regions, 10 of which are Catenulida, difficult to identify for various reasons.

The number of representatives of each genus ever found in each region is given in Table 2 and Fig.2. Species of 181 genera, or 46%, occur in the Palaearctic, 16% in N. America and 37.5% in the rest of the world. To classify the Palearctic species, taxonomists need one genus for every 5.8 species, in North America 4.7 species/genus, in the Neotropic area 4.5 species/genus, 4 in Australia, but one genus for every 3 species in Africa and even less in the other regions. This is of course due to the fact that completely new organisation types are found in those areas which have been studied the least, and the more species get known, the few genera are "needed" and "created" to contain these species. This puts a strong bias in the conclusions when numbers of genera are used as a measurement for biodiversity. Interesting considerations about the pitfalls of measuring biodiversity-using categories above the species level (taxonomic surrogacy) can be found in Bertrand et al. (2006).

Phylogeny

The first comprehensive phylogenetic approach to platyhelminth relationships, based on morphological characters (including ultrastructure) and life histories, was published by Ehlers (1985). The old turbellarian "orders" and "suborders" are now at the same "level" as the former parasitic "classes" (Fig. 3), but

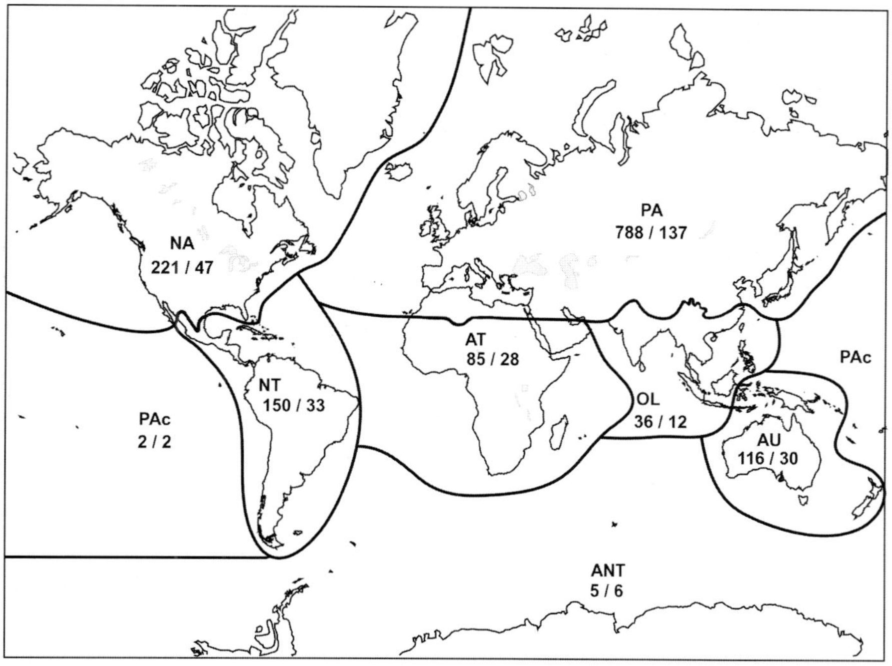

Fig. 2 Species and genus distribution of freshwater platyhelminth per zoogeographic region (species number/genus number). PA—Palaearctic, NA—Nearctic, NT—Neotropical, AT—Afrotropical, OL—Oriental, AU—Australasian, Pac—Pacific & Oceanic Island, ANT—Antarctic

Table 2 Number of genera of which representatives were recorded in the various biogeographical regions

Taxon	PA	NA	NT	AT	OL	AU	PAc	ANT	TOT	OBS
Acoela	2								2	2
Catenulida	9	6	8	5	1	1			10	30
Macrostomida	4	2	2	1	2	1			4	12
Polycladida					1				1	1
Lecithoepitheliata	3	2	3	2		1		1	3	12
Prolecithophora	5	1			1	1			8	8
Proseriata[a]	5	1	3	1				1	9	11
Dalyellioida	14	6	3	4		3			16	30
Typhloplanoida	37	15	6	10	2	2		1	42	73
Temnocephalida	5		1	1	1	10			15	18
Kalyptorhynchia	20	2	1	1	1	1			20	26
Tricladida	33	12	6	3	3	10	2	3	51	72
Total	137	47	33	28	12	30	2	6	181	295
% obs. of total obs.	46.4	15.9	11.2	9.5	4.1	10.2	0.7	2.0	–	–
# species observed	788	221	150	85	36	116	2	5	1303	1404
# species/# genera	5.8	4.7	4.5	3.0	3.0	3.9	1.0	0.8	7.2	–

TOT: number of genera; OBS: total number of observations of those genera; >1: number of genera observed in more than one region. PA: Palaearctic; NA: Nearctic; NT: Neotropical; AT: Afrotropical; OL: Oriental; AU: Australasian; PAc: Pacific & Oceanic Islands; ANT: Antarctic

[a] Including *Bothrioplana semperi* Hofsten, 1907

some important uncertainties remained. In Table 1 we have used these "classic" taxa since these are the names found in the existing literature.

New views on flatworm phylogeny are being developed, based on DNA-sequences. The Platyhelminthes may not be monophyletic and the

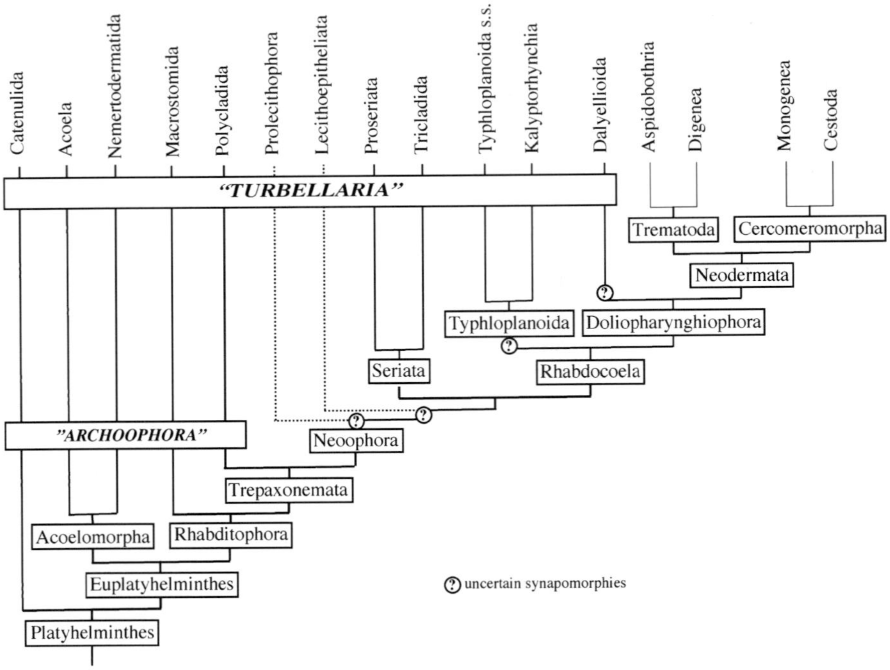

Fig. 3 Phylogenetic relationships of the major platyhelminth taxa according to Ehlers (1985)

Acoelomorpha may not belong to the same clade as the Catenulida + Rhabitophora (see Ruiz-Trillo et al., 2004 and references therein). However, since only two species of acoels have been found in continental waters, we are not considering this matter further.

The phylogenetic relationships of the Platyhelminthes, as they seem to emerge from molecular data, are represented in Fig. 4. This cladogram is a combination of the cladograms of several authors who used various methods of tree building and to assess the support of the clades. Of the marine Retronectidae and Gnosonesimida there are currently no sequences known, and the monophyly of the Catenulida and of the Lecithoepitheliata, as they are defined today on morphological grounds, is still to be confirmed. The monophyly of the freshwater representatives in both these taxa is, however, highly supported. The support for the taxon Trepaxonematida is rather weak and the sistergroup of the Neodermata is still unclear, but is certainly not the "Dalyellioidea", a taxon that even does not seem to exist! The taxon "Typhloplanoidea" does not seem to exist either, and the representatives of these two old groups are distributed over a number of new taxa. Despite remaining questions about flatworm phylogeny, some conclusions on the history of the freshwater turbellarians can be made.

The taxa in bold in the top row in Fig. 4 are all monophyletic and exclusively found in continental habitats. Five out of these six taxa have representatives on all continents, suggesting that their ancestors invaded Pangea before it broke up. There have been at least four major and independent invasions of freshwater: (1) the freshwater Catenulida (i.e. all catenulids except the Retronectidae), (2) the Prorhynchida, (3) the Continenticola (=Tricladida Paludicola + Tricladida Terricola) and (4) Dalyelliidae + Typhloplanidae + Temnocephalida (if their common ancestry is confirmed; if not, then there must have been one or two more independent invasions). The Temnocephalida, ecto-symbionts on crayfish and atyid shrimps, considered a Gondwana taxon by Cannon & Joffe (2001), may have originated later than the other two taxa, but probably not from brackish water "dalyellioids", since these marine species are members of the Neodalyellida which have no close relationship with the Temnocephalida.

The invasion history of the Macrostomida cannot be deduced from this cladogram yet. All freshwater Macrostomida are members of the Macrostomidae and Microstomidae. However, *Macrostomum* and *Microstomum* species are found almost equally abundant in marine and in freshwater habitats and

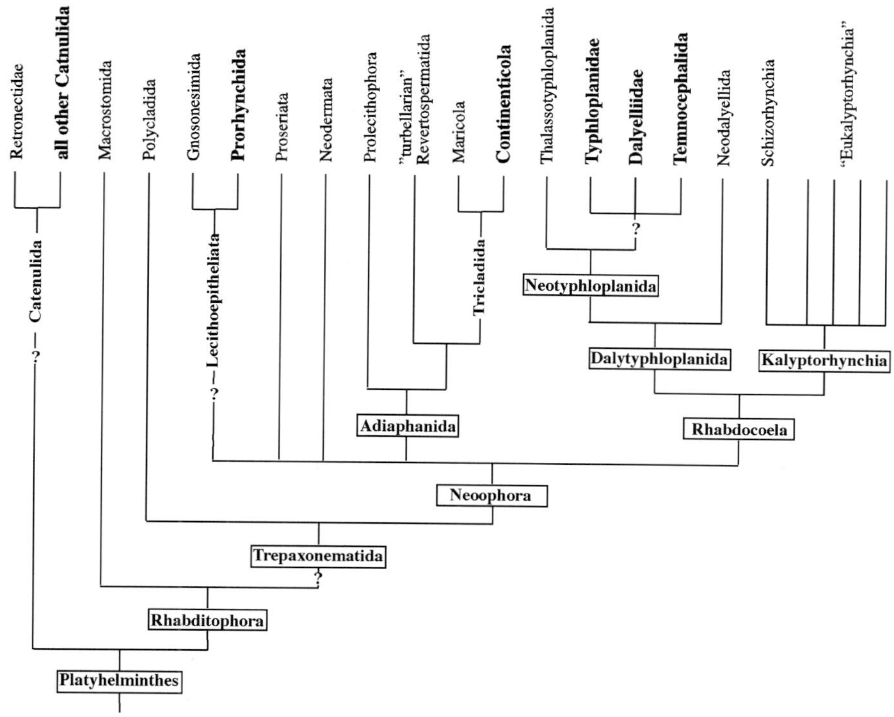

Fig. 4 Phylogenetic relationships of the Platyhelminthes as they appear from 18S rDNA. All named clades are strongly supported, except where indicated by "?". (combined from Baguna et al., 2001; Joffe & Kornakova, 2001; Norén & Jondelius, 2002; Willems et al., 2006)

on all continents. Also the biogeographic history of the freshwater "Eukalyptorhynchia" remains unresolved: representatives of the different "families" (as they are defined today) occur in freshwater.

In the other taxa with representatives in freshwater, independent invasions must have occurred: one single species of Polycladida, some Thalassotyphloplanida and Neodalyellida, and 11 species of Proseriata with different species in the different parts of the world. They all occur in the freshwater zone of rivers and canals connected to the sea.

Conclusions

The number of freshwater flatworms known in the various regions reflects the scientific activity of the past. Although relatively many scientists have been active in Europe, there are still many areas and habitats that have been sampled very poorly, such as the temporary waters around the Mediterranean Sea. Except for lake Baikal and environs, only some old and very scattered data are available for Asia. With these facts in mind, and making a very cautious estimate, the number of species in the Palaearctic must be about 5–10 times higher than known today. Is it realistic then, that only 200–100 or even fewer species occur in the other regions? Certainly not, and without exaggeration it can be said that the number of species of freshwater flatworms is at least one magnitude larger than what is known today.

In view of the above, it is evident that on the basis of such scant and unbalanced information, not many considerations can be made on the distribution of species and higher taxa. Nor can areas with high species richness or endemism be indicated, except perhaps the Baikal Lake with its high sympatric speciation processes.

More sampling in the different parts of the world and further phylogenetic analyses will certainly tell us more about the distribution and the origin in time and space of the freshwater Platyhelminthes. More (young) zoologists should therefore be trained to identify and to describe flatworms. Turbellarians have the reputation to be "difficult" to identify, but they are not, provided one knows how to study them.

References

Artois, T., W. Willems, E. De Roeck, M. Jocqué & L. Brendonck, 2004. Freshwater Rhabdocoela (Platyhelminthes) from ephemeral rock pools from Botswana, with the description of four new species and one new genus. Zoological Science 21: 1063–1072.

Baguna, J., S. Caranza, J. Papas, I. Ruiz-Trillo & M. Riutort, 2001. Molecular taxonomy and phylogeny of the Tricladida. In Littlewood, D. T. J. & R. A. Bray (eds), Interrelationships of the Platyhelminthes. Taylor & Francis, London, 49–56.

Bertrand, Y., F. Pleijel & G. W. Rouse, 2006. Taxonomic surrogacy in biodiversity assessments, and the meaning of Linnaean ranks. Systematics and Biodiversity 4(2): 149–159.

Cannon, L. R. G., 1986. Turbellaria of the World: a Guide to Families and Genera. Queensland Museum, Brisbane, 131.

Cannon, L. R. G. & B. I. Joffe, 2001. The Temnocephalida. In Littlewood, D. T. J. & R. A. Bray (eds), Interrelationships of the Platyhelminthes. Taylor & Francis, London, 83–91.

Damborenea, M. C. & L. R. G. Cannon, 2001. On neotropical Temnocephala (Platyhelminthes). Journal of Natural History 35: 1103–1118.

Ehlers, U., 1985. Das Phylogenetische System der Plathelminthes. Gustav Fischer Verlag, Stuttgart, New York, p 317.

Joffe, B. I. & E. E. Kornakova, 2001. Flatworm phylogeneticist: between molecular hammer and morphological anvil. In Littlewood, D. T. J. & R. A. Bray (eds), Interrelationships of the Platyhelminthes. Taylor & Francis, London, 279–291.

Kawakatsu, M., 1991. History of the study of Turbellaria in Japan. Hydrobiologia 227: 389–398.

Kenk, R., 1989. Revised list of the North American freshwater planarians (Platyhelminthes: Tricladida: Paludicola). Smithsonian contributions. Zoology 476, 10.

Kolasa, J., 2000. The biology and ecology of lotic microturbellarians. Freshwater Biology 44: 5–14.

Marcus, E., 1955. Turbellaria. In Hanström, B., P. Brinck & G. Rudebeck (eds), South African Animal Life—Results of the Lund University Expedition in 1950–1951, vol. I. Almqvist & Wiksell, Stockholm, 101–151.

Marcus, E., 1958. On South American Turbellaria. Anais dal Academia Brasileira de Ciencas 30: 391–417.

Norén, M. & U. Jondelius, 2002. The phylogenetic position of the Prolecitophora (Rhabditophora, 'Platyhelminthes'). Zoologica Scripta 31: 403–414.

Norena, C., C. Damborenea & F. Brusa, 2005. New freshwater interstitial Otoplanidae (Plathelminthes: Proseriata) from the Parana and Uruguay rivers, South America. Journal of Natural History 39: 1457–1468.

Ruiz-Trillo, I., M. Riutort, M. Fourcade, J. Baguña & J. L. Boore, 2004. Mitochondrial genome data support the basal position of Acoelomorpha and the polyphyly of the Platyhelminthes. Molecular Phylogenetics and Evolution 33: 321–332.

Schockaert, E. R., 1996. Turbellarians. In: Hall, G. S. (ed.), Methods for the Examiniation of Organismal Diversity in Soils and Sediments. CAB International, 211–225.

Sewell, K. B. & L. R. G. Cannon, 1998. New temnocephalans from the branchial chamber of Australian *Euastacus* and *Cherax* crayfish hosts. Proceedings of the Linnean Society of New South Wales 119: 21–36.

Sluys, R. & M. Kawakatsu, 2001. Contribution to an inventory of the freshwater planarians of Australia and New Zealand (Platyhelminthes, Tricladida, Dugesiidae), with distribution maps of the species examined. Beaufortia 51: 163–198.

Sluys, R., M. Kawakatsu & R. Ponce de León, 2005. Morphological stasis in an old and widespread group of species: contribution to the taxonomy and biogeography of the genus *Girardia* (Platyhelminthes, Tricladida, Paludicola). Studies on Neotropical Fauna and Environment 40: 155–180.

Timoshkin, O. A., (ed.), 2004. Index of Animal Species Inhabiting Lake Baikal and its Catchment Area, Vol. I, Book 2, Novosibirsk. Russian Academy of Sciences, Nauka, 1344–1491.

Tyler, S., S. Schilling, M. Hooge & L. F. Bush (comp.), 2005. Turbellarian taxonomic database. Version 1.4 http://turbellaria.unimaine.edu.

Willems, W. R., A. Walberg, U. Jondelius, D. T. J. Littlewood, T. Backeljau, E. R. Schockaert & T. J. Artois, 2006. Filling a gap in the phylogeny of flatworms: relationships within the Rhabdocoela (Platyhelminthes), inferred from 18S ribosomal DNA sequences. Zoologica Scripta 35: 1–17.

Young, J. O., 1976. The freshwater Turbellaria of the African continent. Zoologischer Anzeiger 197: 419–432.

Hydrobiologia (2008) 595:49–59
DOI 10.1007/s10750-007-9003-7

FRESHWATER ANIMAL DIVERSITY ASSESSMENT

Global diversity of rotifers (Rotifera) in freshwater

Hendrik Segers

Abstract Rotifera is a Phylum of primary freshwater Metazoa containing two major groups: the heterogonic Monogononta and the exclusively parthenogenetic Bdelloidea. Monogononta contains 1,570 species-level taxa, of which a majority (1,488) are free-living fresh or inland water taxa. Bdelloidea contains 461 "species," only one of which is marine, but with many limnoterrestrial representatives or animals of unknown ecology. Actual numbers may be much higher, considering the occurrence of cryptic speciation in Monogononta and the unsatisfactory nature of taxonomic knowledge. Rotifers, mostly monogononts, occur in all types of water bodies, worldwide. They are particularly diverse in the littoral zone of stagnant waterbodies with soft, slightly acidic water and under oligo- to mesotrophic conditions. The rotifer record is highest in the Northern hemisphere, which may be due to the concentration of studies in those regions. Diversity is highest in the (sub)tropics; hotspots are northeast North America, tropical South America, Southeast Asia, Australia, and Lake Baikal, endemicity is low in Africa (including Madagascar), Europe, the Indian subcontinent, and Antarctica. Although the lack of fossil evidence and of molecular phylogenetic studies are major hindrances, contrasting hypotheses on the origin and evolutionary history of *Brachionus*, *Macrochaetus*, and *Trichocerca* are presented.

Keywords Monogononta · Bdelloidea · Freshwater · Biodiversity · Zoogeography · Review

Introduction

Rotifera (see Wallace et al., 2006 for a recent, comprehensive introduction to the taxon) is a group of primary freshwater invertebrates. Rotifers play a pivotal role in many freshwater ecosystems. They are ubiquitous, occurring in almost all types of freshwater habitat, from large permanent lakes to small temporary puddles, and interstitial and capillary water; from acidic mining lakes to natron lakes and the open ocean, from hyperoligotropic Alpine lakes to sewage ponds. They commonly occur in densities up to 1,000 individuals per liter, and are important filter-feeders on algae and bacteria. Their ubiquity and abundance explain their standing as one of the three main groups of freshwater zooplankton in limnological studies, together with the 'Cladocera' (Anomopoda) and Copepoda, and as organisms used in mass aquaculture. They are permanently and obligatorily connected to aquatic habitats in all active stages, only their resting stages are drought-resistant.

Guest editors: E. V. Balian, C. Lévêque, H. Segers & K. Martens
Freshwater Animal Diversity Assessment

H. Segers (✉)
Belgian Biodiversity Platform, Freshwater Laboratory, Royal Belgian Institute of Natural Sciences, Vautierstraat 29, 1000 Brussels, Belgium
e-mail: Hendrik.Segers@naturalsciences.be

Classically, three groups are recognized within the Phylum Rotifera. The species-poorest is Seisonacea, with only three species living epizootically on marine crustaceans of the genus *Nebalia*. Most well-known and diverse are the predominantly freshwater Bdelloidea and Monogononta. Molecular studies have indicated that a fourth group, Acanthocephala, previously considered a separate Phylum of exclusively endoparasitic organisms, actually belongs to Rotifera (Mark Welch, 2000; Giribet et al., 2000). Little is actually known about the phylogeny of rotifers, due to a lack of modern comprehensive studies (but see Sørensen & Giribet, 2006), and the lack of a robust fossil record.

Rotifers are minute metazoans (50–2,000 μm), characterized by the presence of an anterior ciliated corona, a stiff body wall named lorica bearing variable appendages, and a specialized pharyngeal organ, the mastax, containing hard elements, termed trophi (Fig. 1). Especially, the rotifer's small size, capability of phenotic plasticity and highly adaptable masticatory apparatus are important elements explaining the success of the group. Their propagules consist of single, hard-shelled, and durable encapsulated cysts (monogononts) or anhydrobiotic individuals (bdelloids). These propagules being small and drought-resistant, makes rotifers perfectly adapted to passive, aerial or phoretic dispersal. Monogononts and bdelloids reproduce parthenogenetically. In monogononts, periods of parthenogenetic reproduction are interspersed with sexual phases (heterogony), but bdelloids are unique in being the most diverse group of metazoans in which reproduction is by diploid, mitotic parthenogenesis only. The combination of their high dispersal capacity and their parthenogenetic reproduction, enabling them to establish or renew a population starting off from a single resting stage, and to reach high effective population sizes relatively quickly, makes them theoretically superbly apt (re)colonizers.

The ability of many bdelloids to shift from active to anhydrobiotic stage enables them to live in particularly ephemeral, even predominantly dry conditions such as

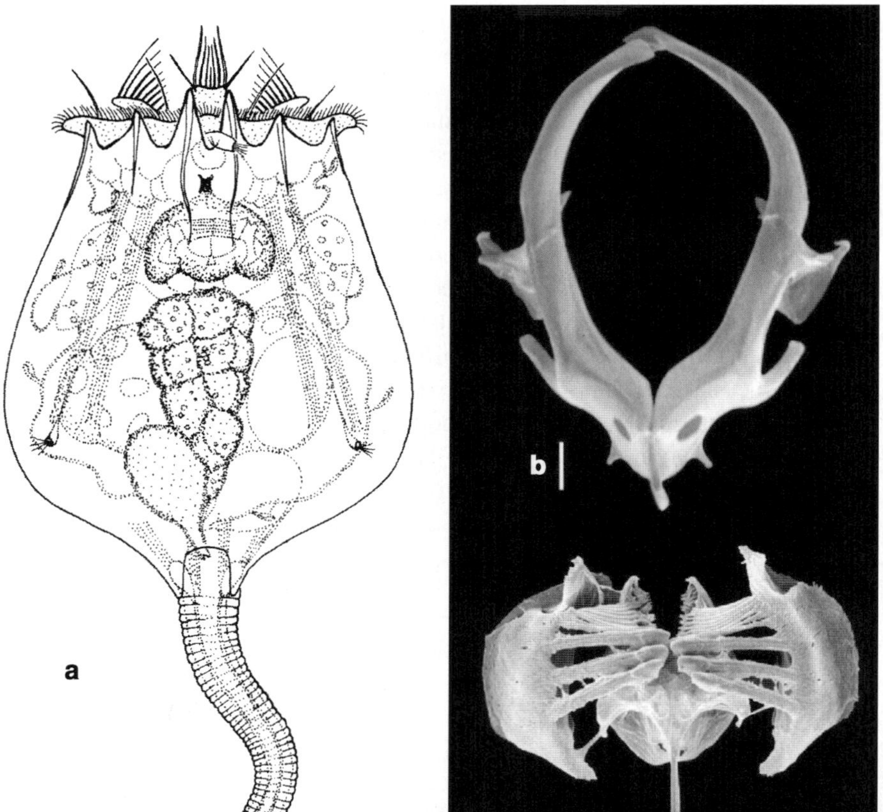

Fig 1 (**a**) Schematic representation of a *Brachionus* rotifer; (**b**) Incudate trophi (*Asplanchna*); (**c**) Malleoramate trophi (*Sinantherina*). Scale bars: 10 μm

lichens or terrestrial mosses. As such, they should probably be considered limnoterrestrial rather then limnetic. Bdelloid rotifers, however, can at present only be identified while alive and need to be examined during feeding and creeping. Their study is, consequently, tedious and very little if any information is available on the ecology of the majority of them. So, notwithstanding that the present project focuses on limnetic representatives of these animals, I include counts of the diversity and distribution of all freshwater bdelloid taxa, as it is not possible to distinguish reliably between the two ecological groups.

Biodiversity of Rotifera

Data collection

Data on which the present analysis is based are restricted to those rotifer taxa that are freshwater or brackishwater and marine. Exclusively marine species have not been included but are listed in the electronic appendix (http://fada.biodiversity.be; see Fontaneto et al., 2006 for a recent review). The taxonomy follows recent taxonomic views as expressed in recent revisions of selected rotifer families (Nogrady et al., 1995; Segers, 1995a, 2003; De Smet, 1996; De Smet & Pourriot, 1997; Nogrady & Segers, 2002), and numerous taxonomic publications. When alternative taxonomies exist, a splitting rather than lumping approach was followed. Species that are insufficiently described and therefore have to be considered *species inquirenda* are not counted. A more complete account on the taxonomic approach is provided in Segers (2007).

Distributional data are based on the literature review of De Ridder (1986, 1991, 1994), De Ridder & Segers (1997), Segers (1995b, 2003) and recent articles (e.g., Jersabek, 2003; Ricci et al., 2003). Rare regional records of species otherwise common in other regions were critically assessed and eventually included only after verifications of published illustrations or material. The data are presented in Segers (2007) and in the electronic appendix (http://fada.biodiversity.be).

Rotifer taxonomy and zoogeography: state of the art

Before analyzing rotifer diversity and distribution, it is necessary to give an account on the limitations of the data. The usual *caveat*, that new species are still to be discovered, applies, but there is more. Rotifer taxonomy is almost exemplary of the taxonomic impediment, as recognized by governments through the Convention on Biological Diversity (see http://www.biodiv.org/programmes/cross-cutting/taxonomy/default.shtml). Serious knowledge gaps exist in the taxonomic system of rotifers and trained taxonomists and curators are (very) few. These deficiencies have a significant impact on our ability to understand the diversity and chorology of these animals. Rotifer taxonomy is all but adequate, an observation that was already made some 25 years ago (Dumont, 1980) but which still holds. Basic, detailed morphological revisions still contribute significantly to our understanding (e.g., Giri & José de Paggi, 2006). Molecular studies with an impact on taxonomy are still scarce. However, the work by Gómez et al. (2002) on the economically important and particularly well-studied *B. plicatilis* O.F. Müller has shown that the taxon, which was long treated as a single but variable species, contains no less then nine different, phylogenetically distinct lineages. Only few of these are morphologically diagnosable (see Ciros-Pérez et al., 2001). Such cryptic speciation is probably common in rotifers, as hinted at by the reproductive isolation of geographically separated, yet morphologically identical strains of *Asplanchna brightwellii* Gosse (see Snell, 1989). These problems are further convoluted in bdelloid rotifers. Here, the difficulties are not only the classic ones hampering rotifer taxonomy (small size of the animals, scarcity of useful morphologic features, high variability: see Ruttner-Kolisko, 1989), but also the practical problem that, to date, only living and actively moving animals can be identified or serve as a basis for taxonomic study. In addition, the animal's unique exclusively parthenogenetic reproduction implies that most species concepts are inapplicable as theoretical framework for their study. Clearly, the counts of rotifer diversity as presented here are tentative and should be interpreted with great caution.

Due to the *caveat* mentioned above, and because identification of rotifers is difficult, rotifer literature is littered with dubious records. Our knowledge on the diversity and distribution of rotifers is moreover biased by the uneven research intensity in different regions (Dumont, 1983). There are only a few rotifer families for which a large number of fairly reliable data is available. These are loricate taxa, which can mostly be identified using external morphology of contracted, fixed material, notably Brachionidae: Pejler (1977) and Dumont (1983), Lecanidae: Segers (1996), and Trichocercidae: Segers (2003).

Genus- and species-level diversity

A total of 1,570 Monogononta and 461 Bdelloidea valid species are presently recognized worldwide (Table 1). Of these, the vast majority (1488 monogononts, 460 bdelloids) are either exclusively freshwater or brackishwater and marine; only 70 described species are exclusively marine (Table 2). The most diverse taxa are Notommatidae, with *Cephalodella* as most speciose genus, the monogeneric Lecanidae, and Dicranophoridae. All of these contain almost exclusively benthic-littoral or psammon-inhabiting species, with a majority inhabiting oligo- to mesotrophic, slightly acidic, soft waters. The same holds for Lepadellidae; *Brachionus* is a notable exception, as most of these prefer alkaline and eutrophic conditions. These preferences are well known and have been commented upon as early as Harring & Myers (1928).

Beres et al. (2005) found that the distribution of genera over families in rotifers is a hollow curve distribution which fits a model given by Hubbell's unified neutral theory of biodiversity (Hubbell, 2001). Basically, this distribution infers that there are relatively numerous taxa containing only one or a few subordinate taxa; that the relative frequency of taxa decreases sharply with increasing number of included subordinate taxa, whereas there are only a few highly diverse taxa (e.g., *Lecane:* 200 species, *Cephalodella*: 159 species). The same seems to hold for the relation between genera and species in Monogononta (Fig. 2), however, it is as yet unclear what this may signify in respect to evolution or biodiversity.

Rotifers, especially monogononts, form a relatively diverse constituent of the fauna of stagnant freshwater ecosystems. Dumont & Segers (1996) calculated that a non-polluted lake with developed weedy littoral would harbour about 150 species in temperate, and up to 250 species in tropical regions. This implies that 7.5–12.5% of all species globally, and ca. one fifth of the regional fauna can be found in a single locality. Myers' (1942) intensive studies on some lakes and ponds in and near the North-American Pocono region (Pennsylvania) yielded 457 Monogononta and 32 Bdelloidea, which constitute more than half of the known Nearctic rotifer fauna in a relatively small region. This remarkably high species diversity, which actually concerns littoral and benthic rotifers, which are mostly present in relatively low numbers, can be ascribed to fine niche partitioning amongst rotifer species in combination with high micro- and macroscale habitat heterogeneity, especially in littoral and benthic environments. On the other hand, local diversity can represent a sizable fraction of regional diversity. This is probably a result of the high (re)colonization and dispersal capacity of rotifers: available niches, even if these are only temporarily present, are relatively quickly filled by recruitment from resting stages that may or may not already be present in the habitat. This situation may be different from that in pelagic habitats, where the presence of a large resting propagule bank produced by locally adapted populations consisting of large numbers of individuals, presents an effective barrier against newly invading genotypes (the Monopolization Hypothesis: De Meester et al., 2002). Alternatively, the observation may be due to a lack of taxonomic resolution in littoral rotifers.

Present distribution and main areas of endemicity

The most diverse and, not coincidently, best-studied region is the Palaearctic, closely followed by the Nearctic region (Map 1). A substantial research effort resulting in a relatively high species record has been devoted to the Neotropical region and, more recently, the Oriental region. There are a fair number of contributions on the Australian and Afrotropical (Ethiopian) regions, but far less on Oceanic islands (see Wallace et al., (2006) for a literature review). That research intensity is largely responsible for this ranking is best illustrated by the regional diversity of taxonomically difficult illoricate taxa such as Dicranophoridae and Notommatidae: the diversity of these in the best studied Palaearctic and Nearctic regions, where most rotifer taxonomists live(d), is almost 7- to 8-fold that of the least studied African region; this is much less so for the relatively easier loricate taxa such as Brachionidae and Lecanidae.

Table 1 Number of genera per family, per region

Number of genera	Palearctic	Afrotropical	Australian	Oriental	Nearctic	Neotropical	Antarctic	Pacific	**Total**[*]
Monogononta									
Asciaporrectidae	1				1				1
Asplanchnidae	3	2	2	3	3	3		1	3
Atrochidae	3	1	2	3	2	1			3
Birgeidae					1				1
Brachionidae	7	7	6	7	7	7	3	1	7
Collothecidae	2	1	2	2	2	2	1	1	2
Conochilidae	1	1	2	1	2	1			2
Dicranophoridae	14	5	8	5	12	6	2	5	19
Epiphanidae	5	4	5	5	5	3	2	4	5
Euchlanidae	4	4	4	5	4	4	1	2	5
Flosculariidae	7	6	7	6	7	7	1	4	7
Gastropodidae	2	2	2	2	2	2			2
Hexarthridae	1	1	1	1	1	1		1	1
Ituridae	1	1	1	1	1	1			1
Lecanidae	1	1	1	1	1	1	1	1	1
Lepadellidae	3	3	3	4	4	4	2	3	4
Lindiidae	1	1	1	1	1	1	1	1	1
Microcodidae	1	1	1	1	1	1	1		1
Mytilinidae	2	2	2	2	2	2	2	1	2
Notommatidae	15	9	11	9	15	10	3	5	18
Proalidae	4	3	3	2	4	3		3	4
Scaridiidae	1	1	1	1	1	1	1	1	1
Synchaetidae	3	3	3	3	4	3		2	4
Testudinellidae	2	2	2	3	2	2		1	3
Tetrasiphonidae	1	1	1	1	1	1			1
Trichocercidae	3	2	3	3	3	2	1	2	3
Trichotriidae	3	3	3	3	3	3		1	3
Trochosphaeridae	3	3	3	3	2	3	1		3
Subtotal:	94	70	80	78	94	76	23	40	108
Bdelloidea									
Adinetidae	2	1	1	1	1	1	1	1	2
Habrotrochidae	3	3	3	1	3	2	1	2	3
Philodinavidae	3	1	2	1	2	2			3
Philodinidae	11	10	10	6	9	9	4	4	12
Subtotal:	19	15	16	9	15	15	6	7	20
Total:	113	85	96	87	109	91	29	47	128

[*] Total number of species includes exclusively marine taxa, not included are Clariaidae (1 species, *Claria segmentata* Kutikova, Markevich & Spiridonov, 1990), and 3 Seisonacea.

Antarctica is a special case; there are quite a few studies but here rotifer diversity is markedly and effectively lower then in other regions (Fig. 3).

Endemicity at higher taxonomic levels is rare in rotifers. There is a single endemic free-living rotifer family, the Nearctic (northeast North American)

Table 2 Number of species-level taxa per family, per biogegraphic region

	PA	NA	NT	AT	OL	AU	PAC	ANT	End.	Cosmo.	World	Mar.
Monogononta[a]												
Asciaporrectidae	3	2	0	0	0	0	0	0	1	0	3	
Asplanchnidae	11	11	10	9	12	9	2	0	2	8	15	
Atrochidae	4	2	1	1	3	2	0	0	0	1	4	
Birgeidae	0	1	0	0	0	0	0	0	1	0	1	
Brachionidae	94	66	71	51	57	58	4	16	94	36	169	1
Brachionus	32	23	32	26	33	34	3	5	29	20	63	
Keratella	21	22	18	15	12	15	0	5	26	7	48	
Notholca	31	13	8	3	2	2	0	6	27	2	40	1
Collothecidae	42	18	15	14	8	12	2	2	24	10	47	
Conochilidae	5	7	5	5	5	6	0	0	1	5	7	
Dicranophoridae[b]	137	93	21	19	15	24	5	6	98	9	181	39
Dicranophorus	36	38	10	12	8	8	1	0	21	7	52	1
Encentrum	64	28	3	2	4	6	1	5	54	1	78	31
Epiphanidae	16	10	10	9	9	8	4	2	4	9	16	
Euchlanidae	19	18	14	15	15	18	3	2	8	11	27	
Flosculariidae	35	38	37	22	23	30	5	2	7	19	50	
Gastropodidae	10	7	8	8	6	7	0	0	2	6	12	
Hexarthridae	11	11	7	8	4	6	3	0	7	4	18	
Ituridae	4	4	4	2	3	5	0	0	0	2	6	
Lecanidae	93	108	94	82	99	61	30	2	81	49	200	
Lepadellidae	95	67	70	54	59	55	18	11	81	37	160	3
Lepadella	66	42	52	39	42	41	11	7	70	25	122	2
Lindiidae	7	11	4	2	3	7	2	1	4	3	13	3
Microcodidae	1	1	1	1	1	1		1	0	1	1	
Mytilinidae	21	10	14	12	12	12	1	2	13	8	29	
Notommatidae	201	165	70	29	48	72	11	11	149	45	277	
Cephalodella	118	79	37	6	26	31	14	8	93	16	159	
Notommata	29	36	12	10	8	14	6	1	25	10	47	
Proalidae	34	34	7	10	7	14	5	0	20	6	47	9
Scaridiidae	3	3	4	4	4	3	1	1	3	2	7	
Synchaetidae	38	26	18	13	15	17	3	0	16	12	45	12
Testudinellidae	19	19	19	18	15	17	1	0	19	9	40	1
Tetrasiphonidae	1	1	1	1	1	1	0	0	0	1	1	
Trichocercidae	50	53	45	39	41	43	18	4	13	34	70	2
Trichotriidae	13	11	15	12	11	10	1	0	10	9	23	
Trochosphaeridae	13	8	1	13	10	13	0	0	5	9	19	
Subtotal	980	805	566	453	486	511	119	63	663	345	1488	70
Bdelloidea												
Adinetidae	17	8	6	7	5	12	1	6	7	5	20	
Habrotrochidae	130	25	37	45	18	53	7	7	75	14	152	
Habrotrocha	108	22	33	39	18	44	6	7	64	*13*	128	
Philodinavidae	3	2	2	1	2	2	0	0	3	0	6	
Philodinidae	220	77	71	85	33	109	6	15	152	41	282	1

Table 2 continued

	PA	NA	NT	AT	OL	AU	PAC	ANT	End.	Cosmo.	World	Mar.
Macrotrachela	75	19	22	31	11	41	3	7	50	*14*	95	
Mniobia	41	11	10	5	0	21		2	29	*2*	49	
Philodina	35	17	14	24	6	18	1	5	28	*10*	50	
Subtotal	370	112	116	138	58	176	14	28	237	60	460	1
Total	1,350	917	682	591	544	687	133	91	900	405	1948	71

PA: Palaearctic; NA: Nearctic; NT: Neotropical; AT: Afrotropical; OL: Oriental; AU: Australasian; PAC: Pacific Oceanic Islands; ANT: Antarctic. End. = Endemics, Cosmo. = Cosmopolites, Mar. = Marine

[a] Excluding Clariaidae, a monospecific family of exclusively parasitic animals living in terrestrial Oligochaeta

[b] Excluding *Albertia* (4 species) and *Balatro* (7species), exclusively endoparasitic in Oligochaeta (both) and gastropods (*Albertia*);

Endemics: present in one region only

Cosmopolites: present in 5 or more regions

Marine: exclusively marine species

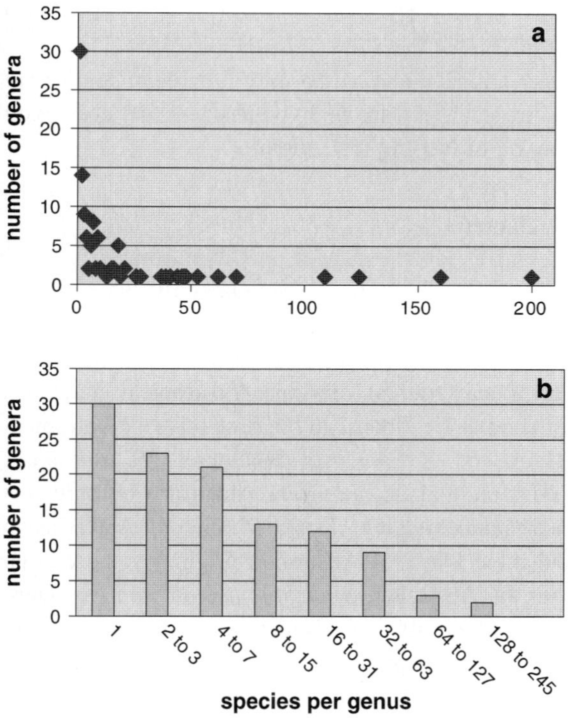

Fig 2 Distribution of rotifer species diversity over different genera. (**a**) normal representation, (**b**) number of species (x-axis) sorted out in octaves

Birgeidae. A number of endemic genera exist: In the Palaearctic these are *Pseudoharringia*, the psammobiotic *Wigrella*, the European Alpine *Glaciera* and the Baikalian *Inflatana*; in the Nearctic (northeast North American) *Rousseletia* and the littoral *Streptognatha*, and, probably, *Pseudoploesoma* (the appurtenance of *P. greeni* Koste to this genus is doubtful: De Smet & Segers, unpublished); in the Oriental region *Pseudoeuchlanis* and *Anchitestudinella;* and the Subantarctic (Kerguelen Island) *Pourriotia*. The biogeographical relevance of these is, however, low: all but *Wigrella* are monospecific, many (*Glaciera, Inflatana, Pseudoeuchlanis, Anchitestudinella* and *Pourriotia*) have only been found once. The fate of *Dorria* is revealing: this monospecific genus was long considered a rare northeast North American endemic taxon, until it was found in southern Australia and on Hawaii (Jersabek, 2003). More reliable, also taxonomically, are Birgeidae, *Streptognatha* and *Pseudoploesoma*; all three of these are northeast North American. This concurs with the main center of endemicity of Trichocercidae (Segers, 2003).

Endemic species occur in all regions and in all but the species-poorest rotifer genera and families. The count of endemics in Table 2, however, underrepresents endemicity and complexity of the distributions of rotifers: quite a few species technically occur in more than one biogeographical region as accepted for this study, yet are clearly restricted to a circumscribed area (e.g., *Keratella kostei* Paggi occurs in Patagonia, the Falkland Islands and South Georgia Island hence both in the Neotropical and Antarctic region) or have far more restricted ranges (e.g., the numerous Baikalian endemics, mostly of *Notholca*). Lecanidae is a

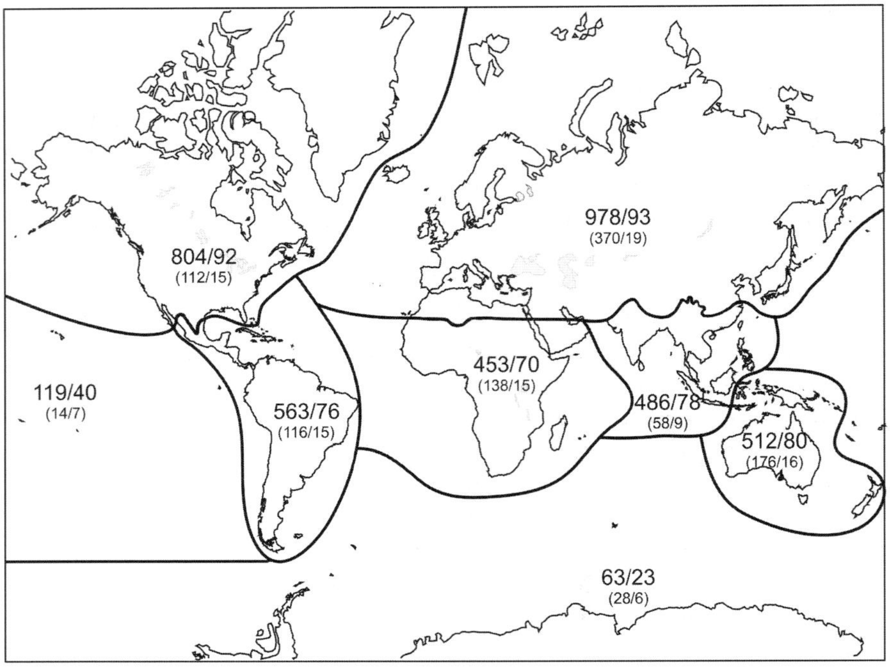

Fig 3 Rotifer diversity in the major biogeographic regions. Number of species/number of genera. Upper: Monogononta, Lower: Bdelloidea. PA—Palaearctic; NA—Nearctic; NT—Neotropical; AT—Afrotropical ; OL—Oriental; AU—Australasia; PAC—Pacific Oceanic Islands; ANT—Antarctic

good illustration of the diversity of distribution patterns (Segers, 1996). Since this 1996 paper, over 30 *Lecane* have been added as valid, either as a result of the application of a less inclusive taxonomic concept or by the description of new species. In general, ranges of *Lecane* have been refined and counts of regional endemicity increased, notwithstanding that some range extensions have been reported. Lecanidae species are predominantly (sub)tropical or warm-water, with numerous regional and local endemics, and some Holarctic, Palaeotropical, Australasian, New World, and Old World taxa illustrating more complex patterns.

Also Brachionidae contains taxa with well-documented ranges (see Pejler, 1977; Dumont, 1983). An update on the distribution of some Brachionidae is as follows:

Anuraeopsis

Of the eight species considered valid here, four are regional endemics. Whereas *A. cristata* Bērziņš, *A. miracleae* Koste and *A. urawensis* Sudzuki are rare, taxonomically difficult and may have been overlooked, the two Neotropical taxa (*A. quadriantennata* (Koste) and *A. sioli* Koste are meaningful, as they are unmistakable and have been recorded repeatedly. As all *Anuraeopsis* species are warm-water animals, and as the only reliable endemics are Neotropical, it can be hypothesized that the taxon may be of Neotropical origin.

Brachionus

This species-rich and predominantly warm-water genus contains 29 endemic (sub)species, most of which are Neotropical (9) or Australian (7). There are only three Oriental, and one Afrotropical endemics. Three taxa are American but probably of Neotropical origin (*B. havanaensis* Rousselet, *B. satanicus* Rousselet and *B. zahniseri* Ahlstrom). *Brachionus dichotomus reductus* Koste & Shiel is Australasian and most likely of Australian origin, by its relation with the Australian *B. dichotomus dichotomus* Shephard. Most of the Neotropical and Australian endemics are phylogenetically and taxonomically distinct. This is much less clear for the Palaearctic and Nearctic endemics, most of which are clear relatives of the *B. plicatilis* complex (*B. asplanchnoides*, *B. ibericus*, *B. spatiosus*). The emerging pattern is one of centered endemicity in South America and Australia, with hardly any endemicity in Africa and the Northern hemisphere. Such a pattern may hint at a late Cretaceous South American-Antarctic-Australian (see Hay et al., 1999), rather than a Gondwanan (Dumont, 1983) origin of the taxon.

Keratella

Within Brachionidae, *Keratella* is the genus with the highest degree of endemicity (52%), and this may even be an underestimate considering the confused taxonomy of a number of species complexes like *Keratella cochlearis*. Endemicity is high in the Eastern Palearctic (*K. mongoliana* Segers & Rong, *K. sinensis* Segers & Wang, *K. trapezoida* Zhuge & Huang, *K. wangi* Zhuge & Huang and *K. zhugeae* Segers & Rong) and Northern Nearctic (*K. armadura* Stemberger, *K. canadensis* Bērziņš, *K. crassa* Ahlstrom, *K. taurocephala* Myers). Here, a Southern hemisphere cold-water faunal component is represented by *K. kostei* Paggi, *K. sancta* Russell (New Zealand, Kerguelen, Macquarie Island) and *K. reducta* (Huber-Pestalozzi) (Cape region, South Africa), amongst others. Considering the relatively small area of southern hemisphere temperate regions, these taxa balance the northern hemisphere temperate *Keratella* fauna. In addition, there are some reliable Australian (e.g., *K. australis* Bērziņš), Oriental (*K. edmondsoni* Ahlstrom), and warm-water Neotropical (*K. nhamundaiensis* Koste) endemics, as well as Palaeotropical (*K. javana* Hauer) and Holarctic (*K. hiemalis* Carlin) taxa. In contrast to *Brachionus*, no clear general pattern emerges in *Keratella*.

Another remarkable genus is *Macrochaetus*. It contains 6 endemics out of 13 species, 4 of which are Neotropical. Three of these are clearly distinct and quite primitive in lacking the elongate dorsal spines typical of the genus. Hence, also *Macrochaetus* could be Neotropical in origin. The surmised origin of *Brachionus* and *Macrochaetus* contrasts with *Trichocerca*, in which a northern hemisphere pre-Pleistocene origin, followed by glacial extinctions in the (west) Palearctic, was postulated to account for an observed lack of endemics in the tropics versus high endemicity in northeast North America (Segers, 2003).

Clearly, and notwithstanding the unsatisfactory nature of our knowledge of their taxonomy, rotifers do exhibit complex and fascinating patterns of diversity and distribution as illustrated in a number of contributions (Green, 1972; Pejler, 1977; De Ridder, 1981; Dumont, 1983; Segers, 1996, 2003). In summary, many species are cosmopolitan, either or not exhibiting latitudinal variation as a result of temperature preferences. Regional differences may result from environmental conditions such as water chemistry. Endemism is real and occurs at diverse geographical scales; more complex patterns exist. Rotifer diversity is highest in the tropics, with endemicity centered in tropical South America and Australia; tropical Africa including Madagascar and the Indian subcontinent are notable for their relatively poor rotifer fauna including few endemics. Hotspots occur in northeast North America, Australia (probably west Australia) and, in contrast to the low endemicity on the Indian subcontinent, Southeast Asia. On a more local scale, Lake Baikal is most noteworthy by its high endemicity; much less is known of other ancient lakes. (Harring & Myers, 1928; Green, 1972; Pejler, 1977, Dumont, 1983; Segers, 1996, 2001, 2003). The remarkable rotifer diversity in northeast North America, in contrast to the low endemicity in European waters is attributed to the presence of glacial refugia in the region during the Pleistocene, at least for *Trichocerca* (Segers, 2003).

Fenchel & Finlay (2004) postulated that small-sized organisms (<1 mm) tend to have a cosmopolitan distribution as a consequence of huge absolute population sizes. At the local scale, their diversity exceeds that of larger organisms yet at the global scale this relation is reversed because endemism is largely responsible for the species richness of large-sized taxa. A latitudinal diversity gradient is absent or weak. Monogonont rotifers appear to comply with this pattern: their local diversity is relatively high compared to the total species diversity of the group, and cosmopolitanism is important. On the other hand, a latitudinal diversity gradient is clearly evident in rotifers (e.g., Green, 1972). Two factors may account for this apparent contradiction: first, the statement that all rotifer resting stages are eminently suited for dispersal may not be correct. Such a generalization is contradicted by the abundance of well-documented cases of locally endemic rotifers. Second, the monopolizing effect of large resting propagule banks may counteract successful colonization.

Human-related issues

Rotifer distribution and diversity is largely influenced in two ways. The most important is that of the decline of the water quality in freshwater ecosystems. As mentioned above, the most diverse rotifer assemblages can be found in soft, slightly acidic, oligo- to mesotrophic waters. These are particularly vulnerable to eutrophication and salinization. Regarding water pollution by pesticides, there are numerous laboratory studies on rotifer ecotoxicology, even using rotifers as test organisms for ecotoxicological assessments. The effects of pollutants on rotifer diversity in nature also has been studied. Rotifers are often less sensitive to insecticides than cladocerans and their sensitivity to specific compounds varies widely. They also exhibit indirect effects from exposure to toxicants, e.g., through reduction of competition from more sensitive organisms or cascading food web effects (see Wallace et al., 2006).

Due to the large dispersal and colonization capacities of many species, rotifers are easily transported to new habitats by man. An illustrative case is that of *Filinia camasecla* Myers, 1938, which was described from the Panama Canal zone; however, the species has subsequently never been found back in the Americas, but has been shown to be a relatively common Oriental species. Several additional instances are known of rotifers being introduced to regions where they did not naturally occur before (e.g., Dartnall, 2005; see Wallace et al., 2006). This phenomenon may have been going on for a long time (see Pejler, 1977) and may be responsible for isolated records of regionally common species outside their natural range. It may, however, have passed unnoticed because of the small size of rotifers and dearth of comprehensive studies. The same reasons explain why rotifers have hardly been used in biodiversity assessments and conservation, notwithstanding their economic relevance in aquaculture.

References

Beres, K. A., R. L. Wallace & H. H. Segers, 2005. Rotifers and Hubbell's unified neutral theory of Biodiversity and Biogeography. Natural Resource Modeling 18(3): 363–376.

Ciros-Pérez, J., A. Gómez & M. Serra, 2001. On the taxonomy of three sympatric sibling species of the *Brachionus plicatilis* (Rotifera) complex from Spain, with the description of *B. ibericus* n. sp. Journal of Plankton Research 23: 1311–1328.

Dartnall, H. J. G., 2005. Are Antarctic planktonic rotifers anthropogenic introductions? Quekett Journal of Microscopy 40: 137–143.

De Meester, L., A. Gómez, B. Okamura & K. Schwenk, 2002. The monopolization hypothesis and the dispersal-gene flow paradox in aquatic organisms. Acta Oecologica 23: 121–135.

De Ridder, M., 1981. Some considerations on the geographical distribution of rotifers. Hydrobiologia 85: 209–225.

De Ridder, M., 1986. Annotated checklist of non-marine Rotifera from African inland waters. Koninklijk Museum voor Miden Afrika, Tervuren, Zoologische Documentatie 21: 123 pp.

De Ridder, M., 1991. Additions to the "Annotated checklist of non-marine rotifers from African inland waters". Revue d'Hydrobiologie tropicale 24(1): 25–46.

De Ridder, M., 1994. Additions II to the "Annotated checklist of non-marine rotifers from African inland waters". Biologisch Jaarboek Dodonaea 61: 99–153.

De Ridder, M. & H. Segers, 1997. Rotifera Monogononta in six zoogeographical regions after publications between 1960 and 1992. Studiedocumenten van het Koninklijk Belgisch Instituut voor Natuurwetenschappen 88: 481 pp.

De Smet, W. H., 1996. Rotifera 4: The Proalidae (Monogononta). In Nogrady T., & H. J. Dumont (eds), Guides to the identification of the microinvertebrates of the continental waters of the World 9. SPB Academic, The Hague, The Netherlands.

De Smet, W. H. & R. Pourriot, 1997. Rotifera 5: The Dicranophoridae (Monogononta) and the Ituridae (Monogononta). In Nogrady T., & H. J. Dumont (eds), Guides to the Identification of the Microinvertebrates of the Continental Waters of the World 12. SPB Academic, The Hague, The Netherlands.

Dumont, H. J., 1980. Workshop on taxonomy and Biogeography. Hydrobiologia 73: 205–206.

Dumont, H. J., 1983. Biogeography of rotifers. Hydrobiologia 104: 19–30.

Dumont, H. & H. Segers, 1996. Estimating lacustrine zooplankton species richness and complementarity. Hydrobiologia 341: 125–132.

Fenchel, T. & B. J. Finlay, 2004. The ubiquity of small species: patterns of local and global diversity. Bioscience 54: 777–784.

Fontaneto, D., W. H. De Smet & C. Ricci, 2006. Rotifers in saltwater environments, re-evaluation of an inconspicuous taxon. Journal of the Marine Biological Association of the United Kingdom 86: 623–656.

Giri, F. & S. José de Paggi, 2006. Geometric morphometric and biometric analysis for the systematic elucidation of *Brachionus caudatus* Barrois and Daday, 1894 (Rotifera Monogononta Brachionidae) forms. Zoologischer Anzeiger 244: 171–180.

Giribet, G., D. L. Distel, M. Polz, W. Sterrer & W. C. Wheeler, 2000. Triploblastic relationships with emphasis on the acoelomates and the position of Gnathostomulida, Cycliophora, Plathelminthes, and Chaetognatha: a combined approach of 18S rRNA sequences and morphology. Systematic Biology 49: 539–562.

Gómez, A., M. Serra, G. R. Carvalho & D. H. Lunt, 2002. Speciation in ancient cryptic species complexes: evidence from the molecular phylogeny of *Brachionus plicatilis* (Rotifera). Evolution 56: 1431–1445.

Green, J. J., 1972. Latitudinal variation in associations of planktonic Rotifera. Journal of Zoology, London 167: 31–39.

Harring, H. K. & F. J. Myers, 1928. The rotifer fauna of Wisconsin. IV. The Dicranophorinae. Transactions of the Wisconsin Academy of Sciences, Arts and Letters 23: 667–808.

Hay, W. W., R. M. DeConto, C. N. Wold, K. M. Wilson, S. Voigt, M. Schulz, A. R. Wold, W.-C. Dullo, A. B. Ronov, A. N. Balukhovsky & E. Söding, 1999. Alternative global Cretaceous paleogeography. In: Barrera, E. & C. C. Johnson (eds), Evolution of the Cretaceous ocean-climate system. Geological Society of America Special Paper 332: 1–47.

Hubbell, S. P., 2001. The Unified Neutral Theory of Biodiversity and Biogeography. Monographs in Population Biology 32. Princeton University press, Princeton and Oxford.

Jersabek, C. D., 2003. Freshwater Rotifera (Monogononta) from Hawai'i—a preliminary checklist. In: Evenhuis, N. L. & L. G. Eldredge (eds), Records of the Hawaii Biological Survey for 2001–2002—Part II: Notes. Bishop Museum occasional papers 74: 46–72.

Mark Welch, D. B., 2000. Evidence from a protein-coding gene that acanthocephalans are rotifers. Invertebrate Biology 119: 17–26.

Myers, F. J., 1942. The rotatorian fauna of the Pocono Plateau and environs. Proceedings of the Academy of Natural Sciences of Philadelphia 44: 251–285, 3 plates.

Nogrady, T., R. Pourriot & H. Segers, 1995. Rotifera 3: The Notommatidae and The Scaridiidae. In Nogrady T. & H. J. Dumont (eds), Guides to the Identification of the Microinvertebrates of the Continental Waters of the World 8. SPB Academic, The Hague, The Netherlands.

Nogrady, T., & H. Segers (eds), 2002. Rotifera 6; The Asplanchnidae, Gastropodidae, Lindiidae, Microcodinidae, Synchaetidae, Trochosphaeridae. In Dumont, H. J. (ed.), Guides to the Identification of the Microinvertebrates of the Continental Waters of the World 18. Backhuys Publishers BV, Dordrecht, The Netherlands.

Pejler, B., 1977. On the global distribution of the family Brachionidae (Rotatoria). Archiv für Hydrobiologie, Beihefte 8: 212–220.

Ruttner-Kolisko, A., 1989. Problems in taxonomy of rotifers, exemplified by the *Filinia longiseta – terminalis* complex. Hydrobiologia 186/187: 291–298.

Ricci, C., R. Shiel, D. Fontaneto & G. Melone, 2003. Bdelloid rotifers recorded from Australia with description of *Philodinavus aussiensis* n.sp. Zoologischer Anzeiger 242: 241–248.

Segers, H., 1995a. Rotifera 2. The Lecanidae (Monogononta). In Nogrady T. & H.J. Dumont (eds), Guides to the Identification of the Microinvertebrates of the Continental Waters of the World 6. SPB Academic, The Hague, The Netherlands.

Segers, H., 1995b. World records of Lecanidae (Rotifera: Monogononta). Studiedocumenten van het Koninklijk Belgisch Instituut voor Natuurwetenschappen 81: 114 pp.

Segers, H., 1996. The biogeography of littoral *Lecane* Rotifera. Hydrobiologia 323: 169–197.

Segers, H., 2001. Zoogeography of the Southeast Asian Rotifera. Hydrobiologia 446/447: 233–246.

Segers, H., 2003. A biogeographical analysis of rotifers of the genus *Trichocerca* Lamarck, 1801 (Trichocercidae, Monogononta, Rotifera), with notes on taxonomy. Hydrobiologia 500: 113–114.

Segers, H., 2007. A global checklist of the rotifers (Phylum Rotifera). Zootaxa 1564: 1–104.

Snell, T. W., 1989. Systematics, reproductive isolation and species boundaries in monogonont rotifers. Hydrobiologia 186/187: 299–310.

Sørensen, M. V. & G. Giribet, 2006. A modern approach to rotiferan phylogeny: Combining morphological and molecular data. Molecular Phylogenetics and Evolution 40: 585–608.

Wallace, R. L., T. W. Snell, C. Ricci, & T. Nogrady, 2006. Rotifera vol. 1: biology, ecology and systematics (2nd edition). In Segers H., & H. J. Dumont (eds), Guides to the Identification of the Microinvertebrates of the Continental Waters of the World, 23. Kenobi Productions, Gent, Belgium and Backhuys Academic Publishing BV. The Hague, The Netherlands.

Hydrobiologia (2008) 595:61–66
DOI 10.1007/s10750-007-9004-6

FRESHWATER ANIMAL DIVERSITY ASSESSMENT

Global diversity of nemerteans (Nemertea) in freshwater

Per Sundberg · Ray Gibson

Abstract Most ribborn worms (phylum Nemertea) are marine and only 22 of the currently named around 1,200 species are known from freshwater habitats (mainly lakes/ponds). They are all free-living benthic forms found in all continents except Antarctica. The vast majority of species have been recorded from the Palearctic region, but this may reflect sampling efforts rather than biogeography.

Keywords Ribbon worms · Freshwater · Diversity · Phylogeny

Introduction

Nemertean worms are typically bilaterally symmetrical, with long, slender, soft, and contractile bodies covered by a ciliated epidermis. Their major morphological feature is an eversible muscular proboscis contained, when retracted, in a dorsal fluid-filled tubular chamber (the rhynchocoel) that extends above the gut. In anoplan nemerteans the proboscis is either unarmed or provided with rhabdites, in enoplan species the structure is armed by one (Monostilifera) or several (Polystilifera) needle-like stylets. With one known exception (the entocommensal hoplonemertean genus *Malacobdella*) nemerteans are carnivorous, either as active predators or as scavengers. Most species within the phylum are dioecious and the exceptions of hermaphroditic species belong almost all to the taxon Monostilifera. Most nemerteans are oviparous; from the species, where the mode of spawning is known it ranges from broadcast release of gametes into the sea, to pseudocopulation with eggs attached to the benthic substratum (Norenburg & Stricker, 2002). Many heteronemerteans are known to have different forms of pelagic larvae, while hoplonemerteans appear to have direct development. A few nemertean species bear living young. It should be pointed, however, that the reproductive biology for the majority of nemerteans is unknown.

Most nemertean species are from marine or estuarine habitats, but some terrestrial forms are known, and a small number of species have been recorded from freshwater environments. These freshwater species are all free-living benthic, like most of all nemerteans (exceptions are a few endoparasitic species and some pelagic species). They are found under rocks and boulders, among algae, and on mud bottoms on all depths from the littoral and down. The phylogenetic position of nemerteans among the metazoans is

Guest editors: E.V. Balian, C. Lévêque, H. Segers & K. Martens
Freshwater Animal Diversity Assessment

P. Sundberg (✉)
Department of Zoology, Göteborg University, P.O. Box 463, Goteborg 405 30, Sweden
e-mail: P.Sundberg@zool.gu.se

R. Gibson
School of Biological and Earth Sciences, Liverpool John Moores University, Byrom Street, Liverpool L3 3AF, UK

enigmatic and unsettled, although the evidence now points in the direction of an affiliation with protostome coelomates rather than having evolved from an acoelomate stock (e.g., Giribet et al., 2000).

Species diversity

Gibson (1995) listed 1,146 species of nemerteans distributed between 250 genera. Several additional taxa have been established subsequently and the current number of named species is estimated at about 1,200–1,250. However, this is certainly an underestimate of the actual number and new genetic evidence (e.g., Strand & Sundberg, 2005a) furthermore shows that sibling and cryptic species are more common than previously recognized. There is no universal agreement on the actual number of genera and species, but current figures indicate that the Anopla accounts for approximately 38% of the known genera and 44% of the named species, and Enopla for 62% and 56%, respectively. The number of known freshwater nemerteans is small; the 22 reported species (Table 1) represent less than 2% of the total number recorded.

The classification of nemertean species into higher taxa is not based within a phylogenetic framework and many groups are clearly nonmonophyletic. Family placement must, therefore, be considered as provisional and viewed with care. Heteronemerteans account for 23% of the known freshwater forms and comprise five monospecific genera, all placed in the family Lineidae. Conversely, the hoplonemertean species are distributed between six genera and three families; *Campbellonemertes* and *Potamonemertes* are placed in the family Plectonemertidae, a taxon that also contains several terrestrial genera, whereas *Koinoporus*, *Limnemertes* and *Prostoma* are united in the family Tetrastemmidae (Moore & Gibson, 1988). Dawydoff (1937) linked *Otonemertes* (a genus which also contains one marine species) to the exclusively interstitial marine and brackish-water genus *Ototyphlonemertes* (Ototyphlonemertidae) but Moore & Gibson's (1985) discussion suggests that the familial placement of *Otonemertes* is particularly uncertain (Fig. 1).

Phylogeny and historical processes

Phylogenetic analyses (e.g., Sundberg et al., 2001; Thollesson & Norenburg, 2003) indicate that nemerteans are ancestrally and primarily a group of marine

Table 1 The freshwater nemerteans of the world (with authors)

Phylum Nemertea
Class Anopla, Subclass Heteronemertea
Amniclineus Gibson & Qi 1991
Apatronemertes Wilfert & Gibson 1974
Apatronemertes albimaculosa Wilfert & Gibson 1974
Planolineus Beauchamp 1928
Planolineus exsul Beauchamp 1928
Siolineus Du Bois-Reymond Marcus 1948
Siolineus turbidus Du Bois-Reymond Marcus 1948 *Yinia* Sun & Lu, 1998
Yinia pratensis Sun & Lu 1998
Class Enopla, Subclass Hoplonemertea, Superorder Monostilifera
Campbellonemertes Moore & Gibson 1972
Campbellonemertes johnsi Moore & Gibson 1972
Koinoporus Sánchez & Moretto 1988
Koinoporus mapochi Sánchez & Moretto 1988
Limnemertes Gibson & Wang 2002
Limnemertes poyangensis Gibson & Wang 2002
Otonemertes Dawydoff 1937
Otonemertes denisi Dawydoff 1937
Potamonemertes Moore & Gibson 1973
Potamonemertes gibsoni Hickman & Moore 1990 Australia
Potamonemertes percivali Moore & Gibson 1973
Prostoma Dugès 1828
Prostoma asensoriatum (Montgomery 1896)
Prostoma canadiensis Gibson & Moore 1978
Prostoma communopore Senz 1996
Prostoma eilhardi (Montgomery 1894)
Prostoma eilhardi eilhardi (Montgomery 1894)
Prostoma eilhardi macradenum Sun and Yin 1995 (Chernyshev et al. 1998, elevate this subspecies to specific rank as *Prostoma macradenum* Sun & Yin 1995)
Prostoma graecense (Böhmig 1892)
Prostoma hercegovinense Tarman 1961
Prostoma jenningsi Gibson & Young 1971
Prostoma kolasai Gibson & Moore 1976
Prostoma ohmiense Chernyshev Timoshkin and Kawakatsu 1998
Prostoma puteale Beauchamp 1932

Several additional taxa reported from freshwater habitats, not included in the list below, are either not nemerteans or are too poorly described to be accepted as valid (Moore & Gibson, 1985). See Gibson (1995) for full bibliographic references to taxa

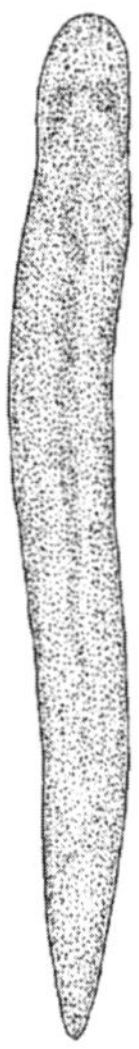

Fig. 1 Habitus of freshwater nemertean *Prostoma graecense* (drawing by R. Gibson)

organisms. Moore & Gibson (1985) and Gibson & Moore (1989) have argued on morphological and physiological grounds that the invasion of freshwater habitats has happened along two distinct routes, either directly via estuarine ancestors, or secondarily via terrestrial/semi-terrestrial relatives. However, when it comes to the hoplonemertean freshwater species, later cladistic analysis did not support this view. Sundberg (1989) instead suggested that the two freshwater genera *Campbellonemertes* and *Potamonemertes* had a common marine ancestor. When it comes to the freshwater genus *Prostoma*, where Moore & Gibson (1985) suggested that they were derived from estuarine/brackish-water species, the phylogenetic analysis based on 18S sequences in Strand & Sundberg (2005b) is inconclusive (Fig. 2). The brackish water species *Cyanophthalma obscura* forms a sister species to the included *Prostoma* species in this analysis, and this clade is in turn a sister to marine species. Thus, the analysis cannot distinguish between whether the most recent common ancestor of *Cyanophthalma* and *Prostoma* was marine or brackish-water. When it comes to the heteronemertean freshwater species there are no phylogenetic analyses testing the hypothesis of Moore & Gibson (1985) that they have occupied the freshwater habitat via an estuarine/brackish-water ancestor.

The information of the distribution of freshwater nemerteans is far too scattered to draw any conclusions, when it comes to historical processes. However, it appears that some records are cases of introduced species (see also below). For example, Strand & Sundberg (2005b) estimated the genetic difference between specimens of *Prostoma graecense* from New Zealand and Sweden to be less than 3% based on mtDNA COI sequences, and around 0.15% based on 18S rRNA sequences. Compared to genetic differentiation within other conspecific nemerteans, and based on analysis of most probable ancestral area, it indicates that this species has been introduced in New Zealand in recent time.

Present distribution and main areas of endemicity (Table 2, Fig. 3)

By far the majority of freshwater nemertean species are known from only single or, at most, very few locations. Two taxa, *Apatronemertes albimaculosa* from freshwater aquarium tanks at the Düsseldorf City Aquarium, Germany, and *Planolineus exsul* from garden ponds at Buitenzorg, Java, were originally described from what were almost certainly artificially introduced specimens and have never been found elsewhere. The remaining heteronemertean species have only been found in single areas, and may be regarded as probably endemic to those regions. *Siolineus turbidus* is known from only four individuals collected in the River Tapajós, a tributary of the Amazonas. The other two *Amniclineus zhujiangensis* (Gibson & Qi, 1991), and *Yinia pratensis* have only been found at their respective locations in the People's Republic of China.

Three of the hoplonemertean species are known only from single locations and are almost certainly endemic to the Australasian region. These species are *Campbellonemertes johnsi* from Campbell Island, *Potamonemertes gibsoni* from Tasmania, and *Potamonemertes percivali* from South Island, New Zealand. *Koinoporus mapochi* has been found at several locations in the central zone of Chile but nowhere else and is probably endemic to this Neotropical region. Two of the other hoplonemertean species are possibly endemic to their regions. These are *Limnemertes poyangensis*, known only from People's Republic of China (Palearctic region), and *Otonemertes denisi* from Tônlé Sab (the Great Lake), Kampuchea (Oriental region).

The remaining, and most diverse, genus of freshwater hoplonemerteans, *Prostoma*, currently contains 11 species; see Gibson & Moore (1976) for a

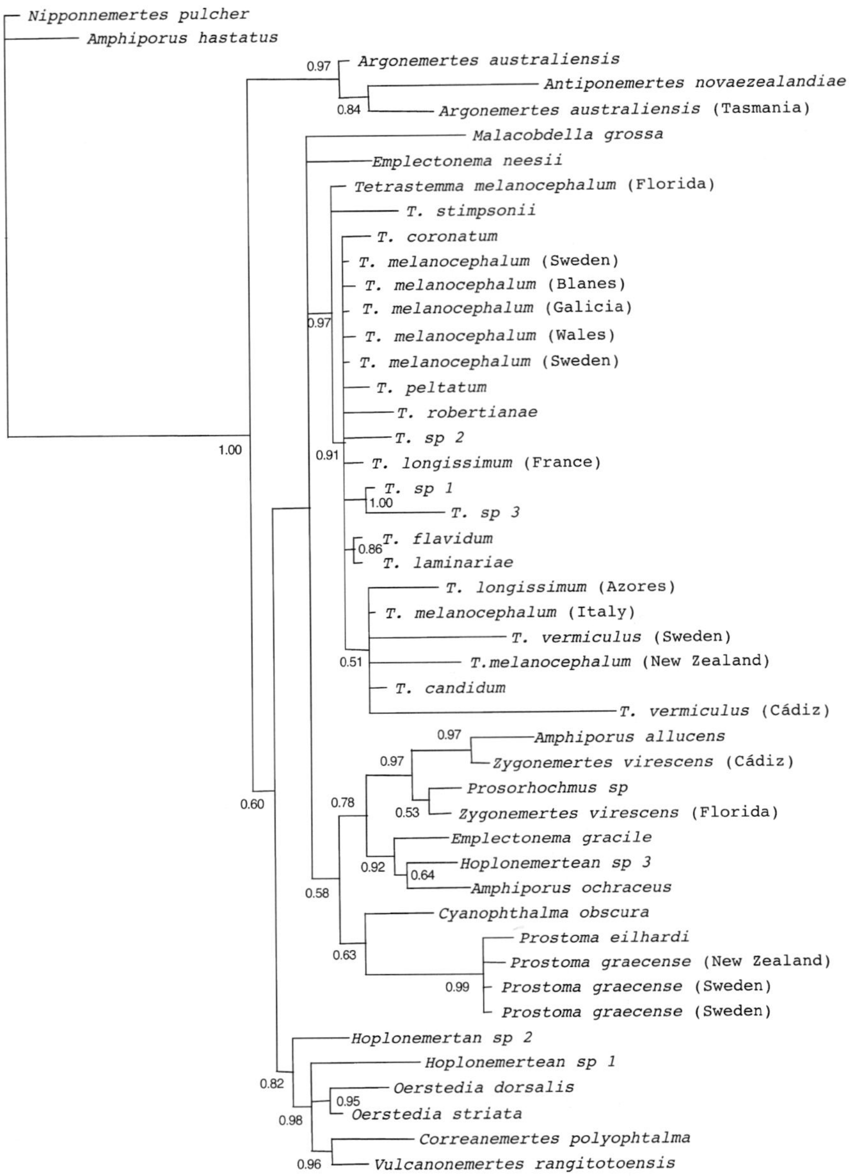

Fig. 2 The phylogeny for a selected number of hoplonemertean taxa to show the position of the genera *Prostoma* (freshwater) and *Cyanophthalma* (brackish water) and sistergroup relationships. The majority rule consensus tree from a Bayesian analysis based on 18S gene sequences is from Strand & Sundberg (2005b). (*T.* stands for *Tetrastemma*)

discussion about the validity of several other supposed species, previously included in this genus. Within the genus *Prostoma* two species, *Prostoma eilhardi* and *Prostoma graecense*, have been reported with a worldwide distribution although the authenticity of the specific identifications cannot always be confirmed from the literature. *Prostoma eilhardi* has been recorded from Europe, Kenya, Rhodesia, southern Africa, South America, St. Vincent, Australia, and New Zealand, whereas *Prostoma graecense* has been found in Europe, Kenya, southern Africa, Australia, New Zealand, Japan, Russia, and South America possibly. The remaining *Prostoma* exhibit very much more restricted distributions. There are two Nearctic species *Prostoma asensoriatum*, from Pennsylvania, USA, and *Prostoma canadiensis*; the latter species was originally found in Lake Huron, Canada, but subsequently recorded from Holland (Moore & Gibson, 1985). Palearctic taxa are *Prostoma communopore* known only from Austria and *Prostoma hercegovinense* found in caves in Bosnia, *Prostoma jenningsi* known from UK, *Prostoma kolasai*, reported from Poland, *Prostoma macradenum* from People's Republic of China, *Prostoma ohmiense* from Japan, and

Table 2 The species of freshwater nemerteans recorded from each of the zoogeographic areas of the world (A: species names listed in ***bold italics*** may be endemic taxa), together with a tabulation of the zoogeographic distribution (B: species number (genus number); PA—Palearctic; NA—Nearctic; NT—Neotropical; AT—Afrotropical ; OL—Oriental; AU—Australasian; PAC—Pacific Oceanic Islands, ANT: Antarctic)

Zoogeographic area	Species recorded
A	
Australasian	***Campbellonemertes johnsi***
	Potamonemertes gibsoni
	Potamonemertes percivali
	Prostoma eilhardi
	Prostoma graecense
Afrotropical	*Prostoma eilhardi*
	Prostoma graecense
Nearctic	*Prostoma asensoriatum*
	Prostoma canadiensis
Neotropical	***Koinoporus mapochi***
	Prostoma eilhardi
	Prostoma graecense
	Siolineus turbidus
Oriental	***Otonemertes denisi***
	Planolineus exsul
Palearctic	***Amniclineus zhujiangensis***
	Apatronemertes albimaculosa
	Limnemertes poyangensis
	Prostoma canadiensis
	Prostoma communopore
	Prostoma eilhardi
	Prostoma graecense
	Prostoma hercegovinense
	Prostoma jenningsi
	Prostoma kolasai
	Prostoma macradenum
	Prostoma ohmiense
	Prostoma puteale
	Yinia pratensis

B

	PA	NA	AT	NT	OL	AU	PAC	ANT	World
Nemertea	14 (5)	2 (1)	2 (1)	4 (3)	2 (2)	5 (3)	0 (0)	0 (0)	22 (12)

Prostoma puteale, found in France and Switzerland. Which, if any, of these species can genuinely be recorded, as endemic is uncertain.

Most freshwater species are known from the Palearctic region (Table 2, Fig. 3); it is, however, important to point out that this probably is a reflection of sampling efforts rather than true geographic distribution of freshwater species. Historically, there is a clear predominance of taxonomists that are interested in nemerteans from this region, which may introduce a bias in the species distribution.

Human related issues

Presently, we know that no freshwater nemertean species that has any economic or medical relevance. One species (*Prostoma jenningsi*) has been listed in

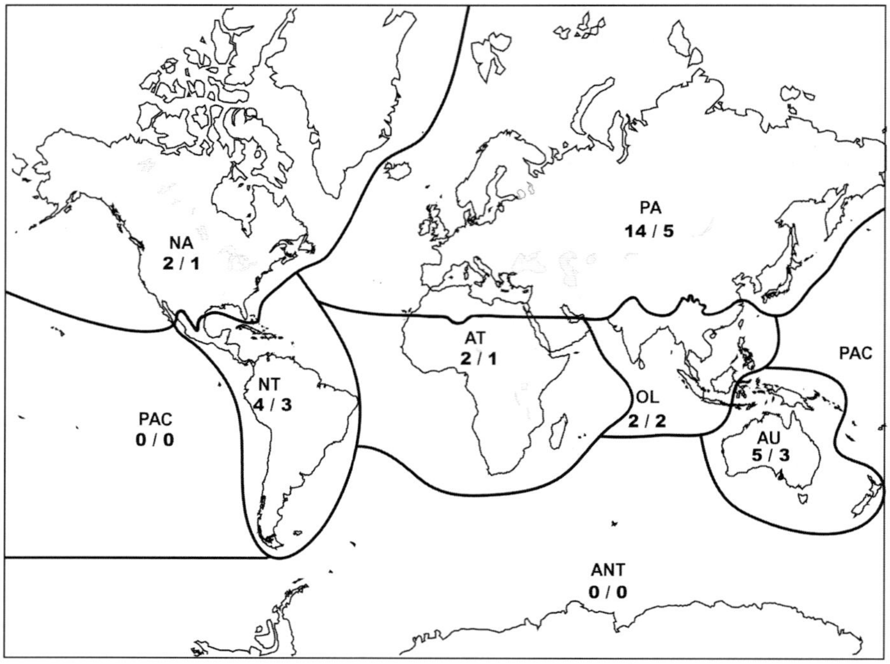

Fig. 3 Map showing designated zoogeographic regions and the number of freshwater nemertean species and genera in each region. PA—Palearctic, NA—Nearctic, NT—Neotropical, AT—Afrotropical, OL—Oriental, AU—Australasian, PAC—Pacific Oceanic Island, ANT—Antarctic

the Red Data Book as a taxon under threat by habitat usage. Other species may also be at risk, but any conclusions can only be conjecture since nemerteans as a group have not been as extensively investigated as many other phyla.

References

Dawydoff, C., 1937. Une Métanémerte nouvelle, appartenant á un groupe purement marin, provenant du Grand Lac du Cambodge. Compte rendu hebdomadaire de Séances de l'Academie des Sciences, Paris 204: 804–806.

Gibson, R., 1995. Nemertean genera and species of the world: An annotated checklist of original names and description citations, synonyms, current taxonomic status, habitats and recorded zoogeographic distribution. Journal of Natural History 29: 271–562.

Gibson, R. & J. Moore, 1976. Freshwater nemerteans. Zoological Journal of the Linnean Society 58: 177–218.

Gibson, R. & J. Moore, 1978. Freshwater nemerteans: New records of *Prostoma* and a description of *Prostoma canadiensis* sp. nov. Zoologischer Anzeiger 201: 77–85.

Gibson, R. & J. Moore, 1989. Functional requirements for the invasion of land and freshwater habitats by nemertean worms. Journal of Zoology, London 219: 517–521.

Giribet, G. G., D. L. Distel, M. Polz, W. Sterrer & W. C. Wheeler, 2000. Triploblastic relationships with emphasis on the acoelomates and the position of Gnathostomulida, Cycliophora, Plathelminthes and Chaetognatha: A combined approach of 18S rDNA sequences and morphology. Systematic Biology 49: 539–562.

Moore, J. & R. Gibson, 1985. The evolution and comparative physiology of terrestrial and freshwater nemerteans. Biological Reviews 60: 257–312.

Moore, J. & R. Gibson, 1988. Further studies on the evolution of land and freshwater nemerteans: Generic relationships among the paramonostiliferous taxa. Journal of Zoology, London 216: 1–20.

Norenburg, J. L. & S. A. Stricker, 2002. Phylum Nemertea. In Young, C., M. Sewell & M. Rice (eds), Atlas of Marine Invertebrate Larvae. Academic Press, 163–177.

Strand, M. & P. Sundberg, 2005a. Delimiting species in the hoplonemertean genus *Tetrastemma* (phylum Nemertea): Morphology is not concordant with phylogeny as evidenced from mtDNA sequences. Biological Journal of the Linnean Society 86: 201–212.

Strand, M. & P. Sundberg, 2005b. Genus *Tetrastemma* Ehrenberg, 1831 (phylum Nemertea)—a natural group? Phylogenetic relationships inferred from partial 18S rRNA sequences. Molecular Phylogenetics and Evolution 37: 144–152.

Sundberg, P., 1989. Phylogeny and cladistic classification of the paramonostiliferous family Plectonemertidae (phylum Nemertea). Cladistics 5: 87–100.

Sundberg, P., J. M. Turbeville & S. Lindh, 2001. Phylogenetic relationships among higher nemertean (Nemertea) taxa inferred from 18S rRNA sequences. Molecular Phylogenetics and Evolution 20: 327–334.

Thollesson, M. & J. L. Norenburg, 2003. Ribbon worm relationships: A phylogeny of the phylum Nemertea. Proceedings of the Royal Society, London. Series B. 270: 407–415.

Hydrobiologia (2008) 595:67–78
DOI 10.1007/s10750-007-9005-5

Global diversity of nematodes (Nematoda) in freshwater

Eyualem Abebe · Wilfrida Decraemer · Paul De Ley

Abstract Despite free-living nematodes being present in all types of limnetic habitats including unfavorable conditions that exclude many other meiobenthic invertebrates, they received less attention than marine and terrestrial forms. Two-fifths of the nematode families, one-fifth of the nearly 1800 genera and only 7% of the about 27,000 nominal species are recorded from freshwater habitats. The Dorylaimia are the most successful in freshwater habitats with nearly two-thirds of all known freshwater nematodes belonging to this subclass. Members of the subclass Enoplia are principally marine though include some exclusively freshwater taxa with extreme endemism. The subclass Chromadoria includes half of the freshwater nematode families and members of the Monhysterida and Plectida are among the most widely reported freshwater nematodes. Studies on freshwater nematodes show extreme regional bias; those from the southern hemisphere are extremely underrepresented, especially compared to European freshwater bodies. The majority of records are from a single biogeographic region. Discussion on nematode endemism is largely premature since apart from Lake Baikal, the nematofauna of ancient lakes as centers of speciation is limited and recent discoveries show high nematode abundance and diversity in cryptic freshwater bodies, underground calcrete formations and stromatolite pools potentially with a high number of new taxa.

Keywords Free-living nematodes · Freshwater nematodes · Nematode biogeography · Distribution · Biodiversity · Global estimate

Guest editors: E.V. Balian, C. Lèvêque, H. Segers & K. Martens
Freshwater Animal Diversity Assessment

Eyualem Abebe (✉)
School of Mathematics, Science and Technology, Department of Biology, Elizabeth City State University, Elizabeth City, NC 27909, USA
e-mail: Ebabebe@mail.ecsu.edu

W. Decraemer
Department of Invertebrates, Royal Belgian Institute of Natural Sciences, Rue Vautier 22, Brussels 1000, Belgium
e-mail: wilfrida.decraemer@naturalsciences.be

P. De Ley
Department of Nematology, University of California, 900 University Avenue Riverside, Riverside, CA 92521, USA
e-mail: paul.deley@ucr.edu

Introduction

Nematodes are the most abundant and arguably the most diverse Metazoa in aquatic sediments. Free-living nematodes are ubiquitous and may be present in all types of limnetic habitats including unfavorable conditions (high temp., acidic, anoxic) that exclude many other meiobenthic invertebrates. Nematode parasites of vertebrates living or frequenting freshwater habitats usually occur only as eggs or within an intermediate host; these are not included here.

However, the insect parasitic mermithids with eggs and different developmental stages (either infective or postparasitic juveniles) and adults in freshwater habitats, are.

Nematodes are generally ranked as a phylum—Nematoda or Nemata. They are unsegmented pseudocoelomates that are typically thread-like. Free-living specimens are, except for representatives of the Mermithidae and Leptosomatidae, under 1 cm in length and usually quite small (0.2–2 mm long). Despite their great diversity in morphology and lifestyle (free-living, parasites of animals and plants), nematodes display a relatively conserved basic body plan that consists of an external cylinder (the body wall with cuticle, epidermis, somatic musculature) and an internal cylinder (the digestive system) separated by a pseudocoelomic cavity that functions as a hydrostatic skeleton. Externally, the body shows little differentiation into sections. The ventral side bears a secretory–excretory pore, the vulva and anus (female) or cloacal opening (male). The lateral sides carry the apertures of the sensory-secretory amphids and may have additional secretory and/or sensory structures. The outer body surface or cuticle may be smooth or ornamented (with transverse striae, punctuations,...) and together with the epidermis functions as a semi-permeable barrier to harmful elements while allowing secretion, excretion, and uptake of various substances. The mouth opening is usually located terminally at the anterior end and surrounded by six lips (basic form) bearing various sense organs which may be papilliform, poriform, or setiform, and which may include the paired amphid openings (mainly in plant-parasitic forms).

Nematodes are in general translucent with much of their internal anatomy observable by light microscopy. Many aquatic species are gland-bearers; they usually possess three epidermal glands in the tail region (caudal glands) mostly ending in a common outlet or spinneret. Secretions of these glands play a role in locomotion and anchoring by allowing temporary attachment of the body to substrates.

All freshwater nematodes, except the adult Mermithidae, possess a continuous digestive tract. The wide diversity of food sources and methods of ingestion is reflected in the structure of the digestive system, especially in the morphology of the buccal cavity and pharynx. Current proposals for dividing nematodes by feeding habit recognize seven types: plant feeders, hyphal feeders, substrate ingesters, bacterial feeders, carnivores, unicellular eukaryote feeders, and animal parasites (Moens et al., 2004). All of these can be found in freshwater habitats; some nematodes may fit in multiple feeding types. The intestine in most freshwater nematodes is a cylindrical tube. In adult Mermithids, however, the intestine is modified into a storage organ or trophosome, separated from the pharynx and rectum.

The central nervous system consists of a nerve ring that usually encircles the pharynx and which connects various ganglia via anteriorly and posteriorly running longitudinal nerves. As noted above, sensory structures (papillae or setae) are concentrated on the anterior end; they function either as mechanoreceptors, chemoreceptors, or a combination of both. In free-living aquatic nematodes, the body may also bear few or numerous somatic sensilla (poriform or setiform). A few Freshwater taxa possess photoreceptor organs such as pigmented areas or ocelli in the pharyngeal region. The secretory–excretory system in most free-living Freshwater taxa consists of a ventral gland or renette cell connected by a duct to the ventral secretory–excretory pore. This system may play a role in excretion of nitrogen in the form of ammonia or urea as well as contributing to osmoregulation and locomotion (Turpeenniemi & Hyvärinene, 1996).

Nematodes are typically amphimictic and have separate males and females. Many species, however, lack males and reproduce either by parthenogenesis or by hermaphroditism (rare among freshwater nematodes, e.g., *Chronogaster troglodytes*). The reproductive system is quite similar in both sexes and generally comprises one or two tubular genital branches. In the female the basic system has two opposed uteri connected to the vagina that opens to the outside via the mid-ventral vulva. Each genital branch consists of a gonad (ovary) and a gonoduct (oviduct and uterus); a spermatheca may be present. Some aquatic species exhibit traumatic insemination whereby the male penetrates the female cuticle with his spicules and releases sperm into the body cavity. A derived system with a single uterus is called monodelphic. The male reproductive system is typically diorchic (with two testes that open into a common *vas deferens*). A part of each *testis* or the anterior part of the *vas deferens* may act as *vesiculum seminalis*. A monorchic condition occurs when only one testis is present after reduction of the posterior (usually) or anterior testis. The copulatory apparatus consists of two sclerotized spicula, rarely fused or reduced to a single spiculum, and a gubernaculum or guiding piece.

Nematode development typically passes through an egg stage and four (occasionally three) juvenile stages with a moult at the end of each stage. During each moult the cuticle is shed and replaced by a new one secreted by the epidermis. In free-living aquatic nematodes, single-celled egg laying appears to be the rule, while mermithids may deposit eggs with fully developed juveniles. The juvenile that hatches from the egg is usually the first stage or J1, although a few groups pass through the first moult before hatching. The generation time of nematodes can, depending on the species concerned, vary from a few days to a year or more. Females are usually oviparous, but in some groups the eggs can hatch inside the body of the female (ovoviviparity). Very little is known about resistant stages, dispersal, and survival of freshwater nematodes.

Species diversity

Estimates of global nematode species diversity have varied widely in the past 15 years, i.e., between one hundred thousand (Andrássy, 1992) and one hundred million (Lambshead, 1993). The current conservative estimate seems to stabilize at about one million species (Lambshead, 2004), a magnitude comparable to estimates for other diverse animal phyla. More than 97% of these potential one million nematode species are currently unknown; the total number currently known to science is close to 27,000 and a large proportion of these are free-living (Hugot, et al., 2001). Some of the reasons for this limited attention include the small size of nematodes and small number of taxonomists unevenly distributed throughout the world. In light of the critical importance of freshwater bodies to humans and the 'International Year of Freshwater' in 2003, it is disheartening to see that nematodes from freshwater habitats have received even less attention than marine or terrestrial forms.

Another factor contributing to the low total number of globally known freshwater nematode species is the relative inaccessibility of taxonomic literature and the possible misidentification of many populations, usually resulting in the creation of "species complexes" with an amalgam of identifying characters (Jacobs, 1984). Two examples are: (1) African populations of *Brevitobrilus* that were considered to belong to *B. graciloides*, later found to comprise more than one species (Tsalolikhin, 1992), and (2) *Monhystera stagnalis*, a species long considered to be ubiquitous with a wide range of morphological characters, might well represent many species (Coomans, pers. comm.). Species complexes mask the true biogeographical and environmental range of individual species within complexes, and discussions on the diversity and biogeography of freshwater nematodes need to be seen within the context of this limitation.

The most recent systematic scheme divides the phylum Nematoda into two classes, three subclasses, 19 orders and 221 families (De Ley and Blaxter, 2004). Andrássy (1999), following a slightly different systematic scheme, provides us with the most recent census of genera of free-living nematodes. He listed a total of 570, 650, and 705 free-living (non-animal parasitic) genera for groups corresponding to De Ley & Blaxter's order Rhabditida, class Chromadorea minus Rhabditida, and Enoplea, respectively.

At family level, both classes Chromadorea and Enoplea, all three sub-classes, two-thirds of the 19 orders, two-fifths of the 221 families, and one-fifth of the nearly 1800 free-living genera have freshwater representatives (Fig. 2). At species level, about 7% of the estimated 27,000 nominal species are considered to be denizens of freshwater habitats (Table 1).

Among the Nematoda, the Dorylaimia are the most successful in freshwater habitats with nearly two-thirds of all known freshwater nematodes belonging to this subclass and 22 of its 26 families having freshwater representatives. Not only are two of its orders, i.e., Dorylaimida and Mononchida, the most common nematodes in freshwater environments with global distribution, but also the zooparasitic Mermithida comprise many species that spend part of their life cycle in freshwater habitats (Fig. 2). Furthermore, Dorylaimia are taxonomically and ecologically diverse, which may suggest an even much larger historical diversity (De Ley et al., 2006).

Dorylaimida are especially species-rich with currently 250 known valid genera and about 2000 species (Peña-Santiago, 2006), of which 80% of the families, more than 40% of the genera and 30% of the species are freshwater and dominate these environments in species diversity except for Antarctica

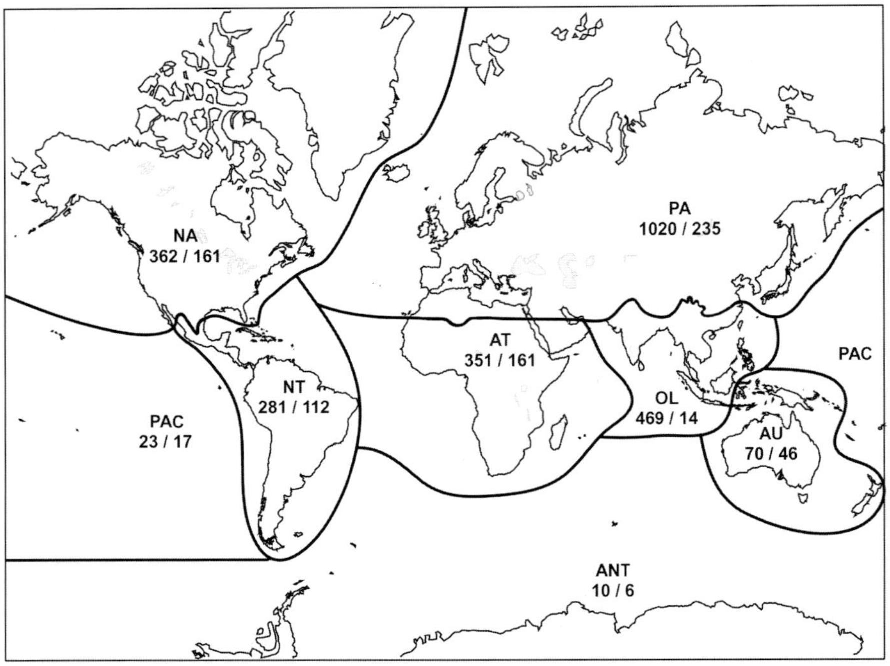

Fig. 1 Distribution of freshwater Nematoda species and genera by zoogeographical region (species number/genus number). PA, Palearctic; NA, Nearctic; NT, Neotropical; AT, Afrotropical; OL, Oriental; AU, Australasian; PAC, Pacific Oceanic Islands, ANT, Antartic

Table 1 Distribution of the number of nematode freshwater species in each biogeographic region. PA: Palearctic, NA: Nearctic, NT: Neotropical, AT: Afrotropical, OL: Oriental, AU: Australasian, PAC: Pacific and oceanic islands, ANT: Antartic

	PA	NA	NT	AT	OL	AU	PAC	ANT	World
Enoplea									
Enoplida	55	5	12	8	5	3	2	1	79
Triplonchida	99	25	26	36	10	6	0	1	140
Dorylaimida	282	116	93	155	186	20	3	0	610
Mononchida	55	36	37	37	27	14	4	0	99
Mermithida	229	63	33	8	164	1	0	0	417
Subtotal	720	245	201	244	392	44	9	2	
Chromadorea									
Chromadorida	29	5	5	6	4	5	1	0	36
Desmoscolecida	4	0	1	1	0	1	4	0	7
Desmodorida	4	3	3	3	0	0	1	1	9
Monhysterida	70	10	12	33	23	5	4	2	114
Araeolaimida	6	3	2	0	3	0	0	0	8
Plectida	54	30	48	35	22	7	4	5	125
Rhabditida	133	66	9	29	25	8	0	0	164
Subtotal	300	117	80	107	77	26	14	8	
Total	1020	362	281	351	469	70	23	10	

(Fig. 2). Many have also successfully adapted to xeric and cryogenic environments, and to moist soils and intermittently drying habitats.

Dorylaimia (except for the zooparasitic Marimermithida) are by large absent from marine environments, hinting at innate physiological constraints that may not be able to address osmotic stress typical of the salty marine environment.

The Mononchida, a less speciose order than Dorylaimida, are also well represented in freshwater

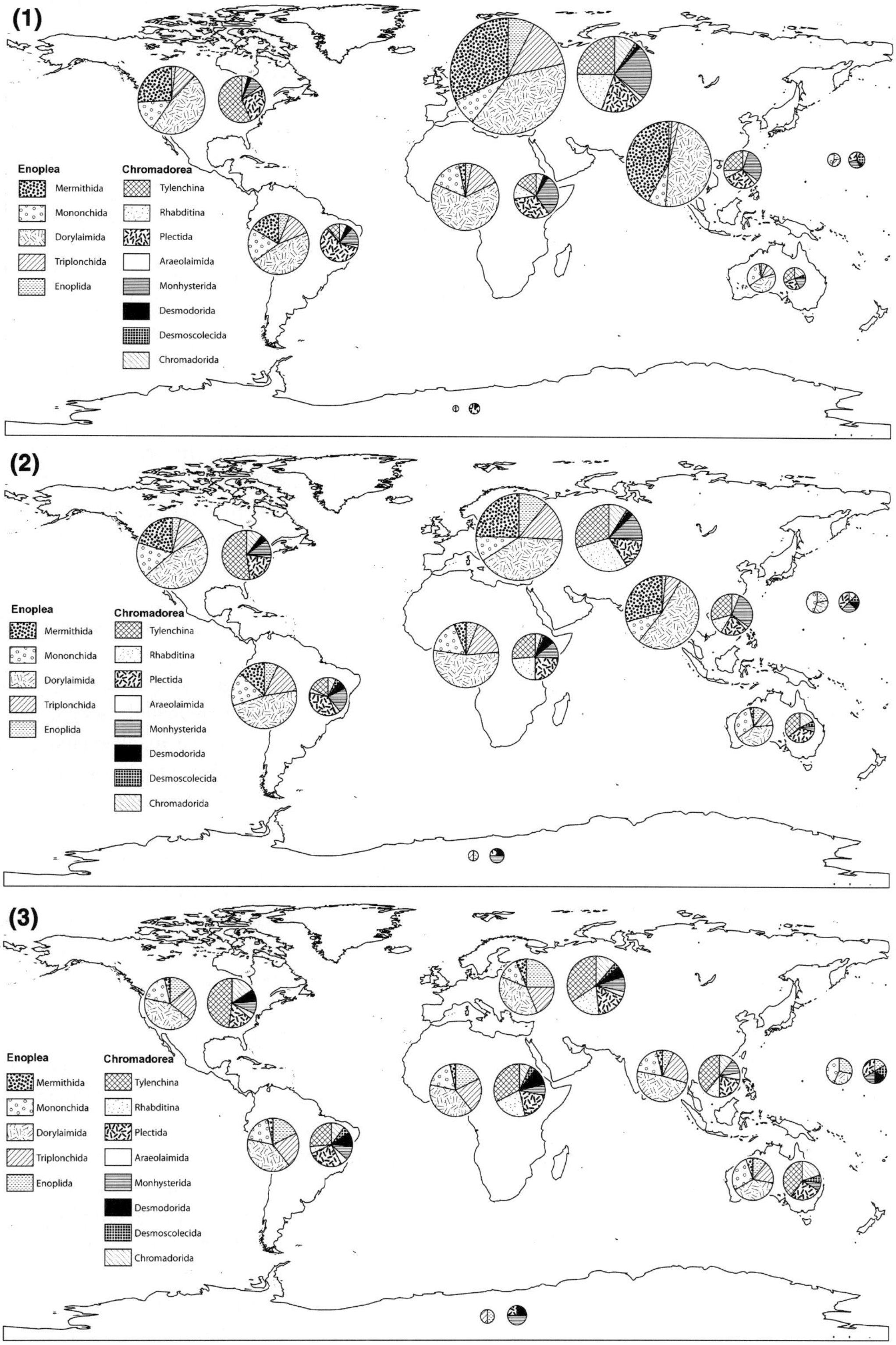

Fig. 2 Proportion of nematod orders in species (**1**), genus (**2**), and family (**3**) numbers per zoogeographic region. In each region: first circle = Enoplea, second circle = Chromadorea

habitats; about 80% of their families and 50% of their genera have freshwater representatives, and about one-quarter of the total 400 species inhabit freshwater environments, many exclusively so (Fig. 2).

The Mermithida have interesting life cycles, many species spending their early juvenile stage as well as their preadult and adult stages in freshwater bodies or sediments. The family of Mermithidae is highly speciose and contributes significantly to freshwater nematode diversity: 16% at genus-level and 23% at species-level (Fig. 2). Despite their diversity, these organisms are encountered infrequently during routine nematode surveys because many mermithid species have a very patchy occurrence, both in space and time.

Members of the subclass Enoplia, comprising the two orders Enoplida and Triplonchida, are principally marine but include important freshwater species. Many are exclusively freshwater taxa with extreme endemism, for example, species in the suborder Tobrilina. About 19% of freshwater nematode families, 15% of genera and 12% of the species belong to Enoplia. Furthermore, this subclass also includes some of the most commonly reported freshwater nematode species. About 50% of the families in the subclass and one quarter of its 700 genera have freshwater representatives. The three genera *Ironus*, *Amphidelus* and *Paramphidelus* are among the most widely reported in freshwater environments.

Triplonchida include the almost exclusively freshwater suborder Tobrilina and the mainly freshwater Tripylina. With its mosaic of large worms of diverse stoma morphology and a largely uncertain systematic position within the Nematoda, the Triplonchida is an important order with close to 150 species reported from freshwater bodies (Zullini, 2006; Fig. 2).

The subclass Chromadoria is the largest of the three subclasses of Nematoda and includes nearly half of the freshwater nematode families in its seven diverse orders (Araeolaimida, Chromadorida, Desmodorida, Desmoscolecida, Monhysterida, Plectida, and Rhabditida). The first four orders are essentially marine with only two species of Araeolaimida, about 2.5% of the species of Desmodorida and Desmoscolecida and about 3.5% of the Chromadorida having been recorded from freshwater habitats. Furthermore, even these low numbers are considered to be overestimates of the actual diversity because the majority of those species reported in these freshwater habitats are also claimed to have been reported in marine habitats (Decraemer and Smol, 2006). Monhysterida and Plectida, on the other hand, are among the most widely reported freshwater nematodes and nearly half of their species are freshwater inhabitants (Fig. 2). These groups include many speciose genera such as *Monhystera* and *Plectus* with wide environmental and zoogeographic ranges, and manifold taxonomic problems.

Among Rhabditida, the suborders Rhabditina and Tylenchina are overall largely terrestrial in their habitat preferences. Both are very diverse groups however, and include many true denizens of freshwater bodies, as well as others that are reported to be accidental occurrences. Usually the Rhabditina only dominate freshwater nematode communities of highly impacted habitats (Zullini, 1988; Bongers, 1990). The Tylenchina, on the other hand, are chiefly parasites of plants and are associated with aquatic plants. As a result they have been the focus of many studies, which probably resulted in a greater effort to record diversity than in most non-parasitic forms (Tables 1, 2, 3).

We do not attempt to estimate the global total number of freshwater nematode species in existence for the simple reason that many inland water bodies are either not sampled at all or are not studied extensively (see also discussion on biogeography below). Existing studies on freshwater nematodes show extreme regional bias; those from the southern hemisphere are extremely underrepresented, especially compared to European freshwater bodies (Fig. 1). The total number of species reported from freshwater environments in Europe is currently nearly 1000. Although very few researchers work on freshwater nematodes, many new species are added every year. There is no reason to expect a different trend in other continents. For instance, we (EA) sampled a number of lakes in the northeastern USA and encountered many new species and genera (unpublished). Consequently, the current total number is primarily a reflection of sampling effort rather than of any genuine differences in regional richness. Nematode biogeography is still in its early stages and in general, distribution of major nematode taxa are discussed per continent. About 53% of the freshwater species are recorded from the Palearctic, more specifically from Europe and Russian territories. Assuming that the total species count will double in

Table 2 Distribution of the number of nematode freshwater genera in each biogeographic region. PA: Palearctic, NA: Nearctic, NT: Neotropical, AT: Afrotropical, OL: Oriental, AU: Australasian, PAC: Pacific and oceanic islands, ANT: Antartic

	PA	NA	NT	AT	OL	AU	PAC	ANT	World
Enoplea									
Enoplida	16	5	5	6	2	3	2	1	19
Triplonchida	22	25	15	13	7	4	0	1	27
Dorylaimida	59	46	45	40	56	12	3	0	103
Mononchida	14	16	14	14	11	9	4	0	20
Mermithida	36	20	5	11	31	1	0	0	52
Subtotal	147	112	84	84	107	29	9	2	
Chromadorea									
Chromadorida	8	5	2	2	2	3	1	0	9
Desmoscolecida	1	0	1	1	0	1	1	0	1
Desmodorida	2	2	3	2	0	0	1	1	5
Monhysterida	11	5	5	6	10	1	2	2	14
Araeolaimida	1	1	0	1	1	0	0	0	1
Plectida	13	10	10	10	7	6	3	1	13
Rhabditida	52	26	21	6	14	6	0	0	55
Subtotal	88	49	42	28	34	17	8	4	
Total	235	161	126	112	141	46	17	6	

Table 3 Distribution of the number of nematode freshwater families in each biogeographic region. PA: Palearctic, NA: Nearctic, NT: Neotropical, AT: Afrotropical, OL: Oriental, AU: Australasian, PAC: Pacific and oceanic islands, ANT: Antartic

	PA	NA	NT	AT	OL	AU	PAC	ANT	World
Enoplea									
Enoplida	8	4	5	5	2	2	2	1	8
Triplonchida	6	6	6	6	6	3	0	1	6
Dorylaimida	12	12	11	11	13	7	2	0	16
Mononchida	4	5	5	5	5	5	3	0	5
Mermithida	2	1	1	1	1	1	0	0	2
Subtotal	32	28	28	28	27	18	7	2	
Chromadorea									
Chromadorida	4	4	2	2	2	3	1	0	4
Desmoscolecida	1	0	1	1	0	1	1	0	1
Desmodorida	2	2	3	2	0	0	1	1	3
Monhysterida	3	2	2	2	2	1	1	2	4
Araeolaimida	1	1	0	1	1	0	0	0	1
Plectida	6	4	5	5	4	4	2	1	6
Rhabditida	18	12	15	6	9	6	0	0	19
Subtotal	35	25	28	19	18	15	6	4	
Total	67	53	56	47	45	33	13	6	

European freshwater bodies, and that seen in the light of many uninventoried ancient lakes in various parts of the world, a roughly similar number of species could be expected from most the other biogeographic regions except for Antarctica, the Oceanic and Pacific Islands and Australia, the global species count from freshwater bodies undoubtedly will be at least about 14,000 species.

Phylogeny and historical processes

In the past 10 years, hypotheses of nematode relationships have become considerably more detailed thanks to the advent of molecular phylogenetics. It is now fairly clear that the old grouping of pseudocoelomates into a phylum Aschelminthes has no phylogenetic basis, and that the closest living relatives of nematodes are probably found in other phyla comprising vermiform moulting animals such as Nematomorpha or Priapulida—but not in ciliated interstitial or aquatic invertebrates such as Turbellaria, Gastrotricha, or Rotifera. The recent proposal of Ecdysozoa (Aguinaldo et al., 1997), an encompassing clade of all moulting invertebrates that would include both Nematoda and Arthropoda, remains much more controversial. Although originally based on molecular data, follow-up analyses based on increasing numbers of molecular loci have produced conflicting results and at this point the molecular and morphological evidence is decidedly ambiguous with regular publication of mutually contradictory studies (see, e.g., Philippe et al., 2005 versus Philip et al., 2005).

Within Nematoda, small subunit ribosomal DNA sequences support three major clades (Aleshin et al., 1998; Blaxter et al., 1998; De Ley & Blaxter, 2004), corresponding largely to the previously recognized subclasses Chromadoria, Enoplia and Dorylaimia (Pearse, 1942; Inglis, 1983)—but not to the traditional classes Secernentea and Adenophorea (Chitwood, 1958). Chromadoria and Enoplia each include various groups of marine, estuarine, and freshwater nematodes, while Dorylaimia are common in freshwater habitats but with very few exceptions unknown from marine or estuarine environments. The relationships between these three clades remain as yet unresolved, a problem that may in part be due to problems with outgroup choice and lingering uncertainty about the exact sister phylum of Nematoda. It is usually assumed that the most recent common ancestor of nematodes was a marine organism, although the lack of resolution for the basal dichotomies in the nematode tree allows for the alternative scenario that nematodes arose in a freshwater environment instead.

The relationships within nematode subclasses, orders, and families are becoming increasingly clear, thanks to a small explosion of phylogenetic studies (e.g., Mullin et al., 2005). Unfortunately, the biogeographical record is, in most cases, far too incomplete to allow for any rigorous analyses of species distribution and the historical processes that have enabled or constrained it. In a few groups of terrestrial nematodes, notably in Leptonchoidea, *Hoplolaimus*, and Longidoridae, patterns of dispersal and vicariance have been detected that reflect limited dispersive abilities and suggest an effective role of oceans as barriers for dispersal between continents of these particular nematodes (Ferris et al., 1976; Topham & Alphey, 1985; Geraert, 1990; Coomans, 1996). No comparable studies exist for freshwater nematodes, however, and it remains unclear to what extent phylogenesis has been driven or constrained by physical barriers and plate tectonics.

Present distribution

Zoogeographic regions and the distribution of nematodes

Current contribution presents a first attempt to summarize and map the biogeographic distribution of freshwater nematode taxa. However, the resulting data have to be interpreted with care.

Nematodes are often microscopic and many have resistant life stages which allow them to take advantage of many effective passive distribution mechanisms through wind, flowing water, and biological agents such as moving animals. Migratory birds typically gravitate toward freshwater sources and, though speculative, may transport resistant stages of nematodes across long distances. In a study of a remote limnetic location in the Galápagos archipelago, Eyualem and Coomans (1995) concluded that ten out of 18 species were cosmopolitan and the remaining six were widely distributed in the Southern hemisphere (two were new records). They argued that the most likely hypothesis to explain the presence of these freshwater nematodes on the Galápagos was through passive and very occasional transport by birds.

Once transported, many freshwater nematodes have special reproductive strategies: parthenogenesis, relatively rapid maturation upon hatching, short generation times and considerable numbers of progeny per female, rendering nematodes efficient colonizers.

Studies of fluctuating environments such as vernal pools and ephemeral water bodies (Hodda et al., 2006) provide excellent examples of the resilience of nematode communities. Their ability to withstand harsh environmental fluctuations allows them to cross barriers that may significantly limit the distribution of larger organisms. Furthermore, the spatial and temporal scales at which evolutionary processes work and diversity hot spots emerge may not be the same for microscopic forms as for larger organisms.

Taxonomical bias

As previously illustrated, records of many species need to be checked for correct identification, a task which is often not possible because no voucher specimens, especially of ecological studies, are stored and literature is not always available. Further, several taxa need revision. Most aquatic genera (including limnetic genera) are either claimed to be ubiquitous or widely distributed (Jacobs, 1984; Eyualem and Coomans, 1995; Michiels and Traunspurger, 2005). As noted above, however, the large majority of species are recorded in the literature from single locations. This apparent contradiction could very well be due to issues of 'doubtful identification' and poor morphological resolution (Jacobs, 1984; Tsalolikhin, 1992; see discussion above). If ubiquity is a general phenomenon in freshwater nematodes, as claimed, we need to confirm it using additional methods than morphology alone.

Distribution versus sampling bias

Although free-living nematodes are present in all types of limnetic habitats, including extreme conditions, discussion of their biogeographic distribution is hampered largely by the regionally biased surveys conducted so far. Some regions are well studied compared to others: for example, the Palearctic region (more specifically Europe and Russian territories) is the most sampled zoogeographic region for freshwater nematodes. Also the more extensive sampling is carried out, the greater the chance that "soil" nematodes are collected from waterlogged habitats and recorded as freshwater nematodes. As a result, the number of freshwater nematodes is biased and in-depth discussions about distribution and endemicity of nematode species are still largely premature.

The recorded limnetic fauna of Antarctica with its extreme environmental conditions is at present restricted to 10 species, 2 species belonging to the Enoplea and 8 to the Chromadorea (5 of these are plectid species). Important orders such as Dorylaimida and Rhabditida have not been reported from antarctic freshwater habitats, although they do occur in antarctic soils. It could well be that some of these species are seasonally aquatic but have not yet been collected at the right moment and in the right places during the brief antarctic summer. Few information is available form the Oceanic and Pacific Islands except for the Solomon Islands and New Caledonia. The freshwater nematofauna of the other biogeographic regions is represented by all orders of the Enoplea and the Chromadorea apart from the mainly marine order Desmoscolecida not observed in Nearctic, Oriental, and Australasian regions. The largely marine order Desmodorida with only a few freshwater taxa has also not been recorded from Australasia and Oriental and Araeolaimida appeared to be absent in the Afrotropical region. In general, the proportion of representatives of the seven orders of the Chromadorea does not vary much between continents; the majority of families belong to the Rhabditida. The largest number of families has been recorded for the Palearctic region with 89% of the total number of freshwater nematode families while Australiasia shows a more aberrant picture on the lower side of the range with 44% representation of freshwater families; the number of species for both biogeographic regions is, respectively, 56% and 3.8% of the total freshwater species.

A closer look at specific groups, for example, the Mononchina, reveals that of the 99 species recorded from freshwater habitats, 58 were recorded from a single biogeographic region, 17 species from 2 regions, 8 species from 4 regions, 4 species from 5 or 6 regions and 7 species were recorded worldwide (except Antarctica). Similar results were found for the typical freshwater taxa within the Tobrilina. Of 100 species, 83 are recorded from a single biogeographic region, 10 species from 2 regions, 4 species from 3 regions and a single species for 4–6 biogeographic regions. No species were recorded for the Antarctic region. Records from one continent are often confined to a single locality.

The effect of bias in sampling effort on the biogeography of nematodes can be most clearly demonstrated by considering the case of Mermithida. The current picture at the genus and species level (Fig. 2) shows a pattern where mermithids are much more diverse in the northern than the southern hemisphere. Consequently, the family Mermithidae is claimed to have a Holarctic distribution (Rubtsov, 1974 in Kozhova et al., 2000). In reality, the current data is inconclusive with regard to global distribution and the observed pattern may simply be an artifact of sampling bias: nematological literature reveals that there has been very little research focus on mermithids in the southern hemisphere and as a result we know little about mermithids of that region (see also discussion above). Furthermore, differences in taxonomic approaches are likely to muddle the picture even further. Thus, the higher number of mermithid species reported from Europe and Asia compared to North America is probably partly due to the tendency of mermithid specialists in North America to be more reluctant in proposing new species or accepting known ones (compare, e.g., Curran & Hominick, 1981 with Rubtsov, 1981).

Endemism

Discussions of nematode endemism are largely premature because of the extremely limited efforts devoted to inventorying freshwater nematodes on a geographic scale, compounded by the almost inevitable operational bias and patchy nature of sampling efforts in hitherto conducted studies. A case in point is the study of nematodes in ancient lakes. These lakes are particularly important in view of the fact that they are natural laboratories for speciation (Rossiter & Kanawabe, 2000) and a deserved focus on their nematofauna may be of paramount significance to our understanding of freshwater nematode biogeography.

Decraemer & Coomans (1994) summarized the nematode species count from the ancient lakes Malawi, Kinneret, Titicaca, Tanganyika, and Ohrid to be 3, 7, 11, 14, and 24, respectively. Lake Baikal is one of the better studied among the ancient lakes with a rich nematofuana of 92 species in 20 families of which about 90% are considered endemic to the lake (Kozhova et al., 2000). Taxonomically, tobrilids disproportionately dominate the community with globally unparalleled diversity making this lake the center of known tobrilid diversity (Tsalolikhin, 1980, 1983). In light of the continuing new species discovery (e.g., Shoshina, 2003) and the effects of sampling effort on species richness estimates (e.g., Michiels & Traunspurger, 2004), the actual nematode species diversity in Lake Baikal could well be much higher than thus far reported. Furthermore, employing taxon delimitation criteria other than morphology may reveal cryptic species. However, no detailed population genetic studies were carried out on freshwater nematodes and only few on free-living marine nematodes (Derycke et al., 2005). The latter molecular study showed a strong genetic differentiation among populations and an inverse correlation between dispersal ability and genetic differentiation. Understanding historical and ecological processes that led to the evolution of a tobrilid flock in lake Baikal, for example, would provide clues in understanding mechanisms of speciation and biogeography of nematodes in other freshwater bodies. This being so, it is plausible that all ancient lakes could similarly be centers of diversity for nematode (sub)groups. Unfortunately nematode communities of many ancient lakes (Lakes Lanao, Victoria, Malawi, Tanganyika, Khubsugul, Biwa, Pannon, and Shanwang) have not been studied in any meaningful way and our knowledge of even these scientifically famous lakes remains extremely patchy. This lack of information is even greater for smaller or much more cryptic freshwater bodies. Recent discoveries of invertebrate hyperdiversity in freshwater habitats such as the underground calcrete formations in western Australia (Leys et al., 2003) or the stromatolite pools in Cuatro Ciénegas, Mexico (Dinger et al. 2005) stand out as locations that are likely to reveal an abundance of nematodes, potentially with a high number of new taxa.

Acknowledgments The first author thanks Dr. William K. Thomas, Hubbard Center for Genome Studies, University of New Hampshire, for allowing him to use the facilities of the center during the preparation of this manuscript, and Drs. István Andrássy (Hungary), Mike Hodda (Australia), Michael Kidd (USA), and Melissa Yoder (USA) for supplying useful literature.

References

Aguinaldo, A. M. A., J. M. Turbeville, L. S. Linford, M. C. Rivera, J. R. Garey, R. A. Raff & J. A. Lake, 1997. Evidence for a clade of nematodes, arthropods and other moulting animals. Nature 387: 489–493.

Aleshin, V. V., O. S. Kedrova, I. A. Milyutina, N. S. Vladychenskaya & N. B. Petrov, 1998. Relationships among nematodes based on the analysis of 18S rRNA gene sequences: molecular evidence for monophyly of chromadorian and secernentian nematodes. Russian Journal of Nematology 6: 175–184.

Andrássy, I., 1999. A census of genera and subgenera of free-living nematodes. Journal of Nematode Morphology and Systematics 2(1): 45–68.

Andrássy, I., 1992. A short census of free-living nematodes. Fundamental and applied Nematologie 15(2): 187–188.

Blaxter, M.L., De Ley, P., Garey, J., Liu, L.X., Scheldemen, P., Vanfleteren, J., Mackey, L.Y., Dorris, M., Frisse, L.M., Vida, J.T., & W.K.Thomas, 1998. A molecular evolutionary framework for the phylum Nematoda. Nature 392: 71–75.

Bongers, T., 1990. The maturity index: an ecological measure of environmental disturbance based on nematode species composition. Oecologia 83: 14–19.

Chitwood, B. G., 1958. The designation of official names for higher taxa of invertebrates. Bulletin of Zoological Nomenclature 15: 860–895.

Coomans, A., 1996. Phylogeny of the Longidoridae. Russian Journal of Nematology 4: 51–60.

Curran, J. & W. M. Hominick, 1981. Description of *Gastromermis metae* sp.n. (Nematoda: Mermithidae) with an assessment of some diagnostic characters and species In *Gastromermis*. Nematologica 27(3): 258–273.

De Ley, P. & M. Blaxter, 2004. A new system for Nematoda: combining morphological characters with molecular trees, and translating clades into ranks and taxa. In Cook, R. & D. J. Hunt (eds), Nematology Monographs and Perspectives 2. Brill Publishing, Leiden, 633–653.

De Ley, P., W. Decraemer & A. Eyualem, 2006. Introduction: summary of present knowledge and research addressing the ecology and taxonomy of freshwater nematodes. In Eyualem Abebe, W. Traunspurger & A. Andrássy (eds), Freshwater Nematodes: Ecology and Taxonomy. CABI Publishing, 3–30.

Decraemer, W. & A. Coomans, 1994. A compendium of our knowledge of the free-living nematofauna of ancient lakes. Archiv für Hydrobiologie Beiheft Ergebnisse. der Limnologie 44: 173–181.

Decraemer, W. & N. Smol, 2006 Order Enoplida. In Eyualem Abebe, W. Traunspurger & A. Andrássy (eds), Freshwater Nematodes: Ecology and Taxonomy. CABI Publishing, 497–573.

Derycke S., T. Remerie, A. Vierstraete, T. Backeljau, J. Vanfleteren, M. Vincx & T. Moens, 2005. Mitochandiral DNA variation and cryptic speciation within the free-living marine nematode *Pellioditis marina*. Marine Ecology, Progress Series 300: 91–103.

Dinger, E. C., A. E. Cohen, D. A. Hendrickson & J. C. Marks, 2005. Aquatic invertebrates of Cuatro Ciénegas, Coahuila, México: natives and exotics. The Southwestern Naturalist 50: 237–281.

Eyualem, A. & A. Coomans, 1995. Freshwater nematodes of the Galápagos. Hydrobiologia 299: 1–51.

Ferris, V. R., C. G. Goseco & J. M. Ferris, 1976. Biogeography of free-living soil nematodes from the perspective of plate tectonics. Science 193: 508–510.

Geraert, E., 1990. Evolution in *Hoplolaimus* (Nematoda, Tylenchida). Nematologica 36: 199–204.

Hodda, M., A. Ocaña & W. Traunspurger, 2006. Nematodes from extreme freshwater habitats. In Eyualem Abebe, W. Traunspurger & A. Andrássy (eds), Freshwater Nematodes: Ecology and Taxonomy. CABI Publishing 179–211.

Hugot J.-P., P. Baujard & S. Morand, 2001. Biodiversity in helminths and nematodes as a field of study: an overview. Nematology 3(3): 199–208.

Inglis, W. G., 1983. An outline classification of the phylum Nematoda. Australian Journal of Zoology 31: 243–255.

Jacobs, L. J., 1984. The free-living inland aquatic nematodes of Africa: a review. Hydrobiologia 113: 259–291.

Kozhova, O. M., E. A. Erbaeva & G. P. Safronov, 2000. The benthic invertebrates of Lake Khubsugul, Mongolia. In Rossiter, A. & H. Kawanabe (eds), Ancient Lakes: Biodiversity, Ecology and Evolution. Advances in Ecological Research volume 31. Academic Press, London, 97–115.

Lambshead, P. J. D., 1993. Recent developments in marine benthic biodiversity research. Oceanis 19: 5–24.

Lambshead, P. J. D., 2004. Marine nematode biodiversity. In Chen, Z. X., S.Y. Chen & D. W. Dickson (eds), Nematology, Advances and Perspectives ACSE-TUP Book Series, 436–467.

Leys, R., C. Watts, S. Cooper & W. Humphreys, 2003. Evolution of subterranean diving beetles (Coleoptera: Dytiscidae: Hydroporini, Bidessini) in the arid zone of Australia. Evolution 57(12): 2819–2834.

Michiels, I. & W. Traunspurger, 2004. A three year study of seasonal dynamics of a zoobenthos community in a eutrophic lake. Nematology 6: 655–669.

Michiels, I. & W. Traunspurger, 2005. Benthic community patterns and the composition of feeding-types and reproductive modes in nematodes. Nematology 7: 21–36.

Moens, T., G. W. Yeates & P. De Ley, 2004. Use of carbon and energy sources by nematodes. In Cook, R. & D. J. Hunt (eds), Nematology Monographs and Perspectives 2. Brill Publishing, Leiden, 529–545.

Mullin, P. G., T. S. Harris & T. O. Powers, 2005. Phylogenetic relationships of Nygolaimina and Dorylaimina (Nematoda: Dorylaimida) inferred from small subunit ribosomal DNA sequences. Nematology 7: 59–79.

Pearse, A. S., 1942. An introduction to parasitology. 375 pp. Springfield: Charles C. Thomas.

Peña-Santiago, R., 2006. Dorylaimida I: Superfamilies Belondiroidea, Nygolaimoidea, and Tylencholaimoidea. In Eyualem Abebe, W. Traunspurger & A. Andrássy (eds), Freshwater Nematodes: Ecology and Taxonomy. CABI Publishing 326–392.

Philip, G. K., C. J. Creevey & J. O. McInerney, 2005. The Opisthokonta and the Ecdysozoa may not be clades: stronger support for the grouping of plant and animal than for animal and fungi and stronger support for the Coelomata than Ecdysozoa. Molecular Biology and Evolution 22: 1175–1184.

Philippe, H., N. Lartillot & H. Brinkmann, 2005. Multigene analyses of bilaterian animals corroborate the monophyly of Ecdysozoa, Lophotrochozoa, and Protostomia. Molecular Biology and Evolution 22: 1246–1253.

Rossiter, A. & H. Kawanabe, 2000. Ancient Lakes: Biodiversity, Ecology and Evolution. Advances in Ecological Research volume 31. Academic Press, London, 624 pp.

Rubtsov, I. A., 1974. Water Mermithidae, Vol. II. Nauka, Leningrad (in Russian).

Rubtsov, I., 1981. New species of mermithids from Estonia and neighbouring areas. IX. The genera *Gastromermis, Brevimermis* and *Eurymermis*. Eesti NSV Teaduste Akademia Toimetised – Izvestia Akademii Nauk Estonskoi SSR. Biologia 3: 192–201.

Shoshina, A. A., 2003. A new trident tobrilid, Setsalia mirabilis gen. et sp.n., from Lake Baikal (Nematoda: Enoplida). Zoosystematica Rossica 12(1): 15–18.

Topham, P. B. & T. J. W. Alphey, 1985. Faunistic analysis of longidorid nematodes in Europe. Journal of Biogeography 12: 165–174.

Tsalolikhin, S. Ya., 1992. Taxonomic notes on African species of the genus Brevitobrilus (Nematoda, Enoplida: Tobrilidae). Zoosystica Rossica 1: 1–15.

Tsalolikhin, S. Ya., 1980. Free-living nematodes of Baikal Lake. Novosibirsk, Nauka, 120 p.

Tsalolikhin S. Ya., 1983. Nematodes of the Families Tobrilidae and Tripylidae of the World Fauna. Leningrad, Nauka, 230 pp.

Turpeenniemi T. A. & H. Hyvärinene, 1996. Structure and role of the renette cell and caudal glands in de nematode Sphaerolaimus gracilis (Monhysterida). Journal of Nematology 28: 318–327.

Zullini A., 1988. The ecology of the Lambro River. Rivista di Idrobiologia 27: 39–58.

Zullini, A., 2006. Order Triplonchida. In Eyualem Abebe, W. Traunspurger & A. Andrássy (eds), Freshwater Nematodes: Ecology and Taxonomy. CABI Publishing, 293–325.

Hydrobiologia (2008) 595:79–83
DOI 10.1007/s10750-007-9112-3

FRESHWATER ANIMAL DIVERSITY ASSESSMENT

Global diversity of hairworms (Nematomorpha: Gordiaceae) in freshwater

George Poinar Jr.

Abstract Freshwater hairworms (Gordiaceae) and marine hairworms (Nectonematoidea) comprise the phylum Nematomorpha. Only the freshwater forms are discussed here. While freshwater hairworms develop as parasites of both aquatic and terrestrial arthropods, they all enter fresh water to mate, oviposit and produce infective stages (preparasitic larvae). The global species diversity of freshwater hairworms based on published descriptions is approximately 326 species and a conservative estimate for the global species diversity in this group would be around 2,000 species.

Keywords Nematomorpha · Hairworms · Gordiaceae

Introduction

There are two classes in the phylum Nematomorpha: the freshwater hairworms or Gordiaceae, comprising a dozen genera and the marine hairworms or Nectonematoidea, with a single genus. The host range, species diversity and distribution of the Nectonematoidea were summarized by Poinar and Brockerhoff (2001) and since all of the five known species are marine, they will not be included here.

While all freshwater hairworms develop as parasites in arthropods, all enter fresh water to mate, oviposit, and infect paratenic hosts. Freshwater hairworm adults, eggs and preparasitic larvae occur in ponds, streams, lakes, and various man-made structures that retain water. The adults range from tan to black in color and from several centimeters to over a meter in length. The body wall is covered with a hard, opaque, multilayered cuticle that varies in texture from smooth to highly sculptured. The surface projections (areoles), warts and/or bristles on the epicuticle as well as the male tail, provide important taxonomic characters. While the anterior end can be attenuated, the tail may be rounded, bilobed, or even trilobed. Hairworm preparasitic larvae are capable of encysting (without development) in a wide range of paratenic hosts, including invertebrates and vertebrates. Development occurs in the body cavity of invertebrates (especially insects) that have ingested paratenic hosts (Poinar, 2001; Schmidt-Rhaesa, 1997; Smith, 2001; Hanelt et al., 2005) Fig. 1.

Guest editors: E. V. Balian, C. Lévêque, H. Segers & K. Martens
Freshwater Animal Diversity Assessment

G. Poinar Jr. (✉)
Department of Zoology, Oregon State University, Corvallis, OR 97331, USA
e-mail: poinarg@science.oregonstate.edu

Species/generic diversity

Some 21 extant and two fossil genera of freshwater hairworms have been described worldwide (Table 1).

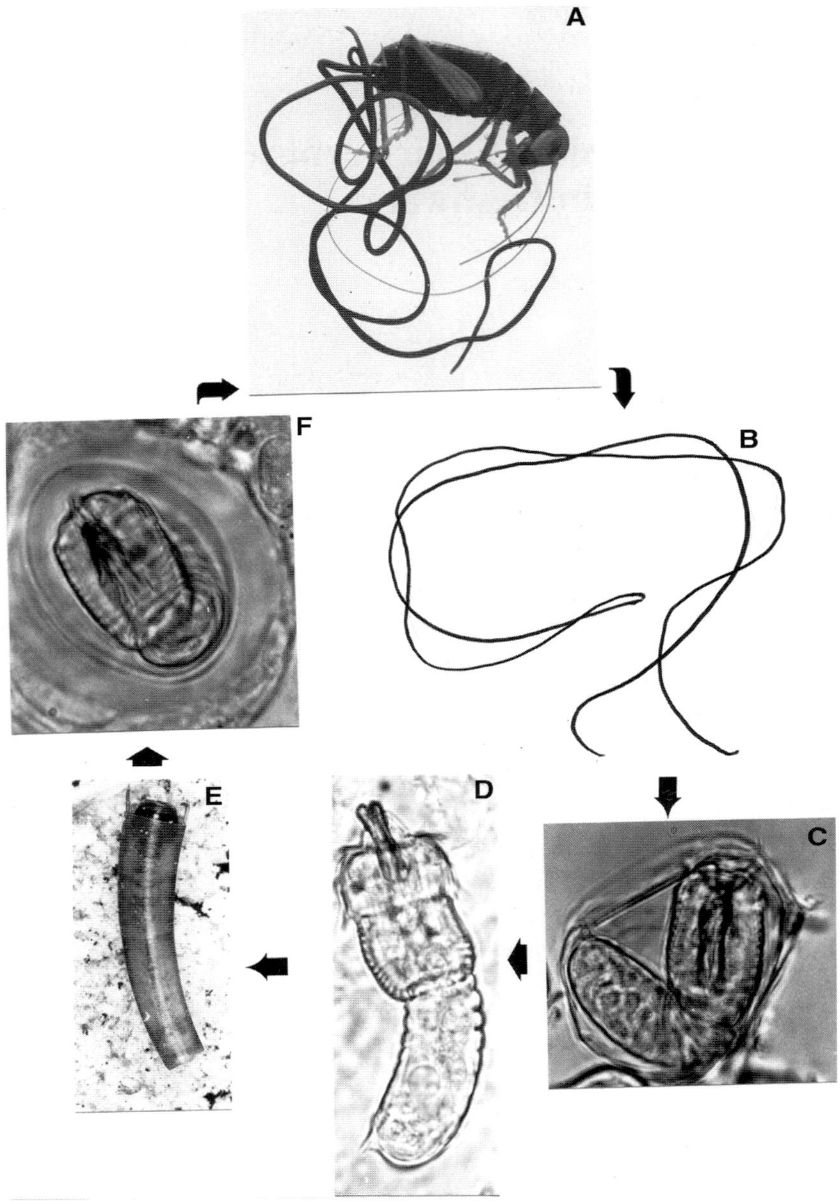

Fig. 1 A typical hairworm life cycle. (**A**). Hairworm adults emerge from an insect host (Orthoptera). (**B**). Hairworms mate and oviposit in freshwater. (**C**). A preparasitic hairworm larva ready to hatch from its egg. (**D**). Using anterior stylets and posterior spines, a larva penetrates the gut of a paratenic host. (**E**). Larvae encyst in the body cavity of paratenic host (here a caddis fly larva). (**F**). A hairworm larva encysted in the tissues of its paratenic host. Development of the hairworm is completed when the paratenic host is eaten, together with the encysted hairworm larvae, by the final (definitive) host. Not drawn to scale

Some of these genera are questionable and further studies are needed to test their reliability. Species diversity is difficult to determine since the intraspecific variation of most species is unknown and some characters now used for species separation may not be reliable. The global species diversity based on published descriptions varies according to authorities, but is around 326 species (Table 2). An estimate for the projected global species diversity in this group would be around 2,000 species (Table 3). The Nematomorpha is an ancient group that probably extends back at least to the Carboniferous, although fossils are extremely rare, the oldest dating from the Early Cretaceous (100 mya) (Poinar & Buckley, 2006) Fig. 2.

Phylogeny and historical processes

The hairworms represent a relict group that is not closely related to any other living phylum. Various authors have attempted to show phylogenetic relationships between hairworms and other 'pseudocoelomate' phyla (summarized by Schmidt-Rhaesa, 1998). However, it is difficult to determine the phylogenetic relationships of a group when there are no completely free-living members and the morphology and development are incompletely known (Bresciani, 1991). Also since the preparastic larvae are quite different from the adults, a phylogenetic position based on morphology will differ depending on the character states chosen. The morphology of adult and preparasitic larval freshwater hairworms is unique and supports the view that hairworms separated from the main body of the Aschelminthes at an early stage. Perhaps molecular studies will reveal if hairworms are related to the kinorhynchs, priapulids, rotifers, or nematodes, the groups considered by most to be their closest relatives.

Table 1 Freshwater hairworm families and genera

Gordiidae May 1919
Acutogordius Heinze, 1952
Gordius L. 1766
Chordodidae May 1919
Beatogordius Heinze, 1934
Chordodiolus Heinz, 1935
Chordodes Creplin, 1847
Cretachordodes Poinar & Buckley, 2006 (fossil genus)
Dacochordodes Căpuse, 1965
Digordius Kirjanova, 1950
Euchordodes Heinze, 1937
Gordionus G. W. Müller, 1927
Lanochordodes Kiryanova, 1950
Neochordodes Carvalho, 1942
Noteochordodes Miralles & Villalobos, 2000
Paleochordodes Poinar, 1999 (fossil genus)
Pantachordodes Heinze, 1954
Parachordodes Camerano, 1897
Paragordius Camerano, 1897
Paragordionus Heinze, 1935
Progordius Kirjanova, 1950
Pseudogordius Yeh & Jordan, 1957
Pseudochordodes Carvalho, 1942
Semigordionus Heinze, 1952
Spinochordodes Kirjanova, 1950

Table 2 Approximate number of hairworm species in geographical regions (total # of species = 326)

Regions	Number of hairworm species	Major references
Africa	64	Camerano (1915); Sciacchitano (1958)
Asia	28	Sciacchitano (1958)
Australia	9	Schmidt-Rhaesa et al. (2003); Schmidt-Rhaesa (2002); Schmidt-Rhaesa & Bryant (2004b)
China	8	Camerano (1915)
Europe	99	Schmidt-Rhaesa (1997)
India	31	Camerano (1915)
Japan	11	Schmidt-Rhaesa (2004a)
Madagascar	5	Sciacchitano (1958)
Malaysia	3	Poinar (2004)
New Zealand	5	Poinar (2006)
North America	18	Poinar & Chandler (2004)
Pacific Islands	13	Sciacchitano (1958); Schmidt-Rhaesa (2003)
South & MesoAmerica	32	Camerano (1915); Sciacchitano (1958); Schmidt-Rhaesa & Menzel (2005)

Table 3 Estimated species diversity of hairworms (total # = 2000)

Zoogeographic Region	Estimated number of species
Palaearctic	250
Nearctic	200
Neotropical	500
Afrotropical	400
Australasian	300
Oriental	300
Pacific Oceanic Islands	50

Present distribution and main areas of endemicity

A worldwide study of the biodiversity of hairworms developing in ground beetles (Carabidae: Coleoptera) showed that at least 70 species of beetles are parasitized by 47 species of hairworms belonging to at least five genera (Poinar et al., 2004). The environmental conditions for hairworm survival are

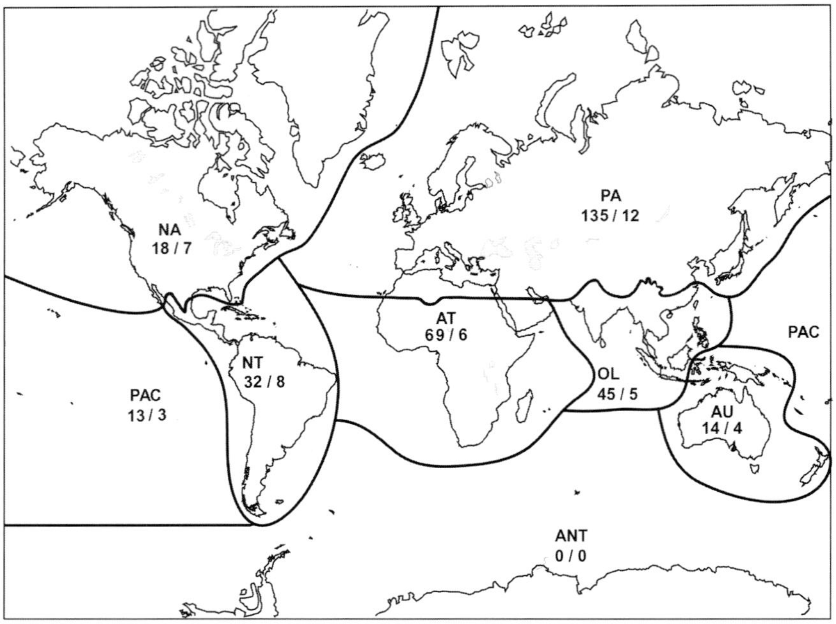

Fig. 2 Biodiversity of hairworm species/genera in each zoogeographic region. Since hairworm genera are widespread, many of the 21 known genera occur throughout the world. There are some 326 described hairworm species. The high level of biodiversity in the Palaearctic probably indicates the higher number of studies undertaken rather than an especially high level of biodiversity. PA—Palaearctic; NA—Nearctic; NT—Neotropical; AT—Afrotropical ; OL—Oriental; AU—Australasian; PAC—Pacific Oceanic Islands; ANT—Antarctic

more complex than that for the host, since most hairworms need both an aquatic paratenic and a terrestrial developmental host.

Human related issues

There is no direct economic or medical importance of freshwater hairworms. Hairworms can serve as natural biological control agents by reducing plant-feeding orthopterans, however they also parasitize predatory insects (Poinar et al., 2004).

Freshwater hairworms are rarely collected since they normally occur in small numbers and are well camouflaged. However in some localities, such as the South Island, New Zealand, one species (*Euchordodes nigromaculatus*) occurs along forest streams in the hundreds at certain times of the year.

The main threats to freshwater hairworms are habitat changes that destroy their hosts and water sources, polluted waters that kill their eggs and newly hatched larvae and predators, parasites and pathogens that attack their paratenic and developmental hosts, and in some cases, hairworms themselves (Poinar, 2001). Most hairworm species probably would be candidates for the Red Data Book species, although a thorough sampling to determine the natural distribution of most species has not been undertaken.

References

Bresciani, J., 1991. Nematomorpha. In Harrison, F. W. & E. E. Ruppert (eds), Microscopic Anatomy of Invertebrates. Wiley-Liss, New York: 197–218.

Camerano, L., 1915. Revisione dei Gordii. Memorie della Reale Academia delle Scienze di Torino. Serie Seconda 66: 1–66.

Hanelt, B., F. Thomas & A. Schmidt-Rhaesa, 2005. Biology of the phylum Nematomorpha. Advances in Parasitology 59: 243–305.

Poinar, Jr., G. O., 2001. Nematoda and Nematomorpha. In Thorp, J. H. & A. P. Covich (eds), Ecology and Classification of North American Freshwater Invertebrates, 2nd ed. Academic Press, San Diego: 255–295.

Poinar, Jr., G. O., 2004. Nematomorpha. In Yule, C. M. & Y. H. Sen (eds), Freshwater Invertebrates of the Malaysian Region. Academy of Sciences Malaysia, Kuala Lumpur: 157–161.

Poinar, Jr., G. O., 2008. Phylum Nematomorpha. In Gordon, D. P. (ed.), The New Zealand Inventory of Biodiversity, Vol. 2. Kingdom Animalia. Chaetognatha, Ecdysozoa, and

Ichnofossils. Canterbury University Press, Christchurch (in press).
Poinar, Jr., G. O. & A. M. Brockerhoff, 2001. Nectonema zealandica n. sp. (Nematomorpha: Nectonematoidea) parasitizing the purple rock crab Hemigrapsus edwardsi (Brachura: Decapoda) in New Zealand, with notes on the prevalence of infection and host defense reactions. Systematic Parasitology 50: 149–157.
Poinar, Jr., G. O., & C. M. Chandler, 2004. Synopsis and identification of North American hairworms (Gordioidea: Nematomorpha). Journal of the Tennessee Academy of Science 79: 1–7.
Poinar, Jr., G. O., J. Rykken & J. Labonte, 2004. *Parachordodes tegonotus* sp. n., (Gordioidea: Nematomorpha), a hairworm parasite of ground beetles (Carabidae: Coleoptera) with a summary of gordiid parasites from Carabidae. Systematic Parasitology 58: 139–148.
Poinar, Jr., G. O. & R. Buckley, 2006. Nematode (Nematoda: Mermithidae) and hairworm (Nematomorpha: Chordodidae) parasites in Early Cretaceous amber. Journal of Invertebrate Pathology 93: 36–41.
Schmidt-Rhaesa, A., 1997. Nematomorpha. In Schwoerbel, J., & P. Zwick (eds), Süsswasserfauna von Mitteleuropa, Vol. 4(4). Gustav Fisher Verlag, Stuttgart: 1–128.
Schmidt-Rhaesa, A., 1998. Phylogenetic relationships of the Nematomorpha a discussion of current hypotheses. Zoologischer Anzeiger 236: 203–216.
Schmidt-Rhaesa, A., 2002. Australian species of *Chordodes* (Nematomorpha) with a description of two new species, remarks on the genus and its life history. Journal of Natural History 36: 1569–1588.
Schmidt-Rhaesa, A., P.-R. Chung & W.-M. Sohn, 2003. *Parachordodes megareolatus*, a new species of horsehair worm (Nematomorpha: gordioida) from Korea. The Korean journal of Systematic Zoology 19: 161–166.
Schmidt-Rhaesa, A., 2003. *Gordionus longistriatus*, a new species of Nematomorpha from Hawaii. Records of the Hawaii Biological Survey for 2003. Bishop Museum Occasional Paper 78: 53–55.
Schmidt-Rhaesa, A., 2004a. Japanese horsehair worms (Nematomorpha) from the Lake Biwa Museum. Species Diversity 9: 97–107.
Schmidt-Rhaesa, A., 2004b. Filling the Gondwana gaps: new species and new reports of *Beatogordius* Heinze, 1934 (Nematomorpha) from Australia and Madagascar. Systematic Parasitology 57: 173–181.
Schmidt-Rhaesa, A. & L. Menzel, 2005. Central American and Caribbean species of horsehair worms (Nematomorpha), with the description of three new species. Journal of Natural History 39: 515–529.
Sciacchitano, I., 1958. Gordioidea del Congo Belga. Annales du Musee Royal du Congo Belge Sciences Zoologiques 67: 7–110.
Smith, D. G., 2001. Pennak's Freshwater Invertebrates of the United States, 4th ed. Porifera to Crustacea. John Wiley and Sons, Inc., New York, 638 pp.

Hydrobiologia (2008) 595:85–91
DOI 10.1007/s10750-007-9006-4

FRESHWATER ANIMAL DIVERSITY ASSESSMENT

Global diversity of gastrotrichs (Gastrotricha) in fresh waters

Maria Balsamo · Jean-Loup d'Hondt · Jacek Kisielewski · Lara Pierboni

Abstract The global diversity of inland water Gastrotricha is poorly known, and information is extremely heterogeneous. Gastrotricha have been studied most widely in Europe and America, whereas data from the other continents are scattered or not even available. This scanty information is related to several reasons, first of which is the technical difficulty in collecting and studying microscopic and soft-bodied species. In addition, the research has been limited mostly to the epibenthos and periphyton in lentic waters, and the gastrotrich taxonomy is still under discussion mainly because of the great intraspecific variability. Three of the five freshwater families are widespread or cosmopolitan, and most genera have been reported from at least two continents. There is strong evidence of a high diversity in genera and species in tropical areas. Nearly a half of the freshwater species are known from only one country or even only from one site, but the insufficient faunistic knowledge does not allow defining them as endemic. The phylogenetic relationships and possible evolutionary trends of inland water species of Gastrotricha are outlined.

Keywords Gastrotricha · Chaetonotida · Freshwater species · Global biodiversity · Geographic distribution

Guest editors: E. V. Balian, C. Lévêque, H. Segers & K. Martens
Freshwater Animal Diversity Assessment

M. Balsamo (✉) · L. Pierboni
Centro Ricerche per la Biodiversità e la Conservazione, Università di Urbino, Urbino, Italy
e-mail: maria.balsamo@uniurb.it

J.-L. d'Hondt
Muséum National d'Histoire Naturelle, Paris, France

J. Kisielewski
Museum and Institute of Zoology, Polish Academy of Science, Warsaw, Poland

Introduction

Gastrotricha are aquatic microinvertebrates composing a constant, important component of the benthic communities in marine and freshwater habitats. In spite of the often high number of populations, gastrotrichs are not yet well known, possibly due to their minute size and body fragility, which make studying them very difficult. The phylum consists of nearly 690 named species, grouped into two orders, Macrodasyida and Chaetonotida, greatly different in morphology, reproductive biology and ecology. Macrodasyida are about 240 worm-like species, all interstitial in marine and estuarine habitats except for two freshwater ones. The roughly 450 species of Chaetonotida are tenpin-shaped, interstitial or epibenthic in marine, brackish, but mainly freshwater habitats (Balsamo & Todaro, 2002; Fig. 1). Most of the 318 inland waters gastrotrich species have been

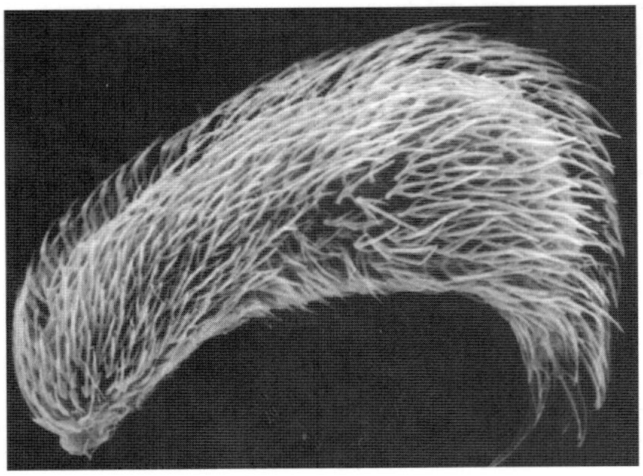

Fig. 1 *Chaetonotus schultzei*, S.E.M

reported from the periphyton and the surface layer of organic sediments in eutrophic, lentic biotopes (Kisielewski, 1998). About 70 species are known from an interstitial habitat, a half of which can be found in lotic waters (see Ricci & Balsamo, 2000). Freshwater Gastrotricha are presently grouped into 21 genera in 5 families of Chaetonotida, and into two genera '*incertae sedis*' of Macrodasyida. Many species have been discovered in the last 50 years, but the list of the species accepted in the current systematization of the group is currently under review following recent extensive and careful faunistic surveys (see for references Schwank, 1990; Kisielewski, 1991, 1998; Naidu & Rao, 2004).

Global species diversity

Past research

First descriptions of freshwater gastrotrich species go back to the XVIII–XIX centuries in detailed studies carried out by famous zoologists, but the systematic research started in late XIX century leading to the first, important monograph by Zelinka (1889) on the European and North American freshwater species. In the beginning of the 1900s scattered faunistic and systematic information was gathered mainly in Europe, but as far away as Africa, India, Ceylon, Tibet, Japan, New Guinea, Jamaica, Paraguay, and America. Since 1970 a significant rise of interest brought about a series of studies mostly in Europe but also in Israel, India, Japan, Korea, US, Argentina and Brazil. Research has touched almost only continental biotopes, but some data are also available for insular fresh waters (Tuscan Archipelago, Italy; Azores, Portugal; Jamaica; see Schwank, 1990; Kisielewski, 1998). The history of the knowledge of brackish-water species is much shorter. Scattered, occasional findings in low-salinity environments have concerned European coastal lagoons, estuaries, deltas, Brazilian mangroves and Amazonian estuaries, even hundreds of kilometres from the sea (see Kisielewski, 1991). Brackish-water gastrotrichs are mostly marine, chaetonotidan species, clearly adapted to great salinity variations. Only a minority of freshwater species can survive salinity, and very few are exclusive to brackish waters and may be endemic of these habitats (Kisielewski, 1991; Tongiorgi et al., 1999).

Estimated global diversity

At a high taxonomical level, the Chaetonotida families Dasydytidae, Neogosseidae and especially Chaetonotidae (subfamily Chaetonotinae) appear to be widely distributed. The other families show a limited distribution: the rare Dichaeturidae have been occasionally found in few European sites, the ditypic Proichthydiidae are only known from South America and Asia, and the subfamily Undulinae (Chaetonotidae) is reported from one site in Amazonia (see Schwank, 1990; Kisielewski, 1991). The only two freshwater monotypic genera of the order Macrodasyida, each recorded in one site, are known only in Europe and South America, respectively (Ruttner-Kolisko, 1955; Kisielewski, 1987). At a super-specific level, almost a half of the genera show an intercontinental distribution. A high diversity of endemic genera in the Brazilian fauna, but not in the European and Levantine ones has been evidenced by Kisielewski (1991). About 1/3 of European species and 1/3–1/2 of South American ones appear cosmopolitan. The distribution of the other species, as well as that of subgenera and genera seems to be restricted to a single continent or, if intercontinental, to the tropical zone. The few studies carried out in the tropical area show a very high generic and specific diversity even of families rare in temperate zones (Kisielewski, 1991). This strongly suggests that freshwater fauna, especially the highly specialized families, will be much richer in the tropical regions. Detailed and reliable faunistic comparisons were made by Kisielewski (1991, 1999), through exhaustive studies carried out in Poland, Brazil and Israel. About 1/3 (33.7%) of the species found in Brazil was known also in Europe, while the percentage of European species was higher (54.5%) in Israel, probably due to the closer zoogeographical relationships and shorter distance between Levant and Europe.

Aspects of phylogeny and evolution

Gastrotricha for a long time have been placed at the base of the 'Aschelminthes', close to Kinorhyncha, Nematoda, Rotifera or Gnathostomulida (see Boaden, 1985). Recent phylogenetic analyses of the protostomes, based on morphological, molecular, developmental and ecological evidences, have agreed on the basal position of the phylum. Gastrotricha have been included in Cycloneuralia, or linked to Gnathostomulida or to Platyhelminthes (see Zrzavy et al., 2002). The phylum has been considered as the sister-group of Ecdysozoa, included in the Spiralia, and in the 'Platyzoa' (see Zrzavy et al., 2002). Despite the key role generally assigned to Gastrotricha in the protostomes phylogeny, their relationships to other lower metazoans are not yet defined (Schmidt-Rhaesa, 2002). The numerous and great differences between the two orders have given rise to hypothesize that they are paraphyletic to Nematoda, or even that they belong to different phyla (Manylov et al., 2004). The monophyly of the orders and of the phylum has been strongly supported by cladistic morphological and molecular (18S rRNA) analyses (Hochberg & Litvaitis, 2000, 2001; Todaro et al., 2006), but the ancestry of each order and the relationships among families are still open questions. Further efforts in detecting possible plesiomorphies useful to clarify internal phylogeny are needed (Ferraguti & Balsamo, 1995; Hochberg & Litvaitis, 2000, 2001). Among Chaetonotida, Neodasyidae and Xenotrichulidae are marine families, Dasydytidae, Neogosseidae and Proichthydiidae are exclusively freshwater, and Dichaeturidae are freshwater with unclear connections with brackish and marine habitats (Kisielewski, 1990). The largest family Chaetonotidae has three marine genera (*Diuronotus, Halichaetonotus, Musellifer*), three freshwater ones (*Arenotus, Polymerurus, Undula*), and five including both marine and freshwater species (*Aspidiophorus, Chaetonotus, Heterolepidoderma, Ichthydium, Lepidodermella*). Hence, most primitive forms were possibly marine, psammic Macrodasyida and Chaetonotida might evolve later, mainly radiating in freshwaters as epibenthic or semipelagic forms. The only two extant freshwater Macrodasyida may represent successful attempts of colonization of inland waters by this marine order. In addition, a few other macrodasyidan species occur in brackish waters, even at salinity, as low as 1‰ (see Kisielewski, 1990). The presence of Macrodasyida far from river mouth, and also in deep beach freshwater springs, suggests a colonization of freshwaters not only through estuarine sediments, but also through water bodies created near beach springs during the marine regression (Kisielewski, 1990). Few, mainly marine Chaetonotida are psammic in brackish waters, but only two species appear exclusive to this habitat, perhaps being rare survivors of the Messinian crisis of the Mediterranean (Tongiorgi et al., 1999). The much greater success of Chaetonotida in colonizing inland waters is proved by the high number of freshwater species, more than 2/3 of the total number of chaetonotidans. The general epibenthic lifestyle is probably related to the organic and muddy nature of these sediments, and appears to have been favoured by particular morphological and biological adaptations [e.g. cuticular sculpturing, parthenogenesis, resting eggs]. Periphytic and semipelagic habitus have possibly developed as adaptations to new, abundant, trophic substrata (vegetation), that are better exploited by this group in lentic waters rather than in lotic ones or in the turbulent, littoral sediments. The primary or secondary presence in the freshwater psammon of the few Chaetonotida cannot be stated with certainty. The radiation of marine Gastrotricha, all interstitial, probably occurred in sandy sediments, whereas the ecological evolutionary trends of inland-water gastrotrichs and the importance of the psammic habitat in this process are still unclear, and could have occurred in different ways in Europe and in South America (see Kisielewski, 1990).

Zoogeography and endemicity

The geographic distribution of the marine gastrotrich fauna is well-known from many world areas (see Hummon, 2001; Naidu & Rao, 2004; Todaro &

Table 1 Total number of inland-water (freshwater + brackish-water) species of Gastrotricha per family and per biogeographical region

Biogeographical Region	PA	NA	NT	AT	OL	AU	PAC	ANT	World
Order/Family									
Chaetonotida	221 + 2	71	91	10	29	8	0	0	316 + 2
Chaetonotidae	192 + 2	60	76	7	25	8	0	0	281
Dasydytidae	21	9	10	0	2	0	0	0	34
Dichaeturidae	3	0	0	0	0	0	0	0	3
Neogosseidae	4	2	4	3	1	0	0	0	8
Proichthydiidae	1	0	1	0	0	0	0	0	2
Macrodasyida	1	0	1	0	0	0	0	0	2
incertae sedis	1	0	1	0	0	0	0	0	2
Total	222 + 2	71	92	10	29	8	0	0	318 + 2

PA: Palearctic; NA: Nearctic; NT: Neotropical; AT: Afrotropical; OL: Oriental; AU: Australasian, PAC: Pacific Oceanic Islands, ANT: Antarctic

Rocha, 2005), whereas that of freshwater species is still very limited, mainly reflecting researchers nationality. Freshwater Gastrotricha have been studied at extremely different levels in the various parts of the world. In Europe, where most research has been done, 213 species have been identified; 92 species have been reported in South America, 71 in North America, 64 in Asia, 10 in Africa, 8 in Oceania and none in Antarctica (Table 1). Some European countries have been the object of special surveys: France (29 spp.; d'Hondt, 1967), Germany (90 spp.; Remane, 1935-36; Schwank, 1990), Italy (92 spp.; Balsamo & Tongiorgi, 1995), Poland (98 spp.; see Kisielewski, 1998), Romania (90 spp.; see Rudescu, 1967); Russia (91 spp; see Tretjakova, 1991), and United Kingdom (58 species; see Martin, 1990). Single or few records are available for other European countries, or even none at all from Portugal [except for Azores] and Netherlands. Data from North America mainly concerns the US (see Weiss, 2001), and Canada (Schwank, 1990), whilst from Central and South America records regard Argentina (see Grosso & Drahg, 1991), Brazil (Kisielewski, 1987, 1991) and French Guyana (d'Hondt et al., 2006). Some information is available for Colombia, Jamaica, Paraguay, Uruguay (see Schwank, 1990). Of the Asian countries only India (see Naidu & Rao, 2004), Israel (Kisielewski, 1999), Japan (see Sudzuki, 1975) and Korea (Lee & Chang, 2000) have been investigated. Few, scattered records are available for Africa and Oceania: New Guinea and Australia (see de Beauchamp, 1932; Hochberg, 2005; Fig. 2). There is no data available for Antarctica and Pacific Islands. This whole picture points out many gaps in the distribution knowledge, as entire world areas have not yet been explored.

Research has especially focused on the epibenthos and periphyton of mesotrophic and eutrophic lentic waters, in which a rich, diversified fauna is known to exist (Kisielewski, 1998). Some special studies have found a few species in sandy and sandy-silty sediments of lentic and lotic fresh waters, in lagoons and estuarine brackish waters, areas generally considered unsuitable for gastrotrichs for various reasons (water turbulence, substrate perturbation, saline excursion etc.) (see Ricci & Balsamo, 2000). We know almost nothing about gastrotrichs from extreme habitats of biogeographic interest, such as inland saline lakes, deep-sea freshwater springs, river springs, warm springs, oasis springs, cave pools and hyporheic waters. A zoogeographical analysis of the inland-water Gastrotricha is at present very difficult due to a heterogeneous faunistic knowledge in different world regions, and a general insufficiency of data. At least a third of the genera and a half of the species known in inland waters have been recorded from only one country, often from only one site (Table 2). The scanty faunistic information from large areas of the world suggests caution in defining these taxa as endemic ones.

Taxonomic descriptions of freshwater species and iconography have been produced by the authors according to personal, not standardized criteria. Permanent slides useful for comparisons are

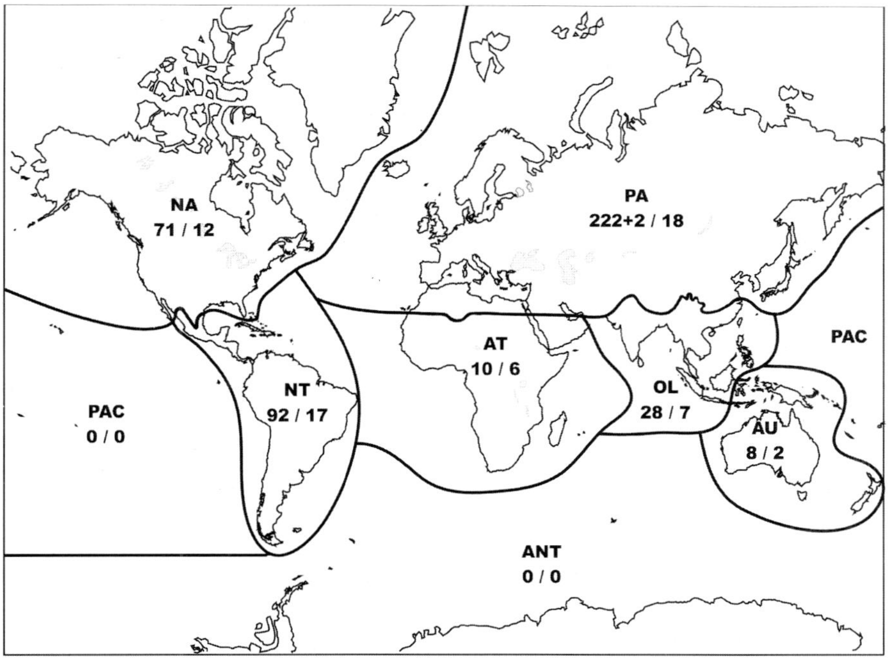

Fig. 2 Total number and zoogeographical distribution of inland water (freshwater + brackish water) species and genera of Gastrotricha. PA, Palearctic; NA, Nearctic; NT, Neotropical; AT, Afrotropical; OL, Oriental; AU, Australasian, PAC, Pacific Oceanic Islands; ANT, Antarctic

Table 2 Total number of inland-water genera of Gastrotricha per family and per biogeographical region

Biogeographical Region	PA	NA	NT	AT	OL	AU	PAC	ANT	World
Order/Family									
Chaetonotida	17	12	16	6	7	2	0	0	21
Chaetonotidae	7	7	8	4	4	2	0	0	9
Dasydytidae	6	4	5	0	2	0	0	0	7
Dichaeturidae	1	0	0	0	0	0	0	0	1
Neogosseidae	2	1	2	2	1	0	0	0	2
Proichthydiidae	1	0	1	0	0	0	0	0	2
Macrodasyida	1	0	1	0	0	0	0	0	2
incertae sedis	1	0	1	0	0	0	0	0	2
Total	18	12	17	6	7	2	0	0	23

relatively few, often with scarce diagnostic value. Most researchers have only worked in one continent; so that an effective comparison of specimens from different continents, but apparently of the same morphospecies, has been possible in very few cases (see Kisielewski, 1991). Even if some molecular studies have tried to shed light on the relationships of and within the phylum, the current taxonomy is still based on morphological and ultrastructural features. Comparisons are difficult, especially with old species, most of which have been insufficiently described and drawn without details that are now required for identification. Many species show a great morphological variability, and several 'forms', or even subspecies, have been described, but the value of these taxa is questionable. All this makes the gastrotrich taxonomy still unreliable, in spite of the recent systematic revisions by Schwank (1990) and Kisielewski (1991, 1998).

Human related issues

As yet there is a no apparent human related issue for freshwater Gastrotricha but they could be used as bioindicators for the quality of the inland waters, as is the case of Macrodasyida in marine habitats. Diversity of freshwater gastrotrich communities is highest in mesotrophic-eutrophic unpolluted lentic waters, although some species appear to be relatively tolerant to changes in some abiotic factors.

Acknowledgements We are greatly indebted to Mitchell J. Weiss for his very careful and complete review of the manuscript, and his constructive criticism and precious collaboration. We thank the anonymous reviewers for their great improvement of the original manuscript.

References

Balsamo, M. & P. Tongiorgi, 1995. Gastrotricha. In Minelli, A., S. Ruffo & S. La Posta (eds), Checklist delle Specie della Fauna d'Italia. Calderini, Bologna 7: 1–11.

Balsamo, M. & M. A. Todaro, 2002. Gastrotricha. In Rundle, S. D., A. Robertson & J. Schmidt-Araya (eds), Freshwater meiofauna: Biology and ecology. Backhuys Publishers, Leiden: 45–61.

de Beauchamp, P., 1932. Scientific results of the Cambridge expedition to the East African lakes, 1930–1931. 6. Rotifères et Gastrotriches. Zoological Journal of the Linnean Society 38: 231–248.

Boaden, P. J. S., 1985. Why is a Gastrotrich? In Conway Morris, S., J. D. George, R. Gibson & H. M. Plat (eds), The Origins and Relationships of Invertebrates. Clarendon Press, Oxford: 248–260.

Ferraguti, M. & M. Balsamo, 1995. Comparative spermatology of Gastrotricha. In Jamieson, B. G. M., J. Ausio & J. L. Justine (eds), Advances in Spermatozoal Taxonomy and Phylogeny. Mémoires du Muséum national d'Histoire naturelle, Paris 166: 105–117.

Grosso, L. E. & F. Drahg, 1991. Gastrotricos dulceacuicolas de la provincia de Tucuman. IV. Acta Zoologica Lilloana 40: 47–51.

Hochberg, R., 2005. First record of *Polymerurus* (Gastrotricha, Chaetonotida) from Australia with the description of a new species from Queensland and of cuticular ultrastructure in *P. nodicaudus*. Invertebrate Biology 124: 119–130.

Hochberg, R. & M. K. Litvaitis, 2000. Phylogeny of Gastrotricha: a morphology-based framework of Gastrotrich relationships. Biological Bulletin 198: 299–305.

Hochberg, R. & M. K. Litvaitis, 2001. Macrodasyida (Gastrotricha): a cladistic analysis of morphology. Invertebrate Biology 120: 124–135.

d'Hondt, J. L., 1967. Documents sur les Gastrotriches dulcicoles des eaux françaises. Annales de Limnologie 3: 381–397.

d'Hondt, J. L., R. Pourriot & R. Rougier, 2006. Nouvelles observations sur les Gastrotriches d'eau douce de Guyane française. Bulletin mensuel de la Société Linnéenne de Lyon 75: 239–245.

Hummon, W.D., 2001. (ed.) Global database for marine Gastrotricha on CD (hummon@ohio.edu). Ohio University Zoological Collections, Athens, U.S.

Kisielewski, J., 1987. Two new interesting genera of Gastrotricha (Macrodasyida and Chaetonotida) from the Brazilian freshwater psammon. Hydrobiologia 153: 23–30.

Kisielewski, J., 1990. Origin and phylogenetic significance of freshwater psammic Gastrotricha. Stygologia 5: 87–92.

Kisielewski, J., 1991. Inland-water Gastrotricha from Brazil. Annales Zoologici (Warsaw) 43: 1–168.

Kisielewski, J., 1998. Brzuchorzeski (Gastrotricha). Fauna Slodkowodna Polski, Zeszyt 31. Windawnictwo Uniwersytetu Lodzkiego, Lódz, 157 pp.

Kisielewski, J., 1999. A preliminary study of the inland-water Gastrotricha of Israel. Israel Journal of Zoology 45: 135–157.

Lee, J. M. & C. Y. Chang, 2000. Freshwater chaetonotid gastrotrichs in Korea. Korean Journal of Systematic Zoology 16: 87–104.

Manylov, O. G., N. S. Vladychenskaya, I. A. Milyutina, O. S. Kedrova, N. P. Korokov, G. A. Dvoryanchikov, V. V. Aleshin & N. B. Petrov, 2004. Analysis of 18S rRNA gene sequences suggests significant molecular differences between Macrodasyida and Chaetonotida (Gastrotricha). Molecular Phylogenetics and Evolution 30: 850–854.

Martin, L. V., 1990. Further observations on Gastrotrichs in Surrey and a provisional British list. Microscopy 36: 415–425.

Naidu, K. V., C. Rao, 2004. The Fauna of India and the adjacent countries—Gastrotricha. Zoological Survey of India, Kolkata, p. 169.

Remane, A., 1935–36. Gastrotricha und Kinorhyncha. In Bronn, H. G. (ed.), Klassen und Ordnungen des Tierreichs, Band 4, Abteilung II, Buch I, Teil 2, Liefrungen 1–2, Akademische Verlagsgesellschaft, Leipzig: 1–385.

Ricci, C. & M. Balsamo, 2000. The biology and ecology of lotic rotifer and gastrotrichs. Freshwater Biology 44: 15–28.

Rudescu, L., 1967. Gastrotricha. Fauna Republicii Socialiste Romania, Vol. 2. Academia Republicii Socialiste Romania, Bucuresti, pp. 289 .

Ruttner-Kolisko, A., 1955. *Rheomorpha neiswestnovae* und *Marinellina flagellata*, zwei phylogenetisch interessante Würmtypen aus dem Süsswasserpsammon. Ősterreichische Zoologischer Zeitschrift 6: 33–69.

Schmidt-Rhaesa, A., 2002. Two dimensions of biodiversity research exemplified by Nematomorpha and Gastrotricha. Integrative and Comparative Biology 42: 633–640.

Schwank, P., 1990. Gastrotricha. In Schwoerbel, J. & P. Zwick (eds), Süsswasserfauna von Mitteleuropa, Band 3, Teil1/2. Gustav Fisher Verlag, Stuttgart, pp. 252.

Sudzuki, M., 1975. A list of Rotifera and Gastrotricha from garden ponds of Tokyo 1974–75 and some notes on *Rhinoglena, Fadeewella, Neogossea*, etc. Proceedings of the Japanese Society of Systematic Zoology 11: 5–12.

Todaro, M. A. & C. E. F. Rocha, 2005. Further data on marine gastrotrichs from the State of São Paulo and the first records from the State of Rio de Janeiro (Brazil). Meiofauna Marina 14: 27–31.

Todaro, M. A., M. J. Telford, A. E. Lockyer & D. T. Littlewood, 2006. Interrelationships of the Gastrotricha and their place among the Metazoa inferred from 18S rRNA genes. Zoologica Scripta 35: 251–259.

Tongiorgi, P., E. Fregni & M. Balsamo, 1999. Gastrotricha from Italian brackish environment with description of a new species of *Chaetonotus*. Journal of the Marine Biological Association, U.K. 79: 585–592.

Tretjakova, E. I., 1991. *Lepidodermella spinifera*, a new species of Gastrotricha with intermediate generic features. Byulletin Moskovogo Obshchestva Ispytatelej Prirody, Otdel Biologiceskij 96: 79–85.

Weiss, M. J., 2001. Widespread hermaphroditism in freshwater gastrotrichs. Invertebrate Biology 120: 308–341.

Zelinka, C., 1889. Die Gastrotrichen. Eine monographische Darstellung ihrer Anatomie, Biologie und Systematik. Zeitschrift für Wissenschaftliche Zoologie 49: 209–384.

Zrzavy, J., 2002. Gastrotricha and metazoan phylogeny. Zoologica Scripta 32:61–82.

Hydrobiologia (2008) 595:93–99
DOI 10.1007/s10750-007-9007-3

Global diversity of bryozoans (Bryozoa or Ectoprocta) in freshwater

Jos A. Massard · Gaby Geimer

Abstract The present study considers 88 bryozoan species occurring in freshwater: 69 phylactolaemate and 19 gymnolaemate species. Roughly 49% of these species are confined to one zoogeographical region. The cosmopolitan status of species like *Fredericella sultana*, *Plumatella repens* or *P. emarginata* has to be reconsidered. Among the Phylactolaemata, which are phylogenetically older than the Gymnolaemata, the gelatinous species (Lophopodidae, Pectinatellidae, Cristatellidae) are more primitive than the branching tubular species (Plumatellidae, Fredericellidae).

Keywords Bryozoa · Ectoprocta · Phylactolaemata · Gymnolaemata · Phylogeny · Zoogeography

Introduction

Bryozoa (Ectoprocta) are small benthic aquatic invertebrates growing on submerged objects as colonies of genetically identical zooids produced by budding (clonal asexual reproduction). They are suspension feeders capturing organic particles with a whorl of ciliated tentacles (lophophore). Whereas earlier estimations amounted to 5600–5700 species, Ryland (2005) estimates that there are more than 8000 extant bryozoan species. Fewer than a hundred species (88 in our checklist) occur in freshwater; most of them belong to the class Phylactolaemata (exclusively living in freshwater), the rest to the class Gymnolaemata (Order: Ctenostomatida) (Fig. 1).

All freshwater and most marine bryozoans are hermaphroditic. Besides sexual reproduction and clonal budding the phylactolaemates reproduce asexually by means of statoblasts (buoyant floatoblasts and fixed sessoblasts), which are very important for species identification. Ctenostomes such as *Paludicella, Victorella, Tanganella, Pottsiella* are known for producing overwintering buds called hibernacula. A new insight into the sexual and asexual reproduction structures of the phylactolaemates is given by Walzl & Wöss (2005) in a recent scanning electron microscopy study of the soft body parts of this group.

All species with branching colonies are strictly sessile, whereas a certain motility has been observed in the gelatinous species *Cristatella mucedo, Pectinatella magnifica, Lophopodella carteri* and *Lophopus crystallinus*. Freshwater bryozoans are not always inconspicuous: giant gelatinous masses of *Pectinatella magnifica*—about 2.5 m long and 0.5 m wide—have been found in Japan; according to T. Wood (pers. comm.) large colonies are also known from

Guest editors: E. V. Balian, C. Lévêque, H. Segers & K. Martens
Freshwater Animal Diversity Assessment

J. A. Massard · G. Geimer
Musée national d'histoire naturelle, 25 rue Munster, Luxembourg 2160, Luxembourg

J. A. Massard (✉)
1A rue des Romains, Echternach 6478, Luxembourg
e-mail: jmassard@pt.lu

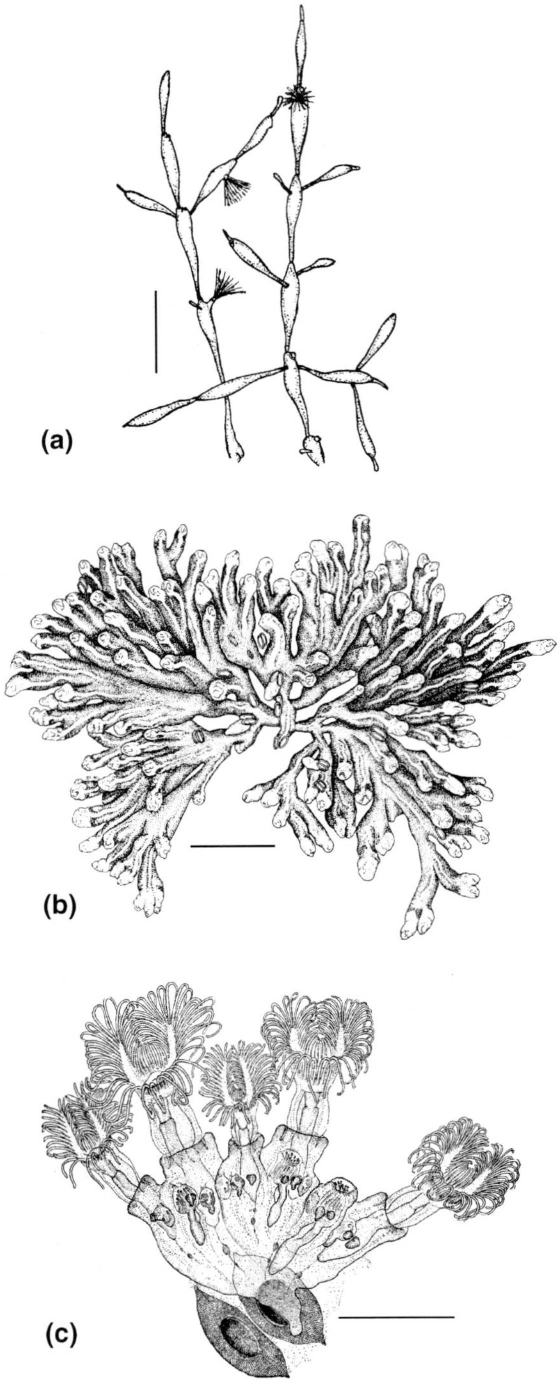

Fig. 1 Habitus of freshwater bryozoans. (**a**) *Paludicella articulata*, portion of a colony, bar = 2 mm (from Prenant & Bobin, 1956). (**b**) *Plumatella casmiana*, young colony, bar = 2 mm (after Rogick, from Geimer & Massard, 1986). (**c**) *Lophopus crystallinus*, zooids of a young colony still attached to the valves of the floatoblast, bar = 1 mm (from Brien, 1960)

both Koreas. *Plumatella fungosa* may form colonies exceeding the size of an adult human head.

Chromosome numbers are known for certain species: *Fredericella sultana*: $2n = 14$; *Plumatella emarginata*: $2n = 14$ or 16; *Cristatella mucedo*: $2n = 16$; *Pectinatella magnifica*: $2n = 18$; *Paludicella articulata*: $2n = 20$ or 22, *Pottsiella erecta*: $2n = 22$ or 24.

Taxonomy and species/generic diversity

Since the publication of the classic works of Prenant & Bobin (1956), Brien (1960) and Wiebach (1960), phylactolaemate taxonomy has advanced considerably (Lacourt, 1968; Mukai, 1999; Wood & Okamura, 2005). The latest revision of phylactolaemate systematics is due to Vinogradov (2004) and still waiting for general acceptance.

The number of species has significantly increased, but proper species discrimination remains problematic and often requires the examination of the floatoblasts and/or sessoblasts by scanning electron microscopy, whose taxonomical potentialities were made out by Wiebach (1975) and Mundy (1980). Geimer & Massard (1986) were "the first to use this tool in a systematic way, clearly distinguishing *Plumatella repens* from *P. fungosa*, and laying to rest the enigmatic *P. coralloides*" (Wood & Okamura, 2005).

Some of the 19 gymnolaemate species in our checklist (Table 1) may be synonymous and need further studies. The most recently discovered species is *Sineportella forbesi* Wood & Marsh, 1996. Just like Wiebach & d'Hondt (1978), we have not included ctenostomate species such as *Bowerbankia*, nor the brackish and estuarine representatives of the cheilostomates (Membraniporidae, Electridae).

Lacourt (1968) lists 32 phylactolaemate species (including 3 species he considers as doubtful), and Bushnell (1973) 39 species. Wood (2002) assumes that there are a total of 77 species (including 24 new species and the confirmation of 14 others). More cautiously Wood & Okamura (2005) state that "the number of phylactolaemate species well exceeds 65, with more expected from Asia and South America". Our own list includes 69 phylactolaemate species (Table 1), but we are aware that some may be doubtful.

Table 1 Numeric data for the different taxa of freshwater Bryozoa

Taxon	Families	Genera	Species
Gymnolaemata Allman, 1856	5	8	19
Victorellidae Hincks, 1880		4	8
Bulbella Braem, 1951			1
Sineportella Wood & Marsh, 1996			1
Tanganella Braem, 1951			2
Victorella Saville Kent, 1870			4
Pottsiellidae Braem, 1940		1	1
Pottsiella Kraepelin, 1887			1
Paludicellidae Allman, 1885		1	2
Paludicella Gervais 1836			2
Arachnidiidae Hincks, 1877		1	1
Arachnoidea (Moore, 1903)			1
Hislopiidae Jullien, 1885		1	7
Hislopia Carter, 1858*			7
Phylactolaemata Allman, 1856	5	16	69
Fredericellidae Hyatt, 1868		2	6
Fredericella Gervais 1838			5
Internectella Gruncharova, 1971			1
Plumatellidae Allman, 1856		9	54
Afrindella Wiebach, 1964			3
Australella Annandale, 1910			1
Gelatinella Toriumi, 1955			1
Hyalinella Jullien, 1885			5
Plumatella Lamarck, 1816			37
Stephanella Oka, 1908			1
Stolella Annandale, 1909			2
Swarupella Shrivastava, 1981			1
Varunella Wiebach, 1974			3
Pectinatellidae Lacourt, 1968		1	1
Pectinatella Leidy, 1851			1
Cristatellidae Allman, 1856		1	1
Cristatella Cuvier, 1798			1
Lophopodidae Rogick, 1935		3	7
Asajirella Oda & Mukai, 1989			1
Lophopodella Rousselet, 1904			5
Lophopus Dumortier, 1835			1
Ectoprocta Nitsche, 1870	10	24	88

*Incl. *Norodonia* Jullien, 1880 and *Echinella* Korotnev, 1901

The most recently described new species are: *Plumatella geimermassardi* Wood & Okamura, 2004; *Plumatella bushnelli* Wood, 2001; *Plumatella mukaii* Wood, 2001; *Plumatella nodulosa* Wood, 2001; *Plumatella similirepens* Wood, 2001.

Phylogeny and historical processes

The most ancient bryozoans are marine stenolaemates from the Lower Ordovician (about 500 million years old). Traces of fossil marine gymnolaemates are present in the Upper Ordivician; there are no fossil records of freshwater gymnolaemates. Fossil phylactolaemates are known only by their statoblasts; among the oldest are plumatellid statoblasts from the Upper Permian in the Asian part of the former U.S.S.R. and possibly pectinatellid statoblasts from the Upper Triassic Molteno Formation in South Africa.

The phylogeny of the gymnolaemate superfamilies with fresh or brackish water representatives (Hislopioidea, Paludicelloidea, Arachnidioidea, Victorelloidea) is illustrated by Todd (2000).

Concerning the phylactolaemates, Toriumi (1956) established the following phylogenetic series based on statoblast structure: *Fredericella, Stephanella, Plumatella/Hyalinella, Gelatinella, Lophopus, Lophopodella, Pectinatella* and *Cristatella.* A similar but more extended phylogenetic tree, additionally based on the number of tentacles, has been published by Lacourt (1968). Using a cladistic approach Mukai (1999) proposes a tree (Fig. 2) where first *Fredericella* and subsequently *Stephanella* separate from the stem of the remaining genera. Then two evolutionary lines appear: (1) *Hyalinella, Lophopus, Lophopodella, Asajirella*; (2) *Gelatinella, Pectinatella, Cristatella.* The genus *Plumatella* and these two monophyletic groups are assumed to be "sister groups that form an unresolved trichotomy".

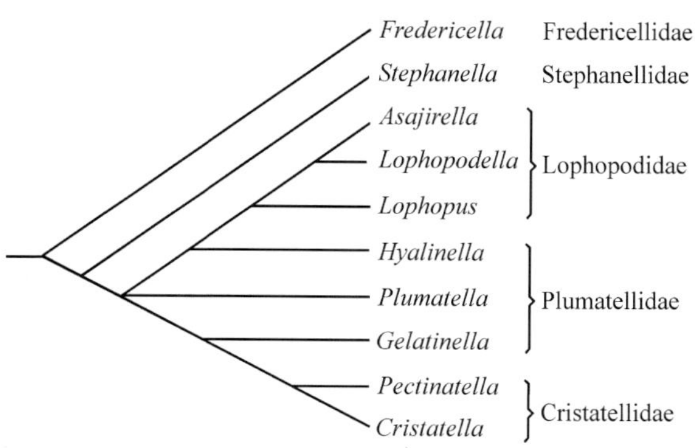

Fig. 2 Phylogeny of the Phylactolaemata proposed by Mukai (1999). The genus *Hyalinella* is here represented by a single species (*H. punctata*) and the genus *Plumatella* includes some allied genera

The phylogenetic tree suggested by distance analysis of 18S rDNA is quite different (Wood & Lore, 2005). At the base of the tree (Fig. 3a) are all globular colonies with large, hooked statoblasts and large lophophores (Lophopodidae, Pectinatellidae, Cristatellidae); at the top of the tree we find the branching, tubular colonies with relatively smaller statoblasts and lophophores (Plumatellidae incl. *Hyalinella punctata*, Fredericellidae). The same study suggests that the phylactolaemates are more closely linked to the phoronids and the brachiopods than to the gymnolaemates (Fig. 3b). This is consistent with studies on NOR-chromosome morphology that suggest a possible derivation of the phylactolaemates from the morphologically similar phoronids (Backus & Banta, 2002).

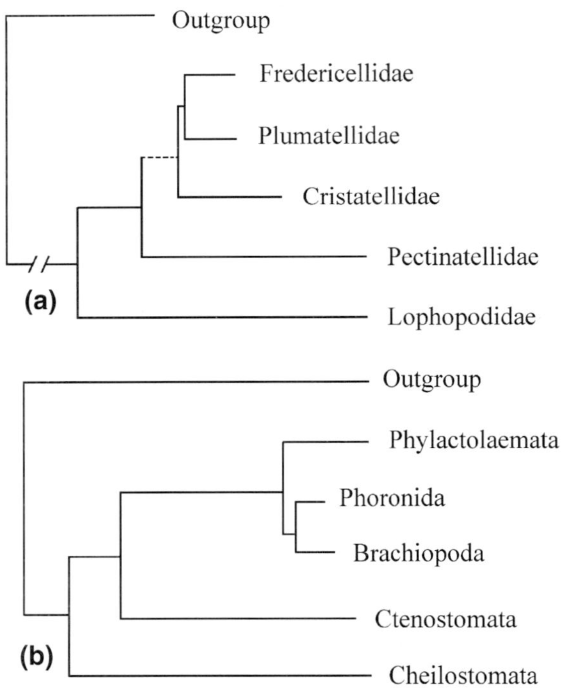

Fig. 3 (**a**) Phylogenetic tree suggested by distance analysis (UPGMA) of 18S rDNA sequence data for 5 families (9 species) of phylactolaemate bryozoans using two phoronid species as the outgroup. Dotted lines have an undefined length (from Wood & Lore, 2005. © Taylor & Francis Group, London). (**b**) Phylogenetic tree suggested by distance analysis (UPGMA) of 18S rDNA sequence data for 14 lophophore-bearing species representing 4 distinct groups, with two cnidarian species serving as the outgroup (from Wood & Lore, 2005. © Taylor & Francis Group, London)

Present distribution and main areas of endemicity

Continental bryozoans are found in all types of waterbodies: ponds, lakes, rivers, estuaries. All the phylactolaemates are freshwater inhabitants. The freshwater gymnolaemates are the ctenostome species: *Paludicella articulata, Pottsiella erecta, Hislopia* spp., *Sineportella forbesi*, *Victorella symbiotica* (Lake Tanganyika) and *Arachnoidea raylankesteri* (Lake Tanganyika, but also brackish lake Birket el Quarum in Eypt). Some other ctenostome species have a preference for brackish water, but may occur also in freshwater: *Victorella continentalis, Bulbella abscondita* (Upper Elbe, Germany), *Tanganella muelleri* (Lake Trasimeno, Italy); *V. bengalensis* is even common in purely freshwaters of Southeast Asia. *Victorella pavida* and *V. bergi* are more tightly bound to brackish water, but traditionally included in the freshwater list (one known inland ocurrence of *V. bergi* is the salt rich Lake Aral in central Asia).

The zoogeographical distribution of freshwater bryozoans has been analysed by Lacourt (1968), Bushnell (1973) and Wood (2002). The checklist compiled for the present article includes a total of 88 species (Tables 2, 3, Fig. 4). A total of 43 species (48.8%) are limited to one zoogeographic region: 12 gymnolaemates (13.6%) and 31 phylactolaemates (35.2%). A total of 22 out of these 43 species are known only from a very restricted area (one or two sites).

The long time accepted cosmopolitan status of species like *Fredericella sultana, Plumatella repens* and *P. emarginata* (Bushnell, 1973) has been challenged (Wood, 2002). Former records of *P. emarginata* may correspond to *P. mukaii*, a new species whose present range includes Japan, Taiwan, Korea, China, India, Indonesia and even Chile, or to *P. reticulata*, another newly described species strongly resembling *P. emarginata*. Former records of *P. repens* may include new species such as *P. nitens, P. nodulosa, P. orbisperma, P. recluse, P. rugosa* and *P. similirepens* (Wood, 2002). Only *P. casmiana* now approaches cosmopolitan status (Wood, 2002), although it is lacking in Australia and South America.

Formerly considered as a holarctic species (Lacourt, 1968, Bushnell, 1973), *Plumatella fungosa* has been identified in New Zealand. *Lophopus crystallinus*, *Cristatella mucedo* and probably also

Table 2 Number of genera recorded in the various zoogeographic regions

Taxon	PA	NA	NT	AT	OL	AU	PAC	World
Gymnolaemata	6	4	4	2	3	2	–	8
Victorellidae	3	2	1	1	1	1	–	4
Pottsiellidae	–	1	1	–	–	–	–	1
Paludicellidae	1	1	1	–	1	1	–	1
Arachnidiidae	1	–	–	1	–	–	–	1
Hislopiidae	1	–	1	–	1	–	–	1
Phylactolaemata	13*	9*	6	8*	14	4	1	16
Fredericellidae	2	1	1	1	2	1	–	2
Plumatellidae	6	4	3	6	9	2	1	9
Pectinatellidae	1	1	1	–	1	–	–	1
Cristatellidae	1	1	–	–	–	–	–	1
Lophopodidae	3	2	1	1	2	1	–	3
Total	19*	13*	10	10*	17	6	1	24

*Including *Stolella*

PA: Palaearctic; NA: Nearctic; NT: Neotropical; AT: Afrotropical; OL: Oriental; AU: Australasian; PAC: Pacific Oceanic Islands

P. fruticosa have kept their holarctic status. *P. fruticosa* reports from India are incorrect (Wood, pers. comm.), and those from Brazil are doubtful. A specimen of *C. mucedo* in the Bryozoa collection at the Zoological Survey of India in Calcutta is not from India but from England (Wood, pers. comm.).

Some species have a disjunct distribution. *Plumatella reticulata* is frequent in North America, but also known from Panama, Israel and Italy. *Plumatella bushnelli* has been recorded in North America and New Zealand. Some of these disjunct distribution patterns are possibly related to the special dispersion means of bryozoans: statoblasts transported by migrating birds (on their feathers and feet, in their guts) or human activity (trade in fish and aquatic plants, shipping traffic, etc.).

There are large gaps in our knowledge of the freshwater bryozoan fauna of Africa, South America, Australia, etc. Even in Europe there remain practically unexplored countries: Portugal, Greece, Albania, etc.

Table 3 Zoogeographical distribution (number of species per family in the different zoogeographic regions)

Taxon	PA	NA	NT	AT	OL	AU	PAC	World
Gymnolaemata	10 (6)	4 (1)	6 (1)	2 (1)	7 (3)	2	0 (0)	19 (12)
Victorellidae	5 (4)	2 (1)	1	1 (1)	2 (1)	1	–	8 (7)
Pottsiellidae	–	1	1	–	–	–	–	1 (0)
Paludicellidae	1	1	2	–	1	1	–	2 (0)
Arachnidiidae	1	–	–	1	–	–	–	1 (0)
Hislopiidae	3 (2)	–	2 (1)	–	4 (2)	–	–	7 (5)
Phylactolaemata	34 (5)	28 (4)	24 (7)	20 (4)	34 (9)	14 (2)	2 (0)	69 (31)
Fredericellidae	4	4	3 (1)	2	4	2	–	6 (1)
Plumatellidae	24[b] (5)	20[b](4)	19[d] (6)	14[b] (3)	25 (8)	11 (2)	2	54 (28)
Pectinatellidae	1	1	1	–	1	–	–	1 (0)
Cristatellidae	1	1	–	–	–	–	–	1 (0)
Lophopodidae	4	2	1	4[c] (1)	4[c] (1)	1	–	7 (2)
Total	44[a] (11)	32[b] (5)	30[d] (8)	22[c] (5)	41[c] (12)	16 (2)	2 (0)	88 (43)

World = total of species per taxon (one given species often occurring in more than one zoogeographic region, the sum of the species numbers of the different regions normally exceeds the total number of species of the family); number in brackets = number of species confined to one zoogeographic region only. PA: Palaearctic; NA: Nearctic; NT: Neotropical; AT: Afrotropical; OL: Oriental; AU: Australasian; PAC: Pacific Oceanic Islands

[a] Including *Fredericella australiensis, Plumatella javanica, Stolella indica* (occurrence/identification to be confirmed)

[b] Including *Stolella indica* Annandale, 1909 (occurrence/identification to be confirmed)

[c] Including *Lophopodella stuhlmanni* Kraepelin, 1914 (doubtful species, considered as *L. carteri* by Toriumi)

[d] Not including *Plumatella fruticosa* (occurrence/identification to be confirmed)

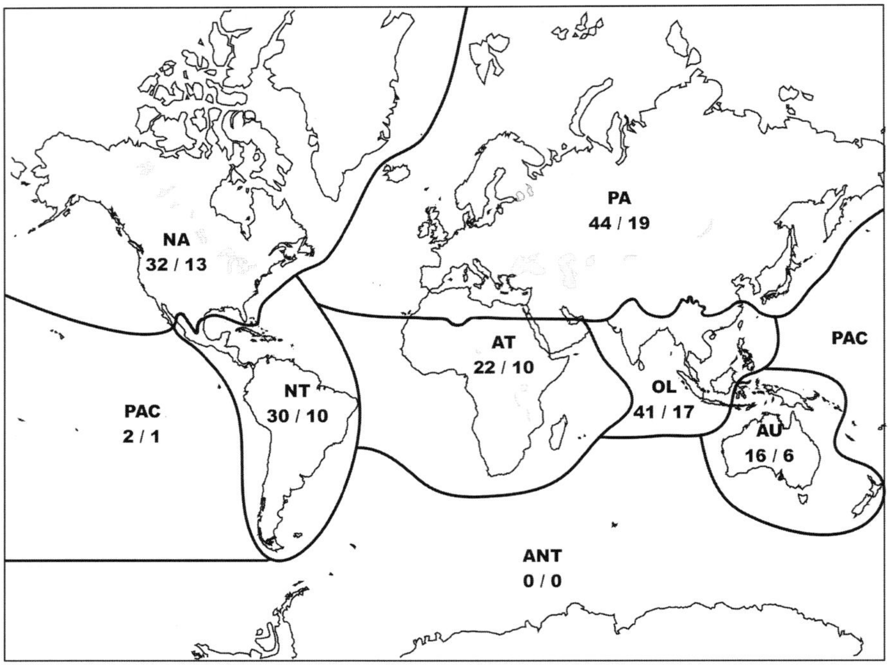

Fig. 4 Zoogeographical distribution map (species number /genus number per region; total: 88/24). PA, Palaearctic; NA, Nearctic; NT, Neotropical; AT, Afrotropical; OL, Oriental; AU, Australasian; PAC, Pacific Oceanic Islands; ANT, Antarctic

Human related issues

A non-protein neurotoxin produced by *Lophopodella carteri* is able to kill fish, probably through inhibition of neurotransmission. Other bioactive components are known from several marine bryozoan species, e.g. bryostatin, an anti-cancer drug produced by *Bugula neritina* (Gymnolaemata: Cheilostomatida).

Freshwater bryozoans are the hosts of the myxozoan parasite *Tetracapsuloides bryosalmonae*, the causitive agent of "proliferative kidney disease" (PKD), a disease of salmonids responsible for economically significant losses in farmed fish and severe reductions in wild fish populations in Europe and North America.

As common fouling animals freshwater bryozoans are occasionally thriving in waterpipes; moreover they may be a nuisance in drinking water treatment stations, wastewater treatment plants and in cooling circuits of thermal or nuclear power stations.

References

Backus, B. T. & W. C. Banta, 2002. NOR-chromosome morphology and evidence for rDNA selection in phylactolaemates. Hydrobiologia 482: 89–95.

Brien, P., 1960. Classe des Bryozoaires. In Grassé, P. (ed.), Traité de Zoologie, 5(2), Paris, Masson: 1054–1335.

Bushnell, J. H., 1973. The freshwater Ectoprocta: a zoogeographical discussion. In Larwood G.P. (ed.), Living and Fossil Bryozoa. Academic Press, London & New York: 503–521.

Geimer, G. & J. A. Massard, 1986. Les Bryozoaires du Grand-Duché de Luxembourg et des régions limitrophes. Travaux Scientifiques du Musée d'Histoire Naturelle de Luxembourg, 7: 1–187.

Lacourt, A. W., 1968. A monograph of the freshwater Bryozoa: Phylactolaemata. Zoölogische Verhandelingen 93: 1–159.

Mukai, H., 1999. Comparative morphological studies on the statoblasts of lower phylactolaemate bryozoans, with discussion on the systematics of Phylactolaemata. Science Reports of the Faculty of Education Gunma University 46: 51–91.

Mundy, S. P., 1980. Stereoscan studies of phylactolaemate bryozoan statoblasts including a key to the statoblasts of the British and European Phylactolaemata. Journal of Zoology 192: 511–530.

Prenant, M. & G. Bobin, 1956. Bryozoaires. Première partie. Entoproctes, Phylactolèmes, Cténostomes. Faune de France, 60: 1–398.

Rogick, M. D., 1941. Studies on freshwater Bryozoa. X. The occurrence of *Plumatella casmiana* in North America. Transactions of the American Microscopical Society, 60: 211–220.

Ryland, J., 2005. Bryozoa: an introductory overview. Denisia 16: 9–20.

Todd, J. A., 2000. The central role of Ctenostomes in bryozoan phylogeny. In Herrera Cubilla, A. & J. B. C. Jackson (eds), Proceedings of the 11th International Bryozoology Association Conference. Balboa (Panama), Smithsonian Tropical Research Institute: 104–135.

Toriumi, M., 1956. Taxonomical study on fresh-water Bryozoa. XVII. General consideration: Interspecific relation of described species and phylogenic consideration. Science Reports of Tohoku University Series IV (Biology) 22: 57–88.

Vinogradov, A. V., 2004. Taxonomical structure of Bryozoans Phylactolaemata. Vestnik Zoologii 38(6): 3–14.

Walzl, M. G. & E. Wöss, 2005. The soft body parts of freshwater bryozoans depicted by scanning electron microscopy. Denisia 16: 49–58.

Wiebach, F., 1960. Bryozoa. In Brohmer, P. et al. (eds), Die Tierwelt Mitteleuropas, 1(8): 1–56, pl. 1–19.

Wiebach, F., 1975. Specific structures of sessoblasts (Bryozoa, Phylactolaemata). In Pouyet, S. (ed.), Bryozoa 1974. Documents des Laboratoires de Géologie de la Faculté des Sciences de Lyon H.S. 3(1): 149–154, pl. 1–3.

Wiebach, F. & J. L. d'Hondt, 1978. Bryozoa. In Illies, J. (ed.), Limnofauna Europaea, 2. Aufl. G. Fischer, Swets & Zeitlinger, Stuttgart, New York, Amsterdam: 492–493.

Wood, T. S., 2002. Freshwater bryozoans: a zoogeographical reassessment. In Wyse Jackson P. N., C. J. Buttler & M. E. Spencer Jones (eds), Bryozoan Studies 2001. Swets & Zeitlinger, Lisse: 339–345.

Wood, T. S. & M. Lore, 2005. The higher phylogeny of Phylactolaemate bryozoans inferred from 18S ribosomal DNA seqences. In Moyano, H. I., J. M. Cancino & P. N. Wyse Jackson (eds), Bryozoan Studies 2004. A.A. Balkema Publishers, Leiden, London: 361–367.

Wood, T. S. & B. Okamura, 2005. A new key to the freshwater bryozoans of Britain, Ireland and Continental Europe, with notes on their ecology. Freshwater biological Association Scientific Publication 63: 1–113.

Hydrobiologia (2008) 595:101–106
DOI 10.1007/s10750-007-9123-0

FRESHWATER ANIMAL DIVERSITY ASSESSMENT

Global diversity of tardigrades (Tardigrada) in freshwater

James R. Garey · Sandra J. McInnes · P. Brent Nichols

Abstract Tardigrada is a phylum closely allied with the arthropods. They are usually less than 0.5 mm in length, have four pairs of lobe-like legs and are either carnivorous or feed on plant material. Most of the 900+ described tardigrade species are limnoterrestrial and live in the thin film of water on the surface of moss, lichens, algae, and other plants and depend on water to remain active and complete their life cycle. In this review of 910 tardigrade species, only 62 species representing13 genera are truly aquatic and not found in limnoterrestrial habitats although many other genera contain limnoterrestrial species occasionally found in freshwater.

Guest editors: E.V. Balian, C. Lévêque, H. Segers & K. Martens
Freshwater Animal Diversity Assessment

Electronic supplementary material The online version of this article (doi:10.1007/s10750-007-9123-0) contains supplementary material, which is available to authorized users.

J. R. Garey (✉) · P. B. Nichols
Division of Cell Biology, Microbiology and Molecular Biology, University of South Florida, Tampa, FL 33620, USA
e-mail: garey@cas.usf.edu

S. J. McInnes
British Antarctic Survey, Natural Environment Research Council, Madingley Road, Cambridge CB3 0ET, UK

Keywords Tardigrada · Biogeography · Phylogeny · Distribution · Diversity

Introduction

Tardigrada is a phylum allied with arthropods. Tardigrades are generally less than 0.5 mm in size, bilaterally symmetrical, and have four pairs of legs. Their biology has been reviewed by Kinchin (1994), Nelson & Marley (2000), and Nelson (2002). Tardigrades are found in freshwater habitats, terrestrial environments, and marine sediments. The tardigrades living in terrestrial environments are the most well-known, where they live in the thin film of water found on mosses, lichens, algae, other plants, leaf litter, and in the soil and are active when at least a thin film of water is present on the substrate. Tardigrades often live alongside bdelloid rotifers, nematodes, protozoans and other animals. Aquatic freshwater tardigrades live upon submerged plants or in the sediment but are not inhabitants of the water column. Some tardigrade species can live in both aquatic freshwater and limnoterrestrial environments. In this article, the term aquatic and/or freshwater will be used to describe tardigrades that live in relatively large bodies of freshwater such as ponds, lakes, streams and rivers. The term limnoterrestrial will be used to describe tardigrades that live in the thin film of water found on mosses, algae and other plants, leaf litter, and soil.

Most tardigrades are gonochoristic with relatively minor sexual dimorphisms that include males being slightly smaller than females. Hermaphrodism and self-fertilization has been documented in only a few, mostly aquatic species. Parthenogenesis is common and can be associated with polyploidy (Bertolani, 2001). Development from egg deposition to hatching can range from 5–40 days. Eutardigrades have direct development but heterotardigrades can display indirect development where first instar larvae lack an anus and gonopore and have fewer claws than adults. Tardigrades become sexually mature after 2–3 molts and molt 4–12 times during a lifetime of 3 or more months. Many tardigrades can undergo various forms of cryptobiosis to enter an environmentally resistant quiescent state. Examples of cryptobiosis include cryobiosis, resistance to freezing (Somme, 1996) and anhydrobiosis, in which internal water is replaced by trehalose to produce a highly resistant tun that can be revived months later (Guidetti & Jönsson, 2002).

Tardigrades have five indistinct segments; a head, three trunk segments each with a pair of lobe-like legs and a caudal segment that contains a fourth pair of legs. The legs of freshwater aquatic and limnoterrestrial tardigrades terminate in claws. The body is covered with a chitinous cuticle that also lines the fore and hind gut. Heterotardigrades are distinguished by cephalic sensory cirri lacking in eutardigrades. Many heterotardigrades are armored by the presence of thick dorsal cuticular plates. Claw structure is important in tardigrade taxonomy (Pilato, 1969). There are numerous major claw types with many recognized variations that distinguish genera. Tardigrades have a complete gut with a complex buccal-pharyngeal apparatus that is also important in taxonomy. The buccal apparatus consists of a mouth, a buccal tube, a muscular sucking pharynx, and a pair of stylets that can extend through the mouth. Most limnoterrestrial and freshwater aquatic tardigrades feed on juices sucked from moss, lichens, algae, and other plants although some tardigrades are carnivorous and consume other mesofauna such as rotifers and nematodes.

Species/generic diversity

Tardigrades are composed of two classes, four orders, at least 90 genera and 900+ species have been described to date. The most complete taxonomic reference for tardigrade species up to 1982 is that of Ramazzotti & Maucci (1983), while Bertolani (1982) focused on aquatic tardigrade species. The number of described tardigrade species has nearly doubled since 1982 (Guidetti & Bertolani, 2005). Tardigrades can be difficult to classify and in some cases the eggs are needed to discriminate among species. The true number of tardigrade species is clearly higher than the 900+ that are currently described. A few species are cosmopolitan, but most tardigrade species appear to be endemic to limited areas. Many other species once thought to be cosmopolitan are now known to be complex species groups (Pilato & Binda, 2001).

Only a few tardigrade taxa are found exclusively in freshwater aquatic habitats in the literature reviewed for this study. Table 1 lists the 62 species of tardigrades known to be exclusively aquatic. Table 2 lists the 13 genera representing five families that contain freshwater aquatic tardigrade species. Only five genera, *Carphania*, *Dactylobiotus*, *Macroversum*, *Pseudobiotus,* and *Thermozodium* were found to be exclusively aquatic in the literature reviewed for this study that included 910 species. Other genera, including *Amphibolus*, *Doryphoribius*, *Eohypsibius*, *Hypsibius*, *Isohypsibius*, *Mixibius*, *Murrayon* and *Thulinius* contain some species that are aquatic. Limnoterrestrial species and genera are listed in Tables 3 and 4 because limnoterrestrial tardigrades are occasionally found in aquatic habitats. The Palaearctic region has the most aquatic genera and species of tardigrades but this is likely to be a sampling artifact due to differences in the intensity of study in that area while the Oceanic Islands have the least.

Little is known of the distribution of freshwater aquatic tardigrades within a habitat. With limnoterrestrial tardigrades microhabitat can be an important factor in distribution. It has been suggested that oxygen tension, pH of the substratum, moisture content of the moss, the thickness of the moss cushion and altitude may all play a role. The extreme patchy distribution of limnoterrestrial tardigrades within seemingly homogeneous habitat has made it difficult to determine which factors cause the unevenness in their distribution. Habitat distribution studies typically do not include enough sampling to test for statistical significance and many of these studies are essentially species lists from different regions (Garey, 2006).

Table 1 Number of freshwater tardigrade species found in biogeographic regions by family. The zeroes represent either a null record (no information) or absence. See Annex 1 in the online supplemental materials for a more detailed listing. PA = Palaearctic, NA = Nearctic, NT = Neotropical, AT = Afrotropical, OL = Oriental, AU = Australasian, PAC = Pacific Oceanic islands, ANT = Antarctic

Families	PA	NA	AT	NT	OL	AU	PAC	ANT	Total freshwater species per genus	Total species per genus
Heterotardigrada										
Oreellidae	0	0	0	0	0	0	0	0	0	2
Carphaniidae	1	0	0	0	0	0	0	0	1	1
Echiniscidae	0	0	0	0	0	0	0	0	0	229
Eutardigrada										
Murrayidae	13	7	3	5	0	1	1	1	19	24
Macrobiotidae	0	0	0	0	0	0	0	0	0	226
Calohypsibiidae	0	0	0	0	0	0	0	0	0	21
Microhypsibiidae	0	0	0	0	0	0	0	0	0	7
Eohypsibiidae	2	1	0	0	0	0	0	0	2	9
Necopinatidae	0	0	0	0	0	0	0	0	0	1
(*Incertae sedis*)	0	0	0	0	0	0	0	0	0	3
Hypsibiidae	33	12	7	8	3	4	1	6	39	368
Milnesiidae	0	0	0	0	0	0	0	0	0	18
sp inquirenda	0	0	0	0	1	0	0	0	1	1
Total	49	20	10	13	4	5	2	7	62	910

Thermozodium esakii is a species of tardigrade reported from a hot spring in Japan and has been proposed to represent a third class of tardigrades known as Mesotardigrada. Neither the type specimens nor locality exist and similar specimens have not been found (Nelson 2002)

Table 2 Number of freshwater tardigrade genera found in biogeographic regions. PA = Palaearctic, NA = Nearctic, NT = Neotropical, AT = Afrotropical, OL = Oriental, AU = Australasian, PAC = Pacific Oceanic islands, ANT = Antarctic

Families	PA	NA	AT	NT	OL	AU	PAC	ANT	Total freshwater genera per family	Total genera per family
Heterotardigrada										
Oreellidae	0	0	0	0	0	0	0	0	0	1
Carphaniidae	1	0	0	0	0	0	0	0	1	1
Echiniscidae)	0	0	0	0	0	0	0	0	0	12
Eutardigrada										
Murrayidae	1	1	1	1	0	1	1	1	3	3
Macrobiotidae	0	0	0	0	0	0	0	0	0	11
Calohypsibiidae	0	0	0	0	0	0	0	0	0	5
Microhypsibiidae	0	0	0	0	0	0	0	0	0	5
Eohypsibiidae	1	1	0	0	0	0	0	0	2	2
Necopinatidae	0	0	0	0	0	0	0	0	0	1
(*Incertae sedis*)	0	0	0	0	0	0	0	0	0	1
Hypsibiidae	3	3	1	1	0	0	0	0	6	20
Milnesiidae)	0	0	0	0	0	0	0	0	0	3
sp inquirenda[a]	0	0	0	0	1	0	0	0	1	1
Biogeographic totals	6	5	2	2	1	1	1	1	13	66

[a] See footnote in Table 1

Table 3 Number of limnoterrestrial tardigrade species found in biogeographic regions by family. The zeroes indicate either a null record (no information) or absence. See Annex 2 in the online supplementary materials for a more detailed listing. PA = Palaearctic, NA = Nearctic, NT = Neotropical, AT = Afrotropical, OL = Oriental, AU = Australasian, PAC = Pacific Oceanic islands, ANT = Antarctic

Families	PA	NA	AT	NT	OL	AU	PAC	ANT	Total species per genus
Heterotardigrada									
Oreellidae	0	0	0	2	0	1	0	1	2
Carphaniidae	0	0	0	0	0	0	0	0	1
Echiniscidae	130	55	31	64	18	37	12	13	229
Eutardigrada									
Murrayidae	6	5	2	1	1	0	0	0	24
Macrobiotidae	104	35	40	47	21	51	13	14	226
Calohypsibiidae	14	7	0	5	1	1	0	2	21
Microhypsibiidae	4	5	1	1	0	1	1	0	7
Eohypsibiidae	8	3	0	0	0	0	0	0	9
Necopinatidae	1	0	0	0	0	0	0	0	1
(*Incertae sedis*)	2	0	1	0	0	1	0	0	3
Hypsibiidae	235	82	35	60	18	43	8	35	368
Milnesiidae	2	2	3	3	1	3	1	1	18
sp inquirenda	0	0	0	0	0	0	0	0	1
Total	506	194	113	183	60	138	35	66	910

See footnote in Table 1

Table 4 Number of limnoterrestrial tardigrade genera found in biogeographic regions. PA = Palaearctic, NA = Nearctic, NT = Neotropical, AT = Afrotropical, OL = Oriental, AU = Australasian, PAC = Pacific Oceanic islands, ANT = Antarctic

Families	PA	NA	AT	NT	OL	AU	PAC	ANT	Total genera per family
Heterotardigrada									
Oreellidae (1)	0	0	0	1	0	1	0	1	1
Carphaniidae (1)	0	0	0	0	0	0	0	0	1
Echiniscidae (12)	9	9	4	8	3	6	3	6	12
Eutardigrada									
Murrayidae (3)	2	1	1	1	1	1	1	1	3
Macrobiotidae (11)	8	5	5	3	4	3	2	2	11
Calohypsibiidae (5)	3	4	0	4	1	1	0	2	5
Microhypsibiidae (2)	2	2	1	1	0	1	1	0	5
Eohypsibiidae (2)	1	1	0	0	0	0	0	0	2
Necopinatidae (1)	1	0	0	0	0	0	0	0	1
(*Incertae sedis*) (1)	1	0	1	0	0	1	0	0	1
Hypsibiidae (20)	13	11	9	11	6	9	7	8	20
Milnesiidae (3)	1	1	1	1	1	3	1	1	3
sp inquirenda[a]	0	0	0	0	0	0	0	0	1
Biogeographic totals:	42	34	22	30	16	26	15	21	66

[a] See footnote in Table 1

Phylogeny and historical processes

Tardigrada is a phylum associated closely with Onychophora and Arthropoda to form Panarthropoda. Like arthropods and nematodes, tardigrades grow through ecdysis and it has been suggested that they belong to a taxon known as Ecdysozoa that contains all molting animals (Aguinaldo et al., 1997). The two groups of tardigrades known today are the heterotardigrades and the eutardigrades and both groups have marine and freshwater members. A recent family level phylogenetic analysis suggests that tardigrades adapted to freshwater aquatic habitats multiple times (Nichols et al., 2006). The present study suggests tardigrades adapted to freshwater aquatic environments at least twice, once among the heterotardigrades in the family Carphaniidae and at least once among the eutardigrades where representatives of three families (Murrayidae, Eohypsibidae, and Hypsibidae) have freshwater aquatic species.

Little is known of the factors that drive change or speciation in tardigrades. Geographic barriers, reproductive biology and substrate quality all are likely involved. It has been suggested that tardigrades evolve slowly (Pilato & Binda, 2001), aided by periods of cryptobiosis, and because of parthenogenesis, new species or populations can readily appear (McInnes & Pugh 1998). There is only weak evidence that anthropogenic forces have an effect on tardigrade evolution although it is clear that tardigrade distribution is affected by pollution (Steiner, 1994; Hohl et al., 2001).

Biogeographical studies

Figure 1 shows the data from Tables 1 and 2 summarized in the form of a biogeographical map. The northern hemisphere appears to have the most diversity, particularly the palaearctic region, which could be due to the more intensive sampling in Europe compared to other regions. Only a few biogeographical studies have been carried out on terrestrial/freshwater tardigrades (e.g., McInnes & Pugh, 1998; Pilato & Binda, 2001). Terrestrial tardigrades appear to be remarkably endemic at the continental level. One study (Pilato & Binda, 2001) found 68% of terrestrial tardigrade species were found in only one biogeographical region while only 6.8% were cosmopolitan. They also found that within

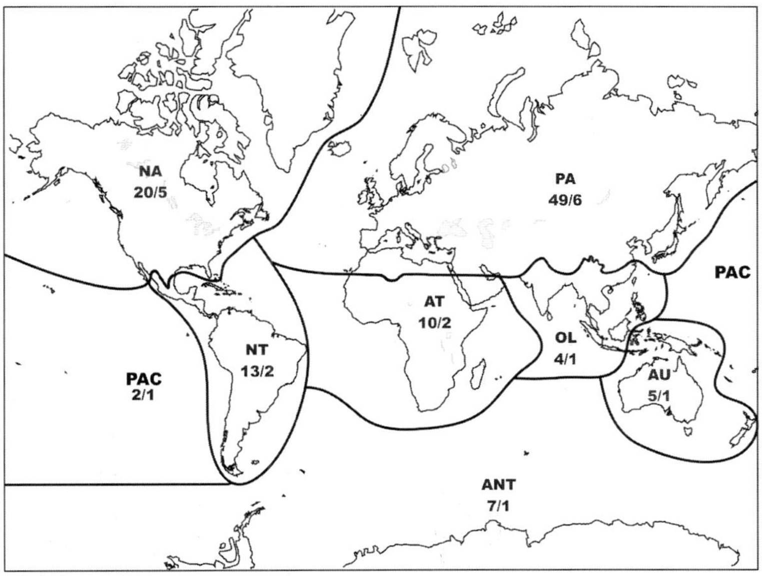

Fig. 1 Summary of the data from Tables 1 and 2 in the context of a biogeographical map. The number preceding the slash represents the number of species that are found exclusively in freshwater aquatic habitats as defined in the text. The number after the slash represents the number of genera with at least one species known to be found exclusively in freshwater aquatic habitats. PA = Palaearctic, NA = Nearctic, NT = Neotropical, AT = Afrotropical, OL = Oriental, AU = Australasian, PAC = Pacific Oceanic islands, ANT = Antarctic

a complex group of species, most often one of the species was cosmopolitan while the other species in the group were endemic to one or a few biogeographical regions. Similar results were found by McInnes & Pugh (1998) where only 22 of the ~800 species considered at that time and 10 of 51 genera were cosmopolitan. They also carried out cluster analyses of tardigrade distribution at the generic and familial level which suggest that 97% of tardigrade species and 82% of genera belong to regional clusters that can be associated with geological events. For example, their cluster analyses show that a laurasian and two gondwanan clusters correlate with the breakup of Pangaea 135 million years ago while two other clusters correspond to the division of East and West Gondwana 65 million years ago.

Economic Importance

Tardigrades have very little economic impact to humans. Their ability to undergo cryptobiosis has created an interest in the medical community and approaches to cell or organ preservation in humans have been tested. Due to the potential medical applications and their pivotal phylogenetic position, branching from the stem lineage that led to arthropods, there has been a renewed interest in the biology of tardigrades at the genomic and proteomic levels. As studies of tardigrade distribution and ecology become more complete they may yet become a useful tool for biogeography (Pilato & Binda, 2001).

References

Aguinaldo, A. M., J. M. Turbeville, L. S. Linford, M. C. Rivera, J. R. Garey, R. A. Raff & J. A. Lake, 1997. Evidence for a clade of nematodes, arthropods, and other moulting animals. Nature 387: 489–493.

Bertolani, R., 1982. 15. Tardigradi. *Guide per il Riconoscimento delle Specie Animali delle Acque Interne Italiane*, Consiglio Nazionale Delle Ricerched, Verona, Italy, pp. 104.

Bertolani, R., 2001. Evolution of the reproductive mechanisms in tardigrades. A review. Zoologischer Anzeiger 240: 247–252.

Garey, J. R., 2006. The biology of tardigrades: and introduction to the 9th international symposium on Tardigrada. Hydrobiologia 558: 1–3.

Guidetti, R. & R. Bertolani, 2005. Tardigrade taxonomy: an updated check list of the taxa and a list of characters for their identification. Zootaxa 845: 1–46.

Guidetti, R. & K. I. Jönsson, 2002. Long-term anhydrobiotic survival in semi-terrestrial micrometazoans. Journal of Zoology, London 257: 181–187.

Hohl, A. M., W. R. Miller & D. Nelson, 2001. The distribution of tardigrades upwind and downwind of a Missouri Coal-burning power plant. Zoolischer Anzeiger 240: 395–401.

Kinchin, I. M., 1994. The Biology of Tardigrades. Portland Press, London.

McInnes, S. J. & P. J. A. Pugh, 1998. Biogeography of limno-terrestrial Tardigrada with particular reference to the Antarctic Fauna. Journal of Biogeography 25: 31–36.

Nelson, D. R., 2001. Tardigrada, In Thorp J. H. & A. P. Covich (eds), Ecology and Classification of North American Freshwater Invertebrates, 2nd edn. Academic Press, New York: 527–550.

Nelson, D. R., 2002. Current status of the Tardigrada: evolution and ecology. Integrative and Comparative Biology 42: 652–659.

Nelson, D. R. & N. J. Marley, 2000. The Biology and ecology of lotic Tardigrada. Freshwater Biology 44:93–108.

Nichols, P. B., D. R. Nelson & J. R. Garey, 2006. A family level analysis of tardigrade phylogeny. Hydrobiologia 558: 53–60.

Pilato, G., 1969. Evoluzione e Nuova Sistemazione degli Eutardigrada. Bolletino di Zoologia 36: 327–345.

Pilato, G. & M. G. Binda, 2001. Biogeography and Limno-terrestrial Tardigrades: are they truly incompatible binomials? Zoologischer Anzeiger 249: 511–516.

Ramazzotti, G. & W. Maucci, 1983. Il Phylum Tardigrada, 3rd edn. (C. W. Beasley, English Translation). Memorie dell'Istituto Italiono di Idrobiologia 41: 1–1012.

Somme, L., 1996. Anhydrobiosis and cold tolerance in tardigrades. European Journal of Entomology 93: 349–57.

Steiner, W. A., 1994. The influence of air pollution on moss-dwelling animals. Revue Suisse de Zoologie 101: 699–724.

Hydrobiologia (2008) 595:107–115
DOI 10.1007/s10750-007-9008-2

Global diversity of polychaetes (Polychaeta; Annelida) in freshwater

Christopher J. Glasby · Tarmo Timm

Abstract A literature review of Polychaeta (Annelida) including Aphanoneura (the oligochaete-like Aeolosomatidae and Potamodrilidae), living in freshwater yielded 168 species, 70 genera and 24 families representing all of the major polychaete clades, but less than 2% of all species. The best-represented families were, in order, Nereididae, Aeolosomatidae, Sabellidae, Spionidae and Histriobdellidae. Fourteen families were represented by a single species and genus. Regions supporting the highest diversity of freshwater polychaetes were in order, Palaearctic, Neotropical, Oriental, Nearctic, Australasian, and Afrotropical. More than half of all species and genera inhabitat lakes and rivers, followed by lagoons/ estuaries, which have a high proportion of euryhaline species, and inland seas. Less common, atypical polychaete habitats include subterranean waters, the hyporheic zone of rivers and plant container habitats (phytotelmata). At least three distinct ecological/ historical processes appear to account for the colonisation of continental waters: invasion of a clade prior to the break-up of Gondwana, as in Aphanoneura, *Namanereis*, *Stratiodrilus*, and *Caobangia*; relatively recent stranding of individual species (relicts); and the temporary visitation of euryhaline species.

Keywords Annelida · Polychaeta · Aphanoneura · Relict · Introduced species · Endemicity · Zoogeography

Guest editors: E. V. Balian, C. Lévêque, H. Segers & K. Martens
Freshwater Animal Diversity Assessment

C. J. Glasby (✉)
Museum & Art Gallery of the Northern Territory, GPO Box 4646, Darwin, NT 0801, Australia
e-mail: chris.glasby@nt.gov.au

T. Timm
Centre for Limnology, Estonian University of Life Sciences, Rannu, Tartumaa 61101, Estonia
e-mail: ttimm@zbi.ee

Introduction

Polychaeta (bristle worms) and Clitellata (oligochaetes and leeches) together comprise the phylum Annelida, or true-segmented worms; however, the taxonomic status of both are currently uncertain, because the Clitellata cluster among the polychaetes making the latter paraphyletic (Rouse & Pleijel, 2001; Struck & Purschke, 2005 and references, therein). Polychaetes have no common morphological features; nevertheless, they can usually be distinguished from Clitellata by the following combination of features: a head with sensory appendages, segmental parapodia bearing numerous chaetae, and most typically they have ciliated pits or patches (nuchal organs) on the back of the head (Glasby et al., 2000; Rouse & Pleijel, 2001). They show sizes from less than a millimetre to over 3 m, although freshwater species tend to be small. There is no typical freshwater polychaete form: motile types tend to be carnivorous or omnivorous and generally have a well-developed head with sensory appendages including eyes, sometimes jaws, and flap-like parapodia which can be highly infused with capillaries that facilitate oxygen exchange (Fig. 1a); sessile types usually live in tubes (soft or calcareous) from which emerge tentacles used in suspension or deposit feeding (Fig. 1b); and commensal or parasitic forms tend to be highly modified for life on their hosts, mostly bivalves and crustaceans (Fig. 1c).

Included among the polychaetes in this review are the enigmatic Aphanoneura, one to several millimetres long, without parapodia but mostly equipped with chaetae arranged in four bundles per segment (Fig. 1d). They usually move by gliding with the help of cilia on the underside of their large prostomium

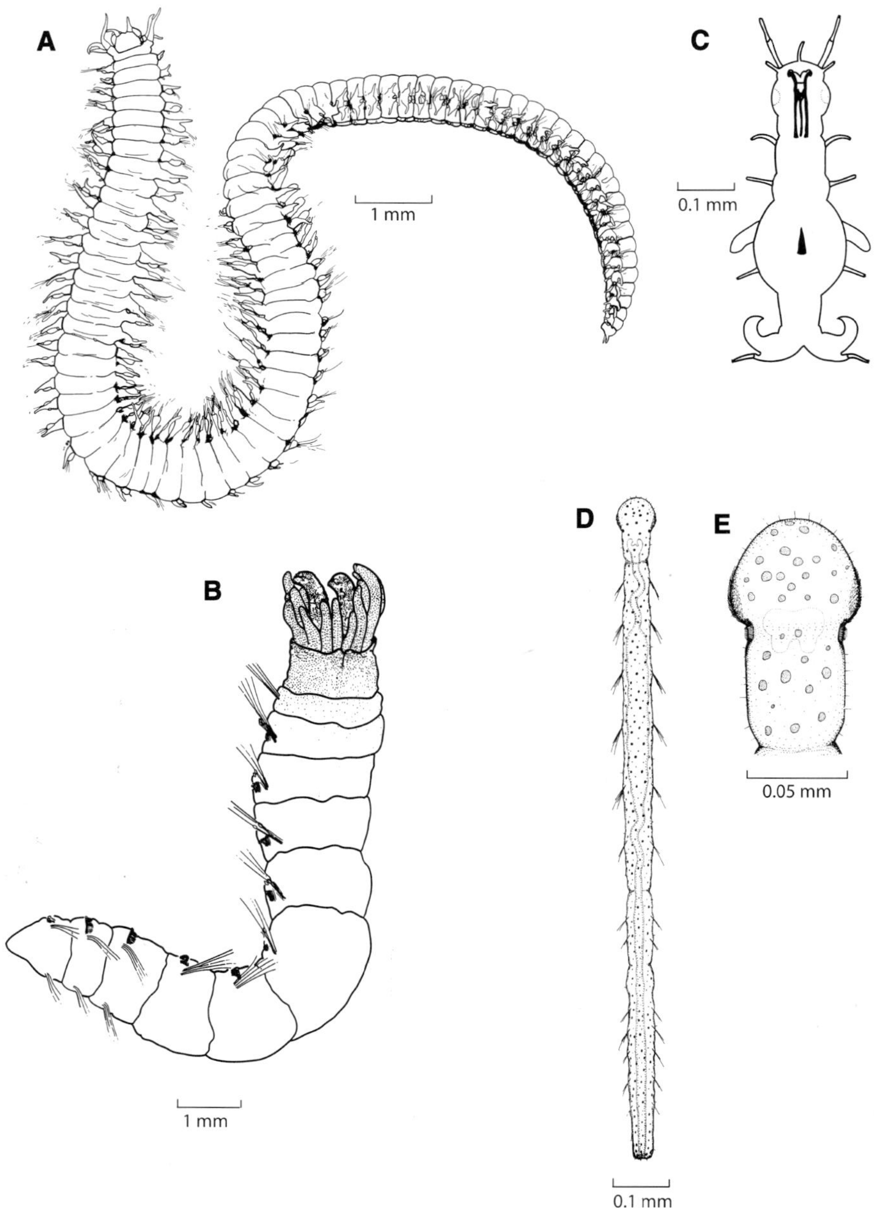

Fig. 1 Habitus of selected freshwater polychaetes. (**A**) *Namanereis cavernicola*, inhabits subterranean aquifers and sinkholes (after Glasby, 1999: Fig. 8c). (**B**) *Manayunkia athalassia*, from saline lake of southern Australia (after Hutchings et al., 1981). (**C**) *Stratiodrilus arreliai* inhabits the branchial lamellae of crabs and crayfish (after Amaral & Morgado, 1997: Fig. 1). (**D**, **E**) *Aeolosoma hemprichi* (after Bunke 1967: Fig. 1a, b). (**D**) Entire body. (**E**) Close up of head

(Fig. 1e). The nominal taxon, as introduced by Timm (1981), comprises two families, Aeolosomatidae and Potamodrilidae; before this, they were treated together as an oligochaete family, Aeolosomatidae. Brinkhurst (1971) first suggested that they do not belong to oligochaetes. Their 'polychaete' nature, as well as the sister-group relationship between Aeolosomatidae and Potamodrilidae, was recently confirmed in gene sequence studies (Struck & Purschke, 2005). Taxonomic reviews are available in Bunke (1967) and Van der Land (1971); some well-known but synonymous names are omitted, as well as numerous species inquirendae or dubiae.

The only previous global review of freshwater polychaetes reported 43 species, comprising 31 purely freshwater species and 12 species found in both fresh and saline waters (Wesenberg-Lund, 1958), although a number of junior synonyms were included. Subsequent reviews are available for Central and South America and the Caribbean (Orensanz, 1977, 1981, 1982); North America (Hartman, 1959; Foster, 1972), South-east Asia (Rouse, 2004), and southern Africa (Day & Day, 2002). The present review includes freshwater species and euryhaline species that have been reported in freshwater. Not counted are the terrestrial species, including *Parergodrilus heideri* Reisinger, *Hrabeiella periglandulata* Pizl & Chalupský, and a few nereidid species rarely encountered in tropical soils. Also excluded are the coastal interstitial species, *Stygocapitella subterranea* Knöllner, *Aeolosoma maritimum* Westheide & Bunke and *A. maritimum dubiosum* Westheide & Schmidt, and species living in marine-dominated subterranean and karst/volcanic habitats such as caves, anchialine ponds and cenotes (sinkholes); this latter habitat supports over 200 polychaete species worldwide (Hartmann-Schröder, 1986).

Diversity

A total of 168 'freshwater' species belonging to 70 genera and 24 families representing all of the major polychaete clades were identified. This includes a few species yet to be formally described. It represents less than 2% of the estimated 9,000 polychaete species, but about 1/3 of the families. The best represented families are in order, Nereididae (55 species, 17 genera), Aeolosomatidae (27/3), Sabellidae (22/8), Spionidae (17/11), Histriobdellidae (10/1), Ampharetidae (6/6), Capitellidae (6/4), Serpulidae (5/2), Nephtyidae (3/1) and Cirratulidae (2/2); the remaining 14 families are represented by a single species and genus (Table 1). Nereididae, Sabellidae and Spionidae are also well represented in marine environments worldwide, while Aeolosomatidae and Histriobdellidae are primarily in freshwater. Histriobdellids are represented in freshwater by the genus *Stratiodrilus*, commensals on the branchial lamellae of crabs and crayfish. Over half of the nereidids belong to a single subfamily, the aptly-named Namanereidinae (Greek, *Nama*, refers to a spring or stream). Most freshwater sabellids belong to the subfamily Fabriciinae, and are either free living (*Monroika* and *Manayunkia*) or bivalve commensals (*Brandika* and *Caobangia*).

Aphanoneura contains 27 valid species of *Aeolosoma*, one species of *Hystricosoma*, one of *Rheomorpha*, and one of *Potamodrilus*; many insufficiently known nominal taxa are omitted, including several subterranean taxa (see lists of species inquirendae and dubiae in Van der Land (1971)). Several species can co-occur in the same benthic habitats; however, the actual diversity at a site often remains obscure without further taxonomic study. The oldest, and therefore, 'most typical' nominal species, *Aeolosoma hemprichi* Ehrenberg, has been extensively recorded from most continents; however misidentifications can be suspected. Supposed 'endemic' species reported from other continents may turn out to be synonyms and some supposedly widely distributed species may be split after subsequent study.

Distribution and endemicity

Freshwater polychaetes are most diverse in the Palaearctic region (67 species representing 32 genera), followed by the Neotropical (53/20), Oriental (48/26), Nearctic (33/22), Australasian (31/15), and Afrotropical regions (12/8) (Table 1; Fig. 2). They rarely occur on oceanic islands (only Nereididae), and are essentially absent from the Antarctic region, except for *Namanereis quadraticeps* Blanchard in Gay, a circum-subantarctic species found in the freshwater seep zones of the upper shores. Although earlier reviews suggested that the Oriental and

Table 1 Numbers of polychaete species and genera (in parentheses) for each major zoogeographical region, arranged alphabetically by family

Taxon	PA	NA	NT	AT	OL	AU	PAC	ANT	World
Aeolosomatidae	25 (3)	8 (1)	14 (1)	3 (1)	5 (1)	2 (1)	1 (1)	–	27 (3)
Ampharetidae	3 (3)	1 (1)	–	–	–	–	–	–	6 (6)
Capitellidae	2 (2)	2 (2)	2 (2)	–	1 (1)	3 (2)	–	–	6 (4)
Cirratulidae	–	2 (2)	–	–	–	–	–	–	2 (2)
Eunicidae	–	–	–	–	1 (1)	–	–	–	1 (1)
Goniadidae	–	–	–	–	1 (1)	–	–	–	1 (1)
Histriobdellidae	–	–	7 (1)	1 (1)	–	2 (1)	–	–	10 (1)
Lumbrineridae	–	–	1 (1)	–	–	–	–	–	1 (1)
Maldanidae	–	–	–	–	1 (1)	–	–	–	1 (1)
Nephtyidae	2 (1)	–	1 (1)	–	2 (1)	–	–	–	3 (1)
Nereididae	8 (4)	10 (8)	21 (9)	5 (3)	17 (9)	15 (5)	9 (3)	1 (1)	55 (17)
Nerillidae	1 (1)	1 (1)	–	–	–	–	–	–	1 (1)
Onuphidae	–	–	–	–	1 (1)	–	–	–	1 (1)
Orbiniidae	–	–	–	–	1 (1)	–	–	–	1 (1)
Paraonidae	1 (1)	–	–	–	–	–	–	–	1 (1)
Phyllodocidae	1 (1)	–	–	–	1 (1)	–	–	–	1 (1)
Pilargidae	–	–	–	–	1 (1)	–	–	–	1 (1)
Pisionidae	–	–	1 (1)	–	–	–	–	–	1 (1)
Potamodrilidae	1 (1)	–	–	–	–	–	–	–	1 (1)
Protodrilidae	1 (1)	–	–	–	–	–	–	–	1 (1)
Sabellidae	8 (3)	2 (2)	–	2 (2)	9 (3)	2 (1)	–	–	22 (8)
Serpulidae	3 (2)	2 (1)	2 (1)	1 (1)	3 (1)	2 (1)	–	–	5 (2)
Spionidae	11 (9)	4 (3)	4 (3)	–	3 (2)	5 (4)	–	–	17 (11)
Sternaspidae	–	–	–	–	1 (1)	–	–	–	1 (1)
Nereidiformia	–	1 (1)	–	–	–	–	–	–	1 (1)
Total	67 (32)	33 (22)	53 (20)	12 (8)	48 (26)	31 (15)	10 (4)	1 (1)	168 (70)

Nereidiformia are Polychaeta *incertae sedis*. PA: Palaearctic; NA: Nearctic; NT: Neotropical; AT: Afrotropical; OL: Oriental; AU: Australasian; PAC: Pacific & Oceanic Islands; ANT: Antarctic

Neotropical regions were the main areas of diversity (e.g., Hartman, 1959; Foster, 1972), they did not include Aphanoneura, a group strongly represented in the Palaearctic. Due to sampling biases and likely taxonomic problems (especially in Aphanoneura), it is not possible to make general statements on large scale endemicity. In particular, the less intensively studied regions (e.g. Afrotropical, eastern Palaearctic) would be expected to show a greater increase in diversity levels following further surveys.

At the regional scale, two areas within the Palaearctic region are notable: Lake Baikal and its tributaries with five endemic species (*Aeolosoma arenicola* Semernoy, *A. singulare* Semernoy, *Manayunkia baicalensis* (Nussbaum), *M. godlewskii* (Nussbaum) and *M. zenkewitschii* Sitnikova) and the Ponto-Caspian region, comprising low saline (0.5–5%) waters of the Black and Caspian Seas, has several characteristic species including *Hypania invalida* (Grube), *Hypaniola kowalewskii* (Grimm), *Parahypania brevicirra* Grimm in Grube, *Manayunkia caspica* Annenkova and *Fabricia stellaris caspica* (Zenkevitsch).

The most common freshwater habitats are lakes and rivers (treated together because of the large number of shared species) with 94 species and 39 genera, followed by coastal lagoons, intermittently isolated lakes and the upper reaches of estuaries with 76 species, 49 genera (Table 2). Inland seas are home to 13 species and 9 genera. Oases/springs and

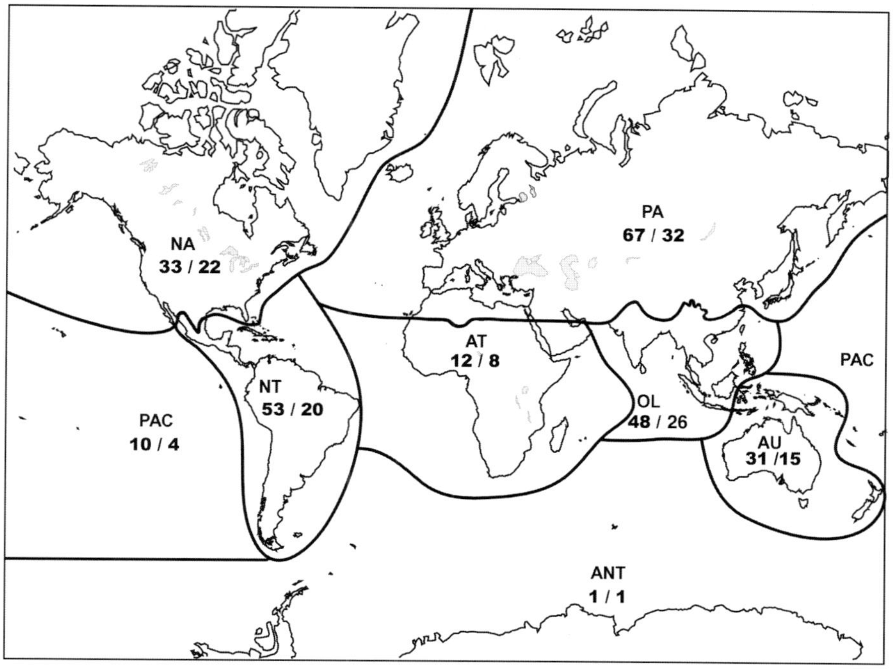

Fig. 2 Numbers of species/genera in each of the major zoogeographical regions (numbers reflect totals in Table 1). Abbreviations used are: PA: Palaearctic; NA: Nearctic; NT: Neotropical; AT: Afrotropical; OL: Oriental; AU: Australasian; PAC: Pacific & Oceanic Islands; ANT: Antarctic

subterranean habitats support a limited polychaete fauna comprising mainly *Namanereis* species (six and seven species, respectively), with most species common to both habitats; the monotypic serpulid, *Marifugia cavatica* Absalon & Hrabě, is apparently restricted to subterranean waters of Croatia. The hyporheic zone is home to the amphi-Atlantic nerillid *Troglochaetus beranecki* Delachaux and the elusive *Namanereis tiriteae* (Winterbourn) from New Zealand and Fiji; the disjunct distribution of both species probably indicates ancient taxa. Plant-associated polychaetes include two species of *Namanereis* and one species of *Namalycastis*, which live in water deposits in the leaf axils (phytotelmata) of suitably formed plants, and some Aphanoneura which live on macrophytes in freshwater lakes, rivers, and streams; however, most Aphanoneura live in sediment (interstitial) and one species is commensal in the branchial chamber of crayfish.

Most freshwater species occur not too far from the sea. Exceptions are the Aphanoneura, the Lake Baikal species, and the river-dwelling species *Namalycastis indica* (Southern) and *Nephtys oligobranchia* Southern found in the Yamuna river about 1600 km from the Ganges delta; even at this distance from the sea the elevation is only 70–100 m above sea level (H. Nesemann, pers. comm.). Other species have been reported at higher elevations, apparently associated with recent tectonic uplift; the most extreme case is that of the nereidid *Lycastoides alticola* Johnson, known only from Sierra de Laguna, Baja California, Mexico, about 2150 m above sea level (Glasby, 1999).

Phylogeny and zoogeography

Although the fossil record for polychaetes is poor, most of the major lineages evidently appeared by the end of the Carboniferous (Rouse & Pleijel, 2001). Surprisingly only a few lineages of Annelida have been successful in colonising freshwater in this time. Apart from the Clitellata, which have had a major radiation on the land, only four other extant annelid clades—Aphanoneura, *Caobangia* (and *Brandika*), *Namanereis* and *Stratiodrilus*—appear to have successfully invaded continental waters either in the Palaeozoic or Mesozoic.

Table 2 Numbers of polychaete species and genera (in parentheses) for each major habitat, arranged alphabetically by family

Family	Lake/River	Estuary/Lagoon	Inland Sea	Oases/Springs	Hyporheic zone	Subterranean water	Phytotelmata
Aeolosomatidae	26 (3)	1 (1)	–	–	–	–	–
Ampharetidae	2 (2)	4 (4)	2 (2)	–	–	–	–
Capitellidae	1 (1)	4 (3)	1 (1)	–	–	–	–
Cirratulidae	–	2 (2)	–	–	–	–	–
Eunicidae	–	1 (1)	–	–	–	–	–
Goniadidae	–	1 (1)	–	–	–	–	–
Histriobdellidae	10 (1)	–	–	–	–	–	–
Lumbrineridae	1 (1)	–	–	–	–	–	–
Maldanidae	–	1 (1)	–	–	–	–	–
Nephtyidae	2 (1)	2 (1)	2 (1)	–	–	–	–
Nereididae	22 (11)	29 (14)	4 (3)	6 (1)	1 (1)	7 (1)	3 (2)
Nerillidae	–	–	–	–	1 (1)	1 (1)	–
Onuphidae	–	1 (1)	–	–	–	–	–
Orbiniidae	–	1 (1)	–	–	–	–	–
Paraonidae	1 (1)	1 (1)	–	–	–	–	–
Phyllodocidae	1 (1)	1 (1)	–	–	–	–	–
Pilargidae	–	1 (1)	–	–	–	–	–
Pisionidae	1 (1)	–	–	–	–	–	–
Potamodrilidae	1 (1)	–	–	–	–	–	–
Protodrilidae	1 (1)	–	–	–	–	–	–
Sabellidae	17 (7)	6 (4)	4 (2)	–	–	–	–
Serpulidae	–	4 (1)	–	–	–	1 (1)	–
Spionidae	7 (6)	14 (9)	–	1 (1)	–	–	–
Sternaspidae	–	1 (1)	–	–	–	–	–
Nereidiformia	1 (1)	–	–	–	–	–	–
Total	94 (39)	76 (49)	13 (9)	7 (2)	2 (2)	9 (3)	3 (2)

Nereidiformia are Polychaeta incertae sedis

A gondwanan radiation is postulated here for the nereidid *Namanereis*, inhabiting subterranean karst environments in New Zealand, Australia, New Guinea, the Arabian Peninsula, Canary Islands and the Caribbean (Fig. 3). Glasby (1999, p. 142) also found support for a southern origin of the group, although a non-gondwanan explanation was suggested at the time. A gondwanan origin has also been postulated for *Stratiodrilus*, most species of which live on freshwater crayfish in Australia, southern South America and Madagascar (Harrison, 1928; Fig. 3). The distribution of both genera supports the idea of a single colonisation of freshwater prior to the break up of Gondwana. The fabriciin sabellids *Caobangia* (and *Brandika*), whose members bore into the shells of freshwater gastropods of South-east Asia and India, also may have a gondwanan origin as parts of South-east Asia are thought to have fragmented from Gondwana before its final breakup (Metcalfe, 1998; Fig. 3). More ancient—possibly Pangaean—radiations may have occurred in the Aphanoneura, and the questionably monotypic genus *Troglochaetus*, which both show amphi-Atlantic distribution patterns. Unlike clitellates, aphanoneurans evidently never have had soil-dwelling ancestors.

Other freshwater taxa belong to several unrelated clades and often occur in isolation from related taxa and their main centre of distribution. Their ancestors may have been stranded during past, relatively recent, climatic (e.g., glaciations and marine transgressions) or eustatic (tectonic uplift) events. A freshwater relict is indicated when there is co-occurrence with unrelated taxa and both show close affinities with marine species in the region (e.g., Croskery, 1978; Schmidt

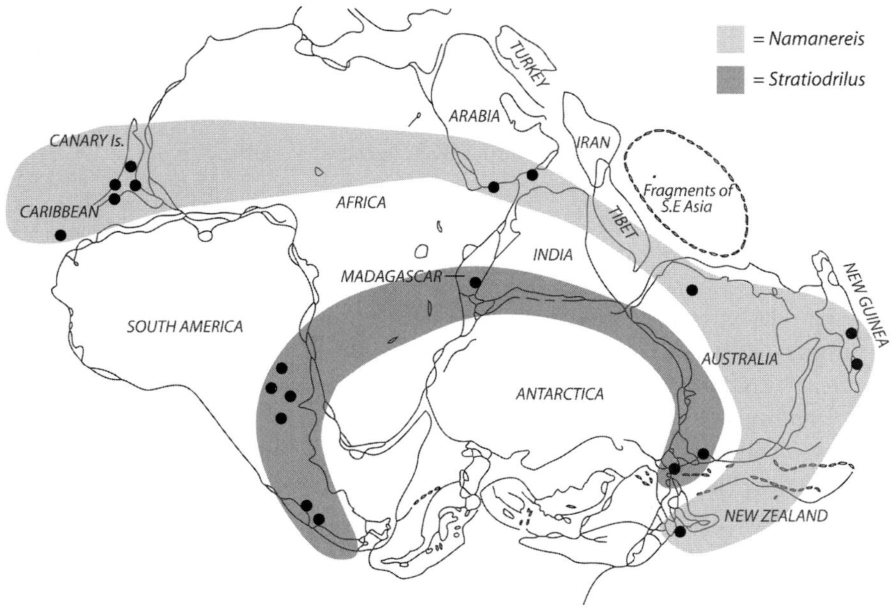

Fig. 3 Distribution of species of *Stratiodrilus* (Histriobdellidae) and *Namanereis* (Nereididae) across Gondwana before its break-up in the mid-Jurassic

& Westheide, 1999). Examples include the Ponto-Caspian sabellids (*Manayunkia* spp.) and ampharetids (*Hypania, Hypaniola, Parhypania*), the cave-dwelling serpulid *Marifugia cavatica* of Croatia, *Hesionides riegerorum* Westheide from a US east coast river, and other species of *Manayunkia* from North America, Lake Baikal and inland saline lakes of Australia (Croskery, 1978; Hutchings et al., 1981; Schmidt & Westheide, 1999).

Freshwater species of a third category are the temporary (either space or time) inhabitants of waters with fluctuating salinity such as estuaries and coastal lagoons, including many Nereididae, Spionidae, and the serpulids, *Ficopomatus* species. At low salinities (below about 10 ppt) and in freshwater, euryhaline species must osmoregulate to maintain correct body volume. Polychaetes living in freshwater also have reproductive modifications that protect their larvae from osmotic stress. These include direct sperm transfer (involving mating) and direct development (i.e., no trochophore stage), either in the body of the adult (e.g., *Hediste limnicola* (Johnson)), or in the parental tube (e.g., *Manayunkia* spp.). Eggs are typically large and yolky. The aeolosomatids undergo asexual reproduction with budding zones (paratomy). Sexual reproduction in all Aphanoneura is performed with copulation, storing the sperm in spermathecae, and laying of single eggs in cocoon-like envelopes. Although most polychaetes have separate sexes, the Aphanoneura are hermaphrodites. Hermaphroditism has also been suggested also for some Namanereidinae (Glasby, 1999) and *Caobangia* (Jones, 1974), but strong evidence is lacking. Reproduction in euryhaline species showing the typical marine pattern of reproduction (broadcast spawning and a pelagic trochophore)—such as *Alitta succinea* (Leuckart; formerly *Neanthes*) and *Ficopomatus enigmaticus* (Fauvel)—is limited to periods of higher salinity, or else the freshwater populations are maintained by recruitment of osmoregulating adults from more saline waters.

Human-related issues

The geographical range of several species has been increased by human activities, both intentionally (e.g., aquaculture) and unintentionally (e.g., interconnection of water basins through canals; shipping activities) (Bij de Vaate et al., 2002). In the Palaearctic, two nereidids, *Hediste diversicolor* and *Alitta succinea*, have been introduced as food for commercially or recreationally exploited fish to the Caspian and Aral Seas (Proskurina, 1980; Khlebovich & Komandentov, 2002). The spionid *Marenzelleria neglecta* Sikorski & Bick was introduced to Europe from North America (Bastrop et al. 1997, as *Marenzelleria* sp. Type II), and the Ponto-Caspian ampharetid *Hypania invalida* (Grube) has successfully spread to the Rhine and Meuse River basins via the Danube (Vanden Bossche et al., 2001; Bij de Vaate et al., 2002) and, together with the zebra mussel, *Dreissena polymorpha*, to the Volga (Shcherbina, 2001). In the Nearctic, the ampharetid *Hobsonia florida* (Hartman), a native of south-eastern USA has been translocated to Oregon (Castillo et al., 2000), the serpulid *Ficopomatus miamiensis* (Treadwell) was introduced from the eastern US to the Gulf of California (Salgado-Barragan et al., 2004) and *Alitta succinea* was intentionally introduced to the Salton Sea, California (Kuhl & Oglesby, 1979).

Few freshwater species have direct economic importance. The Japanese Palolo, *Tylorrhynchus heterochaetus* (Quatrefages) is both a nuisance and of benefit to humans. In Japan the worm causes damage to rice seedlings (Okuda, 1935), whereas in southern China it is eaten either fresh or ground to a fine meal (Chamberlin, 1924). The sabellid *Manayunkia speciosa* (Leidy) is host to *Ceratomyxa shasta* (Noble), a myxosporean parasite of salmonid fishes in the north-western USA (Bartholomew et al., 1997).

One species, the ampharetid polychaete *Alkmaria romijni* Horst, which occurs in sheltered lagoons and estuaries in the UK, is known to be protected by law (UK Wildlife and Countryside Act, 1981) (Gilliland & Sanderson, 2000). Probably other fresh- and brackish-water species will be given similar protection in the future because of their restricted distributions and evolutionary significance.

Acknowledgments CJG would like thank David Karlen and Hasko Nesemann for making available unpublished distribution data and the many colleagues, who provided helpful suggestions in response to a request for information. João Gil and Mary Petersen provided valuable constructive critique in review, and João also supplied literature references and distributional data. We thank Gloria Richards and Sue Dibbs for the preparation of figures.

References

Amaral, A. C. Z. & E. H. Morgado, 1997. *Stratiodrilus* (Annelida: Polychaeta: Histriobdellidae) associated with a freshwater decapod, with the description of a new species. Proceedings of the Biological Society of Washington 110: 471–475.

Bartholomew, J. L., M. J. Whipple, D. G. Stevens & J. L. Fryer, 1997. The life cycle of *Ceratomyxa shasta*, a myxosporean parasite of salmonids, requires a freshwater polychaete as an alternate host. Journal of Parasitology 83: 859–868.

Bastrop, R., M. Roehner, C. Sturmbauer & K. Juerss, 1997. Where did *Marenzelleria* spp. (Polychaeta: Spionidae) in Europe come from? Aquatic Ecology 31: 119–136.

Bij de Vaate, A., K. Jazdzewski, H. A. M. Ketelaars, S. Gollasch & G. Van der Velde, 2002. Geographical patterns in range extension of Ponto-Caspian macroinvertebrate species in Europe. Canadian Journal of Fisheries and Aquatic Sciences 59: 1159–1174.

Brinkhurst, R. O., 1971. Phylogeny and Classification, Part 1. In Brinkhurst, R. O. & B. G. M. Jamieson (eds), Aquatic Oligochaeta of the World. Oliver & Boyd, Edinburgh, 165–177.

Bunke, D., 1967. Zur Morphologie und Systematik der Aeolosomatidae Beddard 1895 und Potamodrilidae nov. fam. (Oligochaeta). Zoologische Jahrbücher, Abteilung für Systematik, Ökologie und Geographie der Tiere 94: 187–368.

Castillo, G. C., H. W. Li & P. A. Rossignol, 2000. Absence of overall feedback in a benthic estuarine community: a system potentially buffered from impacts of biological invasions. Estuaries 23: 275–291.

Chamberlin, R. V., 1924. A new freshwater nereid from China. Proceedings of the Biological Society of Washington 37: 79–82.

Croskery, P., 1978. The freshwater co-occurrence of *Eurytemora affinis* (Copepoda: Calanoida) and *Manayunkia speciosa* (Annelida: Polychaeta): possible relicts of a marine incursion. Hydrobiologia 59: 237–241.

Day, J. H. & J. A. Day, 2002. Polychaeta, Chapter 12. In Day, J. A. & I. J. de Moor (eds), Guides to the Freshwater Invertebrates of Southern Africa. Volume 5: Non-Arthropods. The protozoans, Porifera, Cnidaria, Platyhelminthes, Nemertea, Rotifera, Nematoda, Nematomorpha, Gastrotrichia, Bryozoa, Tardigrada, Polychaeta, Oligochaeta & Hirudinea. Pretoria, Water research Commission, Report No. TT 167/02, 193–202.

Foster, N., 1972. Freshwater Polychaetes (Annelida) of North America. Biota of Freshwater Ecosystems. Identification Manual No. 4. US Government Printing Office, Washington D.C., 15 pp.

Gilliland, P. M. & W. G. Sanderson, 2000. Re-evaluation of marine benthic species of nature conservation importance: a new perspective on certain 'lagoonal specialists' with particular emphasis on *Alkmaria romijni* Horst (Polychaeta: Ampharetidae). Aquatic Conservation: Marine and Freshwater Ecosystems 10: 1–12.

Glasby, C. J., 1999. The Namanereidinae (Polychaeta: Nereididae). Part 1. Taxonomy and phylogeny. Part 2. Cladistic biogeography. Records of the Australian Museum Supplement 25: 1–144.

Glasby, C. J., P. A. Hutchings, K. Fauchald, H. Paxton, G. W. Rouse, C. Watson Russell, & R. S. Wilson, 2000. Class Polychaeta. In Beesley, P. L., G. J. B. Ross & C. J. Glasby (eds), Polychaetes & Allies: The Southern Synthesis. Fauna of Australia. Vol. 4A. Polychaeta, Myzostomida, Pogonophora, Echiura, Sipuncula. CSIRO Publishing, Melbourne, 1–296.

Harrison, L., 1928. On the genus *Stratiodrilus* (Archiannelida: Histriobdella), with a description of a new species from Madagascar. Records of the Australian Museum 16: 116–121.

Hartman, O. 1959. Polychaeta, Chapter 22. In Edmondson, W. T. (ed.), Freshwater Biology, 2nd edn. John Wiley & Sons Inc., 538–541.

Hartmann-Schröder, G., 1986. Polychaeta (incl. Archiannelida). In Botosaneanu, L. (ed.), Stygofauna mundi. A Faunistic, Distributional and Ecological Synthesis of the World Fauna Inhabiting Subterranean Waters (Including the Marine Interstitial). E.J. Brill, Leiden, Netherlands, 210–233.

Hutchings, P. A., P. de Deckker & M. C. Geddes, 1981. A new species of *Manayunkia* (Polychaeta) from ephemeral lakes near the Coorong, South Australia. Transactions of the Royal Society of South Australia 105: 25–28.

Jones, M. L., 1974. On the Caobangiidae, a new family of the Polychaeta, with a redescription of *Caobangia billeti* Giard. Smithsonian Contributions to Zoology 175: 1–55.

Khlebovich, V. V. & A. Y. Komendantov, 2002. Temporary variation of paragnath arrangement in *Hediste diversicolor* and *H. japonica* (Polychaeta, Nereididae). Zoologichesky Zhurnal 81: 503–505.

Kuhl, D. L. & L. C. Oglesby, 1979. Reproduction and survival of the pileworm *Nereis succinea* in higher Salton Sea salinities. Biological Bulletin, Woods Hole 157: 153–165.

Metcalfe, I., 1998. Palaeozoic and mesozoic geological evolution of the SE Asian region: multidisciplinary constraints and implications for biogeography. In Hall, R. & J. D. Holloway (eds), Biogeography and Geological Evolution of SE Asia. Backhuys, Leiden, 25–41.

Okuda, S., 1935. Some lacustrine polychaetes with a list of brackish-water polychaetes found in Japan. Annotationes Zoologicae Japonenses 15: 240–246.

Orensanz, J. M., 1977. Polychaeta. In Hurlbert, S. H. (ed.), Biota acuatica de Sudamerica austral. San Diego State University, San Diego, 97–98.

Orensanz, J. M., 1981. Polychaeta. In Hurlbert, S. H., G. Rodriguez & N. D. dos Santos (eds), Aquatic Biota of Tropical South America. (Being a Compilation of Taxonomic Bibliographies for the Fauna and Flora of Inland Waters of the Tropical Portion of South America). Part 2. Anarthropoda. San Diego University, San Diego, 167–169.

Orensanz, J. M., 1982. Polychaeta. In Hurlbert, S. H. & A. Villalobos-Figueroa (eds), Aquatic Biota of Mexico, Central America and the West Indies (Being a Compilation of Taxonomic Bibliographies for the Fauna and Flora of Island Waters of Mesoamerica and the Caribbean region). San Diego State University, San Diego, 159–161.

Proskurina, Y. S., 1979 [1980]. The state of introduced species in the Aral Sea and the prospects for their spread [English version]. Hydrobiological Journal 15: 30–34.

Rouse, G. W., 2004. Annelida: Polychaeta. In Yule, C. M. & H. S. Yong (eds), Freshwater Invertebrates of the Malaysian region. Academy of Sciences, Malaysia, 194–206.

Rouse, G. W. & F. Pleijel, 2001. Polychaetes. Oxford University Press Inc, New York.

Salgado-Barragan, J., N. Méndez & A. Toledano-Granados, 2004. *Ficopomatus miamiensis* (Polychaeta: Serpulidae) and *Styela canopus* (Ascidiacea: Styelidae), non-native species in Urias estuary, SE Gulf of California, Mexico. Cahiers de Biologie Marine 45: 167–174.

Shcherbina, G. K., 2001. Autoacclimatization of the pontocaspian polychaete *Hypania invalida* from the Upper Volga River basin. Zoologichesky Zhurnal 80: 278–284.

Schmidt, H. & W. Westheide, 1999. Genetic relationships (RAPD-PCR) between geographically separated populations of the "cosmopolitan" interstitial polychaete *Hesionides gohari* (Hesionidae) and the evolutionary origin of the freshwater species *Hesionides riegerorum*. Biological Bulletin (Woods Hole) 196: 216–226.

Struck, T. H. & G. Purschke, 2005. The sister group relationship of Aeolosomatidae and Potamodrilidae (Annelida: "Polychaeta") – a molecular phylogenetic approach based on 18S rDNA and cytochrome oxidase 1. Zoologischer Anzeiger 243: 281–293.

Timm, T., 1981. On the origin and evolution of aquatic Oligochaeta. Eesti NSV Teaduste Akadeemia Toimetised, Bioloogia 30: 174–181.

Vanden Bossche, J.-P., F. Chérot, E. Delooz, F. Grisez & G. Josens, 2001. First record of the Pontocaspian invader *Hypania invalida* (Grube, 1860) (Polychaeta: Ampharetidae) in the River Meuse (Belgium). Belgian Journal of Zoology 131: 183–185.

Van der Land, J., 1971. Family Aeolosomatidae. In Brinkhurst, R. O. & B. G. M. Jamieson (eds), Aquatic Oligochaeta of the World. Oliver & Boyd, Edinburgh, 665–706.

Wesenberg-Lund, E., 1958. Lesser Antillean polychaetes chiefly from brackish waters, with a survey and a bibliography of fresh and brackish water polychaetes. Studies on the Fauna of Curaçao and other Caribbean Islands 8: 1–41.

Hydrobiologia (2008) 595:117–127
DOI 10.1007/s10750-007-9009-1

FRESHWATER ANIMAL DIVERSITY ASSESSMENT

Global diversity of oligochaetous clitellates ("Oligochaeta"; Clitellata) in freshwater

Patrick Martin · Enrique Martinez-Ansemil · Adrian Pinder · Tarmo Timm · Mark J. Wetzel

Abstract Oligochaeta *sensu stricto*, namely clitellates exclusive of branchiobdellids and leeches, occur in marine, estuarine, freshwater and terrestrial environs. About one-third of the almost 5,000 valid species described to date is aquatic. With the exception of some earthworm-like genera (the "megadriles"), aquatic oligochaetes are usually small, ranging from 1 mm to a few centimetres in length (the "microdriles"). Although predominantly terrestrial, 4 of the 14 described megadrile families include species that occur in aquatic or semi-aquatic habitats. The microdriles are fully aquatic, with the exception of the primarily terrestrial family Enchytraeidae, and comprise 13 families. About 1,700 valid species of aquatic oligochaetes are known to date; of these, about 1,100 are freshwater. The most speciose group is the Tubificidae with over 1,000 described species including 582 being considered as freshwater inhabitants. No fewer than 60 species of megadriles are also considered aquatic. Recent years have seen a continuous increase in the number of described species, so that any estimate of the proportion of known freshwater oligochaete species to unknown species would be very imprecise. Molecular studies have recently confirmed the long suspected paraphyly of the Oligochaeta if the group does not include branchiobdellids and leeches, so that Clitellata has become synonymous with "Oligochaeta". The family Capilloventridae has been recently shown to represent a basal clade of Clitellata, supporting an aquatic (freshwater?) origin of the clitellates. In contrast, the adaptation to freshwater of the aquamegadriles is most likely secondary. The Palaearctic region supports the most abundant and diverse freshwater oligochaete fauna, with more than 600 valid species

Guest editors: E. V. Balian, C. Lévêque, H. Segers & K. Martens
Freshwater Animal Diversity Assessment

P. Martin (✉)
Freshwater Biology, Royal Belgian Institute of Natural Sciences, 29 rue Vautier, Brussels 1000, Belgium
e-mail: patrick.martin@sciencesnaturelles.be

E. Martinez-Ansemil
Departamento de Bioloxía Animal, Bioloxía Vexetal e Ecoloxía, Universidade da Coruña, A Coruna 15071, Spain
e-mail: eansemil@udc.es

A. Pinder
Department of Environment and Conservation, Wildlife Research Centre, P.O. Box 51, Wanneroo 6065, Australia
e-mail: adrian.pinder@dec.wa.gov.au

T. Timm
Institute of Agricultural and Environmental Sciences, Centre for Limnology, Estonian University of Life Sciences, Rannu, Tartumaa 61117, Estonia
e-mail: tarmo.timm@emu.ee

M. J. Wetzel
Center for Biodiversity, Illinois Natural History Survey, 1816 South Oak Street, 1021 I-Building MC-652, Champaign, IL 61820, USA
e-mail: mjwetzel@uiuc.edu

described to date; 80% of these are considered endemic. However, it is likely that the apparent concentration of genera and species in the Northern Hemisphere is biassed given the relatively late and still limited interest in the oligochaete fauna of the Southern Hemisphere. Ancient lakes, as well as ground waters, are important centres of endemicity but, except for Lake Baikal, they represent important knowledge gaps. Aquatic oligochaetes perform ecological functions and roles with potentially important repercussions for human health issues. These ecological values of oligochaetes include their importance in aquatic food chains; their impact on sediment structure and water-sediment exchanges; their long history of use in pollution monitoring and assessment; their potential to reduce sludge volumes in sewage treatment systems; and their role as intermediate host for several myxozoan parasites of fishes, including commercially exploited species.

Keywords Freshwater Oligochaeta · Diversity · Phylogeny · Distribution · Endemism

Introduction

The Clitellata Michaelsen, 1919 include all segmented worms (Annelida) that possess a clitellum. This modification of the epidermis develops as a glandular girdle partly behind the female pores and secretes a cocoon in which eggs are laid. Among the annelids, they are commonly distinguished from the Polychaeta by their relative lack of setae (Brinkhurst, 1982a) and other distinctive features such as hermaphroditism, the organisation of the reproductive system and sperm ultrastructure (Purschke et al., 1993; Rouse & Fauchald, 1995; Westheide, 1997).

Clitellates are most often divided into the oligochaetes (sludge worms, earthworms), branchiobdellids (ectoparasites of freshwater crayfish) and leeches (Sawyer, 1986; Rouse & Fauchald, 1995; Brusca & Brusca, 2003). The Oligochaeta have long been suspected, on morphological grounds, to be a paraphyletic group unless it includes branchiobdellids and leeches (Erséus, 1987; Jamieson et al., 1987; Jamieson, 1988; Gelder & Brinkhurst, 1990; Brinkhurst & Gelder, 1991; Ferraguti & Gelder, 1991; Purschke et al., 1993; Brinkhurst, 1994; Brinkhurst, 1999; Ferraguti & Erséus, 1999; Siddall & Burreson, 1996). This paraphyly was recently confirmed by molecular analyses (Martin et al., 2000; Martin, 2001; Siddall et al., 2001) so that Clitellata has become synonymous with "Oligochaeta". No formal revision of the current classification has yet been proposed, however. In this article, we will only consider the Oligochaeta *sensu stricto*: namely, clitellates exclusive of branchiobdellids and leeches (Sket & Trontelj, 2007).

Oligochaetes *s.s.* (Fig. 1) occur in marine, estuarine, freshwater (Balian et al., 2007) and terrestrial environs. About two-thirds of the almost 5,000 valid described species (Erséus, 2005) belong to 'earthworm' families, which vary in length from 2 cm to over 3 m (Avel, 1959). These 'earthworms' are loosely termed "megadriles" (Stephenson, 1930; Brinkhurst, 1982b) and constitute the taxon Crassiclitellata (Jamieson, 1988). Although predominantly terrestrial, 4 of the 14 described megadrile families include species that occur in aquatic or semi-aquatic habitats. They constitute the Aquamegadrili, in contrast to the Terrimegadrili, and consist of the families Almidae, Biwadrilidae, Lutodrilidae and Sparganophilidae (Jamieson et al., 2002).

With the exception of some earthworm-like genera, aquatic oligochaetes are usually very thin and small, ranging from about 1 mm to a few cm in length. They are loosely termed "microdriles" and comprise 13 families. Most microdriles are fully aquatic, with the exception of the Enchytraeidae, a family that is primarily terrestrial; of the 650 described species, 200 are aquatic and 150 marine (Rota, pers. comm.).

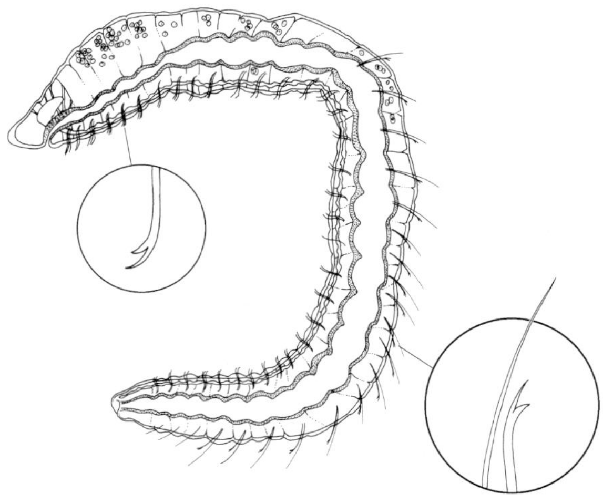

Fig. 1 Habitus Haemonais (Credit: drawing by A. Pinder)

Freshwater oligochaetes occur in a diversity of waterbodies (Timm & Veldhuijzen van Zanten, 2002). Most of them are benthic deposit feeders and burrow in the sediment, ingesting large amounts of particles. A distinctive group of the family Tubificidae, the subfamily Naidinae, is adapted to live on the sediment surface or to swim among the macrovegetation where they feed on algae. Only a few freshwater oligochaetes (e.g., species of *Chaetogaster*) are predatory, but such a behaviour is rare throughout the group. Many oligochaetes carry haemoglobin in the blood and are tolerant of oxygen shortage. Some species have also developed respiratory organs, in the form of long gills or modified posterior end of the body—a probable adaptation to common hypoxic conditions.

All oligochaetes can reproduce sexually, although asexual reproduction is far more common in some genera. The Naidinae, as a whole, reproduce asexually by paratomy (division at special budding zones where regeneration has already begun) or by architomy (fragmentation with subsequent regeneration). The latter process is also common in some lumbriculids and tubificids. Under peculiar environmental conditions, parthenogenesis also occurs in a few species that otherwise most commonly reproduce sexually. There is no larval stage of development (Timm & Veldhuijzen van Zanten, 2002).

Species diversity

About 1,700 valid species of aquatic oligochaetes are known to date; of these, about 600 are marine (Erséus, 2005) and about a hundred exclusively found in groundwater environments (Sambugar et al., 1999; Creuzé des Châtelliers et al. 2007).

The most speciose aquatic oligochaete group is the Tubificidae, which at present includes over 1,000 described species, 582 of which are considered freshwater inhabitants (Table 1). The family is divided into six subfamilies: Tubificinae, Naidinae (formerly Naididae but now treated as a subfamily; Erséus & Gustavsson, 2002), Telmatodrilinae, Rhyacodrilinae, Phallodrilinae and Limnodriloidinae. While most subfamilies have marine representatives, the latter is almost exclusively marine (with the notable exception of *Doliodrilus puertoricensis* Erséus & Milligan, 1988: Martínez-Ansemil et al., 2002). The Phallodrilinae is primarily marine but has several freshwater taxa, most of which inhabit groundwater. Among the tubificid subfamilies, the Naidinae, Tubificinae and Rhyacodrilinae are especially rich in terms of species, comparable to two microdrile families—the Lumbriculidae and the Enchytraeidae (Table 1).

Although megadrile oligochaetes are primarily terrestrial, no fewer than 60 species are considered aquatic or semi-aquatic (Table 1). These occasionally represent a significant biomass of an aquatic site, yet are often overlooked or merely referred to as "earthworms".

One genus, *Metataxis* Righi, 1985, previously thought to belong in the Haplotaxidae, has been shown to be an *incerta sedis* aquatic, megadrile (Brinkhurst, 1988) and for this reason, it has been treated separately from the other aquatic megadriles (Table 1). Some terrimegadrile species, belonging to the families Lumbricidae, Megascolecidae and Ocnerodrilidae, have been repeatedly noted from freshwater environs. These supposedly terrestrial species have been collected in freshwater often enough to suggest that they might be genuine aquatic or semi-aquatic species rather than incidentals.

Among the microdriles, the species richness of the Enchytraeidae is considered to be greatly underestimated. Due to taxonomic difficulties, lack of modern identification guides for most taxa, and few trained/practising systematists, enchytraeids have long been neglected, and rarely are identified, even to genus. We presently recognize 136 nominal species of aquatic enchytraeids, but the proportion of semiaquatic to truly aquatic species is unknown. The enchytraeid systematist Emilia Rota (pers. com.) suggests that as few as 50 enchytraeids may be truly freshwater species. This discrepancy suggests that an accurate estimate of the total number of freshwater enchytraeid species is not possible at this time. The Haplotaxidae, comprised of but 21 recognized species that occur almost exclusively in groundwater habitats, is often neglected for similar reasons and because they are scarce and commonly immature (so are rarely identifiable).

Estimating the potential total number of freshwater oligochaetes is problematic. Knowledge of some biogeographic regions, such as the Nearctic (North America: Kathman & Brinkhurst 1998; Wetzel et al. 2006, 2007) and the Australasian (Pinder et al. 2006) has been well established, while that for other regions (Africa, parts of the Neotropical and Oriental) is still

Table 1 Species diversity of freshwater oligochaetes (Annelida, Clitellata) delineated by primary zoogeographical regions (taxa present in groundwater habitats have also been included between brackets; data and nomenclature as for Septmeber 2005)

		Family	Subfamily	PA	PA without Baikal	NA	NT	AT	OL	AU	ANT	World
Megadrile	Aqua-	Almidae		4	4	1	7	17	15	1	0	41 (1)
Megadrile	Aqua-	Biwadrilidae		1	1	0	0	0	0	0	0	1 (0)
Megadrile	Aqua-	Lutodrilidae		0	0	1	0	0	0	0	0	1 (0)
Megadrile	Aqua-	Sparganophilidae		2	2	12	2	0	0	0	0	14 (0)
Megadrile	Aqua-	Lumbricina fam. (*Metataxis*)		0	0	1 (1)	1	2 (1)	0	0	0	5 (2)
Megadrile	Terri-	Lumbricidae		17	17	4	0	3	0	0	0	29 (9)
Megadrile	Terri-	Megascolecidae		2	2	0	0	0	0	0	0	2 (0)
Megadrile	Terri-	Ocnerodrilidae		1	1	1	0	2	0	0	0	2 (0)
Microdrile		Alluroididae		0	0	0	4	7	0	0	0	11 (0)
Microdrile		Capilloventridae		0	0	0	0	0	0	3 (1)	0	3 (1)
Microdrile		Dorydrilidae		3	3 (3)	0	0	0	0	0	0	3 (3)
Microdrile		Enchytraeidae		63	61 (68)	19	15 (4)	0	3	4	10	136 (68)
Microdrile		Haplotaxidae		12	12 (7)	1 (1)	1	2 (2)	1	6	0	21 (9)
Microdrile		Lumbriculidae		151	107 (44)	57 (10)	0	1	1	2	0	204 (50)
Microdrile		Narapidae		0	0	0	1	0	0	0	0	1 (1)
Microdrile		Opistocystidae		0	0	2	5	1	0	0	0	7 (0)
Microdrile		Parvidrilidae		1	1 (1)	1 (1)	0	0	0	0	0	2 (2)
Microdrile		Phreodrilidae		1	1 (1)	0	4	8 (2)	1	38 (6)	3	50 (10)
Microdrile		Propappidae		3	3 (1)	1	0	0	0	0	0	3 (1)
Microdrile		Tiguassidae		0	0	0	1	0	0	0	0	1 (0)
Microdrile		Tubificidae	Limnodriloidinae	0	0	0	0	1 (1)	0	0	0	1 (1)
Microdrile		Tubificidae	Naidinae	112	85 (41)	74 (20)	101 (24)	56 (3)	80	50 (2)	1	238 (61)
Microdrile		Tubificidae	Phallodrilinae	13	13 (12)	3 (2)	0	2	0	4	0	21 (13)
Microdrile		Tubificidae	Rhyacodrilinae	64	45 (22)	9 (4)	12	10	10	22 (1)	4	102 (24)
Microdrile		Tubificidae	Telmatodrilinae	1	1	2	0	0	1	0	0	3 (0)
Microdrile		Tubificidae	Tubificinae	165	117 (45)	49 (12)	24 (9)	12 (1)	18	16 (2)	0	217 (57)
Microdrile		Tubificidae										582 (156)
		Total		616	476 (245)	238 (50)	178 (37)	124 (9)	130	146 (12)	18	1119 (313)
		Rate of endemism (%)		80	74	56	51	48	27	54 (70)[a]	72	

[a] Rate of endemism in the Australasian region including species considered new to science, but for which descriptions have not yet been completed or published. Subantarctic islands were independently considered with the closer continents (except for Antarctica). PA: Palaearctic; NA: Nearctic; NT: Neotropical; AT: Afrotropical; OL: Oriental; AU: Australasian; ANT: Antarctic

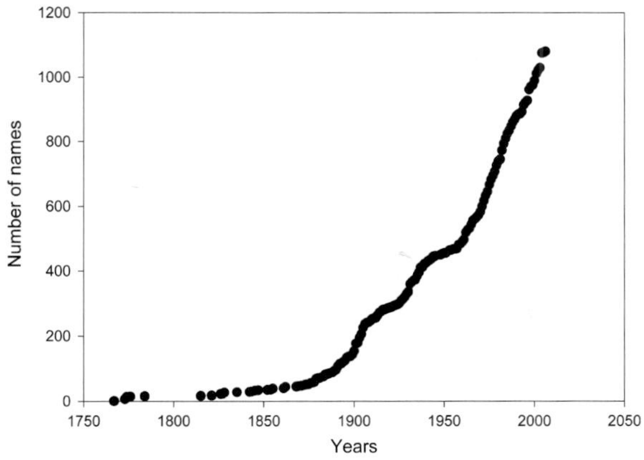

Fig. 2 Evolution of numbers of valid species names of oligochaetes (Annelida, Clitellata) over time

in its infancy. The oligochaete diversity of ground waters is also still poorly known (Juget & Dumnicka, 1986; Giani et al., 2001), especially in Asia. The accumulation of numbers of presently valid names over time shows no trend towards stabilization at a global scale (Fig. 2). Recent years have seen a continuous increase in the number of described species, even in well-studied regions (Palaearctic, Nearctic) (Reynolds & Cook, 1976, 1981, 1989, 1993; Reynolds & Wetzel, 2007) so that any proposed ultimate number would simply be a guess.

Phylogeny and historical processes

The phylogeny of the Oligochaeta has been debated for a long time. As introduced above, the term "Oligochaeta" is itself phylogenetically invalid since it designates a paraphyletic group that does not include all its descendants (leeches and branchiobdellids). Although Oligochaeta is an older name than Clitellata, the latter was explicitly established to include the leeches and branchiobdellids (Michaelsen, 1919) and, hence, should be preferred to the former (Martin, 2001).

In this phylogenetic context, the nature of an ancestor to the Clitellata is inherently problematical, given that there are no fossil remains and most probable outgroups, such as the Polychaeta, share very few characters with clitellates (Jamieson, 1988; Brinkhurst, 1994). The position of the Clitellata in relation to the Annelida remains inconclusive, despite numerous molecular studies addressing the issue (Rota et al., 2001; Erséus, 2005; McHugh, 2005). Placement of the Clitellata within the Polychaeta is not supported, and there are as many candidate sister groups to the Clitellata as there are molecular studies focussed on resolving these relationships.

Based on a particularly defined hypothetical ancestor, it was generally assumed that all oligochaetes *s. s.* could have derived from a segmented worm very close to extant haplotaxids (Brinkhurst, 1984; 1991), thus implying that the split between aquatic and terrestrial groups (the Microdrili and the Megadrili) occurred early in the evolution of the clitellates. The discovery of many marine tubificid species during recent years, belonging to several subfamilies of Tubificidae, prompted Erséus (1987) to propose that oligochaetes could be derived from small forms of marine origin. In contrast, Westheide (1997) and Westheide et al. (1999) postulated a terrestrial origin for the Clitellata, on the basis of functional considerations of clitellate morphological structures.

A recent molecular study (Erséus & Källersjö, 2004) supports an earlier hypothesis (Erséus, 1993; Ferraguti et al., 1996) based on morphological features that Capilloventridae represents a basal clade of Clitellata, and that an aquatic (freshwater?) origin of the clitellates is the most likely. Interestingly, the sister group to the primarily terrestrial megadriles (Crassiclitellata) are the Enchytraeidae, of which two-thirds of the known species are terrestrial. The aquamegadrile families are suspected to have always had an aquatic or amphibious existence. However, molecular studies show that if the monophyletic nature of the Aquamegadrili is not supported, this group does not constitute a basal clade of the Crassiclitellata, which implies that their adaptation to freshwater was secondary (Jamieson et al., 2002).

Present distribution and main areas of endemicity

The Palaearctic region supports the most abundant and diverse freshwater oligochaete fauna, with 616 valid species and 113 genera described to date (or 476 species and 104 genera if the mostly endemic fauna of Lake Baikal is excluded) (Fig. 3; Tables 1, 2). About 80% of Palaearctic species are endemic. The Holarctic region harbours 766 species (626 without Lake Baikal) versus 404 species in all other regions combined.

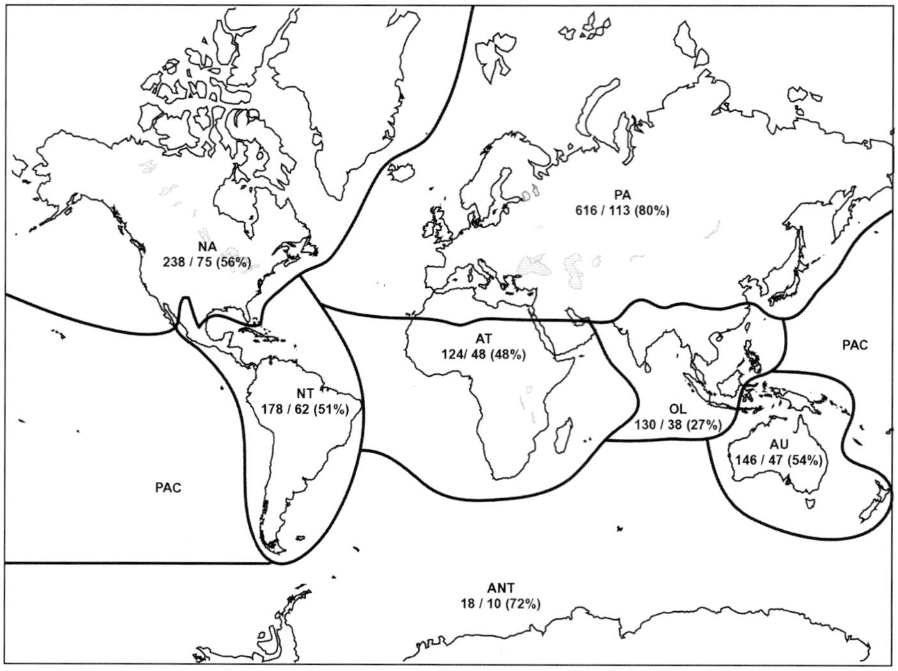

Fig. 3 Species and genus diversity (species number/genus number), and rate of endemism (%) of the freshwater Oligochaeta (Annelida, Clitellata) by primary zoogeographical regions. PA: Palaearctic; NA: Nearctic; NT: Neotropical; AT: Afrotropical ; OL: Oriental; AU: Australasian; PAC: Pacific Oceanic Islands (subantarctic Islands were independently considered with the closer continents except for Antarctica); ANT: Antarctic

The Tubificidae is the only family with a cosmopolitan distribution. The Rhyacodrilinae are present in all regions, including subantarctic islands. Excluding the Antarctic region, the two other numerically most important tubificid subfamilies, the Naidinae and Tubificinae, are recorded from all continents, although their relative contribution to tubificid diversity can change according to the zoogeographical region considered. In particular, the Tubificinae dominate in the Holarctic region (189 spp., vs. 44 spp. in all other regions) whereas the Naidinae are more homogeneously distributed (143 spp., and 156 spp. in the Holarctic, and all other regions, respectively).

The Enchytraeidae is a large and probably cosmopolitan family, but there are no records of freshwater representatives for the Afrotropical region. Given that the family exists in North Africa, and taking into account the taxonomic difficulties associated with this family, it is very likely that freshwater enchytraeids will be found in the Afrotropical region. Enchytraeids are the most diverse group in ground water, but many of them are suspected to be terrestrial and to occur only incidentally in aquifers.

The Lumbriculidae are more limited in distribution, with most species known only from the Holarctic region (204 spp.; Lake Baikal included). Southern occurrences include only two peregrine species, *Lumbriculus variegatus* and *Stylodrilus heringianus*, both of which may have been introduced via shipping (Brinkhurst & Jamieson, 1971). In contrast, the Phreodrilidae have long been known to occur only in the Southern Hemisphere, interpreted in terms of continental drift theory as a relict of an ancient Gondwanan distribution (Brinkhurst & Jamieson, 1971; Giani et al., 1995). The recent accumulation of newly described species from Australia and adjacent areas falls well within this scheme. To date, 49 species are known to occur in the southern regions (Neotropical, Afrotropical, Oriental, Australasia and Antarctica) while only *Astacopsidrilus naceri* Giani & Martin, 1995 is mentioned in the Holarctic region (Morocco). The theory of continental drift was also invoked to explain the observed division of the Alluroididae between the Afrotropical and Neotropical regions, although similarities with the Japanese *Biwadrilus* (Biwadrilidae) give support to a northerly origin (Brinkhurst & Jamieson, 1971).

Seven families only occur in one zoogeographical region. With the exception of the Megascolecidae, these are small, monogeneric families (the Dorydrilidae and the Capilloventridae) or even monospecific families with a (sometimes highly) localized distribution—Biwadrilidae (Japan); Lutodrilidae (Louisiana, USA); Narapidae (interstitial waters of the Parana River, Argentina) and Tiguassidae (Amazonian basin, Brazil)

Table 2 Genus diversity of freshwater oligochaetes (Annelida, Clitellata) delineated by primary zoogeographical regions (a column enumerating taxa restricted to groundwater habitats has also been included)

		Family	Subfamily	PA	PA without Baikal	NA	NT	AT	OL	AU	ANT	Groundwater	World
Megadrile	Aqua-	Almidae		3	3	1	2	3	2	1	0	1	7
Megadrile	Aqua-	Biwadrilidae		1	1	0	0	0	0	0	0	0	1
Megadrile	Aqua-	Lutodrilidae		0	0	1	0	0	0	0	0	0	1
Megadrile	Aqua-	Sparganophilidae		1	1	1	2	0	0	0	0	0	2
Megadrile	Aqua-	Lumbricina fam. (*Metataxis*)		0	0	1	1	1	0	0	0	1	1
Megadrile	Terri-	Lumbricidae		5	5	4	0	3	0	0	0	5	9
Megadrile	Terri-	Megascolecidae		1	1	0	0	0	0	0	0	0	1
Megadrile	Terri-	Ocnerodrilidae		1	1	1	0	2	0	0	0	0	2
Microdrile		Alluroididae		0	0	0	3	4	0	0	0	0	7
Microdrile		Capilloventridae		0	0	0	0	0	0	1	0	1	1
Microdrile		Dorydrilidae		1	1	0	0	0	0	0	0	1	1
Microdrile		Enchytraeidae		8	8	7	11	0	2	4	4	13	18
Microdrile		Haplotaxidae		5	5	1	1	2	1	4	0	3	8
Microdrile		Lumbriculidae		19	17	12	0	1	1	2	0	12	26
Microdrile		Narapidae		0	0	0	1	0	0	0	0	1	1
Microdrile		Opistocystidae		0	0	2	2	1	0	0	0	0	4
Microdrile		Parvidrilidae		1	1	1	0	0	0	0	0	1	1
Microdrile		Phreodrilidae		1	1	0	4	4	1	6	2	5	10
Microdrile		Propappidae		1	1	1	0	0	0	0	0	1	1
Microdrile		Tiguassidae		0	0	0	1	0	0	0	0	0	1
Microdrile		Tubificidae	Limnodriloidinae	0	0	0	0	1	0	0	0	1	1
Microdrile		Tubificidae	Naidinae	22	20	19	16	15	17	11	1	16	28
Microdrile		Tubificidae	Phallodrilinae	4	4	3	0	2	0	2	0	5	9
Microdrile		Tubificidae	Rhyacodrilinae	13	9	4	6	5	6	9	3	5	19
Microdrile		Tubificidae	Telmatodrilinae	1	1	2	0	0	1	0	0	0	2
Microdrile		Tubificidae	Tubificinae	25	24	14	12	4	7	7	0	21	31
Microdrile		Tubificidae		65	58	42	34	27	31	29	4	48	90
		Total		113	104	75	62	48	38	47	10	93	193

Subantarctic islands were independently considered with the closer continents (except for Antarctica). PA: Palaearctic; NA: Nearctic; NT: Neotropical; AT: Afrotropical; OL: Oriental; AU: Australasian; ANT: Antarctic

(Tables 1, 2). The family Dorydrilidae is currently known only from Europe, with records most likely resulting from extensive research focussed on the underground fauna of this region. Interestingly, freshwater Capilloventridae are known only from southeastern Australia, although marine species are present elsewhere. The monogeneric Propappidae, long restricted to the Palaearctic region (Europe and Asia), is now known from North America where it is considered a probable introduction (Coates, pers. comm.).

The presence of freshwater Megascolecidae only in the Palaearctic region is all the more surprising since this primarily terrestrial family is native in all zoogeographical regions (with the exception of the Antarctic) (Jamieson et al., 2002). This observation is not informative, however, in view of uncertainties about how to consider those species occasionally found in aquatic habitats and given that such occurrences are not documented in many regions.

Aquatic and semi-aquatic megadriles are present in all zoogeographical regions (with the exception of the Antarctic). The Almidae are predominantly associated with tropical regions (Oriental, Afrotropical and Neotropical). In contrast, the Sparganophilidae are primarily Nearctic, with occasional occurrences in the Palaearctic and the Neotropical regions. The presence of the Almidae in the Holarctic region is mostly due to one genus, *Criodrilus* Hoffmeister, 1845. As native species, almids are absent from the Australasian region with the single exception of *Glyphidrilus weberi* Horst, 1889—a species that has crossed the Wallace line (Wallace, 1876) and has reached Sulawesi (formerly Celebes). This absence, with other evidence, suggests that the family has had a northern origin (Brinkhurst & Jamieson, 1971).

Lake Baikal is a unique biogeographical region unto itself and a hotspot of oligochaete diversity. About 192 species have been described in this lake so far, of which more than 70% are endemic and species flocks are recognizable or suspected e.g.: *Baikalodrilus* (23 spp.), *Lamprodrilus* (19 spp.), *Stylodrilus* (11 species) and *Isochaetides* (11 spp.) (Semernoy, 2004). Ancient lakes (long-lived waterbodies which have existed for at least 1 Myr; Gorthner, 1994) are important centres of endemicity (Martin, 1996). Unfortunately, they have not yet been comparably studied. For instance, Lake Tanganyika (East Africa) has a similar to age Lake Baikal but yet supports only 16 known species, nine of which are endemic, and no endemic genera (versus 9 in Lake Baikal). This is partly due to an obvious lack of studies, and the oligochaete faunas of the other great African lakes (Lake Malawi, Lake Victoria) are virtually unknown. Recent interest in the great lakes in Africa has revealed a more diverse oligochaete fauna than previously assumed, but better studies of this fauna are still needed.

If only described species are taken into account, 54% of freshwater oligochaete species occurring in the Australasian region are endemic. Numerous species have yet to be described, however, so endemicity may be as high as 70%, making this long-isolated zoogeographical region similar to the Palaearctic in terms of rates of endemicity. The Oriental region is presently represented by surprisingly few endemic species (27%), yet the total number of species is similar to that of some other regions, where the rate of endemicity is much higher. This discrepancy may be the result of a biassed study of the local fauna—one more focussed on areas with high human pressure, and where more banal species live due to their wide ecological valence (see examples in Naidu, 2005). In addition, bias resulting from misidentifications cannot be excluded, given that more than 60% of freshwater oligochaete species occurring in this region are naidines, a group for which the taxonomy is presently deemed unreliable due to a paucity of systematic studies (Brinkhurst & Wetzel, 1984).

Ground waters are important centres of endemicity and refuges for relictual species. Among the 313 nominal species described to date (193 genera), one-third is exclusively found in this environment (stygobionts). The fact that ground waters hold a combination of endemic, sometimes very old and/or thalassoid lineages makes them of the utmost interest as far as biodiversity and conservation topics are concerned.

Human related issues

The oligochaetes have long been known to play a preponderant role in aquatic ecosystems (Giani, 1984). In particular, the impact of benthic oligochaetes on sediment structure and water-sediment exchanges can have important repercussions as far as human issues are concerned. Taking this impact into account is crucial in the study of movement and transport of toxic pollutants in the aquatic ecosystem, since, as a result of the reworking of sediment (bioturbation), some contaminated sediment layers can be re-exposed at the water-sediment interface many years after active discharge of pollutants and deposition have ceased (Golterman et al., 1983).

Aquatic oligochaetes (and in particular, the Tubificidae) have long been associated with polluted waters; perhaps the first mention of oligochaetes and pollution (from foul mud associated with domestic sewage) was by Aristotle [384–322 B.C.] (Thienemann, 1912). Some aquatic oligochaetes are indeed very tolerant to low oxygen levels associated with organically polluted waters and can be abundant in this environment where food supply is abundant and competition absent.

The tolerance of some species to these conditions has led to them being investigated for their potential to reduce sludge volumes in sewage treatment systems (e.g., Wei and Lui (2006). In addition, they are also easy to culture and have rapid and simple life cycles, making them ideal subjects for laboratory sediment toxicity studies.

The cosmopolitan *Tubifex tubifex* is one of the most commonly studied freshwater oligochaetes—both as a health indicator of aquatic environments and as an intermediate host for several myxozoan parasites of fishes. Unfortunately, recent molecular studies have indicated that there may be several cryptic species of *Tubifex* co-occurring in North America and Europe—species that exhibit unique physiological and toxicological responses to the environment as well as resistance to infection by myxozoans (Sturmbauer et al., 1999; Beauchamp et al., 2001, 2006). Such studies are compelling, lending additional support to the importance of taxonomic studies and repercussions of the lack of these on human health related issues.

In spite of these challenges, aquatic oligochaetes have great potential for use in studies relating human health issues to the quality of the freshwater environment, for many reasons, including: their importance in the aquatic food chain; many species are widely distributed and well studied; representatives include freshwater, estuarine, and marine species; as a group, they range from sensitive to insensitive to a wide range of environmental variables; they have a long history of use in pollution monitoring and assessment; and, relevant toxicity and bioaccumulation tests exist (e.g., Reynoldson et al., 1991; Reynoldson & Rodriguez, 1999; Chapman, 2001).

Ackowledgements We gratefully acknowledge Kathryn A. Coates (Bermuda), Steven V. Fend (California, USA), and R. Deedee Kathman (Tennessee, USA) for their assistance in compiling the information for Nearctic species of freshwater oligochaetes; many additional records for Nearctic taxa were secured from the Illinois Natural History Survey Annelida Collection databases. We thank Emilia Rota (Universitá di Siena, Italy) for sharing her knowledge about the Enchytraeidae. Francis Behen (RBINSc, Belgium) assisted the first author in going through the Zoological Records and in feeding his oligochaete database.

References

Avel, M., 1959. Classe des annélides oligochètes. In Grassé, P. P. (ed.), Traité de Zoologie, Vol. 5, No. 1. Paris, 224–462.

Balian, E., H. Segers, C. Lévèque, & K. Martens, 2007. An introduction to the Freshwater Animal Diversity Assessment (FADA) project. Hydrobiologia doi: 10.1007/s10750-007-9235-6

Beauchamp, K. A., R. D. Kathman, T. S. McDowell & R. P. Hedrick, 2001. Molecular phylogeny of tubificid oligochaetes with special emphasis on *Tubifex tubifex* (Tubificidae). Molecular Phylogenetics and Evolution 19: 216–224.

Beauchamp, K. A., M. El-Matbouli, M. Gay, M. P. Georgiadis, B. Nehring & R. P. Hedrick, 2006. The effect of cohabitation of *Tubifex tubifex* (Oligochaeta: Tubificidae) populations on infections to *Myxobolis cerebralis* (Myxozoa: Myxobolidae). Journal of Invertebrate Zoology 91: 1–8.

Brinkhurst, R. O., 1982a. Oligochaeta. In Parker, S. P. (ed.), Synopsis and Classification of Living Organisms, Vol. 1. McGraw-Hill Book Company, New-York, 50–61.

Brinkhurst, R. O., 1982b. Evolution in the Annelida. Canadian Journal of Zoology 60: 1043–1059.

Brinkhurst, R. O., 1984. The position of the Haplotaxidae in the evolution of oligochaete annelids. Hydrobiologia 115: 25–36.

Brinkhurst, R. O., 1988. A taxonomic analysis of the Haplotaxidae (Annelida, Oligochaeta). Canadian Journal of Zoology 66: 2243–2252.

Brinkhurst, R. O., 1991. Ancestors (Oligochaeta). Mitteilungen aus dem Hamburgischen Zoologischen Museum und Institut 88: 97–110.

Brinkhurst, R. O., 1994. Evolutionary relationships within the Clitellata: an update. Megadrilogica 5: 109–112.

Brinkhurst, R. O., 1999. Lumbriculids, branchiobdellidans and leeches: a review of progress. Hydrobiologia 406: 281–290.

Brinkhurst, R. O. & B. G. M. Jamieson, 1971. Distribution and ecology. In Brinkhurst, R. O. & B. G. M. Jamieson (eds), Aquatic Oligochaeta of the World. Oliver & Boyd, Edinburgh, 104–164.

Brinkhurst, R. O. & M. J. Wetzel, 1984. Aquatic Oligochaeta of the world: supplement. A catalogue of new freshwater species, descriptions, and revisions. Canadian Technical Report of Hydrography and Ocean Sciences, No 44.

Brinkhurst, R. O. & S. R. Gelder. 2001. Annelida: Oligochaeta, including Branchiobdellidae. In Thorp, J. H. & and A. P. Covich (eds), Ecology and Classification of North American Freshwater Invertebrates, 2nd edn. Academic Press, San Diego, 431–463.

Brusca, R. C. & G. J. Brusca, 2003. Invertebrates, 2nd edn. Sinauer Associates, Inc., Sunderland, MA.

Chapman, P. M., 2001. Utility and relevance of aquatic oligochaetes in Ecological Risk Assessment. Hydrobiologia 463: 149–169.

Creuzé des Châtelliers, M., P. Martin, J. Juget & M. Lafont, 2007. Status of the Oligochaeta (oligochaetous Clitellata) in the subterranean aquatic environment. Freshwater Biology (in press).

Erséus, C., 1987. Phylogenetic analysis of the aquatic Oligochaeta under the principle of parsimony. Hydrobiologia 155: 75–89.

Erséus, C., 1993. Taxonomy of *Capilloventer* (Capilloventridae), a little-known group of aquatic Oligochaeta, with

descriptions of two new species. Journal of Natural History 27: 1029–1040.
Erséus, C., 1999. *Parvidrilus strayeri*, a new genus and species, an enigmatic interstitial clitellate from underground waters in Alabama. Proceedings of the Biological Society of Washington 112: 327–337.
Erséus, C., 2005. Phylogeny of oligochaetous Clitellata. Hydrobiologia 535/536: 357–372.
Erséus, C. & L. Gustavsson, 2002. A proposal to regard the former family Naididae as a subfamily within the Tubificidae (Annelida, Clitellata). Hydrobiologia 485: 253–256.
Erséus, C. & M. Källerjsö, 2004. 18S rDNA phylogeny of Clitellata (Annelida). Zoologica Scripta 33: 187–196.
Ferraguti, M. & C. Erséus, 1999. Sperm types and their use for a phylogenetic analysis of aquatic clitellates. Hydrobiologia 402: 225–237.
Ferraguti, M. & S. R. Gelder, 1991. The comparative ultrastructure of spermatozoa from five branchiobdellidans (Annelida: Clitellata). Canadian Journal of Zoology 69: 1945–1956.
Ferraguti, M., C. Erséus & A. Pinder, 1996. The spermatozoon of *Capilloventer australis* and the systematic position of the Capilloventridae (Annelida: Oligochaeta). Australian Journal of Zoology 44: 469–478.
Gelder, S. R. & R. O. Brinkhurst, 1990. An assessment of the phylogeny of the Branchiobdellida (Annelida: Clitellata), using PAUP. Canadian Journal of Zoology 68: 1318–1326.
Giani, N., 1984. Contribution à l'étude de la faune d'eau douce et plus particulièrement des Oligochètes. II – Les oligochètes aquatiques : taxinomie, répartition et écologie. Thèse de doctorat d'Etat, Université Paul Sabatier, Toulouse.
Giani, N., P. Martin & J. Juget, 1995. A new species of Phreodrilidae (Oligochaeta), *Astacopsidrilus naceri* sp. nov., from Morocco (North Africa), with notes on the biogeography of the family. Canadian Journal of Zoology 73: 2375–2381.
Giani, N., B. Sambugar, P. Rodriguez & E. Martínez-Ansemil, 2001. Oligochaetes in southern European groundwater: new records and an overview. Hydrobiologia 463: 65–74.
Golterman, H. L., P. G. Sly & R. L. Thomas, 1983. Study of the relationship between water quality and sediment transport. Technical Papers in Hydrobiology, UNESCO.
Gorthner, A., 1994. What is an ancient lake? In Martens K., B. Goddeeris & G. Coulter (eds), Speciation in Ancient Lakes. Archiv für Hydrobiologie Beiheft Ergebnisse der Limnologie 44: 97–100.
Jamieson, B. G. M., 1988. On the phylogeny and higher classification of the Oligochaeta. Cladistics 4: 367–410.
Jamieson, B. G. M., C. Erséus & M. Ferraguti, 1987. Parsimony analysis of the phylogeny of some Oligochaeta (Annelida) using spermatozoal ultrastructure. Cladistics 3: 145–155.
Jamieson, B. G. M., S. Tillier, A. Tillier, J.-L. Justine, E. Ling, K. McDonald & A. F. Hugall, 2002. Phylogeny of the Megascolecidae and Crassiclitellata (Annelida, Oligochaeta): combined versus partitioned analysis using nuclear (28S) and mitochondrial (12S, 16S) rDNA. Zoosystema 24: 707–734.
Juget, J. & E. Dumnicka, 1986. Oligochaeta (incl. Aphaneura) des eaux souterraines continentales. In Botosaneanu, L. (ed.), Sytogofauna Mundi. E.J. Brill., Leiden, 234–244.
Kathman, R. D. & R. O. Brinkhurst, 1998. Guide to the freshwater oligochaetes of North America. Aquatic Resources Center, College Grove, Tennessee, 264 pp
Martin, P., 1996. Oligochaeta and Aphanoneura in ancient lakes: a review. Hydrobiologia 334: 63–72.
Martin, P., 2001. On the origin of the Hirudinea and the demise of the Oligochaeta. Proceedings of the Royal Society of London, Series B 268: 1089–1098.
Martin, P., I. Kaygorodova, D. Yu. Sherbakov & E. Verheyen, 2000. Rapidly evolving lineages impede the resolution of phylogenetic relationships among Clitellata (Annelida). Molecular Phylogenetics and Evolution 15: 355–368.
Martínez-Ansemil, E., N. Giani & B. Sambugar, 2002. Oligochaetes from underground waters of Oman with descriptions of two new species of Phreodrilidae (Oligochaeta): *Antarctodrilus arabicus* n. sp. and *Phreodrilus stocki* n. sp. Contributions to Zoology 71: 147–158.
McHugh, D., 2005. Molecular systematics of polychaetes (Annelida). Hydrobiologia 535/536: 309–318.
Michaelsen, W., 1919. Über die Beziehungen der Hirudineen zu den Oligochäten. Mitteilungen aus dem Naturhistorischen Museum in Hamburg 36: 131–153.
Naidu, K. V., 2005. The Fauna of India and the Adjacent Countries. Aquatic Oligochaeta. Director, Zoological Survey of India, Kolkata.
Pinder, A. M., S. M. Eberhard & W. F. Humphreys, 2006. New phalodrilines (Annelida: Clitellata: Tubificidae) from Western Australian groundwater. Zootaxa 1304: 31–48
Purschke, G., W. Westheide, D. Rohde & R. Brinkhurst, 1993. Morphological reinvestigation and phylogenetic relationships of *Acanthobdella peledina* (Annelida, Clitellata). Zoomorphology 113: 91–101.
Reynolds, J. W. & D. G. Cook, 1976. Nomenclatura Oligochaetologica. A catalogue of names, descriptions and type specimens of the Oligochaeta. University of New Brunswick, Fredericton, New Brunswick.
Reynolds, J. W. & D. G. Cook, 1981. Nomenclatura Oligochaetologica. Supplementum primum. A catalogue of names, descriptions and type specimens of the Oligochaeta. University of New Brunswick, Fredericton, New Brunswick.
Reynolds, J. W. & D. G. Cook, 1989. Nomenclatura Oligochaetologica. Supplementum secundum. A catalogue of names, descriptions and type specimens of the Oligochaeta. New Brunswick Museum Monographic Series (Natural Science) No. 8, Fredericton, New Brunswick.
Reynolds, J. W. & D. G. Cook, 1993. Nomenclatura Oligochaetologica. Supplementum tertium. A catalogue of names, descriptions and type specimens of the Oligochaeta. New Brunswick Museum Monographic Series (Natural Science) No. 9, Fredericton, New Brunswick.
Reynolds, J. W. & M. J. Wetzel, 2007. Nomenclatura oligochaetologica. Supplementum quartum. A catalogue of names, descriptions and type specimens of the Oligochaeta. In prep.
Reynoldson, T. B. & P. Rodriguez, 1999. Field methods and interpretation for sediment bioassessment, Chapter 4. In

Mudroch, A., J. M. Azcue & P. Mudroch (eds), Manual of Bioassessment of Aquatic Sediment Quality. Lewis Publishers, Boca Raton, Florida, 135–175.

Reynoldson, T. B., S. P. Thompson & J. L. Ramsey, 1991. A sediment bioassay using the tubificid oligochaete worm *Tubifex tubifex*. Environmental Toxicology and Chemistry 10: 1061–1072.

Rota, E., P. Martin & C. Erséus, 2001. Soil-dwelling polychaetes: enigmatic as ever? Some hints on their phylogenetic relationships as suggested by a maximum parsimony analysis of 18S rRNA gene sequences. Contributions to Zoology 70: 127–138.

Rouse, G. W. & K. Fauchald, 1995. The articulation of annelids. Zoologica Scripta 24: 269–301.

Sambugar, B., N. Giani & E. Martínez-Ansemil, 1999. Groundwater oligochaetes from Southern-Europe. Tubificidae with marine phyletic affinities: new data with description of a new species, review and consideration on their origin. Mémoires de Biospéologie 26: 107–116.

Sawyer, R. T., 1986. Leech Biology and Behaviour. II. Feeding Biology, Ecology, and Systematics. Clarendon Press, Oxford.

Semernoy, V. P., 2004. Oligochaeta of Lake Baikal. In Timm, T., G. F. Mazepova & O. A. Timoshkin (eds), Guides and Keys to Identification of Fauna and Flora of Lake Baikal. Nauka, Novosibirsk.

Siddall, M. E. & E. M. Burreson, 1996. Leeches (Oligochaeta?: Euhirudinea), their phylogeny and the evolution of life-history strategies. Hydrobiologia 334: 277–285.

Siddall, M. E., K. Apakupakul, E. M. Burreson, K. A. Coates, C. Erséus, S. R. Gelder, M. Källersjö & H. Trapido-Rosenthal, 2001. Validating Livanow: molecular data agree that leeches, branchiobdellidans and *Acanthobdella peledina* form a monophyletic group of oligochaetes. Molecular Phylogenetics and Evolution 21: 346–351.

Sket, B., & P. Trontelj, 2007. Global diversity of leeches (Hirudinea) in freshwater. In Balian, E. V., C. Lévêque, H. Segers & K. Martens (eds), Freshwater Animal Diversity Assessment. Hydrobiologia. doi:10.1007/s10750-007-9010-8

Stephenson, J., 1930. The Oligochaeta. Oxford University Press, Oxford.

Sturmbauer, C., G. B. Opadiya, H. Niederstätter, A. Riedman & R. Dallinger, 1999. Mitochondrial DNA reveals cryptic oligochaete species differing in Cadmium resistance. Molecular Biology and Evolution 16: 967–974.

Thienemann, A. 1912. Aristotles und die Abwasserbiologie. Festschrift Medizinisch-Naturwissenschaftlichen Gesellschaft Münster. Commissionsverlag, Universitäts Buchhandlung Franz Coppenrath Münster.

Timm, T. & H. Veldhuijzen van Zanten, 2002. Freshwater Oligochaeta of North-West Europe. World Biodiversity Database, CD-ROM Series, Biodiversity Center of ETI, Multimedia Interactive Software, Macintosh & Windows Version 1.0, Amsterdam.

Wallace, A. R., 1876. The Geographical Distribution of Animals with a Study of the Relations of Living and Extinct Faunas as Elucidating the Past Changes of the Earth's Surface. Macmillan & Co., London.

Westheide, W., 1997. The direction of evolution within the Polychaeta. Journal of Natural History 31: 1–15.

Westheide, W., D. McHugh, G. Purschke & G. Rouse, 1999. Systematization of the Annelida: different approaches. Hydrobiologia 402: 291–307.

Wei, Y. & J. Liu, 2006. Sludge reduction with a novel combined worm-reactor. Hydrobiologia 564: 213–222.

Wetzel, M. J., S. V. Fend, K. A. Coates, R. D. Kathman & S. R. Gelder, 2006. Taxonomy, systematics, and ecology of the aquatic Oligochaeta and Branchiobdellidae (Annelida, Clitcllata) of North America, with emphasis on the fauna occurring in Florida. A workbook. 10 September. Illinois Natural History Survey, Champaign.

Wetzel, M. J., R. D. Kathman, S. V. Fend & K. A_ Coates, 2007. Classification and Checklist of the Freshwater Oligochaeta occurring in North America North of Mexico. World Wide Web URL: http://www.inhs.uiuc.edu/~mjwetzel/FWOligoNAChklst.html. 15 July 2007.

Hydrobiologia (2008) 595:129–137
DOI 10.1007/s10750-007-9010-8

FRESHWATER ANIMAL DIVERSITY ASSESSMENT

Global diversity of leeches (Hirudinea) in freshwater

Boris Sket · Peter Trontelj

Abstract Leeches (Hirudinea) constitute a relatively small monophyletic group of highly specialized annelids, but may play important roles as invertebrate predators in freshwater, while others are infamous for their ectoparasitic bloodsucking. About 15% of the 680 described species are marine and slightly less have switched to terrestrial life; the rest are freshwater, divided among 91 genera. They are globally distributed on all continents except Antarctica, reaching the highest diversity in the Holarctic region with one-half of all continental species. Known areas of local endemism are the ancient Siberian lake Bajkal and lake Ohrid (about 10 species each) on the Balkan Peninsula, which is an endemicity area in itself. A small number of sanguivorous species known as "medicinal leeches" have played an important role in traditional and modern medicine, most noticeably four *Hirudo* spp. from the Western Palearctic.

Keywords Hirudinea · Leeches · Freshwater · Fauna · Biodiversity · Biogeography

Guest editors: E. V. Balian, C. Lévêque, H. Segers & K. Martens
Freshwater Animal Diversity Assessment

B. Sket (✉) · P. Trontelj
Oddelek za biologijo, Biotehniška fakulteta, Univerza v Ljubljani, Večna pot 111, Ljubljana 1000, Slovenia
e-mail: Boris.Sket@bf.uni-lj.si

Introduction

Leeches (Hirudinea) used to be considered as derivatives of oligochaetes, *Acanthobdella*, with few chaetae still present, being the connecting link. It is now clear that Hirudinea (Achaetobdellae, Euhirudinea, Hirudinida) constitute a monophylum that renders the Oligochaeta paraphyletic (Martin, 2001; Siddall et al., 2001). The Acanthobdellida and the Branchiobdellida (crustacean epizoans or parasites traditionally linked with Oligochaeta) are the two closest relatives of true leeches, the latter being the more likely sister group, according to molecular phylogenetic investigations (Siddall et al., 2001).

Leeches are hermaphroditic annelids with totally reduced both parapodia and chetae, with unpaired male and female genital openings in the region of the glandulose belt called clitellum, and with a sucker on both the anterior and posterior end of the body (Fig. 1). Each somite is superficially divided into usually three to five, but sometimes more than 10 annuli. The coelomic cavity is transformed into a contiguous system of channels. Many leeches are blood-sucking on vertebrates or invertebrates; the others are mainly predators, rarely scavengers. Most inhabit freshwaters, but there are marine and terrestrial species, too. They reproduce by eggs deposited in cocoons secreted by the clitellum. Their ontogeny is direct, without larval stages.

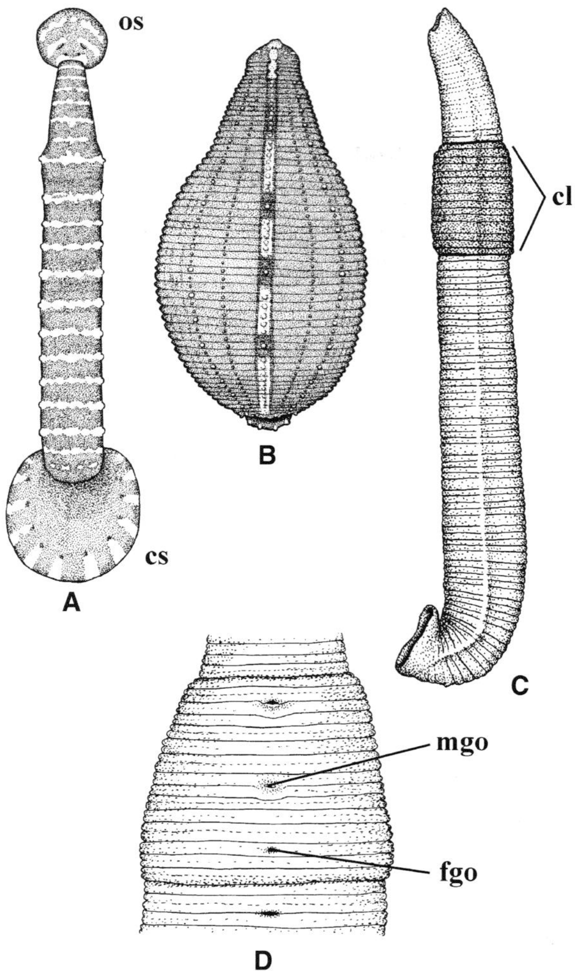

Fig. 1 Habitus of three typical representatives of major hirudinean taxa: (**A**) *Cystobranchus fasciatus*, a fish leech (Piscicolidae), dorsal view; (**B**) *Placobdella costata*, a glossiphoniid, dorsal view; (**C**) – *Barbronia assiuti*, an erpobdelliform, lateral view; (**D**) Ventral view of the clitellum with genital and accessory openings; the latter are found only in some species. **os**—oral sucker, **cs**—caudal sucker, **mgo**—male genital opening; **fgo**—female genital opening. Original drawings are a courtesy of Hasko Nesemann

Species diversity

This compilation was greatly facilitated by some previous local or global monographs, mainly indirectly through Sawyer (1986); beside the latter, such are by Harding & Moore (1927), Klemm (1972), Lukin (1976), Ringuelet (1980), Yang (1996), Nesemann & Neubert (1999), and a series by Soos (1965–1969). Numerous primary sources (not listed in References) were used to complete the work.

Linnaeus (1789) described 14 species of *Hirudo*, classified as Vermes Intestina. The classical taxonomy used to be based on the general shape, color pattern, position and number of eyes, later also on distance between genital openings. Moquin-Tandon (1846) presented anatomical pictures which were later used extensively; so was also the system of annulation with an appropriate coding, proposed by Moore (1898). Pawlowski (1948) added the male atrium shape. However, nobody assessed the variability level of the morphological and anatomical characters. A study of the hirudinean diversity of the former Yugoslavia area, notorious for its complex biodiversity, showed that neither the anatomy of the reproductive apparatus nor the annulation is stable enough to be reliable, as character above the species level (Sket, 1968). This, and some phylogenetic uncertainties, triggered first molecular studies (Trontelj et al., 1996; Siddall & Burreson, 1998) followed by many others that are nowadays yielding surprising solutions predicting discovery of further unexpected errors of the current taxonomy.

However, morphological means allowed us to describe approximately 680 species till now, some 480 of which are freshwater. Bielecki (1997) introduced new detailed morphological and anatomical standards in the systematics of piscicolids (fish leeches) that gave rise to a manifold increase of European fish leech species. This, along with a considerable ‘cryptic’ diversity among European erpobdellids (own studies, unpublished), suggests that a large number of species has still remained unrecognized even within the traditionally best-studied European hirudinean fauna. The cumulative plot of described species (Fig. 2) shows no sign of a plateau.

These facts also imply that leech taxonomy—at the species level and higher—is in a revolutionary phase right now. In the present overview (Table 1), we mainly follow the system of Sawyer (1986) but we could not ignore some novel findings; therefore, some of our solutions are partial and provisional. Too little care has been taken till now to represent genera in molecular studies by their type species and species by their topo-type populations, as these are the only reliable means for assuring the taxonomical identity of objects. Neither the number of families, nor their generic subdivision, are final.

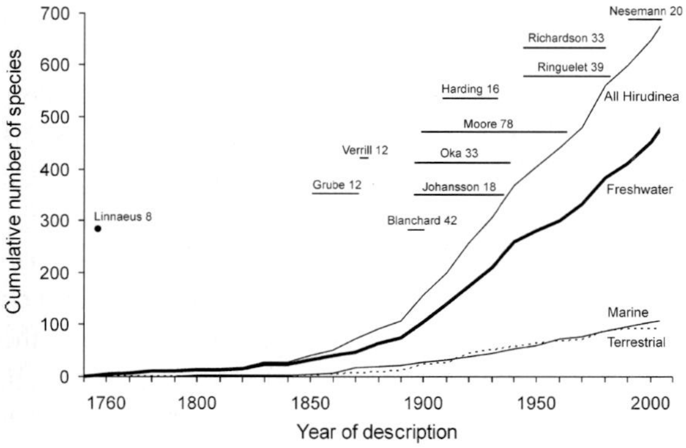

Fig. 2 Cumulative number of described hirudinean species per decade. Authors with significant contributions (if 15 or more species after 1900, 20 or more after 1950) and the phase of their publishing activity are shown

Species diversity is relatively evenly distributed among four major hirudinean clades, the proboscis-bearing Piscicolidae and Glossiphoniidae, the jawed Hirudiniformes, and the predaceous Erpobdelliformes (see Table 1 for classification and species numbers). The latter two groups are sisters, often referred to as arhynchobdellids for their lack of a proboscis. This leaves the small tropical marine and freshwater family Ozobranchidae with an uncertain position among the paraphyletic group of proboscis-bearing leeches, traditionally known as rhynchobdellids.

Piscicolidae are parasites, mainly on fishes, a large part of them being the only marine leeches. Bound to freshwater only, but ecologically very diverse and globally distributed, are the Glossiphoniidae. They feed as parasites on vertebrates and invertebrates; some may be predators if the prey is small enough. Arhynchobdellids live predominantly in freshwaters, but might also be amphibious or terrestrial. Most hirudiniform species are parasitic on vertebrates; some are also predators of small invertebrates, occasionally even scavengers. Erpobdelliforms are exclusively predatory species with an aquatic or amphibious lifestyle. They swallow their prey as a whole using their muscular pharynx. They may occur in high densities and are among the most important invertebrate predators in freshwater communities.

Traditionally a part of the Hirudinea but now recognized as a clade of their own, the

Table 1 Taxonomic overview and diversity of the Hirudinea[a]

Major higher taxon	Family	Freshwater genera	Freshwater species	Terrestrial (T) or Marine (M) species
Acanthobdellida	Acanthobdellidae	1	2	0
"Rhynchobdellida"	Glossiphoniidae	25	208	0
	Piscicolidae	17	57	100 (M)
	Ozobranchidae	2	7	2 (M)
Arhynchobdellida				
Erpobdelliformes	Americobdellidae	0	0	1 (T)
	Erpobdellidae	10[b]	69	0
	Salifidae	7	28	1 (T)
Hirudiniformes	Cylicobdellidae	0	0	28 (T)
	Semiscolecidae	4	13	0
	Haemopidae	3	18	0
	Hirudinidae[c]	17	60	4 (T)
	Macrobdellidae	5	20	0
	Haemadipsidae	0	0	50 (T)
	Xerobdellidae	0	0	8 (T)
Total		91	482	102 M
				92 T

[a] The traditional conception including acanthobdellids but not branchiobdellids. Not considered are species *inquirendae* (of ca 110 species *inquirendae* listed by Soos 1965–1969, only 10 have been classified later and considered here)

[b] Traditional genera are mostly para- or polyphyletic, a final subdivision of the family is not yet established

[c] Including Hirudinariinae, Praobdellinae, Richardsonianinae, Ornithobdellinae; provisionally also *Limnatis*

Acanthobdellida are limited to two species at the boreal fringe of the Holarctic. They parasitize fishes. The larger, more than 100 spp. strong Branchiobdellida have been traditionally discussed with Oligochaeta.

Phylogeny and historical processes

Based on recent molecular phylogenetic hypotheses (e.g., Apakupakul et al., 1999; Trontelj et al., 1999; Borda & Siddall, 2004; Utevsky & Trontelj, 2004), it can be concluded that leeches are primarily and essentially freshwater animals with few switches to marine and terrestrial habitats (Fig. 3). Further, their notorious ectoparasitic bloodsucking might be a sophistication of a less specialized commensalism or parasitism inherited from ancestors shared with branchiobdellids and acanthobdellids. It has been proposed that the first true leech was a proboscis-bearing bloodsucker, and that bloodfeeding facilitated by jaws as in the medicinal leech evolved independently, following a predatory stage (Trontelj et al., 1999; Borda & Siddall, 2004).

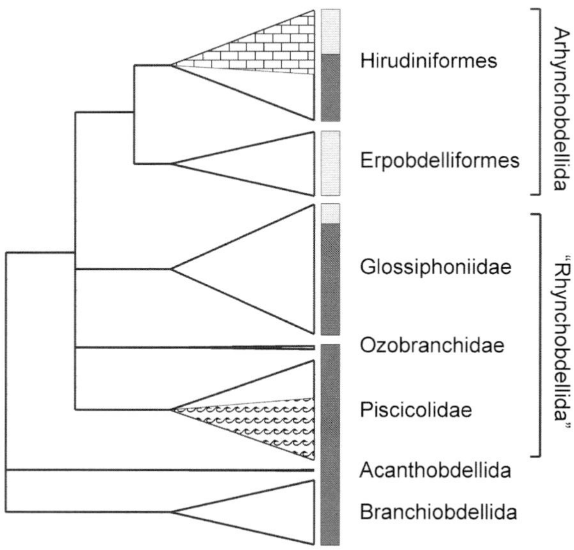

Fig. 3 Simplified phylogenetic tree of main hirudinean and sister taxa, the height of the triangle reflecting the species richness of each clade. Waves represent the share of marine, bricks the share of terrestrial species, while white areas correspond to freshwater (and amphibian) species. The share of terrestrial erpobdelliforms and marine ozobranchids (two species each) is too small to be shown. Dark vertical bars represent ecto-commensal or parasitic feeding, light bars predatory behavior. Only relationships that have consistently received high support in all molecular phylogenetic studies are drawn as resolved

A second focus of current molecular phylogenetic work is on species-level relationships and alpha taxonomy. A number of studies have indicated that not only much of the traditional low-level taxonomy is mislead by highly homoplastic characters, but also a great deal of species diversity remained overlooked or ignored. Perhaps the most striking example of the former case is the family Erpobdellidae, in which, after molecular scrutiny, virtually all characters used for subfamilial and generic subdivision (e.g., annulation, genital anatomy, and color patterns) turned out to be useless for this purpose (Trontelj & Sket, 2000; Siddall, 2002). Lack of reliable taxonomic characters is the main reason for recently discovered cases of cryptic diversity among erpobdellids. Finally, the most famous of all leeches, the European medicinal leech, is represented by at least three species as demonstrated by phylogenetic analyses of nuclear and mitochondrial DNA sequences (Trontelj & Utevsky, 2005). Their coloration pattern, often rejected as too variable, has ultimately proven to be a reliable identification feature.

Present distribution and main areas of endemicity

Global distribution patterns

Erpobdellidae are evidently limited to the Holarctic. The only species in New Zealand can hardly avoid the suspicion of a taxonomical mistake, while in Mexico the group crosses the border of the Neotropical region. Closely related species are widely distributed across the Palearctic, as well as between Palearctic and Nearctic. A particularly high number of erpobdellid species are known from Europe, but the degree of knowledge is territorially biased. The Haemopidae s.str are mainly Nearctic (*Haemopis*) and East Palearctic (*Whitmania*), with a few species of *Haemopis* distributed in Europe. Generally Holarctic, and mainly Western Palearctic, is the whole freshwater section of Piscicolidae. *Myzobdella* (Neotropical) and *Limnotrachelobdella* spp. (E Palearctic) might be secondary invaders from the sea. Limited to the Holarctic are also some genera of Glosiiphoniidae: *Glossiphonia*, *Placobdella* with the majority of species in the Nearctis, and *Torix* limited to the east of the

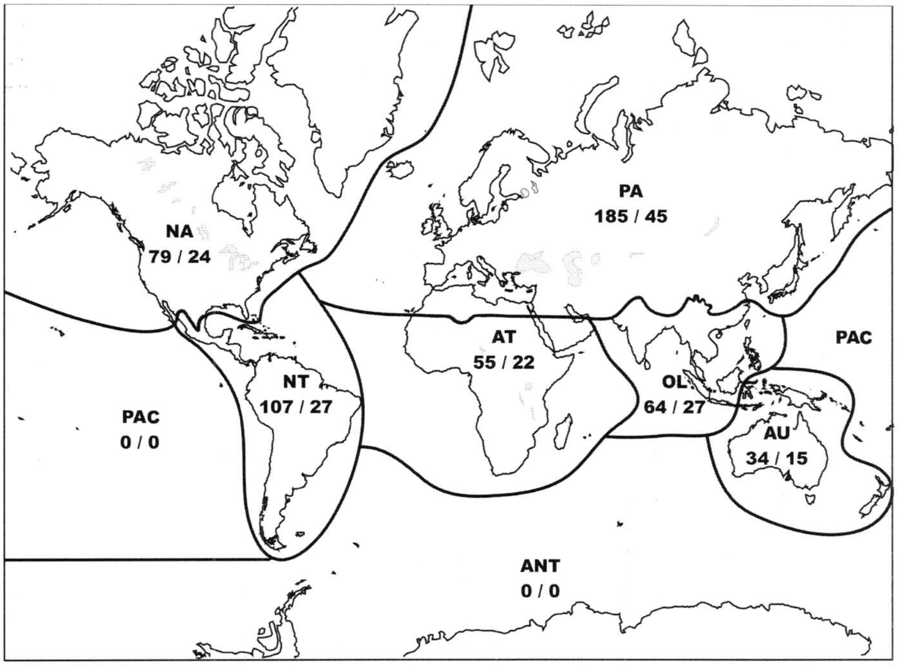

Fig. 4 Total species and genus numbers of Hirudinea per zoogeographic regions (Species number/Genus number). PA—Palearctic, NA—Nearctic, NT—Neotropical, AT—Afrotropical, OL—Oriental, AU—Australasian, PAC—Pacific Oceanic Islands, ANT—Antarctic

Eastern Palearctis. In unison, they make the Holarctic with one-half of all continental leech species the most diverse biogeographic region (Fig. 4, Table 2).

In the Neotropical region, the glossiphoniid genus *Haementeria* is mainly tropical and hardly crossing the northern border of the region. Biogeographically similar, and probably closely related, are the smaller genera *Gloiobdella* and *Adaetobdella*. The genus *Helobdella* has, on the other hand, richly speciated (more than 35 species known) in colder (elevated) regions of South America. It developed a couple of ecologically very successful species, one of which spread widely into the Holarctic region, possibly as an accidental passenger on migrating aquatic birds. Two additional species succeeded to spread out of the Neotropical region by some human means. A similar distribution display the hirudiniform sister families Semiscolecidae and Macrobdellidae, the former being limited to the Neotropical region, the latter with at least one genus each in the Neotropics and the Nearctis.

The Oriental region is still relatively rich in leeches but has virtually no endemic groups. The region can be characterized by some smaller genera with predominantly Oriental species and only slight extrusions into the Eastern Palearctic. Such are the glossiphoniid *Paraclepsis* and hirudiniform *Myxobdella, Poecilobdella* and *Hirudinaria.* Most Oriental groups are in fact Paleotropical, occuring also in the Afrotropical region. Oriental and Afrotropical is the genus *Asiaticobdella*; mainly Oriental, although generally tropical are freshwater Ozobranchidae and the rich glossiphoniid genus *Placobdelloides. Limnatis* is also present outside tropics in southern parts of Europe. The hirudinid *Praobdella* seems to be purely African.

Somehow in the warmer East is also the gravity point of the family Salifidae; its main genera, *Salifa* and *Barbronia*, are both present in the Oriental region and in the eastern Palearctis, spreading slightly to the Afrotropical, the Western Palearctic, and even into the Australian regions. Since some of its species have clearly demonstrated good spreading abilities, it is questionable whether such a distribution pattern is ancient, and if it is natural at all.

The Australasian region, except for the endemic hirudinid subfamily Richardsonianinae (*Bassianobdella, Goddardobdella, Richardsonianus*), is inhabited by very few freshwater leech species. They belong to different genera and might be either results of late natural introductions (e.g., by birds) or even of taxonomical errors. The most numerously represented are the snail leeches *Alboglossiphonia*, a globally distributed genus. Australasia is the domain of the

Table 2 Number of freshwater hirudinean species and genera found in major biogeographical regions

	PA	NA	NT	AT	OL	AU	World
Number of species							
Acanthobdellida							
Acanthobdellidae	2	1	0	0	0	0	2
Rhynchobdellida							
Glossiphoniidae	64	39	69	20	31	13	208
Piscicolidae	40	7	5	2	4	0	57
Ozobranchidae	1	0	1	1	3	1	7
Arhynchobdellida							
Americobdellidae	0	0	0	0	0	0	0
Erpobdellidae	46	18	3	0	2	1	69
Salifidae	10	0	0	8	9	4	28
Cylicobdellidae	0	0	0	0	0	0	0
Semiscolecidae	0	1	13	0	0	0	13
Haemopidae	9	7	1	1	1	0	18
Hirudinidae	15	0	1	18	14	15	60
Macrobdellidae	0	6	14	0	0	0	20
Haemadipsidae	0	0	0	0	0	0	0
Xerobdellidae	0	0	0	0	0	0	0
Total species	187	79	107	50	64	34	482
Number of genera							
Acanthobdellida							
Acanthobdellidae	1	1	0	0	0	0	1
Rhynchobdellida							
Glossiphoniidae	15	11	12	9	10	4	25
Piscicolidae	10	3	2	2	4	0	17
Ozobranchidae	1	0	1	1	1	1	2
Arhynchobdellida							
Americobdellidae	0	0	0	0	0	0	0
Erpobdellidae	6	5	3	0	1	1	10
Salifidae	4	0	0	3	3	2	7
Cylicobdellidae	0	0	0	0	0	0	0
Semiscolecidae	0	1	4	0	0	0	4
Haemopidae	2	1	1	1	1	0	3
Hirudinidae	6	0	1	6	7	7	17
Macrobdellidae	0	2	3	0	0	0	5
Haemadipsidae	0	0	0	0	0	0	0
Xerobdellidae	0	0	0	0	0	0	0
Total genera	45	24	27	22	27	15	91

PA: Palaearctic Region, NA: Nearctic Region, NT: Neotropical Region, AT: Afrotropical Region, OL: Oriental Region, AU: Australasian Region

terrestrial Haemadipsidae, which might be most diverse here, while spreading throughout the Oriental region and even crossing into the SE of the Eastern Palearctic.

The distribution type of some phyletic groups is at present not definable, e.g., the glossiphoniid *Theromyzon* has species distributed across all biogeographical regions except for the Australasian. One could easily explain this by their relation to waterfowl. The explanation of the even distribution of *Alboglossiphonia* throughout all regions is more enigmatic. In oceans, piscicolids are present around all continents and at all latitudes. Previous biogeographical analyses have been published by Soos (1970), Ringuelet (1980), and Sawyer (1986).

Endemicity areas

Discovery of species flocks and therefore of rich endemic faunas can be predicted for the next future. They may occur in SE Europe (own studies) and in the southern United States (according to Govedich et al., 1999). Most ancient lakes have single or no endemic leech species. Likewise, with only two *Caspiobdella* spp., the number of endemics in the (brackish) Caspian is surprisingly low, but a number of European species seems to originate from Ponto-Caspian waters. Richer is Lake Bajkal with three piscicolids (*Baicalobdella torquata, Codonobdella truncata, C. zelenskiji*) and some glossiphoniids (*Baicaloclepsis echinulata, B. grubei, Torix baicalensis* and probably some *Theromyzon* spp.). The Balkan lake of Ohrid harbors the richest known endemic leech fauna (Sket, 1968, 1989) with some endemic glossiphoniids (*Glossiphonia complanata maculosa, G. pulchella*) and a flock of eight erpobdellid species (*'Dina' ohridana* aggregate). The latter are young species according to their mitochondrial DNA divergence (unpublished results), but remarkably differentiated in their body shapes (Sket, 1989). Another probable Lake Ohrid endemic is *Piscicola pawlowskii*.

Ecological specialists

Some species occur in caves (Sket, 1986; unpublished data). These may be generalists, like *Haemopis sanguisuga*, or troglobionts, like *H. caeca* from Dobrogea in Romania. Particularly rich in troglobionts is the family Erpobdellidae. Some described or undescribed cave species are present in southern Europe (N Italy–Balkans–Turkey–Caucasus), their derivative is also the extraordinarily transformed *Croatobranchus mestrovi* form deep caves in the Croatian Dinaric mountains. Some undescribed species, probably erpobdellids, occur in caves of China and the US.

Another habitat less frequently inhabited by leeches are brackish waters, like lakes and lagoons along the SE Indian coast (lake Chilka) with *Pterobdella amara, Aestabdella caeca, Calliobdella olivacea*. Both Caspian piscicolids and the Ponto-Caspian *Archaeobdella esmonti* can be attributed to this group.

Among terrestrial specialists the giant Chilean *Americobdella valdiviana*, representing its own family, is one of the largest leeches (reportedly measuring more than 20 cm, along with two other giants, the Amazonian freshwater species *Haementeria ghiliani* and the Antarctic marine *Megaliobdella szidati*). Ornithobdellinae, a couple of terrestrial hirudinid species, are feeding on sea-birds and can be found in their colonies in Australia and New Zealand. In the Xerobdellidae, *Xerobdella* spp. are terrestrial predators in temperate to alpine climates of Europe, whereas the neotropical *Mesobdella* and *Diestecostoma* spp. are reported as sanguivorous. Entirely terrestrial are also the haematophagous Haemadipsidae and predatory Cylicobdellidae.

Human related issues

Leeches have been intimately connected to humans throughout nearly 2000 years of documented history of Western medicine. While in ancient times the haematophagous medicinal leech (*Hirudo medicinalis*) was considered as panacea, nowadays mainly its bio-active anticoagulant and anti-inflammatory substances are attracting medical and pharmaceutical attention (Sohn et al., 2001; Whitaker et al., 2004). Moreover, the direct therapeutic application of leeches is experiencing a renaissance, albeit for different purposes, e.g., to restore blood circulation after reconstructive surgery or, recently, to treat osteoarthritis (Pilcher, 2004). Although the leeches are now commercially bred in leech farms, the annual consumption will probably never approach the nineteenth century numbers when up to 100 million leeches per year were imported to France alone. It has only recently become clear that most commercially used leeches are not the species officially declared (*H. medicinalis*), but rather its congener *H. verbana* or sometimes *H. orientalis*. Other species, mainly of the SE Asian genus *Hirudinaria* have been exploited medically and are sometimes even offered for sale as genuine medicinal leeches. Even some rhynchobdellids (*Haementeria officinalis* in Mexico, *Placobdella costata* in Krym) have been in medical use.

The "medicinal leech" is protected and/or listed as endangered species in many European countries. It is not clear to what extent the alleged unfavorable conservation status is a consequence of past over-harvesting and how much of it can be contributed to more recent habitat destruction. Moreover, as long as the new taxonomic knowledge is not taken into account, we will not even know which species we are struggling to preserve.

Through centuries of exploitation and translocation the natural distribution of all *Hirudo* spp. was probably substantially affected by humans. More conspicuous, however, are transcontinental introductions, like the one of SE Asian *Hirudinaria manillensis* to the West Indies (Kutschera & Roth, 2006), probably also as a consequence of transport for medical purposes. Accidental transfers of leeches have resulted in several successful invasions of new ranges, most noticeably by the misleadingly named Neotropic glossiphoniid *Helobdella europaea* to Europe and Australia (e.g., Kutschera 2004), or the salifid *Barbronia weberi* from Asia to Europe and Australia. Other non-native *Helobdella* spp. have been reported from Europe, and the Australian *Barbronia arcana* from Mexico (Oceguera-Figueroa et al., 2005). Although *B. weberi* has been characterized as invasive (Govedich et al., 2003), the invasiveness of most non-native leech species does not approach the aggressive nature of some invasive crayfishes and fishes.

Acknowledgments We thank Hasko Nesemann for kindly providing his drawings, and Gregor Bračko for his help with copying out some literature data. This work was in part supported by the Slovenian Research Agency.

References

Apakupakul, K., M. E. Siddall & E. M. Burreson, 1999. Higher level relationships of leeches (Annelida: Clitellata: Euhirudinea) based on morphology and gene sequences. Molecular Phylogenetics and Evolution 12: 350–359.

Bielecki, A., 1997. Fish leeches of Poland in relation to the Palearctic piscicolines (Hirudinea: Piscicolidae: Piscicolinae). Genus 8: 223–375.

Borda, E. & M. E. Siddall, 2004. Arhynchobdellida (Annelida: Oligochaeta: Hirudinida): phylogenetic relationships and evolution. Molecular Phylogenetics and Evolution 30: 213–225.

Govedich, F. R., B. A. Bain, M. Burd & R. W. Davies, 2003. Reproductive biology of the invasive Asian freshwater leech *Barbronia weberi* (Blanchard, 1897). Hydrobiologia 510: 125–129.

Govedich, F. R., D. W. Blinn, R. H. Hevly & P. S. Keim, 1999. Cryptic radiation in erpobdellid leeches in xeric landscapes: a molecular analysis of population differentiation. Canadian Journal of Zoology 77: 52–57.

Harding W. A. & J. P. Moore, 1927. Hirudinea. In Shipley, E. (ed.), The Fauna of British India, including Ceylon and Burma. London.

Klemm, D. J., 1972. Freshwater leeches (Annelida: Hirudinea) of North America. Biota of freshwater ecosystems identification manual no. 8. Water Pollution Control Research Series. U.S. Environmental Protection Agency, Washington, D.C.

Kutschera, U., 2004. The freshwater leech *Helobdella europaea* (Hirudinea: Glossiphoniidae): an invasive species from South America? Lauterbornia 52: 153–162.

Kutschera, U. & M. Roth, 2006. Notes on the ecology of the Asian medicinal leech *Hirudinaria manillensis* (Hirudinea: Hirudinidae). Lauterbornia 56: 9–13.

Linnaeus, C., 1789. Systema Naturae. Vermes.

Lukin, E. I., 1976. Pijavki. Tom I. Pijavki presnyh i solonovatyh vodoemov, Izdatel'stvo "Nauka", Leningrad.

Martin, P., 2001. On the origin of the Hirudinea and the demise of the Oligochaeta. Proceedings of the Royal Society of London Series B—Biological Sciences 268: 1089–1098.

Moore, J. P., 1898. The leeches of the U. S. National Museum. Proceedings US National Museum 21(1160): 543–563.

Moquin-Tandon, A., 1846. Monographie de la famille des Hirudinees. Paris.

Nesemann, H. & E. Neubert, 1999. Annelida: Clitellata: Branchiobdellida, Acanthobdellea, Hirudinea. In: Süßwasserfauna von Mitteleuropa, 6/2. Spektrum Akademischer Verlag, Heidelberg, Berlin.

Oceguera-Figueroa, A., V. León-Règagnon & M. E. Siddall, 2005. Phylogeny and revision of the Erpobdelliformes (Annelida, Arhynchobdellida) from Mexico based on nuclear and mitochondrial gene sequences. Revista Mexicana de Biodiversidad 76: 191–198.

Pawlowski, L. K., 1948. Contributions á la systématique des sangsues dú genre *Erpobdella* de Blainville. Acta zoologica et oecologica Universitatis Lodziensis, sect. III 8: 1–55.

Pilcher, H., 2004. Medicinal leeches: stuck on you. Nature 432: 10–11.

Ringuelet, R. A., 1980. Biogeografia de los hirudineos de America del Sur y de Mesoamerica. Obra del Centenario del Museo de La Plata 6: 1–27.

Sawyer, R. T., 1986. Leech biology and behaviour. Clarendon Press, Oxford.

Siddall, M. E. & E. M. Burreson, 1998. Phylogeny of leeches (Hirudinea) based on mitochondrial cytochrome c oxidase subunit I. Molecular Phylogenetics and Evolution 9: 156–162.

Siddall, M. E., 2002. Phylogeny of the leech family Erpobdellidae (Hirudinea: Oligochaeta). Inveretbrate Systematics 16: 1–6.

Siddall, M. E., K. Apakupakul, E. M. Burreson, K. A. Coates, C. Erséus, S. R. Gelder, M. Källersjö & H. Trapido-Rosenthal, 2001. Validating Livanow: molecular data agree that leeches, Branchiobdellidans and Acanthobdella peledina form a monophyletic group of oligochaetes. Molecular Phylogenetics and Evolution 21: 346–351.

Sket, B., 1968. K poznavanju faune pijavk (Hirudinea) v Jugoslaviji (Zur Kenntnis der Egel–Fauna Jugoslawiens). Razprave SAZU, Cl. IV., 11(4): 127–197.

Sket B., 1986. Hirudinea. In Botosaneanu, L. (ed.), Stygofauna Mundi, pp. 250–253.

Sket, B., 1989. Intralacustrine speciation in the genus *Dina* (Hirudinea, Erpobdellidae) in Lake Ohrid (Yugoslavia). Hydrobiologia 182: 49–58.

Sohn, J. H., H. A. Kang, K. J. Rao, C. H. Kim, E. S. Choi, B. H. Chung & S. K. Rhee, 2001. Current status of the anticoagulant hirudin: its biotechnological production and clinical practice. Applied Microbiology and Biotechnology. 57: 606–613.

Soós, Á., 1965. Identification key to the leech (Hirudinoidea) genera of the world, with a catalogue of the species. I. Family: Piscicolidae. Acta zoologica Academiae scientiarum Hungaricae 11: 417–463.

Soós, Á., 1969. Identification key to the leech (Hirudinoidea) genera of the world, with a catalogue of the species. VI. Family: Glossiphoniidae. Acta zoologica Academiae scientiarum Hungaricae 15: 397–454.

Soós, Á., 1970. A zoogeographical sketch of the fresh-water and terrestrial leeches (Hirudinoidea). Opuscula zoologica Budapest 10(2): 313–324.

Trontelj, P., B. Sket, P. Dovč & G. Steinbrueck, 1996. Phylogenetic relationship in European erpobdellid leeches (Hirudinea: Erpobdellidae) inferred from restriction site data of the 18S ribosomal gene and ITS2 region. Journal of Zoological Systematics and Evolutionary Research 34: 85–93.

Trontelj, P., B. Sket & G. Steinbrück, 1999. Molecular phylogeny of leeches: congruence of nuclear and mitochondrial rDNA data sets and the origin of bloodsucking. Journal of Zoological Systematics and Evolutionary Research 37: 141–147.

Trontelj, P. & B. Sket, 2000. Molecular re-assessment of some phylogenetic, taxonomic and biogeographic relationships between the leech genera *Dina* and *Trocheta* (Hirudinea: Erpobdellidae). Hydrobiologia 438: 227–235.

Trontelj, P. & S. Y. Utevsky, 2005 Celebrity with a neglected taxonomy: molecular systematics of the medicinal leech (genus *Hirudo*). Molecular Phylogenetics and Evolution 34: 616–624.

Utevsky, S. & P. Trontelj, 2004. Phylogenetic relationships of fish leeches (Hirudinea, Piscicolidae) based on mitochondrial DNA sequences and morphological data. Zoologica Scripta 33: 375–385.

Whitaker, I. S., J. Rao, D. Izadi & P. E. Butler, 2004. *Hirudo medicinalis*: ancient origins of, and trends in the use of medicinal leeches throughout history. British Journal of Oral & Maxillofacial Surgery 42: 133–137.

Yang, T., 1996. Annelida Hirudinea. In Fauna Sinica. Science Press, Beijing.

Hydrobiologia (2008) 595:139–147
DOI 10.1007/s10750-007-9011-7

FRESHWATER ANIMAL DIVERSITY ASSESSMENT

Global diversity of freshwater mussels (Mollusca, Bivalvia) in freshwater

Arthur E. Bogan

Abstract The term freshwater bivalve is very inclusive and not very informative. There are representatives of at least 19 families that have at least one representative living in freshwater. This suggests at least 14 different invasions of freshwater. At least nine families have small to large radiations in the freshwater environment: Corbiculidae, Sphaeriidae, Dreissenidae, and the unioniforme families: Hyriidae, Margaritiferidae, Unionidae, Etheriidae, Iridinidae, and Mycetopodidae. The unioniforme families contain at least 180 genera and about 800 species. This order is characterized by the unique parasitic larval stage on the gills, fins or the body of a particular host fish. This order of freshwater bivalves is suffering a very high rate of extinction, with about 37 species considered presumed extinct in North America alone. The level of endangerment and extinction facing these animals is primarily the result of habitat destruction or modification.

Keywords Bivalve · Etheriidae · Extinction · Freshwater mussel · Hyriidae · Iridinidae · Margaritiferidae · Mycetopodidae · Unionidae · Unioniformes

Guest editors: E. V. Balian, C. Lévêque, H. Segers & K. Martens
Freshwater Animal Diversity Assessment

A. E. Bogan (✉)
Research Laboratory, North Carolina State Museum of Natural Sciences, 4301 Reedy Creek Road, Raleigh, NC 27607, USA
e-mail: Arthur.bogan@ncmail.net

Introduction

Freshwater bivalves provide a filtering service in rivers and lakes. Many species are often found in dense aggregations and filter out large quantities of blue–green algae, diatoms, bacteria, fine-particulate organic particles, as well as silt, absorb heavy metals and large organic molecules. All of the taxa included here are obligate freshwater organisms and spend their entire life cycle in freshwater.

Freshwater bivalves are not a monophyletic group and represent at least 19 families in three subclasses of bivalves. Most families are represented by only a few genera or species. Taxa with large radiations in freshwater include the Sphaeriidae, Corbiculidae, and the Order Unioniformes with 6 families, about 180 genera and about 800 species. Bivalves are mollusks without a head have a single foot enclosing the visceral mass, two pair of gills, and the sexes are typically separate. Each individual has two valves surrounding the body composed of calcium carbonate, either as calcite or aragonitic crystal structure. Unioniforme shells have aragonitic crystal structure.

The life history of freshwater bivalves is varied and depends on the family being discussed. Those species from primarily marine bivalve families have veliger or brooded larvae (McMahon & Bogan, 2001). The unioniforme bivalves are unique among bivalves, having an obligate parasitic larval stage on the gills, fins or sides of a host fish (Wächtler et al., 2001).

Shell shape varies among the families reflecting partially their phylogentic history and partially the habitat in which they are living. Byssally attached mussels are often much thinner shelled than those species living buried in cobble and gravel substrates. Many of the species of the Unionforme families have heavy shells with a variety of surface sculpture that aid in stability in the substrate.

Most of the species in this group are infaunal organisms burrowing into substrates varying from sand to cobbles and gravel but a few species exploit the exposed hard surfaces by attaching to hard surfaces with byssal threads like blue mussels and the zebra mussels.

Species/generic diversity

Freshwater bivalves are found in 3 different subclasses, separated into 5 separate orders and divided among 19 families within the Class Bivalvia (Deaton & Greenberg, 1991) (Table 1). There are 206 recognized genera of freshwater bivalves, most families represented by only one to five genera. Species diversity in the Dreissenidae follows Rosenberg & Ludyanskiy (1994). Large bivalve radiations in freshwater have occurred in the Sphaeriidae and the six unioniforme families. The species diversity mirrors the diversity of genera with about 1026 species (Tables 1, 2). Once again the highest diversity is found in the Sphaeriidae and the six unioniforme families. Corbiculidae species are over described based on variable shell form, and indications are that there are only a few species (Brandt, 1974; Morton, 1979; Subba Rao, 1989). Generic and species counts were based on literature for Sphaeriidae (Burch, 1975; Mandahl-Barth 1988; Smith, 1992; Dreher Mansur 1993; Daget, 1998; Korniushin & Glaubrecht, 2002; Lee & Ó Foighil, 2003). Estimates of the generic and specific diversity were more difficult to compile for the unioniforme families, due to the variation in systematic philosophy, lack of overview data for areas of the world. We have chosen to ignore for purposes of this exercise the over-inflation of taxonomic levels by the Russian malacologist of the Starobogatov school. Total genera and species were based on major reviews and localized faunal accounts (Ortmann, 1912; Pilsbry & Bequaert, 1927; McMichael & Hiscock, 1958; Haas, 1969; Brandt, 1975; Liu, 1979; Mandahl-Barth, 1988; Subba Rao, 1989; Smith, 1992; Starobogatov 1995; Bonetto, 1997; Daget, 1998; Turgeon et al., 1998; Bogan and Hoeh, 2000; Smith, 2001; Walker et al., 2001; Huff, et al., 2004).

Phylogeny and historical processes

Our current understanding of the phylogeny of the bivalves is still developing. Higher level phylogenies have been developed for bivalves supporting the subclasses recognized on the basis of morphological characters. However, phylogenetic analyses at the family level are just developing. The overall phylogeny of the Order Unioniformes, a monophyletic group is still in a state of flux. Based on recent DNA analyses, the Margaritiferidae, Unioidae, Mycetopodidae, Iridinidae are all monophyletic. Hyriidae genera from South America and Australasia form monophyletic sister clades, but whose relationships to other unioniforme families is still uncertain (Graf, 2000; Hoeh, et al., 1998, 1999, 2001). Curole and Kocher (2002) based on DNA anlyses suggested the family Margaritiferidae branched off from the Unionidae at a minimum of 230 MYA and estimated the subclass Paleoheterodonta diverged from the rest of Bivalvia at approximately 500 MYA (Middle Cambrian).

Speciation in freshwater bivalves may be driven by separation of stream systems by vicariant events or separate invasions of freshwater. In the Unioniformes speciation may be tied to speciation in host fishes. There has been little discussion of the factors driving speciation in unioniforme bivalves.

Present distribution and main areas of endemicity

Diversity of freshwater bivalves across the main zoogeographic areas is extremely variable (Tables 1, 2; Fig. 1). A total of 19 families with 206 genera and an estimated 1026 species are reported from freshwater. Two main areas of diversity and endemism in freshwater bivalves are the southeastern United States and the Oriental region. This diversity is primarily in

Table 1 Total number of genera in families of freshwater bivalves with representatives found in freshwater

	PA	NA	AT	NT	OL	AU	PAC	ANT	World
Subclass Pteriomorpha									
Order Arcoida									
Arcidae	0	0	0	0	1	0	0	0	1
Order Mytiloida									
Mytilidae	0	0	2	1(I)	1	0	0	0	3 (I)
Subclass Paleoheterodonta									
Order Unioniformes									
Etheriidae	0	0	1	0	0	0	0	0	1
Hyriidae	0	0	0	9	0	8	0	0	17
Iridinidae	0	0	6	0	0	0	0	0	6
Margaritiferidae[a]	3	2	0	0	1	0	0	0	3
Mycetopodidae	0	0	0	12	0	0	0	0	12
Unionidae[b]	26(I)	51(I)	6	20(I)	38(I)	1	0	0	142
Total Unionifomes	29	53	13	41	39	9	0	0	180
Subclass Heterodonta									
Order Veneroida									
Cardiidae	2	0	0	0	0	0	0	0	2
Corbiculidae	1	1(I)	1	2(I)	2	2	0	0	3
Sphaeriidae	4	4	3	5	2	2	2(I)	0	5
Dreissenidae	2	1(I)	1	0	0	0	0	0	3
Solenidae	0	0	0	0	1	0	0	0	1
Donacidae	0	0	2	0	0	0	0	0	2
Navaculidae	0	0	0	0	1	0	0	0	1
Order Myoida									
Corbulidae	1	0	0	0	0	0	0	0	1
Erodonidae	0	0	1	1	0	0	0	0	2
Teridinidae	0	0	0	1	0	0	0	0	1
Subclass Anomalodesmata									
Lyonsiidae	0	0	0	1	0	0	0	0	1
Total	40	59	23	51	47	13	2(I)	0	206

PA, Palaearctic; NA, Nearctic; NT, Neotropical; AT, Afrotropical; OL, Oriental; AU, Australasian; PAC, Pacific Oceanic Islands; ANT, Antarctic

(I) are taxa introduced outside of their native range

[a] The genus *Margaritifera* occurs in three regions

[b] The genus *Unio* occurs in two different regions

the Unionidae. The distribution of unioniforme families does not completely correspond to the standard zoogeographic regions (Fig. 2A–F).

Antarctic area

There are no known modern freshwater bivalves from Antarctica.

Oceanic Islands-Pacific area

There are two genera and two species of Sphaeriidae known as introduced species from Hawaii.

Australasian area

The freshwater bivalve fauna of this region includes representatives of 4 families, 13 genera and 43

Table 2 Total number of species in families of freshwater bivalves with representatives found in freshwater

	PA	NA	AT	NT	OL	AU	PAC	ANT	World
Subclass Pteriomorpha									
Order Arcoida									
Arcidae	0	0	0	0	4	0	0	0	4
Order Mytiloida									
Mytilidae	0	0	2	1(I)	2	0	0	0	5
Subclass Paleoheterodonta									
Order Unioniformes									
Etheriidae	0	0	1	0	0	0	0	0	1
Hyriidae	0	0	0	55	0	28	0	0	83
Iridinidae	0	0	41	0	0	0	0	0	41
Margaritiferidae	6	5	0	0	1	0	0	0	12
Mycetopodidae	0	0	0	39	0	0	0	0	39
Unionidae	86(I)	297(I)	32	85(I)	120	1	0	0	621
Total Unionifomes	92	302	74	179	121	29	0	0	797
Subclass Heterodonta									
Order Veneroida									
Cardiidae	5	0	0	0	0	0	0	0	5
Corbiculidae	[a]	2(I)	2	2(I)	[a]	[a]	0	0	6[a]
Sphaeriidae	34	45(I)	35	41	20	14	2(I)	0	196
Dreissenidae	5	2(I)	1	0	0	0	0	0	5
Solenidae	0	0	0	0	1	0	0	0	1
Donacidae	0	0	2	0	0	0	0	0	2
Navaculidae	0	0	0	0	2	0	0	0	2
Order Myoida									
Corbulidae	1	0	0	0	0	0	0	0	1
Erodonidae	0	0	1	1	0	0	0	0	2
Teridinidae	0	0	0	1	0	0	0	0	1
Subclass Anomalodesmata									
Lyonsiidae	0	0	0	1	0	0	0	0	1
Total	137	351	117	226	150	43	2	0	1026

PA, Palaearctic; NA, Nearctic; NT, Neotropical; AT, Afrotropical; OL, Oriental; AU, Australasian; PAC, Pacific Oceanic Islands; ANT, Antarctic

(I) are taxa introduced outside of their native range

[a] The total number of species in the genus is unkown at this time. The group is over-described based on shell shape variation

species. Diversity in the area is dominated by the Hyriidae with 8 genera and 28 species. Hyriids are restricted to Australia, Tasmania, New Zealand, New Guinea and the Solomon Islands. The number of species of Corbiculidae found in Australia is unknown but represented by an abundance of named shell shapes (Smith, 1992).

Palaearctic area

The diversity found in this region is dominated by the diversity of the Unionidae and Sphaeriidae with the remaining diversity contributed by six other families. Considering the vast area covered by this region, the diversity is not evenly distributed. Western Europe,

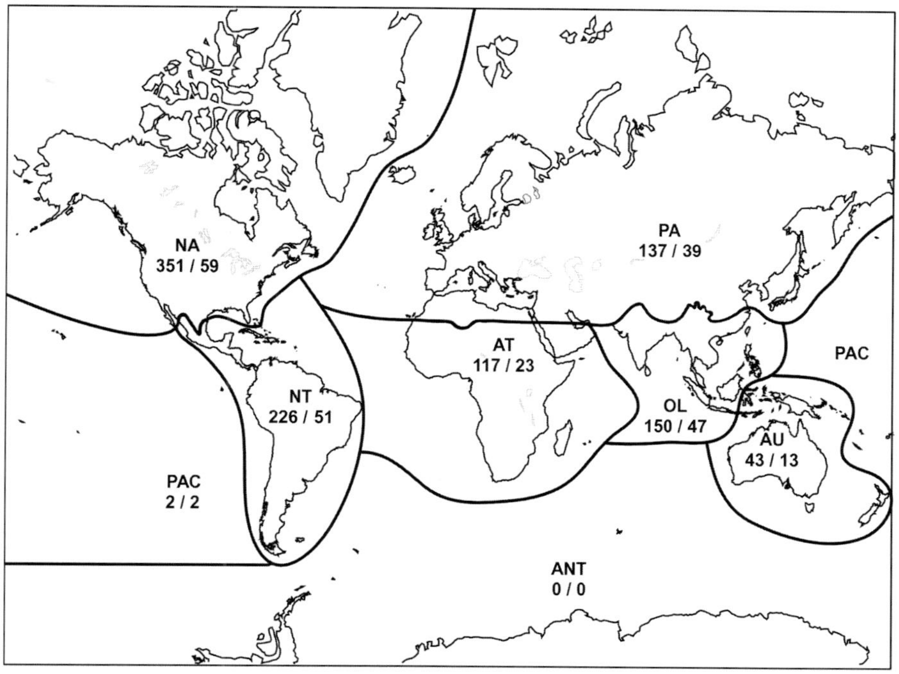

Fig. 1 Distribution of freshwater bivalvia species and genera (SP/GN) per zoogeographic region: ANT, Antarctica; AT, Afrotropical; AU, Australasia; NA, Nearctic; NT, Neotropical; OL, Oriental; PA, Palaearctic; PAC, Pacific Oceanic Islands, ANT, Antarctic

Russia, the trans-Caucasus region and Siberia have a rather limited diversity including representatives of seven families from the area but the greatest diversity in the Unionidae occurs in the eastern region extending from the Amur River basin in the north to southern China, including the Yangtze River basin (Wu, 1998). The diversity of the Unionidae in Western Europe and the region east to the Trans-Caucasus and south to Israel is limited to 6 genera while the Yangtze River basin has 14 genera.

Afro-tropical area

Nine families represented by 23 genera and 117 species are reported from sub-Saharan Africa and the Nile River. Two families with 2 genera and 4 species of freshwater unioniforme bivalves have been reported for Madagascar. Two families, Iridinidae and the Unionidae account for the greatest amount of the generic and species level diversity in this area.

Oriental area

This region's freshwater bivalve fauna is represented by 8 families, 47 genera and 150 species. This fauna can be broken into two separate components, one on the Indian plate extending from extreme eastern Iran east through Pakistan, Afghanistan, India and Bangladesh, and western Myanmar. Southern India is home to two monotypic endemic genera, one a cemented Unionidae (Subba Rao, 1989). The second faunal component extends from Myanmar east down the Malay Peninsula to Java, Borneo, the Philippines, Thailand, Laos, Cambodia, Vietnam, and southern China.

Nearctic area

The Nearctic freshwater bivalve fauna is globally the most diverse with 5 families, 59 genera and 302 species. The greatest diversity of freshwater bivalve genera and species occurs in this area followed by the Oriental region, especially in the Mekong River basin (Brandt, 1974). This diversity is the result of the high level of diversity of the Unionidae of the southeastern United States with 42 genera of and 271 species (e.g., Neves et al., 1998).

Neotropical area

The fauna of this area is diverse with 9 families, 51 genera and 226 species. The freshwater bivalve fauna

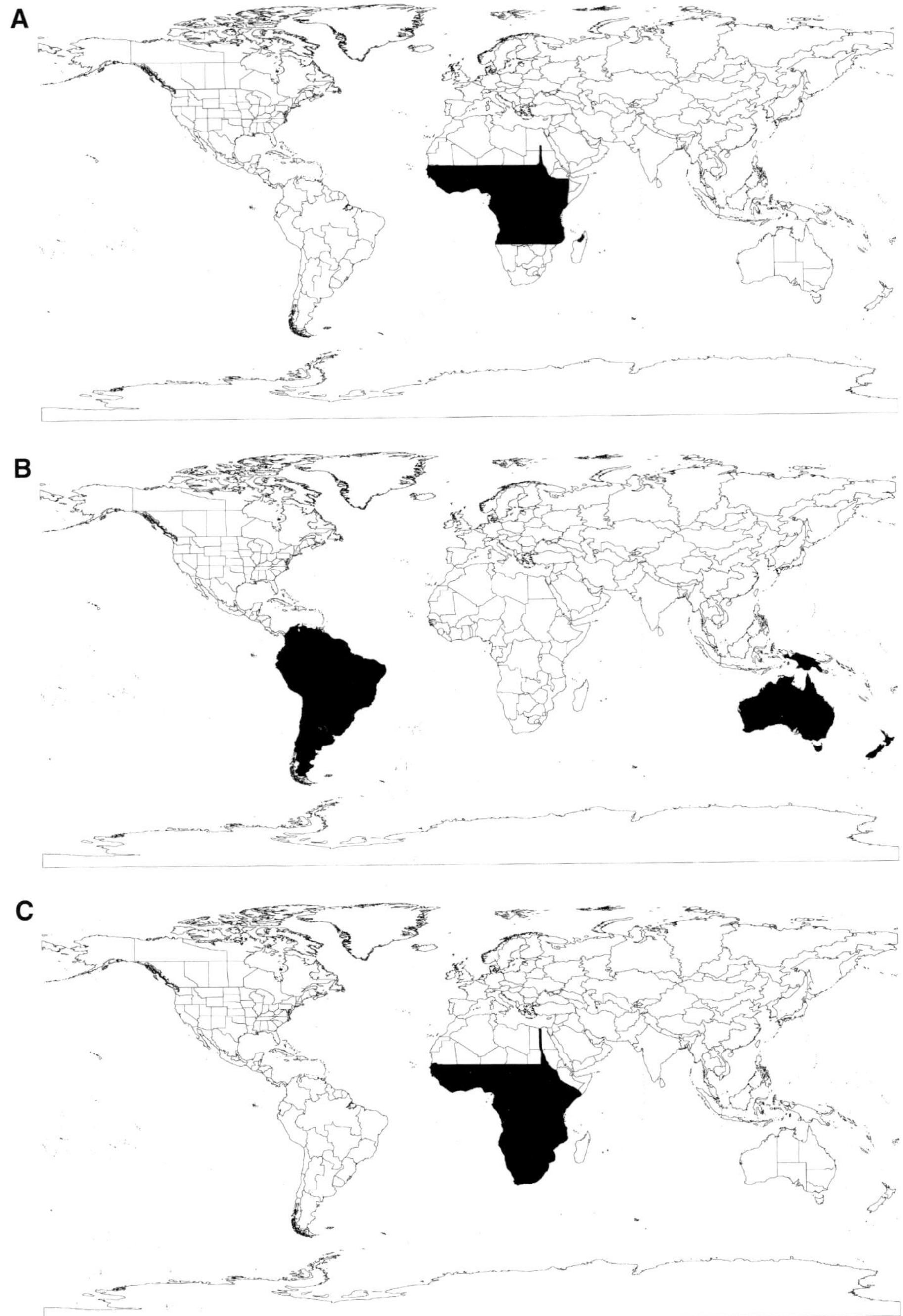

Fig. 2 (**A**) Distribution of Etheriidae, (**B**) Distribution of Hyriidae, (**C**) Distribution of Iridinidae, (**D**) Distribution of Margaritiferidae, (**E**) Distribution of Mycetopodidae, (**F**) Distribution of Unionidae

of this region is poorly known and has not been synthesized recently. The Hyriidae and Mycetopodidae along with the Sphaeriidae account for the majority of the diversity in South America. One genus of Mycetopodidae extends northward through Central America to west Central Mexico. The Unionidae in the Neotropical Area account for 20 genera and 85 species, but are only found in the area from Central Mexico south to Panama and are absent from South America.

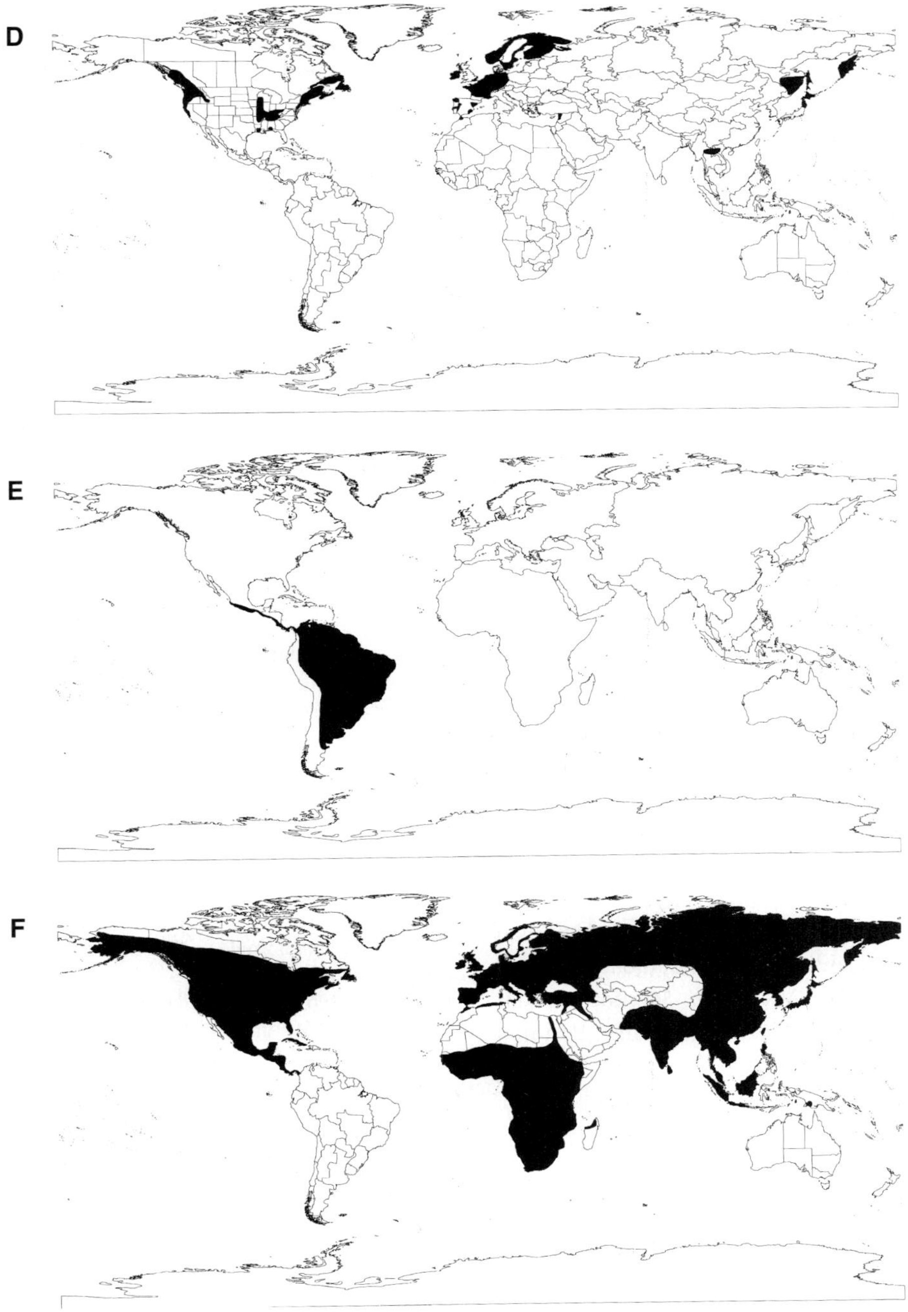

Fig. 2 continued

Human related issues

In various areas of the world, freshwater bivalves are a supplemental food source. The status of freshwater faunas is only incompletely known, but for freshwater mollusks it is declining (Bogan, 1993; Bogan, 1998; Lydeard et al., 2004). This decline is well documented for the very diverse freshwater molluscan fauna of the southeastern United States and suggested for the rest of the world (Bogan 1993; Neves et al., 1998; Lydeard et al., 2004). The consensus is the most dramatic cause of the declines and extinctions of freshwater bivalves is habitat modification and destruction. This can be due to the effects of dams, canalization, changes in water depth, due to flow changes and changes in fine particle deposition (silt and sand). These modifications affect not only the freshwater mussels, but also the fish they rely on for the unioniforme mussel's parasitic life stage. Additional impacts include water withdrawal for industry and irrigation, and pollution, including the creation of impervious areas within the watershed, due to urbanization and road building.

The freshwater bivalve fauna of Africa and South America is poorly known and there is still much confusion around the number of species recognized. As pointed out in Lydeard et al. (2004), the basic surveys of invertebrate animals are "critically important, particularly in poorly inventoried areas, if managers are to determine appropriate locations for conservation efforts." Taxonomic studies go hand in hand with these surveys.

Brackish water bivalves

Many families of marine bivalves have a few representative genera or species that have invaded brackish water but have not made it into freshwater habitats. Representatives of at least 27 bivalve families are found in brackish water: Anomiidae, Arcidae, Cardiidae, Corbiculidae, Corbulidae, Cyrenoididae, Cultellidae, Donacidae, Dreissenidae, Glauconomidae, Gryphaeidae, Isognomonoidae, Limidae, Lyonsiidae, Lucinidae, Mactridae, Mesodesmatidae, Mytilidae, Ostreidae, Pharidae, Pholadidae, Psammobiidae, Tellinidae, Teredinidae, Trapezidae, Ungulinidae, and Veneridae (Deaton & Greenberg, 1991; P. Mikkelsen, Personal communication).

Acknowledgments Paula Mikkelsen provided insight into the variety of families occurring in brackish water. Cynthia M. Bogan and Jamie Smith have kindly read and criticized various versions of this article. Jonathan Raine assisted with the production of the maps.

References

Bogan, A. E., 1993. Freshwater bivalve extinctions: search for a cause. American Zoologist 33: 599–609.

Bogan, A. E., 1998. Freshwater molluscan conservation in North America: problems and practices. In Killeen, I. J., M. B. Seddon & A. M. Holmes (eds), Molluscan Conservation: A Strategy for the 21st Century. Journal of Conchology, Special Publication Number 2, 223–230.

Bogan, A. E. & W. R. Hoeh, 2000. On becoming cemented: evolutionary relationships among the genera in the freshwater bivalve family Etheriidae (Bivalvia: Unionoida). In Harper, E. M., Taylor, J. D. & Crame, J. A. (eds), The Evolutionary Biology of the Bivalvia. Geological Society, London Special Publication 177, 159–168.

Bonetto, A. A., 1997. Las 'ostras de agua dulce' (Muteloidea: Mutelidae). Su taxonomia y distribucion geografica en el conjunto de las Naiades del Mundo. Biociências 5: 113–142.

Brandt, R. A. M., 1974. The non-marine aquatic Mollusca of Thailand. Archiv für Molluskenkunde 105: 1–423.

Burch, J. B., 1975. Freshwater Sphaeriancean Clams (Mollusca: Pelecypoda) of North America. Malacological Publications, Hamburg, Michigan 96 pp.

Curole, J. P. & T. D. Kocher. 2002. Ancient sex-specific extension of the Cytochrome c Oxidase II Gene in bivalves and the fidelity of doubly-uniparental inheritance. Molecular Biology and Evolution 19: 1323–1328.

Daget, J., 1998. Catalogue raisonné des Mollusques bivalves d'eau africains. Backhuys Publishers, Leiden and OSTROM, Paris.

Deaton, L. E. & M. J. Greenberg, 1991. The adaptation of bivalve molluscs to oligohaline and fresh waters: phylogenetic and physiological aspects. Malacological Review 24: 1–18.

Dreher Mansur, M. C., 1993. Morphologische Untersuchungen an *Eupera* Bourguignat, 1854, und *Byssanodonta* Orbigny, 1846. Zur phylogenetischen Systematik der Sphaeriidae und Corbiculidae (Bivalvia, Venerioda). Doctoral dissertation, Eberhard-Karls-Universität, Tübingen, Germany, 125 pp., 152 Figures, 9 Tables.

Graf, D. L., 2000. The Etheriidae revisited: a phylogenetic analysis of hyriid relationships (Mollusca: Bivalvia: Paleoheterodonta: Unionoida*)*. Occasional papers of the Museum of Zoology, The University of Michigan Number 729: 1–21.

Haas, F., 1969. Superfamilia Unionacea. Das Tierreich (Berlin) 88:x + 663 pp.

Hoeh, W. R., M. B. Black, R. Gustafson, A. E. Bogan, R. A. Lutz & R. C. Vrijenhoek, 1998. Testing alternative hypotheses of *Neotrigonia* (Bivalvia: Trigonioida) Phylogenetic relationships using Cytochrome c Oxidase Subunit 1 DNA sequences. Malacologia 40: 267–278.

Hoeh, W. R., A. E. Bogan, K. S. Cummings & S. E. Guttman, 1999. Evolutionary relationships among the higher taxa of freshwater mussels (Bivalvia: Unionoida): inferences on phylogeny and character evolution from analyses of DNA sequence Data. Malacological Review 31: 111–130.

Hoeh, W. R., A. E. Bogan & W. H. Heard. 2001. A phylogenetic perspective on the evolution of morphological and reproductive characteristics in the Unionoida. In Bauer G. & K. Wächtler (eds), Ecology and Evolutionary Biology of Freshwater Mussels, Unionoida. Ecological Studies, Vol. 145. Springer Verlag, 257–280.

Huff, S. W., D. Campbell, D. L. Gustafson, C. Lydeard, C. R. Altaba & G. Giribet, 2004. Investigations into the phylogenetic relationships of freshwater pearl mussels (Bivalvia: Margaritiferidae) based on molecular data: implications for their taxonomy and biogeography. Journal of Molluscan Studies 70: 379–388.

Korniushin, A. V. & M. Glaubrecht, 2002. Phylogenetic analysis based on the morphology of viviparous freshwater clams of the family Sphaeriidae (Mollusca, Bivalvia, Veneroida). Zoologica Scripta 31: 415–459.

Lee, T. & D. Ó Foighil, 2003. Phylogenetic structure of the Sphaeriinae, a global clade of freshwater bivalve mollusks, inferred from nuclear (ITS-1) and mitochondrial (16S) ribosomal gene sequences. Zoological Journal of the Linnean Society 137: 245–260.

Liu, Y. Y., 1979. Freshwater Mollusks of China. Economic Fauna of China. Beijing, China, 134 pp.

Lydeard, C., R. H. Cowie, A. E. Bogan, P. Bouchet, K. S. Cummings, T. J. Frest, D. G. Herbert, R. Hershler, O. Gargominy, K. Perez, W. F. Ponder, B. Roth, M. Seddon, E. E. Strong & F. G. Thompson, 2004. The global decline of nonmarine mollusks. BioScience 54: 321–330.

Mandahl-Barth, G., 1988. Studies on African freshwater bivalves. Danish Bilarziasis Laboratory, Charlottenlund.

McMahon, R. F. & A. E. Bogan. 2001. Mollusca: Bivalvia. In J. H. Thorpe & A. P. Covich. Ecology and Classification of North American Freshwater Invertebrates, 2nd edn. Academic Press, 331–429.

McMichael, D. F. & I. D. Hiscock, 1958. A monograph of the freshwater mussels (Mollusca: Pelecypoda) of the Australian region. Australian Journal of Marine and Freshwater Research 9: 372–508, pls. 1–19.

Morton, B., 1979. *Corbicula* in Asia. In Britton J. C. (ed.), Proceedings, First International *Corbicula* Symposium. Texas Christian University Research Foundation, Fort Worth, Texas, 16–38.

Neves, R. J., A. E. Bogan, J. D. Williams, S. A. Ahlstedt, & P. D. Hartfield. 1997 (March 1998). Status of aquatic mollusks in the southeastern United States: a downward spiral of diversity. In G. W. Benz & D. E. Collins (eds), Aquatic Fauna in Peril: The Southeastern Perspective. Special Publication No. 1, Southeast Aquatic Research Institute, Lenz Design and Communications, Decatur, GA. 554 pp. [Published May 1998], 43–86.

Ortmann, A. E., 1912. Notes upon the families and genera of the Najades. Annals of the Carnegie Museum 8: 222–365.

Pilsbry, H. A. & J. Bequaert, 1927. The aquatic mollusks of the Belgian Congo, with a geographical and ecological account of Congo malacology. Bulletin of the American Museum of Natural History 53: 69–602.

Rosenberg, G. & M. L. Ludyanskiy, 1994. A nomenclatural review of *Dreissena* (Bivalvia: Dreissenidae), with identification of the quagga mussel as *Dreissena bugensis*. Canadian Journal of Fisheries and Aquatic Sciences 51: 1474–1484.

Smith, B. J., 1992.Non-marine Mollusca. In Houston, W. W. K. (ed.), Zoological Catalogue of Australia, Vol. 8, xii. AGPS, Canberra, 405 pp.

Smith, D. G., 2001. Systematic and distribution of the Recent Margaritiferidae. In Harper, E. M., J. D. Taylor & J. A. Crame (eds), The Evolutionary Biology of the Bivalvia. Geological Society, London, Special Publication 177, 33–49.

Starobogatov, Y., 1995. The pearly freshwater mussels (Mollusca, Unionoida, Margaritiferidae) of Russia. In Proceedings of the 9th International Colloquium of the European Invertebrate Survey, Helsinki, 3–4 September 1993. WWF Finland Report No. 7, 109–112.

Subba Rao, N. V., 1989. Handbook Freshwater Molluscs of India. Zoological Survey of India, Calcutta, India, 289 pp.

Turgeon, D. D., J. F. Quinn, Jr., A. E. Bogan, E. V. Coan, F. G. Hochberg, W. G. Lyons, P. Mikkelsen, R. J. Neves, C. F. E. Roper, G. Rosenberg, B. Roth, A. Scheltema, M. J. Sweeney, F. G. Thompson, M. Vecchione & J. D. Williams. 1998. Common and Scientific Names of Aquatic Invertebrates from the United States and Canada: Mollusks. American Fisheries Society Special Publication 26. Second Edition, 536 pp.

Wächtler, K., M. C. Dreher-Mansur & T. Richter, 2001. Larval types and early postlarval biology in Naiads (Unionoida). In Bauer, G. & K. Wächtler (eds), Ecology and Evolution of the Freshwater Mussels Unionoida. Ecological Studies, Vol. 145. Springer-Verlag, Berlin, 95–125.

Walker, K. F., M. Byrne, C. W. Hickey & D. S. Roper. 2001. Freshwater mussels (Hyriidae) of Austalasia. In G. Bauer & K. Wächtler (eds), Ecology and Evolution of Freshwater Mussels, Unionoida. Ecological Studies, Vol. 145. Springer Verlag, 5–31.

Wu, X. P., 1998. Studies on Freshwater Mollusca in Mid-Lower Reaches of Chang Jiang River. Doctoral Dissertation. Academia Sinica, Wuhan, China, 197 pp.

Hydrobiologia (2008) 595:149–166
DOI 10.1007/s10750-007-9012-6

FRESHWATER ANIMAL DIVERSITY ASSESSMENT

Global diversity of gastropods (Gastropoda; Mollusca) in freshwater

Ellen E. Strong · Olivier Gargominy · Winston F. Ponder · Philippe Bouchet

Abstract The world's gastropod fauna from continental waters comprises ~4,000 valid described species and a minimum of 33–38 independent lineages of Recent Neritimorpha, Caenogastropoda and Heterobranchia (including the Pulmonata). The caenogastropod component dominates in terms of species richness and diversity of morphology, physiology, life and reproductive modes and has produced several highly speciose endemic radiations. Ancient oligotrophic lakes (e.g., Baikal, Ohrid, Tanganyika) are key hotspots of gastropod diversity; also noteworthy are a number of lower river basins (e.g., Congo, Mekong, Mobile Bay). But unlike many other invertebrates, small streams, springs and groundwater systems have produced the most speciose associations of freshwater gastropods. Despite their ecological importance in many aquatic ecosystems, understanding of even their systematics is discouragingly incomplete. The world's freshwater gastropod fauna faces unprecedented threats from habitat loss and degradation and introduced fishes and other pests. Unsustainable use of ground water, landscape modification and stock damage are destroying many streams and springs in rural/pastoral areas, and pose the most significant threats to the large diversity of narrow range endemics in springs and ground water. Despite comprising only ~5% of the world's gastropod fauna, freshwater gastropods account for ~20% of recorded mollusc extinctions. However, the status of the great majority of taxa is unknown, a situation that is exacerbated by a lack of experts and critical baseline data relating to distribution, abundance, basic life history, physiology, morphology and diet. Thus, the already considerable magnitude of extinction and high levels of threat indicated by the IUCN Red List of Threatened Species is certainly a significant underestimate.

Keywords Phylogeny · Taxonomy · Biogeography · Endemicity · Radiations · Life history · Fossil record · Biomonitoring · Disease transmission · Conservation

Guest editors: E. V. Balian, C. Lévêque, H. Segers and K. Martens
Freshwater Animal Diversity Assessment

E. E. Strong (✉)
Department of Invertebrate Zoology, Smithsonian Institution, National Museum of Natural History, MRC 163, P.O. Box 37012, Washington, DC 20013-7012, USA
e-mail: StrongE@si.edu

O. Gargominy · P. Bouchet
Muséum National d'Histoire Naturelle, 55 Rue Buffon, 75005 Paris, France

W. F. Ponder
Malacology Section, Aquatic Zoology, Australian Museum, 6 College Street, Sydney, NSW 2010, Australia

Introduction

The Mollusca is an extraordinarily varied phylum—with estimates of 80,000–100,000 described species and total diversity possibly as high as 200,000, they are second only to arthropods in species richness. The largest molluscan classes—Gastropoda and Bivalvia—have repeatedly and successfully colonized continental ("fresh") waters. Freshwater gastropods are found on every continent except Antarctica and in nearly all aquatic habitats including rivers, lakes, streams, swamps, underground aquifers and springs, as well as temporary ponds, drainage ditches and other ephemeral and seasonal waters. Most live submerged, and many are specialized for particular habitats—aquatic vegetation, stones, rocks, wood and other solid surfaces, or soft sediment. Some are amphibious and a few are able to tolerate periods of time out of water (e.g., some Ampullariidae); others are capable of prolonged periods of aestivation in soil during dry periods. Few groups (notably some of the rissooidean families) are found in highly saline inland habitats such as the Caspian Sea or salt lakes in Central Asia, Africa and Australia.

Most freshwater gastropods are micro-herbivorous and/or micro-omnivorous grazers feeding on bacterial films, algae and diatoms, but there are a number of exceptions: the predominantly marine Buccinidae, Marginellidae and Acochlidiida and the entirely freshwater Glacidorbidae are predators; Viviparidae and Bithyniidae are ctenidial suspension feeders at least in part; Ampullariidae are primarily macroherbivorous and are also known to feed on bryozoans and planorbid eggs. There are no pelagic/nektonic or parasitic species, with the great majority being benthic crawlers. A rare exception is the Helicostoidae—a monotypic caenogastropod family of uncertain affinity from China that lives cemented to limestone blocks (Lamy, 1926).

Taxonomic composition

New suites of anatomical, ultrastructural and molecular characters developed in the past 30 years have fuelled a revolution in our understanding of gastropod phylogenetics (Haszprunar, 1988; Ponder & Lindberg, 1997; Colgan et al., 2003; Strong, 2003). Several well supported clades are currently recognized: Caenogastropoda (containing most of the former Mesogastropoda and all the Neogastropoda); its sister group, Heterobranchia (containing the former Opisthobranchia and Pulmonata, as well as a few "mesogastropod" groups); Vetigastropoda (including many of the former Archaeogastropoda) and Neritimorpha (previously a subgroup of "archaeogastropods").

The world's freshwater gastropod fauna is dominated by two main components: the Caenogastropoda and pulmonate heterobranchs. Several additional basal lineages of heterobranchs have also invaded freshwater (Valvatidae, Glacidorbidae, Acochlidiida) as well as some Neritimorpha (Neritiliidae, Neritidae). Only the Viviparoidea, Glacidorboidea and nearly all Hygrophila comprise superfamilial (or above) groupings with members represented exclusively in freshwater. Of the 409 families of Recent gastropods currently recognized (Bouchet & Rocroi, 2005), 26 are composed of taxa that are wholly or mostly restricted to freshwater, four have significant taxonomic representation in freshwater biotopes (Neritidae, Assimineidae, Hydrobiidae, Stenothyridae), and three are marine groups with isolated genera that have invaded freshwater [*Cremnoconchus* (Littorinidae), *Clea* (Buccinidae), *Rivomarginella* (Marginellidae)] (Table 1).

The caenogastropod component of the freshwater fauna represents numerous independent lineages and many separate colonization events. Several clades have produced spectacular endemic radiations, namely Rissooidea (Hydrobiidae s.l., Pomatiopsidae) and Cerithioidea (Pachychilidae, Paludomidae and Pleuroceridae). With the exception of a few parthenogenetic taxa [*Campeloma* (Viviparidae), *Melanoides* (Thiaridae), *Potamopyrgus antipodarum* (Hydrobiidae)], they are exclusively dioecious and reproduction is sexual. Of all the freshwater groups, only the cerithioids are aphallate and transfer sperm using spermatophores; all others use a penis. Most lay egg capsules, and development is intracapsular with embryos emerging as crawling juveniles. A free-swimming dispersal stage is present in some species, particularly those that inhabit the lower reaches of coastal streams, with a free-swimming veliger larva that may develop in the sea (Neritidae, some Thiaridae). However, many species are brooders [all Viviparidae, some Cerithioidea, Rissooidea and Heterobranchia (see below)] and retain their young in brood pouches that represent modifications of the oviduct, mantle cavity or cephalic haemocoel. While it has been suggested that there is a significant selective advantage for parental care and hence brooding among freshwater molluscs (e.g., Köhler et al., 2004), the great majority of freshwater

Table 1 Taxonomic representation and distribution of freshwater gastropods

Taxon	Representation In Freshwater	Habitat
Neritimorpha		
Superfamily Helicinoidea		
Family Neritiliidae	*	Anchialine and coastal running waters
Superfamily Neritoidea		
Family Neritidae	+	Primarily lower reaches of coastal rivers and streams, estuaries
Caenogastropoda		
Architaenioglossa		
Superfamily Ampullarioidea		
Family Ampullariidae	*	Quiet, muddy rivers, lakes, ponds, canals, rice paddies, swamps
Superfamily Viviparoidea		
Family Viviparidae	*	Rivers, lakes, ponds, swamps, canals
Sorbeoconcha		
Superfamily Cerithioidea		
Family Melanopsidae	*	Springs, streams
Family Paludomidae	*	Lakes, rivers, streams (including radiation in Lake Tanganyika)
Family Pachychilidae	*	Lakes, rivers, streams (including radiation in Sulawesi lakes)
Family Pleuroceridae	*	Rivers, streams
Family Thiaridae	*	Rivers, streams
Hypsogastropoda		
Superfamily Littorinoidea		
Family Littorinidae (*Cremnoconchus*)	–	Waterfalls
Superfamily Rissooidea		
Family Amnicolidae	*	Rivers and streams
Family Assimineidae	+, <	Estuaries, freshwater rivers and streams, springs
Family Bithyniidae	*	Quiet muddy rivers, lakes, ponds, canals, swamps
Family Cochliopidae	*	Rivers and streams, swamps, lakes
Family Helicostoidae (*Helicostoa*)	*	Cemented on limestone rocks
Family Hydrobiidae	+	Greatest diversity springs; also streams and rivers, lakes, groundwater systems, caves, estuarine marshes and mudflats
Family Lithoglyphidae	*	Streams, rivers
Family Moitessieriidae	*	Groundwater systems, caves
Family Pomatiopsidae	*/<	Rivers, permanent wetlands, stream edges, some saline springs/lakes.
Family Stenothyridae	+	Rivers, streams, estuarine
Neogastropoda		
Superfamily Buccinoidea		
Family Buccinidae (*Clea*)	–	Lower reaches of rivers
Superfamily Muricoidea		
Family Marginellidae(*Rivomarginella*)	–	Rivers, lakes and canals
Heterobranchia		
Superfamily Glacidorboidea		
Family Glacidorbidae	*	Swamps, lakes, streams

Table 1 continued

Taxon	Representation In Freshwater	Habitat
Superfamily Valvatoidea		
Family Valvatidae	*	Cold, clean lakes rivers, streams
Opisthobranchia		
Acochlidiida		
Superfamily Acochlidioidea		
Family Acochlidiidae	*	Lower reaches of rivers
Superfamily Hedylopsoidea		
Family Tantulidae	*	Lakes
Superfamily Strubellioidea		
Family Strubelliidae	*	Lower reaches of rivers
Pulmonata		
Basommatophora		
Hygrophila		
Superfamily Chilinoidea		
Family Chilinidae	*	On stones and rocks in lakes and running water
Family Latiidae	*	On stones and rocks in running streams and rivers
Superfamily Acroloxoidea		
Family Acroloxidae	*	Lakes (including several Lake Baikal and Lake Ohrid endemics)
Superfamily Lymnaeoidea		
Family Lymnaeidae	*	Flowing rivers and streams, lakes to stagnant ponds, swamps
Superfamily Planorboidea		
Family Planorbidae	*	Low energy temporary and permanent ponds, streams, rivers, springs, lakes
Family Physidae	*	Ponds, wetlands, eutrophic streams, temporary aquatic habitats, springs

Classification follows Bouchet & Rocroi (2005). Note that the higher classification of the Acochlidiida is uncertain. Bouchet & Rocroi (2005) refer to it as "Group Acochlidiacea"; we tentatively use the recently proposed ordinal level name, Acochlidiida. '*' – Wholly/mostly freshwater; '+' – Partly freshwater; '–' – Isolated freshwater; '<' –Amphibious

gastropods are not brooders. Partly as a consequence of their life-history traits, many species are poor dispersers as reflected in high degrees of genetic differentiation between populations (e.g., Ponder & Colgan, 2002). In addition, they are typically habitat specialists, have restricted geographic ranges, long maturation times, low fecundity and are comparatively long lived, rendering them more susceptible to human-mediated threats (e.g., Lydeard et al., 2004; see also below).

The heterobranch component is less diverse and represents relatively few independent colonization events (see below). Heterobranchs are exclusively hermaphroditic and some pulmonates are capable of self-fertilization, although sperm exchange is typical. As in most caenogastropods, development is intracapsular. Brooding is rare and has only been documented in a planorbid limpet (Albrecht & Glaubrecht, 2006), and some glacidorbids (Ponder, 1986; Ponder & Avern, 2000).

Freshwater pulmonates have their greatest diversity primarily in the holarctic, but are distributed worldwide, with some species widely dispersed pests. They are characterized by comparably few, relatively widespread taxa and have produced a few endemic radiations, but never approaching the scale of caenogastropods. Most pulmonates have only limited ability to exploit deeper water habitats because they lack a ctenidium (true molluscan gill) and instead use a thin, vascularized "lung" for gas exchange. However, planorbids have a secondary gill (pseudobranch) and the efficient respiratory pigment haemoglobin so are better equipped to exploit oxygen-depleted environments. Others are associated with lentic habitats, occupying the shallows of lakes and/or temporary or ephemeral bodies of water. Many pulmonates have broad environmental tolerances, tend to be more resistant to eutrophication, anoxia, and brief exposure to air and have short generation times. Nevertheless, there are many exceptions, with some pulmonates having very short ranges including some endemic to (ancient) lakes (Boss, 1978), springs (Brown, 2001; Taylor, 2003) or a short section of a single river (Ponder & Waterhouse, 1997) while others are endangered (e.g., *Camptoceras* in Japan). These traits, together with at least some being capable of self-fertilization, enable many pulmonates to be readily passively dispersed (see below) and some are highly successful colonizers, as reflected in their ability to occupy new or ephemeral habitats (e.g., Økland, 1990) and in comparably less genetic structuring (e.g., Dillon, 2000). This renders many of them more resilient to human-mediated threats and less extinction prone than other freshwater gastropods (Boss, 1978; Davis, 1982; Michel, 1994).

Species diversity

Global patterns of freshwater gastropod species diversity are notoriously difficult to evaluate. The current taxonomy is a complex mixture of taxonomic traditions and practices of numerous generations of workers on different continents (Bouchet, 2006). Early studies of some taxa resulted in the recognition of a few conchologically variable and widespread species, or conversely in the unwarranted enormous inflation of nominal taxa, including species, subspecies and "morphs", particularly so in North America and Europe [e.g., North American Pleuroceridae with over 1,000 nominal taxa and ~200 considered valid (Graf, 2001); Physidae with ~460 nominal taxa, ~80 considered valid (Taylor, 2003); European Lymnaeidae (see below)]. When applied to such complex groups, modern analytical methods incorporating molecular and newly interpreted morphological characters, combined with a new appreciation of ecological and geographical patterns, have led to a more refined understanding of genera and species. Such studies have demonstrated that many currently recognized species are not monophyletic (Minton & Lydeard, 2003; Wethington, 2004) and/or have revealed unrecognized species complexes [e.g., European and North American lymnaeids (Remigio & Blair, 1997; Remigio, 2002); North American pleurocerids (Lydeard et al., 1998); Indonesian pachychilids (von Rintelen & Glaubrecht, 2005)]. Alternatively, some past studies have overindulged in synonymy, for example Hubendick's (1951) major review of world wide Lymnaeidae recognized only 38 valid species and two genera, while recent studies (e.g., Remigio & Blair, 1997; Kruglov, 2005) have indicated that there are several valid genera and a number of additional species, including several synonymized by Hubendick. Morphological studies on large new collections can also reveal significant previously unsuspected diversity, particularly with minute taxa, as for example among Australian glacidorbids and bithyniids (Ponder & Avern, 2000; Ponder, 2004c) and the so-called hydrobioids (see below). There is, nevertheless, a strong bias towards larger sized taxa and towards the developed world, such as North America, Europe, Japan and Australasia. A testament to our incomplete knowledge is that ~45 new freshwater gastropod species are described on average each year, with about 87% from these better studied regions (Bouchet, unpubl. data).

Complicating efforts to evaluate their diversity, it is not feasible to accurately assess genus-level diversity for freshwater gastropods. In the absence of provincial or global revisions at the level of families or superfamilies, generic concepts are often applied locally and vary between regions—some studies employing narrow generic concepts, others very broad ones. In many areas, there are no modern treatments for much of the fauna while in others the faunas are well known and many groups have undergone recent systematic revision using molecular and/or morphological methods. In general terms, the concepts of tropical genera tend to be older and hence broader and more likely polyphyletic. In contrast, genera from many temperate biomes are often more narrowly defined. We believe that species-level data do not suffer so much from geographic differences in historical treatment and conceptual approach.

With the above caveats, the global freshwater gastropod fauna is estimated as approximately 4,000

Table 2 Total number of valid described species of freshwater gastropods arranged by main zoogeographical region; number of introduced species is indicated in parentheses

	PA	NA	NT	AT	OL	AU	PAC	ANT	World
Neritimorpha									
Neritiliidae	4	0	0	2	4	2	3	0	5
Neritidae	45–55	2	~10	14	20–45	~40	42	0	~110
Caenogastropoda									
Ampullariidae	(1)	1 (1)	50–113	28	25 (4)	(1)	0 (4)	0	105–170
Viviparidae	20–25	27	1	19	40–60	19 (1)	0 (2)	0	125–150
Sorbeoconcha									
Melanopsidae	20–50	0	0	0	0	1	2	0	~25–50
Paludomidae	0	0	0	66	28	?	0	0	~100
Pachychilidae	0	0	30–60	22	70–100	43	0	0	165–225
Pleuroceridae	35	156	0	0	4	0	0	0	~200
Thiaridae	20	0	30	34	20–40	20–40	20–35	0	135
Hypsogastropoda									
Littorinidae	0	0	0	0	2	0	0	0	2
Amnicolidae	150–200	19	0	0	0	0	0	0	~200
Assimineidae	0	2	?	11	4	2	0	0	~20
Bithyniidae	45	0	0	34	~25	24	0 (1)	0	~130
Cochliopidae	17	50	176	3	0	0	0	0	246
Helicostoidae	0	0	0	0	1	0	0	0	1
Hydrobiidae	700–750	105	21	13	7	252 (1)	75 (1)	0	~1250
Lithoglyphidae	30	61	?	0			0	0	~100
Moitessieriidae	55	0	0	0	0	0	0	0	55
Pomatiopsidae	17	6	1	10	~130	9	0	0	~170
Stenothyridae	6	0	0	0	~50	~5	0	0	~60
Neogastropoda									
Buccinidae	0	0	0	0	8–10	0	0	0	8–10
Marginellidae	0	0	0	0	2	0	0	0	2
Heterobranchia									
Glacidorbidae	0	0	1	0	0	19	0	0	20
Valvatidae	60	10	0	1	0	0	0	0	71
Acochlidiida									
Acochlidiidae	0	0	0	0	0	2	2	0	4
Tantulidae	0	0	1	0	0	0	0	0	1
Strubelliidae	0	0	0	0	0	1	1	0	1
Pulmonata									
Chilinidae	0	0	~15	0	0	0	0	0	~15
Latiidae	0	0	0	0	0	1	0	0	1
Acroloxidae	40	1	0	0	0	0	0	0	~40
Lymnaeidae	40–120	56	7	2	19	7	5 (2)	0	~100
Planorbidae	100–200	57	59	116	49	43	8 (2)	0	~250
Physidae	15	31	38	(1)	1	(1)	0 (4)	0	~80
Total	1,408–1,711	585	440–533	366	509–606	490–514	154–169	0	3,795–3,972
All red list categories (Excluding LC)	94	215	10	100	2	92	11	0	

PA: Palaearctic, NA: Nearctic, NT: Neotropical, AT: Afrotropical, OL: Oriental, AU: Australasian, PAC: Pacific Oceanic Islands, ANT: Antarctic

valid described species (Table 2). In some cases, the number of species is certainly overestimated, but these are vastly overshadowed by areas of the world yet to be even superficially inventoried with most likely thousands waiting to be discovered (Lydeard et al., 2004), either as entirely new entities or through the recognition of cryptic taxa. The most speciose assemblage by far is the hydrobioids (Rissooidea)—a diversity long masked by their tiny, rather featureless shells and often very restricted ranges. While most families are probably known within 70–90% of actual diversity, the estimated 1,000 species of hydrobioids may represent as little as 25% of their actual diversity as evidenced by the fact that they comprise about 80% of current new species descriptions (compiled 1997–2003; Bouchet, unpubl. data). This suggests that the total number of freshwater gastropods is probably on the order of ~8,000 species.

Phylogeny and historical processes

The phylogenetic framework

In addition to our changing concepts of higher classification and species diversity, the phylogenetic framework for a few freshwater clades has been considerably refined, especially with the use of molecular techniques (see below). However, few comprehensive phylogenies for individual families or the higher taxonomic groupings that contain freshwater taxa have been published to date. For those that have been published, variable taxon sampling, incongruence between morphological and molecular data, compounded by weak support of basal nodes, has often resulted in conflicting interpretations concerning the monophyly and/or affinity of freshwater clades and the number of freshwater invasions [e.g., Neritimorpha (Holthuis, 1995; Kano et al., 2002); Architaenioglossa (Colgan et al., 2003; Simone, 2004); Hygrophila (Barker, 2001; Dayrat et al. 2001); Cerithioidea (e.g., Lydeard et al., 2002); Rissooidea (see below)].

The large assemblage of marine, brackish and freshwater lineages currently placed in the Rissooidea arguably are in the most urgent need of revision. This putative superfamily encompasses the largest and most threatened radiations of freshwater taxa and yet their systematics are just beginning to be clarified. The only phylogenetic analysis encompassing the whole group (Ponder, 1988) requires rigorous testing using molecular data and a substantial sampling of outgroup taxa; results with a small subset of taxa indicate that the rissooideans as presently recognized, are at least diphyletic (Colgan et al., 2007). In the past, all brackish and freshwater members of the group were united in the heterogeneous "Hydrobiidae" (=hydrobioid, or Hydrobiidae s.l.) by some authors, while others recognized different families and even superfamilies. Based on molecular and refined anatomical data, the composition of several monophyletic lineages from within this assemblage has begun to be elucidated (e.g., Amnicolidae, Cochliopidae, Moitessieriidae and Lithoglyphidae) (e.g., Wilke et al., 2001; Hausdorf et al., 2003). Nevertheless, the affinities and composition of many families remain to be more thoroughly evaluated; indeed monophyly of the Hydrobiidae as currently defined is unlikely (Haase, 2005). Additionally, establishing a robust phylogenetic framework for this group will clarify our understanding of their conquest of freshwater. For example, it was estimated that New Zealand "hydrobiids" (=Tateinae, possibly a distinct family; Ponder, unpubl. data) independently conquered freshwater three times (Haase, 2005); it appears that this has happened separately in a number of other hydrobioid groups.

The affinities of valvatids and their allies were long unstable and they were often placed in the wrong higher taxa, in part due to their combination of plesiomorphic and autapomorphic features and small body size (Fig. 1). Detailed anatomical work and refinement of morphological homologies clarified the basal position of valvatoideans in the Heterobranchia and the assemblage of other allied lineages (Haszprunar, 1988; Ponder, 1991; Barker, 2001) with confirmation from molecular studies (Colgan et al., 2003). However, the position of the probably paedomorphic glacidorbids within the Heterobranchia is still disputed (see Ponder & Avern, 2000).

Surprisingly little has been done regarding the phylogenetic relationships of the freshwater pulmonates (Hygrophila), although some families, notably Planorbidae (Morgan et al., 2002; Albrecht et al., 2004), Physidae (Wethington, 2004) and Lymnaeidae (see above) have recently been investigated using mainly molecular data. However, some old

Fig. 1 *Valvata studeri*. Boeters & Falkner, 1998. Size 3 mm. Photo courtesy G. Falkner

classifications remain firmly entrenched. For example, the major group of freshwater limpets, the Ancylidae, was shown by Hubendick (1978) to be almost indistinguishable from Planorbidae, a finding ignored by many subsequent workers outside Europe. Recent molecular analyses have shown that the limpet form has arisen several times within the planorbids (Albrecht et al., 2004), with the typical ancylids nested within that family.

But for many taxa, no modern cladistic and/or taxonomic treatment is available (Chilinoidea, Acochlidiida). In contrast, some freshwater representatives have not been sampled in existing cladistic studies, leaving their systematic affinities unresolved (e.g., *Clea* in the Buccinidae); rarely the taxonomic placement of taxon is unknown (Helicostoidae).

Despite our often limited grasp of phylogenetic relationships, it is clear that gastropods have invaded freshwater biotopes many times. Published estimates, although not comparable as classifications have changed and fossil lineages have been variably included or excluded, range from 6 to 7 (Hutchinson, 1967), or 10 (Taylor in Gray, 1988), to as many as 15 Recent freshwater gastropod colonizations (Vermeij & Dudley, 2000). Based on the current classification (Bouchet & Rocroi, 2005) and our present understanding of gastropod phylogenetic relationships, we estimate that there are a minimum of 33–38 independent freshwater lineages represented among Recent gastropods: in the Rissooidea, there are at least 2 each in Assimineidae and Cochliopidae, 1–2 in Pomatiopsidae, at least 1 each in Stenothyridae, Lithoglyphidae, Moitessieriidae, 1 in Bithyniidae, possibly 1 in Helicostoidae, possibly 6–8 in the Hydrobiidae; 5–6 in the Neritimorpha (Holthuis, 1995); 2–3 in the Cerithioidea (Lydeard et al., 2002); probably 2 each in the "Architaenioglossa" (e.g., Simone, 2004) and the Acochlidiida; and 1 in each of the Litttorinidae, Buccinidae, Marginellidae, Glacidorbidae, Valvatidae and Hygrophila (see Table 1).

The fossil record

While shelled marine molluscs have an excellent fossil record that of freshwater taxa is relatively poor. Fossilization in freshwater habitats is biased towards lowland and lake deposits, with many other habitats that are significant for gastropod diversity represented poorly or not at all (e.g., springs, streams, groundwater). This incomplete record is compounded by the poor preservation potential of the often light, thin shells of many freshwater taxa and acidic environments. Thus, the fossil record for freshwater gastropods is patchy at best and likely to significantly underestimate the age and diversity of freshwater lineages. Moreover, assignments of Palaeozoic fossils to modern freshwater lineages, often based on fragmentary shells, are problematic. Despite these difficulties, most modern groups appear to make their first appearance during the Jurassic or Cretaceous (Tracey et al., 1993), with most families in place by the end of the Mesozoic (Taylor in Gray, 1988; Taylor, 1988). Other elements of apparently more recent marine origin first appear during the Tertiary: chilinids first appear in the Late Paleocene or early Eocene, neritiliids during the Middle Eocene and freshwater buccinids are first known from the Miocene. There is no fossil record for freshwater littorinids or marginellids.

Regardless of their earliest documented occurrence, the cosmopolitan distribution pattern of many lineages indicates their widespread presence in Pangaea long before the break-up of this supercontinent (e.g., Viviparidae). Others are widely distributed on several major continents and have continental biogeographic patterns consistent with a Gondwanan origin (e.g., Pachychilidae—S. America, Africa, Madagascar, Asia; Thiaridae *s.s.*—S. America, Africa, Asia, India, Australia; Ampullariidae—S. America, Africa, S. Asia). Glacidorbidae are found in southern

Australia and Chile (Ponder & Avern, 2000), also suggesting a Gondwanan origin. Those of more recent marine origin occupy more isolated habitats and have not penetrated far inland (*Clea*, *Rivomarginella*, Acochlidiida).

Distribution and main areas of endemicity

Like other freshwater and marine invertebrates, freshwater gastropods present an overall pattern of high diversity in the tropics, with decreasing species richness as well as decreasing endemicity at higher latitudes. There are, however, always exceptions; for example, Tasmania has the most diverse freshwater fauna in Australia, and some groups have low tropical diversity (hydrobioid families, Glacidorbidae). Unlike for land snails, small oceanic islands are noteworthy for generally low levels of freshwater gastropod species richness and endemism (e.g., Starmühlner, 1979), although there are again some exceptions where the number of endemics is surprisingly high [e.g., Lord Howe Island (Ponder, 1982); Viti Levu, Fiji (Haase et al., 2006)].

Of course, both vicariance and dispersal have shaped modern distribution patterns; while vicariance arguably has been dominant in historical contexts, dispersal has certainly played an important role, including via such mechanisms as by animal transport (birds, insects), rafting on aquatic vegetation, marine/brackish larval dispersal phase, stream capture and even by air (e.g., cyclonic storms) (Purchon, 1977). Obviously, the significance and impact of each mechanism is more a function of the individual characteristics of each lineage: life habit (e.g. living on aquatic vegetation vs. attached beneath stones), ecological and physiological tolerances of individuals, mode of respiration, vagility, tolerance to saline water, sexual, reproductive and developmental strategies and ability to withstand desiccation. Such variables differ significantly among species and lineages and, hence, determine local patchiness and geographic range (Purchon, 1977; Davis, 1982; Taylor, 1988; Ponder & Colgan, 2002).

Thus, many apparently ancient freshwater taxa have broad geographic ranges primarily as a result of vicariance modified by dispersal. These lineages mostly belong to higher taxa comprising exclusively freshwater members (Viviparidae, Bithyniidae, Hydrobiidae s.l., Planorbidae and Lymnaeidae); other presumably old lineages are more restricted in geographic range (Glacidorbidae, Chilinidae, Latiidae, Acroloxidae). All are highly modified reflecting the special challenges presented by life in this biotope. Other groups are freshwater remnants of previously euryhaline groups (e.g., Melanopsidae), have euryhaline and/or marine members (e.g., Neritidae, Littorinidae, Stenothyridae, Assimineidae) and/or are amphidromous (some Thiaridae, Neritidae and probably at least some Stenothyridae) with greater opportunities for dispersal and colonization. The presumed most recent colonizers (e.g., Littorinidae, Buccinidae, Marginellidae, some Assimineidae) are characterized by being less highly modified, less speciose and have a more restricted distribution with more or less clear kinship to marine and/or brackish water relatives (e.g., Purchon, 1977). For a summary of continental distribution patterns of freshwater gastropod families and genera, see Bănărescu (1990), although the classification differs from the one adopted here.

At the level of continents, the Palearctic region has the most speciose freshwater gastropod fauna ($\sim$1,408–1,711 valid, described species), with the remaining continental regions of comparable diversity ($\sim$350–600 species). Apart from Africa, most regions have seen marked increases in recent years through the description of the highly endemic hydrobioid faunas (see *Phylogenetic Framework*, above). Surprisingly species-poor are the rivers and streams of South America, particularly of the Amazon basin, which contain, among other things an extraordinary diversity of freshwater fishes; it is not yet clear if this is a sampling/study artefact or an actual pattern. In contrast, groups important from an economic, human health or veterinary perspective (see below) have received considerable attention, even in developing countries.

While a thorough species-level inventory is far from complete, some continental areas stand out for their exceptional diversity and disproportionately high numbers of endemics. Gargominy & Bouchet (1998) identified 27 areas of special importance for freshwater mollusc diversity as key hotspots of diversity with high rates of endemism among freshwater gastropods. Regrettably, most areas important for molluscan diversity have not been recognized by inclusion in the Ramsar List of Wetlands of

International Importance (www.ramsar.org/key_sitelist.htm). Although a number of resolutions have greatly expanded the classification of wetlands currently recognized under the Ramsar typology (Ramsar Convention Secretariat, 2004), few government parties have used these additional criteria to designate sites.

Global hotspots of freshwater gastropod diversity can be broadly classified according to 4 main categories (see Table 3):

1. *Springs and groundwater*. Springs, and sometimes the small headwater streams fed by them, are inhabited by taxa that are typically not found in larger streams or rivers. Single sites usually have low species richness (1–6 species) with populations consisting of 100's, and often 1,000's or even (rarely) millions of individuals. However, as a consequence of spatial isolating mechanisms, spring and headwater habitats regionally support rich assemblages of gastropods dominated primarily by hydrobioids. Similarly, underground aquifers, including underground rivers, are also dominated by hydrobioids with over 300 stygobiont species documented worldwide. As such habitats extend over very small areas, and as most species occur in only a very limited number of sites with single-site endemics commonplace, spring-dwelling gastropods are extremely vulnerable to loss of habitat. Remarkable examples include the artesian springs of the Great Artesian Basin of Australia (Ponder, 2004a); springs and small streams in SE Australia and Tasmania (Ponder & Colgan, 2002) and New Caledonia (Haase & Bouchet 1998); springs and caves in the Dinaric Alps of the Balkans (Radoman, 1983), and other karst regions of France and Spain (Bank, 2004); aquifer-fed springs in Florida, the arid south western United States and Mexico (Hershler, 1998, 1999) (Fig. 2).
2. *Large rivers and their first and second order tributaries*. The Congo (Africa), Mekong (Asia), Mobile Bay basin (North America), Uruguay and Rio de la Plata (South America) are noteworthy for their mollusc faunas that are sometimes extremely speciose, and often do not occur in other types of freshwater habitats (Fig. 2); the Zrmanja in eastern Europe and the coastal rivers of the Guinean region in Africa are also locally important hotspots. The most speciose representatives are usually microhabitat specialists, with highly patchy distributions scattered among the mosaic of microhabitats (flow regimes, sediment type, vegetation) offered by rivers and streams. Habitats of special importance are rapids which are inhabited by species adapted to highly oxygenated water. The gastropods are dominated by the Viviparidae (North America, Eurasia, Oriental region, Australia), Pachychilidae, Pleuroceridae (North America, Japan), Thiaridae (tropical regions), Pomatiopsidae and Stenothyridae (Oriental region); pulmonates are usually only poorly represented (Fig. 3).
3. *Ancient oligotrophic lakes*. Ancient lakes with the most speciose faunas include Lakes Baikal, Ohrid, Tanganyika and the Sulawesi lakes (Fig. 2), with the Viviparidae, Pachychilidae, Paludomidae, Thiaridae and hydrobioid families among the Caenogastropoda and the heterobranch families Planorbidae, Acroloxidae, Ancylidae and Valvatidae best represented. Rissooid and cerithioid lineages predominate among the groups prone to radiate in ancient lakes (Boss, 1978), typically with one clade or the other being dominant, often to the almost complete exclusion of members of the other lineage (e.g., Michel, 1994); Lake Poso (Haase & Bouchet, 2006) and the Malili lakes in Sulawesi are exceptions (Bouchet, 1995). As elsewhere, pulmonates are typically less speciose and have lower rates of endemicity. Planorbids are the most speciose of the pulmonate groups, but tend to be better represented in temperate rather than tropical lakes. Fossil gastropod faunas of long-lived lakes such as the well-known Miocene Lake Steinheim (Janz, 1999) and Plio-Pleistocene Lake Turkana (Williamson, 1981) have been important and influential (but not uncontroversial) models in evolutionary biology for rates and patterns of speciation.
4. *Monsoonal wetlands* and their associated rivers and streams can harbour significant faunas, as for example, in many parts of Asia and northern Australia, which are dominated by Viviparidae, Thiaridae, Bithyniidae, Lymnaeidae and

Table 3 Gastropod species hotspot diversity categorized by primary habitat

Region/Drainage/Basin	Species (endemic)	Dominant taxa
Springs and groundwater		
South western U.S.	~100 (≥58)	Hydrobioid families
Cuatro Cienegas basin, Mexico	12 (9)	Hydrobioid families
Florida, U.S.	84 (43)	Hydrobioid families
Mountainous regions in Southern France and Spain	150 (140)	Hydrobioid families
Southern Alps and Balkans region	220 (200)	Hydrobioid families
Great Artesian basin, Australia*	59 (42)	Hydrobiidae
Western Tasmania, Australia*	206 (191)	Hydrobiidae
New Caledonia	81 (65)	Hydrobiidae
Ancient oligotrophic lakes		
Titicaca	24 (15)	Hydrobioid families, Planorbidae
Ohrid and Ohrid basin	72 (55)	Hydrobioid families, Lymnaeidae, Planorbidae
Victoria	28 (13)	Viviparidae, Planorbidae
Tanganyika*	83 (65)	Paludomidae: 18 endemic genera with important radiation in *Lavigeria*
Malawi	28 (16)	Ampullariidae, Thiaridae
Baikal	147 (114)	Amnicolidae, Lithoglyphidae, Valvatidae, Planorbidae, Acroloxidae
Biwa	38 (19)	endemic subgenus *Biwamelania* (Pleuroceridae), Planorbidae
Inle and Inle watershed	44 (30)	Viviparidae, Pachychilidae, Bithyniidae
Sulawesi lakes	~50 (~40)	Pachychilidae, Hydrobiidae, Planorbidae; 3 endemic genera
Large rivers and their first and second order tributaries		
Tombigbee-Alabama rivers of the Mobile Bay basin	~118 (110)	Pleuroceridae (76 species); 6 endemic genera
Lower Uruguay River and Rio de la Plata, Argentina-Uruguay-Brazil	54 (26)	Pachychilidae
Western lowland forest of Guinea and Ivory Coast	~28 (~19 + 9 near endemic)	*Saulea*(Ampullariidae), *Sierraia* (Bithyniidae), *Soapitia* (Hydrobiidae), *Pseudocleopatra* (Paludomidae)
Lower Zaire Basin	96 (24)	Pachychilidae, Paludomidae, Thiaridae, Bithyniidae, Assimineidae, hydrobioid families; 5 endemic 'rheophilous' genera
Zrmanja	16 (5)	Hydrobioid families
Northwestern Ghats, India	~60 (~10)	2 endemic genera: *Turbinicola* (Ampullariidae), *Cremnoconchus* (Littorinidae)
Lower Mekong River in Thailand, Laos, Cambodia	~140 (111)	Triculinae (Pomatiopsidae) (92 endemic species); Stenothyridae (19 endemic species); Buccinidae; Marginellidae
Monsoonal wetlands		
Northern Australia	56 (13)	Viviparidae, Thiaridae, Bithyniidae, Lymnaeidae, Planorbidae

Data on monsoonal wetlands are included only for Northern Australia; reliable figures for other areas are unavailable. Main source: Gargominy & Bouchet 1998, unpubl. data. Number of endemic species is indicated in parentheses. "*" – Estimate includes undescribed species when such information is available. Note that the hydrobiid fauna of Tasmania is primarily from small groundwater-fed streams, some rivers, caves and a few springs

Planorbidae. For example, according to a recent analysis, the monsoonal rivers and associated wetlands flowing into the Gulf of Carpentaria in northern Australia have 56 species, 13 of which are endemic (Ponder, unpubl. data). Reliable comparative data is not available for other likely similarly diverse areas in e.g., S.E. Asia.

Fig. 2 Hotspots of gastropod diversity. A–H. Springs and groundwater. I–Q. Lakes. R–X. Rivers. Y. Monsoonal wetlands. A: South western U.S.; B: Cuatro Cienegas basin, Mexico; C: Florida, U.S.; D: Mountainous regions in Southern France and Spain; E: Southern Alps and Balkans region; Northern Italy, Austria, former Yugoslavia, Bulgaria, Greece; F: Great Artesian basin, Australia; G: Western Tasmania, Australia; H: New Caledonia. I: Titicaca, Peru-Bolivia; J: Ohrid and Ohrid basin, former Yugoslavia; K: Victoria; Kenya, Sudan, Uganda; L: Tanganyika; Burundi, Tanzania, D.R. Congo; M: Malawi; Malawi, Mozambique; N: Baikal, Russia; O: Biwa, Japan; P: Inle, Burma; Q: Sulawesi lakes, Indonesia. R: Tombigbee-Alabama rivers of the Mobile Bay basin; S: Lower Uruguay River and Rio de la Plata; Argentina, Uruguay, Brazil; T: Western lowland forest of Guinea and Ivory Coast; U: Lower Zaire Basin; V: Zrmanja; W: Northwestern Ghats, India; X: Lower Mekong River; Thailand, Laos, Cambodia. Y: Northern Australia

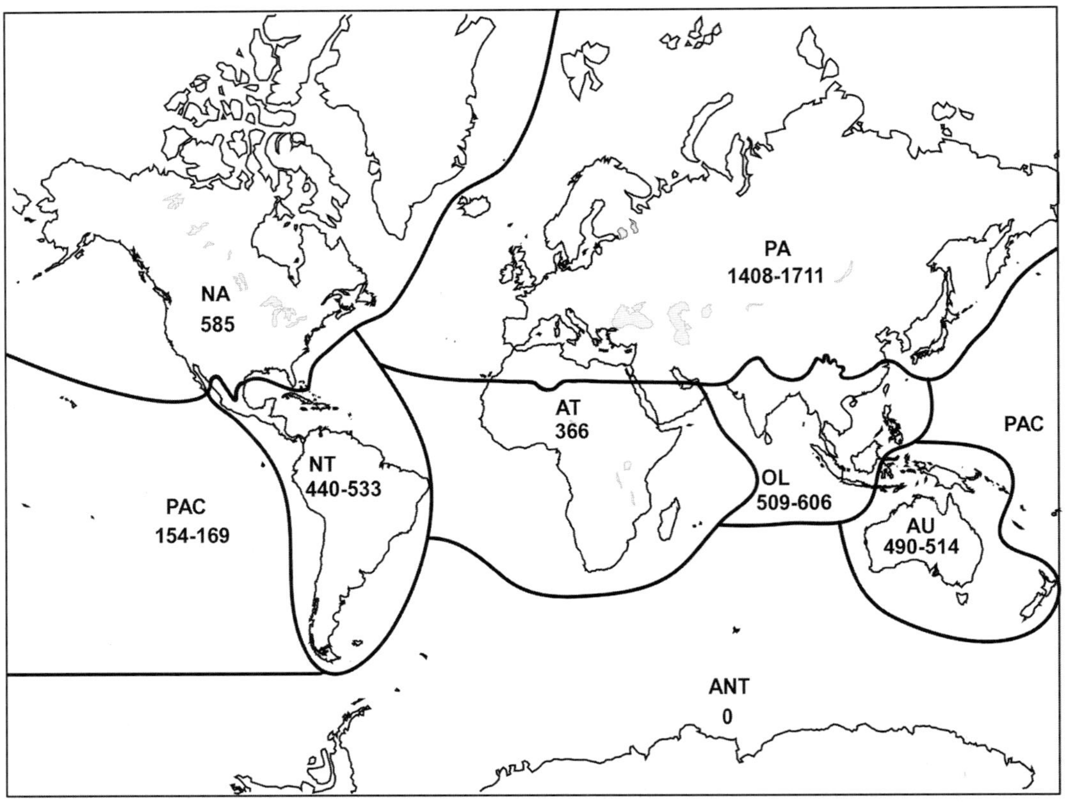

Fig. 3 Distribution of freshwater gastropod species per zoogeographic region. PA—Palaearctic, NA—Nearctic, NT—Neotropical, AT—Afrotropical, OL—Oriental, AU—Australasian, PAC—Pacific Oceanic Islands, ANT—Antarctic

Human related issues

Utility of freshwater gastropods

The potential of freshwater molluscs as indicators is largely unrealized but could be a powerful tool in raising awareness and improving their public image (Ponder, 1994; Seddon, 1998). Their low vagility, adequate size, often large population numbers and the ease of collection and identification of many species render them a useful and practical tool in biomonitoring programs (Chirombe et al. 1997; Langston et al., 1998; Lee et al., 2002). For example, freshwater gastropods are promising tools as pollution indicators through assessments of molluscan community composition and/or biological monitoring programs that rate water quality and status of aquatic biotopes based on invertebrate assemblages. They also have utility in monitoring and assessing the effects of endocrine-disrupting compounds and as monitors of heavy metal contamination (e.g., Salanki et al., 2003; El-Gamal & Sharshar, 2004). Owing to practical considerations (simple anatomy, low cost, fewer ethical issues), freshwater molluscs are also being used in neurotoxicological testing to evaluate the effects of environmental pollutants on neuronal processes and to clarify the mechanisms of action of these substances at the cellular level (Salanki, 2000).

Freshwater gastropods and human health

Some freshwater snails are vectors of disease, serving as the intermediate hosts for a number of infections for which humans or their livestock are definitive hosts. The most significant are snail-transmitted helminthiases caused by trematodes (flukes). At least 40 million people are infected with liver (*Opisthorchis*) and lung flukes (*Paragonimus*) and over 200 million people with schistosomiasis (Peters & Pasvol, 2001) primarily in Africa, Southeast Asia and South America—often with devastating socio-economic consequences. The principal vectors are pomatiopsids and planorbids (schistosomiasis), as well as pachychilids, pleurocerids, thiarids, bithyniids and lymnaeids (liver and lung flukes) (Malek & Cheng, 1974; Davis, 1980; Davis et al., 1994; Ponder et al., 2006). Dam construction has had the adverse effect of enlarging suitable habitat for snail vectors and increasing the prevalence of schistosomiasis (McAllister et al., 2000). Humans are also affected by a number of other infections for which they are accidental hosts, such as angiostrongyliases (nematode infections of rodents and other mammals) which pass through ampullariid intermediate hosts. Ampullariids and pachychilids are often locally harvested as a food resource in Southeast Asia, Philippines and Indonesia furthering the spread of angiostrongyliasis and paragonimiasis, respectively (e.g. Liat et al., 1978).

Exotic freshwater gastropod species

Freshwater snails are routinely inadvertently introduced mainly through the aquarium trade in association with aquatic plants and freshwater fish. Accidental introductions also occur with aquaculture, as fouling organisms on ships and boats and through canals or other modifications of existing waterways (Pointier, 1999; Cowie & Robinson, 2003). The most successful colonizers have been pulmonates (Physidae, Lymnaeidae, Planorbidae) and parthenogenetic species (*Melanoides tuberculata*, *Potamopyrgus antipodarum*), as a single individual is often sufficient to establish a viable population. Introduced taxa tend to flourish in modified environments where they often outnumber native species or are the only ones present.

Although inadvertent introductions are far more common, deliberate introductions have been the most successful and typically the most harmful to native faunas, as a concerted effort is made to ensure their success (Cowie & Robinson, 2003). As with accidental introductions, deliberate introductions have occurred most commonly through the aquarium trade. But freshwater snails have also been introduced intentionally for use as food (Ampullariidae) and as biocontrol agents for invasive aquatic macrophytes (Ampullariidae) and for vectors of disease (see above) (Pointier, 1999; Cowie & Robinson, 2003). Deliberate introductions have been carried out with little or no thought of the impact on native species, rarely with pre-release testing or post-release monitoring of non-target impacts (Cowie, 2001). Consequently, some exotic species (notably *Pomacea canaliculata*) have become serious pests, adversely impacting agriculture (rice, taro production) and/or native faunas and floras through predation and competition (Purchon, 1977; Cowie, 2001).

Threats

Regrettably, only 2% of all mollusc species have had their conservation status rigorously assessed, so current estimates of threat are a severe underestimate (Seddon, 1998; Lydeard et al., 2004). Nevertheless, it is clear that terrestrial and freshwater molluscs arguably represent the most threatened group of animals (Lydeard et al., 2004). Freshwater gastropods, which comprise ~5% of the world's gastropod fauna, face a disproportionately high degree of threat; of the 289 species of molluscs listed as extinct in the 2006 IUCN Red List of Threatened Species (www.redlist.org), 57 (~20%) are gastropod species from continental waters. Terrestrial gastropods, representing ~30% of the world's gastropod fauna, are also facing a major crisis with 197 species listed as extinct (Table 4).

The decline of the world's freshwater gastropod fauna, indeed of freshwater molluscs in general, can be attributed to two main drivers: life-history traits and anthropogenic effects. As described above, in addition to low vagility, the most sensitive species are habitat specialists, have restricted geographic ranges, long maturation times, low fecundity and are comparatively long lived. These traits render them unable to adapt to conspicuous changes in flow regimes, siltation and pollution and unable to effectively compete with introduced species. In many areas, the most significant cause of declines in native snail populations has been dam construction for flood control, hydroelectric power generation, recreation and water storage, which has converted species-rich riffle and shoal habitats into low-energy rivers and pools, greatly reducing and fragmenting suitable habitats and resulting in a cascade of effects both up and downstream (Bogan, 1998; McAllister et al., 2000). This does not always lead to increased numbers of lentic taxa, as changes in flooding regimes can also have adverse impacts on species adapted to such habitats (McAllister et al., 2000). Similarly, the regulation of flow regimes in previously relatively stable habitats may adversely affect species unable to adapt to dramatic changes in water levels and/or velocities. More subtle changes induced as a result of these disturbances also contribute to species declines. For example, a change in the nature of biofilms as a result of altered flow regimes in the Murray – Darling system in Australia has caused the near extinction of riverine viviparids (Sheldon & Walker, 1997).

Threats to spring snails are of a different nature. They are mostly narrow range endemics that can go from unthreatened or vulnerable to extinct without any transitional level of threat, as it may take only one intervention to destroy the only known population of a species. For instance, depletion of ground water for a number of urban and rural uses including water capture for stock, irrigation or mining, spring or landscape modification and trampling by cattle have already destroyed many springs in rural/pastoral areas of Europe, United States and Australia (Sada & Vinyard, 2002; Ponder & Walker, 2003).

Additional sources of habitat degradation, fragmentation and/or loss include gravel mining and other sources of mine waste pollution, dredging, channelization, siltation from agriculture and logging, pesticide and heavy metal loading, organic pollution, acidification, salination, waterborne disease control, urban and agricultural development, unsustainable water extraction for irrigation, stock and urban use,

Table 4 Comparison of rates of threat for groups of molluscs

	~Described valid species diversity	Extinct	Critically endangered	Endangered	Vulnerable	All red list categories (Excluding LC)	Rate of threat
Mollusca		289	265	222	488	2,085	
Gastropoda	~78,000	258	213	194	473	1,882	0.024
Freshwater	~4,000	57	45	62	204	520	0.130
Terrestrial	~24,000	197	166	130	265	1,281	0.053
Marine	~50,000	4	2	3	6	84	0.00168

Source: 2006 IUCN Red List of Threatened Species (www.redlist.org). Rate of threat is estimated from number of Red Listed species (excludes Least Concern) as a percent of estimated currently valid species diversity; does not take into account proportion of species assessed and thus may not accurately reflect relative rate of threat across categories. LC: Least Concern

competition and/or smothering from introduced species (Thomas, 1997; Bogan, 1998; Seddon, 1998; McAllister et al., 2000; Ponder & Walker, 2003). As with damming, it is often not just localized damage, but the cascade of effects both up and downstream that impact a wide range of communities.

Conservation priorities

Despite significant roles in human cultural history, molluscs, as with many invertebrates, have a poor public image (Kay, 1995a). This attitude further impedes allocation of meagre conservation resources in competition with the demands from larger charismatic animals and plants (Ponder, 1995; Wells, 1995; Bouchet & Gargominy, 1998; Seddon, 1998; Lydeard et al., 2004). The often drab-coloured and uncharismatic freshwater species, arguably facing the most serious risks and most deservedly meriting public concern and action, are desperately in need of champions.

Effective management of these threatened resources is often complicated by habitat fragmentation or political obstacles, as large rivers and lakes are often transnational. Although often not feasible, narrow range endemics inhabiting a single stream or spring are best preserved within large protected areas (Ponder, 1995), as many critical sites outside reserves can be so small that they attract little interest from conservation agencies and can suffer from edge effects. Moreover, as noted above, currently recognized species do not necessarily reflect natural evolutionary entities, with clear implications for devising accurate and effective management strategies based on species-targeted approaches. In contrast, habitat-based conservation strategies circumvent many of these problems and may be the preferred option in many circumstances (e.g., Ponder, 2004b).

Major museum collections are a key component of understanding the spatial distribution of species, both past and present (Wells, 1995; Ponder, 1999, 2004b), but much of this information is not yet accessible to the global community via computerized databases. This lack of access hampers or prevents assessment of conservation status by contributors to the Red List—a vital communication tool between scientists and conservation strategists and managers, as well as local or national conservation agencies. Currently, Mollusc Action Plans, as called for by the IUCN Species Survival Commission (SSC), are limited in the extent to which they can offer explicit recommendations in comparison with well-known taxa (e.g., tetrapods) (Bouchet & Gargominy, 1998). With the limitations discussed above, the magnitude of the threat of extinction as indicated by the IUCN Red List of Threatened Species, is certainly a grave underestimate.

Conserving our molluscan resources will effectively require a multiplicity of approaches, including research (systematics, ecology, life history, physiology, morphology, genetics), inventories (distribution, population size, biogeography), enhanced database infrastructure including digitization of significant museum collections, mitigation of human impacts, active intervention to promote recovery (including removal of invasive species, captive breeding programs, re-introduction, restoration of habitat), training in taxonomic expertise and enhanced communication and outreach (Kay, 1995b; Ponder, 1995; Seddon, 1998; McAllister et al., 2000; Lydeard et al., 2004). But considering the already-documented severity of the plight of freshwater gastropods, these strategies cannot proceed in a step-wise linear fashion—by then it will be too late (e.g., Wells, 1995; Lydeard et al., 2004; Ponder, 2004b).

References

Albrecht, C. & M. Glaubrecht, 2006. Brood care among basommatophorans: a unique reproductive strategy in the freshwater limpet snail *Protancylus* (Heterobranchia: Protancylidae), endemic to ancient lakes on Sulawesi, Indonesia. Acta Zoologica 87: 49–58.

Albrecht, C., T. Wilke, K. Kuhn & B. Streit, 2004. Convergent evolution of shell shape in freshwater limpets: the African genus *Burnupia*. Zoological Journal of the Linnean Society 140: 577–586.

Bănărescu, P., 1990. Zoogeography of Fresh Waters, Volume 1: General Distribution and Dispersal of Freshwater Animals. AULA-Verlag, Wiesbaden.

Bank, R., 2004. Fauna Europaea: Gastropoda. www.faunaeur.org.

Barker, G. M., 2001. Gastropods on land: phylogeny, diversity and adaptive morphology. In Barker, G. M. (ed.), Biology of Terrestrial Molluscs. CABI Publishing, Wallingford.

Bogan, A. E., 1998. Freshwater molluscan conservation in North America: problems and practices. Journal of Conchology, Special Publication 2: 223–230.

Boss, K., 1978. On the evolution of gastropods in ancient lakes. In Fretter, V. & J. Peake (eds), Pulmonates, Vol. 2a.

Systematics, Evolution and Ecology. Academic Press, London: 385–428.
Bouchet, P., 1995. A major new mollusc radiation discovered in the ancient lakes of Sulawesi. 12th International Malacological Congress, Vigo, Abstracts: 14–15.
Bouchet, P., 2006. Valid until synonymized, or invalid until proven valid? A response to Davis (2004) on species check-lists. Malacologia 48: 311–320.
Bouchet, P. & O. Gargominy, 1998. Action plan formulation for molluscan conservation: getting the facts together for a global perspective. Journal of Conchology Special Publication 2: 45–50.
Bouchet, P. & J.-P. Rocroi, 2005. Classification and nomenclator of gastropod families. Malacologia 47: 1–397.
Brown, D. S., 2001. Freshwater snails of the genus *Gyraulus* (Planorbidae) in Australia: taxa of the mainland. Molluscan Research 21: 17–107.
Chirombe, L., A. Maredza, G. Chingwena & Y. S. Naik, 1997. Use of molluscs to monitor water pollution. In Proceedings of "Workshop on Medical Malacology in Africa", Harare, Zimbabwe, September 22–26, 1997: 9–28.
Colgan, D. J., W. F. Ponder, E. Beacham & J. M. Macaranas, 2003. Molecular phylogenetic studies of Gastropoda based on six gene segments representing coding or non-coding and mitochondrial or nuclear DNA. Molluscan Research 23: 123–148.
Colgan, D. J., W. F. Ponder, E. Beacham & J. Macaranas, 2007. Molecular phylogenetics of Caenogastropoda (Gastropoda: Mollusca). Molecular Phylogenetics and Evolution 42: 717–737.
Cowie, R. H., 2001. Can snails ever be effective and safe biocontrol agents? International Journal of Pest Management 47: 23–40.
Cowie, R. H. & D. G. Robinson, 2003. Pathways of introduction of nonindigenous land and freshwater snails and slugs. In Ruiz, G. & J. T. Carlton, (eds), Invasive Species: Vectors and Management Strategies. Island Press, Washington, DC: 93–122.
Davis, G. M., 1980. Snail hosts of Asian *Schistosoma* infecting man: evolution and coevolution. Malacological Review Supplement 2: 195–238.
Davis, G. M., 1982. Historical and ecological factors in the evolution, adaptive radiation, and biogeography of freshwater molluscs. American Zoology 22: 375–395.
Davis, G. M., C. E. Chen, Z. B. Kang & Y. Y. Liu, 1994. Snail hosts of *Paragonimus* in Asia and the Americas. Biomedical and Environmental Sciences 7: 369–382.
Dayrat, B., A. Tillier, G. Lecointre & S. Tillier, 2001. New clades of euthyneuran gastropods (Mollusca) from 28S rRNA sequences. Molecular Phylogenetics and Evolution 19: 225–235.
Dillon, R. T. Jr., 2000. The Ecology of Freshwater Molluscs. Cambridge University Press, United Kingdom.
El-Gamal, M. M. & K. M. Sharshar, 2004. Heavy metals partitioning between soft tissues and shells of some freshwater and marine molluscs as accumulation indicators for their monitoring in the environment. Invertebrate Zoology and Parasitology 45: 141–161.
Gargominy, O. & P. Bouchet, 1998. In Groombridge, B. & M. Jenkins (eds), Freshwater Biodiversity: A Preliminary Global Assessment. World Conservation Monitoring Center—World Conservation Press, Cambridge, UK.
Graf, D. L., 2001. The cleansing of the Augean Stables, or a lexicon of the nominal species of the Pleuroceridae (Gastropoda: Prosobranchia) of recent North America, North of Mexico. Walkerana 12: 1–124.
Gray, J., 1988. Evolution of the freshwater ecosystem: the fossil record. Palaeogeography, Palaeoclimatology, Palaeoecology 62: 1–214.
Haase, M., 2005. Rapid and convergent evolution of parental care in hydrobiid gastropods from New Zealand. Journal of Evolutionary Biology 18: 1076–1086.
Haase, M. & P. Bouchet, 1998. Radiation of crenobiontic gastropods on an ancient continental island: the *Hemistomia*-clade in New Caledonia (Gastropoda : Hydrobiidae). Hydrobiologia 367: 43–129.
Haase, M. & P. Bouchet, 2006. The radiation of hydrobioid gastropods (Caenogastropoda, Rissooidea) in ancient Lake Poso, Sulawesi. Hydrobiologia 556: 17–46.
Haase, M., W. F. Ponder & P. Bouchet, 2006. The genus *Fluviopupa* Pilsbry, 1911 from Fiji (Caenogastropoda, Rissooidea). Journal of Molluscan Studies, 72: 119–136.
Haszprunar, G., 1988. On the origin and evolution of major gastropod groups, with special reference to the Streptoneura (Mollusca). Journal of Molluscan Studies, 54: 367–441.
Hausdorf, B., P. Röpstorf & F. Riedel, 2003. Relationships and origin of endemic Lake Baikal gastropods (Caenogastropoda: Rissooidea) based on mitochondrial DNA sequences. Molecular Phylogenetics and Evolution 26: 435–443.
Hershler, R., 1998. A systematic review of the hydrobiid snails (Gastropoda: Rissooidea) of the Great Basin, western United States. Part I. Genus *Pyrgulopsis*. The Veliger 41: 1–132.
Hershler, R., 1999. A systematic review of the hydrobiid snails (Gastropoda: Rissooidea) of the Great Basin, western United States. Part II. Genera *Colligyrus, Fluminicola, Pristinicola* and *Tryonia*. The Veliger 42: 306–337.
Holthuis, B. V., 1995. Evolution between Marine and Fresh-Water Habitats: A Case Study of the Gastropod Suborder Neritopsina. Unpublished Ph.D. thesis, University of Washington.
Hubendick, B., 1951. Recent Lymnaeidae: their variation, morphology, taxonomy, nomenclature and distribution. Kungliga Svenska Vetenskapakademiens Handlingar 3: 1–233.
Hubendick, B., 1978. Systematics and comparative morphology of the Basommatophora. In Fretter V. & J. Peake (eds), Pulmonates, Vol. 2a. Systematics, Evolution and Ecology. Academic Press, London: 1–47.
Hutchinson, G. E., 1967. A Treatise on Limnology: Vol. II, Introduction to Lake Biology and the Limnoplankton. John Wiley & Sons, New York.
Janz, H., 1999. Hilgendorf's planorbid tree—the first introduction of Darwin's theory of transmutation into palaeontology. Paleontological Research 3: 287–293.
Kano, Y., S. Chiba & T. Kase, 2002. Major adaptive radiation in neritopsine gastropods estimated from 28S rRNA sequences and fossil records. Proceedings of the Royal Society, London, Series B 269: 2457–2465.

Kay, E. A., 1995a. Hug a slug—save a snail: a status report on molluscan diversity and a framework for conservation action. In Kay, E. A. (ed.), The Conservation Biology of Molluscs. Proceedings of a symposium held at the 9th International Malacological Congress. Edinburgh, Scotland, 1986: 53–79.

Kay, E. A., 1995b. Actions required for the conservation of molluscan diversity. In Kay, E. A. (ed.), The Conservation Biology of Molluscs. Proceedings of a symposium held at the 9th International Malacological Congress. Edinburgh, Scotland, 1986: 68–79.

Köhler, F., T. Rintelen, A. Meyer & M. Glaubrecht, 2004. Multiple origin of viviparity in Southeast Asian gastropods (Cerithioidea: Pachychilidae) and its evolutionary implications. Evolution 58: 2215–2226.

Kruglov, N. D., 2005. Molluscs of family Lymnaeidae (Gastropoda Pulmonata) in Europe and northern Asia. SGPU. Publishing, Smolensk: 1–507.

Lamy, E., 1926. Sur une coquille enigmatique. Journal de Conchyliologie 70: 51–56.

Langston, W. J., M. J. Bebianno & G. R. Burt, 1998. Metal handling strategies in molluscs. In Langston, W. J. & M. J. Bebianno (eds), Metal Metabolism in Aquatic Environments. Chapman & Hall, London: 219–283.

Lee, L. E. J., J. Stassen, A. McDonald, C. Culshaw, A. D. Venosa & K. Lee, 2002. Snails as biomonitors of oil-spill and bioremediation strategies. Bioremediation Journal 6: 373–386.

Liat, L. B., Y. L. Fong, M. Krishnansamy, P. Ramachandran & S. Mansour, 1978. Freshwater snail consumption and angiostrongyliasis in Malaya. Tropical and Geographical Medicine 30: 241–246.

Lydeard, C., J. H. Yoder, W. E. Holznagel, F. G. Thompson & P. Hartfield, 1998. Phylogenetic utility of the 5′-half of mitochondrial 16S rDNA gene sequences for inferring relationships of *Elimia* (Cerithioidea: Pleuroceridae). Malacologia 39: 183–193.

Lydeard, C., W. E. Holznagel, M. Glaubrecht & W. F. Ponder, 2002. Molecular phylogeny of a circum-global, diverse gastropod superfamily (Cerithioidea: Mollusca: Caenogastropoda) pushing the deepest phylogenetic limits of mitochondrial LSU rDNA sequences. Molecular Phylogenetics and Evolution 22: 399–406.

Lydeard, C., R. H. Cowie, A. E. Bogan, P. Bouchet, K. S. Cummings, T. J. Frest, D. G. Herbert, R. Hershler, O. Gargominy, K. Perez, W. F. Ponder, B. Roth, M. Seddon, E. E. Strong & F. G. Thompson, 2004. The global decline of nonmarine mollusks. BioScience 54: 321–330.

Malek, E. A. & T. C. Cheng, 1974. Medical and Economic Malacology. Academic Press, New York and London.

McAllister, D. E., J. F. Craig, N. Davidson, S. Delany & M. Seddon, 2000. Biodiversity impacts of large dams. A contributing paper to the World Commission on Dams. On-line at: http://www.damsreport.org/docs/kbase/contrib/env245.pdf.

Michel, E., 1994. Why snails radiate: a review of gastropod evolution in long-lived lakes, both Recent and fossil. In Martens, K., B. Goddeeris & G. Coulter (eds), Speciation in Ancient Lakes. Academic Press, London: 285–317.

Minton, R. L. & C. Lydeard, 2003. Phylogeny, taxonomy, genetics and global heritage ranks of an imperiled, freshwater snail genus *Lithasia* (Pleuroceridae). Molecular Ecology 12: 75–87.

Morgan, J. A. T., R. J. DeJong, Y. Jung, K. Khallaayoune, S. Kock, G. M. Mkoji & E. S. Loker, 2002. A phylogeny of planorbid snails, with implications for the evolution of *Schistosoma* parasites. Molecular Phylogenetics and Evolution 25: 477–488.

Økland, J., 1990. Lakes and snails. Environment and Gastropoda in 1,500 Norwegian lakes, ponds and rivers. Universal Book Services/Dr. W. Backhuys. Oegstgeest, The Netherlands.

Peters, W. & G. Pasvol, 2001. Tropical Medicine and Parasitology, 5th edn. Elsevier.

Pointier, J. P., 1999. Invading freshwater gastropods: some conflicting aspects for public health. Malacologia 41: 403–411.

Ponder, W. F., 1982. Hydrobiidae of Lord Howe Island (Mollusca: Gastropoda: Prosobranchia). Australian Journal of Marine and Freshwater Research 33: 89–159.

Ponder, W. F., 1986. Glacidorbidae (Glacidorbiacea : Basommatophora), a new family and superfamily of operculate freshwater gastropods. Zoological Journal of the Linnean Society 87: 53–83.

Ponder, W. F., 1988. The truncatelloidean (=Rissoacean) radiation—a preliminary phylogeny. In Ponder, W.F. (ed.), Prosobranch Phylogeny. Malacological Review, Supplement 4: 129–166.

Ponder, W. F., 1991. Marine valvatoideans-implications for early heterobranch phylogeny. Journal of Molluscan Studies 57: 21–32.

Ponder, W. F., 1994. Australian freshwater Mollusca: conservation priorities and indicator species. Memoirs of the Queensland Museum 36: 191–196.

Ponder, W. F., 1995. The conservation of non-marine molluscs in perspective. In van Bruggen, A. C., S. M. Wells & Th. C. M. Kemperman (eds), Biodiversity and Conservation of the Mollusca, Proceedings of the Alan Memorial Symposium on the Biodiversity and Conservation of the Mollusca, Eleventh International Malacological Congress, Siena, Italy, 1992: 55–67.

Ponder, W. F., 1999. Using museum collection data to assist in biodiversity assessment. In Ponder, W.F. & D. Lunney (eds), The Other 99%. The Conservation and Biodiversity of Invertebrates. Transactions of the Royal Zoological Society of NSW, Mosman: 253–256.

Ponder, W. F., 2004a. Endemic aquatic macroinvertebrates of artesian springs of the Great Artesian Basin—progress and future directions. Records of the South Australian Museum Monograph Series 7: 101–110.

Ponder, W. F., 2004b. Conservation of molluscs and other beasts without backbones; issues, strategies and the role of museum collections. In Killeen, I. J. & M. B. Seddon (eds), Molluscan Biodiversity and Conservation. Journal of Conchology Special Publication number 3: 7–21.

Ponder, W. F., 2004c. Monograph of the Australian Bithyniidae (Caenogastropoda: Rissooidea). Zootaxa 230: 1–126.

Ponder, W. F. & G. J. Avern, 2000. The Glacidorbidae (Mollusca: Gastropoda: Heterobranchia) of Australia. Records of the Australian Museum 52: 307–353.

Ponder, W.F. & D. J. Colgan, 2002. What makes a narrow range taxon? Insights from Australian freshwater snails. Invertebrate Systematics 16: 571–582.

Ponder, W. F. & D. R. Lindberg, 1997. Towards a phylogeny of gastropod molluscs—an analysis using morphological characters. Zoological Journal of the Linnean Society 19: 83–265.

Ponder, W. F., J. Walker & L. Puslednik, 2006. Australian freshwater snails as intermediate hosts for trematodes. National Parasitology Symposium Proceedings. Centre for Infectious Diseases and Microbiology-Public Health, ICPMR, Westmead Hospital, NSW: 57–94.

Ponder, W. F. & K. F. Walker, 2003. From mound springs to mighty rivers: the conservation status of freshwater molluscs in Australia. Aquatic Ecosystem Health & Management 6: 19–28.

Ponder, W.F. & J. Waterhouse, 1997. A new genus and species of Lymnaeidae from the lower Franklin River, south western Tasmania. Journal of Molluscan Studies 63: 441–468.

Purchon, I., 1977. The Biology of the Mollusca. 2nd edn. Pergamon Press, Oxford.

Radoman, P., 1983. Hydrobioidea a superfamily of Prosobranchia (Gastropoda) 1. Systematics Srpska Akademija Nauka i Umetnosti Posebna Izdanja 57: 1–256.

Ramsar Convention Secretariat, 2004. The Ramsar Convention Manual: a Guide to the Convention on Wetlands (Ramsar, Iran, 1971), 3rd edn. Ramsar Convention Secretariat, Gland, Switzerland.

Remigio, E. A., 2002. Molecular phylogenetic relationships in the aquatic snail genus *Lymnaea*, the intermediate host of the causative agent of fascioliasis: insights from broader taxon sampling. Parasitology Research 88: 687–696.

Remigio, E. A. & D. Blair, 1997. Molecular systematics of the freshwater snail family Lymnaeidae (Pulmonata: Basommatophora) utilising mitochondrial ribosomal DNA sequences. Journal of Molluscan Studies 63: 173–185.

Sada, D. W. & G. L. Vinyard, 2002. Anthropogenic changes in historical biogeography of Great Basin aquatic biota. In Hershler, R., D. B. Madsen & D. R. Currey (eds), Great Basin Aquatic Systems History. Smithsonian Contributions to the Earth Sciences. No. 33: 277–295.

Salanki, J., 2000. Invertebrates in neurotoxicology. Acta Biologica Hungarica 51: 287–307.

Salanki, J., A. Farkas, T. Kamardina & K. S. Rozsa, 2003. Molluscs in biological monitoring of water quality. Toxicology Letters 140–141: 403–410.

Seddon, M., 1998. Red listing for molluscs: a tool for conservation? Journal of Conchology Special Publication 2: 27–44.

Sheldon, F. & K. F. Walker, 1997. Changes in biofilms induced by flow regulation could explain extinctions of aquatic snails in the lower River Murray, Australia. Hydrobiologia 347: 97–108.

Simone, L. R. L., 2004. Comparative morphology and phylogeny of representatives of the superfamilies of architaenioglossans and the Annulariidae (Mollusca, Caenogastropoda). Arqivos do Museu Nacional (Rio de Janeiro) 62: 387–504.

Starmühlner, F., 1979. Distribution of freshwater molluscs in mountain streams of tropical Indo-Pacific islands. Malacologia 18: 245–255.

Strong, E. E., 2003. Refining molluscan characters: morphology, character coding and the phylogeny of the Caenogastropoda (Gastropoda). Zoological Journal of the Linnean Society 137: 447–554.

Taylor, D. W., 1988. Aspects of freshwater mollusc ecological biogeography. Palaeogeography, Palaeoclimatology, Palaeoecology 62: 511–576.

Taylor, D. W., 2003. Introduction to Physidae (Gastropoda: Hygrophila), biogeography, classification, morphology. Revista de Biologia Tropical 51, Supplement 1: 1–287.

Thomas, J. D., 1997. Anthropogenic influences on molluscan biodiversity. In Proceedings of "Workshop on Medical Malacology in Africa", Harare, Zimbabwe, September 22–26, 1997: 9–28.

Tracey, S., J. A. Todd & D. H. Erwin, 1993. Mollusca: Gastropoda. In Benton, M. J. (ed.), The Fossil Record. Chapman & Hall, London: 131–167.

Vermeij, G. J. & R. Dudley, 2000. Why are there so few evolutionary transitions between aquatic and terrestrial ecosystems? Biological Journal of the Linnean Society 70: 541–554.

von Rintelen, T. & M. Glaubrecht, 2005. Anatomy of an adaptive radiation: a unique reproductive strategy in the endemic freshwater gastropod *Tylomelania* (Cerithioidea: Pachychilidae) on Sulawesi, Indonesia and its biogeographical implications. Biological Journal of the Linnean Society 85: 513–542.

Wells, S. M., 1995. Molluscs and the conservation of biodiversity. In van Bruggen, A. C., S. M. Wells & Th. C. M. Kemperman (eds), Biodiversity and Conservation of the Mollusca, Proceedings of the Alan Solem Memorial Symposium on the Biodiversity and Conservation of the Mollusca, Eleventh International Malacological Congress. Siena, Italy, 1992: 21–36.

Wethington, A. R. 2004. Phylogeny, taxonomy, and evolution of reproductive isolation in *Physa* (Pulmonata: Physidae). Unpublished Ph.D. Thesis, The University of Alabama, Tuscaloosa, Alabama, USA.

Wilke, T., G. M. Davis, A. Falniowski, F. Giusti, M. Bodon & M. Szarowska, 2001. Molecular systematics of Hydrobiidae (Gastropoda: Rissooidea): testing monophyly and phylogenetic relationships. Proceedings of the Academy of Natural Sciences of Philadelphia 151: 1–21.

Williamson, P. G., 1981. Palaeontological documentation of speciation in Cenozoic molluscs from Turkana Basin. Nature 293: 437–443.

Hydrobiologia (2008) 595:167–176
DOI 10.1007/s10750-007-9119-9

FRESHWATER ANIMAL DIVERSITY ASSESSMENT

Global diversity of large branchiopods (Crustacea: Branchiopoda) in freshwater

Luc Brendonck · D. Christopher Rogers · Jorgen Olesen · Stephen Weeks · Walter R. Hoeh

Abstract With about 500 known species worldwide, the large brachiopods are a relatively small group of primitive crustaceans. With few exceptions they live in temporary aquatic systems that are most abundant in arid and semi arid areas. As many regions remain unexplored and as especially the number of species in clam shrimps and tadpole shrimps is underestimated due to difficult identification, the species list will increase with future surveys. The Branchiopoda are monophyletic, but inter-ordinal relationships, as well as many evolutionary relationships at lower taxonomic levels are still unclear. Ongoing molecular studies will more accurately depict species diversity and phylogenetic patterns. With the exception of some anostracan families, most families are not restricted to the northern or southern hemisphere or specific zoogeographical regions. Large branchiopods are used for the assessment of the quality and function of temporary wetlands. Due to the reduction in number and quality of temporary wetlands, several species became endangered and are red listed by the IUCN.

Keywords Large branchiopod diversity · Anostraca · Spinicaudata · Laevicaudata · Cyclestherida · Notostraca · Temporary pools

Guest editors: E.V. Balian, C. Lévêque, H. Segers & K. Martens
Freshwater Animal Diversity Assessment

Electronic supplementary material The online version of this article (doi:10.1007/s10750-007-9119-9) contains supplementary material, which is available to authorized users.

L. Brendonck (✉)
Laboratory of Aquatic Ecology, KULeuven,
Deberiotstraat 32, Leuven 3000, Belgium
e-mail: Luc.Brendonck@bio.kuleuven.be

D. C. Rogers
EcoAnalysts, Inc., 166 Buckeye St., Woodland, CA 95695, USA

J. Olesen
Zoological Museum, University of Copenhagen,
Universitetsparken 15, Copenhagen Ø 2100, Denmark

S. Weeks
Integrated Bioscience Program, Department of Biology,
The University of Akron, Akron, OH 44325-3908, USA

W. R. Hoeh
Department of Biological Sciences, Kent State University,
Kent, OH, USA

Introduction

The crustacean class Branchiopoda (clam shrimp (Spinicaudata, Laevicaudata, and Cyclestherida), tadpole shrimp (Notostraca), and cladocerans [=Phyllopoda] plus fairy shrimp [=Anostraca]) is a morphologically diverse group of ecologically important, largely freshwater organisms that has a fossil record extending back to the Upper Cambrian (Walossek, 1993). Breeding systems within the class are highly varied, including dioecy, androdioecy, parthenogenesis, and cyclic parthenogenesis (Sassaman, 1995; Weeks et al., 2006a, b). Branchiopoda without cladocerans are commonly known as the "large branchiopods."

Most large branchiopod species rely on seasonal wetlands such as playas, vernal (rain and snow-melt) pools, salt flats, and alkali pans, while other species occur in permanent playas, fishless alkali lakes, and salt lakes (Dumont & Negrea, 2002). These habitats are, for the most part, seasonal pools that are dry for a significant portion of the year, or several years (Dumont & Negrea, 2002).

The vast majority of anostracans are omnivorous filter feeders, indiscriminately filtering particles from the water column with their continually beating thoracopods (Brendonck, 1993a, b). A few species are predators on rotifers, copepods, cladocerans, and other Anostraca (Boudrias & Pires, 2002; Rogers et al., 2006). The tadpole shrimp are predominantly benthic and omnivorous, feeding on detritus and living or dead organisms (Martin, 1992). Clam shrimp typically feed nonselectively on detritus and algae in suspension (Belk, 1982).

Branchiopod crustaceans rely on banks of resting eggs (or "cysts;" Fig. 1) to bridge periods of drought or frost (Brendonck, 1996) and to buffer against the effects of environmental variability. The eggs lay dormant in the substrate until the pool dries and refills during the subsequent rains. In the temporally fluctuating environment of temporary aquatic habitats, usually only part of the dormant eggs, if any,

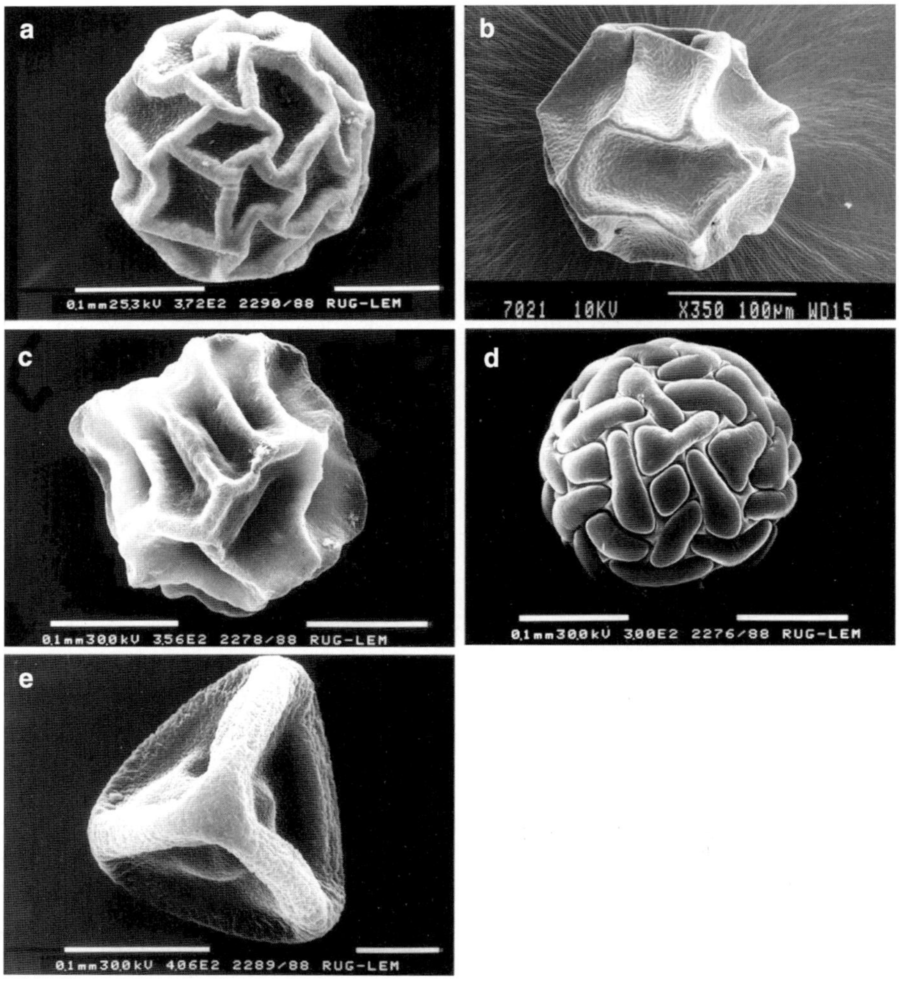

Fig. 1 Various large branchiopod egg morphology: (**a**) *Streptocephalus proboscideus* (Anostraca); (**b**) *Streptocephalus indistinctus* (Anostraca); (**c**) *Eulimnadia cylindrova* (Spinicaudata); (**d**) *Streptocephalus ovamboensis* (Anostraca), (**e**) *Streptocephalus zuluensis* (Anostraca)

hatches during each fill (Hildrew, 1985; Brendonck et al., 1998). This process can lead to the generation of an egg bank that can remain viable for decades or possibly centuries without re-hydration (Belk, 1998), and can harbor potentially great genetic and species diversity (Ellner & Hairston, 1994). The external morphology (shape and ornamentation; Fig. 1) of eggs is an important taxonomic tool (Mura, 1986; Belk, 1989; Martin, 1989; Martin & Belk, 1989; Brendonck et al., 1992; Brendonck & Coomans, 1994a, b). In many cases this allows the assessment of the temporary pool communities even when no water is present.

Large branchiopod crustaceans are arguably the least known of all macroscopic invertebrates in temporary inland waters even though they are the flagship group for these ecosystems and for temporary pool invertebrates in general (Colburn, 2004).

Species/generic diversity

In many areas of the world, the diversity, species composition, and conservation status of large branchiopods remain largely unknown. We therefore expect to discover and describe many new taxa in these largely unexplored areas (e.g., South America, northern, western and eastern Africa, Western Australia, and Asia).

Anostraca

Anostracans (Fig. 2a) are elongated crustaceans, which lack a carapace and are constantly swimming on their backs in the water column. The Anostraca are by far the most taxonomically diverse group at all levels (Tables 1, 2). There are ~300 species on all continents (including Antarctica; Belk & Brtek, 1995, 1997; Dumont & Negrea, 2002), currently arranged in 26 genera and eight families (Tables 1, 2; Weekers et al., 2002, Rogers, 2002b, 2006). Nearly 1/4 of these species are known only from the type localities, or from <3 localities (Belk & Brtek, 1995, 1997). Additionally, there are ~20 undescribed species and at least one undescribed genus.

Typically, anostracan species are differentiated based upon the form and ornamentation of the male second antennae, which are modified into large claspers to grasp the female during copulation (Rogers, 2002a). Linder (1941) demonstrated the importance of copulatory structures, especially the male genitalia, in defining genera. Starting from this base, Belk (1991, 1995), Brendonck (1995), Brendonck & Belk (1997), and Rogers (2002b, 2006) established genital morphology as the defining criteria for anostracan genera. To date, most of the taxonomy and phylogeny has been based on the morphology of these antennal and genital characters.

Notostraca

The order Notostraca (tadpole shrimp; Fig. 2c, d) is composed of the family Triopsidae and includes two genera (*Triops* Schrank, 1803 and *Lepidurus* Leach, 1819; Tables 1, 2) of "living fossils" which have undergone minimal gross morphological change since their divergence over 250 million years ago (Longhurst, 1955).

The Notostraca exhibit plasticity in external morphology, making the demarcation of species on this basis a difficult task (Rogers, 2001). The absence of well-defined criteria allowed taxonomists to describe many 'new species' in such a way that the nominal species of *Triops* and *Lepidurus* amounted to more than 70 in the 1950s (Longhurst, 1955). Linder (1952) and Longhurst (1955) reviewed the alpha taxonomy of the Notostraca and recognized only about 11 species with a wide geographic distribution. Until recently, this classification was extensively used "even attaining the status of dogma" (Sassaman et al., 1997). However, the genetic and molecular studies of Sassaman et al. (1997), Suno-Uchi et al. (1997), King & Hanner (1998), and Murugan et al. (2002) have demonstrated that Linder's and Longhurst's classifications have long obscured understanding of the real species-level diversity of the Notostraca.

Notostracan systematics is further complicated by the discovery of different modes of reproduction (Sassaman & Weeks, 1993; Sassaman, 1995). The determination of the mode of reproduction is important for the understanding of the variety of species.

Suno-Uchi et al. (1997) studied the three morphospecies *Triops cancriformis*, *T. granarius*, and *T. longicaudatus* from Japan, analyzing a segment of the mt16S rRNA gene. They proposed that there

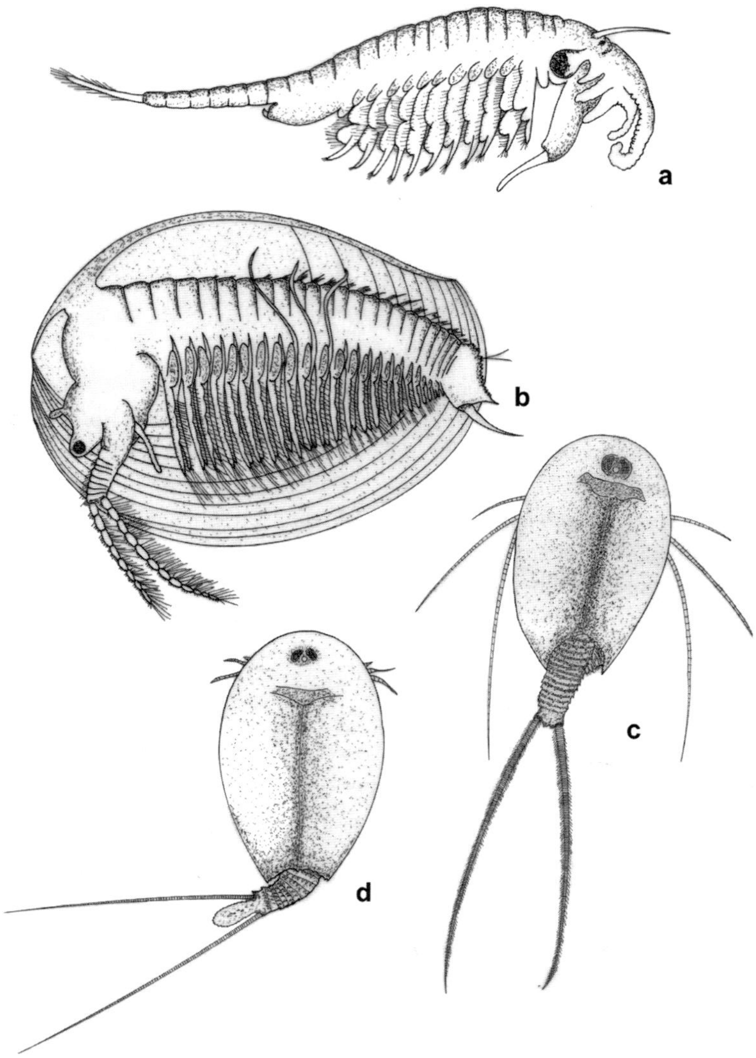

Fig. 2 Habitus of large branchiopods: (**a**) adult male *Chirocephalus diaphanus* (Anostraca), (**b**) adult *Limnadia lenticularis* (Spinicaudata), (**c**) adult *Triops cancriformis* (Notostraca), (**d**) adult *Lepidurus apus* (Notostraca). (Drawings from: Brendonck, 1989)

are four, not three, phylogenetic species. King & Hanner (1998) studied a fragment of the mt12S rRNA gene in four nominal species of *Lepidurus* from Canada and the USA, and deduced that there are five, not four, reproductively isolated species, one of them recently described (Rogers, 2001). Using the mt12S- and mt16S rRNA genes, Murugan et al. (2002) studied five *Triops* populations from México and explored the phylogeny of the genus. Analysis of mt12S rDNA data was in agreement with previous allozyme studies (Sassaman et al., 1997), and showed the nominal (morphological) species *T. longicaudatus* to be a mixture of several species such that, of the seven *Triops* American populations studied, six phylogenetic species can be identified. These molecular data, contrary to a phylogenetic proposal based on morphology (Maeda-Martínez et al., 2000), also indicated that two morphologically and reproductively divergent forms can be grouped into a single monophyletic clade, and that the Old World species *T. cancriformis* may represent a lineage that is independent from other species in that genus (Murugan et al., 2002).

Spinicaudata, Laevicaudata, and Cyclestherida

These small, freshwater crustaceans have laterally compressed bodies enclosed by a bivalved carapace (Fig. 2b). Although clam shrimp are common

Table 1 Species [a] distribution of Anostraca, Notostraca, Spinicaudata and Laevicaudata per family in zoogeopgrahic regions

	PA	NA	NT	AT	OL	AU	PAC	ANT	World
Anostraca[a]	110	64	33	56	11	48	1	1	307
Artemiidae	6	3	2	1	1	1	1	0	9
Parartemiidae	0	0	0	0	0	13+	0	0	13
Branchinectidae	6	24	15	0	0	0	0	1	45
Thamnocephalidae	5	6	16	4	2	32	0	0	62
Streptocephalidae	16	15	0	24	7	2	0	0	56
Branchipodidae	9	0	0	26		0	0	0	35
Tanymastigitidae	8	0	0	0	0	0	0	0	8
Chirocephalidae	61	16	0	1	1	0	0	0	81
Notostraca[a]	7	7	2	1	2	2	0	0	15
Triopsidae	7	7	2	1	2	2	0	0	15
Spinicaudata[a]	~50	~15	~13	~20	~30	23	1	0	~150
Cyzicidae	~20	6	2	~7	~10	10	0	0	~90
Leptestheriidae	~15	1	~5	~8	~3	0	0	0	~37
Limnadiidae	~15	~10	~10	~10	~12	13	1	0	~55
Laevicaudata	8	7	13	4	4	2	0	0	36
Lynceidae	8	7	13	4	4	2	0	0	36

PA, Palaearctic; NA, Nearctic; NT, Neotropical; AT, Afrotropical; OL, Oriental; AU, Australasian; PAC, Pacific Oceanic Islands; ANT, Antarctic

[a] Several taxa are known from more than one region

Table 2 Genus distribution of Anostraca, Notostraca, Spinicaudata and Laevicaudata per family in zoogeopgrahic regions

	PA	NA	NT	AT	OL	AU	PAC	ANT	World
Anostraca	16	11	10	6	6	4	1	1	26
Artemiidae	1	1	1	1	1	1	1	0	1
Parartemiidae	0	0	0	0	0	1	0	0	1
Branchinectidae	1	1	0	1	0	0	0	1	1
Thamnocephalidae	2	3	3	4	2	1	0	0	6
Streptocephalidae	1	1	1	0	1	1	0	0	1
Branchipodidae	2	0	4	0	1	0	0	0	5
Tanymastigitidae	2	0	0	0	0	0	0	0	2
Chirocephalidae	7	5	1	0	1	0	0	0	9
Notostraca	2	2	1	2	1	2	0	0	2
Triopsidae	2	2	1	2	1	2	0	0	2
Spinicaudata	9	5	5	4	6	5	1	0	12
Cyzicidae	3	2	2	1	3	2	0	0	4
Leptestheriidae	3	1	1	1	1	0	0	0	3
Limnadiidae	3	2	2	2	2	3	1	0	5
Laevicaudata	1	2	2	2	1	1	0	0	3
Lynceidae	1	2	2	2	1	1	0	0	**3**

PA, Palaearctic; NA, Nearctic; NT, Neotropical; AT, Afrotropical; OL, Oriental; AU, Australasian; PAC, Pacific Oceanic Islands; ANT, Antarctic

worldwide, they have not been the subjects of intense study. A few investigations have assessed their genetics (Duff et al., 2004), phylogeny and biogeography (Hoeh et al., 2006), and reproductive biology (Weeks et al., 2005, 2006a, b; Scanabissi et al., 2006), but most attention has been devoted to the morphological systematics of these animals and ~116 species are currently recognized. The three orders consist of five extant families and 19 genera (Tables 1, 2). Traditionally the genus *Cyclestheria* was considered to be part of the Spinicaudata, but Olesen (1998) and Spears and Abele (2000) suggested that spinicaudatans are paraphyletic, with respect to the Cladocera.

The systematics within the Spinicaudata has been problematic for centuries, and the principal difficulties are still far from being resolved. Presently, the Spinicaudata is subdivided into three families (Martin & Davis, 2001) (Tables 1, 2), but the monophyly of two of these (Cyzicidae and Leptestheridae) is highly uncertain (the monophyly of the third, Limnadiidae, is strongly supported; Hoeh et al., 2006). Similar problems can be found at both the genus and species levels in all three families. Spinicaudatans are well known for showing great 'morphological plasticity' making species identification exceptionally difficult in some taxa. The usual problems of poor descriptions and lack of type material add to the difficulties.

Belk (1989) re-emphasized the importance of the morphology of the outer covering of the desiccated resting eggs as a taxonomic character.

Phylogeny

The monophyly of the Branchiopoda has been strongly supported by recent phylogenetic analyses (e.g., Spears & Abele, 2000) but inter-ordinal relationships within the Phyllopoda, as well as many evolutionary relationships at lower taxonomic levels throughout the class, have not been clearly elucidated (e.g., Braband et al., 2002).

In last decades there were interesting rearrangements in the systematics of higher taxa of the Branchiopoda (Fryer, 1987; Walossek, 1993; Martin & Cash-Clark, 1995). Clam shrimps, formerly thought to be one order (the "Conchostraca"), have turned out to be paraphyletic with respect to the Cladocera (the water fleas; Olesen, 1998; Spears & Abele, 2000; Braband et al., 2002), and probably also with respect to the Notostraca (tadpole shrimps; Stenderup et al., 2006). The "Conchostraca" are now divided into the Laevicaudata and the Spinicaudata (Fig. 2b) and the Cyclestherida.

Information on phylogenetic relationships in large branchiopods is most advanced, but still fragmentary in anostracans. The anostracan family Streptocephalidae has been subjected to morphological (Hamer et al., 1994a, b) and morphological-cladistic analyses (Maeda-Martínez et al., 1995a, b). To date there are only 5 phylogenetic studies using molecular data that focused on the Anostraca: *Branchinella* (Remigio et al., 2003), *Parartemia* (Remigio et al., 2001), *Streptocephalus* (Daniels et al., 2004), and two studies attempting to resolve familial relationships (Remigio & Hebert, 2000; Weekers et al., 2002) with one study focusing on the ordinal relationships (de Waard et al., 2006). These studies generated more questions about the relationships between the genera (Rogers, 2002b; Weekers et al., 2002).

Present distribution and main areas of endemicity

Large branchiopods have a worldwide distribution (Fig. 3), including the polar areas (e.g., the anostracan *Branchinecta gaini* is rather widespread on the antarctic peninsula), but reach their maximum abundance and species richness in steppes and deserts, where temporary water bodies abound. The vast majority of species are fresh-water; however some species of *Branchinella*, *Streptocephalus*, and *Thamnocephalus*, as well as all species of *Parartemia* and *Artemia*, live in inland saline waters.

Family, generic, and species diversity of large branchiopods at the level of the main zoogeographical areas are presented in Tables 1 and 2. Some anostracan families are endemic to only one of the areas such as Parartemiidae to Australasia and Tanymastigitidae (Anostraca) to the Palearctic. Notostracan and clam shrimp families are always distributed over at least one of the zoogeographical areas. Except for the former families, not one is entirely restricted to the southern or northern hemisphere, which would reflect a Gondwanaland or Laurasia origin, respectively.

Some clam shrimp genera like *Cyzicus* and *Eocyzicus* appear to be Laurasian, with their current

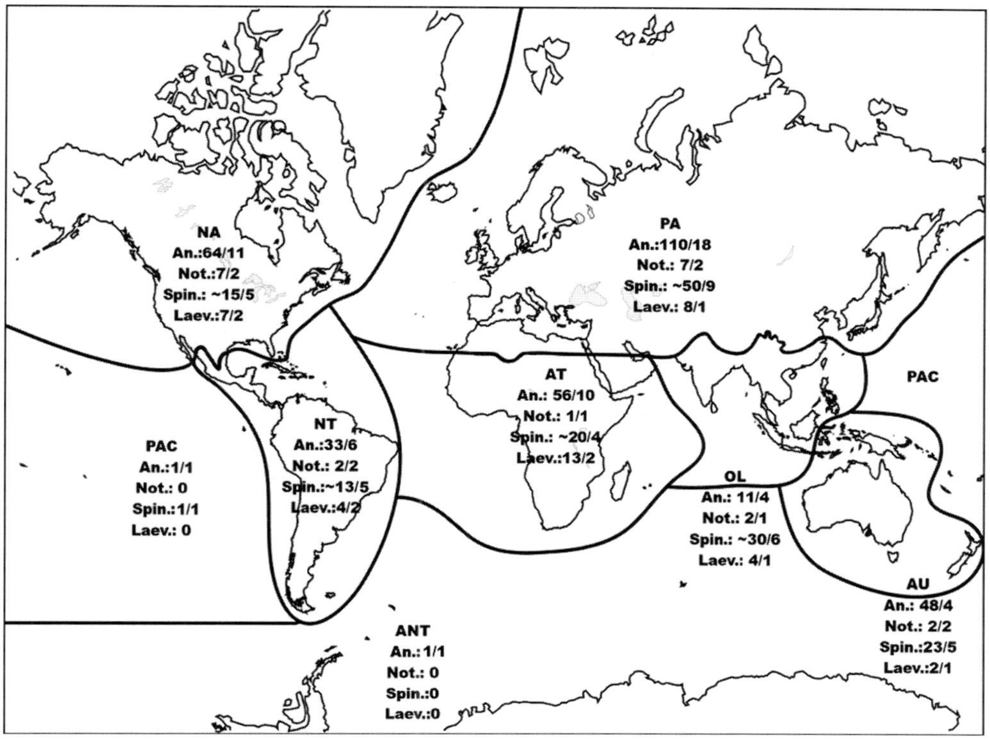

Fig. 3 Species and genus diversity of Anostraca (An), Notostraca (Not), Spinicaudata (Spin) and Laevicaudata (Laev) in each of the seven zoogeographical regions. PA, Palaearctic; NA, Nearctic; NT, Neotropical; AT, Afrotropical; OL, Oriental; AU, Australasian; PAC, Pacific Oceanic Islands; ANT, Antarctic

ranges encompassing Eurasia, Africa, and North America. The genus *Lepthesteriella*, in contrast, seems Gondwanian, and occurs in Africa, Madagascar, and Asia. Lynceids are cosmopolitan, save in Antarctica. *Cyclestheria hislopi* (if representing only one species, which is uncertain) is pantropical.

Human related issues

Worldwide changes in land use, brought about mainly by agriculture and urbanization, has led to a global loss and deterioration of temporary pools. It has been estimated that during the post-war period in England, nearly 40% of ephemeral ponds have disappeared, which is a rate loss of $\sim 1\%$ per year. Several studies indicate a loss of between 50% and 90% of original Californian vernal pools (Bauder, 1986). For most regions of the world, and especially the developing countries, no such figures are available, but are likely to be as dramatic, due to uncontrolled spraying with insecticides, mining, and agricultural activities (Brendonck & Williams, 2000).

Large branchiopods are used in the United States for assessment of ephemeral wetland habitat functions and values (Rogers, 1998), and five US species are currently listed as threatened or endangered in California and Oregon. Another anostracan species has been petitioned for federal endangered status and is protected as a “rare, threatened or endangered species” under the California Environmental Quality Act. Another species was proposed for federal listing in the state of Georgia, four species are proposed for endangered species status in New Mexico, one species is listed as vulnerable in Australia, and one in Brazil. About 32 large branchiopod species are red listed by the IUCN (IUCN, 2000). Furthermore, the world’s first large branchiopod preserve has been developed in Austria (Eder & Hödl, 2002) protecting the anostracan *Chirocephalus shadinii*. Five of the six endangered fairy shrimp in California are endemic to that state, and were only discovered since 1990.

Some large branchiopod species are important economically, including species that are: harvested for aquaculture (*Artemia* spp.), for applications in aquatic toxicology (Brendonck & Persoone, 1993), for human consumption (*Streptocephalus sirindhornae*), or as pets ("Sea Monkeys" *Artemia* "*nyos*") (Sanoamuang et al., 2000).

Some *Triops* forms (Notostraca) are pests in rice fields in at least seven countries on four continents (Grigarick et al., 1961). Alternatively, these shrimp have been used to control weeds in rice fields of Japan (Takahashi, 1977), and also have been proposed as a biological control agent of mosquitoes (Tietze & Mulla, 1991). Given their rapid growth, early maturation, and uniparental reproduction via resting eggs, two *Triops* forms are being studied for their potential use in aquaculture (Obregón-Barboza et al., 2001).

References

Bauder, E. T., 1986. San Diego vernal pools: Recent and projected losses; their condition; and threats to their existence, 1979–1990. Report prepared for Endangered Plant Project, California Department of Fish and Game, Sacramento, CA.

Belk, D., 1982. Branchiopoda. In Parker, S. P. (ed.), Synopsis and Classification of Living Organisms, Vol. 2. McGraw-Hill, New York, 174–180.

Belk, D., 1989. Identifying species in the conchostracan genus *Eulimnadia* by egg shell morphology. Journal of Crustacean Biology 9(1): 115–125.

Belk, D., 1991. Anostracan mating behavior: A case of scramble-competition polygyny. In Bauer, R. T. & J. W. Martin (eds), Crustacean Sexual Biology. Columbia University Press, New York, 111–125.

Belk, D., 1995. Uncovering the Laurasian roots of *Eubranchipus*. Hydrobiologia 298: 241–243.

Belk, D., 1998. Global status and trends in ephemeral pool invertebrate conservation: implications for California fairy shrimp. In Witham, C. W., & E. T. Bauder (eds), 147–150.

Belk, D. & J. Brtek, 1995. Checklist of the Anostraca. Hydrobiologia 298: 315–353.

Belk, D. & J. Brtek, 1997. Supplement to 'checklist of the Anostraca'. Hydrobiologia 359: 243–245.

Boudrias, M. A. & J. Pires, 2002. Unusual sensory setae of the raptorial *Branchinecta gigas* (Branchiopoda: Anostraca). Hydrobiologia 486: 19–27.

Braband, A. R., R. Hiesel & G. Scholtz, (2002). Phylogenetic relationships within the phyllopoda (Crustacea, Branchiopoda) based on mitochondrial and nuclear markers. Molecular Phylogenetics and Evolution 25: 229–244.

Brendonck, L., 1989. A review of the phyllopods (Crustacea: Anostraca, Notostraca, Conchostraca) of the Belgian fauna. Verhandelingen van het symposium "Invertebraten van Belgiё" 1989: 129–135.

Brendonck, L., 1993a. Feeding in the fairy shrimp *Streptocephalus proboscideus* (Frauenfeld) (Branchiopoda: Anostraca). I. Aspects of the feeding biology. Journal of Crustacean Biology 13: 235–244.

Brendonck, L., 1993b. Feeding in the fairy shrimp *Streptocephalus proboscideus* (Frauenfeld) (Branchiopoda: Anostraca). II. Influence of environmental conditions on feeding rate. Journal of Crustacean Biology 13: 245–255.

Brendonck, L., 1995. An updated diagnosis of the branchiopodid genera (Branchiopoda: Anostraca, Branchipodidae) with reflections on the genus concept by Dubois (1988) and the importance of genital morphology in anostracan taxonomy. Archiv für Hydrobiologie 107(2): 149–186.

Brendonck, L., 1996. Diapause, quiescence, hatching requirements: What we can learn from large freshwater branchiopods (Crustacea: Branchiopoda: Anostraca, Notostraca, Conchostraca). Hydrobiologia 320: 85–97.

Brendonck, L. & D. Belk, 1997. On potentials and relevance of the use of copulatory structures in anostracan taxonomy. Hydrobiologia 359: 83–92.

Brendonck, L. & A. Coomans, 1994a. Egg morphology in African Streptocephalidae (Crustacea: Branchiopoda: Anostraca). Part 1: South of Zambezi and Kunene rivers. Archiv für Hydrobiologie/Suppl. 99 (Monographische Beiträge) 3: 313–334.

Brendonck, L. & A. Coomans, 1994b. Egg morphology in African Streptocephalidae (Crustacea: Branchiopoda: Anostraca). Part 2: North of Zambezi and Kunene rivers, and Madagascar. Archiv für Hydrobiologie/Suppl. 99 (Monographische Beiträge) 3: 335–356.

Brendonck, L. & G. Persoone, 1993. Biological/ecological characteristics of large freshwater branchiopods from endorheic regions and consequences for their use in cyst-based toxicity tests. In Soares, A. M. V. M. & P. Calow (eds), Progress in Standardization of Aquatic Toxicity Tests, Chapt. 2. CRC Press, Boca Raton, FL, 7–35.

Brendonck, L. & W. D. Williams, 2000. Biodiversity in wetlands of dry regions (drylands). In Gopal, B., W. J. Junk & J. A. Davis (eds), Biodiversity in Wetlands: Assessment, Function and Conservation, Vol. 1. Backhuys Publishers, Leiden, The Netherlands, 181–194.

Brendonck, L., M. Hamer & A. Thiery, 1992. The occurrence of tetrahedral eggs in the Streptocephalidae (Crustacea:Branchiopoda:Anostraca) with descriptions of a new subgenus: *Parastreptocephalus* and a new species: *Streptocephalus (Parastreptocephalus)zuluensis* Brendonck and Hamer. Journal of Crustacean Biology 12: 282–297.

Brendonck, L., B. J. Riddoch, V. Van de Weghe & T. Van Dooren, 1998. The maintenance of egg banks in very short-lived pools—a case study with anostracans (Branchiopoda). In Brendonck, L., L. De Meester & N. G. Hairston Jr. (eds), Evolutionary and Ecological Aspects of Crustacean Diapause. Archiv für Hydrobiologie (Special issues) 52: 141–161.

Colburn, E. A., 2004. Vernal Pools. Natural History and Conservation. McNaughton & Gunn, Inc., Saline, MI.

Daniels, S. R., M. Hamer & D. C. Rogers, 2004. Molecular evidence suggests an ancient radiation for the fairy shrimp genus *Streptocephalus* (Branchiopoda: Anostraca). Biological Journal of the Linnean Society 82: 313–327.

de Waard, J. R., V. Sacherova, M. E. A. Cristescu, E. Remigio, T. Crease & P. D. N. Hebert, 2006. Probing the relationships of the branchiopod crustaceans. Molecular Phylogenetics and Evolution 39: 491–502.

Duff, R. J., W. R. Hoeh, D. G. Cook & S. C. Weeks, 2004. Isolation and characterization of thirteen polymorphic microsatellite loci from the clam shrimp *Eulimnadia texana* (Crustacea: Spinicaudata). Molecular Ecology Notes 4: 397–399.

Dumont, H. J. & S. Negrea, 2002. Introduction to the Class Branchiopoda. Guides to the Microinvertebrates of the Continental Waters of the World. Backhuys, Leiden.

Eder, E. & W. Hödl, 2002. Large freshwater branchiopods in Austria: Diversity, threats, conservational status. In Escobar-Briones E. & F. Alvarez (eds), Modern Approaches to the Study of Crustacea. Kluwer Academic/Plenum Publishers, New York, 281–289.

Ellner, S. & N.G. Hairston Jr., 1994. Role of overlapping generations in maintaining genetic variation in a fluctuating environment. The American Naturalists 143: 403–417.

Fryer, G., 1987. A new classification of the branchiopod Crustacea. Zoological Journal of the Linnean Society 91: 357–383.

Grigarick, A. A., W. H. Lange & D. C. Finfrock, 1961. Control of the tadpole shrimp, *Triops longicaudatus*, in California rice fields. Journal of Economic Entomology 54: 36–40.

Hamer, M., L. Brendonck, A. Coomans & C. C. Appleton, 1994a. A review of African Streptocephalidae (Crustacea: Branchiopoda: Anostraca) Part 1: South of Zambezi and Kunene rivers. Archiv für Hydrobiologie 3: 235–277.

Hamer, M., L. Brendonck, A. Coomans & C. C. Appleton, 1994b. A review of African Streptocephalidae (Crustacea: Branchiopoda: Anostraca). Part 2: North of Zambezi and Kunene rivers, and Madagascar. Archiv für Hydrobiologie 3: 279–311.

Hildrew, A. G., 1985. A quantitative study of the life history of a fairy shrimp (Branchiopoda: Anostraca) in relation to the temporary nature of its habitat, a Kenyan rainpool. The Journal of Animal Ecology 54: 99–110.

Hoeh, W. R., N. D. Smallwood, D. M. Senyo, E. G. Chapman & S. C. Weeks, 2006. Evaluating the monophyly of *Eulimnadia* and the Limnadiinae (Branchiopoda: Spinicaudata) using DNA sequences. Journal of Crustacean Biology 26(2): 182–192.

International Union of Conservation of Nature and Natural Resources, 2000, 2000. IUCN red list of threatened species, compiled by Craig Hilton-Taylor. IUCN – The World Conservation Union Species Survival Commission.

King, J. L. & R. Hanner, 1998. Cryptic species in a "living fossil" lineage: taxonomic and phylogenetic relationships within the genus *Lepidurus* (Crustacea: Notostraca) in North America. Molecular Phylogenetics and Evolution 10: 23–36.

Linder, F., 1941. Contributions to the morphology and the taxonomy of the Branchiopoda Anostraca. Zoologiska Bidrag fran Uppsala, 10: 101–302.

Linder, F., 1952. Contributions to the morphology and taxonomy of the Branchiopoda, Notostraca, with special reference to the North American species. Proceedings United States National Museum 102: 1–69.

Longhurst, A. R., 1955. A review of the Notostraca. Bulletin (British Alabama Museum of Natural History, Zoology) 3: 3–57.

Maeda-Martinez, A. M., D. Belk, H. Obregon-Barboza & H. J. Dumont, 1995a. Diagnosis and phylogeny of the New World Streptocephalidae (Branchiopoda: Anostraca). Hydrobiologia 298: 15–44.

Maeda-Martinez, A. M., D. Belk, H. Obregon-Barboza & H. J. Dumont, 1995b. A contribution to the systematics of the Streptocephalidae (Branchiopoda: Anostraca). Hydrobiologia 298: 203–232.

Maeda-Martínez, A. M., H. Obregón-Barboza, H. García-Velazco & G. Murugan, 2000. A proposal on the phylogeny and the historical biogeography of the tadpole shrimp *Triops*. Anostracan News 8: 1–4.

Martin, J. W., 1989. *Eulimnadia belki*, a new clam shrimp from Cozumel, Mexico (Conchostraca: Limnadiidae), with a review of Central and South American species of the genus *Eulimnadia*. Journal of Crustacean Biology 9: 104–114.

Martin, J. W., 1992. Branchiopoda. In Harrison, F. W. & A. G. Humes (eds), Microscopic Anatomy of Invertebrates, Crustacea, Vol. 9. Wiley-Liss, New York, 25–224.

Martin, J. W. & D. Belk, 1989. *Eulimnadia ovilunata* and *E. ovisimilis*, new species of clam shrimps (Crustacea, Branchiopoda, Spinicaudata) from South America. Proceedings of the Biological Society of Washington 102(4): 894–900.

Martin, J. W. & C. Cash-Clark, 1995. The external morphology of the onychopod "cladoceran" genus Bythotrephes (Crustacea, Branchiopoda, Onychopoda, Cercopagididae), with notes on the morphology and phylogeny of the order Onychopoda. Zoologica Scripta 24: 61–90.

Martin, J. W. & G. E. Davis, 2001 An Updated Classification of the Recent Crustacea Natural History. Museum of Los Angeles County, Los Angeles.

Mura, G., 1986. SEM morphological survey on the egg shell in the italian Anostracans (Crustacea, Branchiopoda). Hydrobiologia 134: 273–286.

Murugan, G., A. M. Maeda-Martínez, H. Obregón-Barboza & N. Y. Hernández-Saavedra, 2002. Molecular characterization of the tadpole shrimp *Triops* (Branchiopoda: Notostraca) from the Baja California Peninsula, México: New insights on species diversity and phylogeny of the genus. Studies on Large Branchiopod Biology, Hydrobiologia 486: 101–113.

Obregón-Barboza, H., A. M. Maeda-Martínez & G. Murugan, 2001. Reproduction, molting, and growth of two Mexican uniparental forms of the tadpole shrimp *Triops* (Branchiopoda: Notostraca), under a recirculating culture system. Hydrobiologia 462: 173–184.

Olesen, J., (1998). A phylogenetic analysis of the Conchostraca and Cladocera (Crustacea, Branchiopoda, Diplostraca). Zoological Journal of the Linnean Society 122: 491–536.

Remigio, E. A. & P. D. N. Hebert, 2000. Affinities among anostracan (Crustacea: Branchiopoda) families inferred from phylogenetic analyses of multiple gene sequences. Molecular Phylogenetic Evolution 17: 117–128.

Remigio, E. A., P. D. N. Hebert & A. Savage, 2001. Phylogenetic relationships and remarkable radiation in

Parartemia (Crustacea:Anostraca), the endemic brine shrimp of Australia: Evidence from mitochondrial DNA sequences. Biological Journal of Linnean Society 74: 59–71.

Remigio, E. A., B. V. Timms & P. D. N. Hebert, 2003. Phylogenetic systematics of the Australian fairy shrimp genus *Branchinella* based on mitochondrial DNA sequences. Journal of Crustacean Biology 23: 436–442.

Rogers, D. C., 1998. Aquatic macroinvertebrate occurrences and population trends in constructed and natural vernal pools in Folsom, California. In Witham, C. W., E. T. Bauder, D. Belk, W. R. Ferrin Jr. & R. Orduff (eds), Ecology, Conservation, and Management of Vernal Pool Ecosystems—Proceedings from a 1996 Conference. California Native Plant Society, Sacramento, CA, 224–235.

Rogers, D. C., 2001. Revision of the Nearctic *Lepidurus* (Notostraca). Journal of Crustacean Biology 21: 991–1006.

Rogers, D. C., 2002a. Amplexial morphology of selected Anostraca. Hydrobiologia 486: 1–18.

Rogers, D. C., 2002b. A morphological re-evaluation of the anostracan families Linderiellidae and Polyartemiidae, with a redescription of the linderiellid *Dexteria floridana* (Dexter 1956) (Crustacea: Branchiopoda). Hydrobiologia 486: 56–61.

Rogers, D. C., D. Quinney, J. Weaver & J. Olesen, 2006. A new giant species of predatory fairy shrimp from Idaho (Branchiopoda: Anostraca). Journal of Crustacean Biology, 26: 1–16.

Rogers, D. C., 2006. A genus level revision of the Thamnocephalidae (Crustacea: Branchiopoda: Anostraca). Zootaxa (in press).

Sanoamuang, L. A., G. Murugan, P. H. H. Weekers & H. J. Dumont, 2000. *Streptocephalus sirindhornae*, new species of freshwater fairy shrimp (Anostraca) from Thailand. Journal of Crustacean Biology 20: 599–565.

Sassaman, C., 1995. Sex determination and evolution of unisexuality in the Conchostraca. Hydrobiologia 298: 45–65.

Sassaman, C. & S. Weeks, 1993. Sex determination in the conchostracan *Eulimnadia texana*. The American Naturalist 141: 314–328.

Sassaman, C., M. A. Simovich & M. Fugate, 1997. Reproductive isolation and genetic differentiation in North American species of *Triops* (Crustacea: Branchiopoda: Notostraca). Hydrobiologia 359: 125–147.

Scanabissi, F., M. Cesari, S. K. Reed & S. C. Weeks, 2006. Ultrastructure of the male gonad and male gametogenesis in *Eulimnadia texana* (Packard, 1871) (Crustacea, Branchiopoda, Spinicaudata). Invertebrate Biology 125: 117–124.

Spears, T. & L. G. Abele, 2000. Branchiopod monophyly and interordinal phyogeny inferred from 18S ribosomal RNA. Journal of Crustacean Biology 20(1): 1–24.

Stenderup, J. T., J. Olesen & H. Glenner, 2006. Molecular phylogeny of the Branchiopoda (Crustacea)-Multiple approaches suggest a 'diplostracan' ancestry of the Notostraca. Molecular Phylogenetics and Evolution 41: 182–194.

Suno-Uchi, N., F. Sasaki, S. Chiba & M. Kawata, 1997. Morphological stasis and phylogenetic relationships in tadpole shrimps, *Triops* (Crustacea: Notostraca). Biological Journal of the Linnean Society 61: 439–457.

Takahashi, F., 1977. Pioneer life of the tadpole shrimps, *Triops* spp. (Notostraca: Triopsidae). Applied Journal of Entomolgy and Zoology 12: 104–117.

Tietze, N. S. & M. S. Mulla, 1991. Biological control of *Culex* mosquitoes (Diptera: Culicidae) by the tadpole shrimp, *Triops longicaudataus* (Notostraca: Triopsidae). Journal of Medical Entomology 28: 24–31.

Walossek, D., (1993). The Upper Cambrian *Rehbachiella* and the phylogeny of Branchiopoda and Crustacea. Fossils and Strata 32: 1–202.

Weekers, P. H. H., G. Murugan, J. Vanfleteren, D. Belk & H. J. Dumont, 2002. Phylogenetic analysis of anostracans (Branchiopoda: Anostraca) inferred from SSU rDNA sequences. Molecular Phylogenetics and Evolution 25: 535–544.

Weeks, S. C., R. T. Posgai, M. Cesari & F. Scanabissi, 2005. Androdioecy inferred in the clam shrimp *Eulimnadia agassizii* (Spinicaudata: Limnadiidae). Journal of Crustacean Biology 25: 323–328.

Weeks, S. C., C. Benvenuto & S. K. Reed, 2006a. When males and hermaphrodites coexist: a review of androdioecy in animals. Integrative and Comparative Biology 46: 449–464.

Weeks, S. C., T. F. Sanderson, S. K. Reed, M. Zofkova, B. Knott, U. Balaraman, G. Pereira, D. M. Senyo & W. R. Hoeh, 2006b. Ancient androdioecy in the freshwater crustacean *Eulimnadia*. Proceedings of the Royal Society of London, Series B 273: 725–734.

Hydrobiologia (2008) 595:177–184
DOI 10.1007/s10750-007-9013-5

Global diversity of cladocerans (Cladocera; Crustacea) in freshwater

L. Forró · N. M. Korovchinsky · A. A. Kotov · A. Petrusek

Abstract Cladocera is a primarily-freshwater monophyletic group, an important component of the microcrustacean zooplankton. They inhabit most types of continental fresh and saline water habitats, occurring more abundantly in both temporary and permanent stagnant waters. Cladocera is an ancient group of Palaeozoic origin. About 620 species are currently known, but we estimate that the real number of species is 2–4 times higher. A number of currently-recognised widespread species can be expected to harbour extensive cryptic diversity.

Keywords Cladocera · Species richness · Global assessment · Biogeography · Endemicity

Guest editors: E. V. Balian, C. Lévêque, H. Segers & K. Martens
Freshwater Animal Diversity Assessment

L. Forró (✉)
Department of Zoology, Hungarian Natural History Museum, Baross u. 13, 1088 Budapest, Hungary
e-mail: forro@zoo.zoo.nhmus.hu

N. M. Korovchinsky · A. A. Kotov
A.N. Severtsov Institute of Ecology and Evolution, Leninsky prospect 33, 119071 Moscow, Russia

A. Petrusek
Department of Ecology, Faculty of Science, Charles University in Prague, Viničná 7, 12844 Prague 2, Czechia

Introduction

Cladocerans ("water fleas") are primarily-freshwater small-sized (0.2–6 mm, and up to 18 mm in single case of *Leptodora kindtii*) branchiopod crustaceans, inhabiting pelagic, littoral, and benthic zones. Four cladoceran orders are recognised (Fryer, 1987): Anomopoda, Ctenopoda, Onychopoda, and the monotypic Haplopoda (see Fig. 1 A–D for representatives of each order). Most species occur in continental fresh or saline waters, although two ctenopods and several onychopods from the family Podonidae are truly marine, and a few more ctenopod, anomopod and onychopod species occur in brackish waters. Seven known species may be regarded as true inhabitants of subterranean environment, and a few others (of the family Chydoridae) live in semi-terrestrial conditions.

The trunk and appendages of most cladocerans (Anomopoda and Ctenopoda) are enclosed in a bivalved carapace. Tagmosis of the body is obscure (except in *Leptodora kindtii*, the single representative of Haplopoda), and a single eye and ocellus are usually present. Antennules are uniramous, while antennae are biramous (except in females of *Holopedium*), natatory, with 2–4 segments per branch. Four to six pairs of trunk limbs are either mostly similar in shape (Ctenopoda, Onychopoda, Haplopoda) or modified individually for various functions (Anomopoda).

Water fleas are important components of the fauna of fresh waters; they are particularly

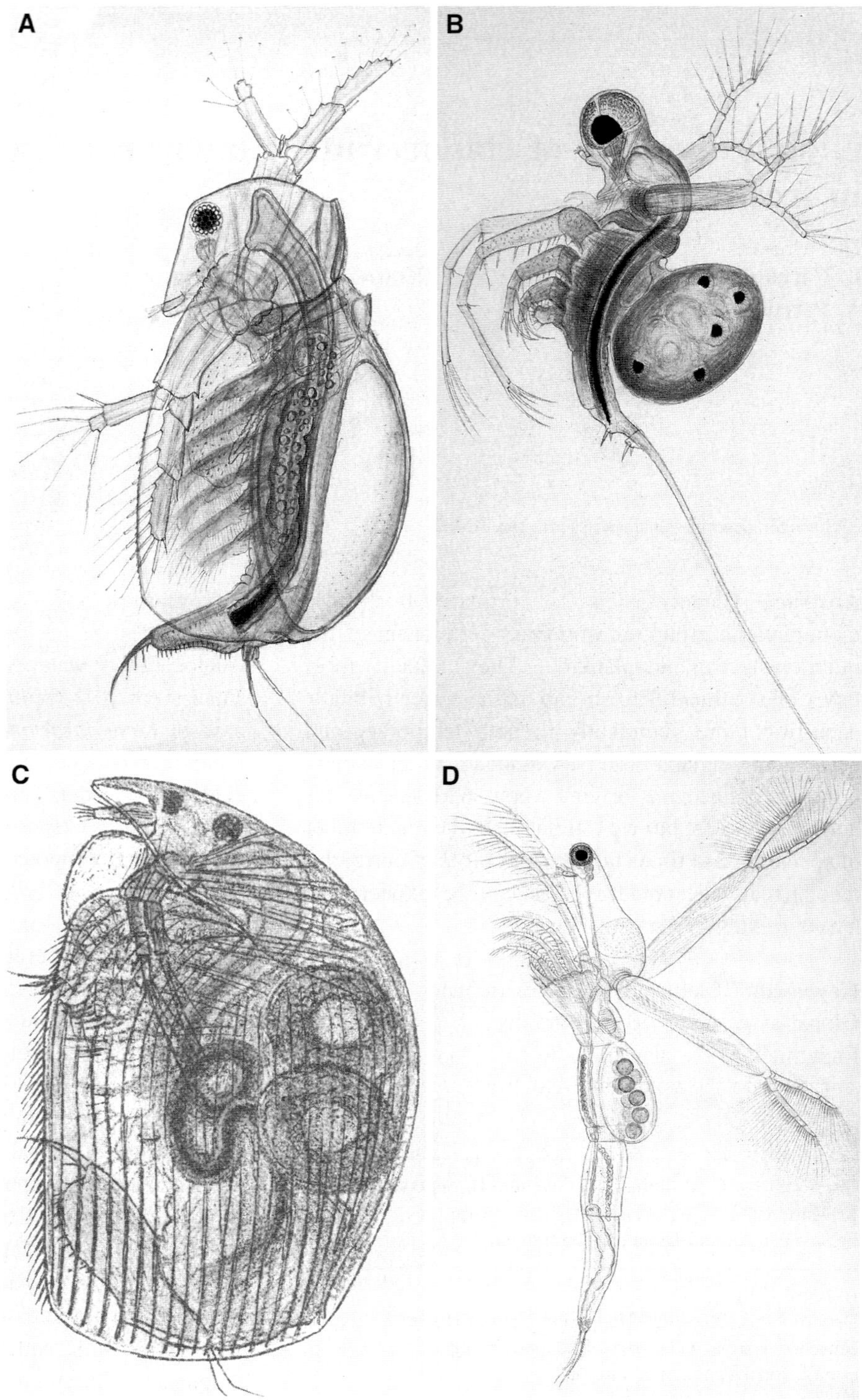

Fig. 1 A—*Sida crystallina* B—*Bythotrephes longimanus* C—*Alona* sp. D—*Leptodora kindti* (original drawings by G O Sars)

significant in the food web of stagnant waters. Most species are filter-feeders; onychopods and haplopods are predatory. They usually reproduce by cyclical parthenogenesis (but asexual lineages are known as well), and populations are mostly dominated by females. Sexual dimorphism is normally rather distinct. Sexually produced diapausing eggs are resistant to desiccation and other unfavourable

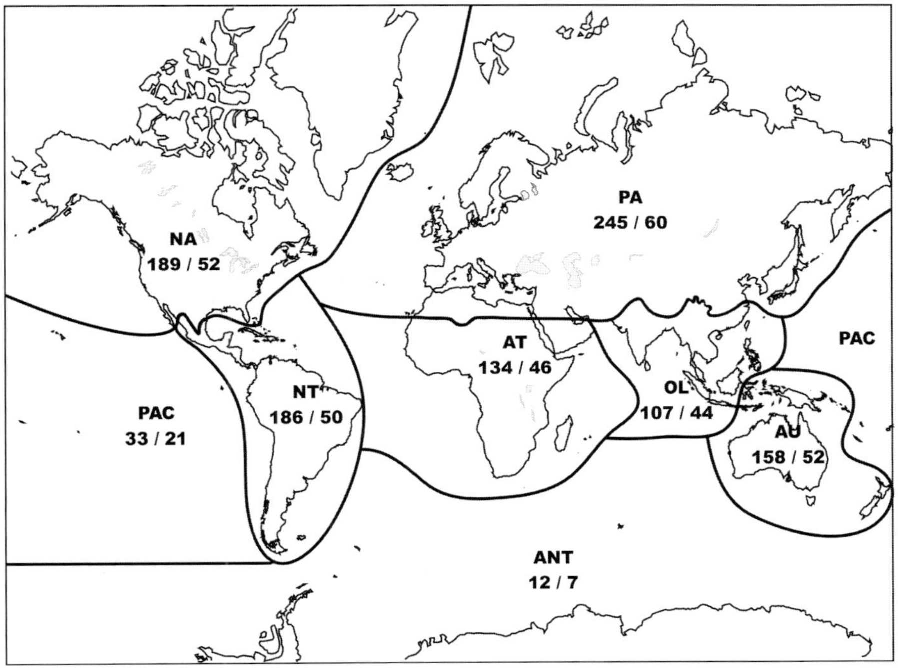

Fig. 2 Global Distribution of species and genus diversity by zoogeographic region (Species Number/ Genus Number). PA—Palaearctic; NA—Nearctic; NT—Neotropical; AT—Afrotropical ; OL—Oriental; AU—Australasian; PAC—Pacific Oceanic Islands; ANT—Antarctic

conditions, and may even survive passage through the digestive track of birds (Figuerola & Green, 2002); thus, they are important propagules for passive dispersal.

The first information on Cladocera date from the 17th century; the history of research has been divided into three to seven major phases (for a detailed discussion see Korovchinsky, 1997; Dumont & Negrea, 2002). An important change of paradigm, characterised by the rejection of the prevailing assumption of cosmopolitanism of cladoceran species, occurred around the 1950–1980s with a new approach to the taxonomy and phylogeny of Chydoridae (Frey, 1959, 1982, 1987). Subsequently, the concept of non-cosmopolitanism has been supported by numerous morphological, as well as molecular, studies. The increasing use of molecular tools in recent years has and will continue to have a strong impact on our understanding of cladoceran diversity, phylogeny and biogeography (e.g., Adamowicz et al., 2004; Cox & Hebert, 2001; Schwenk et al., 2000; Taylor et al., 2002).

Species diversity

The currently accepted number of cladoceran species based on existing descriptions is around 620. The tables summarize the currently known number of species and genera within orders and families of the group (Tables 1, 2, Fig. 2), based on recent major publications (Smirnov, 1992a, 1996; Korovchinsky, 1996, 2004; Orlova-Bienkowskaja, 2001; Dumont & Negrea, 2002; Benzie, 2005; Kotov & Stifter, 2006) and additional published or as yet unpublished sources. The described taxonomic diversity of Cladocera, however, underestimates the reality, and even higher-ranked taxa are still being discovered, e.g., a new family, Dumontiidae (Santos-Flores & Dodson, 2003).

Only about 45–50% of the species may be considered to be more or less well described and valid, while the status of other species is vague, and many of them likely represent cryptic complexes (Korovchinsky, 1996). The families Chydoridae, Daphniidae, Ilyocryptidae, and Sididae have been studied comparatively better. The largest number of valid species is known from Europe, North America, Australia, and South America, and the smallest number from Africa and Southern Asia. This, however, at least partly reflects the intensity of research rather than real patterns of diversity.

Adamowicz & Purvis (2005) estimated three correction factors to extrapolate global branchiopod diversity from the diversity of described species, and

Table 1 Number of Cladocera species currently known in the main biogeographic areas. PA: Palaearctic; NA: Nearctic; NT: Neotropical; AT: Afrotropical; OL: Oriental; AU: Australasian; PAC: Pacific Oceanic Islands; ANT: Antarctic. (numbers in parentheses indicate endemic species)

	PA	NA	NT	AT	OL	AU	PAC	ANT	World
Order Anomopoda	195 (83)	169 (66)	170 (89)	125 (24)	89 (20)	149 (78)	29 (0)	12 (6)	537
Family Daphniidae*	58 (21)	58 (25)	32 (13)	25 (1)	17 (1)	26 (13)	6 (0)	3 (2)	121
Family Moinidae*	13 (6)	7 (2)	10 (5)	10 (1)	3 (0)	7 (3)	4 (0)	0	29
Family Dumontiidae	0	1 (1)	0	0	0	0	0	0	1
Family Ilyocryptidae	11 (3)	10 (2)	9 (4)	8 (3)	5 (3)	5 (3)	1 (0)	1 (0)	28
Family Bosminidae	4 (0)	8 (3)	7 (3)	3 (0)	4 (1)	3 (0)	1 (0)	0	14
Family Acantholeberidae	1 (0)	1 (0)	0	0	0	0	0	0	1
Family Ophryoxidae	3 (1)	3 (1)	0	0	0	0	0	0	3
Family Macrothricidae	16 (10)	10 (5)	21 (12)	12 (2)	12 (4)	20 (9)	2 (0)	3 (1)	60
Family Neothricidae	0	0	0	0	0	3 (3)	0	0	3
Family Eurycercidae	4 (2)	5 (3)	2 (1)	1 (0)	1 (0)	0	0	0	8
Family Chydoridae	85 (40)	66 (24)	89 (51)	66 (17)	48 (11)	85 (47)	15 (0)	5 (3)	269
Order Ctenopoda	17 (5)	18 (7)	16 (9)	9 (0)	15 (4)	9 (5)	4 (1)	0	50
Family Holopediidae	1 (0)	2 (1)	1 (1)	0	0	0	0	0	3
Family Sididae	16 (5)	16 (6)	15 (8)	9 (0)	15 (4)	9 (5)	4 (1)	0	47
Order Haplopoda	1 (0)	1 (0)	0	0	1 (0)	0	0	0	1
Family Leptodoridae	1 (0)	1 (0)	0	0	0	0	0	0	1
Order Onychopoda	32 (31)	1 (0)	0	0	1 (0)	0	0	0	32
Family Polyphemidae	2 (1)	1 (0)	0	0	1 (0)	0	0	0	2
Family Podonidae	17 (17)	0	0	0	0	0	0	0	17
Family Cercopagidae*	13 (13)	0	0	0	0	0	0	0	13
Total	245 (119)	189 (73)	186 (98)	134 (24)	107 (24)	158 (83)	33 (1)	12 (6)	620

* Invasive species not considered

predicted that there are about 2.1 times more branchiopod species in nature than currently known. The overall cladoceran species richness is probably up to 4 times higher than currently known. This is supported by the results of molecular studies. Detailed studies, combining morphological analyses and molecular tools, are especially promising for delineating species boundaries in groups with relatively uniform morphology, fewer qualitative characters, and widespread phenotypic plasticity. Although most molecular analyses have so far focused on a single model genus, *Daphnia*, within a relatively short time this led to the discovery of an unprecedented number of cryptic lineages. According to Hebert & Taylor (1997), the global total for the genus *Daphnia* (including *Daphniopsis*) is likely closer to 200 species instead of 75 included in the last monograph on the genus (Benzie, 2005). Similar patterns of widespread cryptic diversity and high numbers of undescribed lineages can be seen in other groups, e.g., in *Moina* (Petrusek et al., 2004 and unpublished data), *Ilyocryptus* (Kotov & Štifter, 2006) and several genera of the Chydoridae.

Phylogeny and historical processes

Cladocerans probably derived from large bodied branchiopod ancestors. Recent molecular analyzes have suggested two alternative phylogenetic relationships among cladoceran orders. The monophyly of Gymnomera (Haplopoda and Onychopoda) is supported in both cases but one hypothesis suggests a sister relationship between Anomopoda and the remaining three orders (Swain & Taylor, 2003), while the other clusters Anomopoda and Ctenopoda together (De Waard et al., 2006). Other authors (see review in Negrea et al., 1999) have recently proposed alternative hypotheses on ordinal-level relationships

Table 2 Number of Cladocera genera currently known in the main biogeographic areas. PA: Palaearctic; NA: Nearctic; NT: Neotropical; AT: Afrotropical; OL: Oriental; AU: Australasian; PAC: Pacific Oceanic Islands; ANT: Antarctic. (numbers in parentheses indicate endemic genera)

	PA	NA	NT	AT	OL	AU	PAC	ANT	World
Order Anomopoda	44 (4)	43 (3)	44 (3)	42 (1)	36 (1)	48 (11)	19 (0)	7 (0)	76
Family Daphniidae	5 (0)	5 (0)	4 (0)	5 (0)	4 (0)	4 (0)	4 (0)	2 (0)	5
Family Moinidae	1 (0)	2 (0)	2 (0)	2 (0)	2 (0)	2 (0)	1 (0)	0	2
Family Dumontiidae	0	1 (1)	0	0	0	0	0	0	1
Family Ilyocryptidae	1 (0)	1 (0)	1 (0)	1 (0)	1 (0)	1 (0)	1 (0)	1 (0)	1
Family Bosminidae	2 (0)	2 (0)	2 (0)	2 (0)	2 (0)	2 (0)	1 (0)	0	2
Family Acantholeberidae	1 (0)	1 (0)	0	0	0	0	0	0	1
Family Ophryoxidae	1 (0)	2 (1)	0	0	0	0	0	0	2
Family Macrothricidae	6 (0)	6 (0)	7 (2)	5 (0)	4 (0)	5 (0)	1 (0)	1 (0)	11
Family Neothricidae	0	0	0	0	0	1 (1)	0	0	1
Family Eurycercidae	1 (0)	1 (0)	1 (0)	1 (0)	1 (0)	0	0	0	1
Family Chydoridae	26 (4)	22 (1)	27 (1)	26 (1)	22 (1)	33 (10)	11 (0)	3 (0)	49
Order Ctenopoda	7 (1)	7 (0)	6 (0)	4 (0)	6 (0)	4 (0)	2 (0)	0	8
Family Holopediidae	1 (0)	1 (0)	1 (0)	0	0	0	0	0	1
Family Sididae	6 (1)	6 (0)	5 (0)	4 (0)	6 (0)	4 (0)	2 (0)	0	7
Order Haplopoda	1 (0)	1 (0)	0	0	1 (0)	0	0	0	1
Family Leptodoridae	1 (0)	1 (0)	0	0	0	0	0	0	1
Order Onychopoda	8 (5)	1 (0)	0	0	1 (0)	0	0	0	10
Family Polyphemidae	1 (0)	1 (0)	0	0	1 (0)	0	0	0	1
Family Podonidae	5 (3)	0	0	0	0	0	0	0	**7**
Family Cercopagidae	2 (2)	0	0	0	0	0	0	0	2
Total	60 (10)	52 (3)	50 (3)	46 (1)	44 (1)	52 (11)	21 (0)	7 (0)	95

for the Cladocera; these were, however, based only on classical (Fryer, 1987) or cladistic analyses of morphological traits.

Cladocera is an ancient group of Palaeozoic origin (Dumont & Negrea, 2002), but their unambiguous fossil remains are known only from the Mesozoic (Smirnov, 1971, 1992b; Kotov & Korovchinsky, 2006). Recently, Anderson et al. (2004) described crustaceans similar to the Cladocera from the Early Devonian. Molecular phylogenetic data have revealed that the subfamilies of Chydoridae (Anomopoda) were separated in the Middle Palaeozoic (about 400 Myr ago; Sacherová & Hebert, 2003) and representatives of the genus *Daphnia* differentiated at least 200 Myr ago (Colbourne & Hebert, 1996). Any Mesozoic scenarios, such as 'Gondwana-Laurasia' (e.g., Benzie, 2005), are only moderately applicable to cladoceran groups, especially at a generic and subgeneric level. In spite of the general antiquity of Cladocera, radiation within some groups is only recent or even contemporary, e.g., in some Holarctic *Daphnia* and *Bosmina* (Colbourne & Hebert, 1996; Taylor et al., 2002).

Present distribution and main areas of endemicity

The distribution and patterns of endemicity of higher-level taxonomic groups are relatively well-known. The known species diversity, as well as the number of endemic taxa, is nevertheless bound to increase with further faunistic research, especially from non-northern temperate regions, and with the application of detailed morphological and molecular tools to resolve cryptic species complexes. Some endemic species have narrow distributions, and it is therefore likely that many remain overlooked.

The Holarctic cladoceran inland fauna is rich and composed of all four orders. Two orders (Haplopoda and Onychopoda), three families, 13 genera (including those of the Caspian Sea and Lake Baikal), and about 250 known species are endemic for the region. Many taxa are presumably old and phylogenetically divergent, monotypic, or composed of a few species. The Palaearctic taxa are more diverse than those of the Nearctic due to, first of all, the presence of numerous Caspian and Baikalian endemics. Among other zoogeographical regions, Australasia is rich in endemics, represented by one family, one subfamily, one tribe, 11 genera, and 83+ species, while known endemics in Oriental and Neotropical regions are of a lower rank or fewer (one tribe, one genus and 21+ species, and three genera and 98+ species, respectively). The Afrotropical region, though poorly studied, seems to be especially deprived of known higher-level endemic cladoceran taxa, being represented by a single endemic genus and 24+ endemic species.

Cladoceran species richness does not change evenly with latitude but concentrates in the warm temperate to subtropical zone of both hemispheres ($\sim$25–50°, including mountain areas within the true tropics) (Korovchinsky, 2006). In the belt from the Mediterranean through Central Asia including the Pontocaspian region, northern India to East Asia (Amur region and China), five genera and over 100 known endemic species occur, while those in the North Palaearctic do not exceed 55–60. In North America, the area embracing the United States, Mexican plateau, and southern Canada, is inhabited by many endemics, including one family (Dumontiidae), two genera, and over 70 species. Southern Australia, Tasmania, and New Zealand are rich in endemics of high taxonomic rank: one subfamily (Sayciinae), one tribe (Australospilini), 8 genera, and about 80 currently known species compose altogether most of the known Australasian endemics. Only five of them are shared between Australia and New Zealand, which itself has seven endemic species/subspecies. As the African cladoceran fauna has been relatively poorly studied, there are only a few described endemics (one genus and about 10 species), but e.g., endemic *Daphnia* fauna of the Ethiopian biogeographic region is certainly significantly more numerous (Mergeay et al., unpublished). Cladocerans of subtropical and temperate South America include two endemic genera and 17+ endemic species. In total, the species richness of the southern temperate—subtropical zone amounts to more than 100 endemic species. The intermediate tropical zone, from which altogether 163 species are known, is characterized by fewer endemic taxa of comparatively lower taxonomic rank: one tribe (Indialonini) and nine genera (Korovchinsky, 2006).

A bipolar (antitropical) disjunct distribution of faunal complexes and taxa (*Daphnia, Pleuroxus, Tretocephala* etc.), the wide ranges of some species (though some of these likely form species complexes) and the narrow restriction of others, the presence of isolated populations, and concentration of endemics in the warm temperate—subtropical zone of both hemispheres, are typical traits of cladoceran zoogeography.

Such patterns stimulated the analysis of cladoceran faunal formation by the modern version of the concept of 'ejected relicts' instead of vicariance. This hypothesis consider the extant Cladocera as a relict group (Korovchinsky, 2006), whose taxa were widely distributed in the past. Tertiary climatic changes, primarily within the present tropical and boreal latitudes, resulted in mass extinction of their biotas, while the warm temperate—subtropical regions remained comparatively unchanged. Additional factors (e.g., the radiation of freshwater planktivorous fish) could have operated in conjunction with climate changes as well.

While this scenario might be likely for most of cladocerans, molecular data suggest that vicariance processes and allopatric speciation at both the intercontinental level and in regional refugia within continents plays a significant role in shaping species diversity in at least some genera (e.g., *Daphnia*). Sweepstake intercontinental dispersals, followed by a local radiation, seems to have been important factors in augmenting the diversity in the different biogeographic regions. Founder effects coupled with habitat shifts, such as pond-lake transitions (Lynch, 1985) or, possibly, shifts among substrates in littoral groups, are also regarded as potentially important drivers of speciation. Finally, interspecific hybridization and hybrid speciation plays an important role in dynamic young species complexes in *Daphnia*, though reports of other hybridizing cladocerans are scarce (Schwenk & Spaak, 1995).

Human related issues

Cladocerans (especially *Daphnia*) are important model organisms in both basic and applied research, due to their easy culturing, short generation time, and clonal reproduction. Species of *Daphnia* have been widely used in ecological and evolutionary studies (e.g., on trophic interactions, diel vertical migration, interspecific hybridisation, polyploidy and asexuality, host-parasite interactions etc.), and the soon to be available sequence of the whole *Daphnia pulex* s.l. genome will open further research possibilities in genomics and other fields. Cladocerans have also gained certain economic importance as they are also widely used in aquaculture, and large filter-feeding planktonic species have an indirect economic impact as important fish food or phytoplankton-controlling group. These animals as intermediate hosts of some parasites may potentially pose a threat to human health.

A high diversity of cladocerans can be found in the littoral zone of stagnant waters, as well as in temporary water bodies. These habitats are often negatively influenced by human activities, and especially the loss of temporary waters may lead to a decrease of diversity or even local extinction of some species.

Some cladocerans have recently invaded successfully other continents through human-mediated dispersal, and it is likely that this trend will increase. For example, non-indigenous species of *Daphnia* are widespread in Europe, North America or Africa (e.g., Havel et al., 1995; Mergeay et al., 2005), though mostly without a strong ecological impact. The invasion of predatory onychopods (especially *Bythotrephes*) from the Palaearctic into the Laurentian Great Lakes and those of the Canadian Shield, however, have influenced the native fauna significantly (Yan et al., 2002).

Acknowledgements The following funding sources partially supported our research: "Biodiversity" Program of Russian Academy of Sciences (grant 1.1.5) and Russian Foundation for Basic Research (grant 06–04-48624 for AAK and NMK), Czech Ministry for Education (MSM0021620828 supporting AP) and the Hungarian National R&D Programme (contract No: 3B023-04 supporting LF). We would like to thank Nikolai N. Smirnov, Vladimír Kořínek, Dorothy Berner, and Joachim Mergeay for sharing of unpublished data and/or valuable comments on the manuscript and David Hardekopf for language corrections. We are also grateful to an anonymous referee for reading an earlier version of the paper and Sarah Adamowicz for her review.

References

Adamowicz, S. J., P. D. N. Hebert & M. C. Marinone, 2004. Species diversity and endemism in the *Daphnia* of Argentina: a genetic investigation. Zoological Journal of the Linnean Society 140: 171–205.

Adamowicz, S. J. & A. Purvis, 2005. How many branchiopod crustacean species are there? Quantifying the components of underestimation. Global Ecology and Biogeography 14: 455–468.

Anderson, L. I., W. R. B. Crighton & H. Hass, 2004. A new univalve crustacean from the Early Devonian Rhynie chert hot-spring complex. Transactions of the Royal Society of Edinburgh: Earth Sciences 94: 355–369.

Benzie, J. A. H., 2005. The genus *Daphnia* (including *Daphniopsis*) (Anomopoda: Daphniidae). In: Dumont H. J. (ed.), Guides to the Identification of the Microinvertebrates of the Continental Waters of the World 21. Kenobi Productions, Ghent & Backhuys Publishers, Leiden.

Colbourne, J. K. & P. D. N. Hebert, 1996. The systematics of North American *Daphnia* (Crustacea: Anomopoda): a molecular phylogenetic approach. Philosophical Transactions of the Royal Society of London, Series B 351: 349–360.

Cox, A. J. & P. D. N. Hebert, 2001. Colonization, extinction, and phylogeographic patterning in a freshwater crustacean. Molecular Ecology 10: 371–386.

DeWaard, J., V. Sacherová, M. E. A. Cristescu, E. A. Remigio, T. J. Crease & P. D. N. Hebert, 2006. Probing the relationships of the branchiopod crustaceans. Molecular Phylogenetics and Evolution 39: 491–502.

Dumont, H. J. & S. V. Negrea, 2002. Introduction to the class Branchiopoda. In: Dumont H. J. (ed.), Guides to the Identification of the Microinvertebrates of the Continental Waters of the World 19. Backhuys Publishers, Leiden.

Figuerola, J. & A. J. Green, 2002. Dispersal of aquatic organisms by waterbirds: a review of past research and priorities for future studies. Freshwater Biology 47: 483–494.

Frey, D. G., 1959. The taxonomic and phylogenetic significance of the head pores of the Chydoridae (Cladocera). Internationale Revue der gesamten Hydrobiologie 44: 27–50.

Frey, D. G., 1982. Questions concerning cosmopolitanism in Cladocera. Archiv für Hydrobiologie 93: 484–502.

Frey, D. G., 1987. The taxonomy and biogeography of the Cladocera. Hydrobiologia 145: 5–17.

Fryer, G., 1987. Morphology and the classification of the so-called Cladocera. Hydrobiologia 145: 19–28.

Havel, J. E., W. R. Mabee & J. R. Jones, 1995. Invasion of the exotic cladoceran *Daphnia lumholtzi* into North American reservoirs. Canadian Journal of Fisheries and Aquatic Sciences 52: 151–160.

Hebert, P. D. N. & D. J. Taylor, 1997. The future of cladoceran genetics: methodologies and targets. Hydrobiologia 360: 295–299.

Korovchinsky, N. M., 1996. How many species of Cladocera are there? Hydrobiologia 321: 191–204.

Korovchinsky, N. M., 1997. On the history of studies on cladoceran taxonomy and morphology, with emphasis on early work and causes of insufficient knowledge of the diversity of the group. Hydrobiologia 360: 1–11.

Korovchinsky, N. M., 2004. Cladocerans of the Order Ctenopoda of the World Fauna (Morphology, Systematics, Ecology, Biogeography). KMK Publishers, Moscow [In Russian].

Korovchinsky, N. M., 2006. The Cladocera (Crustacea: Branchiopoda) as a relict group. Zoological Journal of the Linnean Society 147: 109–124.

Kotov, A. A. & N. M. Korovchinsky, 2006. First record of fossil Mesozoic Ctenopoda (Crustacea, Cladocera). Zoological Journal of the Linnean Society 146: 269–274.

Kotov, A. A. & P. Štifter, 2006. Ilyocryptidae of the World. In: Dumont H. J. (ed.), Guides to the Identification of the Microinvertebrates of the Continental Waters of the World 22. Kenobi Productions, Ghent & Backhuys Publishers, Leiden.

Lynch, M., 1985. Speciation in the Cladocera. Verhandlungen der internationalen Vereinigung für theoretische und angewandte Limnologie 22: 3116–3123.

Mergeay, J., D. Verschuren & L. De Meester, 2005. Cryptic invasion and dispersal of an American *Daphnia* in East Africa. Limnology & Oceanography 50: 1278–1283.

Negrea, S., N. Botnariuc & H. J. Dumont, 1999. Phylogeny, evolution and classification of the Branchiopoda (Crustacea). Hydrobiologia 412: 191–212.

Orlova-Bienkowskaja, M. Y., 2001. Daphniidae: genus *Simocephalus*. In: Dumont H. J. (ed.), Guides to the Identification of the Microinvertebrates of the Continental Waters of the World 17. Backhuys Publishers, Leiden.

Petrusek, A., M. Černý & E. Audenaert, 2004. Large intercontinental differentiation of *Moina micrura* (Crustacea: Anomopoda): one less cosmopolitan cladoceran?. Hydrobiologia 526: 73–81.

Sacherová, V. & P. D. N. Hebert, 2003. The evolutionary history of the Chydoridae (Crustacea: Cladocera). Biological Journal of the Linnean Society 79: 629–643.

Santos-Flores, C. J. & S. I. Dodson, 2003. *Dumontia oregonensis* n. fam., n. gen., n. sp., a cladoceran representing a new family of 'water-fleas' (Crustacea, Anomopoda) from U.S.A., with notes on the classification of the Order Anomopoda. Hydrobiologia 500: 145–155.

Schwenk, K. & P. Spaak, 1995. Evolutionary and ecological consequences of interspecific hybridization in cladocerans. Experientia 51: 465–481.

Schwenk, K., D. Posada & P. D. N. Hebert, 2000. Molecular systematics of European *Hyalodaphnia*: the role of contemporary hybridization in ancient species. Proceedings of the Royal Society of London, Series B 267: 1833–1842.

Smirnov, N. N., 1971. New species of the genus *Archedaphnia* from the Jurassic deposits of Transbaikalia. Paleontologicheskij Zhurnal 3: 119–121 [In Russian].

Smirnov, N. N., 1992a. The Macrothricidae of the World. Guides to the Identification of the Microinvertebrates of the Continental Waters of the World 1. SPB Academic Publishers, The Hague.

Smirnov, N. N., 1992b. Mesozoic Anomopoda (Crustacea) from Mongolia. Zoological Journal of the Linnean Society 104: 97–116.

Smirnov, N. N., 1996. Cladocera: the Chydorinae and Sayciinae (Chydoridae) of the world. In: Dumont H. J. (ed.), Guides to the Identification of the Microinvertebrates of the Continental Waters of the World 11. SPB Academic Publishers, Amsterdam.

Swain, T. D. & D. J. Taylor, 2003. Structural rRNA characters support monophyly of raptorial limbs and paraphyly of limb specialization in water fleas. Proceedings of the Royal Society of London, Series B 270: 887–896.

Taylor, D. J., C. R. Ishikane & R. A. Haney, 2002. The systematics of Holarctic bosminids and a revision that reconciles molecular and morphological evolution. Limnology and Oceanography 47: 1486–1495.

Yan, N. D., R. Girard & S. Boudreau, 2002. An introduced invertebrate predator (*Bythotrephes*) reduces zooplankton species richness. Ecology Letters 5: 481–485.

Hydrobiologia (2008) 595:185–193
DOI 10.1007/s10750-007-9245-4

Global diversity of ostracods (Ostracoda, Crustacea) in freshwater

Koen Martens · Isa Schön · Claude Meisch · David J. Horne

Abstract There are close to 2,000 subjective species and about 200 genera of Recent non-marine Ostracoda. Together, Cyprididae (1,000 spp.) and Candonidae (c. 550 spp.) represent more than 75% of the extant specific diversity; the remaining 11 families comprise the other 25% of the species. The Palaearctic region has the highest absolute non-marine ostracod diversity, followed by the Afrotropical. The Australian region has the highest relative endemicity. About 90% of the species and 60% of the genera occur in one zoogeographical region only. This means that all the biological mechanisms which lead up to efficient dispersal and which are present in at least part of the non-marine Ostracoda (e.g. brooding, drought-resistant eggs, parthenogenesis) have not induced common cosmopolitan distributions in ostracods. Several habitats are hotspots for ostracod diversity and endemicity. For example, it appears that the ancient lakes hold up to 25% of the total ostracod diversity. Other speciation-prone habitats are groundwater, temporary pools and Australian salt lakes; in the latter two instances, cladogenesis has often been paralleled by gigantism. The present ostracod diversity results from 9 to 12 separate invasions of the non-marine habitat, starting about 400 Myr ago. Genetic diversity can be very different in different species, mostly, but not always, related to reproductive mode.

Keywords Ostracoda · Freshwater · Species · Genera · Ancient lakes

Guest editors: E. V. Balian, C. Lévêque, H. Segers & K. Martens
Freshwater Animal Diversity Assessment

K. Martens (✉) · I. Schön
Freshwater Biology, Royal Belgian Institute of Natural Sciences, Vautierstraat 29, 1000 Brussels, Belgium
e-mail: martens@naturalsciences.be

K. Martens
University of Ghent, Ghent, Belgium

C. Meisch
Musée national d'Histoire Naturelle, Luxembourg, Luxembourg

D. J. Horne
Department of Geography, Queen Mary, University of London, London, UK

Introduction

Mussel-shrimps, or Ostracoda, are small, bivalved Crustacea. Their calcified carapaces have an average length of c. 1 mm and completely envelop the reduced body (Fig. 1). Ostracods are very common in most inland waters, where they abound in the benthic and periphytic animal communities, but they also occur in marine, interstitial and even (semi-) terrestrial environments. Ostracoda are of great interest as a model group in various ecological and evolutionary studies. This is mainly so because the

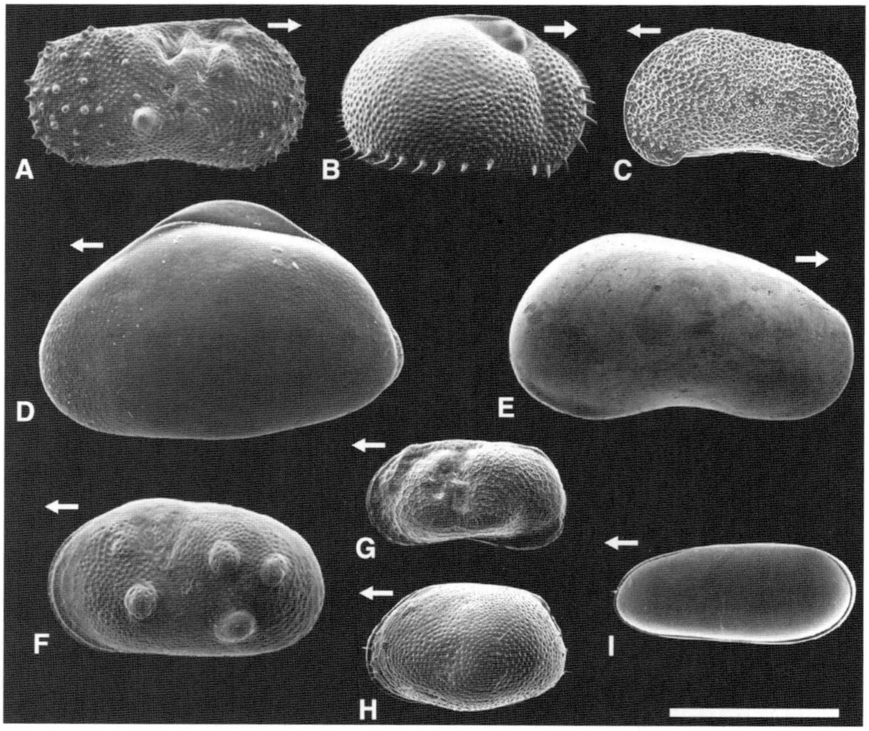

Fig. 1 External views of carapaces of main extant non-marine ostracod groups. A, *Ilyocypris* (Ilyocyprididae, Cypridoidea); B, *Centrocypris* (Notodromadidae, Cypridoidea); C, *Potamocypris* (Cyprididae, Cypridoidea); D, *Cyprinotus* (Cyprididae, Cypridoidea); E, *Candona* (Candonidae, Cypridoidea); F, *Cyprideis* (Cytherideidae, Cytheroidea); G, *Limnocythere* (Limnocytheridae, Cytheroidea); H, *Metacypris* (Limnocytheridae, Cytheroidea); I, *Darwinula* (Darwinulidae, Darwinuloidea). Scale bar = 0.5 mm

calcified valves of non-marine ostracods can be very common in lake sediments and this adds a real-time frame to the evolution of ostracod lineages as well as of their biological traits. At present, ostracods are popular model organisms for research on the evolution of reproductive modes (Martens, 1998a) and as proxies for climate and ecosystem changes (Holmes & Chivas, 2002).

There are three main lineages of recent non-marine ostracods (Fig. 2), all belonging to the Podocopida. All three originated in the Palaeozoic and are between 450 and 360 million years old. The Cytheroidea are mostly marine, but have several non-marine incursions of which the Limnocytheridae are the most common. The Darwinuloidea, with one extant family, are fully non-marine, but have only about 30 extant species. The largest group, the Cypridoidea, comprises 4 families.

In spite of their general presence in aquatic habitats, there still appears to be some aversion towards the study of Ostracoda, when compared to other meiobenthic crustaceans. This has two main reasons. Firstly, correct specific and even generic identification of ostracods generally requires a full dissection. To acquire the skills to do this properly can easily take several months. Secondly, there are almost no identification books or illustrated specific keys (with few exceptions, mostly e.g. Meisch, 2000 for western Europe), so that identification of animals from most zoogeographical regions requires a full set of copies of all (original) descriptions, which are often difficult to track down.

The present article sets out to analyse the extant specific and generic diversity of non-marine Ostracoda, based on literature reviews. However, there are still large numbers of undescribed species, either because certain regions have been unexplored (e.g. the Amazon floodplain), or because known endemic faunas have not yet been described (e.g. Lake Malawi, from which dozens of new, but thus far undescribed, species are known, Martens, 2003), or because many cryptic species remain unrecognized

Fig. 2 Origin of the main ostracod lineages in the Podocopida. Asterisks indicate the three lineage with non-marine radiations. Zenker/no Zenker refers to presence or absence of Zenker organ in males. (After Martens, 1998a)

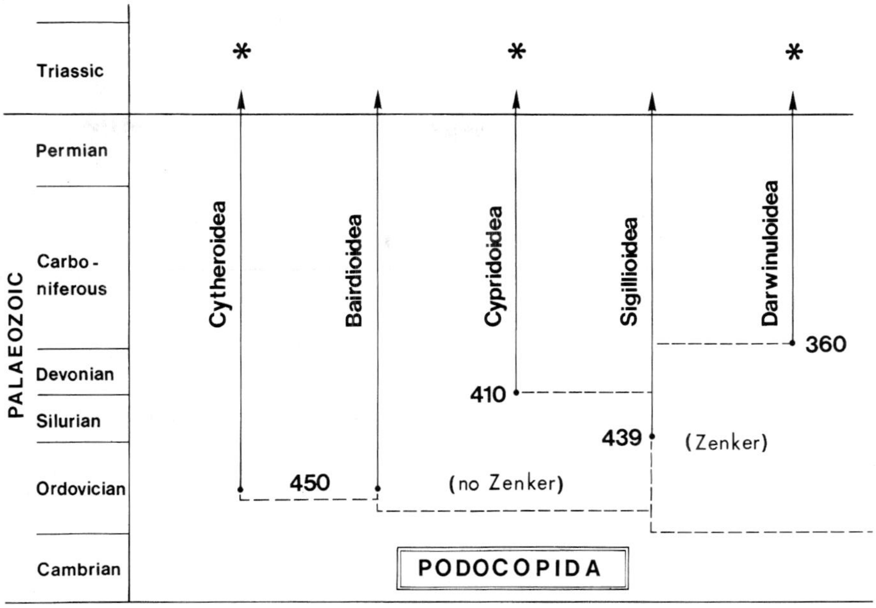

(e.g. in the Lake Baikal *Cytherissa* species flock, Schön & Martens, unpublished). The present survey is therefore only a snapshot in time.

The present data are compiled from the Cologne Database (Kempf, 1980a, b, 1991a, b and subsequent unpublished up dates). These lists include all Phanerozoic non-marine ostracod genera and species, both living and fossil. Moreover, these lists are fully objective, i.e. all published combinations of generic and specific names are included, while also synonymies are listed independently. Extraction of the information needed for the present article, therefore, required four steps. Firstly, Recent taxa had to be separated from fully fossil ones, as the present analyses deal with Recent (extant) taxa only. Secondly, the objective nomenclatorial lists had to be converted into subjective lists, i.e. lists with *real* number of described species in the most recent nomenclature. This is a cumbersome and continuously ongoing process, for which a good deal of the literature has to be consulted, unless revisions and/or checklists are available (e.g. Meisch, 2000 for western Europe, Martens, 1984 for Africa and Martens & Behen, 1994 for South America). Thirdly, the distribution of these subjective taxa over the different zoogeographical regions needed to be plotted. Again, primary literature had to be consulted, but this survey will require constant updating. Finally, rates of endemicity were determined. Endemic here means that the species/genus occurs in one zoogeographical region only. Endemicity can be much narrower (e.g. occurring in one lake only), but for the present analysis the unit of endemicity is the zoogeographical region.

Several other caveats exist:

1. Several (semi-) terrestrial species of Ostracoda are known, which either occur in leaf-litter, or in mosses in splash zones of waterfalls. Although some of these taxa have meanwhile also been found in fully lacustrine conditions (e.g. *Terrestricythere*), we here classify them as 'limno-terrestrial' (see Balian et al., this volume).
2. Also in view of the agreements set for all the chapters in the present volume, non-free-living species were not included in the surveys. Therefore, the c. 200 species of the cytheroid family Entocytheridae were not included in the present list. Entocytheridae are parasitic or commensal on gills and other body parts of other crustaceans, such as Isopoda, Amphipoda and crayfish; they mostly occur in the Holarctic, with highest diversity in the Nearctic, and in the northeastern part of the Neotropical region.
3. Karanovic (2007) described several dozens new species and several new genera from groundwater of the Pilbara (NW Australia). As this document was not available to us during the

present research, these taxa are not included here.

4. Seemingly ad hoc synonyms (e.g. for the candonids of North America) were not included. At a time when increasing numbers of cryptic species are discovered, it seems unwise to lump existing taxa together without checking type specimens, which in many cases are available in public museums.

Diversity and endemicity of non-marine Ostracoda

Species and generic diversity

There are 1,936 subjective aquatic species of extant non-marine Ostracoda species (and 12 limno-terrestrial species—Table 1) and about 189 aquatic genera (and 5 limno-terrestrial—Table 2). Both in species and in genera, the family Cyprididae as it stands today (i.e. including the Cypridopsinae) takes up about half of diversity, with the Candonidae taking about 25% of the total diversity. Of the 11 other families, only the Limnocytheridae with c. 10% of total diversity of genera and less of specific diversity can also be called speciose. All other families are limited to smaller numbers of genera and species.

At a specific level, nearly all families, including the large Cyprididae and Candonidae, have endemicity rates of around 90%, meaning that only about a tenth of all species have intercontinental distributions. Calculated over all known species, close to 94% of all species are thus far known from one zoogeographical region only.

At the generic level, endemicity is of course lower, with about 60% of the genera occurring in one zoogeographical region only. In Cyprididae, c. 60% of all genera are endemic, in Candonidae and in Limnocytheridae almost 75%. Most families occur in all zoogeographical regions, except for Notodromadidae which have thus far not been recorded with certainty from the Neotropical region (Tables 1, 2). The hotspot of diversity of this group is without any doubt in the Oriental region. Darwinulidae have several genera and species with intercontinental

Table 1 Total number of species (endemic species between brackets) of extant non-marine Ostracoda in the zoogeographical provinces

Species	PA	NA	NT	AT	OL	AU	PAC	ANT	World
Cyprididae	206 (163)	154 (101)	169 (137)	317 (292)	154 (132)	106 (99)	3 (0)	2 (2)	998 (926)
Candonidae	333 (306)	101 (74)	40 (36)	52 (52)	17 (14)	35 (35)	0	0	545 (517)
Ilyocyprididae	27 (22)	3 (0)	2 (0)	1 (1)	8 (4)	3 (2)	0	0	33 (29)
Notodromadidae	5 (2)	3 (0)	3 (3)	12 (10)	15 (12)	5 (4)	0	0	36 (31)
Darwinulidae	6 (3)	3 (1)	12 (7)	9 (6)	4 (2)	9 (6)	1 (0)	0	29 (25)
Limnocytheridae	34 (32)	29 (25)	25 (23)	45 (44)	1 (1)	14 (14)	1 (1)	1 (1)	144 (141)
Cytherideidae	60 (58)	8 (4)	10 (8)	19 (18)	0	3 (2)	0	0	93 (90)
Leptocytheridae	19 (19)	1 (1)	0	0	0	1 (1)	0	0	21 (21)
Xestoleberidae	3 (2)	1 (1)	2 (1)	0	0	0	0	0	6 (4)
Cytheruridae	4 (4)	2 (2)	14 (14)	0	0	0	0	0	20 (20)
Loxoconchidae	4 (3)	1 (0)	0	0	0	0	0	0	4 (3)
Hemicytheridae	1 (1)	0	0	0	0	0	0	0	1 (1)
Incertae sedis (*Romeis*)	1 (1)	0	0	0	0	0	0	0	1 (1)
Total	702 (620)	298 (211)	275 (232)	455 (424)	199 (165)	176 (163)	5 (1)	3 (3)	1,936 (1819)
Limno-terrestrial									
Cyprididae	1 (1)	0	2 (2)	4 (4)	0	3 (3)	0	0	10 (10)
Candonidae	1 (1)	0	3 (3)	1 (1)	0	0	0	0	5 (5)
Terrestricytheridae	4 (3)	0	0	0	0	0	2 (0)	0	4 (2)

PA: Palaearctic, NA: Nearctic, NT: Neotropical, AT: Afrotropical, OL: Oriental, AU: Australasian; PAC: Pacific Oceanic Islands, ANT: Antarctic

Table 2 Total number of genera (endemic genera between brackets) of extant non-marine Ostracoda in the zoogeographical provinces

Genera	PA	NA	NT	AT	OL	AU	PAC	ANT	World
Cyprididae	41 (10)	30 (3)	21 (2)	45 (21)	28 (4)	31 (16)	2 (0)	2 (0)	94 (56)
Candonidae	16 (6)	11 (3)	9 (3)	8 (4)	7 (1)	15 (12)	0	0	39 (29)
Ilyocyprididae	1 (0)	1 (0)	1 (0)	1 (0)	1 (0)	1 (0)	0	0	1 (0)
Notodromadidae	3 (0)	3 (0)	2 (0)	3 (1)	5 (1)	2 (0)	0	0	8 (2)
Darwinulidae	4 (0)	3 (0)	5 (0)	4 (0)	4 (0)	3 (0)	1 (0)	0	5 (0)
Limnocytheridae	10 (6)	3 (0)	6 (4)	6 (2)	1 (1)	3 (1)	1 (0)	1 (0)	19 (14)
Cytherideidae	2 (0)	2 (0)	1 (0)	6 (5)	0	1 (0)	0	0	7 (5)
Leptocytheridae	1 (0)	1 (0)	0	0	0	1 (0)	0	0	1 (0)
Xestoleberidae	1 (0)	1 (0)	1 (0)	0	0	0	0	0	1 (0)
Cytheruridae	2 (1)	1 (0)	3 (1)	0	0	0	0	0	4 (2)
Loxoconchidae	3 (2)	1 (0)	0	0	0	0	0	0	3 (2)
Hemicytheridae	1 (1)	0	0	0	0	0	0	0	1 (1)
Incertae sedis (*Romeis*)	1 (1)	0	0	0	0	0	0	0	1 (1)
Total	87 (27)	57 (6)	55 (11)	73 (34)	46 (7)	57 (29)	4 (0)	3 (0)	189 (114)
Limno-terrestrial									
Cyprididae	1 (0)	0	1 (0)	1 (1)	0	1 (1)	0	0	3 (2)
Candonidae	1 (0)	0	2 (1)	1 (1)	0	0	0	0	3 (2)
Terrestricytheridae	1 (0)	0	0	0	0	0	1 (0)	0	1 (0)

PA: Palaearctic, NA: Nearctic, NT: Neotropical, AT: Afrotropical, OL: Oriental, AU: Australasian; PAC: Pacific Oceanic Islands, ANT: Antarctic

distribution, which is surprising for a group without drought-resistant eggs, but the age of the group (see below) and the reproductive mode might at least partly explain this.

Genetic diversity

Little information is available on genetic diversity in non-marine Ostracoda. Rossi et al. (1998) found more than 200 allozyme clones for *Eucypris virens* in Europe, while Schön et al. (2000) found 18% divergence in the mitochondrial COI marker for the same species in a similar number of European populations. The same authors found a much lower genetic variability in *Darwinula stevensoni*: only 7 clones and 3.5% divergence, respectively. These and subsequent authors have correlated this discrepancy with reproductive modes: *E. virens* is a species with mixed reproduction (both sexual and parthenogenetic females exist) with high-standing clonal variability and the ability to generate new genetic variability through (mainly intraspecific) hybridization between males and asexual females. There is also a high incidence of polyploidy in *E. virens*. *Darwinula stevensoni*, on the other hand, is a putative ancient asexual which has an almost identical genotype from northern Europe to South Africa. The presence of several reproductive modes makes Ostracoda an excellent model group for the study of one of the main evolutionary questions: the paradox of sex.

Phylogeny and historical processes

Marine ostracods invaded non-marine habitats from different lineages and at various times. According to the summary of Martens & Horne (in press), the Darwinulidae (one of the least diverse groups to date) most likely are the oldest living non-marine group, as they invaded non-marine habitats in the Devonian (c. 400–370 Myr ago). Limnocytheridae and Cytherideidae followed at the end of the Permian (c. 250 Myr). All four Cypridoidean families most likely invaded non-marine habitats sometime in the Mid to Late Jurassic (c. 175–150 Myr), but it is at present not at all clear if this involved one common ancestor, or if the four lineages derived from different invasions from different marine ancestors. It is thought that the other families, which mostly are largely marine and all cytheroidean (e.g. Xestoleberidae, Leptocytheridae, Loxoconchidae, etc.), only recently invaded non-marine habitats, e.g. in the Neogene or even in the Quaternary (Horne, 2003). Finally, the history of the limno-terrestrial species is unclear as no fossil record is available here. So, depending on whether or not cypridoidean colonization was by 1 or 4 ancestors, there have been between 9 and 12 independent incursions from marine to non-marine habitats during the past 400 million years.

Present distribution

The Zoogeographical Region with the highest specific diversity is the Palaearctic (Fig. 3), with more than 700 species, 80% of these are endemic (of which c. 200 are endemic to Lake Baikal). The Nearctic has only about 300 species, of which 71% are endemic, but c. 200 largely endemic species of the Entocytheridae are not included here, as they are all parasitic/commensal. Nearctic and Palaearctic together hold almost half of all described species. There is also a discrepancy between the Afrotropical (with 455 species (93% endemic) and the Neotropical with c. 275 species (84% endemic). The 176 species (92% endemic) of the Australian region and the c. 200 species of the Oriental region (OL—83% endemic) are with certainty a gross underestimation of the actual diversity. Antarctica and the Pacific regions are virtually unknown and will not further be discussed here.

At the generic level (Fig. 3), the picture is largely congruent. Again, the highest levels of endemicity are in Palaearctic (87, 30% endemic), whereas the Nearctic has only 6 endemic out of 57 genera (c. 10%). Both the Neotropical and the Oriental region have few endemic genera, while the Australian region has the highest percentage of endemic genera (29 out of 57 or about 50%). In the Afrotropical region, about half of the 73 genera are endemic and this number is almost certain to increase through further studies because of the high diversity of the Cypridopsinae.

Cyprididae are most common in the Afrotropical region (Fig. 4). Of the 25 subfamilies (including 3 presently unnamed), 5 are monospecific. Of the remaining 20, the 5 most speciose are the Eucypridinae (78 species mostly in Palaearctic), the Cyprinotinae (129 species, mostly in Afrotropical region), the Herpetocypridinae (151 species, mostly in Palaearctic and Neotropical), the Cypricercinae (171 species, mostly in Afrotropical region and Neotropical) and the Cypridopsinae (202 species, almost half of these in the Afrotropical region).

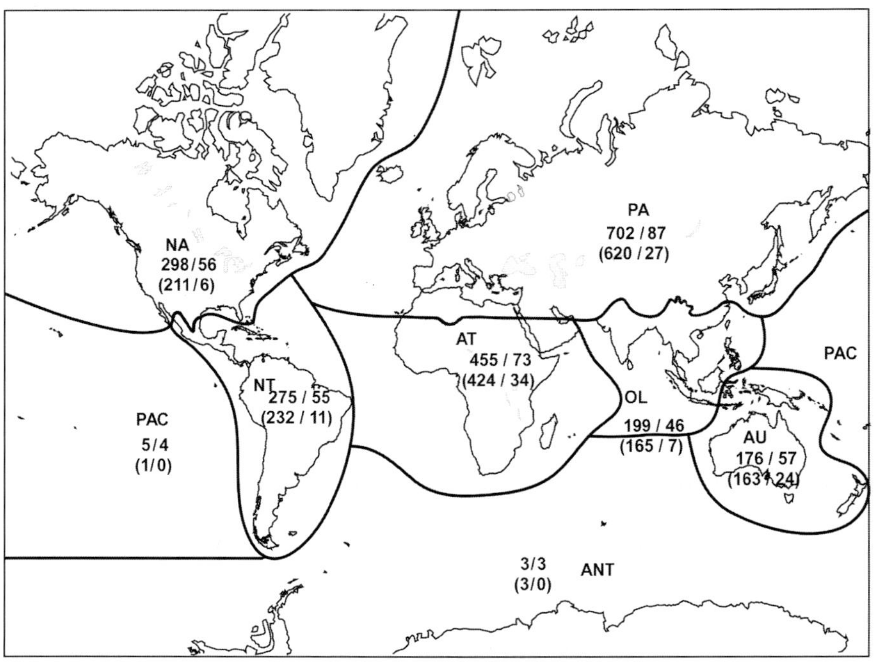

Fig. 3 Diversity and endemicity of ostracods (species/genus numbers and in parentheses endemic species/ endemic genus numbers). PA—Palaearctic, NA—Nearctic, NT—Neotropical, AT—Afrotropical, OL—Oriental, AU—Australasian, PAC—Pacific Oceanic Islands, ANT—Antarctic

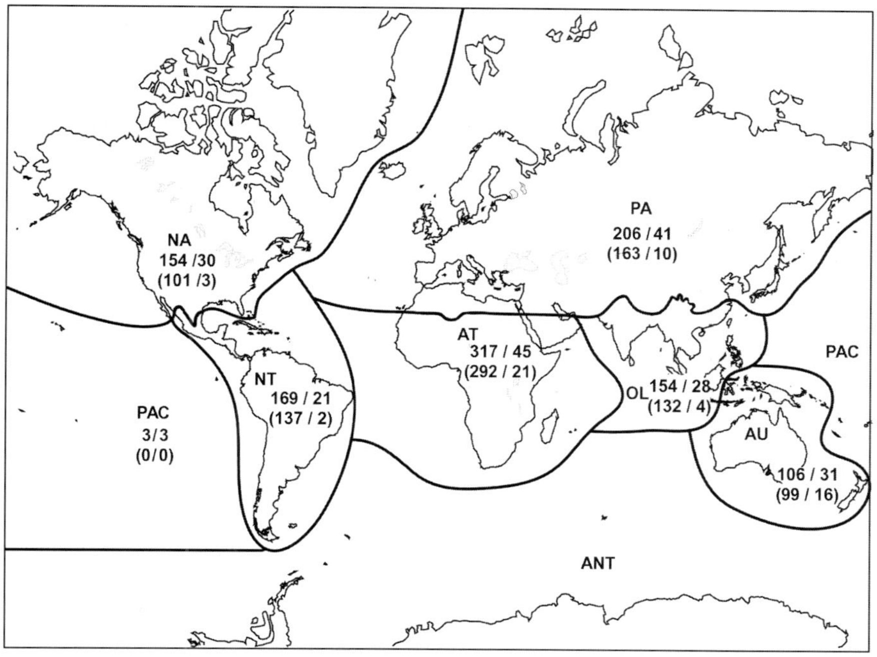

Fig. 4 Diversity and endemicity of Cyprididae (species/genus numbers and in parentheses endemic species/endemic genus numbers). PA—Palaearctic, NA—Nearctic, NT—Neotropical, AT—Afrotropical, OL—Oriental, AU—Australasian, PAC—Pacific Oceanic Islands, and ANT—Antarctic

Most non-marine ostracods either have dry resistant eggs, or are parthenogenetic, or are brooders, or have a mixture of all of these strategies, which are thought to facilitate long-distance dispersal. Indeed, wind, aquatic birds and humans could easily transport such stages (McKenzie, 1986) and in the case of parthenogenetic or gravid brooding females, one specimen is theoretically enough to found a new population. However, global comparison at neither the specific nor the generic levels supports these views. It would thus appear that ostracods are not such good dispersers at all. This is further shown by the existence of habitat-related endemic radiations, like the ones cited below from groundwater, ancient lakes and even temporary pools.

One of the reasons for this is that efficient dispersal does not guarantee the establishment of viable populations. This is, for example, supported by the very low number of species shared between Afrotropical and Neotropical regions (c. 15% shared), even if some taxonomic confusion might introduce some bias in this number. If intercontinental dispersal were indeed important, then the number of shared species would be considerably higher.

Another reason might be that ostracods in general are far less speciation prone than some other groups, and can show morphological stasis over long time spans. Taxa can be quite ancient: the species *Darwinula stevensoni* might be as old as 20–25 Myr (Straub, 1952), the extinct genus *Pattersoncypris* lived 150 Myr ago (Smith, 2000) and closely resembles the present day Cyprinotinae. Horne & Martens (1998) argue that *Cyprois* and *Stenocypris*-like species were already present in the Early Cretaceous (150–100 Myr).

In spite of the preceding arguments, there are some habitat types that hold higher numbers of endemics than others. Ancient lakes, especially Lake Tanganyika and Lake Baikal, hold extensive ostracod radiations, for example, the species flocks in the Cytherideidae and the Candonidae (Mazepova, 1990; Martens, 1994; Wouters & Martens, 2001). Younger ancient lakes such as Lake Titicaca, Lake Ohrid and some further East African lakes have extensive flocks in the Limnocytheridae and the Candonidae (Martens, 1994). With c. 200 species in Lake Baikal, about 100 in Lake Tanganyika and several dozens in other lakes, the ancient lakes are thought to hold 20–25% of the total non-marine ostracod diversity in the world.

Other speciation-prone habitats are those that have no or reduced predation pressure, for example temporary pools. In this case, speciation often coincides with gigantism. In temporary habitats, extensive radiations are known in Megalocypridinae (Afrotropical) and Cypridinae (Afrotropical and

Neotropical). The temporary pool fauna of South West Africa has a generic endemicity equalled only by the East African ancient lakes (Martens, 1998b).

Several ostracod groups in the Australian region are well adapted to lacustrine life in habitats with changing salinities, for example the Mytilocypridinae, the genera *Reticypris* and *Diacypris* and several others.

Subterranean ostracod faunas (Danielopol et al., 1994), finally, are well documented from the West Indies (Broodbakker, 1984). They seem poorly represented in most of Africa, although this could be owing to a lack of study. The recent discovery of dozens of endemic genera and close to 100 endemic species in the Pilbara area of western Australia exemplifies this possibility (Karanovic, 2007).

Finally, limno-terrestrial ostracods were thus far known from a few isolated cases in mainly Afrotropical and Australian regions. Recently, extensive radiations of especially Darwinulidae and Candonidae have been discovered and are being described from South America (Pinto et al., 2005).

Conclusions

With about 2,000 species worldwide, non-marine ostracods are not amongst the most species-rich groups in the freshwaters of the world. Whereas Holarctic faunas are reasonably well documented, southern hemisphere regions remain ill-explored. Especially the ancient lakes of the world (presently accounting for 20–25% of the world's non-marine ostracods) and groundwater faunas in Australia, Africa and South America could comprise significant numbers of presently undescribed taxa. Extant collections hold dozens of undescribed species, but further exploration is vital if non-marine ostracod faunas of the world are to be described with any degree of accuracy. Ostracods are the most common extant arthropod group with the most complete fossil record. Continued documentation of extant distribution patterns will therefore confirm their status as a model group for evolutionary studies.

Acknowledgements Karel Wouters, Hendrik Segers and Estelle Balian (Brussels) suggested improvements on an earlier version of the manuscript. Funding for the present research was received from the Belgian Science Policy (Agreement 'freshwater biodiversity'), the Fonds voor Wetenschappelijk Onderzoek, Vlaanderen (project "Ecological biogeography of freshwater rockpool communities"; coordinator: L. Brendonck, K.U.Leuven) and the EU (RTN Marie Curie Project 'Sexasex', coordinator KM).

References

Balian, E., H. Segers, C. Lévèque & K. Martens, 2008. An Introduction to the freshwater animal diversity assessment (FADA) project. In Balian, E. et al. (eds), Freshwater Animal Diversity Assessment. Hydrobiologia, doi: 10.1007/s10750-007-9235-6

Broodbakker, N. W., 1984. The distribution and zoogeography of freshwater Ostracoda (Crustacea) in the West Indies, with emphasis on species inhabiting wells. Bijdragen tot de Dierkunde 52(2): 25–50.

Danielopol, D. L., P. Marmonier, A. J. Boulton & G. Bonaduce, 1994. World subterranean ostracod biogeography: dispersal or vicariance? Hydrobiologia 287: 119–129.

Holmes, J. A. & A. Chivas (eds), 2002. The Ostracoda: Applications in Quaternary Research. American Geophysical Union, Washington, DC.

Horne, D. J., 2003. Key events in the ecological radiation of the Ostracoda. In Park, L. E. & Smith, A. J. (eds), Bridging the Gap: Trends in the Ostracode Biological And Geological Sciences. The Paleontological Society Papers 9: 181–201.

Horne, D. J. & K. Martens, 1998. An assessment of the importance of resting eggs for the evolutionary success of non-marine Ostracoda (Crustacea). In Brendonck, L., L. De Meester & N. Hairston (eds), Evolutionary and Ecological Aspects of Crustacean Diapause. Advances in Limnology 52: 549–561.

Karanovic, I., 2007. Candoninae Ostracodes from the Pilbara Region in Western Australia. Crustaceana Monographs 7: 432 pp.

Kempf, E., 1980a. Index and Bibliography of non-marine ostracods. 1. Index A. Geologisches Institut, Universität Köln, 35: 188 pp.

Kempf, E., 1980b. Index and Bibliography of non-marine ostracods. 4. Bibliography A. Geologisches Institut, Universität Köln, 38: 186 pp.

Kempf, E., 1991a. Index and Bibliography of non-marine ostracods. 6. Index A, Supplement 1. Geologisches Institut, Universität Köln, working copy 10: 51 pp.

Kempf, E., 1991b. Index and Bibliography of non-marine ostracods. Bibliography C. Geologisches Institut, Universität Köln, working copy, 9: 72 pp.

McKenzie, K. G., 1986: Ostracoda: new aspects of their biogeography. In Gore, R. H. & K. L. Heck (eds), Crustacean Biogeography. A.A. Balkema, Rotterdam, Boston: 257–277.

Martens, K., 1984. Annotated checklist of non-marine ostracods (Ostracoda, Crustacea) from African inland waters. Zoologische dokumentatie van het koninklijk Museum voor Midden Afrika, Tervuren, 20: 51 pp.

Martens, K., 1994. Ostracod speciation in ancient lakes: a review. In Martens, K., Goddeeris, B. & Coulter, G. (eds), Speciation in Ancient Lakes. Advances in Limnology 44: 203–222.

Martens, K. (ed.) 1998a. Sex and Parthenogenesis – Evolutionary Ecology of Reproductive Modes in Non-marine Ostracods. Backhuys Publishers, Leiden, 334 pp.

Martens, K., 1998b. Diversity and endemicity of Recent non-marine ostracods (Crustacea, Ostracoda) from Africa and South America: a faunal comparison. Verhandlungen der Internationalen Vereinigung für Limnologie 26(4): 2093–2097.

Martens, K., 2003. On the evolution of *Gomphocythere* (Crustacea, Ostracoda) in Lake Nyassa/Malawi (East Africa), with the description of five new species. Hydrobiologia 497: 121–144.

Martens, K. & F. Behen, 1994. A checklist of the non-marine ostracods (Crustacea, Ostracoda) from South-American inland waters and adjacent islands. Travaux scientifiques du Musée d'Histoire naturelle de Luxembourg 22: 81 pp.

Martens, K. & D. J. Horne, (in press). Ostracoda. In G. Likkens (ed.), Encyclopedia of Inland Waters. Elsevier Publ.

Mazepova, G., 1990. Rakushkovye ratchki (Ostracoda) Baikala. Nauk. Sib. Otdel. Akad. Nauk. SSSR, Novosibirsk, 470 pp.

Meisch, C., 2000. Freshwater Ostracoda of Western and Central Europe. Spektrum Akademischer Verlag GmbH, Heidelberg, Berlin, 522 p.

Pinto, R. L., C. E. F. Rocha & K. Martens, 2005. On new terrestrial ostracods (Crustacea, Ostracoda) from Brazil, primarily from Sao Paulo State. Zoological Journal of the Linnean Society 145: 145–173.

Rossi, V., I. Schön, R. K. Butlin & P. Menozzi, 1998. Clonal genetic diversity. In Martens, K., (ed.), Sex and Parthenogenesis, Evolutionary Ecology of Reproductive Modes in Non-marine Ostracods. Backhuys Publishers, Leiden: 257–274.

Schön, I., A. Gandolfi, E. Di Masso, V. Rossi, H. I. Griffiths, K. Martens, R. K. Butlin, 2000. Long-term persistence of asexuality through mixed reproduction in *Eucypris virens* (Crustacea, Ostracoda). Heredity 84: 161–169.

Smith, R. J., 2000. Morphology and ontogeny of Cretaceous ostracods with preserved appendages from Brazil. Palaeontology 43: 63–98.

Straub, E. W., 1952. Mikropaläontologische Untersuchungen im Tertiär zwischen Ehingen und Ulm an dem Donau. Geologisches Jahrbuch 66, 433–523.

Wouters, K., K. Martens, 2001. On the *Cyprideis* species flock (Crustacea, Ostracoda) in Lake Tanganyika, with the description of four new species. Hydrobiologia 450: 111–127.

Hydrobiologia (2008) 595:195–207
DOI 10.1007/s10750-007-9014-4

FRESHWATER ANIMAL DIVERSITY ASSESSMENT

Global diversity of copepods (Crustacea: Copepoda) in freshwater

Geoff A. Boxshall · Danielle Defaye

Abstract The zoogeographic distributions of the 2,814 species of copepods reported from freshwater are analysed. Faunal diversity is compared between zoogeographic regions: the Palaearctic region has more than double the species richness of the next most diverse region, the Neotropical. Historical factors affecting levels of diversity are identified. More than 90% of all freshwater copepods are endemic to a single-zoogeographic region and endemic genera occur in all regions except Antarctica. Species that are not endemic to a single region include the highly vagile and cosmopolitan species occurring in four or more regions. The greatest faunal connectivity, as identified by Sørensen's Index, is between Palaearctic and Nearctic regions, and identifies the Holarctic taxa. Key human-related issues, such as the role of copepods as vectors for human parasites and the losses caused by parasitic copepods in commercial aquaculture, are mentioned.

Guest editors: E.V. Balian, C. Lévêque, H. Segers and K. Martens
Freshwater Animal Diversity Assessment

G. A. Boxshall (✉)
Department of Zoology, The Natural History Museum, Cromwell Road, London SW7 5BD, UK
e-mail: G.Boxshall@nhm.ac.uk

D. Defaye
Département Milieux et Peuplements Aquatiques, Muséum national d'Histoire naturelle, CP53, 61 rue Buffon, Paris cedex 05 75231, France

Keywords Zoogeography · Copepods · Freshwater · Endemism · Species richness

Introduction

Copepods occur in almost all freshwater habitats from the largest ancient lakes to subterranean waters, from pools of glacial meltwater to hot springs, and from hypersaline lakes to phytotelmata. They are extremely abundant in freshwater and comprise a major component of most planktonic, benthic and groundwater communities, including semi-terrestrial situations such as damp moss and leaf litter in humid forests. With an estimated 13,000 morphospecies known, the greatest diversity of copepods is found in the marine environment, but approximately 2,814 species inhabit freshwater. Copepods originated in the marine environment: Huys & Boxshall (1991) hypothesized that all ten orders of Copepoda recognized at that time had their origins in the marine hyperbenthic community. Boxshall & Jaume (2000) recognized a minimum of 22 lineages as having independently colonized freshwater, but several other predominantly coastal marine and estuarine taxa have made incursions into freshwater, due to their basic euryhalinity (cf. Lee, 1999). The greatest difficulty in the analysis of global diversity of copepods in freshwater is defining the salinity limits for inclusion. In nature, habitats exhibit a continuum of salinity regimes from fresh to fully marine. Drawing a line through the "brackish" water is a somewhat arbitrary process especially since precise salinity data are often not available for records from estuaries or coastal lagoons. Saline and hypersaline inland continental waters tend to be populated by representatives of typically freshwater higher taxa and are, therefore, included in this analysis.

Most freshwater copepods are free-living, but they have adopted parasitism in numerous independent lineages. About 330 species of copepods in freshwater are parasitic, most commonly on fish hosts, but also on molluscs. Copepods also live as commensal epibionts on freshwater invertebrates, such as crayfishes, medusae and sponges.

It is important to set estimates of regional species richness in their historical context. Nineteenth century species concepts were largely established by and applied by European natural historians who tended to record European-like species from around the world under classical European names. In the second half of the 20th century revisionary studies generated more fine-scale taxonomic resolution and the recognition of numerous species complexes replacing so-called cosmopolitan species. The pioneering work on chydorid cladocerans by Frey (summarised in Frey, 1986) was soon mirrored for copepods, such as *Mesocyclops leuckarti*, but in many copepod taxa such fine-scale resolution has yet to be achieved for morphospecies. In addition, the advent of molecular methodology has not yet fully impacted on species-identity concepts in freshwater copepods. Studies of the genetics of the invasive temorid *Eurytemora affinis* have revealed discordant rates of morphological differentiation, molecular evolution and reproductive isolation, resulting in speciation events that are not accompanied by morphological differentiation in the secondary sexual characters traditionally employed for species discrimination. Instead of being a cosmopolitan euryhaline Holarctic species, *E. affinis* is now recognised as a species complex of genetically divergent and physiologically distinct populations with habitat transitions, such as the invasion of freshwater, entailing rapid evolutionary events (Lee, 1999, 2000).

Species diversity

Copepods from five orders occur in freshwater (Table 1) and this diversity is reflected in the enormous range of life styles adopted, from small-particle feeding to predation and parasitism (Boxshall & Halsey, 2004). In the Calanoida, the Diaptomidae is the dominant family in inland waters in Europe, Asia, North America, Africa and northern low-altitude South America. It does not occur in New Zealand and New Caledonia, and only two species (*Tropodiaptomus australis* and *Eodiaptomus lumholtzi*) have been found in Australia. The family comprises about 440 species in four subfamilies: the two largest comprising 24 species (Paradiaptominae) and about 410 species (Diaptominae). Diaptomids in general are small-particle feeders (Fig. 1A) but the biology, diversity and evolutionary histories of these two subfamilies differ markedly. The Diaptominae is extremely widely distributed and most are planktonic, inhabiting the water column. The Paradiaptominae is restricted to Africa, with outliers in southern Europe and south-western Asia. They are often restricted to extreme habitats, such as temporary pools. The other two subfamilies comprise small-sized species inhabiting subterranean waters.

The Centropagidae contains marine, brackish, freshwater and athalassic saline water species. They are the most important calanoids in continental waters in Australasia. Species of the *Boeckella*-group (comprising at least those approximately 70 species referred to *Boeckella*, *Hemiboeckella* and *Calamoecia*) occur in fresh and athalassic saline waters in Australia, New Zealand, southern and high-altitude South America and New Caledonia (Bayly, 1995; Defaye, 1998; Dussart & Defaye, 2002). A single species, *B. triarticulata* (Fig. 1D), occurs in Mongolia (see records in Dussart & Defaye, 2002) and an introduced population of this species has been reported from Italy. Three centropagid genera, *Limnocalanus*, *Osphranticum* and *Sinocalanus*, are present in continental waters in the Northern Hemisphere. The sole species of *Osphranticum* occurs in freshwater in North America, as far south as Guatemala. *Limnocalanus* species occur in freshwater lakes and in the Caspian Sea. *Sinocalanus* comprises five species found primarily in estuarine waters.

The Temoridae also comprises a mix of coastal marine, estuarine and freshwater taxa. According to Dussart & Defaye (2002), 30 species in three genera, *Epischura*, *Eurytemora* and *Heterocope* Sars, inhabit continental waters in the northern Hemisphere. Most of the *Eurytemora* species are brackish water forms, although *E. velox* and *E. lacustris* are freshwater

Table 1 Number of freshwater species recorded per zoogeographic region

Taxon	PA region	NA region	NT region	AT region	OL region	AU region	PAC	ANT region	World species number
Calanoida									
Diaptomidae	144	77	82	72	92	2	–	–	441
Centropagidae	7	4	22	–	4	47	2	7	79
Temoridae	22	18	–	–	–	–	–	–	30
Aetideidae (*Senecella*)	2	2	–	–	–	–	–	–	2
Harpacticoida									
Canthocamptidae	325	59	109	46	58	51	3	4	627
Parastenocarididae	108	8	65	34	16	2	1	–	249
Ameiridae	97	9	17	9	10	15	6	–	149
Miraciidae (Diosaccidae)	27	–	6	13	3	8	1	–	53
Ectinosomatidae	11	1	–	–	1	–	–	–	13
Phyllognathopodidae	3	2	3	3	3	3	–	1	10
Harpacticidae	4	–	–	–	1	–	–	–	4
Chappuisiidae	2	–	–	–	–	–	–	–	2
Huntemanniidae	3	1	1	1	–	–	–	–	6
Laophontidae	2	3	1	2	2	3	–	–	7
Cletodidae	1	–	–	–	3	–	–	–	3
Darcythompsoniidae	1	–	–	–	–	–	–	–	1
Cyclopoida									
Cyclopidae	337	114	174	167	115	64	16	5	800
Oithonidae	–	2	6	–	2	–	–	–	7
Ozmanidae	–	–	2	–	–	–	–	–	2
Lernaeidae	17	8	10	39	47	2	–	–	114
Ergasilidae	56	26	63	19	23	8	–	–	173
Gelyelloida									
Gelyellidae	2	–	–	–	–	–	–	–	2
Siphonostomatoida									
Lernaeopodidae	31	12	–	–	1	–	–	–	38
Caligidae	1	–	–	–	–	–	–	–	1
Dichelesthiidae	1	1	–	–	–	–	–	–	1
Total	1,204	347	561	405	381	205	29	17	2,814

PA: Palaearctic, NA: Nearctic, NT: Neotropical, AT: Afrotropical, OL: Oriental, AU: Australasian, PAC: Pacific Oceanic Islands, ANT: Antarctic

species. All are restricted to the higher latitudes of the Northern Hemisphere.

Currently placed in the Aetideidae (Boxshall & Halsey, 2004), *Senecella* comprises two species, *S. calanoides* and *S. siberica*, distributed in high latitude North America and northern Siberia, around the shores of the Kara and Laptev Seas. Both occur in brackish coastal waters, as well as in freshwater. The Pseudodiaptomidae and Acartiidae comprise primarily coastal and estuarine species. A few species occur in fresh, brackish and continental waters (Dussart & Defaye, 2002) but neither family is considered further.

The Canthocamptidae is the largest family in the Harpacticoida, comprising about 627 species. It is predominantly freshwater in distribution, but contains fully marine taxa and specialist estuarine forms, such as *Mesochra*. *Cletocamptus* is of uncertain taxonomic

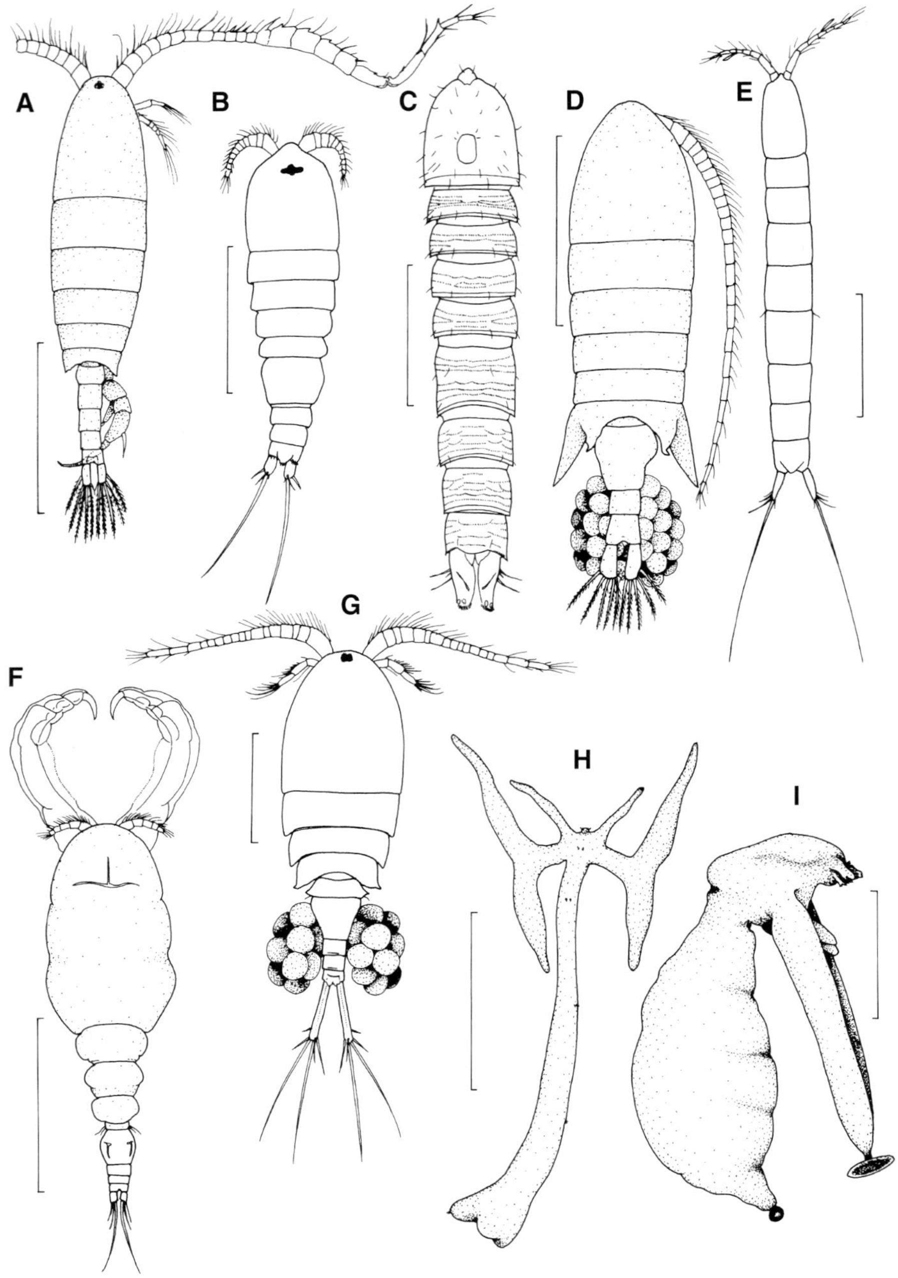

Fig. 1 The diversity of copepods in fresh water. (**A**) Diaptomidae (*Eudiaptomus gracilis*, male); (**B**) Cyclopidae (*Graeteriella unisetigera*, female); (**C**) Canthocamptidae (*Moraria* (*Baikalomoraria*) *brevicauda*, female); (**D**) Centropagidae (*Boeckella triarticulata* female); (**E**) Parastenocarididae (*Parastenocaris roettgeri*, female); (**F**) Ergasilidae (*Acusicola mazatlanesis*, female); (**G**) Cyclopidae (*Cyclops abyssorum*, female); (**H**) Lernaeidae (*Lernaea cyprinacea*, female); (**I**) Lernaeopodidae (*Achtheres percarum*, female). Scale bars: A, G = 0.5 mm, B, C = 0.2 mm, D, I = 1 mm, E = 0.1 mm, F = 0.4 mm, H = 5 mm

affinity and its species occur throughout the salinity range but a few species occur in full freshwater, such as in Lake Titicaca. Virtually all species of Parastenocarididae are found in freshwater, mainly in interstitial habitats. Freshwater Canthocamptidae (Fig. 1C) and the Parastenocarididae (Fig. 1E) both have cosmopolitan distributions, except for the latter's absence from New Zealand (Dussart & Defaye, 1990). The difference in species richness may partly be an artifact, due to incomplete sampling of the groundwater habitats favoured by parastenocaridids. Two smaller families, the Chappuisiidae (two species) and the Phyllognathopodidae (about 10 species) occur exclusively in freshwater but the former is found only in Northern Europe, whereas the latter is extremely widely distributed.

The Ameiridae contains almost 150 freshwater species and is especially diverse in subterranean waters (Rouch, 1986; Galassi, 2001). Freshwater species are found widely across Europe, Asia, North America and Africa, especially northern Africa (Dussart & Defaye, 1990), although this may represent sampling bias. It is one of the more speciose families in freshwater and is dominated by the group of genera related to *Nitokra* and *Nitocrella*. These genera inhabit karstic habitats, calcrete formations and other subterranean waters. Some ameirids have entered into symbiotic relationships with freshwater invertebrates.

The Ectinosomatidae is overwhelmingly marine but contains a few estuarine and freshwater species. The latter belong in six genera: *Pseudectinosoma*, *Halectinosoma*, *Pseudobradya*, *Ectinosoma*, *Rangabradya* and *Arenosetella*. Available information on phylogenetic relationships indicates that there have been several independent colonizations of freshwater within the family, such as that within *Pseudectinosoma* (e.g. Galassi et al., 1999). The Harpacticidae is similar, mainly marine with a few freshwater forms. It contains genera such as *Tigriopus*, which typically live in coastal splash zone pools and are tolerant of extreme variations in salinity, and *Harpacticus* which is common in coastal lagoons. These forms are not treated here as freshwater. The true freshwater clade consists only of *Harpacticella* species, which are found in the Palaearctic and Oriental Regions.

The Miraciidae (=formerly the Diosaccidae) is another predominantly marine family, but contains numerous freshwater forms, especially in the *Schizopera*-group of genera. These freshwater forms are found in Western Eurasia, Australia and Africa, especially in the Rift Valley lakes, where an evolutionary radiation has taken place. There are 10 species of the *Schizopera*-group described from Lake Tanganyika (Coulter, 1991). The speciose marine family Laophontidae contains a small number of brackish and freshwater species: some *Laophonte* species occur in coastal lagoons, and *Troglolaophonte* occurs in a brackish cave. These brackish forms are excluded from the analysis. True freshwater laophontids include the cosmopolitan *Onychocamptus mohammed* and some species of *Heterolaophonte* (e.g. *H. quinquespinosa* and *H. stromi*) that occur in inland saline lakes in southeastern Australia and in high Arctic lakes. The Cletodidae is primarily marine but contains a few euryhaline (brackish to freshwater) taxa, such as *Limnocletodes*. The Huntemanniidae is similar—predominantly marine with a few freshwater and estuarine species. *Nannopus palustris*, for example, inhabits the fresher zones of estuaries across the Holarctic, the Caspian Sea and Lake Kinneret, and *N. perplexus* was described from Lake Tanganyika. The darcythompsoniid *Leptocaris brevicornis* occurs in Lake Kinneret and a lake in El Salvador.

The largest copepod family is the Cyclopidae which comprises over 800 species (Dussart & Defaye, 2006), the great majority belonging to two freshwater subfamilies, Eucyclopinae and Cyclopinae. The Euryteinae contains marine and estuarine species only, and the Halicyclopinae consists predominantly of brackish forms belonging to *Halicyclops* and related genera, with a few freshwater species. The Euryteinae is excluded from subsequent consideration. The Cyclopidae (Fig. 1G) is cosmopolitan in freshwaters and its members exploit a huge variety of habitats, from subterranean waters (Fig. 1B) to ancient lakes, and life styles from small-particle feeders, to predators or even parasites of gastropod molluscs (Boxshall & Strong, 2006).

The Oithonidae contains primarily marine planktonic forms but several species are found in brackish habitats, such as mangrove swamps, and a few occur in freshwater. Some species occur exclusively in freshwater: there is, for example, a small cluster of *Oithona* species in east coast drainages of South America (Rocha, 1986). *Limnoithona* is a brackish-water specialist.

The Ergasilidae has a unique life cycle within the copepods: its developmental stages from nauplius to adult, occurring in the plankton, only after mating do fertilized adult females seek out and infest hosts for the final, parasitic phase of their life cycle. Most ergasilids are known only from their parasitic females and these typically utilize fishes as hosts, with a few on bivalve mollusks, attaching by claw-like antennae (Fig. 1F). Ergasilids are found on and around the coastal margins of all continents excluding Antarctica, and occur in all salinity regimes although the majority can be regarded as freshwater. Estuarine ergasilids commonly utilize hosts, which move freely between salinity regimes.

The Lernaeidae comprises 114 species, all parasites of freshwater fishes, classified in two monophyletic lineages, the subfamilies Lernaeinae and Lamprogleninae (Ho, 1998). All Lamprogleninae are found in the Palaearctic, Oriental and Afrotropical regions. The Lernaeinae is more widely distributed, occurring in all regions except Antarctica, although the Australian fauna comprises only two species. The Ozmanidae consists of two endoparasitic species in freshwater gastropod molluscs in South America.

The only members of the order Siphonostomatoida found in freshwater are parasites of fishes. The *Salmincola*-clade (Fig. 1I) within the Lernaeopodidae comprises 38 species in seven genera, which utilize a wide range of fish hosts (Kabata, 1979). The Caligidae contains over 400 marine species but just a single freshwater species, *Caligus lacustris*, which also occurs in brackish waters and the Black Sea. *Dichelesthium oblongum* parasitises sturgeons and was regarded by Kabata (1979) as a marine species carried into freshwater by the migrations of its host.

Finally, the order Gelyelloida comprises a single family, the Gelyellidae, one genus and two species currently reported only from groundwater habitats of south-western Europe.

Present distribution and main areas of endemicity

Cosmopolitan taxa occurring in freshwater in all zoogeographic regions except the Antarctic include the families Ameiridae, Canthocamptidae, Cyclopidae, Diaptomidae, Ergasilidae, Harpacticidae, Laophontidae, Lernaeidae, Parastenocarididae and Phyllognathopodidae, although the distribution of the Diaptomidae and the Parastenocarididae within the Australasian region is limited. Almost as widely distributed are the Centropagidae, which occur in all regions except the Afrotropical. For these taxa, presence/absence data at the family level are too coarse to reveal distribution patterns and quantitative data, such as species richness per region, and qualitative data at a finer-taxonomic resolution, such as genus or subgenus level, must be considered. The large, cosmopolitan canthocamptid genus *Attheyella* is an excellent example. It has been divided into six subgenera: of these, two (*Attheyella* and *Mrazekiella*) are predominantly Palaearctic in distribution, two (*Chappuisiella* and *Delachauxiella*) are predominantly Neotropical, and one (*Canthosella*) occurs in both Oriental and Neotropical regions (Fig. 2).

Palaearctic region

The Palaearctic region has 1,204 recorded species—more than double the number recorded for any other region (Table 1). The major contributors are the Cyclopidae (28%), especially *Diacyclops*, *Eucyclops*,

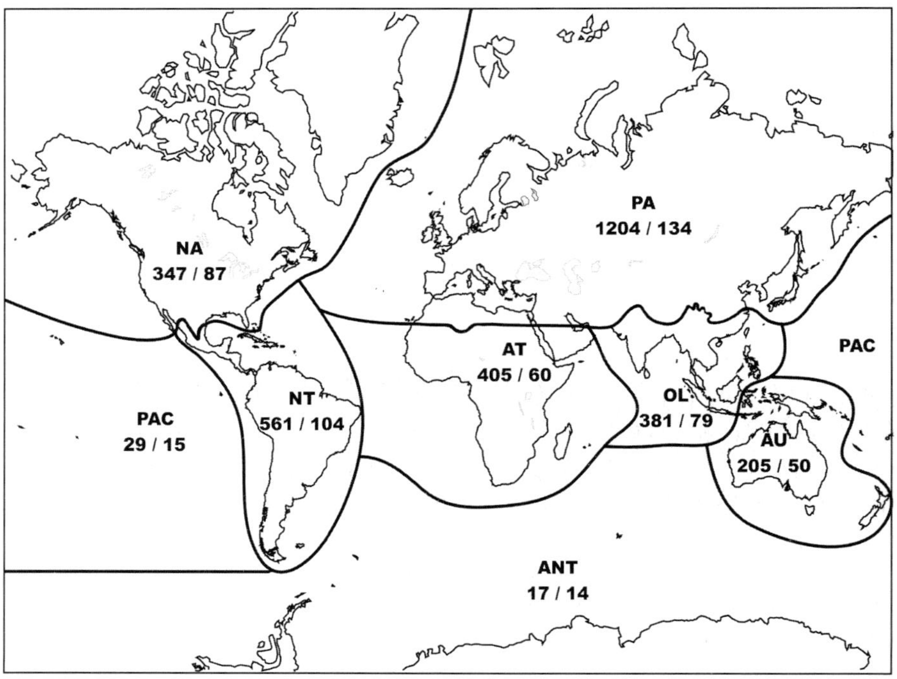

Fig. 2 Geographical distribution of species and genera of freshwater copepods (Species number/ Genus number), coded according to the main zoogeographic regions. PA: Palaearctic, NA: Nearctic, NT: Neotropical, AT: Afrotropical, OL: Oriental, AU: Australasian, PAC: Pacific Oceanic Islands, ANT: Antarctic

Acanthocyclops and *Speocyclops*, the Canthocamptidae (27%) especially *Elaphoidella*, *Moraria*, *Bryocamptus* and *Attheyella* (*Attheyella*) and *Attheyella* (*Mrazekiella*), and the Diaptomidae (12%), especially *Arctodiaptomus*, *Diaptomus*, *Eudiaptomus* and *Hemidiaptomus*. Parasites comprise less than 9% of the total but the Palaearctic also has the greatest species richness, as well as the greatest family-level diversity of fish parasites.

Several small lineages including the freshwater Temoridae, *Senecella*-species, the *Harpacticella*-group and the *Salmincola*-lineage, occur only in the more northerly part of the Northern Hemisphere. The restriction of these relatively small lineages to the northern part of the Holarctic region suggests that the Pleistocene glaciations may have had a profound effect on their distributions. All may be postglacial invasions.

Nearctic region

The Nearctic, with 347 recorded species, exhibits less than one-third of the richness of the Palaearctic (Table 1), and a significant proportion of species is shared between these two regions, i.e. are Holarctic in distribution. At the genus level, there are few Nearctic endemics, since most are Holarctic. The major contributors to the species richness are the Cyclopidae (33%), especially *Diacyclops* and *Acanthocyclops*, the Diaptomidae (22%) especially *Aglaodiaptomus*, *Hesperodiaptomus* and *Leptodiaptomus*, and the Canthocamptidae (17%), especially *Elaphoidella*. The dominant parasitic families are the Ergasilidae (7%) and Lernaeopodidae (3%).

Afrotropical region

Sub-Saharan Africa has a copepod fauna of 405 species (Table 1). The major contributors are the Cyclopidae (41%), especially *Eucyclops* and *Thermocyclops*, and the Diaptomidae (18%) especially *Tropodiaptomus*, followed by the Lernaeidae (10%). The benthic families Canthocamptidae, particularly *Elaphoidella* and *Echinocamptus*, and Parastenocarididae contribute only 11% and 8% to the total, respectively. Centropagids are absent from African inland waters and it has been inferred that Africa was probably never colonized by centropagids (Bayly, 1995; Boxshall & Jaume, 2000). Boxshall & Jaume (2000) also speculated that the ancestors of the subfamily Paradiaptominae were the first calanoids to colonize freshwater on the African plate and that the diaptomines invaded the African continental waters initially from the North, and spread southwards, diversifying and largely displacing the paradiaptomines, except in more extreme habitats.

Neotropical region

The Neotropical region has the second-highest richness, with 561-recorded species (Table 1), although this number is expected to rise with increasing study. The major contributors are the Cyclopidae (31%), especially *Eucyclops*, *Metacyclops* and *Mesocyclops*, the Canthocamptidae (19%), especially *Elaphoidella*, *Attheyella* (*Chappuisiella*) and *A.* (*Delachauxiella*), the Diaptomidae (15%), and Parastenocarididae (12%). Within the Diaptomidae, the most speciose genera are *Notodiaptomus* and *Argyrodiaptomus*, both endemic to the region. Boxshall & Jaume (2000) postulated that the presence of diaptomids at low altitudes in the northern and central parts of South America resulted from a late invasion from North America, occurring after closure of the Panama gap in the Pliocene about 3 mya. After invading from the North, the diaptomids would have spread rapidly, through the highly interconnected, lowland river systems that make South America unique. Boxshall & Jaume speculated that the diaptomids replaced the existing calanoid fauna of *Boeckella*-group centropagids except at high altitudes (i.e. the Andean cordillera) and at high latitudes (e.g. Patagonian), where they are still dominant today. Bayly (1995) argued that the *Boeckella*-group first invaded southern inland continental waters at a time when Australia, New Zealand and South America were still linked but Africa, Madagascar and India had already drifted northwards. Bayly estimated this colonization as occurring between 120 mya and 80 mya.

A small cluster of *Oithona* species has been recorded in Amazonia (Rocha, 1986). The extensive brackish interface between marine and freshwater habitats in the Amazon may have provided the route by which *Oithona* has colonized neotropical freshwaters, but Boxshall & Jaume (2000) noted the possibility that *Oithona* might have arrived during a marine incursion into the Upper Amazon (cf. Lovejoy et al., 1998).

The Ergasilidae comprises 11% of the total richness but Thatcher (1998) calculated that only a tiny proportion of the possible ergasilid fauna of Amazonia is currently known: our knowledge of the Ergasilidae is incomplete. El-Rashidy (1999) showed that most Neotropical freshwater ergasilids probably belong to a single monophyletic lineage which has diversified spectacularly in Amazonia and now encompasses 13 endemic genera. The genus *Acusicola* has subsequently spread northwards through Central America and a single species has penetrated the Nearctic, as far as Texas. This is further evidence that the closure of the Panama gap in the Pliocene permitted an intense period of North/South faunal exchange in both directions (cf. Suárez-Morales et al., 2004).

Oriental region

The diversity of the Oriental region is relatively low, with only 16 families represented by 381 species (Table 1). The major contributors are the Cyclopidae (30%), especially *Mesocyclops* and *Thermocyclops*, the Diaptomidae (24%), especially *Tropodiaptomus* and *Heliodiaptomus*, the Canthocamptidae (15%), especially *Elaphoidella*, and the Lernaeidae (12%). The species richness may be expected to rise substantially since the benthic fauna has been only patchily sampled.

The Oriental region is home to seven genera of Lernaeidae, three of which (*Indopeniculus*, *Pillainus*, and *Indolernaea*) are endemic (Table 2). However, the bulk of the species richness is contributed by *Lamproglena* (22 species) and *Lernaea* (15 species). The former also occurs in the Afrotropical and Palaearctic regions, whereas the latter is cosmopolitan. Approximately 45% of lernaeids occur in the Oriental region and another 35% is reported from the Afrotropical region. According to Ho (1998), this pattern, with 80% of species confined to Asia and Africa, resulted from an explosive cladogenesis on the "Indian Raft", which took place after the ancestral lernaeids colonized the Cyprinidae as hosts.

Australasian region

The Australasian fauna exhibits the lowest richness (205 species) of all regions except the Pacific islands and Antarctica (Table 1). However, despite the aridity of large areas of the Australian continent, which might explain the apparent paucity of species, recent research into the subterranean fauna of arid Western Australia (Karanovic, 2004) has revealed an unexpectedly high diversity and degree of endemism. The major contributors to the overall species richness are the Cyclopidae (31%), especially *Mesocyclops* and *Eucyclops*, the Canthocamptidae (25%) especially *Canthocamptus*, and the Centropagidae (23%). The centropagids have been the focus for much biogeographical research. Bayly (1995) inferred that Australia was inhabited by numerous *Boeckella* species prior to its separation from Antarctica, and contrasted it with *Calamoecia*, which almost certainly evolved after separation.

Australian athalassic saline waters are inhabited by a specialized fauna comprising centropagids (such as *Calamoecia clitellata* and *C. salina*), canthocamptids (such as *Mesochra baylyi*), and cyclopids (such as *Merideicyclops baylyi*). They are likely derived from freshwater ancestors, according to Bayly (1993), who suggested that these halobiont species probably evolved after the mid-Miocene (16 mya) and that their successful colonization was related to the evolution of exceptional cellular tolerance or cellular osmoregulation.

Pacific island region

The Pacific island fauna is low in diversity with only 29 species recorded (Table 1). Calanoids are rare: two centropagids are described from New Caledonia and no diaptomids are known. One brackish ergasilid is reported but no freshwater copepod parasites are known, although a little research has been done on the larger islands. Benthic copepods comprise the bulk of the fauna, but several are vagile species with widespread distributions, such as *Nitokra lacustris*, *Phyllognathopus viguieri*, *Paracyclops chiltoni*, *Mesocyclops aspericornis* and *Tropocyclops prasinus*. The proportion of regional endemics is relatively small.

Antarctic region

Most of the 17 reported species occur on the Subantarctic islands. Pugh et al. (2002) hypothesized that all continental and maritime Antarctic Crustacea represent Holocene immigrants. More specifically, they interpreted the population of the centropagid

Table 2 Number of freshwater genera recorded per zoogeographic region; number in parentheses indicates number of genera that are endemic to the region

Taxon	PA region	NA region	NT region	AT region	OL region	AU region	PAC	ANT region	Total genera number
Calanoida									
Diaptomidae	25 (9)	14 (4)	18 (14)	7 (3)	22 (10)	2 (0)	–	–	60 (40)
Centropagidae	3 (0)	3 (0)	3 (3)	–	2 (0)	7 (2)	2 (1)	3 (0)	8 (5)
Temoridae	3 (0)	3 (0)	–	–	–	–	–	–	4 (0)
Aetideidae (*Senecella*)	1 (0)	1 (0)	–	–	–	–	–	–	1 (0)
Harpacticoida									
Canthocamptidae	23 (11)	9 (1)	12 (3)	6 (1)	9 (2)	8 (2)	1 (0)	3 (0)	38 (21)
Parastenocarididae	2 (1)	1 (0)	6 (5)	1 (0)	1 (0)	1 (0)	1 (0)	–	6 (4)
Ameiridae	6 (1)	5 (0)	4 (0)	3 (0)	3 (0)	6 (2)	1 (0)	1 (0)	9 (4)
Miraciidae (Diosaccidae)	2 (0)	–	1 (0)	1 (0)	2 (1)	1 (0)	1 (0)	–	2 (1)
Ectinosomatidae	5 (1)	1 (0)	–	–	1 (1)	–	–	–	6 (2)
Phyllognathopodidae	1 (0)	1 (0)	2 (1)	1 (0)	2 (1)	1 (0)	–	1 (0)	3 (2)
Harpacticidae	1 (0)	–	–	–	1 (0)	–	–	–	1 (1)
Chappuisiidae	1 (1)	–	–	–	–	–	–	–	1 (1)
Huntemanniidae	2 (0)	2 (0)	1 (0)	1 (0)	1 (0)	–	–	–	2 (0)
Laophontidae	4 (0)	3 (0)	2 (0)	2 (0)	2 (0)	3 (0)	–	1 (0)	6 (0)
Cletodidae	1 (0)	–	–	–	1 (0)	–	–	–	1 (0)
Darcythompsoniidae	1 (0)	–	–	–	–	–	–	–	1 (0)
Cyclopoida									
Cyclopidae	32 (6)	23 (1)	30 (11)	25 (2)	19 (1)	19 (3)	9 (0)	5 (0)	55 (24)
Oithonidae	–	1 (0)	2 (0)	–	1 (0)	–	–	–	2 (0)
Ozmanidae	–	–	1 (1)	–	–	–	–	–	1 (1)
Lernaeidae	4 (0)	1 (0)	6 (5)	7 (3)	7 (3)	1 (0)	–	–	16 (11)
Ergasilidae	8 (1)	5 (0)	16 (13)	6 (1)	5 (0)	3 (1)	–	–	24 (16)
Gelyelloida									
Gelyellidae	1 (1)	–	–	–	–	–	–	–	1(1)
Siphonostomatoida									
Lernaeopodidae	6 (3)	3 (1)	–	–	–	–	–	–	7 (4)
Caligidae	1 (0)	–	–	–	–	–	–	–	1 (0)
Dichelesthiidae	1 (0)	1 (0)	–	–	–	–	–	–	1 (0)
Total	134 (35)	87 (8)	104 (54)	60 (12)	79 (19)	50 (10)	15 (1)	14 (0)	257 (138)

PA: Palaearctic, NA: Nearctic, NT: Neotropical, AT: Afrotropical, OL: Oriental, AU: Australasian, PAC: Pacific Oceanic Islands, ANT: Antarctic

Boeckella poppei inhabiting freshwater lakes in East Antarctica, as a recent anthropogenic introduction, and that of the endemic *Gladioferens antarcticus* as a 'marine interloper'. These interpretations were challenged by Bayly et al. (2003) who presented robust evidence that *B. poppei* has been present in isolated populations in East Antarctica for significant time periods, possibly pre-dating the current interglacial, and that *G. antarcticus* is a true Antarctic endemic whose ancestors have been present in the region, since before Australia separated from Antarctica.

Hotspots

The concept of biodiversity hotspots as areas of exceptional biotic richness set against a background of relatively low diversity has been widely adopted, as a method of identifying priority areas for conservation. Most analyses have identified ancient lakes as hotspots for freshwater biodiversity. For copepods, Lake Baikal with over 120 species, the majority of which are endemic (Boxshall & Evstigneeva, 1994), and Lake Tanganyika with 69 species, of which 34 (49%) are endemic (Coulter, 1991; Boxshall & Strong, 2006), constitute prime hotspots. In both the copepod fauna comprises several species flocks, defined as monophyletic clades endemic to the lake and presumed to have originated via intralacustrine speciation. The age of the lake (i.e. the time available for speciation to have taken place), the heterogeneity of habitats at all scales within the lake, and their biotic complexity are all potential factors affecting the generation and maintenance of high levels of biodiversity in ancient lakes. Boxshall & Evstigneeva (1994) suggested that trophic specialisation was a key driver in the adaptive radiation of the cyclopids in Lake Baikal in contrast to the canthocamptids where they considered differentiation in mating behaviour to be a key driver.

In recent decades, awareness of the diversity of groundwater faunas has increased. Sket (1999) for example, noted that 70% of the 245 harpacticoid copepods and 60% of the 145 cyclopoids recorded in the *Limnofauna Europaea* compendium were stygobites. Culver & Sket (2000) recommended that cave systems inhabited by at least 20 stygobitic species should be designated biodiversity hotspots. Worldwide only 20 karstic systems and two sites in porous aquifers meet this criterion (Culver & Sket, 2000; Danielopol & Pospisil, 2002). The species richness of the Lobau wetland in the Danube flood plain led Danielopol & Pospisil (2002) to refer to it as a biodiversity hotspot. In a sampled area of 0.8 km^2 they discovered 30 copepod species of which 10 were hypogean. The situation is not unique: Karanovic (2004) discovered a subterranean fauna of 31 species (24 new) in the arid Murchison region of western Australia, and Stoch (1995) reported 70 species from the Venezia Giulia (N.E. Italy and southeastern Slovenia) of which 33 (47%) were stygobionts. Galassi (2001) reviewed the patterns of diversity of groundwater copepods, noting that stygobiont copepods inhabit all kinds of aquifers (karstic, fissured and porous). She identified local heterogeneity, patchiness in geomorphological and hydrological characteristics, as well as biotic interactions as causal factors affecting diversity patterns over a range of spatial and temporal scales.

Endemism

The level of endemism in freshwater copepods seems remarkable. The great majority (2,558 out of 2,814 species, or 90.9%) of species occur in a single region. The Neotropical, Afrotropical, Palaearctic and Australasian Regions each display in excess of 80% endemism (Table 3). Nearctic and Oriental Regions display intermediate levels (65–75%) and the Pacific islands and Antarctica both display between 45% and 50%. Species that are not endemic to a single region include highly vagile and cosmopolitan species that occur in four or more regions.

Endemic genera occur in all regions except Antarctica (Table 2), in all orders and across all life styles. The Neotropical is home to the greatest number of endemic genera (Table 2). There is evidence that in the Ergasilidae most of the endemic genera represent a single monophyletic lineage (El-Rashidy, 1999). Similarly, Boxshall & Halsey (2004) recognized the possibility that the cluster of five lernaeid genera endemic to the Neotropics may be closely related (i.e. form a monophyletic group). No phylogeny exists for the Diaptomidae but, building on the speculation (Boxshall & Jaume, 2000) that diaptomines only invaded South America after the closure of the Panama gap, it is possible that the cluster of 14 endemic genera might also form a monophyletic lineage.

Faunal links

Pair-wise analysis of the occurrence of species that inhabit more than one zoogeographic region (Table 4) provides a relative measure of the faunal connectivity between regions. The dominant links as estimated by Sørenson's index (SI) are those between the Palaearctic and Nearctic, between the Nearctic and Neotropical, and between the Oriental and Australasian regions. The Palaearctic to Nearctic link identifies Holarctic taxa and these typically belong to lineages, the distributions of which have been profoundly affected by the Pleistocene glaciations. The Nearctic to Neotropical link in part testifies to the strength of

Table 3 Number of endemic species recorded from each zoogeographic region

Taxon	PA region	NA region	NT region	AT region	OL region	AU region	PAC	ANT region	Total endemic species number
Calanoida									
Diaptomidae	121	65	77	71	81	2	–	–	417
Centropagidae	3	1	14	–	1	45	2	1	67
Temoridae	11	8	–	–	–	–	–	–	19
Aetideidae (*Senecella*)	–	–	–	–	–	–	–	–	
Harpacticoida									
Canthocamptidae	285	46	114	37	40	51	1	1	575
Parastenocarididae	117	9	69	36	16	1	1	–	249
Ameiridae	94	6	11	9	4	9	3	3	139
Miraciidae (Diosaccidae)	25	–	5	11	1	6	1	–	49
Ectinosomatidae	11	–	–	–	1	–	–	–	12
Phyllognathopodidae	1	–	2	2	1	1	–	1	8
Harpacticidae	3	–	–	–	–	–	–	–	3
Chappuisiidae	2	–	–	–	–	–	–	–	2
Huntemanniidae	1	1	–	1	1	–	–	–	4
Laophontidae	2	–	–	–	–	1	–	–	3
Cletodidae	–	–	–	–	2	–	–	–	2
Darcythompsoniidae	1	–	–	–	–	–	–	–	1
Cyclopoida									
Cyclopidae	254	56	130	123	68	43	4	2	680
Oithonidae	–	–	5	–	–	–	–	–	5
Ozmanidae	–	–	2	–	–	–	–	–	2
Lernaeidae	11	7	10	37	42	1	–	–	108
Ergasilidae	50	21	63	17	20	7	1	–	179
Gelyelloida									
Gelyellidae	2	–	–	–	–	–	–	–	2
Siphonostomatoida									
Lernaeopodidae	25	5	–	–	1	–	–	–	31
Caligidae	1	–	–	–	–	–	–	–	1
Dichelesthiidae	–	–	–	–	–	–	–	–	
Total endemic species	1,020	225	502	344	279	167	13	8	2,558
Total species (Table 1)	1,204	347	561	405	381	205	29	17	2,814
Proportion of endemics	84.7%	64.8%	89.5%	84.9%	73.2%	81.5%	44.8%	47.1%	90.9

PA: Palaearctic, NA: Nearctic, NT: Neotropical, AT: Afrotropical, OL: Oriental, AU: Australasian, PAC: Pacific Oceanic Islands, ANT: Antarctic

the faunal exchange that took place between North and South America after the closure of the Panama gap. The Oriental to Australasian link may indicate the main dispersal route of the freshwater colonists of Australia. There is also a relatively strong link between Palaearctic and Oriental regions, which may reflect the weakness of the barrier separating these regions, particularly in central and eastern China. The link between Antarctic and Neotropical regions is strong—10 out of the 17 species present in the Antarctic are shared with the Neotropics. These are mainly centropagids and cyclopids that are distributed from the Falklands to the circum-Antarctic islands. This faunal link is not identified by

Table 4 Faunal links between zoogeographic regions estimated using Sørensen's index, based on species-level data: values exceeding 0.1000 shown in bold

	PA	NA	AT	NT	OL	AU	PAC
NA	**0.1251**						
AT	0.0646	0.0771					
NT	0.0453	**0.1101**	0.0621				
OL	0.0984	0.0769	0.0941	0.0679			
AU	0.0397	0.0507	0.0623	0.0574	**0.1058**		
PAC	0.0162	0.0319	0.0461	0.0407	0.0585	0.0684	
ANT	0.0066	0.0275	0.0190	0.0346	0.0101	0.0270	0.0435

Sørenson's Index, given the paucity of the Antarctic fauna relative to the Neotropical.

Introduced species

There are numerous examples of introduced species from all life styles, planktonic to parasitic. In free-living copepods, at least 18 species have been involved in intercontinental introduction (Reid & Pinto-Coelho, 1994). The translocation of fish stocks for aquaculture has been implicated in the introduction of parasites such as *Paraergasilus longidigitus*, *Neoergasilus japonicus* and *Lernaea cyprinacea* (Fig. 1H) Introductions have also been associated with the transportation of tropical plants for horticulture and agriculture. For example, the presence of the Asiatic centropagid *Boeckella triarticulata* in Italy was attributed to movement of Chinese carp, the introduction of *Mesocyclops ogunnus* to the Cayman Islands and Brazil may have resulted from transportation of organisms for aquaculture, and it has been suggested that *M. pepheiensis* (as *M. ruttneri*), an East-Asian species, was introduced to the southern USA by rice culture. The uncontrolled exchange of ballast water has been implicated in the introduction of estuarine taxa, including *Limnoithona*, *Sinocalanus*, *Pseudodiaptomus* and *Acartia*.

Human related issues

Copepods act as intermediate hosts for a variety of parasites. Species of *Mesocyclops* and *Thermocyclops* are intermediate hosts for guineaworm (*Dracunculus medinensis*), a debilitating nematode parasite. Guineaworm larvae are consumed by copepods and remain in the body cavity until ingested by drinking unfiltered water. Guineaworm is less prevalent than in the past but remains a human health problem, particularly in West Africa and India. Other parasites of man that utilize copepods as intermediate hosts include the fish tapeworm, *Diphyllobothrium latum*.

Parasitic copepods such as *Lernaea cyprinacea* and *Ergasilus sieboldi* can be serious pests, causing mass mortality and significant commercial losses in freshwater aquaculture. They are often found in inconspicuous microhabitats on the host and may be overlooked in fish quarantine checks. Translocation of fish stocks, without sufficient checks, is responsible for the widespread introduction of these pests outside of their natural ranges.

Free-living copepods can be voracious predators and this functional role has been exploited in the use of *Mesocyclops* species as biological control agents against mosquitoes. The spread of diseases such as malaria and dengue fever involves mosquito vectors and *Mesocyclops* have been experimentally trialled as a biocontrol agent for mosquito larvae. The success of these trials has been variable, but in Viet Nam for example, *Mesocyclops* has been used effectively in controlling mosquito larvae at local levels.

Desertification and the overexploitation of ground water are major conservation issues for copepods. Groundwater copepods, especially those with narrow geographical distributions, are at risk from intrusion of saline waters as the groundwater table is lowered.

Acknowledgements We would like to thank Estelle Balian and Koen Martens, for the invitation to contribute to this volume, and Diana Galassi and an anonymous reviewer for their suggested improvements.

References

Bayly, I. A. E., 1993. The fauna of athalassic saline waters in Australia and the Altiplano of South America: comparisons and historical perspectives. Hydrobiologia 267: 225–231.

Bayly, I. A. E., 1995. Distinctive aspects of the zooplankton of large lakes in Australasia, Antarctica and South America. Marine and Freshwater Research 46: 1109–1120.

Bayly, I. A. E, J. A. E. Gibson, B. Wagner & K. M. Swadling, 2003. Taxonomy, ecology and zoogeography of two East Antarctic freshwater calanoid copepod species: *Boeckella poppei* and *Gladioferens antarcticus*. Antarctic Science 15: 439–448.

Boxshall, G. A. & T. D. Evstigneeva, 1994. The evolution of species flocks of copepods in Lake Baikal: a preliminary analysis. In: Martens, K., B. Goddeeris & G. Coulter (eds), Speciation in Ancient Lakes. Archiv für Hydrobiologie, Ergebn Limnology 44: 235–245.

Boxshall, G. A. & S. H. Halsey, 2004. An Introduction to Copepod Diversity. The Ray Society, London, 966 pp.

Boxshall, G. A. & D. Jaume, 2000. Making waves: the repeated colonization of fresh water by copepod crustaceans. Advances in Ecological Research 31: 61–79.

Boxshall, G. A. & E. E. Strong, 2006. An extraordinary shift in life habit in a genus of cyclopid copepods from Lake Tanganyika. Zoological Journal of the Linnean Society 146: 275–285.

Coulter, G., 1991. Lake Tanganyika and its Life. Oxford University Press, Oxford.

Culver, D. C. & B. Sket, 2000. Hotspots of subterranean biodiversity in caves and wells. Journal of Cave and Karst Studies 62: 11–17.

Danielopol, D. L. & P. Pospisil, 2002. Taxonomic diversity of Crustacea Cyclopoida in the Austrian "Danube Floodplain" national park. Vie et Milieu 52: 67–75.

Defaye, D., 1998. Description of the first *Boeckella* (Copepoda, Calanoida, Centropagidae) from New Caledonia. Crustaceana 71: 686–699.

Dussart, B. & D. Defaye, 1990. Répertoire Mondial Crustacés Copépodes des Eaux Intérieures. I. Harpacticoïdes. Crustaceana Supplement 16: 1–384.

Dussart, B. & D. Defaye, 2002. World Directory of Crustacea Copepoda of Inland Waters. I – Calaniformes. Backhuys Publishers, Leiden, 276 pp.

Dussart, B. & D. Defaye, 2006. World Directory of Crustacea Copepoda of Inland Waters. II - Cyclopiformes. Backhuys Publishers, Leiden. 354 pp.

El-Rashidy, H., 1999. Copepods and Grey Mullets (Mugilidae). PhD Thesis, University of London.

Frey, D. G., 1986. The non-cosmopolitanism of chydorid Cladocera; implications for biogeography and evolution. Crustacean Issues 4: 237–256.

Galassi, D. M. P., 2001. Groundwater copepods: diversity patterns over ecological and evolutionary scales. Hydrobiologia 453/454: 227–253.

Galassi, D. M. P., M.-J. Dole-Olivier & P. De Laurentiis, 1999. Phylogeny and biogeography of the genus *Pseudectinosoma* and description of *P. janineae* sp.n. (Crustacea, Copepoda, Ectinosomatidae). Zoologica Scripta 28: 289–303.

Ho, J.-s., 1998. Cladistics of the Lernaeidae (Cyclopoida), a major family of freshwater fish parasites. Journal of Marine Systems 15: 177–183.

Huys, R. & G. A. Boxshall, 1991. Copepod Evolution. The Ray Society, London, 468 pp.

Kabata Z., 1979. Parasitic Copepoda of British fishes. The Ray Society, London, 468 pp.

Karanovic, T., 2004. Subterranean Copepoda from arid Western Australia. Crustacean Monographs 3. Koninklijke Brill, Leiden, 366 pp.

Lee, C. E., 1999. Rapid and repeated invasions of fresh water by the saltwater copepod *Eurytemora affinis*. Evolution 53: 1423–1434.

Lee, C. E., 2000. Global phylogeography of a cryptic species complex and reproductive isolation between genetically proximate 'populations'. Evolution 54: 2014–2027.

Lovejoy, N. R., E. Bermingham & A. P. Martin, 1998. Marine incursion into South America. Nature 196: 421–422.

Pugh, P. J. A., H. J. G. Dartnall & S. J. McInnes, 2002. The non-marine Crustacea of Antarctica and the islands of the Southern Ocean: biodiversity and biogeography. Journal of Natural History 36: 1047–1103.

Reid, J. W. & R. M. Pinto-Coelho, 1994. An Afro-Asian continental copepod, *Mesocyclops ogunnus*, found in Brazil; with a new key to the species of *Mesocyclops* in South America and a review of intercontinental introductions of copepods. Limnologica 24: 359–368.

Rocha, C. E. F., 1986. Freshwater copepods of the genus *Oithona* Baird, 1843 from the Amazonian region (Brazil). Revista de Biologia Tropical 18: 213–220.

Rouch, R., 1986. Copepoda: les Harpacticoïdes souterrains des eaux douces continentales. In Botosaneanu, L. (ed.), Stygofauna Mundi, E.J.Brill, Leiden, 321–355.

Sket, B., 1999. High biodiversity in hypogean waters and its endangerment – the situation in Slovenia, the Dinaric Karst and Europe. Crustaceana 72: 767–779.

Stoch, F., 1995. The ecological and historical determinants of crustacean diversity in groundwaters, or: why are there so many species? Mémoires de Biospéologie 22: 139–160.

Suárez-Morales, E., J. W. Reid, F. Fiers & T. M. Iliffe, 2004. Historical biogeography and distribution of the freshwater cyclopine copepods of the Yucatan Peninsula, Mexico. Journal of Biogeography 31: 1051–1063.

Thatcher, V. E., 1998. Copepods and fishes in the Brazilian Amazon. Journal of Marine Systems 15: 97–112.

Hydrobiologia (2008) 595:209–212
DOI 10.1007/s10750-007-9015-3

FRESHWATER ANIMAL DIVERSITY ASSESSMENT

Global diversity of fishlice (Crustacea: Branchiura: Argulidae) in freshwater

William J. Poly

Abstract The Branchiura of freshwater habitats, consisting of the valid genera *Argulus*, *Chonopeltis*, *Dipteropeltis*, and *Dolops*, presently contains 113 valid species and 12 undescribed species. The entire group is composed of ectoparasitic species that usually live on fishes. The highest diversity of genera and species occurs in the Afrotropical and Neotropical regions. The diversity of the freshwater species surpasses that of species in marine and brackish waters, but this could be due to inadequate study of the fauna of the latter habitats. One species, *Argulus japonicus*, has been introduced from east/southeast Asia to all other continents, except Antarctica. Studies of higher level relationships place the Branchiura with either Pentastomida or Ostracoda. Hypotheses about phylogenetic relationships of either the genera or species in this group have not been proposed.

Keywords Fishlice · Fishlouse · Parasites · Nonindigenous species · Biogeography

Guest editors: E. V. Balian, C. Lévêque, H. Segers & K. Martens
Freshwater Animal Diversity Assessment

The present paper was intended as a very brief overview of the known diversity of the Branchiura in freshwaters only, and regrettably is deficient in mention of much information and many relevant citations pertaining to the group.

W. J. Poly (✉)
Aquatic Biodiversity Research Institute, 11205 Schroeder Road, Saint Marys, OH 45885, USA
e-mail: wpoly@calacademy.org

W. J. Poly
Department of Ichthyology, California Academy of Sciences, Golden Gate Park, San Francisco, CA 94118, USA

Introduction

The subclass Branchiura contains a single family, the Argulidae, and four valid genera: *Argulus* Müller, 1785, *Chonopeltis* Thiele, 1900, *Dipteropeltis* Calman, 1912, and *Dolops* Audouin, 1837. Yamaguti (1963) erected subfamilies within the Argulidae as well as the family Dipteropeltidae; however, neither subfamily designations nor the family group name Dipteropeltidae are used herein. Branchiurans are ectoparasites of fishes primarily but occasionally live on amphibians or invertebrates, and they can move about freely on their hosts (Stuhlmann, 1891; Yamaguti, 1963; Cressey, 1978; Jackson & Marcogliese, 1995; Poly, 2003). Species of Branchiura are known by the common name, fishlouse (plural: fishlice).

Branchiurans are compressed dorsoventrally with a circular to oval shield-like carapace, a pair of compound eyes anteriorly, four pairs of swimming legs on the thorax, and a short, unsegmented abdomen posteriorly (Fig. 1). Total length of adults ranges from a few millimeters to just over 30 mm, and adult females tend

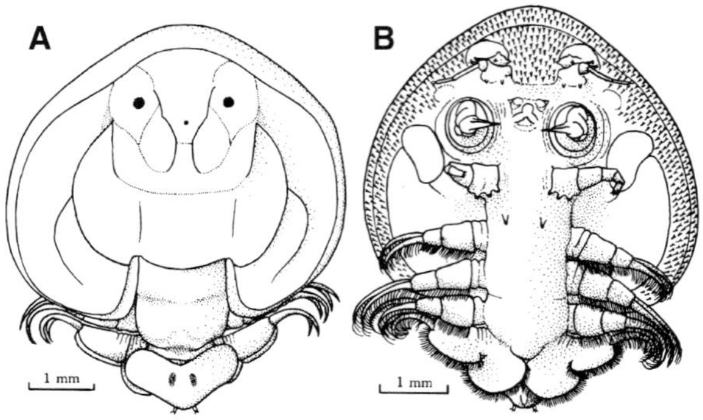

Fig. 1 Typical body form of a branchiuran. Dorsal (**A**) and ventral (**B**) views of the female holotype of *Dolops tasmanianus* Fryer, 1969. Copyright © CSIRO; Reproduced from the Australian Journal of Zoology 17: 49–64 (Fryer, 1969) with permission of CSIRO PUBLISHING, Melbourne, Australia (http:// www.publish.csiro.au/journals/ajz)

to be larger than males. *Argulus*, *Chonopeltis*, and *Dipteropeltis* possess a pair of suction cups (modified first maxillae) in late juvenile and adult stages, whereas *Dolops* retains the larval, claw-like appendages as adults (Fig. 1). On legs 2–4, male *Argulus* spp. and *Chonopeltis* spp. may have secondary sexual modifications that are absent on females. Other differences between the sexes include the presence of a pair of testes in the male's abdomen and a pair of spermathecae in the female's abdomen, a greater number of support rods in suction cups of females, and also usually a higher number of sclerites in the support rods of females (*Argulus*). Eggs often can be seen in the thorax of an adult female, and in some species eggs occur in the thorax and in the lobes of the carapace as well. Shapes of respiratory areas, shapes of carapace and abdomen, number of sclerites in suction cup support rods and number of support rods per suction cup (*Argulus*), features of the second maxillae (especially basal plate), pigmentation, and secondary sexual characters of males are some of the most useful taxonomic characters for distinguishing species (Wilson, 1902, 1944; Sikama, 1938; Meehean, 1940; Poly, 2005).

Mating takes place on the host fish, and later, the female leaves the host to lay eggs, which are attached with an adhesive substance to objects, such as rocks, plants, or sticks. Sperm are stored in the spermathecae of females in *Argulus* and presumably in *Chonopeltis* and *Dipteropeltis*, whereas males of the genus *Dolops* deposit a spermatophore on the females (Carvalho, 1941; Fryer, 1958, 1960). Further information on the ecology and morphology of argulids can be found in Jurine (1806), Clark (1902), Wilson (1902), Calman (1912), Tokioka (1936), Loro (1964), van Niekerk & Kok (1989), Rushton-Mellor & Boxshall (1994), and Van As & Van As (1999).

Species diversity

Dipteropeltis contains one valid species and occurs only in South America, *Chonopeltis* has 14 valid species and is found only in Africa, and *Dolops* holds 13 valid species with all but two species in South America. Species within the three aforementioned genera live in freshwater only. *Argulus* contains about 129 valid species and occurs on or around all continents, except Antarctica, in marine and estuarine (n = 44) and freshwater habitats (n = 85). In total, the diversity of freshwater Branchiura stands at 113 species. In addition, the author is describing at least 12 new species from freshwaters of North America and Australia. The highest diversity of genera and species occurs in the Afrotropical and Neotropical regions (Fig. 2, Table 1). The Nearctic and Oriental regions contain nearly equal numbers of species, whereas the Palearctic and Australasian regions have much less diversity.

Phylogeny and historical processes

Phylogenetic studies concerning the Branchiura pertained to higher level relationships only. Results of several studies supported a sister group relationship between the Branchiura and the Pentastomida or the Ostracoda (discussed in Martin & Davis, 2001; Wingstrand, 1972; Lavrov et al., 2004; Regier et al., 2005 and references therein). Phylogenetic studies of the generic and specific relationships have not been attempted yet. Further detailed taxonomic study of the group will provide much of the data needed for more meaningful and complete phylogenetic analyses.

Present distribution and main areas of endemicity

Biogeographic analyses of the Branchiura were made by Fryer (1969) with regard to *Dolops*, which displays a Gondwanan distribution (South America,

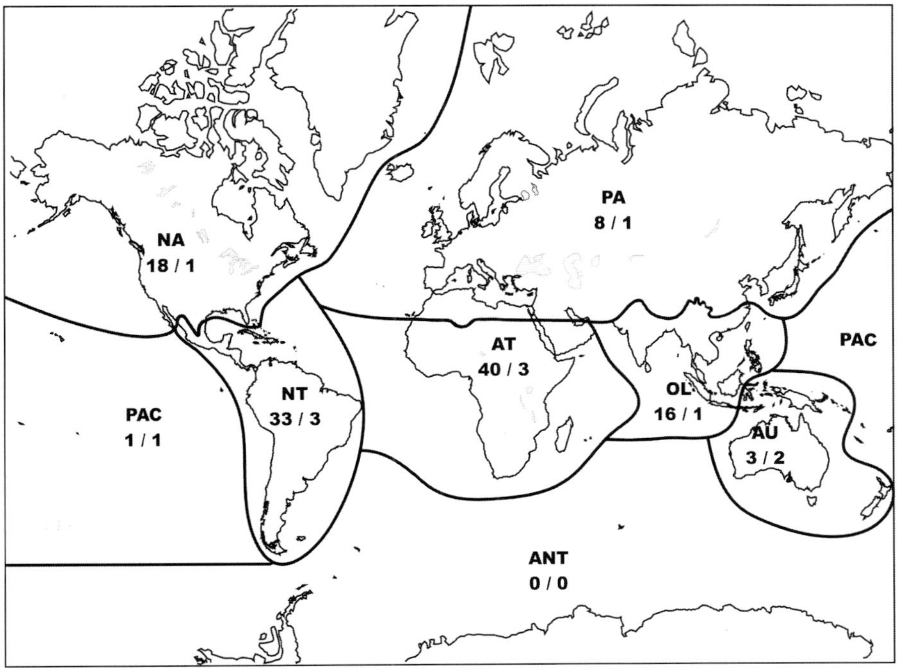

Fig. 2 Distribution of species of Branchiura occurring in freshwater habitats in each of the biogeographic regions of the world. *Argulus japonicus* was included in the number of species for all regions where it is either native (east/southeast Asia) or introduced. PA—Palearctic, NA—Nearctic, NT—Neotropical, AT—Afrotropical, OL—Oriental, AU—Australasian, PAC—Pacific Oceanic Islands, ANT—Antarctic

Table 1 Number of species per genus of Branchiura occurring in freshwater habitats in each of the biogeographic regions of the world

	PA	NA	NT	AT	OL	AU	PAC	ANT	World
Argulus	8	18	21	25	16	2	1		85
Chonopeltis				14					14
Dipteropeltis			1						1
Dolops			11	1		1			13
Total Argulidae	8 (1)	18 (1)	33 (3)	40 (3)	16 (1)	3 (2)	1 (1)	0	113

The Japanese fishlouse, *Argulus japonicus*, was included in all regions where it is either native or introduced, but in the overall total, this species was counted only once. Hence, the total from adding all separate regions is 119, minus 6 occurrences of *A. japonicus* equals 113. Number of genera per region in parentheses. PA: Palearctic, NA: Nearctic, NT: Neotropical, AT: Afrotropical, OL: Oriental, AU: Australasian, PAC: Pacific Oceanic Islands, ANT: Antarctic

Africa, and Tasmania), and the biogeography of African species of *Argulus*, *Chonopeltis*, and *Dolops* was discussed by Fryer (1968). *Chonopeltis* and *Dipteropeltis* are endemic to Africa and South America, respectively. Little else has ever been published on biogeography of this group. *Argulus japonicus* Thiele, 1900 has been introduced from east/southeast Asia to all other continents, except Antarctica (Table 1). Native species do not occur across great distances in most cases, except for some species in the Afrotropical region, such as *Dolops ranarum* (Stuhlmann, 1891) and *Argulus africanus* Thiele, 1900 (Fryer, 1968), and for *Argulus foliaceus* (Linné, 1758) in the Palearctic region. Definitive discussions about species' distributions cannot be made until additional species are described and further collecting and taxonomic study reveal true diversity more accurately.

Human related issues

Branchiurans occasionally attain high densities on their hosts, resulting in fish kills in natural bodies of water (rarely; usually lentic waters) or fish hatchery/aquaculture operations (more commonly). Fishlice also appear at times in large public aquaria, home aquaria, or small outdoor ponds and can contribute to mortality of ornamental fishes.

Acknowledgments The author expresses much appreciation to the organizers of the meeting on Freshwater Animal Diversity Assessment held in Mechelen, Belgium for their careful planning, logistical support, and financial assistance, resulting in an immensely productive and enjoyable session, and for all their efforts to assemble this volume. The author also thanks Geoff Boxshall and an anonymous reviewer for their comments on the manuscript.

References

Calman, W. T., 1912. On *Dipteropeltis*, a new genus of the crustacean order Branchiura. Proceedings of the Zoological Society of London 74: 763–766, 1 pl.

Carvalho, J. de P., 1941. Sôbre *Dipteropeltis hirundo* Calman, Crustáceo (Branchiura) parasito de peixes d'agua doce. Boletins da Faculdade de Filosofia, Ciências e Letras (Universidade de São Paulo, Brasil) 22 (Zoologia no. 5): 265–277.

Clark, F. N., 1902. *Argulus foliaceus*. A contribution to the life history. South London Entomological & Natural History Society (for 1902) 12–21, 2 pls.

Cressey, R. F., 1978. Marine flora and fauna of the northeastern United States. Crustacea: Branchiura. NOAA Technical Report Circular 413: 1–10.

Fryer, G., 1958. Occurrence of spermatophores in the genus *Dolops* (Crustacea: Branchiura). Nature 181: 1011–1012.

Fryer, G., 1960. The spermatophores of *Dolops ranarum* (Crustacea, Branchiura): their structure, formation, and transfer. Quarterly Journal of Microscopical Science 101: 407–432.

Fryer, G., 1968. The parasitic Crustacea of African freshwater fishes; their biology and distribution. Journal of Zoology, London 156: 45–95.

Fryer, G., 1969. A new freshwater species of the genus *Dolops* (Crustacea: Branchiura) parasitic on a galaxiid fish of Tasmania—with comments on disjunct distribution patterns in the southern hemisphere. Australian Journal of Zoology 17: 49–64.

Jackson, C. J. & D. J. Marcogliese, 1995. An unique association between *Argulus alosae* (Branchiura) and *Mysis stenolepis* (Mysidacea). Crustaceana 68: 910–912.

Jurine, A. [Jurine fils], 1806. Mémoire sur l'Argule foliacé (*Argulus foliaceus*). Annales du Muséum d'Histoire Naturelle (Paris) 7: 431–458, 1 pl.

Lavrov, D. V., W. M. Brown & J. L. Boore, 2004. Phylogenetic position of the Pentastomida and (pan)crustacean relationships. Proceedings of the Royal Society of London (B) 271: 537–544.

Loro, R., 1964. Morfologia degli stadi larvali e dell'adulto di *Argulus giordanii*. Archivio di Oceanografia e Limnologia 13: 387–418.

Martin, J. W. & G. E. Davis, 2001. An updated classification of the Recent Crustacea. Natural History Museum of Los Angeles County, Science Series 39: 1–124.

Meehean O. L., 1940. A review of the parasitic Crustacea of the genus *Argulus* in the collections of the United States National Museum. Proceedings of the United States National Museum 88: 459–522.

van Niekerk, J. P. & D. J. Kok, 1989. *Chonopeltis australis* (Branchuira): structural, developmental and functional aspects of the trophic appendages. Crustaceana 57: 51–56.

Poly, W. J., 2003. *Argulus ambystoma*, a new species parasitic on the salamander *Ambystoma dumerilii* from México (Crustacea: Branchiura: Argulidae). Ohio Journal of Science 103: 52–61.

Poly, W. J., 2005. *Argulus yucatanus* n. sp. (Crustacea: Branchiura) parasitic on *Cichlasoma urophthalmus* from Yucatan, Mexico. Gulf and Caribbean Research 17: 1–13.

Regier, J. C., J. W. Shultz & R. E. Kambic, 2005. Pancrustacean phylogeny: hexapods are terrestrial crustaceans and maxillopods are not monophyletic. Proceedings of the Royal Society of London (B) 272: 395–401.

Rushton-Mellor, S. K. & G. A. Boxshall, 1994. The developmental sequence of *Argulus foliaceus* (Crustacea: Branchiura). Journal of Natural History 28: 763–785.

Sikama, Y., 1938. On a new species of *Argulus* found in a marine fish in Japan. Journal of the Shanghai Science Institute, Section III 4: 129–134, pls. XI–XII.

Stuhlmann, F., 1891. Zur kenntniss der fauna central-afrikanischer seen. II. ueber eine neue art der Arguliden-gattung *Gyropeltis*. Zoologische Jahrbücher. Abtheilung für systematik, geographie und biologie der Thiere 6: 152–154.

Tokioka, T., 1936. Larval development and metamorphosis of *Argulus japonicus*. Memoirs of the College of Science, Kyoto Imperial University (Series B) 12: 93–114.

Van As, L. L. & J. G. Van As, 1999. Aspects of the morphology and a review of the taxonomic status of three species of the genus *Chonopeltis* (Crustacea: Branchiura) from the Orange-Vaal and South West Cape river systems, South Africa. Folia Parasitologica 46: 221–228.

Wilson, C. B., 1902. North American parasitic copepods of the family Argulidae, with a bibliography of the group and a systematic review of all known species. Proceedings of the United States National Museum 25: 635–742, 20 pls.

Wilson, C. B., 1944. Parasitic copepods in the United States National Museum. Proceedings of the United States National Museum 94: 529–582, 15 pls.

Wingstrand, K. G., 1972. Comparative spermatology of a pentastomid, *Raillietiella hemidactyli*, and a branchiuran crustacean, *Argulus foliaceus*, with a discussion of pentastomid relationships. Det Kongelike Danske Videnskabernes Selskab Biologiske Skrifter 19: 1–72.

Yamaguti, S., 1963. Parasitic Copepoda and Branchiura of Fishes. Interscience Publishers, John Wiley & Sons, Inc., 1104 pp.

Hydrobiologia (2008) 595:213–218
DOI 10.1007/s10750-007-9016-2

Global diversity of mysids (Crustacea-Mysida) in freshwater

Megan L. Porter · Kenneth Meland · Wayne Price

Abstract In this article we present a biogeographical assessment of species diversity within the Mysida (Crustacea: Malacostraca: Peracarida) from inland waters. Inland species represent 6.7% (72 species) of mysid diversity. These species represent three of the four families within the Mysida (Lepidomysidae, Stygiomysidae, and Mysidae) and are concentrated in the Palaearctic and Neotropical regions. The inland mysid species distributional patterns can be explained by four main groups representing different freshwater invasion routes: (1) Subterranean Tethyan relicts (24 spp.); (2) Autochthonous Ponto-Caspian endemics (20 spp.); (3) *Mysis* spp. 'Glacial Relicts' (8 spp.); and (4) Euryhaline estuarine species (20 spp.). The center of inland mysid species diversity is the Ponto-Caspian region, containing 24 species, a large portion of which are the results of a radiation in the genus *Paramysis*.

Keywords Inland fauna · Freshwater biology · Mysid · Diversity

Introduction

The order Mysida (Crustacea: Malacostraca: Peracarida), first described in 1776 by Müller, contains over 1,000 described species distributed throughout the waters of the world (Wittmann, 1999). Although >90% of mysid species are exclusively marine, the remaining species represent either species from coastal habitats with direct marine connections (e.g., estuaries, coastal rivers, marine caves) or from true invasions of inland freshwaters (Audzijonytė, 2006; Mauchline, 1980). This review deals with the zoogeography and historical processes leading to the current diversity of freshwater mysid species. While most of the true freshwater mysids occur in 'continental' lacustrine and riverine habitats, many of the subterranean mysids occur only from island habitats; for our purposes we have, therefore, chosen to define 'inland' species, as any species with documented populations occurring in freshwaters (salinity <3 g/l). Inland mysids range in size from 3 to 22 mm, and due to the brood pouch present in mature females are

---

Electronic Supplementary Material The online version of this article (doi:10.1007/s10750-007-9016-2) contains supplementary material, which is available to authorized users.

Guest editors: E. V. Balian, C. Lévêque, H. Segers and K. Martens
Freshwater Animal Diversity Assessment

M. L. Porter (✉)
Department of Biological Science, University of Maryland, Baltimore County, 1000 Hilltop Circle, Baltimore, MD 21250, USA
e-mail: porter@umbc.edu

K. Meland
Department of Biology, University of Bergen, Bergen, Norway

W. Price
Department of Biology, University of Tampa, Tampa, FL 33606, USA

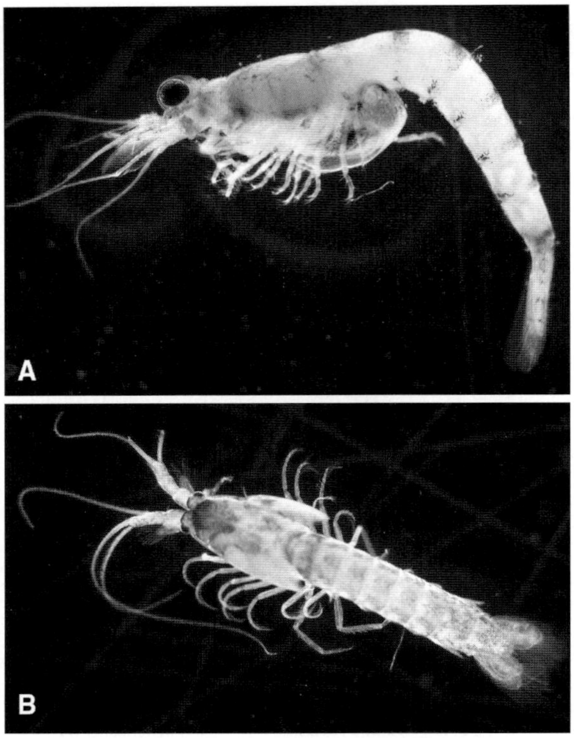

Fig. 1 A—*Taphromysis bowmani*; B—*Spelaeomysis* sp. (Photos by E. Peebles, courtesy of the American Fisheries Society)

often called 'opossum shrimp' (Fig. 1). Most inland species are nektobenthic (although a few species are pelagic), omnivorous suspension-feeders, or scavengers. Some species can be found in large numbers, serving as an important food source for many other organisms.

Species diversity

The Mysida contain ~1,073 species, only 72 of which are documented from inland waters (See supplemental material). Out of the four families within the Mysida *sensu* Martin & Davis (2001), three contain species from inland systems (Tables 1, 2). Two of these families, the Lepidomysidae and the Stygiomysidae, are monogeneric taxa (*Spelaeomysis* and *Stygiomysis*, respectively) containing species adapted to subterranean habitats (e.g., caves, groundwater, wells, and crab burrows). While all of the species in these two families (9 and 7, respectively) are from inland habitats, many are from island or coastal systems with direct connections to marine environments. Nonetheless, 7 Lepidomysidae and 5 Stygiomysidae have made the transition to what can be considered freshwater environments. In comparison, the family Mysidae contains the largest diversity of inland taxa (60 species, 23 genera), with most of this diversity found within the subfamily Mysinae, tribe Mysini (Tables 1, 2).

Zoogeography

The inland mysid fauna are located mainly in the Palaearctic (39 species, 15 genera) and Neotropical (20 species, 6 genera) biogeographical regions (Tables 1, 2; supplementary tables; Fig. 2). Within freshwater habitats, primarily from these two broad regions, four historical diversification patterns can

Table 1 Total number of inland mysid species found in the major geographical regions from each family (**), subfamily (*), and tribe (°) containing at least one inland species. Only species numbers from each family are used to calculate the final totals for each region. See supplementary tables for detailed information. PA: Palaearctic; NA: Nearctic; NT: Neotropical; AT: Afrotropical ; OL: Oriental; AU: Australasian; PAC: Pacific & Oceanic Islands; ANT: Antarctic

	PA	NA	NT	AT	OL	AU	PAC	ANT	Total inland species
**Lepidomysidae	1	1	4	0	2	0	0	0	7
**Stygiomysidae	1	0	4	0	0	0	0	0	5
**Mysidae	37	10	12	1	5	1	0	0	60
*Rhopalophthalminae	0	0	0	0	1	0	0	0	1
*Mysinae	37	10	12	1	4	1	0	0	59
°Heteromysini	0	1	0	0	0	0	0	0	1
°Leptomysini	0	1	1	0	0	1	0	0	2
°Mysini	37	8	11	1	4	0	0	0	56
Total	39	11	20	1	7	1	0	0	72

Note: Several species occur in more than one geographical region, resulting in higher numbers of occurrences (79) than species (72)

Table 2 Total number of inland mysid genera from the major geographical regions from each family (**), subfamily (*), and tribe (°) containing at least one inland genus. Only genera from each family are used to calculate the final totals for each region. See supplementary tables for detailed information. PA: Palaearctic; NA: Nearctic; NT: Neotropical; AT: Afrotropical; OL: Oriental; AU: Australasian; PAC: Pacific & Oceanic Islands; ANT: Antarctic

	PA	NA	NT	AT	OL	AU	PAC	ANT	Total inland genera
**Lepidomysidae	1	1	1	0	1	0	0	0	1
**Stygiomysidae	1	0	1	0	0	0	0	0	1
**Mysidae	13	6	4	1	5	1	0	0	23
*Rhopalophthalminae	0	0	0	0	1	0	0	0	1
*Mysinae	13	6	4	1	4	1	0	0	22
°Heteromysini	0	1	0	0	0	0	0	0	1
°Leptomysini	0	1	1	0	0	1	0	0	2
°Mysini	13	4	3	1	4	0	0	0	19
Total	15	7	6	1	6	1	0	0	25

Note: Several genera occur in more than one geographical region, resulting in higher numbers of occurrences (36) than inland genera (25)

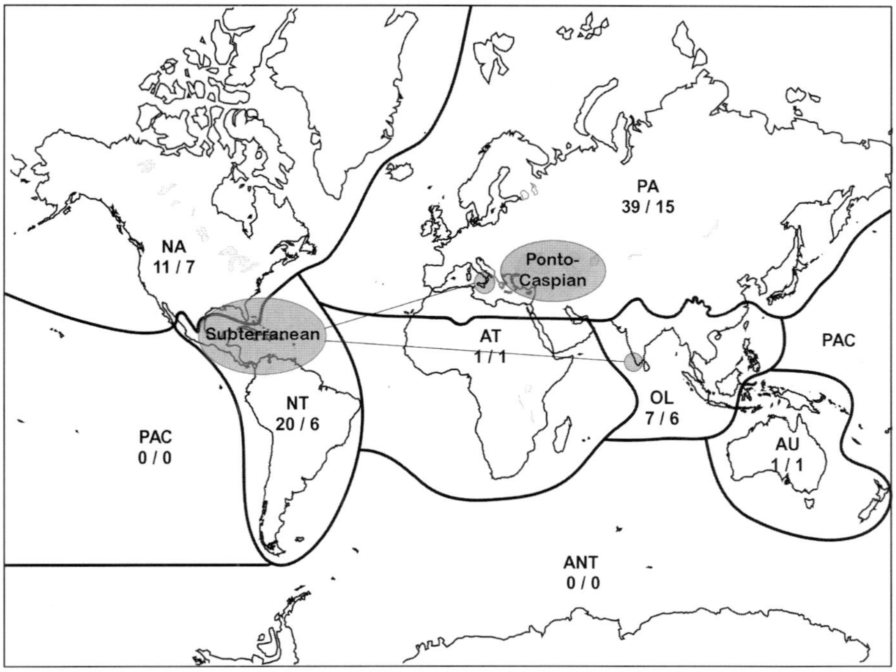

Fig. 2 Biogeographic regions indicating the numbers of inland mysid species and genera (SP/GN) found in each: PA—Palaearctic; NA—Nearctic; NT—Neotropical; AT—Afrotropical; OL—Oriental; AU—Australasian; PAC—Pacific; ANT—Antarctica; Grey circles indicate areas of high biodiversity, i.e., the Ponto-Caspian region and the distribution of Lepidopmysidae and Stygiomysidae continental subterranean species (lines connect disjunct regions of occurrence)

account for a large percentage of mysid species diversity.

Tethyan subterranean relicts

All of the subterranean/groundwater mysid genera considered here (*Antromysis* —6 spp., *Spelaeomysis*—7 spp., *Stygiomysis*—5 spp., and the monotypic *Troglomysis vjetrenicensis*) are found in a distributional pattern suggesting a Tethyan origin, likely colonizing groundwater habitats due to the uplift and stranding of marine ancestors during Miocene regressions of the Tethys and Mediterranean seas (Boxshall & Jaume, 2000). These hypothesized Tethyian dispersal events resulted in ancient mysid genera successfully colonizing early regions of Central America, the Caribbean, and Indian and Mediterranean basins, resulting in widespread genera that were later isolated, and presently represented by species often endemic to a single groundwater system. Additionally, the surface genera *Parvimysis* (2 spp.) and *Surinamysis* (3 spp.), distributed in the Caribbean and South America, are closely related to the genus *Antromysis* and may also be a part of this Tethyan distribution.

Autochthonous Ponto-Caspian endemics

The Ponto-Caspian basin, consisting of the Black, Azov, and Caspian Seas, is composed of inland seas with complex geological histories dating back to the Paratethys Sea (20 Mya), including periods as lacustrine environments (Banarescu, 1991). The Ponto-Caspian mysid fauna (autochthonous + 'glacial relicts', see below) are generally considered the center of inland mysid species diversity. The autochthonous mysids that evolved in these enclosed continental basins, occur in fresh and brackish water portions of the basins including rivers, lakes, and estuaries, and are endemic to one basin or are found in parts of all three (Table 3). Although this fauna consists of seven genera and 20 species, a large portion of the endemic mysid diversity in these basins is the result of a radiation in the genus *Paramysis* (Table 3).

In a phylogeographic study of *Limnomysis benedeni* and six *Paramysis* species across the Ponto-caspian region (Audzijonytė et al., 2006), three main patterns were identified: (1) no deep subdivisions across the entire region, (2) genealogical splits matching geographical borders among basins, and (3) divergent lineages occurring only within the Caspian. The discordant molecular subdivisions among these co-distributed mysid species suggest that the similar zoogeographic patterns were formed at different times (Audzijonytė et al., 2006).

Table 3 List of species endemic to the Black, Azov, and Caspian Seas, including the occurrence in each basin (data from Audzijonytė, 2006). In addition to the species counted here, there are at least four additional endemic Ponto-Caspian mysids (*Diamysis mecznikovi, Hemimysis serrata, Paramysis agigensis, Paramysis pontica*) whose distributions are unclear or do not occur in freshwaters and therefore have not been included in the species list

	Black Sea	Azov Sea	Caspian Sea
'Glacial-relict'			
Mysis amblyops			×
Mysis caspia			×
Mysis macrolepis			×
Mysis microphthalma			×
Autochtonous Ponto-Caspian			
Caspiomysis knipowitschi			×
Diamysis pengoi	×		
Diamysis pusilla			×
Hemimysis anomala	×	×	×
Katamysis warpachowsky	×	×	×
Limnomysis benedeni	×	×	×
Paramysis baeri	×	×	×
Paramysis eurylepis			×
Paramysis grimmi			×
Paramysis incerta			×
Paramysis inflata			×
Paramysis intermedia	×	×	×
Paramysis kessleri	×		×
Paramysis kosswigi	×		
Paramysis kroyeri	×		
Paramysis lacustris	×	×	×
Paramysis loxolepis			×
Paramysis sowinskii		×	×
Paramysis ullskyi	×	×	×
Schistomysis elegans			×

Mysis 'glacial relicts'

The genus *Mysis* comprises 14 species, of which eight have inland freshwater distributions and can be divided into two groups: (1) The *M. relicta* group (*M. relicta, M. diluviana, M. salemaai, M. segerstralei*) with a circumpolar distribution from boreal and subarctic lakes of the previously glaciated areas of Europe and North America (Audzijonytė & Väinölä, 2005), and (2) four Caspian Sea endemics (Table 3). Phylogenetic analyses indicate that the inland *Mysis* species are a monophyletic assemblage, sister to circumarctic marine species (Audzijonytė et al., 2005). The separation of freshwater/continental *Mysis* spp. and circumarctic marine species is estimated to have taken place 3–7 Mya (Audzijonytė et al., 2005), and does not correspond to the timing in general hypotheses of continental invasions, such as mid-Pleistocene glaciation events or mid-Tertiary separation of the Arctic and Caspian basins (see Väinölä, 1995). Furthermore, molecular divergences among the boreal Nearctic (*M. diluviana*) and Palearctic (*M. relicta, M. salemaai, M. segerstralei*) species and the Caspian endemics indicate inland colonizations occurring at different times (Audzijonytė & Väinölä, 2006). *M. relicta* and *M. diluviana* are considered the oldest of *Mysis* freshwater species that independently colonized their respective European and American ranges during early Pleistocene (Väinölä et al., 1994). In contrast, *M. salemaai* and *M. segerstralei* are younger closely related species that have penetrated freshwaters more recently (Audzijonytė & Väinölä, 2006). As for the four endemic Caspian *Mysis* species, small molecular divergences suggest a recent, possibly late Pleistocene, sympatric radiation, possibly driven by adaptation to a deep pelagic habitat by *M. amblyops* and *M. microphthalma* (Väinölä, 1995; Audzijonytė et al., 2005).

Euryhaline estuarine fauna

Most of the remaining mysids (20 spp., 14 genera) are euryhaline species with at least one population occurring in marginal freshwaters. These species have only very recently invaded freshwaters in portions of their distributions.

Phylogeny

The taxonomic position and phylogenetic affiliations within the Mysida are currently under debate. Historically, the Mysida were considered members of the crustacean superorder Peracarida, placed as a sister taxon to the Lophogastrida in the order Mysidacea. More recently, molecular studies have led to raising the Mysida (and Lophogastrida) to ordinal rank (Spears et al., 2005). With respect to those families containing inland fauna, there is also taxonomic and phylogenetic uncertainty. For example, molecular and morphological data show that the subterranean family Stygiomysidae is more closely related to the order Mictacea than to other Mysida families, suggesting that they be removed from the order Mysida and placed within a separate order, Stygiomysida, comprising the families Stygiomysidae and Lepidomysidae (Meland & Willassen, 2007).

Conservation issues

At least 19 inland mysid species are associated with groundwater habitats (caves, wells, and crab burrows) having very limited areas of distribution that are highly susceptible to pollution from the surface. These species in particular are in need of assessment for conservation ranking, as they are often found in aquifers important to local communities as a source of freshwater and may serve as indicators of water quality. Many freshwater mysid species have also served an important role in both US and European fisheries, where they have been introduced into lakes and reservoirs to serve as food for commercially important fish species (Mordukhai-Boltovskoi, 1979; Northcote, 1991).

In contrast, the autochthonous Ponto-Caspian mysids are currently invading aquatic ecosystems of Northern Europe as a result of human activities (De Vaate et al., 2002; Leppakoski et al., 2002). The impact of invasive mysid species on native lacustrine and riverine ecosystems can be large, including a severe reduction in zooplankton abundance, with concomitant negative effects on higher consumers (Spencer et al., 1991; Ketelaars et al., 1999).

Acknowledgments We thank R. Väinölä for advice concerning the Ponto-Caspian mysid fauna and for the helpful suggestions of one anonymous reviewer. *T. bowmani* and *A. almyra* images by E. Peebles from 'Common and Scientific Names of Aquatic Invertebrates from the United States and Canada: Crustaceans' are courtesy of the American Fisheries Society. This work was supported by NSF grant DEB-0206537.

References

Audzijonytė, A., 2006. Diversity and zoogeography of continental mysid crustaceans. Walter and Andrée de Nottbeck Foundation Scientific Reports 28: 1–46.

Audzijonytė, A. & R. Väinölä, 2005. Diversity and distributions of circumpolar fresh- and brackish-water *Mysis* (Crustacea: Mysida): descriptions of *M. relicta* Lovén, 1862, *M. salemaai* n. sp., *M. segerstralei* n. sp. and *M. diluviana* n. sp., based on molecular and morphological characters. Hydrobiologia 544: 89–141.

Audzijonytė, A., & R. Väinölä, 2006. Phylogeographic analyses of a circumarctic coastal and a boreal lacustrine mysid crustacean, and evidence of fast postglacial mtDNA rates. Molecular Ecology 15: 3287–3301.

Audzijonytė, A., M. E. Daneliya, & R. Väinölä, 2006. Comparative phylogeography of Ponto-Caspian mysid crustaceans: isolation and exchange among dynamic inland sea basins. Molecular Ecology 15: 2969–2984.

Audzijonytė, A., J. Damgaard, S.-L. Varvio, J. K. Vainio & R. Väinölä, 2005. Phylogeny of *Mysis* (Crustacea, Mysida): history of continental invasions inferred from molecular and morphological data. Cladistics 21: 575–596.

Banarescu P., 1991. Zoogeography of Fresh Waters, Vol. 2. Distribution and Dispersal of Freshwater Animals in North American and Eurasia. Aula Verlag, Wiesbaden.

Boxshall, G. A. & D. Jaume, 2000. Discoveries of cave misophrioids (Crustacea: Copepoda) shed new light on the origin of anchialine faunas. Zoologischer Anzeiger 239: 1–19.

De Vaate, A. B., K. Jazdzewski, H. A. M. Ketelaars, S. Gollasch, G. Van der Velde, 2002. Geographical patterns in range extension of Ponto-Caspian macroinvertebrate species in Europe. Canadian Journal of Fisheries and Aquatic Sciences 59: 1159–1174.

Ketelaars, H. A. M., F. E. Lambregts-van de Clundert, C. J. Carpentier, A. J. Wagenvoort & W. Hoogenboezem, 1999. Ecological effects of the mass occurrence of the Ponto-Caspian invader, *Hemimysis anomala* GO Sars, 1907 (Crustacea: Mysidacea), in a freshwater storage reservoir in the Netherlands, with notes on its autecology and new records. Hydrobiologia 394: 233–248.

Leppakoski, E., S. Golllasch, P. Gruszka, H. Ojaveer, S. Olenin & V. Panov, 2002. The Baltic: a sea of invaders. Canadian Journal of Fisheries and Aquatic Science 59: 1175–1188.

Martin, J. W. & G. E. Davis, 2001. An updated classification of the recent crustacea. Natural History Museum of Los Angeles County, Science Series 39: 1–124.

Mauchline, J., 1980. The biology of mysids and euphausiids. Advances in Marine Biology 18: 1–680.

Meland, K. & E. Willassen, 2007. The disunity of "Mysidacea" (Crustacea). Molecular Phylogenetics and Evolution. http://dx.doi.org/10.1016/j.ympev.2007.02.009

Mordukhai-Boltovskoi, F. D., 1979. Composition and distribution of Caspian fauna in the light of modern data. Internationale Revue der Gesamten Hydrobiologie 64: 1–38.

Northcote, T. G., 1991. Success, problems, and control of introduced mysid populations in lakes and reservoirs. American Fisheries Society Symposium 9: 5–16.

Spears, T., R. W. DeBry, L. G. Abele & K. Chodyla, 2005. Peracarid monophyly and interordinal phylogeny inferred from nuclear small-subunit ribosomal DNA sequences (Crustacea: Malacostraca: Peracarida). Proceedings of the Biological Society of Washington 118: 117–157.

Spencer, C. N., B. R. McClelland & J. A. Stanford, 1991. Shrimp stocking, salmon collapse, and eagle displacement. Bioscience 41: 14–21.

Väinölä, R., 1995. Origin and recent endemic divergence of a Caspian *Mysis* species flock with affinities to the "glacial relict" crustaceans in boreal lakes. Evolution 49: 1215–1223.

Väinölä, R., B. J. Riddoch, R. D. Ward & R. I. Jones, 1994. Genetic zoogeography of the *Mysis relicta* species group (Crustacea: Mysidacea) in northern Europe and North America. Canadian Journal of Fisheries and Aquatic Science 51: 1490–1505.

Wittmann K. J., 1999. Global biodiversity in Mysidacea, with notes on the effects of human impact. In Schram F. R. & J. C. von Vaupel Klein (eds), Crustaceans and the Biodiversity Crisis. Proceedings of the Fourth International Crustacean Congress, Amsterdam, The Netherlands, July 20–24, 1998, Vol. I. Brill NV, Leiden: 511–525.

Hydrobiologia (2008) 595:219–224
DOI 10.1007/s10750-007-9017-1

Global diversity of spelaeogriphaceans & thermosbaenaceans (Crustacea; Spelaeogriphacea & Thermosbaenacea) in freshwater

D. Jaume

Abstract Spelaeogriphaceans and thermosbaenaceans are two orders of eyeless, unpigmented peracarid crustaceans represented by very few species from subterranean waters. Spelaeogriphaceans occur only in continental waters, either running or still, in limestone or sandstone caves, or in calcrete aquifers. The four species known are limnic except one occurring in slightly brackish water loosely associated with an endorheic basin. The Thermosbaenacea are primarily marine, with only 18 species recorded in limnic conditions or in brackish inland waters whose salinity does not derive from dilution of seawater. They occur in limestone caves, the interstitial medium associated to alluvial deposits, or in thermo-mineral springs. Spelaeogriphaceans are found on the southern continents, in ancient cratons not affected by sea transgressions at least since the Early Cretaceous, when Gondwana started to break-up. The former integration of these terranes into Gondwana suggests that the penetration of spelaeogriphaceans in continental waters took place previous to the fragmentation of this super continent (starting ca. 140 Ma), and that their current distribution pattern was driven by continental drift. The distribution of the Thermosbaenacea matches precisely the area covered by the ancient Tethys Sea or its coastlines. They are most probably relicts of a once widespread shallow-water marine Tethyan fauna stranded in interstitial or crevicular groundwater during marine regressions.

Keywords Freshwater · Global assessment · Species richness · Peracarida · Crustacea

Guest editors: E. V. Balian, C. Lévêque, H. Segers & K. Martens
Freshwater Animal Diversity Assessment

Electronic supplementary material The online version of this article (doi:10.1007/s10750-007-9017-1) contains supplementary material, which is available to authorized users.

D. Jaume (✉)
IMEDEA (CSIC-UIB), Instituto Mediterráneo de Estudios Avanzados, c/Miquel Marquès 21, Esporles, Illes Balears 07190, Spain
e-mail: d.jaume@uib.es

Introduction

Spelaeogriphaceans and thermosbaenaceans are two orders of eyeless, unpigmented peracarid crustaceans represented by very few species from subterranean waters. The Spelaeogriphacea Gordon, 1957 occur only in continental waters, either running or still, in limestone or sandstone caves, or in calcrete aquifers. All species are limnic except one occurring in slightly brackish water loosely associated to an endorheic basin.

Thermosbaenacea Monod, 1927, in contrast, seem to be primarily marine although only 5 out of the 34 species known are euhaline, living in lava tubes or in the interstices between submersed coarse sand grains, volcanic tephra or coral rubble (Wagner, 1994). Most taxa occur in the anchialine environment associated with marine coastal areas, where water salinity varies sharply across the water column and derives from dilution of sea water. Only 18 species have been recorded in limnic conditions or in brackish inland waters whose salinity does not derive from dilution of seawater. They occur in limestone caves, the interstitial medium associated to alluvial deposits or in thermo-mineral springs.

The spelaeogriphacean body is roughly cylindrical, with the cephalothorax incorporating only the first thoracomere (which carries the maxillipeds) and with a short, posteriorly directed carapace covering only part of the second thoracomere dorsally (Fig. 1A). The pereion consists of seven free pereiomeres, each with a pair of similar biramous stenopodial pereiopods, and the pleon comprises six free pleomeres, each with a pair of well-developed biramous pleopods. The telson is free, articulated to the 6th pleomere. The most remarkable autapomorphy of the group is the transformation of the exopods of the posterior pairs of pereiopods into non-setose, respiratory paddles.

Contrary to spelaeogriphaceans, whose inclusion in the Peracarida is undeniable since brooding females display a thoracic ventral marsupium formed by oöstegites (= foliaceous medial extensions of the pereiopodal coxae), the inclusion of the Thermosbaenacea in this group is debatable since here the embryos are carried in a dorsal brood pouch derived from the carapace (Richter & Scholtz, 2001). Apart of this, the rest of features of the thermosbaenacean body plan is roughly similar to the spelaeogriphaceans (Fig. 1B); just notice that the pleopods are now vestigial and present only on pleomeres 1 and 2, and that the condition of the posterior extension of the carapace varies from covering the second thoracomere only (in males and non-brooding females of Thermosbaenidae, Monodellidae and Halosbaenidae) to cover all thoracomeres (Tulumellidae). *Thermosbaena mirabilis* is a highly modified thermosbaenacean that separates from the ordinary morphology of the group in having a pleotelson (formed by the 6th pleomere and telson), and the display of only five pairs of pereiopods.

The mouthparts of spelaeogriphaceans and thermosbaenaceans conform to a functional series of scrapers in an arrangement considered to fit for

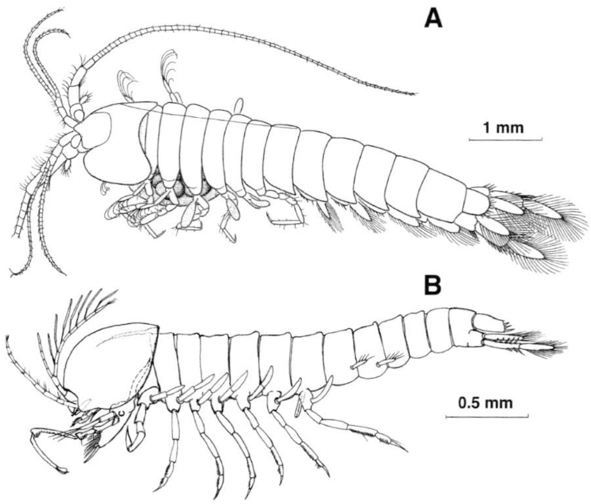

Fig. 1 General aspect of Spelaeogriphacea and Thermosbaenacea. (**A**) *Spelaeogriphus lepidops* Gordon 1957, a spelaeogriphacean from Table Mountain, South Africa (after Gordon, 1960). (**B**) male of *Tethysbaena atlantomaroccana* (Boutin & Cals, 1985), a freshwater thermosbaenacean from Morocco; notice that ornamentation of pereopodal exopods is omitted from figure (after Cals & Boutin, 1985)

Table 1 Global diversity of Spelaeogriphacea

	Distribution	Habitat	Salinity range
Spelaeogriphidae			
Mangkurtu Poore & Humphreys, 1998			
mityula Poore & Humphreys, 1998	NW Australia	Borehole wells	Limnic
kutjarra Poore & Humphreys, 2003	NW Australia	Borehole wells	Oligohaline
Potiicoara Pires, 1987			
brasiliensis Pires, 1987	Brazil	Caves	Limnic
Spelaeogriphus Gordon, 1957			
lepidops Gordon, 1957	South Africa	Caves	Limnic

Table 2 Global diversity of non-marine Thermosbaenacea

	Distribution	Habitat	Salinity range
Thermosbaenidae			
Thermosbaena Monod, 1924			
mirabilis Monod, 1924	Tunisia	Thermo-mineral springs	Presumed oligohaline
Monodellidae			
Tethysbaena Wagner, 1994			
juriaani Wagner, 1994	Dominican Rep.	Anchialine limestone wells and spring	Limnic to oligohaline
gaweini Wagner, 1994	Dominican Rep.	Wells in alluvial sediments	Limnic to oligohaline
haitiensis Wagner, 1994	Haiti	Spring, wells and alluvial sediments	Limnic to oligohaline
juglandis Wagner, 1994	Haiti	Wells in alluvial sediments	Limnic
lazarei Wagner, 1994	Cuba	Cave and river interstitial	Limnic to oligohaline
tinima Wagner, 1994	Cuba	Wells in limestone	Limnic
calsi Wagner, 1994	Saint John; Tortola (British Virgin Islands)	Wells in alluvial deposits	Limnic to oligohaline
relicta (Pór, 1962)	Israel	Thermo-mineral springs	Oligohaline to hyperhaline
somala (Chelazzi & Messana, 1982)	Somalia	Wells in limestone	Oligohaline to polyhaline (evaporites)
atlantomaroccana (Cals & Boutin, 1885)	Morocco	Wells in alluvial deposits	Presumed limnic
tarsiensis Wagner, 1994	Spain	Well in alluvial deposit	Limnic
texana (Maguire, 1965)	Texas (U.S.A.)	Artesian wells	Limnic
vinabayesi Wagner, 1994	Isla Juventud (Cuba)	Cave	Limnic
Halosbaenidae			
Limnosbaena Stock, 1976			
finki (Mestrov & Lattinger-Penko, 1969)	Bosnia-Hercegovina; Italy	Interstitial of river alluvia; caves	Limnic
sp. Wagner, 1994	France	Well	Limnic
Halosbaena Stock, 1976			
tulki Poore & Humphreys, 1992	NW Australia	Calcrete aquifers	Oligohaline (evaporites)
Theosbaena Cals & Boutin, 1985			
cambodjana Cals & Boutin, 1985	Cambodia; Thailand	Caves	Limnic

Salinity tolerance ranges: Limnic: <0.5‰; Oligohaline: 0.5–5‰; Polyhaline: 18–30‰; Hyperhaline: >40‰

scraping small food particles from a substrate (Fryer, 1964).

Species diversity

The Spelaeogriphacea comprises a single family, the Spelaeogriphidae Gordon, 1957 with three genera: *Speleaeogriphus* and *Potiicoara* (both monotypic), and *Mangkurtu*, with two species (Table 1). Three fossil forms from the Carboniferous of Canada, the Upper Jurassic of China, and the Lower Cretaceous of Spain are treated also as spelaeogriphaceans by some authors (Schram, 1974; Shen et al., 1998; 1999). Nevertheless, none of these fossil forms preserves the diagnostic pereopodal exopods, and their body tagmosis and short carapace do not preclude their allocation to other peracaridan orders.

The Thermosbaenacea embraces four families, of which only three include non-marine species (Table 2). The family Thermosbaenidae is monotypic. The Monodellidae includes two genera:

Monodella (monotypic) and *Tethysbaena* (23 species), of which only 13 species of *Tethysbaena* are considered here as non-marine. The Halosbaenidae consists of the truly limnic genera *Limnosbaena* (two species, one of them not formally described) and *Theosbaena* (monotypic), plus the genus *Halosbaena*, with two marine species plus one from non-marine, brackish inland waters. The fourth thermosbaenacean family Tulumellidae, comprising the single genus *Tulumella* (three species), is fully marine. There is no fossil record of the Thermosbaenacea known to date.

Many species of thermosbaenaceans have been reported from oligohaline water in coastal aquifers or anchialine environments only, frequently around haloclines where water salinity changes abruptly from limnic-oligohaline to marine euhaline. These taxa (shown in Appendix Table 1) are not included in the total estimate of non-marine species since none has been reported from pure fresh waters, nor from marine euhaline water either.

Present distribution and historical processes

Living spelaeogriphaceans appear associated with freshwater in southern continents, in ancient cratons not affected by sea transgressions at least since the Early Cretaceous, when Gondwana started to break-up (Fig. 2; Table 3; Appendix Fig. A). *Spelaeogriphus lepidops* is known only from two caves in South Africa excavated in Ordovician quartzites (Gordon, 1957). *Potiicoara brasiliensis*, from two caves in Upper Proterozoic limestone at Mato Grosso do Sul (Brazil; Pires, 1987). And the two species of *Mangkurtu*, from borehole wells in calcrete of Middle to Late Tertiary age on north-western Australia, although these deposits overlie and are in direct contact with Early Cretaceous alluvial conglomerates (Poore & Humphreys, 1998; 2003). The former integration of these terranes in Gondwana suggests that the penetration of spelaeogriphaceans in continental waters took place previous to the fragmentation of this supercontinent (starting ca. 140 Ma), and

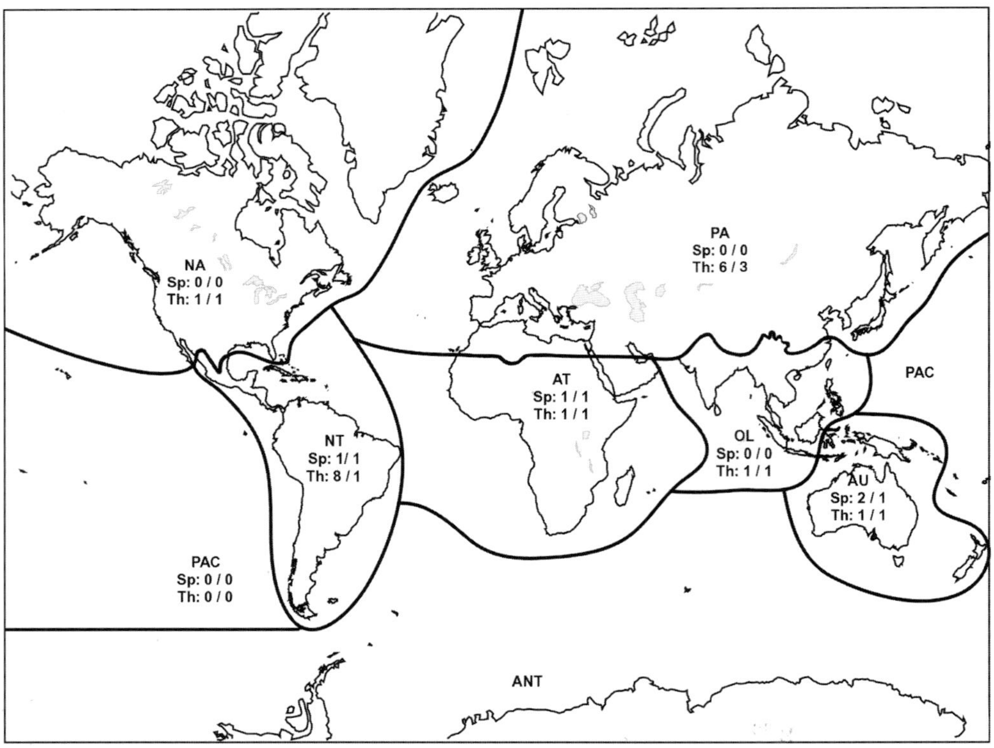

Fig. 2 Global distribution of Spelaeogriphacea and continental water Thermosbaenacea (Species number/Genus number). PA—Palaearctic, NA—Nearctic, NT—Neotropical, AT—Afrotropical, OL—Oriental, AU—Australasian, PAC—Pacific Oceanc Islands, ANT—Antarctic

Table 3 Global and per biogeographic region diversity (species number) of Spelaeogriphacea and non-marine Thermosbaenacea

	PA	NA	NT	AT	OL	AU	PAC	World
Spelaeogriphacea								
Spelaeogriphidae	–	–	1 (1)	1 (1)	–	2 (1)	–	4 (3)
Thermosbaenacea								
Thermosbaenidae	1 (1)	–	–	–	–	–	–	1 (1)
Monodellidae	3 (1)	1 (1)	8 (1)	1 (1)	–	–	–	13 (1)
Halosbaenidae	2 (1)	–	–	–	1 (1)	1 (1)	–	4 (3)
Total Thermosbaenacea	6 (3)	1 (1)	8 (1)	1 (1)	1 (1)	1 (1)	–	18 (5)

In brackets, number of genera. No records of these groups exist from Pacific oceanic islands or Antarctica. PA: Palaearctic, NA: Nearctic, NT: Neotropical, AT: Afrotropical, OL: Oriental, AU: Australasian, PAC: Pacific Oceanc Islands, ANT: Antarctic

that their current distribution pattern was driven by continental drift.

The distribution of the order Thermosbaenacea matches precisely the area covered by the ancient Tethys Sea or its coastlines (Fig. 2, 3; Table 3; see also Appendix Figs A and B). They are most probably relicts of a once widespread shallow-water marine Tethyan fauna, stranded in interstitial or crevicular groundwater during episodes of marine regression. The amphi-Atlantic distribution of *Tethysbaena*, or the so-called "full Tethyan track" displayed by *Halosbaena* (with species in the Caribbean, Canary Islands and Australia) suggest the origin of these genera dates back to at least the final opening of the Atlantic (95 Ma) and the maximum extent of the Tethys sea (120 Ma), respectively. The timing of colonisation of continental waters by thermosbaenacean lineages is probably more recent (all inland water stations fall within areas covered by the sea in Pliocene or more recent times), and the speciation process can have been peripatric.

The distribution of the Thermosbaenidae casts doubts on whether they represent a separate family from its closest relative, the Monodellidae. The single representative of the family dwells in a Tunisian thermo-mineral spring placed in a zone recently covered by the sea (Late Pliocene). This station falls within the area covered by the Monodellidae, which comprises a widespread amphi-Atlantic genus (indicative of the ancient origin for the family; see above) present also in N Africa (*Tethysbaena*), plus a second genus *Monodella*, monotypic and know only from a single-anchialine cave in Italy. The Thermosbaenidae are probably a young,

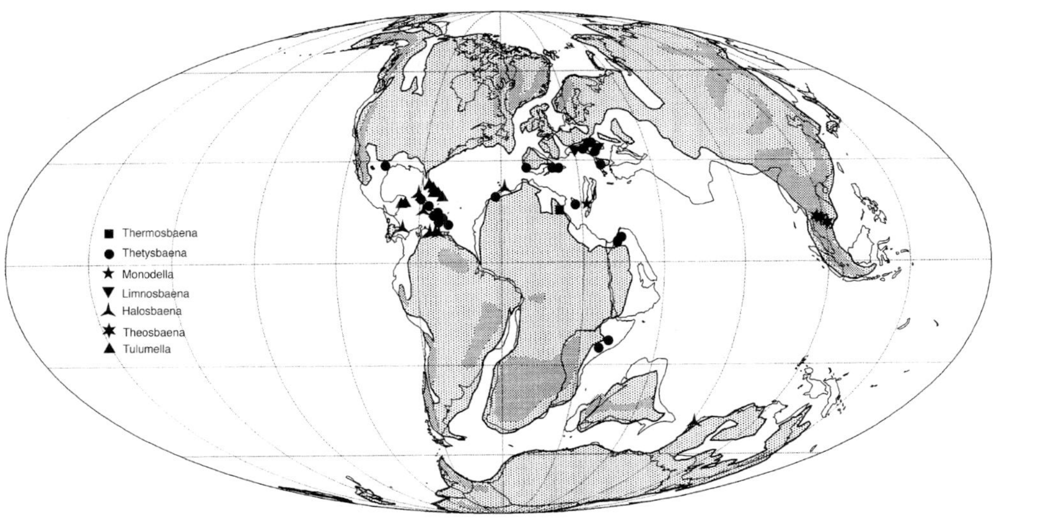

Fig. 3 Translocation of present distribution of Thermosbaenacea (including both marine and continental water taxa) to an Aptian (Lower Cretaceous; 120 Ma) palaeo-coastline map. Shaded areas denote emerged lands

highly modified member of the family Monodellidae adapted to an exceptional habitat (hot springs; see Wagner, 1994: 317).

Acknowledgement This is a contribution to Spanish MEC project CGL2005-02217/ BOS.

References

Boutin, C. & P. Cals, 1985. Importance en biogeography évolutive de la découverte d'un Crustacé phréatobie, *Monodella atlantomaroccana* n. sp. (Thermosbaenacea) dans la plaine alluviale de Marrakech (Maroc atlantique). Comptes Rendus de l'Académie des Sciences, Paris 300 (III, 7): 267–270.

Cals, P. & C. Boutin, 1985. Découverte au Cambodge, domaine ancien de la Téthys orientale, d'un nouveau <<fossile vivant>> *Theosbaena cambodjiana* n.g., n.sp. (Crustacea, Thermosbaenacea). Comptes rendus de l'Académie des sciences, (3) 300(8): 337–340.

Fryer, G., 1964. Studies on the functional morphology and feeding mechanism of *Monodella argentarii* Stella (Crustacea: Thermosbaenacea). Transactions of the Royal Society of Edinburgh 66: 49–90.

Gordon, I., 1957. On *Spelaeogriphus*, a new cavernicolous crustacean from South Africa. Bulletin of the British Museum (Natural History), (Zoology) 5: 31–47.

Gordon, I., 1960. On a *Stygiomysis* from the West Indies, with a note on *Spelaeogriphus* (Crustacea, Peracarida). Bulletin of the British Museum (Natural History), (Zoology) 6: 285–326.

Pires, A. M. S., 1987. *Potiicoara brasiliensis*: a new genus and species of Spelaeogriphacea (Crustacea: Peracarida) from Brazil, with a phylogenetic analysis of the Peracarida. Journal of Natural History 21: 225–238.

Poore, G. C. B. & W. F. Humphreys, 1998. First record of Spelaeogriphacea from Australasia: a new genus and species from an aquifer in the arid Pilbara of Western Australia. Crustaceana 71: 721–742.

Poore, G. C. B. & W. F. Humphreys, 2003. Second species of *Mangkurtu* (Spelaeogriphacea) from north-western Australia. Records of the Western Australian Museum 22: 67–74.

Richter, S. & G. Scholtz, 2001. Phylogenetic analysis of the Malacostraca (Crustacea). Journal of Zoological Systematics and Evolutionary Research 39: 113–136.

Schram, F. R., 1974. Paleozoic Peracarida of North America. Fieldiana Geology 33: 95–124.

Shen, Y.-B., F. R. Schram & R. S. Taylor, 1999. *Liaoningogriphus quadripartitus* (Malacostraca: Spelaeogriphacea) from the Jehol Biota and notes on its paleoecology. Palaeoworld 11: 175–187 (in Chinese).

Shen, Y. -B., R. S. Taylor & F. R. Schram, 1998. A new spelaeogriphacean (Crustacea: Peracarida) from the Upper Jurassic of China. Contributions to Zoology 68: 19–35.

Wagner, H. P., 1994. A monographic review of the Thermosbaenacea (Crustacea: Peracarida). Zoologische Verhandelingen 291: 1–338.

Hydrobiologia (2008) 595:225–230
DOI 10.1007/s10750-007-9018-0

FRESHWATER ANIMAL DIVERSITY ASSESSMENT

Global diversity of cumaceans & tanaidaceans (Crustacea: Cumacea & Tanaidacea) in freshwater

D. Jaume · G. A. Boxshall

Abstract Cumacea and Tanaidacea are marginal groups in continental waters. Although many euryhaline species from both groups are found in estuaries and coastal lagoons, most occur only temporarily in non-marine habitats, appearing unable to form stable populations there. A total of 21 genuinely non-marine cumaceans are known, mostly concentrated in the Ponto-Caspian region, and only four tanaids have been reported from non-marine environments. Most non-marine cumaceans (19 species) belong in the Pseudocumatidae and appear restricted to the Caspian Sea (with salinity up to 13‰) and its peripheral fluvial basins, including the northern, lower salinity zones of the Black Sea (Sea of Azov). There are nine Ponto-Caspian genera, all endemic to the region. Only two other taxa (in the family Nannastacidae) occur in areas free of any marine–water influence, in river basins in North and South America. Both seem able to survive in waters of raised salinity of the lower reaches of these fluvial systems; but neither has been recorded in full salinity marine environments. The only non-marine tanaidacean thus far known lives in a slightly brackish inland spring in Northern Australia. The genus includes a second species, from a brackish-water lake at the Bismarck Archipelago, tentatively included here as non-marine also. Two additional species of tanaidaceans have been reported from non-marine habitats but both also occur in the sea.

Keywords Freshwater · Global assessment · Species richness · Peracarida · Crustacea

Introduction

Comprising about 1,300 and 900 marine species, respectively, the Cumacea and Tanaidacea are only marginal groups in continental waters. Although many euryhaline species in both taxa are found in estuaries and coastal lagoons, most occur only temporarily in non-marine habitats, appearing unable to form stable populations there. Only 21 genuinely non-marine cumaceans are known, most of which occur in the Ponto-Caspian region, whereas just four tanaids have been reported in non-marine habitats. Both groups are orders of peracarid crustaceans that are mainly adapted to a fossorial life-style in non-consolidated marine sediments, especially in deep

Guest editors: E. V. Balian, C. Lévêque, H. Segers & K. Martens
Freshwater Animal Diversity Assessment

D. Jaume (✉)
IMEDEA (CSIC-UIB), Instituto Mediterráneo de Estudios Avanzados, C/ Miquel Marquès 21, Esporles, Illes Balears 07190, Spain
e-mail: d.jaume@uib.es

G. A. Boxshall
Department of Zoology, The Natural History Museum, Cromwell Road, London SW7 5BD, UK
e-mail: g.boxshall@nhm.ac.uk

waters, although they can appear regularly in night-time plankton hauls in shallow waters.

The characteristic body form of a cumacean consists of a large, variably inflated cephalothorax incorporating the first 3 (of 8) thoracic somites, plus an elongate, narrow abdomen terminating in a pair of long and slender uropods. The cephalothorax displays a pair of frontal extensions, the pseudorostral lobes, which converge medially in most instances, whereas its lateral portions act as paired branchial chambers accommodating the respiratory epipodites of the first maxillipeds (see below). All thoracopods except the first, second and eighth are primitively biramous. The first pair (=first maxillipeds) is characteristic, possessing a respiratory coxal epipodite provided with digitiform extensions in addition to a narrow frontal extension, which together with the corresponding cephalothoracic pseudorostral lobe, forms a branchial siphon (exhalant canal) for the corresponding branchial chamber. The abdomen comprises six free pleomeres and a free telson, although in some families the latter is incorporated into the sixth pleomere forming a pleotelson. Apart from the uropods on the last pleomere, there are up to five pairs of pleopods in males, but a maximum of only one pair in females. Reduction in number of pairs of pleopods is common. All these limbs are originally biramous, with a 2-segmented exopod and a unisegmented endopod; the endopod of the uropod can be up to 3-articulate. Cumaceans are primarily deposit feeders, although some are apparently predators of foraminifers and other crustaceans. Most live half-buried in soft sediments.

General morphological characteristics for the order Tanaidacea include: a small cephalothorax incorporating the first two thoracic somites, six free thoracic somites, five abdominal somites bearing pleopods, and a pleotelson with a pair of uropods. All thoracopods except the third (=first pereiopod) of most apseudomorphs, and some other pereiopods of the manca stages of the genus *Kalliapseudes*, are uniramous. The maxillipeds (=first thoracopods) possess a respiratory coxal epipodite, which is concealed under the lateral margin of the cephalothorax (branchial cavity). The second pair of thoracopods is prehensile, displaying a chelate distal portion ("chelipeds"). The pleopods and uropods are basically biramous with 2-segmented exopods and unisegmented endopods, although both rami of the uropods can be multi-articulate, due to the display of cuticular annulations. Tanaidaceans are primarily tube or tunnel dwellers, and are generally considered to be deposit feeders.

Species diversity, distribution and historical processes

Non-marine cumaceans belong to two of the eight recognised families: Pseudocumatidae Sars and Nannastacidae Bate. Most non-marine species (19) are pseudocumatids and their distribution is focused around the Caspian Sea (maximum salinity 13‰) and its peripheral fluvial basins, including the northern, lower salinity zones of the Black Sea (Sea of Azov) (see Tables 1 and 3; Fig. 1). They represent nine genera, all endemic to the region, although the taxonomic status of some genera is equivocal (e.g. *Charsarocuma*; see comments by Sars (1914: 32) on its presumed synonymy with *Schizoramphus*) and their validity should be tested. The natural distribution of these taxa within the Ponto-Caspian region is difficult to ascertain since dispersal via artificial canals and reservoirs, by shipping, or even by deliberate introduction as fish food, may have had a profound effect (see Băcescu & Petrescu, 1999, and references therein). *Stenocuma graciloides* has recently been reported from the Gulf of Finland (Baltic Sea), where it may have been transported by ships passing through the Volga-Baltic waterway from its North Caspian home (Antsulevich, 2005). The presumed deliberate introduction of *Stenocuma gracilis*, *Pterocuma pectinata* and *Schizorhamphus scabriusculus* into the Aral sea, to serve as fish food (Karpevitch, 1960; quoted in Băcescu & Petrescu, 1999), seems not to have succeeded (Nikolay Aladin, pers. comm.). Apart from these Caspian pseudocumatids, only two other taxa (from the Nannastacidae) occur in areas free of any marine-water influence, in river basins in North and South America. Both seem able to survive in waters of raised salinity of the lower reaches of these fluvial systems (see Tables 1 and 3; Fig. 1), but neither has been recorded in full salinity marine environments. These two monotypic genera are endemic to their respective river basins.

Sars (1914) considered that the Caspian Cumacea were derived from a single ancestral form originating from the Mediterranean, probably belonging to the

Table 1 Global diversity of non-marine Cumacea (distribution of Ponto-Caspian taxa after Băcescu (1992) and Antsulevich (2005))

Order Cumacea	Distribution
Family Pseudocumatidae G. O. Sars, 1878	
Genus *Carinocuma* Mordukhai-Boltovskoi & Romanova, 1973	
C. birsteini Mordukhai-Boltovskoi & Romanova, 1973	Caspian Sea
Genus *Caspiocuma* G. O. Sars, 1900	
C. campylaspoides (G. O. Sars, 1897)	Caspian Sea; Volga, Don, Bug and Dniestr river basins
Genus *Charsarocuma* Derzhavin, 1912	
C. knipowitschi Derzhavin, 1912	Caspian Sea; pre-delta region of Volga
Genus *Hyrcanocuma* Derzhavin, 1912	
H. sarsi Derzhavin, 1912	Caspian Sea
Genus *Stenocuma* G. O. Sars, 1900	
S. gracilis (G. O. Sars, 1894)	Caspian Sea; Volga
S. graciloides (G. O. Sars, 1894)	Caspian Sea; Estuaries of Volga, Don, Dniestr and Danube; Black Sea (Azov); Gulf of Finland (Baltic Sea)
S. tenuicauda (G. O. Sars, 1894)	Caspian Sea; Volga
S. diastyloides (G. O. Sars, 1897)	Caspian Sea
S. cercarioides G. O. Sars, 1894	Caspian Sea; Volga, Don, Bug and Dniestr river basins; Black Sea
S. laevis (G. O. Sars, 1914)	Caspian Sea
Genus *Pterocuma* G. O. Sars, 1900	
P. pectinata (Sowinski, 1893)	Caspian Sea; Volga, Danube and Dniestr river basins; Black Sea
P. rostratum (G. O. Sars, 1894)	Caspian Sea; Volga; estuaries of Dniepr, Bug and Danube; Black Sea
P. sowinskyi (G. O. Sars, 1894)	Caspian; Volga; delta of Don; Black Sea
P. grandis G. O. Sars, 1914	Caspian Sea
Genus *Schizorhamphus* Băcescu, 1992	
S. bilamellatus (G. O. Sars, 1894)	Caspian Sea; Volga
S. eudorelloides (G. O. Sars, 1894)	Caspian Sea (up to 264 m depth); river mouths of Danube, Dniestr and Prut; Black Sea
S. scabriusculus (G. O. Sars, 1894)	Caspian Sea; Danube, Dniestr, Bug and Dniepr rivers
Genus *Strauchia* Czerniavsky, 1868	
S. taurica Czerniavsky, 1868	Caspian Sea
Genus *Volgocuma* Derzhavin, 1912	
V. telmatophora Derzhavin, 1912	Caspian Sea; Volga; Black Sea
Family Nannastacidae Bate, 1866	
Genus *Almyracuma* Jones & Burbanck, 1959	
A. proximoculi Jones & Burbanck, 1959	Intertidal freshwater springs at Cape Cod, and limnetic zone of Lower Hudson river (latter 1–30‰; Simpson et al., 1985), NE U.S.A.
Genus *Claudicuma* Roccatagliata, 1981	
C. platense Roccatagliata, 1981	Río de la Plata (Argentina), from Buenos Aires (0.5‰) to Punta del Indio (1.8–7.0‰) (Roccatagliata, 1991)

marine genus *Pseudocuma* Sars, 1865, which includes three Mediterranean species, one of which is also present in the Black Sea. The Caspian genus *Stenocuma* is considered to be a subgenus of *Pseudocuma* by some authors (Băcescu, 1992). Dumont (2000: 186) believed that Caspian cumaceans were derived from ancestral forms that lived in estuaries and tidal zones of rivers that discharged into

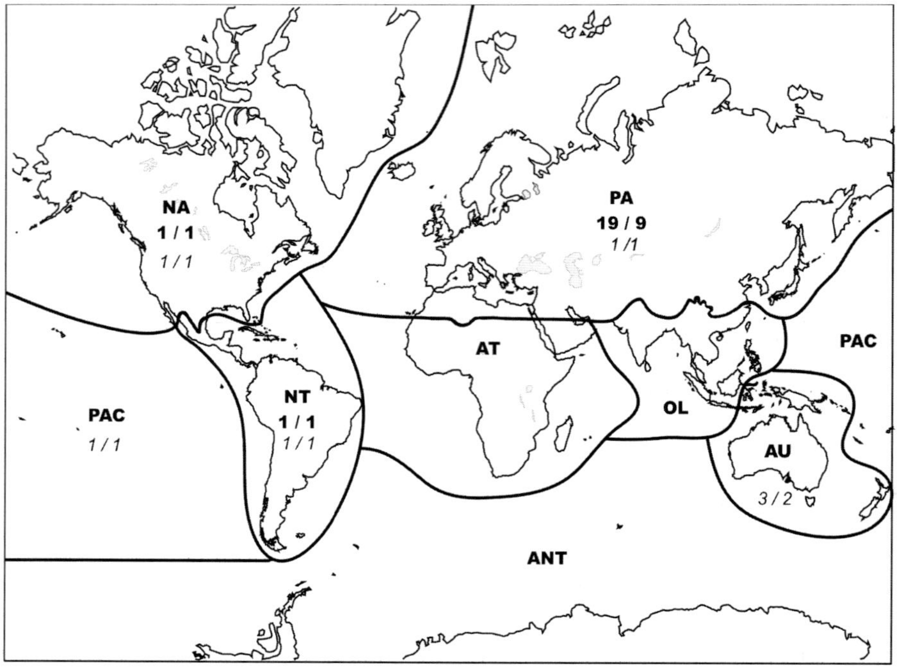

Fig. 1 Total species and genus numbers of Cumacea (Bold) and Tanaidacea (italics) per zoogeographic regions (Species number/ Genus number). PA: Palaearctic, NA: Nearctic, NT: Neotropical, AT: Afrotropical, OL: Oriental, AU: Australasian, PAC: Pacific Oceanic Islands, ANT: Antarctic

the (proto-) Mediterranean before the closure of the Sarmatian Basin, a vanished Miocene brackish lake that covered the entire Ponto-Caspian region from 14.5 to 8.3 Myr ago. Their osmoregulatory abilities would have preadapted them to life in the brackish Sarmatian lake.

The only truely non-marine tanaidacean known is *Pseudohalmyrapseudes aquadulcis* (Parapseudidae) which lives in a slightly brackish inland spring in Northern Australia (see Tables 2 and 3; Fig. 1; Larsen & Hansknecht, 2004). The genus includes a second species, *P. mussauensis*, from a brackish-water lake in the Bismarck Archipelago; this species is tentatively included here as non-marine, since the genus has not been recorded yet in fully marine environments, and Shiino (1965) was rather vague in his description of the salinity regime of the lake where the species was discovered (see Tables 2 and 3; Fig. 1).

Table 2 Global diversity of non-marine Tanaidacea

Order Tanaidacea	Distribution
Family Parapseudidae Gutu, 1981	
Genus *Pseudohalmyrapseudes* Larsen & Hansknecht, 2004	
P. aquadulcis Larsen & Hansknecht, 2004	"Freshwater spring" (but 1.93‰ in salinity), Australian Northern Territory
P. mussauensis (Shiino, 1965)	"Brackish lake", Bismarck Archipelago (Papua New Guinea)
Family Tanaidae Dana, 1849	
Genus *Sinelobus* Sieg, 1980	
S. stanfordi (Richardson, 1901)	Marine, plus freshwater, hypohaline and hypersaline inland waters of Galapagos, Japan, Hong Kong, New Zealand, Australia, Argentina, Kurile Islands, West Indies, Florida and Brazil (see Larsen & Hansknecht, 2004, and references therein)
Family Anarthruridae Lang, 1971	
Genus *Paraleptognathia* Kudinova-Pasternak, 1981	
P. longiremis (Lilljeborg, 1864)	Deep sea plus… Lake Baikal! (Kudinova-Pasternak, 1972) Record requiring confirmation

Table 3 Global and per biogeographic region diversity (species number) of non-marine Cumacea and Tanaidacea

	PA	NA	NT	AT	OL	AU	PAC	World
Nannastacidae		1 (1)	1 (1)					2 (2)
Pseudocumatidae	19 (9)							19 (9)
Total Cumacea	19 (9)	1 (1)	1 (1)					21 (11)
Anarthruridae	1 (1)							1 (1)
Parapseudidae						2 (1)		2 (1)
Tanaidae	1 (1)	1 (1)	1 (1)		1 (1)	1 (1)	1 (1)	1 (1)
Total Tanaidacea	2 (2)	1 (1)	1 (1)		1 (1)	3 (2)	1 (1)	4 (3)

In brackets, number of genera. No records of these groups exist from Antarctica. PA: Palaearctic, NA: Nearctic, NT: Neotropical, AT: Afrotropical, OL: Oriental, AU: Australasian, PAC: Pacific Oceanic Islands, ANT: Antarctic.

Larsen & Hansknecht (2004) suggest that *Pseudohalmyrapseudes* occupies an intermediate position between the euryhaline genus *Halmyrapseudes* Băcescu & Gutu, 1972 and *Longiflagrum* Gutu, 1995, although no phylogenetic analysis was performed to support this suggestion. The Australian species is inferred to have reached the spring it inhabits via the groundwater system, although the possibility of an upstream migration from the ocean cannot be ruled out.

Two other species of tanaidaceans have been reported from non-marine habitats, but both occur also in marine environments. *Sinelobus stanfordi* (Tanaidae), a widely distributed euryharine taxon, has been reported repeatedly from geographically scattered freshwater, hypohaline or hypersaline lakes (Tables 2 and 3; Fig. 1). In addition, there is a doubtful record of the deep-sea trench *Paraleptognathia longiremis* (Anarthruridae) from Lake Baikal (Kudinova-Pasternak, 1972; Tables 2 and 3; Fig. 1). This record requires confirmation and, as Larsen & Hansknecht (2004) point out, the conspecificity of the non-marine populations of these two taxa to their corresponding marine forms should be confirmed, suggesting that the current diversity of non-marine tanaidacean species is underestimated.

Acknowledgement This is a contribution to Spanish MEC project CGL2005-02217/ BOS.

References

Antsulevich, A. E., 2005. First finding of Cumacea crustaceans in the Gulf of Finland. Vestnik Sankt-Peterburgskogo Universiteta: Seriia 3—Biologiia 1: 84–87 (in Russian with English summary).

Băcescu, M., 1992. Cumacea II. Crustaceorum Catalogus 8: 175–468.

Băcescu, M. & I. Petrescu, 1999. Ordre des Cumacés (Cumacea Krøyer, 1846). In Forest, J. (ed.), Traité de Zoologie. Anatomie, Systématique, Biologie. Tome VII (Fascicule III A), Crustacés Péracarides. Mémoires de l'Institut Océanographique, Monaco 19: 391–428.

Czerniavsky, V., 1868. Materialia ad zoographiam Ponticam comparatum. Transactions of the first meeting of Russian naturalists in Saint Petersburg, 1868: 19–136, 8 pls.

Derzhavin, A., 1912. Neue Cumaceen aus dem Kaspischen Meer. Zoologischer Anzeiger 39: 273–284.

Dumont, H. J., 2000. Endemism in the Ponto-Caspian Fauna, with special emphasis on the Onychopoda (Crustacea). Advances in Ecological Research 31: 181–196.

Jones, N. S. & W. D. Burbanck, 1959. *Almyracuma proximoculi* gen. et spec. nov. (Crustacea, Cumacea) from brackish water of Cape Cod, Massachusetts. Biological Bulletin 116: 115–124.

Kudinova-Pasternak, R. K., 1972. Notes about the tanaidacean fauna (Crustacea, Malacostraca) of the Kirmadec Trench. Complex Research of the Nature of the Ocean. Publications Moscow University 3: 257–258.

Larsen, K. & T. Hansknecht, 2004. A new genus and species of freshwater tanaidacean *Pseudohalmyrapseudes aquadulcis* (Apseudomorpha: Parapseudidae), from Northern Territory, Australia. Journal of Crustacean Biology 24: 567–575.

Mordukhai-Boltovskoi, F. D. & N. N. Romanova, 1973. A new genus of Cumacea from the Caspian Sea. Zoologicheskii Zhurnal 52: 429–432.

Roccatagliata, D., 1981. *Claudicuma platensis* gen. et sp. nov. (Crustacea, Cumacea) de la ribera argentina del Río de la Plata. Physis (Buenos Aires) 39: 79–87.

Roccatagliata, D., 1991. *Claudicuma platense* Roccatagliata, 1981 (Cumacea): a new reproductive pattern. Journal of Crustacean Biology 11: 113–122.

Sars, G. O., 1894. Crustacea Caspia, part II. Cumacea. Bulletin de l'Academie Impériale des Sciences de St Petersburg 16: 297–338.

Sars, G. O., 1897. On some additional Crustacea from the Caspian Sea. Annuaire du Musée Zoologique de l'Académie Imperiale des Sciences. St Petersburg 1897: 1–73, 8 pls.

Sars, G. O., 1914. Report on the Cumacea of the Caspian Expedition 1904. Trudy Kaspiĭskoĭ Ékspeditzī 1904 ghoda. 4: 1–34 (in Russian), 1–32 (in English), 12 pls.

Shiino, S. M., 1965. Tanaidacea from the Bismarck Archipelago. Videnskabelige Meddelelser fra Dansk Naturhistorisk Forening i Kjøbenhavn 128: 177–203.

Simpson, K. W., J. P. Fagnani, D. M. DeNicola & R. W. Bode, 1985. Widespread distribution of some estuarine crustaceans (*Cyathura polita*, *Chiridotea almyra*, *Almyracuma proximoculi*) in the limnetic zone of the Lower Hudson River, New York. Estuaries 8: 373–380.

Sowinski, W., 1893. Report on the Crustacea collected by Dr. Ostroumow in the Sea of Azov. Zapiski Kievskago Obshchestva Estestvoispytatelei 14: 289–405, (in Russian).

Hydrobiologia (2008) 595:231–240
DOI 10.1007/s10750-007-9019-z

FRESHWATER ANIMAL DIVERSITY ASSESSMENT

Global diversity of Isopod crustaceans (Crustacea; Isopoda) in freshwater

George D. F. Wilson

Abstract The isopod crustaceans are diverse both morphologically and in described species numbers. Nearly 950 described species (~9% of all isopods) live in continental waters, and possibly 1,400 species remain undescribed. The high frequency of cryptic species suggests that these figures are underestimates. Several major freshwater taxa have ancient biogeographic patterns dating from the division of the continents into Laurasia (Asellidae, Stenasellidae) and Gondwana (Phreatoicidea, Protojaniridae and *Heterias*). The suborder Asellota has the most described freshwater species, mostly in the families Asellidae and Stenasellidae. The suborder Phreatoicidea has the largest number of endemic genera. Other primary freshwater taxa have small numbers of described species, although more species are being discovered, especially in the southern hemisphere. The Oniscidea, although primarily terrestrial, has a small number of freshwater species. A diverse group of more derived isopods, the 'Flabellifera' sensu lato has regionally important species richness, such as in the Amazon River. These taxa are transitional between marine and freshwater realms and represent multiple colonisations of continental habitats. Most species of freshwater isopods species and many genera are narrow range endemics. This endemism ensures that human demand for fresh water will place these isopods at an increasing risk of extinction, as has already happened in a few documented cases.

Keywords Isopoda · Crustacea · Gondwana · Laurasia · Diversity feeding · Reproduction · Habits · Fresh water · Classification

Introduction

The Isopoda are a diverse group of crustaceans, with more than 10,300 species found in all realms from the deepest oceans to the montane terrestrial habitats; approximately 9% of these species live in continental waters. Isopods are thought of as dorsoventrally flattened, and indeed many species fit this morphological stereotype. Diverse taxa found in the deep sea and those found in groundwater habitats depart considerably from this generalised body plan. Palaeontogical and phylogenetic evidence (Brusca & Wilson, 1991; Wilson & Edgecombe, 2003) suggests that the ancestral isopod may have had a narrow

Guest editors: E. V. Balian, C. Lévêque, H. Segers & K. Martens
Freshwater Animal Diversity Assessment

Electronic supplementary material The online version of this article (doi:10.1007/s10750-007-9019-z) contains supplementary material, which is available to authorized users.

G. D. F. Wilson (✉)
Invertebrate Zoology, Australian Museum, 6 College Street, Sydney, NSW 2010, Australia
e-mail: buz.wilson@austmus.gov.au

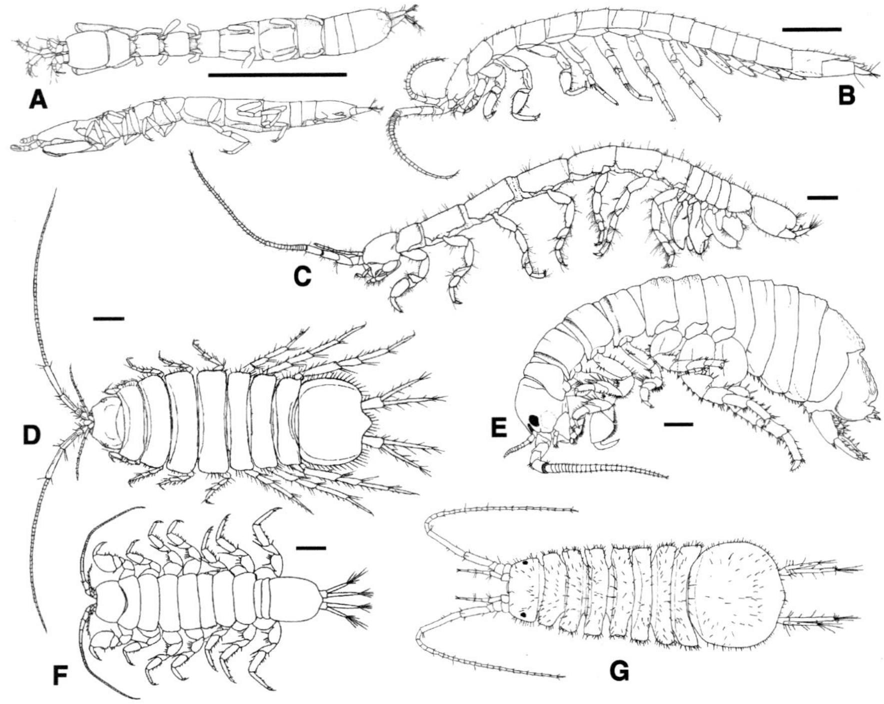

Fig. 1 Freshwater Isopoda, a selection of body forms. (**A**) Microcerberidae sp. (interstitial, Western Australia), dorsal and lateral view; (**B**) *Pygolabis* sp., Tainisopidae (hypogean, Western Australia); (**C**) *Phreatoicoides gracilis*, Hypsimetopidae (epigean, Victoria Australia); (**D**) *Asellus aquaticus*, Asellidae (epigean, Europe, from Sars 1897); (**E**) *Eophreatoicus* sp., Amphisopidae (epigean, Northern Territory Australia); (**F**), *Stenasellus chapmani*, Stenasellidae (hypogean, Indonesia, from Coineau et al. 1994; (**G**) *Heterias* sp., Janiridae (hyporheic & pholoteric, South America; from Bowman et al. 1987). Scale bars 1 mm, except for A, 0.5 mm

vaulted body with legs projecting ventrally (Fig. 1E, *Eophreatoicus*, Amphisopidae). Freshwater taxa include either typical flattened isopods (Fig. 1D, F, G) or narrow body forms (Fig. 1B, C, E), along with a few taxa that are thin and vermiform, often with legs that emerge close to the dorsal surface (Fig. 1A). Other peculiarities of isopods include respiration using their broad posterior limbs (swimming legs or pleopods) with the heart positioned in the posterior part of the body, and biphasic moulting, wherein the back part of the body is cast off before the anterior part. Limb forms are diverse in the isopods, but the first walking leg (second thoracic limb) is modified for grasping in most species.

Feeding

Isopods have a broad range of feeding types from omnivory in Sphaeromatidae to carnivory in the Cirolanidae. Oniscideans and Asellidae are well-known as leaf litter shredders and have bacterial endosymbionts to aid digestion (Zimmer, 2002, Zimmer & Bartholme, 2003). Tainisopidae may be carnivorous scavengers because they can be captured using baited traps. Most freshwater isopods (e.g. Asellota or Phreatoicidea) can be characterised as generalised detritivores-omnivores, but may facultatively choose other items. Phreatoicideans feed on decaying vegetation and roots, or perhaps the microflora and microfauna associated with these substrates (Wilson & Fenwick, 1999), but on occasion will engage in carnivory. Among the 942 described species found in continental waters, the presumptive feeding types (based on extrapolation from taxa where habits are known) are as follows: 3.2% are carnivores, 6.9% scavenger-carnivores, 9.9% ectoparasites, 0.4% herbivores, 6.1% omnivores and the remaining 73.5% are detritivores-omnivores, mostly Asellota and Phreatoicidea.

Reproduction

Isopods, like all peracarid crustaceans, have direct development with the young brooded in a ventral pouch until they are released as small adults. Isopods have internal fertilisation (Wilson, 1991) that occurs prior to the release of embryos into the marsupium, unlike other peracarid crustaceans. Brood sizes range from 4–5 young in tiny interstitial isopods to hundreds in the parasitic forms, and lifetime reproduction may be limited to one or several broods in most species (Johnson et al., 2001). Many isopods, especially the suborders Asellota and Oniscidea, have secondary sexual features for intromission in both males and females that are also useful for systematics. Brooding of the young, direct development and internal fertilisation may be major contributory factors in the high degree of endemism observed in most isopod taxa (Wilson, 1991).

Habitats

Isopods occur in epigean lotic and lentic habitats (e.g. Asellidae like the common European *Asellus aquaticus* and Phreatoicidae in Tasmania), but many live in a variety of subterranean habitats. The Microcerberidae are found interstitially in freshwater or marine sands. Many families are limited to cavernicolous or subterranean habitats, such as Stenasellidae, Microparasellidae, or Tainisopidae. North American and European members of the Asellidae can be both epigean and hypogean (e.g. Turk et al., 1996; Lewis & Bowman, 1981). Some taxa (e.g. Hypsimetopidae or *Heterias*, Janiridae) could best be described as infaunal, living in near subsurface habitats, either burrowing among submerged roots, living in submerged burrows of other animals (pholoteros) or in the subsurface water of streams (hyporheos). A few isopods occur in unusual habitats, such as *Thermosphaeroma thermophilum* in hot springs of the USA southwest. Some oniscideans, which are ordinarily terrestrial, have re-invaded the continental saline waters (e.g. *Haloniscus searlei*) or even normal freshwater (e.g. Trichoniscidae and Styloniscidae). Australian collection records suggest that some Philosciidae and Trichoniscidae may be amphibious (see also Taiti & Humphreys, 2001; Tabacaru, 1999).

Methods

(See additional information on the article webpage). The biodiversity of freshwater isopods is derived from my research on the Phreatoicidea and Asellota, and from the online "World List of Isopoda" (Kensley et al., 2005). The classification is derived from that list (*not* as in Banerescu, 1990), but includes the informal taxon 'Flabellifera' sensu lato (see Wilson, 1999). The World List uses the traditional classification of the 'Flabellifera' that is known to be paraphyletic (Brusca & Wilson, 1991; Wägele, 1989; Tabacaru & Danielopol, 1999). The Microcerberidea includes two families, Microcerberidae and Atlantasellidae (*not* Asellota as in Banarescu, 1990; Wägele, 1983; Jaume, 2001). The peculiar family Calabozidae is classified as Oniscidea owing to its possession of in-group genitalia and coxal plates incorporated into the body (Brusca & Wilson, 1991). Marine species, including those from anchialine cave and marine beach interstitial environments, were filtered out of the downloaded list, either using the type habitat from the list or by consulting the original literature. The data included species from saline continental waters, such as *Haloniscus*. Subspecies records were treated as species-level taxa. Undescribed species (e.g. *Heterias* species) known to me were added to the list where possible, although less than 100 species were added. An estimate of the unknown species was determined for the Phreatoicidea (Wilson in progress; see supplementary information), and information from Gouws et al. (2004, 2005). A diversity estimate for other isopod groups used the simple known to unknown ratio from the Phreatoicidea as applied to the other taxa (Table 1). Although the assumption of similarity between Phreatoicidea and other freshwater isopods has obvious problems, this procedure at least provides an hypothesis for further refinement.

Species diversity

Of the entirely freshwater isopods (marked with an asterisk in Table 1), the Asellota has the most of the 942 described species, with the largest number of species in the family Asellidae, followed by the Stenasellidae. The Phreatoicidea have at least four families with many undescribed species (see Table in supplementary information) that may double the number of described species. Other freshwater families have small numbers of described species, although more species are being discovered as surveys are carried out in the southern hemisphere. The Protojaniridae are tiny and fragile, and may require specialised techniques to recover them from hypogean habitats; 12 species in five genera are described, but more remain to be found. Recently,

Table 1 Species Diversity of Freshwater Isopoda. Estimation method and classification explained in text and supplementary material (see additional information)

Suborder	Family	Species, described and new	Estimated unknown species	Estimated total diversity
PHREATOICIDEA Stebbing, 1893	*Amphisopidae Nicholls, 1943	36	48	84
	*Hypsimetopidae Nicholls, 1943	11	19	30
	*Incertae sedis (*Crenisopus*)	1		1
	*Phreatoicidae Chilton, 1891	49	71	120
	*Ponderellidae Wilson & Keable, 2004	2		2
Subtotal, used for estimates other suborders		99	138	237
Unknown to Known ratio 1.39				
ASELLOTA Latreille, 1803	*Asellidae Rafinesque-Schmaltz, 1815	379	529	908
	Janiridae G. O. Sars, 1897	76	106	182
	Microparasellidae Karaman, 1933	73	102	175
	*Protojaniridae Fresi, Idato & Scipione, 1980	15	21	36
	*Stenasellidae Dudich, 1924	73	102	175
MICROCERBERIDEA Lang, 1961	Microcerberidae Karaman, 1933	21	30	51
ONISCIDEA Latreille, 1803	*Calabozoidae Van Lieshout, 1983	2	3	5
	Philosciidae Kinahan, 1857	1	2	3
	Scyphacidae Dana, 1852	5	7	12
	Trichoniscidae Sars, 1899	1	2	3
'FLABELLIFERA' sensu lato	Aegidae Leach, 1815	1	2	3
	Anthuridae Leach, 1814	19	27	46
	Bopyridae Rafinesque-Schmaltz, 1815	33	46	79
	Cirolanidae Dana, 1852	65	91	156
	Corallanidae Hansen, 1890	4	6	10
	Cymothoidae Leach, 1818	51	72	123
	Entoniscidae Kossmann, 1881	4	6	10
	Idoteidae Samouelle, 1819	4	6	10
	Leptanthuridae Poore, 2001	3	5	8
	Paranthuridae Menzies & Glynn, 1968	1	2	3
	Sphaeromatidae Latreille, 1825	57	80	137
	*Tainisopidae Wilson, 2003	7	10	17
	Total	994	1395	2625

(*Entirely freshwater families)

J. Pérez-Schultheiss in Chile sent specimens of a new protojanirid; another new species is known from northern Australia. The application of "known to unknown" estimates from the Phreatoicidea to the other freshwater isopods results in 62% more than those known, or approximately a total of 2,630 species (Table 1).

Evidence from molecular studies suggest that this estimate could be highly conservative. RAPD (random amplified polymorphic DNA) studies on both species of Asellidae and Stenasellidae (Baratti et al., 1999; Verovnik et al., 2003) have uncovered previously unsuspected diversity in well-known populations of *Stenasellus* and *Proasellus*. Similar results have been obtained from studies of genetic variation using enzymatic loci (*Proasellus*: Ketmaier, 2002) or the mtDNA cytochrome oxidase I gene (CO-I) (*Stenasellus*: Ketmaier et al., 2003). Cryptic species in the epigean phreatoicidean genus *Mesamphisopus* (Gouws et al., 2004, 2005) could include between 15–20 species in South Africa (G. Gouws pers. comm.). Four new Tainisopidae species of the hypogean genus *Pygolabis* (Keable & Wilson, 2006) from the Pilbara region of Western Australia are morphologically similar and difficult to identify, but their species concepts are supported by CO-I studies (C. Francis pers. comm.). These results suggest that freshwater isopods have many more species-level taxa than the current list might suggest.

Phylogeny and historical patterns

Isopods that live in freshwater may be divided into two groups based on their presumptive age and adaptation to fresh water. The first group contains exclusively freshwater higher-level taxa (i.e., families) that are named 'primary freshwater.' Several entirely freshwater families of Asellota, and the entire suborder Phreatoicidea are considered 'primary freshwater.' Both subordinal taxa are ancient, with originations in the Palaeozoic (Wilson, 1999) and both are derived basally in most phylogenies of the Isopoda, owing to their lack of specialised broad coxal plates (Brusca & Wilson, 1991) and other derived features that characterise the remainder of the Isopoda (Wägele, 1989). The 'secondary freshwater' group are those higher level taxa that have members ranging from marine to freshwater habitats. The secondary freshwater groups are in the process of evolving freshwater habits, and have marine representatives at a low taxonomic level (i.e. within the same family or genus).

The transition from marine to freshwater habitats is repeated independently in all isopod groups, and obligate freshwater adaptations appear at different phylogenetic levels. For example, the Palaeophreatoicidae, an extinct Palaeozoic family of the Phreatoicidea, were marine or possibly estuarine, while the fossil *Protamphisopus*, which is classified among the crown group phreatoicideans, appears in freshwater lacustrine Triassic facies (Wilson & Edgecombe, 2003). The Asellota have both freshwater and marine taxa but do not fossilise, so whether the ancestor was freshwater or marine is not directly determinable. Several diverse higher-level asellote taxa are strictly freshwater (Asellidae, Stenasellidae, Protojaniridae). Henry & Magniez (1995) proposed that independent clades of Asellidae evolved from separate marine ancestors. No marine asellid taxa, however, are known in the modern fauna, so this hypothesis remains untested. The Microcerberidae was argued by Wägele et al. (1995) to be primitively freshwater, but this hypothesis is unparsimonious (Wilson, 1996). Wägele (1983) highlighted similarities of Microcerberidae and Atlantasellidae (known only from insular marine caves; Jaume, 2001), such as the coxal plates or spines on the anterior pereonites. These plates, found also in marine microcerberids, are plesiomorphic at the family-level because they are present in both families. Finding interstitial microcerberids with coxal spines on continental Australia (Fig. 1A) suggests that these marine taxa have colonised freshwater independently in different parts of the world. Their freshwater distribution pattern is similar to 'Flabellifera' sensu lato and Oniscidea (Fig. 2).

The remainder of the isopods, the terrestrial isopods (Oniscidea) and the 'Flabellifera' sensu lato, are derived much later in phylogenetic estimates (Wägele, 1989; Brusca & Wilson, 1991; Tabacaru & Danielopol, 1999), and the fossil record of modern families does not begin until the middle to late Mesozoic or later, with few peculiar taxa possibly related to modern families appearing in the Triassic (e.g. Guinot et al., 2005). The Oniscidea, although terrestrial, have their least-derived taxa living on marine seashores, and have a few freshwater taxa. Among these later-derived, secondarily freshwater taxa, some genera may be found in either fresh or saline waters, or are clearly transitional, like the bopyrid genus *Probopyrus* that parasitises members of the estuarine and freshwater decapod family Palaemonidae. The peculiar aquatic family Calabozidae is exclusively freshwater, but may be derived from terrestrial ancestors (Brusca & Wilson, 1991).

The Australian family Tainisopidae, among the higher isopods, lacks known marine or estuarine representatives. All of its species are hypogean, narrow range endemics, and retain isopod plesiomorphies lost by other 'Flabellifera' *sensu lato*. Whether this family should be classified as 'secondary' or 'primary' freshwater depends on its phylogenetic relationships relative to the remainder of the Isopoda. These relationships, however, are still controversial, with two competing subordinal placements in the literature (Wilson, 2003; Brandt & Poore, 2003).

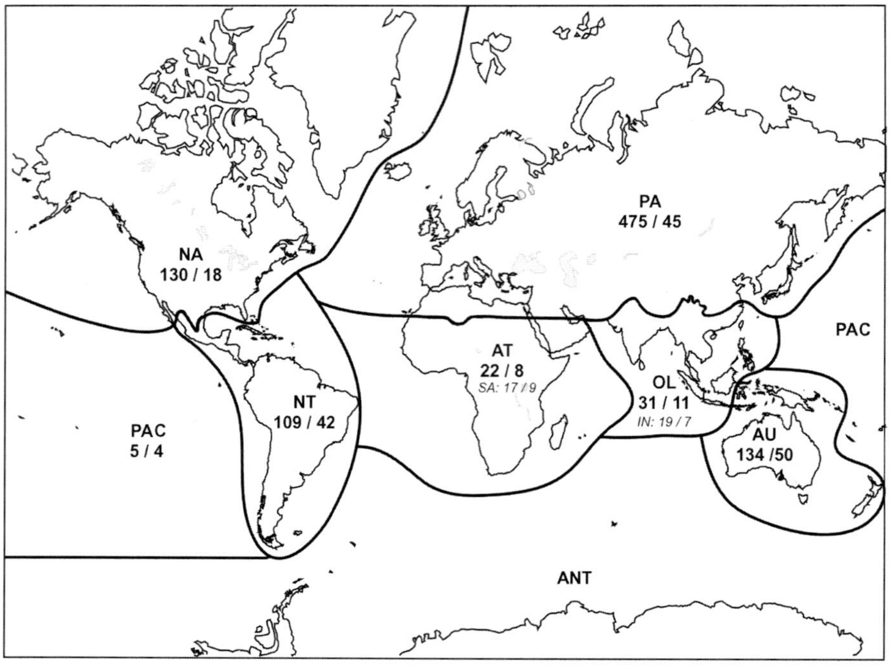

Fig. 2 Distribution of freshwater Isopoda species and genera by zoogeographical region (species number/genus number). Regional abbreviations, with Gondwanan subareas shown separately in parentheses: PA, Palaearctic; NA, Nearctic; NT, Neotropical; AT(SA), Afrotropical (South African); OL (IN), Oriental (Indian); AU, Australasia; PAC, Pacific Oceanic Islands; ANT, Antarctic. Only described species included

Several marine Asellota ancestors may have colonised freshwaters late in Pangean times, and subsequent independent evolution gave rise to two freshwater groups, the Laurasian Asellidae and Stenasellidae and the Gondwanan Protojaniridae (see next section). Where the asellotan genus *Heterias*, found on continents South America, Australia and New Zealand, fits into this picture will ultimately depend on the resolution of the phylogeny of the Janiridae. The presence of *Heterias* and the Protojaniridae on fragments of Gondwana argues for independent freshwater colonisation events, as the two taxa are distinct and not closely related (Wilson, 1987). The ancestral Asellota may have been diverse prior to the break-up of Gondwana, because at least four distinct lineages gave rise to fresh water taxa. If this is the case, then the Asellota has a minimum age of Triassic.

Endemicity and distribution

Except for a few widespread species like *Asellus aquaticus* or epiparasitic *Tachea* species, most species of freshwater isopods are narrow range endemics. Thus all areas in Table 3 are made up of species found only on one landmass, with patterns reflecting, at least partially, sampling effort. Generic endemism (Table 4) is similarly high, with only a few secondary aquatic taxa like *Probopyrus* (Bopyridae) or *Gnorimosphaeroma* (Sphaeromatidae) appearing on more than one continent. Genera counted in Table 4 are mostly unique to fresh water, even in the secondarily aquatic groups. Surprisingly, Australia has the highest number of genera even though the largest number of species is found in the Palaearctic, where more research has been done. This pattern arises because, in addition to different taxonomic styles among asellotan and phreatoicidean workers, speciose genera (such as *Asellus*) are widespread in Eurasia while Australian genera have highly restricted ranges, on scales of 10^{0-2} km. In the northern hemisphere, glaciation, which had much less impact in the south, may have had the dual role of pruning the fauna of rare unique taxa (thus decreasing generic diversity) as well as providing a rich environment for speciation, with multiple opportunities for diversification during the advance and retreat of the ice caps (Magniez, 1974). Several phreatoicidean genera are more widespread, such as the speciose *Colubotelson*, which can be found in most freshwater bodies in Tasmania and upland springs in Victoria, and the genus *Crenoicus*, which is characteristic of highland bogs and springs on the Great Dividing Range. But these taxa are the exception rather than the rule, probably attesting to

Table 2 Geographic partitioning of freshwater isopod species diversity

Suborder region	PA	NA	NT	AT(SA)	OL(IN)	AU	PAC	World
Asellota Latreille, 1803	384	120	7	19(8)	13(4)	6	2	563
'Flabellifera' sensu lato	82	9	96	3(2)	17(13)	25	2	249
Microcerberidea Lang, 1961	8	1	4	(3)	(1)	4		21
Oniscidea Latreille, 1803	1		2			5	1	9
Phreatoicidea Stebbing, 1893				(4)	(2)	94		100
Regional total	475	130	109	22(17)	31(19)	134	5	942

Regional abbreviations, with Gondwanan subareas shown separately in parentheses: PA, Palaearctic; NA, Nearctic; NT, Neotropical; AT(SA), Afrotropical (South African); OL (IN), Oriental (Indian); AU, Australasian; PAC, Pacific Oceanic Islands. Only described species included

Table 3 Geographic partitioning of freshwater species diversity in the suborder Asellota

Family region	PA	NA	NT	AT(SA)	OL(IN)	AU	PAC	World
Asellidae Rafinesque-Schmaltz, 1815	265	112	1		1			379
Janiridae G. O. Sars, 1897	19		1	1		3		24
Microparasellidae Karaman, 1933A	65		3			2	2	72
Protojaniridae Fresi, Idato & Scipione, 1980			2	(8)	(4)	1		15
Stenasellidae Dudich, 1924	35	8		18	12			73
Regional Total	384	120	7	19(8)	13(4)	6	2	563

See Table 2 for regional abbreviations

Table 4 Geographic partitioning of freshwater isopod generic diversity

Suborder region	PA	NA	NT	AT (SA)	OL (IN)	AU	PAC	World
ASELLOTA Latreille, 1803	26	11	5	6(3)	2(2)	3	1	59
'FLABELLIFERA' sensu lato	15	6	32	2(2)	8(4)	13	2	84
MICROCERBERIDEA Lang, 1961	3	1	3	(3)	1	1		12
ONISCIDEA Latreille, 1803	1		2			4	1	8
PHREATOICIDEA Stebbing, 1893				(1)	(1)	29		31
Regional Total	45	18	42	8(9)	11(7)	50	4	194

See Table 2 for regional abbreviations. Gondwanan areas in parentheses

the geological history of Australia as an arid continent, with patchy epigean sources of water and extensive marine transgressions during the Mesozoic era that transformed the continent into a series of smaller islands (Wilson & Johnson, 1999). The secondary freshwater taxa seem to be patchily spread across the continents, representing either sampling effort or pecularities of the region. The Amazon and other major rivers of South America (Neotropics) are significant hot spots for the fish-parasitic group Cymothoidae, with more than 40 species in this family alone. This result may reflect the diversity of their hosts in this region (see Chapt. 43).

The distribution of the freshwater isopods on the continental scale shows significant non-random patterns (Tables 2, 3, 4; Fig. 2) among the more ancient groups. The Asellotan families Asellidae and Stenasellidae conversely show a Laurasian pattern, with no species occurring among known terranes of Gondwana. A single species of *Caecidotea* (Asellidae) has been reported from the highlands of Guatemala (Argano, 1977), but this record may be the southern limit of a Nearctic pattern for the genus. The Stenasellidae have numerous African records, and scattered records among Oriental and southern margins of North America. This pattern appears to be

Tethyan, but more detail is needed on the phylogeny of the Stenasellidae and Asellidae before a historical biogeographical assessment can be made. The Phreatoicidea have a strictly Gondwanan pattern (Wilson & Edgecombe, 2003), and occur only in the terranes of Gondwana, including South Africa, India and Australia-New Zealand. Since the ancestors of the freshwater Asellota and the Phreatoicidea were marine and possibly cosmopolitan in the Palaeozoic, these patterns may be interpreted as the result of colonisation of freshwaters, with subsequent extinction in marine waters. Some Asellota also show a Gondwanan pattern. The family Janiridae, which is known to be non-monophyletic (Wilson, 1994), contains a diverse group of transitional freshwater and marine taxa (best exemplified by the European genus *Jaera*). Among these taxa, the southern hemisphere genus *Heterias* occurs in Australia and in South America (Wilson & Wägele, 1994). *Heterias* species are diverse in southern Australian hyporheos and pholeteros, and recently they have been found in New Zealand by Dean Olsen. The Protojaniridae are strictly freshwater and occur on terranes derived from Gondwana. These observations could be related to rareness and sampling bias, but documented effort in the northern hemisphere argues that this pattern is real. Until recently, protojanirids were only known from Sri Lanka and South Africa, but undescribed species have been found recently in Australia and in Chile Fig. 3.

Human related issues

As more human demand on water resources impacts all parts of the world, we can expect that freshwater isopods will become increasingly at risk of extinction. A point made above, that most species of freshwater isopods are short-range endemics, comes to the foreground in our consideration of human impacts. Wherever small endemic populations of isopods occur, human over-exploitation of water may be a threat to their continued survival. Almost certainly, phreatoicidean species have become extinct owing to water and land use practises in Australia. For example, the artesian spring at the type locality of

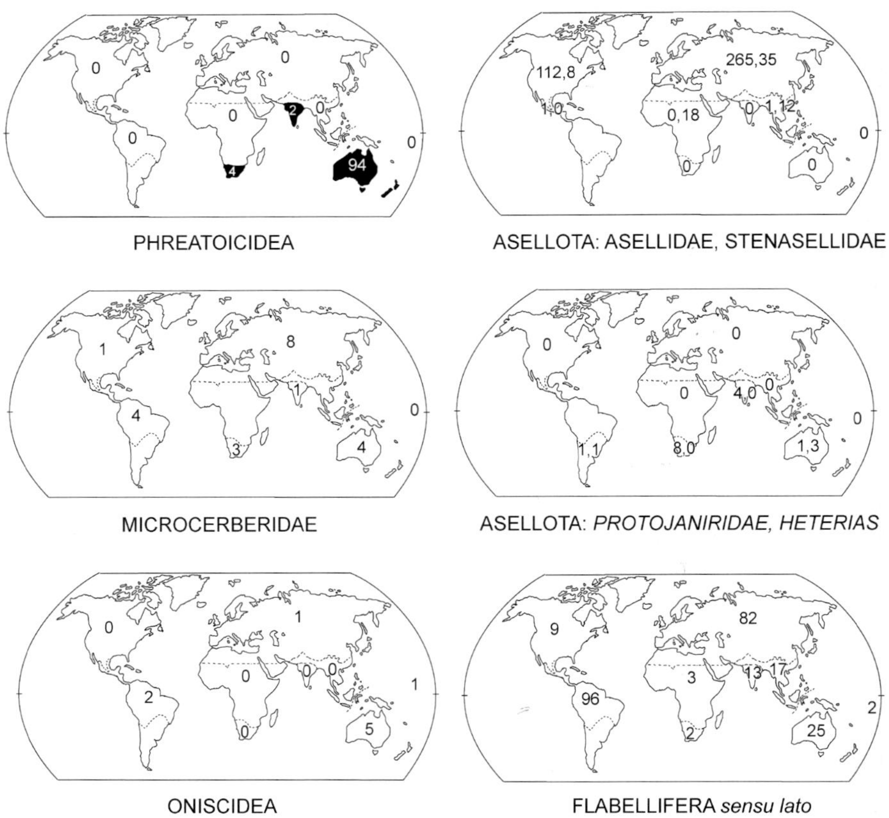

Fig. 3 Global distribution of freshwater isopod diversity. The areas are marked as in Tables 3–5. The darkened areas on the Phreatoicidean map indicate Gondwana biogeographic regions. Not all Asellota shown

Phreatomerus latipes, a bore at Hergott (Maree) in South Australia, has become extinct, presumedly along with the unique population of this isopod species (W. Ponder pers. comm.). Land clearing in the last 200 years along the Great Dividing Range in New South Wales are likely to have been responsible for the extinction of many *Crenoicus* species, by causing the disappearance of the highland springs and *Sphagnum* bogs where they occur. The genus type species, *C. mixtus* is probably extinct because the springs that supplied water to the town of Ballarat (Nicholls, 1944), where this species lived, are now occupied by a large dam and surrounded by a pine plantation. The risks for epigean species are more easily assessed than for the hypogean species, because the latter may be easily collected. Phreatobitic species are only collected from springs, wells, bores and caves, but these animals are clearly adapted to living deep underground in narrow cracks and crevices, where we have little chance to discover their true distribution. As a result, we have great difficulties for assessing the risk to these species where human activities over-exploit the subterranean aquifers. As discussed above, genetic studies show that each restricted aquifer can have an isolated and phylogenetically unique population. As a result, conservation activities for such hypogean species must understand the hydrology of the region, and assessments of their populations must continue while water is being used. To do otherwise is to risk the loss of a substantial component of the regional phylogenetic diversity.

Acknowledgements Information in this article comes from communications and specimens sent by colleagues around the world. In particular, I would like to recognise important contributors of specimens and/or information from particular localities: New South Wales and elsewhere in Australia - W. Ponder; Northern Territory - C. Humphrey; Western Australia - C. Francis, S. Halse & coworkers, S. Eberhard, P. Horwitz and W. Humphreys; South Africa - G. Gouws; Brazil - C. Noro; New Zealand - D. Olsen; Chile - J. Pérez-Schultheiss. I am grateful to Marilyn Schotte (National Museum of Natural History, USA), as the maintainer of the World List of Isopods, which formed the starting point for this article. Helpful advice on the distribution of parasitic taxa was kindly given by J. Markham and J. Shields. Two referees made helpful suggestions for the revision of this article. Research on freshwater isopods at the Australian Museum has been supported by Australian Biological Resources Survey grants to myself, Stephen Keable and Chris Humphrey (Environmental Research Institute of the Supervising Scientist), and a contract from the Department of Conservation and Land Management (Western Australia). Finally, I thank Estelle Balian and Koen Martens for inviting me to the workshop and handling this manuscript.

References

Argano, R., 1977. Asellota del messico meridionale e Guatemala (Crustacea, Isopoda). problemi attuali di scienza e di cultura. Accademia Nazionale dei Lincei, Quaderno N. 171: 101–124.

Banarescu, P. 1990. Zoogeography of Fresh Waters. General Distribution and Dispersal of Freshwater Animals, Vol. 1. Aula-Verlag, Wiesbaden, Germany: 511.

Baratti, M., M. Bazzicalupo, C. de Filippo & G. Messana, 1999. Detection of genetic variability in stygobitic isopods using RAPD markers. Crustaceana (Leiden) 72: 625–634.

Bowman, T. E., R. Prins & J. Arenas, 1987. The occurrence of the freshwater isopod *Heterias* (*Fritzianira*) *exul* in the Lakes region of Chile, with notes on the genus *Heterias* (Asellota: Janiridae). Hydrobiologia 146: 275–281.

Brandt, A. & G. C. B. Poore, 2003. Higher classification of the flabelliferan and related Isopoda based on a reappraisal of relationships. Invertebrate Systematics 17: 893–923.

Brusca, R. C. & G. D. F. Wilson, 1991. A phylogenetic analysis of the Isopoda with some classificatory recommendations. Memoirs of the Queensland Museum 31: 143–204.

Coineau, J.-P., J.-P. Henry, G. Magniez & I. Negoescu, 1994. Isopoda aquatica. In Juberthie C. & V. Decu (eds), Encyclopaedia Biospeologica. Société de Biospéologie Bucarest, 1, 123–140.

Gouws, G., B. A. Stewart & S. R. Daniels, 2004. Cryptic species within the freshwater isopod Mesamphisopus capensis (Phreatoicidea: Amphisopodidae) in the Western Cape, South Africa: allozyme and 12S rRNA sequence data and morphometric evidence. Biological Journal of the Linnean Society 81: 235–253.

Gouws, G., B. A. Stewart & C. A. Matthee, 2005. Lack of taxonomic differentiation in an apparently widespread freshwater isopod morphotype (Phreatoicidea : Mesamphisopidae : *Mesamphisopus*) from South Africa. Molecular Phylogenetics and Evolution 37: 289–305.

Guinot, D., G. D. F. Wilson & F. R. Schram, 2005. Jurassic isopod (Malacostraca: Peracarida) from Ranville, Normandy, France. Journal of Paleontology 79: 954–960.

Henry, J.-P. & G. Magniez, 1995. Nouvelles données sur les Asellidae épigés d'Extrême-Orient (Crustacea, Isopoda, Asellota). Contributions To Zoology (Bijdragen tot de Dierkunde) 65: 101–122.

Jaume, D., 2001. A new atlantasellid isopod (Asellota : Aselloidea) from the flooded coastal karst of the Dominican Republic (Hispaniola): evidence for an exopod on a thoracic limb and biogeographical implications. Journal of Zoology 255: 221–233.

Johnson, W. S., M. Stevens & L. Watling, 2001. Reproduction and Development of Marine Peracaridans. Advances in Marine Biology 39: 105–260.

Keable, S. J. & G. D. F. Wilson, 2006. New species of *Pygolabis* Wilson, 2003 (Isopoda, Tainisopidae) from Western Australia. Zootaxa Zootaxa 1116: 1–27.

Kensley, B. F., M. Schotte & S. Shilling, 2005. World list of marine, freshwater and terrestrial Crustacea Isopoda. National Museum of Natural History Smithsonian Institution: Washington D.C., USA. HTTP://www.nmnh.si.edu/iz/isopod/index.html; accessed 08 September 2005.

Ketmaier, V., 2002. Isolation by distance, gene flow and phylogeography in the *Proasellus coxalis*-group (Crustacea, Isopoda) in Central Italy: allozyme data. Aquatic Sciences 64: 66–75.

Ketmaier, V., R. Argano & A. Caccone, 2003. Phylogeography and molecular rates of subterranean aquatic Stenasellid Isopods with a peri-Tyrrhenian distribution. Molecular Ecology 12: 547–555.

Lewis, J. J. & T. E. Bowman, 1981. The subterranean asellids (*Caecidotea*) of Illinois (Crustacea: Isopoda: Asellidae). Smithsonian Contributions to Zoology 335: 1–66.

Magniez, G., 1974. Données faunistiques et ecologiques sur les Stenasellidae. International Journal of Speleology 6: 1–180.

Nicholls, G. E., 1944. The Phreatoicoidea. Part II. The Phreatoicidae. Papers and Proceedings of the Royal Society of Tasmania 1943: 1–156.

Sars, G. O. 1897. Isopoda. In Crustacea of Norway, part V, VI, Idotheidae, Arcturidae, Asellidae, Ianiridae, Munnidae. Bergen Museum, Norway: 81–116.

Tabacaru, I., 1999. L'adaptation a la vie aquatique d'un remarquable trichoniscide cavernicole, *Cantabroniscus primitivus* Vandel, et le probleme de la monophylie des isopods terrestres. Travaux del'Institut Speologie "Emile Racovitsa" 37–38: 115–132.

Tabacaru, I. & D. L. Danielopol, 1999. Contribution a la connaissance de la phylogénie des Isopoda (Crustacea). Vie et Milieu 49: 163–176.

Taiti, S. & W. F. Humphreys, 2001. New aquatic Oniscidea (Crustacea: Isopoda) from groundwater calcretes of Western Australia. Records of the Western Australia Museum, Supplement 64: 133–151.

Turk, S., B. Sket & S. Sarbu, 1996. Comparison between some epigean and hypogean populations of *Asellus aquaticus* (Crustacea: Isopoda: Asellidae). Hydrobiologia 337: 161–170.

Verovnik, R., B. Sket, S. Prevorcnik & P. Trontelj, 2003. Random amplified polymorphic DNA diversity among surface and subterranean populations of *Asellus aquaticus* (Crustacea: Isopoda). Genetica 119: 155–165.

Wägele, J. W., 1983. On the origin of the Microcerberidae (Crustacea: Isopoda). Zeitschrift fuer Zoologische Systematik und Evolutionsforschung 21: 249–262.

Wägele, J. W., 1989. Evolution und phylogenetisches System der Isopoda. Stand der Forschung und neue Erkenntnisse. Zoologica 140: 1–262.

Wägele, J.-W., N. J. Voelz & J. V. McArthur, 1995. Older than the Atlantic Ocean: Discovery of a fresh-water *Microcerberus* (Isopoda) in North America and erection of *Coxicerberus*, new genus. Journal of Crustacean Biology 15: 733–745.

Wilson, G. D. F., 1987. The road to the Janiroidea: Comparative morphology and evolution of the asellote isopod crustaceans. Zeitschrift für Zoologische Systematik und Evolutionsforschung 25: 257–280.

Wilson, G. D. F., 1991. Functional morphology and evolution of isopod genitalia. In: Bauer R. T. & J. W. Martin (eds), Crustacean Sexual Biology. Columbia University Press, New York/Oxford: 228–245.

Wilson, G. D. F., 1994. A phylogenetic analysis of the isopod family Janiridae (Crustacea). Invertebrate Taxonomy 8: 749–766.

Wilson, G. D. F., 1996. Of uropods and isopod crustacean trees: A comparison of "groundpattern" and cladistic methods. Vie et Milieu 46: 139–153.

Wilson, G. D. F., 1999. Some of the deep-sea fauna is ancient. Crustaceana (Leiden) 72: 1020–1030.

Wilson, G. D. F., 2003. A new genus of Tainisopidae fam. nov. (Crustacea: Isopoda) from the Pilbara, Western Australia. Zootaxa 245: 1–20.

Wilson, G. D. F. & G. D. Edgecombe, 2003. The Triassic isopod *Protamphisopus wianamattensis* (Chilton) and comparison with extant taxa (Crustacea, Phreatoicidea). Journal of Paleontology 77: 454–470.

Wilson, G. D. F. & G. D. Fenwick, 1999. Taxonomy and ecology of *Phreatoicus typicus* Chilton, 1883 (Crustacea, Isopoda, Phreatoicidae). Journal of The Royal Society of New Zealand 29: 41–64.

Wilson, G. D. F. & R. T. Johnson, 1999. Ancient endemism among freshwater isopods (Crustacea, Phreatoicidea). In: Ponder W. F. & D. Lunney (eds), The Other 99%. The conservation and Biodiversity of Invertebrates. Transactions of the Royal Zoological Society of New South Wales, Mossman: 264–268.

Wilson, G. D. F. & J.-W. Wägele, 1994. A systematic review of the family Janiridae (Crustacea: Isopoda: Asellota). Invertebrate Taxonomy 8: 683–747.

Zimmer, M., 2002. Nutrition in terrestrial isopods (Isopoda: Oniscidea): an evolutionary-ecological approach. Biological Reviews 77: 455–493.

Zimmer, M. & S. Bartholme, 2003. Bacterial endosymbionts in *Asellus aquaticus* (Isopoda) and *Gammarus pulex* (Amphipoda) and their contribution to digestion. Limnology and Oceanography 48: 2208–2213.

Hydrobiologia (2008) 595:241–255
DOI 10.1007/s10750-007-9020-6

FRESHWATER ANIMAL DIVERSITY ASSESSMENT

Global diversity of amphipods (Amphipoda; Crustacea) in freshwater

R. Väinölä · J. D. S. Witt · M. Grabowski · J. H. Bradbury · K. Jazdzewski · B. Sket

Abstract Amphipods are brooding peracaridan crustaceans whose young undergo direct development, with no independent larval dispersal stage. Most species are epibenthic, benthic, or subterranean. There are some 1,870 amphipod species and subspecies recognized from fresh or inland waters worldwide at the end of 2005. This accounts for 20% of the total known amphipod diversity. The actual diversity may still be several-fold. Amphipods are most abundant in cool and temperate environments; they are particularly diversified in subterranean environments and in running waters (fragmented habitats), and in temperate ancient lakes, but are notably rare in the tropics. Of the described freshwater taxa 70% are Palearctic, 13% Nearctic, 7% Neotropical, 6% Australasian and 3% Afrotropical. Approximately 45% of the taxa are subterranean; subterranean diversity is highest in the karst landscapes of Central and Southern Europe (e.g., Niphargidae), North America (Crangonyctidae), and Australia (Paramelitidae). The majority of Palearctic epigean amphipods are in the superfamily Gammaroidea, whereas talitroid amphipods (*Hyalella*) account for all Neotropic and much of the Nearctic epigean fauna. Major concentrations of endemic species diversity occur in Southern Europe, Lake Baikal, the Ponto-Caspian basin, Southern Australia (including Tasmania), and the south-eastern USA. Endemic family diversity is similarly centered in the Western Palearctic and Lake Baikal. Freshwater amphipods are greatly polyphyletic, continental invasions have taken place repeatedly in different time frames and regions of the world. In the recent decades, human mediated invasions of Ponto-Caspian amphipods have had great impacts on European fluvial ecosystems.

Keywords Biogeography · Continental invasions · Endemism · Gammaridea · Malacostraca · Species diversity

Guest editors: E. V. Balian, C. Lévêque, H. Segers and K. Martens
Freshwater Animal Diversity Assessment

R. Väinölä (✉)
Finnish Museum of Natural History, University of Helsinki, POB 17, 00014 Helsinki, Finland
e-mail: risto.vainola@helsinki.fi

J. D. S. Witt
Department of Biology, University of Waterloo, Waterloo, ON, Canada N2L 3G1

M. Grabowski · K. Jazdzewski
Department of Invertebrate Zoology and Hydrobiology, University of Lodz, 90-237 Lodz, Poland

J. H. Bradbury
School of Earth and Environmental Sciences, University of Adelaide, Adelaide, SA 5005, Australia

B. Sket
Department of Biology, Biotechnical Faculty, University of Ljubljana, 1001 Ljubljana, Slovenia

Introduction

Amphipods are an order of macroscopic crustaceans of the class Malacostraca. Along with the isopods, cumaceans, mysids, and tanaidaceans they belong to the superorder Peracarida, whose life cycle is characterized by direct development and no independent larval stage. Amphipod females carry their embryos in a brood chamber between the thoracic legs (pereiopods). When released, the juveniles reach maturity after several molts, without any metamorphosis.

The amphipod body is segmented throughout and usually laterally compressed, with a more or less curved or hook-like profile (Fig. 1). Amphipods generally have a pair of sessile lateral compound eyes, but most subterranean species and some that inhabit deep waters are eyeless. Adult body lengths of freshwater species range 2–40 mm, most commonly between 5 and 15 mm.

Most amphipods are marine, but they also inhabit a wide spectrum of freshwater habitats. Freshwater taxa are particularly diversified in relatively cool running waters and subterranean habitats. The number of benthic burrowing (fossorial) species is relatively low, but they may be abundant in large lakes and estuaries. Epibenthic taxa are more diverse, and are usually associated with the littoral vegetation of lakes and rivers and bottom matrix of small streams and springs. Some taxa are nektobenthic, whereas pelagic and parasitic or commensal species are rare in freshwaters, unlike in the seas. There is also a widespread family of terrestrial amphipods (talitrids).

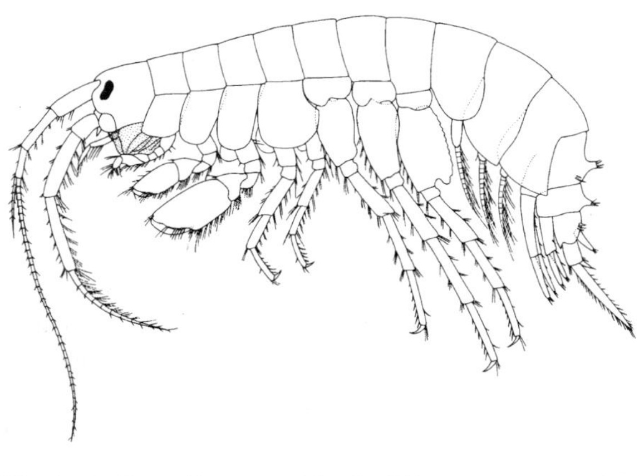

Fig. 1 Outline of an amphipod, *Dikerogammarus oskari* Birstein (size ca. 20 mm; from Barnard & Barnard, 1983). The head is followed by 7 + 3+ 3 pereional + pleonal + urosomal segments. The appendages shown include antennae 1 and 2, mandibular palp, maxillipede, gnathopods 1 and 2 (=pereiopods 1,2), pereiopods 3–7, of which two are directed backwards, three forwards (hence, *amphi podos*, or both legs), three pleopods, and finally three uropods and a terminal telson

Amphipods are among the most diverse hypogean animal groups overall (Sket, 1999), and hypogean taxa account for approximately 45% of freshwater amphipod species. These troglobiotic (stygobiotic) species and races are generally characterized by the loss of eyes and pigmentation, and by elongation of the trunk and/or appendages (troglomorphy). In karst environments or lava fields they inhabit systems of flooded fissures and caverns. Interstitial groundwater taxa are also included in the hypogean component; they are generally slender and small in size. In many areas hypogean taxa can only be accessed through wells and surface springs.

Amphipod feeding habits are diverse; they can be herbivores, detritivores, carnivores, or omnivores. Most subterranean species are supposedly omnivorous, and even when predatory, they indirectly depend on organic debris derived from surface environments. Amphipods can be important in the diets of fish, and frequently serve as intermediate hosts of their parasites. They often play a critical role in aquatic food webs, acting as conduits of nutrients and energy to higher trophic levels.

Species diversity

Data composition

This survey of world amphipod species diversity covers taxa that can complete their life cycles in true fresh waters, and also taxa native to brackish or saline inland water bodies permanently disconnected from the ocean (e.g., the Caspian Sea). We do not include taxa from anchihaline habitats directly connected to the sea, such as marginal marine caves, or those in beach interstitial waters, whereas taxa noted from slightly brackish wells (inland) have been included. Estuarine taxa are also excluded if they do not have landlocked freshwater populations. The interpretation in several cases was necessarily subjective or arbitrary due to incomplete environmental data available. Only taxa properly described by the end of 2005 are counted; no distinction is made between species and subspecies.

The data were compiled in geographical sections by the individual authors (see electronic appendix).

Where available, recent published reviews or web resources covering geographical regions (e.g., Griffiths & Stewart, 2001; Fenwick, 2001; Kamaltynov, 2002; Lowry & Stoddart, 2003) or family or genus level taxa were used in the first place (e.g., Koenemann & Holsinger, 1999; Zhang & Holsinger, 2003; Vonk & Schram, 2005; Fišer et al., 2005). For the remaining hypogean groups, *Stygofauna Mundi* (Botosaneanu, 1986) was taken as an authoritative baseline; Barnard & Barnard's (1983) comprehensive account of freshwater amphipod diversity described prior to 1980 was used as the starting point for others. A listing of new amphipod taxa described in 1974–2004, compiled by Vader (2005a) and based on his continued bibliographical monitoring in *Amphipod Newsletter* vol. 11–30, was used as a principal pointer to newer data.

Numbers of taxa: systematic account

Our total count of continental amphipod species in this inventory is 1,870, of which 145 are listed as non-nominate subspecies (Table 1). This accounts for 20% of the ca. 9,100 amphipods known worldwide according to Vader (2005a, b). The continental taxa are distributed in 53 families and 293 genera (Tables 2, 3), but only 27 families are strictly limited to continental waters. Given some 180 amphipod families recognized globally, the continental and strictly continental families respectively represent 29% and 15% of overall family diversity. The family concepts applied here are in some cases narrower than those in Martin and Davis (2001), including e.g., the new families in Kamaltynov (2002).

Barnard & Barnard (1983) enumerated 1,088 freshwater amphipod species that had been described by June 1979; a decade later the count was at 1,195 (Barnard & Karaman, 1991). Our total continental number represents an increase of 70% over the past 26.5 years, which is somewhat more than that from 6,200 to 9,100 over a similar period for all amphipods (Vader 2005b).

The higher-level systematics of amphipods remains confused, and no convincing phylogenetic hypothesis has been presented (e.g., Barnard & Karaman, 1991; Bousfield & Shih, 1994; Martin & Davis, 2001; Myers & Lowry, 2003). Yet, freshwater amphipods can be broadly discussed in terms of superfamily groups (Table 1; modified from Bousfield, 1983, Bousfield & Shih, 1994), each with a characteristic biogeography and freshwater history. Note however that the superfamily or even family concepts used are not always unequivocal; the divisions are here adopted for pragmatic reasons and are not meant as a taxonomic statement.

Gammaroids (ca. 800 spp. in continental waters) are widespread across the Holarctic, and account for most of the Palearctic epigean diversity. The distribution of the Gammaridae (304) and the genus *Gammarus* itself (204) are similarly broad, centered in Europe but extending to China and North America; they also include taxa in coastal marine waters. Other families and genera are regional, e.g., endemic to the Ponto-Caspian basin (Pontogammaridae etc.), Lake Baikal, or as with Anisogammaridae, to the North Pacific involving both marine, coastal, and inland waters.

Crangonyctoids (341) are a widespread, exclusively freshwater group largely characteristic of subterranean habitats. Different families occupy the Northern and Southern Hemispheres. The Crangonyctidae (209) comprise the majority of North American freshwater amphipod diversity, but some species in the genera *Stygobromus, Synurella and Crangonyx* also occur in Eurasia. About 80% of species are hypogean, while others inhabit small epigean water bodies. Their sister family, the Pseudocrangonyctidae (16), inhabit East Asia (Holsinger, 1994). In the south, the Paramelitidae (69) is shared by Australia and South Africa, and a number of smaller families are endemic to New Zealand, Australia, Madagascar, and South Africa (Fig. 3). Species diversity is lower in the Southern Hemisphere, but much of the Australian diversity remains undescribed (Bradbury, unpubl.).

Niphargoids (319) are the most diverse Palearctic hypogean amphipods, and also include a few epigean taxa. They are distributed through central and particularly southeastern Europe, where they exhibit high levels of endemism in karst systems. *Niphargus* (305) is currently the largest freshwater amphipod genus (Fišer et al., 2005).

Talitroids of the genus *Hyalella* (58) are the only epigean freshwater amphipods in the Neotropics and important in the Nearctic. Talitroids (chiltoniins) are also present in Australasia. The majority of world terrestrial amphipods (about 250) are in the family Talitridae; few of them have entered fresh water.

Table 1 Numbers of described continental amphipod species (and subspecies) native in the major biogeographical regions, listed by family

Superfamily Family	Region							TOT	sspp	Ecology
	PA	NA	NT	AT	OL[e]	AU	PAC			
Crangonyctoidea										
Allocrangonyctidae*		2						2		H
Austroniphargidae*				3				3		E
Crangonyctidae*	25	184						209	6	H(e)
Crymostygiidae*	1							1		H
Neoniphargidae*[a]			2			23		25		H/E
Paracrangonyctidae*						2		2		H
Paramelitidae*				26		43		69		E/H
Perthiidae*						2		2		E
Phreatogammaridae*						4		4		E(h)
Pseudocrangonyctidae*	16							16		H(e)
Sternophysingidae*				8				8		H(e)
Gammaroidea										
Acanthogammaridae*[b]	159							159	10	E
Baikalogammaridae*[b]	1							1		E
Eulimnogammaridae*[b]	122							122	8	E
Macrohectopidae*[b]	1							1		E(pelagic)
Micruropodidae*[b]	55							55	2	E
Pachyschesidae*[b]	16							16		E:p
Pallaseidae*[b]	22							22	1	E
Anisogammaridae	21	9						30		E
Behningiellidae*	4							4		E:p
Caspicolidae*	1							1		E:p
Gammaracanthidae	3	1						3		E
Gammaridae[c]	292	13						304	27	E(h)
Iphigenellidae*	1							1		E:p
Pontogammaridae*	66							66		E
Typhlogammaridae*	8							8		H
Niphargoidea										
Niphargidae*	319							319	80	H(e)
Bogidielloidea										
Artesiidae*		2						2		H
Bogidiellidae s.l.	38	1	34	6	6	1	1[d]	87	3	H
Corophioidea										
Aoridae	2				1			3		E
Corophiidae	10					1	2	13		E
Kamakidae	4							4		E
Photidae					1			1		E
Eusiroidea										
Calliopiidae	3							3		E
Eusiridae	5					4		9		E(h)

Table 1 continued

Superfamily Family	Region							TOT	sspp	Ecology
	PA	NA	NT	AT	OL[e]	AU	PAC			
Hadzioidea										
Hadziidae	14	12	33	3			1	63	4	H
Melitidae	2		1	1	6	15	4	29		H
Metacrangonyctidae	16		2					18		H
Pseudoniphargidae	50							50	1	H
Eoniphargids	2				1			3		H
Liljeborgioidea										
Sebidae		2						2		H
Salentinellidae	16							16	3	H
Lysianassoidea										
Lysianassidae			1					1		H
Uristidae	2							2		E
Oedicerotoidea										
Oedicerotidae	1				1			2	1	E
Paracalliopiidae						4		4		E
Pontoporeioidea										
Pontoporeiidae	2	1						3		E
Talitroidea										
Ceinidae						7		7		E/H
Hyalellidae* (Dogielinotidae)		8	51					58		E
Hyalidae				3				3		E/H
Talitridae	4	1			1	1	2	9		E(h)
Ingolfielloidea										
Ingolfiellidae	9		4	6				19		H
Metaingolfiellidae*	1							1		H
Total	1,319	236	128	56	17[e]	107	9 + 1[d]	1870	145	
	PA	NA	NT	AT	OL	AU	PAC	TOT	sspp	

PA—Palearctic, NA—Nearctic, NT—Neotropical, AT—Afrotropical, OL—Oriental, AU—Australasian, PAC—Pacific Islands: TOT—World). sspp: number of taxa listed as non-nominate subspecies, included in the total count. Ecology: E—epigean, H—hypogean, H(e)—predominantly hypogean (75–95% of described species hypogean), H/E—50–74% of species hypogean, E/H and E(h) correspondingly; p—parasitic or commensal. Asterisks (*) indicate families that live exclusively in fresh/continental waters (including the Ponto-Caspian basin)

[a] Includes falklandiellids (2 spp) of Falkland Islands. [b] Endemic Baikal families [c] Count includes *Sensonator valentiensis* (incertae sedis). [d] Includes the single species recorded from the Antarctic zone (Kerguelen). [e] The OL count excludes southern China, here included in PA instead

A number of additional superfamilies are present on several continents and distributed both in fresh, marine and interface habitats.

Most *bogidielloids* live in hypogean continental waters (89), fewer are marine. The distribution is almost worldwide, but diversity is centered around the Mediterranean and in continental South America, Mexico, and the West Indies. Notably the generic diversity is highest in the Neotropics while species diversity is highest in the Mediterranean area (Koenemann & Holsinger, 1999).

Hadzioids are primarily marine amphipods that have colonized fresh waters in several regions of the world. They are characteristic for the Caribbean region, the coastal Western Palearctic (Hadziidae, Pseudoniphargidae, Metacrangonyctidae), and Wes-

Table 2 Numbers of continental amphipod genera recorded in the major biogeographical regions

Superfamily Family	Region							TOT	sh	Endemism/distribution
	PA	NA	NT	AT	OL[d]	AU	PAC			
Crangonyctoidea										
Allocrangonyctidae*		1						1		
Austroniphargidae*				2				2		Madagascar
Crangonyctidae*	3	5						5	3	
Crymostygiidae*	1							1		Iceland
Neoniphargidae*[a]			2			8		10		Australia[a]
Paracrangonyctidae*						1		1		New Zealand
Paramelitidae*				3		13		16		
Perthiidae*						1		1		Australia
Phreatogammaridae*						1		1		New Zealand
Pseudocrangonyctidae*	2							2		East Asia
Sternophysingidae*				1				1		South Africa
Gammaroidea										
Acanthogammaridae*	39							39		Baikal
Baikalogammaridae*	1							1		Baikal
Eulimnogammaridae*	17							17		Baikal
Macrohectopidae*	1							1		Baikal
Micruropodidae*	5							5		Baikal
Pachyschesidae*	1							1		Baikal
Pallaseidae*	8							8		Baikal
Anisogammaridae	5	1						6		
Behningiellidae*	4							4		Ponto-Caspian
Caspicolidae*	1							1		Ponto-Caspian
Gammaracanthidae	1	1						1	1	"Glacial relict"
Gammaridae[b]	21	1						21	1	
Iphigenellidae*	1							1		Ponto-Caspian
Pontogammaridae*	22							22		Ponto-Caspian
Typhlogammaridae*	5							5		SE Europe
Niphargoidea										
Niphargidae*	9							9		West Palearctic
Bogidielloidea										
Artesiidae*		1						1		Texas
Bogidiellidae s.l.	6	1	16	3	3		1[c]	27	2	
Corophioidea										
Aoridae	1				1			1	1	
Corophiidae	2					1	2	4	1	
Kamakidae	1							1		
Photidae					1			1		
Eusiroidea										
Calliopiidae	2							2		
Eusiridae	2							2		

Table 2 continued

Superfamily Family	Region							TOT	sh	Endemism/distribution
	PA	NA	NT	AT	OL[d]	AU	PAC			
Hadzioidea										
Hadziidae	6	8	12	2			1	28	1	
Melitidae	2		1	1	3	5	4	11	2	
Metacrangonyctidae	3		1					3	1	
Pseudoniphargidae	2							2		
Eoniphargids	1				1			2		
Liljeborgioidea										
Sebidae		1						1		
Salentinellidae	2							2		SW Europe
Lysianassoidea										
Lysianassidae			1					1		
Uristidae	1							1		"Glacial relict"
Oedicrotoidea										
Oedicerotidae	1				1			2		
Paracalliopiidae						1		1		
Pontoporeioidea										
Pontoporeiidae	1	1						2		"Glacial relict"
Talitroidea										
Ceinidae						3		3		
Hyalellidae* (Dogielinotidae)		1	1					1	1	
Hyalidae				1				1		
Talitridae	2	1			[1]	[1]	1	3	1	
Ingolfielloidea										
Ingolfiellidae	1		1	4				5	1	
Metaingolfiellidae*	1							1		Italy
Total	185	23	35	17	10[d]	34	8 + 1[c]	293	16	
	PA	NA	NT	AT	OL	AU	PAC			

See Table 1 legend for general explanation. Sh: shared genera, i.e., those distributed in more than one region. Last column indicates regional endemism or distributional features; "glacial relicts"—Arctic marine genera present in boreal and Arctic lakes and/or the Caspian Sea

[a] Includes falklandiellids (2 spp. on Falkland Islands). [b] Count includes *Sensonator* (incertae sedis). [c] Includes the single species recorded from the Antarctic zone (Kerguelen). [d] The OL count excludes southern China, here attributed to PA (with mainly shared genera)

tern Australia (Melitidae). There are also significant numbers of taxa in subterranean marginal brackish or marine habitats, particularly in anchihaline caves on both coasts of the northern Atlantic and on oceanic islands.

Corophioids are primarily marine, burrowing and tube-building amphipods, which have penetrated epigean brackish (estuaries, lagoons) and freshwater environments repeatedly along continental margins (Kamakidae, Aoridae), and diversified in the Ponto-Caspian waters (Corophiidae).

Ingolfielloids comprise a few species overall (40 in total), but have an exceptional diversity of habitat preferences from the deep sea to high mountain riverine interstitial habitats, as well as other subterranean waters of fresh, brackish and marine continental habitats (Vonk & Schram, 2003, 2005). Ingolfielloids have a specialized vermiform morphology, and most of the 21 freshwater species are small (ca. 2 mm), except for a group of taxa in sub-Saharan Africa (10–20 mm). The distribution is generally spotty, with continental taxa in sub-equatorial Africa,

Table 3 Summary of described systematic diversity of continental amphipods, by major biogeographical regions

	Region								TOT
	PA	NA	AT	NT	OL	AU	PAC	ANT	
Number of species + subspecies	1319	236	56	128	17	107	9	1	1870
(including non-nominate sspp)	(138)	(5)		(2)					(145)
Proportion of world spp. + sspp.	70%	13%	3%	7%	1%	6%	0.5%		100%
Number of genera	185	23	17	35	10	34	8	1	293
Number of families	38	12	8	8	7	12	4	1	53
No. endemic (continental) families	16	2	2	–	–	4	–	–	24
No. species per genus	7	10	3	4	2	3	1	1	6.4
No. freshwater species per family	35	20	7	16	2	9	2	1	35.2
No. genera per family	5	2	2	4	1	3	2	1	5.5
Proportion of hypogean species	39%	70%	46%	61%	77%	53%	55%		46%
No. additional introduced species	3	1							

the Canaries, southern Europe, South America, and the Caribbean.

Several other families are typically marine and occasionally represented in fresh or continental waters by a few species, contributing much to family level diversity, but little to species diversity. These include the families Eusiridae, Hyalidae, Lysianassidae, Oedicerotidae, Pontoporeiidae, Sebidae, Uristidae etc. (see Table 1).

Phylogeny and historical processes

The higher-level classification and phylogenetic relationships among amphipods are in flux and not agreed upon (e.g., Martin & Davis, 2001). At the highest levels, amphipods are variously divided into three or four suborders. All freshwater amphipods are sometimes included in the suborder Gammaridea, which overall comprises the majority amphipod diversity (90–93%). Alternatively, either the ingolfielloids or corophioids are attributed different suborders (Ingolfiellidea, Corophiidea) (Barnard & Karaman, 1991; Myers & Lowry, 2003; Vonk & Schram, 2003). A robust hierarchical classification has been impeded by greatly mosaic (homoplastic) character distributions. Comprehensive phylogenetic analyses do not exist at this time, and the results from preliminary molecular work have been generally inconsistent with the previously proposed schemes (e.g., Englisch et al., 2003, Myers & Lowry, 2003; Macdonald et al., 2005). Amphipod fossils are rare and date no earlier than the Eocene, however the order is considered to be at least 250 Myr old, and on the basis of current biogeographic patterns, had diversified in fresh waters by the time Pangaea broke up 180 Myr ago (e.g., Bousfield, 1983).

Continental invasions

Freshwater amphipods are clearly polyphyletic; their evolution has involved repeated colonizations of continental habitats at different times and in different regions. Three salient historical freshwater patterns are recognizable in view of the continental paleogeographical evolution.

An ancient continental ancestry

An early widespread distribution on ancient landmasses is suggested particularly by the distribution of the Crangonyctoidea (Fig. 3). The nearly pancontinental, exclusively freshwater range suggests a Pangaean continental ancestry. Even the Holarctic family Crangonyctidae (and its individual genera) seem to have occupied Laurasia prior to the separation of Eurasia and North America in the late Mesozoic (Holsinger, 1994). The presence of the Paramelitidae in both Australia and South Africa suggests their ancestors occurred on Gondwanaland before its fragmentation in the Jurassic. The other major superfamily Gammaroidea is also typically continental, but only on land masses formerly part

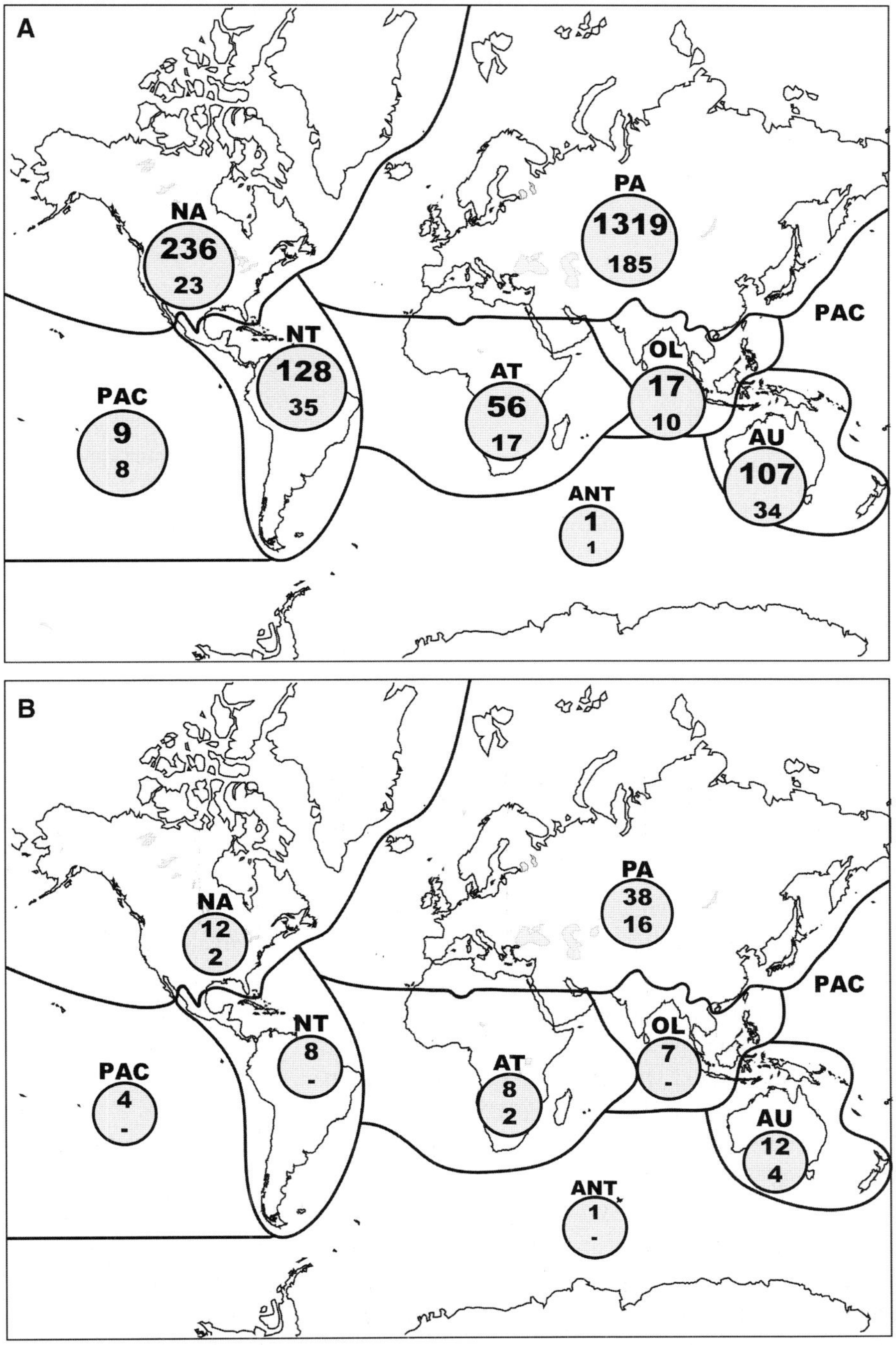

Fig. 2 Distribution of amphipod species, genera, families, and endemic families in inland waters of major biogeographical regions. (**a**) The total number of continental species and subspecies (upper number) and the number of genera (lower number); (**b**) Number of families encountered in continental waters (upper) and the number of continental families endemic to a given region. PA—Palearctic, NA—Nearctic, NT—Neotropical, AT—Afrotropical, OL—Oriental, AU—Australasian, PAC—Pacific Oceanic Islands, ANT—Antarctic

of Laurasia. However, they occur in both marine and inland waters, and continental invasions have taken place repeatedly. Yet the main gammaroid diversity throughout Eurasian continent, including the Gammaridae and the closely associated Baikalian and Ponto-Caspian family complexes, are thought to be of common continental origin and probably secondarily entered coastal seas (Barnard & Barnard, 1983; Macdonald et al., 2005).

Marine relict distributions

Hadzioids have a characteristic 'Tethyan belt' circum-temperate distribution, from Central America to

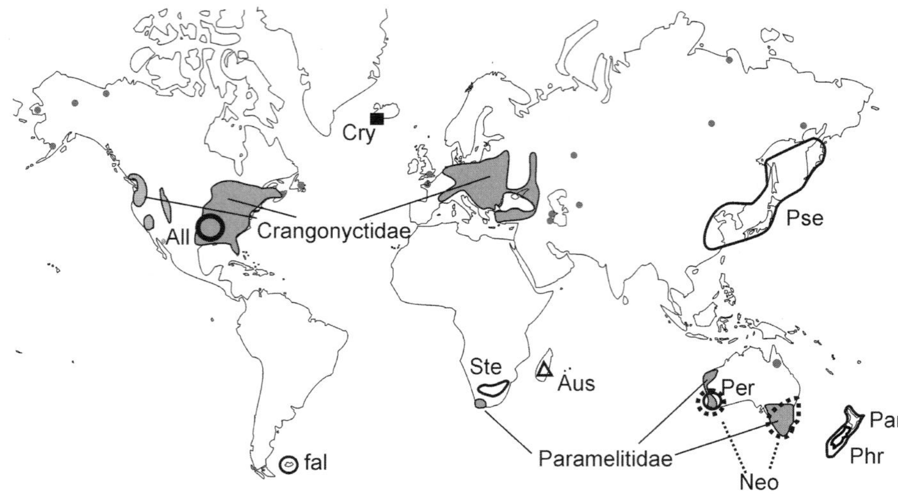

Fig. 3 Family level distributions of crangonyctoid amphipods (schematic). The exclusively freshwater superfamily Crangonyctoidea is thought to have a continental ancestry predating the break-up of Pangaea. The two families Crangonyctidae and Paramelitidae are still present on two continents, other families are endemic to single landmasses: All—Allocrangonyctidae, Aus—Austrocrangonyctidae, Cry—Crymostygiidae, fal—"falklandiellids", Neo—Neoniphargidae, Par—Paracrangonyctidae, Per—Perthiidae, Phr—Phreatogammaridae, Pse—Pseudocrangonyctidae, Ste—Sternophysingidae

the Mediterranean and SE Asia, and occur in both marine and freshwater environments. Their continental distribution in areas covered by shallow marine embayments of the Tethys in the Late Cretaceaous or Tertiary, and on oceanic islands uplifted during the same time frame, provides evidence for independent and old origins in different regions. The numerous hadzioids in marginal marine habitats, particularly anchialine caves, indicate that continental invasion is still an ongoing process (Holsinger, 1994). The bogidielloids have a similar overall distribution, but the question of primary versus secondary invasions is confused by their deep inland penetration of South America (Koenemann & Holsinger, 1999).

Distributional correlations with regional marine regressions suggest similar invasion histories in other, regional groups e.g., in the West Palearctic, but the ancestral relationships mostly remain unclear. Particularly enigmatic are the origins of the almost exclusively freshwater (only secondarily brackish) Niphargidae, diversified in the SE European karst areas that were still submerged in the Eocene, and that of Pseudoniphargidae in the Mediterranean and Central Atlantic islands. The continental vs. marine ancestry of the Ingolfiellidae also remains controversial (Vonk & Schram, 2003).

Recent continental invasions

Several epigean families that ordinarily occur in estuaries and lagoons have occasionally invaded continental waters, including the corophioideans, eusiroids, "glacial relict" pontoporeiids and gammaracanthids. These relatively recent (Neogene) invaders have not yet diversified in fresh waters, but in some cases their ranges are broad (Barnard & Barnard, 1983).

Present distribution and areas of endemism

Inter-continental patterns

Continental amphipods are typical of cool-temperate climates and are notably rare in the tropics. In warmer climates freshwater taxa still occur in subterranean environments where temperatures remain relatively cool. Diversity is also low in high northern latitudes that were directly affected by the Pleistocene glaciations, and have only recently been recolonized. Highest diversities are in the middle latitudes, e.g., the Mediterranean Europe, southern North America, or southern Australia.

The inter-continental distribution of amphipods is remarkably uneven. The Palearctic harbors 70% of described freshwater species, the Nearctic 13%, Neotropics 7%, Australasia 6%, Afrotropics 3% (Table 2, Fig. 2a). Further exploration of the Australian, Nearctic and Neotropic faunas is expected to increase their share substantially. At the family level, the Palearctic harbors 38 families (72% of 53), while the Nearctic and Australasia possess 12 each (22%), and the remaining regions 4–8 (8–15%) (Table 3, Fig. 2b).

Family level endemism is also centered in the Palearctic, with 16 endemic freshwater families (mostly in the ancient lakes), whereas Australasia has 4, and the Nearctic and Afrotropics have each 2 (Table 3, Fig. 2b). Many of these families are local and contain only one or a few species, while others have undergone remarkable regional radiations (e.g., Niphargidae, Pontogammaridae, four Baikalian families). Only 3 of the 27 exclusively freshwater families are distributed on more than one continent (Crangonyctidae, Hyalellidae, Paramelitidae). Genus level inter-continental distributions are rare even in the less strictly continental groups (Table 2). The most widespread single freshwater species is *Gammarus lacustris*, which has a nearly circum-boreal range that also extends into southern Europe and central Asia.

Hotspots

The major centers of regionally endemic species diversity (speciation hotspots) are in Southern Europe, Lake Baikal, and the Ponto-Caspian region, all also rich in endemic families. Additional concentrations occur in Southern Australia (including Tasmania), and in karst regions of the eastern United States. These are treated under the major biogeographical subdivisions themselves below.

Afrotropics (AT)

There are almost no epigean freshwater amphipods in tropical Africa, including the rift valley lakes. However, South Africa harbors two important crangonyctoid families, the epigean Paramelitidae and the hypogean Sternophysingidae (Griffiths & Stewart, 2001), in addition to a peculiar group of large-bodied Ingofiellidae (Vonk & Schram, 2003). Some Bogidiellidae are recorded from NE Africa. Madagascar has an endemic epigean family, the Austroniphargidae, and single representatives of marine families have colonized inland waters of oceanic islands (Table 1).

The Palearctic (PA)

The large fauna of the Palearctic is treated in five (zoo)geographically distinct sections: the hotspots in the ancient Lake Baikal and the Ponto-Caspian, the remaining West and East Palearctic, and northern previously glaciated areas, which may be considered part of a Northen Holarctic province extending to North America.

West Palearctic

This area alone (Europe, Northern Africa, and the Near East) contains nearly half of the world's continental described amphipods (ca. 800 in all if the Ponto-Caspian is included). Hypogean species are in majority (ca. 500), and they also represent a notable number of families. Apart from the most diverse niphargids and pseudoniphargids, these include the similarly endemic salentinellids and typhlogammarids, as well as bogidiellids, hadziids, metacrangonyctids, melitids, gammarids, and crangonyctids. The diversity is concentrated in areas around the Mediterranean and West Atlantic, and in many cases related to the ancient marine regressions of the area. The epigean diversity is dominated by *Gammarus* (ca. 100) and *Echinogammarus* s.l. (60), the latter mainly consisting of localized species in the Mediterranean region.

East Palearctic

This fauna (ca. 150 spp.) is still incompletely characterized. *Gammarus* is the most diverse epigean genus along with the anisogammarid *Jesogammarus* in the east, and accompanied by some recent marginal marine crawl-outs such as kamakids (Barnard & Barnard, 1983). The hypogean fauna includes the endemic Pseudocrangonyctidae along with some bogidiellids along the Pacific margin, and gammarids in the continental interior. Note that the Palearctic species count here encompasses whole China, including some 30 spp. from its southern provinces (particularly Yunnan highland), which more generally would be included in the Oriental region, but in fresh waters are clearly associated with East Asia to the north (Banarescu, 1990).

Lake Baikal

Lake Baikal is the largest and oldest freshwater lake on earth and has an extremely rich endemic fauna; the documented diversity and endemicity are highest in the amphipods (Kozhova & Izmesteva, 1998). There are 363 endemic species and subspecies recorded within the lake itself and further 12 in the downstream watershed, representing 72 genera and 7 endemic families (Kamaltynov, 2002). From molecular data, the true species diversity appears still much higher (Väinölä & Kamaltynov, 1999). The spectacular diversity in size, shape, body armament, color, ecology, habitat, and life style involves many patterns that have parallels in marine systems but not in other fresh waters (e.g., Takhteev, 2000). Still the fauna has no evident recent marine affinities. The whole Baikalian diversity appears to be most closely related to, and derived from the Holarctic continental gammarids, particularly *Gammarus* (Englisch et al., 2003; Macdonald et al., 2005). At any rate the diversity appears relatively old and did not evolve in the current, Pleistocene type of environment (Väinölä & Kamaltynov, 1999). Some of the peculiarities include gigantism, extreme sexual dimorphism, a mysidiform morphology and life-style of a single dominant pelagic species, and a family specialized as brood parasites of other amphipods. A single Baikalian pallaseid species has naturally dispersed to boreal lakes in Northern Europe.

The Ponto-Caspian

The Caspian Sea in fact is the world's largest lake, though brackish (no outlet; salinity <13 ppt). The Ponto-Caspian basin also encompasses the Azov and Black seas (currently connected to the Mediterranean), and was derived from the Neogene epicontinental seas in the area (Paratethys; Dumont, 1998). The indigenous non-marine amphipod fauna comprises some 95 species from several lineages, 40 of them found in the Caspian alone (Mordukhai-Boltovskoi, 1969). Apart from the rich endemic radiation of Pontogammaridae there are three peculiar small families, with, e.g., commensal and parasitic species, and a flock of corophiids. The Caspian is also inhabited by a deepwater community of recent (Pliocene) Arctic marine immigrants. Only a few Ponto-Caspian species naturally occur outside of the region, but many have recently spread with man (see below).

The Holarctic North

The most salient geographical pattern in the diversity of both northern continents is that of a few widespread species in the north, and increasing diversity with narrower ranges toward the south. The previously glaciated, and hence recently (re)colonized regions are dominated by taxa with strong dispersal abilities, such as *Gammarus lacustris* and some North American *Hyalella* taxa, that can be dispersed short distances via waterfowl, and others that dispersed through the periglacial lakes that formed as the ice sheets retreated (*Monoporeia affinis, Gammaracanthus lacustris* and *Pallaseopsis quadrispinosa* in Europe, *Diporeia hoyi* in North America). Remarkably, a few taxa survived the glaciations in subterranean refugia (three *Stygobromus* spp. in North America, *Crymostygius thingvallensis* in Iceland and *Niphargus* spp. on the British Isles.

The Nearctic (NA)

The Nearctic fauna comprises 236 freshwater taxa, of which only 10% occur in regions affected by glaciations. Diversity in non-glaciated regions consists mainly of narrow endemics, often known only from a single locality. Two-thirds of the described Nearctic diversity is subterranean (Table 3).

The majority of described species are crangonyctoids, which have an ancient freshwater ancestry (see above). These are mainly troglobiotic, and even when epigean, they are often troglophilic (Holsinger, 1994; Zhang and Holsinger, 2003). Their diversity is highest in the karst landscapes of Eastern North America, and crangonyctoid taxa west of the Great Plains probably represent a more recent radiation (Holsinger, 1993). The talitroid genus *Hyalella* is also exclusively continental, but of South American origin. In addition to eight formally described North

American *Hyalella* taxa, molecular data have revealed tens of undescribed species. Several of these even occur in formerly glaciated areas (Witt & Hebert 2000), but the most extreme levels of local endemism were found in desert springs in California and Nevada, where most spring populations contain highly divergent lineages (Witt et al., 2006). Extrapolating to similar but unexplored regions, *Hyalella* may indeed be the most diverse North American*Hyalella* amphipod genus, with 500 or more taxa.

Gammaroids exhibit several independent patterns of freshwater colonization. Anisogammarids (*Ramellogammarus*) native to Pacific coastal watersheds may be relatively recent marine crawl-outs. A group of six relatively widespread *Gammarus* species in eastern North America (e.g., *G. fasciatus*) appear to be more ancient. Since *Gammarus* has a Palearctic center of diversity, this trans-Atlantic pattern suggests a secondary colonization of North America from Atlantic marine ancestors. A largely estuarine species, *G. tigrinus*, may represent the most recent example of this crawl-out process. Finally, *G. lacustris* appears to be a recent trans-Beringian colonizer from Eurasia, and *Chaetogammarus ischnus* is a recent human mediated trans-Atlantic invader.

A diverse group of narrowly endemic taxa in areas surrounding the northwest Gulf of Mexico (Texas to NE Mexico) show a true marine relict distribution, resulting from ancient marine regressions (Holsinger, 1994). These include the weckeliid group of the hadziids, along with sebids, bogidiellids, and the *Gammarus pecos* group.

Neotropics (NT)

Associated with the last mentioned element, many hadzioid 'relict' species also occur in fresh or weakly brackish caves and wells of the Caribbean, and many more inhabit coastal and anchihaline (subterranean) environments (Holsinger, 1993). Numerous bogidielloid and a few ingolfielloid species also occur in this region.

The documented South American amphipod diversity is relatively low at the family, genus and species levels. The 13 subterranean bogidiellid species are in 9 genera, but epigean diversity is restricted to the single genus *Hyalella*, widespread across the New World. The 51 described *Hyalella* species undoubtedly underrepresent the actual diversity, particularly in Lake Titicaca, where the genus has undergone an endemic radiation with parallels to other ancient lakes (Dejoux, 1994).

Australasia (AU)

Most Australian species are crangonyctoids, which occur from alpine regions to hypogean habitats in tropical and subtropical areas (Lowry & Stoddart, 2003). Although these are the best studied group in Australasia, the majority of taxa remain undescribed. For example, molecular data have indicated that subterranean aquifers in Western Australia are each inhabited by a unique cryptic lineage, and 200 of these aquifers have yet to be surveyed (Bradbury et al., unpubl. data).

In addition to the paramelitids (shared with Southern Africa) that are relicts of Gondwanaland, two additional crangonyctoid families are endemic to Australia, and another two to New Zealand. The most important factor that has provoked diversification in these groups was the onset of drier conditions beginning in the Eocene. Where marine incursions have occurred, or salinities are relatively high, the crangonyctoids have however been replaced, principally by the Hyalidae.

The remarkable diversity of the Tasmanian genus *Antipodeus* may represent a Pleistocene radiation. This genus dominates the cave faunas of Tasmania, as well as sub-alpine and alpine streams. The Tasmanian neoniphargids, and most of those in mainland Australia are also restricted to previously glaciated alpine regions. The documented New Zealand fauna comprises some 20 freshwater species, but again most of the diversity remains undescribed (Fenwick, 2001, and pers. comm.).

Other regions (PAC, OL, AN)

Freshwater amphipods in the Pacific, Oriental (excluding China, see above) and Antarctic (Kerguelen) regions are typically marine crawl-outs and comprise notably few species but relatively many families (Table 1, Fig. 2). There is however significant diversity in marginal marine habitats (brackish lagoons, anchihaline caves) of these areas.

Human related issues

Threats

The narrow endemism in caves and comparable restricted habitats is a special feature of amphipods by which individual species (and even some genera or families) are extremely vulnerable to extinction through habitat destruction and degradation (e.g., groundwater depletion or pollution) in several regions of the world (e.g., Sket, 1999; Witt et al., 2006).

Recent biotic invasions

Another main human effect on amphipod diversity is through facilitating range expansion across biogeographical boundaries. Inter-continental exchanges of non-marine species are still few (four cases only), but intra-continental invasions enabled by break-up of natural geographical barriers have thoroughly changed freshwater faunas, particularly in Europe. Notably, most of the recent invasions, even intra-continental, have been by taxa tolerant of brackish water also. Particularly, the Ponto-Caspian fauna has long evolved in isolation at a changeable interface of fresh and brackish environments, and thus preadapted to use emerging new dispersal opportunities. The colonizing success of species from the estuaries of major Ponto-Caspian rivers may be partly related to environmental disturbances and pollution in their new territories, creating conditions with high ionic concentrations, and to their natural ability to survive in brackish estuaries and harbors (Bij de Vaate et al., 2002).

The initial invasions by Ponto-Caspian taxa were enabled by the creation of canal networks interconnecting the major eastern and western European river systems since the late 1700s. The process was later enhanced by intentional transfers of potential fish food organisms to hydropower reservoirs, particularly from the Black Sea to the Baltic drainages. In the Soviet Union, 17 amphipod species were used in the transplantations during 1940–1970, among them Ponto-Caspian *Chelicorophium curvispinum*, *Dikerogammarus haemobaphes*, *Pontogammarus robustoides*, *Obesogammarus crassus*, *Chaetogammarus ischnus*, and *C. warpachowskyi* (Jazdzewski, 1980). Still the rate and range of the invasions have dramatically increased since the late 1980s, and in the 2000s many North and Central European river communities are undergoing major change with the aggressive expansion of *D. villosus* (Bij de Vaate et al., 2002). Even a Baikalian littoral species *Gmelinoides fasciatus* (the most eurytopic member of the endemic complex) has recently been established in NE Europe (Panov & Berezina, 2003).

Chaetogammarus ischnus is the single amphipod recently spread to North America along with a more general trans-Atlantic wave of Ponto-Caspian invaders (Vanderploeg et al., 2002). The North American euryhaline *Gammarus tigrinus* in turn was introduced to Britain and then intentionally to Germany in 1957 to replace locally extinct native species (Jazdzewski, 1980), and has since then broadly occupied river, lake, and estuarine habitats in Europe.

Acknowledgments We thank all those who answered our queries, including J. Holsinger, H. Morino, G. Fenwick, and R. Kamaltynov. We are particularly grateful to W. Vader and C. Fišer for their contributions in compiling the data available at their websites. Partial support was provided by a grant from the University of Helsinki Research Funds.

References

Banarescu, P., 1990–1995. Zoogeography of Fresh Waters, Vol. I–III. Aula-Verlag, Wiesbaden.

Barnard, J. L. & C. M. Barnard, 1983. Freshwater Amphipoda of the World. Part I, Evolutionary Patterns: i–xvii, 1–358. Part II, Handbook and Bibliography: xix, 359–830. Hayfield Associates, Mt. Vernon, Virginia.

Barnard, J. L. & G. S. Karaman, 1991. The families and genera of marine gammaridean Amphipoda (except marine gammaroids). Records of the Australian Museum Suppl. 13(1): 1–417.

Bij de Vaate, A., K. Jazdzewski, H. A. M. Ketelaars, S. Gollasch & G. Van der Velde, 2002. Geographical patterns in range extension of Ponto-Caspian macroinvertebrate species in Europe. Canadian Journal of Fisheries & Aquatic Sciences 59: 1159–1174.

Botosaneanu L. (ed.), 1986. Stygofauna Mundi: A Faunistic, Distributional, and Ecological Synthesis of the World Fauna Inhabiting Subterranean Waters (Including the Marine Interstitial). Brill/Backhuys, Leiden.

Bousfield, E. L., 1983. An updated phyletic classification and palaeohistory of the Amphipoda. In Schram, F. R. (ed.), Crustacean Phylogeny. Crustacean Issues 1: 257–277.

Bousfield, E. L. & C. T. Shih, 1994. The phyletic classification of amphipod crustaceans: problems in resolution. Amphipacifica 1(3), 76–134.

Dejoux, C., 1994. Lake Titicaca. Archiv für Hydrobiologie, Ergebnisse der Limnologie 44: 35–42.

Dumont, H. J., 1998. The Caspian Lake: history, biota, structure, and function. Limnology and Oceanography 43: 44–52.

Englisch, U., C. O. Coleman & J. W. Wägele, 2003. First observations on the phylogeny of the families Gammaridae, Crangonyctidae, Melitidae, Niphargidae, Megaluropidae and Oedicerotidae (Amphipoda, Crustacea), using small subunit rDNA gene sequences. Journal of Natural History 37: 2461–2486.

Fenwick, G. D., 2001. The freshwater Amphipoda (Crustacea) of New Zealand: a review. Journal of the Royal Society of New Zealand 31: 341–363.

Fišer, C., B. Sket & P. Trontelj, 2005. *Niphargus* homepage. URL http://www.bf.uni-lj.si/bi/zoologija/cene_fiser/niphargus.

Griffiths, C. I. & B. A. Stewart, 2001: Amphipoda. In Day, J. A., B. A. Stewart, I. J. deMoor & A. E. Louw (eds), Guides to the Freshwater Invertebrates of Southern Africa, Vol. 4: Crustacea III, Bathynellacea, Amphipoda, Isopoda, Spelaeogriphacea, Tanaidacea, Decapoda. WRC Report No TT 141/01. Water Research Commission, Pretoria: 28–49.

Holsinger, J. R., 1993. Biodiversity of subterranean amphipod crustaceans: global patterns and zoogeographic implications. Journal of Natural History 27: 821–835.

Holsinger, J. R., 1994. Pattern and process in the biogeography of subterranean amphipods. Hydrobiologia 287, 131–145.

Jazdzewski, K., 1980. Range extensions of some gammaridean species in European inland waters caused by human activity. Crustaceana (Suppl) 6: 84–107.

Kamaltynov, R. M. 2002 (dated 2001), Amfipody (Amphipoda: Gammaroidea). In Timoshkin, O. A. (ed.), Annotirovannyi Spisok Fauny Ozera Baikal i ego Vodosbornogo Basseina, Vol. I (I). Ozero Baikal (Index to the animal species inhabiting Lake Baikal and its catchment area, Vol. I(I). Lake Baikal). Nauka, Novosibirsk: 572–831.

Koenemann, S. & J. R. Holsinger, 1999. Phylogenetic analysis of the amphipod family Bogidiellidae s. lat., and revision of taxa above the species level. Crustaceana 72: 781–816.

Kozhova, O. M. & L. R. Izmesteva (eds), 1998. Lake Baikal: Evolution and Biodiversity. Backhyus Publishers, Leiden.

Lowry, J. K. & H. E. Stoddart, 2003. Crustacea: Malacostraca: Peracarida: Ampihpoda, Cumacea, Mysidacea. In Beesley, P. L. & W. W. K. Houston (eds), Zoological Catalogue of Australia, Vol 19.2B. CSIRO Publishing, Melbourne, Australia, xii+531 pp.

Macdonald, K. S. III, L. Yampolsky & J. E. Duffy, 2005. Molecular and morphological evolution of the amphipod radiation of Lake Baikal. Molecular Phylogenetics and Evolution 35: 323–343.

Martin, J. W. & G. E. Davis, 2001. An Updated Classification of the Recent Crustacea. Natural History Museum of Los Angeles County, Science Series: Vol. 39, pp. vii–124.

Myers, A. A. & J. K. Lowry, 2003. A phylogeny and a new classification of the Corophiidea Leach, 1814 (Amphipoda). Journal of Crustacean Biology 23: 443–485.

Panov, V. E. & N. A. Berezina, 2003. Invasion history, biology and impacts of the Baikalian amphipod *Gmelinoides fasciatus*. In Leppäkoski, E., S. Gollasch & S. Olenin (eds), Invasive Aquatic Species of Europe. Distribution, Impacts and Management. Kluwer, Dordrecht: 96–103.

Sket, B., 1999. The nature of biodiversity in hypogean waters and how it is endangered. Biodiversity & Conservation 8: 1319–1338.

Takhteev, V. V., 2000. Trends in the evolution of Baikal amphipods and evolutionary parallels with some marine malacostracan faunas. Advances in Ecological Research 31: 196–220.

Vader, W., 2005a. How Many Amphipod Species? Poster Presented at XII International Amphipod Colloquium. Cork, Ireland.

Vader, W., 2005b. New amphipod species described in the period 1974–2004. Amphipod Newsletter 28 (URL http://www.imv.uit.no/amphipod/AN28.pdf).

Väinölä, R. & R. M. Kamaltynov, 1999. Species diversity and speciation in the endemic amphipods of Lake Baikal: molecular evidence. Crustaceana 72: 945–956.

Vanderploeg, H. A., T. F. Nalepa, D. J. Jude, E. L. Mills, K. T. Holeck, J. R. Liebig, I. A. Grigorovich & H. Ojaveer, 2002. Dispersal and emerging ecological impacts of Ponto-Caspian species in the Laurentian Great Lakes. Canadian Journal of Fisheries and Aquatic Sciences 59: 1209–1228.

Vonk, R. & F. R. Schram, 2003. Ingolfiellidea (Crustacea, Malacostraca, Amphipoda): a phylogenetic and biogeographic analysis. Contributions to Zoology 72: 39–72.

Vonk, R. & F. R. Schram, 2005. World catalogue and bibliography of the Ingolfiellidea. URL http://www.science.uva.nl/onderzoek/cepa/ingolfiellideorumcatalogus.htm.

Witt, J. D. S. & P. D. N. Hebert, 2000. Cryptic species diversity and evolution in the amphipod genus *Hyalella* in Central Glaciated North America: a molecular phylogenetic approach. Canadian Journal of Fisheries and Aquatic Sciences 57: 687–698.

Witt, J. D. S., D. L. Threloff & P. D. N. Hebert, 2006. DNA barcoding reveals extraordinary cryptic diversity in an amphipod genus: implications for desert spring conservation. Molecular Ecology 15: 3073–3082.

Zhang, J. & J. R. Holsinger, 2003. Systematics of the freshwater amphipod genus *Crangonyx* (Crangonyctidae) in North America. Virginia Museum of Natural History, Memoir 6, 274 pp.

Hydrobiologia (2008) 595:257–266
DOI 10.1007/s10750-007-9021-5

FRESHWATER ANIMAL DIVERSITY ASSESSMENT

Global diversity of syncarids (Syncarida; Crustacea) in freshwater

A. I. Camacho · A. G. Valdecasas

Abstract Syncarida are malacostracan crustaceans that live exclusively in fresh and brackish waters all over the world. With the exception of a few Anaspidacea that live in free freshwater (streams, ponds, superficial lakes and caves) the great majority inhabit the interstitial groundwater (they are stygobiont that live in groundwater in the wide sense). The Syncarida lack a carapace; have compound eyes (absent in subterranean taxa); the range in body size from 0.55 to 55 mm long and are more or less cylindrical in body shape; they have separate sexes with no free-swimming larval stage. Only the epigean Anaspidaea have coloration. Fossil Syncarida comprises two orders: Palaeocaridacea (five families, 15 genera and 20 species from Europe, USA and Brazil) and Anaspidacea (two monospecific genera from Australia). Anaspidacea also has present-day representatives: five families with 12 genera and 21 living species that live only in the Southern Hemisphere. Bathynellacea, the third order of Syncarida, with no fossil representatives, has two families, with 66 genera and 219 species that are widely distributed throughout all continents, except Antarctica. Since 1950, new species of Bathynellacea have been discovered with regularity, however many countries remain poorly sampled. The accumulation curves for Parabathynellidae, Bathynellidae and the whole of Bathynellacea demostrate that new species descriptions continue to accumulate at a rate that is well beyond the "plateau" level.

Keywords Syncarida · Taxonoic diversity · Phylogeny · Biogeography

Guest editors: E. V. Balian, C. Lévêque, H. Segers & K. Martens
Freshwater Animal Diversity Assessment

Electronic supplementary material The online version of this article (doi:10.1007/s10750-007-9021-5) contains supplementary material, which is available to authorized users.

A. I. Camacho (✉) · A. G. Valdecasas
Departamento de Biodiversidad y Biología Evolutiva, Museo Nacional de Ciencias Naturales (CSIC), C/José Gutiérrez Abascal 2, Madrid 28006, Spain
e-mail: mcnac22@mncn.csic.es

Introduction

Syncarida Packard, 1885 are malacostracan crustaceans that were highly diversified during the Palaeozoic. First known in fossil forms is *Uronectes fimbriatus* (Jordan, 1847), it took 35 years before a living representative was found: *Bathynella natans* Vejdovsky, 1882, in a water well in Prague, and a few years later, *Anaspides tasmaniae* Thompson, 1893 in a stream in Tasmania. Fossil forms were initially linked to isopods and amphipods until Packard described Syncarida in 1885. Later, Calman (1896) gave Syncarida the rank of order and included them in the Eumalacostraca. With the increasing number of species discovered (Chappuis, 1917–1929) the first classifications were proposed (Siewing, 1959; Brooks, 1962). Noodt (1964) provided the phylogenetic foundations for the present-day systematics of the group, erecting the order Stygocaridacea and introducing the first classification of the Bathynellacea Chappuis 1915. This was adopted by Brooks (1969) and completed by Schminke (1973, 1975, 1978a). Schram (1984) advanced several alternative cladograms for the fossil and present day families and Coineau (1996; 1998) synthesized all of these hypotheses and classifications. Currently, fossil and living Syncarida are included in three orders: Palaeocaridacea, that comprises only fossil forms, Anaspidacea with living and fossil representatives, and Bathynellacea with living forms only.

The Syncarida lack a carapace, and the cephalon typically show a mandibular seat. Compound eyes (absent in subterranean taxa) can be either pedunculate or sessile. The body ranges between 0.55 and 55 mm long and is more or less cylindrical, with the pereon comprising eight free somites in members of the orders Bathynellacea and Palaeocaridacea (fossil), and seven free somites in the order Anaspidacea, where the first thoracomere is incorporated into the cephalon to compose a cephalothorax. The pleon comprises six free somites plus a telson in the Anaspidacea, whereas the telson is incorporated into the 6th pleonite to form a pleotelson in the Bathynellacea (Fig. 1). The thoracopods of syncarids (7–8 pairs) are biramous, with the last pair transformed into a copulatory organ in males in some instances, in some cases, the 7th thoracopod is lacking. The pleopods (0–5 pairs) are biramous, uniramous, vestigial or absent; the first two pairs can make up the so-called petasma in males. The uropods are biramous and inserted on the 6th pleonite or on the pleotelson; if foliaceous, they form a caudal fan with the telson (the case of the telson is free); when the telson is incorporated into a pleotelson, the uropod branches are styliform, and there are two furcal branches on the terminal margin of the pleotelson.

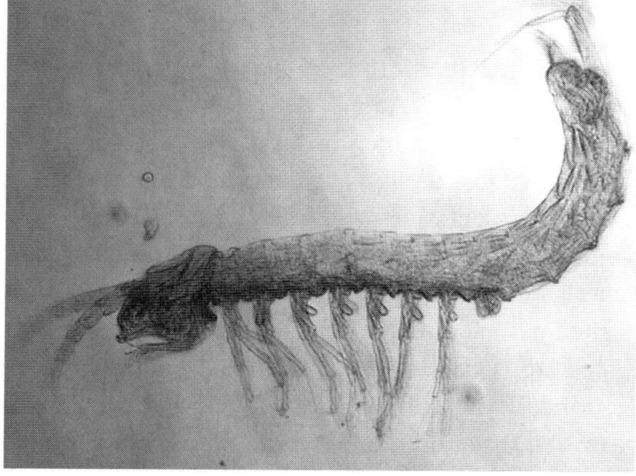

Fig. 1 Bathynellacea habitus

Only the epigean Anaspidacea exhibit coloration, due to the presence of chromatophores.

The fossil syncarids lived in brackish water in estuaries and lagoons from the lower Carboniferous to the Permian (Schram, 1981). Present day representatives live mainly in freshwater; only a few Bathynellacea and Stygocaridacea live in estuaries in oligo to polyhaline water and are euryhaline and eurythermal. Except for a few Anaspididae that live in free freshwater (streams, ponds, superficial lakes and caves) the great majority of syncarids inhabit the interstitial groundwater (they are stygobionts that live in groundwater in the broad sense). The type of substrate used as ranges from coarse to fine, and the depth varies depending on the taxa, wiht the Bathynellacea inhabiting the subterranean aquifers, including two species living in Lake Baikal at depths between 20 and 1440 m.

The interstitial syncarids are detritivores, feeding on bacteria, animal and vegetal remains, and fungi adhered to sand grains; Bathynellidae filter and consume large pieces (Serban, 1980), some Parabathynellidae (*Iberobathynella*) eat harpaticoid copepods (Coineau, 1996), and the epigean Anaspidacea, living in caves, galleries of other crustaceans, or in gravel banks, can be herbivorous (algae), carnivorous or can feed on faecal pellets (*Allanaspides*).

The epigean Anaspidacea swim freely or walk among the vegetation and have an escape reaction known as a caridoid response. Interstitial forms need to be in contact with sand grains (positive thigmotaxis) and move by walking with their thoracopods.

Syncarids have separate sexes though mating has never been observed. Fecundation appears to take place in the oviduct (Smith, 1908) and eggs are deposited one by one. Metamorphosis takes place inside the egg and there are no free-swiming larvae. Forms leaving the egg are similar to the adult but smaller in size and successive moults are necessary to attain adult size. Descriptions of the development of *Anaspides tasmaniae* can be found in Hickman (1937), and of Bathynellacea in Jakobi (1954),

Serban (1972), Schminke (1973) and Coineau (1998). The embryonic stage takes 9 months in *Antrobathynella stammeri* (Coineau, 1998) and life span can be up to 2.5 years in *Iberobathynella* (Coineau, 1998).

Species diversity

Fossil Syncarida comprises two orders: Palaeocaridacea (five families, 15 genera and 20 species from Europe, USA and Brazil), diversified in the Carboniferous, are the most primitive and do not have living representatives. The order Anaspidacea includes two monospecific genera, *Anaspidites* from the Australian Triassic and *Koonaspides* from the Australian lower Cretaceous, both belonging to the family Anaspididae. This order also has living representatives: five families (Table 1a and b; see Table in annex available on the website dedicated to FADA-Chapters) comprising 12 genera and 21 living species.

Bathynellacea, the third order of Syncarida, with no fossil representatives, comprises two families with 66 genera and 219 species: Parabathynellidae, 39 genera and 128 species, and Bathynellidae, 27 genera and 91 species. The latter necessitates a thorough revision.

Since 1950, new species of Bathynellacea have been discovered with regularity. Ten species of Parabathynellidae were known to occur in the Iberian Peninsula as of 1980. Following two decades of extensive work, the figure has risen to 30 species (Camacho 2003c). At present, there are 17 species of Bathynellidae known to occur in France, and five species in the Iberian Peninsula. However, this figure will soon exceed 20 species as we are currently describing 14 additional species. Nonetheless, many countries remain poorly sampled.

Table 1 (a) number of species and subspecies and (b) number of genera of Syncarida in "the Classic Biogeographic Region" of the world (based on Wallace's regions). All the Syncarida are aquatic species

(a)									
SP: Species Number	PA	NA	NT	AT	OL	AU	PAc	ANT	World
Bathynellacea	128	12	21	27	12	20	0	0	219
Parabathynellidae	53	8	17	22	9	19	0	0	128
Bathynellidae	75	4	4	5	3	1	0	0	91
Anaspidacea	0	0	8	0	0	13	0	0	21
Anaspididae	0	0	0	0	0	5	0	0	5
Koonungidae	0	0	0	0	0	3	0	0	3
Psammaspididae	0	0	0	0	0	2	0	0	2
Stygocarididae	0	0	7	0	0	3	0	0	10
Patagonaspididae	0	0	1	0	0	0	0	0	1
SYNCARIDA	128	12	29	27	12	33	0	0	240
(b)									
GN: Genera Number	PA	NA	NT	AT	OL	AU	PAc	ANT	World
Bathynellacea	30	6	14	18	9	6	0	0	66
Parabathynellidae	10	4	11	14	7	5	0	0	39
Bathynellidae	20	2	3	4	2	1	0	0	27
Anaspidacea	0	0	4	0	0	9	0	0	12
Anaspididae	0	0	0	0	0	3	0	0	3
Koonungidae	0	0	0	0	0	2	0	0	2
Psammaspididae	0	0	0	0	0	2	0	0	2
Stygocarididae	0	0	3	0	0	2	0	0	4
Patagonaspididae	0	0	1	0	0	0	0	0	1
SYNCARIDA	30	6	18	18	9	15	0	0	78

PA: Palaearctic; NA: Nearctic; NT: Neotropical; AT: Afrotropical ; OL: Oriental; AU: Australasian; PAc: Pacific & Oceanic Islands; ANT: Antarctic

Thus, the understanding of the global syncarid taxonomic diversity is fraught with uncertainty derived from several sources:

- First, some areas have been poorly sampled, especially Africa, South America and Asia.
- Second, those that have been sampled on a more or less regular basis, probably have not been sampled enough, as demostrated by Camacho & Valdecasas (2003).
- Third, some species are poorly described, and the genera known to date are in urgent need of revision.

(See Fig. 2 for a summary of number of species and genera by zoogeographical region.)

Figure 3a shows three accumulation curves for Parabathynellidae, Bathynellidae and the whole of Bathynellacea. There is no need to fit a function to the raw data to see that new species descriptions still accumulate at a rate that is well beyond the "plateau" level.

A generic diversity was calculated using rarefaction estimation (Simberloff, 1972). This procedure is frequently used in ecology, and takes the number of specimens per species in a collection to estimate the number of species in collections with smaller numbers of specimens. However, a useful taxonomic diversity index can be obtained by rarefacting species into genera, as is frequently done in paleontology (e.g. Favtosky et al., 2004). In our case we took the total Syncarid distribution of species per genera (see Table in annex available on the website dedicated to FADA-Chapters) to estimate generic distribution per biogeographical region, as defined by Wallace (1876). Only two regions show a clear discrepancy between observed and estimated number of genera (Table 2), the Paleartic and the Australasian regions. Rarefaction calculation is highly dependent on the "mother" list, so these figures could change drastically when some regions are better characterized. Figure 3b shows this curve for the whole range of species of Syncarida. It gives the number of genera predicted for a collection of species, if the real "universe" of Syncarida had a similar pattern to the sampling distribution. About 95% limits are also indicated in the figure. In general, the analysis points to the occurrence of undersampling, especially in the Palaearctic and Australasian region.

Finally, we have carried out a resampling scheme, with the whole Syncarid group, dividing the number of species by five year sampling periods, from 1880 to 2005 (55 sampling periods, s.p.), and resampling with replacement one thousand sets of 55 s. p. Results have been organized in 25 classes (Fig 3c). At the 5%

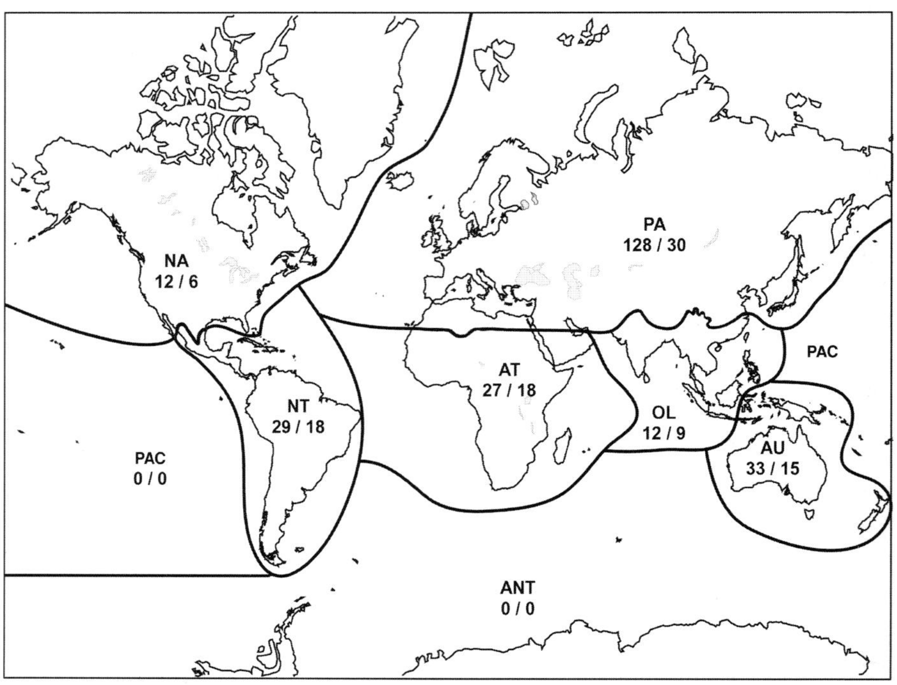

Fig. 2 Distribution of freshwater Syncarida species and genera by zoogeographical region (species number/genus number). PA—Palaearctic; NA—Nearctic; NT—Neotropical; AT—Afrotropical ; OL—Oriental; AU-Australasian; PAC—Pacific & Oceanic Islands; ANT—Antarctic

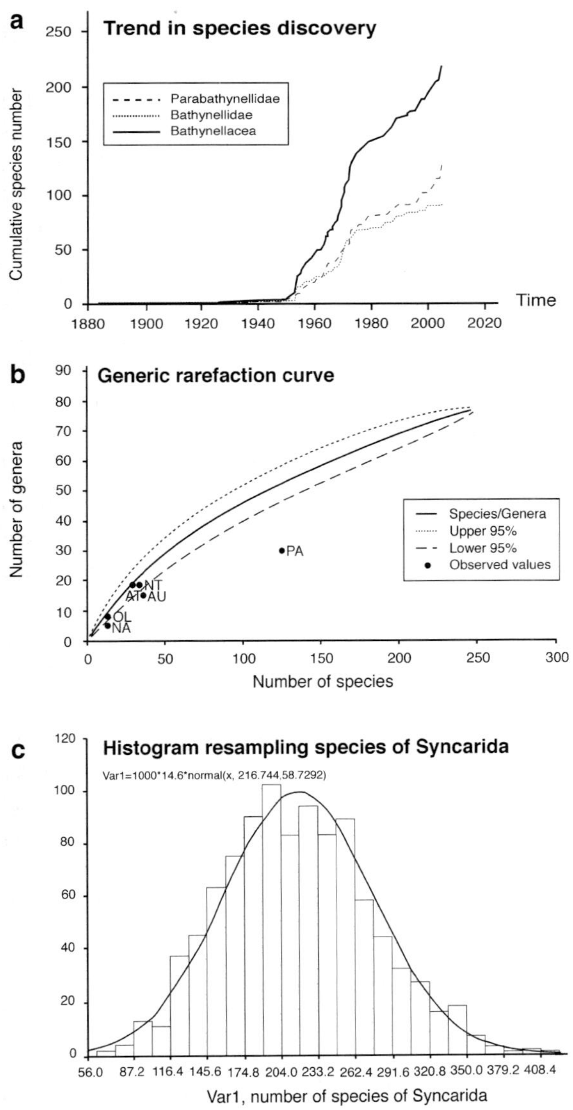

Fig. 3 (**a**) Three accumulation curves for Parabathynellidae, Bathynellidae and the whole Bathynellacea; (**b**) rarefaction curve for all the species of Syncarida (PA = Palaearctic Region; NA = Nearctic Region; AT = Afrotropical Region; NT = Neotropical Region; OL = Oriental Region and AU = Australasian Region). (**c**) resampling curve (see text)

Table 2 Estimated number of Syncarid genera per biogeographical region using rarefaction

	PA	NA	NT	AT	OL	AU
Observed	30	6	18 9	18		15
E (G)	53	10	19	18	10	21
st.dev.(G)	3.1	1.4	2.2	2.1	1.3	2.3

PA: Palaearctic; NA: Nearctic; NT: Neotropical; AT: Afrotropical; OL: Oriental; AU: Australasian

significance level (for a two tail distribution test), the lowest values are 111 species and the highest is 341. Between this range lies the other 95% population figure. The highest number obtained within this 95% is 341, a number of species clearly above the present 240 species. However, this could be a poor estimation of the total Syncarid diversity, as around 100 species are still awaiting description.

Phylogeny and Historical proccesses

The Syncarida were already diversified during the Palaeozoic. They were one of the most important elements of the crustacean communities in brackish and littoral-marginal freshwaters of Laurentia during the Carboniferous. During the Permo-Trias they were present in several areas of the Gondwana, more specifically in Australia and Brazil, where Anaspidacea still live today (Schram, 1981, 1984). Present day distribution includes all continents, except Antarctica. Two alternative hypotheses have been proposed to explain Syncarid distribution on a continental scale.

In the first scenario, the littoral (surface) marine ancestor entered actively, through swimming larvae, into the continental waters and from there, through the hydrographic network of superficial waters into caves, karst and the alluvial porous systems (stygobionts of freshwater origin) (Schminke, 1981). The presence of syncarids in brackish interstitial waters, in this case, would be of secondary derivation. This hypothesis is supported by the presence of Anaspidacea in superficial freshwaters and of rare Bathynellacea in old mountain ranges (e.g. in Spain, see Camacho & Coineau, 1989).

The second scenario addresses the fact that present-day Anaspidacea and Bathynellacea live in areas that have suffered one or several trangression-regression events of the Tethys Sea after the Triassic. The biphase colonization model (Boutin & Coineau, 1990; Notenboom, 1991; Coineau & Boutin, 1992) states that in a first step the littoral marine ancestral syncarid actively colonized the littoral sediments and became interstitial; in a second step, considered passive, during the regression of the Tethys, a part of the littoral interstitial population remained in place and became phreatobitic. Meanwhile, the piezometric level fell and the rest of the population remained littoral as the shore line gradually retreated (thalassostygobiont origin). At the end of the regression the two populations were genetically isolated, and a speciation took place due to a vicariant event (this stage corresponds to the "Regression Model Evolution" of Stock, 1980). Many syncarids inhabit the same habitat as other stygobionts (of undeniable marine origin) that fit this model. The majority of present-day Bathynellacea from littoral waters and many fossil taxa known from littoral facies lend credence to this hypothesis (Boutin & Coineau, 1987; Camacho & Coineau, 1989; Coineau, 1996, 1998; Camacho, 2004; Camacho et al., 2006).

The Syncarida could have evolved by progenesis from an ancestral type of zoea larva. The Anaspidacea exhibit the typical caridoid facies (except for the caparace), and the Stygocarididae and Bathynellacea derive from repeated progenetic events. With the "Zoea theory" Schminke (1981) showed that the Bathynellacea have a larva-like morphology similar to their ancestral form. The five families of the Anaspidacea belong to an evolutive series connecting the surface living forms to the interstitial stygobionts; as they adapted to the interstitial subterranean life, the species of every derived clade looked similar to the juvenile stages of the more primitive sister groups in the adaptive phylogenetic series (Coineau, 1996, 2000).

The traditional phylogenetic tree assumes that the Syncarida are primitive Eumalacostraca (Coineau, 1996; Schram & Hof, 1998; Spears & Abele, 1999). Present day alternatives on syncarid phylogeny are that Syncarida includes the sister groups Bathynellacea and Anaspidacea (Richter & Scholtz, 2001) or that the Bathynellacea are the ancestor of the Malacostraca, including the Anaspidacea (Lange & Schram, 1999). In a recent phylogenetic analysis (Camacho et al., 2002) based on molecular data derived from the 16S rDNA of different Malacostraca, including a species of Bathynellacea and five species of Anaspidacea, Bathynellacea remained in a basal position and clearly split off from the Syncarida, but lay within the Eumalacostraca. Serban (1972) has previously advocated for a superorder Podophallocarida for the Bathynellacea outside of the Syncarida. More studies are clearly needed to resolve this question.

The phylogenetic relationships among the families of the Syncarida have been explored by several authors (Schminke, 1975, Schram, 1984; Coineau, 1996). Figure 4 shows a cladogram with the more commoly accepted relationships (Coineau, 1996).

Other phylogenetic studies within Bathynellacea and Anaspidacea include Grosso & Peralta (2002), Schminke (1973, 1975), Schram (1984), Coineau (1996), Camacho (1987, 2003c), Camacho et al. (1997, 2000), Guil and Camacho (2001) and Cho (1995, 2001), however, there is no global study that focuse on the relationships among all genera of both families.

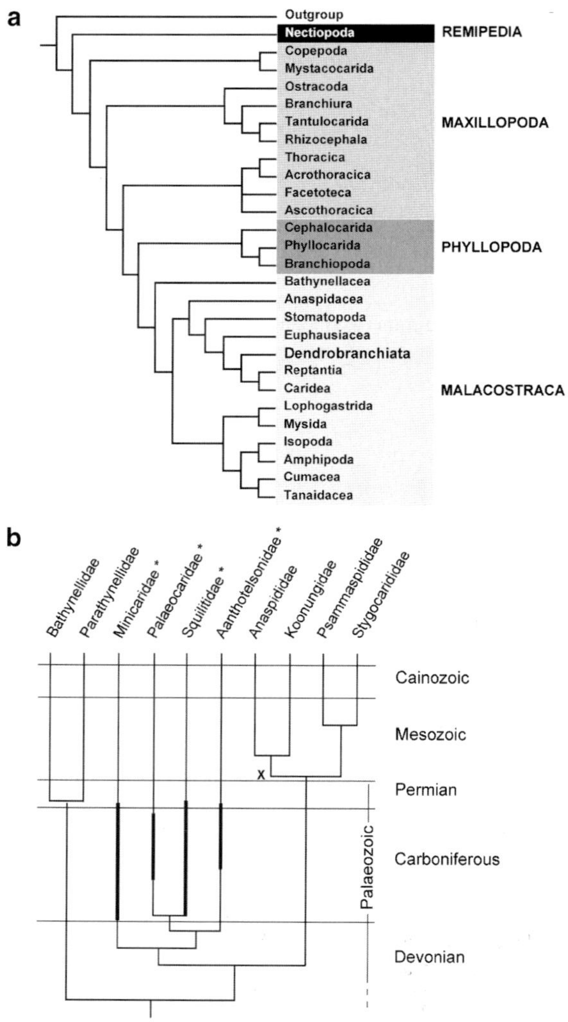

Fig. 4 (**a**) Crustacean phylogeny inferred by morphology (slightly modified from Lange & Schram, 1999); (**b**) Phylogenetic relationship of Syncarida (after Schminke, 1975; Schram, 1984; Coineau, 1996)

Present distribution and main areas of endemicity

The distribution of both fossil and living Syncarida can be explained by a double vicariant biogeographic model, taking into account plate tectonics and the evolution of the Tethys Sea during the Mesozoic and the Cenozoic. The extinct taxa, endemic to the Laurentia in the Carboniferous, could have dispersed through the Pangea, since its split during the Permo-Trias. The fragmentation of Pangea during the Mesozoic could have produced a discontinuity in the distribution (Schram, 1977), and the Syncarida of each area could have evolved independently. The Anaspidacea, living only in the Southern Hemisphere

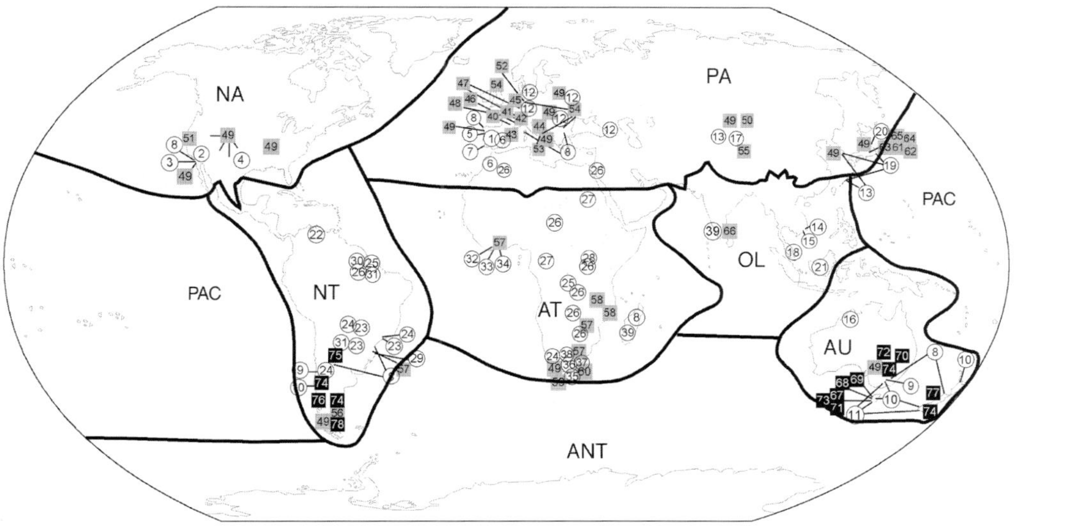

Fig. 5 World distribution of Syncarida genera per biogeographical area (based on Wallace's Region). **White circle = Parabathynellidae**: 1. *Iberobathynella* Schminke, 1973; 2. *Californibathynella* Camacho et Serban, 1998; 3. *Califobathynella* Cho, 1997; 4. *Texanobathynella* Delamare et al., 1975; 5. *Guadalopebathynella* Camacho et Serban, 1998; 6. *Paraiberobathynella* Camacho et Serban, 1998; 7. *Hexaiberobathynella* Camacho et Serban, 1998; 8. *Hexabathynella* Schminke, 1972; 9. *Chilibathynella* Noodt, 1963; 10. *Atopobathynella* Schminke, 1973; 11. *Notobathynella* Schminke, 1973; 12. *Parabathynella* Chappuis, 1926; 13. *Eobathynella* Birstein et Ljovuschkin, 1964; 14. *Paraeobathynella* Camacho, 2005; 15. *Sketinella* Camacho, 2005; 16. *Kimberleybathynella* Cho et al., 2005; 17. *Issykkulibathynella* Serban, 1984; 18. *Batubathynella* Schminke, 1973; 19. *Allobathynella* Morimoto et Miura, 1957; 20. *Nipponbathynella* Schminke, 1973; 21. *Sabahbathynella* Schminke, 1988; 22. *Psalidobathynella* Schminke, 1979; 23. *Leptobathynella* Noodt, 1965; 24. *Parvulobathynella* Schminke, 1973; 25. *Thermobathynella* Capart, 1951; 26. *Cteniobathynella* Schminke, 1973; 27. *Nilobathynella* Dumont, 1984; 28. *Heterodontobathynella* Schminke, 1973; 29. *Brasilibathynella* Jakobi, 1958; 30. *Odontobathynella* Delamare et Serban, 1979; 31. *Noodtibathynella* Schminke, 1973; 32. *Haplophallonella* Serban et Coineau, 1975; 33. *Lamtobathynella* Serban et Coineau, 1987; 34. *Acantobathynella* Coineau, 1967; 35. *Ctenophallonella* Coineau et Serban, 1978; 36. *Racovitzaibathynella* Serban et Coineau, 1994; 37. *Afrobathynella* Schminke, 1976; 38. *Nunubathynella* Schminke, 1976; 39. *Habrobathynella* Schminke, 1973. **Grey square = Bathynellidae**: 40. *Gallobathynella* Serban et al., 1971; 41. *Clamousella* Serban, 1989; 42. *Vejdovskybathynella* Serban et Leclerc, 1984; 43. *Hispanobathynella* Serban et al., 1971; 44. *Meridiobathynella* Serban et al., 1971; 45. *Pseudobathynella* Serban et al., 1971; 46. *Vandelibathynella* Serban et al., 1971; 47. *Delamareibathynella* Serban, 1989; 48. *Parameridiobathynella* Serban et Leclerc., 1984; 49. *Bathynella* Vejdovsky, 1882; 50. *Baikalobathynella* Birstein et Ljovuschkin, 1967; 51. *Pacificabathynella* Schminke et Noodt, 1988; 52. *Pseudantrobathynella* Schminke, 1988; 53. *Sardobathynella* Serban, 1973; 54. *Antrobathynella* Serban, 1966; 55. *Tianschanobathynella* Serban, 1993; 56. *Austrobathynella* Delamare Deboutteville, 1960; 57. *Nannobathynella* Noodt, 1969; 58. *Agnatobathynella* Schminke, 1980; 59. *Transvaalthynella* Serban et Coineau, 1975; 60. *Transkeilthynella* Serban et Coineau, 1975; 61. *Uenobathynella* Serban, 2000; 62. *Parauenobathynella* Serban, 2000; 63. *Paradoxibathynella* Serban, 2000; 64. *Morimotobathynella* Serban, 2000; 65. *Nihobathynella* Serban, 2000; 66. *Serbanibathynella* Reddy et Schminke, 2005. **Black square = Anaspididae Thomson, 1884**: 67. *Anaspides* Thomson, 1894; 68. *Paranaspides* Smith, 1908; 69. *Allanaspides* Swain et al., 1970. Koonungidae Sayce, 1880: 70. *Koonunga* Sayce, 1907; 71. *Micraspides* Nicholls,1931. Psammaspididae Schminke, 1974: 72. *Psammaspides* Schminke, 1974; 73. *Eucrenonaspides* Knott et Lake,1980; Stygocarididae Noodt, 1963: 74. *Stygocaris* Noodt, 1963; 75. *Parastygocaris* Noodt, 1963; 76. *Oncostygocaris* Schminke, 1980; 77. *Stygocarella* Schminke, 1980. Patagonaspididae Grosso et Peralte, 2002: 78. *Patagonaspides* Grosso et Peralta, 2002

(in South America and in the Australo-Tasmano-New Zealand area) (see Fig. 5) would reflect a Gondwanian distribution, similar to the distribution of the fossil Malacostraca of the Permo-Trias.

The Palaearctic is the best known region for this fauna and here the distribution of Bathynellidae is broader than that of the Parabathynellidae (Fig. 5). Bathynellidae is more diversified and widely distributed in the northern hemisphere (21 genera and 82 species) than in the southern hemisphere (7 genera and 9 species). Parabathynellidae are widely distributed and diversified through the temperate and tropical regions. North America is the region with the lowest number of genera (6) and species (7) known.

Within Bathynellidae, the genus *Bathynella* Vejdovsky, 1882, has more than 50 species and subspecies, but only a few have been properly studied. The Asian species and a few others should be revised as some of them may belong to different genera (Serban, 2000). According to Serban (2000), *Bathynella* does not occur in Asia. In the European context, it spreads to southeast Europe, but is not present in the Iberian Peninsula. There is only one case of a genus of this family that is found in two very distinct areas, *Nannobathynella* being present with one species each in Brazil and the Ivory Coast.

Within the family Parabathynellidae there are a few genera with several species distributed through very different biogeographic regions; *Chilibathynella* and *Atopobathynella* being found in Chile and in Australia; *Parvulobathynella, Cteniobathynella* and *Thermobathynella* live in South America and Africa; and *Habrobathynella*, that was known only from Africa and has been found recently in India. The distribution of Iberobatinellids is amphi-Atlantic, and is related to the opening of the North Atlantic Ocean. The genera *Iberobathynella*, *Guadalopebathynella* and *Hexaiberobathynella* are restricted to the Iberian Peninsula and the Balearic islands, while some related genera such as *Texanobathynella* and *Californibathynella* live in North America and *Paraiberobathynella* in the Iberian Peninsula and North Africa (Coineau, 1998; Camacho et al. 2000).

The genus *Hexabathynella* is the only one displaying a worldwide distribution. However, the majority of the species has been found in just one locality or in localities of a limited area (Camacho, 2003a). The majority of genera live in very limited areas, although this can be, due to the temporaly and spatialy sporadic nature of the collecting. For this reason, it does not seem appropiate to designate them endemics. When sampling is done intensively and extensively in different seasons over different years and with different sampling techniques in different habitats (caves, springs, wells and the interstitial environment associated with streams), previously restricted species have been shown to have a wider distribution, e.g. *Iberobathynella imuniensis*, *Paraiberobathynella fagei* or *Hexaiberobathynella mateusi* (Camacho, 2003c).

Figure 5 shows the general distribution of the genera of known Syncarida represented in the classic biogeographical areas (based on the Wallace's Region).

Acknowledgements We gratefully acknowledge the assistance of C. Puch, J.M. Becerra and M.L. Peláez. This study was supported by projects EVK2-CT-2001-00121 (PASCALIS) and Convenio Junta de Castilla y León- CSIC (2002–2004).

References

Main References (See complete list of references - auxiliary references – in annex available on the website dedicated to FADA-Chapters)

Boutin, C., & N. Coineau, 1987. *Iberobathynella* (Crustacea: Syncarida: Bathynellacea) sur le continent africain. Implications paléobiogéographiques. Comptes Rendus de l' Académie des Sciences, Paris 304: 355–358.

Boutin, C. & N. Coineau, 1990. "Regression model", "Modèle Biphase" d'évolution et origine des micro-organismes stygobies interstitiels continentaux. Revue de Micropaléontologie 33 (3–4): 303–322.

Brooks, H. K., 1962. On the fossil Anaspidacea, with a revision of the classification of the Syncarida. Crustaceana 4: 229–242.

Brooks, H. K., 1969. Syncarida. In Moore, R. C. (ed.), Treatise on Invertebrate Paleontology, Part R, Arthropoda 4, Vol. 1. Geological Society of America and University of Kansas, Lawrence: 345–359.

Calman, W. T., 1896. On the genus Anaspides and its affinities with certain fossil Crustacea. Transaction of the Royal Society of Edimburgh 38: 787–802.

Camacho, A. I., 1987. La Familia Parabathynellidae en la Península Ibérica: Taxonomía, Filogenia y Biogeografía. Tesis Doctoral (unpublished), Universidad Autonoma de Madrid, 890 pp.

Camacho, A. I., 2003a. Historical biogeography of *Hexabathynella*, a cosmopolitan genus of groundwater Syncarida (Crustacea, Bathynellacea, Parabathynellidae). Biological Journal of the Linnean Society 78: 457–466.

Camacho, A. I., 2003c. An overview of the distribution of the Parabathynellidae (Crustacea Syncarida Bathynellacea) on the Iberian Peninsula and Balearic Islands. Graellsia, 59(1): 63–78.

Camacho, A. I., 2004. An overview of *Hexabathynella* (Crustacea, Syncarida, Parabathynellidae) with the description of a new species. Journal of Natural History 28: 1249–1261.

Camacho, A. I. & N. Coineau, 1989. Les Bathynellacea d'Espagne: Répartition et Biogéographie. Mémoires de Biospéléologie XVI(43): 111–124.

Camacho, A. I., E. Bello, & G. F. Estabrook, 1997. A statistical approach to the evaluation of characters to estimate evolutionary relationships among the species of the aquatic subterranean genus *Iberobathynella* (Crustacea, Syncarida). Biological journal of the Linnean Society 60: 221–241.

Camacho, A. I. & A. G. Valdecasas, 2003. Evaluating extinction in rare habitats: an essay. Graellsia 59(2–3): 409–414.

Camacho, A. I., E. Serban & N. Guil, 2000. Phylogenetical review and biogeographic remarks on the interstitial and subterranean freshwater iberobathynells (Crustacea, Syncarida, Parabathynellidae). Journal of Natural History 34: 563–585.

Camacho A. I., I. Rey, B. A. Dorda, A. Machordom & A. G. Valdecasas, 2002. A note on the systematic position of the Bathynellacea (Crustacea, Malacostraca) using molecular evidence. Contribution to Zoology 71(4): 123–129.

Camacho, A. I., T. Torres, E. Ortiz, C. Puch & A. G. Valdecasas, 2006. Small-scale biogeographical pattern in groundwater Crustacea (Syncarida, Parabathynellidae). Biodiversity and Conservation, 15: 3527–3541.

Coineau, N., 1996. Sous-Classe des Eumalacostracés (Eumalacostraca Grobben, 1892). Super-Ordre des Syncarides (Syncarida Packard, 1885). In Forest, J. (ed.), Traité de Zoologie, Crustacés VII(2): 897–954.

Coineau, N., 1998. Syncarida. In Juberthie, C. & V. Decu (eds), Encyclopaedia Biospeologica, Tome II, Moulis, Bucarest, Société Biospéologie: 863–876.

Coineau, C., 2000. In Wilkens, H., D.C. Culver & W.F. Humphreys (eds), Subterranean Ecosystems, Ecosystems of the world 30, Elsevier : 189–210.

Coineau, N. & C. Boutin, 1992. Biological processes in space and time: colonization. evolution and speciation in interstitial stygobionts. In Camacho A. I. (ed.), The Natural History of Biospeleology. Monografías del M.N.C.N. 7(CSIC), Madrid: 423–451.

Chappuis, P. A., 1929. Anaspidacea. In Kükenthal-Krumbach, Handb Zoology 3(6): 594–606.

Cho, J. L., 1995. Systematik und Biogeographie von *Hexabathynella* Schminke, 1973 sowie ein Beitrag zur Taxonomie der "Leptobathynellinae" Noodt, 1964 (Bathynellacea, Syncarida, Malacostraca). Kiel, Doctoral Thesis (unpublished).

Cho, J. L., 2001. Phylogeny and zoogeography of three new species of the genus *Hexabathynella* (Crustacea, Malacostraca, Bathynellacea) from North America, Zoologica Scripta 30(2), 145–157.

Fastovsky, D. E., Y. Huang, J. Hsu, J. Martin-McNaughton, P. M. Sheehan & D. B. Weishampel, 2004, Shape of Mesozoic dinosaur richness: Geology 32: 877–884.

Grosso, L. E., & M. Peralta, 2002. *Patagonaspides* gen.n.; *P. sandroruffoi* sp.n. (Crustacea, Syncarida). First phreatobite species of a new anaspidacean family discovered in Patagonia with cladistic analisis of Stygocaridinea (Anaspidacea). Bulletino del Museo Cívico di Storia Naturale di Verona 26: 105–118.

Guil, N., & A. I. Camacho, 2001. Historical Biogeography of *Iberobathynella* (Crustacea, Syncarida, Bathynellacea), an aquatic subterranean genus of Parabathynellids, endemic to the Iberian Peninsula. Global Ecology and Biogeography 10: 487–501.

Hickman, V. V., 1937.The embriology of the Syncarid Crustacean *Anaspides tasmaniae*. Papers and Proceedings of the Royal Society of Tasmania: 1–36.

Jakobi, H., 1954. Biologie, Entwicklungsgeschichte und Systematik von *Bathynella natans* Vejd. Zoologisches Jahrbuch, Systematic 83(1/2): 1–62.

Jordan, H., 1847. Entdeckung fossiler Crustaceen im Saarbrücken'schen Steinkohlengebirge. Verhandlungen des natur historischen Vereines preussischen Rheinlande 4: 89–92.

Lange, S. & F. Schram, 1999. Evolución y filogenia de los crustáceos. Boletín de la Sociedad Entomológica Aragonesa 26: 235–254.

Noodt, W., 1964. Natürliches System und Biogeographie der Syncarida. Gewässer und Abwässer 37/38: 77–186.

Notenboom, J., 1991. Marine regressions and the evolution of groundwater dwelling amphipods (Crustacea). Journal of Biogeography 18: 437–454.

Richter, S., & G. Scholtz, 2001. Phylogenetic analysis of the Malacostraca (Crustacea). Journal of Zoological Systematics and Evolutionary Research 39: 113–136.

Schram, F. R., 1977. Paleozoogeography of the late Paleozoic and Triassic Malacostraca. Systematic Zoology 26: 367–379.

Schram, F. R., 1981. On the classification of Eumalacostraca. Journal of Crustacean Biology 1(1): 1–10.

Schram, F. R., 1984. Fossil Syncarida. Transaction of the San Diego Society of Natural History 20: 189–246.

Schram, F. R., & C. H. J. Hof, 1998. Fossils and the interrelationships of Major Crustacean groups. In: Edgecombe, G. D. (ed.) Arthropod Fossils and Phylogeny. Columbia University Press, New York: 233–302.

Schminke, H. K., 1973. Evolution, System und Verbreitungsgeschichte der Familie Parabathynellidae (Bathynellacea, Malacostraca). Mikrofauna Meeresboden 24: 1–192.

Schminke, H. K., 1975. Phylogenie und Verbreitungsgeschichte der Syncarida (Crustacea, Malacostraca). Verhandlungen Zoologische Gesellschaft, Bochum: 384–388.

Schminke, H. K., 1978a. Die phylogenetische Stellung der Stygocarididae (Crustacea: Syncarida). Unter besonderer Berücksichtingung morphologischer Ähnlichkeiten mit Larvenformen der Eucarida. Zeitschrift für zoologische Systematik und Evolutionforschung 16: 225–239.

Schminke, H. K., 1981. Adaptation of Bathynellacea (Crustacea, Syncarida) to the life in the interstitial ("Zoea Theory"). International Revue Gesamter Hydrobiologie 66(4): 576–637.

Serban, E., 1972. Bathynella (Podophallocarida, Bathynellacea). Travaux de l'Institut de Spéologie "Émile Racovitza" 11: 1–398.

Serban, E., 1980. La mandibule et l'individualisation des ensembles évolutifs majeurs dans l'ordre des Bathynellacea (Malacostraca: Podophallocarida). Bijdragen tot de Dierkunde 50(1): 155–189.

Serban, E., 2000. *Uenobathynella* n.g., *Parauenobathynella* n.g., *Morimotobathynella* n.g., *Nihobathynella* n.g. et *Paradoxibathynella* n.g., Bathynellinae du Japón (Bathynellidae, Bathynellacea, Podophallocarida). Travaux de l'Institut de Spéologie Émile Racovitza 36: 3–61.

Siewing, R., 1959. Syncarida. In Bronn, H. G. (ed.), Klassen und Ordnungen des Tierreichs. Akademische Verlagsgesellsch. Geest & Portig, Leipzig, Thomson, 5(1): 121; 4(2): 1–121.

Simberloff, D., 1972. Models in biogeography. In: Schopf, J. M. (ed.), Models in Paleobiology. Freeman, Cooper & Company, San Francisco: 160–191.

Smith, G., 1908. On the Anaspidacea. Living and Fossil. Quaternaly Journal of Microscopy Sciences 53: 489–578.

Spears, T. & L. G. Abele, 1999. Crustacean phylogeny inferred from 18s rDNA. In: Fortey R. A. & R. H. Thomas (eds), Arthropod relationship. Chapman and May, London: 160–187.

Stock, J. H., 1980. Regression model evolution as exemplified by genus *Pseudoniphargus* (Amphipoda). Bijdragen tot Dierkunde 50(1): 105–144.

Wallace, A. R. 1876. The geographical distribution of animals. Harper, New York, xxiii +503, xi + 553 pp (reprinted 1962, Hafner, New York).

Hydrobiologia (2008) 595:267–273
DOI 10.1007/s10750-007-9022-4

FRESHWATER ANIMAL DIVERSITY ASSESSMENT

Global diversity of crabs (Aeglidae: Anomura: Decapoda) in freshwater

Georgina Bond-Buckup · Carlos G. Jara · Marcos Pérez-Losada · Ludwig Buckup · Keith A. Crandall

Abstract The freshwater anomuran crabs of the family Aeglidae are all restricted to southern South America occurring in Chile, Brazil, Bolivia, Uruguay, Paraguay, and Argentina. The family consists of a single genus, *Aegla*, containing 63 currently described species. There are another 5–10 known yet undescribed species to complement this diversity. The aeglids occur in freshwater lakes, streams, rivers, and in caves with freshwater. The origin of the family appears to be from marine ancestors from the Pacific invading streams in Chile about 75 mya radiating both in Chile and again on the eastern side of the Andes, particularly in Brazil. Of the 63 species, 23 or 36.5% are considered under threat and are in need of conservation action.

Keywords Conservation · Freshwater biology · Crab · Decapoda · Anomura · Aeglidae · Diversity · South America

Guest editors: E. V. Balian, C. Lévêque, H. Segers & K. Martens
Freshwater Animal Diversity Assessment

G. Bond-Buckup · L. Buckup
Departamento de Zoologia, Universidade Federal Do Rio Grande Do Sul, Porto Alegre, Brazil

G. Bond-Buckup
e-mail: gbond_buckup@yahoo.com.br

C. G. Jara
Instituto de Zoología, Universidad Austral de Chile, Casilla 567, Valdivia, Chile

M. Pérez-Losada · K. A. Crandall
Department of Biology, Brigham Young University, Provo, Utah, 84602-5181, USA

K. A. Crandall (✉)
Monte L. Bean Life Science Museum, Brigham Young University, Provo, Utah, 84602-5181, USA
e-mail: keith_crandall@byu.edu

Introduction

The Aeglidae are the most abundant and widely distributed freshwater decapod "crabs" in southern South America. Unlike true brachyuran crabs, however, in aeglids the fifth pair of pereiopods is reduced in size, lacking walking capacity (Lopretto, 1978; Martin & Abele, 1988); they also possess tiny chelae with which they groom branchiae and eggs attached to the pleopods of females and the underside of the abdomen (Martin & Felgenahuer, 1986) (Fig. 1, inset). All aeglids are primarily aquatic and occur in lakes, streams, and caves, at depths of down to 320 m in Chilean lakes (Jara, 1977), and at altitudes of up to ~3,500 m in northeastern Argentinean cordilleras (Bond-Buckup & Buckup, 1994). Aeglids are the only anomuran family restricted to the Neotropical region of South America. Taxonomically, aeglids are included within the anomuran superfamily Galatheoidea, but there is some morphological evidence (e.g., gill structure and caparace sutures) and molecular

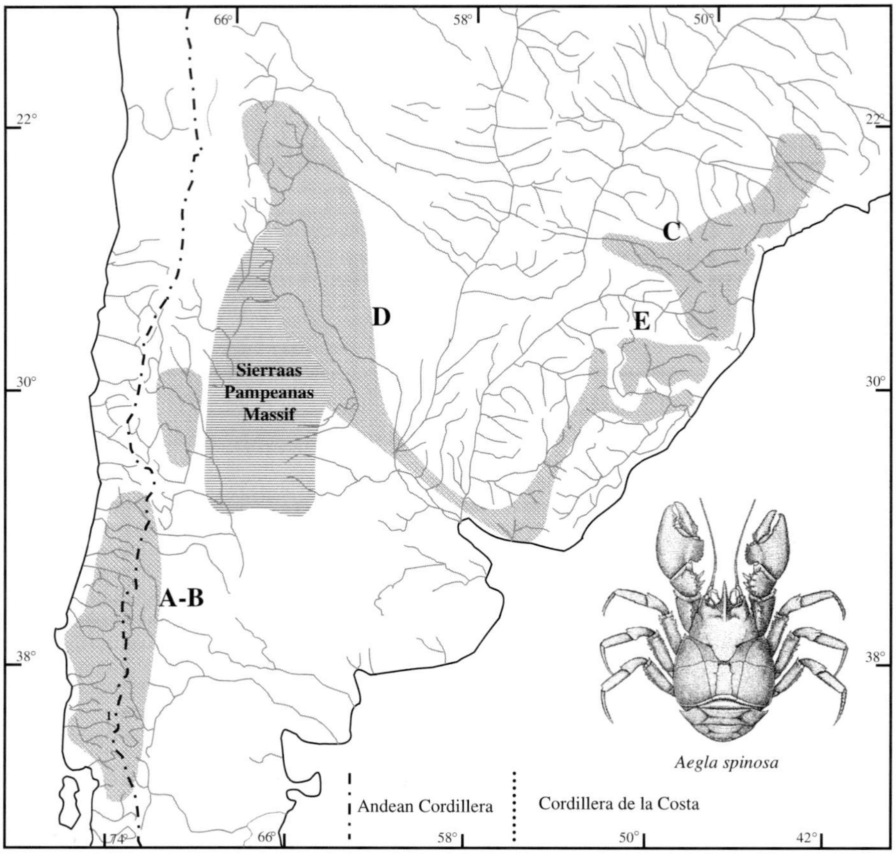

Fig. 1 Map of southern South America indicating the major river systems and distribution of the major clades (cf. Fig. 2) of species diversity in the aeglid crabs. A drawing of *Aegla spinosa* from Bond-Buckup and Buckup (1994) is shown as an inset

data that suggest the Aeglidae should be in their own superfamily (Martin & Abele, 1986; Pérez-Losada etal., 2002b; Tudge & Scheltinga, 2002). From a conservation perspective, several of the known species are very restricted in distribution, and they and their habitats are considered threatened (Pérez-Losada etal., 2002a). From an ecological perspective, aeglids are unique because they are the only anomuran family entirely restricted to freshwater habitats.

The adult size of aeglids does not surpass 60 mm carapace length (CL), and color varies according to the substrate, from greenish brown to almost black; in some lacustrine populations blue, yellow, orange, and red specimens are also sporadically found (Jara, 1989). *Aegla* are dioecious; males lack pleopods, and genital pores open on the coxa of fifth pair of pereiopods at the tip of membranous tubes; female genital pores open on the coxa of third pereopods (Martin & Abele, 1988). Spawning occurs chiefly during the autumn, but some species spawn continuously throughout the year (Bahamonde & López, 1961; Rodrigues & Hebling, 1978; Bueno & Bond-Buckup, 2000). The incubation period lasts between four and eight months, so that an adult female produces no more than one clutch of eggs per annum. Fecundity depends on the size of females, ranging between 120 eggs at 12.5 mm CL to 400 at 22.5 mm CL female in *A. laevis* (Bahamonde & López, 1961); between 115 eggs at 14.99 mm CL to 368 at 19.18 mm CL female in *A. leptodactyla* (Buckup personal observation); and between 699 eggs at 29 mm to 1043 at 33 mm CL female in *A. rostrata* (Jara, 1977). Egg size ranges between 1.00 and 1.37 mm diameter (Jara, 1977). Recruitment normally occurs once a year, in springtime (August to October) (Bahamonde & López, 1961; López 1965; Bueno & Bond-Buckup, 2000), but for *A. longirostri* and *A. castro*, it seems that two recruitment periods exist (Swiech-Ayoub & Masunari, 2001; Delevati etal., 2005). *Aegla* species lack larval stages;

Table 1 Species from the freshwater anomuran crab family Aeglidae including the countries within which they are distributed

Species	Distribution	Habitat
Aegla abtao Schmitt, 1942	Chile	River
Aegla affinis Schmitt, 1942	Chile, Argentina	River
Aegla alacalufi Jara & López, 1981	Chile	Small streams
Aegla araucaniensis Jara, 1980	Chile	River
Aegla bahamondei Jara, 1982	Chile	River
Aegla camargoi Buckup & Rossi, 1977	Brazil	River
Aegla castro Schmitt, 1942	Brazil	River
Aegla cavernicola Turkay, 1972	Brazil	Cave
Aegla cholchol Jara & Palacios, 1999	Chile	River
Aegla concepcionensis Schmitt, 1942	Chile	Small streams
Aegla denticulata denticulate Nicolet, 1849	Chile	River
Aegla denticulata lacustris Jara, 1989	Chile	Lake
Aegla expansa Jara, 1992	Chile	Small stream
Aegla franca Schmitt, 1942	Brazil	River
Aegla franciscana Buckup & Rossi, 1977	Brazil	River
Aegla grisella Bond-Buckup & Buckup, 1994	Brazil	River
Aegla hueicollensis Jara & Palacios, 1999	Chile	River
Aegla humahuaca Schmitt, 1942	Argentina	Headwater & Lake
Aegla inconspicua Bond-Buckup & Buckup, 1994	Brazil	River
Aegla inermis Bond-Buckup & Buckup, 1994	Brazil	River
Aegla intercalata Bond-Buckup & Buckup, 1994	Argentina	River
Aegla itacolomiensis Bond-Buckup & Buckup, 1994	Brazil	River
Aegla jarai Bond-Buckup & Buckup, 1994	Brazil	River
Aegla jujuyana Schmitt, 1942	Brazil	River
Aegla laevis laevis Latreille, 1818	Chile	River
Aegla laevis talcahuano Schmitt, 1942	Chile	River
Aegla lata Bond-Buckup & Buckup, 1994	Brazil	River
Aegla leptochela Bond-Buckup & Buckup, 1994	Brazil	Cave
Aegla leptodactyla Buckup & Rossi, 1977	Brazil	River
Aegla ligulata Bond-Buckup & Buckup, 1994	Brazil	River
Aegla longirostri Bond-Buckup & Buckup, 1994	Brazil	River
Aegla manni Jara, 1980	Chile	Small streams
Aegla marginata Bond-Buckup & Buckup, 1994	Brazil	River
Aegla microphthalma Bond-Buckup & Buckup, 1994	Brazil	Cave
Aegla neuquensis Schmitt, 1942	Chile	River
Aegla odebrechtii Müller, 1876	Brazil	River
Aegla obstipa Bond-Buckup & Buckup, 1994	Brazil	River
Aegla occidentalis Jara etal., 2003	Chile	River & Lake
Aegla papudo Schmitt, 1942	Chile	River
Aegla parana Schmitt, 1942	Brazil	River
Aegla parva Bond-Buckup & Buckup, 1994	Brazil	River
Aegla paulensis Schmitt, 1942	Brazil	River
Aegla perobae Hebling & Rodrigues, 1977	Brazil	Headwater Cave
Aegla pewenchae Jara, 1994	Chile	River
Aegla plana Buckup & Rossi, 1977	Brazil	River

Table 1 continued

Species	Distribution	Habitat
Aegla platensis Schmitt, 1942	Brazil, Argentina, Paraguay, Uruguay	River
Aegla prado Schmitt, 1942	Brazil, Uruguay	Swamp & Lake
Aegla ringueleti Bond-Buckup & Buckup, 1994	Argentina	River
Aegla riolimayana Schmitt, 1942	Chile, Argentina	River
Aegla rossiana Bond-Buckup & Buckup, 1994	Brazil	River
Aegla rostrata Jara, 1977	Chile	Lake
Aegla sanlorenzo Schmitt, 1942	Argentina	River
Aegla scamosa Ringuelet, 1948	Argentina	River
Aegla schmitii Hobbs III, 1979	Brazil	River
Aegla septentrionalis Bond-Buckup & Buckup, 1994	Argentina, Bolivia	Headwater
Aegla serrana Buckup & Rossi, 1977	Brazil	River & Headwater
Aegla singularis Ringuelet, 1948	Argentina, Brazil	River
Aegla spectabilis Jara, 1986	Chile	River
Aegla spinipalma Bond-Buckup & Buckup, 1994	Brazil	River
Aegla spinosa Bond-Buckup & Buckup, 1994	Brazil	River
Aegla strinatii Turkay, 1972	Brazil	River
Aegla uruguayana Schmitt, 1942	Argentina, Uruguay, Brazil	River
Aegla violacea Bond-Buckup & Buckup, 1994	Brazil	River

offspring hatch as juvenile "crabs" of 1.13 - 1.58 mm CL (Bond-Buckup etal., 1999) that remain with the mother for 3–4 days before living independently (López etal., 2004). Population density varies widely, reaching up to 250 individuals/m^2 (Bahamonde & López, 1961) in highly productive conditions.

The ecological role of *Aegla* species has not been assessed but their omnivorous diet includes periphyton, decaying allochtonous vegetable matter, aquatic invertebrates (Bahamonde & López, 1961; Burns, 1972; Lara & Moreno, 1995, Castro-Souza & Bond-Buckup, 2004), and fine particulate organic matter (Isler, 1988). Additionally, they constitute a relevant dietary item for the non-native rainbow trout (*Oncorhynchus mykiss*) in Chile and South Brazil and brown trout (*Salmo trutta fario*) in Chile (Burns, 1972; Arenas, 1978; Buckup, personal observation), and for the Chilean and Brazilian species of river otter (*Lontra provocax*) (Medina, 1998).

Species diversity

The present Aeglidae belong to a single genus, *Aegla* Leach, 1820, consisting of 63 described species (Bond-Buckup & Buckup, 1994; Bond-Buckup, 2003) (Table 1); including, newly described species based on recent molecular phylogenetic analyses (Jara etal., 2003). By our count, there are at least six additional species waiting to be described. Of these species, 57 are found mainly in rivers, only two in lakes, and four in cave habitats.

Phylogeny and historical processes

Ortmann (1902) proposed that aeglid species from Chile represented the more primitive forms of the genus. However, Schmitt (1942) hypothesized that *Aegla* from the Atlantic side of South America were more primitive, and species ranging in the Chilean streams were more derived. Recent estimates of phylogenetic relationships based on a variety of molecular data support the Pacific origin hypothesis (Pérez-Losada etal., 2004) (Fig. 2) and suggest that the group, as a whole originated around 75 mya. The western *Aegla* species radiated, approximately, 40–45 mya (clades A and B—Fig. 2), but the speciation of the central and eastern taxa took place later, around 23–35 mya (clades C-E in Fig. 2) (see Pérez-Losada etal., 2004 for more detailed discussions on the phylogenetic relationships among all the aeglid species).

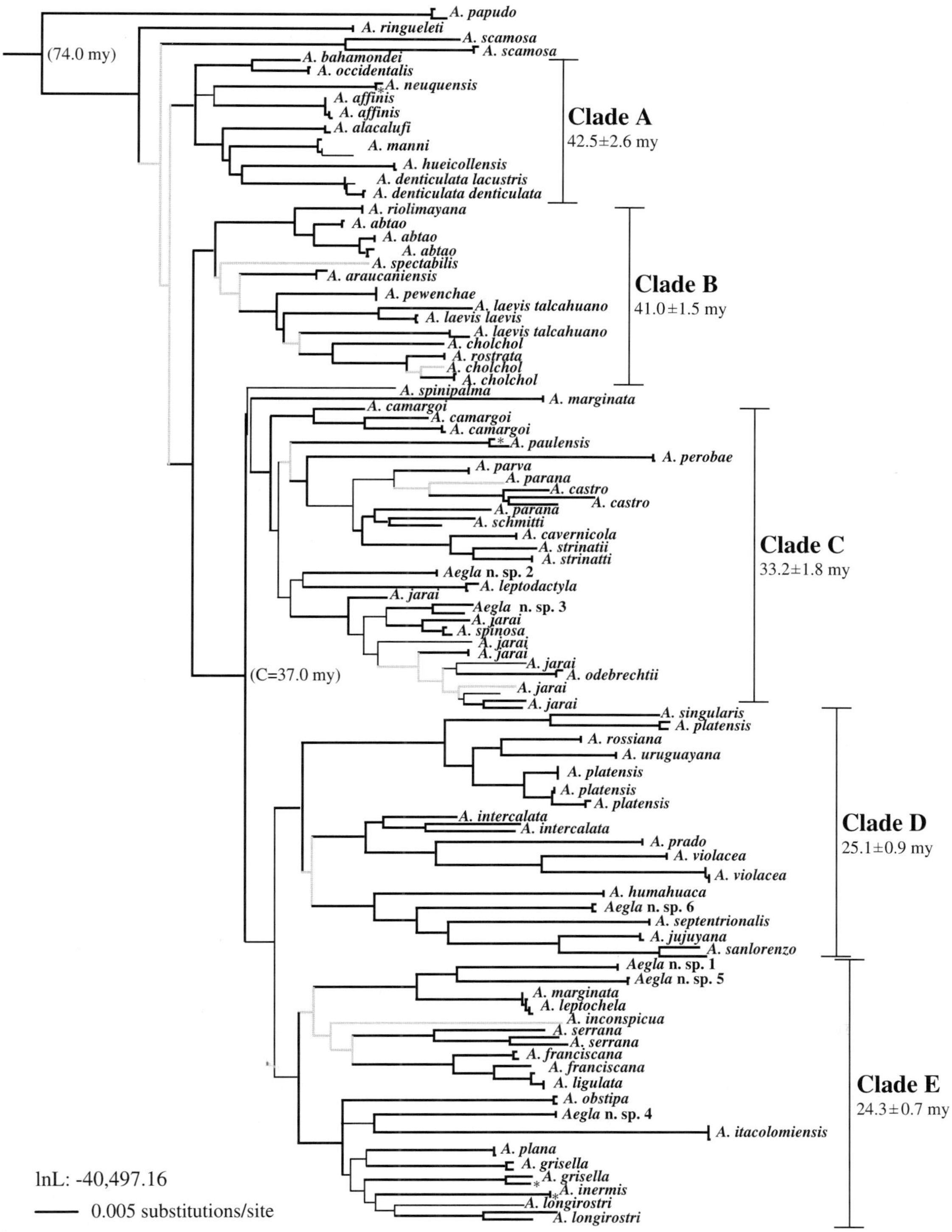

Fig. 2 Bayesian and maximum likelihood estimate of phylogenetic relationships among the aeglid species using the GTR+Γ+I model of evolution. Clade support is graphically indicated as follows: ▬ bp ≥ 70% and p*P* ≥ 0.95, ▬ 50% ≤ bp < 70% and/or 0.75 ≤ p*P* < 0.95, and — bp < 50% and/or p*P* < 0.75. Branch lengths are shown proportional to the amount of change along the branches in the maximum likelihood tree with estimated divergence dates shown for the root and major clades (see Pérez-Losada etal., 2004, for details)

Present distribution and areas of endemicity

The aeglids are distributed in southern South America (Fig. 1) including Chile, Brazil, Argentina, Uruguay, Bolivia, and Paraguay. Thus, all 63 species are contained within the Neotropical region. There are 16 species endemic to Chile, seven species endemic to Argentina and 36 species endemic to southern Brazil (Bond-Buckup etal. 2003). They occur in all the main rivers of southern South America, except the most southern Patagonian drainages.

Human related issues

Most *Aegla* species have very narrow distributional areas and are therefore of significant conservation concern. Of the 63 species currently recognized in the genus, using the IUCN Red List criteria (IUCN, 2001), we recognize 23 species or 36.5 % as under threat or endangered, mainly due to their narrow distributions and the rapid degradation of the freshwater habitats they occupy.

In Chile, aeglids are recognized as an important food item for exotic salmonid species (*Salmo trutta fario* and *Oncorhynchus mykiss*) which support a fast-growing fly-fishing sport industry, especially in the Llanquihue Lake region and South Brazilian Highlands. On the negative side, the fast-growing fruit and wine producing industries in Central Chile constitute a threat for conservation of aeglids, and benthic river fauna in general, due to the widespread use of biocides, which likely accumulate and have an impact in rivers and streams. Threats for conservation of aeglids also derive from silvicultural practices related to the establishment of extensive plantations of pine and *Eucalyptus*, mainly along the Coastal Cordillera and south Brazil. Additionally, it must be pointed out that the exaggerated use of pesticides in the widespread apple tree and potatoes cultivation and mainly, the hog raising activities along the majority of the South Brazilian Rivers, are an important menace to the native populations of aeglids, particularly in the states of Santa Catarina and Rio Grande do Sul, in Brazil.

Acknowledgements We thank Estelle Balian and Koen Martens for their invitation to participate in this exciting project and for their helpful comments on our paper. We also thank Darren Yeo and an anonymous reviewer for their helpful comments to improve the paper. Our work was supported by the Conselho Nacional de Desenvolvimento Científico e Tecnológico (CNPq) (National Research Council, Brazil - GBB) and NSF grants OISE-0530267 and EF-0531762.

References

Arenas, J. N., 1978. Análisis de la alimentación de *Salmo gairdneri* Richardson en el Lago Riñihue y Río San Pedro, Chile. *Medio Ambiente* **3**: 50–58.

Bahamonde, N. & M. T. López, 1961. Estudios biológicos en la población de *Aegla laevis laevis* (Latreille) de El Monte. Investigaciones Zoológicas Chilenas 7: 19–58.

Bond-Buckup, G., L. Buckup, P. B. Araujo, 2003. Crustáceos. In: Fontana, C.F., Bencke, G.A., Reis, R.E. (org.) *Livro Vermelho da Fauna Ameaçada de Extrinção no Rio Grande do Sul*. Porto Alegre: EDIPUCRS, 2003. pp. 73–83.

Bond-Buckup, G., 2003. Familia Aeglidae. In Melo, G. A. S. (ed.), Manual de identificação dos crustacea Decapoda, de água doce do Brasil. Editora Loyola, São Paulo: 21–116.

Bond-Buckup, G. & L. Buckup, 1994. A família Aeglidae (Crustacea, Decapoda, Anomura). Arquivos de Zoologia 32: 159–347.

Bond-Buckup G., A. P. Bueno, K. A. Keinecke, 1999. Morphological characteristics of juvenile specimens of *Aegla* (Decapoda, Anomura, Aeglidae). Proceedings of the Fourth International Crustacean Congress, 1998, 372–381. Koninklijke Brill NV, Leiden, 1999.

Bueno, A.P. & G. Bond-Buckup, 2000. Dinâmica populacional de *Aegla platensis* Schmitt (Crustacea, Decapoda, Aeglidae). Revista Brasileira de Zoologia 17: 43–49.

Burns, J. W., 1972. The distribution and life history of South American freshwater crabs (*Aegla*) and their role in trout streams and lakes. Transactions of the American Fisheries Society 101: 595–607.

Delevati Colpo, K., L. Oliveira Ribeiro & S. Santos, 2005. Population biology of the freshwater anomuran Aegla longirostri (Aeglidae) from South Brasilian streams. Journal of Crustacean Biology 25: 495–499.

Castro-Souza, T. & G. Bond-Buckup, 2004. The trophic niche of two sympatric *Aegla* Leach species (Crustacea, Aeglidae) in a tributary of hydrographic basin of Pelotas river, Rio Grande do Sul, Brazil. Revista Brasileira de Zoologia 21(4): 805–813.

IUCN , 2001. IUCN Red List Categories: Version 3.1. IUCN Species Survival Commission, Gland, Switzerland.

Isler, M. L., 1988. Alimentación natural, conducta alimentaria y preferencia dietaria en *Aegla denticulata* Nicolet, 1849 (Crustacea: Decapoda: Anomura: Aeglidae). Unpublished Thesis, Universidad Austral de Chile, Valdivia, Chile 50 pp.

Jara, C. G., 1977. *Aegla rostrata* n.sp., (Decapoda, Aeglidae), nuevo crustáceo dulceacuícola del Sur de Chile. Studies on Neotropical Fauna and Environment 12: 165–176.

Jara, C., 1980. Dos nuevas especies de *Aegla Leach* (Crustacea, Decapoda, Anomura) del sistema hidrográfico del Río Valdivia. Anales del Museo de Historia Natural de Valparaíso 13: 255–266.

Jara, C., 1982. *Aegla bahamondei*, new Species (Crustacea: Decapoda: Anomura) from the coastal mountain range of Nahuelbuta, Chile. Journal of Crustacean Biology 2: 232–238.

Jara, C., 1992. *Aegla expansa*, new species (Crustacea: Decapoda: Anomura: Aeglidae), from the lower Bío-Bío River Basin, Concepción, Chile. Gayana (Zoología) 56(1–2): 49–57.

Jara , C. G., 1989. *Aegla denticulata lacustris*, new subspecies, from Lake Rupanco, Chile (Crustacea: Decapoda: Anomura: Aeglidae). Proceedings of the Biological Society of Washington 102: 385–393.

Jara, C. G. & M. T. López, 1981. A new species of freshwater crab (Crustacea: Anomura: Aeglidae) from Insular South Chile. Proceedings of the Biological Society of Washington 94(1): 34–41.

Jara, C. & V. L. Palacios, 1999. Two new species of *Aegla* Leach (Crustacea: Anomura: Aeglidae). Proceedings of the Biological Society of Washington 122: 106–109.

Jara ,C.G., M. Pérez-Losada & K.A. Crandall, 2003. *Aegla occidentalis* (Crustacea: Decapoda: Aeglidae), a new species of freshwater crab from the Nahuelbuta Coastal Range, Chile. Proceedings of the Biological Society of Washington 116: 933–942.

Lara, G. & C. Moreno, 1995. Effects of predation of *Aegla abtao* (Crustacea, Aeglidae) on the spatial distribution pattern and abundance of *Diplodon chilensis* (Bivalvia, Hyriidae) in Lake Panguipulli, Chile. Revista Chilena de Historia Natural 68: 123–129.

López, M. T., 1965 Estudios biológicos en *Aegla odebrechtii paulensis* Schmitt (Crustacea, Decapoda, Anomura). Boletim de Zoologia da Facultade de Ciências e Letras de Sao Paulo 25: 301–314.

López, L .S., V. Viau, M. Lavolpe, G. Bond-Buckup & E. M. Rodriguez, 2004. Juvenile hatching and maternal care in Aegla uruguayana (Anomura, Aeglidae). Journal of Crustacean Biology 24: 309–313.

Lopretto, E. C., 1978. Estructura exoesqueletaria y miológica del quinto par de pereiópodos del macho de la familia Aeglidae (Crustacea Anomura). Limnobios 1: 284–298.

Martin, J. W. & L. G. Abele, 1986. Phylogenetic relationships of the genus *Aegla* (Decapoda, Anomura, Aeglidae), with comments on anomuran phylogeny. Journal of Crustacean Biology 6: 576–616.

Martin, J.W. & L.G Abele, 1988. External morphology of the genus *Aegla* (Decapoda, Anomura, Aeglidae). Smithsonian Contributions to Zoology 453: iv+46 pp.

Martin, J. W. & B. E. Felgenhuer, 1986. Grooming behaviour and the morphology of grooming appendages in the endemic South American crab genus *Aegla* (Decapoda, Anomura, Aeglidae). Journal of zoology (London, England : A) 209: 213–224.

Medina, G., 1998. Seasonal variations and changes in the diet of southern river otter in different freshwater habitats in Chile. Acta Theriologica 43: 285–292.

Ortmann A.E., 1902. The geographical distribution of freshwater decapods and its bearing upon ancient geography. Proceedings of the American Philosophical Society 41: 267–400.

Pérez-Losada, M., G. Bond-Buckup, C. G. Jara & K. A. Crandall, 2004. Molecular systematics and biogeography of the Southern South American freshwater "crabs" *Aegla* (Decapoda: Anomura: Aeglidae) using multiple heuristic tree search approaches. Systematic Biology 53: 767–780.

Pérez-Losada, M, C. G. Jara, G. Bond-Buckup & K.A. Crandall, 2002a. Conservation phylogenetics of Chilean freshwater crabs *Aegla* (Anomura, Aeglidae): Assigining priorities for aquatic habitat protection. Biological Conservation 105: 345–353.

Pérez-Losada, M, C. G. Jara, G. Bond-Buckup, M. L. Porter & K. A. Crandall, 2002b. Phylogenetic position of the freshwater Anomuran family Aeglidae. Journal of Crustacean Biology 22: 670–676.

Rodrigues, W. & J. N. Hebling, 1978. Estudos biológicos em *Aegla perobae* Hebling & Rodrigues , 1977 (Decapoda, Anomura). Revista Brasileira de Biologia 38: 383–390.

Schmitt W (1942) The species of Aegla, endemic South American freshwater crustaceans. Proceedings of the United States National Museum 91: 431–520.

Swiech-Ayoub, B. P. & S. Masunari, 2001. Biologia reprodutiva de Aegla castro Schmitt (crustacea, Anomura, Aeglidae) no Buraco do Padre, Ponta Grossa, Paraná, Brasil. Revista Brasileira de Zoologia, Curitiba 18(suppl.3): 1019–1030.

Tudge ,C. C., D. M. Scheltinga, 2002. Spermatozoal morphology of the freshwater anomuran *Aegla longirostris* Bond-Buckup and Buckup, 1994 (Crustacea: Decapoda: Aeglidae) from South America. Proceedings of the Biological Society of Washington 115: 118–128.

Hydrobiologia (2008) 595:275–286
DOI 10.1007/s10750-007-9023-3

Global diversity of crabs (Crustacea: Decapoda: Brachyura) in freshwater

Darren C. J. Yeo · Peter K. L. Ng · Neil Cumberlidge · Célio Magalhães · Savel R. Daniels · Martha R. Campos

Abstract An assessment of the global freshwater crab diversity is presented. A total of 1,476 species in 14 families are currently known from all zoogeographical regions (except Antarctica), including 1,306 species in eight exclusively freshwater families (Pseudothelphusidae, Trichodactylidae, Potamonautidae, Deckeniidae, Platythelphusidae, Potamidae, Gecarcinucidae and Parathelphusidae). Estimates of true freshwater crab diversity including likely numbers of undescribed taxa suggest that the field remains largely in a "discovery" phase. Main ideas on the origins, diversification, and phylogeny of true freshwater crabs are briefly discussed. The economic importance of freshwater crabs is also highlighted.

Keywords Global assessment · Freshwater crab · Diversity · Crustacea · Decapoda · Brachyura · Species estimates

Guest editors: E. V. Balian, C. Lévêque, H. Segers & K. Martens
Freshwater Animal Diversity Assessment

D. C. J. Yeo (✉) · P. K. L. Ng
Department of Biological Sciences, National University of Singapore, 14 Science Drive 4, Singapore 117543, Republic of Singapore
e-mail: darrenyeo@nus.edu.sg

N. Cumberlidge
Department of Biology, Northern Michigan University, Marquette, MI 49855, USA

C. Magalhães
Instituto Nacional de Pesquisas da Amazônia, Caixa postal 478, Manaus 69011-970 Amazonas, Brazil

S. R. Daniels
Department of Botany and Zoology, University of Stellenbosch, Private Bag X1, Matieland 7602, South Africa

M. R. Campos
Instituto de Ciencias Naturales, Universidad Nacional de Colombia, Apartado Aéreo, Bogota 103698, Colombia

Introduction

Of the more than 6,700 known species of brachyuran crabs, over 1,300 are true freshwater crabs. True freshwater crabs are regarded as those that have adopted freshwater, semi-terrestrial or terrestrial modes of life, and are characterized by their ability to complete their life cycle independently of the marine environment. These crabs are currently assigned to eight exclusively freshwater families—Pseudothelphusidae and Trichodactylidae (Mexico, Central and South America), Potamonautidae (Africa and Madagascar), Deckeniidae and Platythelphusidae (East Africa), Potamidae (North Africa, southern Europe, Asia), Gecarcinucidae (Seychelles, Asia), and Parathelphusidae (Asia, Australasia) (Martin & Davis, 2001). Wholly or primary freshwater taxa undergo direct development in which the large, yolky eggs hatch directly into juvenile crabs. Crabs found in freshwater also include numerous euryhaline species or secondary freshwater species from primarily marine brachyuran families (e.g. Sesarmidae, Varunidae, Hymenosomatidae). Although these freshwater species are fully adapted to freshwater/terrestrial living, most do not have direct development in their life cycle (though highly abbreviated development occurs in some) and most possess one or more larval stages. The diversity of these taxa is also assessed in this chapter (Tables 1 and 2), but the emphasis will be placed on the true freshwater crabs.

Freshwater crabs belong to the Order Decapoda, the crustacean group that also includes lobsters, prawns, crayfish and hermit crabs, which share the characteristic presence of five pairs of thoracic legs (pereiopods). In freshwater crabs the first pereiopods are modified as pincers (chelipeds), and the remaining four pairs are relatively unspecialised walking legs (Fig. 1). The general body plan of freshwater crabs consists of a head, thorax and abdomen, with the head and thorax (cephalothorax) covered by a broad carapace, and the abdomen reduced, flattened and flexed under the thoracic sternum. In adults, the male abdomen is slim and narrow, and is either triangular or T-shaped, while the female abdomen is broad and round and covers nearly the entire thoracic sternum. Adult males bear two pairs of abdominal appendages (pleopods) that are modified into copulatory structures known as gonopods. Gonopod structure is taxonomically important, especially because the external morphology of freshwater crabs tends to be rather conservative (see Ng, 1988; Cumberlidge, 1999, for details).

Freshwater crabs are found in the tropics and subtropics in most parts of the world, and occur in a wide variety of aquatic and terrestrial habitats. These decapods are present in almost all freshwater bodies, from clear, fast-flowing montane streams to sluggish lowland rivers and streams, as well as in peat and freshwater swamps, stagnant ponds and rice fields, and even in pools in tree holes and leaf axils. A fair number are also adapted to live in caves. Among the primarily aquatic freshwater crabs, some (e.g. potamids) are entirely adapted to living in fresh water, and are not thought to be able to survive for long in salt water, while others (e.g. parathelphusids) are more tolerant of saline conditions, and can survive immersion in salt water for short periods of time. Terrestrial species may occur well away from permanent freshwater sources, either moving among the forest floor litter or, in some cases, even climbing trees (Ng, 1988; Ng & Tay, 2001; Cumberlidge et al., 2005). These freshwater crab species do not require regular immersion in fresh water and can obtain water either from food, from drinking dew or casual water, or by capillary or osmotic uptake from moist substrata. In the present chapter, such species are also categorised as "freshwater-dependent" species (see later; Table 1).

Freshwater crabs are primarily nocturnal, preferring to remain hidden during the day in sheltered places and foraging mostly at night. They are mostly omnivorous scavengers, mainly feeding on plant matter, but some are opportunistic carnivores, feeding either on live prey such as fish and prawns or on dead animals that they encounter (Ng, 1988; unpub.), and cannibalism is not uncommon (unpub.). Crabs themselves also constitute an important food resource for many species of fishes, birds, caymans, turtles and mammals (see Ng, 1988; Magalhães, 2003).

Diversity and endemicity

Known global diversity

There are currently a total of 238 genera and 1,476 species of known freshwater crabs from 14 families (including 1,306 true freshwater crab species in eight families: Pseudothelphusidae, Trichodactylidae, Potamonautidae, Deckeniidae, Platythelphusidae, Potamidae, Gecarcinucidae and Parathelphusidae) (as of 1st August 2006). The species and genus diversity of primary as well as secondary freshwater species are listed by zoogeographical regions (sensu Cox, 2001) in Tables 1 and 2, respectively. The total number of species or genera of certain families listed in the last column of Tables 1 and 2, respectively, do not tally with the sum of species or genera from each zoogeographical region because some taxa occur in more than one region (see later, Tables 1 and 2 footnotes).

Table 1 also lists separately the number of "freshwater-dependent" species (WDpt). In the context of the present volume, our assessment of the freshwater crab species required grouping them into two broad ecological categories to reflect their different habitat preferences and degrees of their association with freshwater habitats, viz., "real aquatic" species and "freshwater-dependent"

Table 1 Global species diversity of freshwater crabs

Family	PA		NA		NT		AT		OL		AU		PAC		World[c]	
	Total FW	WDpt	Total FW	WDpt	Total FW	WDpt	Total FW	WDpt	Total FW	WDpt	Total FW	WDpt	Total FW	WDpt	Total FW	WDpt
[a]Potamidae[d]	86	–	–	–	–	–	3	1	435	58	–	–	–	–	509	59
[a]Potamonautidae	–	–	–	–	–	–	115	9	–	–	–	–	–	–	115	9
[a]Deckeniidae	–	–	–	–	–	–	2	2	–	–	–	–	–	–	2	2
[a]Platythelphusidae	–	–	–	–	–	–	9	–	–	–	–	–	–	–	9	–
[a]Parathelphusidae[e]	4	–	–	–	–	–	–	–	243	14	52	8	–	–	298	22
[a]Gecarcinucidae	–	–	–	–	–	–	8	6	36	3	–	–	–	–	44	9
[a]Pseudothelphusidae	–	–	16	–	262	3	–	–	–	–	–	–	–	–	278	3
[a]Trichodactylidae	–	–	–	–	51	–	–	–	–	–	–	–	–	–	51	–
[b]Gecarcinidae[f]	–	–	3	3	9	9	9	9	8	8	6	6	8	8	20	20
[b]Hymenosomatidae[g]	1	–	–	–	1	–	1	–	8	–	9	–	3	–	22	–
[b]Ocypodidae	–	–	–	–	–	–	–	–	2	–	–	–	–	–	2	–
[b]Sesarmidae[h]	2	–	–	–	14	14	1	1	72	68	15	13	7	7	101	95
[b]Goneplacidae	–	–	–	–	–	–	–	–	–	–	2	–	2	–	4	–
[b]Varunidae[i]	4	–	–	–	3	–	1	–	14	–	5	–	4	–	21	–
Total[j]	97	–	19	3	340	26	149	28	818	151	89	27	24	15	1476	219

Total FW: Aquatic + "freshwater dependent" species. WDpt: "freshwater-dependent" species (see text under "Known global diversity" for definition). PA: Palaearctic, NA: Nearctic, NT: Neotropical, AT: Afrotropical, OL: Oriental, AU: Australasian, PAC: Pacific Oceanic Islands

[a] True (or primary) freshwater crab family

[b] Secondary freshwater crab family

[c] "World" refers to the actual total number of freshwater species, which may not necessarily be the sum of the totals for each zoogeographical region as some species occur in more than one region (see text)

[d] Potamidae—15 species with Palaearctic/Oriental distributions

[e] Parathelphusidae—1 species with Palaearctic/Oriental distribution

[f] Gecarcinidae—13 species with various overlapping distributions (3 Nearctic/Neotropical + 4 Afrotropical/Neotropical + 4 Afrotropical/Oriental/Australasian/Pacific + 2 Oriental/Australasian/Pacific)

[g] Hymenosomatidae—1 species with Neotropical/Australasian distribution

[h] Sesarmidae—6 species with various overlapping distributions (1 Palaearctic/Oriental/Australasian + 1 Afrotropical/Oriental/Australasian + 1 Oriental/Australasian + 2 Oriental/Australasian/Pacific + 1 Australasian/Pacific)

[i] Varunidae—7 species with various overlapping distributions (3 Palaearctic/Oriental + 1 Palaearctic/Afrotropical/Oriental/Australasian/Pacific + 3 Oriental/Australasian)

[j] Total—43 species with various overlapping distributions (see above for details)

Table 2 Global genus diversity of freshwater crabs

Family	PA	NA	NT	AT	OL	AU	PAC	World[c]
[a]Potamidae[d]	9	–	–	2	72	–	–	78
[a]Potamonautidae	–	–	–	12	–	–	–	12
[a]Deckeniidae	–	–	–	1	–	–	–	1
[a]Platythelphusidae	–	–	–	1	–	–	–	1
[a]Parathelphusidae[e]	1	–	–	–	35	9	–	42
[a]Gecarcinucidae	–	–	–	4	10	–	–	14
[a]Pseudothelphusidae[f]	–	2	41	–	–	–	–	41
[a]Trichodactylidae	–	–	15	–	–	–	–	15
[b]Gecarcinidae[g]	–	2	3	4	4	4	4	6
[b]Hymenosomatidae[h]	1	–	1	1	3	2	2	6
[b]Ocypodidae	–	–	–	–	1	–	–	1
[b]Sesarmidae[i]	1	–	3	1	7	4	3	10
[b]Goneplacidae[j]	–	–	–	–	–	1	2	2
[b]Varunidae[k]	2	–	2	1	7	4	2	9
Total[l]	14	4	65	27	139	24	13	238

PA: Palaearctic, NA: Nearctic, NT: Neotropical, AT: Afrotropical, OL: Oriental, AU: Australasian, PAC: Pacific Oceanic Islands

[a] True (or primary) freshwater crab family

[b] Secondary freshwater crab family

[c] "World" refers to the actual total number of freshwater genera, which may not necessarily be the sum of the totals for each zoogeographical region as some genera occur in more than one region (see text)

[d] Potamidae—5 genera with Palaearctic/Oriental distributions

[e] Parathelphusidae—3 genera with various overlapping distributions (1 Palaearctic/Oriental + 2 Oriental/Australasian)

[f] Pseudothelphusidae—2 genera with Nearctic/Neotropical distributions

[g] Gecarcinidae—6 genera with various overlapping distributions (1 Nearctic/Neotropical + 1 Nearctic/Afrotropical/Neotropical/Oriental/Australasian/Pacific + 1 Afrotropica/Neotropical + 2 Afrotropical/Oriental/Australasian/Pacific + 1 Oriental/Australasian/Pacific)

[h] Hymenosomatidae—2 genera with different overlapping distributions (1 Palaearctic/Oriental/Pacific + 1 Neotropical/Oriental/Australasian)

[i] Sesarmidae—4 genera with various overlapping distributions (1 Palaearctic/Oriental/Australasian + 1 Afrotropical/Oriental/Australasian/Pacific + 2 Oriental/Australasian/Pacific)

[j] Goneplacidae—1 genus with Australasian/Pacific distribution

[k] Varunidae—6 genera with various overlapping distributions (1 Palaearctic/Oriental + 1 Palaearctic/Afrotropical/Oriental/Australasian/Pacific + 3 Oriental/Australasian + 1 Oriental/Pacific)

[l] Total—29 genera with various overlapping distributions (see above for details)

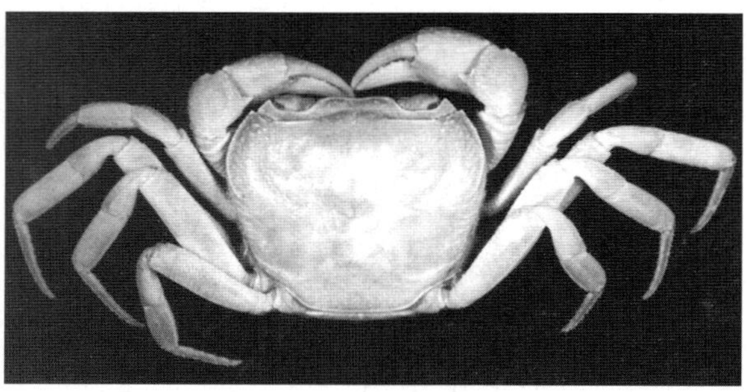

Fig. 1 Habitus

species. "Real aquatic" species are ones that are dependent on freshwater habitats to complete their life cycle i.e. part or all of the life cycle occurs in the water. Thus, adopting a practical approach, this study has included under "real aquatic" species all fully aquatic as well as semi-terrestrial species that are generally found in and around (or associated with) traditional freshwater environments (streams, rivers, lakes, ponds, swamps). Under "freshwater-dependent" species (WDpt), this work has included the more terrestrial species in which the adults are not primarily found in and around (or associated with) traditional freshwater environments, but are nevertheless dependent on wet/humid terrestrial environments for survival e.g. tree-climbing crabs, forest floor dwellers, dry cave dwellers. These include many so-called terrestrial crabs that have juvenile stages that can occur in water.

Estimated global biodiversity

Two methods are proposed here for estimating the global diversity of true freshwater crabs:

(1) Based on extrapolation. Estimated global diversity of true freshwater crabs: 2,155 species

Yeo & Ng (1999) used the species number per unit area of Thailand (1.8×10^{-4} species/km^2) as a reference for estimating the fauna for the whole of Indochina. These authors applied this ratio to Indochina [=Thailand, Laos, Cambodia, Vietnam and Myanmar] (1,939,320 km^2), and obtained a working estimate of 349 species for this region. There are currently 212 species known from Indochina (Yeo & Ng, unpub.) which gives an approximate ratio of actual to estimated species of 1:1.65. The 13 species of freshwater crabs known from Madagascar give a similar estimate (Cumberlidge & Sternberg, 2002). Although simplistic, no similar objective estimates have been attempted thus far. Applying a ratio of actual to estimated species of 1:1.65 to the known true freshwater crab global diversity from the previous section (1,306 species), gives an estimated global diversity of 2,155 species. Given that Indochina lies in one of the most species-rich areas of the global range of freshwater crabs, this figure may tend to be an overestimate. Considering the large numbers of undescribed species known and/or likely to be discovered in the near future, freshwater crab taxonomy must be regarded as still being in its "discovery" phase. The cumulative curves of new species described over time for the two South American crab families suggests that the diversity of the Pseudothelphusidae is still far from being well-known because the curve (Fig. 2a) is still ascending. On the other hand, the curve for the trichodactylids (Fig. 2b) seems to have already reached an asymptote, suggesting at least for this group, the discovery phase is ending. The several phases of tectonic uplift that affected most of the western and northern

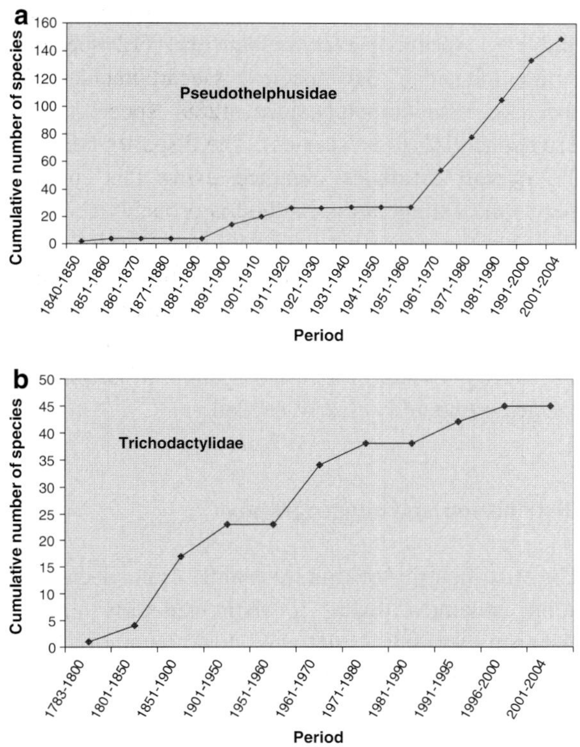

Fig. 2 Cumulative number of South American species of: (**a**) Pseudothelphusidae described since 1840; and (**b**) Trichodactylidae described since 1783

margins of South America produced many vicariance events (Lundberg et al., 1998) that could account for the high diversification of the pseudothelphusids along the Andes. Pseudothelphusids are usually distributed in mountainous regions with restricted distributional ranges and there still are several unexplored areas in the Andes, Guyana and the Central Brazilian Massifs from where new taxa are still being found. Species of trichodactylids usually have extensive ranges along the relatively uniform, tectonically stable lowlands of the continent's huge hydrographic basins, have not speciated as much as the pseudothelphusids, and the number of new species of trichodactylids still awaiting discovery is expected to be low.

(2) Based on numbers of as yet undescribed species known. Estimated global diversity of true freshwater crabs: 1,430 species.

This estimate is based simply on the total number of described and undescribed species of true freshwater crabs known to the authors. The breakdown of these estimates by family is as follows: Potamidae

(570 species); Potamonautidae (135 species); Deckeniidae (3 species); Platythelphusidae (12 species); Parathelphusidae (310 species); Gecarcinucidae (60 species); Pseudothelphusidae (289 species) and Trichodactylidae (51 species). The disparity between the overall estimates obtained using this method versus the extrapolation method is probably accentuated by the fact that some of our family level estimates here are conservative. Reality is probably somewhere in between, which means that there are at least 128–846 more species yet to be described/ discovered. A more accurate system of estimating species numbers is clearly needed.

Distribution and zoogeography

The vast majority of true freshwater crab species are point endemics owing to their generally limited dispersal abilities, relatively low fecundity, and stenotopic habits. Most genera of true freshwater crabs are endemic to their respective zoogeographical regions (sensu Cox, 2001; see below).

The distribution of freshwater crab diversity across the main zoogeographical regions (sensu Cox, 2001) adopted in this volume is illustrated in Table 1 (species), Table 2 (genera), and Fig. 3 (total and true freshwater crabs). It should be noted, however, that the phylogeographical patterns of some taxa are not always reflected by this categorisation. One such anomaly is with the family Potamidae, in which the two subfamilies have relatively distinct distributions: Potaminae is clearly a Palaearctic group with the main diversity in southern Europe/North Africa/Near East/Middle East, while Potamiscinae is an Oriental group with the main diversity in East and Southeast Asia. These two groups are only linked by the potamids of the northwestern Oriental Region, where their distributions overlap around parts of Myanmar and northeastern India (Yeo & Ng, 2003). Following strictly the zoogeographical regions of Cox (2001), however, a significant proportion of East Asian

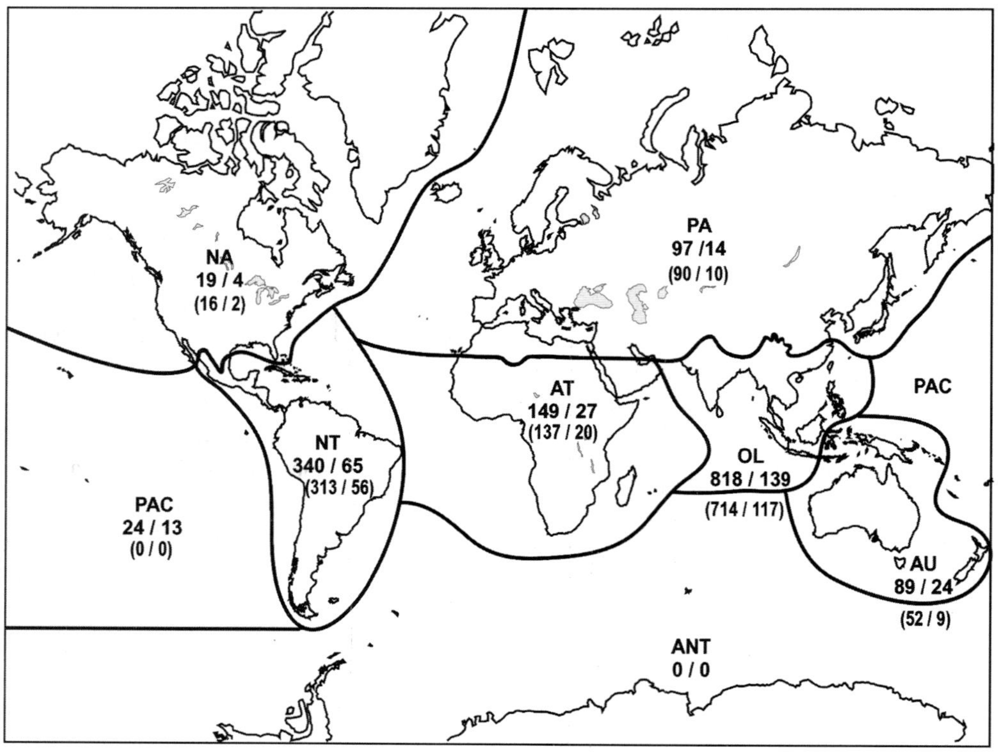

Fig. 3 Map of zoogeographical regions sensu Cox (2001) showing total freshwater crab distribution (Species number/ Genus number). True freshwater crab distribution is shown in parentheses. (PA—Palaearctic; NA—Nearctic; AT—Afrotropical; NT—Neotropical; OL—Oriental; AU—Australasian; PAC—Pacific Oceanic Islands; ANT—Antartica)

potamiscines will fall into the Palaearctic region (together with potamines) instead.

Furthermore, a small group (two genera with three species) of potamids on the island of Socotra off the horn of Africa that is clearly affiliated with the Paleartctic potamines falls misleadingly under the Afrotropical region instead. Similarly, the definition in this volume of the Neotropical region as "excluding highlands of Mexico" and the Nearctic as "including highlands of Mexico", ends up assigning pseudothelphusids found in the Mexican highlands to the Nearctic region despite their Neotropical affinities.

In addition, not all taxa are restricted to a single zoogeographical region (e.g. *Aparapotamon grahami* and *Geothelphusa* spp. [Potamidae] occur in both Palaearctic and Oriental regions; and *Parathelphusa* and *Sundathelphusa* [Parathelphusidae] occur in both Oriental and Australasian regions). The families containing species and/or genera that have distributions which overlap adjacent zoogeographical regions are: Potamidae, Parathelphusidae, Pseudothelphusidae, Gecarcinidae, Hymenosomatidae, Sesarmidae, Goneplacidae and Varunidae. Because of this, as mentioned earlier, the "World" totals listed in the last column of Tables 1 and 2 for these families are less than the sum of the number of taxa from each zoogeographical region on the map (Fig. 3) or in the corresponding row of each table (see Tables 1 and 2 footnotes).

Major historic processes leading to global biodiversity patterns

Sternberg et al. (1999) summarised the hypotheses for the origin and diversification of the true freshwater crabs into the polyphyletic, archaic and phylogenetic schools. The polyphyletic school (e.g. Bott, 1955; Pretzmann, 1973) considered that the freshwater crab families originated from a number of different marine ancestors; in this case, morphological similarities would be the result of convergence, not common ancestry.

In the archaic population school, vicariance has been suggested as the key mechanism and the breakup of Gondwanaland a key historic process. The pseudothelphusoid and gecarcinucoid freshwater crabs share a two-segmented bilobed mandibular palp (a presumptive synapomorph) and freshwater crab taxa with this character are found today in the majority of Gondwanan fragments (South America, Africa, Madagascar, the Seychelles, India, Southeast Asia, and Australasia). Crabs are postulated to have had a Gondwanan origin with present day distribution patterns resulting from the breakup of the supercontinent (Rodríguez, 1986; Ng & Rodríguez, 1995; Ng et al., 1995; Yeo & Ng, 1999). There is, however, no paleontological support for this view and there is incongruence with regard to what is known about Brachyuran evolution. Daniels et al. (2006), however, argue that the two-segmented bilobed mandibular palp may be a convergent character, and the pseudothelphusoid and gecarcinucoids may not be that closely related. Bănărescu (1990) proposed long distance transoceanic dispersal as an alternative mechanism to explain the largely insular distribution of the Parathelphusidae in Sundaic Southeast Asia. However, this has been challenged by Ng & Rodríguez (1995), who argued that Bănărescu's ideas made dubious assumptions about the ecology and origins of the Parathelphusidae (and about the phylogeny of all freshwater crabs). The dominance of gecarcinucoid crabs in the Indian peninsula (and the absence of potamids) could also be explained as the result of their long isolation on a Gondwanan continental fragment before it collided with continental Asia (where potamids are found in large numbers).

In contrast to the above hypotheses, the phylogenetic school (Sternberg et al., 1999, following Colosi's (1921) ideas) suggested that the pseudothelphusoid, gecarcinucoid, and potamoid freshwater crabs form a monophyletic group that may have had a more recent, post-Gondwanan origin. Here the present global distribution pattern is thought to be the result of colonisation of the tropical continental margins from a common ancestral marine group consisting of a monophyletic thoracotreme clade widely distributed along littoral areas of the southern Tethys Sea during the Cretaceous that eventually gave rise to the modern families after independent diversification into the freshwater environments. This post-Gondwanan (Cretaceous) transoceanic dispersal hypothesis was most recently supported by molecular evidence presented by Daniels et al. (2006) based on work carried out with emphasis on Afrotropical freshwater crabs. Another hypothesis recently proposed by

Klaus et al. (2006) suggested that gecarcinucoid crabs originated in Africa and reached South Asia via transoceanic dispersal and a series of theorised "stepping-stone" islands. The available evidence might now suggest a mixture of vicariance and dispersal, although consensus has still to be reached in this complex issue. Clearly, there are many ideas and hypotheses being proposed, a sure sign that there is increasing interest in using freshwater crabs for biogeographic studies.

The current Eurasian distribution of the Potamidae shows that one subfamily (Potaminae) occurs in western Eurasia (in North Africa, southern Europe, Socotra, the Middle East and the Himalayas), and one subfamily (the Potamiscinae) is in Southeast Asia, China and Japan. There is reason to believe that this distributional pattern may be the result of dispersal from a continental Asian origin (Yeo & Ng, 2003). A trend is apparent in the relative distributions of these two subfamilies that suggests potamids may have spread westwards into Eurasia (as potamines) and southwards into insular Southeast Asia (as potamiscines) from a continental Asian origin. Additional circumstantial evidence for this trend is seen in the distinct decline in potamid diversity westward from Southeast Asia, whereby southern Europe has only one genus (*Potamon*) and East Asia has some 40 genera. The distributional trend shown by potamines and potamiscines corresponds to a similar trend shown by freshwater crab superfamilies discussed by Yeo & Ng (1999).

Rodríguez (1982, 1986) explained the current distribution of Neotropical pseudothelphusids by vicariance events and secondary dispersion. In his hypothesis, based on Rosen's (1976) model for Caribbean biogeography, an ancestral group that occupied the Proto-Antillean archipelago, characterised by a plesiomorphic character of the third maxilliped (presence of long exognath), gave rise to two different groups after the Caribbean Plate drifted northeastward between Central and South America during the late Mesozoic and Cenozoic: the Epilobocerinae in the Antilles and the Strengerianini in northern Colombia. Based on sympatries and geographic morphoclines in somatic and gonopodal characters, Rodríguez (1986) distinguished three distinct chorological series that would have radiated from a dispersal centre in northern Colombia towards Central America and Mexico (Pseudothelphusini and Potamocarcinini), northern and eastern South America (Kingsleyini) and the southern Andes (Hypolobocerini).

The South American trichodactylids are phylogenetically separate from all other freshwater crab families which points to an independent invasion of this habitat by the group's supposedly marine portunoid ancestors (Rodríguez, 1992; Sternberg et al., 1999; Martin & Davis, 2001). The morphological cladistic analysis of Sternberg et al. (1999) identified grapsids as probable sister taxa to the non-trichodactylid freshwater crabs, which contradicted the assertions of earlier authors (Bott, 1955) who suggested ancestry from marine crab groups such as the Xanthoidea or the Portunoidea. There is, however, still some uncertainty about freshwater crab origins and relationships.

In addition to vicariance and dispersal, distributional limits of true freshwater crabs are also influenced by a host of other factors that interact with these two key processes. These include abiotic factors such as climate, hydrology, topography and altitude as well as biotic factors such as habitat vegetation and inter-specific competition (see Rodríguez, 1986; Ng, 1988; Barbaresi & Gherardi, 1997; Cumberlidge, 1999; Dai, 1999; Rodríguez & Suárez, 2004; Magalhães et al., 2005; Marijnissen et al., 2005).

Phylogeny

The recent surge in alpha taxonomic descriptions of freshwater crabs has been accompanied by an increase in interest in their phylogenetic relationships and higher taxonomy. Most studies in the late 20th century have accepted the traditional morphology-based classification system proposed by Bott (1970) comprising three superfamilies and eleven families. This has been challenged over the last decades by some workers who questioned the superfamily system and synonymised three families, namely Sundathelphusidae, now a junior synonym of Parathelphusidae (Ng, 1988; Chia & Ng, 1998); and Isolapotamidae and Sinopotamidae, both now junior synonyms of Potamidae (Ng, 1988; Dai, 1999; Yeo & Ng, 2003). Brandis (2002) had recently argued for the revalidation of the families Isolapotamidae and Sinopotamidae for two apparently discreet monophyletic potamid taxa; however, this was challenged by Yeo & Ng (2003) who suggested that Brandis' (2002) groupings might instead be infra-familial clades within the Potamidae (subfamily Potamiscinae).

Morphological cladistic studies (e.g. Sternberg et al., 1999; Cumberlidge, 1999; Sternberg & Cumberlidge, 1999, 2001) argue for recognising the remaining eight families in two main lineages. These authors argued that the true freshwater crabs are paraphyletic and consist of two distinct lineages: (1) the monophyletic Trichodactylidae in the predominantly marine superfamily Portunoidea and (2) a monophyletic group consisting of all remaining freshwater crab families assigned to three superfamilies, viz., Potamoidea (Potamidae, Potamonautidae, Deckeniidae, Platythelphusidae), Gecarcinucoidea (Parathelphusidae, Gecarcinucidae) and Pseudothelphusoidea (Pseudothelphusidae). Various authors have expressed doubts about the existing family classification and it is clear that there is still much to be done before a reasonable consensus can be reached. Most recently, Daniels et al. (2006) have suggested that some of these families may be artificial, while Brandis (2002), and Klaus et al. (2006) and Cumberlidge et al. (2007) have each offered different systems of higher classification. We have erred on the side of "conservativeness" in the system adopted here. The differences between workers will only encourage more morphological and molecular work to help resolve the conflicting views to the classification of these animals. Currently, there are few other molecular phylogenies of freshwater crabs and those available have a limited geographical scope (e.g. Japan: Segawa, 2000; Taiwan: Shih et al., 2004, 2007; South Africa: Daniels et al., 2002; India and Sri Lanka: Bossuyt et al., 2004; East Africa: Marijnissen et al., 2006; Malay Peninsula: Yeo et al., 2007). The status of the phylogenetic relationships of the freshwater crabs is, therefore, still controversial in the face of incongruent morphological and molecular studies and is the subject of ongoing work by several research groups including the present authors. For convenience, we follow the higher taxonomy of the freshwater crabs proposed by Martin & Davis (2001).

Bott (1955) estimated that the age of freshwater crabs was about 65 million years, with an origin between the end of the Cretaceous and the beginning of the Tertiary for the African Potamonautidae. Later, in agreement with Rodríguez (1986) that New World pseudothelphusoids and Old World gecarcinucoids share a synapomorphy (a bilobed mandibular palp), Ng et al. (1995) suggested that the freshwater crabs possessing this synapomorph were at least 120 million years old by inference, corresponding to the timing of the split between South America and Africa. More recently, Daniels et al. (2006) estimated that Afrotropical crabs radiated some 75.03–78.6 million years ago. The fossil evidence, however, does not support any of the above proposed age estimates, with the oldest known fossil from the Upper Tertiary in northern India and Europe being not much older than 23 mya (Bott, 1955; Glaessner, 1969). Ng et al. (1995), however, warned against making firm conclusions based on this dearth of fossils, commenting that, "the rarity and difficulty of forming (and finding) freshwater fossils is well known".

One of the key processes driving freshwater crab diversification is likely to be allopatric speciation resulting from geographic isolation. This is helped by the relatively low fecundity and poor dispersal abilities of freshwater crabs; and is often facilitated by the habitat heterogeneity and numerous ecological niches and microhabitats afforded by the complicated topography and hydrology of their

Fig. 4 Freshwater crabs (*Somanniathelphusa dangi*: Parathelphusidae) being sold as food in a rural market in northern Vietnam

environments. Nevertheless, the significance of sympatric speciation cannot be discounted, especially for lacustrine species (e.g. *Platythelphusa* in the African Rift Valley Lakes (Cumberlidge et al., 1999; Marijnissen et al., 2006); and *Parathelphusa* and allies in the Sulawesi lakes (Chia & Ng, 2006)). All these factors have led to a high degree of endemism in freshwater crabs.

Human-related issues

Economic and medical uses

Freshwater crabs are an important protein source and are consumed in many parts of the world. Ng (1988) noted that in Thailand, large potamids and parathelphusids are occasionally eaten by locals. Yeo & Ng (1998) commented that potamids are important in the diet of rural and hill tribes of northern Vietnam (Fig. 4). Freshwater crabs are also consumed for purported medicinal and tonic properties, including treatment of stomach ailments and physical injuries (Dai, 1999). In South America, indigenous groups use freshwater crabs, particularly large pseudothelphusids, for food (Finkers 1986).

Medically, freshwater crabs are important because they are intermediate hosts to the parasitic lung fluke, *Paragonimus* (Platyhelminthes) which causes paragonimiasis. This dangerous disease affects humans when they consume infected crabs (Ng, 1988; Dai, 1999; Cumberlidge, 1999). Rodríguez & Magalhães (2005) listed the pseudothelphusid species reported as hosts for *Paragonimus* and discussed its occurrence in the neotropics. Although proper cooking would kill the parasite, many of the more rural communities prefer to consume freshwater crabs half-raw (Ng, 1988; Dai, 1999).

Freshwater crabs are sometimes sold in the aquarium trade. These are usually the more colourful Indochinese potamids e.g. *Demanietta khirikhan, Pudaengon arnamicai, Terrapotamon abbotti*, although less gaudy parathelphusids like *Heterothelphusa fatum* are also sold. The trichodactylid crab, *Dilocarcinus pagei*, is captured for bait in game fishing of large catfishes in the Pantanal Matogrossense, a swampy area in the Paraguay River basin (Magalhães, 2000). The impact of these activities on the freshwater crab fauna, however, is low as demand is low and collection irregular.

Threats and conservation issues

Based on a recent assessment of the conservation status of Malaysian freshwater crabs, a few patterns emerged (Ng & Yeo, 2007). The restricted distributions of most of the freshwater crab species pose serious problems for conservation. Fortunately, for the time being, the species with the most restricted distributions are often those that inhabit offshore islands or mountains, areas that are generally less immediately impacted by man. The loss of natural forest as a result of land development and agriculture has generally affected lowlands more severely. However, many lowland species (e.g. *Parathelphusa maculata*) may have survived because they have relatively wider distributions. Specialist species (e.g. obligate cavernicoles like *Cerberusa caeca*), and highland taxa (e.g. *Johora grallator*) are at higher risk because they have a restricted range and are less tolerant of habitat changes. A similar study was made of the Sri Lankan fauna by Bahir et al. (2005), and of the Tanzanian and Southern African freshwater crabs by Reed & Cumberlidge (2006) and Cumberlidge & Daniels (2007).

The conservation of freshwater crabs will depend primarily on preserving large enough natural forest areas to maintain the good water quality of the original streams. Thus, the need to establish more nature reserves and national parks, together with careful planning and development is imperative.

Acknowledgements This study has benefited from the help and advice from many colleagues. We thank Tohru Naruse and Paul Clark for their many helpful comments and suggestions which have helped improve the manuscript. Support from the Belgium Biodiversity Platform, the Belgian Science Policy, the Royal Belgium Institute of Natural Sciences, and the Department of Biological Sciences, National University of Singapore, is gratefully acknowledged.

References

Bahir, M. M., P. K. L. Ng, K. Crandall & R. Pethiyagoda, 2005. A conservation assessment of the freshwater crabs of Sri Lanka. Raffles Bulletin of Zoology Supplement 12: 121–126.

Bănărescu, R. M., 1990. Zoogeography of Freshwaters. 1. General distribution and dispersal of freshwater animals. AULA-Verlag, Weisbaden, 1–511.

Barbaresi, S. & F. Gherardi, 1997. Italian freshwater decapods: exclusion between the crayfish *Austropotamobius pallipes* (Faxon) and the river crab *Potamon fluviatile* (Herbst). Bulletin française de la Pêche et de la Piscicolture 347: 731–747.

Bossuyt, F., M. Meegaskumbura, N. Beenaerts, D. J. Gower, R. Pethiyagoda, K. Roelants, A. Mannaert, M. Wilkinson, M. M. Bahir, K. Manamendra-Arachchi, P. K. L. Ng, C. J. Schneider, O. V. Oommen & M. C. Milinkovitch, 2004. Local endemism within the western Ghats-Sri Lanka biodiversity hotspot. Science 306: 479–481.

Bott, R., 1955. Die Süsswasserkrabben von Afrika (Crustacea, Decapoda) und ihre Stammesgeschichte. Annales du Musee Royal du Congo belge 1: 209–352, Pls. 1–30, Figs. 1–103.

Bott, R., 1970. Die Süsswasserkrabben von Europa, Asien, Australien und ihre Stammesgeschichte. Eine Revision der Potamoidea und Parathelphusoidea (Crustacea, Decapoda). Abhandlungen der Senckenbergischen Naturforschenden Gesellschaft, Frankfurt 526: 1–338, Pls. 1–58.

Brandis, D., 2002. On the taxonomic status and biogeography of the Isolapotamidae Bott, 1970 (Decapoda, Brachyura). Journal of Natural History, London 36: 1291–1339.

Chia, O. K. S. & P. K. L. Ng, 1998. Is Sundathelphusidae Bott, 1969 a valid taxon? A cladistic appraisal. Proceedings and Abstracts of the 4th International Crustacean Congress, Amsterdam: 72.

Chia, O. K. S. & P. K. L. Ng, 2006. The freshwater crabs of Sulawesi, with descriptions of two new genera and four new species (Crustacea: Decapoda: Brachyura: Parathelphusidae). Raffles Bulletin of Zoology 54: 383–428.

Colosi, G., 1921. La distribuzione geografica dei Potamonidae. Rivista di Biologia 3: 294–301.

Cox, C. B., 2001. The biogeographic regions reconsidered. Journal of Biogeography 28: 511–523.

Cumberlidge, N., R. v. Sternberg, R. Bills & H. A. Martin, 1999. A revision of the genus *Platythelphusa* A. Milne-Edwards, 1887, from Lake Tanganyika, East Africa (Decapoda: Potamoidea: Platythelphusidae). Journal of Natural History 33: 1487–1512.

Cumberlidge, N., 1999. The Freshwater Crabs of West Africa: Family Potamonautidae. Institut de Recherche pour le Développement, Collection Faune et Flore Tropicales no. 36, Paris, 1–382.

Cumberlidge, N. & S. R. Daniels, 2007. A conservation assessment of the freshwater crabs of southern Africa (Brachyura: Potamonautidae). African Journal of Ecology (in press).

Cumberlidge, N., S. R. Daniels & R. v. Sternberg, 2007. A revision of the higher taxonomy of the Afrotropical freshwater crabs (Decapoda: Brachyura) with a discussion of their biogeography. Biological Journal of the Linnean Society (in press).

Cumberlidge, N., D. B. Fenolio, M. E. Walvoord & J. Stout, 2005. Tree-climbing crabs (Potamonautidae and Sesarmidae) from phytotelmic microhabitats in rainforest canopy in Madagascar. Journal of Crustacean Biology 25: 302–308.

Cumberlidge, N. & R. v. Sternberg, 2002. The freshwater crabs of Madagascar (Crustacea, Decapoda, Potamoidea). Zoosystema 24: 41–79.

Dai, A. Y., 1999. Fauna Sinica (Arthropoda, Crustacea, Malacostraca, Decapoda, Parathelphusidae, Potamidae). Editorial Committee of Fauna Sinica, Academia Sinica, Science Press, Beijing, 501 pp, 238 figs., 30 pls.

Daniels, S. R., N. Cumberlidge, M. Pérez-Losada, S. A. E. Marijnissen & K. A. Crandall, 2006. Evolution of Afrotropical freshwater crab lineages obscured by morphological convergence. Molecular Phylogenetics and Evolution 40: 227–235.

Daniels, S. R., B. A. Stewart, G. Gouws, M. Cunningham & C. A. Matthee, 2002. Phylogenetic relationships of the southern African freshwater crab fauna derived from multiple data sets reveal biogeographic patterning. Molecular Phylogenetics and Evolution 25: 511–523.

Finkers, J., 1986. Los Yanomami y su sistema alimenticio (Yanomami Nii Pë). Vicariato Apostólico de Puerto Ayacucho, Puerto Ayacucho. Monografía n° 2, 262 pp.

Glaessner, M. F., 1969. Decapoda. In Moore, R. C. (ed), Treatise on Invertebrate Palaeontology. Part R. Arthropoda 4. Geological Society of America, 2: R399–R533.

Klaus, S., C. D. Schubart & D. Brandis, 2006. Phylogeny, biogeography and a new taxonomy for the Gecarcinucoidea Rathbun, 1904 (Decapoda: Brachyura). Organisms, Diversity and Evolution 6: 199–217.

Lundberg, J. G., L. G. Marshall, J. Guerrero, B. Norton, M. C. Malabarba & F. Wesselingh, 1998. The Stage for Neotropical Fish Diversification: A History of Tropical South American Rivers. In Malabarba, L. R., R. E. Reis, R. P. Vari, C. A. S. Lucena & Z. M. S. Lucena (eds), Phylogeny and Classification of Neoptropical Fishes. EDIPUCRS, Porto Alegre, 12–48.

Magalhães, C., 2000. Diversity and abundance of decapod crustaceans in the Rio Negro basin, Pantanal, Mato Grosso do Sul, Brazil. In Chernoff, B., L. E. Alonso, J. R. Montambault & R. Lourival (eds), A Biological Assessment of the Aquatic Ecossystems of the Pantanal, Mato Grosso do Sul, Brazil. Conservation International, Washington, DC, 56–62. Bulletin of Biological Assessment 18.

Magalhães, C., 2003. Famílias Pseudothelphusidae e Trichodactylidae. In Melo, G. A. S. (ed), Manual de Identificação dos Crustacea Decapoda de Água Doce do Brasil. Editora Loyola, São Paulo, 143–287.

Magalhães, C., F. A. Abrunhosa, M. de Oliveira Pereira & M. A. Melo, 2005. New records of *Fredius denticulatus* (H. Milne-Edwards, 1853) and *F. reflexifrons* (Ortmann, 1897), and the eastern limits of the distribution of pseudothelphusid crabs (Crustacea: Decapoda) in Brazil. Acta Amazonica 35: 93–96.

Martin, J. W. & G. E. Davis, 2001. An updated classification of the recent Crustacea. Natural History Museum of Los Angeles County Science Series 39, 124 pp.

Marijnissen, S. A. E., E. Ellinor Michel, S. R. Daniels, D. Erpenbeck, S. B. J. Menken & F. R. Schram, 2006. Molecular evidence for recent divergence of Lake

Tanganyika endemic crabs (Decapoda: Platythelphusidae). Molecular Phylogenetics and Evolution 40: 628–634.

Marijnissen, S. A. E., S. Lange & N. Cumberlidge, 2005. Revised distribution of the African freshwater crab genus *Deckenia* Hilgendorf, 1868 (Brachyura, Potamoidea, Deckeniidae). Crustaceana 78: 889–896.

Ng, P. K. L., 1988. The Freshwater Crabs of Peninsular Malaysia and Singapore. Department of Zoology, National University of Singapore, Shinglee Press, Singapore, i–viii, 1–156, Figs. 1–63, 4 colour plates.

Ng, P. K. L & G. Rodríguez, 1995. Freshwater crabs as poor zoogeographical indicators: a critique of Bănărescu (1990). Crustaceana 68: 636–645.

Ng, P. K. L., Z. Števčić & G. Pretzmann, 1995. A revision of the family Deckeniidae Ortmann, 1897 (Crustacea: Decapoda: Brachyura: Potamoidea), with description of a new genus (Gecarcinucidae: Gecarcinucoidea) from Seychelles, Indian Ocean. Journal of Natural History 29: 581–600.

Ng, P. K. L. & F. W. M. Tay, 2001. The freshwater crabs of Sri Lanka (Decapoda: Brachyura: Paratelphusidae). Zeylanica 6: 113–199.

Ng, P. K. L. & D. C. J. Yeo, 2007. Malaysian freshwater crabs: conservation prospects and challenges. In Chua, L. S. L., L. G. Kirton & L. G. Shaw (eds), Status of Biological Diversity in Malaysia and Threat Assessment of Plant Species in Malaysia. Proceedings of the Seminar and Workshop. 28–30 June 2005, 95–120.

Pretzmann, G., 1973. Grundlagen und Ergebnisse der Systematik der Pseudothelphusidae. Zeitschrift fuer Zoologische Systematik und Evolutionsforschung 11: 196-218.

Reed, S. K. & N. Cumberlidge, 2006. Taxonomy and biogeography of the freshwater crabs of Tanzania, East Africa (Brachyura: Potamoidea: Potamonautidae, Platythelphusidae, Deckeniidae). Zootaxa 1262: 1–139.

Rodríguez, G., 1982. Les crabes d'eau douce d'Amerique. Famille des Pseudothelphusidae. ORSTOM, Paris, 224 pp, Faune Tropicale 22.

Rodríguez, G., 1986. Centers of radiation of freshwater crabs in the neotropics. In Gore, R. H. & K. L. Heck (eds), Biogeography of Crustacea, Crustacean Issues 3: 51–67.

Rodríguez, G., 1992. The Freshwater Crabs of America. Family Trichodactylidae. Office de la Recherche Scientifique d'Outre Mer (ORSTOM), Paris, 200 pp.

Rodríguez, G. & C. Magalhães, 2005. Recent advances in the biology of the Neotropical freshwater crab family Pseudothelphusidae (Crustacea, Decapoda, Brachyura). Revista brasileira de Zoologia 22: 354–365.

Rodríguez, G. & H. Suárez, 2004. A revision of the freshwater crabs of the family Pseudothelphusidae (Decapoda: Brachyura) from Peru with notes on the southern limits of the family. Amazoniana 18: 11–28.

Rosen, D. E., 1976. A vicariance model of Caribbean biogeography. Systematic Zoology 24: 431–464.

Segawa, R., (2000). Molecular phylogenetic study of potamoid crabs in Ryukyu Islands. Kaiyo Monthly 32: 241–245.

Shih, H.-T., S.-H. Fang & P. K. L. Ng, 2007. Phylogeny of the freshwater crab genus *Somanniathelphusa* Bott (Decapoda : Parathelphusidae) from Taiwan and the coastal regions of China, with notes on their biogeography. Invertebrate Systematics 21: 29–37.

Shih, H.-T., P. K. L. Ng & H.-W. Chang, 2004. The systematics of the genus *Geothelphusa* (Crustacea, Decapoda, Brachyura, Potamidae) from southern Taiwan: A molecular appraisal. Zoological Studies 43: 561–570.

Sternberg, R. v. & N. Cumberlidge, 1999. A cladistic analysis of the genus *Platythelphusa* A. Milne-Edwards, 1887 from Lake Tanganyika, East Africa (Decapoda: Potamoidea: Platythelphusidae) with comments on the phylogenetic position of the group. Journal of Natural History 33: 493–511.

Sternberg, R. v. & N. Cumberlidge, 2001. Notes on the position of the true freshwater crabs within the Brachyrhynchan Eubrachyura (Crustacea: Decapoda: Brachyura). Hydrobiologia 449: 21–39.

Sternberg, R. v., N. Cumberlidge & G. Rodríguez, 1999. On the marine sister groups of the freshwater crabs (Crustacea: Decapoda). Journal of Zoological Systematics and Evolutionary Research 37: 19–38.

Yeo, D. C. J. & P. K. L. Ng, 1998. Freshwater crabs of the *Potamon tannanti* species group (Crustacea, Decapoda, Brachyura, Potamidae) of northern Indochina. Raffles Bulletin of Zoology 46: 627–650.

Yeo, D. C. J. & P. K. L. Ng, 1999. The state of freshwater crab taxonomy in Indochina (Decapoda, Brachyura). In Schram, F. R. & J. C. von Vaupel Klein (eds), Crustaceans and the Biodiversity Crisis, Proceedings of the Fourth International Crustacean Congress, 1998, vol I, 637–646.

Yeo, D. C. J. & P. K. L. Ng, 2003. Recognition of two subfamilies in the Potamidae Ortmann, 1896 (Brachyura, Potamidae), with a note on the genus *Potamon* Savigny. Crustaceana 76: 1219–1235.

Yeo, D. C. J., H.-T. Shih, R. Meier & P. K. L. Ng, 2007. Phylogeny and biogeography of the freshwater crab genus *Johora* (Crustacea: Brachyura: Potamidae) from the Malay Peninsula, and the origins of its insular fauna. Zoologica Scripta 36: 255–269.

Hydrobiologia (2008) 595:287–293
DOI 10.1007/s10750-007-9024-2

FRESHWATER ANIMAL DIVERSITY ASSESSMENT

Global diversity of shrimps (Crustacea: Decapoda: Caridea) in freshwater

S. De Grave · Y. Cai · A. Anker

Abstract Freshwater caridean shrimps account for approximately a quarter of all described Caridea, numerically dominated by the Atyidae and Palaemonidae. With the exception of Antarctica, freshwater shrimp are present in all biogeographical regions. However, the Oriental region harbours the majority of species, whilst the Nearctic and western Palaearctic are very species-poor. Many species are important components of subsistence fisheries, whilst the Giant River Prawn forms the basis of an extensive aquaculture industry. A total of 13 species are threatened or endangered, with one species formally extinct.

Guest editors: E. V. Balian, C. Lévêque, H. Segers and K. Martens
Freshwater Animal Diversity Assessment

S. De Grave (✉)
Oxford University Museum of Natural History, Parks Road, Oxford OX1 3PW, UK
e-mail: sammy.degrave@oum.ox.ac.uk

Y. Cai
Biodiversity Centre, National Parks Board, 1 Cluny Road, Singapore 259569, Republic of Singapore
e-mail: caiyixiong@yahoo.com

A. Anker
Smithsonian Tropical Research Institute, Naos Unit 0948, APO AA, Miami, FL 34002-0948, USA
e-mail: ankera@si.edu

Keywords Caridea · Decapoda · Freshwater · Biodiversity

Introduction

The suborder Caridea presently consists of approximately 2,500 described species in 31 families. They occur in all aquatic habitats on the planet, from pelagic marine species and those dwelling in anchialine caves through to freshwater species. A total of 655 freshwater species (just over a quarter of all described carideans) are presently known. Amongst the freshwater families, the two most speciose are the near exclusively freshwater Atyidae and the Palaemonidae (Subfamily Palaemoninae), which also have brackish water and marine representatives. Freshwater shrimps occur in a vast range of habitats, from torrential mountain streams down to sluggish, oligohaline waters. Freshwater stygobiont taxa are well represented, with many more taxa occasionally found in subterraneous habitats (see Holthuis, 1986). Atyidae are characterised by unique feeding chelipeds, with the complex brushes on the first and second pereiopods (Fig. 1A) filtering out suspended matter or sweeping up microbial films (Fryer, 1977). Members of the Palaemonidae exhibit a wide variety in habitus, from the unspecialised habitus of *Palaemonetes* species through to the males of *Macrobrachium*, which have unusually enlarged second chelae (Fig. 1B), used in agonistic interactions and

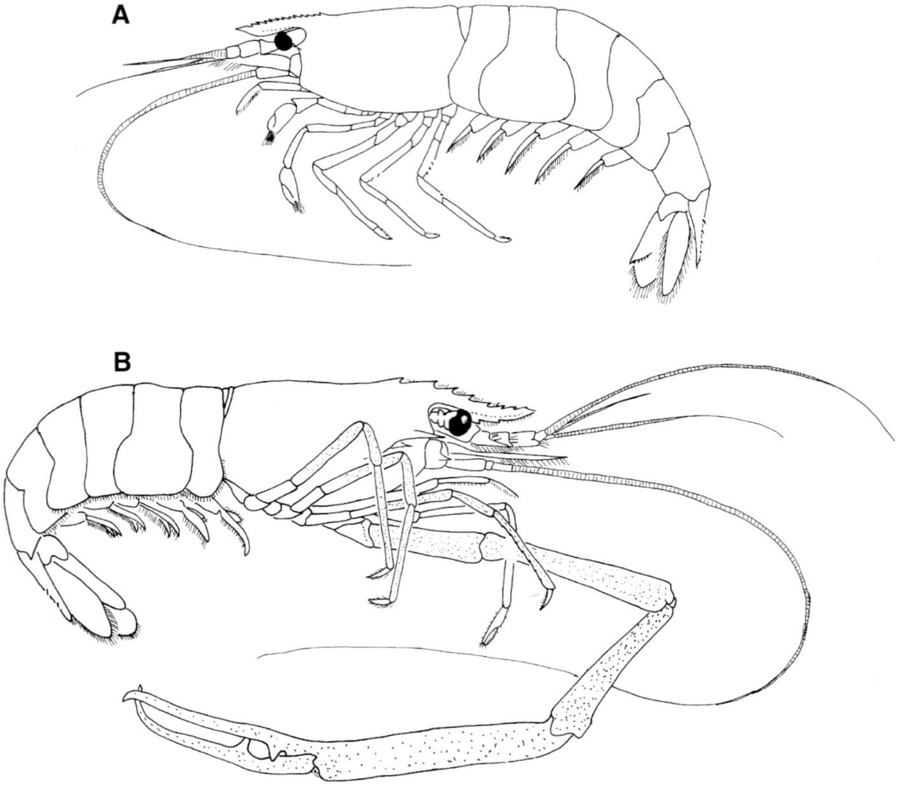

Fig. 1 Habitus of (**A**) *Caridina weberi* De Man, and (**B**) *Macrobrachium lar* (Fabricius)

predation. Many freshwater taxa are important as sources for artisanal fisheries (Holthuis, 1980), with one species, *Macrobrachium rosenbergii* (De Man), being an important aquaculture species. Freshwater species are also becoming increasingly popular in the aquarium trade. Abbreviated larval development occurs in many species of Atyidae and freshwater Palaemoninae, with members of *Neocaridina* having completely lost larval development, hatching as post larvae. Although abbreviated larval development allows these species to complete their life cycle in freshwater, many other species of both families complete part of their larval cycle in brackish water or even in fully marine conditions.

Species diversity

Freshwater species of carideans belong to eight families/subfamilies, numerically dominated by the Atyidae, with 359 species/subspecies (Table 1). Although this family is considered in many textbooks as restricted to freshwater habitats, some anchialine genera are known (e.g. *Antecaridina*, *Halocaridina*, *Typhlatya*), whilst juveniles of *Atya* have been found under fully marine conditions in Atlantic waters. Although the most speciose genus *Caridina* occurs in six biogeographic regions, many genera and species are either only known from their type locality or have a narrow geographical distribution (e.g. *Lancaris* is restricted to Sri Lanka, see Cai & Bahir, 2005). Some species are morphologically adapted to live in fast-flowing water, such as the Caribbean *Atya scabra* (Leach), which lives beneath rocks under waterfalls and in rapids, whilst other species, such as many *Caridina* species are adapted to lakeshore weed beds, usually displaying a more gracile habitus. Cave dwelling taxa are well represented with many exclusively stygobiont genera. Of particular ecological interest are the only two freshwater commensal species (a widespread mode of life in marine shrimp species): *Limnocaridina iridinae* Roth-Woltereck from the mantle cavity of a unionid clam from Lake Tanganyika (Roth-Woltereck, 1958) and a *Caridina* species from Lake Towuti in Sulawesi living with freshwater sponges (Cai, pers. obs.).

The second most speciose family is the Palaemonidae (Table 1), with many more marine and brackish

Table 1 Geographical distribution of freshwater Caridea described up to December 2005

Family/Subfamily	PA	NA	NT	AT	OL	AU	PAC	ANT	World
Species + subspecies									
Alpheidae	–	–	1	1	2	–	–	–	4
Atyidae	31	4	19	59	209	55	13	–	359
Desmocarididae	–	–	–	2	–	–	–	–	2
Kakudicarididae	–	1	–	–	–	2	–	–	3
Palaemonidae	13	12	83	28	138	30	12	–	276
Euryrhynchinae	–	–	4	2	–	–	–	–	6
Typhlocaridinae	3	–	–	–	–	–	–	–	3
Xiphocarididae	–	–	2	–	–	–	–	–	2
Total	47	17	109	92	349	87	25	–	655
Genera									
Alpheidae	–	–	1	1	1	–	–	–	2
Atyidae	9	2	5	11	13	11	4	–	35
Desmocarididae	–	–	–	1	–	–	–	–	1
Kakudicarididae	–	1	–	–	–	2	–	–	3
Palaemonidae	4	2	9	2	7	2	2	–	13
Euryrhynchinae	–	–	1	2	–	–	–	–	3
Typhlocaridinae	1	–	–	–	–	–	–	–	1
Xiphocarididae	–	–	1	–	–	–	–	–	1
Total	14	5	17	17	21	15	6	–	59

Note that the family Typhlocarididae is divided into two subfamilies: Typhlocaridinae and Euryrhynchinae. PA: Palaearctic, NA: Nearctic, NT: Neotropical, AT: Afrotropical, OL: Oriental, AU: Australasian, PAC: Pacific Oceanic Islands, ANT: Antarctic

water species known than there are freshwater taxa, all of the latter being restricted to the subfamily Palaemoninae. The numerically dominant genus is *Macrobrachium*, restricted to fresh and brackish water except for the enigmatic, single record of an undescribed species from Canadian waters. Other species-rich genera are *Palaemonetes*, a taxonomically poorly defined world-wide genus, and *Palaemon*. Some species of *Palaemonetes* are exclusively freshwater, such as the North American *Palaemonetes kadakiensis* Rathbun, but several estuarine species can tolerate fully freshwater conditions. Several species of *Palaemon* have also been recorded from marine, brackish and freshwater environments, e.g. *Palaemon concinnus* Dana (see Marquet, 1991).

Fewer species are known in the near exclusively freshwater Typhlocarididae (one species being anchialine). The Typhlocarididae comprises two subfamilies, the monogeneric stygobiont Typhlocaridinae, and the Euryrhynchinae with representatives in South America and West Africa. Although the Alpheidae are one of the most species-rich shrimp families, a few freshwater species are known. Some members of the genus *Potamalpheops* and *Alpheus cyanoteles* Yeo & Ng occur in freshwater, with several other species either tolerating seasonally fresh water or occurring in oligohaline waters.

It is difficult to estimate the true species richness of freshwater shrimps, as every year new taxa continue to be described, mainly in the two most numerically dominant genera, *Caridina* and *Macrobrachium*. As a result, species discovery curves are not flattening out (Fig. 2), and it can be expected that many more species await discovery. New genera also continue to be erected, for instance for morphologically disparate species previously placed in *Caridina* (e.g. the genera *Lancaris*, *Sinodina*, *Paracaridina*). Genetic studies have only recently started in freshwater shrimps, with for instance the work of Baker et al. (2004) highlighting the presence of several cryptic lineages in Australian *Paratya*, some of which may well represent true species.

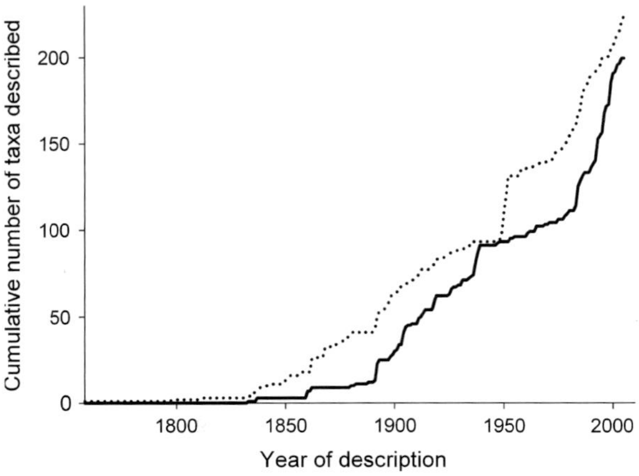

Fig. 2 Rates of discovery for the two most species-rich genera of freshwater shrimp, *Caridina* (full line) and *Macrobrachium* (dotted line)

Phylogeny and historical processes

In common with the remainder of the Caridea, very little can be said on the phylogeny of freshwater shrimps, mainly due to the paucity of higher level cladistic and genetic studies, with the study of Pereira (1997) being the sole exception for freshwater taxa. This study clearly highlighted the non-monophyletic status of many of the palaemonid genera, recently also suggested by genetic work (Murphy & Austin, 2005).

It is however evident that, at a minimum, three invasions of freshwater must have taken place, as palaemonoid, atyoid and alpheoid shrimps are not closely related. Several lines of evidence, including the occurrence of freshwater fossils of Cretaceous age (Beurlen, 1950) testify that Atyidae are ancient inhabitants of freshwater, having diverged early from an ancestral marine stock (Fryer, 1977). The origin of *Macrobrachium* has been suggested to be in the late Oligocene or early Miocene (Short, 2004; Murphy & Austin, 2005), although there are evidence that multiple invasions of freshwater are involved (Pereira, 1997; Murphy & Austin, 2005).

Present distribution and main areas of endemicity

With the exception of Antarctica, freshwater carideans shrimps are present in all the main biogeographical regions (Fig. 3). The Oriental region harbours three times as many species as the next most species-rich provinces: Neotropical, Afrotropical and Australasian. The Nearctic region harbours the lowest number of taxa, primarily due to the absence of

Fig. 3 Geographical distribution of species/subspecies and genera described up to December 2005 (Species number/Genus number), coded according to the main zoogeographic regions. PA: Palaearctic, NA: Nearctic, NT: Neotropical, AT: Afrotropical, OL: Oriental, AU: Australasian, PAC: Pacific Oceanic Islands, ANT: Antarctic

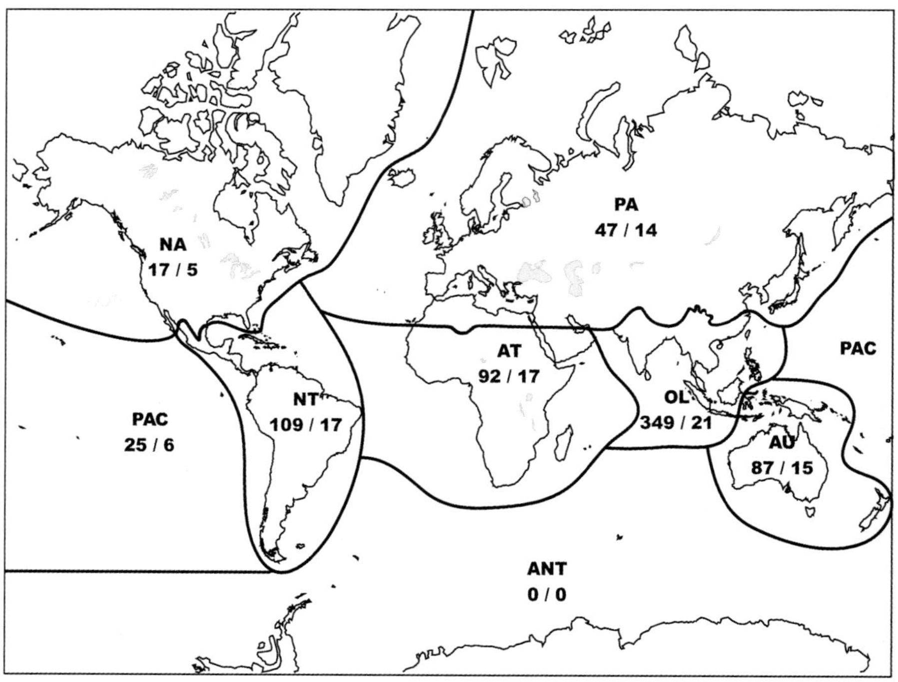

Table 2 Geographical distribution of troglobitic freshwater species of Caridea described up to December 2005

Family/Subfamily	PA	NA	NT	AT	OL	AU	PAC	ANT	World
Atyidae	13	2	–	10	6	11	1	–	43
Kakudicarididae	–	1	–	–	–	–	–	–	1
Palaemonidae	–	2	15	–	4	1	–	–	22
Typhlocaridinae	3	–	–	–	–	–	–	–	3
Total	16	5	15	10	10	12	1	–	69

PA: Palaearctic, NA: Nearctic, NT: Neotropical, AT: Afrotropical, OL: Oriental, AU: Australasian, PAC: Pacific Oceanic Islands, ANT: Antarctic

Caridina and the low number of *Macrobrachium* present. Although the Palaearctic region harbours 47 taxa, there is a marked discrepancy in species composition between the western and eastern Palaearctic, with no overlapping species distributions.

Interestingly, the number of troglobitic taxa is roughly similar across several of the main biogeographic regions (Table 2), although within each region they are not uniformly distributed. Within the Afrotropical region for instance, all except two species (*Caridina lovoensis* Roth-Woltereck from Zaire and *Caridina lanzana* Holthuis from Somalia) occur in Madagascar (Holthuis, 1986).

On a family level, the Atyidae are present in five biogeographic regions, but with very few representatives in the western Palaearctic, Neotropical or Nearctic regions. The only representatives of this family in the western Palaearctic are the genera *Atyaephyra*, *Typhlatya*, *Troglocaris* and *Dugastella*, with in addition *Caridina nilotica* occurring in Egypt. The atyid fauna of the Nearctic is very impoverished, with only four species in two genera (*Palaemonias*, *Syncaris*), whilst the Neotropical fauna consists of 19 species, primarily of the genus *Atya*.

The family Palaemonidae also occurs in all six biogeographic regions, but with distinct generic level differences (Table 1). *Macrobrachium* exhibits its highest diversity in the Oriental region (123 species), with far fewer species occurring in the Neotropical (53 species) and the Australasian regions (28 species). Only three species occur in the Nearctic region, whilst the eastern Palaearctic only harbours four species and the Pacific region 10.

The Desmocarididae are restricted to West Africa (Powell, 1977), in contrast the Euryrhynchinae occur in Brazil (*Euryrhynchus*) and west Africa (*Euryrhynchina*, *Euryrhynchoides*) (Powell, 1976). The Xiphocarididae are restricted to the northern part of the Neotropical region, chiefly being distributed in the Caribbean islands; whilst the Typhlocaridinae only occur in the western Palaearctic. Perhaps the most disjunct distribution on a family level is exhibited by the Kakaducarididae, with one troglobitic species, *Calathaemon holthuisi* (Strenth), from Texas, and two epigean species in Australia (*Kakaducaris glabra* Bruce, *Leptopalaemon gagudjui* Bruce & Short) (see Bruce, 1993).

The majority of genera are restricted to a single biogeographical region, with some notable exceptions, primarily those genera with marine or anchialine representatives. For instance, *Potamalpheops* occurs worldwide, but the only freshwater representatives are found in Mexico, West Africa and Singapore (Anker, 2005). Similarly, *Typhlatya* is distributed on both sides of the Atlantic, although freshwater species are only found in Europe (Jaume & Bréhier, 2005). *Caridina*, *Macrobrachium*, *Neocaridina* and *Paratya* have representatives in adjacent parts of the Afrotropical/Oriental and/or Australasian/Pacific regions. The genus *Atya* has representatives on either side of the Atlantic (Hobbs & Hart, 1982), whilst the troglobitic genus *Parisia* occurs in Madagascar, Australia and the Philippines.

On a species level, few taxa are present in more than one biogeographical region. A suite of nine species is distributed from East Africa (Afrotropical) across the Oriental region through to the Australasian/Pacific regions. However, the taxonomic status of many of these populations may be in doubt. In addition, this includes species such as *Palaemon debilis* Dana and *P. concinnus*, both also recorded from marine and estuarine habitats (Chace & Bruce, 1993), and it remains unclear if these taxa are present in freshwater throughout their range. In addition, two species of *Atya* (*Atya gabonensis* Giebel, *Atya scabra*) are known from both sides of the Atlantic;

whilst there is some species overlap between the eastern Palaearctic and the north-eastern Oriental region.

On a more regional scale, several biodiversity hotspots emerge. Madagascar harbours 26 species of Atyidae, many of which are troglobitic; with just over three quarters of all species being endemic to the island. Of the 28 species of atyids reported from Sulawesi, half of them are endemic to the island group, with a further 14 undescribed species recently discovered from either cave habitats or in the ancient lakes of the Malii group. Although the Chinese fauna can still be considered imperfectly known, 136 taxa are known to occur, many of which are endemic to Hunan province (Liang, 2004). As the majority of newly described Atyidae are from China, many more species may await discovery. Although the total species richness of Lake Tanganyika and the Amazon region is relatively low, both areas are rich in endemic genera (*Limnocaridina, Caridella* and *Atyella* in Lake Tanganyika, *Pseudopalaemon* and *Euryrhynchus* in the Amazon).

Human related issues

Holthuis (1980) listed 16 species of Atyidae and 61 species of freshwater Palaemonidae as either of commercial interest or forming an important component of subsistence fishing, especially in South America, Africa, the Far East and many Pacific islands. In addition, he noted that many, unidentified and probably mixed populations of small species of *Caridina* are heavily fished in the Philippines, China, Sulawesi, Madagascar and parts of the Ganges Delta. In the Philippines, these are eaten fresh or salted and made into a fermented product, called alamang.

The Giant River Prawn, *Macrobrachium rosenbergii*, is not only commercially fished within its native range (India to northern Australia), but also

Table 3 Species afforded special protection by IUCN or state legislation

Species	Distribution	IUCN status	Other
Creaseria morleyi (Creaser)	Mexico	–	Threatened
Macrobrachium acherontium Holthuis	Mexico	–	Special protection afforded
Macrobrachium villalobosi Hobbs	Mexico	–	Special protection afforded
Neopalaemon nahuatlus Hobbs	Mexico	–	In danger of extinction
Palaemonias alabamae Smalley	USA	Endangered	Listed as endangered in the Endangered Species Act, 1975
Palaemonias ganteri Hay	USA	Endangered	Listed as endangered in the Endangered Species Act, 1975
Palaemonetes antrorum *Benedict*	USA	Endangered	–
Palaemonetes cummingi Chace	USA	Vulnerable	listed as threatened in the Endangered Species Act, 1975
Stygiocaris lancifera Holthuis	Australia	–	Likely to become extinct. Listed in the Wildlife Conservation (specially protected fauna) Notice 2003
Syncaris pasadenae (Kingsley)	USA	Extinct	–
Syncaris pacifica (Holmes)	USA	Endangered	listed as endangered in the US Endangered Species Act, 1975
Troglocaris anopthalmus (Kollar)	Former Yugoslavia and Italy	Vulnerable	–
Troglomexicanus perezfarfanteae (Villalobos)	Mexico	–	In danger of extinction

Three subspecies of *T. anopthalmus* are known, which are not distinguished in the IUCN list. The other category refers to local or state protection categories

forms the basis of intensive prawn farming since the 1960s, both in its native range and outside (e.g. Alaska, USA, Nicaragua). The annual production is estimated to be around 200,000 tonnes, worth about a billion dollars. Recently Chinese prawn farms have also begun culturing *Macrobrachium nipponense* (De Haan), whilst minor quantities of *Macrobrachium malcomsoni* (H. Milne Edwards) are commercially farmed in India.

Several species are now available in the freshwater aquarium trade, some of which are being specifically cultured for this purpose (Werner, 2003). At least 18 species of *Caridina*, *Neocaridina*, *Atya*, *Atyopis*, *Atyoida* and *Macrobrachium* are now regularly offered for sale, mainly Asiatic species.

A total of 13 species have been either listed in the IUCN Red Data Book (2004 edition) or are otherwise afforded local or state protection (Table 3). With the exception of *Syncaris pasadenae* (Kingsley), all these taxa are stygobionts, many only known from a single cave or cave system, with these systems being under threat of human encroachment and ground water pollution. *Syncaris pasadenae* is the only shrimp species presently considered as extinct. Formally it occurred in a few streams in the Los Angeles area, but the species has not been found since 1933, despite extensive searches (Martin & Wicksten, 2004).

References

Anker, A., 2005. Presence of the alpheid shrimp genus *Potamalpheops* Powell, 1979 (Crustacea: Decapoda: Caridea) in South Asia, with description of a new species from Sri Lanka. The Raffles Bulletin of Zoology, Supplement 12: 31–37.

Baker, A. M., D. A. Hurwood, M. Krogh & J. M. Hughes, 2004. Mitochondrial DNA signatures of restricted gene flow within divergent lineages of an atyid shrimp (*Paratya australiensis*). Heredity 93: 196–207.

Beurlen, K., 1950. Alguns restos de crustáceos decapods d'água doce fósseis no Brazil. Anais da Academia brasiliera de ciencias 22: 453–459.

Bruce, A. J., 1993. *Kakaducaris glabra* gen. nov., sp. nov., a new freshwater shrimp from the Kakadu National Park, Northern territory, Australia (Crustacea: Decapoda: Palaemonidae), with the designation of a new subfamily Kakaducaridinae. Hydrobiologia 268: 27–44.

Cai, Y. & M. M. Bahir, 2005. *Lancaris*, a new genus of freshwater shrimp from Sri Lanka (Crustacea: Decapoda: Atyidae). The Raffles Bulletin of Zoology Supplement 12: 39–46.

Chace, F. A. Jr. & A. J. Bruce, 1993. The caridean shrimps (Crustacea: Decapoda) of the Albatross Philippine expedition 1907–1910, part 6: superfamily Palaemonoidea. Smithsonian Contributions to Zoology 543: 1–152.

Fryer, G., 1977. Studies on the functional morphology and ecology of atyid prawns of Dominica. Philosophical Transactions of the Royal Society of London, B 277: 57–129.

Hobbs, H. H. J & C. W. Hart Jr., 1982. The shrimp genus *Atya* (Decapoda: Atyidae). Smithsonian Contributions to Zoology 364: 1–143.

Holthuis, L. B., 1980. FAO Species Catalogue. Vol. 1. Shrimps and Prawns of the World. An Annotated Catalogue of Species of Interest to Fisheries. FAO, Rome.

Holthuis, L. B., 1986. Decapoda. In Botosaneanu, L. (ed.), Stygofauna Mundi. A Faunistic, Distributional, and Ecological Synthesis of the World Fauna Inhabiting Subterranean Waters (including the marine interstitial). E.J. Brill, Leiden, 589–615.

Jaume, D. & F. Bréhier, 2005. A new species of *Typhlatya* (Crustacea: Decapoda: Atyidae) from anchialine caves on the French Mediterranean coast. Zoological Journal of the Linnean Society 144: 387–414.

Liang, X.-Q., 2004. Fauna Sinica. Invertebrata Vol. 36. Crustacea Decapoda Atyidae. Science Press, Beijing.

Marquet, G., 1991. Freshwater crustaceans from French Polynesia: taxonomy, distribution and biomass (Decapoda). Crustaceana 61: 125–140.

Martin, J. W. & M. K. Wicksten, 2004. Review and redescription of the freshwater atyid shrimp genus *Syncaris* Holmes, 1900, in California. Journal of Crustacean Biology 24: 447–462.

Murphy, N. P. & C. M. Austin, 2005. Phylogenetic relationships of the globally distributed freshwater prawn genus *Macrobrachium* (Crustacea: Decapoda: Palaemonidae): biogeography, taxonomy and the convergent evolution of abbreviated larval development. Zoologica Scripta 34: 187–197.

Pereira, G., 1997. A cladistic analysis of the freshwater shrimps of the family Palaemonidae (Crustacea, Decapoda, Caridea). Acta Biologica Venezuelica 17(Suppl.): 1–69.

Powell, C. B., 1976. Two new freshwater shrimps from West Africa: the first euryrhynchids (Decapoda: Palaemonidae) reported from the Old World. Revue de Zoologie Africaine 90: 884–902.

Powell, C. B., 1977. A revision of the African freshwater shrimp genus *Desmocaris* Sollaud, with ecological notes and description of a new species (Crustacea, Decapoda, Palaemonidae). Revue de Zoologie Africaine 91: 649–674.

Roth-Woltereck, E., 1958. *Limnocaridina iridinae* n. sp., eine interessante garnele aus dem Tanganjika-See (Decapoda Atyidae). Zoologischer Anzeigner 161: 188–192.

Short, J. W., 2004. A revision of the Australian river prawns, *Macrobrachium* (Crustacea: Decapoda: Palaemonidae). Hydrobiologia 525: 1–100.

Werner, U., 2003. Shrimps, Crayfishes, and Crabs in the Freshwater Aquarium. Aqualog, Rodgau.

Hydrobiologia (2008) 595:295–301
DOI 10.1007/s10750-007-9120-3

Global diversity of crayfish (Astacidae, Cambaridae, and Parastacidae—Decapoda) in freshwater

Keith A. Crandall · Jennifer E. Buhay

Abstract The freshwater crayfishes are distributed across all but the Indian and Antarctic continents with centers of diversity in the southeastern Appalachian Mountains in the Northern Hemisphere and in south–east Australia in the Southern Hemisphere. There are currently over 640 described species of freshwater crayfishes with an average of 5–10 species still being described each year. Freshwater crayfishes can serve as keystone species in aquatic habitats, but a few species are also significantly invasive and can cause impressive damage to the fragile freshwater habitat. Crayfishes inhabit caves, burrows, streams, lakes and strong burrowers can even be found in terrestrial habitats where they have burrowed to the water table or where rainfall is sufficiently abundant to provide the needed moisture. The freshwater crayfishes, like the habitats in which they are encountered, are generally endangered to some degree and conservation efforts would do well to focus on them as key elements of the freshwater ecosystem.

Guest editors: E. V. Balian, C. Lévêque, H. Segers & K. Martens
Freshwater Animal Diversity Assessment

K. A. Crandall (✉) · J. E. Buhay
Department of Biology, Brigham Young University, Provo, UT 84602-5181, USA
e-mail: keith_crandall@byu.edu

K. A. Crandall
Monte L. Bean Life Science Museum, Brigham Young University, Provo, UT 84602-5181, USA

Keywords Crayfish · Biodiversity · Phylogeny · Conservation

Introduction

Freshwater crayfishes are a highly diverse group of decapod crustaceans (over 640 species) with two centers of diversity, one in the Southern Appalachian Mountains of the southeastern United States (Northern Hemisphere center) and one in south–east Australia (Southern Hemisphere center). Crayfishes occupy four main habitat types; primary burrowers (those crayfish who spend their entire life cycle in borrows – indeed, primary burrowers can burrow down to the water table and are not restricted to freshwater but are essentially terrestrial), stream-dwellers, pond/lake/large river dwellers (including secondary burrowers who do require connectivity of burrows with freshwater), and stygobitic species (obligate cave-dwellers). Each habitat type has distinctive morphological adaptations for those ecosystems (Fig. 1). For example, primary burrowers typically have a vaulted carapace to accommodate larger gill surface area and robust pinchers for digging and burrow protection. Stream specialists typically have large abdomens for swimming and are highly intolerant of low oxygen content in the water. Cave species have the typical suite of cave-adapted morphologies including loss of pigmentation, loss of eyes, elongated antennae, and elongated limbs. Most

Fig. 1 *Orconectes incomptus* (Hobbs & Barr 1972), a cave adapted freshwater crayfish from the Cumberland Plateau escarpment in northern Tennessee, USA (photo by Jennifer E. Buhay©)

crayfish typically has a life span of about 2 years (although some surface and cave species can live beyond 20 years). They reproduce sexually, although hermaphroditic specimens and even parthenogenetic specimens are known to occur (Scholtz et al. 2003). Freshwater crayfishes are omnivorous and typically nocturnal. They are voracious eaters and can be extremely destructive when introduced to non-native habitats. A few freshwater crayfish are particularly invasive (e.g., *Orconectes rusticus*) and/or were distributed in the aquaculture trade (e.g., *Procambarus clarkii*). These introduced species wreak havoc on natural ecosystems to which they are not native, but the vast majority of freshwater crayfishes are extremely narrowly distributed and hence are endangered due mainly to the destruction of the freshwater ecosystems in general. In this article, we review the freshwater crayfish diversity as it is currently understood and emphasize both the phylogenetic diversity and conservation status of this interesting group of freshwater crustaceans.

Species diversity

There are over 640 described species of freshwater crayfish worldwide (Fig. 2; Table 1). Taxonomically, they are organized into two superfamilies, the Northern Hemisphere Astacoidea and the Southern Hemisphere Parastacoidea. The Astacoidea contain two families, the Cambaridae (with by far the most species diversity, containing over 420 species contained within 12 genera) and the Astacidae with six genera and 39 species (Hobbs 1989). The Parastacoidea are composed of a single family, Parastacidae, consisting of 15 genera and over 170 species (Table 1).

Phylogeny and historical processes

The freshwater crayfish are a monophyletic group and a sister taxon to the clawed lobsters from the superfamily Nephropoidea (Crandall et al. 2000) and together the Nephropoidea, Astacoidea, and Parastacoidea make up the Astacidea (Fig. 3). While much work has been done in terms of testing the monophyly and sister relationships among the crayfish and their allies, relatively little work has been published on the phylogenetic relationships among the genera of freshwater crayfish. To a large extent, this is reflective of the fact that initial studies have shown that many of the genera (especially in the family Cambaridae) do not form monophyletic groups, thereby requiring relatively complete taxonomic sampling to perform reasonable phylogenetic analyses (Sinclair et al. 2004). This is a difficult task given the large number of species of freshwater crayfish.

Given the geographic distribution of this group (Fig. 2) and the strong support for a monophyletic origin, the crayfishes must have originated in Pangaea by the Triassic period (185–225 mya). The separation of the two crayfish superfamilies represents the splitting of Pangaea into northern (Laurasia) and southern (Gondwana) land-masses around 185 mya. This separation is clearly seen in the estimates of crayfish phylogenetic relationships (Fig. 3). The antiquity of the crayfishes is supported by recent fossil evidence from Colorado and Utah with fossil crayfish and burrows associated with Permian and Early Triassic (265 mya) deposits (Hasiotis and Mitchell 1993). Furthermore, the phylogenic connection of the Southern Hemisphere crayfishes represented in southern South America, Madagascar, New Zealand, and Australia corresponds to the distribution patterns of the predatory dinosaur group Abelisauridae (Sampson et al. 1998). Thus the crayfishes offer further support for the hypothesis suggesting extended contact between these land-masses and the antiquity of the freshwater crayfish lineage (Crandall et al. 2000; Hobbs 1988; Scholtz and Richter 1995).

Table 1 Total number of described freshwater crayfish species by taxon and distribution in the geographic regions of focus. PA: Palaearctic; NA: Nearctic; NT: Neotropical; AT: Afrotropical; OL: Oriental; AU: Australasian; PAC: Pacific Oceanic Islands; ANT: Antarctic

Family	Genus	PA	NA	AT	NT	OL	AU	World
Astacidae		*31*	*8*	*0*	*0*		*0*	*39*
	Astacus	5	0	0	0		0	5
	Atlantoastacus	8	0	0	0		0	8
39 species	Austropotamobius	7	0	0	0		0	7
	Caspiastacus	2	0	0	0		0	2
	Pacifastacus	0	8	0	0		0	8
	Pontastacus	9	0	0	0		0	9
Cambaridae		*7*	*374*	*0*	*48*		*0*	*423*
	Barbicambarus	0	1	0	0		0	1
	Bouchardina	0	1	0	0		0	1
445 species	Cambarellus	0	8	0	9		0	17
	Cambaroides	7	0	0	0		0	7
	Cambarus	0	97	0	0		0	95
	Distocambarus	0	5	0	0		0	5
	Fallicambarus	0	18	0	0		0	18
	Faxonella	0	4	0	0		0	4
	Hobbseus	0	7	0	0		0	7
	Orconectes	0	91	0	0		0	89
	Procambarus	0	140	0	39		0	177
	Troglocambarus	0	2	0	0		0	2
Parastacidae		*0*	*0*	*9*	*16*		*151*	*176*
	Astacoides	0	0	9	0		0	9
	Astacopsis	0	0	0	0		3	3
175 species	Cherax	0	0	0	0		45	45
	Engaeus	0	0	0	0		39	39
	Engaewa	0	0	0	0		5	5
	Euastacus	0	0	0	0		43	43
	Geocharax	0	0	0	0		2	2
	Gramastacus	0	0	0	0		1	1
	Ombrastacoides	0	0	0	0		11	11
	Paranephrops	0	0	0	0		2	2
	Parastacus	0	0	0	8		0	8
	Samastacus	0	0	0	1		0	1
	Spinastacoides	0	0	0	3		0	3
	Tenuibranchiurus	0	0	0	1		0	1
	Virilastacus	0	0	0	3		0	3
Total		*38*	*382*	*9*	*64*		*151*	*638*

Present distribution and areas of endemicity

The most species rich family of freshwater crayfish, the Cambaridae, are distributed in North America east of the Rocky Mountains, north into southern Canada and south through Mexico and, surprisingly, in Asia (Table 1). The Asian endemic genus *Cambaroides* is a bit of an enigma both biogeographically

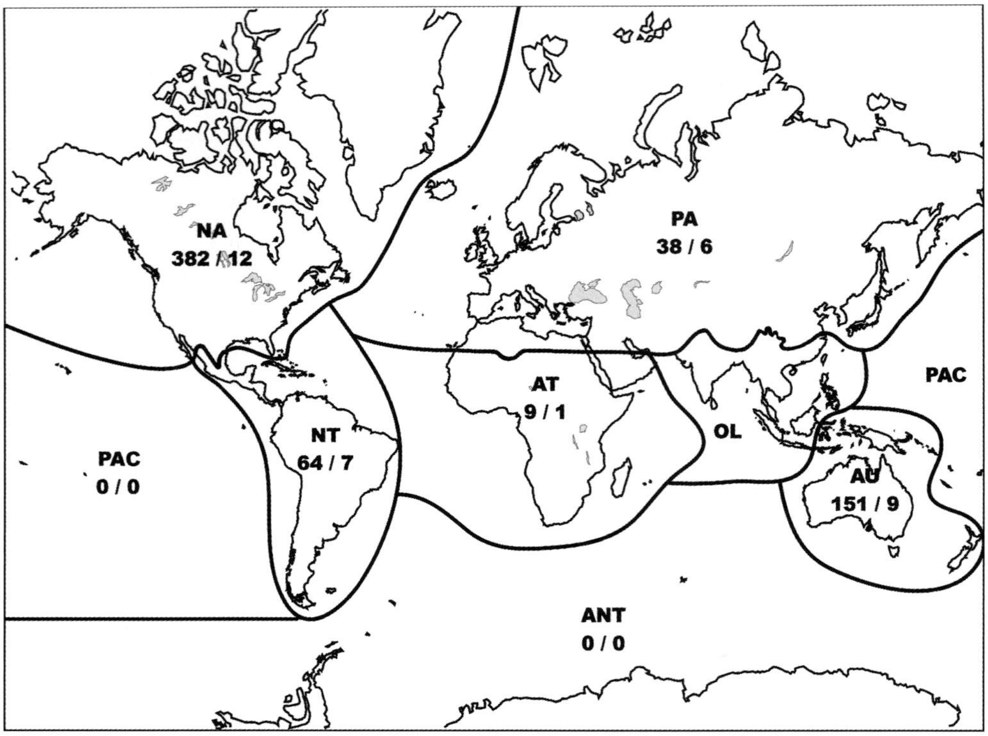

Fig. 2 Geographic distribution of the global freshwater crayfish diversity (species number/genus number). PA—Palaearctic; NA—Nearctic; NT—Neotropical; AT—Afrotropical; OL—Oriental; AU—Australasian; PAC—Pacific Oceanic Islands; ANT—Antarctic

as well as phylogenetically. Some phylogenetic analyses place it weakly with the European species (which makes more biogeographic sense), but a good sampling of *Cambaroides* to place into a robust phylogenetic analysis has yet to occur. The Astacidae are distributed west of the Rocky Mountains (mainly in the Pacific Northwest) and in Europe (Table 1). In the superfamily Parastacoidea, 10 of the 15 genera are found in Australia. The remaining genera are distributed in southern South America (three endemic genera with 12 species distributed in southern Chile, Uruguay, and southern Brazil), New Zealand (with one endemic genus and two described species), and Madagascar (with one endemic genus and nine described species).

The greatest species diversity in the freshwater crayfishes occurs in the southern Appalachian Mountains of the southeastern United States (Table 1). This region is also home to a number of other highly endemic and highly endangered stream species including freshwater fishes, salamanders, snails, and mussels. Based on phylogenetic results and biogeographic analyses, much of the species diversity in the Cambaridae is of relatively recent origin (compared to the Parastacidae) and seems to have been driven by isolating effects of pre- and post-Pleistocene river drainage changes (Crandall and Templeton 1999). Population genetic studies of species groups are beginning to unravel the evolutionary mechanisms associated with driving this amazing diversity (Buhay and Crandall 2005; Fetzner and Crandall 2003).

The Southern Hemisphere crayfishes have a center of diversity in south–east Australia. These taxa (at least the genus-level divisions) appear to be much older in origin relative to the Northern Hemisphere taxa. These genera appear to form well-defined monophyletic groups (e.g., Schull et al. 2005) as a result. The species richness of many of these genera appears to be the result of vicariance and dispersal with isolation events (e.g., Ponniah and Hughes 2004; Schull et al. 2005).

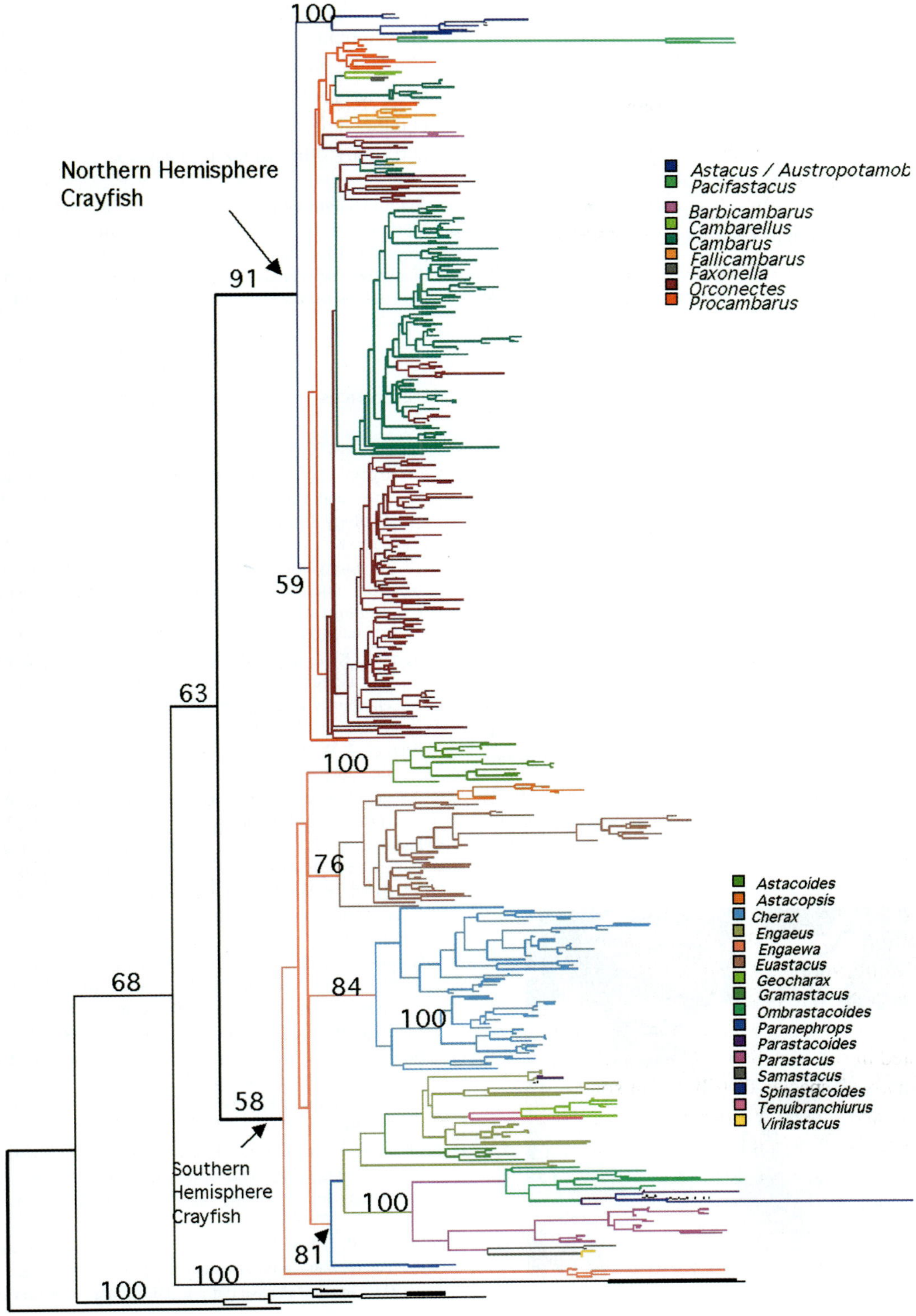

Fig. 3 Phylogenetic relationships among the global freshwater crayfish showing the strongly supported division between the two superfamilies of the Northern Hemisphere (Astacoidea) and the Southern Hemisphere (Parastacoidea). (Sinclair et al. 2004)

Human related issues

Freshwater crayfish are a much sought after food item in many cultures (Fig. 4) and a significant aquacultural commodity. Unfortunately for many of the charismatic species (e.g., the Tasmanian Giant Lobster, *Astacopsis gouldi*), over-harvesting coupled with degradation of habitat has resulted in the endangerment of many crayfish species (e.g., Horwitz 1994). Other areas, such as Madagascar, support a sustainable harvest of crayfish, yet these populations are typically imperiled by stream degradation (Jones et al. 2005). While all 640+ species of freshwater crayfish have yet to be categorized for conservation status relative to the IUCN Red List Criteria (IUCN 2001), a large number of species already occur on that list. Taylor et al. (1996) found that over 50% of the United States species of freshwater crayfish are endangered or threatened to some degree. Yet, surprising, very few of these species are under national protection with a few more under local (state, provincial) protection. For example, in the United States, 20 species are known from less than five localities (with 15 known only from a single locality) and well over 210 species are considered endangered or threatened, but only four species are listed by the federal government on the Endangered Species List. Native crayfish play an essential role in their native habitat and may act as keystone species for the freshwater ecosystems where they occur. Yet this biodiversity is at great risk of loss if national governments (especially in the US and Australia where the bulk of the species diversity occurs) do not take appropriate measures to protect the highly endangered freshwater ecosystems that house these and many other amazing species.

Fig. 4 Crayfish are staple in many cultures both as a food source and source of income; this example is from a roadside crayfish stand in Madagascar with both *Astacoides madagascarensis* and *Astacoides betsiloensis* for sale (Photo by Julia Jones© 1999 used with permission)

Acknowledgments We thank Estelle Balian and Koen Martens for their invitation to participate in this exciting project and for their helpful comments on our article. We also thank the anonymous reviewers for their helpful comments to improve the article. Our work was supported by NSF grant EF-0531762.

References

Buhay, J. E. & K. A. Crandall, 2005. Subterranean phylogeography of freshwater crayfishes shows extensive gene flow and surprisingly large population sizes. Molecular Ecology 14: 4259–4273.

Crandall, K. A., D. J. Harris & J. W. Fetzner, 2000. The monophyletic origin of freshwater crayfishes estimated from nuclear and mitochondrial DNA sequences. Proceedings of the Royal Society of London, Series B 267: 1679–1686.

Crandall, K. A. & A. R. Templeton, 1999. The zoogeography and centers of origin of the crayfish subgenus *Procericambarus* (Decapoda: Cambaridae). Evolution 53: 123–134.

Fetzner, J. W. Jr. & K. A. Crandall, 2003. Linear habitats and the nested clade analysis: An empirical evaluation of geographic vs. river distances using an Ozark crayfish (Decapoda: Cambaridae). Evolution 57: 2101–2118.

Hasiotis, S. T. & C. E. Mitchell, 1993. A comparison of crayfish burrow morphologies: Triassic and Holocene fossil, paleo- and neo-ichnological evidence, and the identification of their burrowing signatures. Ichnos 2: 291–314.

Hobbs, H. H. Jr., 1988. Crayfish distribution, adaptive radiation and evolution. In Holdich, D. M. & R. S. Lowery (eds), Freshwater Crayfish: Biology, Management and Exploitation. Timber Press, Portland, OR: 52–82.

Hobbs, H. H. Jr., 1989. An illustrated checklist of the American Crayfishes (Decapoda: Astacidae, Cambaridae, and Parastacidae). Smithsonian Contributions to Zoology 480: 1–236.

Hobbs, H. H. Jr. & T. C. Barr, Jr., 1972. Origins and affinities of the troglobitic crayfishes of North America (Decapoda: Astacidae). II. Genus Orconectes. Smithsonian Contributions to Zoology 105: 1–84.

Horwitz, P., 1994. Distribution and conservation status of the Tasmanian giant freshwater lobster *Astacopsis gouldi* (Decapoda: Parastacidae). Biological Conservation 69: 199–206.

IUCN. 2001. IUCN Red List Categories: Version 3.1. IUCN Species Survival Commission, Gland, Switzerland.

Jones, J. P. G., F. B. Andriahajaina, N. J. Hockley, A. Balmford & O. R. Ravoahangimalala, 2005. A multidisciplinary approach to assessing the sustainability of freshwater crayfish harvesting in Madagascar. Conservation Biology 19: 1863–1871.

Ponniah, M. & J. Hughes, 2004. The evolution of Queensland spiny mountain crayfish of the genus *Euastacus*: I. Testing vicariance and dispersal with interspecific mtDNA. Evolution 58: 1073–1085.

Sampson, S. D., L. M. Witmer, C. A. Forster, D. W. Krause, P. M. O'Connor, P. Dodson & F. Ravoavy, 1998. Predatory dinosaur remains from Madagascar: Implications for the Cretaceous Biogeography of Gondwana. Science 280: 1048–1051.

Scholtz, G., A. Braband, L. Tolley, A. Reimann, B. Mittmann, C. Lukhaup, F. Steuerwald & G. Vogt, 2003. Ecology: Parthenogenesis in an outsider crayfish. Nature 421: 806.

Scholtz, G. & S. Richter, 1995. Phylogenetic systematics of the reptantian Decapoda (Crustacea, Malacostraca). Zoological Journal of the Linnean Society 113: 289–328.

Schull, H. C., M. Perez-Losada, D. Blair, K. Sewell, E. A. Sinclair, S. Lawler, M. Ponniah & K. A. Crandall, 2005. Phylogeny and biogeography of the freshwater crayfish *Euastacus* (Decapoda: Parastacidae) based on nuclear and mitochondrial DNA. Molecular Phylogenetics and Evolution:accepted.

Sinclair, E. A., J. W. Fetzner Jr., J. Buhay & K. A. Crandall, 2004. Proposal to complete a phylogenetic taxonomy and systematic revision for freshwater crayfish (Astacida). Freshwater Crayfish 14: 1–9.

Taylor, C. A., M. L. Warren Jr., J. F. Fitzpatrick Jr., H. H. Hobbs III, R. F. Jezerinac, W. L. Pflieger & H. W. Robison, 1996. Conservation status of crayfishes of the United States and Canada. Fisheries 21: 25–38.

Hydrobiologia (2008) 595:303–315
DOI 10.1007/s10750-007-9025-1

Global diversity of water mites (Acari, Hydrachnidia; Arachnida) in freshwater

Antonio Di Sabatino · Harry Smit · Reinhard Gerecke · Tom Goldschmidt · Noriko Matsumoto · Bruno Cicolani

Abstract The Hydrachnidia (water mites) represent the most important group of the Arachnida in fresh water. Over 6,000 species have been described worldwide, representing 57 families, 81 subfamilies and more than 400 genera. The article analyzes extant water mite diversity and biogeography. Data on distribution and species richness of water mites are substantial but still far from complete. Many parts of the world are poorly investigated, Oriental and Afrotropical regions in particular. Moreover, information among different freshwater habitats is unbalanced with springs and interstitial waters disproportionately unrepresented. Therefore, more than 10,000 species could be reasonably expected to occur in inland waters worldwide. Based on available information, the Palaearctic region represents one of the better investigated areas with the highest number of species recorded (1,642 species). More than 1,000 species have been recorded in each of the Neotropical (1,305 species) and Nearctic regions (1,025 species). Known species richness is lower in Afrotropical (787 species) and Australasian (694 species) regions, and lowest in the Oriental region (554 species). The total number of genera is not correlated with species richness and is distinctly higher in the Neotropical (164 genera); genus richness is similar in the Palaearctic, Nearctic and Australasian regions (128–131 genera) and is lower in the Afrotropical and Oriental regions with 110 and 94 genera, respectively. A mean number of about three genera per family occur in the Palaeartic, Nearctic and Oriental while an average of more than four genera characterizes the families of Australasian and Afrotropical regions and more than five genera those of the Neotropical. Australasian fauna is also characterized by the highest percentage of endemic genera (62%), followed by Neotropical (50.6%) and Afrotropical (47.2%) regions. Lower values are recorded for the Palaearctic (26.9%), Oriental (24.4%) and Nearctic

Guest editors: E. V. Balian, C. Lévêque, H. Segers & K. Martens
Freshwater Animal Diversity Assessment

A. Di Sabatino (✉) · B. Cicolani
Dipartimento di Scienze Ambientali,
University of L'Aquila, Via Vetoio,
L'Aquila 67100, Italy
e-mail: adisab@univaq.it

H. Smit
Zoological Museum, University of Amsterdam,
Plantage Middenlaan 64, 1018 DH Amsterdam,
The Netherlands

R. Gerecke
Biesingerstr. 11, D-7024 Tübingen, Germany

T. Goldschmidt
Zoologisches Institut, University of Karlsruhe,
Kornblumenstr. 13, D-76128 Karlsruhe, Germany

N. Matsumoto
Division of Biological Sciences,
Graduate School of Science, Hokkaido University,
Sapporo, Japan

(21.4%). The Palaearctic and Nearctic have the highest faunistic similarity, some minor affinities are also evident for the generic diversification of Holarctic and Oriental families. The faunas of Southern Hemisphere bioregions are more distinct and characterized by the presence of ancient Gondwanan clades with a regional diversification particularly evident in the Neotropics and Australasia. This scenario of water mite diversity and distribution reflect the basic vicariance pattern, isolation, phylogenetic diversification, recent climatic vicissitudes and episodes of dispersal between adjacent land masses together with extant ecological factors can be evoked to explain distribution patterns at a global scale.

Keywords Biodiversity · Biogeography · Freshwaters · Global assessment · Water mites · Hydrachnidia

Introduction

The Hydrachnidia (water mites), also called Hydrachnellae, Hydracarina, Hydrachnida, represent the most important group of the Arachnida in freshwater. Originating from terrestrial ancestors, they have colonized all kinds of freshwater habitats. Water mites are highly diversified both in lotic and lentic habitats, as well as in springs and interstitial waters (Di Sabatino et al., 2000, 2003; Smith et al., 2001). Hydrachnidia belong to the cohort Parasitengona (Actinedida), a group whose species are characterized by a complex life cycle involving a heteromorphic parasitic larva, two pupa-like resting stages (proto- and tritonymph) and free-living predacious deutonymphs and adults. The resting stages provide an adaptation for avoiding unfavourable conditions in unstable environments, and larval parasitism on flying insects confers substantial advantages ensuring dispersal and rapid exploitation of new habitats (Smith et al., 2001).

Water mites are characterized by bright colours and a highly diversified morphology. Plesiotypically, body shape is globular but it may also be flattened dorso-ventrally or laterally, or elongated into a worm-like form (Fig. 1). Length ranges from 0.2 mm up to 10 mm, although most species are between 0.5 and 1.5 mm. As in all Acari, the body of a water mite is divided into two principal units, gnathosoma and idiosoma. The gnathosoma is a complex of trophic and sensory structures composed of a sclerotized capsule (capitulum) and two pairs of appendages (palps and chelicerae). The idiosoma, or body proper, may be soft-skinned or the integumental muscle-attachment sites are transformed to more or less extended sclerotized plates with tendencies to develop complete sclerotization. The idiosoma is also characterized by the presence of series of defensive glands and mechanoreceptive slit organs. The ventral side includes four pairs of sclerotized coxal plates (insertion points for legs and leg muscles), the genital field and the opening of the excretory system. The four pairs of legs of adults are six-segmented and usually bear one pair of terminal claws. Size and chaetotaxy of leg segments are modified in relation to modes of locomotion and reproduction. For more detailed information see Di Sabatino et al. (2000, 2002), Smith et al. (2001) and references therein.

Water mites are grouped into eight superfamilies with more than 50 families. Over 420 genera and about 6,000 species are described (Viets, 1987; Smit, in http//www.watermite.org). However, this number likely greatly underestimates global richness of water mites. For example, about 5,500 species are calculated to occur in the Neotropics alone (Goldschmidt, 2002).

Species diversity

The most recent catalogue of water mites (Viets, 1987) included more than 5,000 described species in 50 families and 310 genera. Since then, our knowledge on taxonomy, diversity and distribution has significantly improved. More than 1,000 species have been described and 65 new genera, nine subfamilies (three of which were reinstated) and ten new families (six of which were reinstated) were established (Table 1).

Most of these new species have come from the southern hemisphere. More than 230 species have been added to the list of the Neotropical freshwater fauna with about 140 additional species recorded for South America and 100 for Central America. Australasian is the second region in terms of new species discovered after 1986, with 115 species from Australia, 35 from New Zealand and 34 from western Pacific islands. A considerable number of species were also reported from the still poorly known freshwater fauna of the Eastern Palaearctic with more

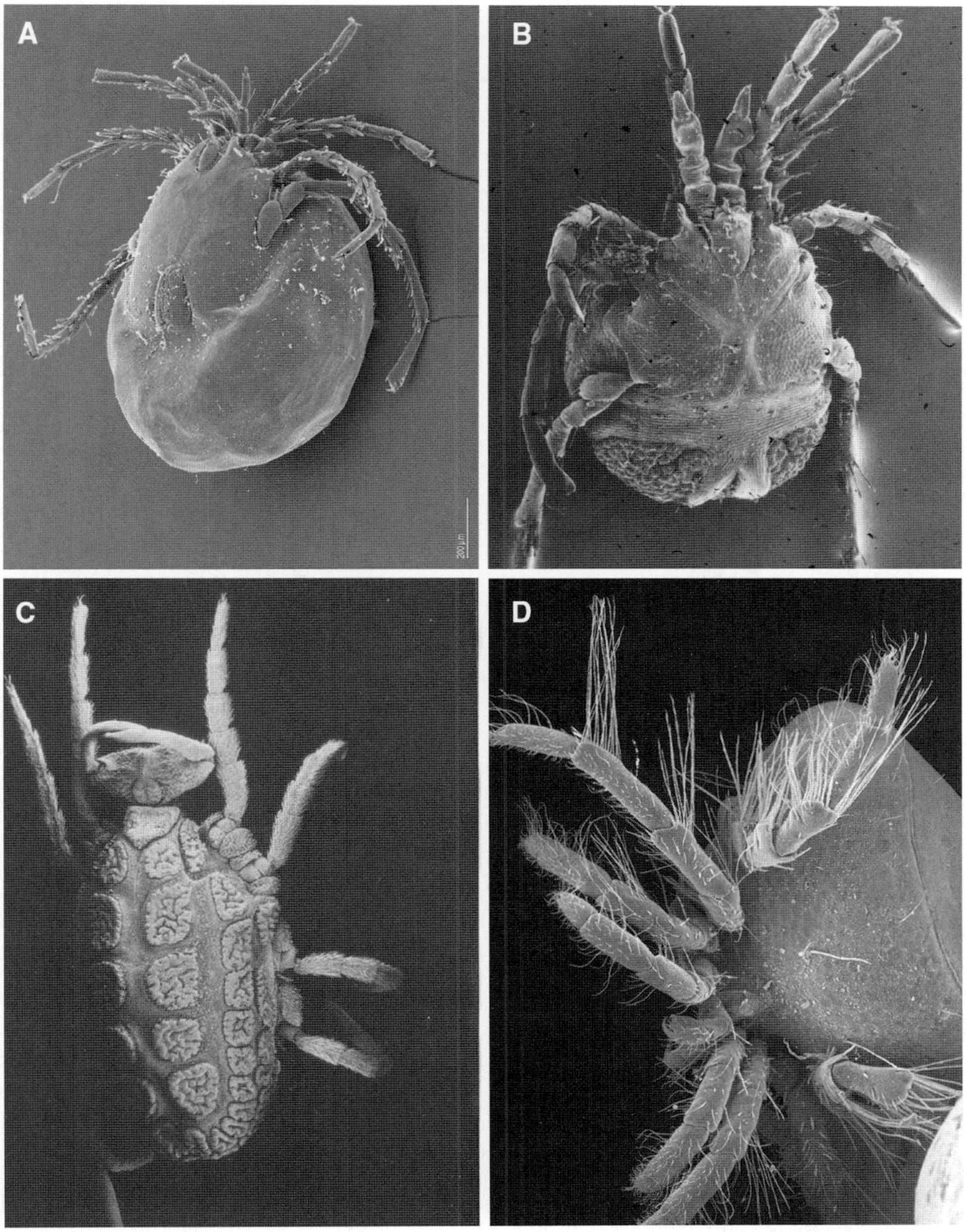

Fig. 1 Scanning electron microphotographs of some water mite genera. (**A**)—*Lebertia* (Lebertiidae), ventral view. A widespread genus presents in all bioregions except Australasian. (**B**)—*Feltria* (Feltriidae), ventral view. A genus particularly diversified in the Northern Hemisphere, not reported from Afrotropical and Australasian. (**C**)—*Apheviderulix*, dorsal view. The only genus of the enigmatic family Apheviderulicidae with distribution limited to South Western Palaearctic and South Western North America. (**D**)—*Harpagopalpus* (Harpagopalpidae), frontal view. The few species of this genus are only known from the Afrotropical and Oriental regions. (Photos (**A**) and (**D**) are from Gerecke; (**B**) from collection Carl Bader, Basel; (**C**) from Gerecke et al. (1999)

than 150 species described from China and Russia. More than 100 species were also added to the faunal lists of North America and Europe. At higher taxonomic levels, new families were described from the western Palaearctic, Nearctic, Oriental and Australasian, with Australia and USA being the countries with the highest number of new genera.

These considerations suggest that the actual number of water mite species in the world is large but cannot be readily estimated. A species-accumulation

Table 1 New water mite taxa described after 1986 (data from Smit in http//www.watermite.org)

	W Palaearctic	E Palaearctic	Nearctic	Neotropical	Afrotropical	Oriental	Australasian
Families	1[a]	–	2[a]	–	–	1	1
Subfamilies	–	–	3[b]	2	1	–	1[b]
Genera	4	2	17	9	6	6	23
Species	118	181	110	239	49	133	184

[a] The family Aphevideruliciidae is distributed throughout N-America and W-Palaearctic

[b] The subfamily Cyclomomoniinae is represented both in Nearctic and Australia

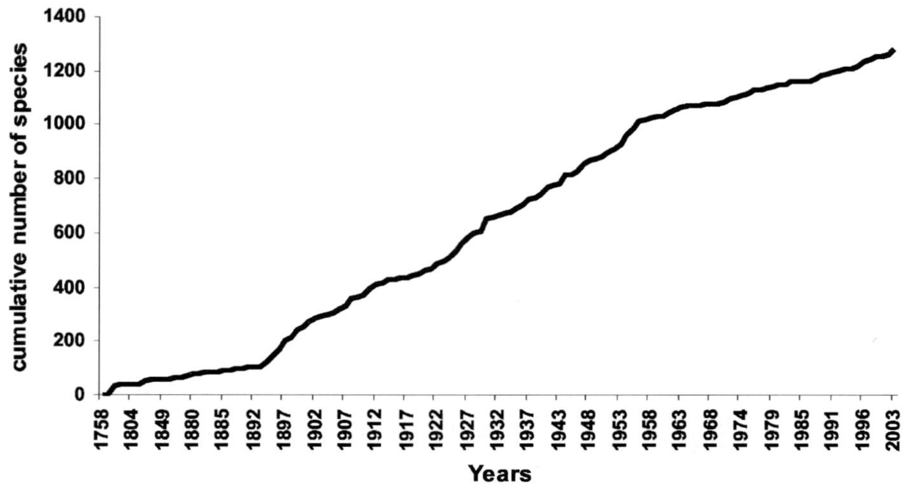

Fig. 2 Cumulative numbers of water mite species described from the Western Palaearctic region (data from Gerecke in http// www. watermite.org)

curve calculated for the western Palaearctic, one of the better investigated areas of the world, indicates that the number of new species discovered is still increasing and there is no indication that it is reaching a plateau (Fig. 2).

Data on present distribution and species richness of water mites are substantial but still far from complete (Table 2). Many parts of the world, Afrotropical and Oriental regions in particular, are poorly investigated. Moreover, information on distribution and species diversity among different freshwater habitats is unbalanced, with particularly scanty documentation of taxa that dwell in springs or hyporheic-interstitial habitats. Therefore, more than 10,000 species could be reasonably expected to occur in inland waters world-wide.

Table 2 Taxonomic diversity of water mites at global and biogeographic region scales

	PA	NA	NT	AT	OL	AU	World
Families	42	44	31	28	33	30	57
Subfamilies	40	69	55	44	51	50	82
Genera	131	131	164	110	94	128	428
Species	1,642	1,025	1,305	787	554	694	>6,000

PA: Palaearctic; NA: Nearctic; NT: Neotropical; AT: Afrotropical; OL: Oriental; AU: Australasian

Phylogeny and historical processes

Hydrachnidia probably originated during the Mid-Palaeozoic era (400 Mya) from terrestrial Parasitengona-like ancestors, but see also Smith & Cook (1999) for an alternative hypothesis. From plesiotypic unstable and intermittent habitats (temporary pools, seepage areas and small springs) they invaded all kind of freshwater habitats. The only two fossil records from Tertiary deposits refer to larval specimens of highly derivative modern clades (Poinar, 1985). No attempt has been made so far to produce a hypothesis about the age and phylogenetic relationships of the major clades with the support of molecular data. Differentiation of extant superfamilies started from early Mesozoic

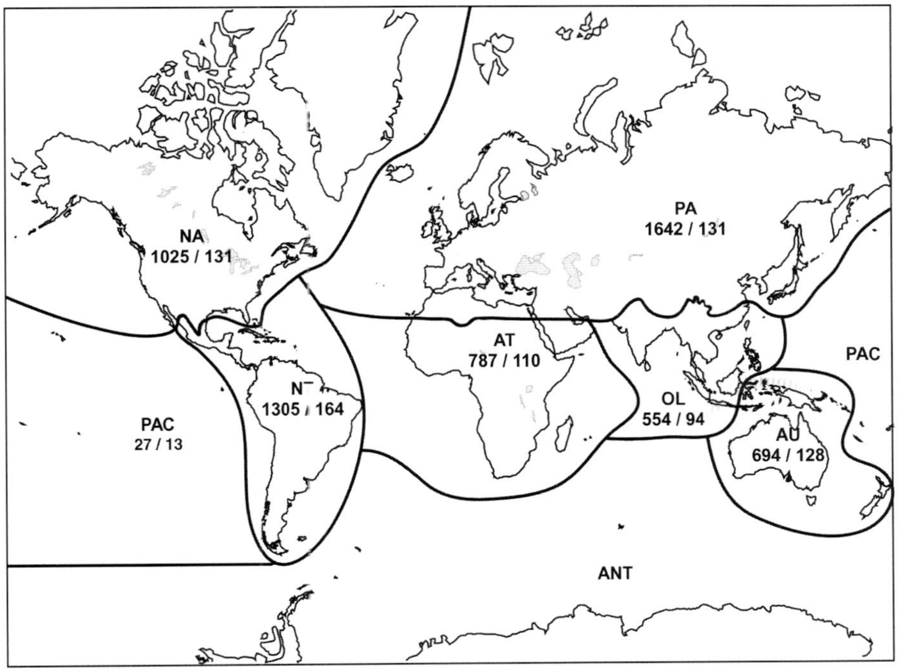

Fig. 3 Distribution of freshwater mite (Hydrachnidia) species and genera (SP/GN) per zoogeographic region: PA—Palaearctic; NA—Nearctic; NT—Neotropical; AT—Afrotropical; OL—Oriental; AU—Australasian; PAC—Pacific Oceanic Islands; ANT—Antarctic

(240 Mya), Stygothrombioidea, Hydrovolzioidea, Hydrachnoidea, Eylaoidea and Hydryphantoidea are by most authors considered "lower water mites". Members of these superfamilies share plesiotypic features such as aerial larvae bearing six-segmented legs and parasitism of a wide variety of insect hosts. However, also within these groups, numerous adaptations to particular environmental conditions are realized. Probably from hydryphantoid-like ancestors, the "higher water mites" (Lebertioidea, Hygrobatoidea and Arrenuroidea) evolved and radiated, perhaps in part as a response to the diversification of nematoceran insect hosts. The systematics of higher groups is still in a state of flux. For alternative systems of superfamilial and familial classifications see Tuzovskij (1987), Witte (1991), Harvey (1998) and Panesar (2004).

Present distribution and main areas of endemicity

The currently available information does allow us to present a preliminary scenario on water mite diversity at a global scale (Fig. 3).

Palaearctic

Due to a long tradition of European water mite research and more recent intensive field-work in southern and central Europe, the Western Palaearctic is one of the better investigated areas of the world (Gerecke & Lehmann, 2005—see also Gerecke in http//www.watermite.org). A total of about 1,600 species in 42 families, 40 subfamilies and 131 genera have been documented. More than 1,100 species (114 genera and 40 families) are recorded from the Western Palaearctic, most of them (about 800 species) from countries surrounding the Mediterranean Sea (Di Sabatino & Gerecke, unpublished database). From central Europe a recent list reports 620 species (Gerecke & Lehmann, 2005). Data from the Eastern Palaearctic are more scanty and mostly refer to some Russian territories (Sokolow, 1940; Tuzovskij, 1997) and China (Jin, 1997). Ongoing research and unpublished records suggest that more than 3,000 species are potentially present in this zoogeographic region.

Nearctic

About 1,700 species are estimated to occur in North America, north of Mexico (Smith et al., 2001). So far, 1,025 species have been reported, in 44 families, 69 subfamilies and 131 genera. Over 800 species (100 genera, 35 families) occur in Canada (Smith et al., 1998). Distributional data and check-lists are also available for some Canadian ecoregions (Smith et al., 1998) and North America (Mitchell, 1954; Habeeb, 1967).

Afrotropical

Older data (Viets, 1970; Van Rensburg, 1974) and more recent papers on the distribution of Afrotropical water mites, record about 800 species belonging to 110 genera, 44 subfamilies and 28 families from this region (Gerecke, unpublished). Liberia (180 species), South Africa (160 species) and Cameroon (150 species) are the countries best investigated (Cook, 1966; Viets, 1970; Harrison, 2000). From Madagascar, a country known as particularly rich in species and endemic taxa, only 65 water mite species have been documented so far, with the fauna of springs, running waters and subterranean habitats mostly unknown (Gerecke, 2004). The real diversity is likely to be one order of magnitude higher (Goldschmidt & Gerecke, 2003).

Neotropical

About 1,300 species in over 160 genera, 55 subfamilies and 31 families have been recorded from Central and South America, with large differences in the state of knowledge of different regions and a high percentage of new species discovered also in the better investigated areas (Goldschmidt, 2002; Rosso de Ferradas & Fernandez, 2005). The number of known species ranges from more than 260 from Southern Brazil (Lundblad, 1941, 1942, 1943a, 1943b, 1944), 220 from Southern Mexico (Cook, 1980) and about 180 from Central Chile (Cook, 1988). Conversely, only 5 species are reported from Bolivia and no data exist from Jamaica, Nicaragua or French Guiana.

Oriental

In the Oriental region more than 500 species in 94 genera, 51 subfamilies and 33 families have been found. Our knowledge of the region is far from complete, many areas have hardly been explored (e.g. many Indonesian islands, Southeast Asia, Himalayas, Philippines) but see also Wiles (2004) and references therein. A more detailed documentation exists only from India with 230 species recorded, in 26 families and 61 genera (Cook, 1967).

Australasian

One of the most distinct water mite faunas in the world has evolved in Australasia. About 780 species in 128 genera, 50 subfamilies and 30 families have been found. Australia (24 families, 89 genera, 440 species) and New Zealand (19 families, 52 genera, 143 species) are the best examined countries of the region (Cook, 1983, 1986; Harvey, 1998; Smit, 2005), but our knowledge is still insufficient. New Guinea is very poorly studied, and hardly any data are available from the adjacent islands (e.g. Solomon Islands, Bismarck Archipelago, New Britain).

Oceanic islands/pacific

Only eight families, 13 genera and 27 species are known from Pacific islands (Smit, 2005). Three species are reported from Vanuatu, 12 from Fiji and 14 from the Caroline Islands. Only two species are known from Western Samoa and Hawaii. Water mites are absent from the Galapagos Islands (Gerecke et al., 1996).

Antarctic

No water mites have been reported so far from this region. Their absence is confirmed by recent studies from sub-Antarctic islands (Pugh & Dartnall, 1994; Dartnall, 2005).

Endemism and faunistic similarity at family and subfamily level

A preliminary analysis of the extant differential distribution of families and their generic diversification (Table 3) shows that 19 of the 57 known families (33%) occur in all bioregions, three families (5%) have a near cosmopolitan distribution being absent only from Australasia, 12 (21%) are restricted to two bioregions and 11 are endemic to single regions (Table 4). About 75% of the families occur both in the Palaearctic and Nearctic, 30–33 families (more than 50%) have been documented from Australasian, Oriental and Neotropical regions, and only 27 (47%) are represented in the Afrotropical region.

Table 3 Genera richness of water mite families in each biogeographic region

Family	PA	NA	NT	AT	OL	AU	World
Stygothrombiidae	1	1	0	0	0	0	1
Hydrovolziidae	2	1	0	1	1	0	3
Acherontacaridae	1	1	0	0	1	0	1
Hydrachnidae	1	1	1	1	1	1	1
Limnocharidae	4	3	3	2	2	3	6
Eylaidae	1	1	2	1	1	1	2
Apheviderulicidae	1	1	0	0	0	0	1
Piersigiidae	1	1	0	0	1	2	3
Hydryphantidae	27	23	12	9	13	10	50
Hydrodromidae	1	1	1	1	1	1	1
Teratothyadidae	0	0	0	1	1	0	2
Rhynchohydracaridae	0	1	3	0	0	0	3
Ctenothyadidae	0	0	0	0	1	1	2
Thermacaridae	1	1	1	0	1	0	1
Zelandothyadidae	0	0	0	0	0	2	2
Stygotoniidae	0	0	0	0	0	1	1
Sperchontidae	2	2	4	1	1	1	4
Teutoniidae	1	1	0	0	0	0	1
Rutripalpidae	1	1	0	0	0	0	1
Anisitsiellidae	8	5	11	9	16	6	33
Lebertiidae	2	3	2	1	1	0	4
Acucapitidae	0	0	0	0	1	0	1
Oxidae	2	2	3	3	2	2	4
Torrenticolidae	4	4	5	3	5	1	6
Pontarachnidae	0	0	0	1	0	1	1
Limnesiidae	3	9	20	4	3	4	27
Omartacaridae	0	1	1	0	1	1	2
Hygrobatidae	5	6	33	19	5	24	77
Ferradasiidae	0	0	1	0	0	0	1
Unionicolidae	3	4	8	11	4	6	18
Feltriidae	1	1	1	0	1	0	1
Pionidae	9	9	4	2	3	6	14
Astacocrotonidae	0	0	0	0	0	1	1
Wettinidae	1	1	0	1	0	3	5
Frontipodopsidae	1	1	1	2	1	1	2
Aturidae	19	16	26	18	14	29	62
Lethaxonidae	1	2	1	1	1	0	2
Mideidae	2	1	0	0	0	0	2
Gretacaridae	0	0	0	0	0	1	1
Momoniidae	4	3	4	1	2	6	13
Mideopsidae	1	1	3	3	2	4	8
Nudomideopsidae	1	3	0	0	0	2	3
Uchidastygacaridae	2	3	0	0	0	1	4
Kantacaridae	1	0	0	0	0	0	1
Nipponacaridae	1	0	0	0	0	0	1
Neoacaridae	1	2	1	0	0	0	2
Bogatiidae	1	1	0	0	0	0	2
Chappuisididae	2	1	1	0	0	0	2
Krendowskiidae	1	2	3	1	1	0	4
Acalyptonotidae	1	2	0	0	0	0	2
Athienemanniidae	4	4	1	3	1	4	16
Harpagopalpidae	0	0	0	1	1	0	1
Hungarohydracaridae	3	0	1	1	2	0	4
Arenohydracaridae	0	1	1	0	0	0	1
Amoenacaridae	0	1	0	0	0	0	1
Laversiidae	0	1	0	0	0	0	1
Arrenuridae	2	1	5	8	2	2	12
Total genera	131	131	164	110	94	128	428
Total families	42	44	31	28	33	30	

PA: Palaearctic; NA: Nearctic; NT: Neotropical; AT: Afrotropical; OL: Oriental; AU: Australasian

The distribution of families shows that the Australasian fauna is particularly distinct (Fig. 4). The Palaearctic and Nearctic (Holarctic realm) have the highest-mutual affinity and are linked, at a low level of similarity, with the Neotropical, Afrotropical and Oriental regions, with the Neotropical distinctly set off from the Afrotropical and Oriental.

An analysis of the regional endemism of water mite taxa (Table 4) shows that four families are endemic to the Australasian region, a further two are exclusive to the Palaearctic, Nearctic and Neotropical, a single family is limited to the Oriental and no endemic families are known from the Afrotropical region. At the subfamily level, the Neotropical region is distinctive with 12 exclusive subfamilies, followed by Palaearctic (4), Nearctic (3) and Australasian (3). A last consideration can be made on the distribution of taxa exclusive for pairs of adjacent biogeographic regions (Fig. 5). Some taxa possibly reflecting an older Laurasian or North-Pangean distribution have their geographical range limited to the Holarctic while two subfamilies are limited to the Palaearctic and Oriental regions. In the Southern Hemisphere, the Neotropical shows a closer affinity with the Nearctic (two families and five subfamilies) than with the Afrotropical region

Table 4 Water mite families and subfamilies endemic to each biogeographic region

	PA	NA	NT	AT	OL	AU
Family	Kantacaridae	Laversiidae	Ferradasiidae		Acucapitidae	Gretacaridae
	Nipponacaridae	Amoenoacaridae	Arenohydracaridae			Astacocrotonidae
						Stygotoniidae
						Zelandothyadidae
Subfamily	Stygolimnesiinae	Cowichaniinae	Epallagopodinae	Omanohydracarinae	Nicalimnesiinae	Schminkeinae
	Bogatiinae	Chimerathyadinae	Tyrrelliinae		Bharatoniinae	Guineaxonopsinae
	Balcanohydracarinae	Stygameracariinae	Ankelothyadinae			Notomundamellinae
	Tsushimacarinae		Neotorrenticolinae			
			Cubanohydracarinae			
			Rheolimnesiinae			
			Psammolimnesiinae			
			Mixolimnesiinae			
			Rhynchohydracarinae			
			Plaumanniinae			
			Santiagoacarinae			
			Apeltosperchontinae			

PA: Palaearctic; NA: Nearctic; NT: Neotropical; AT: Afrotropical; OL: Oriental; AU: Australasian

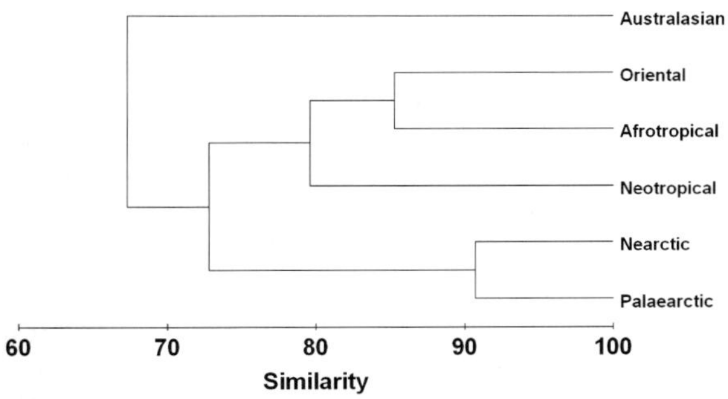

Fig. 4 Dendrogram of faunistic similarity between bioregions based on presence–absence of water mite families (Sørensen index)

(one subfamily). Two families and two subfamilies have an Afrotropical–Oriental distribution and only a single family and subfamily are exclusive to the Australasian–Oriental regions.

Endemism and faunistic similarity at generic and species level

When described species are considered, the Palaearctic is the most species-rich region (27% of the known species) followed by Neotropical (22%) and Nearctic (17%). A lower richness is recorded from

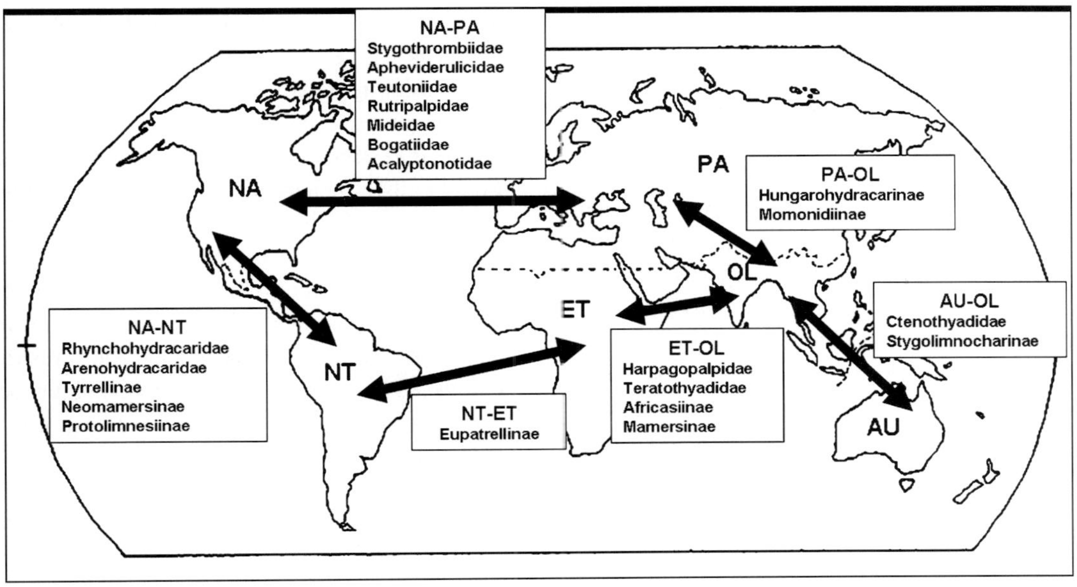

Fig. 5 Water mite families and subfamilies with distributions actually limited to adjacent biogeographic regions. PA—Palaearctic; NA—Nearctic; NT—Neotropical; ET—Afrotropical; OL—Oriental; AU—Australasian; PAC—Pacific Oceanic Islands; ANT—Antarctic

Table 5 Species richness of water mite superfamilies in each biogeographic region

Superfamily	PA	NA	NT	AT	OL	AU	World
Stygothrombioidea	4	6	0	0	0	0	10
Hydrovolzioidea	16	6	0	1	7	0	30
Eylaioidea	59	36	35	16	10	20	176
Hydrachnoidea	82	26	3	24	15	16	171
Hydryphantoidea	183	65	51	35	38	38	410
Lebertioidea	375	141	144	150	142	63	1,015
Hygrobatoidea	634	473	789	395	240	434	2,965
Arrenuroidea	289	272	273	166	102	123	1,230
Total species	1,642	1,025	1,305	787	554	694	6,007

PA: Palaearctic; NA: Nearctic; NT: Neotropical; AT: Afrotropical; OL: Oriental; AU: Australasian

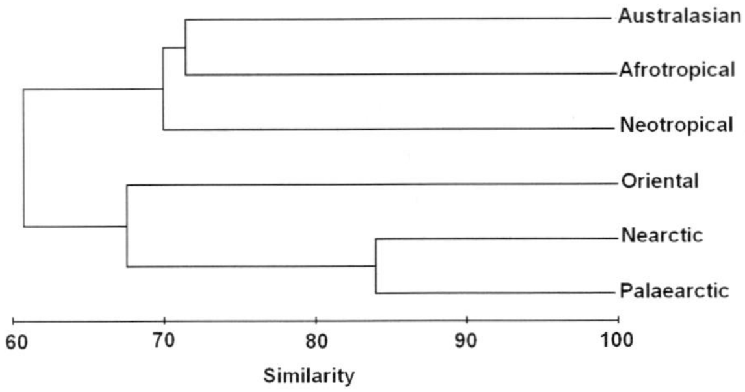

Fig. 6 Dendrogram of faunistic similarity between bioregions based on generic richness of water mite families (Bray-Curtis index)

Table 6 Numbers of endemic water mite families, subfamilies, genera and percentage of endemic genera in each biogeographic region

	PA	NA	NT	AT	OL	AU
Endemic families	2	2	2	–	1	4
Endemic subfamilies	4	3	12	1	2	3
Endemic genera	34	28	83	52	23	80
Percent endemic genera	26.9	21.4	50.6	47.2	24.4	62.0

PA: Palaearctic; NA: Nearctic; NT: Neotropical; AT: Afrotropical; OL: Oriental; AU: Australasian

Australasian and Afrotropical regions (11–13%) followed by the Oriental region with only 9% of described species (Table 5).

The total number of genera (Table 3) is not correlated with species richness and is distinctly higher in the Neotropical (164 genera); genus richness is similar in the Palaearctic, Nearctic and Australasian regions and is lower in Afrotropical and Oriental regions. Moreover, as reported in other groups (Ephemeroptera, this volume), Northern and Southern Hemispheres are characterized by a different mean number of genera per family: about three genera per family occur in the Palaeartic, Nearctic and Oriental regions while an average of more than four genera characterizes the families of Australasian and Afrotropical regions and more than five genera those of the Neotropical.

Compared to the presence–absence of families, a quantitative analysis of the richness of genera among families (Fig. 6) leads to slightly different results. The regions are here grouped into two clusters, the first including again the Palaearctic and Nearctic, but linked at low level of similarity with the Oriental; the second is formed by Neotropical and Afrotropical–Australasian regions.

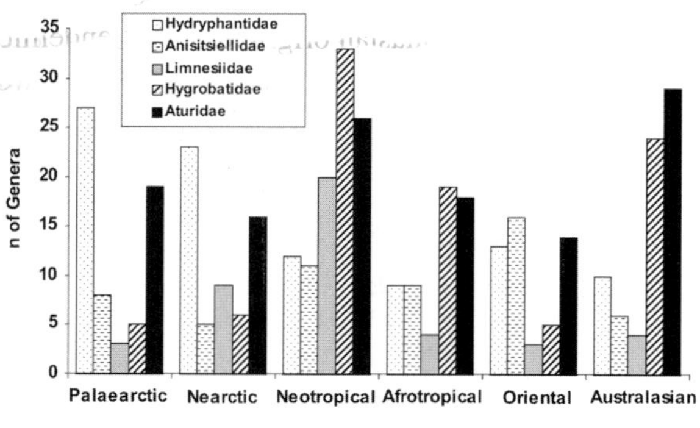

Fig. 7 Generic diversity of some cosmopolitan water mite families in each bioregion

This pattern seems mainly determined by the generic diversification of some cosmopolitan families (Fig. 7). The Holarctic and Oriental regions are characterized by the diversification of hydryphantids and the Oriental furthermore by the massive presence of anisitsiellid genera. On the other hand, Hygrobatidae, Limnesiidae and Aturidae are more diversified in the Neotropics, Afrotropical and Australasian regions with the Neotropical being unique because of the extreme diversification of Limnesiidae.

The percentages of endemic genera (Table 6) follow the same pattern described above for the generic diversification of the most cosmopolitan families, with regions of the Southern Hemisphere having a high degree of endemism with about 50% of the genera exclusive for each area. Also in this case, Australasia is distinctive with 62% of the genera known only from this region. Conversely, in the Palaearctic, Nearctic and Oriental regions the percentages of endemic genera are markedly lower (between 21 and 27%).

At species level, the Nearctic is characterized by the dominance of lentic adapted species of the genera *Arrenurus*, *Unionicola* and *Piona*, which represent about 30% of the North American water mite fauna; also remarkable is the diversification of the lotic-adapted genera *Feltria* (37 species) and *Aturus* (49 species). The Palaearctic fauna is mostly dominated by lotic species of the genera *Atractides*, *Lebertia* and *Sperchon* (more than 25% of the whole fauna) and by lentic-adapted species of the genera *Hydrachna*, *Piona* and *Eylais*. In the Neotropical, species of the genera *Koenikea, Limnesia, Corticacarus* and *Torrenticola* represent about 40% of the Neotropical mite fauna. The Afrotropical water mite fauna is mostly represented by species of the genera *Arrenurus, Neumania, Hygrobates, Torrenticola* and *Monatractides* (more than 40% of the whole fauna). The Oriental region is characterized by a relatively large number of *Monatractides, Torrenticola* and *Atractides* species, while the genus *Lebertia* is little represented. Almost 30% of the Australasian fauna is represented by species of endemic (*Procorticacarus*) and Gondwanan genera (*Australiobates, Aspidiobates*), and by species of the more widespread genera *Limnesia, Axonopsella, Koenikea* and *Frontipoda*. The latter two genera have the highest-species richness in the austral part of the world. A number of near-cosmopolitan genera are absent in the Australasian region i.e. *Lebertia, Atractides* and *Torrenticola*.

Global distribution patterns

From the data presented above we can conclude that the extant distribution of water mites is in good agreement with the basic vicariance pattern proposed by Smith & Cook (1999). A differential distribution of water mite fauna had been established before the break-up of Pangea with a distinction between Northern and Southern Pangean clades. After the separation, Laurasia and Gondwanaland were both characterized by the presence of a distinct temperate and tropical fauna.

In Laurasia a prevalent temperate fauna evolved and diversified from the original stock with tropical clades mainly confined to the southern part of the supercontinent. These elements may have persisted until today as localized endemism in non-glaciated areas of both North America and southern Eurasia (possible extant representatives could be the family Apheviderulicidae and the genera *Javalbia, Utaxatax* and *Momonisia*). Palaearctic and Nearctic endemics are mainly localized in Eastern Eurasia and Western North America which represented the western and eastern borders of Laurasia. Areas of particular interest with endemic species and genera are also localized in southwestern North America and in the Mediterranean area in southern Europe.

In the Southern Hemisphere the situation is more complex. Gondwanaland was subjected to multiple fragmentation and extensive migration (drifting) of continental blocks. India separated first and moved largely northward experiencing dramatic climatic change. Most of the original fauna disappeared and only a few Gondwanan elements persist today (Teratothyadidae, Harpagopalpidae). The extant Oriental fauna is mostly dominated by immigrant Southern Palaearctic elements and by some Australasian taxa.

The water mite fauna of currently southern bioregions is mainly derived from ancient tropical Gondwanan clades. The long-term isolation and phylogenetic diversification of most of these clades in Australia and New Zealand could explain the distinctive fauna of Australasian. Remnants of older temperate elements could have persisted at high elevation or high latitude in South America, Australia and New Zealand.

The separation of the Oriental and the Australasian regions has always been of great interest to science. Many boundary lines have been proposed between these two faunal regions (for a review see Simpson, 1977 and Cox, 2001). When one considers the water mite fauna, Lydekker's Line (Lydekker, 1896) seems to match most closely the boundary of the Australasian fauna. Lydekker's Line corresponds with the edge of the continental shelf of New Guinea and Australia. East of this line a typical Australasian fauna is found, with characteristic genera including *Australiobates* and *Procorticacarus*. Sulawesi has a predominantly Asian fauna, with member of the family Torrenticolidae dominating. Moreover, species of *Atractides*, a genus absent from Australasia, have been reported from Sulawesi. No records have been published from the intermediate area between Sulawesi and Australia.

In the Pacific, species of islands north of the Equator are of Asian origin, while those south of the Equator are of Australasian origin. Only one endemic genus is known i.e. *Fijilimnesia* from Fiji. Two *Aspidiobates* species, a genus with a Gondwanan distribution, are found on Vanuatu, which has never been a part of Gondwana. The number of species decreases from west to east, and from Western Samoa eastward streams are devoid of water mites (Smit, 2005).

The fauna of South America also conserves a typical mixture of tropical and temperate elements. The recent establishment of the Central American land bridge did not alter the original composition of the fauna except for the northernmost part and the Caribbean area. Conversely, a number of taxa migrated northward and invaded North America (Smith & Cook, 1999; Smith et al., 2001).

These considerations suggest that the actual scenario of water mite diversity and distribution reflect essentially the basic vicariance pattern. Isolation, phylogenetic diversification, recent climatic vicissitudes and episodes of dispersal between adjacent land masses together with extant ecological factors can be evoked to explain distribution patterns at a global scale.

Human related issues

Water mites represent a robust component of aquatic communities in terms of both abundance and species richness. Their high specialization to particular microhabitats and the strength of biotic interactions as predators (deutonymphs and adults) and parasites (larvae) are important features not shared by other components of macroinvertebrate communities. Perhaps because of this specificity and interconnectedness, they reflect natural changes and human induced modifications of freshwater communities (see Di Sabatino et al., 2000, 2002 and references therein). Water mites are mostly absent from springs, running waters, ponds and lakes whose natural sediment structure has been destroyed and/or water quality reduced by organic waste waters, heavy metals or other poisonous compounds. Moreover, due to the high number of taxa specialized for life in spring habitats (Di Sabatino et al., 2003) or in the upper part of the interstitial zone, their study offers an important but seldom exploited tool for understanding and monitoring human impacts in groundwaters (Boulton et al., 2003) and groundwater–surfacewater interactions.

Acknowledgements We wish to thank H. Proctor (Alberta, Canada) and three anonymous reviewers for their helpful comments on an earlier draft of this article. ADS wish to thank P. Vignini (L'Aquila, Italy) for her valuable help during the preparation of the manuscript. ADS and BC were in part financially supported by the Italian MIUR (ex 60%).

References

Boulton, A. J., W. F. Humphreys & S. M. Eberhard, 2003. Imperilled subsurface waters in Australia: biodiversity, threatening processes and conservation. Aquatic Ecosystem Health & Management 6: 41–54.

Cook, D. R., 1966. The water mites of Liberia. Memoirs of the American Entomological Institute 6: III+418 pp.

Cook, D. R., 1967. Water mites from India. Memoirs of the American Entomological Institute 9: III+411 pp.

Cook, D. R., 1980. Studies on neotropical water mites. Memoirs of the American Entomological Institute 31: V+645 pp.

Cook, D. R., 1983. Rheophilic and hyporheic water mites of New Zealand. Contributions of the American Entomological Institute 21: II+224 pp.

Cook, D. R., 1986. Water mites from Australia. Memoirs of the American Entomological Institute 40: IV+568 pp.

Cook, D. R., 1988. Water mites from Chile. Memoirs of the American Entomological Institute 42: 356.

Cox, C. B., 2001. The biogeographic regions reconsidered. Journal of Biogeography 28: 511–523.

Dartnall, H. J. G., 2005. Freshwater invertebrates of subantarctic South Georgia. Journal of Natural History 39: 3321–3342.

Di Sabatino, A., R. Gerecke & P. Martin, 2000. The biology and ecology of lotic water mites (Hydrachnidia). Freshwater Biology 44: 47–62.

Di Sabatino, A., P. Martin, R. Gerecke & B. Cicolani, 2002. Hydrachnidia (Water mites). In Rundle S., A. Robertson & J. Schmid-Araya (eds), Freshwater Meiofauna: Biology and Ecology. Backhuys Publisher, Leiden, 105–133.

Di Sabatino, A., B. Cicolani & R. Gerecke, 2003. Biodiversity and distribution of water mites (Acari, Hydrachnidia) in spring habitats. Freshwater Biology 48: 2163–2173.

Gerecke, R., 2004. The water mites of Madagascar (Acari: Hydrachnidia): a revised list completed by original material conserved at the Muséum national d'Histoire naturelle, Paris. Zoosystema 26: 393–418.

Gerecke, R. & E. O. Lehmann, 2005. Towards a long term monitoring of Central European water mite faunas (Acari: Hydrachnidia and Halacaridae)—considerations on the background of data from 1900 to 2000. Limnologica 35: 45–51.

Gerecke, R., S. B. Peck & H. E. Pehofer, 1996. The invertebrate fauna of the inland waters of the Galápagos Archipelago (Ecuador)—a limnological and zoogeographical summary. Archiv für Hydrobiologie, Supplement 107: 113–147.

Gerecke, R., I. M. Smith & D. R. Cook, 1999. Three new species of *Apheviderulix gen. nov.* and proposal of Apheviderulicidae *fam. nov.* (Acari: Hydrachnidia: Eylaoidea). Hydrobiologia 397: 133–147.

Goldschmidt, T., 2002. The biodiversity of Neotropical water mites. In Bernini, F., R. Nannelli, G. Nuzzaci & F. de Lillo (eds), Acarid Phylogeny and Evolution. Adaptations in mites and ticks. Kluwer Academic Publishers, 91–99.

Goldschmidt, T. & R. Gerecke, 2003. Zur Kenntnis der Wassermilbenfauna (Acari, Hydrachnidia) in Fließgewässern und Quellen auf Madagaskar. Tagungsbericht DGL Braunschweig 2002: 755–760.

Habeeb, H., 1967. A check list of North American water-mites. Leaflets of Acadian Biology 43: 1–8.

Harrison, A.D., 2000. The Water Mites (Acari: Hydrachnellae) of South Africa. http://www.rhodes.ac.za/academic/departments/zooento/Martin/amites.html

Harvey, M. S., 1998. The Australian Water Mites. A Guide to Families and Genera. Monographs on Invertebrate Taxonomy 4. CSIRO Publishing, Collingwood.

Jin, D., 1997. Hydrachnellae—morphology, systematics. A primary study of Chinese fauna. Guizhou Science and Technology Publishing House.

Lundblad, O., 1941. Die Hydracarinenfauna Südbrasiliens und Paraguays. Erster Teil. Kungliga Svenska Vetenskapsakademiens Handlingar 19: 1–183.

Lundblad, O., 1942. Die Hydracarinenfauna Südbrasiliens und Paraguays. Zweiter Teil. Kungliga Svenska Vetenskapsakademiens Handlingar 20: 1–175.

Lundblad, O., 1943a. Die Hydracarinenfauna Südbrasiliens und Paraguays, Dritter Teil. Kungliga Svenska Vetenskapsakademiens Handlingar 20: 1–148.

Lundblad, O., 1943b. Die Hydracarinenfauna Südbrasiliens und Paraguays, Vierter Teil. Kungliga Svenska Vetenskapsakademiens Handlingar 20: 1–171.

Lundblad, O., 1944. Die Hydracarinenfauna Südbrasiliens und Paraguays, Fünfter Teil. Kungliga Svenska Vetenskapsakademiens Handlingar 20: 1–182.

Lydekker, R., 1896. A Geographical History of Mammals. University Press, Cambridge.

Mitchell, R. D., 1954. Check list of North American watermites. Fieldiana, Zoology 35: 27–70.

Panesar, A., 2004. Evolution in water mites (Hydrachnellae, Actinedida, Acari). A revision of the Anisitsiellidae Koenike, 1910. Bonner Zoologische Monographien 52: 1–144.

Pugh, P. J. A. & H. J. G. Dartnall, 1994. The Acari of fresh- and brackish water habitats in the Antarctic and sub-Antartctic regions. Polar Biology 14: 401–404.

Poinar, G. O., 1985. Fossil evidence of insect parasitism by mites. International Journal of Acarology 11: 37–38.

Rosso de Ferradas B. & H. R. Fernandez, 2005. Elenco y biogeografia de los Acaros acuaticos (Acari, Parasitengona, Hydrachnidia) de Sudamerica. Graellsia, 61: 181–224.

Simpson, G. H., 1977. Too many lines; the limits of the Oriental and Australian zoogeographic regions. Proceedings of the American Philosophical Society 121: 107–120.

Smit, H., 2005. A review of the water mite fauna from the Australasian and Pacific region (Acari: Hydrachnidia). In Weigmann, P., G. Alberti, A. Wohltmann & S. Ragusa (eds), Acarine Biodiversity in the Natural and Human Sphere. Phytophaga 14: 525–530.

Smith, I. M. & D. R. Cook, 1999. An assessment of global distribution patterns in water mites (Acari: Hydrachnidia) In Mitchell R., G. Needham & W. C. Welbourn (eds), Acarology IX, Vol. 2, Symposia. Ohio Biological Survey. Columbus, Ohio, 523–527.

Smith, I. M., D. R. Cook & B. P. Smith, 2001. Water mites (Hydrachnida) and other arachnids. In Thorp J. H. & A. P. Covich (eds), Ecology and Classification of North American Freshwater Invertebrates (2nd edition). Academic Press, San Diego, California, 551–659.

Smith, I. M., E. E. Lindquist & V. Behan-Pelletier, 1998. Mites (Acari). In Smith, I. M. & G. G. Scudder (eds), Assessment of Species Diversity in the Montane Cordillera Ecozone. Burlington: Ecological Monitoring and Assessment Network. Available at: http://www.naturewatch.ca/eman/reports/publications/99_montane/mites/mites06.html

Sokolow, I., 1940. Hydracarina (1.re partie: Hydrachnellae). In: Sernow D., &T. Stackelberg (eds), Faune de l'URSS. Arachnides. Institute Zoology Academy of Science. URSS, Moscow, Leningrad, 5: 1–511.

Tuzovskij, P. V., 1987. Morphology and Postembryonal Development of Water Mites. Nauka, Moskwa: 1–176.

Tuzovskij, P.V., 1997. Hydrachnidia. In Tsalolikhin, S. J. (ed.), Key to Freshwater Invertebrates of Russia and Adjacent Lands. Institute Zoology Academy of Science, URSS 3: 13–35.

Van Rensburg, C. A., 1974. A checklist of the Ethiopian water mites with notes on their distribution. Wetenskaplike Bydraes van die Potchefstroomse Universiteit vir C.H.O. Reeks B, Natuurwetenskappe 68, pp. 1–51.

Viets, K. O., 1970. Unser Zuwachs an Kenntnissen über die aus Afrika bekannten Wassermilben (Hydrachnellae, Acari) (mit Anhang: Limnohalacaridae). Hydrobiologia, 35: 65–126.

Viets, K. O., 1987. Die Milben des Süβwassers (Hydrachnellae und Halacaridae [part.], Acari). 2. Katalog. Sonderbände des Naturwissenschaftlichen Vereins in Hamburg 8: 1–1012.

Wiles, P. R., 2004. Water mites (Acari: Hydrachnidia) from South-East Asia, Thailand and Sulawesi Tenggara (Indonesia): descriptions of new species and new records. Journal of Natural History 38: 2153–2165.

Witte, H., 1991. Indirect sperm transfer in prostigmatic mites from a phylogenetic viewpoint. In: Schuster R., P. W. Murphy (eds), The Acari. Reproduction, Development and Life-history Strategies. Chapman and Hall, London, 137–176.

Hydrobiologia (2008) 595:317–322
DOI 10.1007/s10750-007-9026-0

Global diversity of halacarid mites (Halacaridae: Acari: Arachnida) in freshwater

Ilse Bartsch

Abstract Halacarid mites have successfully invaded the sea and approximately 56 species colonized the freshwater. Invasion from the sea into continental waters probably started in the Mesozoic or Pre-Mesozoic and went on in the following epochs. The number of genera (14) and species (34) recorded from the Palaearctic is remarkably higher than that of other geographical regions. These numbers do not imply that the Palaearctic is a centre of origin, they reflect the sampling activity rather than reliable data on diversity.

Keywords Halacaridae · freshwater · distribution · phylogeny · endemicity

Introduction

Most representatives of the family Halacaridae (Order Prostigmata) are marine, to date slightly more than 1,000 marine species are described versus about 56 from continental freshwater. Halacarid mites, marine, as well as freshwater taxa, are benthic throughout their life; they are unable to swim; plankton stages are not known.

Halacarids are small-sized, rarely more than 500 μm in length, and in general less conspicuous than the majority of water mites, they are never as intensely coloured as many hydrachnids and never as heavily sclerotized, as oribatids and mesostigmatids. In halacarid mites, legs I and II are directed forward, legs III and IV backward, the epimeres of the two anterior pairs of legs by a gap separated from the posterior epimeres. The legs of adults are six-segmented, the palps four-segmented. The idiosoma generally bears four dorsal plates, though in some species the plates may be fused or reduced.

Freshwater halacarids are found in subterranean and surface waters; they live in springs, wells, the hyporheic zone of rivers and flocculent ooze of lakes, in artificial filters, sandy deposits, amongst colonial organisms, gill chambers, mosses and vascular plants, in humic as well as in brackish coastal waters. Many of the freshwater halacarid species proved to be euryoecious and very tolerant towards short-term physical and chemical fluctuations in their environment. A stable and predictable habitat and successful competition seem to be more important for colonization than environmental parameters.

The mites run through one larval and two nymphal stages before the final moult to adults. The fecundity is low, often a single generation per year and less than 50 offsprings per female. Adults and juveniles live in the same substratum. Migration, dispersal or resting

Guest editors: E. V. Balian, C. Lévêque, H. Segers & K. Martens
Freshwater Animal Diversity Assessment

I. Bartsch (✉)
Forschungsinstitut Senckenberg, Deutsches Zentrum für marine Biodiversitätsforschung, Notkestr. 85, Hamburg 22607, Germany
e-mail: bartsch@meeresforschung.de

stages are not known. Reproduction can be dioecious or parthenogenetic.

Halacarid mites are often classified as predators or scavengers, feeding on small metazoans and protozoans, but some species feed on algal and plant cells, and probably bacteria and fungi are used by halacarid mites as well. One species is expected to be a parasite.

Species diversity

The first record of a marine mite ('insecta marina') from the seashore was published more than 200 years ago (Baster, 1758: pl.10, Fig. 1). A century later, Gosse (1855) introduced the genus *Halacarus* and Murray (1877) established the family Halacaridae for 'mites living habitually under the sea'. The first halacarid mite from freshwater, *Leptognathus violaceus*, was described by Kramer (1879). Viets (1927, 1933) separated between marine and freshwater genera and set up the family Porohalacaridae Viets 1933 for freshwater species with external genital acetabula, in contrast to the Halacaridae which included marine species with internal genital acetabula. Since of numerous discrepancies between genera belonging to the Porohalacaridae on the one hand and characters shared with marine genera, the division based on presence or absence of external genital acetabula was abandoned (Newell, 1947; Petrova, 1974; Bartsch, 1989b, 1996). Nowadays, freshwater halacarid genera are incorporated in halacarid subfamilies, which

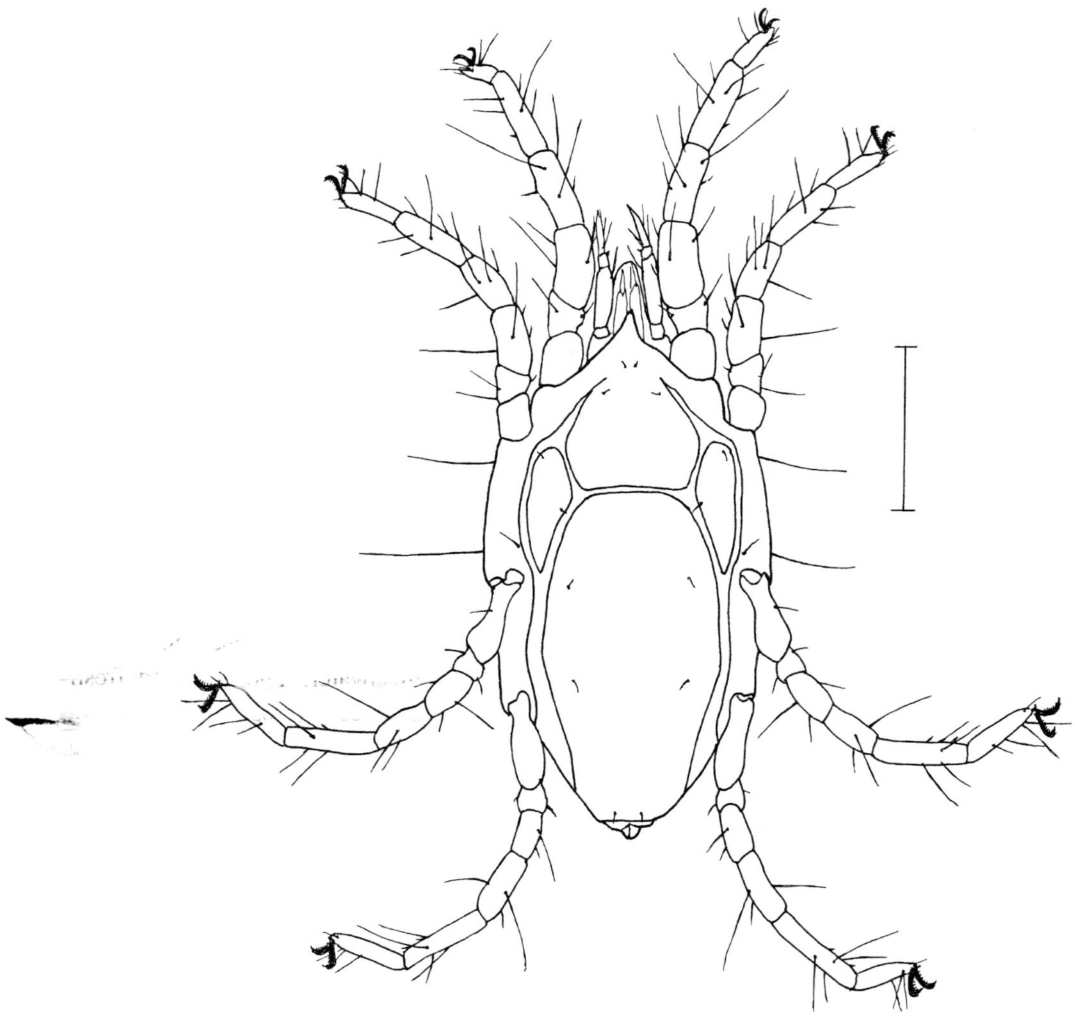

Fig. 1 Habitus of *Lobohalacarus*. Scale = 100 μm

Table 1 Freshwater Halacaridae (Acari), genera of and number of species recorded from each zoogeographical region

	PA	NA	NT	AT	OL	AU	PAC	ANT	World
Astacopsiphagus	0	0	0	0	0	1	0	0	1
Acarothrix	0	1	0	0	0	0	0	0	1
Copidognathus	4	1	0	0	0	0	0	0	5
Caspihalacarus	1	0	0	0	0	0	0	0	1
Halacarellus	1	0	0	0	0	0	0	0	1
Hamohalacarus	0	1	0	0	0	0	0	0	1
Himejacarus	1	0	0	0	0	0	0	0	1
Limnohalacarus	2	1	1	4	2	2	0	0	13
Lobohalacarus	2	1	2	1	0	2	1	1	6
Lohmannella	5	0	0	0	0	0	0	0	5
Porohalacarus	3	1	0	0	0	1	0	0	3
Parasoldanellonyx	3	1	0	0	0	0	0	0	3
Peregrinacarus	0	0	1	0	0	0	0	1	2
Porolohmannella	1	1	0	0	0	0	0	0	1
Ropohalacarus	1	1	0	0	0	0	0	0	1
Soldanellonyx	8	3	0	2	1	1	1	0	9
Stygohalacarus	1	0	0	0	0	0	0	0	1
Troglohalacarus	1	0	0	0	0	0	0	0	1
Total	34	12	4	7	3	7	2	2	56

PA: Palaearctic region; NA: Nearctic region; NT: Neotropical region; AT: Afrotropical region; OL: Oriental region; AU: Australasian region; PAC: Pacific Oceanic Islands; ANT: Antarctic region

include either fresh or mostly marine species, some of the freshwater halacarid mites are assigned to else marine genera.

Approximately 56 halacarid species (Table 1) are known from freshwater habitats. This figure includes species of genera in general restricted to fresh or diluted brackish (0.5–5‰) water, as well as limnic representatives of genera widespread in marine habitats. Not included are species commonly found in refreshed brackish water (about 0.5‰ or less) but most abundant in coastal waters (salinity 5–30‰). Subspecies are not mentioned separately. Some of the species, as well as genera are in need of re-examination to prove their identity. Not included in the table (Table 1) and map (Fig. 2) are two species found in aquaria in Europe, but most likely being introduced from overseas.

Phylogeny and historical process

Arachnids have an evolutionary age dating back to the Cambrian and probably all major orders of Acari have existed, since the Silurian (Vitzthum, 1943). The mites in the sea, the halacarid mites, are expected to have evolved from aquatic or semi-aquatic prostigmatid ancestors which colonized the sea shores. The taxon most likely exists since the Pre-Mesozoic or Mesozoic (Bartsch 1982, 1989a, 1996). Freshwater halacarids are polyphyletic, descendants of marine or brackish water ancestors (Petrova, 1974; Bartsch 1982, 1996). For halacarid mites, living in a wide range of substrata, both an epigean and hypogean way was open to migrate from the sea into freshwater. The increased freshwater influx and hence need for osmoregulation was met by either moving the genital acetabula to an external position or increasing the genital acetabula or epimeral pores. Both are osmoregulatorily active areas, and an external position and/or enlargement will be of competitive advantage in a life in diluted brackish or freshwater.

Several of the present-day freshwater genera and even species may have invaded continental waters in the Mesozoic or even Pre-Mesozoic (Bartsch, 1996). These probably first immigrants are today restricted

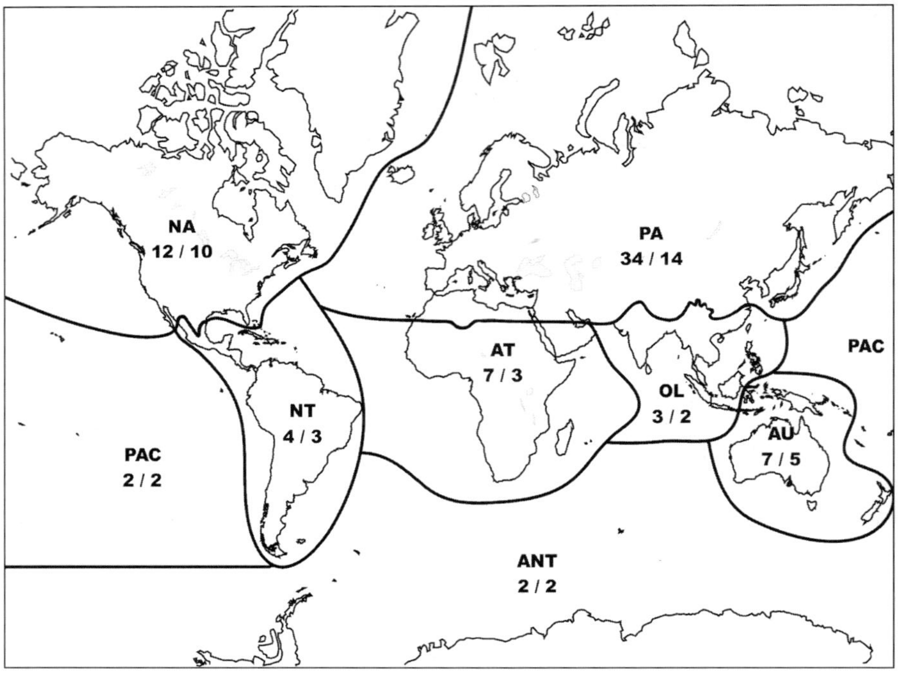

Fig. 2 Diversity of freshwater Halacaridae (Acari). Number of species/ Number of genera recorded from each zoogeographical region. PA: Palaearctic region; NA: Nearctic region; NT: Neotropical region; AT: Afrotropical region; OL: Oriental region; AU: Australasian region; PAC: Pacific Oceanic Islands; ANT: Antarctic region

to freshwater areas, none is directly allied to a marine genus, they all have external genital acetabula and, compared to marine forms, often a reduced number of setae. A remarkable euryvalence of these genera and species is demonstrated by the wide spectrum of habitats in which they are found.

The process of invading freshwater was not restricted to the Pre-Mesozoic and Mesozoic. Some of the present-day freshwater genera demonstrate a close similarity to marine genera. Their speciation assumedly is correlated with geological events. Halacarid populations in land-locked, refreshing water bodies survived in that their genital acetabula became enlarged and moved to an external position. These genera may date back to the Tertiary or Late Mesozoic; they are found in a restricted geographical area.

Amongst the halacarid mites present in freshwater there are representatives of otherwise marine genera, e.g. of the genera *Copidognathus*, *Halacarellus* and *Lohmannella*, which differ from congeners by their enlarged epimeral pores or genital acetabula. Many of these species live in areas, which in the preceding one to five million years have been connected with the sea.

The process of marine halacarid mites penetrating into freshwater habitats is still going on, several species are frequently found both in very diluted brackish or freshwater, as well as on the seashore at up to 30‰. The other way, freshwater species or genera getting adapted to life in sea, will occur, too.

Present distribution and main areas of endemicity

Compared to other geographic areas, the number of genera and species (Tables 1 and 2) recorded from the Palaearctic is surprisingly high. The data do not imply that the Palaearctic is a centre of origin, instead they mirror the poor sampling activity in the other areas. Europe is most intensely studied with regard to its halacarid fauna. The two species mentioned from the area Antarctica are from sub-Antarctic islands, though true Antarctic records are expected to be found, too.

The palaearctic freshwater halacarid fauna includes widely spread, as well as endemic genera and species. An example of a widespread species is *Soldanellonyx monardi*, with records from the Palaearctic (Europe and Asia), Nearctic (North America), Afrotropics (Kenya), Orients (Java), Neotropics (Falklands), Australasia (Australia), and Oceanic islands in the Pacific (Hawaii). *Porohalacarus alpinus*, *Lobohalacarus weberi* and *Porolohmannella violacea*, too, are expected to be found in all

Table 2 Freshwater Halacaridae (Acari), subfamilies and number of genera recorded from each zoogeographical region

	PA	NA	NT	AT	OL	AU	PAC	ANT	World
Astacopsiphaginae	0	0	0	0	0	1	0	0	1
Copidognathinae	1	2	0	0	0	0	0	0	2
Halacarinae	5	2	1	1	0	2	1	2	6
Limnohalacarinae	5	4	2	2	2	2	1	0	6
Lohmannellinae	1	0	0	0	0	0	0	0	1
Porolohmannellinae	1	1	0	0	0	0	0	0	1
Ropohalacarinae	1	1	0	0	0	0	0	0	1
Total	14	10	3	3	2	5	2	2	18

PA: Palaearctic region; NA: Nearctic region; NT: Neotropical region; AT: Afrotropical region; OL: Oriental region; AU: Australasian region; PAC: Pacific Oceanic Islands; ANT: Antarctic region

geographical areas. Genera and species with a very restricted distribution assumedly have invaded freshwater recently, correlated with geological events, rather than to represent relicts of a former pangaean fauna. Examples are *Caspihalacarus*, *Stygohalacarus,* and *Troglohalacarus*, and the freshwater representatives of the marine genera *Copidognathus* (four species), *Halacarellus* (one species) and *Lohmannella* (four species). Records of these genera and species are from an area once being covered or connected with the western part of the Tethys, the Proto-Mediterranean and Paratethys, a water body that, due to rising mountain chains and changes in sea level was splitted into numerous basins, each with its own development, e.g. drastic freshwater input, and hence a change in its fauna and flora.

Similarly, in other regions coastal basins have been isolated from the sea and adaptation to the new environment resulted in speciation. Freshwater representatives of the marine genus *Copidognathus* are not restricted to Europe and are expected to be found in all continents. Ancestors of the eastern Australian endemic genus *Astacopsiphagus*, a genus found amongst the gills of parastacid crayfish, are expected to have been associated with the crayfish hosts while these were living in coastal lagoons and together with their hosts became adapted to freshwater.

Halacarid mites have a very low dispersal ability, there are no migration or resting stages, and the fecundity is low, still many genera and even species are spread worldwide. Halacarids are known to be generalists with immense ecological tolerance as to the microhabitat and its fluctuations of chemical and physical parameters. This tolerance is often higher than that of other small co-occurring arthropods, mites can survive short-term deteriorating conditions. Dispersal could be via slow migration both along epigean and hypogean passages, by extreme floods and storms, as well as by vectors such as insects, birds, mammals and, more recent, by Man. Being non-spezialized, halacarid mites transported into a new region had and have a chance to build up a population.

References

Bartsch, I., 1982. Halacariden (Acari) im Süßwasser von Rhode Island, USA, mit einer Diskussion über Verbreitung und Abstammung der Halacaridae. Gewässer und Abwässer 68/69: 41–58.

Bartsch, I., 1989a. Marine mites (Halacaroidea: Acari): a geographical and ecological survey. Hydrobiologia 178: 21–42.

Bartsch, I., 1989b. Süßwasserbewohnende Halacariden und ihre Einordnung in das System der Halacaroidea (Acari). Acarologia 30: 217–239.

Bartsch, I., 1996. Halacarids (Halacaroidea, Acari) in freshwater. Multiple invasions from the Paleozoic onwards? Journal of Natural History 30: 67–99.

Baster, J., 1758. Observationes de corallinis, iisque insidentibus polypis, aliisque animaliculis marinis. Philosophical Transactions of the Royal Society 50: 258–280.

Gosse, P. H., 1855. Notes on some new or little-known marine animals. Annals and Magazine of Natural History, Ser. 2 16(91): 27–36.

Kramer, P., 1879. Ueber die Milbengattungen *Leptognathus* Hodge, *Raphignathus* Dug., *Caligonus* Koch und die neue Gattung *Cryptognathus*. Archiv für Naturgeschichte 45: 142–157.

Murray, A., 1877. Economic Entomology. Aptera. South Kensington Museum Science Handbooks. Chapman & Hall, London.

Newell, I. M., 1947. A systematic and ecological study of the Halacaridae of eastern North America. Bulletin of the Bingham Oceanographic Collection 10: 1–232.

Petrova, A., 1974. Sur la migration des halacariens dans les eaux douces et la position systématique des halacariens et limnohalacariens. Vie et Milieu, Série C 24: 87–96.

Viets, K., 1927. Die Halacaridae der Nordsee. Zeitschrift für wissenschaftliche Zoologie 130: 83–173.

Viets, K., 1933. Vierte Mitteilung über Wassermilben aus unterirdischen Gewässern (Hydrachnellae et Halacaridae, Acari). Zoologischer Anzeiger 102: 277–288.

Vitzthum, H., 1943. Acarina. In H. G. Bronn's Klassen und Ordnungen des Tierreichs 5(IV, 5). Akademische Verlagsgesellschaft, Leipzig, 1011.

Hydrobiologia (2008) 595:323–328
DOI 10.1007/s10750-007-9027-z

Global diversity of oribatids (Oribatida: Acari: Arachnida)

Heinrich Schatz · Valerie Behan-Pelletier

Abstract Oribatid mites are primarily terrestrial. Only about 90 species (less than 1% of all known oribatid species) from 10 genera are truly aquatic, with reproduction and all stages of their life cycle living in freshwater. Adaptation to aquatic conditions evolved independently in different taxa. However, many terrestrial species can also be found in aquatic habitats, either as chance stragglers from the surrounding habitats, or from periodic or unpredictable floodings, where they can survive for long periods. In spite of their low species richness aquatic oribatids can be very abundant in different freshwater habitats as in lentic (pools, lakes, water-filled microhabitats) or flowing waters (springs, rivers, streams), mainly on submerged plants. The heavily sclerotized exoskeletons of several species enables subfossil or fossil preservation in lakes or bog sediments.

Guest editors: E. V. Balian, C. Lévêque, H. Segers & K. Martens
Freshwater Animal Diversity Assessment

H. Schatz (✉)
Institut für Zoologie/Ökologie, Leopold-Franzens-Universität, Technikerstr. 25, 6020 Innsbruck, Austria
e-mail: heinrich.schatz@uibk.ac.at

V. Behan-Pelletier
Biodiversity Program, Agriculture and Agri-Food Canada, K.W. Neatby Bldg., Ottawa, ON, CanadaK1A 0C6
e-mail: behanpv@agr.gc.ca

Keywords Acari · Oribatida · Aquatic distribution · Global diversity

Introduction

Compared to larger, "easily visible" arthropods, the minute oribatid mites are still poorly known, although they are among the most abundant and diverse of the mesofauna. Up to 500.000 individuals per m^2 can be found in forest litter, representing over 100 species. They play an important role in decomposition of plant litter, in nutrient cycling, in soil formation, and in distribution of fungal spores (e.g. Schneider et al., 2005). Oribatida can also be abundant in aquatic habitats, but species richness is very low (Fernandez & Athias-Binche, 1986).

All oribatid mite species have a life-cycle comprising a prelarval, a larval, three nymphal stages (proto-, deuto-, tritonymph), and adult (Walter & Proctor, 1999). The proportion of oribatid fauna that is sexual is considerably lower in aquatic systems than in soil or litter (Behan-Pelletier & Eamer, 2007, in press). Many species, especially those in macropyline superfamilies are parthenogenetic (Norton & Palmer, 1991; Norton et al., 1993; Maraun et al., 2003). In addition, most species in the aquatic brachypyline superfamily Hydrozetoidea are parthenogenic. Some sexual species do occur in the genus *Hydrozetes* and these show sexual dimorphism (Grandjean, 1948). Some aquatic taxa (e.g. *Trhypochthoniellus*, *Hydrozetes*) are highly variable in their morphology. This has led to separation of species in the past which are regarded as forms today (Deichsel, 2005; Weigmann, 2006). Where mating occurs in Oribatida it is always by indirect sperm transfer (Walter & Proctor, 1999). The life cycle of Oribatida is mainly characterized by slow growth, low reproductive potential, iteroparity and long adult life span. Synchronization with the annual climatic cycle is frequent, at least in temperate zones, but life cycles can be prolonged in species at high latitude or high elevation (Schatz, 1985; Norton, 1994; Søvik, 2004).

Most oribatid species have specific habitat preferences. The vast majority lives in terrestrial habitats, such as plant litter, soil, suspended soil, mosses and lichens. They are particularly species rich and abundant in humid habitats, such as forest soils and moorland. In addition, a range of markedly xerobiont species, occur in semi-deserts, in arid meadows, on rocks and on the bark and leaves of trees and shrubs (Walter & Proctor, 1999). A small number of oribatid mites are bound to various aquatic habitats including springs, seepages, temporary and permanent pools, reed belts around lakes, rivers, streams, phytotelmata and other water-filled microhabitats, as well as the brackish and marine sublittoral and littoral (Behan-Pelletier & Eamer, 2007, in press). They do not swim and live mainly on water plants, or in stream or lake sediment. Although the aquatic Oribatida encompass only a small number of species, abundance can be high, for example, *Hydrozetes lemnae* can reach population densities of a mean of 1600 per 200 cc^{-2} of water (Fernandez & Athias-Binche, 1986). A review on possible adaptation of oribatid mites to the aquatic environment and of their morphological, behavioural and physiological modifications was published recently (Behan-Pelletier & Eamer, 2007, in press). Many terrestrial oribatid mite species recorded from aquatic habitats are chance stragglers from surrounding habitats. These can survive for short to long periods in submerged conditions (see examples in Behan-Pelletier & Eamer, 2007, in press). The same is true for terrestrial species in floodplains, subject to periodic or unpredictable floodings (Adis & Messner, 1991; Weigmann 1997, 2005). In contrast, many "aquatic" species can withstand short to prolonged dry periods (Behan-Pelletier & Eamer, 2007, in press).

Species diversity

Up to date, almost 10.000 oribatid species are described (Schatz, 2002, 2005; Subias, 2004); however, estimates of the world fauna range from 50,000 to 100,000 species (Colloff & Halliday, 1998; Schatz, 2002). Less than 1% of all described species are known to be aquatic. Piffl (1978) gave a short overview of European aquatic Oribatida. According to him, only the genus *Hydrozetes* can be designated as truly aquatic, in addition to a number of species from different families, which were found on submerged plants. Recently, Behan-Pelletier & Eamer (2007, in press) considered 15 genera in 11 families to be represented in freshwater. Weigmann & Deichsel (2006) present 17 oribatid species from Central Europe that live exclusively or regularly in freshwater habitats.

If the term "aquatic" is restricted to taxa which reproduce in water, with all stages of their life-cycle living in freshwater habitats or at its margins, only few oribatid genera (and not always all included species) can be called truly aquatic: *Mucronothrus*, *Trhypochthoniellus*, *Aquanothrus*, *Chudalupia*, *Tegeocranellus*, *Hydrozetes*, *Limnozetella*, *Limnozetes* (Fig. 1), *Heterozetes*, *Zetomimus* (87 species from 7 families, see Tables 1, 2). Numerous species of other genera are also freshwater inhabitants but they seem to need saturated air to reproduce rather than water *per se*, and are also found in wet moss and sodden organic debris: these include some or all species of *Platynothrus*, *Trhypochthonius*, *Mainothrus*, *Malaconothrus*, *Trimalaconothrus*, *Astegistes*, *Naiazetes*,

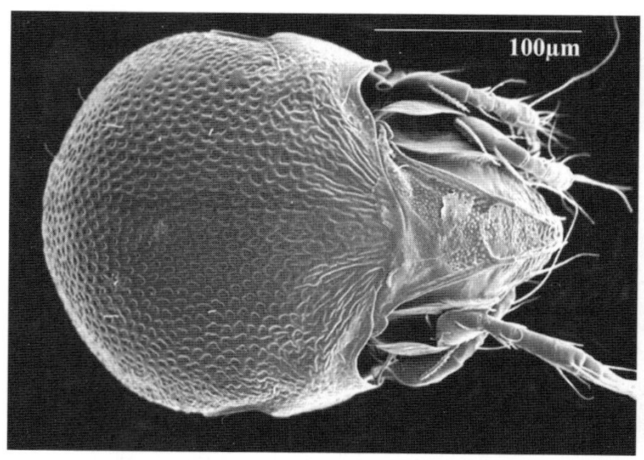

Fig. 1 Habitus of Oribatida: *Limnozetes guyi* dorsal

Table 1 Species number of aquatic Oribatida

SP: Species Number	PA	NA	NT	AT	OL	AU	PAc	ANT	World
Mucronothridae	1	2	1			1			2
Ameronothridae				1		1			2
Trhypochthonellidae	2	2	1	1	2	1	1		4
Tegeocranellidae	3	4	3	2	3	2	1		16
Hydrozetidae	9	8	9	3	3	1	1		27
Limnozetidae	6	9	2		3				17
Zetomimidae	6	6	5		1	1			18
Total	27	31	21	7	12	7	3	0	86

PA: Palaearctic; NA: Nearctic; NT: Neotropical; AT: Afrotropical; CL: Oriental; AU: Australasian; PAc: Pacific & Oceanic Islands; ANT: Antarctic

Table 2 Genus number of aquatic Oribatida

GN: Genus Number	PA	NA	NT	AT	OL	AU	PAc	ANT	World
Mucronothridae	1	1	1			1			1
Ameronothridae				1		1			2
Trhypochthonellidae	1	1	1	1	1	1	1		1
Tegeocranellidae	1	1	1	1	1	1	1		1
Hydrozetidae	1	1	1	1	1	1	1		1
Limnozetidae	1	1	1		2				2
Zetomimidae	2	2	2		1	1			2
Total	7	7	7	4	6	6	3	0	10

PA: Palaearctic; NA: Nearctic; NT: Neotropical; AT: Afrotropical; OL: Oriental; AU: Australasian; PAc: Pacific & Oceanic Islands; ANT: Antarctic

Ceratozetes, *Edwardzetes*, *Sphaerozetes*, *Minguezetes*, *Allogalumna*, *Orthogalumna* and others. Additionally, littoral species in the ameronothroid families Ameronothridae (*Ameronothrus*), Fortuyniidae and Selenoribatidae tolerate submergence in salt or freshwater, but are not active when submerged, and thus cannot be considered truly aquatic or semi-aquatic (Behan-Pelletier & Eamer, 2007, in press).

Empty exoskeletons of drowned terrestrial brachypyline mites from surrounding habitats are often found in aquatic samples (Schatz & Gerecke, 1996). These exoskeletons might be preserved under certain circumstances, especially in peatland, lake and bog sediments, forming subfossil or fossil taphocenoses (Erickson, 1988; Krivolutsky et al., 1990; Solhøy & Solhøy, 2000). These and the subfossils of aquatic Oribatida, may contribute important information on environmental changes (Schelvis, 1990; Erickson, 1996; Solhøy, 2001; Erickson et al., 2003).

The majority of aquatic Oribatida from the genera mentioned above is recorded from the Palaearctic and Nearctic Regions (Holarctic Region), which are better investigated than most tropical regions (Fig. 2). No true aquatic species were, hitherto, recorded from the Subantarctic and Antarctic area, although two hemi-edaphic species were found in aquatic moss in South Georgia (Pugh, 1996). Also, more than half of Ameronothridae species (some of which are semi-aquatic) were described from the Antarctic and surrounding islands.

Phylogeny and historical processes

An overview on the phylogeny and classification of oribatid mites based on morphological data is given in Weigmann (2006). The suborder includes a paraphyletic group of five cohorts, also known as

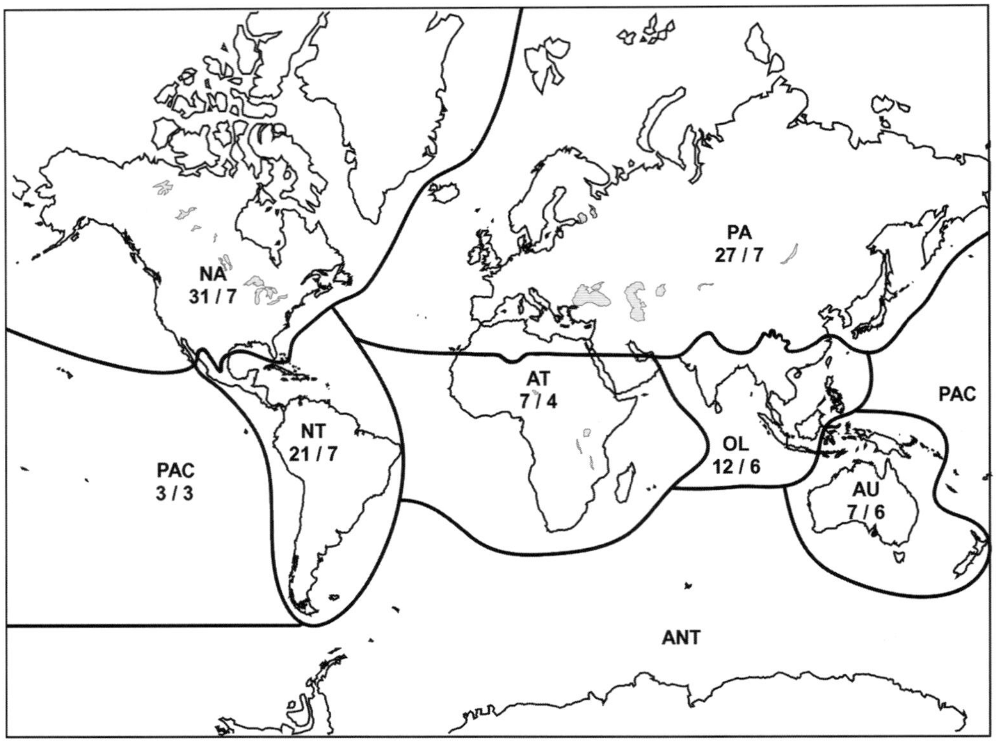

Fig. 2 Diversity of freshwater Oribatida (Acari) species and genera (Species number/Genus number) per zoogeographic region: PA, Palaearctic; NA, Nearctic; NT, Neotropical; AT, Afrotropical; OL, Oriental; AU, Australasian; Pac, Pacific & Oceanic Islands; ANT, Antarctic

"Macropylina" or "Lower Oribatida" (Balogh & Mahunka, 1979), and the probably monophyletic lineage Brachypylina. The oribatid cohort Astigmatina is not considered herein (see Norton, 1994, 1998), though it includes species known to be aquatic (Bücking et al., 1998). Maraun et al. (2004) present phylogenetic relationships and radiation of oribatid mites using DNA characters. The neighbour-joining tree of the investigated oribatid species is in general agreement with trees based on morphology (see Grandjean, 1954, 1965, 1969, 1970; Norton, 1984; Haumann, 1991).

Oribatid mites are known as fossils at least since the Devonian Period (Norton et al., 1988a), and possibly earlier (Bernini et al., 2002), and have been recorded from palaeozoic coal-swamp forests (Labandeira et al., 1997). Fossil aquatic *Hydrozetes* species are reported from the Jurassic Period in Sweden (Sivhed & Wallwork, 1978), from the Paleocene in Canada (Baker & Wighton, 1984), from the Pliocene in Siberia (Druk, 1982), and together with *Limnozetes* species in numerous records from the Pleistocene (Krivolutsky et al., 1990). Although the aquatic *Mucronothrus nasalis* lacks fossil evidence, Hammer (1965) suggested that this is a very old species, predating the breakup of the supercontinent Pangaea about 200 million year ago. The distribution of this species is worldwide but discontinuous, and seems to be limited by temperature to cold bogs or cold spring-fed waters (Norton et al., 1988b).

References

Adis, J. & B. Messner, 1991. Langzeit-Überflutungsresistenz als Überlebensstrategie bei terrestrischen Arthropoden – Beispiele aus zentralamazonischen Überschwemmungsgebieten. Deutsche Entomologische Zeitschrift, Neue Folge 38: 211–223.

Baker, G. T. & D. C. Wighton, 1984. Fossil aquatic Oribatid mites (Acari, Oribatida, Hydrozetidae, Hydrozetes) from the Paleocene of South-Central Alberta, Canada. Canadian Entomologist 116: 773–775.

Balogh, J. & S. Mahunka, 1979. New taxa in the system of the Oribatida (Acari). Acta Zoologica Academiae Scientiarum Hungaricae 71: 279–290.

Behan-Pelletier, V. M. & B. Eamer, 2007. Aquatic Acari: ecology, morphology ad behaviour. In Morales-Malacara,

J. B., V. M. Behan-Pelletier, E. Ueckermann, T. M. Pérez, E. Estrada, C. Gispert & M. Badii (eds), Acarology XI: Proceedings of the International Congress. Instituto de Biología UNAM; Facultad de Ciencias, UNAM; Sociedad Latinoamericana de Acarología, México (in press).

Bernini, F., G. Carnevale, G. Bagnoli & S. Stouge, 2002. An early Ordovician oribatid mite (Acari: Oribatida) from the island of Öland, Sweden. In Bernini, F., R. Nannelli, G. Nuzzaci & E. De Lillo (eds), Acarid Phylogeny and Evolution: Adapation in Mites and Ticks. Proceedings of the IV Symposium of the European Association of Acarologists, Siena 2000. Kluwer Academic Publishers, Dordrecht, Boston, London: 45–47.

Bücking, J., H. Ernst & F. Siemer, 1998. Population dynamics of phytophagous mites inhabiting rocky shores – K-strategists in an extreme environment? In Ebermann, E. (ed.), Arthropod Biology: Contributions to Morphology, Ecology and Systematics. Biosystematics and Ecology Series 14: 93–143.

Colloff, M. J. & R. B. Halliday, 1998. Oribatid Mites: a Catalogue of the Australian Genera and Species. Monographs on Invertebrate Taxonomy, Vol. 6. CSIRO Publishing, Melbourne, 224 pp.

Deichsel, R., (2005): A morphometric analysis of the parthenogenetic oribatid mites *Hydrozetes lacustris* and *Hydrozetes parisiensis* – sister species or morphotypes? Phytophaga 14(2004): 377–382.

Druk, A. Ja., 1982. Beetle mites of certain types of bogs in the Moscow Region. In Soil Invertebrates of the Moscow Region. Nauka Publishers, Moscow: 72–77 (in Russian).

Erickson, J. M., 1988. Fossil oribatid mites as tools for Quarternary palaeoecologists: preservation quality, quantities, and taphonomy. Bulletin of the Buffalo Society of Natural Sciences 33: 207–226.

Erickson, J. M., 1996. Can Palaeacarology contribute to global change research? In Mitchell, R., D. J. Horn, G. R. Needham & W. Calvin Welbourn (eds), Acarology IX – Proceedings. Ohio Biological Survey, Columbus, Ohio, vol. 1: 533–537.

Erickson, J. M., R. B. Platt Jr. & D. H. Jennings, 2003. Holocene fossil oribatid mite biofacies as proxies of palaeohabitat at the Hiscock site, Byron, New York. Bulletin of the Buffalo Society of Natural Sciences 37: 176–189.

Fernandez, N. A. & F. Athias-Binche, 1986. Analyse demographique d'une population d'Hydrozetes lemnae Coggi, Acarien Oribate infeode a la lentille d'eau *Lemna gibba* L. en Argentine. I. Methodes et techniques, demographie d' *H. lemnae* comparaisons avec d'autre Oribates. Zoologisches Jahrbuch Systematik 113: 213–228.

Grandjean, F., 1948. Sur les *Hydrozetes* (Acariens) de l'Europe occidentale. Bulletin du Museum national d' Histoire naturelle 20(2): 328–335.

Grandjean, F., 1954. Essai de classification des Oribates (Acariens). Bulletin de la Societé Zoologique de France 78(1953): 421–446.

Grandjean, F., 1965. Complément à mon travail de 1953 sur la classification des Oribates. Acarologia 7: 713–734.

Grandjean, F., 1969. Considérations sur le classement des Oribates. Leur division en 6 groupes majeurs. Acarologia 11: 127–153.

Grandjean, F., 1970. Stases. Actinopiline. Rappel de ma classification des Acariens en 3 groupes majeurs. Terminologie en soma. Acarologia 11: 796–827.

Hammer, M., 1965. Are low temperatures a species-preserving factor? Acta Univ. Lundensis, Section II 2:1–10.

Haumann, G., 1991. Zur Phylogenie primitiver Oribatiden, Acari: Oribatida. dbv Verlag für die Technische Universität Graz, Acari: 237 pp.

Krivolutsky, D. A., A. Ja. Druk, I. S. Ejtminaviciute, L. M. Laskova & E. Karppinen, 1990. Fossil Oribatid Mites. Mokslas Publishers, Vilnius: 109 pp. (in Russian).

Labandeira, C. C., T. L. Phillips & R. A. Norton, 1997. Oribatid mites and the decomposition of plant tissues in paleozoic coal swamp forests. Palaios 12: 319–353.

Maraun, M., M. Heethoff, S. Scheu, R. A. Norton, G. Weigmann & R. H. Thomas, 2003. Radiation in sexual and parthenogenetic oribatid mites (Oribatida, Acari) as indicated by genetic divergence of closely related species. Experimental and Applied Acarology 29: 265–277.

Maraun, M., M. Heethoff, K. Schneider, S. Scheu, G. Weigmann, J. Cianciolo, R. H. Thomas & R. A. Norton, 2004. Molecular phylogeny of oribatid mites (Oribatida, Acari): evidence for multiple radiations of parthenogenetic lineages. Experimental and Applied Acarology 33: 183–201.

Norton, R. A., 1984. Monophyletic groups in the Enarthronota (Sarcoptiformes). In Griffiths, D. A. & C. E. Bowman (eds), Acarology VI. Horwood, Chichester, Vol. 1: 233–240.

Norton, R. A., 1994. Evolutionary aspects of oribatid mite life histories and consequences for the origin of the Astigmata. In Houck, M. (ed.), Mites. Ecological and Evolutionary Analyses of Life-history Patterns. Chapman and Hall, New York: 99–135.

Norton, R. A., 1998. Morphological evidence for the evolutionary origin of Astigmata (Acari: Acariformes). Experimental Applied Acarology 22: 559–594.

Norton R. A., P. M. Bonamo, J. D. Grierson & W. A. Shear, 1988a. Oribatid mite fossils from a terrestrial Devonian deposit near Gilboa, New York. Journal of Paleontolology 62: 259–269.

Norton, R. A., J. B. Kethley, D. E. Johnston & B. M. OConnor, 1993. Phylogenetic perspectives on genetic systems and reproductive modes of mites. In Wrensch, D. L. & M. A. Ebbert (eds), Evolution and Diversity of Sex Ratio in Insects and Mites. Chapman and Hall, New York: 8–99.

Norton, R. A. & S. C. Palmer, 1991. The distribution, mechanisms and evolutionary significance of parthenogenesis in oribatid mites. In Schuster, R. & P. W. Murphy (eds), The Acari – Reproduction, Development and Life-history Strategies. Chapman and Hall, London, New York: 107–136.

Norton, R. A., D. D. Williams, I. D. Hogg & S. C. Palmer, 1988b. Biology of the oribatid mite *Mucronothrus nasalis* (Acari: Oribatida: Trhypochthoniidae) from a small coldwater springbrook in Eastern Canada. Canadian Journal of Zoology 66: 622–629.

Piffl, E., 1978. Oribatei. In Illies, J. (ed.), Limnofauna Europaea, 2nd Ed., Fischer, Stuttgart, New York: 182–183.

Pugh, P. J. A., 1996. Edaphic oribatid mites (Cryptostigmata: Acarina) associated with an aquatic moss on sub-Antarctic South Georgia. Pedobiologia 40: 113–117.

Schatz, H., 1985. The life cycle of an alpine Oribatid mite, *Oromurcia sudetica* Willmann. Acarologia 26: 95–100.

Schatz, H., 2002. Die Oribatidenliteratur und die beschriebenen Oribatidenarten (1758–2001) – Eine Analyse. Abhandlungen und Berichte des Naturkunde Museums Görlitz 74: 37–45.

Schatz, H., 2005. Diversity and global distribution of oribatid mites – evaluation of the present state of knowledge. Phytophaga 14(2004): 485–500.

Schatz, H., & R. Gerecke, 1996. Hornmilben (Acari, Oribatida) aus Quellen und Quellbächen im Nationalpark Berchtesgaden (Oberbayern) und in den Südlichen Alpen (Trentino – Alto Adige). Berichte des naturwissenschaftlich-medizinischen Vereins Innsbruck 83: 121–144.

Schelvis, J., 1990. The reconstruction of local environments on the basis of remains of oribatid mites (Acari: Oribatida). Journal of Archaeological Science 17: 559–572.

Schneider, K., K. Renker, S. Scheu & M. Maraun, 2005. Feeding biology of oribatid mites: a mini review. Phytophaga 14(2004): 247–256.

Sivhed, U., & J. A. Wallwork, 1978. An early jurassic Oribatid mite from Southern Sweden. Geologiska Föreningen i Stockholm Förhandlingar 100: 65–70.

Solhøy, T., 2001. Oribatid mites. In Smol, J. P., J. B. Birks & W. M. Last (eds), Tracking Environmental Change Using Lake Sediments, Vol. 4. Zoological Indicators. Kluwer Academic Publishers, Dordrecht, The Netherlands: 81–104.

Solhøy, I. W., & T. Solhøy, 2000. The fossil oribatid mite fauna (Acari, Oribatida) in late glacial and early holocene sediments in Krakenes Lake, Western Norway. Journal of Paleolimnology 23: 35–47.

Søvik, G., 2004. The biology and life history of arctic populations of the littoral mite Ameronothrus lineatus (Acari, Oribatida). Experimental and Applied Acarology 34: 3–20.

Subias, L. S., 2004. Listado sistemático, sinonímico y biogeográfico de los Ácaros Oribátidos (Acarifomes, Oribatida) del mundo (1748–2002). Graellsia 60: 3–305.

Walter, D. E. & H. C. Proctor, 1999. Mites. Ecology, Evolution and Behaviour. CABI Publishing, Wallingford, New York, Sydney: 322 pp.

Weigmann, G., 1997. Die Hornmilben-Fauna (Acari, Oribatida) in Auenböden des Unteren Odertals. Faunistisch-ökologische Mitteilungen 7: 319–333.

Weigmann, G., 2005. Recovery of the oribatid mite community in a flood plain after decline due to long time inundation. Phytophaga 14(2004): 201–208.

Weigmann, G., 2006. Hornmilben (Oribatida). Die Tierwelt Deutschlands, 76. Teil. Goecke & Evers, Keltern, 520 pp.

Weigmann, G. & R. Deichsel, 2006. Acari: Limnic Oribatida. In Gerecke, R. (ed.), Chelicerata: Araneae, Acari I. Süßwasserfauna von Mitteleuropa, Vol. 7/2-1. Spektrum, München: 89–112.

Hydrobiologia (2008) 595:329–338
DOI 10.1007/s10750-007-9116-z

FRESHWATER ANIMAL DIVERSITY ASSESSMENT

Global diversity of springtails (Collembola; Hexapoda) in freshwater

Louis Deharveng · Cyrille A. D'Haese · Anne Bedos

Abstract Water-dependency appeared independently in several clades of the class Collembola, which is basically of terrestrial origin according to recent phylogenetic analyses. Though moderately diversified (less than 8,000 species), Collembola are among the most numerous terrestrial arthropods in wetland communities, with a small number of species living on the surface of water. Many species are dependent on water-saturated atmosphere of caves, and on snow or ice in high mountains. A total of 525 water-dependent species have been recognized, of which 103 are linked to free freshwaters and 109 to anchialine or marine waters. Many interstitial species are also dependent to an unknown extent on water saturation in the deep layers of the soil. The numbers provided here are underestimates, as Collembola are extremely poorly known outside the Holarctis, and the ecology of described species usually not documented. However, a general biogeographical pattern is emerging from available data. The most remarkable feature is that about 15% of the fauna is water-dependent in the holarctic region, compared to 4% in the tropics and southern hemisphere.

Guest editors: E. V. Balian, C. Lévêque, H. Segers & K. Martens
Freshwater Animal Diversity Assessment

Electronic supplementary material The online version of this article (doi:10.1007/s10750-007-9116-z) contains supplementary material, which is available to authorized users.

L. Deharveng (✉) · C. A. D'Haese · A. Bedos
UMR5202 CNRS, Origine, Structure et Evolution de la Biodiversité, Muséum National d'Histoire Naturelle, CP50 bat. Entomologie, 45 rue Buffon, Paris 75005, France
e-mail: deharven@mnhn.fr

Keywords Collembola · Wet habitats · Biogeography

Introduction

The springtails (Collembola) are small hexapods measuring 0.2–8 mm, of elongated to globular habitus (Fig. 1). Many species have a ventral jumping organ on abdominal segments III and IV named the furca, which is reduced and even lacking in several groups. Some species have a life cycle of less than 2 weeks (*Folsomia candida*), while others need several months to reach the adult stage. They have adapted to almost all terrestrial habitats, and are numerically dominant or subdominant in invertebrate communities from seashore to the highest altitudes, from deep caves to the forest canopy, from Antarctica glaciers to tropical deserts.

Collembola are basically terrestrial animals. But many species are more or less connected to freshwater or marine water. On the terrestrial side of the water/terrestrial ecotones, Collembola are found in large number (Deharveng & Bedos, 2004). They are often among the most abundant arthropods both in

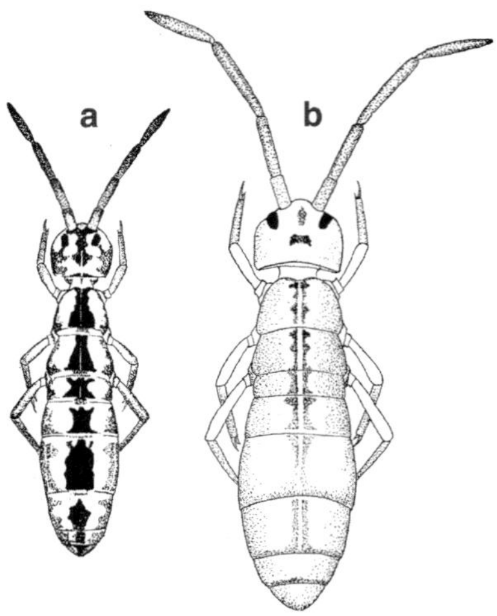

Fig. 1 *Isotomurus alticolus* (Carl) from Austria, non-cave (**a**) and cave (**b**) specimens (after Stach, 1947)

density (Petersen & Luxton, 1982) and in species number (Stork, 1988; Stork & Blackburn, 1993), even if their participation to the biomass remains low (Petersen, 1994). Though no Collembola are adapted to life in water, many have developed successful mechanisms to exploit water surface or water film in extremely humid environments. First, Collembola have a strongly water-repellant body cuticle which allows them to float on the surface of water (Noble-Nesbitt, 1963). Thus, flooding acts as a major dispersal agent for springtails. Second, claws and mucro are often strongly modified in hydrophilous Collembola, especially among neustonic species. Claws tend to become thin and elongated compared to those of non-hydrophilous relatives. This trend is observed in many species living in habitats where free water is abundant (along streams, around ponds, in caves), whatever the group (Christiansen, 1961; Deharveng, 1988; D'Haese, 2003). Claw elongation is considered an adaptive character as it appears independently in phyletically remote lineages. Christiansen (1965) investigated its functional basis, and showed that it allowed cave Collembola to walk on water. The same is probably true for non-cave species, which exhibit the same modification of the claw morphology. Another morphological modification limited to neustonic Symphypleona and Neelipleona, consists in a considerable enlargement of the mucro.

Water-dependent Collembola may be allocated to four main categories according to the conditions of water availability, and the relative importance of non-hydric factors of their habitats.

(1) Epigean hydrophilous species. They live on the surface of freshwater, on soil or on low vegetation close to water.
(2) Cryophilous species. They are active on, under or at the edge of permanent snow fields or glaciers. They are also present on or near small water puddles which forms at the edge of snow fields or on ice surface during summer. Permanently cold habitats occur either at high elevation in temperate and tropical mountains, or at high latitudes (Antarctic and Arctic regions). Collembola are relatively diversified in such extreme habitats, where they tend to represent the largest component of the invertebrate fauna (Cassagnau, 1961).
(3) Cave hydrophilous species. They are strictly associated with water-saturated atmosphere of large subterranean voids (caves and "Milieu Souterrain Superficiel", Juberthie, 2000). Collembola represent the most diversified and numerous invertebrates in these subterranean habitats. Thibaud (1970) showed experimentally that several cave Hypogastruridae cannot survive more than a few hours below 96% of relative humidity. Collembola only found on or very close to free water (waterfall, underground stream, film of water on rocks) with extremely elongate appendages and claws are the most sensitive (Sket, 2004).
(4) A fourth category, interstitial hydrophilous species, is likely to exist inside humid sandy sediments, at the level of water table. However, ecological data are lacking to ascertain whether some of the numerous minute interstitial species which have been discovered these last two decades (Thibaud & Christian, 1997) may be considered as hydrophilous, and they will not be dealt with here.

The ecological categories defined above are more or less relevant to anchialine or brackish water fauna as well. The tolerance to freshwater of most euryhaline Collembola is unknown but a few *Pseudanurida* and *Yuukianura* are found both near marine water and near inland streams.

All ecological categories listed above concern Collembola which may live at the surface of water or ice, but there also are records of species collected underwater, including sea water (Jacquemart & Jacques, 1980). Some records are probably pollution of water samples by epigean species. But some species (not necessarily hydrophilous) have been experimentally shown to be able to survive submersion in sea water for days (Coulson et al., 2002). This phenomenon is more documented among freshwater species. Some Onychiuridae, Tomoceridae, Entomobryidae (Ruppel, 1953) and Symphypleona (Falkenhan & Fox, 1932; Delamare Deboutteville, 1952) may live a few days underwater. Thibaud (1970) tested different Hypogastruridae species in this respect. The animals were able to move and survive under water up to 40 days after hatching and up to 36 days for adults (in *Typhlogastrura balazuci*, a cave-dwelling species), but were usually unable to moult (though moulting of *Ceratophysella armata* was obtained by Britt (1951)). However, there is no evidence so far that the total life cycle of any species may occur underwater. In the field, eggs may hatch after long period of flooding (Beck, 1976; Tamm, 1984). To which extent this ability may contribute to explain the rapid recover of some riparian communities after flooding, as documented in Russell et al. (2004) and Russel & Griegel (2006), remains to be explored.

Taxonomic diversity

With 7,761 species and 643 genera (Janssens, 2006), including 2,000 species in Europe (Fauna Europaea, 2004), Collembola are not among the most speciose groups of living invertebrates. However, being among the less known, the group experiences a high annual rate of species richness increase (Deharveng, 2004), and its relative contribution to global biodiversity is growing more than that of most other invertebrates, with a total number of species possibly as high as 50,000 according to Hopkin (1997, 1998).

Collembola are distributed in four orders (Poduromorpha, Entomobryomorpha, Symphypleona and Neelipleona) and 30 families (Deharveng, 2004). The proportion of hydrophilous species is difficult to establish, because ecological and biological data on most species, especially in the tropics, are lacking. An analysis based on 3,788 species (half the total number of Collembolan species) for which a minimal information on habitats was available or was extrapolated from our knowledge of the concerned genera, allows to recognize 109 marine species, 414 freshwater species in three of the four categories defined above (103 epigean hydrophilous, 26 cryophilous, 285 cave hydrophilous), and 2 marine species which are also freshwater hydrophilous. Tables 1 and 2 reflect the general pattern of hydrophilous taxonomic richness in the group across the four orders of Collembola and the different biogeographical regions on Earth. The contribution of the different orders to the biodiversity of each habitat is quite uneven, and the same degree of unevenness is retrieved inside the orders themselves. Among Entomobryomorpha for instance, Isotomidae bring about half of the total number of hydrophilous Collembola, including most of the cryophilous ones, but it has practically no cave species. Conversely, Entomobryidae, Oncopoduridae and Tomoceridae, all three Entomobryomorpha families, contribute significantly to cave biodiversity, but nothing for other habitats. Among biogeographical regions, striking patterns are emerging. Northern temperate and boreal regions (Nearctic and Palaearctic) have about 15% of hydrophilous species in their fauna, while tropical and southern hemisphere regions have 1–6% (4% when pooled together).

However, these values of species richness are rapidly changing. In Europe, we begin to realize that the classical *Isotomurus palustris* Mueller, 1776, the most abundant "species" of holarctic temperate wetlands, is probably a cluster of closely related forms that are awaiting description (Cassagnau, 1987; Deharveng & Lek, 1993; Carapelli et al., 2001). Hydrophilous species of several other genera have been overlooked in the literature (*Pseudisotoma microchaeta* Cassagnau, for instance) or were recently discovered in Europe (*Isotomiella hygrophila* Sterzynska & Kaprus, 2001). According to Murphy (1990), the Sminthurididae (genera *Sminthurides*, *Sphaeridia* and their allies) which are dominant in damp habitats of the tropics may also encompass a large diversity of undescribed species (Murphy, 1966). This points to the necessity of a taxonomic re-appraisal of existing species, as well as increased sampling effort.

Origin and phylogeny of Collembola

Collembola provided the first known hexapods fossils from the early Devonian (Hirst & Maulik, 1926;

Table 1 Number of Collembolan species in the different regions of the world (total, WDpt or freshwater-dependent species, including Epig or epigean hydrophilous species, see Introduction) and the four orders of Collembola

Number of Species	PA			NA			AT			NT			OL		
	Total	WDpt	Epig	Total	WDpt	Epig	Total	WDpt	Epig	Total	WDpt	Epig	Total	WDpt	Epig
Entomobryomorpha	928	151	33	168	22	13	147	1	1	245	8	3	319	24	5
Neelipleona	6	3	1	4	0	0	3	0	0	4	0	0	2	0	0
Poduromorpha	999	145	16	190	17	12	97	2	1	247	10	4	203	1	0
Symphypleona	169	39	18	54	10	7	49	3	3	59	10	10	41	9	8
Total	2,102	338	68	416	49	32	296	6	5	555	28	17	565	34	13
Number of Species	AU			PAC			ANT			Total					
	Total	WDpt	Epig	Total	WDpt	Epig	Total	WDpt	Epig	Total	WDpt	Epig			
Entomobryomorpha	113	4	4	107	2	2	23	0	0	1,711	190	43			
Neelipleona	2	0	0	4	0	0	1	0	0	10	3	1			
Poduromorpha	91	0	0	84	0	0	26	1	1	1,677	157	19			
Symphypleona	57	2	2	30	1	1	6	0	0	390	64	40			
Total	263	6	6	225	3	3	56	1	1	3,788	414	103			

Values derived from a subset of 3,788 species of Collembola. The main bias is the inclusion of Nepal in the palaearctic region, whereas its Collembolan fauna is relatively well documented and mostly of tropical or tropical montane affinities. PA—Palaearctic; NA—Nearctic; NT—Neotropical; AT—Afrotropical; OL—Oriental; AU—Australasian; PAC—Pacific Oceanic Islands; ANT—Antarctic

Table 2 Number of Collembolan genera in the different regions of the world and the four orders of Collembola

Number of Genera	PA		NA		AT		NT		OL		AU		PAC		ANT		Total	
	WDpt	Epig	WDpt	Epig	WDpt	Epig	WDpt	Epig	WDpt	Epig	WDpt	Epig	WDpt	Epig	WDpt	Epig	WDpt	Epig
Entomobryomorpha	31	13	10	5	1	1	4	1	6	1	1	1	1	1	0	0	33	13
Neelipleona	1	1	0	0	0	0	0	0	0	0	0	0	0	0	0	0	1	1
Poduromorpha	35	6	7	4	2	1	8	4	1	0	0	0	0	0	1	1	35	7
Symphypleona	4	4	5	4	2	2	3	3	3	2	2	2	1	1	0	0	9	8
Total	71	24	22	13	5	4	15	8	10	3	3	3	2	2	1	1	78	29

Cassagnau, 1990). However, phylogenetic analyses cannot integrate this information because the known fossils are in too poor condition, and we have to rely entirely on extant species to reconstruct evolutionary history of the group. Collembola were long considered as a group of Insects. Actually, they represent a class in their own, of controversial affinities among Arthropods. A detailed phylogeny of Collembola based on morphological and molecular characters has been produced by D'Haese (2002, 2003, Fig. 2), who tested the hypothesis of an aquatic origin for the group. Physiological data suggest that Collembola might have evolved from marine ancestors (haemolymph with high osmotic pressure and mainly composed of inorganic salts), but this is still a matter of debate in the broader context of hexapod origin (Little, 1983, 1990). The secondary transition from terrestrial to freshwater habitats is physiologically more convincing. The phylogeny of D'Haese (2003) shows that life on water was not ancestral but derived, supporting a terrestrial (edaphic) origin for the springtails. Among terrestrial Collembola, those lacking molar plate on the mandible (Neanuridae, Brachystomellidae, Odontellidae) are supposed to ingest the water film which surrounds soil particles, rich in bacteria and Protozoa ("suspension feeders", Adams & Salmon, 1972). This aquatic type of feeding, considered as derived phylogenetically, has however rarely led to adaptation to aquatic life. Actually, only a few Collembolan lineages developed adaptations to aquatic life, and these adaptations probably arose independently. But they have been surprisingly successful. *Podura aquatica* Linné, 1758, the only species of the family Poduridae, is extremely frequent and abundant in northern temperate countries. The genus *Isotomurus* has undergone a tremendous diversification in the Holarctic region, colonizing all kinds of damp habitats. *Sminthurides*, often very abundant near and on freshwater worldwide, exhibits an intermediate level of species diversification.

Geographical distribution

Hydrophilous Collembola are present worldwide (Thibaud & Massoud, 1986; Christiansen & Snider, 1996). However, the scarcity of information about this category of springtails prevents any sound

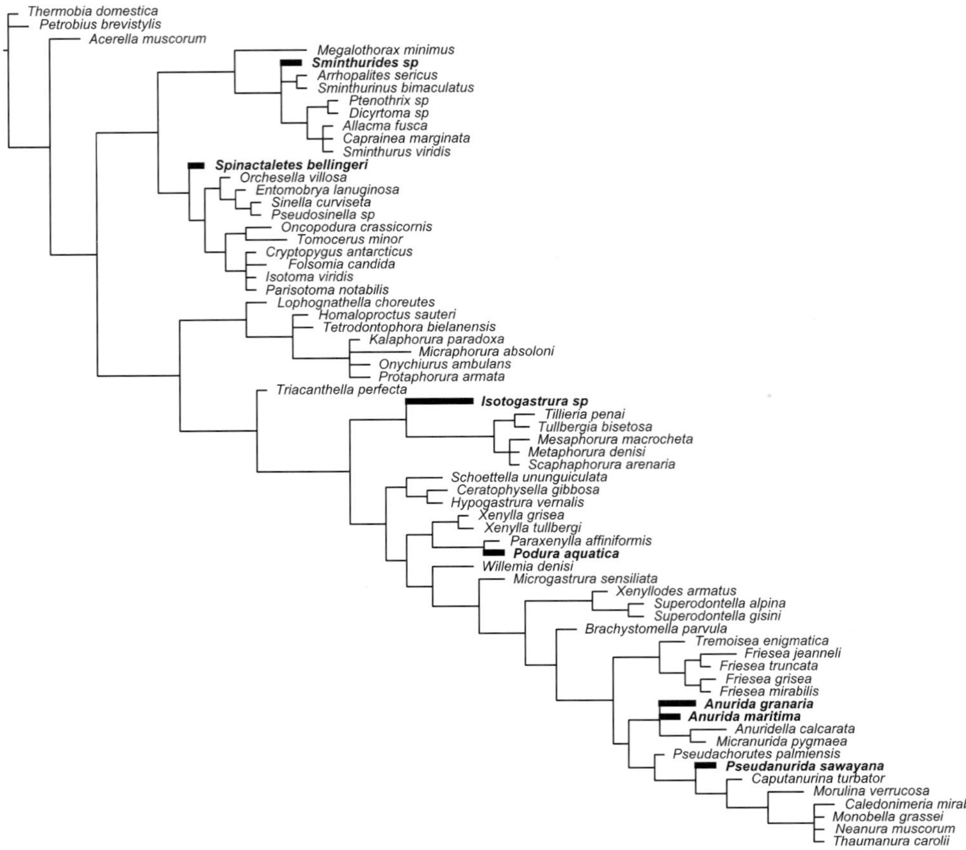

Fig. 2 A phylogeny of Collembola based on morphological data after D'Haese (2003). Water-adapted species are indicated in bold (4 marine species, and 3 freshwater-dependent species: *Sminthurides* sp., *Podura aquatica* and *Anurida granaria*).

biogeographical analysis. Data for tropical Africa are limited to Gambia (Murphy, 1960). For tropical Asia, almost all original information comes from Singapore (Murphy, 1990), with a few from Indonesia (Handschin, 1931). The work of Deharveng & Bedos (2004) on aquatic Collembola of Malaysia is a simple synthesis of existing literature data. Publications not specifically devoted to water-dependent Collembola, especially the excellent taxonomic works of Yosii (1971) on tropical Asia, remain elusive about the ecology of species. Many distributional data for the new world are found in the works of Christiansen & Bellinger (1998), and Mari Mutt & Bellinger (1990), but species ecology is not documented. The comprehensive study of Fjellberg (1985) about Alaskan Poduromorpha, more informative in this respect, is geographically more limited. The information available about Australian and Subantarctic Collembola rarely include data about the dependence of species on water (Deharveng, 1981; Greenslade, 1994). In Europe, several important works deal with Collembolan fauna of wetlands (e.g. Strenzke, 1955; Palissa, 2000), and may be used for regional biogeography but some are taxonomically outdated. Detailed regional fauna (Fjellberg, 1994; Jordana et al., 1997), local ecological studies (Deharveng & Lek, 1995; Lek-Ang et al., 1999; Sterzynska & Ehrnsberger, 1999; Lek-Ang & Deharveng, 2002; Russell et al., 2004; Russel & Griegel, 2006) as well as the classical works of Stach (for instance on Isotomidae in 1947, on Sminthuridae in 1956) and Gisin (1960) provide sound ecological information on water-dependent species. Yosii published many articles (1977) on Japanese Collembola, with very limited information or no data on species ecology. The synopses of the Palaearctic fauna (Zimdars & Dunger, 1994; Bretfeld, 1999; Potapov, 2001; Thibaud et al., 2004) put together a huge amount of ecological

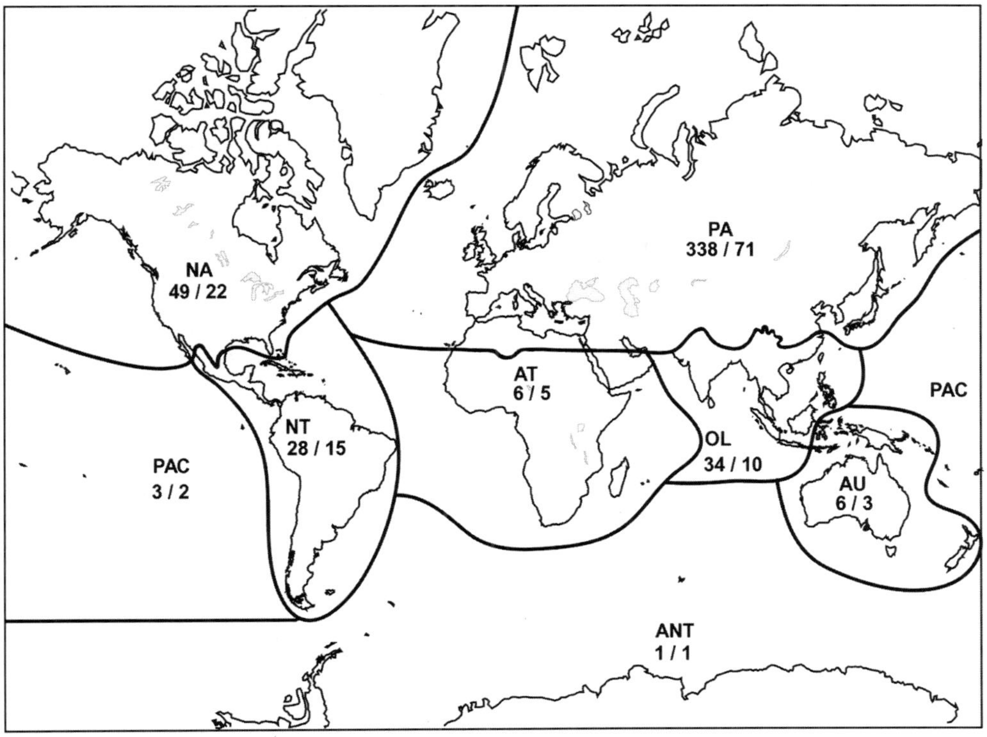

Fig. 3 Global distribution of freshwater-dependent species of Collembola identified from a subset of 3,788 species (species number/genera number). PA—Palaearctic; NA—Nearctic; NT—Neotropical; AT—Afrotropical; OL—Oriental; AU—Australasian; PAC—Pacific Oceanic Islands; ANT—Antarctic

as well as distributional information, and represent the richest source of information for the old world (Fig. 3).

In spite of this uneven information, available data point to a surprising contrast between tropical and non-tropical regions in each ecological category. Two of the main genera of tropical, non-interstitial, fully marine Collembola (*Pseudanurida*, *Oudemansia*) are widely distributed across at least two continents, but are absent in temperate and boreal regions, with the noticeable exception of the Koreo-Japanese area. Conversely, *Friesea*, *Anuridella*, *Halisotoma*, well-represented in the holarctic region, are rare or absent in marine habitats of the tropics. Similar contrast is retrieved among interstitial forms (Thibaud & Christian, 1997), some of which are probably hydrophilous as well. Among epigean hydrophilous species, *Isotomurus* is dominant in non-tropical regions while *Sminthurides* is more diversified in the tropics and the southern hemisphere. Among cave restricted species the contrast is also striking between temperate regions, where Onychiuridae and Hypogastruridae are well diversified, and tropical regions, where these genera are virtually absent, but Paronellidae very diversified in caves.

Aquatic Collembola include a few widespread (like *Podura aquatica*) and many endemic species. Endemism patterns are extremely contrasted among regions, habitats and taxa, in terms of proportion of endemics in the fauna and range size of the species. Considering the species with a known range smaller than 500,000 sq km as endemic, and keeping in mind that distributional information is still poor for ascertain endemic status of many species in the tropics, a few general features nevertheless emerge from available data.

Possibly as a result of quaternary glaciations, hydrophilous endemics seem to be absent in boreal and northern temperate regions of Europe. In contrast, Mediterranean regions exhibit a high richness in endemics, many of them undescribed (unpublished data). At the world scale, high endemicity areas for hydrophilous Collembola are difficult to spot with precision given the huge geographical gaps in our knowledge. However, supra-specific diversity peaks in at least three areas: the Far-East of the Palaearctic region (Japan, Korea, Primorye), the Pyreneo-Cantabric range and the Dinarides. The Far-East Palaearctic hot-spot reflects the mixing of mostly-tropical lineages with palaearctic ones. The two other regions, located in western Europe, are world hot-spots for cave biodiversity, perhaps as a result of an history of available habitats and high productivity levels longer than in surrounding regions (Culver et al., 2006).

The link between habitats and endemicity is also obvious. Endemicity among marine hydrophilous Collembola is lower than among freshwater ones. Marine Collembola have in particular several species widespread in Europe or in large regions of the world. The less fragmented structure of littoral habitats may explain this pattern. Not surprisingly, endemism rates vary also greatly in the recognized categories of freshwater Collembola, in relation to the proneness of habitats to fragmentation. Several monospecific and phyletically isolated genera exist among cave hydrophilous springtails (*Gnathofolsomia* in the Austrian Alps, *Bessoniella* and *Ongulogastrura* in the Pyrenees, *Ongulonychiurus* in the Cantabrian range, are the most remarkable). Subterranean habitats have also many endemic species belonging to more widespread genera. Cryophilous springtails have only one endemic genus (*Gnathisotoma*, restricted to the Pyrenees) with two species; all other endemics of this ecological category belong to non-endemic genera. No genus is known as endemic of epigean wet habitats. Most are widespread, with either one or a few species of very large distribution (*Podura* and *Pseudobourletiella*, mostly in northern hemisphere, for instance), or several largely distributed species (*Sminthurides*), or many closely related and often endemic species (*Isotomurus*). Many endemic species are expected in this last genus, as a result of a recent re-appraisal of its taxonomic forms (Carapelli et al., 2001).

References

Adams, E. C. G. & J. T. Salmon, 1972. The Mouth-parts and Feeding Methods of *Brachystomella parvula* (Schaeffer) (Collembola: Brachystomellidae). Transactions of the Royal Entomological Society, London 124(3): 269–286.

Beck, L., 1976. Zum Massenwechsel der Makro-Arthropodenfauna des Bodens in Überschwemmungswäldern des Zemtralen Amazonasgebietes. Amazoniana 6: 1–20.

Bretfeld, G., 1999. Symphypleona. In Dunger, W. (ed.), Synopses on Palaearctic Collembola, Vol. 2. Staatliches Museum für Naturkunde Görlitz 71: 1–318.

Britt, N. W., 1951. Observations on the Life-history of the Collembolen *Achorutes armatus*. American Microscopic Society Transactions 70:119–132.

Carapelli, A., F. Frati, P. P. Fanciulli & R. Dallai, 2001. Taxonomic Revision of 14 South-western European Species of *Isotomurus* (Collembola, Isotomidae), with Description of Four New Species and the Designation of the Neotype for *I. palustris*. Zoologica Scripta 30: 115–143.

Cassagnau, P., 1961. Ecologie du sol dans les Pyrénées Centrales. Les Biocénoses de Collemboles. Hermann, Paris, 265 pp.

Cassagnau, P., 1987. A Propos des Types de Coloration chez *Isotomurus Palustris* (Müller) (Collemboles). Revue d'Ecologie et de Biologie du Sol 24: 85–89.

Cassagnau, P., 1990. Des Hexapodes Vieux de 400 Millions D'années: Les Collemboles. I. Biologie et Évolution. L'Année Biologique 29: 1–37.

Christiansen, K. A., 1961. Convergence and Parallelism in Cave Entomobryinae. Evolution 15(3): 288–301.

Christiansen, K. A., 1965. Behavior and Form in the Evolution of Cave Collembola. Evolution 19: 529–537.

Christiansen, K. A. & P. F. Bellinger, 1998. The Collembola of North America, North of Rio Grande. Grinnell College, Grinnell, Iowa, 1520 pp.

Christiansen, K. A. & R. J. Snider, 1996. Aquatic Collembola. In Merritt, R. W. & K. W. Cummins (eds), An Introduction to the Aquatic Insects of North America. Kendall/Hunt Publ., Iowa: 113–123.

Coulson, S. J., I. D. Hodkinson, N. R. Webb & J. A. Harrison, 2002. Survival of Terrestrial Soil-dwelling Arthropods on and in Seawater: Implications for Trans-oceanic Dispersal. Functional Ecology 16(3): 353–356.

Culver D. C., L. Deharveng, A. Bedos, J. J. Lewis, M. Madden, J. R. Reddell, B. Sket, P. Trontelj & D. White, 2006. The Mid-Latitude Biodiversity Ridge in Terrestrial Cave Fauna. Ecography 29: 120–128.

Deharveng, L., 1981. Collemboles des Iles Subantarctiques de l'Océan Indien. Biologie des sols, CNFRA 48: 33–108.

Deharveng, L., 1988. Collemboles Cavernicoles VII.—*Pseudosinella bessoni* n.sp. et note sur L'évolution Morphologique de la Griffe Chez les *Pseudosinella*. Revue Suisse de Zoologie 95(1): 203–208.

Deharveng, L., 2004. Recent Advances in Collembolan Systematics. Pedobiologia 48: 415–433.

Deharveng, L. & A. Bedos, 2004. Insecta: Collembola. In Yule, C.M. & H. S. Yong (eds), Freshwater Invertebrates of the Malaysian Region. Academy of Sciences, Malaysia: 384–393.

Deharveng, L. & S. Lek, 1993. Remarques sur la Morphologie et la Taxonomie du Genre *Isotomurus* Börner, 1903 et Description de Deux Espèces Nouvelles de France (Collembola: Isotomidae). Annales de la Société entomologique de France 29: 245–259.

Deharveng, L. & S. Lek, 1995. High Diversity and Community Permeability: The Riparian Collembola (Insecta) of a Pyrenean Massif. Hydrobiologia 312: 59–74.

Delamare Deboutteville, C., 1952. Données Nouvelles sur la Biologie des Animaux Cavernicoles. Notes biospéologiques 7: 15–20.

D'Haese, C. A., 2002. Were the First Springtails Semi-aquatic? A Phylogenetic Approach by Means of 28S rDNA and Optimization Alignment. Proceedings of the Royal Society, London B 269: 1143–1151.

D'Haese, C. A., 2003. Morphological Appraisal of Collembola Phylogeny with Special Emphasis on Poduromorpha and a Test of the Aquatic Origin Hypothesis. Zoologica Scripta 32: 523–586.

Falkenhan, H.-H. & I. Fox, 1932. Biologische Beobachtungen an *Sminthurides aquaticus* (Collembola). Zeitschrift für wissenschaftliche zoologie 141(4): 525–580.

Fauna Europaea Web Service, 2004. Fauna Europaea version 1.1, Available online at http://www.faunaeur.org.

Fjellberg, A., 1985. Arctic Collembola I. Alaskan Collembola of the Families Poduridae, Hypogastruridae, Odontellidae, Brachystomellidae and Neanuridae. Entomologica Scandinavica Suppl. 21: 1–126.

Fjellberg, A., 1994. The Collembola of the Norwegian Arctic Islands. Meddelelser 133: 1–57.

Gisin, H., 1960. Collembolenfauna Europas. Museum d'Histoire Naturelle, Genève, 312 pp.

Greenslade, P. J., 1994. Collembola. In Houston, W. W. K. (ed.), Zoological Catalogue of Australia, Vol. 22. Protura, Collembola, Diplura. CSIRO, Melbourne, Australia: 19–138.

Handschin, E., 1931. XLVI. Hygrophile Collembolen aus Niederländisch-Indien. Archiv für Hydrobiologie. Tropische Binnengewässer suppl. 9: 472–490.

Hirst, S. & S. Maulik, 1926. On some Arthropod Remains from the Rhynie Chert (Old Red Sandstone). Geological Magazine 63: 69–71.

Hopkin, S. P., 1997. Biology of the Springtails (Insecta: Collembola). Oxford University Press, 330 pp.

Hopkin, S. P., 1998. Collembola: The Most Abundant Insects on Earth. Antenna 22: 117–121.

Jacquemart, S. & J.-M. Jacques, 1980. A Propos d'un Collembole Entomobryen à la fois Marin et Désertique. Annales de la Société Royale Zoologique de Belgique 109 (1): 9–18.

Janssens, F., 2006. Checklist of the Collembola: Taxonomical Speciation Rate. http://www.collembola.org/doc/history.htm.

Jordana, R., J. I. Arbea, C. Simon & M. J. Luciañez, 1997. Collembola Poduromorpha. In Ramos, M. A. et al. (eds), Fauna Iberica, Vol. 8. Museo Nacional de Ciencas Naturales. CSIC, Madrid: 1–807.

Juberthie, C., 2000. The Diversity of the Karstic and Pseudokarstic Hypogean Habitats in the World. In Wilkens, H., D. C. Culver & W. F. Humphreys (eds), Ecosystems of the World 30, Subterranean Ecosystems. Elsevier, Amsterdam: 17–39.

Lek-Ang, S. & L. Deharveng, 2002. Seasonal Variations of Collembola (Hexapoda) Assemblages in Damp Riparian Habitats. Archiv für Hydrobiologie 153: 635–655.

Lek-Ang, S., L. Deharveng & S. Lek, 1999. Predictive Models of Collembolan Diversity and Abundance in a Riparian Habitat. Ecological Modelling 120: 247–260.

Little, C., 1983. The Colonisation of Land. Origins and Adaptations of Terrestrial Animals. Cambridge University Press, Cambridge.

Little, C., 1990. The Terrestrial Invasion. An Ecophysiological Approach to the Origins of Land Animals. Cambridge University Press, Cambridge.

Mari Mutt, J. A. & P. F. Bellinger, 1990. A catalog of the Neotropical Collembola, Including Nearctic Areas of Mexico. Fauna & Flora Handbook 5. Sandhill Crane Press, Gainesville, Florida, 237 pp.

Murphy, D. H., 1960. Collembola Symphypleona from the Gambia, with a Note on the Biogeography of Some Characteristic Savanna Forms. Proceedings of the Zoological Society of London 134: 557–594.

Murphy, D. H., 1966. Taxonomy and Bionomics of Eight Species of Sphaeridia (Collembola, Sminthuridae) in a Singapore Flood-plain. Revue d'Ecologie et de Biologie du Sol 3: 65–96.

Murphy , D. H., 1990. "Walkers on Water" an Account of the Pleuston of Singapore. Essays in Zoology: 153–168.

Noble-Nesbitt, P. J., 1963. Transpiration in *Podura aquatica* L. (Collembola) and the Wetting Properties of its Cuticle. Journal of Experimental Biology 40: 681–700.

Palissa A., 2000. Collembola. In Susswasserfauna von Mitteleuropa, Band 10. Gustav Fischer Verlag, 166 pp.

Petersen, H., 1994. A Review of Collembolan Ecology in Ecosystem Context. Acta Zoologica Fennica 95: 11–118.

Petersen, H. & M. Luxton, 1982. A Comparative Analysis of Soil Fauna Populations and Their Role in Decomposition Processes. Oikos 39: 287–388.

Potapov M., 2001. Isotomidae. In Dunger, W. (ed.), Synopses on Palaearctic Collembola, Vol. 3. Staatliches Museum für Naturkunde Görlitz 73: 1–603.

Ruppel, H., 1953. Physiologische Untersuchungen uber die Bedeutung des Ventraltubusund die Atmung der Collembolen. Zoologische Jahrbuch 2: 537–542.

Russell, D. J. & A. Griegel, 2006. Influence of Variable Inundation Regimes on Soil Collembola. Pedobiologia 50(2): 165–175.

Russell, D. J., A. Hauth & O. Fox, 2004. Community Dynamics of Soil Collembola in Floodplains of the Upper Rhine Valley. Pedobiologia 48(5–6): 527–536.

Sket, B., 2004. The Cave Hygropetric: A Little Known Habitat and Its Inhabitants. Archiv für Hydrobiologie 160: 413–425.

Stach, J., 1947. The Apterygotan Fauna of Poland in Relation to the World-fauna of this Group of Insects. Family: Isotomidae. Acta Monographica Museum Historia Naturalis, Krakow, 488 pp.

Stach, J., 1956. The Apterygotan Fauna of Poland in Relation to the World-fauna of This Group of Insects. Family: Sminthuridae. Acta Monographica Museum Historia Naturalis, Krakow, 287 pp.

Sterzynska, M. & R. Ehrnsberger, 1999. Diversity and Structure of Collembola Communities in Wetlands. In Tajovsky, K. & V. Pizl (eds), Soil Zoology in Central Europe. T. J. Posekany, Ceske Budejovice, Czech Republic: 325–334.

Stork, N. E., 1988. Insect Diversity: Facts, Fictions and Speculation. Biological Journal of the Linnean Society 35: 321–337.

Stork, N. E. & T. M. Blackburn, 1993. Abundance, Body Size and Biomass of Arthropods in Tropical Forest. Oikos 67: 483–489.

Strenzke, K., 1955. Thalassobionte und Thalassophile Collembola. Die Tierwelt der Nord und Ostsee 36: 1–52.

Tamm, J. C., 1984. Surviving Long Submergence in the Egg Stage—a Successful Strategy of Terrestrial Arthropods Living on Flood Plains (Collembola, Acari, Diptera). Oecologia 61(3): 417–419.

Thibaud, J. M., 1970. Biologie et Écologie des Collemboles Hypogastruridae Édaphiques et Cavernicoles. Mémoires du Museum National d'Histoire Naturelle, Paris, Série A, Zoologie 61: 83–201.dfs.

Thibaud, J. M. & E. Christian, 1997. Biodiversity of Interstitial Collembola (Insecta) in Sand Sediments. European Journal of Soil Biology 33: 123–127.

Thibaud, J. M. & Z. Massoud, 1986. Insecta: Collembola. In Botosaneanu L. (ed.), Stygofauna Mundi. Brill/Backhuys, Leiden: 616–617.

Thibaud J. M., H. J. Schulz & M. M. da Gama Assalino, 2004. Hypogastruridae. In Dunger, W. (ed.), Synopses on Palaearctic Collembola, Vol. 4. Staatliches Museum für Naturkunde Görlitz 75: 1–287.

Yosii, R., 1971. Collembola of Khumbu Himal. Khumbu Himal 4: 80–130.

Yosii, R., 1977. Critical Check List of the Japanese Species of Collembola. Contributions from the Biological Laboratory, Kyoto University 25: 141–170.

Zimdars, B. & W. Dunger, 1994. Tullbergiinae. In Dunger, W. (ed.), Synopses on Palaearctic Collembola, Vol. 1. Staatliches Museum für Naturkunde Görlitz 68: 1–71.

Hydrobiologia (2008) 595:339–350
DOI 10.1007/s10750-007-9028-y

Global diversity of mayflies (Ephemeroptera, Insecta) in freshwater

Helen M. Barber-James · Jean-Luc Gattolliat · Michel Sartori · Michael D. Hubbard

Abstract The extant global Ephemeroptera fauna is represented by over 3,000 described species in 42 families and more than 400 genera. The highest generic diversity occurs in the Neotropics, with a correspondingly high species diversity, while the Palaearctic has the lowest generic diversity, but a high species diversity. Such distribution patterns may relate to how long evolutionary processes have been carrying on in isolation in a bioregion. Over an extended period, there may be extinction of species, but evolution of more genera. Dramatic extinction events such as the K-T mass extinction have affected current mayfly diversity and distribution. Climatic history plays an important role in the rate of speciation in an area, with regions which have been climatically stable over long periods having fewer species per genus, when compared to regions subjected to climatic stresses, such as glaciation. A total of 13 families are endemic to specific bioregions, with eight among them being monospecific. Most of these have restricted distributions which may be the result of them being the relict of a previously more diverse, but presently almost completely extinct family, or may be the consequence of vicariance events, resulting from evolution due to long-term isolation.

Keywords Ephemeroptera · Bioregion · Endemicity · Diversity · Distribution · Introgression

Guest editors: E. V. Balian, C. Lévêque, H. Segers & K. Martens
Freshwater Animal Diversity Assessment

H. M. Barber-James (✉)
Department of Freshwater Invertebrates, Albany Museum, Grahamstown, South Africa
e-mail: H.James@ru.ac.za

H. M. Barber-James
Department of Entomology, Rhodes University, Grahamstown, South Africa

J.-L. Gattolliat · M. Sartori
Museum of Zoology, Lausanne, Switzerland

M. D. Hubbard
Laboratory of Aquatic Entomology, Florida A & M University, Tallahassee, FL, USA

Introduction

The Ephemeroptera (mayflies) are an ancient lineage of insects, dating back to the late Carboniferous or early Permian periods, some 290 mya. It is thought that they attained their highest diversity during the Mesozoic (Brittain & Sartori, 2003). They are clearly the most primitive and ancient of the extant insect groups (Edmunds & McCafferty, 1988). The relationship of Ephemeroptera with other modern winged insects is still a subject of debate. Together with the Odonata, mayflies were traditionally placed in the Paleoptera, which was considered the sister group of all other extant primarily winged orders (Kukalová-Peck, 1991). More recently, it was suggested that

Ephemeroptera per se are the sister group of Odonata + Neoptera. This is based on a number of morphological features unique to mayflies, as well as on recent DNA-based phylogeny (Wheeler et al., 2001; Ogden & Whiting, 2003).

The nymphal stage of mayflies (Fig. 1) is the dominant life history stage, and is always aquatic. The nymphs undergo a series of moults as they grow, the precise number being variable within a species, depending on external factors, such as temperature, food availability and current velocity (Brittain & Sartori, 2003). Ranges from 10 to 50 instars have been reported (Ruffieux et al., 1996). Typically, nymphs have up to seven pairs of abdominal gills, usually three caudal filaments, and mouthparts generally adapted for collector/gatherer and deposit feeding. A few species are predaceous and some are scrapers. Certain groups are burrowers, and have variously developed mandibular tusks and frontal processes to loosen the substrate, and flattened legs for digging. Burrowers usually have feathery gills, which are folded over the abdomen and used to create a current through their burrow. Mayfly nymphs colonize all types of freshwaters but are more diversified in running waters than in lakes or ponds. A couple of species can even be found in brackish waters.

Mayflies undergo hemimetabolous metamorphosis, having a unique maturation stage between the nymph and adult, the subimago. Subimagos appear superficially similar to the adults, but are sexually immature. Their wings and abdomens are covered with small water-resistant microtrichia, which help them to leave the water after moulting from the final instar nymph (Edmunds & McCafferty, 1988). Except for a few exceptions, such as female Polymitarcyidae and Palingeniidae (which are mature as subimagos), most adults have transparent wings and glossy abdomens, having shed the subimaginal cuticle, and males have extended forelegs for grasping the female during mating. Usually, mayfly adults live from a few hours to a few weeks depending on the species. Many species have male mating swarms forming at dawn or dusk. Females have various methods of oviposition and the number of eggs laid varies according to species and size of female and eggs (Sartori & Sartori-Fausel, 1991; Brittain and Sartori, 2003). Length and number of life cycles per year depend largely on geographic locality and size of the species, with large burrowers in temperate climates taking over 2 years to mature, while tropical species may have several generations in a year.

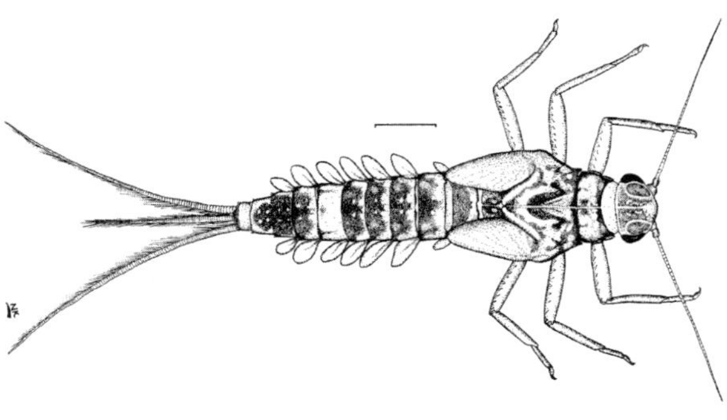

Fig. 1 Typical mayfly nymph (Baetidae)

Species and generic diversity

Ephemeroptera are represented by 42 families (Tables 1, 2), with a little over 3,000 described species (Table 1) in ca. 400 genera (Table 2). Studies of their taxonomy are still in progress and numerous unknown species and genera await description, mainly in tropical areas. This synthesis includes studies up to October 2005. The supraspecific nomenclature has been the subject of numerous changes over the last few years, with a great increase in the number of recognized genera and families. Some of these changes are due to the fact that more and more phylogenetic studies are now being undertaken, leading to more monophyletic clades (Brittain & Sartori, 2003).

Species and generic diversity is presented in Fig. 2. Clearly, collecting effort is reflected in the diversity pattern seen. The Holarctic Realm exhibits the highest species diversity and is also where the fauna is the best known. The faunas of the Afrotropical and Oriental Realms are probably underestimated because large areas are still "terrae incognitae" (e.g., Central Africa, parts of South America and Southeast Asia). A recent 80 km^2 survey of the mayfly fauna of a lowland tropical forest in Borneo led to the discovery of at least ten new genera and tens of new species (Sartori et al., 2003). Mayflies are unknown from Antarctica. The only significant Pacific components are located in New Caledonia,

Table 1 Number of Ephemeroptera species per realm, as of October 2005

Family	PA	NA	NT	AT	OL	AU	World
Acanthametropodidae	1	2	0	0	0	0	3
Ameletidae	20	34	0	0	2	0	56
Ameletopsidae	0	0	3	0	0	4	7
Ametropodidae	1	2	0	0	0	0	3
Arthropleidae	1	1	0	0	0	0	2
Vietnamellidae	0	0	0	0	6	1	7
Baetidae	156	153	161	185	139	39	833
Baetiscidae	0	12	0	0	0	0	12
Behningiidae	4	1	0	0	1	0	6
Caenidae	54	27	35	44	28	12	200
Chromarcyidae	0	0	0	0	2	0	2
Coloburiscidae	0	0	1	0	0	5	6
Coryphoridae	0	0	1	0	0	0	1
Dipteromimidae	2	0	0	0	0	0	2
Ephemerellidae	63	71	0	0	21	0	155
Ephemeridae	27	14	3	13	31	0	88
Ephemerythidae	0	0	0	5	0	0	5
Euthyplociidae	0	1	7	8	3	0	19
Heptageniidae	322	129	3	20	35	0	509
Ichthybotidae	0	0	0	0	0	2	2
Isonychiidae	6	16	1	0	7	0	30
Leptohyphidae	0	38	100	0	0	0	138
Leptophlebiidae	62	87	212	52	59	136	608
Machadorythidae	0	0	0	1	0	0	1
Melanemerellidae	0	0	1	0	0	0	1
Metretopodidae	4	9	0	0	0	0	13
Neoephemeridae	3	4	0	0	4	0	11
Nesameletidae	0	0	1	0	0	6	7
Oligoneuriidae	12	8	20	11	0	0	51
Oniscigastridae	0	0	2	0	0	6	8
Palingeniidae	10	2	0	3	15	4	34
Polymitarcyidae	5	7	56	4	12	0	84
Potamanthidae	8	5	0	0	10	0	23
Prosopistomatidae	5	0	0	4	8	2	19
Pseudironidae	0	1	0	0	0	0	1
Rallidentidae	0	0	0	0	0	1	1
Siphlaenigmatidae	0	0	0	0	0	1	1
Siphlonuridae	23	26	0	0	0	0	49
Siphluriscidae	1	0	0	0	0	0	1
Teloganellidae	0	0	0	0	1	0	1
Teloganodidae	0	0	0	8	5	0	13
Tricorythidae	0	0	0	32	1	0	33
Total	790	650	607	390	390	219	3046

PA: Palaearctic, NA: Nearctic, NT: Neotropical, AT: Afrotropical, OL: Oriental, AU: Australasian

Table 2 Number of Ephemeroptera genera per realm, as of October 2005

Family	PA	NA	NT	AT	OL	AU	World
Acanthametropodidae	1	2	0	0	0	0	2
Ameletidae	2	1	0	0	1	0	2
Ameletopsidae	0	0	2	0	0	2	4
Ametropodidae	1	1	0	0	0	0	1
Arthropleidae	1	1	0	0	0	0	1
Vietnamellidae	0	0	0	0	1	1	2
Baetidae	16	23	29	41	21	11	97
Baetiscidae	0	1	0	0	0	0	1
Behningiidae	2	1	0	0	1	0	3
Caenidae	2	4	5	8	5	3	17
Chromarcyidae	0	0	0	0	1	0	1
Coloburiscidae	0	0	1	0	0	2	3
Coryphoridae	0	0	1	0	0	0	1
Dipteromimidae	1	0	0	0	0	0	1
Ephemerellidae	8	10	0	0	8	0	18
Ephemeridae	1	3	1	3	2	0	6
Ephemerythidae	0	0	0	1	0	0	1
Euthyplociidae	0	1	3	1	1	0	5
Heptageniidae	14	16	3	2	11	0	31
Ichthybotidae	0	0	0	0	0	1	1
Isonychiidae	1	1	1	0	1	0	1
Leptohyphidae	0	4	9	0	0	0	10
Leptophlebiidae	9	10	45	17	19	50	131
Machadorythidae	0	0	0	1	0	0	1
Melanemerellidae	0	0	1	0	0	0	1
Metretopodidae	2	2	0	0	0	0	3
Neoephemeridae	3	1	0	0	1	0	3
Nesameletidae	0	0	1	0	0	2	3
Oligoneuriidae	3	2	6	3	0	0	11
Oniscigastridae	0	0	1	0	0	2	3
Palingeniidae	4	1	0	1	2	1	8
Polymitarcyidae	1	3	3	4	2	0	7
Potamanthidae	2	1	0	0	2	0	3
Prosopistomatidae	1	0	0	1	1	1	1
Pseudironidae	0	1	0	0	0	0	1
Rallidentidae	0	0	0	0	0	1	1
Siphlaenigmatidae	0	0	0	0	0	1	1
Siphlonuridae	2	4	0	0	0	0	4
Siphluriscidae	1	0	0	0	0	0	1
Teloganellidae	0	0	0	0	1	0	1
Teloganodidae	0	0	0	5	2	0	7
Tricorythidae	0	0	0	5	1	0	5
Total	78	94	112	93	84	78	405

Total number mentioned. PA: Palaearctic, NA: Nearctic, NT: Neotropical, AT: Afrotropical, OL: Oriental, AU: Australasian. The last column differs from the total of all realms because some genera may be found in several realms

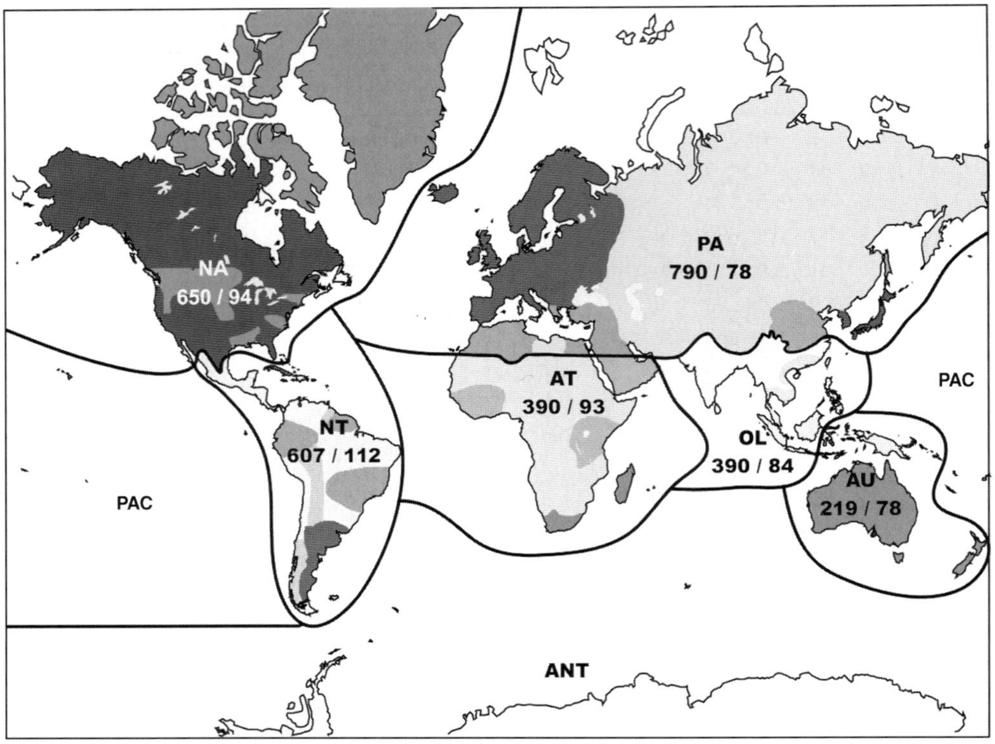

Fig. 2 Ephemeroptera diversity: number of species/number of genera per realm as of October 2005. *Dark color* indicates well-known fauna, *medium color* indicates data available, *pale* indicates paucity of data. PA: Palaearctic, NA: Nearctic, NT: Neotropical, AT: Afrotropical, Au: Australasian (including the Pacific realm PAC). ANT: Antarctic

comprising 18 genera and 37 species of Leptophlebiidae, all strictly endemic (Peters et al. 1978; Peters and Peters, 1980, 1981a, b). They have been considered as part of Australasia in the following discussion (Fig. 2).

The generic diversity (Table 2) reflects a different pattern to species diversity (Table 1) when viewed by bioregion. For example, the Palearctic realm has the highest species diversity but the lowest generic diversity. As a whole, the Northern Hemisphere (Holarctic Realm) possesses fewer genera than the Southern Hemisphere. The diversity among families is very heterogeneous. Within the families, 40% (17 families) are monogeneric, suggesting that they represent isolated and relict branches of formerly more diversified lineages, or relatively young and recently evolved groups. They are fairly evenly distributed between the Southern and Northern Hemispheres. Approximately three quarters of the species (Table 1) and genera (Table 2) belong to only five families. The Leptophlebiidae is the most diverse mayfly family at the generic level, while the Baetidae have more species in fewer genera. In Australasia and the Neotropics, Leptophlebiidae represent 62% and 35% of their mayfly species, respectively, while they range from 8% to 15% in other bioregions. In contrast, Baetidae are more evenly distributed among the bioregions, making up 20–25% of the species, with the exception of the Afrotropical and Oriental regions, where they represent 47% and 36%, respectively. The Heptageniidae have the third highest diversity. They are notably more diverse in the Northern Hemisphere, and contribute to 41% and 20% of the Palearctic and the Nearctic mayfly species, respectively. They constitute a minor group in other bioregions, but are absent from Australia. Although distributed worldwide, the Caenidae play a less important role in mayfly biodiversity (less than 8%) in all regions except in the Afrotropics (11%), although it should be noted that in terms of biomass they may at times exceed any other group. The Ephemerellidae, although absent from several regions, contribute to 11% of the diversity in the Nearctic, 8% in the Palaearctic and 5% in the Oriental (Tables 1, 2).

Twelve genera encompass one third of the total mayfly species richness worldwide (Table 3). They have a wide range extension, being present in at least two bioregions, with the exception of two genera which are restricted to the Americas. On the other hand, 37% (149) of the genera are monospecific with a strong proportion biased towards ancient Gondwanan relicts of the South Hemisphere. They contribute 27% and 22% of the generic richness of the Neotropics and Australasia, but only 7% and 11% of the Palearctic and Nearctic, respectively.

One may question why one area may be more diverse than another at species level and yet have fewer genera, or the converse. Past changes in the earth's climate and the availability of suitable habitats may have a major role in influencing the patterns seen today. Populations may have been evolving in isolation in some bioregions (such as the Afrotropics) for longer than in others (such as the Oriental Region) resulting in extinction of species but evolution of more genera through time. When comparing the mean number of species between different realms, the effect of different climatic histories in different areas becomes apparent. For example, there is a mean of three species per genus in Australasia, which has been stable over millions of years, compared to ten in the Palearctic, which has been affected by orogenesis and recent glaciation.

A number of authors have recently produced a series of chapters, which make useful supplementary reading, summarizing the history and extent of knowledge of Ephemeroptera biodiversity and systematics (see Alba-Tercedor 2001 and subsequent authors in the same series). A still growing number of catalogs are available and can be reached through the website "Ephemeroptera Galactica" and subsequent links (Hubbard, 2006).

Phylogeny and historical processes

The Ephemeroptera are the oldest extant order of winged insects, and the taxa present today represent the surviving branches of evolution. It is likely that a large degree of extinction of the order has occurred and some basal lineages have disappeared. The first comprehensive work on the systematics of mayflies was that of Eaton (1883–1888). His division of the mayflies into sections constituted the basis of the modern classification and remained almost unchanged for a century except for the hierarchical ranks of the sections.

Edmunds (1962) made the first attempt to reconstruct the phylogeny of the order, then McCafferty & Edmunds (1979) divided the mayflies in two suborders, Pannota and Schistonota. This reconstruction was widely used and constituted the basis for McCafferty's (1991) higher classification, where the Schistonota concept was considered paraphyletic, and replaced by three new suborders, Retracheata, Setisura and Pisciforma. Recently, Kluge (2004) proposed revisions at different levels of the phylogeny of

Table 3 The twelve most diverse genera among Ephemeroptera, with number of described species and main distribution

Genus	Family	Total spp.	Distribution
Baetis Leach	Baetidae	158	Cosmopolitan except South America
Caenis Stephens	Caenidae	135	Cosmopolitan except Australasia
Rhithrogena Eaton	Heptageniidae	134	Holarctic and neotropical
Epeorus Eaton	Heptageniidae	71	Holarctic and oriental
Cloeon Leach	Baetidae	66	Cosmopolitan except neotropics
Thraulodes Ulmer	Leptophlebiidae	61	Panamerican
Ephemera Linnaeus	Ephemeridae	59	Holarctic, oriental, afrotropical
Ecdyonurus Eaton	Heptageniidae	59	Holarctic and neotropical
Pseudocloeon Klapálek/*Labiobaetis* N. & K.	Baetidae	58	Cosmopolitan except neotropics
Paraleptophlebia Lestage	Leptophlebiidae	56	Holarctic and oriental
Ameletus Eaton	Ameletidae	55	Holarctic and oriental
Tricorythodes Ulmer	Leptohyphidae	54	Panamerican
Total		966 (32%)	

the mayflies and summarized them in the first comprehensive modern treatment including all mayfly lineages (except Leptophlebiidae and Baetidae). Unfortunately, the approach of Kluge's work remains difficult for most workers because of the use of non-ranking nomenclature (including plesiomorphons). McCafferty's (updated with his subsequent partial reconstructions, quoted below) and Kluge's systems are roughly congruent (Fig. 3). They divide the order into four main lineages: (1) Carapacea (McCafferty)/Posteritorna (Kluge); (2) Furcatergalia; (3) Setisura; (4) Pisciforma (McCafferty)/Tridentisata (Kluge). Although the four main lineages are generally accepted, the relationships between and among them, the placement of some taxa and the rankings remain problematic. The most confusing lineages are the Siphlonuroidea, the Ephemerelloidea and the Ephemeroidea. Since the appearance of McCafferty's (1991) system, about ten cladistic studies have been undertaken on family and superfamily relationships helping to better understand aspects of mayfly phylogeny, but none of them have included the entire order (e.g., Kluge et al., 1995; McCafferty & Wang, 2000; Kluge, 2003; Molineri & Dominguez, 2003; McCafferty, 2004; Wang & McCafferty, 2004). A phylogeny based on molecular data was proposed and will greatly help to understand the higher relationships within the order (Ogden & Whiting, 2005); note that this was constructed before the publication of McCafferty's (2004) reclassification of the Scapphodonta, which now includes the family Palingeniidae. It is likely that the details of phylogenies will continue to be debated in the light of further new information in the future.

Distribution and endemicity

Historically, mayflies were considered as having very low-dispersal ability with oceans or mountain ranges constituting barriers to dispersal. Therefore, the present distribution was believed to be explained only by vicariance, radiation, and extinction events and was a reflection of geological events, especially

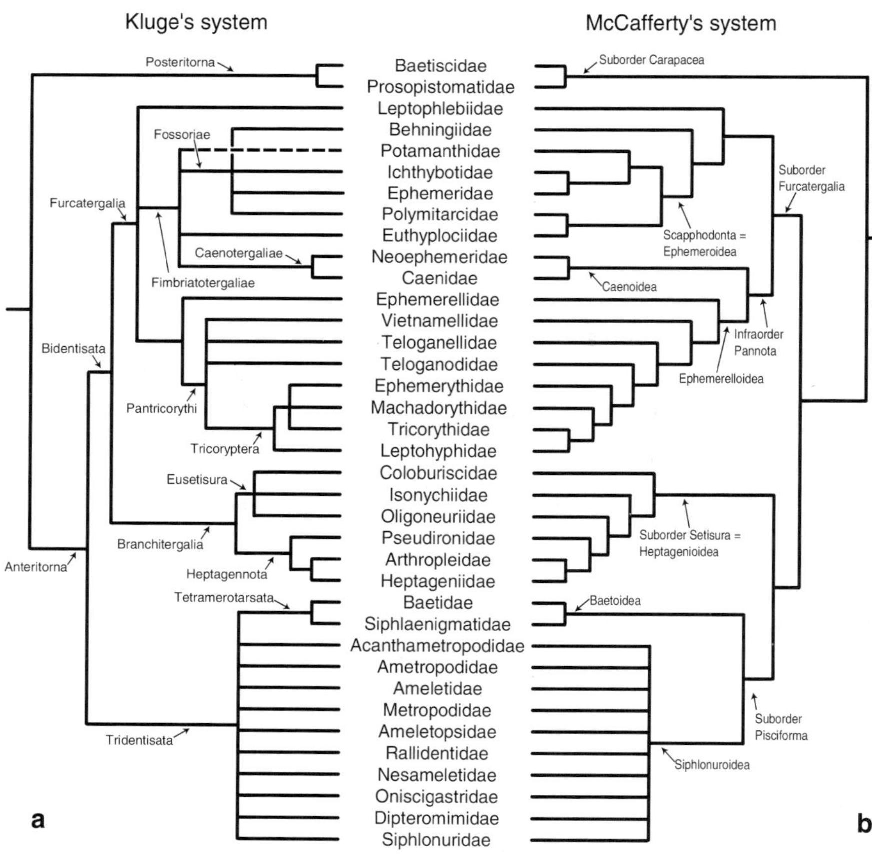

Fig. 3 Comparison of the two most recent phylogenetic reconstructions for the Ephemeroptera order, **a**. Kluge, **b**. McCafferty (from Ogden & Whiting (2005) with permission)

the break-off and drifting of the continents (Edmunds, 1972, 1975). Interchanges were considered to be restricted to regions directly in touch (McCafferty, 1998). Recent evidence has shown that the power of dispersal of mayflies had been greatly underestimated. The Baetidae in the South East Indian Ocean illustrate this case: the molecular reconstructions of the Afrotropical Baetidae phylogeny shows that there are strong affinities between Madagascar and continental Africa and very low affinities with other realms (opposing reliance upon the tectonic plate theory alone) and that transoceanic dispersal had occurred between Madagascar and continental Africa in both directions until recently (Monaghan et al., 2005). Simultaneously, the discovery of two species on the recent volcanic island of la Réunion indicates unambiguously that mayflies can disperse over more than 700 km (Gattolliat, 2004). Consequently, vicariance, radiation, and dispersal must be taken into account to explain present faunal composition and distribution.

Ephemeroptera have a worldwide distribution, being absent only from Antarctica and some remote oceanic islands (Hubbard, unpubl. results). Only three of the 42 families possess such a distribution (Baetidae, Caenidae, and Leptophlebiidae, Tables 1, 2). The origin of the Leptophlebiidae and Baetidae is probably Pangean, but the former has greater diversity and presents the highest degree of endemicity in the Neotropics and Australasia, and the latter in the Neotropical and Afrotropical regions. Caenidae are mainly diversified at the specific level in the Palearctic realm and at the generic in the Australasian, Afrotropical and Neotropical realms. It seems evident in regard to the almost cosmopolitan distribution of some genera included in these families (*Baetis* and *Cloeon* for the Baetidae, *Choroterpes* for the Leptophlebiidae and *Caenis* for the Caenidae) that dispersal events among regions must have taken place after the initial splitting of the Gondwanan plates. Heptageniidae, Ephemeridae, and Polymitarcyidae have a similar distribution except that they do not enter the Australasian realm, suggesting they have a Laurasian origin and reached the Southern Hemisphere well after the Gondwanan separation. This is emphasized by the fact that Heptageniidae only enter the Neotropics in Central America and are absent from South America. The monogeneric family Prosopistomatidae is present in all regions except the Americas, where it is replaced in the Nearctic by its sister family Baetiscidae. This distribution can be hypothetically explained by a relatively recent origin of the family (possibly Afrotropical or Oriental) and a secondary colonization of the other regions. Although this chapter has not discussed fossil mayflies, these do shed an interesting light on some of the distributions. For example, a fossil prosopistomatid has been described from Burmese amber (Sinitshenkova, 2000), dating at about 100 million years (Cruickshank & Ko, 2003), indicating the long-term presence of the family in the Oriental region. Staniczek et al. (2002) report a fossil Baetiscidae from Baltic amber, indicating a once more widespread Laurasian distribution of this family.

Another important aspect to consider at this point is the effect of extinction on the mayfly fauna. Although background extinction takes place more or less constantly (e.g., Raup, 1994), mass extinction has had a notable effect on the composition of the mayfly families and genera. McCafferty (1990, 1991) points out that the Mesozoic mayfly fauna was considerably different compared to the Cenozoic fauna, following the mass Cretaceous-Tertiary (K-T) extinctions.

Looking at extant distributions, five families have a strict Holarctic distribution (Acanthametropodidae, Ametropodidae, Arthropleidae, Metretopodidae, Siphlonuridae) and five others (Ameletidae, Behningiidae, Ephemerellidae, Neoephemeridae, Potamanthidae) extend their Holarctic range (through the Palearctic) to the Oriental realm (Tables 1, 2). It seems reasonable to conclude that these all have a Laurasian origin. Four families are amphinotic (Ameletopsidae, Coloburiscidae, Nesameletidae, Oniscigastridae) which confirm that South America, Australia, and New Zealand share the same Gondwanan origin. Two families, Teloganodidae and Tricorythidae are spread through the Oriental and Afrotropical bioregions, including Madagascar, possibly indicating a more tropical Gondwanan origin.

The distribution of the Palingeniidae is puzzling, as they are absent from the Neotropics and Continental Africa, but present in Madagascar. A once worldwide distribution with a complete extinction in South America and continental Africa seems quite unlikely. A Lemurian (Madagascar and Deccan plate) origin with subsequent colonization of the Palearctic and introgression to Papua New Guinea could be more appropriate but does not explain adequately the presence of *Pentagenia* in the Nearctic. However, as

Edmunds (1972) stated, with the immensity of time, improbable dispersals can become probable.

Considering endemicity, 13 families are endemic to specific bioregions, with eight among them being monospecific. Most of these have restricted distributions and greatly contribute to the distinctness of the fauna in that region (e.g., the monospecific Rallidentidae and Siphlaenigmatidae for New Zealand; Dipteromimidae for Japan; Coryphoridae in the Amazonian basin). The restricted size of these families can be explained by two possible processes; they can be the relict of a previously more diversified but presently almost completely extinct family, in other words, refugial (e.g., Siphluriscidae in China), while other families, especially those endemic to islands, may be the consequence of vicariance events, resulting from evolution due to long-term isolation (Fig. 4).

In contrast to the widely distributed genera (Table 3), it is worth noting that 79% of the 405 known genera are endemic to a single realm. No genera exhibit an amphinotic distribution, attesting to the ancient splitting of Gondwanaland. For others, it has to be mentioned that 21 genera (5%) are Panamerican, whereas 13 (3%) are Holarctic. The extant distribution of some genera can be explained, as for the families, by introgression into an adjacent bioregion: from Palearctic to Oriental (e.g., *Torleya*) or from Oriental to Palearctic (e.g., *Baetiella*). A single genus may present a disjunctive distribution, e.g., *Habrophlebiodes* (Leptophlebiidae) being represented in the eastern Nearctic and Oriental by four species in each realm. It has been suggested that this vicariant distribution could be associated with the Arcto-Tertiary forest that covered most of the Northern Hemisphere during the Early Tertiary into the Pleistocene (Peters, 1988).

Of the 3,046 species reported here, just over 60 (2%) are distributed among two bioregions. The main patterns are:

- Circumpolar (Arctic) species, (10 species, such as some Baetidae, Ephemerellidae and Siphlonuridae);
- Panamerican species with a Neotropical origin, with introgression into the Nearctic (32 species, mainly Baetidae, Leptophlebiidae and Tricorythidae);

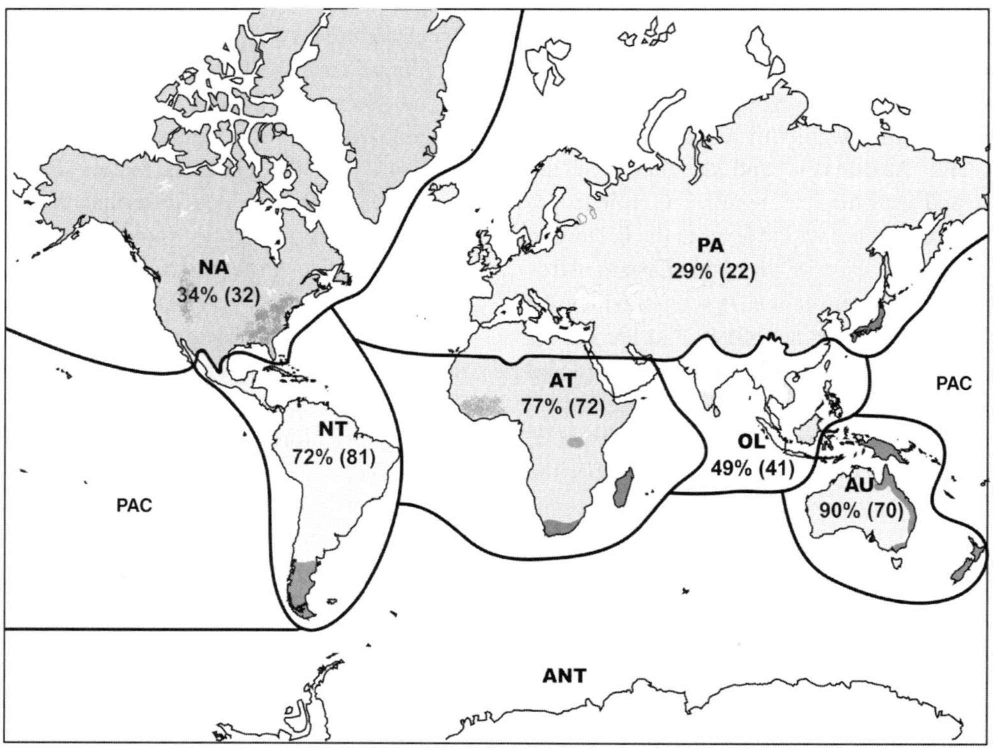

Fig. 4 Ephemeroptera endemism: percentage of endemic genera, number of endemic genera in brackets. Dark color indicates high level of endemicity

- Panamerican species with a Nearctic origin, with introgression into the Neotropics (8 species, mainly Heptageniidae);
- Palearctic species with introgression into the Oriental realm (6 species, mainly Baetidae and Heptageniidae);
- Oriental species with introgression into the Palearctic (4 species, mainly Ephemerellidae);
- Transpacific distribution (2 species within Baetidae and Heptageniidae seem to have a transpacific distribution, being present in the Eastern Palearctic and Western Nearctic).

Southern and Northern Hemisphere regions not only differ in the generic richness, but also notably in the degree of endemism (Fig. 4). The different faunas from the Southern Hemisphere possess a high degree of generic endemicity (comprising between 72% and 90%). Moreover, most of the centers of endemicity are in the Southern Hemisphere (Southern South America, Southeastern South Africa, Madagascar, Eastern coast of Australia, New Zealand, New Caledonia). Affinities between Southern Hemisphere regions are low. The Australasian fauna is the most distinctive. With a generic degree of endemicity of 90%, it shares only a few cosmopolitan genera with other regions, although intermittent stepping-stone archipelagos have been available intermittently for dispersal between the Oriental and Australasian realm (McCafferty, 1999). This can explain the presence of the Prosopistomatidae in northern Australia, and on several of the islands between Australia and Asia, these having migrated southward during times of lower sea level.

The Neotropical fauna also has low affinities with other Southern hemisphere regions. It seems much closer to the Nearctic fauna. This is mainly due to the formation of the Central American landbridge during the Pliocene allowing introgression (or Great American Interchange) from both sides during the Pleistocene and Holocene (McCafferty, 1998; but see Savage et al., 2005 for an alternate view). Africa has been separated for so long that it has little in common with South America or Australasia, except for some of the globally dispersed families such as Baetidae and Leptophlebiidae. The only remnants of cold adapted taxa in this area may be some of the endemic genera of the Leptophlebiidae (e.g., *Aprionyx*, and *Castanophlebia*), in the southern and south-western tip of Africa. Amphinotic families, such as Nesameletidae, encompass vicariant genera restricted to Southern South America (*Metamonius*), New Zealand (*Nesameletus*) and Australia (*Ameletoides*). The sharing of fauna between these two regions is well documented (e.g., Alexander, 1929; Brundin, 1966; Edmunds, 1972, 1975), and indicates that there was once an important center of evolution in the south, with Antarctica being a part of that center.

Contrary to the Southern Hemisphere, Northern Hemisphere bioregions possess a low level of endemicity, both at family and generic levels. Only Japan (with the endemic family Dipteromimidae), and China (with the monotypic family Siphluriscidae), as well as some parts of North America (with Baetiscidae (once more widespread, discussed above) and Pseudironidae) can be considered as important centers of endemicity. Most of the taxa present a wide distribution at the same latitude.

The Oriental realm presents an intermediate situation between the Northern and Southern Hemispheres. About half of the genera are endemic to this region. This is clearly more than the Holarctic fauna but it is still lower than regions in the Southern Hemisphere, even if some new endemic genera are still undescribed (mainly from Malaysia and Indonesia). The Oriental region shows important affinities with both the Holarctic and Afrotropical realms; this is probably partially due to the dual origin of the region, with India belonging to Gondwana and the remaining part of Laurasian origin.

Table 4 summarizes affinities of the mayfly fauna between realms both at the family and the generic levels.

Important differences in the diversity and endemicity between Northern and Southern Hemispheres have been elucidated. These can be explained by several different hypotheses:

- The Gondwanan fauna was originally much more diverse than the Laurasian fauna;
- A higher level of extinction occurred in the Northern Hemisphere (due for instance to the climate changes in the Pleistocene);
- In the Northern Hemisphere, bioregions are oriented East–West where similar latitudes (especially in the Palearctic) imply much more homogenous environments. In the Southern Hemisphere, bioregions are oriented North–South

Table 4 Sörensen's index of similarities between the different realms at the family level (above) and generic level (below). The higher the value, the more similar are the faunas

	Neotropical	Nearctic	Palaearctic	Oriental	Australasian	Afrotropical
Neotropical		0.51	0.42	0.44	0.48	0.53
Nearctic	0.30		0.84	0.65	0.22	0.49
Palaearctic	0.06	0.42		0.67	0.29	0.50
Oriental	0.06	0.24	0.43		0.36	0.65
Australasian	0.01	0.06	0.06	0.09		0.37
Afrotropical	0.03	0.10	0.18	0.20	0.07	

meaning they are much more subject to climatic changes;

- The main islands suitable for vicariance and radiation events of the mayflies are present in the Southern Hemisphere (Madagascar, New Zealand, Australia, New Guinea, and Borneo).

Human impacts

Deforestation is one of the primary threats to mayfly biodiversity and conservation in the tropics (e.g., Madagascar, Borneo) (Benstead et al., 2003; Benstead & Pringle, 2004; Dudgeon, 2000a, b), whereas pollution (Rosenberg & Resh, 1993) and/or building and reshaping of the banks leading to a lack of connectivity with the floodplain (Buijse et al., 2002) or habitat fragmentation (Zwick, 1992) are the main causes in temperate areas.

Many studies have been carried out in recent year on the effects of climate changes on mayflies. Clearly, climate changes are affecting the behavior and ultimately the ecology of some mayflies: for example, small increases in temperature (3°C) over the short term cause early emergence (McKee & Atkinson, 2000), possibly when the flow regime is not right for successful egg laying (Harper & Peckarsky, 2005). Climate changes alter precipitation pattern, leading to greater flood magnitude and frequency in certain rivers. This results in changes in ecological structure and function, and loss of diversity through too frequent scouring (Poff, 2002). Beniston (2006) reviews climatic changes in the Alps, and notes that with the warming trend observed, glaciers have considerably reduced in volume and area over the last 150 years. With the continuing trend of temperature increase, the proportion of glacial melt and snow melt waters will change and lead to drastic changes in macroinvertebrate communities, including mayflies (Brown et al., 2003, 2006).

Stream acidification is another negative human effect. In Europe acidification is still having a negative impact on invertebrate communities: in the Vosges mountains (France), Guerold et al. (2000) found a high reduction in diversity for many aquatic species and that Ephemeroptera totally disappeared from some streams.

Very few mayflies have been listed under the IUCN red listing criteria, although this is probably because of a lack of knowledge of specific ecologies, rather than an indication of the state of the fauna. The 1998 Red List of the German fauna (Malzacher et al., 1998) lists a much larger number of mayflies, probably because the environmental requirements are better known there than most places.

Some species are thought to have undergone recent extinction: *Pentagenia robusta* (USA) (Palingeniidae), *Isonychia diversa* (USA) (Isonychiidae), *Siphlonurus luridipennis* (USA) (Siphlonuridae), *Ephemera compar* (USA) and *Ephemera mooiana* (South Africa) (Ephemeridae), all inhabitants of large rivers, though it is possible some may yet be rediscovered. The largest and one of the most vulnerable European mayfly, *Palingenia longicauda* (Palingeniidae) has been recently added to the Bern Convention of the Council of Europe (Sartori & Landolt, 1998); formerly distributed in large rivers throughout Europe, it is now restricted to the Tisza watershed in Hungary. Many more species need to be added to this list, as more and more habitat destruction takes place.

Acknowledgments Thanks to two referees for useful input to this article, and to N. Köhly for the illustration of the nymph for Fig. 1.

References

Alba-Tercedor, J., 2001. Introduction to the panel discussion "The status and knowledge of Ephemeroptera". In Dominguez, E. (ed.), Trends in Research in Ephemeroptera and Plecoptera. Kluwer Academic/Plenum Publishers, New York, 1–2.

Alexander, C. P., 1929. Diptera of Patagonia and South Chile. Part 1. Crane flies. British Museum of Natural History, London, 240 pp.

Beniston, M., 2006. Mountain weather and climate: a general overview and a focus on climatic change in the Alps. Hydrobiologia 562: 3–16.

Benstead, J. P., P. H. de Rham, J.-L. Gattolliat, F. M. Gibon, P. V. Loiselle, M. Sartori & J. S. Sparks, 2003. Conserving Madagascar's freshwater biodiversity. Bioscience 53: 1101–1111.

Benstead, J. P. & C. M. Pringle, 2004. Deforestation alters the resource base and biomass of endemic stream insects in eastern Madagascar. Freshwater Biology 49: 490–501.

Brittain, J. E. & M. Sartori, 2003. Ephemeroptera. In Resh, V. H. & R. T. Cardé (eds), Encyclopedia of Insects. Academic Press, Amsterdam, 373–380.

Brown, L. E., D. M. Hannah & A. M. Milner, 2003. Alpine stream habitat classification: an alternative approach incorporating the role of dynamic water source contributions. Arctic Antarctic and Alpine Research 35: 313–322.

Brown, L. E., A. M. Milner & D. M. Hannah, 2006. Stability and persistence of alpine stream macroinvertebrate communities and the role of physicochemical habitat variables. Hydrobiologia 560: 159–173.

Brundin, L., 1966. Transantarctic Relationships and Their Significance as Evidenced by Chironomid Midges. Almqvist & Wiksell, Stockholm, 1–472.

Buijse, A. D., H. Coops, M. Staras, L. H. Jans, G. J. Van Geest, R. E. Grift & B. W. Ibelings, 2002. Restoration strategies for river floodplains along large lowland rivers in Europe. Freshwater Biology 47: 889–907.

Cruickshank, R. D. & K. Ko, 2003. Geology of an amber locality in Hukawng Valley, Northern Myanmar. Journal of Asian Earth Sciences 21: 441–455.

Dudgeon, D., 2000a. The ecology of tropical Asian rivers and streams in relation to biodiversity conservation. Annual Review of Ecology and Systematics 31: 239–263.

Dudgeon, D., 2000b. Riverine biodiversity in Asia: a challenge for conservation biology. Hydrobiologia 418: 1–13.

Eaton, A. E., 1883–1888. A revisional monograph of recent Ephemeridae or mayflies. Transactions of the Linnean Society of London, 2nd Series, Zoology 3: 1–352.

Edmunds, G. F., 1962. The principles applied in determining the hierarchic level of the higher categories of Ephemeroptera. Systematic Zoology 11: 22–31.

Edmunds, G. F., 1972. Biogeography and evolution of Ephemeroptera. Annual Review of Entomology 17: 21–42.

Edmunds, G. F., 1975. Phylogenetic biogeography of mayflies. Annals of the Missouri Botanical Garden 62: 251–263.

Edmunds, G. F. & W. P. McCafferty, 1988. The mayfly subimago. Annual Review of Entomology 33: 509–529.

Gattolliat, J.-L., 2004. First reports of the genus *Nigrobaetis* Novikova & Kluge (Ephemeroptera: Baetidae) from Madagascar and La Réunion with observations on Afrotropical biogeography. Revue Suisse de Zoologie 111: 657–669.

Guerold, F., J.-P. Boudot, G. Jacquemin, D. Vein, D. Merlet & J. Rouiller, 2000. Macroinvertebrate community loss as a result of headwater stream acidification in the Vosges Mountains (N-E France). Biodiversity and Conservation 9: 767–783.

Harper, M. P & B. L. Peckarsky, 2005. Emergence cues of a mayfly in a high-altitude stream ecosystem: potential response to climate change. Ecological Applications 16: 612–621.

Hubbard, M. D., 2006. Available on internet at http://www.famu.org/mayfly/.

Kluge, N. J., 2003. System and phylogeny of Pinnatitergaliae (Ephemeroptera). In Gaino E. (ed.), Research Update on Ephemeroptera and Plecoptera. University of Perugia, Italy, 145–152.

Kluge, N. J., 2004. The Phylogenetic System of Ephemeroptera. Kluwer Academic Publishers, Dordrecht, 1–442.

Kluge, N. J., D. Studemann, P. Landolt & T. Gonser, 1995. A reclassification of Siphlonuroidea (Ephemeroptera). Mitteilungen der Schweizerischen Entomologischen Gesellschaft 68: 103–132.

Kukalová-Peck, J., 1991. Fossil history and the evolution of hexapod structures. In Naumann, I. D. (ed.), The Insects of Australia, 2nd edn., Vol. 1. CSIRO, Melbourne University Press, Australia, 141–179.

Malzacher, P., U. Jacob, A. Haybach & H. Reusch, 1998. Rote Liste der Eintagsfliegen (Ephemeroptera). In Naturschutz, B. F. (ed.), Rote Liste gefährdeter Tiere in Deutschland, Bonn, 264–267.

McCafferty, W. P., 1990. Chapter 2. Ephemeroptera. In Grimaldi, D. A. (ed.), Insects from the Santana Formation, Lower Cretaceous, of Brazil. Bulletin America Museum of Natural History, 20–25.

McCafferty, W. P., 1991. Toward a phylogenetic classification of the Ephemeroptera (Insecta): a commentary on systematics. Annals of the Entomological Society of America 84: 343–360.

McCafferty, W. P., 1998. Ephemeroptera and the great American interchange. Journal of the North American Benthological Society 17: 1–20.

McCafferty, W. P., 1999. Biodiversity and biogeography: examples from global studies of Ephemeroptera. In Proceedings of the Symposium on Nature Conservation and Entomology in the 21st Century. The Entomological Society of Korea, 3–22.

McCafferty, W. P., 2004. Higher classification of the burrowing mayflies (Ephemeroptera: Scapphodonta). Entomological News 115: 84–92.

McCafferty, W. P. & G. F. Edmunds, 1979. The higher classification of the Ephemeroptera and its evolutionary basis. Annals of the Entomological Society of America 72: 5–12.

McCafferty, W. P. & T.-Q. Wang, 2000. Phylogenetic systematics of the major lineages of Pannote mayflies (Ephemeroptera, Pannota). Transactions of the American Entomological Society 126: 9–101.

McKee, D. & D. Atkinson, 2000. The influence of climate change scenarios on populations of the mayfly *Cloeon dipterum*. Hydrobiologia 441: 55–62.

Molineri, C. & E. Dominguez, 2003. Nymph and egg of *Melanemerella brasiliana* (Ephemeroptera: Ephemerelloidea: Melanemerellidae), with comments on its systematic position and the higher classification of Ephemerelloidea. Journal of the North American Benthological Society 22: 263–275.

Monaghan, M. T., J.-L. Gattolliat, M. Sartori, J.-M. Elouard, H. M. James, P. Derleth, O. Glaizot, F. de Moor & A. P. Vogler, 2005. Trans-oceanic and endemic origins of the small minnow mayflies (Ephemeroptera, Baetidae) of Madagascar. Proceedings of the Royal Society B 272: 1829–1836.

Ogden, T. H. & M. F. Whiting, 2003. The problem with "the Paleoptera Problem:" sense and sensitivity. Cladistics – the International Journal of the Willi Hennig Society 19: 432–442.

Ogden, T. H. & M. F. Whiting, 2005. Phylogeny of Ephemeroptera (mayflies) based on molecular evidence. Molecular Phylogenetics and Evolution 37: 625–643.

Peters, W. L., 1988. Origins of the North American Ephemeroptera fauna, especially the Leptophlebiidae. Memoirs of the Entomological Society of Canada 144: 13–24.

Peters, W. L. & J. G. Peters, 1980. The Leptophlebiidae of New Caledonia (Ephemeroptera). Part II. Systematics. Cahiers O.R.S.T.O.M., Series Hydrobiologie 13: 61–82.

Peters, W. L. & J. G. Peters, 1981a. The Leptophlebiidae: Atalophlebiinae of New Caledonia (Ephemeroptera). Part III – Systematics. Revue d'Hydrobiologie Tropical 14: 233–243.

Peters, W. L. & J. G. Peters, 1981b. The Leptophlebiidae: Atalophlebiinae of new Caledonia (Ephemeroptera). Part IV – systematics. Revue d'Hydrobiologie Tropical 14: 245–252.

Peters, W. L., J. G. Peters & G. F. Edmunds, 1978. The Leptophlebiidae of New Caledonia (Ephemeroptera). Part I. Introduction and systematics. Cahiers de l'ORSTOM, série Hydrobiologie 12: 97–117.

Poff, N. R., 2002. Ecological response to and management of increased flooding caused by climate change. Philosophical Transactions of the Royal Society of London A 360: 1497–1510.

Raup, D. M., 1994. The role of extinction in evolution. Proceedings of the National Academy of Sciences of the United States of America 91: 6758–6763.

Rosenberg, D. M. & V. Resh, 1993. Freshwater Biomonitoring and Benthic Macroinvertebrates. Chapman & Hall, New York, 1–488.

Ruffieux, L., M. Sartori & G. L'Eplattenier, 1996. Palmen body: a reliable structure to estimate the number of instars in *Siphlonurus aestivalis* (Eaton) (Ephemeroptera: Siphlonuridae). International Journal of Insect Morphology and Embryology 25: 341–344.

Sartori, M., P. Derleth & J. L. Gattolliat, 2003. New data about the mayflies (Ephemeroptera) from Borneo. In Gaino, E. (ed.), Research Update on Ephemeroptera and Plecoptera. Perugia, University of Perugia, Italy, 403–406.

Sartori, M. & P. Landolt, 1998. Memorandum concernant la candidature de *Palingenia longicauda* (Olivier, 1791) (Insecta Ephemeroptera) à son inscription en annexe de la Convention de Berne. Strasbourg, Document T-PVS (98) 15, Council of Europe, 1–9.

Sartori, M. & A. Sartori-Fausel, 1991. Variabilité de la durée du stade subimaginal et de la fécondité chez *Siphlonurus aestivalis* (Eaton) (Ephemeroptera; Siphlonuridae). Revue Suisse de Zoologie 98: 717–723.

Savage, H. M., R. W. Flowers & V. W. Porras, 2005. Rediscovery of *Choroterpes atramentum* in Costa Rica, type species of *Tikuna* new genus (Ephemeroptera: Leptophlebiidae: Atalophlebiinae), and its role in the "Great American Interchange". Zootaxa 932: 1–14.

Sinitshenkova, N. D., 2000. The first fossil prosopistomatid mayfly from Burmese amber (Ephemeroptera: Prosopistomatidae). Bulletin of the Natural History Museum, London (Geology) 56: 25–28.

Staniczek, A. H., T. Bechly & G. Bechly, 2002. First fossil record of the mayfly family Baetiscidae from Baltic amber (Insecta: Ephemeroptera). Stuttgarter Beiträge zur Naturkunde B (Geologie und Paläontologie) 322: 1–11.

Wang, T. Q. & W. P. McCafferty, 2004. Heptageniidae (Ephemeroptera) of the world. Part I: phylogenetic higher classification. Transactions of the American Entomological Society 130: 11–45.

Wheeler, W. C., M. Whiting, Q. D. Wheeler & J. M. Carpenter, 2001. The phylogeny of the extant hexapod orders. Cladistics – The International Journal of the Willi Hennig Society 17: 113–169.

Zwick, P., 1992. Stream habitat fragmentation – a threat to biodiversity. Biodiversity and Conservation 1: 80–97.

Hydrobiologia (2008) 595:351–363
DOI 10.1007/s10750-007-9029-x

Global diversity of dragonflies (Odonata) in freshwater

Vincent J. Kalkman · Viola Clausnitzer · Klaas-Douwe B. Dijkstra · Albert G. Orr · Dennis R. Paulson · Jan van Tol

Abstract Larvae of almost all of the 5,680 species of the insect order Odonata (dragonflies and damselflies) are dependent on freshwater habitats. Both larvae and adults are predators. The order is relatively well studied, and the actual number of species may be close to 7,000. Many species have small distributional ranges, and are habitat specialists, including inhabitants of alpine mountain bogs, seepage areas in tropical rain forests, and waterfalls. They are often successfully used as indicators for environmental health and conservation management. The highest diversity is found in flowing waters in rain forests of the tropics, the Oriental and Neotropical regions being the most speciose. This paper discusses diversity, summarises the biogeography of dragonflies in the different biogeographical regions and gives the total number of species and genera per family per biogeographical region. Examples are given of areas of particular diversity, in terms of areas of endemism, presence of ancient lineages or remarkable recent radiations but no well-based review of areas with high endemism of dragonflies is available so far. The conservation status of dragonflies is briefly discussed. Species confined to small remnants of forest in the tropics are most under threat of extinction by human activities.

Keywords Odonata · Dragonflies · Diversity · Endemicity · Biogeography · Conservation

Guest editors: E. V. Balian, C. Lévêque, H. Segers and K. Martens
Freshwater Animal Diversity Assessment

V. J. Kalkman (✉) · K.-D. B. Dijkstra · J. van Tol
National Museum of Natural History Naturalis, PO Box 9517, 2300 RA Leiden, The Netherlands
e-mail: kalkman@naturalis.nnm.nl

K.-D. B. Dijkstra
e-mail: dijkstra@naturalis.nnm.nl

J. van Tol
e-mail: tol@naturalis.nnm.nl

V. Clausnitzer
Gräfestraße 17, 06110 Halle/Saale, Germany
e-mail: violacl@t-online.de

A. G. Orr
CRC-TREM, AES, ENS, Griffith University, Nathan, QLD Q4111, Australia
e-mail: agorr@bigpond.com

D. R. Paulson
Slater Museum of Natural History, University of Puget Sound, Tacoma, WA 98416, USA
e-mail: dpaulson@ups.edu

Introduction

With 5,680 extant species, dragonflies are a relatively small order of insects. Their size and colour and their diurnal and often conspicuous behaviour make them a popular group for both professional and amateur entomologists.

Dragonflies are among the most ancient of winged insects, dating back well into the Permian (Grimaldi & Engel, 2005). They include the largest insect that ever lived, the griffenfly *Meganeuropsis permiana* Carpenter, with a wingspan of c. 70 cm. Dragonflies are recognised by their long, slender abdomen; large globular eyes, often making up a large portion of the head; short antennae; and long wings, which have a conspicuous nodus and usually a pterostigma. They possess a unique mechanism of indirect sperm transfer: sperm are produced in the testes situated at the abdomen tip, but the secondary copulatory organs that transfer them to the female lie on the ventral side of the abdomen base. Sperm must be transferred externally to this organ before copulation. This copulatory organ is used not only to inseminate, but also to remove the sperm of the female's previous mates. Sperm competition in Odonata was first reported by Waage (1979) and stimulated numerous studies, making dragonflies one of the most studied animal groups in terms of reproductive behaviour. Another unique feature of odonates is the strongly modified labium of the larva, which can be extended at great speed to seize prey.

The extant dragonflies are divided into two suborders, the Zygoptera or damselflies and the Anisoptera or true dragonflies (Fig. 1). Until recently a third suborder, Anisozygoptera, was recognised,

Fig. 1 The damselfly *Pseudagrion kersteni* and the dragonfly *Orthetrum julia* sitting on the same perch. (photo: Viola Clausnitzer)

with two extant species from Japan and the eastern Himalayas. The Anisozygoptera, which have some features recalling Zygoptera, are now often included in Anisoptera (as done here), or combined with them under the new name Epiprocta. Zygopterans have a broad head with widely separated eyes and similar fore and hind wings. Most species rest with wings closed. The larvae are slender and rely mainly on two or three caudal gills for respiration. Anisoptera are on average larger and more robust than Zygoptera. Their hind wings are distinctly broader at their base than the fore wings and in most families the eyes touch on top of the head. At rest most species spread their wings. The larvae are typically much sturdier than those of Zygoptera and lack caudal gills: oxygen is absorbed through gills in the rectum. A general outline of odonate diversity is given by Silsby (2001). A checklist of all dragonflies including synonyms and references is found on http://www.odonata.info (van Tol, 2005)

Dragonfly larvae live in freshwater environments and only a few species tolerate brackish conditions, two of which even live in salt marshes and mangroves. Both running and standing waters are used, while a few species are semi-terrestrial or inhabit water held in tree holes, leaf axils and other phytotelmata. Many species have small distributional ranges, and are habitat specialists, including inhabitants of alpine mountain bogs, seepage areas in tropical rain forests, and waterfalls. Larvae prey on all kinds of small animals up to the size of tadpoles and small fish. Larvae take from a few weeks to 7 years to develop. Emergence takes place above water on plants or on the shore, after which most species leave the water edge to mature. Males return to the water to search for females or to establish territories. Females often only return to mate and to oviposit. Information on the life history and behaviour of odonates is thoroughly summarised in Corbet's (1999) review of odonate behaviour and ecology.

Species diversity

Information on the number of species of odonates is derived from the Global Species Database Odonata prepared for the Catalogue of Life (van Tol, 2005). Taxa were assigned to one or more of the biogeographical regions based on the authors information and several key references (e.g., Lieftinck, 1949; Watson et al., 1991; Westfall & May, 1996; Okudaira et al., 1999; Needham et al., 2000; Tsuda, 2000; Wang, 2000). Subspecies were not considered. Table 1 enumerates the number of genera and species per family for each biogeographical region. Family-level classification of odonates is poorly resolved, although most families are broadly accepted. The most recent contribution to the higher classification of dragonflies was published by Rehn (2003). With the advent of molecular techniques, revision of family-level classification may be expected.

In total 5,680 species of Odonata are known, 2,739 belonging to the suborder Zygoptera (19 families) and 2,941 to the suborder Anisoptera (12 families). Table 1 and Fig. 2 show that the tropics support by far the most species of dragonflies. Besides higher diversity at the species level, the number of families is also much greater in the tropics (Fig. 3). Twelve of the 31 families are restricted mostly to running waters within tropical forest habitats. The two largest families (Coenagrionidae and Libellulidae) are believed to be relatively recent (Rehn, 2003). Almost all ubiquitous species belong to these two families and they dominate in unshaded habitats with stagnant water (both artificial and natural, e.g., savannas). Both families include species with the greatest migratory capacity, including those with distributions spanning more than one continent and almost all species found on isolated islands.

It is estimated that between 1,000 and 1,500 species of dragonflies await description (Table 1). If this is true, the actual number of extant species may be close to 7,000. The Oriental, Australasian and especially the Neotropical regions hold the highest number of undescribed species. In the latter, new species are still discovered more rapidly than descriptions are published (Paulson, 2004). The fauna of Africa is relatively well known and relatively depauperate. Overall the families Platystictidae, Protoneuridae, Gomphidae and Corduliidae are believed to hold relatively many undescribed species. They are typically inconspicuous odonates with small ranges, often confined to seepages or small runnels in tropical forest. Gomphidae, Corduliidae and also Chlorogomphidae in Southeast Asia are difficult to collect as they spend little time at the waterside. The two largest families, Coenagrionidae and Libellulidae, are relatively well known, because most species are conspicuous and many favour open habitats, although in absolute numbers they still represent a large proportion of species to be described. This is especially so for the Coenagrionidae in South America. Since 1970 an average of 38 species have been described annually (Fig. 4). With an undiminished rate of description an estimated 95% of all species will be described in 2030.

Processes influencing diversity of dragonflies

Factors influencing the distribution of dragonfly diversity can be divided into historical (geological) and ecological factors. Both determine current species diversity, while composition at family and genus level is predominantly determined by the first. Dragonflies are an ancient group, and present-day distribution reflects the distribution of the families before the break-up of Pangaea and subsequent supercontinental schisms. However, more well-founded phylogenetic reconstructions are needed before a satisfactory synthesis of this subject can be written.

Today's patterns of dragonfly diversity correspond largely with the present climatological zones. Temperature accounts for a sharp increase of diversity from the poles to the equator, while precipitation obscures this pattern by reducing diversity in areas of low precipitation, resulting in 'gaps' in diversity. Diversity of tropical odonates is at least partly explained by the high diversity of aquatic habitats in tropical forests (Orr, 2006), especially in montane areas (Oppel, 2005). Mountains not only provide a greater contemporary diversity of habitats, but also a greater potential for survival in regional refugia. The relative long-term stability of forest habitats (also in the short term, the limited seasonality), which provides opportunities for animals with a specialist lifestyle, might also explain the high diversity of tropical odonates.

Speciation events in dragonflies can be directly linked to isolation events in the geological past such as Andean orogeny (De Marmels, 2001) and isolation in refugia in southern Europe during the ice ages (Sternberg, 1998). Polhemus (1997) showed how a single coenagrionid founder evolved into 22 species of *Megalagrion* on the Hawaiian Islands. Speciation was not only promoted by isolation after the colonisation of newly formed volcanic islands, but also by

Table 1 (a) Number of species per family per biogeographical region. (b) Number of genera per family per biogeographical region

	PA	NA	NT	AT	OL	AU	PAC	ANT	World
(a)									
Aeshnidae	57	42	127	44	138	76	13	–	441
Amphipterygidae	–	–	3	2	5	–	–	–	10
Austropetaliidae	–	–	7	–	–	4	–	–	11
Calopterygidae	37	8	61	20	60	4	–	–	171
Chlorocyphidae	3	–	–	41	80	15	–	–	135
Coenagrionidae	95	101	370	197	185	170	88	–	1084
Cordulegastridae	18	9	1	–	27	–	–	–	49
Chlorogomphidae	5	–	–	–	40	–	–	–	41
Corduliidae	20	50	37	17	57	54	12	–	239
Dicteriadidae	–	–	2	–	–	–	–	–	2
Epiophlebiidae	1	–	–	–	1	–	–	–	2
Euphaeidae	11	–	–	–	65	1	–	–	69
Gomphidae	127	100	273	149	358	42	–	–	966
Hemiphlebiidae	–	–	–	–	–	1	–	–	1
Isostictidae	–	–	–	–	–	41	5	–	46
Lestidae	17	19	38	17	39	29	3	–	148
Lestoideidae	2	–	–	–	4	9	–	–	13
Libellulidae	121	107	352	245	190	184	32	–	1012
Macromiidae	6	9	2	37	50	16	–	–	122
Megapodagrionidae	2	–	130	38	28	57	5	–	260
Neopetaliidae	–	–	1	–	–	–	0	–	1
Perilestidae	–	–	18	1	–	–	–	–	20
Petaluridae	1	2	1	–	–	6	–	–	10
Platycnemididae	23	–	–	33	130	37	–	–	210
Platystictidae	–	1	42	–	119	29	1	–	191
Polythoridae	–	–	58	–	–	–	–	–	58
Protoneuridae	1	3	94	37	57	52	–	–	245
Pseudolestidae	7	–	–	–	15	–	–	–	22
Pseudostigmatidae	–	–	18	1	–	–	–	–	19
Synlestidae	6	–	1	10	17	8	–	–	37
Synthemistidae	–	–	–	–	–	35	9	–	43
Total	560	451	1636	889	1665	870	168	0	5680
Undescribed	75–100	5–10	400–500	100–125	300–400	175–250	30–40	0	1085–1425
(b)									
Aeshnidae	14	13	15	6	18	19	7	–	48
Amphipterygidae	–	–	2	1	1	–	–	–	4
Austropetaliidae	–	–	2	–	–	2	–	–	8
Calopterygidae	8	3	3	3	10	1	–	–	16
Chlorocyphidae	3	–	–	3	14	4	–	–	18
Coenagrionidae	15	15	38	15	23	24	12	–	90
Cordulegastridae	3	1	1	–	5	–	–	–	5
Chlorogomphidae	1	–	–	–	1	–	–	–	1
Corduliidae	6	8	2	6	7	16	3	–	41
Dicteriadidae	–	–	2	–	–	–	–	–	2

Table 1 continued

	PA	NA	NT	AT	OL	AU	PAC	ANT	World
Epiophlebiidae	1	–	–	–	1	–	–	–	1
Euphaeidae	5	–	–	–	12	1	–	–	12
Gomphidae	33	14	26	20	43	9	–	–	92
Hemiphlebiidae	–	–	–	–	–	1	–	–	1
Isostictidae	–	–	–	–	–	11	1	–	12
Lestidae	3	2	2	1	5	3	3	–	8
Lestoideidae	1	–	–	–	1	2	–	–	3
Libellulidae	31	27	44	53	56	45	16	0	143
Macromiidae	2	1	2	1	2	2	–	–	4
Megapodagrionidae	2	–	14	6	10	6	3	–	39
Neopetaliidae	–	–	1	–	–	–	–	–	1
Perilestidae	–	–	2	1	–	–	–	–	3
Petaluridae	1	2	1	–	–	2	–	–	5
Platycnemididae	4	–	–	9	8	11	–	–	25
Platystictidae	–	1	1	–	5	2	1	–	6
Polythoridae	–	–	8	–	–	–	–	–	8
Protoneuridae	1	2	14	4	8	1	–	–	25
Pseudolestidae	1	–	—	–	3	–	–	–	3
Pseudostigmatidae	–	–	5	1	–	–	–	–	6
Synlestidae	2	–	1	2	2	3	–	–	8
Synthemistidae	–	–	–	–	–	4	1	–	4
Total	137	89	186	132	235	169	47	0	642

habitat specialisation (stagnant water, seepage, phytotelmata and swift streams) within an island. Speciation has also been promoted by the isolation of patches of tropical forest due to climatological factors (Dijkstra & Clausnitzer, 2006). Large river systems such as the Amazon and Congo basins, forming an ever-changing mosaic of land and water, probably also facilitated speciation, but distribution patterns in these regions are known insufficiently to verify this hypothesis.

A brief outline of odonate biodiversity within the biogeographical regions

Palaearctic

Large parts of the Palaearctic are relatively species poor when compared with areas at the same latitude in North America. Europe for instance has only slightly more than half the number of species of Texas. Exceptions are Japan, Korea and the part of China included in the Palaearctic. The faunal diversity in these areas is at least as high as in North America and is far richer than in Europe. In China the Palaearctic fauna merges gradually into the Oriental fauna. This transition zone is very rich compared to the other parts of the Palaearctic and harbours many species not occurring elsewhere in the Palaearctic. The large differences in diversity between different parts of the Palaearctic are largely due to the advance of glaciers during the Pleistocene ice ages, which impoverished the fauna in the western two-thirds of the Palaearctic. Here the main mountain ranges and seas run east–west (e.g., the Mediterranean Sea, the Pyrenees, Alps and Himalayas) thus forming a barrier for northern species retreating southwards. Similar factors also apply today as Oriental species can easily penetrate into the Palaearctic, but northward expansion of African and Oriental species into the western Palaearctic is hampered by the same barriers as those limiting southward retreat in the past. The ice ages also promoted speciation by isolating species in various refugia, especially evident in Europe. Most Palaearc-

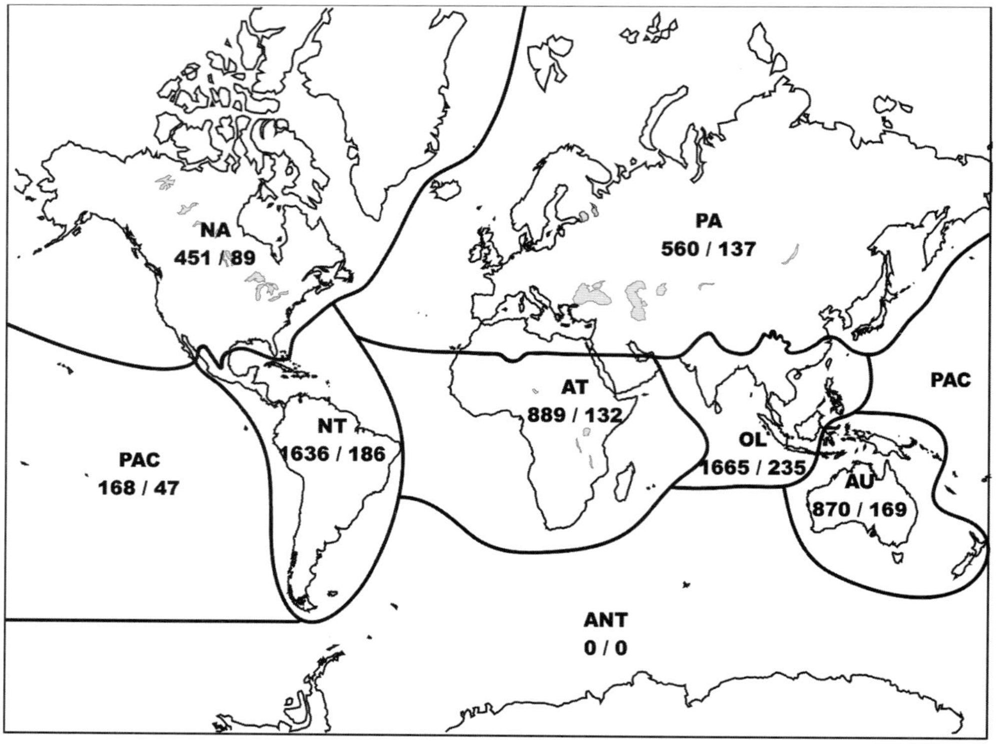

Fig. 2 Diversity of dragonflies per biogeographical region (species number/genus number). PA—Palaearctic, NA—Nearctic, NT—Neotropical, AT—Afrotropical, OL—Oriental, AU—Australasian, PAC—Pacific Oceanic Islands, ANT—Antarctic

tic species with a more northern distribution are widespread, several ranging from Europe to eastern Russia or even into the Nearctic. Whether after the ice ages these wide-ranging species colonised the eastern Palaearctic from the western Palaearctic or vice versa is still a point of debate (Kosterin, 2005).

Nearctic

The dragonfly fauna of the Nearctic is richer than that of most of the Palaearctic. As in the Palaearctic, the eastern part of the Nearctic is richer than the western part, and most eastern states in the USA have larger species lists than all of Europe. This is presumably because the humid East has had a continuous connection with the wet tropics to the south, and numerous tropical species have moved into south-eastern USA, while the West has gone through arid periods when odonate dispersal was interrupted and aquatic faunas were presumably extirpated by glaciation. The species of the wet forests on the west coast of Mexico are restricted from advancing northward by thorn forest and then desert, but species of the moister uplands of the Mexican Plateau have also moved north into the southwestern states. Thus the latter region is a centre of diversity and endemism in North America, as are the north-eastern and south-eastern coastal plains, Allegheny-Appalachian

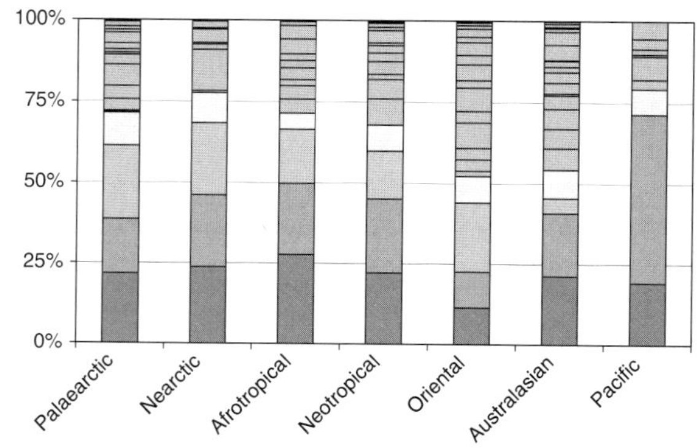

Fig. 3 Percentages of species belonging to a family for the seven different biogeographical regions. The four largest families are at the bottom with from bottom to top: Coenagrionidae, Libellulidae, Gomphidae and Aeshnidae

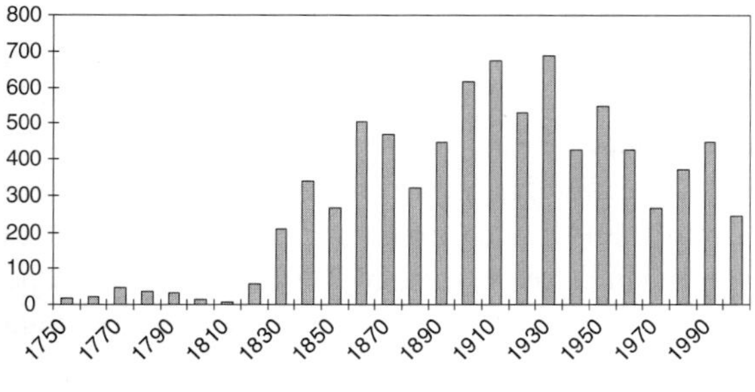

Fig. 4 Rate of description of new taxa in Odonata

uplands, Ozarks, Great Plains, and Pacific coast. Stream-dwelling gomphids are especially likely to show restricted distributions and diversification, and they comprise the largest odonate family in the eastern Nearctic. However, many odonate species, both northern and southern, are wide-ranging over the entire moist eastern half or all across the continent. Others are restricted to the West, often both arid and humid parts of it, as ultimately it is the presence of water bodies that determines their distributions. Some of even the largest odonate families appear to show different origins in the Nearctic, for example coenagrionids and libellulids mostly from the tropics and gomphids and corduliids mostly from northern latitudes. There is a substantial boreal fauna; Canada holds 208 species, but many of them are restricted to the southeastern border region, including tropical genera such as *Hetaerina*, *Argia*, and *Pantala* (Catling et al., 2005). Special features of the Nearctic include the presence of two petalurids, a Pacific Northwest montane species (*Tanypteryx hageni*) with nearest relative in Japan and a southeastern lowland species (*Tachopteryx thoreyi*); certain genera (*Tanypteryx*, *Lanthus*, *Stylogomphus* and *Hagenius*) that show a distinct relationship between eastern Asia and eastern Nearctic; and a very recent radiation of *Enallagma* (Brown et al., 2000).

Afrotropical

It is notable that, among tropical faunas, the Afrotropical fauna is relatively poor and its composition is nearest that of the Holarctic, with few families and a large proportion of Coenagrionidae and Libellulidae (Dijkstra & Clausnitzer, 2006). This may be explained by the relatively unstable climatological history of the continent, which favoured species capable of colonising recent or temporary habitats. The extent of tropical forest in Africa is believed to have contracted substantially during periods with a cooler and drier climate. As a consequence the 'old' African fauna seems to be largely gone, although relicts remain in isolated areas that were apparently more stable. Examples are the genera *Pentaphlebia* (Amphipterygidae) and *Nubiolestes* (Perilestidae) in the Cameroon highlands and *Coryphagrion* (Pseudostigmatidae) in the East Coast forests, which all have their only relatives in tropical America. The families Synlestidae and Megapodagronidae, which have a global but rather fragmented distribution, are largely restricted to South Africa and Madagascar, respectively. On the other hand, the present-day extent of forest and other tropical habitats, such as the continent's famous savannahs, has allowed remarkable speciation in a few genera (e.g., *Chlorocypha*, *Pseudagrion*, *Paragomphus*, *Phyllomacromia*, *Orthetrum* and *Trithemis*). These groups often have strong Asian affinities, suggesting palaeotropic faunal exchange followed by rapid radiation in periods with a more favourable climate. A small but interesting element in the fauna of the eastern coast and Indian Ocean islands are genera of probably Papuan-Australian origin (*Hemicordulia*, *Teinobasis*), that probably reached Africa by wind-aided trans-oceanic dispersal. The highest odonate diversity, as well as the greatest number of range-restricted species, is found in the Guineo-Congolian forest, which stretches from Senegal to western Kenya. The richest area is the Cameroon highlands and the surrounding Lower Guinea lowland forest. The Upper Guinean forest, Congo Basin and Albertine Rift are other core areas within this forest belt. Outside it, coastal East Africa (including the Eastern Arc Mountains), the Ethiopian highlands and South Africa are notable for their endemism. Although the approximately 175 odonate species of Madagascar include distinctly Afrotropical elements, 60% of Anisoptera and almost 95% of Zygoptera species are endemic. Endemism and diversity is greatest on the island's wet eastern coast.

Oriental

The Oriental region is, together with the Neotropical region, by far the most species-rich of the eight regions recognised here. In China the Oriental and Palaearctic faunas merge gradually along a climatic gradient. The Chlorogomphidae and Euphaeidae are largely confined to the Oriental region although both have outlying species occurring in the Palaearctic, and several families such as the Chlorocyphidae, Platycnemididae, Platystictidae and Pseudolestidae are exceptionally well represented. Within the region, several loosely defined subregions, each with a characteristic dragonfly fauna, may be recognised: i.e., the Indian subcontinent, Sundaland, the Philippines, and the main landmass of southeast and east Asia (including tropical and subtropical China, but excluding the Malaya peninsula). The latter subregion exhibits the highest diversity in both species and genera of the entire Oriental region, presumably owing to its large area, numerous mountain ranges intersected by major rivers, and mosaic of forest types. Particularly speciose is the area including the north of Thailand, Laos and Vietnam together with tropical China, recognised by some as distinct faunistic sub-region (van Tol & Rozendaal, 1995; Wilson & Reels, 2003). Within the Indian sub-region the greatest number of species and endemics occurs in tropical forest refugia. Richest are the tropical and subtropical forests to the south of the Himalayas, including Sikkim, North Bengal and the Khasi Hills, with other centres of diversity in the Western Ghats and Nilgiris and the wet south-western and central part of Sri Lanka (Lahiri, 1989; Bedjanič, 2004). Extensive semi-arid parts of the subcontinent host a depauperate and unexceptional fauna. Present-day Sundaland is divided into several large land masses which were contiguous as recently as 8,000 years ago when sea levels were lower. Highest levels of endemicity and species richness occur in north Borneo among forest stream dwellers in montane and mixed dipterocarp forest, but Java, Sumatra and the Malay peninsula all host distinctive faunas. The fauna of the Philippines has a high number of endemics (more than 60% of the named species) sharing elements with both the Oriental and the Australasian fauna. Its numerous islands have facilitated speciation, resulting in a high number of endemic species in genera such as *Drepanosticta*, *Amphicnemis*, *Teinobasis*, *Risiocnemis* and *Oligoaeschna* (Hämäläinen & Müller, 1997).

Australasian region

The Australasian dragonfly fauna is very distinct with a strong representation of small families either largely confined to the region or showing a relict distribution. For several families a large percentage of the world fauna is restricted to the Australasian region: Austropetaliidae (36%), Isostictidae (89%), Lestoideidae (69%), Petaluridae (60%) and Synthemistidae (81%). Hemiphlebiidae and Cordulephyinae (Corduliidae) are both endemic for continental Australia. The Austropetaliidae are only shared with the southern Andes and are therefore believed to be of Gondwanian origin. The Petaluridae and the Synlestidae are good examples of families showing a relict distribution. The majority of dragonflies of the Australian continent occur in the eastern Great Dividing Range and in the adjacent narrow coastal strip to the east of this, and in the wetter parts of the southwest. Greatest diversity is to be found in the north-eastern wet tropics of Queensland. The dry interior of the continent has a depauperate fauna of widespread eurytopic species. The New Zealand fauna is poor with only 17 species (Rowe, 1992) including two species of Petaluridae. New Guinea is very species-rich with a high percentage of endemics, owing to the perhumid tropical conditions and a highly dissected, mountainous topography that creates numerous isolated stream systems, each including a wide altitudinal range. New Guinea was formed during the mid-late Caenozoic when the northward moving Australian plate collided with island arcs to the north, resulting in massive uplifting and orogenesis. The island arcs were part of a complex archipelago that probably played a part in faunal exchange between the Oriental region and the Australasian region, resulting in unexpected affinities between the Philippines and New Guinea (van Tol & Gassmann, 2005). New Guinea and Australia were connected as recently as 8,000 years ago and generally have strong biogeographical affinities. They share a high diversity of Megapodagrionidae, Isostictidae and Synthemistidae. However considering their shared history the differences in the dragonfly fauna is remarkable. Especially striking is the absence of Platystictidae and Platycnemididae in Australia and conversely the virtual lack of Gomphidae and Brachytroninae (Aeshnidae) in New Guinea (Lieftinck, 1949). The Moluccas and Lesser Sundas (Nusa Tenggara) have probably never been connected to either the Oriental region or the Australasian region. The islands of Lesser Sundas have a drier climate than the rest of the Indonesian archipelago but their faunas are generally commensurate with island area. Most of the species on these islands, including the many endemics, are of Oriental origin (Lieftinck, 1953). The Moluccan fauna is largely derived from New Guinea, is rather depauperate, and is perhaps most notable for its lack of the genus *Neurobasis* (Calopterygidae), present in New Guinea, Sulawesi and the Philippines. The island of Sulawesi was formed by the collision of several elements of Laurasian origin and Gondwanian origin. The dragonfly fauna is therefore a blend of species of Australasian and Oriental origin, although the latter dominate (van Tol & Gassmann, 2005). No current review of Sulawesi dragonflies is available, but it is known that the fauna is less species-rich than might be expected (van Tol, 1987) The family Chlorocyphidae shows exceptional higher-level diversity, as does Borneo, which perhaps dates back to the most recent connection of the two land masses 42 mya.

Pacific

As might be expected, the Pacific is species poor. Species present can be divided into two groups: those with a very small area of distribution, being often confined to a single island or island group, and highly vagile eurytopic species which occur on most Pacific islands, and which generally also occur throughout much of the Oriental or Australasian regions (or both). Even in Hawaii this phenomenon occurs, although the widespread species originate from the Americas. Both the widespread species and the endemics belong mainly to the Coenagrionidae and the Libellulidae. In the Coenagrionidae the colonisation of an island or group of islands was often followed by speciation events leading to a group of closely related species. This has occurred on Hawaii (*Megalagrion*), Pohnpei (*Teinobasis*), Fiji (*Nesobasis* and *Melanesobasis*) and Samoa (*Pacificagrion* and *Amorphostigma*). An exception to this pattern is New Caledonia, which drifted away from continental Australia at the end of the Cretaceous, and is moderately species rich. It has an interesting fauna showing distinct affinities with Australia and New Guinea and has numerous endemic species and several endemic genera (Davies, 2002).

Neotropical

Although North and South America have numerous genera and species in common, this is primarily because the boundary between them is political rather than biogeographical. Nevertheless, the two faunas are quite distinct, with a strong faunal break at middle elevations around the Mexican Plateau, many Nearctic species in temperate habitats on that plateau, and tropical species surrounding it in the lowlands (Paulson, 1982). Dispersal was apparently much greater from south to north when Panama emerged in the Pliocene to provide a pathway between the continents, and that dispersal continues today. The Polythoridae, Dicteriadidae and Neopetaliidae are endemic to the region, the latter confined to the southern Andes while the former two are distributed in the tropics. Largely confined to this region are the Austropetaliidae, Perilestidae and Pseudostigmatidae. The latter includes 18 species of very elongate spider-eating, phytotelmata-breeding damselflies which are among the most remarkable odonates. Significant regions of odonate diversification include the Mexican Plateau, Chiapas to Honduras highlands, Costa Rica-Panama highlands, northern Andes, eastern Andean foothills, tepuis of the Guyana Shield, Guyana lowlands, Atlantic forests of Brazil, Rio Paraná basin, and southern Andes. In the last, Gondwanian groups, including the Austropetaliidae, Neopetaliidae, Petaluridae, and *Gomphomacromia*, are prominent. This leaves the huge Amazon basin, poorly known but presumably with its own regions of endemism. The Neotropical fauna equals that of the Oriental region in both modern (species) and ancient (family) diversity. The complexity of the mountain ranges extending from Mexico to Chile and the varied climates along their length have produced a great variety of odonate habitats, as well as providing repeated opportunities for speciation, and *Argia*, with 108 named species, is the star of this show. Other characteristic neotropical genera that have diversified widely in the region include *Heteragrion*, *Palaemnema*, *Acanthagrion*, *Telebasis*, *Phyllogomphoides*, *Progomphus*, *Erythrodiplax* and *Micrathyria*. High biodiversity is the rule for all of the countries in this region, but nevertheless, the Neotropical fauna is the least known in the world. The highest known local diversity of odonates is in South America, with 186 species at a single site in southern Peru. Much of the fauna of the West Indies comes from adjacent Mexico and South America, but the large Greater Antillean islands have numerous endemics, including *Hypolestes* of poorly known affinities.

Table 2 Examples of areas with a high number of endemic dragonflies

Biogeographical region	Name of area	Number of species	Number of endemic species	Estimate of endemism (%)
Afrotropical	Ethiopia	96	12	13
	South Africa	160	30	19
	Madagascar	175	135	77
Oriental	Taiwan	142	21	15
	Borneo	272	124	46
	Hainan	127	20	16
	Sri Lanka	116	53	46
Palaearctic	Northwest Africa (Morocco, Algeria and Tunisia)	70	4	6
	Japan	215	74	34
Neotropical	Cuba	80	5	6
	Costa Rica	265	32	12
	Venezuela	489	90	18
Australasian	Sulawesi	124	55	44
	New Zealand	17	10	59
Pacific	Hawaii	36	26	72
	New Caledonia	55	22	40

Antarctica

No species are known from this region and it is unlikely that any species of dragonfly will reproduce there although it is not impossible that some species might be found as vagrants.

Areas of endemicity

No well-based review of areas with high endemism of dragonflies is available. However, this is intended in the near future as part of a Global Dragonfly Assessment. Regional projects to identify areas of endemism carried out so far include an analysis of endemism in freshwater biotas partly based on Zygoptera for New Guinea and on Zygoptera and Anisoptera in eastern Africa (Polhemus et al., 2004; Darwall et al., 2005) and are presently being prepared for southern and western Africa by the IUCN Freshwater Biodiversity Assessment Programme. Some examples of levels of endemism in different regions are summarised in Table 2, showing large inter-area differences between areas in absolute and relative numbers of endemic species. Species in the temperate region have wide distributions and the percentage of localised species is low, whereas faunas with both high absolute and relative endemism are mainly found in moist tropical forests. Although at present there is no sound basis for identifying the most important areas of endemism, it goes without question that the faunas of the islands of New Guinea, Sulawesi, Sri Lanka and Madagascar are exceptionally rich in endemics (see Table 2). It is noteworthy that the percentage of endemic Zygoptera is almost always much higher than the percentage of endemic Anisoptera. Examples of this are Madagascar (60% in Anisoptera, 95% in Zygoptera), the Philippines (31%, 86%) and Sri Lanka (30%, 68%).

Human-related issues

Dragonflies have little economic value, although they are used as food and as magical or medicinal resources at a local scale, and to an unknown extent may influence populations of disease vectors. The group features prominently in nature management in the temperate regions of the world (Westfall & May, 1996; Kosterin et al., 2004; Sahlén et al., 2004) and they are often used as indicators for environmental health and conservation management (Corbet, 1999). Their sensitivity to structural habitat quality (e.g., forest cover, water chemistry) and amphibious habits make dragonflies well suited for use in evaluating environmental change in the long term (biogeography, climatology) and in the short term (conservation biology), both above and below the water surface (e.g., Clark & Samways, 1996; Sahlén & Ekestubbe, 2001; Clausnitzer, 2003; Foote & Hornung, 2005; Osborn, 2005). Dragonflies are often used in both fundamental and applied research because of the relative ease with which they can be observed and their broad array of behaviours. In many regions reliable identification literature is available, so species can be determined fairly easily by the layman. This enables mapping schemes conducted by volunteers, facilitating the use of distributional data on dragonflies in management. From a global perspective, odonates are among the best known insect groups with respect to taxonomy and distribution, and, apart from butterflies, probably no other group of insects receives so much attention from the general public and has so many organisations devoted to its study. An overview of the conservation and research status of the world's Odonata can be found in Clausnitzer & Jödicke (2004).

Many species in the temperate region have shown a dramatic decline in distribution and abundance since the second half of the 20th century (Westfall & May, 1996; Sahlén et al., 2004; Inoue, 2004). This has been caused by habitat destruction, eutrophication, acidification and pollution of aquatic habitats in general, and the canalisation of streams and rivers. Most of these species are not under immediate threat of extinction as they have wide ranges. A marked exception comes from the Ogasawara Islands, Japan, where five endemic species are on the verge of extinction due to the introduction in the 1980s of an *Anolis* lizard (Karube, 2005). At least in parts of Europe, some of the species considered threatened recovered since the 1990s as result of improved water management. Recently it has become evident that many dragonflies of temperate regions are responding, both in distribution and phenology, to global climate change (Ott, 2001). The ranges of common and widespread southern species are expanding in Europe but there is as yet no strong evidence that northern species are decreasing as a result of the rising temperatures, as might be expected.

Most of the world's dragonflies are restricted to the tropics, especially to forest, where the diversity of the group is greatest. Tropical species of open landscapes are generally better able to colonise new habitats than species confined to forest, and therefore have wider ranges on average and seem to be less influenced by habitat alteration. A clear exception is a number of South African endemics which declined due to shading of their habitat by invasive alien trees (Kinvig & Samways, 2000), and recovered after removal of the trees (Samways, 2004). Destruction of tropical forest is probably the most important threat to global odonate diversity, potentially resulting in the extinction of numerous species. Unfortunately these species are often poorly known, making it difficult to say whether a species is genuinely rare or merely overlooked. Evaluating the conservation status of most naturally rare species is hardly possible. Examples of data deficiency are known from Africa (e.g., Dijkstra & Clausnitzer, 2006), South America (Paulson, 2006), the Oriental region (Orr, 2004) and New Guinea. More fieldwork is thus essential to establish the true ranges of these species and to determine areas of endemism within larger tropical forest areas. There is, however, no doubt that species confined to small remnants of forest in areas under high human pressure are endangered. Examples of such sites include many of the Philippine islands, Hawaii, the small pockets of forest in the Eastern Arc Mountains of East Africa and the Caribbean islands of Cuba, Hispaniola and Jamaica, but a well-founded overview of threatened areas of high importance for dragonflies is wanting. Especially susceptible are species depending on forest on small islands such as those of the Seychelles (Samways, 2003). Here the disappearance of forest-cover not only results in alteration of the habitat but also may change precipitation patterns.

Dragonflies have shown to be useful for nature management and conservation, and recently an increased effort is being made to make information on dragonflies available for both scientists and policymakers. Important initiatives taken are the update of the IUCN red list, the 'Pan-Africa Freshwater Biodiversity Assessment' started by the IUCN (Darwall et al., 2005), which includes dragonflies among other taxa, and the 'Global Dragonfly Assessment' initiated in 2005. The latter should hopefully result in a more detailed overview of the areas of endemism and conservation priority in the coming years.

References

Bedjanič, M., 2004. Odonata fauna of Sri Lanka: research and threat status. In Clausnitzer, V. & R. Jödicke (eds), Guardians of the Watershed. Global Status of Dragonflies: Critical Species, Threat and Conservation. International Journal of Odonatology 7: 279–294.

Brown, J. M., M. A. McPeek & M. L. May, 2000. A phylogenetic perspective on habitat shifts and diversity in the North American *Enallagma* damselflies. Systematic Biology 49: 697–712.

Catling, P. M., R. A. Cannings & P. M. Brunelle, 2005. An annotated checklist of the Odonata of Canada. Bulletin of American Odontology 9: 1–20.

Clark, T. E. & M. F. Samways, 1996. Dragonflies (Odonata) as indicators of biotope quality in the Kruger National Park, South Africa. Journal of Applied Ecology 33: 1001–1012.

Clausnitzer, V. & R. Jödicke (eds), 2004. Guardians of the watershed. Global status of dragonflies: critical species, threat and conservation. International Journal of Odonatology 7: 385–398.

Clausnitzer, V., 2003. Dragonfly communities in coastal habitats of Kenya: indication of biotope quality and the need of conservation measures. Biodiversity and Conservation 12: 333–356.

Corbet, P. S., 1999. Dragonflies Behaviour and Ecology of Odonata. Harley Books, Essex, England.

Darwall, W., K. Smith, T. Lowe & J.-C. Vié, 2005. The Status and Distribution of Freshwater Biodiversity in Eastern Africa. IUCN SSC Freshwater Assessment Programme. IUCN, Gland, Switzerland and Cambridge, UK.

Davies, D. A. L., 2002. The odonate fauna of New Caledonia, including the description of a new species and a new subspecies. Odonatologica 31: 229–251.

De Marmels, J. C., 2001. Revision of *Megapodagrion* Selys, 1886 (Insecta, Odonata: Megapodagrionidae). Dissertation, Universität Zürich.

Dijkstra, K.-D. B. & V. Clausnitzer, 2006. Thoughts from Africa: how can forest influence species composition, diversity and speciation in tropical Odonata? In Cordero Rivera, A. (ed.), Forest and Dragonflies. Pensoft Publishers, Sofia.

Foote, A. L. & C. L. R. Hornung, 2005. Odonates as biological indicators of grazing effects on Canadian prairie wetlands. Ecological Entomology 30: 273–283.

Grimaldi, D. & M. S. Engel, 2005. Evolution of the Insects. Cambridge University press, New York.

Hämäläinen, M. & R. A. Müller, 1997. Synopsis of the Philippine Odonata, with lists of species recorded from forty islands. Odonatologica 26: 249–315.

Inoue, K., 2004. Critical species of Odonata in Japan. In Clausnitzer, V. & R. Jödicke (eds), Guardians of the Watershed. Global Status of Dragonflies: Critical Species, Threat and Conservation. International Journal of Odonatology 7: 311–324.

Karube, H., 2005. Why are endemic Odonates endangered in oceanic islands Ogasawara? Abstract Book of the 4th WDA International Symposium of Odonatology. Ponteverda, Spain.

Kinvig, R. G. & M. J. Samways, 2000. Conserving dragonflies (Odonata) along streams running through commercial forestry. Odonatologica 29: 195–208.

Kosterin, O. E., 2005. Western range limits and isolates of eastern odonata species in Siberia and their putative origins. Odonatologica 34: 219–242.

Kosterin, O. E., E. I. Malikova & A. Yu. Haritonov, 2004. Critical species of Odonata in the Asian part of the former USSR and the Republic of Mongolia. In Clausnitzer, V. & R. Jödicke (eds), Guardians of the Watershed. Global Status of Dragonflies: Critical Species, Threat and Conservation. International Journal of Odonatology 7: 341–370.

Lahriri, A. R., 1989. On the status of rare Indian odonate species. Advances in Odonatology 4: 53–56.

Lieftinck, M. A., 1949. The dragonflies (Odonata) of New Guinea and neighbouring islands. Part VII. Results of the Third Archbold expedition 1938–1939 and of the Le Roux Expedition 1939 to Netherlands New Guinea (II. Zygoptera). Nova Guinea (N.S.) 5: 1–271.

Lieftinck, M. A., 1953. The Odonata of the island Sumba with a survey of the dragonfly fauna of the Lesser Sunda Islands. Verhandlungen der Naturforschenden Gesellschaft Basel 64: 118–228.

Needham, J. G., M. J. Westfall & M. L. May, 2000. Dragonflies of North America. Scientific Publishers, Gainesville.

Okudaira, M., M. Sugimura, S. Ishida & K. Kojima, 1999. Dragonflies of the Japanese Archipelago in Color. Hokkaido UP.

Oppel, S., 2005. Habitat associations of an Odonata community in a lower montane rainforest in Papua New Guinea. International Journal of Odonatology 8: 243–257.

Orr, A.G., 2004. Critical species of Odonata in Malaysia, Indonesia, Singapore and Brunei. In Clausnitzer, V. & R. Jödicke (eds), Guardians of the Watershed. Global Status of Dragonflies: Critical Species, Threat and Conservation. International Journal of Odonatology 7: 371–384.

Orr, A. G., 2006. Odonata in Bornean tropical rain forest formations: diversity, endemicity and implications for conservation management. In Cordero Rivera, A. (ed.), Forest and Dragonflies. Pensoft Publishers, Sofia.

Osborn, R., 2005. Odonata as indicators of habitat quality at lakes in Louisiana, United States. Odonatologica 34: 259–270.

Ott, J., 2001. Expansion of Mediterranean Odonata in Germany and Europe: consequences of climate changes. In Walther, G.-R., C. A. Burga & P. J. Edwards (eds), 'Fingerprints' of Climate Change: Adapted Behaviour and Shifted Species Ranges. Kluwer Academic/Plenum Publishers, New York, Boston, Dordrecht, London, Moscow: 89–111.

Paulson, D. R., 1982. Odonata. In Hurlbert, S. H. & A. Villalobos-Figueroa (eds), Aquatic Biota of Mexico, Central America and the West Indies. San Diego State University, San Diego.

Paulson, D. R., 2004. Critical species of Odonata in the Neotropics. In Clausnitzer, V. & R. Jödicke (eds), Guardians of the Watershed. Global Status of Dragonflies: Critical Species, Threat and Conservation. International Journal of Odonatology 7: 163–188.

Paulson, D., 2006. The importance of forest to Neotropical dragonflies. In Cordero Rivera, A. (ed.), Forest and Dragonflies. Pensoft Publishers, Sofia.

Polhemus, D. A., 1997. Phylogenetic analysis of the Hawaiian Damselfly Genus *Megalagrion* (Odonata: Coenagrionidae): Implications for biogeography, ecology, and conservation biology. Pacific Science 51: 395–412.

Polhemus, D. A., R. A. Englund & G. R. Allen, 2004. Freshwater biotas of New Guinea and nearby islands: analysis of endemism, richness, and threats. Bishop Museum Technical Report 31.

Rehn, A. C., 2003. Phylogenetic analysis of higher-level relationships of Odonata. Systematic Entomology 28: 181–239.

Rowe, R., 1992. The Dragonflies of New Zealand. Oxford University Press.

Sahlén, G. & K. Ekestubbe, 2001. Identification of dragonflies (Odonata) as indicators of general species richness in boreal forest lakes. Biodiversity and Conservation 10: 673–690.

Sahlén, G., R. Bernard, A. C. Rivera, R. Ketelaar & F. Suhling, 2004. Critical species of Odonata in Europe. In Clausnitzer, V. & R. Jödicke (eds), Guardians of the Watershed. Global Status of Dragonflies: Critical Species, Threat and Conservation. International Journal of Odonatology 7: 385–398.

Samways, M. J., 2003. Conservation of an endemic odonate fauna in the Seychelles Archipelago. Odonatologica 32: 177–182.

Samways, M. J., 2004. Critical species of Odonata in southern Africa. In Clausnitzer, V. & R. Jödicke (eds), Guardians of the Watershed. Global Status of Dragonflies: Critical Species, Threat and Conservation. International Journal of Odonatology 7: 255–262.

Silsby, J., 2001. Dragonflies of the World. Natural History Museum/CSIRO.

Sternberg, K., 1998. The postglacial colonization of Central Europe by dragonflies, with special reference to southwestern Germany (Insecta, Odonata. Journal of Biogeography 25: 319–337.

Tsuda, S., 2000. A Distributional List of World Odonata. Private Publication, Osaka.

van Tol, J., 2005. Global Species Database Odonata. http://www.odonata.info (5 August 2005) (also available via www.species2000.org).

van Tol, J. & D. Gassmann, 2005. Zoogeography of freshwater invertebrates of Southeast Asia, with special reference to Odonata. In Renema, W. (ed.), Biogeography, Time and Place: Distributions and Islands. Springer.

van Tol, J. & F. G. Rozendaal, 1995. Records of Calopterygoidea from Vietnam, with description of two new species (Zygoptera: Amphipterygidae, Calopterygidae, Chlorocyphidae, Euphaeidae). Odonatologica 24: 89–107.

van Tol, J., 1987. The Odonata of Sulawesi (Celebes), Indonesia an introduction. Advances in Odonatology 3: 147–155.

Waage, J. K., 1979. Dual function of the damselfly penis: sperm removal and transfer. Science 203: 916–918.

Wang, L.-J., 2000. Dragonflies of Taiwan. Jemjem Calendar, Taipei.

Watson, J. A. L., G. Theischinger & H. M. Abbey, 1991. The Australian Dragonflies. CSIRO, Canberra and Melbourne.

Westfall, M. J. & M. L. May, 1996. Damselflies of North America. Scientific Publishers, Gainesville.

Wilson, K. D. P. & G. T. Reels, 2003. Odonata of Guangxi Zhuang Autonomous Region, China, I. Zygoptera. Odonatologica 32: 237–279.

Hydrobiologia (2008) 595:365–377
DOI 10.1007/s10750-007-9031-3

FRESHWATER ANIMAL DIVERSITY ASSESSMENT

Global diversity of stoneflies (Plecoptera; Insecta) in freshwater

Romolo Fochetti ·
José Manuel Tierno de Figueroa

Abstract Plecoptera, or stoneflies, is a small order of hemimetabolous insects: according to our data, more than 3,497 species have been described so far in the world. The total number of species has enormously increased in the last 30 years (2,000 species estimated in 1976) and, if the trend continues, then it will nearly double in the near future. The order is divided into the suborders Arctoperlaria and Antarctoperlaria, and includes 16 families: 12 arctoperlarian and 4 antarctoperlarian. The Arctoperlaria account for a total number of 3,179 species, and Antarctoperlaria, only 318 species. The total number of genera is 286. We give in this article the estimated number of species for each family. The fauna and diversity of stonefly in North America (650 species reported) and Europe (426 species) are best known. Nevertheless, in the last 25 years, a mean of 2.6 Plecoptera species per year were described in Europe. Stonefly-faunas of Australia (191 species, Tasmania included) and New Zealand (104 species) are relatively well-known, while our knowledge of the Plecoptera of Central and South America (95 and 378 species respectively) is poor and still not representative of the real diversity. Africa has a reduced stonefly fauna (126 species). Asian stonefly diversity (approximately 1,527 species) is much greater than that of Europe or North America despite the fact that, except for Japan and Asiatic Russia that have been well studied, our knowledge of the remaining Asiatic areas is extremely poor. Even though our data indicate the Holarctic Region as the diversity hot-spot for the order, the analysis of the specific diversity divided by family suggests also an important role of tropical stoneflies.

Keywords Plecoptera · Distribution · Diversity · Zoogeographical regions

Guest editors: E.V. Balian, C. Lévêque, H. Segers & K. Martens
Freshwater Animal Diversity Assessment

R. Fochetti (✉)
Dipartimento di Scienze Ambientali, Università della Tuscia, Largo dell'Università snc, 01100 Viterbo, Italy
e-mail: fochetti@unitus.it

J. M. Tierno de Figueroa
Departamento de Biologia Animal, Facultad de Ciencias, Universidad de Granada, 18071 Granada, Spain

Introduction

Plecoptera is a small order of hemimetabolous insects, commonly called stoneflies, with more than 3,497 described species (this article). Stoneflies are distributed over all continents except Antarctica, and constitute a significant ecological component of running water ecosystems. Their ecological requirements greatly limit the dispersal capacity of the nymphs and, because adults have reduced flight ability, stoneflies show a high percentage of endemism. Due to the growing pollution and alteration of water courses and to the high stenoecity of Plecoptera, numerous stonefly species are reduced to small isolated populations and many others have already gone extinct. Stoneflies are probably one of the most endangered groups of insects (Fochetti & Tierno de Figueroa, 2004).

Plecoptera can be easily recognized by several morphological characters: soft body, three segmented tarsi, elongate filiform antennae, mandibulated mouthparts, two compound eyes, two or three ocelli, two usually long cerci, 10-segmented abdomen with vestiges of the eleventh segment. Adults have two pair of membranous large wings (sometimes reduced or absent), and subequal fore and hind wings (hind wings slightly wider) that fold horizontally over and around abdomen when at rest (hence the name: *plecos* = folded; *pteros* = wings) (Fig. 1A). Nymphs are similar to adults (Fig. 1B), with a closed tracheal system with or without filamentous gills. When present, gills are located on different parts of the body.

Plecoptera nymphs are aquatic and live mainly in cold, well-oxygenated running waters, although some species can also be found in lakes. Few species are adapted to terrestrial life in Sub-Antarctic areas: Hynes (1976) reported the tendency of southern hemisphere nymphs to leave the water. However, the increasing number of stoneflies described from the tropics and their high rate of endemicity can modify the common belief that Plecoptera are cold-water specialists, and suggest instead that the true hot-spot for Plecoptera diversity are tropical areas (Zwick, pers. com.). The life cycle of stoneflies lasts for one or more years, but there are also bi- or tri-voltine species. Nymphal or egg diapause is not uncommon. The nymphs can moult up to 33 times before emerging. They feed on animal or vegetable matter as collectors, scrapers, shredders or predators.

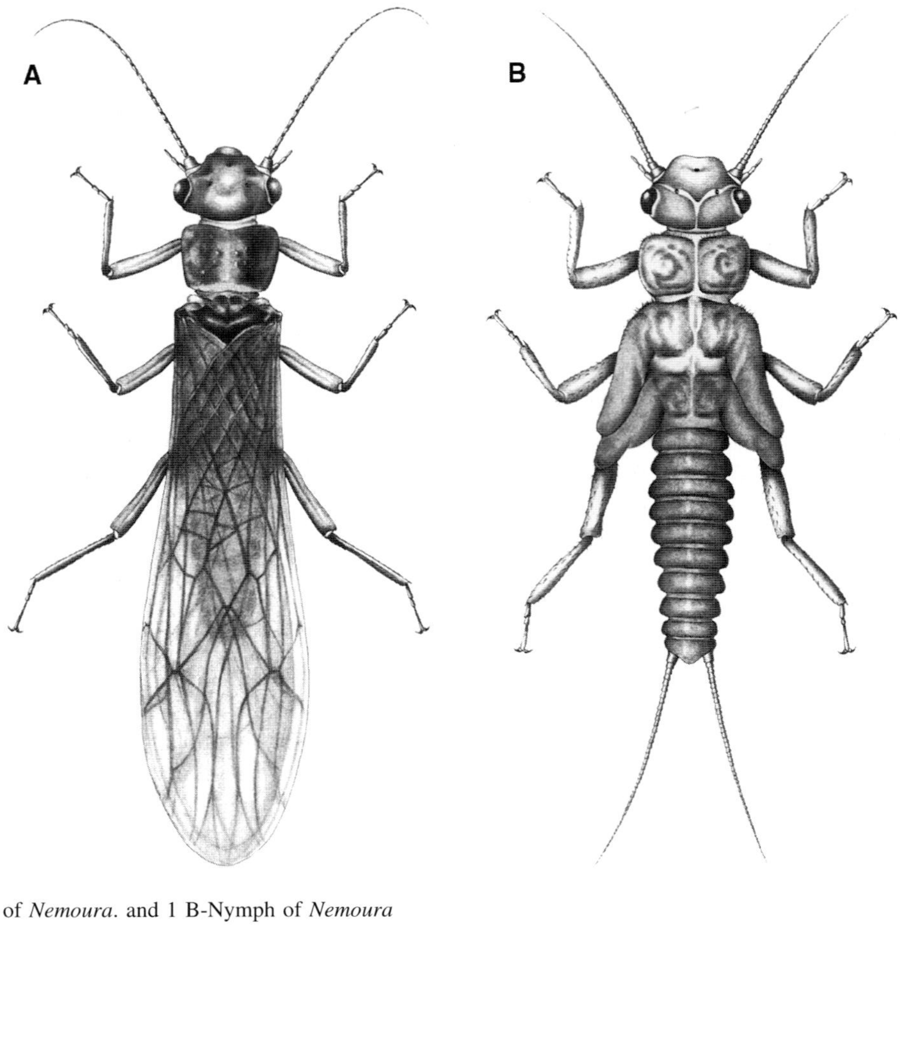

Fig. 1 A-Adult of *Nemoura*. and 1 B-Nymph of *Nemoura*

Adults are terrestrial or sub-aerial, with the exception of *Capnia lacustra* Jewett, whose adults have been collected only at a depth of 60–80 m in the Lake Tahoe (USA) (Jewett, 1963), and the adult females of *Zapada cinctipes* (Bank) that can remain under water for a long time (20–60 min) for ovipositing (Tozer, 1979). Adults lifespan is short, from a few days to weeks; they display complex behaviours related to mate searching (such as the use of intersexual vibrational communication in Arctoperlarian stoneflies) and oviposition (Stewart, 1994). The adult feeds on pollen, lichens, cyanobacteria, etc., although the adults of the largest species do not seem to feed (Tierno de Figueroa & Fochetti, 2001). Females produce 25–3,000 eggs, and lay their egg-clutches, or rarely each egg separately, on the water. Parthenogenesis and ovoviviparity have been also described for some species.

Species diversity

According to our data, approximately 3,500 species of Plecoptera have been reported so far in the world (Tables 1, 2). They are divided as: North America 650 species (Stark & Baumann, 2005), Central America 95 species, South America 378 species (Heckman, 2003), Europe 426 species (Fochetti & Tierno de Figueroa, 2004), Africa 126 species (31 from Mediterranean Africa, with only 12 endemic species, 50 from tropical Africa, 33 from South Africa and 12 from Madagascar), Asia approximately 1,527 species [(about 784 from South-East Asia = 31 Taiwan, 32 Korea, 25 Philippines, 39 Borneo, 36 Indonesia, 17 Malaysia, 29 Thailand, 20 Vietnam, two Laos, three Myanmar, 28 Bhutan, two Bangladesh, 10 Sri Lanka, 112 India, 63 Nepal, 20 Pakistan, nine Afghanistan, 306 Japan, no species from Cambodia; Sivec & Yang 2001), China 350 (?) species (deduced from Yuzhou & Junhua, 2001, who report 231 species of Perlidae), West Asia about 114 species (two Syria, two Israel, 25 Iran, 15 Lebanon, 70 Asiatic Turkey), Asiatic Russia 279 species (179 Nemouroidea and about 100 Perloidea). Although a small number of species are shared by several countries, we consider the total sum as a good approximation)], Australia 191 species (Michaelis & Yule, 1988), New Zealand 104 species (McLellan, 2006).

Very few species are shared among zoogeographical regions (mainly between Nearctic and Palaearctic and between Palaearctic and Oriental) or among continents, i.e. at borders between Africa and Europe or at the borders between Europe and Asia. Australia and New Zealand are obviously isolated; deserts or dry lands prevent stonefly migration from North to Central and South America, and from the Palaearctic to the Ethiopic Regions. Thus, the total number of described species should not be too far from the reported one.

In comparison to the papers by Hynes (1976) and Zwick (1980) who reported about 2,000 described stonefly species, the total number has enourmously increased in the last 25 years and, if the trend continues then it will nearly double in the near future.

The fauna and diversity of stonefly in North America and Europe are best known. Nevertheless, new species have been described from these areas at a high rate: in the last 25 years, a mean of 2.6 Plecoptera species per year were described in Europe (Fochetti & Tierno de Figueroa, 2005). Stonefly-faunas of Australia (Tasmania included) and New Zealand are relatively well-known, while our knowledge of the Plecoptera of Central and South America is poor and still not representative of the real diversity.

Asian stonefly diversity is much greater than that of Europe or North America. This is true despite the fact that, except for Japan and Asiatic Russia that have been well studied, our knowledge of the remaining Asiatic areas is extremely poor. Our greatest lack of knowledge and the highest uncertainty regard Asiatic faunas occurs in the Oriental Region. For instance, according to Yuzhou & Junhua (2001) over 230 species of Perlidae, the most speciose family in the country, have so far been described from China. The identity of many of these species is questionable and cannot be checked because most of the types were lost or destroyed; nonetheless, intensive future research will certainly lead to the discovery of hundreds of new species in this country. In fact, a few intensive collecting trips to Taiwan almost doubled its stonefly-fauna (Sivec & Yang, 2001).

In Africa, a reduced number of species were collected, but we do not expect a great increase in Plecoptera diversity in the future. The results of recent studies on the systematics of the tropical tricky

Table 1 Number of Plecoptera species in each family and each continent

	NAm	CAm	SAm	Eu	AF	AS[a]	NZ	AUS
Eustheniidae			2				4	15
Diamphipnoidae			5					
Austroperlidae			4				1	10
Gripopterygidae			73				70	134
Pteronarcyidae	10	1				2		
Peltoperlidae	22					24		
Styloperlidae						8		
Perlodidae	125	3		62	3	118		
Perlidae	82	71	277	18	55	546		
Chloroperlidae	95	2		19	1	89		
Scopuridae						8		
Taeniopterygidae	34			41	2	65		
Notonemouridae			17		43		29	32
Nemouridae	71	9		132	8	413		
Capniidae	156	8		21	4	126		
Leuctridae	55	1		133	10	128		
Total	650	95	378	426	126	1,527	104	191

[a] Asiatic Russia and China data have been arbitrarily subdivided (NAm = North America; CAm = Central American; SAm = South America; Eu = Europe; AF = Africa; AS = Asia; NZ = New Zealand; AUS = Australia)

Table 2 Number of Plecoptera species in each family and each zoogeographical region

	PA[a]	NA	NT	AT	OL[b]	AU	World
Eustheniidae			2			19	21
Diamphipnoidae			5				5
Austroperlidae			4			11	15
Gripopterygidae			73			204	277
Pteronarcyidae		10	1				13
Peltoperlidae		22					46
Styloperlidae							8
Perlodidae	63	125	4				311
Perlidae	18	82	348	52			1,049
Chloroperlidae	20	95	2				206
Scopuridae							8
Taeniopterygidae	42	34					142
Notonemouridae			17	43		61	121
Nemouridae	136	71	9				633
Capniidae	21	156	8				315
Leuctridae	138	55	1				327
Total	438[a]	650	474	95	350?	295	3,497

[a] Palaearctic Asia not included. PA: Palaearctic Region, NA: Nearctic Region, NT: Neotropical Region, AT: Afrotropical Region, OL: Oriental Region, AU: Australasian Region. [b] The distribution of Oriental Plecoptera cannot be unequivocally established

genus *Neoperla* might increase the species list and some new species could be described from Mediterranean countries (Morocco, Algeria, Tunisia), but the total stonefly diversity will not increase to the level of the other continents.

As we expected, the analysis of Plecoptera diversity and distribution shows their well-known ecological requirements as: they are absent from arid or semi-arid areas everywhere in the world. There are almost no records from West Australia, no records at all for Saudi Arabia, Yemen, Oman, Emirates, Kuwait and Iraq in Asia, and stonefly diversity in Africa is greatly limited by deserts and dry lands.

The order is divided into the suborders Arctoperlaria and Antarctoperlaria (see the phylogeny chapter below), and includes 16 families: 12 arctoperlarian and four antarctoperlarian. The Arctoperlaria account for a total number of 3,179 species, and Antarctoperlaria, only 318 species. They are listed in Table 2 with the estimated number of species for each family. The total number of genera is 286 (Table 3). Even though Table 1 is incomplete because it lacks details on the distribution of Chinese families and Far East Russia species, Perlidae is the most diverse family, with more than 1,000 described species. Since the last estimate (Zwick, 1980), the number of perlid species almost doubled, due to the increase of faunistic research in Asia and, above all, to the improved knowledge of the systematics of the tropical genera *Neoperla* and *Anacroneuria*. The family Perlidae spreads over Holarctic, Oriental, Afrotropical and Neotropical Regions. The most diverse perlid areas are Asia, and Central and South America. Presently the Chinese perlid species represent about 25% of the world Perlidae, making China one of the most important areas of the world for perlid diversity (Yuzhou & Junhua, 2001). The Neotropical perlid fauna currently includes more than 350 species: *Anacroneuria* is the dominant and most diverse stonefly genus, (Stark, 2001) with almost 300 described species.

Perlodidae presently includes about 300 species, mainly in the Holarctic region. The most speciose perlodid genus, *Isoperla*, has a Holarctic distribution and has more than 100 species, a third of the total. A similar distribution is shown by the ancient genus *Arcynopteryx* [Aubert (1959) stated it is an Ercinic relict] and by the genus *Diura*. *A. compacta* (McLachlan) is distributed in North America and Europe, as well as *D. bicaudata* (Linnaeus) and *D. nanseni* (Kempny).

Chloroperlidae include more than 200 species mainly in the Holarctic region. The genera *Sweltsa* and *Alloperla* have a Nearctic and East Palaearctic distribution and includes almost 80 species. In Europe, only four genera and 19 species are present.

Taeniopterygidae is a family present in the Nearctic and Palaearctic Regions, with almost 150 species. The genus *Taeniopteryx* (20 species) is present in Europe and North America (even though the morphology of the nymphs slightly differs in the two continents), as well as the genus *Oemopteryx*. The genus *Doddsia* is shared by North America and Japan. The West Palaearctic genus *Brachyptera* includes more than 30 species.

Nemouridae was considered the most diverse family, with almost 400 described species (Zwick, 1980). The number currently is 633 species, but it is now the second most diverse family due to the enormous increase of Perlidae. The family has an Holarctic distribution with some genera reaching the Oriental Region; the genera *Amphinemura* (Holarctic; more than 100 species), *Nemoura* (Holarctic; 175 species) and *Protonemura* (Palaearctic; 135 species) account for 2/3 of the whole family. *Nemoura arctica* Esben-Petersen has a circum-arctic distribution.

Capniidae and Leuctridae include a similar numbers of species, 311 and 321 respectively, both with mainly Holarctic distribution and a strict systematic affinity. The most diverse genera are the Holarctic *Capnia* (110 species), *Allocapnia* (Nearctic; 43 species), *Mesocapnia* (Nearctic and East Palaearctic, 19 species) among Capniidae, and *Leuctra* (Holarctic, reaching Northern India; 190 species) and *Rhopalopsole* (East Palaearctic and Oriental; 20 species) among Leuctridae. The genus *Megaleuctra* presently includes six North American and one North Korea species; it is also known the fossil record in North America and in Baltic amber.

Notonemouridae (121 species), is considered the sister taxon of Nemouridae (but see below), and thus is an Arctoperlarian family, but shows a disjunct South African (included Madagascar), South American and Australian (included New Zealand) distribution. The family may not be monophyletic (Zwick, 1990) and its biogeography is not fully understood yet. *Austrocercella* (15 species) is the most diverse genus.

Among the remaining Arctoperlaria, the small family (eight species) of wingless stoneflies Scopuridae is known only from Japan and Korea, and the family Pteronarcyidae (13 species) is present in North America and East Asia. Peltoperlidae has a disjunct North American and East Asian distribution (44 species). Styloperlidae a recently instituted family (eight species), erected from aberrant Peltoperlidae, is presently known only from China and Taiwan.

Among Antarctoperlaria, Gripopterygidae is the most diverse family, with 277 described species. It has a disjunct South American and Australian (included New Zealand) distribution, but no genera are shared between the two continents. *Dinotoperla* (33 species), *Leptoperla* (28 species), *Riekoperla* (26 species) are the most diverse Australian genera, *Zelandobius* (27 species) is the most diverse genus of New Zealand. The remaining Antarctoperlaria families are scarcely diversified: Diamphipnoidae (five species) is a small South American family, Eustheniidae includes two genera in South America (two species) one genus in New Zealand (four species) and three genera in Australia (15 species), while Austroperlidae includes 10 species in Australia (five genera), a single species, *Austroperla cyrene* Tylliard, in New Zealand, and four species (three genera) in South America.

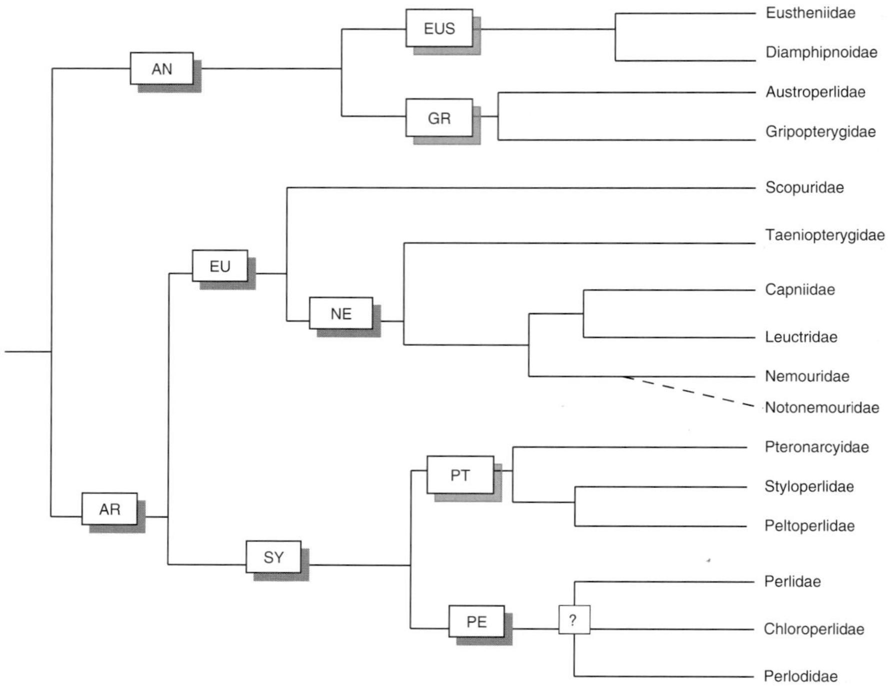

Fig. 2 Plecoptera phylogeny according to Zwick (2000). AN: Antarctoperlaria; AR: Arctoperlaria; EUS: Eusthenioidea; GR: Gripopterygoidea; EU: Euholognatha; NE: Nemouroidea; SY: Systellognatha; PT: Pteronarcyoidea; PE: Perloidea. (redrawn from Zwick, 1990)

Among Antarctoperlaria Eustheniidae are predators, as are Perloidea in the Northern hemisphere Arctoperlaria; Austroperlidae are shredders as are the Arctoperlaria Pteronarcyidae. Many Gripopterygidae are scrapers, as are Brachypterainae in the Arctoperlaria fauna; other Gripopterygidae genera are ecologically similar to genera and families of Nemouroidea.

Phylogeny and historical processes

Stoneflies are Pterygota, Neoptera, Exopterygota insects. Although they are usually included among Polyneoptera (possibly they are the sister-group of the remaining Polyneoptera), the relationships of Plecoptera with other orders of this subdivision are poorly understood (Gullan & Cranston, 2005). Among Polyneoptera, Plecoptera can be identified by a combination of mostly primitive characters. Moreover, only a few derived characters support the Plecoptera monophyly (Zwick, 2000): in the adult, gonads form loops with the anterior ends medially fused; in the male, there is a complex arrangement of two superimposed seminal vesicles each of which forms an anterior loop; in the nymph, strong oblique intersegmental ventrolongitudinal muscles are present that support laterally undulating swimming (this last character also occurs in Zygoptera). Two other probably derived characters have been reported for Plecoptera include the absence of an ovipositor (although it can secondarily appear in some taxa) and the existence of an accessory circulatory organ (cercus heart) in Perlidae and Perlodidae, that can be present in the other families as well.

The order includes 16 families whose relationships have been studied by several authors: we will follow the more recent and widely accepted classification by Zwick (2000) (Fig. 2), which recognizes two large groups: Antarctoperlaria, present only in the Southern Hemisphere, and Arctoperlaria, distributed mostly in the Northern Hemisphere. The first taxon includes four families, the latter includes 12 families belonging to two subgroups (Systellognatha and Euholognatha) with six families each one.

The monophyly of Antarctoperlaria is supported by morphological characters, while monophyly of Arctoperlaria is supported by ethological and morphological characters related to mate finding: the intersexual vibrational communication (drumming, tremulation, rubbing or combinations of them) and/or the existence of structures associated to it in the male (lobes, hammers, etc.) that are widely present in all families of this group except Scopuridae, and absent in Antarctoperlaria (Zwick, 2000). Among the existing families, only Notonemouridae seems to be a polyphyletic assemblage (McLellan, 2000).

According to Rohdendorf & Rasnitsyn (1980), recent stoneflies derived from Plecopteroidea from the Permian Period (including three orders: fossils Protoperlaria and Paraplecoptera, possibly not monophyletic, and Plecoptera). The oldest known Plecoptera fossils are from the early Permian Period, 258–263 m.y. ago (Sinitshenkova, 1997). According to this author, the origin and evolution of the group is linked to temperate areas; during the Mesozoic Period and the role of this group in lenthic ecosystems was much more important than it has been afterwards.

According to Illies (1965), Banarescu (1990) and Zwick (1990, 2000) Arctoperlaria and Antarctoperlaria originated as independent lines at the splitting of Pangea and the subsequent separation of Gondwanaland and Laurasia, at the end of Triassic Period (Fig. 3). Antarctoperlaria possibly began their diversification before the continents separated, producing some sister-groups distributed in South America, Australia and New Zealand. The absence of Antarctoperlaria from South Africa and India may be interpreted as a later extinction event. These lands became warmer and drier during their northward journey after separation from Antarctica and Australia. Before the isolation of Madagascar, Africa was for sure colonized by one family of Antarctoperlaria, the Notonemouridae (Zwick, 1990).

The Arctoperlaria families mainly have aHolarctic distribution (with the exception of Notonemouridae which are exclusive of the Southern Hemisphere), with taxa (even species) in common between North America and East Asia. The European stonefly-fauna shares only a few genera with Asia and very few with North America. This may be due to Pleistocenic events, such as the temporal separation of Europe and Asia by the Turgai Strait, the extinctions that occurred in Europe (families Pteronarcyidae and Peltoperlidae) and the faunistic interchanges through the Bering Strait. The presence of arctoperlarian taxa in the Southern Hemisphere could be due to invasions from the Northern Hemisphere (Zwick, 1981). The distribution of the very distinct *Notonemoura* group excludes recent dispersal. It appears that the present distribution results from dispersal across Antarctica (Zwick, 1981, 1990). Nevertheless, Zwick (2000) reported two alternative hypotheses: extinction of Antarctoperlaria in the Northern Hemisphere, and of Arctoperlaria in the Southern one. These hypotheses are based on the fact that, although stoneflies are an ancient insect order, their extant families may not necessarily be old, revealing recent and numerous speciation and extinction events, more than disperssive processes. Moreover, the scarce flight ability of these insects makes improbable long distance

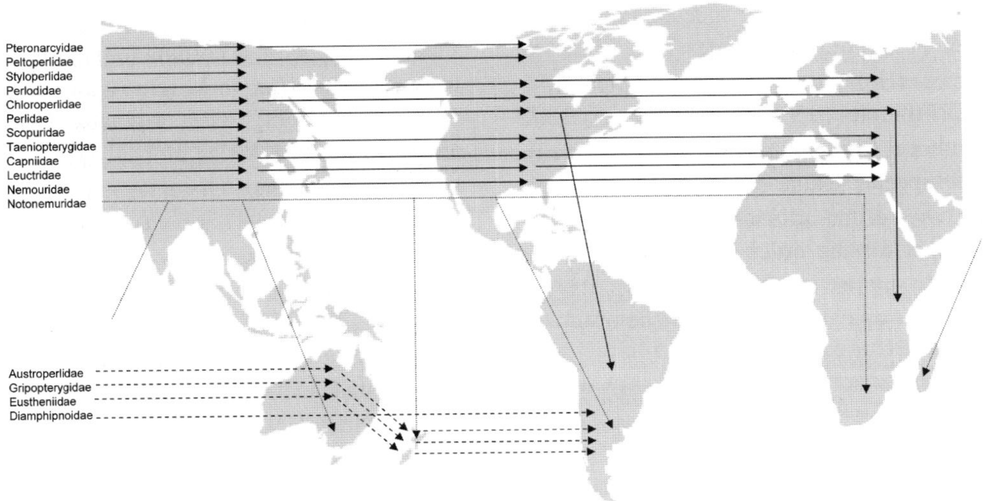

Fig. 3 Plecoptera families distribution; Arctoperlaria = continuous line; Antarctoperlaria = dotted line (modified from Zwick, 1990)

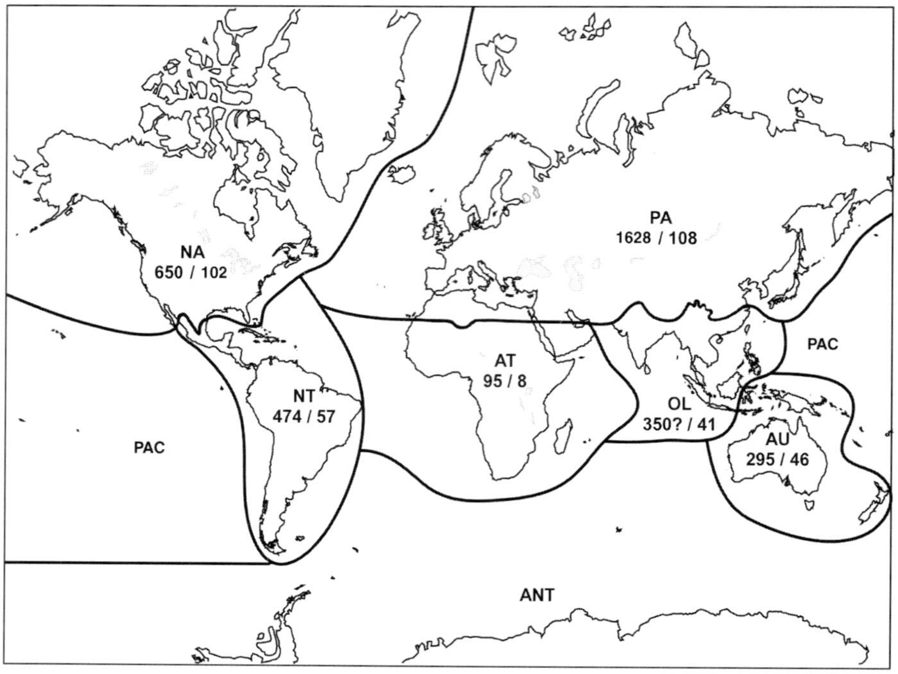

Fig. 4 Distribution of freshwater Plecoptera species and genera by zoogeographical region (species number/genus number). PA—Palaearctic; NA—Nearctic; NT—Neotropical; AT—Afrotropical; OL—Oriental; AU—Australasian; PAC—Pacific & Oceanic Islands, ANT: Antarctic

dispersal (Zwick, 2000). Also, according to Sinitshenkova (1997), the Northern and Southern faunas were isolated from the late Permian Period, when the Pangea still existed.

Present distribution and main areas of endemicity

The diversity for the various zoogeographical regions can be only roughly estimated; it should be approximately as follows: Nearctic 650 species, Palaearctic 1,628 species (it arbitrarily includes also Southern Chinese species) (Holarctic 2,278), Neotropical 474, Afrotropical (Ethiopic) 95 species, Oriental 350? species (Southern Chinese species not included), Australian 295 species (Table 3, Fig. 4). Even though these data indicate the Holarctic Region as the diversity hot-spot for the order, the analysis of the specific diversity divided by family suggests also an important role of tropical stoneflies (see below).

The Afrotropical Region (Ethiopian)

As reported by Zwick (1998), although Africa (as well as India) was part of Gondwana, its plecopterafauna lacks members of the Gondwanian Antarctoperlaria. As stated before, Anctartoperlaria must have been present in Gondwana before Africa (and India) split from it; climatic changes are thought to have caused their disappearance from these lands (Banarescu, 1990; Zwick, 2003). Only two Arctoperlaria families (Perlidae and Notonemouridae) are present today in the Afrotropical Region, and with a low specific diversity; one of them (Perlidae) is absent in Madagascar. Only the genus *Neoperla* Needham occurs throughout tropical Africa, including South Africa, with an uncertain number of species. They vary, according to the different authors, from only one [*Neoperla spio* (Newman)] to at least 50 (Zwick, 1998). The originally Oriental genus *Neoperla* reached Africa probably via the same route as the mammalian Ungulata during the Tertiary Period (Zwick, 1998). About 31 species of Notonemuridae, belonging to six genera (*Aphanicerca* Tillyard; *Aphanicercella* Tillyard; *Balinskycercella* Stevens & Picker; *Afronemura* Illies; *Aphanicercopsis* Barnard, and *Desmonemoura* Tillyard) besides two yet undescribed *Neoperla* species are present in South Africa (Villet, 2000). Only the endemic Notonemourid genus *Madenemura* Paulian, with the subgenera (considered as genera by some authors) *Madenemura* and *Tsaranemura* (Paulian, 1951), is present in Madagascar. Unfortunately, only the

Table 3 Number of Plecoptera genera in each family and each zoogeographical region and total genera number

	PA	NA	NT	AT	OL	AU	World
Eustheniidae			2			3	5
Diamphipnoidae			2				2
Austroperlidae			3			6	9
Gripopterygidae			24			25	49
Pteronarcyidae	1	2	1				2
Peltoperlidae	4	6			4		11
Styloperlidae					2		2
Perlodidae	29	30	2		1		51
Perlidae	20	15	11	1	24		52
Chloroperlidae	14	12	2		2		17
Scopuridae	1						1
Taeniopterygidae	11	6					12
Notonemouridae			4	7 (8)		12	23
Nemouridae	8	12	2		6		21
Capniidae	13	10	3				17
Leuctridae	7	9	1		2		12
Total	108	102	57	8(9)	41	46	286

PA: Palaearctic, NA: Nearctic, NT: Neotropical, AT: Afrotropical, OL: Oriental, AU: Australasian

females of some species were described; the disappearance of holotypes and the lack of description of some material from collections make it difficult to estimate the total number of species living in Madagascar (Elouard, pers. com.). The totally endemic notonemourid fauna of Madagascar together with the absence of *Neoperla*, may be explained by the early isolation of Lemuria.

The Oriental Region

Seven Arctoperlaria families are present in the Oriental Region, including 41 genera. The absence of Antarctoperlaria in India may be explained by the same reasons of its absence in Africa (see above). One family, Styloperlidae, composed of two genera and eight species endemic from Taiwan and South China, is endemic to this region. Peltoperlidae (with four Oriental genera) has a Nearctic-Asian distribution. The remaining families with Oriental representatives are also widely distributed in the Holarctic Region; only two of them are well-diversified in the Oriental Region: Perlidae (with 24 genera, 12 of which endemic from the Oriental Region) and Nemouridae (with six genera). Perlodidae (one genus), Chloroperlidae (two genera) and Leuctridae (two genera) complete the list. Unfortunately, the poor and fragmentary knowledge of this region makes it difficult to estimate the real diversity and the exact number of Plecoptera species that must be much larger than what we know. In fact, the Chinese stonefly fauna is almost unknown and hundreds of species are probably still undescribed (Sivec & Yang, 2001). These authors estimate approximately 350 Plecoptera species in the countries forming the Oriental Region, Southern China excluded.

The Oriental stonefly fauna was colonized from close Palaearctic areas (Zwick, 1980), as suggested by the decrease in species numbers towards the south. The first penetration was by members of three families—Nemouridae, Peltoperlidae and Perlidae—that extended down into the rain forest of southeast Asia, thus having somewhat overcome the usual aversion of the order for warm climates.

The Nearctic Region

The Nearctic Region includes 650 described species, distributed in 102 genera and nine families. Seven of these families (Capniidae, Leuctridae, Nemouridae, Taeniopterygidae, Chloroperlidae, Perlidae, Perlodidae) and 10 genera (*Capnia*, *Leuctra*, *Amphinemura*, *Nemoura*, *Oemopteryx*, *Taeniopteryx*, *Agnetina*, *Isoperla*, *Arcynopteryx*, *Diura*) also occur in the Palaearctic Region. There is a substantial division of eastern and western North America genera; only a few of them, in fact, transgress the barrier formed by the central prairies region and the Mississippi valley. Twelve of the western genera are shared with the Russian Far East or Japan, suggesting a recent invasion from the west. The genus *Anacroneuria* invaded from the south, bringing species into Central America, Texas and Arizona. This occurred after the formation of the Isthmus of Panama, during the Pliocene Period, about 4 millions years ago. Very few species have been able to disperse south into Central America, which has a mainly Neotropical stonefly-fauna.

According to Hynes (1988) North America has four groups of stoneflies with different origins:

(1) The original eastern fauna that was present before the final break from Europe, a great many of which may have evolved initially in North America;
(2) Invasions from the west, possibly over a long period, while the Cordillera was forming;
(3) Invasion of one genus from the south after the formation of the isthmus;
(4) Invasion by several species from the Bering Strait within the last few thousand years. These species may have survived the Pleistocene in the Alaskan refugium.

The Neotropical Region

Central America

The stonefly-fauna of Central America is poorly diversified. Deserts and dry areas prevent the immigration of the rich North American Plecoptera fauna: in fact less than 25 "Nearctic" species are known from Central America. They came during the Pleistocene Period, when stoneflies were able to cross the discontinuity between the southern Rocky Mountains and the Sierra Madre Occidental and disperse into northern, central and southern Mexico (Sargent et al., 1991). On the other hand the "tropical" Plecoptera fauna is also scarce: only the genus *Anacroneuria* (Perlidae) has been able to colonize from the South, with a few scattered species across the countries (34 in Mexico, eight in Nicaragua, 12 in Costa Rica, 15 in Panama, two in Trinidad). However, much research is needed to get a reliable knowledge of Central America stonefly-fauna.

South America

The stonefly fauna of South America is a complex mixture of arctoperlarian and antarctoperlarian species included in six families: Perlidae, Notonemouridae, Diamphinoidae, Eustheniidae, Austroperlidae, and Gripopterygidae (Tables 1, 2, 3). Among Antarctoperlaria, Eustheniidae (two species) is confined to southern South America, with two monospecific genera. The sister family, Diamphipnoidae (five species), is found only in central Chile, with two genera. Austroperlidae (four species) also occurs only in Chile, with three genera. The family Gripopterygidae (73 species), the largest family of the suborder (24 genera, 10 of which are monospecific) is more widespread. Unlike the former families, the latter is also present in subtropical, but not in tropical, South America. Only two families of Arctoperlaria are found in this region, Perlidae and Notonemouridae. Perlidae is the most diverse family (277 species), including 11 genera: the very speciose genus *Anacroneuria*, restricted to Central America and the tropical South America, includes almost 220 species. *Kempnya* (29 species) and *Macrogynoplax* (12 species) are diversified genera as well, while the remaining genera have few species. Finally, Notonemouridae (17 species), the only Plecoptera family with a Gondwanian distribution (not circumantarctic like the Antarctoperlaria), occurs in southern South America with four genera. However, it must be reminded that the monophyly of this family is still questioned (Zwick, 2000).

It must be noted that, considering the number of species, the true hot-spot of this area (that is almost coincident with the Neotropical Region), is the tropical belt. However, this is due to the presence of several speciose genera (especially the genus *Anacroneuria*). Considering generic diversity, the subtropical and temperate areas are more diversified, with many genera including few species.

The Palaearctic Region

The Palaearctic Region which includes Europe, Northern Africa and a great part of Asia (except Southeast), includes ten families and more than 1,600 Plecoptera species.

Seven families (Perlidae, Perlodidae, Chloroperlidae, Taeniopterygidae, Nemouridae, Capniidae and Leuctridae), 35 genera and 426 species constitute the Plecoptera fauna of the European continent (Fochetti & Tierno de Figueroa, 2004). Thirty-seven percent of the European genera are monospecific, whereas the four more diversified ones (*Leuctra, Protonemura, Nemoura* and *Isoperla*) include approximately 70% of the European stonefly species. This fauna is relatively well known from a taxonomical and faunistical point of view. Nevertheless, new species continue to be described.

Although the European Plecoptera fauna is diverse, considerable extinctions probably occurred during Pleistocenic glaciations, because the most important European mountain chains are E–W oriented: when ice advanced southwards they hindered the movement of fauna to the South. Moreover, there have been limited relationship between European and Asian stonefly-faunas, as reported above. Thus, faunistic interchanges were only possible through Asia Minor and the mountain systems, whereas the Central European plains acted as a barrier, and only a few taxa could move across the Polar Circle (such as *Diura* Billberg or *Arcynopteryx* Klapálek) (Zwick, 1980).

High numbers of species exist in the Mediterranean peninsulas (particularly the Western ones) including many endemic species. This is due to the presence of mountain systems, which represent an optimal environment for stoneflies, and to the effect of glaciations, that favoured isolation and speciation processes. Also many species occur in Central Europe mountainous areas (Alpine or Carpathic) where cold streams are abundant. Nevertheless, the post-glacial invasion by widespread species has lowered the number of endemic species. On the contrary, Northern Europe has low species richness. Finally, no species or a few number of them are present in the Atlantic islands as a consequence of the low dispersion capacity of stoneflies.

Northern Africa Plecoptera are included in the same seven families present in Europe (Perlidae, Perlodidae, Chloroperlidae, Taeniopterygidae, Nemouridae, Capniidae and Leuctridae) and in the same genera, with only one non-European genus: *Afroperlodes* Miron & Zwick. Some species are shared with Southern Europe or are very closely related. Currently 15 genera and about 31 species have been reported for this area (Sánchez-Ortega & Azzouz, 1998), approximately 1/3 of which are endemic. The similarity with Southern Europe and particularly with the Southern Iberian Peninsula may be explained by the existence in the past of a land bridge connecting the two areas. In fact, the Betic System (Spain) and the Rif Mountain (North West Africa) were separated only in the Lower Pliocene, when the Gibraltar Strait was formed (Sánchez-Ortega & Azzouz, 1998).

All Arctoperlaria families, except Notonemouridae and Styloperlidae, are present in Palaearctic Asia. The family Scopuridae is endemic from this region, and has a small distribution area and includes eight species belonging to the genus *Scopura* Uéno, which only occurs in Japan and Korea. Pteronarcyidae (only one genus with Asiatic species) and Peltoperlidae (four Palaearctic genera) have a Nearctic-Asian distribution. The remaining families show a wide Holarctic distribution and are well represented in the Asian part of the Palaearctic Region, some of them reaching the Oriental Region.

East Palaearctic (Asian) Plecoptera diversity is much greater than all the remaining continents, despite the poor and fragmentary knowledge of this area: the exact number of Asian Plecoptera species is difficult to estimate, but hundreds of new species will probably be described in the future. Japan with 306 species is an exception, because stoneflies from this country have been well studied, while the Chinese stonefly fauna is almost unknown (Sivec & Yang, 2001). Zhiltzova (1997, 2001) reported 350 species in the former USSR (also including the European part), and noted (1) a marked impoverishment of the fauna towards the North; (2) a high diversity of species in the Caucasus (3) a certain scarcity of Central Asia fauna, with a high degree of endemisms at the specific level concentrated in the Himalayan region; (4) a rather poor fauna in the East Siberian area, even though rich in genera number and endemic species, and (5) a very rich fauna in the South Russian Far East. Finally, the stonefly-fauna of South-West part of Asia is partly unknown; due to the general dry climate of the area, new species may be described in the future only from mountain systems.

From a zoogeographical point of view, the distribution of Pteronarcyidae, Peltoperlidae and several taxa within other families demonstrates that these evolved in the area that encompasses East and North Asia as well as North America, when the two continents were still joined together (Banarescu, 1990). The presence of the genus *Perlodes* Banks only in Europe and Japan, may be explained by its extinction in Siberia during the Ice Age (Banarescu, 1990).

The Australian Region

Australia currently includes 191 species from 26 genera and four families. Gripopterygidae is shared with South America while Notonemouridae occurs in both South America and South Africa. All 191 Plecoptera species and 24 of the 26 Australian genera are endemic (Theischinger 1991). Tasmanian Plecoptera includes 47 species, 41 of which are endemic to the island. Plecoptera can be found throughout Australia, except the Northern Territory. Their distribution in Western and South Australia is restricted, due to the arid conditions. The genera *Crypturoperla, Tasmanoperla* (Austroperlidae), *Tasmanocerca* (Notonemouridae) and *Cardioperla* (Gripopterygidae) are limited to Tasmania. Many genera in all families are confined to mainland Australia. Only the family Gripopterygidae occurs in Western Australia.

New Zealand has 104 described species (McLellan, 2006), and the same four families as Australia. All genera are endemics to this island system, except *Notonemoura* and *Stenoperla*, that are shared with Australia, even though species of this genera are endemic in both countries. A considerable proportion of the entire stonefly-fauna is found in the high precipitation region of the Southern Alps and its subsidary mountain ranges (McLellan, 2000), and endemisms are most numerous in South Island.

At least one species of *Neoperla* occurs in New Guinea, while no Plecoptera at all are present in New Caledonia.

The total Plecoptera species of the Australian Region is 295. This number is remarkably small if compared to the Holarctic and Oriental faunas, which are of entirely different composition, sharing no families with Australia (Zwick, 1981).

Human related issues

Stoneflies represent a very important component of rivers and streams, both as biomass and as diversity of ecological roles, acting as primary or secondary consumers and as prey for other macroinvertebrates and fishes, including those of economic importance. From a scientific point of view, they have been used as biogeographical indicators and in evolutionary researches (for instance in the insect wings and respiratory proteins evolution). Their importance as human food is scarce, although some big species are included in the diet of some cultures. Stoneflies have been used for centuries in the sport of fly-fishing, and fishermen have a good general knowledge of them. Plecoptera are frequently employed as bioindicators of water quality: the highest values as indicators of good ecological and environmental quality are normally assigned to this group. Stoneflies do not cause economic damages to human crops although some damage to fruit trees and ornamental plants by adult Taeniopterygidae have been reported.

Unfortunately, water quality degradation and physical alteration of streams and rivers have caused a decrease in the number of Plecoptera species, particularly those inhabiting lowland rivers of industrialized countries (such as many Central European ones), or highly populated countries like the Asiatic ones. *Taeniopteryx araneoides* (Klapálek) and *Oemopteryx loewi* (Albarda), once common in large Central Europe rivers, are now extinct (Zwick, 2004). These are among the very few documented cases of extinction in insects. Many undescribed species, mainly from Asia, will probably go extinct before they are discovered (Sivec & Yang, 2001). Nevertheless, few Plecoptera species are included in the "official" lists of threatened species. Only one species, the chloroperlid *Alloperla roberti* Surdick endemic from Illinois (USA), is reported as extinct in the IUCN Red List of Threatened Species, (IUCN, 2004).

Acknowledgements We would like to thank P. Zwick (Schlitz, Germany) and Ian McLellan (Westport, New Zealand) for providing material and useful suggestions. B. P. Stark (Mississippi, U.S.A.) and K. W. Stewart (Texas, U.S.A.) commented and improved an early version of the paper and are heartly thanked.

References

Aubert, J., 1959. Plecoptera. Insecta Helvetica 1: 1–140.

Banarescu, P. 1990. Zoogeography of Fresh Water. General Distribution and Dispersal of Freshwater Animal, Vol. 1. Aula-Verlag, Wiesbaden.

Fochetti, R. & J. M. Tierno de Figueroa, 2004. Plecoptera. Fauna Europaea Web Service. http://www.faunaeur.org (Fauna Europaea project, ref. EVR1-CT-1999–2001).

Fochetti, R. & J. M. Tierno de Figueroa, 2005. I Plecotteri europei: distribuzione e stato di conservazione. XX Congresso Italiano di Entomologia. Assisi 13–18.VI.2005.

Gullan, P. J. & P. S. Cranston, 2005. The Insects. An outline of Entomology. 3rd ed. Blackwell Publishing Ltd, Oxford.

Heckman, C. W., 2003. Encyclopedia of South American Aquatic Insects: Plecoptera. Kluwer Academic Publishers, Dordrecht.

Hynes, H. B. N., 1976. Biology of Plecoptera. Annual Review of Entomology 21: 135–153.

Hynes, H. B. N., 1988. Biogeography and origins of the North American stoneflies. Memoires Entomological Society Canada 144: 31–37.

Illies, J., 1965. Phylogeny and zoogeography of the Plecoptera. Annual Review of Entomology 10: 117–140.

IUCN, 2004. 2004 IUCN Red List of Threatened Species. http://www.redlist.org.

Jewett, S. G., 1963. A stonefly aquatic in the adult stage. Science 139: 484–485.

McLellan, I. D., 2000. A revision of *Cristaperla* (Plecoptera: Notonemouridae) and some comments on Notonemouridae and its generic groups. New Zealand Journal of Zoology 27: 233–244.

McLellan, I. D. 2006. Endemism and biogeography of New Zealand Plecoptera (Insecta). Illiesia 2: 15–23. http://www2.pms-lj.si/illiesia/papers.html.

Michaelis F. & C. M. Yule, 1988. Plecoptera. In Walton, D. W. & W. W. K. Houston (eds), Ephemeroptera, Megaloptera, Odonata, Plecoptera, Trichoptera. Zoological Catalogue of Australia, Vol. 6. Canberra, AGPS xi 315 pp (133–176), (Updated by Andrew A. Calder, CSIRO Entomology, Canberra, Australia, 2002).

Paulian, R., 1951. Faunes des eaux douces de Madagascar I. Plécoptères et Mégaloptères. Mémoires Institut Scientifique Madagascar 6: 53–62.

Rohdendorf, B. B. & A. P. Rasnitsyn, 1980. [Historical Development of Insect Class]. Trudy Paleontologischeskogo Instituta 178: 1–267 (in Russian).

Sánchez-Ortega, A. & M. Azzouz, 1998. Faunistique et phénologie des Plécoptères (Insecta, Plecoptera) du Rif marocain (Afrique du nord). Relations avec les autres aires de la región méditerranéenne occidentale. Mitteilungen der Schweizerischen Entomologischen Gesellschaft 71: 449–461.

Sargent, B. J., R. W. Baumann & B. C. Kondratieff, 1991. Zoogeographic affinities of the Nearctic stonefly fauna of Mexico. Southwestern Naturalist 36: 323–331.

Shinitshenkova, N. D. 1997. Paleontology of stoneflies. In Landolt, P. & M. Sartori (eds), Ephemeroptera and Plecoptera, Biology-Ecology-Systematics. MTL Fribourg, 561–565.

Sivec, I. & P. Yang, 2001. Stoneflies of Taiwan within the oriental stonefly fauna diversity. In Domínguez E. (ed.), Trends in Research in Ephemeroptera and Plecoptera. Kluwer Academic/Plenum Publishers, New York, 401–404.

Stark B. P., 2001. A synopsis of the Neotropical Perlidae. In Domínguez E. (ed.), Trends in Research in Ephemeroptera and Plecoptera. Kluwer Academic/Plenum Publishers, New York, 405–422.

Stark, B. P. & R. W. Baumann, 2005. North American Stonefly (Plecoptera) complete list. http://www.mlbean.byu.edu/plecoptera/list.asp.

Stewart, K. W., 1994. Theoretical considerations of mate finding and other adult behaviors of Plecoptera. Aquatic Insects 16: 95–104.

Theischinger G., 1991. Plecoptera (Stoneflies). In CSIRO (ed.), The Insects of Australia. A textbook for students and research workers. Vol. 1, 2nd ed. Melbourne University Press, Melbourne, 311–319.

Tierno de Figueroa, J. M. & R. Fochetti, 2001. On the adult feeding of several European stoneflies (Plecoptera). Entomological News 112: 130–134.

Tozer, W., 1979. Underwater behavioural thermoregulation in the adult stonefly, *Zapada cinctipes*. Nature (London) 281: 566–567.

Villet, M. H. 2000. The stoneflies (Plecoptera) of South Africa. http://www.ru.ac.za./academic/departments/zooento/Martin/plecoptera.html.

Yuzhou, D. & H. Junhua, 2001. Progress on taxonomic study of the family Perlidae from China. In Domínguez E. (ed.), Trends in Research in Ephemeroptera and Plecoptera. Kluwer Academic/Plenum Publishers, New York, 369–375.

Zhiltzova, L. A., 1997. Zoogeographic features of the Euholognatha fauna (Plecoptera) of Russia and adjacent territories (within the limits of the former USSR). In Landolt, P. & M. Sartori (eds), Ephemeroptera and Plecoptera: Biology-Ecology-Systematics. MTL Fribourg, 186–192.

Zhiltzova, L. A., 2001. Plecoptera fauna of Capniidae of Russia and adjacent territories (within the limits of the former USSR). In Domínguez E. (ed.), Trends in Research in Ephemeroptera and Plecoptera. Kluwer Academic/Plenum Publishers, New York, 423–429.

Zwick, P. 1980. Plecoptera (Steinfliegen). In Handbuch der Zoologie, Walter de Gruyter, Berlin, 1–115.

Zwick, P., 1981. Plecoptera. In: Keast A. (ed.) Ecological Biogeography of Australia, Vol. 2. The Hague, Junk, 1171–1182.

Zwick, P., 1990. Transantarctic relationships in the Plecoptera. In Campbell I.C. (ed.), Mayflies and Stoneflies, Life History and Biolgy. Kluwer Academic Publishers, London, 141–148.

Zwick, P. 1998. The African *Neoperla* (Plecoptera: Perlidae) revisited: towards an overdue revision. XIII International Symposium on Plecoptera, IX International Conference on Ephemeroptera. Tucumán (Argentina), 92.

Zwick, P., 2000. Phylogenetic System and Zoogeography of the Plecoptera. Annual Review of Entomology 45: 709–746.

Zwick, P., 2003. Biogeographical patterns. In Resh V. H. & R. T. Cardé, (eds), Encyclopedia of Insects. Academic Press, Elsevier Science, USA, 94–103.

Zwick, P., 2004. Key to the West Paleartic genera of stoneflies (Plecoptera) in the larval stage. Limnologica 34: 315–348.

Hydrobiologia (2008) 595:379–391
DOI 10.1007/s10750-007-9033-1

FRESHWATER ANIMAL DIVERSITY ASSESSMENT

Global diversity of true bugs (Heteroptera; Insecta) in freshwater

John T. Polhemus · Dan A. Polhemus

Abstract The aquatic and semi-aquatic Heteroptera, consisting of the infraorders Leptopodomorpha, Gerromorpha, and Nepomorpha, comprise a significant component of the world's aquatic insect biota. Within these three infraorders as a whole there are currently 23 families, 343 genera and 4,810 species group taxa considered valid, of which 20 families, 326 genera and 4,656 species inhabit freshwater. In addition, more than 1,100 unequivocally diagnosed species remain to be described. Aquatic Heteroptera occur on all continents except Antarctica, and are most numerous in the tropical regions, although there are many distinctly cold-adapted genera. Overall species richness is highest in the Neotropical and Oriental regions, which harbor 1,289 and 1,103 species, respectively. In comparison to these core tropical regions, species richness is significantly lower in the Afrotropical (799 species), Australasian (654 species), Palearctic (496 species), Nearctic (424 species) and Pacific (37 species) regions. Aquatic Heteroptera are notable for utilizing an exceptionally broad range of habitats, from marine and intertidal to arctic and high alpine, across a global altitudinal range of 0–4,700 m. Species may be found in almost every freshwater biotope, and many exhibit striking morphological adaptations to their aquatic environment, making them excellent subjects for ecological and biogeographic studies.

Keywords Heteroptera · Aquatic · Diversity · Richness · Distribution · Endemism

Guest editors: E. V. Balian, C. Lévêque, H. Segers & K. Martens
Freshwater Animal Diversity Assessment

J. T. Polhemus (✉)
Colorado Entomological Institute, 3115 S. York St., Englewood, CO 80113, USA
e-mail: jtpolhemus@msn.com

D. A. Polhemus
Department of Entomology, Smithsonian Institution, 10th & Constitution NW, Washington, DC 20560, USA

Present Address:
D. A. Polhemus
Department of Natural Sciences, Bishop Museum, 1525 Bernice St., Honolulu, HI 96817, USA
e-mail: bugman@bishopmuseum.org

Introduction

The aquatic and semi-aquatic Heteroptera, referred to herein subsequently as "aquatic Heteroptera", are composed of three monophyletic infraorders, the Gerromorpha, Nepomorpha, and Leptopodomorpha, with worldwide distribution. Because of their general abundance in many freshwater systems, coupled with unusual morphological specializations for exploitation of specialized microhabitats, the group has long attracted the interest of aquatic entomologists, and has a relatively mature taxonomy (Polhemus & Polhemus, in press). The majority of species occurring in Europe, North America, and Australia have been now described, and significant recent progress has also been made in documenting the aquatic Heteroptera biotas of tropical South America, Africa, Asia and the Pacific. Although many taxa still remain to be formally named from these latter regions, the biotas have in many cases been moderately well collected and informally diagnosed to the morphospecies level.

The analysis of global patterns of taxonomic richness in aquatic Heteroptera presented herein follows the guidelines developed in a workshop entitled "A global assessment of animal diversity in freshwater" convened in Mechelen, Belgium, October 13–16, 2005. The term "freshwater" excludes species inhabiting dry terrestrial or marine habitats, but can include marine species that also inhabit essentially freshwater, e.g., several species of the typically marine gerrid genus *Halobates*. It was decided to include as freshwater species two different categories of taxa, defined as: (1) "real aquatic species" with all or part of their lifecycle in aquatic habitats (in or on the water), and (2) "water dependent species" with a close/specific dependence on aquatic habitats. Under these guidelines all Leptopodomorpha (the shorebugs) are deemed to be water dependent (WDpt), and all Gerromorpha (water striders; which live on the water, but only rarely in the water) and all Nepomorpha (water bugs; which live mostly in the water, except for two littoral families) are deemed to be real aquatic species. In order to present a balanced picture of the aquatic and semi-aquatic Heteroptera of the world, we give both the entire global fauna, and the freshwater component of the global fauna in Table 1 according to the guidelines stated above. Table 2 gives the distribution of the freshwater species, and Table 3 the genera, assigned to the biogeographical regions developed during the workshop. Table 4 gives the estimated number of species that are known to us as morphospecies but as yet undescribed, to provide a more complete picture of the total global fauna as presently known. We also discuss below the "excluded marine" species and the "Holarctic" species.

Table 1 Global freshwater aquatic Heteroptera diversity—Fall 2005

Family	Genera	Species
Leptopodomorpha		
Aepophilidae	0 (1)	0 (1)
Omaniidae	0 (2)	0 (6)
Leptopodidae	8 (10)	32 (39)
Saldidae	25 (29)	322 (335)
Total	33 (42)	354 (381)
Gerromorpha		
Gerridae	62 (67)	700 (751)
Hebridae	9 (9)	221 (221)
Hermatobatidae	0 (1)	0 (9)
Hydrometridae	7 (7)	125 (126)
Macroveliidae	3 (3)	3 (3)
Mesoveliidae	12 (12)	46 (46)
Paraphrynoveliidae	1 (1)	2 (2)
Veliidae	58 (61)	896 (962)
Total	152 (161)	1993 (2,120)
Nepomorpha		
Aphelocheiridae	1 (1)	78 (78)
Belostomatidae	9 (9)	160 (160)
Corixidae	35 (35)	607 (607)
Gelastocoridae	3 (3)	111 (111)
Helotrephidae	21 (21)	180 (180)
Naucoridae	37 (37)	391 (391)
Nepidae	15 (15)	268 (268)
Notonectidae	11 (11)	400 (400)
Ochteridae	3 (3)	68 (68)
Pleidae	3 (3)	38 (38)
Potamocoridae	2 (2)	8 (8)
Total	140 (140)	2309 (2309)
Total	326 (343)	4656 (4810)

Summary excludes fossils; (x) total genus/species number of aquatic Heteroptera including species of marine and xeric habitats

General aspects of aquatic heteroptera

Heteroptera are hemimetabolous insects, typically developing via a series of 5 nymphal instars. The body consists of 3 distinct parts (although the head and thorax are closely adjoined in Pleidae and Helotrephidae), with mouthparts specialized for piercing and sucking (except in the Corixidae). All aquatic families are predaceous (except once again for the Corixidae, which are omnivores), with their prey consisting of any organism that can be subdued by injection of a venom consisting of toxins and proteolytic enzymes. In aquatic systems devoid of large fishes, aquatic Heteroptera may sometimes represent the top predators in the trophic chain; this

Table 2 Global distribution of aquatic Heteroptera species diversity

	PA	NA	NT	AT	OL	PAC	AU	WORLD
Leptopodomorpha (all are water dependent; WDpt 100%)								
Leptopodidae	7	1	1	13	6	0	4	32
Saldidae	147	70	41	28	22	13	23	344
Total	154	71	42	41	28	13	27	376
Gerromorpha (most are water surface dwellers; WDpt 2%)								
Gerridae	51	47	141	66	287	8	113	712
Hebridae	16	15	31	77	76	0	8	223
Hydrometridae	6	6	37	31	30	4	15	129
Macroveliidae	0	2	1	0	0	0	0	3
Mesoveliidae	2	3	15	5	9	2	13	49
Paraphrynoveliidae	0	0	0	2	0	0	0	2
Veliidae	44	31	290	158	199	5	176	903
Total	119	104	515	339	601	19	325	2021
Nepomorpha (most are sub-surface dwellers; WDpt 7%)								
Aphelocheiridae	19	0	0	6	47	0	6	78
Belostomatidae	4	17	111	23	9	0	5	169
Corixidae	140	136	152	111	77	0	46	662
Gelastocoridae	2	7	48	2	9	1	47	116
Helotrephidae	0	0	10	31	111	0	12	164
Naucoridae	6	29	186	67	74	0	36	398
Nepidae	7	13	93	84	48	0	23	268
Notonectidae	36	35	96	85	75	3	92	422
Ochteridae	3	6	16	6	15	0	29	75
Pleidae	6	6	12	4	9	1	6	44
Potamocoridae	0	0	8	0	0	0	0	8
Total	223	249	732	419	474	5	302	2404
Total species richness by region	496	424	1289	799	1103	37	654	4801

PA, Palaearctic; NA, Nearctic; NT, Neotropical; AT, Afrotropical; OL, Oriental; AU, Australasian; PAC, Pacific Oceanic Islands; ANT, Antarctic

is particularly true for some of the larger taxa in the familes Belostomatidae and Nepidae. Members of this group also exhibit considerable variation in body size, ranging from <1 mm (the genus *Micronecta* in the Corixidae) to >110 mm (the genus *Lethocerus* in the Belostomatidae).

Across the aquatic Heteroptera as a whole the morphology of the included families and genera is extremely variable, with many demonstrating striking morphological adaptations to particular microhabitats (Fig. 1-1–1-4). The Hydrometridae, or water measurers, have extremely elongate bodies and legs, allowing them to walk in stilt-like fashion across the surfaces of ponds and stream pools. The Helotrephidae, or BB Bugs, are by contrast compact, globular bugs that swim upside down, inhabiting quiet waters of lotic and occasionally lentic ecosystems. The Nepidae, or water scorpions, have a siphon to pierce the water surface for air, and bodies that resemble sticks and leaves, allowing them to remain motionless as sit-and-wait predators along pond margins, waiting for prey to come within reach. In the Notonectidae, or backswimmers, the genus *Anisops* utilizes haemoglobin to bind or release oxygen, allowing individuals to attain neutral buoyancy and remain quietly suspended at any depth in the water column. Aphelocheiridae and some Naucoridae use a plastron of ultramicroscopic hairs to create a physical gill form of respiration, obviating the

Table 3 Global distribution of aquatic Heteroptera generic diversity

	PA	NA	NT	AT	OL	PAC	AU	WORLD
Leptopodomorpha (all are water dependent; WDpt 100%)								
Leptopodidae	5	1	1	5	3	0	1	16
Saldidae	12	11	6	8	7	1	6	51
Total	17	12	7	13	10	1	7	67
Gerromorpha (most are water surface dwellers; WDpt 8%)								
Gerridae	6	8	20	17	31	3	20	104
Hebridae	2	3	3	2	6	0	3	19
Hydrometridae	1	1	4	2	2	3	1	14
Macroveliidae	0	2	1	0	0	0	0	3
Mesoveliidae	1	1	4	3	5	2	4	20
Paraphrynoveliidae	0	0	0	1	0	0	0	1
Veliidae	3	4	14	17	18	2	20	78
Total	13	19	46	42	62	10	48	239
Nepomorpha (most are sub-surface dwellers; WDpt 4%)								
Aphelocheiridae	1	0	0	1	1	0	1	4
Belostomatidae	3	3	6	4	3	0	2	21
Corixidae	13	18	14	7	5	0	5	62
Gelastocoridae	1	2	3	1	1	1	1	10
Helotrephidae	0	0	2	6	14	0	1	23
Naucoridae	3	4	13	9	13	0	8	50
Nepidae	2	3	4	6	6	0	5	26
Notonectidae	4	3	4	5	6	3	6	31
Ochteridae	1	1	2	1	1	0	2	8
Pleidae	2	2	2	1	1	1	1	10
Potamocoridae	0	0	2	0	0	0	0	2
Total	30	36	52	41	51	5	32	247
Total genus richness by region	60	67	105	96	123	16	87	

PA, Palaearctic; NA, Nearctic; NT, Neotropical; AT, Afrotropical; OL, Oriental; AU, Australasian; PAC, Pacific Oceanic Islands; ANT, Antarctic

need to surface for air and thus allowing them to hunt for prey beneath stones in swift running waters. The Belostomatidae, or giant water bugs (Fig. 1-3), are fierce predators with powerful venoms, the largest capable of killing sizable fish, and in some cases even birds. In the Veliidae, the genus *Phoreticovelia*, or Zeus Bugs, has phoretic males that are fed by secretions from the anterior thorax of the females (Arnqvist et al., 2003). The Saldidae, or shore bugs (Fig. 1-1), feed largely on sub-surface prey in littoral habitats that they locate via chemoreceptive setae on the antennae (Polhemus, 1985). The above are just a few representative examples of the great ecological and morphological plasticity exhibited by these insects.

Modern water bug lineages can be traced back by fossil evidence to the Mesozoic, with some families, such as Belostomatidae, occurring in the Triassic, 180 Mya (Polhemus, 2000). Cladistic evidence indicates that they are most likely derived from terrestrial ancestors. The age of the group is reflected by the fact that numerous sister genera and species are separated by modern oceans; examples include Japan/Mexico (*Speovelia* in the Mesoveliidae), South America/Africa (*Platyvelia*/*Angilia*, in the Veliidae), and South America/Australasia (*Metrobates*/*Metrobatoides*, in the Gerridae). Vicariance, orogeny, and tectonic fusion or fragmentation are the most likely drivers of diversification, a prime example being the geologically young, mountainous, and tectonically

Table 4 Estimated number of undescribed aquatic Heteroptera species and global distribution

Family	Total genera	Total species	PA	NA	NT	AT	OL	PAC	AU
Leptopodomorpha									
Aepophilidae	1	0	0	0	0	0	0	0	0
Omaniidae	2	0	0	0	0	0	0	0	0
Leptopodidae	10	14	0	0	0	0	8	0	6
Saldidae	29	96	5	5	25	8	8	25	20
Total	42	110	5	5	25	8	16	25	26
Gerromorpha									
Gerridae	67	86	6	0	15	15	30	0	20
Hebridae	9	96	6	0	20	20	30	0	20
Hermatobatidae (m)	1	(3)	0	0	0	0	(1)	(1)	(1)
Hydrometridae	7	26	3	0	8	8	4	0	3
Macroveliidae	3	2	0	0	2	0	0	0	0
Mesoveliidae	12	26	0	0	4	4	10	0	8
Paraphrynoveliidae	1	0	0	0	0	0	0	0	0
Veliidae	61	316	0	0	66	70	40	15	125
Total	161	552	15	0	115	117	114	15	176
Nepomorpha									
Aphelocheiridae	1	31	3	0	0	15	10	0	3
Belostomatidae	9	24	0	0	10	3	8	0	3
Corixidae	35	66	0	6	10	20	15	0	15
Gelastocoridae	3	16	0	0	6	0	2	0	8
Helotrephidae	21	52	0	0	6	6	30	0	10
Naucoridae	37	81	0	0	26	20	20	0	15
Nepidae	15	50	0	0	10	10	20	0	10
Notonectidae	11	55	0	0	10	15	15	0	15
Ochteridae	3	63	0	0	20	3	20	0	20
Pleidae	3	10	0	0	5	0	0	0	5
Potamocoridae	2	4	0	0	4	0	0	0	0
Total	140	452	3	6	107	92	140	0	104
Total	343	1114	23	11	247	217	270	40	306

PA, Palaearctic; NA, Nearctic; NT, Neotropical; AT, Afrotropical; OL, Oriental; AU, Australasian; PAC, Pacific Oceanic Islands; ANT, Antarctic. (Genera given are existing genera; species are undescribed.) This summary does not include fossils, or marine (m) species

complex island of New Guinea, where nearly every major river basin displays significant local endemicity (Polhemus, 1996; Polhemus & Polhemus, 1998).

The phylogeny of the Heteroptera as a whole and of the individual infraorders within it has been presented in a series of cladograms by Schuh & Slater (1995, p. 5). These cladistic interpretations, which were based primarily on morphological characters, were derived from the works of various authors, as follows: Leptopodomorpha (Schuh & Polhemus, 1980; Polhemus, 1985; see Schuh & Slater, 1995, p. 134); Gerromorpha (Andersen, 1982; see Schuh & Slater, 1995, p. 84); Nepomorpha (Mahner, 1993; see Schuh & Slater, 1995, p. 110). Although a limited amount of molecular data contributed to the cladograms presented by Schuh & Slater (1995, p. 5), there are now a number of ongoing studies testing the current phylogenies with more extensive molecular data sets (e.g., Hebsgaard et al., 2004).

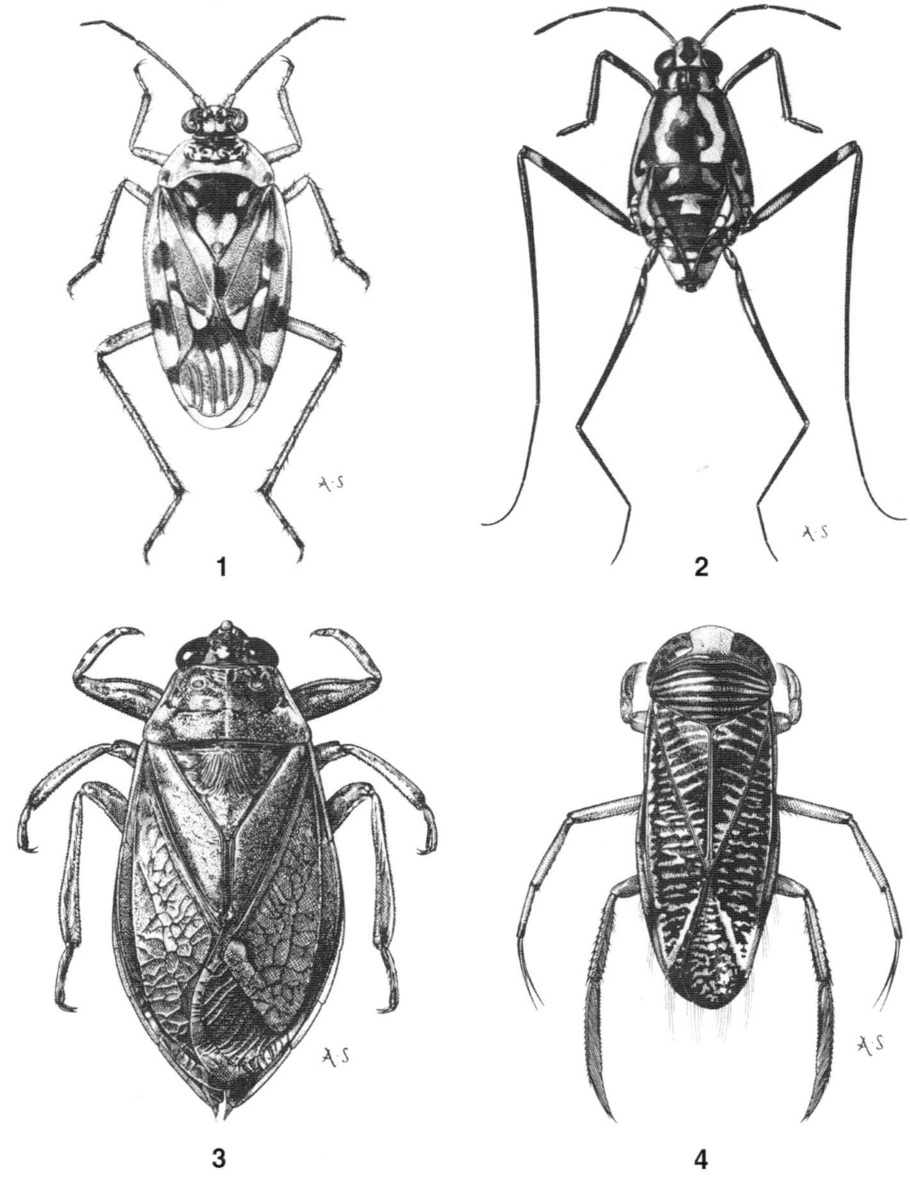

Fig. 1 Exemplar taxa of aquatic Heteroptera in various infraorders. 1-1, *Pentacora signoreti* (Guerin) (Saldidae: Leptopodomorpha); 1-2, *Trepobates becki* Drake & Harris (Gerridae: Gerromorpha); 1-3, *Belostoma bakeri* Montandon (Belostomatidae: Nepomorpha); 1-4. *Sigrara mckinstryi* Hungerford (Corixidae: Nepomorpha)

Species diversity

Over the past 40 years, the senior author has compiled a manuscript catalog of all taxonomic names in the aquatic Heteroptera, including synonyms, based primarily on a comprehensive library of the taxonomic literature, consisting of approximately 11,000 references containing the original descriptions of virtually all species as well as their synonyms. In the early 1990s, the junior author facilitated the entry of this manuscript catalog into a flat file database at the Smithsonian Institution, producing an electronic world checklist of all valid names in the three infraorders under study. This database was subsequently refined by additional exhaustive literature searches to detect names published in obscure journals, as well as previously overlooked synonyms and homonyms. Finally, the most recent literature

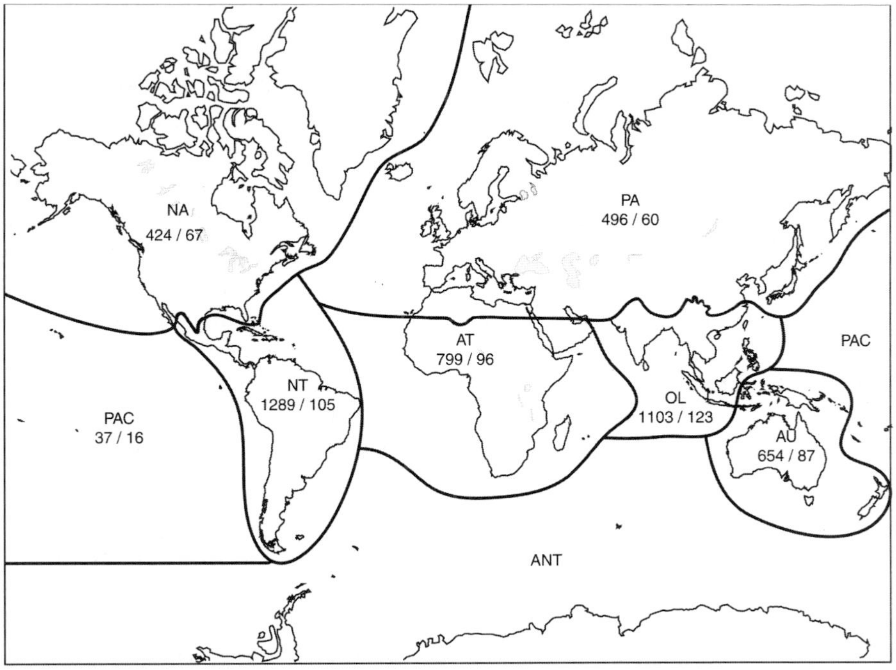

Fig. 2 Total richness of aquatic Heteroptera species in individual biogeographic regions (see Table 2 for additional details on levels of family and genus level richness). PA, Palaearctic; NA, Nearctic; NT, Neotropical; AT, Afrotropical; OL, Oriental; AU, Australasian; PAC, Pacific Oceanic Islands; ANT, Antarctic

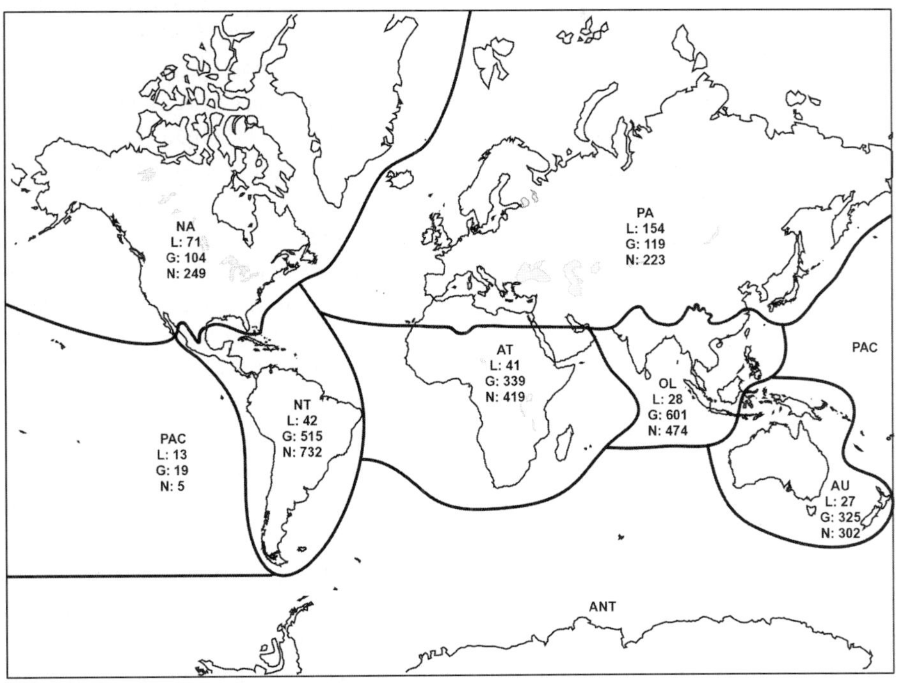

Fig. 3 Total richness of Leptopodomorpha (L), Gerromorpha (G), Nepomorpha (N) species in individual biogeographic regions (see Table 2 for additional details on levels of family and genus level richness). PA, Palaearctic; NA, Nearctic; NT, Neotropical; AT, Afrotropical; OL, Oriental; AU, Australasian; PAC, Pacific Oceanic Islands; ANT, Antarctic

was integrated to bring the database up to date as of 15 November 2005. Parts of this work have been incorporated into regional or world faunal catalogs (Schuh et al., 1987; Henry & Froeschner, 1988; Aukema & Rieger, 1995).

Based on this data analysis, the currently described world biota of aquatic Heteroptera across all three infraorders consists of 23 families, 343 genera and 4,810 species (see Table 1, which includes all species group taxa). Of these, 20 families, 326 genera and 4,656 species are considered to inhabit freshwater (also given in Table 1). We estimate, based on known unequivocally diagnosed but still undescribed morphospecies held in major collections, that a minimum of 1,100 species still await formal taxonomic description (Table 4). Allowing for still further discoveries in remote and undersurveyed areas, and for eventual resolution of varying taxonomic interpretations regarding narrow versus broad species concepts, we would estimate that the total world biota of aquatic Heteroptera probably lies in the range of 7,000 species.

Present distribution

Aquatic Heteroptera occur on all continents except Antarctica, and are most numerous in the tropical regions, although there are many distinctly cold-adapted genera, particularly in the Saldidae and Corixidae. Diversity is highest in first to third order streams (i.e., the smallest streams in a given drainage network), and lowest in large rivers and lakes. A high percentage of tropical endemics are found associated with rheocrenes or headwater streamlets, often in proximity to waterfalls or other habitats with high structural complexity (see Polhemus et al., 1992). As a whole, aquatic Heteroptera are notable for utilizing an exceptionally broad range of aquatic ecosystems, from marine and intertidal to arctic and high alpine, across a global altitudinal range of 0–4,700 m. As such, they are excellent subjects for comparative biogeographical and ecological studies.

Using the world checklist (Table 1), refined to include only freshwater species (also in Table 1), all included species were assigned to a biogeographic region or regions (Table 2) using the criteria specified by Royal Belgium Institute of Natural Sciences. The following biogeographic regions were recognized for the purposes of this analysis (Fig. 2): Nearctic (NA), Neotropical (NT), Afrotropical (AT), Australasian (AU), Oriental (OL), Palaearctic (PA), Pacific (PAc), and Antarctic (ANT). There are no Heteroptera known from the Antarctic region, therefore it was omitted from our tables. Based on this integration of taxonomic and distributional information, it was then possible to determine the total number of species occurring in each major biogeographic region, in terms of both raw species numbers, and representation by individual infraorders, familes and genera (see Tables 2 and 3 and Fig. 3).

Marine and xerophilous taxa have been excluded from this analysis (see Table 1). In the Leptopodomorpha, two marine families (Aepophilidae and Omaniidae), and 4 intertidal genera of Saldidae with 13 species are excluded, along with two genera and 7 species of xerophilous Leptopodidae. In the Gerromorpha, one family (Hermatobatidae) plus 8 marine genera and 126 marine species are excluded, but the estuarine genera that occupy the ecotone from mixohaline to limnetic waters are included in the freshwater totals. While the primarily marine gerrid genus *Halobates* is excluded, with a total of 46 species, three species have been added to the total of freshwater species, because two of these (*H. acherontis* J. Polhemus and *H. robinsoni* Andersen) occupy freshwater rivers in northern Australia, and one (*H. murphyi* J. & D. Polhemus) inhabits Papuan estuaries with mixohaline to limnetic waters.

It should be noted that the total numbers of species and genera from Tables 2 and 3 (distribution) will generally exceed the numbers shown in Table 1 (summary of freshwater aquatic Heteroptera global richness), because certain species and genera occur in more than one biogeographic region as defined herein.

Although aquatic Heteroptera are distributed across all of Earth's major non-polar biogeographic regions, including even the most remote islands of the Indian, Atlantic and Pacific oceans, there are obvious differences in levels of species richness in regard to both geographic regions and the distributions of individual families. Interestingly, there are only 19 Holarctic species in four families: Saldidae (12), Corixidae (5), Gerridae (1), and Veliidae (1); therefore the species totals for individual regions provide a reasonable indication of local biogeographic species richness and endemism. The following synopsis examines these patterns on a continent-by-continent basis (see Table 2).

Palearctic Region (PA)

The Palearctic Region has 16 families containing 496 species. There is a high richness of cold-adapted Corixidae and Saldidae, and a low richness of Veliidae (except Veliinae), and Naucoridae. The Gelastocoridae and Helotrephidae are absent.

Nearctic Region (NA)

The Nearctic Region, with 16 families and 424 species, is nearly equivalent in species richness to the Palearctic. There is a high richness of Corixidae and Saldidae, which each have a number of cold-adapted genera restricted to this region, and a low richness of certain Veliidae (Veliinae). The Helotrephidae and Aphelocheiridae are absent.

Afrotropical Region (AT)

There are 18 families containing 799 species in the Afrotropical Region, with a particularly high richness of Nepidae, Aphelocheiridae, and certain Veliidae (Microveliinae, Rhagoveliinae). Throughout Africa two speciose genera of Nepidae (*Laccotrephes* and *Ranatra*) dominate. The Aphelocheiridae and Veliidae (especially Rhagoveliinae) exhibit notable insular species richness and endemism in Madagascar, with many undescribed species (see Table 4). In contrast to the Neotropical and Australasian regions, there is a low richness of Gelastocoridae, with only two species, and Gerridae show only modest diversification.

Oriental Region (OL)

The Oriental Region has 17 families represented by 1,103 species, rivaling the Neotropical Region in terms of total species richness. There is a particularly high richness of Gerridae, Naucoridae and Helotrephidae, with many undescribed species in these families. By contrast, there is a low richness of Gelastocoridae, and certain Veliidae (Rhagoveliinae). Distinctive groups confined to this region include the Eotrechinae (Gerridae) and Cheirochelini (Naucoridae).

Neotropical Region (NT)

The Neotropical Region, with 18 families and 1,289 species, is the world's richest in terms of described species. The already high richness of Veliidae, Belostomatidae, Naucoridae and Gelastocoridae will be further amplified when the known undescribed species are published. Similarly, the known but undescribed Saldidae and Ochteridae will more than double the present numbers recorded for this region. The family Aphelocheiridae is absent, and the family Helotrephidae has limited representation.

Australasian Region (AU)

The Australasian Region has 17 families containing 654 species, with high richness of Gerridae, Gelastocoridae, Ochteridae, Notonectidae and certain Veliidae (Microveliinae, Rhagoveliinae). The Helotrephidae are absent throughout most of this region, except for Sulawesi and the Lesser Sunda Islands. There are a large number of undescribed Veliidae in this region, most of them in the sub-families Microveliinae and Rhagoveliinae on New Guinea, which exhibits almost total endemism in regard to its water bug fauna at the species level. New Guinea also has high endemism at the generic level, with seven endemic genera of Veliidae (Polhemus & Polhemus, 2005), and six endemic genera of Naucoridae. The extensive water bug fauna of Australia, which is also highly endemic, has recently been monographed by Andersen & Weir (2004).

Pacific Region (PAc)

The Pacific Region has 8 families containing 37 species. We would note that a distinctive eastward extension of this biotic region, not originally considered within the geographic constraints of the workshop that coordinated this analysis, consists of the Insular Pacific with its numerous high islands and atolls (Polhemus, 1996). Although small in total land area, this sub-region spans nearly half the globe and has high endemicity and certain localized foci of high species richness. Islands included in this region include the Fiji, Tonga, Samoa, Caroline, Marshall, Gilbert, Society, Austral, Gambier, Tuamotu, Marquesas and Hawaiian archipelagoes. The faunas of these isolated island groups tend to be composed of species derived from a discrete suite of taxonomic lineages, most notably Veliidae (Microveliinae), Gerridae (Gerrinae), Mesoveliidae, Saldidae, Hydrometridae, and Notonectidae (Anisopinae), with progressive attenuation in an eastward direction from Australia and the Philippines to Hawaii. Although the Pacific Region currently appears depauperate, there are a substantial number of undescribed taxa awaiting publication.

Although held in the Pacific Region, the oldest and largest Fijian island of Viti Levu has a diverse topography and large rivers of continental magnitude, and thus might arguably be included in the Australasian region, similar to New Caledonia, under the criteria of the current analysis.

Antarctic Region

No aquatic Heteroptera are currently recorded from Antarctica, although biogeographic evidence indicates that during the Early Tertiary this continent may have provided a biogeographic corridor for the interchange of certain groups (Gelastocoridae, trepobatine Gerridae) between Australia and South America (J. Polhemus & D. Polhemus, 2002).

Areas of endemicity

Based on our studies and those of many colleagues worldwide, we believe that the main areas of endemicity for aquatic Heteroptera on a world basis are as follows (the following listing does not imply any form of ranking, and there may be others): (1) Madagascar; (2) New Guinea; (3) Indochina; (4) the Malay Archipelago; (5) Australia; (6) tropical central and west Africa; (7) the Guiana Shield of northern South America; (8) the Atlantic rainforests of eastern South America. These areas are largely congruent with the global biodiversity "hotspots" defined by Mittermeier et al. (1998) and are dicussed in detail further below:

(1) Madagascar has an overwhelming preponderance of endemic species, and 10 specialized endemic genera, however, most other genera are shared with mainland Africa. There are a substantial number of species awaiting description in the Veliidae (Microveliinae, *Rhagovelia*), Aphelocheiridae and Naucoridae (*Temnocoris*).

(2) New Guinea has a higher percentage of endemic genera than any other large landmass, primarily concentrated in the Gerromorpha and Nepomorpha (Naucoridae). We estimate the species endemism for this very large island at well over 90%. There are substantial numbers of Veliidae (Rhagoveliinae and Microveliinae), Gerridae (*Tenagogonus*), Notonectidae (*Enithares*) and Ochteridae (*Ochterus*) still awaiting description, all of them endemic.

(3) Indochina is extremely rich in aquatic Heteroptera, and contains a number of endemic genera, although most of its genera are shared with the Australasian and Palearctic regions. The enormous number of endemic species is due to the geological age of the region, the complex topography created by the accretion of the Indian Plate to southern Asia, and region's proximity to the western Austalasian Region and the remainder of Asia, which has resulted in high lineage diversity. A number of intensive studies are presently focused on this area, resulting in a continuing flood of new taxonomic descriptions, mostly of newly discovered endemics.

(4) The Malay Archipelago (defined as extending from the Malay Peninsula to Timor, and excluding New Guinea; see above) is rich in insular endemic species, but has few endemic genera, instead sharing most of its genera with mainland tropical Asia. Our knowledge of this area is still limited; a number of brief baseline surveys have been undertaken but a comprehensive knowledge of the aquatic Heteroptera on a region-wide basis is far from complete, and further discoveries of endemic taxa or even genera, particularly at higher elevations, are nearly certain. A recent compedium of

knowledge to date for the water bugs (excluding shore bugs) of this region is now available (Chen et al., 2005), providing a useful basis for further studies.

(5) Australia is a huge but still under-surveyed continent, especially in its tropical regions. Of the 54 genera of water bugs found on the continent, 14 are endemic, and it will be surprising to see further additions. Most of the species are endemic, and others are shared only with immediately neighboring areas, particularly New Guinea and Timor. About 25% of the Australian species are marine, and some of these latter previously thought to be endemic have been recently found on the islands bordering the northern margin of the Coral Sea, thousands of kilometers to the northeast (Polhemus & Polhemus, 2006). A comprehensive treatment of the known Australian water bugs (excluding shore bugs) is now available (Andersen & Weir, 2004), which will facilitate recognition and treatment of new discoveries.

(6) Tropical central and west Africa has been partly surveyed over time, but in a geographically haphazard fashion, thus is still poorly understood. The known fauna has a number of endemic genera, and a high proportion of endemic species. The few taxa that are known from the plateaus of Angola, which lies on the periphery of this region, are also mostly endemic. Chronic political instability and poor infrastructure make prospects for further field work in this region uncertain at best.

(7) The Guiana Shield is a very rich region for aquatic Heteroptera, due to its great geological age, equatorial location, and high topographic complexity. The region shares most of its genera with the remainder of tropical South America, but has many locally endemic species, a large number of which are yet undescribed. The eastern slopes of the Andes, extending south and west of the Guiana Plateau, also exhibit high richness and endemicity, and might logically be included as an extension of this region bounding the western limits of the Amazon basin. Further collections are badly needed throughout this entire area, but are complicated by remote and complex terrain, and social factors such as the narcotics trade.

In addition to these large-scale areas of endemism and richness, there are many other more localized areas of endemism for aquatic Heteroptera found in all major biogeographic regions (for instance, the Great Basin of the western United States; D. Polhemus & J. Polhemus, 2002). In many cases these are internal sub-divisions of the larger areas discussed above; for instance, a recent study identified 40 areas of endemism for aquatic Heteroptera in New Guinea alone (Polhemus et al., 2004).

Under-collected areas (black holes)

The following are areas that appear to be of high importance but which lack adequate surveys, and as such represent "black holes" of aquatic Heteroptera knowledge (the following listing does not imply any form of ranking): (1) the sub-tropical plateaus of Angola and adjacent southwest Africa; (2) portions of the Malay Archipelago, notably the Lesser Sunda Islands, the Moluccas, and the southeastern mountains of Borneo; (3) portions of interior Indochina, particularly Cambodia (Cardamom Mountains) and Laos; (4) The southeastern margin of the Himalayan uplift, including northern Burma, Assam, southwest China and eastern Tibet; (5) the mountains of Ethiopia and the adjacent Horn of Africa; (6) tropical and sub-tropical West Africa; (7) central tropical Africa, particularly Lake Tanganyika and associated drainages, which are a renowned center of endemism for other freshwater groups; (8) the Atlantic coastal rainforests of Brazil from Rio de Janeiro north to the mouth of the Amazon; (9) many Pacific islands.

For most of these areas we have a few scattered samples, which are sufficient to underscore the importance of these areas and the need for additional surveys (for instance, two endemic genera of Naucoridae are known from Lake Tanganyika). Additional areas that could be added here might include inner Tibet and far western China, although much of this area is cold, dry and likely depauperate for aquatic Heteroptera.

Summary of general biogeographic patterns

Based on the current analysis, the following general biogeographic patterns have emerged:

(1) The Corixidae and Saldidae exhibit a distinctly higher species richness in the Northern Hemisphere in comparison to tropical regions.
(2) Among the Gerromorpha, there is a disproportionate richness of Veliidae in the Neotropical region, and of Gerridae in the Oriental region. In many cases it appears that certain endemic genera are filling ecologically equivalent roles in the two regions (i.e., the gerrid genus *Metrocoris* in the Oriental region may be ecologically equivalent to the veliid genus *Rhagovelia* in the Neotropical region).
(3) The family Helotrephidae is entirely confined to land masses derived from Gondwanaland. The Gelastocoridae and Ochteridae show a nearly similar pattern, except for presumably recent dispersal into the southern section of the Nearctic region via the Mesoamerican land bridge.
(4) The family Aphelocheiridae is confined to the Eastern Hemipshere, with no representation in the Neotropical and Nearctic regions.
(5) Cosmopolitan groups with at least some representation in all non-polar biogeographic regions include Gerridae, Veliidae, Mesoveliidae, Hydrometridae, Hebridae, Pleidae, Belostomatidae, Nepidae, Notonectidae, Saldidae and Leptopodidae.

Conservation

Aquatic Heteroptera are hardy and adaptable insects, and based on current information it appears that most species are not at risk. There are a few documented cases in Europe, North America, and Japan where taxa with particularly circumscribed ranges have become endangered through loss of habitat. This trend seems to be most pronounced in the Saldidae (*Salda morio* Zetterstedt not seen in the Netherlands since 1918, when its peat bog habitat largely disappeared; Aukema, 2003), Naucoridae (*Ambrysus amargosus* LaRivers, endemic to small springs in Ash Meadows, Nevada, threatened by habitat destruction; Polhemus, 1994) and Aphelocheiridae (*Aphelocheirus kawamurae* Matsumura now thought to be extinct because of alteration of the river system from Biwa-ko Lake, Japan; Tomokuni et al., 1995). For tropical regions, our documentation of species ranges and ecologies is so incomplete that no solid conclusions can be drawn in regard to these topics, but potential threats exist from both human alteration of landscape ecologies (such as conversion of rain forests to plantations or cattle pastures) and from ecosystem transformations driven by global climate change.

Many species of aquatic Heteroptera are also useful in terms of global or regional conservation planning for preservation of freshwater biodiversity, due to the largely non-overlapping nature of the broad regional biotas, coupled with a high degree of localized endemism within regions (Polhemus et al., 2004). Certain species are also effective biological controls of mosquitoes and other pest flies because they feed on the immatures, and other species have been documented as natural enemies of rice pests. The various families also display differing tolerances to water pollution, and as a result have potential utility as bioindicators of water quality (Jansson, 1977). In light of this, their preservation is in the best interests of human societies as a whole.

Acknowledgements This project benefited greatly from the efforts of Gerry Louton and his staff at the National Museum of Natural History, Smithsonian Institution, who entered the original manuscript world catalog of aquatic Heteroptera into electronic format. We also wish to thank the Smithsonian Institution's Conservation Research Center, the World Wildlife Fund, Conservation International, the Wildlife Conservation Society, and the National Science Foundation, all of which facilitated surveys of aquatic Heteroptera in remote areas of the world and thereby contributed to the quality of the analysis presented herein. This is a research contribution from Colorado State University, where JTP is an adjunct faculty member, and with which the Colorado Entomological Institute is affiliated.

References

Andersen, N. M., 1982. The semiaquatic bugs (Hemiptera, Gerromorpha). Phylogeny, adaptations, biogeography, and classification. Entomonograph 3: 1–455.

Andersen, N. M. & T. A. Weir, 2004. Australian water bugs. Their biology and identification (Hemiptera-Heteroptera, Gerromorpha & Nepomorpha). Entomonograph 14: 344 pp., 8 col. pls., text figs., maps.

Arnqvist, G., T. M. Jones & M. A. Elgar, 2003. Reversal of sex roles in nuptial feeding. Nature 424: 387.

Aukema, B., 2003. Recent changes in the Dutch Heteroptera fauna (Insecta: Hemiptera). Proceedings of the 13th International colloqium European Invertebrate Survey, September 2001, pp. 39–52.

Aukema, B. & C. Rieger (Eds) Catalogue of the Heteroptera of the Palearctic Region. Netherlands Entomological Society, Amsterdam, Vol. 1, xxvi + 222 pp.

Chen, P. P., N. Nieser & H. Zettel, 2005. The aquatic and semi-aquatic bugs (Heteroptera: Nepomorpha & Gerromorpha) of Malesia. Fauna Malesiana Handbook 5: 1–546.

Hebsgaard, M. B., N. M. Andersen & J. Damgaard, 2004. Phylogeny of the true water bugs (Nepomorpha: Hemiptera-Heteroptera) based on 16S rDNA and morphology. Systematic Entomology 29: 488–508.

Henry, T. J. & R. C. Froeschner, 1988. Catalog of the Heteroptera or True Bugs of Canada and the Continental United States. Brill, Leiden, New York, i–xix, 958 pp.

Jansson, A., 1977. Micronectae (Heteroptera, Corixidae) as indicators of water quality in two lakes in southern Finland. Annales Zoologici Fennici 14: 118–124.

Mahner, M., 1993. Systema Cryptoceratum Phylogeneticum (Insecta Heteroptera). Zoologica, Stuttgart, Heft 143. ix + 302 pp.

Mittermeier, R. A., N. Meyers, J. B. Thomsen, G. A. B. da Fonseca & S. Olivieri, 1998. Biodiversity hotspots and major tropical wilderness areas: Approaches to setting conservation priorities. Conservation Biology 12(3): 516–520.

Polhemus, J. T., 1985. Shore Bugs (Heteroptera, Hemiptera; Saldidae). A World Overview and Taxonomy of Middle American Forms. The Different Drummer, Englewood, Colorado, v + 252 pp.

Polhemus, D. A., 1994. Conservation of aquatic insects: Worldwide crisis or localized threats? American Zoologist 33: 588–598.

Polhemus, D. A., 1996. Island arcs and their influence on Indo-Pacific biogeography (pp. 51–66). In: Keast, A. & S. L. Miller (eds), The Origin and Evolution of Pacific Island Biotas, New Guinea to Eastern Polynesia: Patterns and Processes. SPB Publishing, Amsterdam, viii + 531 pp.

Polhemus, J. T., 2000. North American Mesozoic aquatic Heteroptera (Insecta, Naucoroidea, Nepoidea) from the Todilto Formation, New Mexico (pp. 29–40). In Lucas, S. G. (ed.), New Mexico's Fossil Record 2. New Mexico Museum of Natural History and Science Bulletin No. 16, vi + 284 pp.

Polhemus, D. A., R. A. Englund & G. Allen, 2004. Freshwater biotas of New Guinea and nearby islands: Analysis of endemism, richness and threats. Pacific Biological Survey Contribution 2004-004: 1–68 for Conservation International, Washington, DC.

Polhemus, D. A., J. Maciolek & J. Ford, 1992. An ecosystem classification of inland waters for the tropical Pacific. Micronesica 25: 155–173.

Polhemus, D. A. & J. T. Polhemus, 1998. Assembling New Guinea: 40 million years of island arc accretion as indicated by the distributions of aquatic Heteroptera (Insecta) (pp. 327–340). In Hall, R. & J. D. Holloway (eds), Biogeography and Geological Evolution of SE Asia. Backhuys Publishers, Leiden, ii + 417 pp., 167 figs., 12 col. pls.

Polhemus, D. A. & J. T. Polhemus, 2002. Basins and Ranges: The Biogeography of Aquatic True Bugs in the Great Basin, Vol. 33. Smithsonian Institution Contributions to the Earth Sciences, 235–254.

Polhemus, J. T. & D. A. Polhemus, 2002. The Trepobatinae (Gerridae) of New Guinea and surrounding regions, with a review of the world fauna. Part 6. Phylogeny, biogeography, world checklist, bibliography and final taxonomic addenda. Insect Systematics and Evolution 33: 253–290.

Polhemus, D. A. & J. T. Polhemus, 2005. Two new genera and thirty new species of Microveliinae (Heteroptera: Veliidae) from the East Papua Composite Terrane, far eastern New Guinea. Tijdschrift voor Entomologie 147: 113–189.

Polhemus, J. T. & D. A. Polhemus, 2006. The marine Heteroptera of far eastern New Guinea and adjacent archiplelagoes (Insecta: Gerromorpha). Denisia 19: 927–982.

Polhemus, J. T. & D. A. Polhemus (in press). Global trends in the description of aquatic and semiaquatic Heteroptera species, 1758–2004. Tijdschrift voor Entomologie 150(2).

Schuh, R. T., B. Galil & J. T. Polhemus, 1987. Catalog and bibliography of Leptopodomorpha (Heteroptera). Bulletin of the American Museum of Natural History 185(3): 243–406.

Schuh, R. T. & J. T. Polhemus, 1980. Analysis of taxonomic congruence among morphological, ecological, and biogeographic data sets for the Leptopodomorpha (Hemiptera). Systematic Zoology 29: 1–26.

Schuh, R. T. & J. A. Slater, 1995. True Bugs of the World (Hemiptera: Heteroptera). Classification and Natural History. Cornell University Press, Ithaca and London, xii + 336 pp.

Tomokuni, M., M. Sato, N. Ichikawa, Y. Araki & S. Nishu, 1995. Two Japanese *Aphelocheirus* species (Naucoridae), *A. kawamurae* presumably extinct and *A. nawae* still surviving in Hyogo Prefecture. Rostria 44: 21–25.

Hydrobiologia (2008) 595:393–407
DOI 10.1007/s10750-007-9113-2

FRESHWATER ANIMAL DIVERSITY ASSESSMENT

Global diversity of caddisflies (Trichoptera: Insecta) in freshwater

F. C. de Moor · V. D. Ivanov

Abstract The not yet uploaded Trichoptera World Checklist (TWC) [http://entweb.clemson.edu/database/trichopt/search.htm], as at July 2006, recorded 12,627 species, 610 genera and 46 families of extant and in addition 488 species, 78 genera and 7 families of fossil Trichoptera. An analysis of the 2001 TWC list of present-day Trichoptera diversity at species, generic/subgeneric and family level along the selected Afrotropical, Neotropical, Australian, Oriental, Nearctic and Palaearctic (as a unit or assessed as Eastern and Western) regions reveals uneven distribution patterns. The Oriental and Neotropical are the two most species diverse with 47–77% of the species in widespread genera being recorded in these two regions. Five Trichoptera families comprise 55% of the world's species and 19 families contain fewer than 30 species per family. Ten out of 620 genera contain 29% of the world's known species. Considerable underestimates of Trichoptera diversity for certain regions are recognised. Historical processes in Trichoptera evolution dating back to the middle and late Triassic reveal that the major phylogenetic differentiation in Trichoptera had occurred during the Jurrasic and early Cretaceous. The breakup of Gondwana in the Cretaceous led to further isolation and diversification of Trichoptera. High species endemism is noted to be in tropical or mountainous regions correlated with humid or high rainfall conditions. Repetitive patterns of shared taxa between biogeographical regions suggest possible centres of origin, vicariant events or distribution routes. Related taxa associations between different regions suggest that an alternative biogeographical map reflecting Trichoptera distribution patterns different from the Wallace (The Geographical Distribution of Animals: With a Study of the Relations of Living and Extinct Faunas as Elucidating the Past Changes of the Earth's Surface, Vol. 1, 503 pp., Vol. 2, 607 pp., Macmillan, London, 1876) proposed biogeography patterns should be considered. Anthropogenic development threatens biodiversity and the value of Trichoptera as important functional components of aquatic ecosystems, indicator species of deteriorating conditions and custodians of environmental protection are realised.

Guest editors: E. V. Balian, C. Lévêque, H. Segers & K. Martens
Freshwater Animal Diversity Assessment

F. C. de Moor (✉)
Department of Freshwater Invertebrates,
Makana Biodiversity Centre, Albany Museum,
Grahamstown 6139, South Africa
e-mail: F.deMoor@ru.ac.za

F. C. de Moor
Department of Zoology and Entomology, Rhodes University, Grahamstown 6139, South Africa

V. D. Ivanov
Department of Entomology, Faculty of Biology,
St. Petersburg State University, Universitetskaya nab. 7/9,
St. Petersburg 19034, Russia

Keywords Caddisflies · Biogeography · Fossil-record · Endemism · Environmental protection

Introduction

The order Trichoptera (caddisflies) comprises a group of holometabolous insects closely related to the Lepidoptera. Together the two orders form the superorder Amphiesmenoptera. Adult Trichoptera range in size over two orders of magnitude, from minute with a wing span of less than 3 mm, to large with a wing span approaching 100 mm. Some species have striking colours and wing patterns but they generally range in colour from dull yellow through grey, or brown to black. They are moth-like insects with wings covered by hairs, not scales as in Lepidoptera. Adults have prominent, and in some species exceptionally long, antennae (more than double the length of the forewing). With some exceptions they have well-developed maxillary and labial palps, but never the coiled proboscis that characterises most adult Lepidoptera.

Trichoptera larvae are probably best known for the transportable cases and fixed shelters that many, though not all, species construct. Silk has enabled Trichoptera larvae to develop an enormous array of morphological adaptations for coping with life in almost any kind of freshwater ecosystem (Wiggins, 1996, 2004). Larvae can be distinguished from all other insects with segmented thoracic legs by the presence of a pair of anal prolegs, each with a single curved terminal claw and very short, sometimes almost invisible, antennae consisting of a single segment. The trichopteran pupa is exarate and covered by a semitransparent pupal integument and if fully developed reveals the pharate adult inside. The pupa usually possesses a pair of strong functional mandibles, non functional in the adult, and the abdomen has a number of segments adorned with characteristic sclerotised, dorsal hook-bearing plates. The larval and pupal stages of Trichoptera are, with a few exceptions, entirely dependent on an aquatic environment and are usually abundant in all freshwater ecosystems, from spring sources, mountain streams, large rivers, the splash zones of waterfalls and marshy wetlands, along shorelines and in the depths of lakes, to temporary waters. Certain species are tolerant of high salinities and species in one family, the Chathamiidae, have managed to colonise tidal pools along the sea shore in New Zealand and eastern Australia; some species inhabit the brackish inshore waters of the Baltic and White seas.

The phylogeny of Trichoptera has been studied intensively with explicit methods for 50 years (Ross, 1956, 1964, 1967; Weaver, 1984, 1992a, 1992b; Weaver & Morse, 1986; Wiggins & Wichard, 1989; Wiggins, 1992, 2004; Frania & Wiggins, 1997; Ivanov, 1997, 2002; Morse 1997; Kjer et al., 2001, 2002) (Fig. 1). Morphological, molecular and behavioural features of the adults, larvae and pupae have been used to assess specific and higher taxonomic relationships and form the basis of the hierarchical classification system developed. Subdivision into two suborders—Annulipalpia and Integripalpia—is accepted here, because of their strong support from recent phylogenetic studies. Four families—Hydrobiosidae, Hydroptilidae, Glossosomatidae and Rhyacophilidae—sometimes included in a controversial third group ("Spicipalpia"), remain uncertain in their placement. A detailed comprehensive review of ordinal, familial and infrafamilial phylogenies was provided by Morse (1997, 2003).

Species diversity

Fischer (1960–1973) produced a world catalogue that recorded 5,546 species. The recently updated TWC currently records 12,627 species (Morse, personal communication, July 2006, see also Morse, 1999, 2003). These species are arranged in 610 genera and 46 extant families. In addition, 488 species and 78 genera in seven families are known only from fossil records. New species continue to be described at a considerable rate and it seems—particularly from ongoing studies in the Neotropics, Madagascar, humid regions of Africa, south-east Asia, China and the Phillipines—that the prediction of Schmid (1984), Flint et al. (1999) and Morse (personal communication, 2005), although considered an overestimate by Malicky (1993), that there are in excess of 50,000 species may be closer to the actual figure. If these estimates are correct, this leads to the assumption that only around 20–25% of the world species of Trichoptera have been described.

Species recognition is based primarily on morphological features of the adults, strongly influenced

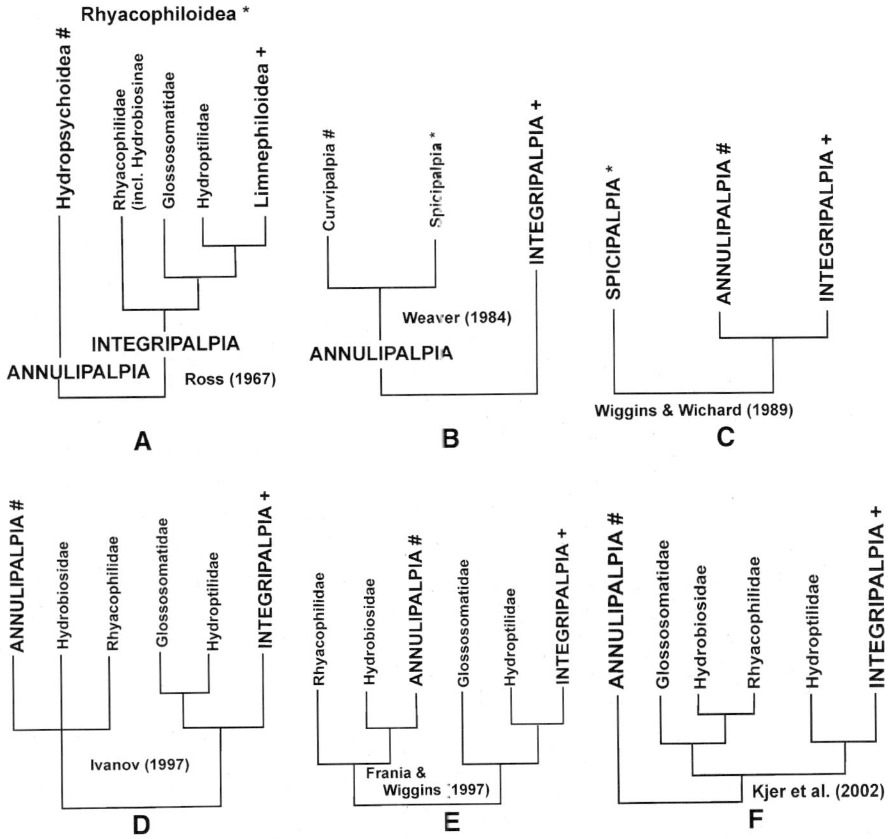

Fig. 1 Six contemporary hypotheses of subordinal relationships of the Trichoptera. Equivalent taxonomic units are indicated by like symbols (e.g. Ross' Hydropsychoidea = Weaver's Curvipalpia = Wiggins & Wichard's Annulipalpia). **A**. From Ross (1967). **B**. From Weaver (1984). **C**. From Wiggins & Wichard (1989), based on pupation only (Wiggins, 1992). **D**. From Ivanov (1997). **E**. Strict consensus of five trees from Frania & Wiggins (1997; Figs. 24, 25). **F**. Simplified phylogram from differentially weighted parsimony analysis of combined data from Kjer et al. 2002. Spicipalpia as used here includes the families Rhyacophilidae, Hydrobiosidae, Glossosomatidae and Hydroptilidae

by detailed studies of the external genitalia of adult male Trichoptera. More recently, molecular sequences in selected RNA and mitochondrial DNA segments have been used to assess species diversity and phylogenetic relationships (Kjer et al., 2001, 2002). The identification of cuticular hydrocarbons in adult Trichoptera presents a further technique that can be used to discern species (Nishimoto et al., 2002). These techniques offer new insights into species diversity and also diversity within species, making it possible to recognise two different species that are morphologically indistinguishable but show considerable genetic diversity, thus making the identification of "sibling" or "aphanic" species possible (Steyskal, 1972).

Generally the world distribution of Trichoptera is considered in a common framework of regions proposed for vertebrates and terrestrial arthropods (Wallace, 1876). The zoogeography of amphibiotic orders, including Trichoptera, differs sufficiently from this to suggest that a different regional classification should be used. In order to present conformity of data to enable comparison with all freshwater fauna reviewed in the other articles in this volume but with one modification, separating the east and west Palaearctic Regions for synthesis, the selection of biogeographical regions for assessing Trichoptera distribution patterns in this article has followed the six major biogeographical regions according to Wallace (1876) (Table 1). A synthesis of the number of genera and species based on the earlier edition of the TWC (last updated 8 January 2001) reveals a total of 11,532 extant species in 620 genera and 94 sub-genera. More than half of these

Table 1 The number of extant genera and species (in parentheses) recorded in Trichoptera families in the major biogeographical regions of the world

Family taxa	W. PA	E. PA	PA	NA	AT	NT	OL	AU	World total
Annulipalpia									
Philopotamidae	5 (51)	5 (41)	7 (91)	5 (56)	4 (87)	4 (257)	10 (346)	5 (60)	17 (886)
Stenopsychidae	–	1 (14)	1 (14)	–	1 (1)	1 (3)	1 (64)	1 (10)	3 (89)
Hydropsychidae	6 (120)	11 (69)	11 (179)	17 (165)	13 (148)	16 (355)	24 (489)	17 (87)	49 (1,409)
Dipseudopsidae	1 (1)	3 (3)	3 (4)	1 (5)	4 (48)	–	4 (47)	2 (3)	6 (104)
Polycentropodidae	8 (88)	9 (38)	11 (118)	8 (77)	7 (20)	7 (173)	10 (230)	8 (42)	23 (656)
Ecnomidae	2 (10)	1 (5)	2 (14)	1 (3)	3 (80)	1 (35)	1 (120)	2 (78)	6 (327)
Xiphocentronidae	–	1 (3)	1 (3)	2 (8)	1 (2)	3 (47)	5 (76)	–	7 (133)
Psychomyiidae	5 (103)	6 (28)	6 (130)	4 (18)	3 (16)	–	7 (234)	2 (5)	8 (400)
"*Spicipalpia*"									
Rhyacophilidae	2 (120)	2 (110)	3 (221)	2 (127)	–	–	3 (350)	1 (1)	4 (696)
Hydrobiosidae	–	1 (2)	1 (2)	1 (5)	–	23 (168)	1 (31)	27 (183)	50 (384)
Glossosomatidae	3 (78)	5 (63)	6 (135)	5 (85)	1 (4)	14 (160)	6 (125)	1 (22)	22 (530)
Hydroptilidae	11 (181)	10 (61)	15 (236)	19 (295)	13 (142)	33 (498)	17 (318)	21 (224)	68 (1,679)
Integripalpia									
Oeconesidae	–	–	–	–	–	–	–	6 (19)	6 (19)
Brachycentridae	2 (30)	5 (28)	6 (56)	5 (37)	–	–	2 (22)	–	7 (112)
Phryganopsychidae	–	1 (2)	1 (2)	–	–	–	1 (2)	–	1 (3)
Lepidostomatidae	6 (25)	9 (55)	12 (79)	3 (75)	3 (37)	1 (18)	23 (187)	–	30 (389)
Pisuliidae	–	–	–	–	2 (15)	–	–	–	2 (15)
Rossianidae	–	–	–	2 (2)	–	–	–	–	2 (2)
Kokiriidae	–	–	–	–	–	1 (1)	–	4 (7)	6 (8)
Plectrotarsidae	–	–	–	–	–	–	–	3 (5)	3 (5)
Phryganeidae	9 (26)	7 (27)	10 (44)	7 (21)	–	–	5 (19)	–	14 (77)
Goeridae	7 (24)	3 (20)	8 (44)	2 (6)	1 (1)	–	5 (110)	1 (2)	12 (160)
Uenoidae	1 (6)	2 (6)	3 (12)	5 (51)	–	–	2 (15)	–	7 (78)
Apataniidae	2 (31)	15 (69)	15 (97)	5 (34)	–	–	5 (60)	–	18 (185)
Limnephilidae	50 (388)	29 (167)	64 (514)	39 (222)	–	10 (45)	17 (102)	1 (3)	95 (861)
Tasimiidae	–	–	–	–	–	2 (2)	–	2 (6)	4 (9)
Odontoceridae	1 (3)	2 (9)	3 (12)	6 (12)	–	2 (25)	4 (41)	2 (4)	12 (103)
Atriplectididae	–	–	–	–	1 (1)	1 (1)	–	1 (1)	4 (5)
Limnocentropodidae	–	1 (1)	1 (1)	–	–	–	1 (14)	–	1 (15)
Philorheithridae	–	–	–	–	–	2 (5)	–	6 (15)	8 (23)
Molannidae	2 (6)	2 (7)	2 (10)	2 (7)	–	–	2 (19)	–	3 (34)
Calamoceratidae	1 (2)	5 (11)	6 (13)	3 (5)	1 (5)	2 (39)	3 (46)	1 (25)	9 (125)
Leptoceridae	14 (127)	13 (102)	18 (212)	8 (116)	18 (302)	12 (143)	16 (597)	18 (207)	48 (1,549)
Sericostomatidae	5 (50)	1 (2)	6 (52)	3 (15)	5 (12)	5 (16)	2 (4)	–	19 (97)
Beraeidae	5 (45)	2 (2)	6 (47)	1 (4)	1 (1)	–	–	–	7 (52)
Anomalopsychidae	–	–	–	–	–	2 (22)	–	–	2 (22)
Helicopsychidae	1 (5)	1 (2)	1 (7)	1 (10)	1 (13)	1 (62)	1 (55)	2 (52)	2 (194)
Chathamiidae	–	–	–	–	–	–	–	2 (5)	2 (5)
Helicophidae	–	–	–	–	–	5 (13)	–	3 (8)	8 (21)
Calocidae	–	–	–	–	–	–	–	7 (20)	7 (20)

Table 1 continued

Family taxa	W. PA	E. PA	PA	NA	AT	NT	OL	AU	World total
Conoesucidae	–	–	–	–	–	–	–	12 (42)	12 (42)
Barbarochthonidae	–	–	–	–	1 (1)	–	–	–	1 (1)
Antipodoeciidae	–	–	–	–	–	–	–	1 (1)	1 (1)
Hydrosalpingidae	–	–	–	–	1 (1)	–	–	–	1 (1)
Petrothrincidae	–	–	–	–	2 (6)	–	–	–	2 (6)
Total genera	149	145	229	157	87	148	169	143	619
Total species	1,520	947	2,349	1,461	944	2,100	3,723	1,140	11,532

The Palaeartic region has been divided into eastern and western regions but is also recorded as a single region for comparative purposes. Zoogeographic regions: PA, Palaearctic; NA, Nearctic; NT, Neotropical; AT, Afrotropical; OL, Oriental; AU, Australasian

known species were recorded from only two regions, the Oriental and Neotropical Regions (Fig. 2). This indicates a high capacity for supporting large numbers of different species in tropical ecosystems, lower rates of species extinctions during the most recent glaciations, and probably a significantly higher rate of speciation in these two regions than in the other regions. This is borne out by the large proportion of the recorded world species in widely distributed genera such as *Chimarra* (35% and 40%), *Orthotrichia* (47% and 27%), *Oecetis* (40% and 7%) and *Setodes* (71% and 0%) found, respectively, in these two regions.

The highest species diversity is recorded in the Oriental Region. With more than 3,700 species, it contains more than double the recorded species for each of the other regions, except the Neotropics. Without exception, all eight families of the suborder Annulipalpia attain their greatest species richness in the Oriental Region. The family Rhyacophilidae and

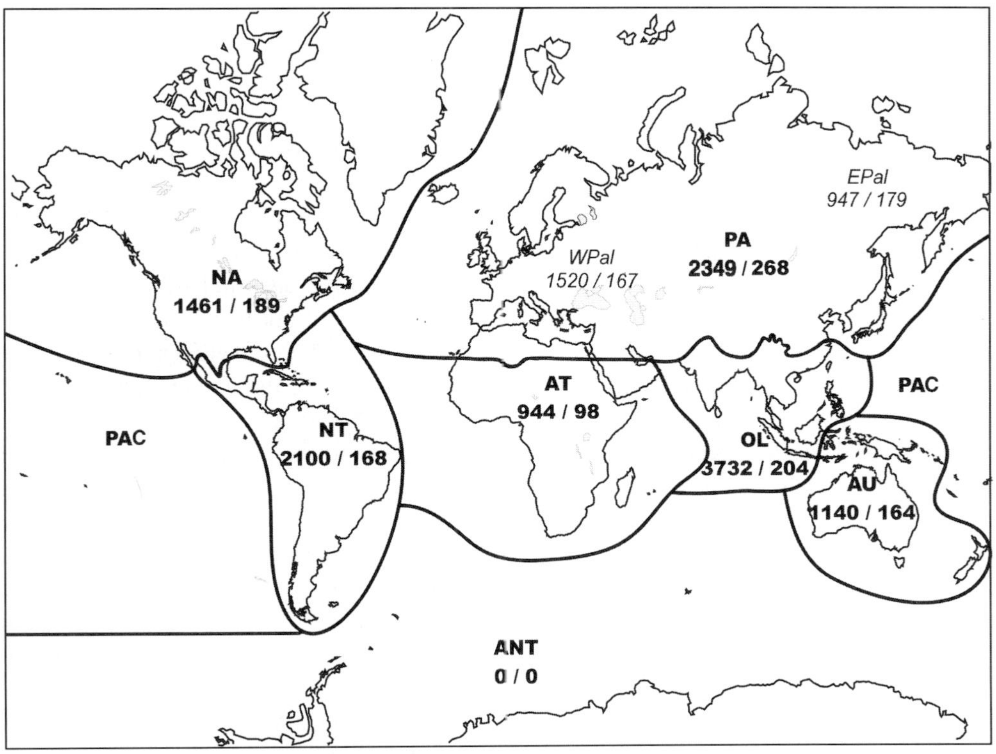

Fig. 2 The current number of species/genera plus subgenera for each of the seven major biogeographical regions. AT, afrotropical; numbers for AU, Australasian include Pacific Oceanic Islands (PAC); PA (WPal), West Palaearctic; PA (EPal), East Palaearctic; NA, Nearctic; NT, neotropical; OL, Oriental

the integripalpian families Lepidostomatidae, Goeridae, Calamoceratidae and Leptoceridae also record their highest number of species in this region. With the large number of species previously described from India by Schmid & Mosely (see expanded reference list on web site), this region also records the highest density of species per unit area at 1.6 species per kilohectare (Morse, 2003). The Neotropical Region records the greatest number of species in the families Hydroptilidae and Glossosomatidae (Table 1). There are no Rhyacophilidae in this region but Hydrobiosidae, confined mostly to southern Patagonia and Chile, are second in species richness after the Australian Region. The West Palaearctic Region records the greatest number of integripalpian species in the families Limnephilidae, Sericostomatidae and Beraeidae.

The distribution of species in the 45 families of Trichoptera is also very uneven with the five most species-rich families comprising 55% of the recorded species. Nineteen families, all in the suborder Integripalpia, comprise fewer than 30 species per family. Ten genera out of the 620 world genera account for 3,299 species, representing 29% of the world total. The numbers of genera within families is also very unevenly distributed. The family Rhyacophilidae, with 696 species, comprises only four extant genera, with 93% of the family's species in the genus *Rhyacophila*. The closest sister family—the Hydrobiosidae—includes 384 species classified in 50 genera, with the most species-rich genus, *Atopsyche*, split into four subgenera containing 30% of the known species in the family. In the family Hydroptilidae there are 20 monobasic genera and a single genus *Hydroptila* records 375 species, contributing 22% of the species in that family. Within the suborder Annulipalpia, the family Philopotamidae (with 17 world genera) has 61% of its species recorded in three subgenera of the genus *Chimarra*. The family Hydropsychidae (with 49 world genera) contains 36% of 1,409 recorded species in two genera (*Cheumatopsyche* and *Hydropsyche*).

Molecular techniques, and more detailed morphological and cladistic techniques have revealed that many of the presently classified large genera are paraphyletic or even polyphyletic. Consequently, some genera need refining to represent monophyletic lineages. Thus the estimations on abundances in generic and higher-level classifications are rather tentative. Larger families like Hydropsychidae, Limnephilidae and Rhyacophilidae await revisions to provide a more reliable basis for determining zoogeographic distribution patterns and phylogenies.

The regional biogeographic diversity of the species and genera recorded in each of the 45 extant families (Table 1) represents considerable underestimates for regions like the Afrotropical and Oriental realms where large numbers of species have been described recently or are awaiting description. Studies in Madagascar (Gibon et al., 2001) reveal that there are at least a further 416 undescribed Madagascan species. Based on the current database and considering only described and recorded species there was a 13% increase in the number of known world species between 2001 and 2006.

A number of regional species lists, catalogues and atlases of Trichoptera, including web sites such as the Trichoptera World Checklist (TWC) (Morse, 1999 http://entweb.clemson.edu/database/trichopt/index.htm), Fauna Europaea (http://www.faunaeur.org) and Checklist of the New Zealand Trichoptera (Ward, 2003 http://www.niwa.co.nz/ncabb) can be consulted for an understanding of Trichoptera diversity.

Historical processes and phylogeny

If representatives from fossil families in the Permian suborder Protomeropina (=Permotrichoptera), which are part of the ancestral Amphiesmenoptera lineage, are considered not to belong in the direct lineage to Trichoptera, then the earliest records of recognisable Trichoptera—in the extinct families Prorhyacophilidae and Necrotauliidae and species recognisable as belonging to the extant family Philopotamidae—are from the Middle and Late Triassic times around 230 mybp (Morse, 1997; Ivanov & Sukatsheva, 2002). It is assumed that all the continents were united in a supercontinent Pangea with a remarkably homogeneous biota, emphasised by the indistinctness of floristic boundaries, recorded throughout (Eskov & Sukatsheva, 1997). This suggests that relatively uniform climatic conditions existed and allowed rapid dispersion of insect groups all over Pangea. The extinct family Necrotauliidae were considered to be the dominant Triassic and Jurassic Trichoptera until the recent revision of old collections by Ansorge (2002) revealed large numbers of Lepidoptera among them. The overall number of Trichoptera recorded from the Triassic is low, both in species and specimen numbers. The earliest aquatic stages of Trichoptera are dated to Late Jurassic times. Trichoptera diversity increased in the Jurassic, with the Lower and Middle Jurassic deposits revealing a number of extinct families (Ivanov & Sukatsheva, 2002). Major phylogenetic differentiation in Trichoptera appeared in the Late Jurassic and Cretaceous (Ivanov & Sukatsheva, 2002) and the biogeographical patterns of those times can therefore help us to understand the present-day peculiarities of Trichoptera distribution.

The earliest Philopotamidae were discovered from late Triassic deposits in the then tropical belt of what would constitute the present day North America and Western Europe (Eskov & Sukatsheva, 1997). Middle Jurassic fossil sediments in Angaraland (which included present-day Siberia) record the earliest species that can be placed in the Integripalpia. The origins of the earliest Rhyacophilidae (Middle Jurassic) and Polycentropodidae (Upper Jurassic) are also recorded from this region. The Rhyacophilidae penetrated the tropical realm in the Early Cretaceous but Polycentropodidae are not recorded there until the Caenozoic from Oligocene-Miocene Dominican Amber. The oldest supposed Hydroptilidae fossils (larval or pupal cases) were found in the Upper Jurassic of Siberia (Ivanov & Sukatsheva, 2002). Presumably the place of origin for this family was somewhere within the non-tropical Old World areas. Originally the species in this family were phytophagous and found in lotic ecosystems. The Hydroptilidae were preadapted to survive in warm waters because of their small size and larval hypermetamorphosis (with very tiny caseless younger instars). Since low oxygen in overheated, organic-rich waters is the most important limiting factor for apneustic immature stages of aquatic insects in the tropics, members of this family are well adapted to tropical situations. Their small size also makes it easy to survive in hygropetric ecosysytems (in a thin film of water over stones, in waterfalls and rapids). The Hydroptilidae show remarkable speciation in the tropics and the diverse S. American fauna clearly demonstrates this (Flint et al., 1999). It is assumed that there were several independent invasions of Hydroptilidae from north to south (from N. America to S. America, from Europe to Africa, and from Asia to Australia). Some species of Hydroptilidae are readily dispersed by wind, and this manner of dispersal is possibly responsible for the peculiar patchy distribution pattern shown by some species. So although the origin of the above family was not considered as tropical, it was adapted to readily invade the tropics and subsequently, with isolation and speciation, developed a diverse fauna (Eskov et al., 2004).

It appears likely that the climate during the Jurassic and Cretaceous was sub-tropical to warm temperate throughout most of the landmasses, without the climatic extremes of the present-day tropical deserts and rainforests. There are notably no Jurassic fossil records from any region of Gondwana (Eskov & Sukatsheva, 1997). Hydrobiosidae appear to have originated in the tropical Jurassic belt (in present day-western Mongolia) and from there spread into more temperate regions. During the Late Jurassic, Leptoceridae were found in the extratropical, warm temperate latitudes of Laurasia (England and Siberia); they dispersed in the Early Cretaceous across other landmasses including Gondwana (Brazil). In contrast, the families Calamoceratidae and Phryganeidae originated in the Early Cretaceous in higher latitudes in the Northern Hemisphere and dispersed to lower latitudes later. The early Cretaceous reveals rapid progress and diversification in Trichoptera case constructions (Ivanov & Sukatsheva, 2002 and the references therein) reflecting extensive speciation. In the Late Cretaceous the Sericostomatidae and Hydroptilidae appear for the first time in deposits in high latitudes in the Northern Hemisphere. Between the early and late Cretaceous the extinction of many of the earlier taxa and dispersal of the taxa described above led to a complete change in pattern of overall Trichoptera diversity. This was caused largely by general transformation of the freshwater ecosystems through the proliferation of angiosperms which resulted in additions of large quantities of foliage debris in surface waters, leading to eutrophication and oxygen depletion (Eskov & Sukatsheva, 1997; Ivanov & Sukatsheva, 2002). During the Cretaceous the breakup of Gondwana further facilitated the isolation of populations of Trichoptera on the newly formed southern continents.

Caenozoic fossil resins and a few poorly studied sedimentary depositional sites reveal a domination of recent Trichoptera at the level of families, a few extinct genera and many extinct species generally related to the extant ones (Ulmer, 1912; Ivanov & Sukatsheva, 2002). The most recent fossil Trichoptera (Middle Miocene Dominican amber) show no significant difference from the Holocene fauna for the same area, indicating that most major evolutionary and dispersal events, at least for these tropical areas, happened before the Miocene (Ivanov & Sukatsheva, 2002). The main feature of the Eocene Baltic amber Trichoptera is the total absence of the large modern family Limnephilidae and the relative paucity of the generally-abundant Hydropsychidae while the family Polycentropodidae is extremely diverse. Based on a number of Caenozoic amber fossils, Limnephilidae are believed to have originated in North America and subsequently spread out across Angaraland/Angarida via the Beringian land bridge into Siberia and Europe (Ivanov & Sukasheva, 2002). Fossil *Wormaldia* species are closely related to extant North American species. In contrast to its present-day paucity in Europe, Caenozoic amber-fossil records of Ecnomidae suggest the previous diversity of this family in the Palaearctic. Similarly, fossil evidence shows the presence of Stenopsychidae and tentatively-identified Dipseudopsidae, now absent from Europe.

Although there were earlier general classifications of subordinal taxa within the Trichoptera, the first hypotheses—linking phylogeny to the dispersal of Trichoptera—to assess phylogenetic relationships were put forward by Ross (1967). He proposed a number of distribution patterns for the explanation and support of his phylogenies but most of his dispersal schemes had no palaeontological evidence. Historical biogeography has been used to identify tracks of phylogenetic relationships across recognised biogeographic regions. This more rigorous testable method produces reduced area cladograms and has been used to identify repeated patterns of biogeographic vicariance events to explain present day distribution patterns in Leptoceridae (Yang & Morse, 2000; Morse & Yang, 2002).

Present distribution and main areas of endemism

The present-day distribution of Trichoptera is nearly cosmopolitan, with only the Polar Regions and small islands remote from continents being excluded. The larvae are almost always aquatic and the adults seldom move far from the water-source on which they are dependent for production of future generations.

The origin and early diversification of Trichoptera are currently considered to have occurred in the early Mesozoic prior to the breakup of Gondwana (see discussion above, Morse, 1997; Ivanov, 2002). Wiggins (1984) noted the uneven distribution pattern of Trichoptera families in the world, with distinct northern and southern hemisphere differences being particularly discernible in the Integripalpia. This pattern reflects the Mesozoic split of the land masses separated by the Tethys Ocean in the North and a series of epicontinental seas in the South. Long isolation between the northern and southern continents could have led to the parallel evolution of separated ancestors of each of the major phylogenetic lineages, so there are southern counterparts of many northern families, forming pairs or triads: Phryganeidae (northern)–Plectrotarsidae (southern); Lepidostomatidae (northern, Laurasian-Oriental)–Oeconesidae (southern, Gondwana-Australian)–Pisuliidae (southern, Gondwana-African); Rhyacophilidae (northern)–Hydrobiosidae (southern). The Hydrobiosidae are especially notable: the family originated in the northern landmasses, then penetrated the southern continents where it evolved into several species-rich lineages, while it became extinct on the northern continents except for a few secondarily migrated species of the genera *Apsilochorema* in the East Palearctic and *Atopsyche* in the West Palearctic regions (Schmid, 1989).

Note should be taken that the present-day distribution of Trichoptera presents a snap-shot in a geological time scale of a continuously changing pattern driven by two major processes; a slow process of evolution, and a more rapid process of vicariance and dispersal moving and mixing of the different faunal elements. The Pleistocene has seen several periods of glaciation when cooling and increased aridity caused rainforests to be reduced to small isolated patches and the great lakes in Africa nearly dried out. This increased aridity would have reduced the suitability of large areas for colonisation by Trichoptera and would have created many small refuge areas for both warm and cold adapted species. The glacial periods were followed by interglacials when temperatures were often considerably warmer than present-day ones. Ocean levels rose by more than 10 m higher than present-day levels and large areas became suitable for colonisation by Trichoptera.

Mobility and dispersal capacity differ from species to species. A number of species in the genus *Drusus* are endemic to specific mountain valleys in the Balkans, whereas others are widespread over Europe and Asia (Malicky, 1979, 1983). The radiation of certain genera resulting in the formation of large concentrations of endemic species in small regions—as seen in the genus *Drusus* in mountain valleys in the Balkans, Apataniidae in Lake Baikal, and *Athripsodes* in the Cape Fold mountains of South Africa—can be considered to have been a result of recent speciation events with ensuing limited dispersal. This was described by Mayr (1942) as "explosive speciation" that resulted in the formation of flocks of closely related species and was inferred for the genus *Drusus* by Malicky (1979). Malicky (1983) proposes a Dinodal biome for explaining the restricted distribution of specialised mountain-stream Trichoptera that could have survived several glacial and interglacial epochs because mountain stream conditions during these periods were relatively stable when compared to lowland areas that show much larger fluctuations of temperature regimes.

The present day distribution of Hydroptilidae shows that generic and species proliferation in this family has occurred mostly in the tropical regions. All fossil records of hydroptilids from amber resins date back to the late Cretaceous from regions that were in the warm to hot belts of the corresponding epochs (Eskov et al., 2004). They are not found in any of the fossil amber from the cool Sibero-Canadian palaeofloristic region (Meyen, 1987). Hydroptilidae do also not track a temperate Gondwana distribution which is revealed by the paucity of genera in the temperate areas of the Australian and Neotropical Regions and the lack of any transoceanic relationships in this family. Hydroptilidae are conservative as regards their dispersal capacity this is borne out by the present-day restriction of the genera *Agraylea* and *Palaeagapetus* to the Holarctic region since the late Cretaceous, as evidenced from fossil resins (Eskov et al., 2004).

Knowledge on the world distribution of Trichoptera is unevenly skewed, with some regions very well known and others hardly explored, a measure of the present state of knowledge is nevertheless presented. The database prepared from the TWC (Morse, 1999) is summarised in Table 1. The strength of association of the 714 genera and subgenera of Trichoptera between each of the seven selected biogeographical regions was assessed through a two-way regional comparison using Sorensen's coefficient of biotic similarity [$SC = 2a/(2a + b + c)$, where a is the number of genera/subgenera common between two regions, b the number of genera/subgenera unique to first region, c the number of genera subgenera unique to second region] (Table 2).

The highest value for Sorensen's coefficient, and thus the greatest regional generic/subgeneric similarity, is between the Oriental and East Palaearctic Regions which shared 111 taxa. This is followed by the East Palaearctic and West Palaearctic Regions sharing 85 taxa. Only one other association (East Palaearctic and Nearctic) is above 0.4 with 63 taxa shared between these two regions.

The seven selected regions for the TWC, essentially represent an artificially imposed biogeography for the Trichoptera. There is clearly a temperate (Chile and Patagonia) and tropical (Brazil and Argentina) Gondwana region for the Neotropical realm, as indicated by many of the Trichoptera families. The Nearctic also shows a strong link on the western side with the Eastern Palaearctic. The similarity of NW American and NE Asian faunas supports the concept of " Angarida/Angaraland" or "Beringia" as a special faunistic region that existed in the past. This was evidently an area of rapid faunistic exchanges in times immediately preceding

Table 2 Sorensen's coefficient produced from a two-way analysis of the relative affinities of 714 Trichoptera genera/subgenera for the seven major biogeographical regions covered

	AT	WP	EP	NA	NT	OL
WP	0.23					
EP	0.28	0.49				
NA	0.17	0.35	0.46			
NT	0.12	0.09	0.10	0.32		
OL	0.33	0.34	0.58	0.31	0.12	
AU	0.20	0.14	0.19	0.15	0.11	0.22

AT, Afrotropical; WP, West Palaearctic; EP, East Palaearctic; NA, Nearctic; NT, Neotropical; OL, Oriental; AU, Australasian

glaciation or shortly after glacial meltdown, when the ocean level was sufficiently low to expose the Beringian landbridge along the Bering Strait (Ross, 1967; Levanidova, 1982; Wiggins & Parker, 2002). The corresponding link between the East Nearctic and West Palearctic is less evident because the continental break has led to an increasing distance and longer period of isolation across the Atlantic Ocean. The recent glaciation events also greatly altered the European and American faunas. The most evident present-day faunal relations are in the Arctic regions revealing a Circumboreal type of distribution. There is significant asymmetry in Trichoptera distribution in northern Europe showing distinct penetration of the cold-adapted species from Asia to the boreal regions spreading from the Urals to Fennoscandia (Spuris, 1986, 1989).

The present-day distribution of families in the seven selected biogeographical regions gives a somewhat less refined analysis but adequately reveals major patterns. There are six endemic families in the Australian Region making it the region with the highest rate of endemism (Fig. 3). The Afrotropical Region comes second with four endemic families, three of which are restricted to the south-western region of South Africa or Madagascar and are considered to be relict populations from temperate Gondwana (Scott & de Moor, 1993). Species in the remaining family, Pisuliidae, are specialised ecological shredders that have populations confined to patches of coastal or montane rainforest in the central and southern half of the African continent and in rain forests in Madagascar. The Neotropical and Nearctic Regions each have one endemic family. All 12 families endemic to any region have low numbers of species; the highest number recorded is 22 species for the Anomalopsychidae. Six additional families share their distribution over two biogeographical regions. Four families are shared between the Neotropical and Australian Regions and two families have species in the East Palaeartic and Oriental Regions. The family Atriplectididae has representative species in the Australian, Afrotropical and Neotropical Regions (Fig. 3, Table 1).

An assessment of the number of endemic genera/subgenera in each region also shows the highest endemism to be in the Australian Region with a figure of 73%, followed by the Neotropical Region with 69% (Fig. 3). The Afrotropical Region (43%) and West Palaearctic Region (41%) also have quite high endemism at generic/subgeneric levels. The Oriental Region with the highest number of genera/subgenera (204) has a similar number of endemics to the West Palaearctic Region, but this only represents 33% of its generic/subgeneric component.

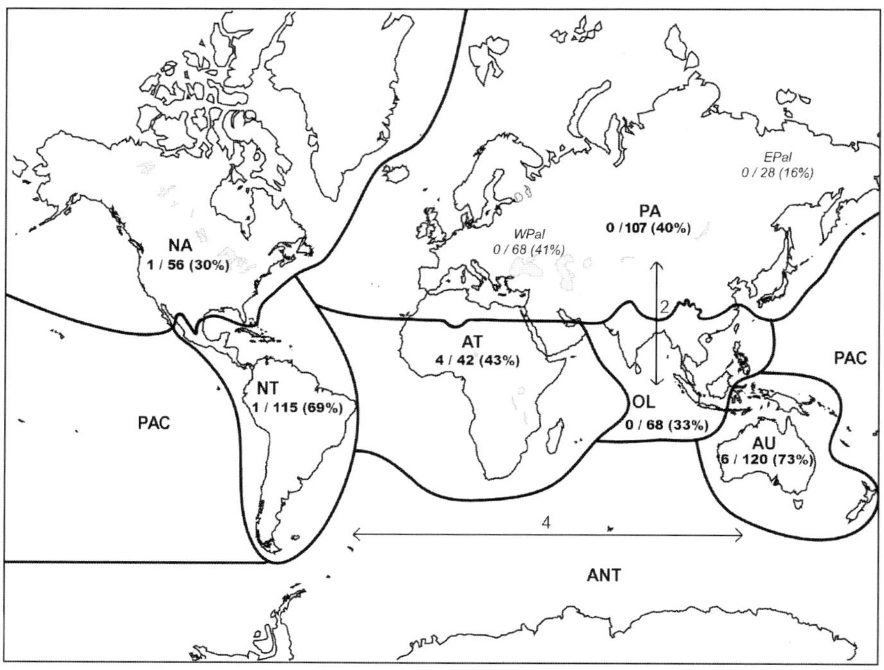

Fig. 3 The number of recorded endemic families and genera plus subgenera (End. Fam./End gen. (% of total genera)) for each major biogeographical region. Arrowed figures indicate the number of families found in two biogeographical regions (two families are common to EPal and OL and four families are common to NT and AU)

One striking aspect is that the areas with the greatest species diversity are not necessarily those with the greatest higher-level taxon diversity. The Neotropical Region, although second highest in species diversity, is only fourth in generic/sub-generic diversity after the Nearctic and East Palaearctic. The subfamily Leptocerinae is most diverse in generic diversity in the Afrotropical Region, but records the highest number of species in the Oriental Region (de Moor 2002a). These discrepancies might indicate rapid speciation in recent times after the formation of rainforest ecosystems in the Late Caenozoic.

Areas with great concentrations of endemic species or with high numbers of species (hotspots of endemism or hotspots of species diversity) can be identified as being in temperate and tropical mountain regions or in areas of high rainfall. The northern Andes, the Amazon basin and mountains of southern and southeastern Brazil are represented in the Neotropics (Flint et al., 1999); the mountain fold belt of the south western and southern Cape in South Africa, the central African highlands and West Africa are represented in the Afrotropics. Numerous endemics are found in the Mediterranean area including Turkey, the southern Caucasus and the Alps (Malicky, 1983, 1986; Malicky & Sipahiler, 1993) as well as the Iberian peninsula in WP. There are smaller centres of endemism in Middle Asia (from Pamir to the Tyan-Shan mountain ranges); in the Altai Mountains, mountain ridges of the Far East and around Lake Baikal. A large amount of endemism occurs in Korea, China and Japan, although the Chinese biogeography is too poorly studied to provide significant detail of smaller regions. Large numbers of local endemic species are found throughout SE Asia, with significant concentrations along the Himalayan slopes and the meridional mountain ridges of Indochina, and also on the islands of Indonesia. The endemism here correlates with the monsoon intensity providing the humid conditions necessary for Trichoptera development.

Repetitive patterns of shared taxa between biogeographical regions suggest possible centres of origin, routes of dispersal or vicariant events. These are apparent from the database of generic/subgeneric distribution and species abundance prepared from the TWC (see internet documents and Morse, 1999). In summarising the foregoing discussions the following biogeographical patterns are proposed:

1. Temperate Gondwana link between Australian and Neotropical Regions with some relict survivors at superfamily level (Sericostomatoidea) also found in the southernmost part of Africa and Madagascar. Several families and closely related genera show this pattern.
2. A shared distribution pattern between the Oriental and Holarctic Regions. This has two alternative routes of dispersal both indicated by species abundances of closely related taxa; the first via the East Palaeartic to the Nearctic Region, and the second to the West Palaearctic Region (the Rhyacophilidae indicate both these patterns).
3. A tropical Gondwana link between Africa, Madagascar and India (genera of Hydropsychidae, Hydroptilidae and Dipseudopsidae). There are also a few instances where Brazilian linkages with West Africa can be discerned with some taxa even extending to the Australian region (family Atriplectididae, genus *Ecnomus* and some Leptocerinae and Hydroptilidae genera).
4. A two-way exchange of Neotropic and Nearctic faunas, indicated by some shared Hydroptilidae and Leptoceridae genera.
5. A more recent link between the Australian and Oriental Region with a number of taxa crossing the Wallace/Weber lines. This is mostly of tropical origin into temperate Australia, but there are some temperate Triplectidinae that have apparently moved into the Oriental realm more recently.

The discovery of Eocene Baltic Amber fossil records of Helicophidae and psyllobetine Hydrobiosidae—groups presently restricted to Australia and New Zealand (Australian Region) and Chile (Neotropical Region)—suggests a very different alternative hypothesis. Possibly these presently-restricted, temperate Gondwana taxa (Southern Hemisphere) had a much wider distribution, previously extending into the Northern Hemisphere, and became extinct in most of their former range (Botosaneanu & Wichard, 1983). The Northern Hemisphere continents in the last 3 million years (Pleistocene) have been exposed to glaciation events which have largely left the southern hemisphere continents unscathed. It is likely that the glaciations led to the local and regional extinction of many widespread taxa that were unable to disperse. Relict populations probably survived in more hospitable climates. Baltic and Dominican Amber fossils also reveal that certain taxa (Leptoceridae: Triplectidinae) were more widespread than indicated by present-day records (Botosaneanu & Wichard, 1981). Some populations of the more plesiomorphic Philopotamidae and Hydrobiosidae—found presently only in temperate Gondwana regions—may therefore also represent a relict fauna of previously widespread taxa that became extinct in other regions in the Northern Hemisphere (Eskov, 1984).

From the above discussion it is suggested that an alternative map highlighting regions on present day continents reflect a somewhat different biogeographical pattern to the generally accepted seven biogeographical Regions. Detailed distribution patterns of Trichoptera and possibly other amphibiotic insects cannot be expounded here, but what is clear is that Gondwana needs to be divided into several more closely related biogeographical regions reflecting a temperate southern region incorporating Australia, New Zealand, Patagonia and southern Chile and possibly the southern tip of Africa and southernmost part of Madagascar; and a tropical region incorporating parts of the Neotropical, African, Madagascar and the Indian-part of the Oriental regions. The proposed map highlighting distinct biogeographic regions and indicating close relationships illustrates these patterns (Fig. 4).

The Northern Holarctic Region also presents a different pattern with at least five clearly discernible divisions (Fig. 4). Thus, the distribution of Hydrobiosidae in S. America shows great extratropical diversity in Patagonia similar to that in New Zealand and Southern Australia, whereas the slopes of the Andes further to the north have increasingly less diversity (Schmid, 1989). The distribution of Trichoptera in Central Asia shows a distinct gap in Eastern Iran and Western Afghanistan between the generally western European fauna of Iran and the Oriental fauna of Afghanistan (Malicky, 1986). The distribution of the "Beringian" species encompasses Eastern Siberia, but generally avoids Western Siberia (Levanidova, 1982).

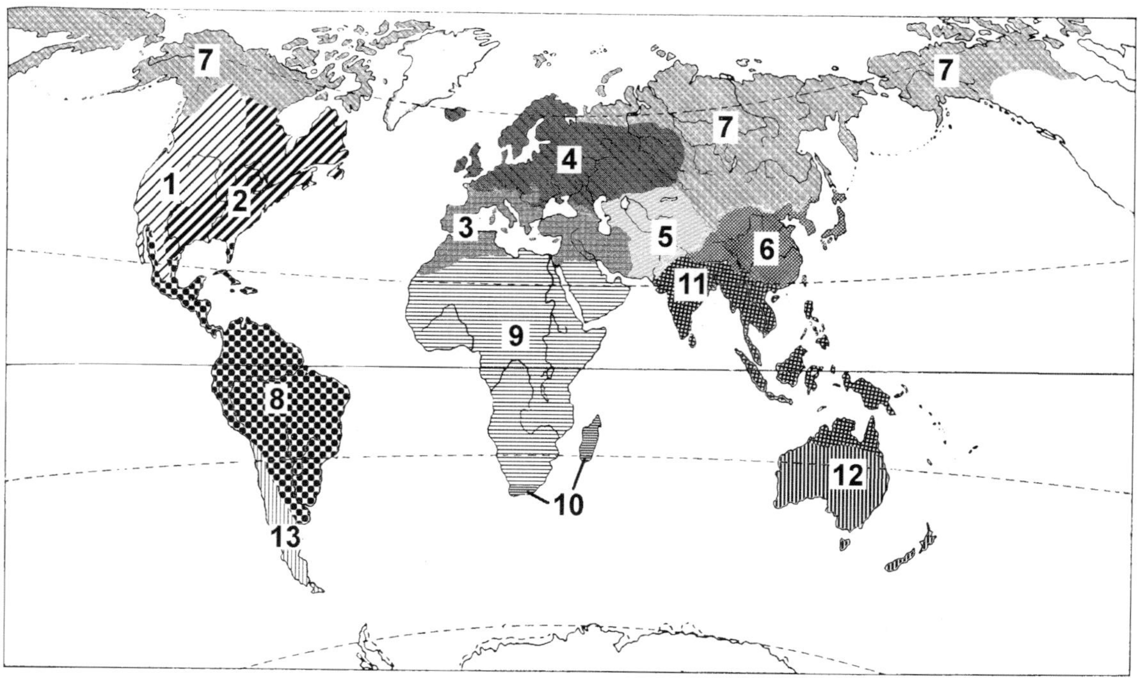

Fig. 4 Preliminary outline for Trichoptera biogeographical regions. 1. West Nearctic, 2. East Nearctic, 3. West Palaearctic, 4. Boreal Palaearctic, 5. South Palaearctic, 6. East Palaearctic, 7. Beringian, 8. Neotropical, 9. Afrotropical, 10. Cape South African, 11. Oriental, 12. Australasian, 13. Patagonian. Similar colours indicate close biogeographical relationships

Human related issues

Trichoptera are abundant in all types of natural aquatic ecosystems. They are apneustic and rely on dissolved oxygen for respiration. With a high diversity of species with both case- and shelter-constructing larvae, they are useful indicators of organic pollution. Trichoptera have been used extensively in biomonitoring assays with indicator species, selected communities or assemblages of species or more broadly based family level identifications of species being used to assess the health status of aquatic ecosystems. Trichoptera, in conjunction with other aquatic insect orders, have also been used to assess aquatic biodiversity EPT (Ephemeroptera, Plecoptera and Trichoptera) or ETS (Ephemeroptera, Trichoptera, Simuliidae: Diptera) and habitat diversity (Hannaford & Resh, 1995; Hewlett, 2000; de Moor, 2002b).

From an ecological perspective, Trichoptera are important processors of organic matter and provide a keystone taxon in the development of the river continuum concept (RCC) (Vannote et al., 1980). As processors of organic matter, collectively known as the functional feeding groups (FFG) of animals, they display the full array of feeding modes (Cummins, 1973). In lotic water filter-feeding, shelter-constructing species are important predators of blackfly larvae and help keep population levels of pest species at acceptable levels (de Moor, 1992). In some instances excessive construction of shelters by collector/filter-feeding larvae have caused the clogging of pipes under bridges and storm-water pipes, creating problems when flooding occurs (McCafferty, 1981). Trichoptera larvae, pupae and adults also form an important link in the food chain and they have also been used extensively by trout fishing enthusiasts as models for "flies" (McCafferty, 1981).

Development and man's activities have considerably reduced Trichoptera species diversity and the need for remedial measures to prevent further damage to ecosystems and loss of Trichoptera species have been discussed by Botosaneanu (1981) and Ivanov (2000). Red data lists for certain countries have been produced and can be used to manage ecosystems to protect vulnerable or rare species of Trichoptera. Anthropogenic development threatens biodiversity and the value of Trichoptera as important functional components of aquatic ecosystems, indicator species of deteriorating conditions and custodians of environmental protection are realised.

Acknowledgements We would like to thank Dr. Koen Martens and Ms. Estelle Balian for organising and facilitating the workshop where discussions of the project and the sharing of the views of all the disciplines presented by the invited coordinators was made possible. The Royal Belgian Institute for Natural Sciences is thanked for providing the funding for air fare and accommodation that allowed F. C. de Moor to attend the meeting in Mechelen in October 2005. Prof. John Morse (Clemson University, USA) and Dr. Alice Wells (Australian Biological Resources Survey) are thanked for their critical evaluation of the manuscript and for providing information and thanks are also due to two anonymous reviewers for suggestions and corrections. Irene de Moor is thanked for editing the manuscript in various stages. V. D. Ivanov is thankful to the Russian Federal Programme for support of the leading Scientific Schools project NSH-7130.2006.4. The Directorate of Eastern Cape Museums is thanked for supporting F. C. de Moor in the research into aquatic biodiversity at the Albany Museum.

References

Ansorge, J., 2002. Revision of the "Trichoptera" described by Geinitz and Handlirsch from the Lower Toarcian of Dobbertin (Germany) based on new material. In Mey, W. (ed.), Proceedings of the 10th International Symposium on Trichoptera. Nova Supplementa Entomologica. Goecke & Evers, Keltern, 15: 55–74.

Botosaneanu, L., 1981. Ordo Trichoptera et Homo insapiens. In Moreti, G. P. (ed.) Proceedings of the 3rd International Symposium on Trichoptera. Dr W. Junk Publishers, The Hague, 11–19.

Botosaneanu, L. & W. Wichard, 1981. Trichoptera from the Baltic Amber. In Moreti, G. P. (ed.) Proceedings of the 3rd International Symposium on Trichoptera. Dr W. Junk Publishers, The Hague, 31–32.

Botosaneanu, L. & W. Wichard, 1983. Upper Cretaceous Siberian and Canadian amber caddisflies (Insecta: Trichoptera). Bijdragen tot de Dierkunde 53: 187–217.

Cummins, K. W., 1973. Trophic relations of aquatic insects. Annual Review of Entomology 18: 183–206.

de Moor, F. C., 1992. Parasites, generalist and specialist predators and their role in limiting the population size of blackflies and in particular *Simulium chutteri* Lewis (Diptera: Simuliidae) in the Vaal River, South Africa. Annals of the Cape Provincial Museums (Natural History) 18(13): 271–291.

de Moor, F. C., 2002a. An assessment of the global distribution of Leptocerinae (Trichoptera) and use of larval characters for determining phylogenetic relationships. In Mey, W. (ed.), Proceedings of the10th International Symposium on Trichoptera. Nova Supplementa Entomologica. Goecke & Evers, Keltern, 15: 293–308.

de Moor, F. C., 2002b. Shortcomings and advantages of using rapid biological assessment techniques for the purpose of characterising rivers in South Africa. Verhandlungen,

Internationale Vereinigung fur theoretische und angewandte Limnologie. Stuttgart 28: 651–662.
Eskov, K. Yu., 1984. Continental drift and the problems of historical biogeography. In Tchernov, Yu. I. (ed.), Faunogenesis and phylocenogenesis. Nauka Publishing House, Moscow, 24–92, (in Russian).
Eskov, K. Yu. & I. D. Sukatsheva, 1997. Geographical distribution of the Paleozoic and Mesozic caddisflies (Insecta: Trichoptera). In Holzenthal, R. W. & O. S. Flint (eds), Proceedings of the 8th International Symposium on Trichoptera. Ohio Biological Survey, Columbus, Ohio, 95–98.
Eskov, K. Yu., V. D. Ivanov, I. D., Sukatsheva, & A. Wells, 2004. Geographic history of the family Hydroptilidae (Trichoptera). In Materialy. II Vserossiyskiy simpozium po amfibioticheskim i vodnym nasekomym. Voronezh, 40–48 (in Russian).
Fischer, F. C. J., 1960–1973. Trichopterorum Catalogus. Amsterdam: Nederlandsche Entomologische Vereeniging, Amsterdam, 15 Vols. & index.
Flint, O. S., Jr., R. W. Holzenthal & S. C. Harris, 1999. Catalog of the Neotropical Caddisflies (Insecta: Trichoptera). Ohio Biological Survey, Columbus, Ohio, iv + 239 pp.
Frania, H. E. & G. B. Wiggins, 1997. Analysis of morphological and behavioural evidence for the phylogeny and higher classification of Trichoptera (Insecta). Royal Ontario Museum Life Sciences Contributions 160. ROM Publications in Life Sciences, Ontario, 68 pp.
Gibon, F.-M., D. Randriamasimanana, & Z. Randriambelo, 2001. Ordre Trichoptera. In Elouard, J.-M. & F.-M. Gibon (eds), Biodiversité et biotypologie des eaux continentales de Madagascar, 141–195.
Hannaford, M. L. & V. H. Resh, 1995. Variability in rapid-bioassessment surveys and habitat assessments in a Northern California stream. Journal of the North American Benthological Society 14: 430–439.
Hewlett, R., 2000. Implications of taxonomic resolution and sample habitat for stream classification at a broad geographic scale. Journal of the North American Benthological Society 19: 352–361.
Ivanov, V. D., 1997. Rhyacophiloidea: A paraphyletic taxon. In Holzenthal, R. W. & O. S. Flint (eds), Proceedings of the 8th International Symposium on Trichoptera. Ohio Biological Survey, Columbus, Ohio, 189–193.
Ivanov, V. D., 2000. Changes of fauna and the problems of Trichoptera protection. In Fauna, problems of ecology, ethology and physiology of the amphibiotic and aquatic insects of Russia. Materials of VI All-Russian Trichopterological Symposium and I All-Russian Symposium on the amphibiotic and aquatic insects. Voronezh State University Voronezh, 19–23.
Ivanov, V. D., 2002. Contribution to the Trichoptera phylogeny: New family tree with consideration of Trichoptera-Lepidoptera relations. In Mey, W. (ed.), Proceedings of the 10th International Symposium on Trichoptera. Nova Supplementa Entomologica. Goecke & Evers, Keltern, 15: 277–292.
Ivanov, V. D. & I. D. Sukatsheva, 2002. Trichoptera (Phryganeida). In Rasnitsyn, A. P. & L. J. Quicke (eds), History of Insects. Kluwer Academic Publishers, Dordrecht, Boston, London, 199–220.
Kjer, K. M., R. J. Blahnick & R. W. Holzenthal, 2001. Phylogeny of Trichoptera (Caddisflies): Characterization of signal and noise within multiple datasets. Systematic Biology 50(6): 781–816.
Kjer, K. M., R. J. Blahnick & R. W. Holzenthal, 2002. Phylogeny of caddisflies (Insecta, Trichoptera). Zoologica Scripta 31: 83–91.
Levanidova, I. M., 1982. Amphibioticheskie nasekomye gornykh oblastey Dal'nego Vostoka SSSR. Faunistika, Ecologiya, Zoogeograpfiya Ephemeroptera, Plecoptera i Trichoptera. Leningrad, Nauka. Amphibiotic insects of the mountain regions of the Far East of the USSR. Faunistics, ecology, zoogeography of Ephemeroptera, Plecoptera, and Trichoptera, (in Russian).
McCafferty, W. P., 1981. Aquatic Entomology. The Fishermen's and Ecologists' Illustrated Guide to Insects and Their Relatives. Jones & Bartlett Publishers, Inc., Boston, xv + 448 pp.
Malicky, H., 1979. Discussion generale, 2. Limnofaune. In Zoogeography and ecology of Greece and neighbouring regions. Biologia Gallo-Hellenica, 8: 379–386.
Malicky, H., 1983. Chorological patterns and biome types of European Trichoptera and other freshwater insects. Archiv fur Hydrobiologie, 96(2): 223–244.
Malicky, H., 1986. Die Köcherfliegen (Trichoptera) des Iran und Afghanistans. Zeitschrift der Arbeitsgemeinschaft Österreichs Entomologen, 38: 1–16.
Malicky, H., 1993. First speculation on the size of areas and the number of species of caddisflies (Trichoptera) in southeastern Asia. In Otto, C. (ed.), Proceedings of the 7th International Symposium on Trichoptera, 92.
Malicky, H. & F. Sipahiler, 1993. Köcherfliegen (Trichoptera) aus der Türkei, mit bemerkungen zu weiteren mediterranen Köcherfliegen. – Mitteilungen der Schweizerischen entomologischen gesellschaft, 66: 457–478.
Mayr, E., 1942. Systematics and the Origin of Species. Columbia University Press, New York.
Meyen, S. V., 1987. Fundamentals of Palaeobotany. Chapman and Hall, London, 432 pp.
Morse, J. C., 1997. Phylogeny of Trichoptera. Annual Review of Entomology 42: 427–450.
Morse, J. C. (ed.), 1999. Trichoptera World Checklist. http://entweb.clemson.edu/database/trichopt/index.htm, effective 22 May 1999, updated 28 July 2000, 8 January 2001.
Morse, J. C., 2003. Trichoptera (Caddisflies). In Resh, V. H. & R. T. Carde (eds), Encyclopedia of Insects. Elsevier, 1145–1151.
Morse, J. C. & L. Yang, 2002. Phylogeny, classification, and historical biogeography of world species of Mystacides (Trichoptera: Leptoceridae), with a new species from Sri Lanka. In Mey, W. (ed.) Proceedings of the10th International Symposium on Trichoptera. Nova Supplementa Entomologica. Goecke & Evers, Keltern, 15: 173–186.
Nishimoto, H., M. Nakabou & K. Tanida, 2002. Identification of Japanese *Cheumatopsyche* (Trichoptera: Hydropsychidae) by analysis of cuticular hydrocarbons. In Mey, W. (ed.), Proceedings of the 10th International Symposium on Trichoptera. Nova Supplementa Entomologica. Goecke & Evers, Keltern, 15: 205–210.
Ross, H. H., 1956. Evolution and Classification of the Caddisflies. University of Illinois Press, Urbana.

Ross, H. H., 1964. Evolution of caddisworm cases and nets. American Zoologist 4: 209–220.

Ross, H. H., 1967. The evolution and past dispersal of the Trichoptera. Annual Review of Entomology 12: 169–206.

Schmid, F., 1984. Essai d'evaluation de la faune mondiale des Trichoptères. In Morse, J. C. (ed.), Proceedings of the 4th International Symposium on Trichoptera. Dr. W. Junk Publishers, Series Entomologica 30. The Hague, 337.

Schmid, F., 1989. Les Hydrobiosides (Trichoptera, Annulipalpia). Bulletin de l'Institut Royal des Sciences Naturelles de Belgique, Entomologie, Supplément 59. 154 pp.

Scott, K. M. F. & F. C. de Moor, 1993. Three recently erected Trichoptera families from South Africa, the Hydrosalpingidae, Petrothrincidae and Barbarocthonidae (Integripalpia: Sericostomatoidea). Annals of the Cape Provincial Museums (Natural History) 18(14): 293–354.

Spuris, Z., 1986. Origin of the caddis flies fauna of the East Baltics. Latvijas Entomologs 29: 45–59.

Spuris, Z., 1989. Synopsis of the fauna of the Trichoptera of the USSR. Riga Zinatne, 89 p.

Steyskal, G. C., 1972. The meaning of the term 'sibling species'. Systematic Zoology 21(4): 446.

Ulmer, G., 1912. Die Trichopteren des Baltischen Bernsteins. Beitrage zur Naturkunde Preussens. Physikalisch-Okonomischen Gesellschaft zu Konigsberg 10: 1–380.

Vannote, R. L., G. W. Minshall, K. W. Cummins, J. R. Sedell & C. E. Cushing, 1980. The river continuum concept. Canadian Journal of Fisheries and Aquatic Science 37: 130–137.

Wallace, A. R., 1876. The Geographical Distribution of Animals: With a Study of the Relations of Living and Extinct Faunas as Elucidating the Past Changes of the Earth's surface, Vol. 1, 503 pp., Vol. 2, 607 pp., Macmillan, London.

Ward, J. B., 2003. Checklist of the New Zealand Trichoptera (caddis). National Center for aquatic biodiversity and security. http://www.niwa.co.nz/ncabb.

Weaver, J. S., III, 1984. The evolution and classification of Trichoptera, with a revision of the Lepidostomatidae and a North American synopsis of this family. PhD thesis, Clemson University, Clemson, SC.

Weaver, J. S., III, 1992a. Remarks on the evolution of Trichoptera: A critique of Wiggins and Wichard's classification. Cladistics 8: 171–180.

Weaver, J. S., III, 1992b. Further remarks on the evolution of Trichoptera: A reply to Wiggins. Cladistics 8: 187–190.

Weaver, G. B., & J. C. Morse, 1986. Evolution of feeding and case-making behaviour in Trichoptera. Journal of the North American Benthological Society 5: 150–158.

Wiggins, G. B., 1984. Trichoptera, some concepts and questions. In Morse, J. C. (ed.), Proceedings of the 4th International Symposium on Trichoptera. Dr. W. Junk Publishers, Series Entomologica 30, The Hague, 1–12.

Wiggins, G. B., 1992. Comments on the phylogeny of pupation behaviour in Trichoptera: A response to weaver. Cladistics 8: 181–185.

Wiggins, G. B., 1996. Larvae of the North American Caddisfly Genera (Trichoptera), 2nd edn. University of Toronto Press, Ontario, 457 pp.

Wiggins, G. B., 2004. Caddisflies, the Underwater Architects. University of Toronto Press, Toronto, Buffalo, London, 292 pp.

Wiggins, G. B. & C. R. Parker, 2002. Caddisflies (Trichoptera) of the Yukon, with analyis of the Beringian and Holarctic species of North America. In Danks, H. V. & J. A. Downes (eds), Insects of the Yukon. Biological survey of Canada (Terrestrial Arthropods), Ottawa, 787–866.

Wiggins, G. B. & W. Wichard, 1989. Phylogeny of pupation in Trichoptera, with proposals on the origin and higher classification of the order. Journal of the North American Benthological Society 8: 260–276.

Yang, L. & J. C. Morse, 2000. Leptoceridae (Trichoptera) of the People's Republic of China. Memoirs of the American Entomological Institute 64(1–7): 1–311.

Hydrobiologia (2008) 595:409–417
DOI 10.1007/s10750-007-9035-z

FRESHWATER ANIMAL DIVERSITY ASSESSMENT

Global diversity of dobsonflies, fishflies, and alderflies (Megaloptera; Insecta) and spongillaflies, nevrorthids, and osmylids (Neuroptera; Insecta) in freshwater

Matthew R. Cover · Vincent H. Resh

Abstract The insect orders Megaloptera and Neuroptera are closely related members of the superorder Neuropterida, a relict lineage of holometabolous insects that also includes the Raphidoptera. Megaloptera, composed of the families Sialidae and Corydalidae (including subfamilies Chauliodinae and Corydalinae), has fully aquatic larvae that occur in a wide variety of lotic and lentic habitats, including temporary streams. In total, 2 of 17 families of Neuroptera have aquatic larvae: Nevrorthidae live in the benthos of fast-flowing streams and Sisyridae reside on freshwater sponges. A third family of Neuroptera, Osmylidae, contains some water-dependent species that reside under leaves and rocks along the margins of waterbodies. We recognize 328 extant, described species of Megaloptera (composed of 116 species of Chauliodinae, 131 species of Corydalinae, and 81 species of Sialidae) and 73 species of aquatic Neuroptera (composed of 12 species of Nevrorthidae and 61 species of Sisyridae). Additionally, we estimate that 45 species of Osmylidae are water-dependent, although the ecology of this group is poorly understood. Chauliodinae and Corydalidae are both found in the New World, the Oriental region, and South Africa, but are absent from Europe, the Middle East, Central Asia, tropical Africa, and boreal regions. Chauliodinae is quite speciose in Australia, whereas Corydalinae is absent. Sialidae is most speciose in temperate regions, and is absent from tropical Africa and portions of the Oriental region. Sisyridae and Osmylidae are nearly cosmopolitan, but the relict family Nevrorthidae is limited to Japan, the Mediterranean, and Australia. The discovery of many new species in recent years, particularly among Corydalidae in the Neotropics and China, suggests that our knowledge of aquatic neuropterid diversity is far from complete.

Keywords Biodiversity · Species richness · Megaloptera · Neuroptera · Neuropterida

Guest editors: E. V. Balian, C. Lévêque, H. Segers and K. Martens
Freshwater Animal Diversity Assessment

M. R. Cover (✉) · V. H. Resh
Department of Environmental Science, Policy, & Management, University of California, Berkeley, 137 Mulford Hall, Berkeley, CA 94720-3114, USA
e-mail: mcover@nature.berkeley.edu

Introduction

The insect orders Megaloptera and Neuroptera (sensu stricto), along with Raphidoptera, are closely related members of the superorder Neuropterida, a relict lineage of holometabolous insects (Grimaldi & Engel, 2005). Neuropterida are primarily a terrestrial group; only the 2 families of Megaloptera and 2 of 17 families of Neuroptera have species with aquatic larval stages. Neuropterida contains remarkably heterogeneous taxa, but the relatively low-species richness ($\sim$6,500 species across three orders) and relict distributions suggest that the group was much more speciose and cosmopolitan in the past (Aspöck, 2002).

A general account of the biology of the Megaloptera and/or Neuroptera is given by New and Theischinger (1993) and New (1986, 1989). The biology of only a few species of Megaloptera and aquatic Neuroptera have been described in detail, with most of the information published on species from Europe, North America, and Japan.

The Megaloptera is composed of two extant families, the Corydalidae and the Sialidae (alderflies). Corydalidae is further subdivided into the subfamilies Corydalinae (dobsonflies) and Chauliodinae (fishflies), although some have regarded these taxa as distinct families (e.g., Aspöck et al., 1980; Theischinger, 1983). Dobsonfly larvae are commonly known as hellgrammites (although this term is sometimes applied to larval Chauliodinae or, in a more general sense, to all large benthic insects used by anglers as bait). Megalopteran larvae have been reported in a wide variety of aquatic habitats. Most Corydalidae are found in fast-flowing riffles under gravel and cobble, but some corydalids are found on snags and sand in low-gradient streams. Some species of Chauliodinae have been reported from lentic habitats, including ponds and swamps (e.g., Dolin & Tarter, 1981). Sialidae are also found in a wide range of freshwater habitats, both lotic and lentic, and are usually associated with soft sediments. Megalopteran larvae have also been reported from several remarkable locations that suggest adaptations to otherwise inhospitable habitats. For example, the larvae of both Sialidae and Chauliodinae have been reported in water-filled vegetation, including tree holes and the purple pitcher plant (*Sarracenia purpurea*) (Fashing, 1994; Hamilton et al., 1996; Pittman et al., 1996). Five species of fishflies from California (USA) representing three genera (*Neohermes*, *Protochauliodes*, and *Dysmicohermes*), and two species of dobsonflies (*Corydalus luteus* and *C. affinis*) in the Neotropics have been found in dry streambeds (Evans, 1972; Contreras-Ramos, 1998). Larvae of *Neohermes* dig burrows beneath large rocks as temporary streams in California dry up, where they remain for 6 months or more until water begins to flow. Many fishfly larvae possess a pair of highly modified abdominal spiracles to facilitate aerial respiration (Fig. 1), whereas dobsonfly larvae possess ventral tufts of tracheal gills. Co-occurring species of Megaloptera can be ecologically separated by differences in larval habitat related to substrate or flow conditions (Evans, 1972; Hayashi, 1989).

Megaloptera larvae are generalist predators of small invertebrates, and have also been observed to scavenge and cannibalize. Larvae of *Protohermes grandis* remained relatively stationary under stones in riffles for many days at a time, presumably acting as sit-and-wait predators (Hayashi & Nakane, 1989).

The life cycles of Megaloptera are usually 1–2 years, although Chauliodinae in cold or seasonally dry streams may live up to 5 years (Evans, 1972). Larvae occupying permanent water bodies crawl to the banks of the stream to build pupal chambers in the soil or leaf litter, while larvae in temporary streams pupate under large stones after streamflow ceases.

Adult Megaloptera live for approximately 1 week and do not feed, although some have been reported to drink water or sugar solutions. Adult Corydalidae are large and conspicuous, especially male dobsonflies in the genera *Corydalus* and *Acanthacorydalis* that possess enormous scimitar-like mandibles. These mandibles are of unknown function, but are likely a secondary sexual characteristic selected for male–male competition or female choice. Females oviposit

Fig. 1 (**A**) Habitus of freshwater Megaloptera: *Neohermes filicornis*, (**B**) Habitus of freshwater Neuroptera: *Sisyra vicaria*

on branches, logs, or stones overhanging the stream, so that freshly hatched larvae fall directly into the water.

The Neuroptera (lacewings) are primarily a terrestrial order. Larval neuropterans all posses sucking mouthparts composed of modified maxillae and mandibles (Fig. 1). Two families have aquatic larvae: Sisyridae (spongillaflies) and Nevrorthidae (often incorrectly spelled Neurorthidae).

Sisyridae larvae are obligate predators of freshwater sponges (Spongillidae), and possibly Bryozoans as well (Parfin & Gurney, 1956; Steffan, 1967; Resh, 1976). Like the Megaloptera, eggs are laid in small clusters on vegetation overhanging the water. Hatchling larvae enter the water and swim until they find a suitable host, where they remain until the end of the third instar (New, 1986). Pupation occurs on vegetation, and adults remain near the water for mating and oviposition. Sisyrids are usually univoltine, although multiple cohorts can be present (Resh, 1976).

Nevrorthidae, once considered part of Sisyridae, is now recognized as a distinct family (Zwick, 1967) or even suborder (i.e., Nevrorthiformia; Aspöck et al., 2001). Nevrorthid larvae have been found in the benthos of fast-flowing streams and are believed to be fully aquatic (Zwick, 1967), although the habits and morphology have not been described for many species (Dudgeon, 1999).

Although little is known about the ecology of the neuropteran family Osmylidae, species in the subfamilies Osmylinae and Kempyninae have been strongly associated with freshwater habitats (New, 1986). The most well-studied species is *Osmylus fulvicephalus*, which is widespread in Europe. Larvae are found under stones, moss, and leaf litter along the margins of streams, where they probe the mud in search of small arthropods (Ward, 1965). Although not considered aquatic or even subaquatic, diapausing larvae on the stream bank can survive immersion by rising floodwaters. Several other species of Osmylinae and Kempyninae have been found in wet or moist zones along streams (New, 1986). For the purposes of this review of freshwater biodiversity, we consider Osmylinae and Kempyninae to be water-dependent because they may be restricted to the margins of aquatic habitats. Other osmylids are known to prefer terrestrial habitats, including the Australian subfamilies Porisminae and Eidoporisminae, which are found under the bark of Eucalyptus trees located far from water (New, 1986). The speciose subfamily Stenosmylinae most likely has terrestrial larvae as well (New, personal communication).

The objectives of this chapter are to enumerate the known worldwide species richness of Megaloptera and aquatic Neuroptera, examine patterns of biogeography and endemicity, and relate these patterns to what is known of the phylogeny of the groups. The discovery of many new species in recent years, particularly among the Corydalidae in the Neotropics and China, suggests that our knowledge of aquatic neuropterid diversity is far from complete.

Species diversity

For the Megaloptera we recognize 328 extant, described species, composed of 116 species of Chauliodinae, 131 species of Corydalinae, and 81 species of Sialidae (Table 1). For the Neuroptera we recognize 73 aquatic species, including 12 species of Nevrorthidae and 61 species of Sisyridae, and 45 species of water-dependent Osmylidae. We have included all described species in the published literature, based on the latest taxonomic revisions (e.g., Contreras-Ramos, 1998; Liu & Yang, 2006a, b), species lists (e.g., Penny et al., 1997; Contreras-Ramos, 1999; Oswald, 2006), and distributional studies (e.g., Bowles, 2006). We excluded, however, several unnamed species that are only known from female specimens. Species richness values are undoubtedly underestimates of true richness, especially for the tropical and subtropical taxa.

The past quarter century has seen the fastest rate of new species descriptions of Megaloptera, most notably in China (by C. Yang, D. Yang, and X. Liu), Australia (by G. Theischinger), and the Neotropics (by A. Contreras-Ramos). Although the Megaloptera of the Holarctic are believed to be well documented, further work in tropical areas, especially south-east Asia and portions of South America, is expected to increase the known richness of Megaloptera to at least 370 species, and possibly more than 400 species. There is evidence that approximately 25 undescribed species of Megaloptera are present in China (X. Liu, personal communication) and perhaps 15 additional species are present in the Neotropics (A. Contreras-Ramos, personal communication).

Table 1 Species richness of Megaloptera and aquatic Neuroptera worldwide and by biogeographic region

	PA		NT		NA		AT		OL		AU		PAC		ANT		Total freshwater species	
	FW	WDpt	FW	WDpt	FW	WDpt	FW	WDpt	FW	WDpt	FW	WDpt	FW	WDpt	FW	WDpt	FW	WDpt
Chauliodinae	11	0	6	0	18	0	8	0	56	0	23	0	0	0	0	0	116	0
Corydalinae	5	0	54	0	4	0	1	0	71	0	0	0	0	0	0	0	131	0
Sialidae	28	0	10	0	24	0	1	0	8	0	4	0	0	0	0	0	81	0
Total Megaloptera	44	0	70	0	46	0	10	0	135	0	27	0	0	0	0	0	328	0
Nevrorthidae	9	0	0	0	0	0	0	0	1	0	2	0	0	0	0	0	12	0
Sisyridae	9	0	25	0	6	0	8	0	8	0	6	0	1	0	0	0	61	0
Osmylidae	0	16	0	4	0	0	0	0	0	9	0	15	0	0	0	0	0	45
Total Neuroptera	34	16	29	4	6	0	8	0	9	9	23	15	1	0	0	0	118	45

"WDpt" represents the number of water-dependent species. "FW" refers to the total number of freshwater including fully aquatic + water dependent species. PA: Palaearctic; NA: Nearctic; NT: Neotropical; AT: Afrotropical; OL: Oriental; AU: Australasian; PAC: Pacific & Oceanic Islands, ANT: Antarctic

China and south-east Asia are major centers of species richness of Corydalinae and Chauliodinae (Table 1, Fig. 2). In China alone, there are 40 described species of Corydalinae and 41 species of Chauliodinae, the majority of these from the Oriental region of China. Additional species have been

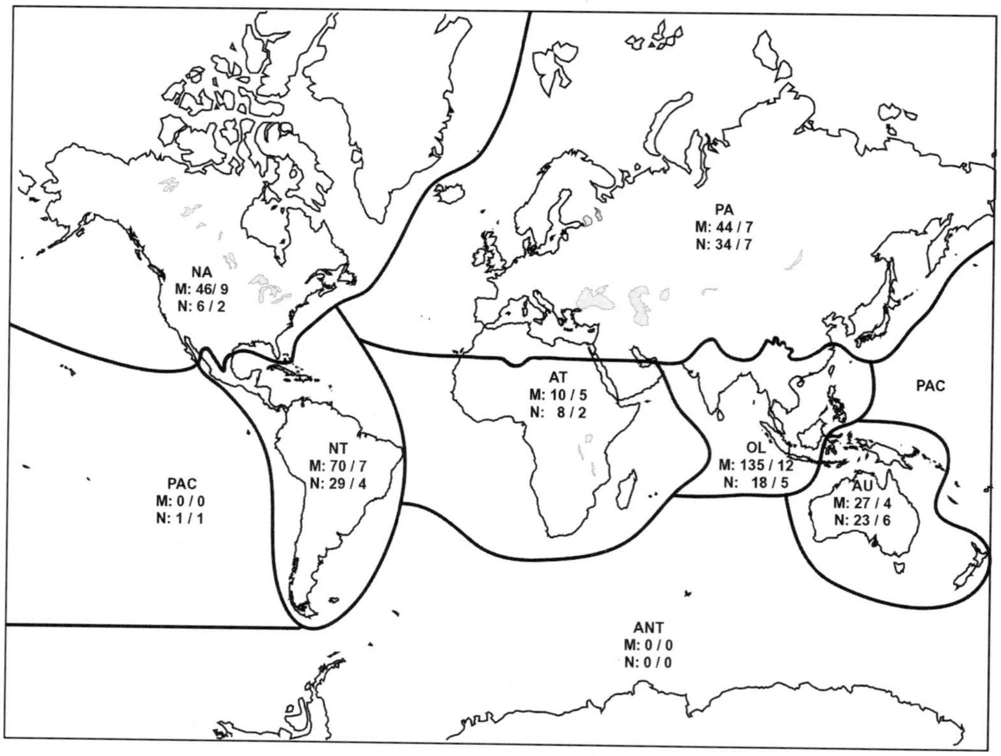

Fig. 2 Distribution of freshwater (aquatic + water dependent) Megaloptera (M) and Neuroptera (N) species and genera by zoogeographical region (species number/genus number). PA—Palaearctic; NA—Nearctic; NT—Neotropical; AT—Afrotropical; OL—Oriental; AU—Australasian; PAC—Pacific & Oceanic Islands, ANT—Antarctic

identified but not yet described, with over 90 species of Corydalidae having been found in China (X. Liu, personal communication). China has been well surveyed relative to the rest of the Oriental region, so given the high endemicity and restricted distributions of the Chinese species we expect that many undescribed, endemic species also occur in India and Southeast Asia.

A second region of high diversity in Corydalinae is Amazonia and the Andes Mountains. For example, 18 species of Corydalinae are known from Venezuela and 16 species are known from Brazil. Chauliodinae is rather sparse in the Neotropics, being restricted to Chile (with one possible undescribed species in Brazil, A. Contreras-Ramos, personal communication), but is much more speciose in Australia (23 species) and the Pacific Coast of the US (11 species). Generic richness follows roughly the same trends as species richness, although the Nearctic (6 genera) is the richest bioregion with respect to the Chauliodinae (Table 2).

Sisyridae and Nevrorthidae have received little recent attention, and the extent of our knowledge of the diversity of these families is unclear. The Sisyridae are particularly speciose in the Neotropics, with 15 species in Brazil. The Nevrorthidae have an extremely disjunct distribution, being limited to the Mediterranean, Japan, Taiwan, and Australia, and are clearly a relict taxonomic group.

The number of water-dependent Osmylidae is highly uncertain because the larvae and biology of many genera are completely unknown. If all members of the Osmylinae and Kempyninae occur along stream margins, the 45 described species in these subfamilies would be categorized as water-dependent osmylids. Water-dependent osmylids are particularly speciose in Australia (13 endemic species) and east Asia (16 species) (New, 1983).

Phylogeny

The Neuropterida are a primitive lineage possessing many apparently pleisomorphic characters of holometabolous insects, such as mobile pupae in Megaloptera and Raphidoptera. The monophyly of Neuropterida is generally accepted based upon inconspicuous characters such as thoracic and abdominal sutures and the musculature of the ovipositor (Kristensen, 1999; Aspöck, 2002). Monophyly is also supported by molecular (18S rDNA) data (Whiting, 2002) and combined morphological/molecular analyses (Wheeler et al., 2001). Neuropterida is hypothesized to comprise the sister group to Coleoptera, based on morphologic characters (i.e., the absence of cruciate cervical muscles, modification of the female terminalia, and characters of the hind-wing-base; Kristensen, 1999; Hornschemeyer, 2002) and some molecular studies (e.g., Wheeler et al., 2001). Other molecular data, however, does not support this hypothesis because the monophyly of Coleoptera has been questioned (Whiting, 2002).

The monophyly of each of the three orders of Neuropterida is generally accepted, although there is less certainty about the Megaloptera than the other two orders. Autapomorphies of Neuroptera are quite conspicuous, and include complex sucking mouthparts and closure of the midgut in the larvae.

Table 2 Generic richness of Megaloptera and aquatic or water-dependent Neuroptera worldwide and by biogeographic region

	PA	NA	NT	AT	OL	AU	PAC	ANT	Total freshwater genera
Chauliodinae	2	6	3	3	4	2	0	0	15
Corydalinae	3	2	3	1	5	0	0	0	9
Sialidae	2	1	1	1	3	2	0	0	7
Total Megaloptera	7	9	7	5	12	4	0	0	31
Nevrorthidae	2	0	0	0	1	1	0	0	3
Sisyridae	1	2	2	2	2	2	1	0	4
Osmylidae	4	0	2	0	1	3	0	0	7
Total Neuroptera	7	2	4	2	5	6	1	0	14

PA: Palaearctic; NA: Nearctic; NT: Neotropical; AT: Afrotropical; OL: Oriental; AU: Australasian; PAC: Pacific & Oceanic Islands, ANT: Antarctic

Raphidoptera are united by many adult characters such as elongation of the female ovipositor, elongate prothorax, and aspects of wing venation. Autapomorphies of the Megaloptera are far from definite, although larval tracheal gills (Aspöck, 2002) and characters of the ovaries (Büning, 1998) have been proposed. Some analyses have suggested that Megaloptera is paraphyletic with respect to Raphidoptera (Achtelig, 1967; S. Winterton, personal communication).

There is considerable controversy surrounding the interordinal relationships within the Neuropterida. Traditionally, Megaloptera has been considered to be a sister group to the Raphidioptera (Achtelig, 1978; Kristensen, 1981; Afzelius & Dallai, 1988), suggesting the independent origin of aquatic larvae in the Megaloptera and Neuroptera (Grimaldi & Engel, 2005). This relationship has been supported by analysis of 18S and 28S rDNA sequences (Whiting et al., 1997; Whiting, 2002). A sister group relationship between Megaloptera and Neuroptera, first suggested by Boudreaux (1979), has been supported by several recent phylogenetic analyses (Aspöck et al., 2001; Aspöck, 2002). Harring & Aspöck (2004) found support for (Megaloptera + Neuroptera) + Raphidoptera using four different genetic markers (16S rDNA, 18S rDNA, COX3 DNA, and COX3 amino acids). These studies propose that aquatic larvae are a synapomorphy of Megaloptera + Neuroptera. A joint morphological/molecular analysis of hexapod phylogeny settled on a trichotomy of Megaloptera, Neuroptera, and Raphidoptera (Wheeler et al., 2001)

Within Megaloptera, the Corydalinae and Chauliodinae are commonly classified as a monophyletic group, with Sialidae as a sister taxa (Davis, 1903; New & Theischinger, 1993; Aspöck et al., 2001). This is supported by several obvious character differences, such as the absence of ocelli in adult Sialidae and the presence of a single long-caudal filament in larval Sialidae (as opposed to 2 anal prolegs in Corydalinae and Chauliodinae). In an analysis of 24 morphological characters, Contreras-Ramos (2004) recently suggested that Sialidae and Chauliodinae share several specialized traits. Characters that are present in Corydalinae but absent in Sialidae and Chauliodinae, including complete mandible dentition, 5-segmented maxillary palps, more than 7Rs branches, and larval tracheal gills, can be interpreted as pleisomorphic in Megaloptera (Contreras-Ramos, 2004). This hypothesis, which proposes that Corydalinae is sister to all other Megaloptera and that Sialidae is the lineage with the most autapomorphies, should be further examined with phylogenetic studies.

Within Neuroptera, the Nevrorthidae, Osmylidae, and Sisyridae are not believed to be closely related. Based on a cladistic analysis of morphologic characters, Aspöck et al. (2001) hypothesize that the Nevrorthidae constitute a basal suborder, the Nevrorthiformia, which retained the aquatic larvae of the ancestor Megaloptera-Neuroptera. The stem-species of the two other sub-orders, Myrmeleontiformia + Hemerobiiformia, is believed to have had terrestrial larvae, suggesting that the Sisyridae acquired fully aquatic larvae secondarily (Aspöck et al., 2001). Further phylogenetic work on the relationships within the Neuroptera is needed to clarify these relationships.

Biogeography and endemicity

Even at the family level, the Megaloptera exhibit discontinuous distributions. Chauliodinae and Corydalinae are both found in the New World, the Oriental region, and South Africa, but are absent from Europe, the Middle East, Central Asia, tropical Africa, and boreal regions. Chauliodinae is quite speciose in Australia, whereas Corydalinae is absent. Sialidae species are most common in temperate regions, and are absent from tropical Africa and portions of the Oriental region. Nearly all genera of Megaloptera exhibit circumscribed distributions, with most being confined to one biogeographic region. No genus occurs in more than one tropical bioregion (e.g., *Sialis* in the Oriental region, *Corydalus* in the Neotropics, etc.).

Megaloptera are generally believed to be poor at dispersal, which may be related to their relict or limited distributions. For example, all larvae are dependent on freshwater, and do not appear to actively enter the drift. Adults have large bodies, are regarded as clumsy flyers, are short-lived, and usually are found near water bodies. Thus, the Megaloptera likely expand their ranges very slowly, and current distributions probably reflect a long history of tectonic and climatic processes. Contreras-

Ramos (1998) suggests that diversification occurs relatively slowly in the Corydalidae because of inherent attributes such as the simple life history strategy (non-specialized habitat) and the external transfer of sperm, which leads to a lack of specialization in their genitalia.

In a recent cladistic analysis, Liu & Yang (2006c) described two main clades within the Chauliodinae: (1) the Asian genera plus the Nearctic genera *Chauliodes* and *Nigronia*, and (2) *Protochauliodes* + *Archichauliodes* + *Neohermes*. The analysis did not include the three Afrotropical genera and three Nearctic genera, however. Unique among the Corydalidae in being found in more than one temperate bioregion, *Protochauliodes* is found in the western United States, Chile, Australia, and New Zealand, and *Archichauliodes* is found in Chile, Australia, and New Zealand. *Chauliodes* and *Nigronia*, closely related to the Asian genera in Liu & Yang's (2006c) phylogeny, are restricted to eastern North America. Liu & Yang's (2006c) results thus suggest that an ancestral fishfly was probably present throughout much of Pangaea, and an early vicariance event well before the breakup of the supercontinent separated the fishflies into an eastern/northern clade (the Asian genera + *Chauliodes* + *Nigronia*) and a western/southern clade (*Protochauliodes* + *Archichauliodes* + *Neohermes*). The current distribution of *Protochauliodes*, *Archichauliodes*, and *Neohermes* corresponds to the western and southern coastal areas of Pangaea. The regions where Chauliodinae are now absent, namely eastern South America, northern and central Africa, and Europe, were located in the interior of Pangaea, which is believed to have been very arid. These inferences, plus the sister-group relationships *Nigronia* + *Ctenochauliodes* and *Chauliodes* + *Anachauliodes* (Liu & Yang, 2006c), suggest that the current recognized genera were fairly well established prior to the breakup of Pangaea. Regions of present-day diversity and endemicity include China ($\sim$35 endemic species) and Australia (25 endemic species). Further work on the phylogeny of the fishflies will undoubtedly improve our understanding of the biogeography of the group.

The biogeography of the Corydalinae has been addressed in systematic accounts by Glorioso (1981), Penny (1993), and Contreras-Ramos (1998). The present distribution can be explained by an ancestral dobsonfly that occurred in southern Gondwana (what is now southern South America, South Africa, and India), with massive radiation and speciation in East Asia following the collision of India and Asia in the early Cenozoic (Penny, 1993). Alternatively, an earlier vicariance event could have separated the Asian lineage from South America and Africa, although this would probably require an early cosmopolitan distribution and subsequent extirpation from the Holarctic (Penny, 1993). Contreras-Ramos (1998) suggests that the South African genus *Chloroniella* (1 sp.) is the most basal dobsonfly, and was likely separated from the Asian and New World lineages during or prior to the breakup of Africa and South America in the Cretaceous. The dobsonfly genera *Protohermes* and *Neurhermes* represent a hypothesized ancestral Asian lineage, whereas *Acanthacorydalis, Neoneuromus*, and *Neuromus* represent a second Asian lineage. All five genera are present today in both India and China. Aside from minor range extensions to the eastern Palaearctic (northern China, Japan, and Korea), all of the species and genera of Oriental Corydalinae are endemic to the region. Likewise, most species of Neotropical Corydalinae are endemic; only four species extend into the Nearctic. *Platyneuromus*, the hypothesized ancestral New World genus, is distributed in Central America and southern Mexico. Its range must have overlapped with the basal members of the other Neotropical dobsonfly genera, *Chloronia* and *Corydalus*, which are endemic to southeastern Brazil. *Platyneuromus* split from *Chloronia* + *Corydalus* during an early vicariance event, which was likely the separation of North America and South America during the breakup of Pangaea (Contreras-Ramos, 1998). The expansion of *Corydalus* into Central and North America is likely a recent affair, perhaps during the Miocene (Contreras-Ramos, 1998).

The biogeography and evolutionary processes of the Sialidae is not as well understood as for the Corydalidae because little is known about the phylogeny of the group. The Sialidae, which are distributed throughout the Holarctic, exhibit a more Laurasian distribution than the Corydalidae. *Sialis* is the most widespread genus of Megaloptera, occurring throughout the Nearctic, Palaearctic, and Oriental region. It is unclear whether *Protosialis*, which is endemic to the Neotropics, represents a recent advance from the Nearctic or a relict distribution. *Sialis jianfengensis*, recently described from China, bears a close resemblance in several key characters to *Protosialis* (Liu & Yang, 2006a). *Indosialis*, known from two species in the Oriental region, is inferred to be a basal sialid and closely related to *Protosialis* (Liu et al., 2006). The presence of endemic genera in Australia and South America suggests that the ancestral Sialidae could have been cosmopolitan. Biogeographical patterns in the Sialidae may be obscured, however, because the family needs revision at the genus level (New & Theischinger, 1993).

The Sisyridae is the most cosmopolitan group of the aquatic Neuroptera. *Climacia* is restricted to the New World, with the greatest diversity in South America. In contrast, *Sisyra* is fairly cosmopolitan. Like the Sialidae, a cladistic analysis of the Sisyridae would undoubtedly help clarify the biogeography of the taxa. Nevrorthidae exhibits an extremely disjunct distribution, with species endemic to Japan (5 species), the Mediterranean (4 species), and Australia (2 species).

The Osmylidae are fairly cosmopolitan but absent from the Nearctic. Over one-half of the Osmylidae are endemic to Southeast Asia, although this is primarily the result of speciation of a single genus (*Spilosmylus*) that is terrestrial (New, 2003). The water-dependent subfamily Kempyninae has a southern distribution (Australia, New Zealand, and southern South America). The genus *Kempynus* is present in both Australia and South America, while *Australysmus* and *Clydosmylus* are both endemic to southeastern Australia. The Osmylinae are distributed throughout central, southern, and eastern Asia, with a few species as far west as Europe and North Africa. *Osmylus* is the most speciose genus of Osmylinae, while *Grandosmylus* is endemic to central Asia and *Lahulus* is confined to the Himalayas.

Acknowledgments We thank Xingyue Liu for supplying information on the Megaloptera of China, including species distributions and richness estimates. Timothy New, Oliver Flint, and Norman Penny provided insight into the aquatic status of the Osmylidae. Atilano Contreras-Ramos provided information on the Neotropical Megaloptera. John Oswald's Bibliography of the Neuropterida (URL: http://entowww.tamu.edu/research/neuropterida/bibhome.html) and Index to the Neuropterida Species of the World (URL: http://insects.tamu.edu/research/neuropterida/neur_sp_index/ins_search.html) were especially helpful for locating references and compiling the species information. Comments from two anonymous reviewers helped to improve the manuscript.

References

Achtelig, M., 1967. Über die anatomie des kopfes von *Raphidia flavipes* stein und die verwandtschaftsbeziehungen der Raphidiidae zu den Megaloptera. Zoologische Jahrbücher (Abteilung für Anatomie und Ontogenie) 84: 249–312.

Achtelig, M., 1978. Entwicklung und morphologie der inner und ausseren weiblichen Genitalorgane der Kamelhalsfliegen (Neuropteroidea: Raphidoptera). Entomologica Germanica 4: 140–163.

Afzelius, B. A. & R. Dallai, 1988. Spermatozoa of Megaloptera and Raphidoptera (Insecta, Neuropteroidea). Journal of Ultrastructure and Molecular Structure Research 101: 185–191.

Aspöck, U., 2002. Phylogeny of the Neuropterida (Insecta: Holometabola). Zoologica Scripta 31: 51–55.

Aspöck, H., U. Aspöck & H. Hölzel, 1980. Die Neuropteran Europas: Eine zusammenfassende Darstellug der Systematik, Okologie und Chlorologie der Neuropteroidea (Megaloptera, Raphidoptera, Planipennia) Europas. Goecke and Evers, Krefeld, Germany.

Aspöck, U., J. D. Plant & H. L. Nemeschkal, 2001. Cladistic analysis of Neuroptera and their systematic position within Neuropterida (Insecta: Holometabola: Neuropterida: Neuroptera). Systematic Entomology 26: 73–86.

Boudreaux, H. B., 1979. Arthropod Phylogeny, with Special Reference to Insects. Wiley, New York, NY.

Bowles, D. E., 2006. Spongillaflies (Neuroptera: Sisyridae) of North America with a key to the larvae and adults. Zootaxa 1357: 1–19.

Büning, J., 1998. The ovariole: structure, type and phylogeny. In Harrison F. W. & M. Locke (eds), Microscopic Anatomy of Invertebrates 11C, Insecta. Wiley-Liss, London: 897–932.

Contreras-Ramos, A., 1998. Systematics of the Dobsonfly Corydalus (Megaloptera: Corydalidae). Thomas Say Publications in Entomology: Monographs, Entomological Society of America, Lanham, Maryland.

Contreras-Ramos, A., 1999. List of species of Neotropical Megaloptera (Neuropterida). Proceedings of the Entomological Society of Washington 101: 274–284.

Contreras-Ramos, A., 2004. Is the family Corydalidae (Neuropterida, Megaloptera) a monophylum? Denisia 13: 135–140.

Davis, K. C., 1903. Sialididae of North and South America. Bulletin of the New York State Museum 68: 442–486.

Dolin, P. S. & D. C. Tarter, 1981. Life history and ecology of *Chauliodes rastricornis* Rambur and *C. pectinicornis* (Linnaeus) (Megaloptera: Corydalidae) in Greenbottom Swamp, Cabell County, West Virginia. Brimleyana 7: 111–120.

Dudgeon, D., 1999. Tropical Asian Streams: Zoobenthos, Ecology, and Conservation. Hong Kong University Press, Hong Kong.

Evans, E. D., 1972. A study of the Megaloptera of the Pacific coastal region of the United States. Ph.D. dissertation. Oregon State University, Corvalis.

Fashing, N. J., 1994. A novel habitat for larvae of the fishfly *Chauliodes pectinicornis* (Megaloptera: Corydalidae). Banisteria 3: 25–26.

Glorioso, M. J., 1981. Systematics of the dobsonfly subfamily Corydalinae (Megaloptera: Corydalidae). Systematic Entomology 6: 253–290.

Grimaldi, D. & M. S. Engel, 2005. Evolution of the Insects. Cambridge University Press, New York, NY.

Hamilton, R., M. Whitaker, T. C. Farmer, A. A. Benn & R. M. Duffield, 1996. A report of Chauliodes (Megaloptera: Corydalidae) in the purple pitcher plant, *Sarracenia purpurea* L. (Sarraceniaceae). Journal of the Kansas Entomological Society 69: 257–259.

Harring, E. & U. Aspöck, 2004. Phylogeny of the Neuropterida: a first molecular approach. Systematic Entomology 29: 415–430.

Hayashi, F., 1989. Microhabitat selection by the fishfly larva, *Parachauliodes japonicus*, in relation to its mode of respiration. Freshwater Biology 21: 489–496.

Hayashi, F. & M. Nakane, 1989. Radio tracking and activity monitoring of the dobsonfly larva, *Protohermes grandis* (Megaloptera: Corydalidae). Oecologia 78: 468–472.

Hornschemeyer, T., 2002. Phylogenetic significance of the wing-base of the Holometabola (Insecta). Zoologica Scripta 31: 17–29.

Kristensen, N. P., 1981. Phylogeny of insect orders. Annual Review of Entomology 26: 135–157.

Kristensen, N. P., 1999. Phylogeny of endopterygote insects, the most successful lineage of living organisms. European Journal of Entomology 96: 237–253.

Liu, X. & D. Yang, 2006a. Revision of the genus *Sialis* from Oriental China (Megaloptera: Sialidae). Zootaxa 1108: 23–35.

Liu, X. & D. Yang, 2006b. Revision of the fishfly genus *Ctenochauliodes* van der Weele (Megaloptera: Corydalidae). Zoologica Scripta 35: 473–490.

Liu, X. & D. Yang, 2006c. Phylogeny of the subfamily Chauliodinae (Megaloptera: Corydalidae), with description of a new genus from the Oriental realm. Systematic Entomology 31: 652–670.

Liu, X., D. Yang & F. Hayashi, 2006. Discovery of *Indosialis* from China, with description of one new species (Megaloptera: Sialidae). Zootaxa 1300: 31–35.

New, T. R., 1983. A revision of the Australian Osmylidae: Kempyninae (Insecta: Neuroptera). Australian Journal of Zoology 31: 393–420.

New, T. R., 1986. A review of the biology of Neuroptera Planipennia. Neuroptera International, Supplemental Series 1: 1–57.

New, T. R., 1989. Planipennia, Lacewings. Handbuch der Zoologie, Vol. 4 (Arthropoda: Insecta), Part 30. Walter de Gruyter, Berlin.

New, T. R., 2003. The Neuroptera of Malesia. Brill, Leiden.

New, T. R. & G. Theischinger, 1993. Megaloptera, Alderflies and Dobsonflies. Handbuch der Zoologie, Vol. 4 (Arthropoda: Insecta), Part 33. Walter de Gruyter, Berlin.

Oswald, J. D., 2006. Index to the Neuropterid species of the world. Version 1.00. URL: http://entowww.tamu.edu/research/neuropterida/neur_sp_index/ins_search.html.

Parfin, S. I. & A. B. Gurney, 1956. The spongilla-flies, with special reference to those of the western hemisphere (Sisyridae, Neuroptera). Proceedings of the United States National Museum 105: 421–529.

Penny, N. D., 1993. The phylogenetic position of *Chloroniella peringueyi* (Megaloptera: Corydalidae) and its zoogeographic significance. Entomological News 104: 17–30.

Penny, N. D., P. A. Adams & L. A. Stange, 1997. Species catalog of the Neuroptera, Megaloptera, and Raphidioptera of America North of Mexico. Proceedings of the California Academy of Sciences 50: 39–114.

Pittman, J. L., T. S. Turner, L. Frederick, R. L. Petersen, M. E. Poston, M. Mackenzie, R. M. Duffield, 1996. Occurrence of alderfly larvae (Megaloptera) in a West Virginia population of the purple pitcher plant, *Sarracenia purpurea* L. (Sarraceniaceae). Entomological News 107: 137–140.

Resh V. H., 1976. Life cycles of invertebrate predators of freshwater sponge. In Harrison F. W. & R. R. Cowden (eds) Aspects of Sponge Biology, Academic Press, New York, New York: 299–314.

Steffan, A. W., 1967. Ectosymbiosis in aquatic insects. In Henry S. M. (ed.), Symbiosis, Vol. II. Academic Press, New York.

Theischinger, G., 1983. The adults of the Australian Megaloptera. Aquatic Insects 5: 77–98.

Ward, P. H., 1965. A contribution to the knowledge of the biology of *Osmylus fulvicephalus* (Scopoli 1763) (Neuroptera, Osmylidae). Entomologist's Gazette 16: 175–182.

Wheeler, W. W., M. Whiting, Q. D. Wheeler & J. M. Carpenter, 2001. The phylogeny of the extant hexapod orders. Cladistics 17: 113–169.

Whiting, M. F., 2002. Phylogeny of the holometabolous insect orders: molecular evidence. Zoologica Scripta, 31: 3–15.

Whiting, M. F., J. C. Carpenter, Q. D. Wheeler & W. C. Wheeler, 1997. The Strepsiptera problem: phylogeny of the holometabolous insect orders inferred from 18S and 28S Ribosomal DNA sequences and morphology. Systematic Biology 46: 1–68.

Zwick, P., 1967. Beschreibung der aquatischen Larve von *Neurorthus fallax* (Rambur) und Errichtung der neuen Planipennierfamilie Neurorthidae fam. nov. Gewasser und Abwasser 44/45: 65–86.

Hydrobiologia (2008) 595:419–442
DOI 10.1007/s10750-007-9117-y

FRESHWATER ANIMAL DIVERSITY ASSESSMENT

Global diversity of water beetles (Coleoptera) in freshwater

M. A. Jäch · M. Balke

Abstract The global diversity of True Water Beetles, False Water Beetles and Phytophilous Water Beetles (sensu Jäch, 1998a. In Jäch & Ji (eds), Water Beetles of China, Vol. II. Zoologisch-Botanische Gesellschaft in Österreich and Wiener Coleopterologenverein, Wien: 25–42.) is assessed. Facultative Water Beetles, Parasitic Water Beetles and Shore Beetles (sensu Jäch, 1998a. In Jäch & Ji (eds), Water Beetles of China, Vol. II. Zoologisch-Botanische Gesellschaft in Österreich and Wiener Coleopterologenverein, Wien: 25–42.) are here classified as "paraquatic" and are thus not included in the assessment. It is estimated that about 18,000 species of aquatic Coleoptera are roaming the earth at present. About 12,600 (70%) of these are already described (deadline: October 2005). About 30 beetle families have aquatic representatives, and in 25 of these families at least 50% of the species are to be considered as aquatic. Six families are supposed to include 1,000 or more aquatic species: Dytiscidae (3,908 described species/5,000 estimated), Hydraenidae (1,380/2,500), Hydrophilidae (1,800/2,320), Elmidae (1,330/1,850), Scirtidae (900/1,700) and Gyrinidae (750/1,000). Scirtidae and Hydraenidae are regarded as the least explored families, the number of described species in each of these two families probably will be almost doubled in the future. The Palearctic (ca. 3,350 described species/ca. 3,900 estimated), the Neotropical (2,510/3,900) and the Afrotropical Region (2,700/3,750) harbour almost the same number of water beetle species, followed by the Oriental (2,200/3,580) and the Australian/Pacific Realm (1,340/2,100). The Nearctic (1,420/1,550) is by far the poorest region in terms of water beetle diversity.

Keywords Insecta · Coleoptera · Water beetles · Global diversity

Guest editors: E.V. Balian, C. Lévêque, H. Segers & K. Martens
Freshwater Animal Diversity Assessment

M. A. Jäch (✉)
Natural History Museum, Burgring 7, Wien, 1010, Austria
e-mail: manfred.jaech@nhm-wien.ac.at

M. Balke
Zoologische Staatssammlung, Münchhausenstraße 21, München,81247, Germany

Introduction

Beetles are holometabolous insects, normally with adecticous, exarate pupae. Adults are characterized by a strongly sclerotized body with the forewings hardened into elytra, which serve to protect the more delicate hind wings, as well as the dorsal surface of the hind two thoracic segments (pterothorax) and abdomen. Other derived characteristics of adult beetles are: presence of a gula; antenna primarily with 11 articles; hind wings folded under elytra;

thorax broadly connected with abdomen, so that the primary functional units of body are head, prothorax, pterothorax-abdomen, rather than the more typical head, thorax, abdomen of many other insect orders; genitalia retracted into abdomen.

Beetles represent the world's most speciose animal order. Although about 400,000 species have been described until today, some biodiversity experts estimate that millions of species may actually roam the earth. Beetles occur on all continents[1] except on Antarctica itself, although many species live on the Subantarctic Islands (e.g., Kerguelen, Campbell).

Beetles represent one of the largest orders of "aquatic" animals. However, the majority of Coleoptera is terrestrial, only a minor percentage can be regarded as "aquatic" (definition given below). According to Crowson (1981: 429) water has been "invaded" at least 10 times independently during the evolution of Coleoptera, but this process in fact seems to have happened more than 20 times, and the water's edge has been "approached" even more often. Water beetles do therefore not represent a single monophyletic clade. Accordingly, their behavioural and morphological adaptations to the aquatic environment are exceptionally diverse. The smallest water beetles are less than 1 mm, the largest ones are about 5 cm long.

In the "great three" aquatic insect orders (Ephemeroptera, Plecoptera and Trichoptera) there is usually a fully submerged, long-lived larval stage and a shorter-lived, fully terrestrial adult stage. This type of life cycle is comparatively scarce in beetles (e.g., Scirtidae, Psephenidae). The life histories of water-associated beetles are in fact extremely manyfold, differing greatly between families. In contrast to most representatives of other higher taxa the ecology of many species of beetles cannot be simply defined as "aquatic" or "terrestrial". In Coleoptera aquatic and terrestrial behaviour very often grade almost imperceptibly into each other at the water's edge. The difficulties in the ecological classification are related mainly to the following factors: (1) amount of time spent in contact with water, (2) degree of submergence, (3) degree of water dependance, (4) motivation for getting into contact with water (food, refuge, etc.). These factors tend to be most variable and they are displayed in various combinations, differing greatly between, and often within beetle families. Even some of the most typical aquatic families have fully terrestrial representatives (e.g., three genera of Dytiscidae) and many of the typical terrestrial families (e.g., Lampyridae) have at least some representatives that spend much time under water. The habitat preference of a beetle may not only differ between the stages (larva, pupa, adult) but it may even be different from one larval instar to another, as in Hydraenidae (Jäch et al., 2005a) or Lampyridae (Ho et al., 1998; Jeng et al., 2003), where a gradual change from aquatic to riparian behaviour is not exceptional in the larval stage. In certain cases behaviour may even deviate between populations of the same species, depending on habitat availability; in *Ochthebius haberfelneri* Reitter (Hydraenidae), *Aphodius alternatus* Horn, *Dyscinetus morator* F. (Scarabaeidae) or *Dryops nitidulus* (Heer) (Dryopidae), aquatic and fully terrestrial populations/specimens have been recorded. Furthermore, many beetle species are very small (>2 mm), being therefore able to live more or less exactly at the land-water margin. It is often not possible to decide whether a certain species is to be regarded as "aquatic" or "terrestrial" unless the biology of adults, pupae and all larval instars has been studied very carefully, which is usually very difficult in most of the tiny representatives. First instars of some riparian beetles (e.g., Sphaeriusidae) are so small that even a very thin film of water is sufficient to keep them partly submerged. The larval instars of most species of tiny beetles are still undescribed today! Of the 1,420 species of Hydraenidae, larvae are known from only about a dozen species, which is about 1%!

In order to overcome these classificatory difficulties, Jäch (1998a) defined six ecological groups based on more than 40 families of beetles more or less strongly associated with aquatic habitats: (1) "True Water Beetles" (at least partly submerged for most of the time of their adult stage), (2) "False Water Beetles" (submerged for most of the time of their larval stage, adults always predominantly terrestrial), (3) "Phytophilous Water Beetles" (living and feeding on water plants (mono- or oligophagous), submerged for at least some time in any developmental stage), (4) "Parasitic Water Beetles" (like Phytophilous Water Beetles, but their hosts are aquatic mammals), (5) "Facultative Water Beetles" (actively submerged

[1] In this chapter "all continents" refers to North America, South America, Europe, Africa, Asia and Australia.

(occasionally or regularly) or actively dwelling on the water surface (occasionally or regularly) for a limited period of time, e.g., while hunting, feeding, seeking refuge, etc., during any of their developmental stages in at least one population) and (6) "Shore Beetles" (riparian, living close to the water's edge during all their developmental stages, not entering water voluntarily). With a few exceptions, the amount of time spent in contact with water is gradually decreasing from category 1–6.

True Water Beetles and False Water Beetles are generally regarded as "aquatic" because they are fully submerged at least in the larval stage. However, in Phytophilous Water Beetles, Parasitic Water Beetles and Facultative Water Beetles the amount of time spent in contact with water and the degree of submergence and/or water dependance are most variable. It is often impossible to decide whether a certain species is to be regarded as "aquatic" or "riparian/terrestrial" unless the biology of adults and all larval instars of all populations has been studied very carefully. The delimitation of Facultative Water Beetles from Shore Beetles and strictly terrestrial ones poses a major problem in connection with the global assessment of the number of freshwater species. River margins, stream banks and lake shores are home to thousands of beetle species. However, the life histories of Facultative Water Beetles, Shore Beetles and truly terrestrial beetles intergrade smoothly and so far no attempt has ever been made to provide exact definitions to distinguish between "aquatic", "paraquatic" (Facultative Water Beetles, Parasitic Water Beetles, Phytophilous Shore Beetles, Shore Beetles) and "purely terrestrial" species. No global assessment of Shore Beetles has ever been carried out. This is most regrettable from the conservation point of view, because river banks are greatly suffering from destruction world-wide. River margins are among the most vulnerable habitats in Europe. Their fauna is thus severely threatened (Jäch et al., 2005b). A global assessment of riparian arthropods (especially Araneae, Orthoptera, Blattodea, Dermaptera, Hemiptera, Hymenoptera, Coleoptera, Diptera) is dearly needed.

It should be noted here that there is no species of Coleoptera (except maybe some stygobiontic ones) that does not voluntarily leave water for at least some hours (e.g., for pupation and/or dispersal flight) during its life. No water beetle has so far been confirmed to be ecologically 100% "aquatic" as, for example, are whales or most fishes.

The terms "amphibious", "amphibiotic", "semi-aquatic" and "semi-terrestrial" were often used in connection with water beetles and other animals. However, it should be stressed that these terms have been variously defined. According to Schaefer & Tischler (1983) amphibious is the same as semi-terrestrial: "denoting an animal, which due to its organisation has to spend certain phases of its life under water and others on land, like dragonflies, mayflies, caddisflies or amphibia". According to Webster's Encyclopedic Unabridged Dictionary of the English Language there is a difference between amphibiotic: "living on land during the adult stage and in water during a larval stage", and amphibious: "living or able to live both on land and in water". Jäch et al. (2005a) defined amphibious as "being able to live below and above the water surface in the same developmental stage". The term semi-aquatic (not found in Schaefer & Tischler, 1983) is defined in Webster's as "partly aquatic; growing or living close to water and sometimes found in water". According to Torre Bueno (1989) semi-aquatic species are "living in wet places or partially in water (Borror et al.)"! It is recommended to avoid all these four special terms or to provide unambiguous definitions when using them.

Aquatic beetles are found to live in almost all kinds of aquatic habitats, such as rivers, springs, lakes, ditches, puddles, phytotelmata, seepages, ground water. They are known to survive trapped in ice. Salinity is also not a limiting factor for some species of water beetles—according to Gerdes et al. (1985) *Ochthebius corrugatus* Rosenhauer can cope with a salinity up to 250‰. However, beetles do not inhabit the oceans[2], although numerous species live at their shores, where they can be found in hypersaline rock pools of the supralittoral, i.e., the spray (or splash) zone slightly above the intertidal zone. In contrast to other insects, water beetles prefer small, richly vegetated ditches. In larger lakes, they prefer the swampy margins, as for instance the reed belt of the Central European Neusiedler See, where water beetle biomass is probably higher than anywhere on earth.

[2] The margins of the Baltic Sea, where some species of water beetles (e.g., *Macroplea mutica*) may regularly be encountered, are rather brackish than truly marine.

Most water beetles are benthic organisms, Gyrinidae adults (and some Staphylinidae) are neuston dwellers, which can glide on the water surface; other species (e.g., some Hydraenidae, Hydrophilidae and Lampyridae) can be found "walking" upside-down on the underside of the surface film, and some are living exclusively on water plants or on their mammal hosts. A considerable number of water beetles are able to swim and dive. Many species, especially those which live in well-oxygenated running waters (Elmidae, Hydraenidae adults) stay submerged for most of their life and breathe by means of a microplastron (a very thin layer of air, held by a dense coating of hydrofuge setae). Some Elmidae were encountered down to a depth of more than 10 m below the water surface.

Apart from typical aquatic biotopes containing water of atmospheric origin, beetles can also live in watery liquids of biogenic origin, e.g., in flowing tree sap (hygroarboreal habitat), which is the preferred choriotope of many Diptera, Nitidulidae (Tree Sap Beetles) and Nosodendridae (Wounded Tree Beetles), where they quite often can get totally submerged. Flowing tree sap and rain water may gather in a natural cavity to form a phytotelma, and accordingly, the habitat distinction between flowing tree sap and phytotelmata is gradual. Other liquid substances, like fresh cow dung, are generally not classified as aquatic habitats, although they are practically nothing else but a watery fluid of biogenic origin containing a high percentage of organic particles and it is therefore not surprising that hydrophilids are a major component of the cow dung fauna. However, coprophilous beetles shall not be included in this survey. Also not included are beetles living on snow (e.g., Cantharidae larvae). Especially in early spring, when the surface of the snow melts, nivicolous insects are indeed walking in a very thin layer of water. Finally, it should be mentioned that cave beetles often need 100% humidity and therefore also strongly depend on H_2O, at least in its gaseous phase.

Comprehensive up-to-date regional country-level information on water beetle species has for instance been published for the Netherlands (Drost et al., 1992), Scandinavia (Nilsson, 1996), China (Jäch & Ji, 1995, 1998, 2003), Malaysia (Balke et al., 2004a), Singapore (Hendrich et al., 2004), and South Carolina (Ciegler, 2003).

A key as well as colour illustrations for many families of water beetles was provided by Jäch & Balke (2003) and Balke et al. (2004a); detailed accounts on the morphology of most water beetle families can be found in the Handbook of Zoology (Beutel & Leschen, 2005a).

Species diversity

For practical reasons, the term "aquatic" is herein used strictly in connection with True Water Beetles, False Water Beetles and Phytophilous Water Beetles (sensu Jäch, 1998a), even if they spend much of their individual life on land.

Facultative Water Beetles, Parasitic Water Beetles and Shore Beetles (sensu Jäch, 1998a) are here considered as paraquatic. Their ecology is inseparably connected with aquatic biotopes although they spend most part of their life on the shores. Paraquatic beetles are just briefly treated herein; they are not included in the assessment. Also not included in this assessment are the few water beetles that occur in hypersaline marine rock pools (e.g., some specialized ochthebiine hydraenids) and Shore Beetles living at sea coasts.

Chelonariidae were repeatedly regarded as aquatic (e.g., Ciegler, 2003). However, numerous authors (e.g., Brown, 1972; Spangler, 1980a; Ivie, 2002; Beutel & Leschen, 2005b) have pointed out that this family is not associated with water.

The order Coleoptera is comprised of four suborders, three of which have aquatic representatives: Myxophaga (77 described species, ca. 90% aquatic), Adephaga (ca. 30,000 described species, ca. 18% aquatic), and Polyphaga (ca. 370,000 described species, ca. 1.25% aquatic). While truly terrestrial species are an exception in the Myxophaga, eight of the 11 extant families of Adephaga are regarded as predominantly aquatic (Gyrinidae, Haliplidae, Meruidae, Noteridae, Amphizoidae, Aspidytidae, Hygrobiidae, Dytiscidae), and only 13 of the ca. 150 recognized families of the large suborder Polyphaga are regarded as "predominantly aquatic" (Helophoridae, Epimetopidae, Hydrochidae, Spercheidae, Hydrophilidae, Hydraenidae, Scirtidae, Elmidae, Dryopidae, Lutrochidae, Psephenidae, Cneoglossidae, Eulichadidae). Larvae of the scirtoid family Decliniidae are still unknown and it cannot be excluded that they are aquatic like scirtid larvae.

In addition to these 25 typically aquatic families, another 12 families (1 adephagan and 11 polyphagan) have at least one representative living in (or in very close association with) a freshwater habitat, but the majority of their members is exclusively terrestrial: Carabidae (Adephaga), Leiodidae, Staphylinidae, Scarabaeidae, Ptilodactylidae, Lampyridae, Nitidulidae, Monotomidae, Chrysomelidae, Nanophyidae, Erirhinidae, Curculionidae (Polyphaga). Larvae of some species of Limnichidae are probably also aquatic (W. D. Shepard, pers. comm.) but detailed ecological studies are still lacking.

Nosodendridae, a small family known to occur in tree sap, are not included in this assessment. However, it should be kept in mind, that further research might reveal, that some nosodendrids are well able to live in phytotelmata like the nitidulid *Amphicrossus japonicus* Reitter (see below).

Typical Shore Beetles (sensu Jäch, 1998a) are found in numerous families, e.g., Carabidae (Adephaga), Lepiceridae (Myxophaga), Helophoridae, Hydrophilidae, Georissidae, Histeridae, Ptiliidae, Leiodidae (Cholevinae), Staphylinidae, Micropeplidae, Scarabaeidae, Elateridae, Limnichidae, Heteroceridae, Lampyridae, Latridiidae, etc. (Polyphaga).

Species described after 2005 are not included in this assessment.

Predominantly aquatic families

At least 50% of the species of each of the 25 families in this group are aquatic. All the typical water beetle families are included here. Ecologically, they are mostly True Water Beetles, like Predaceous Diving Beetles (Dytiscidae), Whirligig Beetles (Gyrinidae), Water Scavenger Beetles (Hydrophilidae) and Riffle Beetles (Elmidae), or, to a lesser percentage, False Water Beetles, e.g., Water Penny Beetles (Psephenidae).

Adephaga

1. Gyrinidae. About 750 described species in 13 genera. Gyrinids occur on all continents. Adults and larvae of all species are strictly aquatic. The majority of the species lives in running water. – Literature: Beutel & Roughley (2005), Mazzoldi (1995).

2. Haliplidae. About 200 described species in five genera are found on all continents, but they are more diverse in the northern temperate regions. Larvae and adults of all species are truly aquatic. They live mainly in stagnant water, some species prefer lotic habitats. This family is remarkably well explored; new discoveries are rather exceptional. – Literature: Vondel (2005).

3. Meruidae. Monogeneric family described from Venezuela in 2005. The single known species lives at the gravelly margins (interstitial) of mountain streams. – Literature: Beutel et al. (2006), Spangler & Steiner (2005).

4. Noteridae. About 250 species in three subfamilies and 14 genera have been described so far. The family occurs on all continents. Adults and larvae are aquatic; *Noterus* pupates under water in air-filled cocoons. Noteridae are commonly found in stagnant water between roots of water plants. The monogeneric Phreatodytinae (six species, all restricted to Japanese groundwater habitats) are considered as distinct family by certain authors. – Literature: Nilsson (2005a).

5. Amphizoidae. Monogeneric family with five described species, known only from North America and China. Larvae and adults of all species are aquatic living in rather fast flowing rivers. – Literature: Nilsson (2005b).

6. Aspidytidae. Monogeneric family with two species from South Africa and China. Larvae and adults live in seepages; pupae unknown. – Literature: Balke et al. (2003, 2005), Nilsson (2005c).

7. Hygrobiidae (Paelobiidae). Monogeneric family with six described species, occurring in Europe, China and Australia. Larvae and adults of all species are aquatic (stagnant water). The discovery of new species is not to be expected. – Literature: Nilsson (2005d).

8. Dytiscidae. With almost 4,000 described species in 175 genera, this is the most speciose water beetle family; it occurs on all continents. A total of 10 subfamilies (Agabinae: 388 spp., Colymbetinae: 130 spp., Copelatinae: 568 spp., Coptotominae: 5 spp., Dytiscinae: 377 spp., Hydrodytinae: 4 spp., Hydroporinae: 2,012 spp., Laccophilinae: 400 spp., Lancetinae: 8 spp., Matinae: 8 spp.) are presently recognized. Larvae and adults of almost all species are aquatic; they live in a wide variety of freshwater habitats: stagnant water, running water, groundwater (as deep as 30 m underground), seepages, phytotelmata; only five species are known to be fully terrestrial (humicolous). – Literature: Larson et al. (2000), Nilsson (2001).

Myxophaga

1. Lepiceridae. Monogeneric family with two New World species. Adults of *Lepicerus inaequalis* are usually found on sand banks very close to streams. Adults have never been found in the water, but assuming that the still undescribed larvae live in sand well saturated with water (interstitial), they might get submerged regularly and should thus be able to breathe under water (Shepard, pers. comm.). The second species, *L. bufo*, is also found in sandy habitats along stream courses, but often well away from the water and is thus classified as paraquatic. Photographs of lepicerid habitats are found at http://www.amatl.net/lepi.html. – Literature: Arce-Pérez (1997), Reichardt (1976).

2. Torridincolidae. About 31 tiny species in seven genera occurring in South America (three genera), Africa (incl. Madagascar) (three genera) and Palearctic Asia (one genus). An undescribed species is known from the Philippines. The species are found in mountain streams; many of them have a preference for hygropetric habitats. So far known, all Torridincolidae are aquatic in all three developmental stages and thus they are the most strongly water-associated beetle family. Undescribed species are known from Paraguay, Venezuela, China and the Philippines. – Literature: Endrödy-Younga (1997a), Jäch (1998b), Spangler (1980b).

3. Hydroscaphidae. About 21 tiny species in three genera occurring on all continents except Australia. One species, *Hydroscapha natans*, is recorded from two biogeographical regions. Adults and larvae of all species are aquatic preferring seepages (hygropetric habitats), hot springs, or the interstitial of gravel banks of streams and rivers. – Literature: Jäch (1995a), Löbl (1994, 2003a).

4. Sphaeriusidae. Monogeneric family with 23 tiny species occurring on all continents. Several species are reportedly terrestrial (humicolous); other species (e.g., the type species *Sphaerius acaroides* Waltl) occur at the margins of running and/or stagnant water, often in wet sand very close to the water surface. As it can be assumed that they get submerged regularly, they are here classified as True Water Beetles. Numerous undescribed species are deposited in various museums. A precise evaluation of the percentage of aquatic, riparian and fully terrestrial species is currently not possible, because the ecology of most species is still unknown. However, it is assumed that the majority of the species is aquatic. – Literature: Arce-Pérez (1997), Beutel & Raffaini (2003), Endrödy-Younga (1997b), Hall (2003), Löbl (1995, 2003b).

Polyphaga

1. Helophoridae[3]. Monogeneric family with about 185 species, more or less confined to the Holarctic Realm. Adults of most species are considered truly aquatic (about 75% living in stagnant water, R. Angus, pers. comm.); several species are usually encountered slightly above the water line and therefore seem to be facultatively aquatic or riparian; only few species are strictly terrestrial. Larvae of *Helophorus* are riparian or strictly terrestrial (humicolous). This family is remarkably well explored; new discoveries are not common, but molecular studies might yield new siblings. – Literature: Angus (1992), Hansen (1999).

2. Epimetopidae[3]. About 29 species in three genera occurring in the New World (*Epimetopus*), Africa (*Eupotemus*) and Asia (*Eumetopus*). Adults of all species are probably aquatic (sandy margins of lentic habitats), ecology of larvae unknown. – Literature: Hansen (1999), Jäch (2002), Ji & Jäch (1998), Skale & Jäch (2003).

3. Hydrochidae[3]. Monogeneric family with about 180 species; hydrochids occur on all continents. All species are truly aquatic, living in well-vegetated stagnant water and/or at the edges of very slowly flowing water. Unfortunately, taxonomic research in this family has been badly corrupted by a single unqualified author, and it is therefore most difficult to keep track with the number of ill-defined new taxa and resulting synonymies being published each year, a thorough revision is dearly needed. – Literature: Hansen (1999), Short & Hebauer (2005).

4. Spercheidae[3]. Monogeneric family with 18 species. The genus occurs on all continents. Larvae

[3] Helophoridae, Epimetopidae, Hydrochidae and Spercheidae are treated as subfamilies of Hydrophilidae in Lawrence & Newton (1995), but they are generally accepted as valid families by most water beetle specialists (see Hansen, 1999).

and adults generally live in stagnant water. – Literature: Hansen (1999), Short & Hebauer (2005).

5. Hydrophilidae. About 2,652 species in 174 genera. Hydrophilids occur on all continents. They comprise four subfamilies: (1) Horelophinae: monotypical, New Zealand, adults obviously riparian or hygropetric; (2) Horelophopsinae: monogeneric, two species, New Guinea (Yapen Island) and Japan (Ryukyu Archipelago), the Japanese species is obviously aquatic, the ecology of the Yapen species is not exactly known; (3) Hydrophilinae: 1,740 species in 66 genera, on all continents, adults and larvae of most species are living in stagnant water, running water, in phytotelmata or seepages, numerous species are reportedly riparian or terrestrial (humicolous); (4) Sphaeridiinae: 909 species in 106 genera, most of which are terrestrial; only seven genera include aquatic representatives. In total, about 70% of Hydrophilidae are aquatic. – Literature: Hansen (1991, 1997, 1999), Short & Hebauer (2005).

6. Hydraenidae. About 1,420 species in about 40 genera. Hydraenids are encountered on all continents and inhabit even some Subantarctic Islands, where only few insects are able to cope with the hostile climatic conditions. While adults of most species are aquatic (stagnant water, running water, seepages), many are riparian or strictly terrestrial and a few species are known to live exclusively in hypersaline marine rock pools. Hydraenid larvae are usually riparian or terrestrial, only the first instar of some species is aquatic. The Subantarctic species are poorly studied. They are, however, not associated with freshwater. This family probably contains the largest number of undescribed species, and it is the only family in which more than 1,000 species are expected to be still undescribed. Two new genera and 95 new species of Hydraenidae have been described between January 2004 and September 2005. Even in the comparatively well explored Europe, about 40 (!) new species of Hydraenidae were described in the last 10 years (1995–2005). Several hundred new species are housed in the collection of the Vienna Natural History Museum. – Literature: Hansen (1998), Jäch et al. (2000).

7. Scirtidae. About 900 species in 30 genera. Scirtids occur on all continents. Larvae are usually aquatic, although there are reports about scirtid larvae found in wet soil and on rotten logs. Imagos are generally terrestrial, but adults of *Hydrocyphon* are occasionally collected under water; pupae of *Hydrocyphon* are also reported to be aquatic. Scirtid larvae are found in running water (about 20%), in stagnant water, phytotelmata, and in groundwater. The ecology of scirtids is most poorly studied. Since there is no detailed information on the presumably terrestrial species, all Scirtidae are here provisionally classified as aquatic (False Water Beetles). – Literature: Klausnitzer (2004), Lawrence (2005), Yoshitomi & Satô (2005).

8. Elmidae. About 1,330 species in 146 genera. Elmids occur on all continents. Two subfamilies are presently recognized: Larainae (26 genera, 130 species), and Elminae (120 genera, 1,200 species). However, the phylogeny of elmids has never been carefully studied, therefore major changes concerning the generic and tribal concept are to be expected. At present, there is a proportionally high amount of genera (seven times as many as in Hydraenidae, a family with a comparable number of described species). Adults and larvae of all species are considered to be aquatic, however, adults of many Larainae species often can be encountered below or a little above the water line or in spray zones of water falls and cascades. Members of this family are generally living in lotic habitats, very few species are encountered at lake shores or in ponds. Dozens of undescribed species are deposited in various museum collections. – Literature: Brown (1981a, 1981b), Kodada & Jäch (2005a).

9. Dryopidae. About 300 species in 33 genera. Dryopids occur in all biogeographical regions, but they are absent from the Australian continent. Larvae are generally riparian or terrestrial; adults of about 75% of the species are regarded as aquatic (lotic and lentic habitats), the remaining ones are riparian or terrestrial (humicolous, arboricolous). The ecology of numerous species living close to the water margins has never been studied thoroughly, and some species of the genus *Dryops* classified here as aquatic may in fact turn out to be riparian. Dozens of undescribed species are deposited in the Vienna Natural History Museum. – Literature: Brown (1981a), Kodada & Jäch (2005b).

10. Lutrochidae. About 15 species, all confined to the New World. Larvae and adults are reported to be aquatic in lotic habitats (riparian gravel, emergent rocks or submerged wood). Lutrochidae are generally regarded as True Water Beetles; however, like in dryopids, thorough ecological studies are still wanting and it cannot be excluded that some species are in fact facultatively aquatic or riparian. Adults occasionally leave the water. So far only one genus is recognized; several undescribed species and at least one new genus are deposited in various museum collections. The family is in need of a taxonomic revision. – Literature: Brown & Murvosh (1970), Ide et al. (2005).

11. Psephenidae. About 272 species in 35 genera. Psephenids occur on all continents. They are comprised of four subfamilies: Eubrianacinae, Eubriinae, Psepheninae, Psephenoidinae. Larvae are always aquatic, almost exclusively occurring in running water; with few exceptions, adults and pupae are strictly terrestrial. – Literature: Arce-Pérez & Shepard (2001), Brown (1981a), Lee et al. (2005).

12. Cneoglossidae. Monogeneric family with eight Neotropical species. Adults are terrestrial. The larva of *Cneoglossa edsoni* was found "inside submerged rotting brushwood, in small shallow streams with moderate to fast running water". Nothing is known about the larval behaviour of the remaining species, which are herein tentatively classified as aquatic. – Literature: Costa et al. (2005).

13. Eulichadidae. About 21 species in two genera occurring in North America and Asia. Larvae live in streams, but adults are strictly terrestrial. A revision of the genus *Eulichas* with description of several new species is in preparation (J. Hájek, pers. comm.). – Literature: Ivie (2005), Jäch (1995b).

Not predominantly aquatic families

The 12 families listed in this group are primarily terrestrial. Most of these families are very well-known and speciose (e.g., Ground Beetles (Carabidae), Rove Beetles (Staphylinidae), Leaf Beetles (Chrysomelidae), Snout Beetles (Curculionidae)). However, they contain a small percentage of aquatic members. The aquatic representatives of these families probably do not include True Water Beetles, only few can be regarded as False Water Beetles (e.g., some Lampyridae and some Ptilodactylidae) and Parasitic Water Beetles. Most species are to be classified as Phytophilous Water Beetles or Facultative Water Beetles; some Phytophilous Water Beetles can dive very well and may spend most of their life under water. However, the degree of submergence and/or water dependance is extremely variable in Phytophilous Beetles.

Adephaga

1. Carabidae. More than 20,000 species. A few of the many riparian members in this family can be classified as Facultative Water Beetles: Adults and larvae of *Carabus clathratus* L., *C. variolosus* F., *C. menetriesi* Hummel (Carabinae) are reported to hunt for snails, crustaceans, insect larvae, tadpoles and small fish under water, thus getting fully submerged; other species, e.g., *Oodes helopioides* (F.) (Oodinae) and *Chlaenius* spp. (Chlaeniinae) are reported to stay under water for some time to seek refuge; specimens of *Brachygnathus* sp. (Panagaeinae[4]) from Paraguay were collected between submerged roots, ca. 15 cm below the water surface; some riparian species of Cicindelinae are known to snatch tadpoles from the shallow water, but they do not get fully submerged. *Hydrotrechus cantabricus* Carabajal et al., a subterranean Trechinae from Spain, is described as being permanently in contact with a layer of water and it may therefore represent a True Water Beetle, however, more ecological research is necessary to confirm this assumption. – Literature: Carabajal et al. (2000), Jäch (1998a), Klausnitzer (1996), Wachmann et al. (1995).

Polyphaga

1. Leiodidae. About 3,000 species and about 250 genera in six subfamilies: Camiarinae, Catopocerinae, Leiodinae, Coloninae, Cholevinae (=Catopinae, incl. Bathysciinae), Platypsyllinae (=Leptininae) (classification acc. to Lawrence & Newton, 1995); at least two of these subfamilies have paraquatic representatives. The platypsylline *Platypsyllus castoris* Ritsema is regarded as a Parasitic Water Beetle. It lives in the fur of beavers (*Castor fiber* L.) and thus gets submerged frequently. A similar ecology is recorded from *Silphopsyllus desmanae* Olsufiew which is known to live on the water mole (*Desmana moschata* L.). The Bosnian cave beetle *Hadesia*

[4] Panagaeinae are included in Harpalinae in Lawrence & Newton (1995).

vaseceki (Cholevinae) was regarded as strictly aquatic by early authors. However, Remy (1940) pointed out that *Hadesia* never gets submerged voluntarily, although it is "extremely hygrophilous", which was confirmed by M.E. Schmid (pers. comm.), who observed several specimens in their natural environment. However, *Hadesia vaseceki* is evidently brushing subterranean seepages for food and therefore it can be classified at least as facultatively aquatic. Another closely related European cave beetle genus, *Antroherpon* Reitter, was reported to live close to subterranean streams. – Literature: Absolon (1915), Chappuis (1927), Jäch (1998a).

2. *Staphylinidae*. More than 30,000 species in ca. 30 subfamilies (classification acc. to Lawrence & Newton, 1995). Although this family contains hundreds of riparian and intertidal species, hardly any of these can be classified as aquatic in the strict sense, although members of the subfamily Steninae can glide ("skim") over the water surface at an enormous speed and are therefore probably the fastest "swimming" beetles!

Lesteva spp. (Omaliinae) are often found on the underside of stones in streams, being thus fully submerged. It is not known for how long they enter the water and whether they do it to search for food or to take refuge. About 100 species have been described in this genus. The subfamily Steninae includes two genera: *Stenus* (ca. 2,100 species), and *Dianous* (ca. 200 species). All species of *Dianous* and hundreds of species of *Stenus* live very close to the edges of freshwater habitats (lentic and lotic). At least one species, *Stenus fornicatus* Stephens, was reported to stay submerged voluntarily. Many *Dianous* and *Stenus* have especially water repellent tarsal ventral surfaces and are able to move very skillfully on the surface film of the water, for which they can use three different modes of locomotion, of which "expansion skating" (releasing an abdominal secretion) is by far the most effective one. No comprehensive field research on the behaviour of Steninae has been carried out and we know very little about the amount of time spent on the water and the motivation to do so (refuge, hunting). Therefore at present it cannot be determined how many of the stenines can be classified as "aquatic" (True Water Beetles) or as "paraquatic" (Facultative Water Beetles, Shore Beetles). Without doubt, there are several hundreds of paraquatic species in this subfamily. A few species of Staphylininae (*Hesperus kovaci* Schillhammer, *Odontolinus* sp.) feed on aquatic dipterous larvae in phytotelmata. Specimens of *Trichocosmetes norae* Schillhammer reportedly sit on leaves above the water surface, and then drop onto the water surface in order to escape enemies. – Literature: Betz (2002), Schillhammer (2002).

3. *Scarabaeidae*. About 28,000 species in 13 subfamilies (classification acc. to Lawrence & Newton, 1995). The facultatively aquatic Rice Beetle, *Dyscinetus morator* (Dynastinae), can spend several hours under water while escaping disturbances or while feeding on water plants (e.g., *Eichhornia crassipes*). The Rice Beetle obviously is not monophagous, it also feeds on terrestrial plants, such as carrots and radishes. Therefore it cannot be classified as Phytophilous Water Beetle. *Aphodius alternatus* (Aphodiinae) usually lives in moist habitats along freshwater margins in North America; however, several populations evidently have adopted truly aquatic habits in grassland vernal pools; hence, this species has to classified as facultatively aquatic. – Literature: Buckingham & Bennet (1989), Rogers (1997).

4. *Ptilodactylidae*. About 500 species in more than 30 genera and five subfamilies (acc. to Lawrence & Newton, 1995). Ptilodactylids occur on all continents, but only introduced synanthropic terrestrial species have been reported from Europe. Larvae of the subfamily Anchytarsinae (on all continents), Areopidinae (one species from North America) and some Cladotominae (e.g., *Drupeus hygropetricus* Lee et al.) are known to be aquatic (or at least facultatively aquatic). They are found between gravel or on submerged wood in lotic habitats or in seepages. Adults are strictly terrestrial. Since the aquatic behaviour has not been confirmed for all species of Anchytarsinae and comprehensive taxonomic revisions have not been carried out, the numbers given in the assessment are rather arbitrary. – Literature: Aberlenc & Allemand (1997), Lawrence & Stribling (1992), Lucht (1998), Stribling (1986).

5. *Lampyridae*. Almost 2,000 species in over 90 genera and eight subfamilies. Larvae of comparatively few species are truly aquatic or facultatively aquatic, living in running or stagnant water. Adults are generally terrestrial. Truly aquatic larval stages were confirmed for six species of the genus *Luciola*. Facultatively aquatic species (e.g., *Pyractomena*

lucifera Melsheimer, *Pristolycus kanoi* Nakane) are found near seepages, creeks, swamps, marine rock pools, etc. – Literature: Buschman (1984), Downie & Arnett (1996), Jeng et al. (2003).

6. Nitidulidae. About 3,000 species in seven subfamilies (classification acc. to Lawrence & Newton, 1995). Nitidulids occur on all continents. Numerous species live in tree sap. Adults of one species, *Amphicrossus japonicus*, are here classified as facultatively aquatic. They are found in "bamboo sap" in Malaysian rainforests and also in phytotelmata, where they can stay submerged for a long time (D. Kovac, pers. comm.).

7. Monotomidae. More than 200 species in two subfamilies, occurring on all continents. Monotomids usually live under bark. *Cyanostolus aeneus* (Richter) was reported to be aquatic, living on submerged wood in streams and rivers. However, no detailed studies have been carried out to determine the amount of time spent in contact with water and the degree of submergence and/or water dependance. Specimens collected recently in Austria were found under the bark of a submerged piece of wood, about 10 cm above the water line (R. Schuh, pers. comm.). *Cyanostolus aeneus* is hence classified as facultatively aquatic (paraquatic). – Literature: Peacock (1978).

8. Chrysomelidae. About 46,000 species in 20 subfamilies (classification acc. to Jolivet & Verma, 2002). Numerous species are regarded as Phytophilous Water Beetles. At least one subfamily, Donaciinae, is obligatorily associated with aquatic plants, with larvae and pupae generally submerged; adults of two genera, *Macroplea* and *Neohaemonia*, are often found under water. Several members of Alticinae[5], Chrysomelinae, Galerucinae and Hispinae live on emergent aquatic plants, usually above the water surface; many of them are capable of walking on the water surface and may survive under water for a long time. Several Central American hispines are found in phytotelmata (water-filled floral bracts). – Literature: Konstantinov (2003), Mohr (1960), Seifert (1982).

9. Nanophyidae[6]. About 300 species in 29 genera (classification acc. to Alonso-Zarazaga & Lyal, 1999). Several species of the genus *Nanophyes* Schönherr are regarded as Phytophilous Water Beetles; they are feeding in the stems of aquatic plants (*Alternanthera*, *Ludwigia*). – Literature: Lawrence & Britton (1994), Sankaran (1972).

10. Erirhinidae[7]. About 300 species in numerous genera and two subfamilies (classification acc. to Alonso-Zarazaga & Lyal, 1999). About 10 genera are known to include Phytophilous Water Beetles. – Literature: Anderson (2002).

11. Curculionidae. The world's most speciose animal family with probably more than 60,000 described species in 16 subfamilies (classification acc. to Alonso-Zarazaga & Lyal, 1999). About seven genera in two subfamilies (Bagoinae, Ceutorhynchinae) are known to include Phytophilous Water Beetles. Most species of Bagoinae are exclusively aquatic, but some were captured on land (M. Alonso-Zarazaga, pers. comm.). Many aquatic weevils can dive very well and spend most of their live under water. – Literature: Angus (1966), Caldara & O'Brien (1998), Colonnelli (2004), Klausnitzer (1996).

Shore beetles/phytophilous shore beetles

When talking about freshwater beetles one must not ignore all those many species that live at the margins of aquatic habitats. They are strongly depending on the microclimate, substrate and the food web of their aquatic environment. Any environmental influence (e.g., water pollution, power plants, drought) effecting the truly aquatic species will have more or less the same effect on the Shore Beetles, although they hardly get into contact with water actively! Their ecology is inseparably connected with the aquatic biotope although they are not part of it physically. In contrast to Facultative Water Beetles they do not enter water voluntarily.

Shore Beetles are often difficult to distinguish from truly terrestrial beetles (especially humicolous ones—living in damp places).

About 22 families containing Shore Beetles (from freshwater shores and sea shores) were listed by Jäch (1998a): Sphaeriusidae, Carabidae, Helophoridae,

[5] Alticinae are included in Galerucinae in Lawrence & Newton (1995).

[6] Nanophyidae are regarded as a subfamily of Brentidae in Lawrence & Newton (1995).

[7] Erirhinidae are regarded as a subfamily of Curculioninae in Lawrence & Newton (1995).

Georissidae, Hydrophilidae, Histeridae, Hydraenidae, Ptiliidae, Leiodidae (Cholevinae), Staphylinidae, Scarabaeidae, Dryopidae, Limnichidae, Heteroceridae, Elateridae, Lampyridae, Phycosecidae, Melyridae, Monotomidae, Tenebrionidae, Salpingidae, Anthicidae. Several additional families, e.g., Lepiceridae, Agyrtidae, Micropeplidae, Ptilodactylidae, Byrrhidae, Lampyridae and Latridiidae must be added to that list.

Altogether, there are certainly several thousand species of Shore Beetles world-wide. For reasons explained above (see Introduction), exact numbers cannot be provided at this point. In Carabidae and Staphylinidae there is a rather high number of riparian representatives, but no detailed assessments were carried out in these two families so far. Georissidae (monogeneric, about 70 species) are generally humicolous (wet sand banks, wet meadows). Heteroceridae (five genera, more than 200 species) are noteworthy because they are exclusively riparian living at muddy margins of running as well as stagnant waters.

Limnichidae (about 40 genera and almost 400 species) are here tentatively listed as a Shore Beetle family. All species are living at wet places. Most representatives are found at the shores of freshwater habitats, often very close to the water's edge and it seems that quite a number of species of Limnichidae (e.g., *Caccothryptus*, *Pseudeucinetus*) might in fact be at least facultatively aquatic. Almost nothing is known about larval behaviour.

Numerous Phytophilous Beetles (numerous Chrysomelidae; numerous Curculinoidea; *Telmatophilus* spp. (Cryptophagidae); *Phalacrus caricis* (Phalacridae)) live on aquatic or riparian plants (e.g., *Carex*, *Typha*, *Phragmites*) but hardly ever get in contact with water. They are therefore classified as paraquatic (Phytophilous Shore Beetles).

Conclusions

Only six of the "aquatic" families treated herein are supposed to include at least 1,000 species of water beetles: Dytiscidae, Hydraenidae, Hydrophilidae, Elmidae, Scirtidae and Gyrinidae (Table 1, Fig. 4).

Without any doubt the Predacious Diving Beetles or Dytiscidae are the world's most speciose water beetle family. With an estimated total of 4,800 species, there are about as many dytiscids as in the next two families (Hydraenidae and Hydrophilidae) taken together. Hydrophilidae are presently regarded as the second most speciose family. However, with the inclusion of the estimated undescribed species, hydraenids will soon overtake hydrophilids in the diversity ranking (Fig. 4). Each of these two families probably comprises distinctly more than 2,000 species. Elmidae and Scirtidae follow next with somewhat less than 2,000 species. Gyrinidae represent a family of medium diversity, with an estimated 1,000 species. All the following families have less than 1,000 aquatic representatives.

With an estimated 1,120 undescribed species (45% of the total number of species estimated), Hydraenidae are definitely one of the least explored families. This is mainly due to their small size (the largest species is only 3.3 mm long), their cryptic habits (many species live in tiny rain forest puddles, where they are difficult to detect) and the enormous degree of local endemism (see below). With an estimate of 800 undescribed species scirtids represent even a higher percentage of undiscovered species: ca. 47%. This can be credited to the lack of specialists (at present there are not more than three scientists contributing to scirtid taxonomy regularly), and to the fact, that the short-lived adults of this family are rarely collected. With about 18% undescribed species, Dytiscidae are comparatively well studied. With approximately 520 undescribed species, elmids (28% undescribed) and hydrophilids (22% undescribed) will also contribute considerably to the increase of the global water beetle fauna. Among the less speciose water beetle families, Dryopidae, with an estimated 43% undescribed species, are noteworthy.

About 12,600 (ca. 3%) of the known species of Coleoptera are regarded as "aquatic". Adding estimations of undescribed aquatic species it can be assumed, that actually about 18,000 water beetle species roam the earth.

Phylogeny and historical process

Coleoptera first occur in fossil records of the Lower Permian of Moravia and the Ural. These beetle ancestors were certainly terrestrial.

Water beetles do not form a monophyletic group but rather occur in three of the four different suborders of the Coleoptera where they have adopted aquatic lifestyles independently. A comprehensive account of the general biology and evolutionary trends of water beetles was published by Crowson (1981: Chap. 13). The sturdy exoskeleton of adults, in connection with a rather well sealed subelytral space that may contain an air reservoir, was most likely functioning as major preadaptation of Coleoptera for the invasion of water. It is obviously this subelytral space that allows even many terrestrial beetles to survive under water for some time. Several aquatic lineages gave rise to secondarily terrestrial forms, such as the hydraenid *Edaphobates puetzi* Jäch & Díaz living in Chinese *Rhododendron*-forests or the blind and wingless dytiscid *Typhlodessus monteithi* Brancucci, which was collected from leaf litter on a New Caledonian mountain top.

Based on morphological analyses and according to recent DNA sequence data (Hughes et al., 2006) Myxophaga are the sister group of Polyphaga. Their age could be mid Triassic.

Molecular data suggest that the typically aquatic adephagan beetle families have entered the aquatic environment only once (e.g., Shull et al., 2001), while morphological data rather suggest two or three independent invasions (Beutel et al., 2006). In any case, Gyrinidae (known from Lower Permian deposits) take a basal position within Hydradephaga. Beutel (1997) suggested that the Adephagan ancestor was carnivorous as both adult and larva, and the aquatic medium was invaded by terrestrial species dwelling at river and/or pond margins. Dated molecular phylogenies remain to be presented, but the wider distribution of several genera and multiple subfamilies in different families suggest that diversification cannot be pinned down to only one tectonic event such as the breakup of Gondwana. Major driving forces in the diversification of the larger families include the adaptation to adult surface dwelling, which led to a speciose group (Gyrinidae) after a new habitat has been conquered, as well as evolution of the simultaneous stroke of adult middle and hind legs in Noteridae and Dytiscidae for advanced swimming behaviour (Ribera et al., 2002; Balke et al., 2005).

The major aquatic groups in the large suborder Polyphaga belong to the Staphyliniformia/Scarabaeiformia (e.g., Hydraenidae, Hydrophilidae) and the Elateriformia (e.g., Elmidae, Psephenidae). Hydraenidae are the sister of the terrestrial Ptiliidae, as proven by morphological and molecular data (Beutel & Leschen, 2005c; Caterino et al., 2005). Hydrophiloidea are very probably the sister of a terrestrial clade as well, probably Histeroidea or Scarabaeidae (Beutel & Leschen, 2005c; Vogler, 2005). The enormous diversity of hydraenids (probably the second largest water beetle family on earth) can obviously be explained by their minute body size enabling them to inhabit even very small aquatic biotopes (e.g., tiny puddles, interstitial of gravel banks), and by the development of an antimicrobial "exocrine secretion delivery system" (ESDS), being composed of exocrine glands interacting with various cuticular structures (especially on underside of head and prothorax). Polyphaga probably have originated in the mid Triassic. Considering the high number of families with single aquatic representatives in different genera (see above under "Not predominantly aquatic families") we can assume that the invasion of the aquatic medium has occurred about 20 times in Polyphaga. Curculionoidea, for instance, must have invaded water independantly at least four times, and in Lampyridae aquatic (or paraquatic) behaviour might be a separate evolutionary process in almost every second of the aquatic species.

The biogeographic history of the different groups largely remains to be investigated. However, it more and more emerges that dispersal must have played an important role in the evolution of dytiscid diversity. For example, many Australian and Pacific species, even morphologically highly modified ones, appear to be of comparably recent origin as inferred from DNA sequence divergences (Balke et al., in prep.). As another example, the diving beetle subfamily Copelatinae, with 568 known species, has a pantropical core range, which might have been formed by dispersal rather than ancient plate tectonic events (Balke et al., 2004b).

Distributions, main areas of endemism

The delimitations of the biogeographic regions in this chapter mainly follow the World Catalogue of Insects, Vols. 1–3, 7 (Adephaga, Hydrophiloidea, Hydraenidae) and the Catalogue of Palaearctic Coleoptera, Vol. 3 (Dryopoidea). However, it should be noted, that the biogeographical limits in the World Catalogue of Insects, Vols. 1–2 (Hydraenidae, Hydrophilidae) differ from Vols. 3, 7 (Adephaga). Weber's Line is here chosen as border between the Oriental and the Australian/Pacific Region. At least for the predominantly aquatic families, Mexico is here entirely regarded as Neotropical.

Subantarctic species are included in the geographically closest realm in this assessment (species from South Georgia have for instance been included in the Neotropical Region).

Water beetles are found in all biogeographic regions, from Ellesmere Island in the north to Tierra del Fuego in the south, and they are encountered even on many of the remotest Pacific Islands. And although, on average, diversity is greater in humid climates, some of the desert regions are quite rich in water beetles (see below).

Water beetles display their greatest diversity in the tropics. Haliplidae, Amphizoidae and Helophoridae are the only exception to this rule. Haliplidae are distinctly more diverse in the Holarctic Realm than in any of the tropical regions, and although most tropical countries are still rather poorly examined we do not expect a significant increase of species in this family. Helophoridae are even more strongly confined to the northern temperate regions; a mere 0.7% of the species are found outside the Holarctic Region! All five known species of Amphizoidae are Holarctic.

In contrast to the statement by Lévêque et al. (2005), most aquatic beetles are not "cosmopolitan, or widespread". Although there is quite a number of wide-spread Holarctic species, only very few species are distributed in more than two realms, e.g., *Rhantus suturalis*. In fact, the diversity and the degree of local endemism is extremely high in almost all water beetle families, especially among those living in running waters in warmer climates.

Dytiscidae. In Europe, the Iberian Peninsula must be regarded as a hot spot of endemism, with 34 out of about 164 Iberian species (ca. 20%) only occurring in Spain and Portugal (Ribera, 2000), but levels of endemism are even higher on the Macaronesian Islands, Madeira (50%) and the Canary Islands (30–40%, depending on taxonomy) (Balke & Hendrich, 1989; Balke et al., 2002). The larger tropical Islands such as Borneo (endemism: 63%) and especially New Guinea (>70%) feature most remarkable diving beetle radiations with very high levels of regional endemism (Balke et al., 2002) where, as far as currently known, species turnover might be high even between neighbouring valleys. In that respect, New Guinea might prove one of the most important global water beetle hot spots where total species numbers remain hard to estimate even at the beginning of the 21st century. Australia, with ca. 90% endemic species, is home to the world's largest assemblage of groundwater beetles, with most of the more than 50 known groundwater dytiscid species occurring in calcrete aquifers in the desert (Leys et al., 2003; Pain, 2005). New Caledonia and Fiji were recently screened for diving beetles, and both islands revealed an unexpectedly diverse fauna, with levels of endemism around 90%.

Hydraenidae (Fig. 1). The presumably second largest water beetle family is unchallenged with regard to endemism. Most of the hydraenid species from warmer climates (arid or tropical) are to be regarded as SORD (species of restricted distribution). The highest species diversities are probably found in tropical/subtropical montane forests. The Hydraenidae of some presumptive biodiversity hot spots, such as Borneo and New Guinea, are still very poorly studied. Only two species of *Hydraena* have been described so far from New Guinea and none from Borneo, although at least 100 undescribed species collected in these two islands during a few water beetle expeditions are deposited in the Vienna Natural History Museum. More than 100 species of *Hydraena* have been described from Turkey and about 60 species from Greece, many of them restricted to a single island, a single mountain range or even to a single streamlet (e.g., *Hydraena pangaei*, *H. gynaephila*). The exceptional diversity of hydraenids can very probably be credited to their small size and their limited dispersal abilities. Paradoxically, even larger rivers have been confirmed as potent dispersal barriers for hydraenids: *Hydraena alpicola* and *H. saga*, two very closely related mountain stream dwelling species, are for instance effectively separated from each other by the Danube River in Lower Austria. Very little is known about the species diversity of hydraenids in montane areas of tropical Africa and tropical South America.

Hydrophilidae. The main areas of endemism are obviously in the tropics but accurate data are still lacking. As in dytiscids and hydraenids, hydrophilids from tropical Africa and tropical South America are very poorly known.

Elmidae. Among the larger water beetle families, only elmids are more or less exclusively confined to running water. Due to their flight abilities, they are less diverse than Hydraenidae, especially in arid

Fig. 1 Habitus of *Hydraena* s.str. (Hydraenidae), undescribed species from China. With a total of about 570 described species *Hydraena* represents the most speciose water beetle genus of the world. With more than a thousand species still awaiting description it may well be the most successful aquatic genus on earth. The antimicrobial "exocrine secretion delivery system" (ESDS), being composed of exocrine glands interacting with various cuticular structures (especially on underside of head and prothorax) is nowhere as specialized as in this genus, which is obviously the key to its "success"

climates. Their greatest diversity is found in montane areas of tropical and subtropical rainforests. In Southeast Asia, Borneo seems to be a biodiversity hot spot. Tropical Africa and tropical South America are still very inadequately explored. Even North America, which has been quite thoroughly studied with regard to most water beetles, is in need of a modern elmid revision. Some of the wide-spread West-Palearctic elmids might in fact represent complexes of closely related species, which should be tested by molecular analyses.

Scirtidae. Very little can at present be said about the main areas of endemism of this very poorly studied family. Without doubt, they are most diverse in tropical countries (lowland as well as montane areas).

Gyrinidae. In the Oriental Realm, Borneo obviously must be regarded as a biodiversity hot spot for this typically tropical family. Africa is comparatively well explored, but tropical South America and New Guinea may still harbour notable numbers of undescribed species.

Counting the presently described species, the Palearctic Region houses the highest number of water beetles species (Table 2, Fig. 2). Although comprehensive water beetle surveys are still lacking for large parts of the Neotropical and Afrotropical Realms, it is estimated (after including the undescribed species) that the Palearctic (ca. 3,350 described species/ca. 3,900 estimated total), the Neotropical (2,510/3,900), and the Afrotropical Region (2,700/3,750) harbour more or less the same number of water beetle species, followed by the Oriental (2,200/3,580) and the Australian/Pacific Realm (1,340/2,100). Undoubtedly, the Nearctic (1,420/1,550) is by far the poorest region in terms of water beetle diversity (Fig. 3).

Within the Palearctic Region, the Mediterranean countries and Anatolia are to be regarded as biodiversity hot spots, at least for certain families. In the comparatively well-explored Oriental Region, Borneo was found to be a hot spot of paramount significance for many water beetle families.

One of the world's most comprehensive biodiversity projects was aimed at the exploration of the water beetle fauna of China, which was virtually unknown before this survey (CWBS)[8] got started in 1993. More than 500 sampling stations have been investigated since then and several hundred new species have been detected, many of which were described in a comprehensive three-volume monograph (Jäch & Ji, 1995, 1998, 2003), compiled by 50 authors from 18 countries. The southeastern parts of China (attributed to the Oriental Region or to the Palearctic Region acc. to different specialists) were found to be extremely diverse with a very high degree of local endemism. Even in the Special Adminstrative Region of Hongkong, where stagnant water habitats have been drained (growing rice is not allowed) and where numerous streams are frequently sprayed to kill mosquitos, about 70 species of water beetles are known meanwhile, some of which seem to be endemic to the former British crown colony (see Jäch, 2004).

[8] See http://www.nhm-wien.ac.at/nhm/2Zoo/coleoptera/publications/chinaindex.html

Table 1 Global assessment of water beetle species; AE = estimated number of aquatic species; + = plus 1–10. Estimations include assumed number of undescribed species and possible new faunal records. PA: Palearctic, NA: Nearctic, AT: Afrotropical, NT: Neotropical, OL: Oriental, AU: Australian, PAC: Pacific

Family/Regions	PA	NA	NT	AT	OL	AU + PAC	Total (all regions)
	AE	AE	AE	AE	AE	AE	AE
Carabidae	?1+	0	0	0	0	0	?1+
Gyrinidae	50+	50+	280	250	300	70	1,000
Haliplidae	61	65+	39+	26+	20+	15	220
Meruidae	0	0	1+	0	0	0	1+
Noteridae	30+	16	93+	95+	30+	8+	270
Amphizoidae	2	3	0	0	0	0	5
Aspidytidae	1+	0	0	1+	0	0	2+
Hygrobiidae	1	0	0	0	1	4	6
Dytiscidae	1,050	500	1,200	1,200	600	600	4,800
Lepiceridae	0	0	1+	0	0	0	1+
Torridincolidae	1+	0	40	6+	1+	0	60
Hydroscaphidae	9+	1	6+	1+	5+	0	40
Sphaeriusidae	20	3+	20	2+	40	2+	100
Helophoridae	155+	40+	4+	3+	6+	0	200
Epimetopidae	0	4	19+	2+	8+	0	50
Hydrochidae	30	30+	50	60	25	50	250
Spercheidae	5+	0	2+	9+	6+	2+	25
Hydrophilidae	400	220	700	550	630	300	2,320
Hydraenidae	800	100	300	450	500	350	2,500
Scirtidae	250	130	250	250	500	300	1,700
Elmidae	340	120	360	400	350	220	1,850
Dryopidae	90	20+	100	70	70	1+	350
Lutrochidae	0	3+	30	0	0	0	30
Ptilodactylidae	7+	3+	8+	30+	4+	5+	?100
Psephenidae	92+	15+	70	50	130	14+	370
Cneoglossidae	0	0	?8+	0	0	0	?8+
Eulichadidae	11+	1	0	0	30	0	30
Lampyridae	6+	0	0	0	1+	0	6+
Chrysomelidae (Donaciinae)	70+	60+	5+	12+	27+	2+	180
Chrysomelidae (other subfamilies)	70	10+	200	20	100	40	440
Nanophyidae	0	1+	0	0	1+	1	3+
Erirhinidae	120	73+	50	120	50	50	470
Curculionidae (Bagoinae)	140	45	0	100	90	40	415
Curculionidae (Ceutorhynchinae)	50	30	1+	10	1+	1+	100
Total	3,900	1,550	3,900	3,750	3,580	2,100	18,000

Human related issues and conservation

Relationships between water beetles and mankind are more diverse than usually expected.

Due to their ability to produce an audible sound, Squeak Beetles, *Hygrobia hermanni* (F.) (Hygrobiidae), were sold as childrens pets on British markets in the past (Wesenberg-Lund, 1943). Still today *Cybister* spp. (Dytiscidae) are sold in Hong Kong pet shops for use in the aquarium (Jäch & Easton, 1998), a formerly common practice in Europe as well (Wesenberg-Lund, 1943), which has unfortunately been forgotten.

Table 2 Global assessment of water beetle species; A = number of described species of True Water Beetles, False Water Beetles and Phytophilous Water Beetles (sensu Jäch, 1998a); T = total number of species (aquatic, paraquatic and terrestrial). Total number of not predominantly aquatic families not provided for biogeographical regions. PA: Palearctic, NA: Nearctic, NT: Neotropical, AT: Afrotropical, OL: Oriental, AU: Australian, PAC: Pacific

Family/Regions	PA		NA		NT		AT		OL		AU + PAC		Total (all regions)	
	T	A	T	A	T	A	T	A	T	A	T	A	T	A
Carabidae		?1		0		0		0		0		0	20,000	?1
Gyrinidae	50	50	50	50	200	200	220	220	210	210	40	40	750	750
Haliplidae	61	61	65	65	39	39	26	26	20	20	15	15	204	204
Meruidae	0	0	0	0	1	1	0	0	0	0	0	0	1	1
Noteridae	30	30	16	16	93	93	95	95	29	29	8	8	250	250
Amphizoidae	2	2	3	3	0	0	0	0	0	0	0	0	5	5
Aspidytidae	1	1	0	0	0	0	1	1	0	0	0	0	2	2
Hygrobiidae	1	1	0	0	0	0	0	0	1	1	4	4	6	6
Dytiscidae	953	953	488	488	743	743	1,060	1,060	534	532	449	446	3,913	3,908
Lepiceridae	0	0	0	0	2	1	0	0	0	0	0	0	2	1
Torridincolidae	1	1	0	0	24	24	6	6	0	0	0	0	31	31
Hydroscaphidae	9	9	1	1	6	6	1	1	5	5	0	0	21	21
Sphaeriusidae	8	4	3	3	1	1	2	2	7	?7	2	2	23	?19
Helophoridae	150	155	42	40	4	4	3	3	6	6	0	0	185	178
Epimetopidae	0	0	4	4	19	19	2	2	8	8	0	0	29	29
Hydrochidae	22	22	30	30	42	42	40	40	13	13	36	36	180	180
Spercheidae	5	5	0	0	2	2	9	9	6	6	2	2	18	18
Hydrophilidae	510	380	243	200	639	570	593	450	687	460	362	210	2,652	1,800
Hydraenidae	700	692	85	85	140	138	230	215	125	120	140	130	1,420	1,380
Scirtidae	200	200	100	100	100	100	50	50	250	250	200	200	900	900
Elmidae	240	240	100	100	260	260	320	320	200	200	150	150	1,330	1,330
Dryopidae	70	69	20	20	70	60	40	35	80	30	4	1	280	200
Lutrochidae	0	0	3	3	12	12	0	0	0	0	0	0	15	15
Ptilodactylidae		?7		?3		?8		?30		?4		?5	500	?56
Psephenidae	92	92	15	15	36	36	14	14	105	105	14	14	272	272
Cneoglossidae	0	0	0	0	8	?8	0	0	0	0	0	0	8	?8
Eulichadidae	11	11	1	1	0	0	0	0	19	19	0	0	19	19
Lampyridae		6		0		0		0		1		0	2,000	6
Chrysomelidae (Donaciinae)		70		60		5		12		27		2	166	166
Chrysomelidae (other subfamilies)		50		10		100		10		70		20	46,000	260
Nanophyidae		0		?1		0		0		?1		1	300	3+
Erirhinidae		90		73		35		50		15		20	300	283
Curculionidae (Bagoinae)		115		35		0		40		50		27	267	260
Curculionidae (Ceutorhynchinae)		29		13		1		2		0		1	1,316	42
Total		3,346		1,419		2,508		2,693		2,189		1,334		12,604

Pemberton (1990) reports about the use of *Cybister* sp. for a kind of lottery in Korea.

Dytiscids were often blamed for causing considerable harm to fish fry (Wesenberg-Lund, 1943), but there are few actual studies on that subject, and more research is needed to assess potential harm (as well as benefits) of water beetles to aquaculture (Vazirani, 1972). Dytiscids are predators of mosquitos and may play an important role in controlling them. Their actual importance is not yet understood, but at least

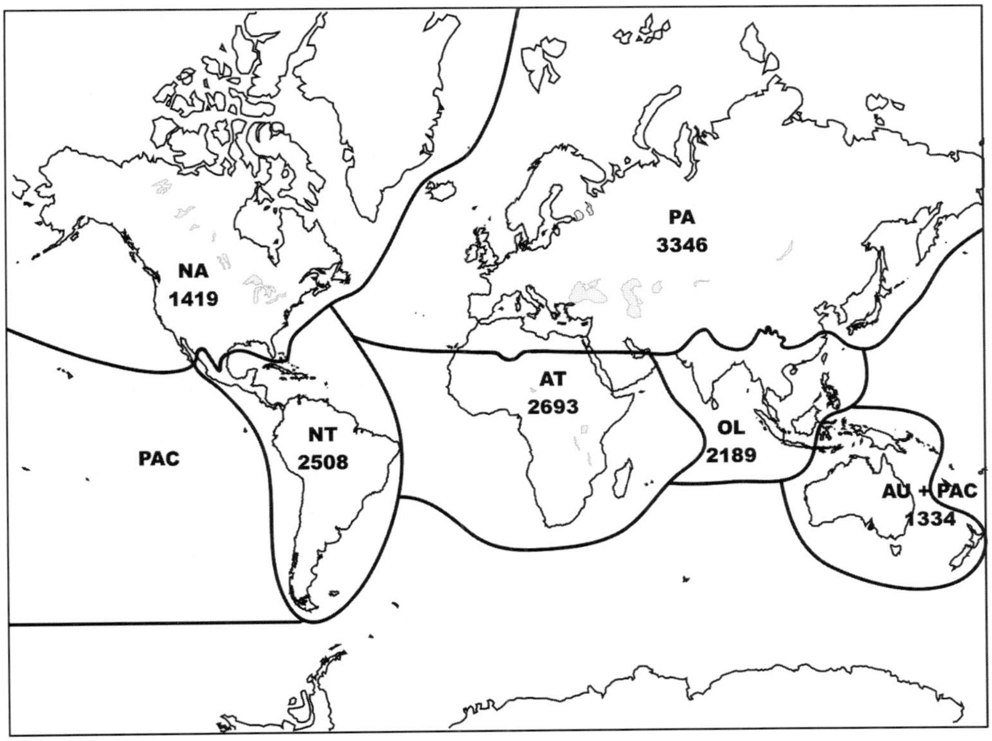

Fig. 2 Global diversity of water beetles by zoogeographic regions. Borders between realms arbitrary. Species from the Antarctic Region (ANT) and the Pacific Region (PAC) have been included to adjacent realms. AU: Australia, AT: Afrotropical, NA: Nearctic, NT: Neotropic, OL: Oriental, PA: Palearctic, PAC: Pacific

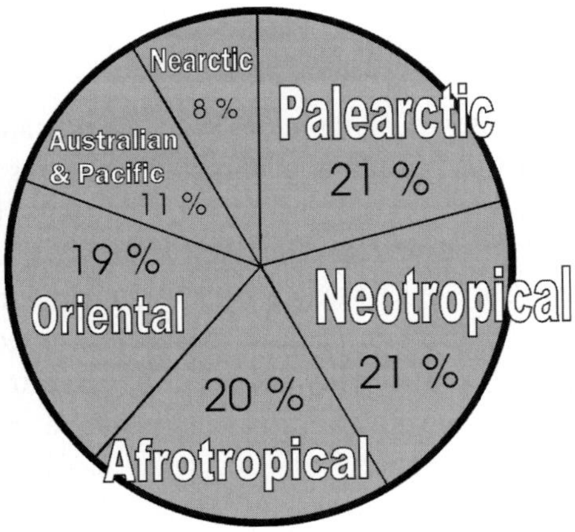

Fig. 3 Estimation of global diversity of water beetles by biogeographic regions

some attention is nowadays paid to that problem (Mogi et al., 1999).

Apart from these marginal relationships between manhood and water beetles, the applied relevance of aquatic Coleoptera can be summarized in three main groups.

Environmental indication

Water beetles, especially Elmidae, are increasingly gaining recognition as biological indicators for (1) water quality (saprobial index), (2) habitat types, (3) biological functionality and (4) species and habitat conservation (red lists) (see Moog, 2003; Jäch et al., 2005b). A comprehensive list of saprobic valencies of the Austrian elmids was published by Moog & Jäch (2003).

The IUCN Red List of Threatened Species (see http://www.redlist.org/) contains also a number of water beetles. However, this list has obviously not been compiled by water beetle experts and it is therefore dearly in need of revision. Two New Caledonian dytiscids, *Rhantus alutaceus* and *R. novaecaledoniae*, both listed as extinct in the 2004 IUCN Red List of Threatened Species, definitely suffer from ongoing habitat loss, but were otherwise still collected during expert fieldwork recently. On the other hand, some species, which have repeatedly been recorded as being probably extinct, e.g.,

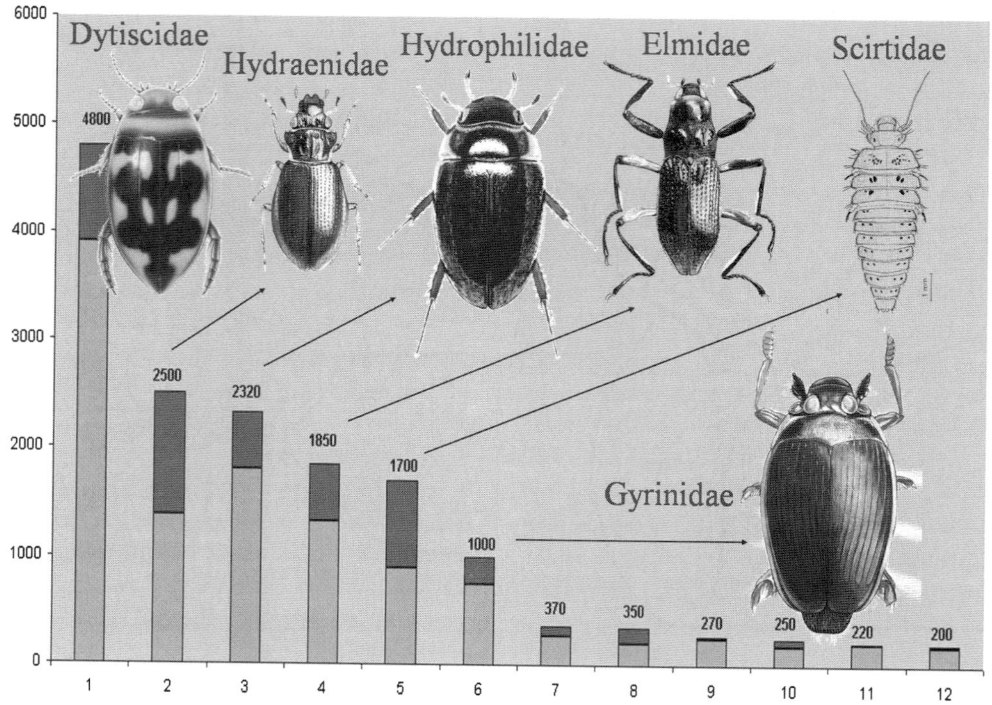

Fig. 4 Global diversity of water beetle families; number of described species (pale grey/blue) and estimation of undescribed (dark grey/purple) species in 12 beetle families; paraquatic and terrestrial members not included; 1—Dytiscidae, 2—Hydraenidae, 3—Hydrophilidae, 4—Elmidae, 5—Scirtidae, 6—Gyrinidae, 7—Psephenidae, 8—Dryopidae, 9—Noteridae, 10—Hydrochidae, 11—Haliplidae, 12—Helophoridae

Hygrobia davidi from China and *Hydraena sappho* from Greece (see Audisio et al., 1996), have not been included in the IUCN Red List so far. Numerous coleopterists (e.g., P. Audisio, M. Hess, U. Heckes, M.A. Jäch, A. Komarek, H. Schillhammer, H. Schönmann) have vainly searched for the latter species in the last two decades.

In general, water beetle communities all around the world suffer from desertification, irrigation, eutrophication induced by lifestock, man, and agriculture, as well as overall loss of primary habitats. Species living in aquatic habitats surrounded by tropical rain forests seem to be especially vulnerable to deforestation, which is devastatingly effecting chemical and physical characters of freshwater biotopes. Furthermore, in numerous water beetles at least one stage of the life cycle is spent outside the water. Therefore, any disturbance of the immediate surroundings of an aquatic habitat must be considered a major threat for its water beetle communities.

Pest and pest control

Several species of Phytophilous Water Beetles are used to control water plants that are regarded as pests while others are destructive to plants of economic importance.

The South American *Agasicles hygrophila* Selman & Vogt (Chrysomelidae: Alticinae) was introduced to several countries on various continents to control Alligatorweed (*Alternanthera*). The Neotropical *Cyrtobagous salviniae* (Erirhinidae) is used as control agent for the Water Fern (*Salvinia*) and has been introduced, for instance, to India, Thailand and Australia. Two species of *Neochetina*, *N. eichhorniae* and *N. bruchi* (Erirhinidae), natives of South America, were released in the USA and Asia where they are most effective in reducing the flowering and potential growth of waterhyacinth (*Eichhornia crassipes*), the world's most important aquatic weed spreading at an alarming rate. The Waterlettuce Weevil, *Neohydronomus affinis* (Erirhinidae), from South America was released in Florida and Australia for biological control of *Pistia stratiotes*. Species of *Nanophyes* (Nanophyidae) are feeding in the stems of invasive plant pests, such as the Water Primrose (*Ludwigia*).

The Rice Water Weevil, *Lissorhoptrus oryzophilus* Kuschel (Erirhinidae), is a most destructive pest of rice (*Oryza sativa*). It originates from America, where it is a particularly severe pest in Louisiana. Larvae of *L. oryzophilus* are responsible for the main damage as they feed on leafs for a short period and then crawl down to the roots; they have paired dorsal hooks to pierce the roots for obtaining oxygen. The fourth larval instar forms a mud-coated cocoon attached to the roots. Serious crop losses are reported in all countries where this species occurs (up to 60% in Japan). *Lissorhoptrus oryzophilus* has been spread accidentally to all continents except Africa and Australia. To Europe (Italy) it was introduced rather recently (Caldara et al., 2004). *Lissorhoptrus oryzophilus* is not only a pest of rice, but it also attacks many other wild grasses and sedges (Poaceae and Cyperaceae) which serve as alternative hosts for adults in or near rice fields, rendering pest management most difficult. Furthermore, resistance to some insecticides has also been reported. In Asia and in California only parthenogenetic females are known. Other Erirhinidae, e.g., *Lissorhoptrus simplex* (Say) and *Afroryzophilus djibai* Lyal, are causing heavy damages too.

Cuisine and medicine

The medical/nutritive relationships between water beetles and mankind have obviously been more diverse in the past, but they have survived to some extent. Even today, various species are used for human consumption (both as medicine and as confection). Eating water beetles is still practised in Madagascar, New Guinea and Asia, for instance in China, where two genera, *Hydrophilus* Geoffroy (Hydrophilidae) and *Cybister* Curtis (Dytiscidae) are offered for sale in Guangdong markets and restaurants. Remarkably, *Cybister* is more expensive than *Hydrophilus* (Jäch, 2003). Ding et al. (1997) published a list of medicinal insects in China and mentioned that *Cybister* was consumed in order to improve blood circulation. According to Hoffmann (1947) water beetles are considered as an anti-diuretic.

More than a century ago, *Austrelmis condimentarius* (Philippi) (Elmidae), was used as seasoning for food in South America, and it was reported to have considerable commercial value (Philippi, 1864; Netolizky, 1916).

In the early 19th century, *Aulonogyrus strigosus* F. (Gyrinidae), was roasted and eaten by Australian aborigines (Mjöberg, 1916; Ochs, 1924) and approximately at the same time, gyrinids were used in Europe as an aphrodisiac for cows and mares (Netolizky, 1916, 1919; Ochs, 1966).

Rubbing Gyrinidae (*Dineutus*, *Gyrinus*, etc.) and Dytiscidae (*Hydaticus*, *Rhantus*) onto young girls breasts for stimulating their growth is a unique traditional practice still alive and wide-spread in East Africa (Kutalek & Kassa, 2005). In their prothoracic defence glands dytiscids produce steroids resembling human hormones! However, there is insufficient evidence to conclude that the chemical defence mechanism of the water beetles is in fact responsible for a possible growth of the breasts.

Acknowledgements The following persons are sincerely thanked for providing useful information and comments: M. Alonso-Zarazaga, R. Angus, R. Caldara, E. Colonnelli, M.-L. Jeng, J. Kodada, A. Komarek, A.S. Konstantinov, D. Kovac, C.-F. Lee, P. Mazzoldi, A.N. Nilsson, I. Ribera, W. Rücker, R. Schuh, W.D. Shepard, C.H.S. Watts and H. Yoshitomi. Special thanks are due to M. Alonso-Zarazaga, R. Angus, R. Caldara, E. Colonnelli, M.-L. Jeng, A. Komarek, A.S. Konstantinov, C.-F. Lee, P. Mazzoldi, A.N. Nilsson and H. Yoshitomi for providing updated species numbers from their personal databases.

References

Aberlenc, H.-P. & R. Allemand, 1997. Acclimatation en France de Ptilodactyla exotica, espèce à moers anthrophiles (Coleoptera, Ptilodactylidae). Bulletin de la Société Entomologique de France 102(2): 93–102.

Absolon, K., 1915. Bericht über höhlenbewohnende Staphyliniden der dinarischen und angrenz enden Karstgebiete. Koleopterologische Rundschau 4: 132–151.

Alonso-Zarazaga, M. A. & C. H. C. Lyal, 1999. A world catalogue of families and genera of Curculionoidea (Insecta: Coleoptera). (Excepting Scolytidae and Platypodidae). Entomopraxis, Barcelona: 315 pp.

Anderson, R. S., 2002. II. Erirhininae Schoenherr 1825. In Arnett, R. H. Jr. et al. (eds), American beetles, Vol. II. CRC Press, Boca Raton, London, New York, Washington: 730–732.

Angus, R. B., 1966. A note on the swimming of Bagous limosus (Gyll.) (Col. Curculionidae). Entomologist's Monthly Magazine 101: 202.

Angus, R. B., 1992. Süßwasserfauna von Mitteleuropa. Bd. 20: Insecta. Coleoptera. 10. Hydro philidae. 2. Helophorinae. Spektrum Akademischer Verlag, Heidelberg: XII+144 pp.

Arce-Pérez, R., 1997. Sinopsis del suborden Myxophaga (Coleoptera) de México. Dugesiana 4(2): 41–50.

Arce-Pérez, R. & W. D. Shepard, 2001. Sinopsis de la familia Psephenidae (Coleoptera: Dryop oidea) de Norte y Centroamérica. Folia Entomológica Mexicana 40(3): 397–406.

Audisio, P., A. De Biase & M. A. Jäch, 1996. Greek species of the genus Hydraena (Subgenus Haenydra Rey) (Coleoptera, Hydraenidae). Aquatic Insects 18(2): 65–90.

Balke, M. & L. Hendrich, 1989. Verbreitung, Lebensweise, Taxonomie und Historie der Dytisciden der Ilha da Madeira (Coleoptera: Dytiscidae). Boletin do Museu Municipal do Funchal XLI 212: 55–83.

Balke, M., L. Hendrich, P. Mazzoldi & O. Biström, 2002. Borneodessus zetteli, new genus and new species, notes on biogeography and a checklist of Dytiscidae from Borneo (Insecta: Coleoptera: Dytiscidae). Journal of Natural History 36: 963–978.

Balke, M., M. A. Jäch & L. Hendrich, 2004a. Insecta: Coleoptera. In Yule, C. M. & H. S. Yong (eds), Freshwater Invertebrates of the Malaysian Region. Academy of Sciences Malaysia, Kuala Lumpur: 555–609.

Balke, M., I. Ribera & R. G. Beutel, 2003. Aspidytidae: On the discovery of a new beetle family: detailed morphological analysis, description of a second species, and key to fossil and extant adephagan families (Coleoptera). In Jäch, M. A. & L. Ji (eds), Water Beetles of China, Vol. III. Zoologisch-Botanische Gesellschaft & Wiener Coleopterologenverein, Wien: 53–66.

Balke, M., I. Ribera & R. G. Beutel, 2005. The systematic position of Aspidytidae, the diversification of Dytiscoidea (Coleoptera, Adephaga) and the phylogenetic signal of third codon positions. Journal for zoological systematics and evolutionary research 43: 223–242.

Balke, M., I. Ribera & A. P. Vogler, 2004b. MtDNA phylogeny and biogeography of Copelatinae, a highly diverse group of tropical diving beetles (Dytiscidae). Molecular Phylogenetics and Evolution 32: 866–880.

Betz, O., 2002. Performance and adaptive value of tarsal morphology in rove beetles of the genus Stenus (Coleoptera, Staphylinidae). The Journal of Experimental Biology 205: 1097–1113.

Beutel, R. G., 1997. Über Phylogenese und Evolution der Coleoptera (Insecta). Abhandlungen des Naturwissenschaftlichen Vereins Hamburg (NF) 31: 1–164.

Beutel, R. G., M. Balke & W. J. Steiner, 2006. The systematic position of Meruidae (Coleoptera, Adephaga) based on a cladistic analysis of morphological characters. Cladistics 22: 102–131.

Beutel, R. G. & R. A. B. Leschen (eds), 2005a. Handbuch der Zoologie/Handbook of Zoology, Vol. IV, Arthropoda: Insecta, Part 38, Coleoptera, Beetles, Vol. I: Morphology and Systematics. de Gruyter, Berlin: xii+578 pp.

Beutel, R. G. & R. A. B. Leschen, 2005b. 10.18. Chelonariidae, Blanchard, 1845. In Beutel, R. G. & R. A. B. Leschen (eds), Handbook of Zoology, Vol. IV (Part 38), Coleoptera, Beetles, Vol. I: Morphology and Systematics (Archostemata, Adephaga, Myxophaga, Polyphaga partim). Walter de Gruyter, Berlin: 543–546.

Beutel, R. G. & R. A. B. Leschen, 2005c. Phylogenetic analyses of Staphyliniformia (Coleoptera) based on characters of larvae and adults. Systematic entomology 30(4): 410–548.

Beutel, R. G. & G. B. Raffaini, 2003. First record of Sphaeriusidae for Argentina (Coleoptera: Myxophaga). Koleopterologische Rundschau 73: 1–6.

Beutel, R. G. & R. E. Roughley, 2005. 7.1. Gyrinidae, Latreille, 1810. In Beutel, R. G. & R. A. B. Leschen (eds), Handbook of Zoology, Vol. IV (Part 38), Coleoptera, Beetles, Vol. I: Morphology and Systematics (Archostemata, Adephaga, Myxophaga, Polyphaga partim). Walter de Gruyter, Berlin: 55–64.

Brown, H. P., 1972. Aquatic dryopoid beetles of the United States. Biota of Freshwater Ecosystems, Identification Manual 6: 1–82.

Brown, H. P., 1981a. A distributional survey of the world genera of aquatic dryopoid beetles (Coleoptera: Dryopidae, Elmidae, and Psephenidae sens.lat.). Pan-Pacific Entomologist 57(1): 133–148.

Brown, H. P., 1981b. Key to the world genera of Larinae (Coleoptera, Dryopoidea, Elmidae), with descriptions of new genera from Hispaniola, Colombia, Australia, and New Guinea. Pan-Pacific Entomologist 57(1): 76–104.

Brown, H. P. & C. M. Murvosh, 1970. Lutrochus arizonicus new species, with notes on ecology and behaviour (Coleoptera, Dryopoidea, Limnichidae). Annals of the Entomological Society of America 63(4): 1030–1035.

Buckingham, G. R. & C. A. Bennet, 1989. Dyscinetus morator (Fab.) (Coleoptera: Scarabaeidae) adults attack waterhyacinth, Eichhornia crassipes (Pontederiaceae). The Coleopterists Bulletin 43(1): 27–33.

Buschman, L. L., 1984. Biology of the firefly Pyractomena lucifera (Coleoptera: Lampyridae). Florida Entomologist 67(4): 529–542.

Caldara, R. & O'Brien, C. W. 1998. Systematics and evolution of Weevils of the genus Bagous. VI. Taxonomic treatment of the species of the Western Palearctic Region (Coleoptera Curculionidae). Memorie della Società entomologica italiana 76 [1997]: 131–347.

Caldara, R., L. Diotti & R. Regalin, 2004. First record for Europe of the rice water weevil, Lissorhoptrus oryzophilus Kuschel (Coleoptera, Curculionoidea, Erirhinidae). Bollettino di Zoologia Agraria e di Bachicoltura, Serie II, 36(1): 165–171.

Carabajal, E., J. García & F. Rodríguez, 2000. Descripción de un nuevo género y una nueva especie de Trechini (Coleoptera: Caraboidea: Trechidae) de la cordillera Cantábrica. Elytron 13 [1999]: 123–131.

Caterino, M. S., T. Hunt & A. P. Vogler, 2005. On the constitution and phylogeny of Staphyliniformia (Insecta: Coleoptera). Molecular Phylogenetics and Evolution 34: 655–672.

Chappuis, P. A., 1927. Die Tierwelt der unterirdischen Gewässer. In A. Thienemann (eds), Die Binnengewässer III. Schweizerbart, Stuttgart: 175 pp., 4 plts.

Ciegler, J., 2003. Biota of South Carolina, Vol. III: Water Beetles of South Carolina (Coleoptera: Gyrinidae, Haliplidae, Noteridae, Dytiscidae, Hydrophilidae, Hydraenidae, Scirtidae, Elmidae, Dryopidae, Limnichidae, Heteroceridae, Psephenidae, Ptilodactylidae, and Chelonariidae). Clemson University, Clemson: 210 pp.

Colonnelli, E., 2004. Catalogue of Ceutorhynchinae of the world, with a key to genera (Insecta: Coleoptera: Curculionidae). Argania editio, Barcelona: 124 pp.

Costa, C., S. A. Vanin & S. Ide, 2005. 18.8. Cneoglossidae Guérin-Méneville, 1843. In Beutel, R. G., & R. A. B. Leschen (eds), Handbook of Zoology, Vol. IV (Part 38), Coleoptera, Beetles, Vol. I: Morphology and Systematics (Archostemata, Adephaga, Myxophaga, Polyphaga partim). Walter de Gruyter, Berlin: 533–536.

Crowson, R. A., 1981. The biology of Coleoptera. Academic Press, London: 802 pp.

Ding, Z., Y. Zhao & X. Gao, 1997. Medicinal insects in China. Ecology of Food and Nutrition 36: 209–220.

Downie, N. M. & R. H. Arnett Jr., 1996. The beetles of northeastern North America, Vol. I, II. The Sandhill Crane Press, Gainseville: 1721 pp.

Drost, M. B. P., H. P. P. J. Cuppen, E. J. Nieukerken van, M. Schreijer (eds), 1992. De waterkevers van Nederland. Utrecht: Uitgeverij K.N.N.V., 279 pp.

Endrödy-Younga, S., 1997a. Active extraction of water-dissolved oxygen and descriptions of new taxa of Torridincolidae (Coleoptera: Myxophaga). Annals of the Transvaal Museum 36(24): 313–332.

Endrödy-Younga, S., 1997b. Microsporidae (Coleoptera: Myxophaga), a new family for the African continent. Annals of the Transvaal Museum 36(23): 309–311.

Gerdes, G., J. Spira & C. Dimentman, 1985. 15. The fauna of the Gavish Sabkha and the Solar Lake – a comparative study. In Friedman, G. M. & W. E. Krumbein (eds), Ecological Studies. Vol. 53: Hypersaline Ecosystems. Springer Verlag, Berlin: 322–345.

Hall, W. E. 2003. Sphaeriusidae (Coleoptera). In M. A. Jäch, & L. Ji (eds), Water Beetles of China, Vol. III. Wien: Zoologisch-Botanische Gesellschaft & Wiener Coleopterologenverein: 37–41.

Hansen, M. 1998. Hydraenidae (Coleoptera). In Hansen, M. (ed.), World Catalogue of Insects, Vol. I. Stenstrup: Apollo Books: 168 pp.

Hansen, M., 1997. A new subfamily for a remarkable new genus and species of Hydrophilidae from New Guinea (Coleoptera: Hydrophilidae). Annales Zoologici (Warszawa) 47: 107–110.

Hansen, M., 1991. The Hydrophiloid Beetles. Phylogeny, classification and a revision of the genera (Coleoptera, Hydrophiloidea). Biologiske Skrifter 40: 1–368.

Hansen, M. 1999. Hydrophiloidea (Coleoptera). In Hansen, M. (ed.), World Catalogue of Insects, Vol. II. Stenstrup: Apollo Books: 416 pp.

Hendrich, L., M. Balke & C. M. Yang, 2004. Aquatic Coleoptera of Singapore: species richness, ecology and conservation. The Raffles Bulletin of Zoology 52(1): 97–145.

Ho, J.-Z., J.-S. Ju & J.-C. Ju, 1998. Discovery of the aquatic larvae of Luciola substriata. Nature Conservation Quarterly 22: 47–51 [in Chinese].

Hoffmann, W. E., 1947. Insects as human food. Proceedings of the Entomological Society of Washington 49(9): 231–237.

Hughes, J., S. J. Longhorn, K. Theodorides, A. de Riva, M. Mejia-Chang, P. G. Foster & A. P. Vogler, 2006. Dense taxonomic EST sampling and its applications for molecular systematics of the Coleoptera (Beetles). Molecular Biology and Evolution 23(2): 268–278.

Ide, S., C. Costa & S. A. Vanin, 2005. 18.4. Lutrochidae Kasap & Crowson, 1975. In Beutel, R. G. & R. A. B. Leschen (eds), Handbook of Zoology, Vol. IV (Part 38), Coleoptera, Beetles, Vol. I: Morphology and Systematics (Archostemata, Adephaga, Myxophaga, Polyphaga partim). Walter de Gruyter, Berlin: 508–512.

Ivie, M. A. 2002. 50. Chelonariidae Blanchard 1845. In Arnett, R. H., M. C. Thomas, P. E. Skelley & J. H. Frank (eds), American Beetles, Vol 2. CRC Press, Boca Raton: 139–141.

Ivie, M. A., 2005. 18.11. Eulichadidae Crowson, 1973. In Beutel, R. G. & R. A. B. Leschen (eds), Handbook of Zoology, Vol. IV (Part 38), Coleoptera, Beetles, Vol. 1: Morphology and Systematics (Archostemata, Adephaga, Myxophaga, Polyphaga partim). Walter de Gruyter, Berlin: 547–551.

Jäch, M. A., 1995a. Hydroscaphidae (Coleoptera). In Jäch, M. A. & L. Ji (eds), Water Beetles of China, Vol. I. Zoologisch-Botanische Gesellschaft in Österreich and Wiener Coleopterologenverein, Wien: 33–34.

Jäch, M. A., 1995b. Eulichadidae: synopsis of the species of the genus Eulichas Jacobson from China, Laos and Vietnam (Coleoptera). In Jäch, M. A. & L. Ji (eds), Water Beetles of China, Vol. I. Zoologisch-Botanische Gesellschaft in Österreich and Wiener Coleopterologenverein, Wien: 359–388.

Jäch, M. A., 1998a. Annotated check list of aquatic and riparian/littoral beetle families of the world (Coleoptera). In Jäch, M. A. & L. Ji (eds), Water Beetles of China, Vol. II. Zoologisch-Botanische Gesellschaft in Österreich and Wiener Coleopterologenverein, Wien: 25–42.

Jäch, M. A., 1998b. Torridincolidae: I. First record of Torridincolidae from China (Coleoptera). In Jäch, M. A. & L. Ji (eds), Water Beetles of China, Vol. II. Zoologisch-Botanische Gesellschaft in Österreich and Wiener Coleopterologenverein, Wien: 51–52.

Jäch, M. A., 2002. First record of Epimetopidae in Laos (Coleoptera: Epimetopidae). Koleoptero logische Rundschau 72: 161–164.

Jäch, M. A., 2003. Fried water beetles Cantonese style. American Entomologist 49(1): 34–37. http://www.nhm-wien.ac.at/nhm/2Zoo/coleoptera/publications/183.pdf.

Jäch, M. A., 2004. The China Water Beetle Survey (CWBS) – a biodiversity project of the superlative celebrates its 10th anniversary. Procupine 30: 5–7.

Jäch, M. A. & M. Balke, 2003. Key to adults of Chinese water beetle families (Coleoptera). In Jäch, M. A & L. Ji (eds), Water Beetles of China, Vol. III. Zoologisch-Botanische Gesellschaft & Wiener Coleopterologenverein, Wien: 21–36.

Jäch, M. A., R. G. Beutel, J. A. Delgado & J. A. Díaz, 2005a. 11.1. Hydraenidae Mulsant, 1844. In Beutel, R. G. &R. A. B. Leschen (eds), Handbook of Zoology, Vol. IV (Part 38), Coleoptera, Beetles, Vol. I: Morphology and Systematics (Archostemata, Adephaga, Myxophaga, Polyphaga partim). Walter de Gruyter, Berlin: 224–251.

Jäch, M. A., R. G. Beutel, J. A. Díaz & J. Kodada, 2000. Subgeneric classification, description of head structures, and world check list of Hydraena Kugelann (Insecta: Coleoptera: Hydraenidae). Annalen des Naturhistorischen Museums in Wien 102B: 177–258.

Jäch, M. A., F. Dietrich & B. Raunig, 2005b. Rote Liste der Zwergwasserkäfer (Hydraenidae) und Krallenkäfer (Elmidae) Österreichs (Insecta: Coleoptera). In Zulka, R. M. (ed.), Rote Listen gefährdeter Tiere Österreichs. Checklisten, Gefährdungsanalyse, Handlungsbedarf. Part 1: Säugetiere, Vögel, Heuschrecken, Wasserkäfer, Netzflügler, Schnabelfliegen, Tagfalter (Grüne Reihe des Lebensministeriums, Vol. 14/1). Bundesministerium für Land- und Forstwirtschaft, Umwelt und Wirtschaft, Wien: 211–284.

Jäch, M. A. & E. R. Easton, 1998. Water beetles of Macao (Coleoptera). In Jäch, M. A. & L. Ji (eds), Water Beetles of China, Vol. II. Zoologisch-Botanische Gesellschaft in Österreich and Wiener Coleopterologenverein, Wien: 43–50.

Jäch, M. A. & L. Ji (eds), 1995. Water Beetles of China, Vol. I. Zoologisch-Botanische Gesellschaft in Österreich and Wiener Coleopterologenverein, Wien: 410 pp.

Jäch, M. A. & L. Ji (eds), 1998. Water Beetles of China, Vol. II. Zoologisch-Botanische Gesellschaft in Österreich and Wiener Coleopterologenverein, Wien: 371 pp.

Jäch, M. A. & L. Ji (eds), 2003. Water Beetles of China, Vol. III. Zoologisch-Botanische Gesellschaft in Österreich and Wiener Coleopterologenverein, Wien: VI+572 pp.

Jeng, M.-L., J. Lai & P.-S. Yang, 2003. Lampyridae: A synopsis of aquatic fireflies with description of a new species (Coleoptera). In Jäch, M. A. & L. Ji (eds), Water Beetles of China, Vol. III. Zoologisch-Botanische Gesellschaft & Wiener Coleopterologenverein, Wien: 539–562.

Ji, L. & M. A. Jäch, 1998. Epimetopidae: Synopsis of the genus Eumetopus Balfour-Browne (Coleoptera). In Jäch M. A. & L. Ji (eds), Water Beetles of China, Vol. II. Zoologisch-Botanische Gesellschaft in Österreich and Wiener Coleopterologen verein, Wien: 195–205.

Jolivet, P. & K. K. Verma, 2002. Biology of Leaf Beetles. Intercept Limited, Andover: 332 pp.

Klausnitzer, B. 1996. Käfer im und am Wasser. Westarp Wissenschaften, Magdeburg. Spektrum Akademischer Verlag, Heidelberg (Die Neue Brehm-Bücherei; Bd. 567), 200 pp.

Klausnitzer, B., 2004. Eine neue Gattung der Familie Scirtidae (Insecta: Coleoptera). Entomologische Abhandlungen 62(1): 77–82.

Kodada, J. & M. A. Jäch, 2005a. 18.2. Elmidae Curtis, 1830. In Beutel, R. G., & R. A. B. Leschen (eds), Handbook of Zoology, Vol. IV (Part 38), Coleoptera, Beetles, Vol. 1: Morphology and Systematics (Archostemata, Adephaga, Myxophaga, Polyphaga partim). Walter de Gruyter, Berlin: 471–496.

Kodada, J. & M. A. Jäch, 2005b. 18.3. Dryopidae Bilberg, 1820 (1817). In Beutel, R. G. & R. A. B. Leschen (eds), Handbook of Zoology, Vol. IV (Part 38), Coleoptera, Beetles, Vol. 1: Morphology and Systematics (Archostemata, Adephaga, Myxophaga, Polyphaga partim). Walter de Gruyter, Berlin: 496–508.

Konstantinov, A. S., 2003. Chrysomelidae: Aquatic leaf beetles of China (Coleoptera). In Jäch, M. A. & L. Ji (eds), Water Beetles of China, Vol. III. Zoologisch-Botanische Gesellschaft & Wiener Coleopterologenverein, Wien: 563–572.

Kutalek, R. & A. Kassa, 2005. The use of gyrinids and dytiscids for stimulating breast growth in East Afica. Journal of Ethnobiology 25(1): 115–128.

Larson, D. J., Y. Alarie & R. E. Roughley, 2000. Predaceous diving beetles (Coleoptera: Dytiscidae) of the Nearctic region, with emphasis on the fauna of Canada and Alaska. NRC Research Press, Ottawa: 982 pp.

Lawrence, J. F., 2005. 15.4. Scirtidae Fleming, 1821. In Beutel, R. G. & R. A. B. Leschen (eds), Handbook of Zoology, Vol. IV (Part 38), Coleoptera, Beetles, Vol. 1: Morphology and Systematics (Archostemata, Adephaga, Myxophaga, Polyphaga partim). Walter de Gruyter, Berlin: 443–450.

Lawrence, J. F. & E. B. Britton, 1994. Australian Beetles. Melbourne University Press, 192 pp.

Lawrence, J. F. & A. F. Newton, 1995. Families and subfamilies of Coleoptera (with selected genera, notes, references and data on family-group names). In Pakaluk, J. & S. A. Ślipiński (eds), Biology, phylogeny and classification of Coleoptera: Papers celebrating the 80th birthday of Roy A. Crowson. Muzeum i Instytut Zoologii PAN, Warszawa: 779–1006+48.

Lawrence, J. F. & J. B. Stribling, 1992. A new genus of Ptilodactylidae (Coleoptera: Elateriformia) from North Queensland, with description of the presumed larva. Journal of the Australian Entomological Society 31(1): 19–27.

Lee, C.-F., M. A. Jäch & R. G. Beutel, 2005. 18.7. Psephenidae Lacordaire, 1854. In Beutel, R. G. & R. A. B. Leschen (eds), Handbook of Zoology, Vol. IV (Part 38), Coleoptera, Beetles, Vol. 1: Morphology and Systematics (Archostemata, Adephaga, Myxophaga, Polyphaga partim). Walter de Gruyter, Berlin: 521–533.

Lévêque, C., E. V. Balian & K. Martens, 2005. An assessment of animal species diversity in continental waters. Hydrobiologia 542: 39–67.

Leys, R., S. J. B. Cooper, C. H. S. Watts & W. F. Humphreys, 2003. Evolution of subterranean diving beetles (Coleoptera, Dytiscidae, Hydroporini, Bidessini) in the arid zone of Australia. Evolution 57: 2819–2834.

Löbl, I., 1994. Les espèces asiatiques du genre Hydroscapha Leconte (Coleoptera, Hydroscaph idae). Archives de Sciences, Genève 47(1): 15–34.

Löbl, I., 1995. New species of terrestrial Microsporus from the Himalaya (Coleoptera: Microspor idae). Entomologische Blätter 91(3): 129–138.

Löbl, I, 2003a. Family Hydroscaphidae LeConte, 1874. In Löbl, I. & A. Smetana (eds), Catalogue of Palaearctic Coleoptera, Vol. 1, p. 35. Stenstrup: Apollo Books: 25.

Löbl, I., 2003b. Family Sphaeriusidae Erichson, 1845. In Löbl, I. & Smetana A. (eds), Catalogue of Palaearctic Coleoptera, Vol. 1, p. 35. Apollo Books, Stenstrup: 25–26.

Lucht, W., 1998. (Familie: Ptilodactylidae). In Lohse, G. A. & B. Klausnitzer (eds), Die Käfer Mitteleuropas XV (4th suppl.). Goecke & Evers, Krefeld: 236.

Mazzoldi, P., 1995. Gyrinidae: Catalogue of Chinese Gyrinidae. In Jäch, M. A. & L. Ji (eds), Water Beetles of China, Vol. I. Zoologisch-Botanische Gesellschaft and Wiener Coleopterologenverein, Wien: 155–172.

Mjöberg, E., 1916. Results of Dr. E. Mjöberg's Swedish Scientific Expedition to Australia 1910–1913. 9. Cicindelidae, Gyrinidae, Lucanidae, Paussidae. Arkiv för Zoologi 10(10): 1–16.

Mogi, M., T. Sunahara & M. Selomo, 1999. Mosquito and aquatic predator communities in ground pools on lands deforested for rice field development in Central Sulawesi,

Indonesia. Journal of the American Mosquito Control Association 15(2): 92–97.
Mohr, K.-H., 1960. Erdflöhe. Neue Brehm Bücherei, Wittenberg: 261 pp.
Moog, O. (ed.), 2003. Fauna Aquatica Austriaca, edition 2002. Wien: Wasserwirtschaftskataster, Bundesministerium für Land- und Forstwirtschaft, Umwelt und Wasserwirtschaft. http://www.wassernet.at/article/archive/5726/.
Moog, O. & M. A. Jäch, 2003. Elmidae. In Moog, O. (ed.): Fauna Aquatica Austriaca, edition 2002. Wien: Wasserwirtschaftskataster, Bundesministerium für Land- und Forstwirtschaft, Umwelt und Wasserwirtschaft. http://www.wassernet.at/article/archive/5726/.
Netolizky, F. 1916. Insekten als Heilmittel. Pharmazeutische Post: 1–45.
Netolizky, F. 1919. Käfer als Nahrungs- und Heilmittel. Koleopterologische Rundschau 7(9–10): 121–129; 8(1–3): 21–26; 8(4–6): 47–60.
Nilsson, A. (ed.), 1996. Aquatic Insects of North Europe. A Taxonomic Handbook, Vol. 1. Apollo Books, Stenstrup: 274 pp.
Nilsson, A. N. 2001. Dytiscidae (Coleoptera). In World Catalogue of Insects 3. Stenstrup: Apollo Books, 395 pp.
Nilsson, A. N. 2005a. Family Noteridae (Coleoptera, Adephaga). In Nilsson, A. N., Vondel & B. J. van 2005 (eds), Amphizoidae, Aspidytidae, Haliplidae, Noteridae and Paelobiidae (Coleoptera, Adephaga). World Catalogue of Insects, Vol. 7. Apollo Books, Stenstrup: 87–158.
Nilsson, A. N. 2005b. Family Amphizoidae (Coleoptera, Adephaga). In Nilsson, A. N. & Vondel, B. J. van 2005 (eds), Amphizoidae, Aspidytidae, Haliplidae, Noteridae and Paelobiidae (Coleoptera, Adephaga). World Catalogue of Insects, Vol. 7. Apollo Books, Stenstrup: 15–18.
Nilsson, A. N. 2005c. Family Aspidytidae (Coleoptera, Adephaga). In Nilsson, A. N. & Vondel, B. J. van 2005 (eds), Amphizoidae, Aspidytidae, Haliplidae, Noteridae and Paelobiidae (Coleoptera, Adephaga). World Catalogue of Insects, Vol. 7. Apollo Books, Stenstrup: 19.
Nilsson, A. N. 2005d. Family Paelobiidae (Coleoptera, Adephaga). In Nilsson, A. N. & van B. J. Vondel, 2005 (eds), Amphizoidae, Aspidytidae, Haliplidae, Noteridae and Paelobiidae (Coleoptera, Adephaga). World Catalogue of Insects, Vol. 7. Apollo Books, Stenstrup: 154–163.
Ochs, G., 1924. Über neue und interessante Gyriniden aus dem British Museum in London. Entomologische Blätter für Biologie und Systematik der Käfer 20: 228–244.
Ochs, G., 1966. Vom Geruch der Taumelkäfer. Entomologische Blätter 62(1): 14–16.
Pain, S., 2005. Life underground, down under. New Scientist 2511: 28–33.
Peacock, E. R., 1978. Notes on the subgenera of European Rhizophagidae (Coleopetra). Entomo logica Scandinavica 9: 236–239.
Pemberton, R. W., 1990. The Korean Water Beetle Game. Pan-Pacific Entomologist 66: 172–174.
Philippi, R. A., 1864. Ein Käferchen, daß als Gewürz dient. Stettiner Entomologische Zeitung 25: 93–96.
Reichardt, H., 1976b. Revision of the Lepiceridae (Coleoptera, Myxophaga). Papéis Avulsos de Zoologia 30(3): 35–42.
Remy, P., 1940. Sur le mode de vie des Hadesia dans la grotte Vjetrenica (Col. Bathysciinae). Revue Francaise d'Entomologie VII(1): 1–8.
Ribera, I., 2000. Biogeography and conservation of Iberian water beetles. Biological Conservation 92: 131–150.
Ribera, I., R. G. Beutel, M. Balke & A. P. Vogler, 2002. Discovery of Aspidytidae, a new family of aquatic Coleoptera. Proceedings of the Royal Society, London B 269: 2351–2356.
Rogers, C., 1997. Aphodius alternatus Horn (Aphodiinae: Scarabaeidae), first record of a semi aquatic scarab beetle. Pan-Pacific Entomologist 73(2): 135–136.
Sankaran, T., 1972. Nanophyes sp. nr. nigritulus Boh. (Col., Curculionidae), a promising biocontrol agent against water-primrose, Ludwigia adscendens (L.). Commonwealth Institute of Biological Control, Bangalore, India, Hara: 8 pp.
Schaefer, M. & W. Tischler, 1983. Wörterbücher der Biologie. Ökologie. Gustav Fischer, Jena: 354 pp.
Schillhammer, H., 2002. Three new Oriental species of Hesperus Fauvel (Coleoptera: Staphylin idae). Koleopterologische Rundschau 72: 127–135.
Seifert, R. P., 1982. Neotropical Heliconia insect communities. The Quaterly Review of Biology 57: 1–28.
Short, A. E. Z. & F. Hebauer, 2005. World Catalogue of Hydrophiloidea – additions and corrections, 1 (1999–2005) (Coleoptera). Koleopterologische Rundschau 76: 315–359.
Shull, V. L., A. P. Vogler, M. D. Baker, D. R. Maddison & P. M. Hammond, 2001. Sequence alignment of 18S Ribosomal RNA and the basal relationships of Adephagan beetles: evidence for monophyly of aquatic families and the placement of Trachypachidae. Systematic Biology 50: 945–969.
Skale, A. & M. A. Jäch, 2003. A new species of Eumetopus Balfour-Browne, 1949 from Nepal (Insecta: Coleoptera: Epimetopidae). In Hartmann, M., & H. Baumbach (eds), Biodiversität und Naturausstattung im Himalaya. Verein der Freunde & Förderer des Naturkundemuseums Erfurt e.V., Erfurt: 195–196, XVI plts.
Spangler, P. J., 1980a. Chelonariid Larva, Aquatic or Not? The Coleopterists Bulletin 34(1): 105–114.
Spangler, P. J., 1980b. A new species of Ytu from Brazil (Coleoptera: Torridincolidae). The Coleopterists Bulletin 34(2): 145–158.
Spangler, P. J. & W. E. Steiner, 2005. A new aquatic beetle family, Meruidae, from Venezuela. Systematic Entomology 30(3): 339–357.
Stribling, J. B., 1986. Revision of Anchytarsus Coleoptera Dryopoidea and a key to the new world genera of Ptilodactylidae. Annals of the Entomological Society of America 79(1): 219–234.
Torre-Bueno, J. R. de la, 1989. The Torre-Bueno glossary of Entomology. The New York Entomological Society, New York: 840 pp.
Vazirani, T. G., 1972 [?]. Aquatic insects in relation to fisheries in India. Cheetal, Journal of the Wildlife and Fisheries Service of India 17(2): 14–24.
Vogler, A. P., 2005. 3. Molecular systematics of Coleoptera: What has been achieved so far? In Beutel, R. G., & R. A. B. Leschen (eds), Handbook of Zoology, Vol. IV (Part 38), Coleoptera, Beetles, Vol. 1: Morphology and Systematics

(Archostemata, Adephaga, Myxophaga, Polyphaga partim). Walter de Gruyter, Berlin: 17–22.
Vondel, B.J. van, 2005. Family Haliplidae (Coleoptera, Adephaga). In Nilsson, A. N. & Vondel, B. J. van (eds), Amphizoidae, Aspidytidae, Haliplidae, Noteridae and Paelobiidae (Coleoptera, Adephaga). World Catalogue of Insects, Vol. 7. Apollo Books, Stenstrup: 20–86.
Wachmann, E., R. Platen & D. Barndt, 1995. Laufkäfer. Beobachtung – Lebensweise. Naturbuch Verlag, Augsburg: 295 pp.
Wesenberg-Lund, C., 1943. Biologie der Süßwasserinsekten. Springer, Berlin: 682 pp.
Yoshitomi, H. & M. Satô, 2005. Scirtidae of the Oriental Region, Part 8. Entomological Review of Japan 60(2): 153–206.

Hydrobiologia (2008) 595:443–445
DOI 10.1007/s10750-007-9133-y

Global biodiversity of Scorpionflies and Hangingflies (Mecoptera) in freshwater

Leonard C. Ferrington Jr.

Abstract Seven species and one sub-species of Mecoptera are aquatic as larvae and pupae. All aquatic species are classified in two genera of the family Nannochoristidae and have very restricted geographic ranges, with three species confined to extreme southern South America, three species and one sub-species confined to New South Wales or Tasmania in Australia, and one species confined to the South Island of New Zealand.

Keyword Nannochoristidae

Introduction

The Mecoptera are represented by nine families (Fig. 1), four of which are restricted to the southern Neotropical and Australasian regions (Riek, 1954). Immature stages of many species are unknown or poorly described, but known larvae of one family, Nannochoristidae, inhabit flowing water and can be collected in springs and smaller streams (Byers, 1989; Pilgrim, 1972; Winterbourn & Gregson, 1989).

Guest editors: E. V. Balian, C. Lévêque, H. Segers & K. Martens
Freshwater Animal Diversity Assessment

L. C. Ferrington Jr. (✉)
Department of Entomology, University of Minnesota, Saint Paul, MN 55108-6125, USA
e-mail: ferri016@umn.edu

Adults of aquatic species are not abundant, but when collected are commonly found near the natal habitat as a consequence of their low vagility. Adults of some species in other genera [e.g., *Apteropanorpa tasmanica* Carpenter (Apteropanorpidae), *Merope tuber* Newman (Meropeidae)] are also predictably encountered on or among riparian or aquatic vegetation, mosses or bryophytes, and the unknown immature stages, when discovered, may conform to the definition of "aquatic" used in this publication. Consequently, the list of taxa provided here represents a minimum estimate for aquatic species of Mecoptera.

Species diversity

Two genera consisting of seven species and one sub-species comprise the family Nannochoristidae (Table 1). Nannochoristidae are unusual among Mecoptera in that they have an elateriform larva, with slightly reduced sclerotization of abdominal segments, that probably represent morphological adaptations to aquatic habits. Larvae of *Microchorista philpotti* (Tillyard), have been recovered from fine organic deposits in areas shielded from current in small woodland streams of South Island, New Zealand (Winterbourn, 1982; Winterbourn & Gregson, 1989). However, larvae of an unidentified second species have been collected repeatedly from lateral

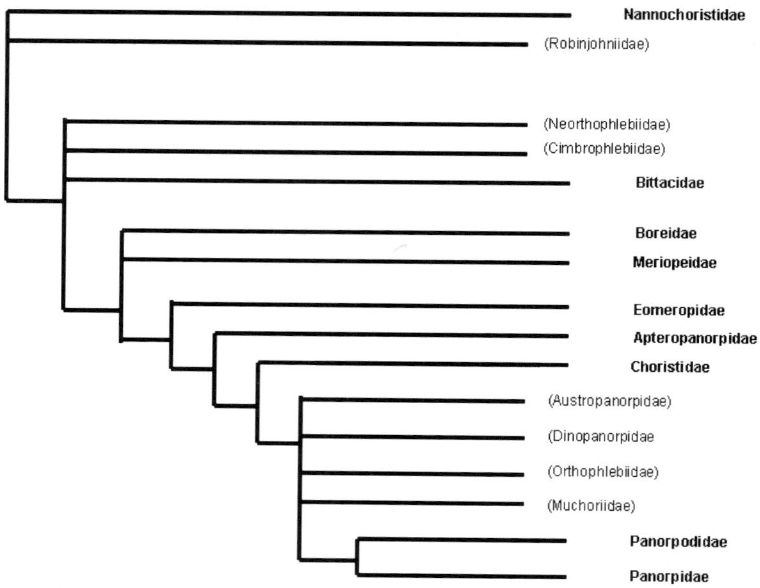

Fig. 1 Phylogenetic relationships of Mecoptera families (modified from Willmann 1987). Extant families shown in bold, extinct families in parenthesis

margins of riffle habitats in fast-flowing streams in Tasmania (Ferrington, unpublished data). Larvae of *M. philpotti* are predatory, feeding on larval Chironomidae and other aquatic Diptera (Winterbourn, 1982). It has been assumed that larvae of all Nannochoristidae species share similar aquatic and trophic habits (Hynes, 1984).

Phylogeny and historical processes

Two fundamental competing hypotheses related to the phylogenetic position of Nannochoristidae exist in recent literature. An earlier phylogeny of Mecoptera proposed by Willmann (1987) is shown in Fig. 1. According to this hypothesis, based primarily on external and internal morphological characters, the Nannochoristidae plus the fossil family Robinjohniidae form the suborder Nannomecoptera, which is the primitive sister group to the suborder Pistillifera, containing all the other families of Mecoptera (Willmann, 1987). An alternative hypothesis more recently proposed by Whiting (2002) was developed by integrating DNA sequence data. Although phylogenetic analyses of single genes across Mecoptera and Siphonaptera was not sufficient to unambiguously resolve the position of Nannochoristidae, a single fully resolved topology was obtained when four genes were analyzed. The analysis supports the view that Mecoptera as defined by Willmann (1987) is paraphyletic, but with two major lineages. The total evidence molecular tree based on 18S + 28S + EF1-α + COII with all characters weighted equally resulted in two clades, where the Nannochoristidae plus Boreidae and Siphonaptera form the basal group to all remaining families of Mecoptera.

Table 1 Aquatic Mecoptera and their known distributions

Species	Author	Distribution
Microchorista philpotti	(Tillyard) 1917	New Zealand, South Island
Nannochorista andina	Byers 1989	Argentina, Chile
Nannochorista dipteroides	Tillyard 1917	Australia: Tasmania
Nannochorista dipteroides eboraca	Tillyard 1917	Australia: New South Wales
Nannochorista edwardsi	Kimmins 1929	Argentina, Chile
Nannochorista holostigma	Tillyard 1917	Australia: Tasmania
Nannochorista maculipennis	Tillyard 1917	Australia: Tasmania
Nannochorista neotropica	Navás 1928	Argentina, Chile

Present distribution and main areas of endemicity

Present distributions of aquatic Nannochoristidae are restricted to southern South America (three species), southeastern Australia, including Tasmania (three species and one sub-species) and the South Island of New Zealand (Byers, 1989; Riek, 1954; Winterbourn & Gregson, 1989). This distribution is in concordance with an historical Gondwanaland distribution and subsequent diversification since the close of the Triassic Period. No fossil species of *Nannochorista* are known, thus the ages of species lineages cannot be confirmed. However, two fossil species of Nannochoristidae recovered from upper Permian strata of eastern Australia, *Nannochoristella reducta* Reik and *Neochoristella optata* Riek, date the family to approximately 260 million years bp (Riek, 1953).

Human related issues

Adults of Nannochoristidae do not bite or otherwise serve as vectors of disease, are not strong fliers and usually remain near the natal habitats. Consequently there are no known negative interactions with humans. Adults and immatures provide essential roles in ecosystem processes such as nutrient cycling and energy flow.

References

Byers, G. W., 1989. The Nannochoristidae of South America (Mecoptera). The University of Kansas Science Bulletin 54(3): 25–34.

Hynes, H. B. N., 1984. The relationships between the taxonomy and ecology of aquatic insects. Chapter 2, pp. 9–23. In: Resh, V. H. & D. M. Rosenberg (eds), The Ecology of Aquatic Insects. Praeger Publishers, New York, NY: 625 pp.

Kimmins, D. E., 1929. Some new and little known Argentine Neuroptera. Revista de la Sociedad Entomológica Argentina 2: 187–192.

Navás, L., 1928. Insectos neotropicas. Revista Chilena de Historia Natural 32: 106–128.

Pilgrim, R. L. C., 1972. The aquatic larva and the pupa of *Choristella philpotti* Tillyard, 1917 (Mecoptera: Nannochoristidae). Pacific Insects 14: 151–168.

Riek, E. F., 1953. Fossil mecopteroid insects from Upper Permian of New South Wales. Records Australian Museum 23: 55–87.

Riek, E. F., 1954. The Australian Mecoptera or Scorpion-flies. Australian Journal of Zoology 2: 143–168.

Tillyard, R. J., 1917. Studies in Mecoptera. 1. The new family Nannochoristidae. Proceedings of the Linnean Society of New South Wales 42: 284–301.

Whiting, M. F., 2002. Mecoptera is paraphyletic: multiple genes and phylogeny of Mecoptera and Siphonaptera. Zoologica Scripta 31: 93–104.

Willman, R., 1987. The phylogenetic system of the Mecoptera. Systematic Entomology 12: 519–524.

Winterbourn, M. J., 1982. The invertebrate fauna of a forest stream and its association with fine particulate matter. New Zealand Journal of Marine and Freshwater Research 16: 271–281.

Winterbourn, M. J. & K. L. D. Gregson, 1989. Guide to the aquatic insects of New Zealand, Revised Edition. Bulletin of the Entomological Society of New Zealand 9: 1–95.

Hydrobiologia (2008) 595:447–455
DOI 10.1007/s10750-007-9130-1

FRESHWATER ANIMAL DIVERSITY ASSESSMENT

Global diversity of non-biting midges (Chironomidae; Insecta-Diptera) in freshwater

Leonard C. Ferrington Jr.

Abstract Chironomidae are common inhabitants of most aquatic habitats, and often dominate aquatic insect communities in both abundance and species richness. Species occur in all continents, including Antarctica, and most major oceanic islands that have been investigated. The family is divided into 11 subfamilies and 22 nominal tribes. Although individual species occur in a wide range of habitats from terrestrial to fully aquatic, a total of 339 genera and 4,147 species are unambiguously aquatic in their immature stages. Greatest species and generic richnesses occur in the Palaearctic Region and Nearctic Region, respectively, but this pattern may largely reflect historical patterns of past taxonomic research efforts.

Keywords Chironomidae · Diversity · Zoogeography · Endemicity · Historical processes

Introduction

Chironomidae is a family of aquatic flies (Diptera) with world-wide distribution. It is the most widespread of all aquatic insect families, with individual species occurring from Antarctica at 68° S latitude (*Belgica antarctica* Jacobs) and sub-antarctic islands (*Parochlus steinenii* (Gerke)) (Edwards & Usher, 1985, Sugg et al., 1983) to Lake Hazen at 81° N on Ellesmere Island (Oliver & Corbet, 1966). They also exhibit extreme elevational ranges, occurring in a glacial-melt stream at 5,600 m in the Himalaya Mountains (Koshima, 1984) to more than 60 m depths in Lake Hovsgol (Hayford & Ferrington, 2006) and >1,000 m depths in Lake Baikal (Linevich, 1971). They are among the most tolerant of aquatic insects to water and air temperatures, with larvae of *Paratendipes thermophilus* Townes occurring in hot springs at temperatures of 38.8°C (Hayford et al., 1995) and adults of *Diamesa mendotae* Muttkowski able to depress their freezing point and survive air temperatures less than –20°C (Carrillo et al., 2004, Bouchard et al., 2006). Although most species appear to be univoltine to trivoltine in seasonal environments (Tokeshi, 1995), life cycle strategies of individual species can also be extreme, with *Apedilum elachistus* Townes maturing from egg to adult in less than 7 days (Nolte, 1996) in rock pool habitats in the Brazilian Pantanal. By contrast, Butler (1982) proposed a 7-year life cycle for two species of *Chironomus* in Alaskan tundra ponds.

Chironomidae are divided into 11 subfamilies and nominally into 22 tribes (not including five additional provisional tribes resulting from dividing Orthocladiini as proposed by Sæther 1979). Overviews of 10

Guest editors: E.V. Balian, C. Lévêque, H. Segers & K. Martens
Freshwater Animal Diversity Assessment

L. C. Ferrington Jr. (✉)
Department of Entomology, University of Minnesota, Saint Paul, MN 55108-6125, USA
e-mail: ferri016@umn.edu

subfamilies are provided in Cranston (1995a), the phylogeny of subfamilies was cladistically analyzed by Sæther (2000b), and zoogeographical patterns were summarized by Ashe et al. (1987) and Sæther (2000a). A list of generic and subgeneric names and synonyms for Chironomidae was published by Ashe (1983). Table 1 lists the subfamilies in approximate phylogenetic order and tribes alphabetically, based primarily on publications by Brundin (1966, 1983), Brundin and Sæther (1978), Sæther (1977, 1979, 1989, 2000b) and Spies (2005). No world catalogs or checklists have been published recently, however regional catalogs that are somewhat dated but have been valuable sources of distributional data for this article are: Ashe & Cranston (1990) for the Palaearctic Region; Sæther & Spies (2004) for Europe; Oliver et al. (1990) for the Nearctic Region; Spies & Reiss (1996) for the Neotropical Region (and Mexico); Freeman & Cranston (1980) for the Afrotropical Region; Cranston & Martin (1989) for the Australian and Oceanic regions and the Antarctic; and Sublette & Sublette (1973) for the Oriental Region. All new taxonomic papers post-dating the respective regional catalogs (and published as of 2006) have been reviewed and incorporated into the totals provided in Table 2 (see Figs. 1, 2).

When developing estimates of richness for aquatic species of Chironomidae, several problematic issues arise. Notwithstanding difficulties associated with differing species interpretations and incorporating new publication data into this overview, which have required continuous refinement of the richness estimates, a more difficult issue deals with a workable definition of "aquatic". Our working group has discussed this issue and have labored to develop a standardized definition (Lévêque et al., 2005). Among Chironomidae it is the larval and pupal stages that typically are constrained to aquatic habitats while adults are aerial and often collected at substantial distances from potential natal habitats. Species descriptions have, by tradition, been based primarily on adults, and knowledge of immature stages of species is disparate among tribes or even among species within a genus. Some genera are known to have terrestrial immature stages and by our workable definition are excluded from the totals. However, in other genera some species may be clearly aquatic as immatures, others recorded only from terrestrial habitats, and still others unknown as immatures. Cases like this make application of our criterion subjective. In these cases, I have opted not to count species unless (1) the collection records specifically say they are recorded from in or near aquatic habitats or (2) unless I have firsthand knowledge of them occurring in one or more of the types of "aquatic conditions" defined by Lévêque

Table 1 Subfamilies and tribes of Chironomidae

Telmatogetoninae	
Usambaromyiinae	
Aphroteniinae	
Chilenomyiinae	
Podonominae	
	Boreochlini
	Podonomini
Tanypodinae	
	Anatopynyiini
	Coelopyiini
	Coelotanypodini[a]
	Macropelopiini
	Natarsiini
	Pentaneurini
	Procladiini
	Tanypodini
Buchonomyiinae	
Diamesinae	
	Boreoheptagyiini[b]
	Diamesini
	Harrisonini
	Heptagyiini[b]
	Lobodiamesini
	Protanypodini[b]
Prodiamesinae	
Orthocladiinae	
	Corynoneurini[c]
	Metriocnemini[c]
	Orthocladiini[c]
Chironominae	
	Chironomini
	Pseudochironomini
	Tanytarsini

[a] Please refer to Spies (2005) for information about validity of this tribe name

[b] Please refer to comments by Spies (2005) for further information about this spelling

[c] Please refer to opinion by Spies (2005) for comments on usage and/or recommendations related to these tribes

Table 2 Genera and species of aquatic Chironomidae by biogeographic region (Fig. 1)

	PA	NA	NT	AT	OL	AU	PAC	ANT	World
Total Genera	181	211	154	104	105	116	29	6	339
Total Species	1,321	1,092	618	406	359	471	155	9	4,147

Continental distribution of freshwater lizards. PA = Palaearctic, NA = Nearctic, NT = Neotropical, AT = Afrotropical, OL = Oriental, AU = Australasian, PAC = Pacific Oceanic islands, ANT = Antarctic

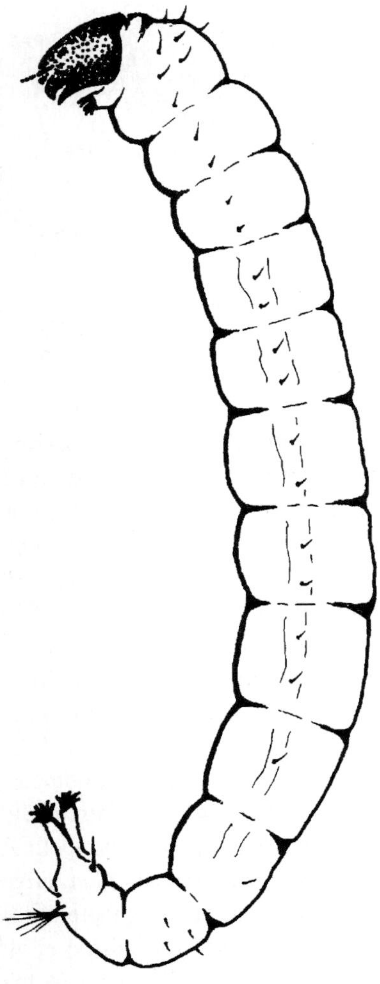

Fig. 1 A chironomid larvae

et al. (2005). However, I have summarily included other species in the genus that are not described as immatures in the totals when all taxa that are known as immatures have been collected from an aquatic habitat. Consequently, the figures provided in this article may be disputed as overly inflated. Without full knowledge, however, of the biology of larvae it is therefore debatable if a general consensus can be obtained for the number of described aquatic species of Chironomidae. However, review of collection records and species accounts confirms a total of 339 genera and 4,147 species can be considered unambiguously aquatic in their immature stages.

Species/generic diversity

When sampled comprehensively, species richness of Chironomidae is usually among the highest of aquatic insect families detected in most aquatic settings, often approaching 80 or more species and occasionally exceeding 100 species per site. Coffman (1989) summarized 152 species richness estimates as a function of stream order and concluded that average richness increased with increasing stream order up to third order, then leveled off or decreased in higher order rivers. Mean species richness (and range of estimates) varied from: 26 (10–64) for first order streams; 44 (13–144) for second order; 63 (25–157) for third order; 51 (25–83) for fourth order; 47 (11–86) for fifth order; 47 (10–99) for sixth order; 45 (12–148) for seventh order and higher.

Although predicted to have lower species richness, some studies of springs, spring runs and intermittent streams show remarkably high values, especially in heterogeneous spring settings or when studied over periods with differing precipitation and hydrologic conditions (e. g., Chou et al., 1999). Ferrington et al. (1995) reported 66 species for an isolated spring in the high plains of western Kansas, but concluded that 48 species were restricted to only one of the five habitats occurring in the spring source. Springs occurring over broad geographic areas show considerable individual variation in richness, however cumulative richness can exceed 100 or more species (e.g., Blackwood et al., 1995; Ferrington, 1998; Lindegaard, 1995).

Patterns of lotic chironomid richness across biotopes were discussed by Coffman (1989). Factors identified as contributing to regional species richness

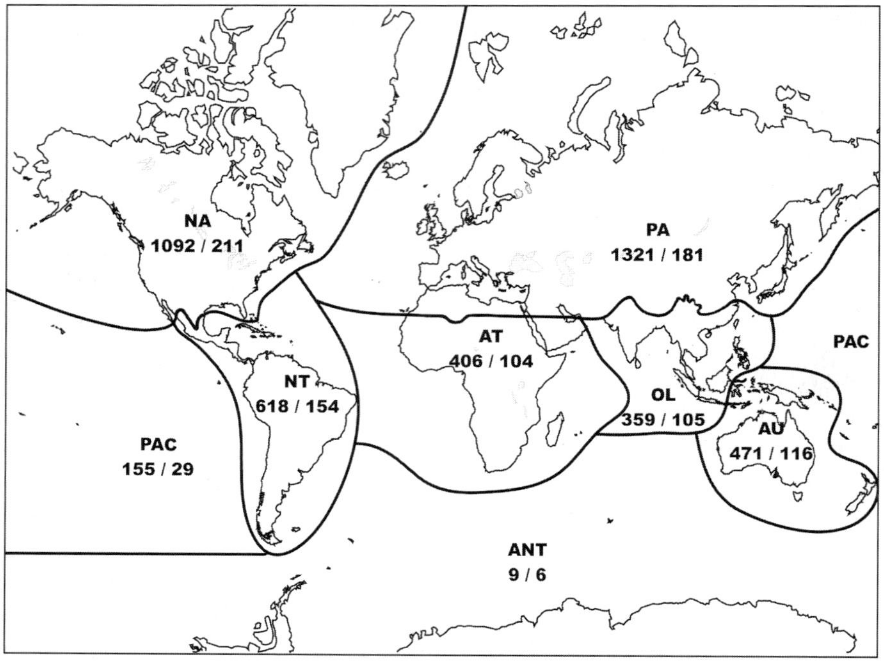

Fig. 2 Global distribution of species and genus diversity of Chironomidae by zoogeographic region (Species number/genus number). PA—Palaearctic; NA—Nearctic; NT—Neotropical; AT—Afrotropical ; OL—Oriental; AU—Australasian; PAC—Pacific Oceanic Islands; ANT—Antarctic

included (1) ecological heterogeneity, (2) size of stream, (3) altitude, (4) latitude, and (5) "biogeographical potential". Consideration of these factors led him to predict highest species richness would occur in 3rd order, temperate, lowland streams in the Holarctic Region. However, in subsequent papers Coffman and de la Rosa (1998) reported 250 species from three streams in Costa Rica, and Coffman et al. (1992) documented 299 species in 31 West African streams, including 175 species in a large river. Consequently, it is likely that regional species richness in some tropical streams approaches or exceeds richness in temperate lowland streams.

Species richness of lakes is generally lower than average stream estimates. However, some estimates for large, heterogeneous systems reach or exceed richness for streams (e.g., Reiss (1968) detected 184 species in Bodensees). When a large array of lake types are considered, the cumulative number of species detected can also be high. For example Ruse (2002) detected 275 species in 30 lakes in England and Wales.

Phylogeny and historical processes

A phylogeny of subfamilies was developed by Sæther (1977). Based strongly on considerations of the structure of female genitalia, the Telmatogetoninae were placed as the primitive sister group to all remaining subfamilies of Chironomidae, a position differing from Brundin's (1966) view of Telmatogetoninae as the apomorph sister group of Diamesinae. In a subsequent paper, Sæther (1979) provided an historical review of earlier classifications and proposed relationships, and additional arguments for the basal placement of Telmatogetoninae. This placement has been criticized by Ashe et al. (1987), based on re-interpretation of homologies of female genitalia and the predominantly marine habits of larvae. However, no compelling arguments have been proposed for an alternative placement of Telmatogetoninae, although preliminary molecular evidence presented by Cranston et al. (2000) appears to support a basal placement of the subfamily.

Chilenomyiinae with one described species, *Chilenomyia paradoxa* Brundin, was considered by Ashe et al. (1987) as a logical replacement for Telmatogetoninae as the primitive sister to all other chironomid subfamilies. Brundin (1983) proposed a placement of Chilenomyiinae as sister group to all subfamilies less the Telmatogetoninae.

The subfamily Usambaromyiinae, with one species known from Tanzania, is considered to be the sister group of Tanypodinae + Aphroteniinae + Podonominae (=semifamily Tanypodoinae of Sæther, 1983) (Andersen and Sæther, 1994).

The monophyly of Tanypodinae + Aphroteniinae + Podonominae is well supported morphologically and generally accepted. Within this clade Tanypodinae form the sister subfamily to Aphroteniinae.

Sæther (1983) proposed the semifamily Chironomoinae to include the remaining subfamilies Buchonomyiinae, Diamesinae, Prodiamesinae, Orthocladiinae and Chironominae. The placement of Buchonomyiinae within the semifamily was disputed by Murray and Ashe (1985), based on re-interpretation of morphology of male genitalia and new characters from the female genitalia, who proposed placement within the semifamily Tanypodoinae. The latter placement appears to be favored by Cranston (1995a), who considered the apparent lack of premandibles in early instar larvae combined with a labral rod and SII setae on pedestals as characters supporting placement in Tanypodoinae.

Sæther (2000b) performed a cladistic analysis of the subfamilies and (using parsimony) concluded the results supported previous findings, placing Buchonomyiinae and Chilenomyiinae basally within the Chironomoinae, and Usambaromyiinae as the sister group of the Tanypodoinae.

Other subfamilies assigned to Chironomoinae are generally considered monophyletic and appropriately placed, with the relationships following Sæther (1977). Cranston (1995a), however, speculated on possible paraphyly of both Diamesinae and Prodiamesinae.

Zoogeography and Endemicity

Estimates of aquatic species richness by region are presented in Table 2. Based on published data the Chironomidae appear to be more species rich in the northern hemisphere than the southern hemisphere, and with greatest richnesses in northern mid-latitudes. The latter pattern does not conform to the more generalized trend of greatest richness in the tropics for many groups of organisms, and Coffman (1989) has argued that a variety of factors contribute to the higher richness of lotic chironomids in mid-latitudes (see earlier text). However, it can also be expected that the apparent patterns of species richness of Chironomidae are largely the result of differing efforts to describe local faunas rather than actual biological patterns, and more recent studies of Costa Rican and West African streams support this conclusion (Coffman and de la Rosa, 1998; Coffman et al., 1992).

Several taxonomic levels of chironomids show strong, but alternative, patterns of endemism and/or zoogeographical distributions. At the subfamily level, the southern hemisphere patterns of endemism for Podonominae and Aphroteniinae conform to patterns of transantarctic diversification and vicariance (Brundin, 1966). The Heptagyiae group (Diamesinae), defined by Brundin (1966) as consisting of five genera, also conforms to this pattern. Other conspicuous examples of endemism at the genus or sub–genus level, but on more restrictive geographic scales, are discussed in Sæther (2000a) and Sæther & Ekrem (2003). Based on phylogenetic hypotheses derived from morphological and limited distributional data, several Gondwanian (early Cretaceous), Laurasian (Quaternary Ice Age) and Inabrezian distributions were postulated. These conclusions provide the basis for developing hypotheses that can be tested when better information about immature stages can be generated. Another example of endemism, but with very strong support from morphological and molecular data, was presented by Papoucheva et al. (2003) for species flocks of *Sergentia* species in Lake Baikal, which they conclude date to tectonic activities 25–27 MY ago. On a more restrictive scale, the distributional pattern of the two described species of *Oliveridia*, with *O. tricornis* widespread in lakes at high latitudes in North America, Greenland, Iceland, northern Norway and *O. hugginsi*, by contrast, limited to one watershed in extreme southern Kansas (Ferrington & Sæther, 1987), strongly suggest post-glacial events since the Kansan Glaciation as contributing to the disjunct distributions of these two species. Vicariance events with intervening dispersal during climatic optima were argued by Willassen and Cranston (1986) as contributing to range extension of the cold-water adapted genus *Diamesa* into Africa and the present day distribution of species within montane habitats.

Hot spots

Areas where large numbers of undescribed species are known to occur include north western Costa Rica, the

Brazilian lowlands, West Africa and high elevation streams in the steppes on the eastern edge of the Andes in Patagonia. In a more general sense, habitats that are widespread geographically but are likely to have high concentrations of undescribed species include: (1) intermittent and ephemeral aquatic habitats; (2) streams and lakes of most extreme low latitudes of major north/south orientated mountain chains (example, southern Rockies in Arizona, New Mexico and southern Appalachians in North America); (3) high-elevation springs and seeps; and (4) oceanic archipelagos. Timing of field work to correspond to early spring collecting soon after ice melt in mid-latitude streams and lakes, will also likely result in undescribed species even in areas where the summer emerging fauna is relatively well known. Similarly, collecting during winter in streams that are strongly dominated by groundwater inputs and remain free of ice and relatively warm in winter may also provide substantial numbers of undescribed species.

Human-related issues

Chironomidae have both positive and negative interactions with humans. Negative interactions include species that are considered as nuisance species (Ali, 1995), species that act as known or suspected sources of environmental allergens (Cranston, 1995b), and species that host and transport pathogenic bacteria (Broza and Halpern, 2001).

Species with dense populations and synchronized emergences from water bodies in highly urbanized landscapes are often considered to be nuisance species when attracted as adults by light to shops, restaurants, homes, or other places where people congregate such as sports stadiums and outdoor banquet or concert halls. Adults can stain buildings as meconium is produced and deposited, and can clog screened air intakes to furnaces (personal observation), air conditioner cooling systems and compressors, or indoor building environment systems. Estimates of maintenance costs and business losses in Florida amounted to millions of dollars annually (Anonymous, 1977) and control efforts were estimated at approximately one million dollars in Venice in the late 1980s/early 1990s (Ali et al., 1992). Possible consequences to the aviation industry have been proposed by Barbato et al. (1990). Larval stages have often been considered as nuisances by infesting drinking-water supply systems (Berg, 1996; Langton et al., 1988), by damaging rice seedlings (Darby, 1962; Ferrarese 1993), or by grazing on apical meristem or leaf-mining in other ornamental aquatic hydrophytes (Tokunaga and Koroda, 1936).

Hemoglobin produced by larvae for respiration, but carried over to the adult, has been shown to have allergenic properties among atopic individuals (Cranston, 1995b; Miyamoto, 1988) that may be of worldwide significance (Sublette and Sublette, 1988). Symptoms can range from mild urticaria, pruritis and conjunctivitis to more severe reactions varying from allergenic rhinitis to asthmatic symptoms or even leading to anaphylaxis.

Occupational allergic reactions are known among aquarists, owners of pet fish stores and employees of fish-food production facilities (Baur, 1982, 1992; Fuchs and Kleinhaus, 1982). In most cases the allergen is known or suspected to be associated with freeze-dried larvae incorporated into the commercial fish food (Cabrerizo Ballesteros et al., 2006). Contact dermatitis is also reported in a person handling bloodworms (Brasch et al., 1992).

Vibrio cholerae, the causative agent of Cholera, has been isolated from chironomid egg masses by Broza and Halpern (2001). They concluded the gelatinous matrix was the sole source of carbon for the bacteria and considered masses to act as natural reservoirs of the bacteria. More recently, Broza et al. (2003) have isolated two non-pathogenic serotypes of *V. cholerae* from adults and have proposed chironomids as a potential air born dispersal mechanism.

Positive interactions with humans have not been extensively quantified, however both adults and immatures provide essential roles in ecosystem processes such as nutrient cycling and energy flow. Consequently chironomids, and other aquatic invertebrates, contribute to ecosystem goods and services upon which human society depends. Other more readily conceptualized interactions benefiting humans are summarized by Armitage (1995) and include serving as food for freshwater commercial and sport fisheries (e.g., Mackay, 1979, Rasmussen, 1990), food for other species of invertebrates and amphibians (e.g., Avery, 1968; Johnson, 1985), and as ingredients in production of insect flour (Bergeron et al., 1988). Mating swarms of adults attract swifts and swallows (Lack and Owen, 1955; St Louis et al.,

1990), at least some species of bats intermittently consume large quantities of adults (Griffith and Gates, 1985), and larvae can be important in diets of migratory waterfowl during breeding seasons (e.g., Austin et al., 1990). Chironomids are also a major component of some live and freeze-dried commercial tropical fish foods (Armitage, 1995).

Larvae, pupae, and even eclosing adults have served as models and mimics for fly fishing. Publications explaining the concept of "match the hatch" provide directions on how to determine if chironomids are emerging and how to tie or select flies that match particular life stages or colors of chironomids. A few random web searches yielded an impressive array of fly-types such as "*Brassie, Griffith's gnat, midge, olive midge, blood midge, disco midge, black palomino midge, AK's midge larva, paramidge, midge pupa, ascending midge pupa, blood midge pupa, hatching midge pupa, parachute Adams, serendipity*", attesting to the interest and enthusiasm of some fly-fishers to mimic the habitus of chironomids.

References

Ali, A., 1995. Nuisance, economic impact and possibilities for control, In: Armitage P. D., P. S. Cranston & L. V. C. Pinder (eds), The Chironomidae: Biology and Ecology of Non-biting Midges. Chapman & Hall, London, UK, 339–364, Chapter 13.

Ali, A., L. Barbato, G. Ceretti, S. Della Sala, R. Riso, G. Marchese & F. D'Andrea, 1992. Efficacy of two temephos formulations against *Chironomus salinarius* (Diptera: Chironomidae) in the saltwater lagoon of Venice, Italy. Journal of the American Mosquito Control Association 8: 353–356.

Andersen, T. & O. A. Sæther, 1994. *Usambaromyia nigrala* gen. n., sp. n., and Usambaromyiinae, a new subfamily among the Chironomidae (Diptera). Aquatic Insects 16: 21–29.

Anonymous, 1977. Economic Impact Statement, Blind Mosquito (Midge) Task Force. Report by Sanford Chamber of Commerce, Seminole County, Florida.

Armitage, 1995. Chironomidae as food. In: Armitage P. D., P. S. Cranston & L. V. C. Pinder (eds), The Chironomidae: Biology and Ecology of Non-biting Midges. Chapman & Hall, London, UK, 423–435, Chapter 17.

Ashe, P., 1983. A catalogue of chironomid genera and subgenera of the world including synonyms (Diptera: Chironomidae). Entomologica Scandinavica Supplement 17: 1–68.

Ashe, P. & P. S. Cranston, 1990. Chironomidae. In: Soos A. & L. Papp (eds), Catalogue of Palaearctic Diptera. Elsevier Science Publishers, Amsterdam, The Netherlands, 113–355.

Ashe, P., D. A. Murray & F. Reiss, 1987. The zoogeographical distribution of Chironomidae. Annales de Limnologie 23: 27–60.

Austin, J. E., J. R. Serie & J. H. Noyes, 1990. Diet of canvasbacks during breeding. Prairie Naturalist 22(3): 171–176.

Avery, R. A., 1968. Food and feeding relations of three species of *Triturus* (Amphibia: Urodela) during the aquatic phases. Oikos 19: 408–412.

Barbato, L. C., L. A. Filia, B. Maraga, O. Pancino & E. Tsuroplis, 1990. Problematiche relative al controllo dei Chironomidi nella Laguna di Venecia. In: D'Andrea F. & G. Marchese (eds), *Chironomidi*, Culicidi, Simulidi—Aspetti Sanitari ed Ecologici. Regione Veneto, ULSS 16, S. I. P., Venezia, Italy, 113–127.

Baur, X., 1982. Chironomid hemoglobin. A major allergen for humans. Chironomus 2: 24–25.

Baur, X., 1992. Chironomid midge allergy. Japanese Journal of Immunology 41: 81–85.

Berg, M. B., 1996. Infestation of enclosed water supplies by chironomids: Two case studies. In: P. Cranston (ed.). Chironomids: From *Genes to Ecosystems*. CSIRO Publications, East Melbourne, Australia, 241–246.

Bergeron, D., R. J. Bushway, F. L. Roberts, I. Kornfield, J. Okedi & A. A. Bushway, 1988. The nutrient composition of an insect flour sample from Lake Victoria, Uganda. Journal of Food Composition and Analysis 1: 371–377.

Blackwood, M. A., S. M. Hall & L. C. Ferrington Jr, 1995. Emergence of Chironomidae (Diptera) from springs in the central high plains of the United States. Journal of the Kansas Entomological Society 68: 132–151.

Bouchard, R. W., Jr, M. A. Carrillo & L. C. Ferrington Jr, 2006. Lower lethal temperature for adult male *Diamesa mendotae* Muttkowski (Diptera: Chironomidae), a winter-emerging aquatic insect. Aquatic Insects 28: 57–66.

Brasch, J., H. Brüning & E. Paulke, 1992. Allergic contact dermatitis from chironomids. Contact Dermatitis 26: 317–320.

Broza, M. & M. Halpern, 2001 Chironomid egg masses and *Vibrio cholerae* . Nature 412: 40.

Broza, M., M. Halpern, H. Gancz & Y. Kashi, 2003. Do chironomids play a key role in onset of Cholera epidemics. Abstract on-line at: http://www.entomology.umn.edu/midge/XVSymposium.htm.

Brundin, L., 1966. Transantarctic relationships and their significance, as evidenced by chironomid midges with a monograph of the subfamilies Podonominae and Aphroteniinae and the austral Heptagyiae. Kunglica Svenska Vetenskapsakademiens Handlungar 11: 1–472 + plates.

Brundin, L., 1983. *Chilenomyia paradoxa* gen. n., sp. n. and Chilenomyiinae, a new subfamily among Chironomidae (Diptera). Entomologica scandinavica 14: 33–45.

Brundin, L & O. A. Sæther, 1978. *Buchonomyia burmanica* sp. n. and Buchonomyiinae, a new subfamily among the Chironomidae (Diptera). Zoologica Scripta 7: 269–275.

Butler, M. G., 1982. A 7-year life cycle for two *Chironomus* species in arctic Alaskan tundra ponds (Diptera: Chironomidae). Canadian Journal of Zoology 60: 58–70.

Cabrerizo Ballesteros S., M. de Barrio, M. L. Baeza, M. Rubio Sotés, 2006. Allergy to chironomid larvae (red midge larvae) in non professional handlers of fish food. Journal of Investigational Allergology and Clinical Immunology 16: 63–68.

Carrillo, M. A., C. A. Cannon & L. C. Ferrington Jr, 2004. Effect of sex and age on supercooling point of winter-active *Diamesa mendotae* Muttkowski (Diptera: Chironomidae). Aquatic Insects 26: 243–251.

Chou, R. Y. M., L. C. Ferrington Jr, B. L. Hayford & H. M. Smith, 1999. Composition and phenology of Chironomidae (Diptera) from an intermittent stream in Kansas. Archives für Hydrobiology 147: 35–64.

Coffman, W. P., 1989. Factors that determine the species richness of lotic communities of Chironomidae. Acta Biologica Debrecina Oecologica Hungarica 3: 95–100.

Coffman, W. P. & C. de la Rosa, 1998. Taxonomic composition and temporal organization of tropical and temperate species assemblages of lotic Chironomidae. Journal of the Kansas Entomological Society 71: 388–406.

Coffman, W. P., C. de la Rosa, K. W. Cummins & M. A. Wilzbach, 1992. Species richness in some Neotropical (Costa Rica) and Afrotropical (West African) lotic communities of Chironomidae. Netherlands Journal of Aquatic Ecology 26: 229–237.

Cranston, P. S., 1995a. Systematics. In: Armitage P. D., P. S. Cranston & L. V. C. Pinder (eds), The Chironomidae: Biology and Ecology of Non-biting Midges. Chapman & Hall, London, UK, 31–61, Chapter 3.

Cranston, P. S., 1995b. Medical significance. In: Armitage P. D., P. S. Cranston & L. V. C. Pinder (eds), The Chironomidae: Biology and Ecology of Non-biting Midges. Chapman & Hall, London, UK, 365–384, Chapter 14.

Cranston, P. S. & J. Martin, 1989. Family Chironomidae. In: Evenhuis N. L. (ed.), Catalogue of the Diptera of the Australasian and Oceanian Regions. Bishop Museum Press, Honolulu, Hawaii, 252–274.

Cranston, P. S., O. A. Sæther & L. Cook, 2000. Molecular evidence supports, fails to refute, or provoke new morphological assessment in Chironomidae (Diptera). Abstract, XXI International Congress of Entomology, 919.

Darby, R. E., 1962. Midges associated with California rice fields, with special reference to their ecology (Diptera: Chironomidae). Hilgardia 32: 1–206.

Edwards, M. & M. B. Usher, 1985. The winged Antarctic midge *Parochlus steinenii* (Gerke) (Diptera: Chironomidae), in the South Shetland Islands. Biological Journal of the Linnaean Society 26: 83–93.

Ferrarese, U., 1993. Chironomids of Italian rice fields. Netherlands Journal of Aquatic Ecology 26: 341–346.

Ferrington, L. C., Jr, 1998. Generic composition of the chironomid fauna in springs of North America. In: L. Botosaneanu (ed.). Studies in Crenobiology: The Biology of Springs & Springbrooks. Backhuys Publishers, Leiden, The Netherlands, 141–155.

Ferrington, L. C., Jr. & O. A. Sæther, 1987. The male, female, pupa and biology of *Oliveridia hugginsi* n. sp. (Diptera: Chironomidae Orthocladiinae) from Kansas. Journal of the Kansas Entomological Society 60: 451–461.

Ferrington, L. C., Jr, R. G. Kavanaugh, F. J. Schmidt & J. L. Kavanaugh, 1995. Habitat separation among Chironomidae (Diptera) in Big Springs. Journal of the Kansas Entomological Society 68: 152–165.

Freeman, P. & P. S. Cranston, 1980. Family Chironomidae. In: Crosskey R. W. (ed.), Catalogue of the Diptera of the Afrotropical Region. British Museum (Natural History), London, England, 175–202.

Fuchs, T. & D. Kleinhaus, 1982. Fischfutter-allergie: Sensibilisierung gegen Chironomiden-Larven (Zuckmückenlarven). Allergologie 5: 81–82.

Griffith, L. A. & J. E. Gates, 1985. Food habits of cave-dwelling bats in the Central Appalachians. Journal of Mammology 66: 451–460.

Hayford, B. & L. C. Ferrington Jr, 2006. Distribution of Chironomidae (Diptera) in Lake Hövsgöl, Mongolia. In Goulden, C. E., T. Sitnikova, J. Gelhaus, B. Boldgiv (eds), The Geology, Biodiversity, Ecology of Lake Hövsgöl (Mongolia). Backhuys Publishers, Leiden, The Netherlands, 433–452, Chapter 26.

Hayford, B. L., J. E. Sublette & S. J. Herrmann, 1995. Distribution of chironomids (Diptera: Chironomidae) and ceratopogonids (Diptera: Ceratopogonidae) along a Colorado thermal spring effluent. Journal of the Kansas Entomological Society (Special Publication 1) 68: 77–92.

Johnson, J. H., 1985. Diel feeding ecology of the nymphs of *Aeshna multicolor* and *Lestes unguiculatus* (Odonata). Freshwater Biology 15: 749–756.

Koshima, S., 1984. A novel cold-tolerant insect found in a Himalayan glacier. Nature 310: 225–227.

Lack, D. & D. F. Owen, 1955. The food of the swift. Journal of Animal Ecology 24: 120–136.

Langton, P. H., P. S. Cranston & P. D. Armitage, 1988. The parthenogenetic midge of water supply systems *Paratanytarsus grimmii* Schneider (Diptera: Chironomidae). Bulletin of Entomological Research 78: 317–328.

Lévêque, C., E. V. Balian & K. Martens, 2005. An assessment of animal species diversity in continental waters. Hydrobiologia 542: 39–67.

Lindegaard, C., 1995. Chironomidae (Diptera) of European cold springs and factors influencing their distribution. Journal of the Kansas Entomological Society 68: 108–131.

Linevich, A. A., 1971. The Chironomidae of Lake Baikal. Limnologica (Berlin) 8: 51–52.

Mackay, A. P., 1979. Trophic dependencies of some larval Chironomidae (Diptera) and fish species in the River Thames. Hydrobiologia 62: 241–247.

Miyamoto, T. (ed.), 1988. International Symposium on Mite and Midge Allergy. Special report of the Ministry of Education, Science and Culture, University of Tokyo, Japan, 372 pp.

Murray, D. A. & P. Ashe, 1985. A description of the adult female of *Buchonomyia thienemanni* Fittkau and a reassessment of the phylogenetic position of the subfamily Buchonomyiinae. Spixiana Supplement 11: 149–160.

Nolte, U., 1996. From egg to adult in less than seven days: *Apedilum elachistus* Chironomidae. In: P. Cranston (ed.), Chironomids: From Genes to Ecosystems. CSIRO Publications, East Melbourne, Australia, 177–184.

Oliver, D. R. & P. S. Corbet, 1966. Aquatic habitats in a high arctic locality: the Lake Hazen camp study area, Ellesmere Island, N. W. T. Defense Research Board of Canada, Report #26: Operation Hazen.

Oliver, D. R., M. E. Dillon & P. S. Cranston, 1990. A catalog of Nearctic Chironomidae. Research Branch Agriculture Canada, Publication 1857/B.

Papoucheva, E., V. Proviz, C. Lambkin, B. Goddeeris & A. Blinov, 2003. Phylogeny of the endemic Baikalian *Sergentia* (Chironomidae, Diptera). Molecular Phylogenetics and Evolution 29: 120–125.

Rasmussen, K., 1990. Some positive and negative effects of stocking whitefish on the ecosystem redevelopment of Hjarbaek Fjord, Denmark. Hydrobiologia 200: 593–602.

Reiss, F., 1968. Ökologische und systematische Untersuchungen an Chironomiden (Diptera) des Bodensees: Ein Beitrag zur lakustrischen Chironomidenfauna des nördlichen Alpenvorlandes. Archives für Hydrobiologie 64: 176–323.

Ruse, L., 2002. Chironomid pupal exuviae as indicators of lake status. Archives für Hydrobiologie 153: 367–390.

Sæther, O. A., 1977. Female genitalia in Chironomidae and other Nematocera: morphology, phylogenies, keys. Bulletin of the Fisheries Research Board of Canada 197: 1–209.

Sæther, O. A., 1979. Hierarchy of the Chironomidae with emphasis on the female genitalia (Diptera). Entomologica scandinavica Supplement 10: 17–26.

Sæther, O. A., 1983. The canalized evolutionary potential: inconsistencies in phylogenetic reasoning. Systematic Zoology 32: 343–359.

Sæther, O. A., 1989. Phylogenetic trends and their evaluation in chironomids with special reference to orthoclads. Acta Biologica Debrecina Oecologica Hungarica 2: 53–75.

Sæther, O. A., 2000a. Zoogeographical patterns in Chironomidae (Diptera). Internationale Vereingung für Theoretische und Angewandte Limnologie 27: 290–302.

Sæther, O. A., 2000b. Phylogeny of the subfamilies of Chironomidae (Diptera) Systematic Entomology 25: 393–403.

Sæther, O. A. & T. Ekrem, 2003. Biogeography of Afrotropical Chironomidae (Diptera), with speciel reference to Gondwanaland. Cimbebasia 19: 175–190.

Sæther, O. A. & M. Spies, 2004. Family Chironomidae (Diptera), on-line @ http://www.faunaeur.org.

St Louis, V. L., L. Breebaart & J. C. Barlow, 1990. Foraging behavior of tree swallows over acidified nonacidic lakes. Canadian Journal of Zoology 68: 2385–2392.

Spies, M., 2005. On selected family group names in Chironomidae (Insecta, Diptera), and related nomenclature. Zootaxa 894: 1–12.

Spies, M. & F. Reiss, 1996. Catalog and bibliography of Neotropical and Mexican Chironomidae (Insecta, Diptera). Spixiana Supplement 22: 61–119.

Sugg, P., J. S. Edwards & J. Baust, 1983. Phenology and life history of *Belgica Antarctica*, an Antarctic midge (Diptera: Chironomidae). Ecological Entomology 8: 105–113.

Sublette, J. E. & M. S. Sublette, 1973. Family Chironomidae. In Delfinado M. & E. D. Hardy (eds), Catalogue of the Diptera of the Oriental Region, Part 1. University Press, Honolulu, Hawaii, 389–422.

Sublette, J. E. & M. Sublette, 1988. An overview of the potential for Chironomidae as a worldwide source for potent allergens. In: Miyamoto T. (ed.), International Symposium on Mite and Midge Allergy. Special report of Ministry of Education, Science and Culture, University of Tokyo, Japan, 190–231.

Tokeshi, M., 1995. Life cycles and population dynamics. In: Armitage P. D., P. S. Cranston & L. V. C. Pinder (eds), The Chironomidae: Biology and Ecology of Non-biting Midges. Chapman & Hall, London, UK, 225–268, Chapter 10.

Tokunaga, M. & M. Kuroda, 1936. *Stenochironomus* midge from Japan (Diptera), with notes on controlling methods of a leaf mining midge. Transactions of the Kansai Entomological Society 7: 1–6.

Willassen, E. & P. S. Cranston, 1986. Afrotropical montane midges (Diptera: Chironomidae, *Diamesa*). Zoological Journal of the Linnaean Society 87: 91–123.

Hydrobiologia (2008) 595:457–467
DOI 10.1007/s10750-007-9131-0

FRESHWATER ANIMAL DIVERSITY ASSESSMENT

Global diversity of craneflies (Insecta, Diptera: Tipulidea or Tipulidae *sensu lato*) in freshwater

Herman de Jong · Pjotr Oosterbroek · Jon Gelhaus · Herbert Reusch · Chen Young

Abstract The Tipulidae s.l.—craneflies—are one of the largest groups of the Diptera containing over 15,270 valid species and subspecies. The immatures of the majority of species live in aquatic or semiaquatic habitats. Some aquatic species live entirely submerged and lack functional spiracles, others come to the surface to take oxygen by using spiracles positioned at the end of the abdomen. Semiaquatic species occur in a wide range of habitats. The semiterrestrial and terrestrial larvae live in environments that are moist or at least humous. All adult craneflies are terrestrial. Conflicting hypotheses on the phylogenetic position of the Tipuloidea within the Diptera continue to exist: some authors consider them to represent one of the oldest lineages of the Diptera, others suppose a close relationship to the Brachycera, the true flies. Current systematic knowledge of the Tipuloidea indicates that the Palaearctic region contains the highest number of genus-group taxa, while the Neotropical region has the highest number of species and subspecies. The Afrotropical and Australasian regions are relatively poor respectively in genera and subgenera and in species and subspecies. The oldest fossils that represent the Tipuloidea date back to the Lower Triassic at about 240 million years. Present-day general distribution patterns of many higher taxa of Tipuloidea probably have a Pangean or Gondwanan origin.

Keyword Freshwater · Biodiversity · Tipulidae · Diptera · Zoogeography · Review

Guest editors: E. V. Balian, C. Lévêque, H. Segers & K. Martens
Freshwater Animal Diversity Assessment.

Electronic supplementary material The online version of this article (doi:10.1007/s10750-007-9131-0) contains supplementary material, which is available to authorized users.

H. de Jong (✉) · P. Oosterbroek
Section Entomology, Zoological Museum Amsterdam, Plantage Middenlaan 64, Amsterdam 1018 DH, The Netherlands
e-mail: Hjong@science.uva.nl

J. Gelhaus
The Academy of Natural Sciences, 1900 Benjamin Franklin Parkway, Philadelphia, PA 19103-1195, USA

H. Reusch
BAL – Bureau of Applied Limnology and Landscape Ecology, Wellendorf 30, 29562 Suhlendorf, Germany

C. Young
Carnegie Museum of Natural History, 4400 Forbes Avenue, Pittsburgh, PA 15213, USA

Introduction

Systematic position

The Tipulidae *sensu lato*, also known as the Tipuloidea and vernacular, as crane flies, constitute one of the largest groups of the Diptera, with more than 15,270 currently recognized valid species and subspecies that are contained in 525 genera and subgenera

(Oosterbroek, 2005). The Tipuloidea, as they will be referred to in the remainder of this text, occur worldwide, ranging from the arctic to equatorial forests and from the intertidal zone to over 5,600 m in certain high mountain ranges (Alexander & Byers, 1981). They belong to the lower Diptera, but their phylogenetic position within the lower Diptera has not yet been established satisfactorily. Traditionally, the Tipuloidea were regarded as representing one of the most primitive lineages of the Diptera, if not the most primitive lineage. Authors supporting this hypothesis are, among others, Wood & Borkent (1989) and Michelsen (1996). An opposing hypothesis claims that the Tipuloidea belong to the more derived lower Diptera and are closely related to the Brachycera, the higher flies. This hypothesis, which is based on a digital phylogenetic analysis of all then available evidence, was first proposed by Oosterbroek & Courtney (1995). Ongoing research, including phylogenetic analyses of fossil information, aims to solve this controversy. The monophyly of the Tipuloidea is supported by a sequence of apomorphies associated with the different life stages (Wood & Borkent, 1989; Oosterbroek & Theowald, 1991, Oosterbroek & Courtney, 1995). Currently, the Tipuloidea are considered to contain the families Cylindrotomidae, Limoniidae, Pediciidae, and Tipulidae *sensu stricto*.

Habitus

Adult Tipuloidea can be recognized by the presence of two complete anal veins in the wing, the lack of ocelli, and the presence of a V-shaped transverse suture on the mesothorax. They are typically slender-bodied flies, with long antennae, wings, legs, and abdomen (Fig. 1). The legs fall off easily. Sizes range from very small species with a wing length of about 2 mm (e.g., *Tasiocera*) up to large species with wing lengths up to 40 mm (e.g., *Holorusia* and *Leptotarsus*). Larvae of the Tipuloidea can be recognized by the hemicephalous, retractible head capsule, and the usually metapneustic respiratory system (rarely apneustic).

Morphology

Adult

Head. Usually with short and inconspicuous rostrum in the Cylindrotomidae, Limoniidae and Pediciidae, rostrum were well-developed in Tipulidae *s.str.* and often with a nasus at tip. Rostrum elongate and slender in *Elephantomyia*, *Helius*, *Toxorhina*, in some species as long as head and thorax combined; mouthparts elongate in *Geranomyia*, forming long rostrum-like structure. Mouthparts are usually well-developed, palpi generally five-segmented, although the number of segments can be reduced. Ultimate palpal segment short in most Cylindrotomidae, Limoniidae and Pediciidae, elongate and slender in Tipulidae *s.str.* Antennae usually long and slender, consisting of 14–16 segments remarkably elongate—up to four times the body length, e.g., in certain species of *Hexatoma*, *Rhabdomastix*, *Leptotarsus*, and *Megistocera*. Flagellomeres usually cylindrical in both sexes, but pectinate in (males of) e.g., *Dicranomyia* (*Zelandoglochina*), *Clytocosmus*, *Ctenophora* and allies, *Elnoretta*, *Platyphasia*, and *Ptilogyna s.l.* Compound eyes large, often touching or nearly touching on ventral side of head. Eyes usually bare, but provided with macrotrichia in between ommatidia in most Pediciidae (absent in *Nipponomyia*). Ocelli absent.

Fig. 1 *Tipula (Acutipula) maxima* Poda, 1761, male. European species with larva in marshy soil or freely aquatic in small rivulets, adult in wet woodland along streams. Germany, photo Jürgen Peters

Thorax with well-developed mesothorax; prothorax and metathorax small were reduced to strip-like elements. Legs long and slender, tips of tibiae without spurs or with one or two spurs. Tarsal claws simple or toothed with one tooth or several teeth, arolium between claws. Halter with long shaft and distinct knob. Wing usually well-developed but sometimes reduced in female or both sexes, especially in species living in isolated localities like islands and mountain ranges, or in cold-adapted species. Venation variable, characteristic for major taxonomic units. Always two complete anal veins are present (excepting *Stibadocerella*, the species of which have one complete anal vein).

Abdomen long or very long, usually slender. External sexual organs are at the tip of the abdomen. Terminalia of male with structural adaptations of the last tergites, sternites and their genital processes. Structures of male terminalia considered species-specific. Female terminalia generally forming an ovipositor consisting of sclerotized, bladelike cerci and hypogynial valves, usually less characteristic for individual species, but showing differences at level of higher taxonomic groups.

Larva

Elongate, cylindrical, tapering toward anterior tip, posterior end often more or less truncate, with flattened spiracular disc. Head capsule usually distinct, retracted into anterior thoracic segments, deeply incised ventrally and often dorso-laterally more reduced. Abdominal segments smooth, or with transverse rows of hairs. Creeping welts or fleshy projections sometimes present. Spiracular disc usually glabrous and surrounded by lobe-like projections. Membranous anal lobes usually present.

Pupa

Pupa obtect, elongate. Eyes prominent. Mesothoracic horns usually simple, ranging from short sessile to very elongate. Antennal sheaths long. Tarsal sheaths arranged side by side, not superimposed, lateral tarsal sheaths partly covered by wing sheaths. Abdomen parallel-sided or almost so, more or less smooth, or provided with welts or spines; prominent marginal or other abdominal spines especially in Limnophilinae and Tipulidae s.str. Abdominal end truncate in male, pointed in female. Anal segment usually with spines.

Identification

There are many publications available for the identification of adult Tipuloidea, although they are rather scattered in the literature. Good starting points are Alexander & Byers (1981) for the Nearctic region, Dienske (1987), Mannheims (1951–1968), Peus (1952), Savchenko (1961–1983 (Fauna SSSR), 1982–1986 (Fauna Ukrainy), 1983, 1989), Savchenko & Krivolutskaya (1976), and Theowald (1973–1980) for the Palaearctic region, Oosterbroek (1998) for Malaysia, Alexander (1929) for the Neotropical region, Dobrotworsky (1968–1974) and Theischinger (1996) for Australia, and Alexander (1956, 1964) and Wood (1952) for the Afrotropical region.

The identification of pre-adult Tipuloidea can be attempted with the aid of Alexander & Byers (1981), Byers (1996), and Gelhaus (1986, 2000, 2002) for the Nearctic region, Brindle (1960, 1967), Brinkmann (1997), Gelhaus & Byers (1994), Hofsvang (1997), Peus (1952), Reusch & Oosterbroek (1997), and Theowald (1957, 1967) for the Palaearctic region, Young (2004) for Malaysia, and Wood (1952) for the Afrotropical region.

Life history

As in all holometabolic insects, the Tipuloidea have four life stages: egg, larva, pupa and adult. The egg stage usually lasts 6–14 days and is followed by four larval stages of variable duration. The pupal stage lasts 5–12 days and the adult usually is short-lived. The majority of species living in temperate environments are univoltine or bivoltine, i.e., they have one or two generations per year. The life cycle of others can be as short as six weeks, or, in the case of some arctic species, can last up to five years. The long life cycle of arctic species principally is caused by the shortness of the arctic summer, rather than by the low temperatures (Pritchard, 1983).

Current main habitats

Adults of the Tipuloidea are exclusively terrestrial, whereas the preadult stages of the majority of species are aquatic or semiaquatic. Adult crane flies usually occur in moist environments in the damp vegetation along the borders of lakes and streams. Certain species or higher taxonomic groups are adapted to open meadows, fairly dry rangelands, and even deserts.

Larvae of crane flies occupy a wide variety of habitats, ranging from strictly aquatic to entirely terrestrial. Fully aquatic larvae can occur in freshwater, especially in rapidly flowing streams, or in brackish water or intertidal zones. Some of these species live entirely submerged, are without functional spiracles, and take oxygen by gas-exchange through the cuticle. Others come to the surface for oxygen by using the spiracles positioned at the posterior end of the body.

A number of aquatic species occur in the benthic zone (epibenthic to hyporheic) of fast flowing rivers and streams. Specialized aquatic species occur on wet cliff and rock faces, piles or bridge piers in or beneath a scum of algal growth, mosses, or liverworts. Others are marine and live on rocks or earth in algal mats that become submerged by the tide. Species are known from cold springs, while others live in phytotelmata, such as water-filled tree-holes, or water-containing axils of leaves of such plants as Bromeliaceae and Liliaceae.

Semi-aquatic species occur in wet or saturated mats or cushions of mosses and liverworts, in leaf drift and rich organic mud along the edges of streams, lakes, ponds, and other waterbodies. Semi-terrestrial and terrestrial species are known from moist to saturated humous soils in woods, meadows, and open lands. Some species live in decaying plant material in various stages of putrefaction, in fungi that are often in advanced stages of decay, and in dead wood, ranging from wet, decayed wood to relatively sound wood. A few species are leafminers, some live in dry moss cushions, and others can be found in relatively dry soil (Alexander, 1931; Alexander & Byers, 1981).

The majority of larvae feed on decaying plant material and its associated microflora (algae and fungi), others on mosses and liverworts. Species of the Pediciinae (Pediciidae) and most Limnophilinae (Limoniidae) are predacious.

Pupation usually takes place in dryer places near the larval habitats.

Species diversity

Numbers of species and subspecies

An overview of the genera, subgenera and number of species and subspecies of the Tipuloidea for each of the biogeographical regions is given in Tables 1, 2. The information is derived from Pjotr Oosterbroek's *Catalogue of the Craneflies of the World*, the contents of which is based on the examination of virtually the entire world literature dealing with the taxonomy, systematics, faunistics and biogeography of the Tipuloidea. The classification employed in the tables is in accordance with the classification used in the

Table 1 Numbers of species and subspecies of the families of the Tipuloidea according to biogeographic region

Family	PA	NA	NT	AT	OL	AU	Total world # valid species
Cylindrotomidae	27	8	1	0	31	8	71
Limoniidae	1579	885	2728	1038	2324	2114	10430
Pediciidae	202	144	13	0	132	5	489
Tipulidae *sensu stricto*	1280	573	805	339	925	385	4188

Higher level classification of the Tipuloidea according to Oosterbroek (2005). All data derived from Oosterbroek (2005)

PA: Palaearctic; NA: Nearctic; NT: Neotropical; AT: Afrotropical; OL: Oriental; AU: Australasian

Table 2 Numbers of the genera and subgenera of the families of the Tipuloidea according to biogeographic region

Family	PA	NA	NT	AT	OL	AU	Total world # valid genera
Cylindrotomidae	6	4	1	0	7	2	9
Limoniidae	137	104	121	90	121	116	314
Pediciidae	17	13	3	0	17	1	26
Tipulidae *sensu stricto*	45	38	36	23	45	30	115

Higher level classification of the Tipuloidea according to Oosterbroek (2005). All data derived from Oosterbroek (2005)

PA: Palaearctic; NA: Nearctic; NT: Neotropical; AT: Afrotropical ; OL: Oriental; AU: Australasian

Catalogue of the Craneflies of the World. The catalogue is available on-line via http://www.science.uva.nl/zma/.

Table 2 shows that the Cylindrotomidae contain 71 valid species and subspecies, the Limoniidae 10,430 species-group taxa, the Pediciidae 489, and the Tipulidae *s.str.* 4,188. As these numbers show, differences in species richness between the major groups are considerable. The numbers of species and subspecies included in the various genera and subgenera of the Tipuloidea also show striking differences. A number of genera are monotypic, e.g., in Cylindrotomidae: *Stibadocerina*, Limoniidae: *Aymaramyia*, *Gonomyopsis*, *Jivaromyia*, *Quathlambia*, etc., Pediciidae: *Savchenkoiana*, Tipulidae s.str.: *Austrotipula*, *Elnoretta*, *Euvaldiviana* etc. Other genera and subgenera hold hundreds of species, e.g., Limoniidae: *Cheilotrichia*, *Erioptera*, *Gonomyia*, *Idiocera*, *Molophilus*, *Ormosia*, *Teucholabis*, *Austrolimnophila*, *Gynoplistia*, *Hexatoma*, *Dicranomyia*, *Geranomyia*, *Limonia*, Pediciidae: *Dicranota*, Tipulidae s.str.: *Dolichopeza*, *Nephrotoma*, *Tipula*. These divergent numbers could be related to disparities in the ages of the higher taxa and the speciation rates in the different taxonomic groups. Otherwise, variance could be an artifact of the current traditional classification with more phylogenetic studies needed to distinguish between natural and apparent species diversity.

More than 11,000 of the presently known 15,250 plus species and subspecies of the Tipuloidea were described between 1910 and 1981 by a single author, the renowned Charles P. Alexander, and certainly many more species still await description. Virtually all currently known species and subspecies are recognized on morphological traits, and re-examination of taxon clusters containing cryptic species using molecular techniques is expected to increase the number of recognized species considerably.

Phylogeny and historical processes

Fossil record

The oldest fossils of Tipuloidea, and of Diptera as such, date back to the Lower Triassic Longtanian of the Grés à Voltzia Formation of Arzviller in France, which is dated at 240 Mya (Krzeminski & Krzeminska, 2003). Other Triassic finds include fossils from the Ladinian-Carnian Madygen Formation in Kyrgyzstan, dated at 230–225 Mya and the Norian-Rhaetian Tologoi Formation in Kazakhstan, dated 216–208 Mya (Shcherbakov et al. 1995). Since the Triassic, a continuous record of fossil Tipuloidea is available through the different geological epochs. The literature dealing with the taxonomy of fossil Diptera up to the early 1990's is summarized in the *Catalogue of the fossil flies of the World* by Neal L. Evenhuis (1994).

Phylogeny

Published hypotheses about the phylogeny of the Tipuloidea include Stary (1992) and Oosterbroek & Theowald (1991). Stary (1992) studied the relationships at the family level within the Tipuloidea, using characters of the adults. He concluded that the monophyletic Limoniidae are the sistergroup of a clade containing the Pediciidae, Cylindrotomidae, and Tipulidae *s.str.*, of which the Pediciidae are the sister group of the Cylindrotomidae + Tipulidae *s.str.* Oosterbroek & Theowald (1991) studied the then available information on pre-adult Tipuloidea at the genus level. According to their results, the Pediciidae, Cylindrotomidae and Tipulidae are monophyletic groups, while the Limoniidae are paraphyletic. Part of the Limoniidae are considered to form a monophyletic clade together with the Cylindrotomidae and Tipulidae, which are regarded as sistergroups (Fig. 2).

The results of these studies by Stary and Oosterbroek & Theowald were primarily based on information on species occurring in the northern hemisphere, and their hypotheses need to be tested by incorporating additional taxa of the Southern hemisphere.

Present distribution and main areas of endemism

Distribution

Figure 3 plots the number of genera and subgenera and the number of species and subspecies of the Tipuloidea for each of the biogeographic regions.

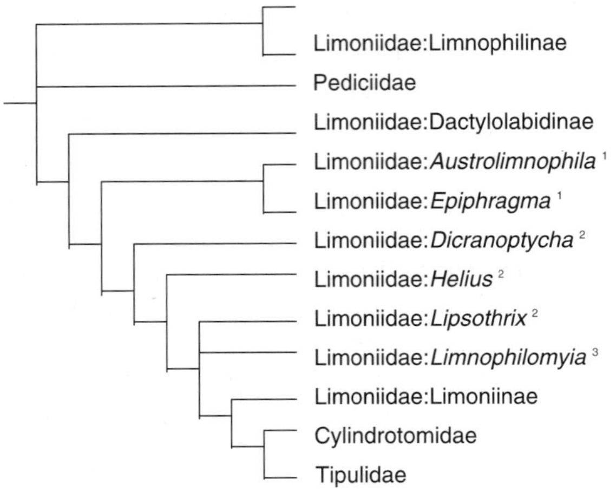

Fig. 2 Cladogram showing the relationships of the subfamilies and families of the Tipuloidea, modified from Oosterbroek & Theowald (1991). 1: Currently assigned to Limnophilinae; 2: currently assigned to Limoniinae; 3: currently assigned to Chioneinae. (see Table 1)

According to the present classification, these numbers are the following: Palaearctic region, 205 genera and subgenera, 3088 species and subspecies; Oriental region, 190 and 3412; Nearctic region, 159 and 1610; Neotropical region, 161 and 3547; Afrotropical region, 113 and 2512; Australasian region, 149 and 1377.

Our present understanding of the Tipuloidea shows that the Palaearctic region is the richest region for the taxa at the genus-group level (205), while the Neotropical region contains the highest number of species and subspecies (3547). The lowest number of genera and subgenera is found in the Afrotropical region (113), and the Australasian region has the lowest number of species and subspecies (1377).

Although the current classification of the Tipuloidea is based primarily on typological taxon concepts and the monophyly of the majority of higher taxa still has to be substantiated, a few preliminary conclusions can be drawn from the information contained in Table 1.

Endemicity and distribution patterns

The greater majority of the species and subspecies of Tipuloidea are restricted to a single biogeographic region and thus underscore the endemicity of these areas, and neighboring biogeographic regions usually share the presence of only a small percentage of species and subspecies. The Palaearctic and Nearctic regions have 88 species and subspecies in common, the Palaearctic and Oriental regions 176, the Nearctic and Neotropical regions 48, and the Oriental and Australasian regions 44. Other combinations of biogeographic regions with shared species or subspecies are the Palaearctic and Neotropical regions (3 species-group taxa), the Palaearctic and Afrotropical

Fig. 3 Numbers of species and subspecies/genera and subgenera of the Tipuloidea for the biogeographic regions. PA—Palaearctic; NA—Nearctic; NT—Neotropical; AT—Afrotropical ; OL—Oriental; AU—Australasian; PAC—Pacific Oceanic Islands; ANT—Antarctic

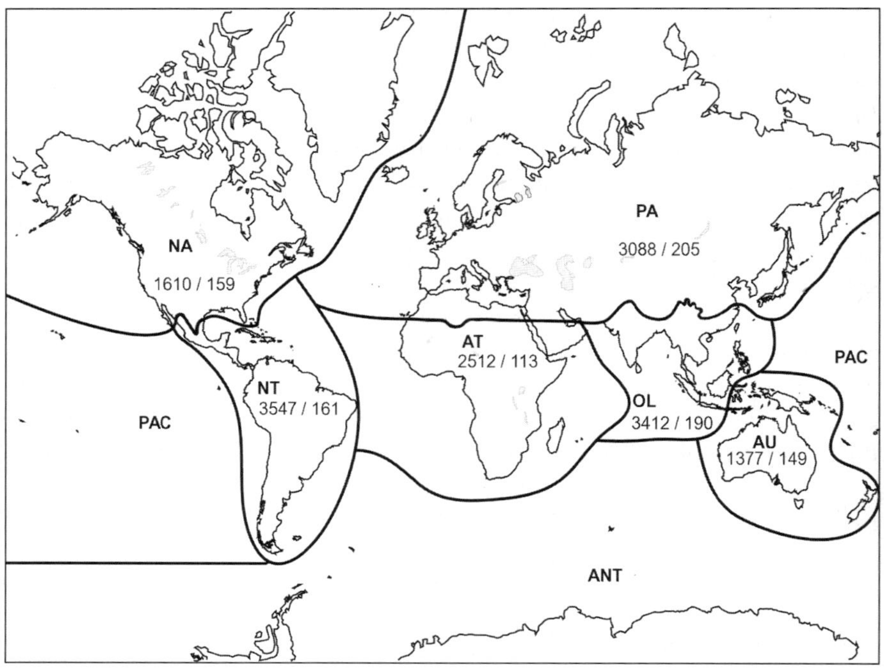

regions (5), the Palaearctic and Australasian regions (9), the Nearctic and Afrotropical regions (2), the Nearctic and Oriental regions (6), the Nearctic and Australian regions (3), the Neotropical and Afrotropical regions (3), the Neotropical and Oriental regions (2), the Neotropical and Australasian regions (3), the Afrotropical and Oriental regions (9), and the Afrotropical and Australasian regions (5). In most of these cases, the taxa in common are ubiquitous species and subspecies such as *Atypophthalmus (Atypophthalmus) umbratus*, *Symplecta (Trimicra) pilipes pilipes*, *Tipula (Tipula) oleracea*, and *Conosia irrorata irrorata*, i.e., taxa of which the range probably or certainly was accidentally extended by man.

Looking at the distribution of the genera and subgenera of the Tipuloidea over the six biogeographic regions recognized, a set of unique distribution patterns can be distinguished (Table 2). Regarding number of endemic genera and subgenera, the Neotropical and Australasian regions are the richest with 61 endemic genus level taxa each. The Afrotropical region contains 35 endemic genera and subgenera, the Palaearctic 29, and the Oriental and Nearctic regions each 25. Other unique distribution patterns with a high frequencies of repetition are the combinations of Holarctic: Palaearctic—Nearctic (25), Eurasia: Palaearctic—Oriental (25), Palearctic—Oriental—Nearctic (33) (*i.e.*, in addition to the occurrence of the combinations Palaearctic—Nearctic and Palaearctic—Oriental), Cosmopolitan: Palaearctic—Oriental—Nearctic—Neotropical—Australasian—Afrotropical (22), Oriental—Australasian (16), Palaearctic—Oriental—Nearctic—Neotropical (13), Neotropical—Australasian (12). All other distribution patterns occur with a frequency of fewer than 10 times (Table 3).

Palaeogeography

As stated above, the first evidence of Diptera in the fossil record date back to the Lower Triassic (240 Mya), during which the terrestrial domain of the earth consisted of a single landmass, Pangaea. The oldest fossil Diptera include species that definitely represent ancestral lineages of the present-day Tipuloidea. Recent distributions of the genera and subgenera of the Tipuloidea to some extent reflect the palaeaogeographic history of the earth from Triassic times onward.

Pangaea existed from the Upper Carboniferous (320 Mya) to the Upper Jurassic (180–160 Mya), when it split into the Northern continent Laurasia and the Southern continent Gondwana.

Since its origin in the Upper Jurassic, Laurasia fragmented into four main landmasses—corresponding to the present-day western Nearctic, eastern Nearctic, western Palaearctic, and eastern Palaearctic—that collided in various combinations during different geological periods (cf. Enghoff, 1995).

Gondwana started to fragment at about 150–130 Mya when Africa drifted away from the remainder of Gondwana. Since then a sequence of vicariance events led to a fragmentation history that can be summarized in the following parenthetical notation: (Africa (India (New Zealand (Australia (South America Antarctica))))).

Pangaea, Laurasia and Gondwana patterns

The Northern and Southern hemispheres share the presence of 166 genus-group taxa. Widespread genera such as *Baeoura*, *Rhabdomastix*, and *Styringomyia* take basic positions in the cladograms by Oosterbroek & Theowald (1991), which could reflect their relatively old age and, with some reservation, their distribution could be called Pangaean. Worldwide distributions are also found in relatively derived taxa, such as *Antocha*, *Molophilus*, and *Orimarga*. More detailed historical biogeographic studies are needed to distinguish between global distributions in the Tipuloidea that should be regarded as Pangaean or that result from historical processes that date back to more recent geological times.

The number of genera and subgenera that exclusively occur in the northern hemisphere (the Palaearctic and Nearctic regions and the combination of both), is 79, the number of genera and subgenera occurring exclusively in the Southern hemisphere (the Oriental, Neotropical, Afrotropical, and Australian regions and all combinations of these) is 221. Both these figures indicate the relative isolation of these two composite areas and could be correlated with the break-up of Pangaea into Laurasia and Gondwana.

The Neotropical and Australasian regions share 12 genus-group taxa. The shared occurrence of genera and subgenera in the Neotropical and Australasian faunas

Table 3 Frequency of distribution patterns of the genera and subgenera of the Tipuloidea across biogeographic region

x = presence of genus/subgenus

PA	NA	NT	AT	OL	AU	Frequency of pattern	Count of regions involved
0	0	x	0	0	0	61	1
0	0	0	0	0	x	61	1
0	0	0	x	0	0	35	1
x	x	0	0	x	0	33	3
x	0	0	0	0	0	29	1
x	x	0	0	0	0	25	2
0	x	0	0	0	0	25	1
x	0	0	0	x	0	25	2
0	0	0	0	x	0	25	1
x	x	x	x	x	x	22	6
0	0	0	0	x	x	16	2
x	x	x	0	x	0	13	4
0	0	x	0	0	x	12	2
x	0	0	x	x	x	9	4
0	x	x	0	0	0	7	2
x	x	0	0	x	x	7	4
x	x	x	x	x	0	6	5
x	x	0	x	x	0	6	4
x	0	x	x	x	x	5	5
0	0	0	x	x	0	5	2
x	x	0	x	x	x	4	5
x	x	x	x	0	0	4	4
x	0	0	x	x	0	4	3
x	x	x	0	x	x	3	5
0	x	x	x	x	x	2	5
0	x	x	0	0	0	2	2
x	0	x	0	0	0	2	2
x	0	0	x	0	0	2	2
0	0	x	x	x	x	2	4
0	0	x	x	0	0	2	2
x	x	0	0	x	x	1	4
x	x	0	x	0	0	1	3
x	x	x	0	0	0	1	3
0	x	0	0	x	0	1	2
0	x	x	x	0	x	1	4
0	x	x	0	0	x	1	3
0	x	x	x	0	0	1	3
x	0	x	0	x	x	1	4
x	0	x	0	x	0	1	3
x	0	x	x	0	0	1	3
0	0	x	0	x	x	1	3
0	0	x	x	0	x	1	3

Higher level classification of the Tipuloidea according to Oosterbroek (2005). All data derived from Oosterbroek (2005)

PA: Palaearctic; NA: Nearctic; NT: Neotropical; AT: Afrotropical ; OL: Oriental; AU: Australasian

of Tipuloidea, as such led Alexander (1929) to suggest a former Antarctic land connection. A closer look at the taxa involved reveals more intricate patterns: four genus-group taxa are present in New Zealand, Australia, and the Neotropical region, four taxa are exclusively found in New Zealand and the Neotropical region, and four taxa are exclusively shared by Australia and the Neotropical region. Most of the Neotropical species concerned occur in the south of that region. In 1960 Hennig published an analysis of the distribution patterns of Diptera occurring in the southern hemisphere, including the Tipuloidea. Lack of information on the phylogenetic relationships of the taxa involved impeded an actual association of distribution patterns with the geological events that caused the break-up of Gondwana. Hennig suggested a range of possible studies on promissing and rewarding groups of Tipuloidea occurring in the area, but since then only a few of such studies have been carried out (cf. De Jong, 1989).

The many endemic genus-group taxa in the Afrotropical region and the relatively few taxa this region shares with other biogeographical regions substantiate the relatively isolated position of the continent. Only a few patterns indicate an exclusive affinity of the Afrotropical region with one or more of the other regions of the Southern hemisphere. Most of the inter-regional patterns involving the Afrotropical region suggest a relationship with one or more of the biogeographical regions from the northern hemisphere. The relatively isolated position of the Afrotropical region probably reflects its early separation from the remainder of Gondwana.

Taxonomic diversification patterns at the species-group level in the Tipuloidea often can be correlated with Tertiary palaeaogeographic and climatic processes (e.g., De Jong, 1998; Oosterbroek, 1980; Oosterbroek & Arntzen, 1992; Tangelder, 1988; Theowald, 1984).

Human related issues

Research

Due to their high species-richness, the Tipuloidea are an excellent study group for looking at patterns of local and regional diversity and endemism. Many species are specialized and have restricted habitat requirements and can, therefore, be used to indicate habitat quality and to identify regions of priority for conservation efforts (Falk, 1991; Oosterbroek, 1994). Tipuloidea species are often extremely abundant and they play key roles in many ecosystems, including aquatic systems, either as consumers or as prey for a wide variety of predators ranging from fish, amphibia, birds and mammals, to spiders and predacious insects.

Tipulid larvae are a component in aquatic biomonitoring—as a group they are considered sensitive to moderately sensitive to human perturbations (e.g., rating of 3 on a 1–10 scale in the Hilsenhoff index, where 10 is highly tolerant of pollution).

Pest species

A few species of Tipuloidea are of economic importance in that their larvae feed on roots of crops or on seedlings of field crops. By their sheer numbers, some of these species can become serious pests (Blackshaw & Coll, 1999). The larva of *Tipula (Yamatotipula) aino* is a chronic aquatic pest of rice seedlings.

References

Alexander, C. P., 1929. Crane-flies. Diptera of Patagonia and South Chile 1: 1–240, pl. 1–12.

Alexander, C. P., 1931. Deutsche Limnologische Sunda-Expedition. The crane-flies (Tipulidae, Diptera). Archiv für Hydrobiologie, Supplement 9, Tropische Binnengewässer 2: 135–191, pl. 2.

Alexander, C. P., 1956. Tipulidae. Ruwenzori Expedition 1934–35 1(7): 129–380, pl. 27, 28.

Alexander, C. P., 1964. Diptera (Nematocera): Tanyderidae, Ptychopteridae, Tipulidae. South African Animal Life 10: 229–441.

Alexander, C. P. & G. W. Byers, 1981. Tipulidae. In McAlpine, J. F., B. V. Peterson, G. E. Shewell, H. J. Teskey, J. R. Vockeroth & D. M. Wood (eds), Manual of Nearctic Diptera, Vol. 1. Research Branch, Agriculture Canada, Monograph 27: 153–190.

Blackshaw, R. P. & C. Coll, 1999. Economically important leatherjackets of grassland and cereals: biology, impact and control. Integrated Pest Management Reviews 4: 143–160.

Brindle, A., 1960. The larvae and pupae of the British Tipulinae (Diptera: Tipulidae). Transactions of the Society of British Entomology 14: 63–114.

Brindle, A., 1967. The larvae and pupae of the British Cylindrotominae and Limoniidae (Diptera, Tipulidae). Transactions of the Society of British Entomology 17: 151–216.

Brinkmann, R., 1997. Diptera Cylindrotomidae. In Nilsson A. (ed.), Aquatic Insects of North Europe, a Taxonomic Handbook, Vol. 2. Apollo Books, Stenstrup: 99–104.

Byers, G. W., 1996. Tipulidae. In Merritt R. W. & Cummins K. W. (eds), An Introduction to the Aquatic Insects of North America. Kendall/Hunt Publishing Company, Dubuque, Iowa: 549–570.

de Jong, H., 1998. In search of historical biogeographic patterns in the western Mediterranean terrestrial fauna. Biological Journal of the Linnean Society 65: 99–164.

de Jong, H., 1989. Revision of the South American taxa *Elnoretta* Alexander, *Euvaldiviana* Alexander and *Valdiviana* Alexander (Diptera: Tipulidae), with a discussion on their phylogeny. Systematic Entomology 14: 243–257.

Dienske, J. W., 1987. An illustrated key to the genera and subgenera of the western Palaearctic Limoniidae (Insecta: Diptera), including a description of the external morphology. Stuttgarter Beiträge zur Naturkunde, Serie A (Biologie) 409: 1–52.

Dobrotwosky, N. V. 1968–1974. The Tipulidae (Diptera) of Australia. Australian Journal of Zoology 16: 459–494; 495–510; Supplement 1: 1–41; Supplement 5: 21–45; Supplement 16: 1–9; 11–43; Supplement 25: 1–14; 15–30; 31–62; 63–75.

Enghoff, H., 1995. Historical biogeography of the Holarctic: area relationships, ancestral areas, and dispersal of non-marine animals. Cladistics 11: 223–263.

Evenhuis, N. L., 1994. Catalogue of the Fossil Flies of the World (Insecta: Diptera). Backhuys Publishers, Leiden: 1–600.

Falk, S., 1991. A review of the scarce and threatened flies of Great Britain (Part 1). Research and Survey in Nature Conservation 39: 1–192.

Gelhaus, J. K., 1986. Larvae of the crane fly genus *Tipula* in North America (Diptera: Tipulidae). The University of Kansas Science Bulletin 53: 121–182.

Gelhaus, J. K. 2000. Manual for the identification of aquatic crane fly larvae for Northwest North America: 1–147. The Northwest Biological Assessment Workgroup's Taxonomic Workshop on Diptera, Ellensburg, Washington.

Gelhaus, J. K., 2002. Manual for the identification of aquatic crane fly larvae for Southeastern Uniited States: 1–205. Carolina Area Benthological Workshop, Durham, North Carolina.

Gelhaus, J. K. & G. W. Byers, 1994. Tipulidae. In Morse, J. C., L.Yang & L. Tian (eds), Aquatic Insects of China Useful for Monitoring Water Quality: 438–458.

Hennig, W., 1960. Die Dipteren-fauna von Neuseeland als systematisches und tiergeografisches Problem. Beiträge zur Entomologie 10: 221–329.

Hofsvang, T., 1997. Diptera Tipulidae, crane flies. In Nilsson A. (ed.), Aquatic Insects of North Europe, a Taxonomic Handbook, Vol. 2. Apollo Books, Stenstrup: 93–98.

Krzeminski, W. & E. Krzeminska, 2003. Triassic Diptera: descriptions, revisions and phylogenetic relations. Acta Zoologica Cracoviensia 46: 153–184 Supplement.

Mannheims, B. 1951–1968. Tipulidae. Fliegen der Palaearktischen Region 3(5) 1: 1–320.

Michelsen, V., 1996. Neodiptera: new insights into the adult morphology and higher level phylogeny of Diptera (Insecta). Zoological Journal of the Linnean Society 117: 71–102.

Oosterbroek, P., 1980. The western palaearctic species of *Nephrotoma* Meigen, 1803 (Diptera, Tipulidae), Part 5, phylogeny and biogeography. Beaufortia 29: 311–393.

Oosterbroek, P., 1994. Biodiversity in the Mediterranean region. In Forey, P. L., C. J. Humphries & R. I. Vane-Wright (eds), Systematics Conservation Evaluation. Systematics Association Special, Vol. 50. Clarendon Press, Oxford: 289–307.

Oosterbroek, P., 1998. The families of Diptera of the Malay Archipelago. Fauna Malesiana Handbooks 1: i-xii, 1–227.

Oosterbroek, P. 2005. Catalogue of the Craneflies of the World, version 17.05.2005. The catalogue is on-line available via http://www.science.uva.nl/zma/.

Oosterbroek, P. & J. W. Arntzen, 1992. Area-cladograms of Circum-Mediterranean taxa in relation to Mediterranean palaeogeography. Journal of Biogeography 19: 3–20.

Oosterbroek, P. & G. Courtney, 1995. Phylogeny of the nematocerous families of Diptera (Insecta). Zoological Journal of the Linnean Society 115: 267–311.

Oosterbroek, P. & B. Theowald, 1991. Phylogeny of the Tipuloidea based on characters of larvae and pupae (Diptera, Nematocera), with an index to the literature except Tipulidae. Tijdschrift voor Entomologie 134: 211–267.

Peus, F., 1952. Cylindrotomidae. Fliegen der Palaearktischen Region 3(5) 3: 1–80.

Pritchard, G., 1983. Biology of Tipulidae. Annual Review of Entomology 28: 1–22.

Reusch, H. & P. Oosterbroek, 1997. Diptera Limoniidae and Pediciidae, short-palped crane flies. In Nilsson A. (ed.), Aquatic Insects of North Europe, a Taxonomic Handbook, Vol. 2. Apollo Books, Stenstrup: 105–132.

Savchenko, E. N., 1961–1983. Crane flies (Diptera, Tipulidae. Fauna SSSR, Diptera 2 (N.S.) 79: 1–488; 89: 1–503; 105: 1–282; 127: 1–585.

Savchenko, E. N., 1982–1986. Limoniid flies. Fauna Ukrainy 14 (2): 1–340; 14 (3): 1–335; 14 (4): 1–180.

Savchenko, E. N., 1983. Limoniidae of South Primorye. Akademia Nauk Ukrainian SSR. Kiev: 1–156.

Savchenko, E. N., 1989. Limoniidae fauna of the USSR. Akademia Nauk Ukrainian SSR, Kiev: 1–377.

Savchenko, E. N. & G. O. Krivolutskaya, 1976. Limoniidae of the South Kuril Islands and South Sakhalin. Akademia Nauk Ukrainian SSR. Kiev: 1–160.

Shcherbakov, D. E., E. D. Lukashevich, & V. A. Blagoderov, 1995 Triassic Diptera and initial radiation of the order. International Journal of Dipterological Research 6: 75–115.

Stary, J., 1992. Phylogeny and classification of Tipulomorpha, with special emphasis on the family Limoniidae. Acta Zoologica Cracoviensia 35: 11–36.

Tangelder, I. R. M., 1988. The biogeography of the Holarctic *Nephrotoma dorsalis* species-group (Diptera, Tipulidae). Beaufortia 38: 1–35.

Theischinger, G., 1996. The Limoniinae (Diptera: Tipulidae) of Australia. VIII. Preliminary key to the genus-group taxa. Stapfia 44: 117–144.

Theowald, B., 1957. Die Entwicklungsstadien der Tipuliden (Diptera, Nematocera) insbesondere der west-palaearktischen Arten. Tijdschrift voor Entomologie 100: 195–308.

Theowald, B., 1967. Familie Tipulidae (Diptera, Nematocera). Larven und Puppen. Bestimmungsbücher zur Bodenfauna Europas 7: 1–100.

Theowald, B., 1973–1980. Tipulidae. Fliegen der Palaearktischen Region 3(5) 1: 321–538.

Theowald, B., 1984. Taxonomie, Phylogenie und Biogeographie der Untergattung *Tipula (Tipula)* Linnaeus, 1758 (Insecta, Diptera, Tipulidae). Tijdschrift voor Entomologie 127: 33–78.

Wood, H. G., 1952. The crane-flies of the South-West Cape. Annals of the South African Museum 39: 1–327.

Wood, D. M., & A. Borkent, 1989. Phylogeny and classification of the Nematocera. In McAlpine, J. F. & D. M. Wood (eds), Manual of Nearctic Diptera, Vol. 3. Research Branch, Agriculture Canada, Monograph 32: 1333–1370.

Young, C. W., 2004. Insecta: Diptera, Tipulidae. In Yule, C. M. & Y. H. Sen (eds), Freshwater Invertebrates of the Malaysian Region: 775–785.

Hydrobiologia (2008) 595:469–475
DOI 10.1007/s10750-007-9114-1

Global diversity of black flies (Diptera: Simuliidae) in freshwater

Douglas C. Currie · Peter H. Adler

Abstract Black flies are a worldwide family of nematocerous Diptera whose immature stages are confined to running waters. They are key organisms in both aquatic and terrestrial ecosystems, but are perhaps best known for the bloodsucking habits of adult females. Attacks by black flies are responsible for reduced tourism, deaths in wild and domestic birds and mammals, and transmission of parasitic diseases to hosts, including humans. About 2,000 nominal species are currently recognized; however, certain geographical regions remain inadequately surveyed. Furthermore, studies of the giant polytene chromosomes of larvae reveal that many morphologically recognized species actually consist of two or more structurally indistinguishable (yet reproductively isolated) sibling species. Calculations derived from the best-known regional fauna—the Nearctic Region—reveal that the actual number of World black fly species exceeds 3,000.

Keywords Simuliidae · Lotic · Filter-feeders · Bloodsuckers · Sibling species · Bioindicators

Guest editors: E.V. Balian, C. Lévêque, H. Segers & K. Martens
Freshwater Animal Diversity Assessment

Electronic supplementary material The online version of this article (doi:10.1007/s10750-007-9114-1) contains supplementary material, which is available to authorized users.

D. C. Currie (✉)
Department of Natural History, Royal Ontario Museum, Toronto, ON, Canada
e-mail: dcurrie@zoo.utoronto.ca

P. H. Adler
Department of Entomology, Clemson University, Clemson, SC, USA
e-mail: padler@clemson.edu

Introduction

Black flies (Simuliidae) represent a relatively small and structurally homogeneous family of nematocerous Diptera. They are most closely related to the Ceratopogonidae, Chironomidae, and Thaumaleidae, which collectively form the culicomorphan superfamily Chironomoidea. The 2000 nominal species of black flies are worldwide in distribution, occurring on all continents except Antarctica. They also populate most major archipelagos except Hawaii, the Falkland Islands, and isolated desert islands.

As with other holometabolous insects, black flies pass through four stages to complete their development: egg, larva, pupa, and adult. The first three stages are confined to running waters which, depending on species, can range in size from tiny headwaters to large rivers. Eggs are deposited on a variety of submerged or emergent substrata, or are simply dropped into the water where they settle into the sediments. Larvae (Fig. 1, top) are sausage-shaped organisms with a well-sclerotized external head capsule that typically bears a pair of labral fans.

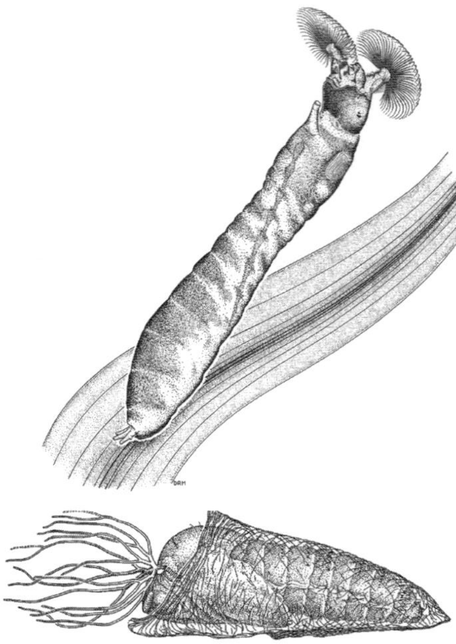

Fig. 1 Simuliidae Habitus: larva (top) (source: Currie, 1986), pupa (bottom) (source: Peterson, 1981)

The thorax bears a single-ventral prothoracic proleg, and the last abdominal segment terminates with a posterior proleg that serves as an attachment organ. Larvae typically pass through seven instars to reach maturity. The fully mature larva (pharate pupa) spins a variously shaped silken cocoon in which to pupate. The pupa (Fig. 1, bottom) is remarkably uniform in shape and reflects the shape of the future adult. A pair of spiracular gills arises from the anterolateral corners of the thorax and typically projects anteriorly or anterodorsally. The gill consists of a number of slender filaments, or in some species it is tubular, club shaped, or spherical. The adults are small, hunchbacked flies with cigar-shaped antennae. Their wings are broad at the base with darkly sclerotized anterior veins and weakly sclerotized posterior veins. Both sexes require sugars (nectar, honeydew) as a source of energy for flight and other metabolic needs. Females of most species require blood from homeothermic hosts (mammals or birds) to develop their eggs. During blood feeding, black flies can transmit parasitic disease agents to their hosts. Some species (ca. 2.4% of the world fauna) have mouthparts that are feebly developed and unable to cut flesh (Crosskey, 1990). These obligatorily autogenous species develop their eggs in the absence of blood.

Black flies are key organisms in both aquatic and terrestrial ecosystems, especially in the boreal biome of the Nearctic and Palearctic Regions (Malmqvist et al., 2004). Larvae occur in huge numbers under favorable conditions, attaining population densities of up to a million individuals/m^2. Under such densely packed conditions, they are an important source of food for many invertebrate (e.g., plecopteran) and vertebrate predators (e.g., salmonids). The filter-feeding habit of larvae plays a role in the processing of organic matter in streams. Fine particulate organic matter, and even dissolved organic matter, is removed from the water column and, as a consequence of the larva's low-digestion efficiency, is egested as nutritious fecal pellets. The pellets sink rapidly to the streambed where they serve as food for members of the collector–gatherer functional feeding guild of invertebrates. Were it not for the black fly larva's ability to assimilate such fine particles, much of the organic matter entrained in the water column would be transported downstream. The importance of fecal pellets in streams was highlighted by a study for a single river in Sweden that showed the average daily transport of fecal pellets reached a staggering 429 tonnes (dry mass) past an imaginary line across the river (Malmqvist et al., 2001). This massive amount of recycled organic matter provides crucial fodder for invertebrates and microorganisms, and has the potential to fertilize river valleys (Malmqvist et al., 2004).

Species & generic diversity

As one of the most prominent members of the benthic community and the second most important group of medically important insects, the family Simuliidae is among the most completely known groups of freshwater arthropods. The Simuliidae, along with the Culicidae in part, are unique among freshwater organisms in that their taxonomy has been greatly aided by band-by-band analyses of the giant polytene chromosomes in the larval silk glands. These giant chromosomes have routinely allowed the discovery of morphologically indistinguishable species. Despite these taxonomic promoters, we suspect that a significant proportion of the world's simuliid fauna remains undiscovered.

Approximately 2000 nominal species of extant black flies in 26 nominal genera are recognized as valid through mid-2006 (Crosskey & Howard, 2004; Adler & Crosskey, unpublished) (Tables 1, 2). We base our estimates of the absolute number of simuliids on mentations derived from one of the best-known regional faunas—the Nearctic fauna (Adler et al., 2004). Out of the 256 species of Nearctic black flies, 186 have had some level of chromosomal screening, albeit minimal in some cases. As a result of these chromosomal studies, 60 species (32.2%) were added to the Nearctic fauna. If we use 32.2% as a minimal estimate of hidden biodiversity (i.e., sibling species), the Nearctic fauna will increase by 23 species (i.e., 32.2% of the 70 chromosomally unscreened species), yielding a total of 279 species. Each of the remaining major regional faunas also would increase by 32.2% of the total number of nominal species known from each region, yielding species counts of 283 (Afrotropical), 258 (Australasian), 469 (Neotropical), 424 (Oriental), 73 (Pacific), and 924 (Palearctic).

The estimate derived for hidden biodiversity is a minimum value. Although giant chromosomes provide powerful *prima facie* evidence of reproductive isolation among species when two opposite and fixed chromosomal sequences are present in sympatry, they cannot always be used to reveal good species. Since Y-chromosome differences occur in the heterozygous condition—male black flies are typically XY—the giant chromosomes cannot always provide a distinction between two possibilities: separate species or a Y-chromosome polymorphism within one species. Given the high frequency of cases where different Y-chromosomes occur among putatively single species, we expect that the number of additional sibling species is considerable. Similarly, giant chromosomes cannot always reveal homosequential sibling species, which are not only morphologically indistinguishable, but also have the same chromosomal banding sequences. These species are real (e.g., Henderson, 1986), but too little prospecting has been done to provide an estimate of their numbers among the Simuliidae. We suspect that the number of homosequential sibling species would easily balance the number of potential synonyms, and for purposes of predicting the absolute number of species we here consider them to cancel one another.

Table 1 Numbers of simuliid species in each biogeographical region for the family, subfamilies, and tribes

	PA	NA	AT	NT	OL	AU	PAC	ANT	World*
Simuliidae	699	256	214	355	321	195	55	2	2000
Parasimuliinae	0	4	0	0	0	0	0	0	4
Simuliinae	699	252	214	355	321	195	55	2	1996
Prosimuliini	77	62	0	0	0	0	0	0	136
Simuliini	622	190	214	355	321	195	55	2	1860

PA: Palaearctic; NA: Nearctic; NT: Neotropical; AT: Afrotropical ; OL: Oriental; AU: Australasian; PAC: Pacific Oceanic Islands; ANT: Antarctic

* Grand totals reflect not only species endemic to a region but also shared species

Table 2 Numbers of simuliid genera in each biogeographical region for the family, subfamilies, and tribes

	PA	NA	AT	NT	OL	AU	PAC	ANT	World*
Simuliidae	12	13	2	10	1	2	1	1	26
Parasimuliinae	0	1	0	0	0	0	0	0	1
Simuliinae	12	12	2	10	1	2	1	1	25
Prosimuliini	6	4	0	0	0	0	0	0	6
Simuliini	6	8	2	10	1	2	1	1	20

PA: Palaearctic; NA: Nearctic; NT: Neotropical; AT: Afrotropical ; OL: Oriental; AU: Australasian; PAC: Pacific Oceanic Islands; ANT: Antarctic

* Grand totals reflect not only species endemic to a region but also shared species

Estimating the number of morphologically distinct species yet to be discovered is more problematic. In areas such as western Europe and North America, the number is probably quite low; only one new morphologically distinct species has been found in the Nearctic Region and none in western Europe in the past four years. On the other hand, in poorly explored hot spots, such as the Himalayas, Cambodia, Peru, and the interior of Irian Jaya, the number is likely to be large; for example, 71 morphologically distinct species were found on five major Indonesian islands in the past four years (Takaoka, 2003). We suggest that a 5% increase in the number of morphologically distinct species above the currently recognized number of nominal species in the Nearctic and Palearctic Regions is reasonable, yielding an increase of 13 and 35 species, respectively. For the Afrotropical, Australasian, Neotropical, Oriental, and Pacific Regions, we estimate a 25% increase in species numbers, yielding increases of 54, 49, 89, 80, and 14 species, respectively. Considering the number of potentially undiscovered morphospecies and sibling species, we suspect that the total number of black flies in the world is more than 3000, with regional contributions to this grand total as follows: 337 (Afrotropical), 2 (Antarctic), 307 (Australasian), 292 (Nearctic), 558 (Neotropical), 504 (Oriental), 87 (Pacific), and 959 (Palearctic).

Phylogeny and historical processes

The earliest definitive simuliid fossils date to late Jurassic times (Kalugina, 1991; Currie and Grimaldi, 2000); however, the fossil record of related families suggests that black flies must have originated considerably earlier. The family Simuliidae, therefore, likely had a Pangean, or effectively Pangean, origin.

There are no comprehensive phylogenies of the world Simuliidae—at least, none that are reconstructed in an explicitly phylogenetic framework. Adler et al. (2004) provided a cladistic analysis of the genera and subgenera of Holarctic black flies, and Moulton (2000) provided an interpretation of suprageneric relationships based on his analysis of molecular sequence data. Both studies recognized a two-subfamily system (Parasimuliinae and Simuliinae), of which the latter was divided into two tribes (Prosimuliini and Simuliini); this system is the one followed here. Members of the subfamily Parasimuliinae exhibit a relict distribution, occurring only in the coastal mountains of western North America. Within the subfamily Simuliinae, members of the tribe Prosimuliini occur only in the Nearctic and Palearctic Regions, whereas the most plesiomorphic members of the tribe Simuliini occur in the Afrotropical, Australasian, and Neotropical Regions. These distributions suggest that the two tribes originated in Laurasia and Gondwanaland, respectively (Currie, 1988).

Present distribution and main areas of endemicity

The Palearctic Region has by far the largest number of described species (Fig. 2), although the dubious taxonomic practices of earlier specialists suggest that many synonyms probably exist among the names of currently recognized species. Nonetheless, significant parts of the Palearctic Region (e.g., Himalayas) remain inadequately surveyed and the current prediction of 959 species might not be unreasonable. The Nearctic Region—the most completely surveyed faunal area—has far fewer than half that number of predicted species. The Neotropical and Oriental Regions have roughly equal numbers of species (558 and 504 species, respectively), although the latter region has not received the same intensity of taxonomic scrutiny as the former.

South Asia is in particular need of study. Relatively few taxonomic studies have been conducted in the area, consisting mainly of isolated species descriptions. The simuliid fauna of the Afrotropical Region has received considerable attention, but most of this effort has been directed toward vectors (e.g., *Simulium damnosum* complex) of the causal agent of human onchocerciasis. The Australasian Region has not been intensively studied since the early 1970s (e.g., Mackerras & Mackerras, 1952; Dumbleton, 1973). Indeed, the 71 species recently described from Irian Jaya, Maluku, and Sulawesi (Takaoka, 2003) represent more than 36% of all the morphospecies currently recognized from Australasia. More intensive surveys in Irian Jaya, Papua New Guinea, and Western Australia will increase the number of species for the region. The simuliid faunas of Antarctica (Crozet Islands) and the Pacific Oceanic Islands have been well surveyed (e.g., Craig et al., 2003; Craig & Joy, 2000); yet, additional species continue to be

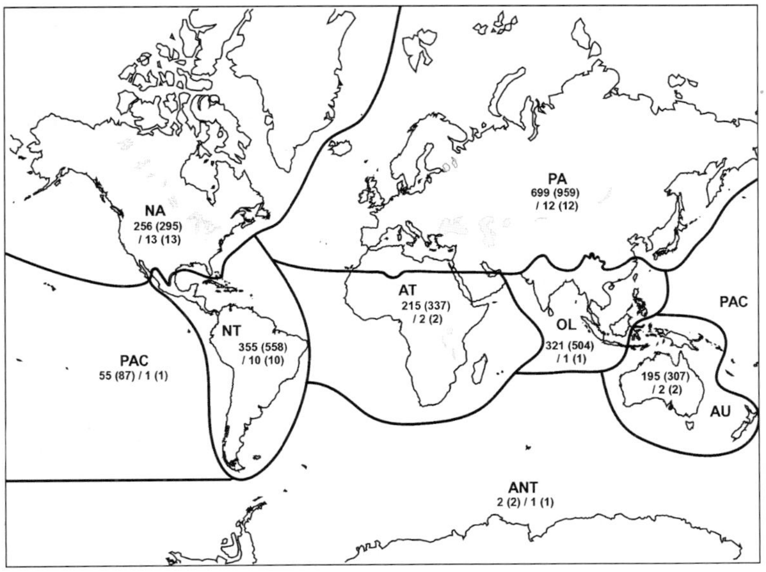

Fig. 2 Known and Estimated (numbers in parentheses) diversity of black flies in each zoogeographic region (Species / Genera). PA, Palaearctic; NA, Nearctic; NT, Neotropical; AT, Afrotropical; OL, Oriental; AU, Australasian; PAC, Pacific Oceanic Islands; ANT, Antarctic

discovered in the Pacific Region as new collections are made.

The Nearctic and Oriental Regions have the fewest endemic taxa among the major zoogeographic areas, perhaps owing in part to their close association with the Palearctic Region (Tables 1, 2). The Neotropical Region has by far the largest number of endemic genus-group taxa; however, a number of currently recognized 'valid' names (e.g., *Kempfsimulium*, *Pedrowygomyia*) undoubtedly will fall into synonymy, as phylogenetic relationships become better understood. In contrast, several additional genus-group taxa will have to be recognized for species that are currently assigned to the Australian "*Paracnephia*." The Afrotropical Region, with its 11 unique genus-group taxa, is second only to the Neotropical Region in terms of endemicity. Antarctica (Crozet Islands) and the Pacific Oceanic Islands have one and two endemic genus-group taxa, respectively.

In terms of faunal similarities, the Nearctic and Palearctic Regions share by far the greatest number (18) of genus-group taxa (Table 3). The Nearctic and Neotropical Regions share six genus-group taxa, as do the Palearctic and Oriental Regions. At the other end of the spectrum, Antarctica (Crozet Islands), with its one endemic genus (*Crozetia*), exhibits no clear relationship with any other zoogeographical region (Craig et al., 2003) (Table 3).

Human-related issues

The bloodsucking habits of female simuliids are responsible for considerable deleterious effects on

Table 3 Genus-group taxa of Simuliidae shared among zoogeographic regions

	PA	NA	NT	AT	OL	AU	PAC	ANT
PA	–							
NA	18	–						
NT	2	6	–					
AT	2	2	0	–				
OL	6	4	1	2	–			
AU	3	2	0	2	4	–		
PAC	1	0	0	0	1	2	–	
ANT	0	0	0	0	0	0	0	–

PA = Palearctic, NA = Nearctic, NT = Neotropical, AT = Afrotropical, OL = Oriental, AU = Australasian, PAC = Pacific Oceanic Islands, ANT = Antarctic

humans and their economic welfare (Crosskey, 1990). Reduced tourism, deaths of domesticated birds and mammals, and transmission of parasitic disease organisms are but a few of the myriad medical and socioeconomic impacts associated with black flies. Human onchocerciasis (river blindness) is the most pressing health-related issue, with up to 18 million people infected in parts of Africa and South and Central America. The causative organism, *Onchocerca volvulus*, is transmitted exclusively by black flies—predominantly members of the *Simulium damnosum* complex in Africa and members of the *Simulium* subgenera *Aspathia* and *Psilopelmia* in South and Central America (Crosskey, 1990). In addition to being the sole vectors of the disease agent of river blindness, black flies are pests of humans due to their swarming and bloodsucking behavior. However, no species of black fly feeds exclusively on humans, and relatively few species include humans among their hosts. Massive outbreaks of anthropophilic species, nonetheless, can have a great impact on tourism and other forms of human activity.

Black flies are also responsible for transmitting parasitic disease organisms, such as filarial worms, protozoans, and arboviruses to a wide variety of domesticated animals (Adler, 2005). Massive attacks by livestock pests such as *Cnephia pecuarum*, *Simulium colombaschense*, *Simulium luggeri*, and *Simulium vampirum* have caused mortality in cattle, horses, mules, pigs, and sheep. Deaths in such instances are typically attributed to toxic shock (simuliotoxicosis) from the salivary injections of many bites. Sublethal attacks can have an economic impact through unrealized weight gains, reduced milk production, malnutrition, impotence, and stress-related phenomena (Adler et al., 2004). The effects of black flies on wild birds and mammals are inadequately studied, but are likely to be as great as those reported for domesticated animals.

Black flies have a negative reputation because of the bloodsucking habit of the females. On a more positive note, the adults provide food for predators, such as birds and odonates, and promote conservation by deterring people from inhabiting or developing wilderness areas. The immature stages not only play a dominant role in lotic communities by processing organic matter, but also are sensitive to anthropogenic inputs and are thus excellent barometers of water quality. *Simulium maculatum* (Meigen), for example, once widespread in central Europe, was extirpated from many large rivers because of pollution (Zwick & Crosskey, 1981).

Where pest species persist or thrive in the face of human activity, various means have been used to control their populations. Historically, chlorinated hydrocarbons such as DDT (dichlorodiphenyltrichloroethane) and methoxychlor were used against black fly larvae. Neither compound was specific to black flies, and both were susceptible to resistance problems. DDT was discontinued in the early 1970s because of its devastating impact on the environment (i.e., bioaccumulation), and methoxychlor and the organophosphate Abate fell into disfavor because of resistance and nonspecificity (Adler et al., 2004). Currently, the biological control agent *Bacillus thuringiensis* variety *israelensis* (*Bti*)—a naturally occurring bacterium—is the remedy of choice against black flies worldwide. Unlike its chemical predecessors, *Bti* has an excellent host specificity, is highly toxic to larval black flies, is safe for humans, and is relatively inexpensive.

References

Adler, P. H., 2005. Black flies, the Simuliidae. In W. C. Marquardt (ed.), Biology of Disease Vectors, 2nd edition. Elsevier Academic Press, San Diego, CA: 127–140.

Adler, P. H., D. C. Currie & D. M. Wood, 2004. The Black Flies (Simuliidae) of North America. Cornell University Press, Ithaca, NY: 941.

Craig, D. A., D. C. Currie & P. Vernon, 2003. *Crozetia* (Diptera: Simuliidae): redescription of *Cr. crozetensis*, *Cr. seguyi*, number of larval instars, phylogenetic relationships and historical biogeography. Zootaxa 259: 1–39.

Craig, D. A. & D. A. Joy, 2000. New species and redescriptions in the central-western Pacific subgenus *Inseliellum* (Diptera: Simuliidae). Annals of the Entomological Society of America 93: 1236–1262.

Crosskey, R. W., 1990. The Natural History of Blackflies. John Wiley, Chichester, U.K.: 711.

Crosskey, R. W. & T. M. Howard, 2004. A Revised Taxonomic and Geographical Inventory of World Blackflies (Diptera: Simuliidae). The Natural History Museum, London. http://www.nhm.ac.uk/research-curation/projects/blackflies/ (accessed 28 September 2006).

Currie, D. C., 1986. An annotated list of and keys to the immature black flies of Alberta (Diptera: Simuliidae). Memoirs of the Entomological Society of Canada 134: 1–90.

Currie, D. C., 1988. Morphology and systematics of primitive Simuliidae (Diptera: Culicomorpha). Ph.D. thesis, University of Alberta, Edmonton. 331 pp.

Currie, D. C. & D. Grimaldi, 2000. A new black fly (Diptera: Simuliidae) genus from mid Cretaceous (Turonian) Amber of New Jersey. In D. Grimaldi (ed.), Studies on Fossils in Amber, with Particular Reference to the Cretaceous of New Jersey. Backhuys Publishers, Leiden, The Netherlands: 473–485.

Dumbleton, L. J., 1973. The genus *Austrosimulium* Tonnoir (Diptera: Simuliidae) with particular reference to the New Zealand fauna. New Zealand Journal of Science 15(1972): 480–584.

Henderson, C. A. P., 1986. Homosequential species 2*a* and 2*b* within the *Prosimulium onychodactylum* complex (Diptera): temporal heterogeneity, linkage disequilibrium, and Wahlund effect. Canadian Journal of Zoology 64: 859–866.

Kalugina, N. S., 1991. New Mesozoic Simuliidae and Leptoconopidae and blood-sucking origin in lower dipterans. Paleontologichesky Zhurnal 1991: 69–80. [In Russian with English summary].

Mackerras, I. M. & M. J. Mackerras, 1952. Notes on Australasian Simuliidae (Diptera). III. Proceedings of the Linnean Society of New South Wales 77: 104–113.

Malmqvist, B., R. S. Wotton & Y. Zhang, 2001. Suspension feeders transform massive amounts of seston in large northern rivers. Oikos 92: 35–43.

Malmqvist, B., P. H. Adler, K. Kuusela, R. W. Merritt & R. S. Wooton, 2004. Black flies in the boreal biome, key organisms in both terrestrial and aquatic environments: a review. Écoscience 11: 187–200.

Moulton, J. K., 2000. Molecular sequence data resolves basal divergences within Simuliidae (Diptera). Systematic Entomology 25: 95–113.

Peterson, B. V., 1981. Simuliidae. In J. F. McAlpine, B. V. Peterson, G. E. Shewell, H. J. Teskey, J. R. Vockeroth, & D. M. Wood (eds.), Manual of Nearctic Diptera, vol. 1, Monograph No. 27. Research Branch, Agriculture Canada, Ottawa: 355–391.

Takaoka, H., 2003. The Black Flies (Diptera: Simuliidae) of Sulawesi, Maluku and Irian Jaya. Kyushu University Press, Fukuoka, Japan: 581.

Zwick, H. & R. W. Crosskey, 1981 ["1980"]. The taxonomy and nomenclature of the blackflies (Diptera: Simuliidae) described by J. W. Meigen. Aquatic Insects 2: 129–173.

Hydrobiologia (2008) 595:477–487
DOI 10.1007/s10750-007-9037-x

FRESHWATER ANIMAL DIVERSITY ASSESSMENT

Global diversity of mosquitoes (Insecta: Diptera: Culicidae) in freshwater

Leopoldo M. Rueda

Abstract Mosquitoes that inhabit freshwater habitats play an important role in the ecological food chain, and many of them are vicious biters and transmitters of human and animal diseases. Relevant information about mosquitoes from various regions of the world are noted, including their morphology, taxonomy, habitats, species diversity, distribution, endemicity, phylogeny, and medical importance.

Keywords Mosquitoes · Culicidae · Diptera · Freshwater · Diversity

Introduction

Mosquitoes are important groups of arthropods that inhabit freshwater habitats. Their role in the ecological food chain is well recognized by many aquatic ecologists. They are prominent bloodsuckers that annoy man, mammals, birds and other animals including reptiles, amphibians, and fish. They are probably the most notoriously undesirable arthropods, with respect to their ability to transmit pathogens causing human malaria, dengue, filariasis, viral encephalitides, and other deadly diseases. They are also known for being irritating biting pests. Sometimes, their nuisance bites are so severe that they make outdoor activities almost impossible in many parts of the world. Many large coastal areas are made unbearable by salt marsh mosquitoes, and the real estate development and the tourism industry are also seriously affected. More than a hundred species of mosquitoes are capable of transmitting various diseases to humans and other animals. *Anopheles* mosquitoes, for example, solely transmit malaria. It is undoubtedly the most serious arthropod vector-borne disease affecting humans. About 90% of all malaria deaths in the world occur in Africa, south of the Sahara. This is because the majority of infections in Africa are caused by *Plasmodium falciparum*, the most dangerous of the four human malaria parasites. It is also due to the fact that efficient malaria vectors (e.g., *Anopheles gambiae* Giles) are widespread in Africa and are very difficult to control. In many parts of the world, the indirect effect of malaria and mosquito-borne diseases in lowered vitality and susceptibility to other non-vector-borne diseases accounted for more deaths as well as reduced production following work losses.

This chapter presents the current information on the morphology, taxonomy, habitats, species diversity, phylogeny, distribution, endemicity, and medical importance of the various groups of mosquitoes

Guest editors: E. V. Balian, C. Lévêque, H. Segers and K. Martens
Freshwater Animal Diversity Assessment

L. M. Rueda (✉)
Walter Reed Biosystematics Unit, Department of Entomology, WRAIR, MSC MRC 534 Smithsonian Institution, 4210 Silver Hill Road, Suitland, MD 20746, USA
e-mail: ruedapol@si.edu

worldwide. It is not the intention of this work to comprehensively review all major topics mentioned above. Selected references are included, and should be consulted for further reading.

General morphology

Mosquitoes, like any arthropods, are bilaterally symmetrical. The adult mosquito (Fig. 1A) is covered with an exoskeleton, and its body is divided into three principal regions: the head, thorax, and abdomen. The head is ovoid in shape, with large compound eyes. It bears five appendages, which consist of two antennae, two maxillary palpi and the proboscis. The thorax, the body region between the head and abdomen, is composed of three segments, the prothorax, mesothorax, and metathorax. Each segment has a pair of jointed legs; in addition, the mesothorax has a pair of functional wings, and the metathorax, a pair of wings represented by knobbed structures called halteres. The abdomen is composed of 10 segments, of which the three terminal segments are specialized for reproduction and excretion. Apparently, the mosquito adults resemble Chironomidae, Dixidae, Chaoboridae, and other Nematocera, which like mosquitoes have aquatic immature stages. Mosquitoes, however, are distinguished from such similar looking dipterous flies by the presence of scales on the wing veins and wing margins, and by their forwardly projecting long proboscis that is adapted for piercing and sucking. In contrast to an adult, the larva (Fig. 1B) is largely composed of soft, membranous tissues in the thorax and abdomen, and hardened, sclerotized plates in the

Fig. 1 (**A**) Mosquito adult female, *Anopheles sinensis*, lateral view. (**B**) Mosquito larvae, *An. sinensis*, dorsal view. (**C**) Mosquito habitat, creek. (**D**) Mosquito habitat, irrigation ditch

head. This allows for the swimming movements and doubling of the body when cleaning the mouth or palatal brushes.

Detailed morphological descriptions and glossaries of the adult, pupa, larva, and egg of mosquitoes are found in several publications (e.g., Harbach & Knight 1980, 1981; Darsie & Ward 2005). Most taxonomic keys to identify mosquitoes are based on morphological characters (e.g., Belkin 1962; Harrison & Scanlon 1975; Huang 1977; Rueda et al. 1997; Rueda 2004; Rattanarithikul et al. 2005). For additional list of identification keys and references, you may visit the WRBU website, http://wrbu.org/.

Biology

Mosquitoes have a holometabolous type of development; that is, having four distinct stages in their life cycle: egg, larva, pupa, and adult. Larvae and pupae of mosquitoes require an environment with standing or flowing water for proper development. The female adult lays either single eggs (e.g., *Aedes, Anopheles*) or in clusters (e.g., *Culex, Culiseta*), up to several hundred at a time, on the surface of the water, on the upper surface of floating vegetation, along the margins of quiet water pools, on the walls of artificial containers or in moist habitat subject to flooding. The larvae (called *wrigglers*) undergo shedding (or *molting*) of the skin (or *exuviae*) four times before becoming pupa. Larvae of most species usually filter out and feed on organic matter and other microorganisms, in the water for about 1–3 weeks or longer depending on the water temperature. Larvae of mosquito predators (e.g., *Toxorhychites*, *Lutzia*) feed on larvae of other mosquitoes. In some predatory species, the first instar is a filter feeder, and the predaceous feeding structures are not developed until the second instar. The pupae (called *tumblers*), or resting stage, appear after the fourth larval molt. Unlike larvae, pupae do not feed, and live for 1–3 days before becoming adults. Only adult female mosquitoes bite humans and animals. Male mosquitoes feed primarily on flower nectars, while the females require the blood meal to produce viable eggs. Some species (*anthropophilous*) prefer to feed on man, while others (*zoophilous*) feed in nature on animals (including mammals and birds) other than humans. Females of *Toxorhynchites* and other mosquitoes do not feed on blood. There are some species that readily feed on fish exposed to air, reptiles, amphibians, and insect larvae (Harris et al. 1969; Harwood & James 1979). Some autogenous females can also produce viable eggs, even without blood meal. Females typically feed every 3–5 days, and in a single feeding a female usually engorges more than its own weight of blood. Some species of mosquitoes (e.g., *Anopheles*) prefer to feed at dusk, twilight or nighttime, while others bite mostly during the daytime (e.g., *Aedes*). Other species exhibit seasonal switching of hosts that provides a mechanism for transmitting diseases from animals to humans (called *zoonotic disease transmission*). Diapause (i.e., hibernation, aestivation, over wintering) occurs in various life stages, e.g. as eggs in *Psorophora* and most *Aedes*; as larvae in *Coquillettidia*, some *Culiseta, Mansonia, Orthopodomyia, Toxorhynchites, Wyeomyia*, some *Aedes*; as adults, often fertilized females, in *Uranotaenia*, most Culex, some *Anopheles* and other *Culiseta*; and as either eggs or larvae in *Culiseta morsitans* (Theobald) (Stojanovitch & Scott 1997).

The stimuli or factors that attract mosquitoes to a human or animal host are complex and are not fully understood. Mosquitoes, like other biting arthropods, use visual, thermal, and olfactory stimuli to locate a host. Olfactory cues may be the most important as a mosquito nears the host but visual stimuli seem important for in flight orientation, particularly over wide ranges. For daytime biters, movement of the host may initiate orientation toward that person or animal. Out of more than 300 compounds that are released from the human body as by-products of metabolism, more than 100 volatile compounds can be detected from human breath. Carbon dioxide is released primarily from the breath and the skin, and can be detected by mosquitoes. Carbon dioxide and octenol are common attractants that are used in monitoring and surveillance of mosquitoes in their habitats (Rueda et al. 2001). The antennae of mosquitoes have chemoreceptors that are stimulated by lactic acid, but can be inhibited by repellents. Repellents (e.g., deet or *N,N*-diethyl-3-methyl-benzamide) are effective personal protective measures against biting insects to reduce or prevent transmission of vector-borne diseases (Rueda et al. 1998).

Habitats

Mosquitoes have diverse habitats that allow them to colonize different kinds of environments. The immature stages of mosquitoes are found in a variety of aquatic habitats (Table 1, Fig. 1C, D), e.g., ponds, streams, ditches, swamps, marshes, temporary and permanent pools, rock holes, tree holes, crab holes, lake margins, plant containers (leaves, fruits, husks, tree holes, bamboo nodes), artificial containers (tires,

Table 1 Common habitats of the mosquito larvae

Habitats*	Examples of mosquitoes	Remarks
1. Flowing streams	*Culex fuscocephala, gelidus; Anopheles kochi; An. spp.*	Include creeks, drainage and irrigation ditches
2. Ponded streams	*An. kochi; Cx. annulus, bitaeniorhynchus; Lutzia fuscanus*	Include flooded stream beds, Chlorophyta-rich habitats, polluted ponds
3. Lake edges	*An. farauti, maculipennis An. quadrimaculatus, pseudopunctipennis Cx. annulirostris, squamosus*	Margins of lakes
4. Swamps and marshes	*An. farauti, gambiae, kochi, punctulatus An. sinensis; An. spp. Cx. annulus, bitaeniorhynchus, gelidus, sitiens Cx. tritaeniorhynchus; Lutzia fuscanus*	Include coastal marshes, mangrove swamps, irrigated fields
5. Shallow permanent ponds	*Aedes longirostris, An. kochi, sinensis Cx. gelidus, tritaeniorhynchus Mansonia uniformis, Mimomyia chamberlaini*	Include fishponds, duckweed ponds
6. Shallow temporary pools	*Ae. communis, excrucians, hexodontus, impiger An. dirus*	Include snowmelt pools
7. Intermittent ephemeral puddles	*An. gambiae, kochi, punctulatus Cx. annulus, fuscocephala, tritaeniorhynchus*	Common in road construction sites resulting from rainy season downpours
8. Natural containers (plant origin)	*Aedes (Aedimorphus, Finlaya, Stegomyia,) spp., Anopheles spp., Armigeres spp., Culex spp., Ficalbia spp., Haemagogus sp., Orthopodomyia spp. Sabethes spp., Toxorhynchites spp., Tripteroides spp. Uranotaenia spp., Wyeomyia spp.*	Include tree holes, internodes, leaf axils, flower bracts, fronds, nuts, pods, pitchers [Graminae(bamboo), Pandanaceae (screwpines), Palmae (palms), Agavaceae (Dracaena), Araceae (taro), Musaceae (banana, abaca), Bromeliaceae (bromeliads, pineapples), Cytanaceae (rafflesias), Nepenthaceae (climbing pitcher plants), Sarraceniaceae (terrestial pitcher plants)]
9. Natural containers (animal and other origins)	*Aedes (Cancraedes, Geoskusea, Levua, Lorrainea) Ae. (Rhinoskusea, Skusea, Stegomyia) spp. Anopheles spp., Culex spp., Culiseta spp. Deinocerites spp., Eretmapodites spp. Uranotaenia spp.*	Include shells of snails, clams, arboreal ant nests, crab holes
10. Artificial containers	*Aedes spp., Culex spp., Toxorhynchites spp.*	Include tires, cans, flower vases, bottles, tanks, troughs, drums, gutters, etc.

* Adapted from Laird (1988) and other references

tin cans, flower vases, bird feeders), and other habitats (Laird 1988; Rueda et al. 2005, 2006). The enormous importance of diverse habitats on the increasing populations of mosquitoes has been well recognized by aquatic ecologists and public health personnel. Furthermore, knowledge of larval habitats is important in determining vector control, as well as for disease prevention purposes. It is extremely necessary in designing effective vector control programs. The most practical way to reduce a local population of pestiferous mosquitoes is to eliminate their habitats as much as possible, particularly sources of standing water, such as old discarded tires, clogged gutters, stumped tree holes, etc. Larval habitats that are not possibly eliminated can be modified (e.g., cleaning clogged ditches, open water management in salt marshes). Appropriate methods can be applied such as using biological control agents (predatory fish, microbials), selected insecticides for permanent habitats (ponds, lakes), or other environmental modification techniques to control breeding of mosquitoes.

Table 2 Number of genera by subfamily/tribe in each zoogeographical region

Subfamily/Tribe	Number of genera[a]						
	PA	NA	NT	AT	OL	AU	World
Subfamily Anophelinae	1	1	2	1	2	2	3
Subfamily Culicinae	18	12	22	14	23	20	39
Tribe Aedeomyiini	0	0	1	1	1	1	1
Tribe Aedini	5	4	3	2	8	5	12
Tribe Culicini	2	2	3	1	1	1	4
Tribe Culisetini	1	1	1	1	1	1	1
Tribe Ficalbiini	2	0	0	2	2	2	2
Tribe Hodgesiini	0	0	0	1	1	1	1
Tribe Mansoniini	2	1	2	2	2	2	2
Tribe Orthopodomyiini	1	1	1	1	1	1	1
Tribe Sabethini	3	1	9	1	4	4	13
Tribe Toxorhynchitini	1	1	1	1	1	1	1
Tribe Uranotaeniini	1	1	1	1	1	1	1
Total	19	13	24	15	25	22	42

PA, Palaeractic; NA, Nearctic; NT, Neotropical; AT, Afrotropical; OL, Oriental; AU, Australasian (including South Pacific Islands)

[a] Based on multiple searches of the Walter Reed Biosystematics Unit web site in December 2005

Species diversity

Mosquitoes belong to the family Culicidae, order Diptera, class Insecta (Hexapoda), phylum Arthropoda. There are two recognized subfamilies, the Anophelinae and Culicinae. Subfamily Culicinae has 11 tribes: Aedeomyiini (*Aedeomyia*), Aedini (*Aedes, Armigeres, Ayurakitia, Eretmapodites, Haemagogus, Heizmannia, Opifex, Psorophora, Tanakius, Udaya, Verrallina, Zeugnomyia*), Culicini (*Culex, Deinocerites, Galindomyia, Lutzia*), Culisetini (*Culiseta*), Ficalbiini (*Ficalbia, Mimomyia*), Hodgesiini (*Hodgesia*), Mansoniini (*Coquillettidia, Mansonia*), Orthopodomyiini (*Orthopodomyia*), Sabethini (*Isostomyia, Johnbelkinia, Limatus, Malaya, Maorigoeldia, Onirion, Runchomyia, Sabethes, Shannoniana, Topomyia, Trichoprosopon, Tripteroides, Wyeomyia*), Toxorhynchitini (*Toxorhynchites*), and Uranotaeniini (*Uranotaenia*). There are about 3,500 species and subspecies, under 140 subgenera in 42 genera of mosquitoes worldwide (Walter Reed Biosystematics Unit 2001). In Culicinae, tribe Sabethini (13) has the greatest number of genera, followed by Aedini (12) and Culicini (4). Tribes Ficalbiini and Mansoniini each have two genera, and the remaining tribes each with one genus. The Oriental region (OL, 25) has the greatest number of genera, followed by the Neotropical region (NT, 24), and the Australasian region (22) (Table 2, Fig. 2).

The diversity of the mosquito species varies among different geographical regions of the world. The greatest diversity of mosquito species is found in the NT (31% of total known species; 1069/3492), followed by the Oriental (30%), Afrotropical (22%), and Australasian (22%) regions. The Nearctic region (5%) has the lowest species diversity (Table 3, Fig. 2). In the NT, the greatest number of species in Culicinae is found under tribe Culicini, followed by Aedini and Sabethini. Only one species in Aedeomyiini is known, but none of Ficalbiini and Hodgesiini from the NT. In the OL, Aedini has the greatest number of species, followed by Culicini and Sabethini. Orthopodomyiini, Hodgesiini, Culisetini, and Aedeomyiini have lower number of species in the OL. In the Afrotropical, Palearctic, Nearctic, and Australasian region, Aedini and Culicini have the greatest number of species. Aedomyiini and Orthopodomyiini have the lowest number of species in the

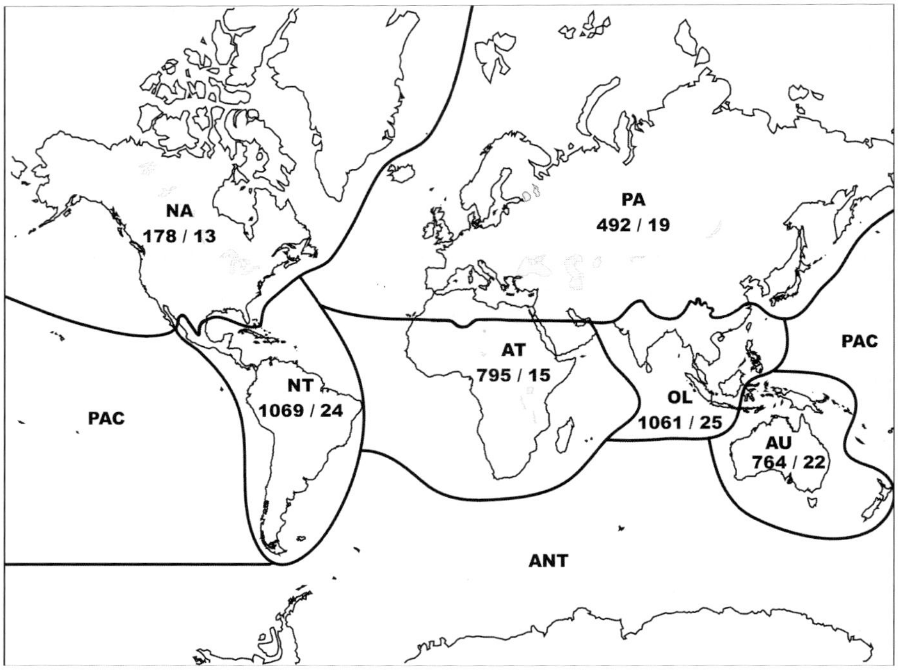

Fig. 2 Number of species and genera in each zoogeographical region (from Tables 2 and 3). PA, Palaearctic; NA, Nearctic; NT, Neotropical; AT, Afrotropical; OL, Oriental; AU, Australasian region including South Pacific Islands (PAC: Pacific Oceanic Islands; ANT: Antarctic)

Table 3 Number of species by subfamily/tribe in each zoogeographical region

Subfamily/tribe	Number of species[a]						
	PA	NA	NT	AT	OL	AU	World
Subfamily Anophelinae	94	21	125	148	168	95	489
Subfamily Culicinae	398	157	944	647	893	669	3003
Tribe Aedeomyiini	0	0	1	3	1	2	6
Tribe Aedini	217	96	258	300	407	302	1262
Tribe Culicini	96	35	360	169	185	155	813
Tribe Culisetini	16	8	8	7	4	15	40
Tribe Ficalbiini	8	0	0	34	15	13	53
Tribe Hodgesiini	0	0	0	4	6	5	11
Tribe Mansoniini	7	2	27	26	18	26	81
Tribe Orthopodomyiini	3	3	7	7	7	1	25
Tribe Sabethini	19	4	223	6	135	89	409
Tribe Toxorhynchitini	10	5	24	22	35	15	88
Tribe Uranotaeniini	22	4	36	69	80	46	215
Total	492	178	1069	795	1061	764	3492

PA, Palaeractic; NA, Nearctic; NT, Neotropical; AT, Afrotropical; OL, Oriental; AU, Australasian (including South Pacific Islands)

[a] Based on multiple searches of the Walter Reed Biosystematics Unit web site in December 2005

Afrotropical and Australasian regions. In the Nearctic region, Mansoniini has the lowest number of species, and none is known of Aedeomyiini, Ficalbiini, and Hodgesini. In the Palearctic region, Orthopodomyiini has the lowest number of species, but none is known of Aedeomyiini and Hodgesiini.

Although mosquito taxonomists use new methods of computerized and molecular analyses and new data sets to address the phylogeny and classification of mosquitoes, many problems arise to change the classification and nomenclature. Due to the importance of mosquitoes in disease transmission, many technical issues need to be resolved to facilitate communication and information exchange among public health practitioners, medical entomologists, parasitologists, epidemiologists, ecologists, and other interested groups. In this article, the taxonomic classification of the family Culicidae as given by Knight & Stone (1977) is followed. Mosquitoes are documented in world catalogs (Knight & Stone 1977 and its three supplements) that are updated regularly (Walter Reed Biosystematics Unit 2001).

Phylogeny

The classification of mosquitoes, Family Culicidae, consists of two subfamilies, the Anophelinae and

Culicinae (Harbach & Kitching 1998, 2005). Subfamily Anophelinae has three genera: *Anopheles* Meigen (cosmopolitan in distribution), *Bironella* Theobald (Australasian), and *Chagasia* Cruz (Neotropical). Subfamily Culicinae has 39 genera (Table 2). Phylogenetic analysis is needed within the family Culicidae. Detailed observations still need to be made that many hypotheses of relationship must be regarded only as a stimulus for further investigation. The phylogeny of anopheline mosquitoes has been examined using morphological characters (Sallum et al. 2000; Harbach & Kitching 1998, 2005) and nuclear, ribosomal and mitochondrial DNA sequences (Krzywinski et al. 2001a, b, Sallum et al. 2002). Based on cladistic analysis of morphological data (Harbach & Kitching 1998), Anophelinae is a monophyletic clade comprised of *Chagasia* in a sister-group relationship to *Bironella* and *Anopheles*. However, separate analyses based on both morphological and molecular data (Sallum et al. 2000, 2002, respectively) suggest that subgenus *Anopheles* is a paraphyletic assemblage relative to *Bironella*. Recent study (Harbach & Kitching 2005) based on morphological data analysis using implied weighting supported the monophyly of Anophelinae, the basal position of *Chagasia*, the monophyly of subgenera *Cellia, Kertezia* and *Nyssorhynchus*, and the sister relationship of *Kertezia* and *Nyssorhynchus*. *Bironella, Lophopodomyia*, and *Stethomyia* are firmly nested within subgenus *Anopheles*.

Mosquito taxon sampling and under representation of the species and groups in the analyses of morphological and molecular data clearly affect the outcome of a phylogenetic reconstruction. In addition, the sequence-based phylogenies are heavily compromised by the selection of DNA fragments, interpretations of gene structure and homology, alignment and sequencing error and choice of phylogenetic method (Harbach & Kitching 2005). Due to the medical importance of *Anopheles* in malaria transmission, many taxonomists, medical entomologists, mosquito control specialists and preventive medicine personnel are very likely to prefer the practice of recognizing subgenera rather than raising those groups to numerous genera.

The phylogeny of tribe Aedini is still not stabilized, despite recent attempts by Reinert et al. (2004). In their cladistic analysis, they used the morphological characters of less than 10% (119/1239) of the species in the tribe Aedini. They tried to translate the results of their cladistic analysis directly into a reclassification of genera. They proposed an elevation of 32 subgenera to generic status, even though most of these genera are poorly defined and difficult or impossible to identify, which results in too much confusion (Savage 2005).

Some fossil mosquitoes have been traced back to the Eocene period (38–54 million years before the present, BP). Most fossil discoveries of mosquitoes have been from the Oligocene period (26–38 million years BP) (Lutz 1985) by which time genera *Aedes, Culex*, and probably *Mansonia*, can be recognized. Although Bode (1953) described new genera and species from immature forms as possibly belonging to the Culicidae, they are difficult to ascertain or confirm. Those interested in the phylogeny and paleontology of mosquitoes should consult Edwards (1923), Christophers (1933), Statz (1944), Wood & Borkent (1989), Harbach & Kitching (1998), and Poinar et al. (2000).

Distribution and endemicity

Mosquitoes have an almost world wide distribution, being found throughout the tropics and temperate regions. They are absent only from a few islands and Antarctica. They can thrive in a variety of habitats with fresh water, brackish water, or any water (clear, turbid or polluted) except in marine habitats with high-salt concentration. Although there are about 3,500 known species and subspecies, there are probably more than 1,000 species that have yet to be found and described. The biodiversity of mosquitoes is very evident, with many genera having worldwide distribution and some genera with limited or endemic distribution. For example, the genera, i.e., *Anopheles, Aedes, Coquillettidia, Culex, Culiseta, Lutzia, Orthopodomyia, Toxorhynchites*, and *Uranotaenia*, have at least one species found in all five regions of the world. About 36% of the 42 known genera are endemic in four regions. In the NT, there are nine endemic genera (*Chagasia, Galindomyia, Isostomyia, Johnbelkinia, Limatus, Onirion, Sabethes, Shannoniana*, and *Trichoprosopon*); in the OL, three endemic genera (*Ayurakitia, Udaya*, and *Zeugnomyia*); in the Australasian region, including South Pacific Islands (AU), two endemic genera

(*Maorigoeldia* and *Opifex*); and in the Afrotropical or Ethiopian region (ET), one endemic genus (*Eretmapodites*). There are 47 species under uncertain subgenera in four genera, namely *Aedes* (1 species in uncertain genus in AU), *Culex* (7 species in NT, 1 in AU), *Wyeomyia* (37 species in NT), and *Culiseta* (1 species in AU). The NT has 58% and 57% of the 81 known subgenera and 42 genera, respectively, worldwide; Nearctic region, 23% and 31%; OL, 52% and 60%; Afrotropical region, 33% and 38%; Australasian region (including South Pacific Islands), 53% and 55%; and Palearctic region, 39% and 45%.

In all five regions, subfamily Culicinae (81–88%) has greater number of species than subfamily Anophelinae. In the subfamily Culicinae, tribe Aedini (46–61%) has the greatest number of species in all regions, except in the NT. Species in all 10 tribes of the subfamily Culicinae are found in the Australasian, Oriental, and Afrotropical regions. Species of tribe Hodgesiini have not been reported from the Nearctic, Palearctic, and NTs, while species of the tribe Aedeomyiini have not been reported from the Palearctic region. Many countries or island groups in the Oriental, Australasian, and NTs have endemic species of mosquitoes. For example, about 40% of the mosquito fauna in the Philippines are found only in that country, and such endemicity is conspicuous especially for the genera *Tripteroides*, *Zeugnomyia*, and *Aedes* (*Finlaya*) and *Armigeres* (*Armigeres*) (Tsukamoto et al. 1985). In the South Pacific Islands, about 84% of the indigenous species are endemic to the islands (Belkin 1962). The endemism of the mosquito fauna in the South Pacific Islands and other parts of the Australasian region, for example, is not confined to the specific level. In addition to two endemic genera (*Opifex*, *Maorigoeldia*), there are several endemic species groups in *Uranotaenia*, *Culex*, *Coquillettidia*, *Aedes*, and *Tripteroides* from this region. Belkin (1962) compared the mosquito fauna of the South Pacific Islands with the other regions of the world. He noted that there are some affinities of South Pacific fauna with the Oriental, Nearctic, Palaearctic and Afrotropical regions, and parts of the NT (South Chile areas). Due to the relatively poor knowledge of numerous mosquito groups, it is very difficult to ascertain the exact affinities of mosquito fauna from various regions. An updated list of endemic species for each country and the general distribution record of each species can be found in the online systematic catalog of Culicidae (Walter Reed Biosystematics Unit 2001).

Human related issues

Medical importance

Many members of the mosquito genera such as *Anopheles, Aedes, Culex,* and *Mansonia* commonly bite humans and animals, and are involved in the transmission of infectious diseases as principal, secondary or bridge vectors. They are a nuisance to humans and other animals. Cutaneous responses to mosquito bites range from common localized wheal- and -flare reactions to delayed bite papules, and anaphylaxis. Bite reactions are the result of sensitization to mosquito salivary antigens, which lead to the formation of specific IgE and IgG antibodies. In extreme cases, severe itching can lead to secondary infection through scratching of the skin. In heavily infested areas, particularly during peak season, they cause annoyance to humans, and can affect the agriculture and tourism industry along coastal areas. These mosquitoes might also be important in the epidemiology of various pathogenic organisms in those areas.

Mosquitoes are important vectors of organisms causing diseases to humans and animals, particularly malaria (plasmodia, i.e., *P falciparum, Plasmodium vivax, Plasmodium malariae, Plasmodium ovale*), lymphatic filariases (filarial worms, i.e., *Wuchereria, Brugia*), arboviral encephalitides (viruses, i.e., dengue, yellow fever, West Nile, Japanese, Eastern Equine, others) (Harwood & James 1979; Peters 1992; Service 1993; Centers for Disease Control and Prevention 2005). Malaria is a very serious disease, affecting about 40% of the world's population, mostly in the tropical and subtropical areas of the world. Despite the heavy use of pesticides to control *Anopheles* vectors of malaria during the 20th century, more than 1 million deaths and over 300 acute malaria cases are still reported annually in the world, with greatest number of deaths occurring in Africa (World Health Organization 2005). Antimalarial drugs have been available for more than 50 years, but effective vaccines are still needed to help boost the campaign against the malarial disease. Important malaria vectors are found in various regions of the world such as NT (*e.g.*, *An. albimanus, An. albitarsis, An. darlingi, An. aquasalis, An. pseudopunctipennis*), Afrotropical region (*An. gambiae, An. funestus, An. arabiensis*), OL (*An. dirus, An. minimus, An. flavirostris*), Palearctic region (*An. lesteri, An. sinensis*), and Australasian region (*An. farauti, An. punctulatus, An. koliensis*) (Peters 1992).

Dengue is another serious arboviral disease of Asia, South and Central America, and Africa. Although it has a low mortality, it has very debilitating symptoms. *Aedes aegypti* and *Ae. albopictus* are the common vectors of dengue. They can easily adapt and proliferate in new areas, resulting in the wide spread of dengue worldwide. For example, in 2003, Brazil reported more than 350,000 human cases and 42 deaths. In 2004, Indonesia had a dengue outbreak that caused more than 54,000 cases and over 600 deaths, while Venezuela reported 11,600 cases. Yellow fever is another important disease in Africa and South and Central America, with about 200,000 cases and 30,000 deaths in 33 countries every year. It is transmitted primarily by *Ae. aegypti*. Most countries have strict regulations and requirements for yellow fever vaccination prior to entering the country that may help in minimizing disease infections among tourists or travelers. West Nile virus (WNV) has spread from its origin in Africa (Uganda) in 1937 into Europe and west and central parts of Asia. In 1999, it first appeared in North America (New York) with 62 cases and 7 human deaths. It is widespread in the United States, Canada, Mexico, and the Caribbean Islands. In the United States, more than 43 species of mosquitoes have tested positive for WNV. *Culex pipiens* complex is the most common mosquito group associated with human and horse infections. Furthermore, significant population growth, demographic movement to urban residential area, and an increase in tourism-based facilities are deemed major factors involved in mosquito-borne disease resurgence in many parts of the world.

Food chain

Many people think that mosquitoes would be one of the last groups of wildlife that they would consider saving. Mosquitoes can be unbearable, biting viciously in forest shades or near salt-water mangroves, especially during the rainy season. Even though mosquitoes are annoying to most people, they play an important role in the food chain, particularly in the everglades and other freshwater habitats. For instance, mosquito larvae are usually at the base of the food chain. In freshwater areas, they are a primary source of food for small fish such as the mosquito fish (*Gambusia*), which in turn are food for medium size fish such as the blue gill and brim. These fish are food for still larger fish, such as large mouth and garfish. The bass and the gar are a food source for the alligator, birds, and humans. Wading birds (i.e., egrets, spoon bills, and wood storks) also benefit from the mosquito food chain. As the mosquito ponds dry up between rains, the mosquito fish and killifish become trapped by the thousands. Wading birds congregate at the receding ponds and feed on concentrated, readily available fish. Furthermore, without mosquito larvae, local fishermen and fish sportsmen in many coastal areas would have less fish to catch (U.S. Department of the Interior 2005).

Acknowledgments I thank R. C. Wilkerson, T. Gaffigan, J. Pecor, T. Anderson, and staff of the Walter Reed Biosystematics Unit for curating specimens, maintaining the WRBU mosquito catalog websites, and providing help during the preparation of the manuscript. Special thanks go to M. A. Sallum, R. Sithiprasasna, and Y. M. Huang for reviewing the manuscript and providing invaluable advice; to C. R. Summers and B. P. Rueda for providing helpful comments and support. This manuscript was prepared under a Memorandum of Understanding between the Walter Reed Army Institute of Research and the Smithsonian Institution, with institutional support provided by both organizations. The opinions and assertions contained herein are those of the author and are not to be construed as official or reflecting the views of the Department of the Army or the Department of Defense.

References

Belkin, J. N., 1962. The Mosquitoes of the South Pacific (Diptera, Culicidae). Univ. California Press, Berkeley, Vol. 1, 608 pp., Vol. 2, 412 pp.

Bode, A., 1953. Die Insektenfauna des ostniedersaechsichen oberen Lias. Palaeontographica 103A, 1–375.

Centers for Disease Control and Prevention. 2005. Mosquito-borne diseases. National Center for Infectious Diseases, Centers for Disease Control, Atlanta, Georgia. http://www.cdc.gov/ncidod/diseases/list_mosquitoes.htm.

Christophers, S. R., 1933. The Fauna of British India, Including Ceylon and Burma. Diptera. Family Culicidae. Tribe Anophelini, Vol. IV. Taylor & Francis, London, 371 pp.

Darsie, R. F. & R. A. Ward, 2005. Identification and Geographical Distribution of the Mosquitoes of North

America, North of Mexico. University Press of Florida, Gainesville, Florida, 384 pp.

Edwards, F. W., 1923. Oligocene mosquitoes in the British museum; with a summary of our present knowledge concerning fossil Culicidae. Quarterly Journal of the Geological Society, London 79: 139–155.

Harbach, R. E. & I. J. Kitching, 1998. Phylogeny and classification of the Culicidae (Diptera). Systematic Entomology 23: 327–370.

Harbach, R. E. & I. J. Kitching, 2005. Reconsideration of anopheline mosquito phylogeny (Diptera: Culicidae: Anophelinae) based on morphological data. Systematics and Biodiversity 3: 345–374.

Harbach, R. E. & K. L. Knight, 1980. Taxonomists' Glossary of Mosquito Anatomy. Plexus Publishing, Inc. Marlton, NJ, 415 pp.

Harbach, R. E. & K. L. Knight, 1981. Corrections and additions to taxonomists' glossary of mosquito anatomy. Mosquito Systematics 13: 201–217.

Harris, P., D. F. Riordan & D. Cooke, 1969. Mosquitoes feeding on insect larvae. Science 164: 184–185.

Harrison, B. A. & J. E. Scanlon, 1975. Medical entomology studies—II. The Subgenus *Anopheles* in Thailand (Diptera: Culicidae). Contributions of the American Entomological Institute.(Ann Arbor) 12: 1–307.

Harwood, R. F. & M. T. James, 1979. Entomology in Human and Animal Health. Macmillan Publishing Co., New York, 548 pp.

Huang, Y. M., 1977. Medical entomology studies. VII. The subgenus *Stegomyia* of *Aedes* in Southeast Asia. II.- The edwardsi group species. III. The w-albus group species. (Diptera: Culicidae). Contributions of the American Entomological Institute (Ann Arbor) 14: 1–111.

Knight, K., A. Stone, 1977. A Catalog of the Mosquitoes of The World (Diptera: Culicidae). Thomas Say Foundation, Vol. 6. College Park, Maryland, 611 pp.

Krzywinski, J., R. C. Wilkerson & N. Besansky, 2001a. Evolution of mitochondrial and ribosomal gene sequences in Anophelinae (Diptera: Culicidae): implications for phylogeny reconstruction. Molecular Phylogenetics and Evolution 18: 479–487.

Krzywinski, J., R. C. Wilkerson & N. Besansky, 2001b. Towards understanding Anophelinae (Diptera: Culicidae) phylogeny: insights from nuclear single copy genes and weight of evidence. Systematic Biology 50: 540–556.

Laird, M., 1988. The Natural History of Larval Mosquito Habitats. Academic Press, London, 555 pp.

Lutz, M., 1985. Eine fossile Stechmucke aus dem Unter-Oligozen von Cereste, Frankreich (Diptera, Culicidae). Palaeontologische Zeitschrift 59: 269–276.

Peters, W., 1992. A Colour Atlas of Arthropods in Clinical Medicine. Wolfe Publishing, London, 304 pp.

Poinar Jr. G. O., T. J. Zavortink, T. Pike & P. A. Johnston, 2000. *Paleoculicis minutus* (Diptera: Culicidae) n. gen., n. sp., from Cretaceous Canadian amber, with a summary of described fossil mosquitoes. Acta Geologica Hispanica 35: 119–128.

Rattanarithikul, R., B. A. Harrison, P. Panthusiri & R. E. Coleman, 2005. Illustrated keys to the mosquitoes of Thailand 1. Background; geographic distribution; lists of genera, subgenera, and species; and a key to the genera. Southeast Asian Journal of Tropical Medicine and Public Health 36: 1–80.

Reinert, J. F., R. E. Harbach & I. J. Kitching, 2004. Phylogeny and classification of Aedini (Diptera: Culicidae), based on morphological characters of all life stages. Zoological Journal of the Linnaean Society 142: 289–368.

Rueda, L. M., 2004. Pictorial keys for the identification of mosquitoes (Diptera: Culicidae) associated with dengue virus transmission. Zootaxa 589: 1–60.

Rueda, L. M., S. A. Stockwell, J. E. Pecor & T. Gaffigan, 1997. Key to the Mosquito Genera of the World (INTKEY Module). The Diptera Dissimenation Disk -vol. 1, North American Dipterists Society, Washington, DC [CD].

Rueda, L. M., L. C. Rutledge & R. K. Gupta, 1998. Effect of skin abrasions on the efficacy of the repellent deet against *Aedes aegypti*. Journal of the American Mosquito Control Association 14: 178–182.

Rueda, L. M., B. Harrison, J. Brown, P. B. Whitt, R. Y. Harrison & R. Gardner, 2001. Evaluation of 1-octen-3-ol, carbon dioxide and light as attractants for mosquitoes with two distinct habitats in North Carolina. Journal of the American Mosquito Control Association 17: 61–66.

Rueda, L. M., M. Iwakami, M. O'Guinn, M. Mogi, B. F. Prendergast, I. Miyagi, T. Toma, J. E. Pecor & R. C. Wilkerson, 2005. Habitats and distribution of *Anopheles sinensis* and associated *Anopheles* Hyrcanus Group in Japan. Journal of the American Mosquito Control Association 21: 458–463.

Rueda, L. M., H. C. Kim, T. A. Klein, J. E. Pecor, C. Li, R. Sithiprasasna, M. Debboun & R. C. Wilkerson, 2006. Distribution and larval habitat characteristics of *Anopheles* Hyrcanus Group and related mosquito species (Diptera: Culicidae) in South Korea. Journal of Vector Control 31: 199–206.

Sallum, M. A. M., T. R. Schultz & R. C. Wilkerson, 2000. Phylogeny of Anophelinae (Diptera: Culicinae) based on morphological characters. Annals of the Entomological Society of America 93: 745–775.

Sallum, M. A. M., T. R. Schultz, P. G. Foster, K. Aronstein, R. A. Wirtz & R. C. Wilkerson, 2002. Phylogeny of Anophelinae (Diptera: Culicidae) based on nuclear ribosomal and mitochondrial DNA sequences. Systematic Entomology 27: 361–382.

Savage, H. M., 2005. Classification of mosquitoes in tribe Aedini (Diptera: Culicidae): Paraphylyphobia, and classification versus cladistic analysis. Journal of Medical Entomology 42: 923–927.

Service, M. W., 1993. Mosquitoes (Culicidae). In: Lane R. P., & Crosskey R. W. (eds), Medical Insects and Arachnids. Chapman & Hall, London, 120–240.

Statz, G., 1944. Plagegister aus dem Reinischen Braunkohlewnwalde. Rheinischer naturfreund 5: 1–16.

Stojanovitch, P. & H. G. Scott, 1997. Illustrated Key to The Adult Male Mosquitoes of America (North of Mexico). Harold G. Scott, Louisiana, 121 pp.

Tsukamoto, M., I. Miyagi & T. Toma, 1985. A revised checklist of the Philippine mosquitoes. Tropical Biomedicine 2: 149–160.

U.S. Department of the Interior, 2005. Everglades. National Park Service, USDI, Homestead, Florida. http://www.nps.gov/ever/current/fbn97-a.html.

Walter Reed Biosystematics Unit, 2001. Systematic Catalog of Culicidae. Smithsonian Institution, Washington, DC, USA. http://wrbu.org.

Wood, D. M. & A. Borkent, 1989. Phylogeny and classification of the Nematocera. In McAlpine J. F. & D. M. Wood (eds.), Manual of Nearctic Diptera, Vol. 3. Research Branch, Agriculture Canada, Monograph 32, 1333–1370.

World Health Organization, 2005. Malaria in Africa. Roll Back Malaria, World Health Organization, Geneva, Switzerland. http://www.rbm.who.int/cmc_upload/0/000/015/370/RBMInfosheet_3.html.

Hydrobiologia (2008) 595:489–519
DOI 10.1007/s10750-007-9127-9

FRESHWATER ANIMAL DIVERSITY ASSESSMENT

Global diversity of dipteran families (Insecta Diptera) in freshwater (excluding Simulidae, Culicidae, Chironomidae, Tipulidae and Tabanidae)

Rüdiger Wagner · Miroslav Barták · Art Borkent · Gregory Courtney · Boudewijn Goddeeris · Jean-Paul Haenni · Lloyd Knutson · Adrian Pont · Graham E. Rotheray · Rudolf Rozkošný · Bradley Sinclair · Norman Woodley · Tadeusz Zatwarnicki · Peter Zwick

Abstract Today's knowledge of worldwide species diversity of 19 families of aquatic Diptera in Continental Waters is presented. Nevertheless, we have to face for certain in most groups a restricted knowledge about distribution, ecology and systematic, particularly in the tropical environments. At the same time we realize a dramatically decline or even lack of specialists being able, having the time or the opportunity to extend or even secure the present information. The respective families with approximate numbers of aquatic species are: Blephariceridae (308), Deuterophlebiidae (14), Nyphomyiidae (7), Psychodidae (∼2.000), Scatopsidae (∼5), Tanyderidae (41), Ptychopteridae (69), Dixidae (173), Corethrellidae (97), Chaoboridae (∼50), Thaumaleidae (∼170),

Guest editors: E. V. Balian, C. Lévêque, H. Segers & K. Martens
Freshwater Animal Diversity Assessment

R. Wagner (✉)
University of Kassel, Kassel, Germany
e-mail: Ruediger.Wagner@uni-kassel.de

M. Barták
Czech University of Agriculture, Praha, Czech Republic

A. Borkent
Salmon Arm, British Columbia, Canada

G. Courtney
Iowa State University, Ames, USA

B. Goddeeris
Royal Belgian Institute of Natural Sciences, Brussels, Belgium

J.-P. Haenni
Muséum d'histoire naturelle, Neuchâtel, Switzerland

L. Knutson
Salita degli Albito 29 (Via Aragonese), Gaeta, LT 04024, Italy

A. Pont
Goring-on-Thames, Reading RG8 0EP, UK

G. E. Rotheray
National Museums of Scotland, Edinburgh, UK

R. Rozkošný
Institute of Botany and Zoology, Masaryk Univerity, Brno, Czech Republic

B. Sinclair
Ontario Plant Laboratories, Canadian Food Inspection Agency, Ottawa, Canada

N. Woodley
USDA, c/o Smithsonian Institution, Washington, DC, USA

T. Zatwarnicki
Department of Zoology, Academy of Agriculture, Wroclaw, Poland

P. Zwick
Schwarzer Stock 9, 36110 Schlitz, Germany

Ceratopogonidae (~6.000), Stratiomyidae (~43), Empididae (~660), Lonchopteridae (2), Syrphidae (~1.080), Sciomyzidae (~190), Ephydridae (~1.500), Muscidae (~870). Numbers of aquatic species will surely increase with increased ecological and taxonomical efforts.

Keywords Aquatic Diptera · Continental waters · Diversity · Distribution · State of knowledge

Foreword

With few exceptions it is difficult to extract from the numerous Diptera families those with aquatic or 'water dependent' species. A key problem for most groups is the lack of knowledge on larval ecology and morphology of many taxa on the one hand, and the great ecological plasticity on the other hand. In general, the majority of 'water dependent' larvae and pupae live in moist to wet grounds (providing substratum, shelter and food) in the surrounding of springs, streams, rivers, ponds lakes or in wetlands where they may occupy a multitude of spatially and temporally variable habitats.

A restricted amount of families show unquestionable aquatic life cycles, i.e. Chaoboridae and Corethrellidae, that are almost exclusively found in the pelagic zone of standing water bodies, or in the families of Blephariceromorpha whose larvae live almost only in torrential mountain streams. Recently, a few Diptera families have been reported from aquatic habitats, e.g. Lonchopteridae (Vaillant, 2002), Scatopsidae (Haenni, Vaillant pers. comm.), and Bibionidae (own observation). Another example is the family Dolichopodidae; several thousand species have been described until now. Some species exhibit aquatic development of larvae; however, larval ecology of most species remains totally unknown. A compilation of 'water dependent other Diptera' (to our best knowledge) is given in Table 1.

Some families were left out of this assessment because we could not find available specialists to deal with the worldwide diversity of these groups (e.g. Athericidae, Rhagionidae, Tabanidae). This is a strikingly clear indication of a dramatic loss or even lack of taxonomic specialists—not only in aquatic Diptera—all over the world.

Authors have tried to provide most complete up to date information, however, in many cases knowledge and existing databases are still far from being complete.

Introduction

Family Blephariceridae (P. Zwick)

For a long time, the slender, long-legged net-winged midges (wing lengths between 4 and 12 mm) were first known by the name Blepharoceridae in the 1840s. Blepharicerid larvae and pupae were only discovered in 1881. They inhabit fast flowing, often torrential waters, from sea-level to high mountains.

Larvae attach to smooth rock substrata with their suckers, head upstream, some resisting speeds of flow over 1 m s^{-1}. Locomotion is by successive detachment of suckers and is slow, except during sideways (!) escape gate.

Distribution on inhabited lands is not uniform, for ecological reasons, plains and terrain without rock substrata cannot be colonized. Blephariceridae are also absent from sandstone and other coarse grained rocks to which suckers cannot attach. Most species occur in permanent streams, but some taxa with long egg diapause survive or even specialize (*Dioptopsis*) in intermittent water courses.

Family Deuterophlebiidae (G. Courtney & R. Wagner)

Deuterophlebiidae (mountain midges) is one of the most specialized dipterous insects. The larvae and pupae of these small flies (<4 mm) are aquatic, found mostly in cold torrential streams, but ranging from small high-gradient creeks to large low-gradient rivers (Turner et al., 1986; Courtney, 1991a). Larvae and pupae are restricted to riffle habitats where they adhere to larger stones. Among the structural and ecological adaptations are eversible larval prolegs and flattened streamlined pupae. Adults have comparatively large wings and males extremely long antennae. Adults that have vestigial mouthparts, live for only a few hours (Courtney, 1991a, 1994a). Univoltine, synchronous life cycles are typical of most Nearctic species but bivoltine, asynchronous

Table 1 Compilation of 'water dependent' Diptera families

Suborder: Nematocera	Suborder: Brachycera
Infraorder: Blephariceromorpha	Infraorder: Tabanomorpha
Superfamily: Blephariceroidea	Superfamily: Tabanoidea
Family: Blephariceridae	Family: Tabanidae
Family: Deuterophlebiidae	Family: Rhagionidae
Superfamily: Nymphomyioidea	Superfamily: Stratiomyoidea
Family: Nymphomyiidae	Family: Stratiomyidae
Superfamily: Bibionoidea	Infraorder: Asilomorpha
Family: Bibionidae	Superfamily: Empidoidea
Infraorder: Psychodomorpha	Family: Empididae
Superfamily: Psychodoidea	Family: Dolichopodidae
Family: Psychodidae	Infraorder: Muscomorpha
Family: Scatopsidae	Section: Aschiza
Infraorder: Ptychopteromorpha	Superfamily: Platypezoidea
Family: Tanyderidae	Family: Lonchopteridae
Family: Ptychopteridae	Superfamily: Syrphoidea
Infraorder: Culicomorpha	Family: Syrphidae
Superfamily: Culicoidea	Section: Schizophora
Family: Dixidae	Subsection: Acalyptratae
Family: Corethrellidae	Superfamily: Lauxanioidea
Family: Chaoboridae	Family: Sciomyzidae
Family: Culicidae	Superfamily: Sphaeroceroidea
Superfamily: Chironomoidea	Fam: Heleomyzidae
Family: Thaumaleidae	Fam: Sphaeroceridae
Family: Simulidae	Superfamily: Ephydroidea
Family: Ceratopogonidae	Family: Ephydridae
Family: Chironomidae	Subsection: Calyptratae
	Superfamily: Muscoidea
	Family: Muscidae

life cycles were also reported by Courtney (1991a); seasonal, habitat and reproductive isolation of sympatric species occurs North American and Himalayan species (Courtney, 1991a, 1994a).

Family Nymphomyiidae
(G. Courtney & R. Wagner)

This peculiar Diptera family was discovered by Tokunaga as recently as 1930, when six specimens were collected along a torrential mountain stream near Kyoto, Japan. Nymphomyiidae were established as a new family based on a single species, *Nymphomyia alba* (Tokunaga 1932, 1935a, b). Phylogenetic relationships have been discussed controversial, but recent studies (Wood & Borkent, 1989; Oosterbroek & Courtney, 1995) suggest the Nymphomyiidae are sister group to the superfamily Blephariceroidea (Blepahriceridae + Deuterophlebiidae). Only few observations of larval and adult biology exist (Cutten & Kevan, 1970; Harper & Lauzon, 1989; Courtney 1994b). Larvae and pupae are cold stenothermous, found mostly in small perennial streams and cold headwaters (Courtney & Jedlicka, 1997). Details of their life cycle and habits were summarized by Courtney (1994b). Adults apparently do not feed and their life span is very short. One or two generations per year were observed in some species. Although some authors have suggested that eggs or pupae pass the winter (Cutten & Kevan, 1970; Back & Wood, 1979), most data on Nearctic nymphomyiids indicate that the larva is the overwintering stage (Harper & Lauzon, 1989; Courtney 1994b).

Psychodidae (R. Wagner)

Adult Psychodidae are small to medium sized Nematocera, the hump-backed body is covered by a dense vestiture of hairs and setae.

Larvae of Phlebotominae and Bruchomyiinae are terrestrial. Probably all Trichomyiinae larvae are xylophageous. Aquatic in the strict sense are Horaiellinae, Sycoracinae and most Psychodinae. The general appearance differs strongly between subfamilies. Larvae of the Old World Sycoracinae are small and asselliform. *Horaiella* larvae have only one curious ventral sucker. The head of Psychodinae larvae is strongly sclerotized, non-retractile. Thoracic and abdominal segments divided into 26 or 27 ring-shaped pseudosegments (annuli). They are more or less heavily sclerotized, with dorsal plates. Arrangement of hairs and setae on these plates are specific in most cases.

Scatopsidae (J.-P. Haenni)

Scatopsidae are minute to small, rather stoutly built, generally blackish midges. Both sexes are holoptic (except in few genera), with typical eye-bridge above antennae, ocelli present and palpi one-segmented. Larvae peripneustic, with the last abdominal segment bearing a pair of large spiracles generally placed at the apex of more or less elongate posterior processes. Larvae of only a few genera are known, with species-specific or at least genus-specific chaetotaxy. Larvae of most genera are terrestrial, saprophageus, living in a wide variety of organic matter in all degrees of decomposition, consequently often in liquid or semi-liquid media. Few larvae are aquatic living under the surface of a thin water film among water-logged dead tree-leaves, while few others are dendrolimnobiontic (Haenni & Vaillant 1990, 1994). Pupae are typically enclosed in last larval skin.

Family Tanyderidae (R. Wagner)

Tanyderidae (primitive crane-flies) are mid-sized to large nematocerans, resembling Tipulidae, wing with five radial veins and only one anal vein reaching the margin of the variously patterned wings. Larvae of five genera are known, they occur in two types of habitats, the hyporheic zone of cobble and sand bottom streams or in the outer layers of submerged rotting wood (Exner & Craig, 1976; Krzeminski & Judd, 1997). They prefer unpolluted mountainous streams. Males sometimes swarm in the evening, at day they hide in the bank vegetation.

Family Ptychopteridae (P. Zwick)

The family Ptychopteridae comprises the genus *Ptychoptera* in the Ptychopterinae, and *Bittacomorpha* and *Bittacomorphella* in the subfamily Bittacomorphinae. Numerous differences in structural detail among the subfamilies exist in all life stages. The best general account of the family remains Peus (1958).

Ptychopteridae have petiolate slender wings 5–12 mm long. The family name refers to sharp longitudinal folds in the membrane that may at first glance be taken for veins. *Ptychoptera* has two of these, the more slender Bittacomorphinae with narrower wings only one. Eggs are laid at the edge of soft aquatic sediments in which larvae and pupae live. However, they breathe atmospheric oxygen with long respiratory tubes. From personal observations of European fauna, habitat specificity seems to be high, and habitats may be small. Preferred habitats range from pristine spring seeps in some species to anoxic sediments with H_2S in others. Studied European *Ptychoptera* (Hodkinson, 1973; Hansen, 1979; Wolf & Zwick, 2001; PZ personal data on all seven German species) are seasonal and univoltine, the last larval instar overwinters. *Bittacomorpha* can be aseasonal and plurivoltine (Bowles, 1998). Adult life seems to last only for some days. Adult habits are poorly known.

Family Dixidae (R. Wagner)

The Greek word Dixos means bifurcate and refers to the two forked veins of the wing. Adults are small, frail, and they do not feed. They remain near their biotops (streams, ponds), and rest in the vegetation. Males of some species form swarms. Eggs are deposited in masses at the water's edge. The life cycle includes four larval instars and pupa. Larvae are pale greyish filter-feeders that lie on the water surface (meniscus-midges). They were attributed to the '*Fauna Hygropetrica*', i.e. the microhabitat where a thin water film flows permanently (or intermittently) over emergent structures. Pupae are much less active than the larvae; they prefer drier microhabitats. *Dixa* larvae prefer stream banks, *Dixella* larvae are found at the edge of standing waters. Some species are restricted to bog or mesotropic lakes and are appropriate bioindicators.

Life history studies are from Wood (1934; afrotropical *Dixa bicolor*, pupal stage 3–4 days). Peach & Fowler (1986) describe all instars of the palaearctic *D. autumnalis* (life cycle about 60 days). Peters & Adamski (1982) give an outline of larval morphology (Nearctic *Dixella nova*). Early cytogenetic studies are from Frizzi et al. (1966). Disney (1999) provided an outstandingly nice compendium of west-palaearctic Dixidae that can be used worldwide as a base of knowledge.

Family Corethrellidae (A. Borkent & R. Wagner)

Corethrellidae are close relatives of Culicidae and Chaoboridae, and at one time were placed in the latter family. The predaceous larvae are restricted to bodies of small standing water, in various phytothelmata such as treeholes, epiphytes and leaf axils or in small isolated ground pools, with these sometimes at the margins of springs or lakes. The larvae feed on a variety of small invertebrates (e.g. crustaceans, nematodes, mosquitoes). The female adults feed on the blood of male frogs and are attracted by their calls.

Family Chaoboridae (R. Wagner & Goddeeris)

Chaoboridae or phantom midges are medium-sized, non-biting Nematocera. The larvae live in standing waters; they possess hydrostatic air bladders (except *Eucorethra*) and they are predators of small arthropods, oligochaetes, rotifers, but also unicellular algae (except the filter feeder *Australomochlonyx*). In *Chaoborus*, larvae are mostly planctonic, transparent and they display circadian vertical migration. Pupae are free living in the pelagial. Some species (*Mochlonyx* and *Cryophila*) live in small, often temporary water bodies. Eggs of species from intermittent biotops overwinter and are resistant to desiccation (Montshadsky, 1960).

Family Thaumaleidae (B. J. Sinclair & R. Wagner)

Thaumaleidae are small, stout, brownish nematocerous Diptera, very uniform in appearance, sometimes confused with Psychodidae or Simuliidae. Both sexes are holoptic and no ocelli are present. They are considered members of the '*Fauna Hygropetrica*', i.e. larvae are restricted to vertical, thin water films flowing over rocks alongside waterfalls and torrents, and hold on to the substrate with anterior and posterior prolegs (Boussy et al., 1994). Larvae are categorized as scrapers, grazing on diatoms. Rapid larval movement is unique and diagnostic, traversing with quick zigzag strokes. Pupae are free-living in the hygropetric zone, bearing leaf-shaped appendages in some Southern Hemisphere species. Parasitism by *Entomacis* sp. (Hymenoptera: Proctotrupiodea: Diapriidae) upon immature *Austrothaumalea denticulata* Theischinger is the only known example (Sinclair, 2000b).

Family Ceratopogonidae (A. Borkent)

The Ceratopogonidae are an extremely diverse and abundant group of flies, commonly known as biting midges or no-see-ums. Biting midge adults are relatively easy to recognize to family. Adult females of four genera suck blood from vertebrates but most of the remaining 99 genera suck blood from other insects. The immatures occur in a wide array of environments where there is at least some moisture, such as rotting vegetation or manure, semiaquatic to fully aquatic habitats, including rivers and lakes. They are most common, both in terms of individuals and in numbers of species, in smaller aquatic habitats such as epiphytic plants, treeholes, pools, rice fields and the margins of marshes, pools and running water. The Ceratopogonidae, in nearly every regard, are one of the most poorly understood of all aquatic groups. Numerous surveys indicate that the group is abundant in many habitats, but because the larvae and pupae cannot be identified even to the generic level with any confidence, they remain largely unstudied.

Family Stratiomyidae
(R. Rozkošný & N. E. Woodley)

Adults of Stratiomyidae range from 2.0 to 26.0 mm in length. They belong to the suborder Brachycera. Larval habitats are very diverse, with both terrestrial and aquatic taxa. The typical aquatic larvae, with the exception of *Ptecticus* (Sarginae) and semi-aquatic larvae of *Beris* (Beridinae) and *Nemotelus* (Nemotelinae), are characterized by a more or less lengthened anal segment (breathing tube) (Rozkošný, 1982–1983).

Hymenopteran parasitoids of aquatic stratiomyid larvae have been known for a long time; reviews are given by McFadden (1967) and Rozkošný (1982).

Family Empididae (B. J. Sinclair)

The Empidoidea (excluding the Dolichopodidae) or dance flies, is a morphologically diverse group, with wing length ranging in size from less than two to over 12 mm. The aquatic larvae are apneustic (except *Oreogeton*) and characterized by paired prolegs or raised pads, partially sclerotized head capsule, with paired metacephalic rods. The aquatic empidoids are predaceous as adults and larvae (Steyskal & Knutson, 1981; Sinclair, 1995; Sinclair & Harkrider, 2004).

In the Northern Hemisphere, surveys of aquatic habitats (mostly running water) that include detailed species lists of empidoids, have been conducted in Canada (Harper, 1980; Landry & Harper, 1985), Germany (Wagner & Gathmann, 1996) and a recent short-term study in the Himalayas (Wagner et al., 2004).

Family Lonchopteridae (M. Bartak)

Lonchopteridae are small, strongly setose yellow to brownish black Brachycera. Both sexes are dichoptical, with conspicuously pointed wing with sexually dimorphic venation. Antennae small, with a rather elongate subapical arista (detailed description and illustrations in e.g. de Meijere, 1906; Smith, 1969; Baud, 1973; Peterson, 1987; Drake, 1996). Larvae live in wet microenvironment of decaying organic matter, among dead leaves, in plant debris, under stones on shores of water courses or they are semiaquatic. They are saprophagous, microphagous or presumably, mycetophagous. Puparium resembles larva. For more details about biology consult Baud (1973), Bährmann & Bellstedt (1988) and Drake (1996).

Family Syrphidae (G. E. Rotheray)

Syrphidae are small to large (4–25 mm long), slender to robust Lower Cyclorrhaphan flies. Many species resemble noxious Hymenoptera in a general way having white, yellow, orange or red markings but some are highly specific mimics of bees or wasps. Adults are often abundant and occur at low and high altitudes and they are found in a wide range of habitats, from deserts to rain forests. They are frequent flower visitors and important pollinators (Thompson & Rotheray, 1998). Larvae are unusually diverse in form and habit with saprophagous, predatory, parasitic and phytophagous species. Information on larva three stages, morphology and life cycle is available from Rotheray & Gilbert (1999), Rotheray et al. (2000), and Rotheray (2003). Larval morphology is modified in various ways for an aquatic way of life (Varley, 1937; Hartley, 1958, 1961; Rotheray & Gilbert, 1999; Rotheray et al., 2000).

Away from ponds and lakes, aquatic breeding sites used by syrphids include a wide range of phytotelmata i.e. water-based container habitats (Frank, 1983; Kitching, 2000) such as tree holes, water tanks of Neotropical bromeliads (Bromeliaceae) and pitchers of oriental pitcher-plants (Nepenthaceae). Some species are coastal in distribution such as the common European syrphid, *Eristalinus aeneus* (Hartley, 1961).

Family Sciomyzidae (R. Rozkošný & L. Knutson)

The Sciomyzidae represent a family of acalyptrate flies which belong, with several related families, to the superfamily Sciomyzoidea. According to recent authors the family includes three subfamilies Huttonininae, Salticellinae and Sciomyzinae, the latter with tribes Sciomyzini and Tetanocerini, but there may be reasons to reinstate a further subfamily (Sepedoninae) for distinctly apotypic genera around *Sepedon* (cf.

Knutson & Ghorpadé, 2004). Sciomyzid larvae feed as aquatic predators or terrestrial parasitoids (and/or predators) of snails and pea mussels, a few attack slugs, aquatic oligochaetes or consume snail eggs. Only two species are known to live as scavengers and feed on various dead invertebrates in addition to living terrestrial snails. The Sciomyzidae thus have colonized a great variety of ecosystems from different types of chiefly stagnant waters, moist habitats of semi-aquatic forms, mesophytic woods and even some xerothermic sites. Information on the development, food preference, ecology, behaviour and biology of Sciomyzidae is unusually rich. In the last 50 years several hundreds of specialized papers and comprehensive reviews have been published, including cladistic analyses of the family (Marinoni & Mathis, 2000; Barker et al., 2004) and evolutionary scenario based on a revised classification of behavioural groups (Knutson & Vala, 2002). Special papers dealing with morphology and biology of immature stages were summarized by e.g. Berg & Knutson (1978), Knutson & Vala (2002) and Barker et al. (2004) and comprehensive information devoted particularly to regional aquatic forms was given by Knutson (1981—Neotropical Region, 1982—Mexico, Central America and West Indies), Rivosecchi (1984—Italy), Rozkošný (1997b, 2002—Europe), Knutson & Ghorpadé (2004—Oriental Region) and Sueyoshi (2005—Japan).

Family Ephydridae (T. Zatwarnicki)

Adults are small to moderate-sized flies, typical length 1.0–8.0 mm, sometimes up to 12.0 mm. Adults of shore flies usually occur in close association with moist substrates and may often be observed on seashores, and along lake shores, rivers, streams, ponds, marshy habitats and also on the water surface of pools. Many adults are known to be polyphagous, feeding on yeast, various algae, and other microscopic organisms. There are few non-specialized predators, capturing and feeding on small insects. Most larvae are aquatic or semi-aquatic and occur in many different microhabitats. Some are uniquely adapted to such inhospitable habitats as inland alkaline or saline lakes, hot springs, crude oil pools and maritime habitats. The most wide-spread larvae develop in semi-liquid media, faeces, or moist shore mud. The majority of shore-flies feeds on various micro-organisms, e.g. algae, protozoa, and bacteria, and/or utilize detritus. Others are leaf-miners of limnic emergent macrophytes, active predators or parasitoids of spiders and frog eggs. A large group of scavengers develops in decomposing organic matter, e.g. carcasses of small animals, carrion and faeces.

Family Muscidae (A. Pont)

Muscidae are small to medium-sized flies (wing-length 1.5–20.0 mm) belonging to the superfamily Muscoidea of the series Calyptrata. Adult flies may be recognised by the absence of strong meral setae, an incomplete anal vein (A_1), the absence of a true dorsal seta on hind tibia, and, in the female sex, the absence of postabdominal spiracles. Larvae are typical maggots, though there are some variations from this ground-plan among aquatic forms. Adults are found in all zoogeographic regions, and in all biotopes except for the most arid. Many species are dark and drab in general appearance, but some tropical genera contain more colourful and patterned species. Larvae are mainly terrestrial, but genera in two subfamilies contain species with aquatic larvae. These may be found around lakes and ponds, rivers and streams, living in mud, wet sand and in the water among mosses, vascular plants and algae. They live as predators of other small aquatic invertebrates, and the adults are also predaceous, mainly on other small insects.

Species Diversity

Family Blephariceridae (P. Zwick)

A forthcoming world catalogue (Hogue & Zwick, in prep.) recognizes 308 species in 27 genera (as of 2004) which are assigned to four tribes in two subfamilies. All taxa are strictly aquatic.

Family Deuterophlebiidae
(G. Courtney & R. Wagner)

The family contains a single genus, *Deuterophlebia*, with fourteen described species, six in western North

America and eight from east and central Asia. Although the immature stages of several Himalayan species have been described, some remain unnamed until properly associated with adults of named species (Courtney, 1994a). The Himalayas and other parts of Asia may harbour additional new species, but it seems likely that the number of undiscovered species will be few.

Family Nymphomyiidae (G. Courtney & R. Wagner)

Seven species in one genus are known to date. Since the first discovery during the past 40 years additional species have been found, in eastern Canada (Ide, 1965), eastern Himalayas (Cutten & Kevan, 1970), Far East Russia (Rohdendorf & Kalugina, 1974), Sakhalin Island (Makarchenko, 1979), Hong Kong, and the south-eastern US (Courtney, 1994b). Larvae discovered recently in Nepal (Courtney, 1998) remain unassociated with adults and, therefore, unnamed.

Psychodidae (R. Wagner)

Six subfamilies are recognized to date, but larval development of Phlebotominae and Bruchomyiinae is obviously terrestrial; larvae of Trichomyiinae as far as we know live in dead wood. Three subfamilies include aquatic or semi-aquatic species.

The monogeneric subfamily Horaiellinae includes four known species. *Horaiella* is not well known but it seems to be restricted to the Himalayas and the mountainous areas of SE Asia and China, where larvae are found in unpolluted mountain streams.

More than 20 Sycoracinae (monogeneric) species are known from the Palaearctic, Afrotropic and Australian regions. The occurrence of Sycoracinae in the Nearctic and in the Neotropical regions is unlikely. All *Sycorax* species are aquatic.

A total of 2,000 species of Psychodinae in approximately 100 genera are currently described but species number is still increasing. Particularly the Neotropical, Afrotropical and Oriental regions are hardly discovered. The description of new species genera or even higher taxa is very probable. Most Psychodinae are aquatic and almost all are water dependent (Table 2).

The current numbers likely underestimate real diversity that is expected to be ten times higher, main issues concern nomenclatural problems and lack of phylogenetic analyses. However, there are several fine revisions of some groups and some phylogenetical attempts (among others Hogue, 1973; Vaillant, 1971–1983; Duckhouse, 1990, 1991).

Scatopsidae (J.-P. Haenni)

About 350 species in 32 genera have been described worldwide, but an aquatic development is known from only few species. But this figure is only provisional since faunas of large regions remains practically unstudied. Aquatic or semi-aquatic larvae are known till now only from five Palaearctic species of the following genera: *Ectaetia*, with *E. platyscelis* (Loew), *Holoplagia*, with *H. richardsi* (Edwards), both in water-filled tree-holes in France (Haenni & Vaillant, 1994); *Reichertella*, with *R. geniculata* (Zetterstedt) among water-logged dead tree leaves (Europe, Haenni, unpublished); finally in an unidentified genus of Scatopsinae, with two different species in little permanent springs and subsequent brooklets in forested areas of France (Haenni & Vaillant, 1994). There is little doubt that aquatic representatives will also be discovered in other zoogeographic regions.

Family Tanyderidae (R. Wagner)

Tanyderidae include more than 40 species in ten extant genera (Krzeminski & Judd, 1997—Table 2). Several are known only from larva and remain unnamed (Judd, homepage).

Family Ptychopteridae (P. Zwick)

Less than a 100 species were named from the Palaearctic, Oriental and Afrotropical Regions, and from Madagascar. Species numbers in Table 2 follow the taxonomic interpretation of regional catalogues and lists (e.g. Zwick, 2004). Many more species may await discovery, even in well-studied Europe new species were recently found.

Table 2 Number of aquatic and water dependent (FW) species per zoogeographic region

	PA	NA	NT	AT	OL	AU	PAC	ANT	World FW Species
Blephariceridae	124	34	76	28	22	19	0	0	322
Edwardsininae	0	0	10	8	0	19	0	0	37
Blepharicerinae	124	34	66	20	22	0	0	0	285
Deuterophlebiidae[1]	8	6	0	0	[1]	0	0	0	14
Nymphomyiidae	3	2	0	0	2	0	0	0	7
Psychodidae	692	114	362	159	279	370	62	0	1,988
Horaiellinae	0	0	0	0	4	0	0	0	4
Sycoracinae	11	0	(1)	4	(1)	5	0	0	22
Psychodinae	681	114	361	155	274	365	62	0	1,917
Scatopsidae	5	?	?	?	?	?	?	?	5
Tanyderidae[2]	9	6	3	1	[3]	22	0	0	41
Ptychopteridae	27	18	?	9	?	15	?	0	69
Dixidae	67	51	17	8	21	16	1	0	180
Corethrellidae[3]	2	7	69	6	4	12		0	97
Chaoboridae[3]	10	15	11	8	6	7		0	50
Thaumaleidae	87	28	6	2	16	33	1	0	174
Ceratopogonidae[3,4]	1,028[b]	600	1,066	622[a]	521[c]	761[d]			5,600
Stratiomyiidae	240	181	252	131	56	76		0	889
Empidoidea (exclusive Dolichopodidae)	296	125	52	51	87	58	2	0	671
"Oreogetoninae"	5	8	0	0	0	10	0	0	23
Ceratomerinae	0	0	1	0	0	34	0	0	35
Clinocerinae	200	58	8	30	37	9	2	0	344
Hemerodromiinae	77	58	43	21	45	5	0	0	299
Trichopezinae	14	1	0	0	5	0	0	0	19
Lonchopteridae	2	?	?	?	?	?	?	0	2
Syrphidae[3]	354	402	257	165	202	68		0	1,341
Sciomyzidae	52	82	30	7	10	8		0	154
Ephydridae	352	329	317	218	113	104	49	2	1,251
Discomyzinae	43	22	25	40	13	6	4	0	120
Hydrelliinae	63	73	54	93	40	22	4	0	312
Gymnomyzinae	75	52	58	36	34	28	6	0	232
Ilytheinae	35	38	50	18	8	8	4	0	127

Table 2 continued

	PA	NA	NT	AT	OL	AU	PAC	ANT	World FW Species
Ephydrinae	136	144	130	31	18	40	31	2	460
Muscidae	128	45	105	140	91	137	28	0	599
Mydaeinae	1	1	0	0	1	0	0	0	1
Coenosiinae Limnophorini	115	39	104	139	89	137	26	0	580
Coenosiinae Coenosiini	12	5	1	1	1	0	2	0	18

PA = Palaearctic, NA = Nearctic, NT = Neotropical, AT = Afrotropical, OL = Oriental, AU = Australasian, PAC = Pacific Oceanic islands, ANT = Antarctic

[1] Species mentioned in Oriental Catalog, but attributed to PA

[2] Protoplasa fitchii and Protoplasa fitchii ssp. Carolinensis, [ ] Palaearctic species in Oriental region

[3] Data in AU include PAC

[4] Ceratopogonidae in the biogeographical regions of the world, catalogues and recent additions. Numbers for the Nearctic and Neotropical Regions, and the total are current. Others are dated and need revision: [a]from Wirth et al. (1980); [b]from Remm (1988), cataloged to 1982; [c]from Wirth (1973), notably out of date; [d]from Debenham (1989). Most species are water dependent

Family Dixidae (R. Wagner)

More than 170 species in seven genera have been described. Species diversity is highest in sources and in headwater streams, river banks are hardly populated. Low species numbers in regions depend on inadequate exploration. Most species are known from the Palaearctic and Neartic regions. *Dixa* and *Dixella* are mainly distributed in the northern hemisphere with relations to Afrotropical and Oriental regions. Most Oriental species have been described from the Himalayas. Only few species are known from the Neotropic and Australian regions. It is unlikely that the present generic concept will remain the same in future.

Family Corethrellidae (A. Borkent & R. Wagner)

The family Corethrellidae is monogeneric with a nearly pantropical distribution. A total of 97 extant species are recognized but many more remain undescribed. The genus is generally restricted to subtropical and tropical climates but some species are known from as far north as southern Canada and eastern Siberia and as far south as New Zealand. A world catalogue of fossil and extant Corethrellidae (and Chaoboridae) was provided by Borkent (1993). A world revision is near completion (Borkent, in prep.).

Family Chaoboridae (R. Wagner & Goddeeris)

Chaoboridae include about 50 extant species in six genera and two subfamilies (Borkent, 1993). *Eucorethra* is monobasic in the Nearctic, the also monobasic *Promochlonyx* and *Australomochlonyx* are endemic to Australia (Colless, 1986). *Mochlonyx* has Holarctic distribution, *Cryophila* is Palaearctic, *Chaoborus* is cosmopolitan (Saether, 2002).

Family Thaumaleidae (B. J. Sinclair & R. Wagner)

Some 170 species of Thaumaleidae in eight genera are known and typically restricted to small distributional areas.

Family Ceratopogonidae (A. Borkent)

The Ceratopogonidae are an extremely diverse, worldwide group, with 5,598 validly named species, placed in 103 genera, in four subfamilies (Borkent & Wirth, 1997). However, large areas of the planet have been poorly sampled and some, like the otherwise species rich South American Andes, have been barely collected at all. It is reasonable to estimate that there are currently at least 15,000 morphologically distinct species on our planet. Most species are assumed to be aquatic or water dependant, but the ecology of many species remains unknown.

Identification guides to genera and species are listed by Borkent & Spinelli (2000; area south of the US); Borkent & Grogan (in press; Nearctic); Boorman (1997; Palaearctic); de Meillon & Wirth (1991); and the other relevant catalogs.

Family Stratiomyidae
(R. Rozkošný & N. E. Woodley)

More than 2,650 species are known from all the main biogeographical regions (Woodley, 2001). Aquatic or at least semi-aquatic larvae are known in 75 spp. (though specific diagnostic characters are sometimes unknown) and presumed in 889 species. The majority of descriptions are based on larvae originating from the Nearctic and the western part of the Palaearctic Regions. The number of aquatic larvae known from other regions is only very limited.

Family Empididae (B. J. Sinclair)

The majority of empidoids breed in damp soil and rotting wood, with larval development in aquatic habitats mostly restricted to two subfamilies, Clinocerinae and Hemerodromiinae. Although larvae are only known for about 1% of the species, the taxa included in Table 2 are classified as aquatic based on where adults are most often collected (e.g. emergent rocks, riparian vegetation), congeneric larvae have been collected elsewhere from aquatic habitats, or adults have been found in emergence traps set over aquatic habitats. Some 660 species in 26 genera are known, with several species found in more than one biogeographical region, e.g. Holarctic species (*Clinocera stagnalis*, *C. nivalis*).

Many species remain undescribed from most regions of the world, especially the Southern Hemisphere, (e.g. *Proagomyia* and *Asymphyloptera*). In North America, the genera *Proclinopyga* and *Dolichocephala* require taxonomic revision. Large numbers of undescribed species of *Hemerodromia* remain unexamined worldwide.

Family Lonchopteridae (M. Bartak)

About 49 species (and two subspecies) are known worldwide all of them cited mostly in a single genus. *Lonchoptera*, but sometimes in combination with other nominal genera (e.g. *Neolonchoptera*, *Dipsa*, *Spilolonchoptera*), but their taxonomic status remains unclear. An aquatic development is known from only two or three palaearctic species.

Geographic ranges are mostly restricted to relatively small areas except broadly distributed Palaearctic species (*L. fallax*), cosmopolitan *L. bifurcata*, Palaearctic + Oriental *L. lutea* and Holarctic *L. uniseta*). Recently two Palaearctic species were mentioned to have water dependent larvae; these were found in rheo- and limnocrenic environments and in madicolous habitats along streams (Vaillant, 2002).

Family Syrphidae (G. E. Rotheray)

About 6,000 species of Syrphidae are known occurring on all continents except the Antarctic. Three subfamiles and about 177 genera are recognised: Microdontinae, Syrphinae and Eristalinae (Thompson & Rotheray 1998). So far as is known microdontines are myrmecophilous (ant associated) and do not include aquatic species and only a few Syrphinae are aquatic. Most aquatic syrphids belong to the Eristalinae. However not all members of this subfamily are aquatic (Rotheray & Gilbert, 1999; Stahls et al., 2003). More than 1.000 species in more than 90 genera are aquatic or water-dependant.

Family Sciomyzidae (R. Rozkosny & L. Knutson)

The known world fauna of Sciomyzidae embraces 533 described species (Rozkošný 1995; Vala et al. in prep.). However, there is a large number of distinguished but still undescribed species and the tropical faunas appear insufficiently investigated. A number of species with aquatic and at least semi-aquatic larvae may be estimated to about 280 spp., though the number of species with known biology of larvae belonging to these ecological groups reach, for the time being, 156 species.

The known aquatic forms are dominant in the subfamily Sciomyzinae (521 spp.) where differences in the larval morphology of the two tribes have been found. Semi-aquatic larvae of Sciomyzini chiefly are parasitoids and predators of exposed aquatic snails whereas aquatic larvae of Tetanocerini mainly include (at least in the last, third instar) overt predators of aquatic molluscs and aquatic oligochaetes.

Family Ephydridae (T. Zatwarnicki)

About 1,800 valid species in 127 genera (five treated also as subgenera) are distributed in all zoogeographic regions and on most oceanic islands, except continental Antarctica (Mathis & Zatwarnicki, 1998). Although inadequately known, larvae of following genera and tribes are not strictly aquatic: parasitoids [*Rhynchopsilopa* and *Trimerina* (Psilopini)], scavengers [Discomyzini (Discomyzinae)], macrophyte leaf-miners [Hydrelliini (Hydrelliinae)], parasitoids or carcasses feeders [Gastropini, Hecamedini, Gymnomyzini (Gymnomyzinae)], and soil algae feeders [*Garifuna*, *Nostima* and *Philygria* (Hyadinini, Ilytheinae)]. Members of five subfamilies (89 genera; 1,251 species) are aquatic: Discomyzinae (120 species), Hydrelliinae (312), Gymnomyzinae (232), Ilytheinae (127) and Ephydrinae (460).

Family Muscidae (A. Pont)

There are probably some 4,500 described species of Muscidae, but the biology and immature stages of only a relatively small number (less than 10%) are known (Skidmore, 1985). The larvae of the more basal subfamilies are terrestrial (Achanthipterinae, Atherigoninae, Azeliinae, Muscinae, Phaoniinae, Cyrtoneurininae), but include a few that live in water accumulations in tree holes, bamboos, etc. The truly aquatic and subaquatic larvae belong to the more apical subfamilies: Mydaeinae (*Graphomya* Robineau-Desvoidy), Coenosiinae tribe Limnophorini (*Spilogona* Schnabl, *Lispoides* Malloch, *Xenomyia* Malloch, *Limnophora* Robineau-Desvoidy, *Lispe* Latreille), and Coenosiinae tribe Coenosiini (*Lispocephala* Pokorny, *Schoenomyza* Haliday). Several hundred species of eight genera are aquatic or water dependant. The Limnophorini are the most abundant and diverse of the aquatic groups, although some of the known larvae are terrestrial not aquatic. What is known about the predatory activity of the adults has been summarised by Werner & Pont (2005): species of *Lispe* are predaceous mainly on the adults and immature stages of mosquitoes (Culicidae) and swarming midges (Chironomidae), whilst species of *Limnophora* and *Xenomyia* are predaceous mainly on adults and immature stages of black flies (Simuliidae). When more is known about the biology of Muscidae in tropical regions, especially in the Neotropical and Afrotropical regions, it is certain that other genera will also be found to have aquatic larvae.

Phylogeny and historical processes

Family Blephariceridae (P. Zwick)

Families Blephariceridae, Deuterophlebiidae and Nymphomyiidae together form the monophyletic Blephariceromorpha (Courtney, 1991b) all of which develop exclusively in streams.

The monophyly of the blepharicerid subfamilies and of the tribes Paltostomatini and Apistomiini is well established (Zwick, 1977; Stuckenberg, 2004); monophyly of Blepharicerini is weakly supported. The Edwardsininae are apparently Gondwanan relicts, with *Paulianina* in Madagascar and *Edwardsina* in Andean South America and south-eastern Australia.

The area of origin of the Blepharicerinae is not clear, extant representatives occur world-wide. The only fossil reliably assigned to the family comes from Far Eastern Siberia and is a close relative and possible ancestor of extant local genera (Lukashevich & Shcherbakov, 1997).

Tribal assignment of *Hapalothrix* (Europe) and *Neohapalothrix* (Central to East Asia) is doubtful. They share some derived characters with Paltostomatini, but also some with Blepharicerini. Additional investigation and re-evaluation of character expressions are needed, also because genera *Blepharicera* and *Tianschanella* (Tien-Shan) lack structures distinctive of other Blephariceridae.

The case of *Hapalothrix* and *Neohapalothrix* is decisive for interpretations of evolution and distributional history of the Blepharicerinae. If these two genera should eventually prove to be Blephariceridae, Paltostomatini would be endemic, and disjunct, in the Afrotropical (only *Elporia*) and Neotropical regions (several endemic genera).

Sister-group to the Paltostomatini is the Apistomyiini (Stuckenberg, 2004) which have their most ancient representatives on New Zealand and New Caledonia. More advanced genera occur in the Oriental region and in East Australia (not Tasmania) (Zwick, 1977, 1998). Genus *Apistomyia* attains its greatest diversity in the Oriental region. It extends into eastern Australia and apparently profits from aerial distribution, with outlying species on Bougainville, Taiwan, and islands in the European Mediterranean region.

Family Deuterophlebiidae (G. Courtney & R. Wagner)

Deuterophlebiidae + Blephariceridae form a monophyletic group (Blephariceroidea) supported by several synapomorphies (Wood & Borkent, 1989; Courtney, 1991b; Oosterbroek & Courtney, 1995). No fossil records of the group are available. The Nearctic fauna is not monophyletic at least two successive invasions into North America are assumed (Courtney, 1994a). Dispersal along Beringia was assumed because during the Pleistocene this area provided ideal conditions to cool adapted taxa; however, a mid Tertiary transgression was more probable than a Pleistocene invasion (Courtney, 1994a). Transatlantic dispersal (until 20–35 million years ago (mya)) is highly improbable.

Family Nymphomyiidae (G. Courtney & R. Wagner)

Some early workers considered the Nymphomyiidae the most primitive Diptera family (e.g. Ide, 1965; Cutten & Kevan, 1970). Courtney (1994b) suggested a relation to the Culicomorpha. Recent analyses (e.g. Oosterbroek & Courtney, 1995) recognised it as sister-group of the Blephariceroidea.

Hoffeins & Hoffeins (1995) found fossil European Nymphomyiidae in the Eocene Baltic and Bitterfeld amber that were described by Wagner et al. (2000). With the discovery of this species, a gap in the distribution pattern of extant Nymphomyiidae was closed. It is probable that the Nymphomyiidae colonised the eastern Nearctic Region via the Thule landbridge approximately 25–30 mya. In Europe, Nymphomyiidae probably became extinct with the Pleistocene climatic alteration. They survived only at some distance from the borders of glaciation with sufficient environmental conditions in areas with torrential streams.

Psychodidae (R. Wagner)

The entire system of Psychodidae sensu lato needs urgent revision. Existing phylogenetical analyses are contradictory even at the subfamily level. Comparison of the extant and Baltic amber Psychodinae faunas shows great differences even on the subfamily and generic level, probably as a consequence of glaciation. On the other hand, psychodids in Caribbean amber are quite similar to the extant fauna, at least on generic level.

Scatopsidae (J.-P. Haenni)

The older still undescribed fossil Scatopsidae date back to the Cretaceous (Siberian and Canadian ambers), and even to the early Cretaceous from Mongolia (Kovalev, 1986), although their relation to the extinct family Protoscatopsidae still must be investigated. Four species from Paleocene/Eocene Baltic amber are the older described taxa (Meunier, 1907) but several additional undescribed species are known (Haenni, unpubl.). Scanty information on fossil history of the family is summarized in Haenni (1997).

Family Tanyderidae (R. Wagner)

Since 1919 Tanyderidae have been given family rank, before they had been included into Tipulidae or Ptychopteridae. There is still some debate about the sister-group of Tanyderidae in the phylogenetical system.

Family Ptychopteridae (P. Zwick)

Generic relationships and distributions suggest a Holarctic origin of the family and its subsequent spread into Africa. Palaeontological evidence is not in conflict with this interpretation. Assignment of several European Liassic fossils to the family is doubtful (Peus, 1958). Ptychopterid pupae were recorded from the Mesozoic of Siberia (Lukashevich, 1995). A Tertiary *Bittacomorphella* from Colorado (Alexander, 1927, 1981) and European Tertiary *Ptychoptera* resemble extant forms (Peus, 1958).

Bittacomorpha appears more derived than *Bittacomorphella.* For further inferences, relations between individual species would have to be known but are not. Many individual species have been compared with some other for selected diagnostic characters, but a comparative morphological study permitting phylogenetic analyses was never made. Distributional pathways of the Ptychopteridae remain therefore unknown. *Paraptychoptera* Tonnoir is a monophyletic endemic European clade and was assigned subgeneric rank within *Ptychoptera*, in order not to turn *Ptychoptera* s. str. paraphyletic (Zwick & Starý, 2003).

Family Dixidae (R. Wagner)

Dixidae are placed in the superfamily Culicoidea, but were also considered a subfamily of the Culicidae. Hennig (1966) mentioned Dixidae from Baltic amber described by Loew and Meunier; he described three additional species in the genus *Paradixa* Tonnoir, 1924 (a synonymy of *Dixella* Dyar & Shannon, 1924). But there are earlier remnants of Dixidae or related taxa (Hennig, l.c.). The Baltic amber species are not very different from extant Palaearctic species, concerning the figured male genitalia.

Dixa and *Dixella* seem to be of northern hemisphere origin with radiation into the adjacent Afrotropical and Oriental regions. *Dixina*, *Neodixa* and *Nothodixa* are exclusively distributed in the southern hemisphere. *Nothodixa* occurs in the Neotropical and Australian regions and probably is a Gondwanan element. Climatic changes from Tertiary to the present may have affected at least the northern hemisphere Dixidae. Probably the warmer postglacial climate and an increasing number of permanently running waters may have led to an increased number of *Dixa* species.

Family Corethrellidae (A. Borkent & R. Wagner)

Corethrellidae have traditionally been placed as subfamily in the related Chaoboridae, but are now recognized as the sister group of Culicidae + Chaoboridae (Wood & Borkent, 1989). In the world catalogue of fossil and extant Corethrellidae and Chaoboridae Borkent (1993) provides substantial information and references on both groups. Fossils are known from various ambers, including one from Lebanese amber, 121 million years old.

Family Chaoboridae (R. Wagner & B. Goddeeris)

Chaoboridae and Culicidae are the sister-group of the Corethrellidae (Wood & Borkent, 1989; Saether, 1997). Together with Dixidae they form the superfamily Culicoidea of the infraorder Culicomorpha. Two synapomorphies, (1) precocious development of adult eyes in the larva, (2) articulate, membranous anal paddles in pupae indicate the monophyly of Chaoboridae and Culicidae (Wood & Borkent, 1989). Within Chaoboridae Eucorethrinae are the plesiomorphic sister group of Chaoborinae (features in Saether, 1970, 1992). Infrafamiliar classification remains unsolved. Only the position of the monobasic *Eucorethra* in its own subfamily is generally accepted. A number of fossil Chaoboridae has been described. The group probably diverged in the Upper Jurassic (Refs. in Saether, 2002). Borkent (1993) presented a world catalogue of fossil and extant Corethrellidae and Chaoboridae with substantial references on both groups.

Since then several new species have been described. Several fossil taxa were grouped in the subfamily Chironomapterinae that is probably paraphyletic (Borkent, 1993). Higher classification needs new studies with classical and molecular methods.

Family Thaumaleidae (B. J. Sinclair & R. Wagner)

Hennig (1973) assigned the Thaumaleidae to the Culicomorpha, and an assumed phylogenetic relationship with Chironomoidea, although they appear to be a somewhat isolated group. This classification has been followed ever since (e.g. Wood & Borkent, 1989; Oosterbroek & Courtney, 1995). Relations among most genera have yet to be clearly analysed, although it appears the Southern Hemisphere species form a monophyletic clade (Sinclair, unpubl. data). Kovalev (1990) described *Mesothaumalea fossilis* from the late Jurassic or early Cretaceous ($\sim$110 to 130 mya), which represents the only known fossil thaumaleid.

Family Ceratopogonidae (A. Borkent)

The basics of Ceratopogonidae phylogeny are reasonably well understood and the family has one of the best fossil records of any group of insect. The relationships between the four subfamilies are well established and the early lineages within these groups at least partially understood. The Leptoconopinae are the sister group of all remaining Ceratopogonidae and the Forcipomyiinae + Dasyheleinae are the sister group of the Ceratopogoninae. There remains a great need to interpret the relationships between the genera in the tribes Ceratopogonini, Heteromyiini, Sphaeromiini, Palpomyiini and Stenoxenini. A molecular study supported the relationships previously indicated by morphological and fossil studies.

Ceratopogonidae are an ancient family, likely arising in the Jurassic. Remarkably, two extant genera, *Leptoconops* and *Austroconops*, the only members of the Leptoconopinae, are present in Lebanese amber, 121 million years old (Borkent & Craig, 2004). Other extinct genera have been described and abundant fossils from a variety of ambers of different ages show a strong congruence with the cladistic relationships based on morphological analysis (Borkent, 2000). Successively younger ambers include successively younger lineages.

The habit of adult females sucking blood from vertebrates is a plesiotypic feature within the family and is homologous with vertebrate feeding in the related families Simuliidae, Corethrellidae and Culicidae. Similarly, the occurrence of Ceratopogonidae in small aquatic habitats is a plesiotypic feature, shared with at least the early lineages of all other Culicomorpha (Borkent & Craig, 2004). Those Ceratopogonidae which are in large rivers and lakes represent derived lineages.

Family Stratiomyidae
(R. Rozkošný & N. E. Woodley)

The first comprehensive phylogenetic information concerning the Stratiomyidae on a world level, a definition of the Xylomyidae as a sister-group of Stratiomyidae and a cladistic analysis of all 12 subfamilies were presented by Woodley (1995, 2001). Subfamilies with aquatic larvae are all in clade six which includes the Rhaphiocerinae, Stratiomyinae and Nemotelinae. The recently discovered larva of *Raphiocera* in a semi-aquatic situation seems to point to an aquatic or semi-aquatic existence for the subfamily. In the Nemotelinae only semi-aquatic larvae of *Nemotelus* are widely distributed, (though aquatic larvae are not excluded in *Brachycara* spp. occurring in littoral marine habitats). Thus, current information indicates that the aquatic lifestyle has evolved once at clade six in the Stratiomyidae (Woodley, 2001), with a few convergent species in other clades such as two *Beris* spp. and *Ptecticus flavifemoratus* discussed above.

The primitive Brachycera probably arose during the Triassic because flies with well-developed brachycerous characters are known from the lower Jurassic. However, no fossils that can be assigned to any extant families are known from these early periods (Woodley, 1989). Only a small number of fossils belonging undoubtedly to the Stratiomyidae are included in a catalogue by Evenhuis (1994). Out of seven species with presumed aquatic larvae, five lived apparently in the Oligocene and were found in Europe or USA, one is from the Eocene or Oligocene (England) and one from the Oligocene or Miocene (Dominican Republic). No reliable information about the time when stratiomyid larvae colonized water environment is available.

Family Empididae (B. J. Sinclair)

Fossils with empidoid-like venation are known from the upper Jurassic, with empidoid subfamilies present by the early Cretaceous (Grimaldi, 1999; Grimaldi & Cumming, 1999). In fact, the Empidoidea are among the best known lineages from the Cretaceous (Grimaldi, 1999). Divergence time estimates for the Empidoidea range between 144 and 163 mya

(Wiegmann et al., 2003). During the past few years, there have been attempts to establish a stable phylogeny using molecular characters (Collins & Wiegmann, 2002a; Moulton & Wiegmann, 2004). A quantitative phylogeny of the empidoids by Sinclair & Cumming (2006), based on morphological characters has also helped to stabilize classification.

Family Lonchopteridae (M. Bartak)

The origin and relationships of the family Lonchopteridae are not known certainly, hence, taxonomic position remains unclear. The family Lonchopteridae is usually placed in the paraphyletic basal cyclorrhaphan taxon "Aschiza" (e.g. Peterson, 1987) of the infraorder Muscomorpha. Within Aschiza, it is either placed to the superfamily Platypezoidea (as sister group to Opetiidae—Collins & Wiegmann, 2002b) or it forms a single taxon on superfamily (Lonchopteroidea) or higher taxonomic levels (e.g. Griffiths, 1972). No reliable fossils are described in details except recent treatments of *Lonchopterites ptisca* and *Lonchopteromorpha asetocella* by Grimaldi & Cumming (1999).

Family Syrphidae (G. E. Rotheray)

Phylogenetic relationships between Syrphidae are under assessment and although some clades are well supported, a consensus has yet to be reached (Rotheray & Gilbert, 1999; Stahls et al., 2003). The earliest fossil syrphids are aged at about 120–130 million years, the time when the supercontinent was breaking apart (Grimaldi & Cumming, 1999). During the break up, basal syrphid lineages probably became separated in South America, South Africa and possibly Australia. When syrphids reached the Palearctic probably from South Africa, diversity rose and lineages spread east into the Oriental and perhaps across the Bering Strait into the Nearctic. Others spread into the Australiasian region. In the other direction, lineages also spread west from the Palearctic into first the Nearctic and from the Nearctic, a subset of lineages crossed into South and Central America and diversified in the Neotropics (Vockeroth, 1969; Thompson, 1972).

Family Sciomyzidae (R. Rozkosny & L. Knutson)

The potential evolution of malacophagy in Diptera, probable origin of Sciomyzidae, their ecological specialisation and generalisation as well as subsequent radiation and some further aspects of sciomyzid evolution are discussed in detail by Knutson & Vala (2002) and Barker et al. (2004). The Sciomyzidae probably evolved from a dryomyzid-like ancestor during the Lower Cretaceous. This generally adapted, saprophagous, acalyptrate ancestor had probably developed a requirement for a diet rich in proteins as a base for a subsequent great variety of malacophagous behaviour. A probable biology of sciomyzid ancestors may be demonstrated by the extant *Atrichomelina pubera* (NA). Its larvae are capable to live as saprophages, predators and parasitoids and their feeding habits are dependant on circumstances.

Very probably the specialised forms developed in different microhabitats, from the original damp situations to the almost strictly aquatic forms on one side and to the terrestrial, hygrophilous and even xerothermic forms, on the other side. This specialisation was, however, markedly influenced by the availability of molluscs (or other invertebrates) as a suitable source of food.

According to the cladistic morphoanalysis of the Sciomyzidae presented by Marinoni & Mathis (2000) and Barker et al. (2004), *Salticella* is at the base of the cladogram and it is more closely related to the Sciomyzini than to the Tetanocerini. Also the monophyly of both tribes of Sciomyzinae was confirmed but *Eutrichomelina* (NT) was transferred from the Sciomyzini to the Tetanocerini. *Renocera* and *Anticheta*, which share some intermediate larval characters, are placed at the base of the Tetanocerini and the genera around *Sepedon* (forming a potential subfamily Sepedoninae) form the most specialised group of the family.

Fossil records of Sciomyzidae (13 described species in five genera, see Vala et al. in prep.) are relatively rich in the framework of acalyptrate flies. All are restricted to the Tertiary. Four genera are extinct and two extant, all species are known from the Eocene/Oligocene and Miocene, and many from the Baltic amber. Species of *Sciomyza* and *Tetanocera* were apparently numerous already in that time. It is not excluded that some fossils from the Upper Jurassic/Lower Cretaceous of Spain belong also to this family (Evenhuis, 1994). Unfortunately, no information on feeding habits or ecological requirements of immature stages of fossil forms is available.

Family Ephydridae (T. Zatwarnicki)

Ephydridae is a family of the Ephydroidea (=Drosophiloidea) within the cyclorrhaphous Schizophora. They are related to Risidae (also treated as a sublineage within shore flies) and Diastatidae due to the possession of a female ventral receptacle (Hennig, 1973). The family is divided into five subfamilies; Discomyzinae and Hydrellinae are one evolutionary line, and Gymnomyzinae, Ilytheinae and Ephydridae are a second morphologically more advanced line (Zatwarnicki, 1992). Probably ancestors fed on decomposing organic material, parasitoids, leaf miners and larvae living in carcasses of small animals developed later. Predators and the use of microorganisms and/or detritus in water and mud evolved independently. Few fossils (4 genera) are known from the Oligocene and related ages.

Family Muscidae (A. Pont)

For at least a century there has been little dispute over the definition and scope of the family Muscidae, and the only fundamental change has been the removal of the subfamily Fanniinae to a separate family Fanniidae. The phylogenetic classification of the family was dealt with comprehensively by Hennig (1965) and, whilst some details have changed since 1965, there has been no new overall review. Some cladistic lineages were outlined by de Carvalho (1989), and a cladistic analysis of the tribe Coenosiini was made by Couri & Pont (2000). A few species, but none belonging to the aquatic groups, have been described from Dominican amber, 15–20 mya (Pont & Carvalho, 1997). A cladistic analysis of the Limnophorini is urgently needed.

Present Distribution and Main Areas of Endemicity

Family Blephariceridae (P. Zwick)

Blephariceridae occur on all continents except Antarctica (Table 2), and on many continental islands, but also on the Oceanic islands of St. Vincent and Bougainville.

Blephariceridae are often disjunctly distributed. Distinctness of the separate subfaunas in the Andes and the Brazilian Shield, respectively, or the Rocky Mountains and the Appalachians (only *Blepharicera*), respectively, suggests long lasting separations maintained by present ecological conditions. The four western Nearctic genera are shared with East Asia and provide evidence of past Trans-Beringian connections. The presence of net-winged midges in the East of Australia and their absence from the rest of the continent has parallels among other stream fauna and is probably due to past and present ecological conditions. The relictual Edwardsininae are well represented in Tasmania, Victoria and New South Wales but lack further north. Conversely, Apistomiini which seem to be immigrants from the tropical north during relatively recent land connections with Papua New Guinea occur all along the Australian East coast, but are absent from Tasmania.

The holarctic fauna is sharply divided into a western part including the Caucasus and adjacent highlands in Iran (endemic genera *Dioptopsis, Liponeura, Hapalothrix*) and an eastern part extending west to Kazachstan and Afghanistan (endemic genera *Asioreas, Horaia, Neohapalothrix, Tianschanella*).

The distribution of West Palaearctic species reflects Pleistocene impact. Net-winged midges are absent from suitable habitats in the British Isles and Scandinavia, and there are only five species in the Alps and the same species plus two more in mountains north of the Alps. The largest number of West Palaearctic species occurs on mountains in the Mediterranean area, each of the Mediterranean peninsulas and Anatolia harbour a high endemic diversity.

East Palaearctic species may range from Kazachstan to Kamchatka and Japan, respectively, but most Japanese species are endemic. The genera *Agathon, Bibiocephala, Philorus* are shared between the East Palaearctic and the West Nearctic regions, *Philorus* extends also into the Oriental region.

The most widespread northern hemisphere genus is *Blepharicera* whose three species groups are very distinct. One occurs in eastern North America, another in the Oriental region where it overlaps with the third which is also widespread in the Palaearctic region and western North America where the most ancient representatives of this third group occur.

Family Deuterophlebiidae
(G. Courtney & R. Wagner)

The extant species are restricted to the northern temperate regions of the earth, although Kennedy (1973) mentioned "one undescribed collection from South America". The distribution is amphi-pacific and appears relictual, but there are no species with a Holarctic distribution (Courtney, 1994a). Six species are known from the western Neartic region, and eight described and perhaps some undescribed species in the Palaearctic region (including the Himalayas). Within the Palaearctic and Nearctic fauna there are two widespread species and several species known from only a few localities. Courtney (1990, 1994a) provided an extensive account of mountain midges with distribution data. The Cascade- and Coast Ranges (North America), Himalayas, Japan and the Korean peninsula appear as main areas of deuteroblebiid endemicity.

Family Nymphomyiidae
(G. Courtney & R. Wagner)

The distribution of extant Nymphomyiidae seems to be restricted to the Holarctic and Oriental Regions (Table 2). Two geographical and phylogenetic 'lines' can be distinguished (Courtney, 1994b): the '*N. alba*-group', consisting of *N. alba* + (*N. rohdendorfi* + *N. levanidovae*) in the eastern Palaearctic Region, and the '*N. walkeri*-group' (*N. walkeri* + *N. dolichopeza*) in the eastern Nearctic Region, including also (*N. brundini* + *N. holoptica*) from the Oriental Region.

Psychodidae (R. Wagner)

Psychodidae occur on all continents except Antarctica (Table 2). Several species (mainly Psychodini and *Clogmia albipunctata*) have been transported by man with organic material (e.g. vegetables) from continent to continent and this is still the case for some other species. Recently, species with larvae developing in Brazilian Bromeliaceae have been detected in Sweden passively transported within these ornamental plants. Our attempt to describe distribution patterns indicates that the tribe Psychodini is distributed worldwide partly due to passive transport by man. Pericomini and Paramormiini are particularly distributed in the northern hemisphere (Holarctic elements), Paramormiini with few more genera in the neighbouring Afrotropical and Oriental regions. Maruinini and Mormiini are probably of southern origin but with large expansion into the northern hemisphere. Setomimini are most abundant in the Neotropical region. Endemicity is often related to specific habitat requirements for water dependant taxa therefore several genera are endemic to individual realms. In general, Psychodid larvae occur in almost all types of wetlands, in springs, streams and along rivers. Even Bromeliads and other small water bodies may be inhabited by specialized taxa. Endemicity is higher in mountainous areas under moderate and tropical climate but endemics are also present in wetlands in the tropics.

Scatopsidae (J.-P. Haenni)

This family of tiny midges is represented in all zoogeographic regions with about 350 described species, but water dependant scatopsids have been recorded only from the Palaearctic region, with five species in three genera (plus one unidentified genus). Their larvae live under the surface of a thin water film among water-logged tree-leaves or are dendrolimnobiontic (Haenni & Vaillant, 1990, 1994), while larvae of most genera are terrestrial, saprophagous, living in a wide variety of organic matter in all degrees of decomposition, consequently often in liquid or semiliquid media. Due to fragmentary present knowledge, no inference may be made upon the real distribution and areas of endemicity of genera with aquatic representatives.

Family Tanyderidae (R. Wagner)

Most genera are endemic to the southwestern Nearctic region, southern Neotropic, South Africa, Australia or New Zealand (Table 2). Only *Protanyderus* has a wider distribution in the Holartic and the adjacent parts of the Oriental region, and *Radinoderus* in the Australian and Oceanian regions. Genus *Mischoderus* is endemic in New Zealand, *Nothoderus* in Australia, *Radinoderus* in Australia and south-east Asian Islands. *Protanyderus* occurs in the Palaearctis and W-Nearctis, *Protoplasa* in the eastern Nearctic. (D. Judd presents substantial information at: http://mgd.nacse.org/cgi-bin/sqml2.0/judd/Pictdb.qml?qml_screen=Picturelist&none=)

Family Ptychopteridae (P. Zwick)

Many species are known only from types, their actual distribution remains unknown. There are no examples of confirmed narrow regional endemism. Instead, ranges of individual species may be very large. For example, *Ptychoptera contaminata* (L.) ranges from Great Britain to Kazachstan (leg. Devyatkov, P.Z. collection), and *P. hugoi* Tjeder from Mongolia to North Sweden (Krzeminski & Zwick, 1993). Several North American and African species also seem to be widespread.

Family Dixidae (R. Wagner)

Faunas of the western and eastern borders of the Paleartic and Nearctic regions are quite well known. More than 40 species occur in the West-Palaearctic, and about a dozen in Japan. Spreading of these taxa into Central Asia remains unknown, but so far four endemic species have been mentioned from that area. Two species are of Holarctic distribution (*Dixella naevia* Peus, N-Europe, N-Russia and Alaska, *Dixella dyari* (Garrett), Sweden, Alaska, Canada). Information on the other biogeographic regions is fragmentary, in particular concerning tropical South America, Africa, Southeast Asia, and partly Australia. At least two species are endemics on oceanic islands (Canary, Madeira). Many species still remain undiscovered in the tropics and in the mountain ranges all over the world in general. Certainly species numbers mentioned represent at most 15–20% of the world biodiversity of Dixidae.

Family Corethrellidae (A. Borkent & R. Wagner)

Out of the 97 extant species known, more than 2/3 is from the Neotropical realm (Table 2). However, the Neotropical Region has been far more extensively sampled and it is likely that the genus will be diverse in other areas, especially in south-east Asia and New Guinea. Two species occur in both the Nearctic and Neotropical Regions and one Japanese species occurs in the Ryukyu Islands in the Oriental Region. Otherwise there are no species occurring in more than one region.

Family Chaoboridae (R. Wagner & Goddeeris)

Most Chaoboridae (even species) have large distribution areas that cover continents, or even one or two biogeographical regions. Two genera are restricted to Australia, one to the Nearctic region. Five species (*C. crystallinus*, *C. flavicans*, *M. fuliginosus*, *M. velutinus*, *C. nyblaei*) are distributed in the Nearctic and Palaearctic regions (Holarctic). *C. festivus* occurs in the Nearctic and Neotropic regions, and *C. queenslandensis* the Australian and Oriental regions.

Family Thaumaleidae
(B. J. Sinclair & R. Wagner)

The degree of endemicity in Thaumaleidae is particularly high, because adults are weak fliers and larvae are restricted to their small, local habitats. Consequently, Thaumaleidae are particularly suitable for biogeographical studies. For example, only six of over 80 species of west-Palaearctic Thaumaleidae are distributed over wide areas within Europe. Most are restricted to islands, mountainous areas or even individual massifs. Most mountainous areas of the earth remain undiscovered. Adult Thaumaleidae are inconspicuous, larvae and pupae are not striking, and many species are on the wing for only a short period. The total amount of species is assumed to be several times higher than the number of species known today.

The most substantial knowledge exists for the west-Palaearctic (European) (Vaillant, 1977; Wagner, 2002) and west- and east-Nearctic Thaumaleidae (Arnaud & Boussy, 1994; Sinclair, 1996). *Thaumalea* is widely distributed in the W Palaearctic, *Andropropsopa* is Holarctic, *Protothaumalea* is endemic in the SW-Palaearctic, and *Trichothaumalea* is widely disjunct with species in the eastern and western Nearctic and in Japan (Sinclair & Saigusa, 2002). Only about five European species are distributed over greater areas, the others are restricted to mountainous regions (e.g. Alps, Pyrenees, Balkans). Only *Th. verralli* is distributed in two biogeographical regions (amphi-Atlantic). Sinclair (1996) suggested that its presence in North America was human assisted; larvae may have been transported over centuries step-by-step in water barrels from NE Europe via Iceland and Greenland to Newfoundland. Neotropical and Australian genera *Niphta*, *Oterere* and *Austrothaumalea* have close affinities to *Afrothaumalea* and all four are unequivocal Gondwanan elements (Sinclair & Stuckenberg, 1995). Up to date publications on Thaumaleidae of the Neotropics (Edwards 1932), Australia and New Zealand (McLellan, 1983, 1988; Theischinger, 1986, 1988) and South Africa (Sinclair & Stuckenberg, 1995) form a good basis for future research.

Family Ceratopogonidae (A. Borkent)

Members of the family occur on all continents other than Antarctica, from sea level to as high as 4,200 m (in Tibet). They occur within 150 km of permanent polar ice in the north and have been recorded from most subantarctic islands. A few genera are particularly good dispersers and there are representatives of the family on every island of even moderate size (Borkent, 1991). Although as a family, Ceratopogonidae are very broadly distributed, many species are known from restricted areas. Lowland, continental species tend to have broader distributions within various regions but at higher elevations, especially above 1,500 m in the tropics, species are increasingly endemic and it is common to have species restricted to a given mountain range. Likewise, many species on continental islands and distant volcanic islands are endemic. In the Palaearctic 1028 species have been recorded, different numbers occur in the Nearctic (600), Afrotropical (622), Neotropical (1,066), Oriental (521) and Australasian (761) regions.

Family Stratiomyidae
(R. Rozkošný & N. E. Woodley)

Stratiomyids are found throughout the world but they are particularly diverse in tropical areas. Almost 1,000 species are known from the Neotropical Region but many fewer species have been recorded in the Palaearctic (426), Nearctic (267), Afrotropical (387), Australasian (407) and Oriental (321) regions. It is evident that especially in the last region many new species await description as indicated by a set of recent studies.

Genus distribution

Forms with confirmed and presumed aquatic larvae occur in all regions (Table 3), though some genera with numerous species predominate in the Palaearctic Region (*Oxycera*, *Nemotelus*, *Stratiomys*) or in the Australasian Region (*Odontomyia*). The species-rich genus *Nemotelus* is evidently absent in the Oriental and Australasian Regions, *Oxycera* in the Neotropical and Australasian Regions, and also *Stratiomys* in the latter region. Some genera occur almost exclusively in the Nearctic Region (*Caloparyphus*), in the Nearctic and Neotropical Regions (*Euparyphus*, *Myxosargus*, *Anoplodontha*, *Hedriodiscus*, *Hoplitimyia*, *Psellidotus*) or only in the Neotropical Region (*Anopisthocrania*, *Chloromela*s, *Metabasis*, *Glariopsis*, *Glaris*, *Pachyptilum*, *Promeranisa*, *Rhingiopsis*, *Stratiomyella*, *Zuerchermyia*).

The described aquatic larvae from the Palaearctic Region (38 spp.) are treated by Vaillant (1951, 1952), Rozkošný (1982–1983, 1997a, 2000) and Rozkošný & Baez (1986). Only three aquatic larvae are described from the Afrotropical Region (Lachaise & Lindner, 1973; Kühbander, 1985). No aquatic larva has been described from the Australasian Region, although species of *Odontomyia* that occur there are almost certainly aquatic. Described aquatic larvae of the Nearctic Region (24 spp.) are summarized by McFadden (1967) and added by Sinclair (1989).

More information on the distribution of genera in individual realms will be presented on the homepage.

Family Empididae (B. J. Sinclair)

Aquatic Empididae are found on all continents except Antarctica (Table 3). The genus *Wiedemannia* predominates in terms of number of species in the Palearctic Region, especially Europe, in contrast to North America, where *Trichoclinocera* has greater diversity and is more numerous and widespread. These two genera are primarily confined to the Northern Hemisphere, with an endemic species group of *Wiedemannia* known from widely disjunct

Table 3 Number of aquatic and water dependent (FW) genera per zoogeographic region

	PA	NA	NT	AT	OL	AU	PAC	ANT	World FW genera
Blephariceridae	12	4	5	2	4	9	0	0	27
Edwardsininae	0	0	1	1	0	1	0	0	2
Blepharicerinae	12	4	4	1	4	8	0	0	25
Deuterophlebiidae	1	1	0	0	0	0	0	0	1
Nymphomyiidae	1	1	0	0	1	0	0	0	1
Psychodidae	50	15	37	21	24	22	9	0	102
Horaiellinae	0	0	0	0	1	0	0	0	1
Sycoracinae	1	0	(1)	1	(1)	1	0	0	3
Psychodinae	49	15	36	20	22	21	9	0	98
Scatopsidae	4	?	?	?	?	?	?	0	4
Tanyderidae	1	3	3	1	(1)	4	0	0	10
Ptychopteridae	2	3	?	1	2	?	?	0	3
Dixidae	2	3	4	1	2	5	–	0	8
Corethrellidae[3]	1	1	1	1	1	1		0	1
Chaoboridae[3]	3	3	1	1	1	3		0	6
Thaumaleidae[4]	4	3	3	1	2	3	1	0	8
Ceratopogonidae	?	?	?	?	?	?	?	?	?
Stratiomyiidae	16	13	38	16	13	7		0	64
Empidoidea (exclusive Dolichopodidae)	18	13	9	7	11	8	3	0	26
"Oreogetoninae"	1	1	0	0	0	1	0	0	2
Ceratomerinae	0	0	1	0	0	2	0	0	2
Clinocerinae	12	7	4	6	8	3	2	0	16
Hemerodromiinae	3	4	4	1	2	2	1	0	4
Trichopezinae	2	1	0	0	1	0	0	0	2
Lonchopteridae	1	?	?	?	?	?	?	0	1
Syrphidae[3]	37	24	33	12	9	10		0	53
Sciomyzidae	16	14	12	3	5	4	1	0	30
Ephydridae	52	50	46	40	27	33	15	2	93
Discomyzinae	4	5	5	4	3	3	2	0	11
Hydrelliinae	13	7	7	13	8	8	3	0	21
Gymnomyzinae	10	12	11	8	6	8	3	0	17
Ilytheinae	7	6	7	5	3	4	4	0	10
Ephydrinae	18	20	16	10	7	10	3	2	34
Muscidae	6	7	6	7	6	6	0	0	19
Mydaeinae	1	1	1	1	1	1	0	0	1
Coenosiinae Limnophorini	5	4	6	3	2	4	2	0	13
Coenosiinae Coenosiini	4	4	2	1	1	0	1	0	5

PA = Palaearctic, NA = Nearctic, NT = Neotropical, AT = Afrotropical, OL = Oriental, AU = Australasian, PAC = Pacific Oceanic islands, ANT = Antarctic

afromontane regions of Africa (Sinclair, 2003) and *Trichoclinocera* distributed as far south as Java (Sinclair & Saigusa, 2005). *Bergenstammia*, *Phaeobalia* and *Clinocerella* do not occur outside of Europe, whereas a series of species of *Kowarzia* extend from Europe to South Africa (Sinclair, 1999).

In the Southern Hemisphere these northern clinocerine genera are mostly absent and replaced by the

Ceratomerinae, and the clinocerine genera, *Asymphyloptera* and *Proagomyia*. Many aquatic species remain undescribed from New Zealand and this habitat is largely unexplored in Chile.

Clinocera is worldwide in distribution, although only a few species groups occur in the Southern Hemisphere (Sinclair, 1995, 2000a). In the Hemerodromiinae, *Chelifera* and *Hemerodromia* are very widespread and not confined to a specific biogeographic region. In contrast, *Metachela* is a Northern Hemisphere taxon and *Neoplasta* is possibly confined to the New World.

The degree of endemicity or "aquatic empidoid hotspots" based on the number of species (incl. undescribed species) can be identified and include for example, New Zealand (*Ceratomerus*—some 40 spp., *Hydropeza*—10 spp.), southeastern Australia (*Ceratomerus*—17 spp., *Clinocera*—7 spp.), western North America (*Clinocera*—+30 spp., other aquatic empidids—+60 spp.), Europe (*Wiedemannia*—+65 spp.) and Himalayas (+45 spp.).

Family Lonchopteridae (M. Bartak)

The main distribution centre of the family is probably the area of south-east Asia. From that area 31 species are known and additional taxa await description. In Europe 13 species occur (four of them restricted to the northern or southern parts), four are known from the Nearctic (one Holarctic and one cosmopolitan), five from the Afrotropis (one cosmopolitan) and 16 species from the Orientalis (one occurring simultaneously in the Palaearctic region), and only a single cosmopolitan, *L. bifurcata* is known from the Neotropical and Australasian regions. Aquatic species are only known from Europe.

Family Syrphidae (G. E. Rotheray)

As stated in the phylogeny section, explanation of distribution and biogeographic patterns remain provisional. At genus-level the greatest region of diversity is the Palaearctic followed by the Oriental. However in terms of species, the richest region is the Neotropical. The Afrotropical and Australasian regions are relatively less biodiverse and include a mix of endemic and more widespread groups.

Family Sciomyzidae (R. Rozkošný & L. Knutson)

Out of 533 described species in 61 genera distributed throughout all the main biogeographical regions aquatic or semi-aquatic larvae are presumed in about 280 species and 31 genera. Most aquatic species are restricted to a single Region (except for holarctically distributed species), but several are slightly to moderately invasive into one or two neighbouring regions and a few are broadly distributed across two regions. The number of aquatic species with the known larvae being found in each region are: Palaearctic (54), Nearctic (79), Afrotropical (6), Australian and Oceanic (8), Neotropical (28), and Oriental (10). The fauna of the northern hemisphere including the Palaearctic and the Nearctic species is apparently more rich (or better known) than that of southern hemisphere.

In the Palearctic, nine genera are apparently endemic but of them aquatic larvae were proved only in *Ilione*, *Pherbina* and *Psacadina*. Some Palaearctic species reach some parts of the Oriental Region (e.g. *Pherbellia cinerella*, *P. nana*, *Hydromya dorsalis*, *Ilione turcestanica*, *Sepedon sphegea*).

Of genera with known aquatic larvae only *Hedria* and *Hoplodictya* are apparently of Nearctic origin, though two of the five species of the latter also colonize some parts of Central America. Species of *Dictya* are distributed very asymmetricaly, embracing 34 Nearctic, eight Neotropical and one Palaearctic species. The North American populations of *Sepedon spinipes* are considered to represent a separate subspecies (*S. spinipes americana*).

Out of ten Afrotropical genera five are endemic but only *Sepedonella* has aquatic larvae. The Afrotropical complex of *Sepedon* species (all with presumably aquatic larvae) is surprisingly rich (42 spp.) also in comparison with any other region and includes a substantial part of Afrotropical species at all (63 spp.). Only *Hydromya dorsalis* penetrate from the Palaearctic and, on the other hand, only *Sepedon hispanica* and *S. ruficeps* reach the southern areas of the West Palaearctic. There is no species common with the Neotropical or Oriental Regions.

The Australasian Region with its 39 sciomyzid species and eight known aquatic larvae appears to be almost as poor as the Oriental Region. Out of six endemic species of the widely distributed *Dichetophora*, *D. australis* seems to be confined to Tasmania and some *Sepedon* species are common with the Oriental Region. A totally endemic fauna with 25 species in four endemic genera is known from New Zealand, where the larvae of *Eulimnia* and five of eight reared species of *Neolimnia* has been proved as aquatic. All these forms very probably developed in a long geographical isolation. The Oceanic sciomyzid fauna is scarce (*Pherbellia dives* and *Sepedon lata* on the Fiji Is., *S. plumbella* on some islands). No original species of Sciomyzidae is known from the Hawaiian Is.

A compartively high number of endemic genera are known in South America. Of 22 genera 13 are confined to the Neotropical region and nine of them include aquatic larvae. Aquatic larvae apparently predominate at least in the endemic genera *Dictyodes*, *Guatemalia*, *Perilimnia*, *Protodictya*, *Sepedonea*, *Shannonia* and in *Thecomyia*. Some species of distinctly Nearctic origin penetrate from the North (species of *Dictya*, *Ditaeniella*, *Hoplodictya*, *Pherbellia*, *Sepedon*, *Tetanocera*) and Neotropical species have been regularly recorded in southern parts of the Nearctic Region (*Pherbellia guttata* and *Sepedomerus macropus*).

The Oriental fauna of Sciomyzidae is only insufficiently investigated because only ten aquatic larvae from the genera *Ditaeniella*, *Hydromya*, *Ilione* and *Sepedon* are known. Some clearly endemic Oriental species belong to the broadly distributed genera and additional aquatic larvae may be presumed especially in the Oriental species of *Pteromicra*, *Dichetophora*, *Sepedon* and *Tetanocera*. Typical Oriental species that are widely distributed throughout the region belong to *Sepedon* (*S. chalybeifrons*, *S. ferruginosa*, *S. lobifera* and *S. senex*). *Pteromicra leucodactyla* is probably endemic on Taiwan. Some species penetrate from the Palaearctic, other widely distributed species are also known from the Australasian region (e.g. *Sepedon plumbella*).

Family Ephydridae (T. Zatwarnicki)

Some genera, especially those with numerous species are either widespread all over the world or absent in only one or two zoogeographic regions (perhaps because of poor collecting). There are genera restricted to the New World [*Cressonomyia*, *Dimecoenia* (also Tristan da Cunha), *Lipochaeta*, *Hydrochasma*], the Old World [*Dryxo*, *Cerobothrium*, *Donaceus*; Holarctic: *Axysta*, *Calocoenia*, *Coenia*, *Dichaeta*, *Eutaenionotum*, *Gymnoclasiopa*, *Lamproscatella*, *Paracoenia*, *Rhinonapaea*, *Schema*, and *Thinoscatella*] or to the Southern Hemisphere [*Oedenopiforma*, *Eleleides* and *Ephydrella*]. A few are distributed in two or three regions: *Diclasiopa*, *Haloscatella* (New Zealand, probably taxonomic artefact), and *Philotelma* in the Holarctic and Afrotropical realms; *Asmeringa*, *Homalometopus* and *Scoliocephalus* in the Afrotropic and Palaearctic regions; *Halmopota* and *Omyxa* in the Palaearctic and Oriental, and *Papuama* in Australasia and the Oriental realms. Twenty seven genera are limited to one zoogeographic region: *Afrolimna*, *Corythophora*, *Isgamera* and *Psilopoidea* (Afrotropical), *Parydroptera*, *Subpelignus* and *Teichomyza* (introduced to Neotropics) (Palearctic), *Callinapaea*, *Cirrula*, *Thiomyia*, *Tronamyia* and *Trimerinoides* (Nearctic), *Austrocoenia*, *Dagus*, *Diedrops*, *Pelinoides*, *Physemops*, *Notiocoenia*, *Paraglenanthe*, *Pectinifer* and *Peltopsilopa* (Neotropical), *Psilephydra* and *Sinops* (Oriental), *Galaterina*, *Parahyadina* and *Tauromima* (Australian and Oceanian), *Amalopteryx* and *Synhoplos* (remote islands—Falkland and Sub-Antarctic). Most species are distributed in one or two realms and only a few are subcosmopolitic: *Discocerina obscurella* (Fallén), *Hydrellia griseola* (Fallén) and *Scatella stagnalis* (Fallén) (Mathis & Zatwarnicki, 1998).

For historical reasons the Palaearctic region is the best known (325 species: Cogan, 1984), the majority of species are widespread. Large numbers of endemic species are known from well studied countries: Italy (14: Canzoneri & Meneghini, 1983), and Japan (30: Myiagi, 1977; probably artificial). Wirth (1965) mentioned in the catalogue 329 species from Nearctic 38 of which are restricted to the Western USA. In the Afrotropis (218 species) endemic species are predominantly restricted to groups of off-shore islands: Madagascar (16 endemic species of 32), Cape Verde Islands (five of 16) and Seychelles Islands (seven of 39, Mathis & Zatwarnicki, 2003). The fauna of Oriental region for sure contains more than the 113 species recorded; some islands, particularly the Philippines (five of 22) and Taiwan (six of 39) bear endemic species. Characteristic feature of 317 species in the Neotropis is the narrow distribution, both on the continent, e.g. Argentina (25 species: Lizarralde de Grosso, 1989), Chile (29 species) and on the islands: Juan Fernandez Islands (21 of 25 species), and the West Indies (20 of 41). In Australasia and on Oceanic Islands (146 species) there are several areas with significant rate of endemism, like New Guinea (4 indigenous species of 18), French Polynesia (six of 12), Samoa (four of six), Hawaii (19 of 22: Tenorio, 1980), and New Zealand (11 of 18).

Family Muscidae (A. Pont)

Spilogona species are confined to temperate and montane habitats and are especially abundant at high latitudes of the Holarctic and southern Neotropics. *Lispe* and *Limnophora* are predominantly subtropical to tropical groups, though *Lispe* is poorly represented in the Neotropics. *Xenomyia* is exclusively Afrotropical. *Lispoides* is entirely New World, and is best developed in the southern Neotropics, and *Schoenomyza*, with a single Old World species, follows the same pattern. *Graphomya* appears to be an ancient genus, found in all zoogeographic regions but with only a small number of species. *Lispocephala* is found in all regions except the Neotropics. Many species are widespread and occur in more than one zoogeographic region, especially Palaearctic/Nearctic (*Spilogona*), Nearctic/Neotropical (*Limnophora*) and Afrotropical/Oriental (*Lispe*, *Limnophora*). Some areas show a high level of endemicity, such as the island of New Guinea (*Limnophora*, 87 species (Shinonaga, 2005)), the Hawaiian Islands (*Lispocephala*, 105 species (Hardy, 1981)), and New Zealand (*Spilogona* s. l., circa 200 species, R. A. Harrison, pers.comm).

Human related issues

Psychodidae (R. Wagner)

Phlebotomine sand-flies are of basic medical interest. The main sand-fly born disease (leishmaniasis) occurs in the Old World largely outside the tropics. Females of *Horaiella* have functional mouthparts, but no information on hosts is available. Females of Old World Sycoracinae bite amphibia and reptilia and transmit diseases of frogs and lizards. A few Psychodinae larvae and adults live in sewage treatments. The cosmopolitan 'trickling filter fly' (*Tinearia alterana*, Say) may cause several diseases: such as asthmatic attacks in sewage treatment workers, and the larvae cause myiasis (Europe, Jordan, Japan). *T. alternata* (Say), *Psychoda albipennis* Zetterstedt, *P. cinerea* Banks and other *Psychoda* species have been quoted as members of an eusynanthropic, exophilous Diptera fauna.

Family Dixidae (R. Wagner)

No relation to any human disease is known. Dixid larvae are extremely sensitive indicators of the presence of surfactant or oil-borne pollutants in streams (Thomas, 1979). Larval mortality increased with decreasing surface tension of water (Fowler, Withers & Dewhurst, 1997).

Family Chaoboridae (R. Wagner & B. Goddeeris)

Swarms of *Chaoborus edulis* Edwards, emerging from the large East African lakes may be so dense that they pose a risk of suffocation. People living in the surroundings of the African lakes Victoria and Nyassa/Malawi are known to collect the huge swarms of adult chaoborids for food.

Family Ceratopogonidae (A. Borkent)

The female adults of a few species in a given region are nasty biters that pester humans and domestic animals. Because of their generally small size, some of these can pass through screens and mesh that keeps other biting pests outside and these can make life insufferable. Worldwide, Ceratopogonidae female adults of three genera *Leptoconops*, *Forcipomyia (Lasiohelea)* and *Culicoides*, transmit harmful diseases or parasites including 66 viruses, 15 species of protozoa and 26 species of filarial nematodes, to a diversity of vertebrate hosts (Borkent, 2004).

Biting midges also provide important services in ecological systems. Many species, especially in the genera *Forcipomyia*, *Atrichopogon* and *Dasyhelea* are important pollinators of such plants as cacao (without them we would not enjoy chocolate!) and rubber trees, and the larvae of many are significant predators of other organisms in aquatic habitats. The adults of most biting midges actually suck blood from other insects.

Family Syrphidae (G. E. Rotheray)

Most syrphids are beneficial or benign as far as humans are concerned. Direct benefits include pollination of crops (Gilbert, 1986), predation of pests such as greenfly and blackfly (Rotheray, 1989). Some aquatic species are used to indicate water quality and others are being investigated for use in clearing up water polluted by organic waste. A few species are pests in bulbs (*Merodon* and *Eumerus*) and there are occasional reports of aquatic larvae causing health problems (Whittington & Rotheray, 1997). In the UK, a programme of work is underway to conserve two endangered saproxylic syrphids in one of the first attempts at species-level conservation (see Malloch Society website: http://www.mallochsociety.org.uk).

Family Sciomyzidae (R. Rozkosny & L. Knutson)

Some freshwater predaceous larvae appear to be suitable potential biocontrol agents of aquatic snails that are intermediate hosts of parasitic flatworms causing schistosomiasis of man and fasciolasis of cattle and sheep. Some species have been thus introduced to Pacific areas as new natural enemies or to support the native populations of sciomyzids.

The most recent review of Sciomyzidae as biological control agents are those of Barker et al. (2004) and Knutson & Vala (2006).

Family Ephydridae (T. Zatwarnicki)

The leaf or stem mines of *Hydrellia*, particularly *H. griseola* become extensive enough to damage crops of watercress, rice, barley, and other irrigated cereals (Grigarick, 1959). On the other hand the same species had been used as biocontrol agent against noxious weeds that invade wetlands in the USA (Foote, 1995). A *Notiphila* species in Japan (Koizumi, 1949) and *Ephydra macellaria* Egger in Egypt (Andres, 1913) have been reported to damage rice. *Psilopa leucostoma* Meigen is a leaf miner affecting sugar beet production in the USA (Landis et al., 1967). *Teichomyza fusca* Macquart incidentally causes myiasis (Goetghebuer, 1928).

Family Muscidae (A. Pont)

Mosquitoes (Culicidae) and black flies (Simuliidae) are among the most important groups of biting insects from the point-of-view of human and animal health. The species of Limnophorini, both as adults and as larvae, are known to act as significant predators of both these groups. The dipteran predators of black flies have been reviewed by Werner & Pont (2003), and the same authors have recently discussed the adult predaceous behaviour of the entire tribe Limnophorini (Werner & Pont, 2005). Apart from some work on the genus *Xenomyia* in Africa, almost all that

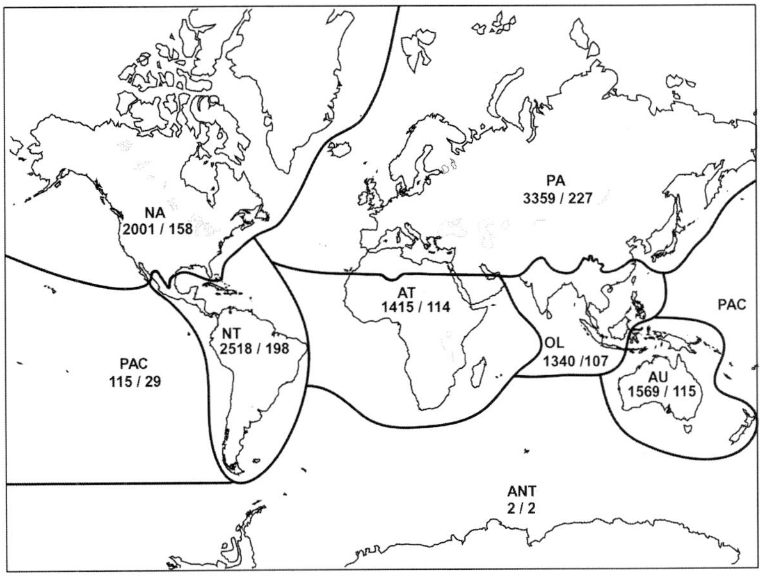

Fig. 1 Distribution of Dipteran species and genera (Species number/Genus number) by zoogeographic regions (excluding Diptera families: Simulidae, Culicidae, Chironomidae, Tipulidae and Tabanidae). PA—Palaearctic; NA—Nearctic; NT—Neotropical; AT—Afrotropical; OL—Oriental; AU—Australasian; PAC—Pacific Oceanic Islands; ANT—Antarctic

is known about predation by the Limnophorini is based on work in Europe. This would be a worthwhile field for further investigation anywhere in the world (Fig. 1).

Acknowledgement Family Stratiomyidae (R. Rozkošný & N. E. Woodley) & Family Sciomyzidae (R. Rozkošný & L. Knutson): These studies were supported by the Ministry of Education of the Czech Republic (Programme MSM0021622416).

References

Alexander, C. P., 1927. Diptera. Fam. Ptychopteridae. Genera Insectorum 188: 1–12 +1 plate.

Alexander, C. P., 1981. Ptychopteridae. In McAlpine, J. F., B. V. Peterson, G. E. Shewell, H. J. Teskey, J. R. Vockeroth & D. M. Wood (eds), Manual of Nearctic Diptera, Vol. 1, Agriculture Canada, Monograph 27: 325–328.

Andres, A., 1913. Note preliminare sur un ravageur du riz. Bulletin de la Société Entomologique d'Egypte 3: 40-42.

Arnaud, P. H. & I. A. Boussy, 1994. The adult Thaumaleidae (Diptera: Culicomorpha) of Western North America. Myia 5: 41–152.

Back, C. & D. M. Wood, 1979. *Palaeodipteron walkeri* (Diptera: Nymphomyiidae) in northern Quebec. The Canadian Entomologist 111: 1287–1291.

Bährmann, R. & R. Bellstedt, 1988. Beobachtungen und Untersuchungen zum Vorkommen der Lonchopteriden auf dem Gebiet der DDR, mit einer Bestimmungstabelle der Arten. Deutsch entomologische Zeitschrift, Neue Folge 35: 265–279.

Barker, G. M., L. V. Knutson, J.-C. Vala, J. B. Coupland & J. K. Barnes, 2004. Overview of the biology of marsh flies (Diptera: Sciomyzidae), with special reference to predators and parasitoids of terrestrial gastropods. In Barker, G. M. (ed.), Natural Enemies of Terrestrial Molluscs. CAB International, Wallingford: 159–225.

Baud, F., 1973. Biologie et cytologie de cinq espèces du genre *Lonchoptera* Meig. (Dipt.) don't l'une est parthénogénétique et les autres bisexuées avec quelques remarques d'ordre taxonomique. Revue suisse de Zoologie 80: 473–515.

Berg, C. O. & L. Knutson, 1978. Biology and systematics of the Sciomyzidae. Annual Review of Entomology 23: 239–258.

Boorman, J., 1997. Family Ceratopogonidae. In Papp, L. & B. Darvas (eds), Contributions to a Manual of Palaearctic Diptera (With Special Reference to Flies of Economic Importance), Vol. 2. Science Herald, Budapest, Hungary: 349–368.

Borkent, A., 1991. The Ceratopogonidae (Diptera) of the Galápagos Islands, Ecuador with a discussion of their phylogenetic relationships and zoogeographic origins. Entomologica Scandinavica 22: 97–122.

Borkent, A., 1993. A world catalogue of fossil and extant Corethrellidae and Chaoboridae (Diptera), with a listing of references to keys, bionomic information and description of each known life stage. Entomologica Scandinavica 24: 1–24.

Borkent, A., 2000. Biting midges (Ceratopogonidae: Diptera) from Lower Cretaceous Lebanese amber with a discussion of the diversity and patterns found in other ambers. In Grimaldi, D. (ed.), Studies on Fossils in Amber, with Particular Reference to the Cretaceous of New Jersey. Backhuys Publishers, Leiden, The Netherlands: 355–452.

Borkent, A., 2004. 10. Ceratopogonidae. In Marquardt, W. C. (ed.), Biology of Disease Vectors, 2nd edn. Elsevier Academic Press: 113–126.

Borkent, A. & D. A. Craig, 2004. *Austroconops* Wirth and Lee, a Lower Cretaceous genus of biting midges yet living in Western Australia: a new species, first description of the immatures and discussion of their biology and phylogeny (Diptera: Ceratopogonidae). American Museum Novitates 3449: 1–67.

Borkent, A. & G. R. Spinelli, 2000. Catalog of New World Biting Midges South of the United States (Diptera: Ceratopogonidae). Contributions on Entomology, International 4: 1–107.

Borkent, A. & W. L. Grogan, (in press). Catalog of the New World Biting Midges North of Mexico (Ceratopogonidae: Diptera). Myia.

Borkent, A. & W. W. Wirth, 1997. World Species of Biting Midges (Diptera: Ceratopogonidae). Bulletin of the American Museum of Natural History 233: 257.

Boussy, I. A., J. M. Gillespie & P. H. Arnaud, 1994. External structures of larval *Thaumalea buckae* Arnaud and Boussy (Diptera, Thaumaleidae). Myia 5: 195–201.

Bowles, D. A., 1998. Life History of *Bittacomorpha clavipes* (Fabricius) (Diptera: Ptychopteridae) in an Ozark Spring, U.S.A Aquatic Insects 20: 29–34.

Canzoneri, S. & D. Meneghini, 1983. Ephydridae e Canaceidae. In Calderini (ed.), Fauna d'Italia, Vol. 20, Bologna: 1–337.

de Carvalho, C. J. B., 1989. Classificação de Muscidae (Diptera): uma proposta através da análise cladística. Revista brasileira de Zoologia 6: 627-648.

Cogan, B. H., 1984. Family Ephydridae. In Soós, A. & L. Papp (eds), Catalogue of Palaearctic Diptera 10: 126–176.

Colless, D. H., 1986. The Australian Chaoboridae (Dipetra). Australian Journal of Zoology, Supplementary Series 124: 1–66.

Collins, K. P. & B. M. Wiegmann, 2002a. Phylogenetic relationships and placement of the Empidoidea (Diptera: Brachycera) based on 28S rDNA and EF–1 sequences. Insect Systematics and Evolution 33: 421–444.

Collins, K. P. & B. M. Wiegmann, 2002b. Phylogenetic relationships of the lower Cyclorrhapha (Diptera: Brachycera) based on 28S rDNA sequences. Insect Systematics and Evolution 33: 445–456.

Couri, M. S. & A. C. Pont, 2000. Cladistic analysis of Coenosiini (Diptera: Muscidae: Coenosiinae). Systematic Entomology 25: 373–392.

Courtney, G. W., 1990. Revision of Nearctic mountain midges (Diptera: Deuterophlebiidae). Journal of Natural History 24: 81–118.

Courtney, G. W., 1991a. Life history patterns of Nearctic mountain midges (Diptera: Deuterophlebiidae). Journal of the North American Benthological Society 10: 177–197.

Courtney, G. W., 1991b. Phylogenetic analysis of the Blephariceromorpha, with special reference to mountain midges (Diptera: Deuterophlebiidae). Systematic Entomology 16: 137–172.

Courtney, G. W., 1994a. Revision of Palaearctic mountain midges (Diptera: Deuterophlebiidae), with phylogenetic and biogeographic analyses of world species. Systematic Entomology 19: 1–24.

Courtney, G. W., 1994b. Biosystematics of the Nymphomyiidae (Insecta: Diptera): life history, morphology, and phylogenetic relationships. Smithsonian Contributions to Zoology 550: 1–41.

Courtney, G. W. & L. Jedlicka, 1997. Family Nymphomyiidae. In Papp, L. & B. Darvas (eds), Contributions to a Manual of Palaearctic Diptera. Science Herald, Budapest: 21–27.

Courtney, G. W., 1998. First records of the Nymphomyiidae (Diptera) in Nepal. Proceedings of the Entomological Society Washington 100: 595–597.

Cutten, F. E. A. & D. K. McE. Kevan, 1970. The Nymphomyiidae (Diptera) with special reference to *Palaeodipteron walkeri* Ide and its larva in Quebec, and a description of a new genus and species from India. Canadian Journal of Zoology 48: 1–24.

De Meillon, B. & W. W. Wirth, 1991. The genera and subgenera (excluding *Culicoides*) of the Afrotropical Biting Midges (Diptera: Ceratopogonidae). Annals of the Natal Museum 32: 27–147.

Disney, R. H. L., 1999. British Dixidae (Meniscus midges) and Thaumaleidae (Trickle Midges): Keys with ecological notes. Freshwater Biological Association, Ambleside, Publication No. 56, 128 pp, Kendal, Cumbria.

Drake, C. M., 1996. The larva and habitat of *Lonchoptera nigrociliata* (Diptera: Lonchopteridae). Dipterists Digest 3: 28–31.

Duckhouse, D. A., 1990. The Australian genera of Pericomoid Psychodidae (Diptera) and the status of related Enderlein genera in the Neotropics. Invertebrate Taxonomy 3: 721–746.

Duckhouse, D. A., 1991. A revision of Australopapuan and New Caledonian *Brunettia* (Diptera, Psychodidae). Invertebrate Taxonomy 4: 973–1030.

Edwards, F. W., 1932. Diptera of Patagonia and South Chile II. Nematocera, Thaumaleidae: 111–115.

Evenhuis, N. L., 1994. Catalog of the fossil flies of the world (Insecta: Diptera). Backhuys Publishers, Leiden, 600 pp. Carvalho, C. J. B. de, M. S. Couri, A. C. Pont, D. Pamplona & S. M. Lopes, 2005. A Catalogue of the Muscidae (Diptera) of the Neotropical Region. Zootaxa 860: 1–282.

Exner, K. & D. A. Craig, 1976. Larvae of Alberta Tanyderidae (Diptera: Nematocera). Quaestiones Entomologicae 12: 219–237.

Foote, B. A., 1995. Biology of Shore Flies. Annual Revue of Entomology 40: 417–442.

Fowler, J. A., I. D. Withers & F. Dewhurst, 1997. Meniscus midges (Diptera: Dixidae) as indicators of surfactant pollutants. Entomologiste 116: 24–27.

Frank J. H., 1983. Bromeliad phytotelmata and their biota, especially mosquitoes. In Frank, J. H. & L. P. Lounibos (eds), Phytotelmata: Terrestrial plants as hosts for aquatic insect communities. Plexus Press, Medford, New Jersey: 101–128.

Frizzi, G., C. Contini & M. Mameli, 1966. Ulteriori ricerche citogenetiche sui Dixidae della Sardegna. Atti Associatione Genetica Italiana 11: 286–291. Pavia.

Gilbert, F. S., 1986. Hoverflies. Naturalists' Handbooks 5, 66 pp.

Goetghebuer, M., 1928. Note sur un cas de myiase intestinale. Bulletin et Annales de la Société Entomologique de Belgique 68: 237–239.

Griffiths, G. C. D., 1972. The phylogenetic classification of Diptera Cyclorrhapha, with special reference to the structure of the male postabdomen. Series entomologica 8. Jung, Hague.

Grigarick, A. A., 1959. Bionomic of the rice leaf miner, *Hydrellia griseola* (Fallén), in California (Diptera: Ephydridae). Hilgardia 29: 1–80.

Grimaldi, D., 1999. The co-radiations of pollinating insects and angiosperms in the Cretaceous. Annals of the Missouri Botanical Garden 86: 373–406.

Grimaldi, D. & J. Cumming, 1999. Brachyceran Diptera in Cretaceous ambers and Mesozoic diversification of the Eremoneura. Bulletin of the American Museum of Natural History 239: 1–124.

Haenni, J.-P., 1997. 2.12. Family Scatopsidae. In Papp, L. & B. Darvas (eds), Contributions to a Manual of Palaearctic Diptera (with special reference to flies of economic importance), Vol. 2. Nematocera and Lower Brachycera. Science Herald, Budapest: 255–272.

Haenni, J.-P. & F. Vaillant, 1990. First evidence of Scatopsidae (Diptera, Nematocera) with aquatic larvae. Second International Congress of Dipterology, Bratislava, August 27–September 1st, 1990, abstract book: 314.

Haenni, J.-P. & F. Vaillant, 1994. Description of dendrolimnobiontic larvae of Scatopsidae (Diptera) with a review of our knowledge of the preimaginal stages of the family. Mitteilungen der Schweizerischen Entomologischen Gesellschaft 67: 43–59.

Hansen S. B., 1979. Livscyklus og vækst hos arter af *Ptychoptera* (Diptera, Nematocera) in en dansk bæk. Entomologisk Meddellelser 47: 33–38.

Hardy, D. E., 1981. Insects of Hawaii. A manual of the insects of the Hawaiian Islands, including an enumeration of the species and notes on their origin, distribution, hosts, parasites, etc. Vol. 14. Diptera: Cyclorrhapha IV, series Schizophora, section Calyptratae. The University Press of Hawaii, Honolulu, 491 pp.

Harper, P. P., 1980. Phenology and distribution of aquatic dance flies (Diptera: Empididae) in a Laurentian watershed. American Midland Naturalist 104: 110–117.

Harper, P. P. & M. Lauzon, 1989. Life cycle of the nymph fly *Palaeodipteron walkeri* Ide 1965 (Diptera: Nymphomyiidae) in the White Mountains of southern Quebec. The Canadian Entomologist 121: 603–607.

Hartley J. C., 1958. The root-piercing spiracles of the larva of *Chrysogaster hirtella* Loew (Diptera: Syrphidae). Proceedings Royal Entomological Society of London (A) 33: 81–87.

Hartley J. C., 1961. A taxonomic account of the larvae of some British Syrphidae. Proceedings of the Zoological Society of London 136: 505–573.

Hennig, W., 1965. Vorarbeiten zu einem phylogenetischen System der Muscidae (Diptera: Cyclorrhapha). Stuttgarter Beiträge zur Naturkunde 141: 1–100.

Hennig, W., 1966. Dixidae aus dem Baltischen Bernstein, mit Bemerkungen über einige andere fossile Arten aus der Gruppe Culicoidea (Diptera Nematocera). Stuttgarter Beiträge zur Naturkunde 153: 1–16.

Hennig, W., 1973. Diptera (Zweiflügler). In Helmcke, J.-G., D. Starck & H. Wermuth (eds), Handbuch der Zoologie. IV. Band: Arthorpoda – 2. Hälfte: Insecta, 2. Teil/31. Walter de Gruyter, Berlin, New York: 1–200.

Hodkinson, I. D., 1973. The immature stages of *Ptychoptera lenis lenis* (Diptera: Ptychopteridae) with notes on their biology. Canadian Entomologist 105: 1091–1099.

Hoffeins, C. & H. W. Hoffeins, 1995. Erstfund von Nymphomyiidae im Baltischen Bernstein. Fossilien 4/95: 207–210.

Hogue, C. L., 1973. A taxonomic review of the genus *Maruina* (Diptera: Psychodidae). Natural History Museum, Los Angeles County, Science Bulletin 17: 1–69.

Ide, F. P., 1965. A fly of the archaic family Nymphomyiidae (Diptera) from North America. The Canadian Entomologist 97: 496–507.

Judd, D.: Family Tanyderida (Diptera: Ptychopteromorpha) – http://mgd.nasce.org/DJFlies/index.html.

Kennedy, H. D., 1973. Family Deuterophlebiidae. In Delfinado M. D. & D. E. Hardy (eds), Catalog of the Diptera of the Oriental Region, Vol. 1. Suborder Nematocera. University of Hawaii Press: 260.

Kitching R. L., 2000. Food webs and container habitats: the natural history and ecology of phytotelmata. Cambridge University Press, 431 pp.

Knutson, L., 1981. Sciomyzidae. In Hurlbert, S. H., G. Rodrigues & N. D. Santos (eds), Aquatic Biota of Tropical South America, Part 1: Arthropoda. San Diego, State University, San Diego, California: 310–312.

Knutson, L., 1982. Sciomyzidae. In Hurlbert, S. H. & A. Villalobos-Figueroa (eds), Aquatic Biota of Mexico, Central America and the West Indies. San Diego State University, San Diego, California: 467–469.

Knutson, L. & K. Ghorpadé, 2004. Family Sciomyzidae. In Yule, C. M. & H. S. Yong (eds), The Freshwater Invertebrates of the Malaysian Region. National Academy of Sciences, Kuala Lumpur: 831–844.

Knutson, L. & J. C. Vala, 2002. An evolutionary scenario of Sciomyzidae and Phaeomyiidae (Diptera). Annales de la Société Entomologique de France (nouvelle série) 38: 145–162.

Knutson, L. & J. C. Vala, 2006. Biology of snail-killing Sciomyzidae. University Press, Cambridge (in press).

Koizumi, K., 1949. Description of a new species of *Notiphila*, a serious pest of rice roots (Diptera, Ephydridae). Matsumushi 3: 91–93.

Kovalev, V. G., 1986. Insects in the Early Cretaceous ecosystems of Western Mongolia. Transactions of the Palaeontological Institute of the USSR 28: 127–154 [in Russian].

Kovalev, V. G., 1990. Geological history and systematic position of the Thaumaleidae. Entomological Review 69(4): 121–131. (Translation from: Entomologicheskoe obozrenie 4 1989: 798–808).

Krzeminski, W. & P. Zwick, 1993. New and Little Known Ptychopteridae (Diptera) from the Palaearctic Region. Aquatic Insects 15: 65–87.

Krzeminski, W. & D. D. Judd, 1997. Family Tanyderidae. In Papp, L. & B. Darvas (eds), Contibutions to a Manual of Palaearctic Diptera, Vol. 2. Science Herald, Budapest: 281–289.

Kühbander, M., 1985. Beschreibung der Larve von *Odontomyia smaragdifera* (Lindner, 1938) aus Kenia und Anmerkungen zur Biologie (Diptera, Stratiomyidae). Entomofauna 6: 417–423.

Lachaise, D. & E. Lindner, 1973. Les Diptères des savanes tropicales préforestières de Lamto (Cote-d'Ivoire). I. – Note écologique sur les Stratiomyidae de Lamto et description d'une espèce nouvelle: *Odontomyia magnifica*. Annales de la Société Entomologique de France (nouvelle série) 9: 593–608.

Landis, B. J., R. L. Wallis & R. D. Redmond, 1967. *Psilopa leucostoma*, a new leaf miner of sugar beets in the United States. Journal of Economic Entomology 60: 115–118.

Landry, B. & P. P. Harper, 1985. The aquatic dance fly fauna of a subarctic river system in Québec, with the description of a new species of *Hemerodromia* (Diptera: Empididae). The Canadian Entomologist 117: 1379–1386.

Lizarralde de Grosso, M., 1989. Ephydridae de la Republica Argentina (Insecta – Diptera). Serie Monographica y Didactica No 3, Facultad de Ciencias Naturales e Instituto Miguel Lillo, Universidad Nacional de Tucumán: 1–93.

Lukashevich, E. D., 1995. First pupae of the Eoptychopteridae and Ptychopteridae from the Mesozoic of Siberia (Insecta: Diptera). Palaeontological Journal 29: 164–171.

Lukashevich, E. D. & D. E. Shcherbakov, 1997. A first find of net-winged midges (Blephariceridae, Diptera) in the Mesozoic. Neues Jahrbuch für Geologie und Palaeontologie, Monatshefte 11: 639–646.

Makarchenko, E. A., 1979. *Nymphomyia rohdendorfi* sp.n., a new representative of archaic insects (Diptera, Nymphomyiidae) from the Upper Kolyma Flow. Zoologicheskii Zhurnal 58: 1070–1073.

Marinoni, L. & W. N. Mathis, 2000. A cladistic analysis of Sciomyzidae Fallén (Diptera). Proceedings of the Biological Society of Washington 113: 162–209.

Mathis, W. N. & T. Zatwarnicki, 1998. 3.49. Family Ephydridae. In Papp, L. & B. Darvas (eds), Contributions to a Manual of Palaearctic Diptera (with special reference to flies of economic importance), Vol. 3, higher Brachycera. Science Herald, Budapest: 537–570.

Mathis, W. N. & T. Zatwarnicki, 2003. Shore flies of the Republic of Seychelles (Diptera: Ephydridae). Annales Zoologici 53(4): 585–650.

McFadden, M. W., 1967. Soldier fly larvae in America North of Mexico. Proceedings of the United States National Museum, 121(No. 3569): 1–72.

McLellan, I. D., 1983. New diagnosis for genus *Austrothaumalea*, and redescription of *A. neozealandica* (Diptera: Thaumaleidae). New Zealand Journal of Zoology 10: 267–270.

McLellan, I. D., 1988. A revision of New Zealand Thaumaleidae (Diptera: Nematocera) with description of new species and a new genus. New Zealand Journal of Zoology 15: 563–575.

Meunier, F., 1907. Beitrag zur Fauna der Bibioniden, Simuliden und Rhyphiden des Bernsteins. Jahrbuch der Königlich

Preussischen Geologischen Landesanstalt und Bergakademie Berlin 24: 391–404, Tafel 17.

Meijere, J. C. H., 1906. Die Lonchopteren des palaearktischen Gebietes. Tijdschrift voor Entomologie 49: 44–98.

Miyagi, I., 1977. Ephydridae (Insecta: Diptera). Fauna Japonica: 1–113; 49 plates.

Mondchadsky, A., 1960. Intraspecific relations of predaceaous larvae of the subfamily Chaoborinae (Diptera, Culicidae). Report 1 (In Russian with English summary). Entomological Review, Washington 38: 453–462.

Moulton, J. K. & B. M. Wiegmann, 2004. Evolution and phylogenetic utility of CAD (rudimentary) among Mesozoic-aged Eremoneuran Diptera (Insecta). Molecular Phlyogenetics and Evolution 31: 363–378.

Oosterbroek, P. & G. W. Courtney, 1995. Phylogeny of the nematocerous families of Diptera (Insecta). Zoological Journal of the Linnean Society 115: 267–311.

Peach, W. J. & J. A. Fowler, 1986. Life cycle and laboratory culture of *Dixella autumnalis* Meigen (Dipt., Dixidae). Entomologist's Monthly Magazine 122: 59–62.

Peters, T. M. & D. Adamski, 1982. A description of the larva of *Dixella nova* (Walker) (Diptera; Dixidae). Proceedings of the Entomological Society of Washington 82: 521–528.

Peterson, B. V., 1987. 49. Lonchopteridae. In McAlpine, J. F. (ed.), Manual of Nearctic Diptera, Vol. 3, Monograph 32. Research Branch, Agriculture Canada, Ottawa: 675–680.

Peus, F., 1958. 10a. Tanyderidae 10b. Liriopeidae. In Lindner, E. (ed.), Die Fliegen der paläarktischen Region 3, Lieferung 200: 1–44.

Pont, A. C. & C. J. B. de Carvalho, 1997. Three species of Muscidae (Diptera) from Dominican amber. Studia Dipterologica 4: 173–181.

Rivosecchi, L., 1984. Famiglia Sciomyzidae. In Rivosecchi, L. (ed.), Ditteri (Diptera). No. 28. Guide per il Riconoscimento delle Specie Animali delle Acque Interne Italiane. Consiglio Nazionale delle Richerche AQ/I/206. Valdonega, Verona: 142–152.

Rohdendorf, B. B. & N. S. Kalugina, 1974. The finding of peculiar Diptera of the family Nymphomyiidae in Primorje Territory. Revue d'Entomologie de URSS 1974: 686–694 (in Russian).

Rotheray, G. E., 1989. Aphid Predators. Naturalists' Handbooks 11, 66 pp.

Rotheray, G. E., 2003. The predatory larvae of two *Nepenthosyrphus* species living in pitcher plants (Diptera, Syrphidae). Studia dipterologica 10: 219–226.

Rotheray, G. E. & F. Gilbert, 1999. Phylogeny of Palaearctic Syrphidae (Diptera): evidence from larval stages. Zoological Journal of the Linnean Society 127: 1–112.

Rotheray, G. E., M. Zumbado, E. G. Hancock & F. C. Thompson, 2000. Remarkable aquatic predators in the genus *Ocyptamus* (Diptera, Syrphidae). Studia dipterologica 7: 385–398.

Rozkošný, R., 1982–1983. A biosystematic study of the European Stratiomyidae (Diptera). Vols. 1 and 2, Dr. W. Junk Publishers, The Hague: 401 and 431 pp.

Rozkošný, R., 1995. World distribution of Sciomyzidae based on the list of species (Diptera). Studia Dipterologica 2: 221–238.

Rozkošný, R., 1997a. Diptera Stratiomyidae, soldier flies. In Nilsson, A. N. (ed.), Aquatic Insects of North Europe – A Taxonomic Handbook, Vol. 2. Apollo Books, Stenstrup: 321–332.

Rozkošný, R., 1997b. Diptera Sciomyzidae, Snail killing flies. In Nilsson, A. N. (ed.), Aquatic Insects of North Europe – A Taxonomic Handbook, Vol. 2. Apollo Books, Stenstrup: 363–381.

Rozkošný, R., 2000. Insecta: Diptera: Stratiomyidae. In Schwoerbel, J. & P. Zwick (eds), Süßwasserfauna von Mitteleuropa, Vol. 21/18, Spektrum Verlag, Heidelberg-Berlin: 1–110.

Rozkošný, R., 2002. Insecta: Diptera: Sciomyzidae. In Schwoerbel, J. & P. Zwick (eds), Süßwasserfauna von Mitteleuropa, Vol. 21/23, Spektrum Verlag, Heidelberg-Berlin: 15–122.

Rozkošný, R. & M. Baez, 1986. Larvae of the Canary Islands Stratiomyidae (Diptera). Vieraea 16: 341–349.

Saether, O. A., 1970. Nearctic and Palaearctic *Chaoborus* (Diptera: Chaoboridae): Bulletin, Fisheries Research Board of Canada 174, VII + 57 pp.

Saether, O. A., 1992. Resdescription of *Cryophila lapponica* Bergroth (Diptera: Chaoboridae) and the phylogenetic relationships of chaoborid genera. Aquatic Insects 14: 1–21, 193–194.

Saether, O. A., 1997. 2.18 Family Corethrellidae. In Papp, L. & B. Darvas (eds), Contributions to a Manual of Palaearctic Diptera (with Special Reference to the Flies of Economic Importance). Science Herald, Budapest: 319–323.

Saether, O. A., 2002. Insecta: Diptera: Chaoboridae. In Schwoerbel, J. & P. Zwick (eds), Süßwasserfauna von Mitteleuropa, Vol. 21, Insecta. Diptera. 10/11. Chaoboridae und Thaumaleidae Spektrum Verlag, Heidelberg-Berlin: 1–38.

Shinonaga, S., 2005. Studies on the genus *Limnophora* from New Guinea and Solomon Islands (Diptera, Muscidae). Japanese Journal of Systematic Entomology 11: 73–152.

Sinclair, B. J., 1989. The biology of *Euparyphus* Gerstaecker and *Caloparyphus* James occurring in madicolous habitats of eastern North America, with descriptions of adult and immature stages (Diptera: Stratiomyidae). Canadian Journal of Zoology 67: 33–41.

Sinclair, B. J., 1995. Generic revision of the Clinocerinae (Empididae), and description and phylogenetic relationships of the Trichopezinae, new status (Diptera: Empidoidea). The Canadian Entomologist 127: 665–752.

Sinclair, B. J., 1996. A review of the Thaumaleidae (Diptera: Culicomorpha) of eastern North America, including a redefinition of the genus *Androprosopa* Mik. Entomologica Scandinavica 27: 361–376.

Sinclair, B. J., 1999. Review of the Clinocerinae of southern Africa (Diptera: Empididae). Annals of the Natal Museum 40: 103–125.

Sinclair, B. J., 2000a. Review of the genus *Clinocera* from Australia and New Zealand (Diptera: Empididae: Clinocerinae). Invertebrate Taxonomy 14: 347–361.

Sinclair, B. J., 2000b. Immature stages of Australian *Austrothaumalea* Tonnoir and *Niphta* Theischinger (Diptera: Thaumaleidae). Australian Journal of Entomology 39: 171–176.

Sinclair, B. J., 2003. Southern African Empidoidea (Diptera) – phylogenetic patterns and biogeographic implications. Cimbebasia 29: 205–213.

Sinclair, B. J. & J. M. Cumming, 2006. The morphology, higher-level phylogeny and classification of the Empidoidea (Diptera). Zootaxa 1180: 1–172

Sinclair, B. J. & J. R. Harkrider, 2004. The immature stages and rearing observations of the aquatic dance fly, *Roederiodes wirthi* Coquillett (Diptera: Empididae: Clinocerinae), with taxonomic notes on the genus. Studia Dipterologica 11: 51–61.

Sinclair, B. J. & T. Saigusa, 2002. A new species of the seepage midge genus *Trichothaumalea* Edwards from Japan (Diptera: Thaumaleidae). Insect Systematics and Evolution 33: 175–184.

Sinclair, B. J. & T. Saigusa, 2005. Revision of the *Trichoclinocera dasyscutellum* group from East Asia (Diptera: Empididae: Clinocerinae). Bonner Zoologische Beiträge 53(2004): 193–209.

Sinclair, B. J. & B. R. Stuckenberg, 1995. Review of the Thaumaleidae of South Africa. Annals of the Natal Museum 36: 209–214.

Skidmore, P., 1985. The biology of the Muscidae of the world. Series Entomologica 29: 550.

Smith, K. G. V., 1969. Diptera, Lonchopteridae. In Handbooks for the identification of British Insects Volume 10: 1–9.

Stahls, G., H. Heikki, G. Rotheray, J. Muona & F. Gilbert, 2003. Phylogeny of Syrphidae (Diptera) inferred from combined analysis of molecular and morphological characters. Systematic Entomology 28: 433–450.

Steyskal, G. C. & L. V. Knutson, 1981. Empididae [Chapter 47]. In McAlpine, J. F., B. V. Peterson, G. E. Shewell, H. J. Teskey, J. R. Vockeroth & D. M. Wood (eds), Manual of Nearctic Diptera, Vol. 1. Agriculture Canada Monograph 27: 607–624.

Stuckenberg, B. R., 2004. Labial morphology in Blephariceridae (Diptera: Nematocera): a new interpretation with phylogenetic implications, and a note on colocephaly. African Invertebrates 45: 223–236.

Sueyoshi, M., 2005. Sciomyzidae. In Kawai, T. & K. Tanida (eds), Aquatic Insects of Japan: Manual with Keys and Illustrations. Tokai University Press, Kanagawa: 1229–1256.

Tenorio, J. A., 1980. Family Ephydridae. In Hardy D. E. & M. D. Delfinado (eds), Insects of Hawaii. Diptera: Cyclorrhapha III, series Schizophora section Acalypterae, exclusive of family Drosophilidae, Vol. 13. The University Press of Hawaii: 251–351.

Theischinger, G., 1986. Australian Thaumaleidae (Insecta: Diptera). Records of the Australia Museum 38: 291–317.

Theischinger, G., 1988. *Austrothaumalea bickeli* spec. nov., a new thaumaleid from Australia (Insecta: Diptera: Thaumaleidae). Stapfia 17: 211–213.

Thomas, A., 1979. Diptères torrenticoles peu connus. VI. Les Dixidae du sud-ouest de la France (Nematocera) (*Dixa puberula* Loew, 1849. Écologie, microhabitat et intérêt pratique pour le dépistage des pollutions per les stations touristiques de montagne). Bulletin de la Societé d'Histoire Naturelle de Toulouse 115: 242–268.

Thompson, F. C., 1972. A contribution to a generic revision of the neotropical Milesiinae (Diptera, Syrphidae). Arquivos de Zoologica, Sao Paulo 23: 73–215.

Thompson, F. C. & G. E. Rotheray, 1998. Family Syrphidae. In Papp, L. & B. Darvas (eds), Contributions to a Manual of Palaearctic Diptera (with Special Reference to Flies of Economic Importance), Vol. 3. Science Herald, Budapest, Hungary: 81–140.

Tokunaga, M., 1932. A remarkable dipterous insect from Japan, *Nymphomyia alba*, gen. et sp. nov. Annotationes Zoologicae Japonenses 13: 559–569.

Tokunaga, M., 1935a. On the pupae of the Nymphomyiid fly. Mushi 8: 44–52.

Tokunaga, M., 1935b. A morphological study of a nymphomyid fly. The Philippine Journal of Science 56: 127–214.

Turner, W. J., J. M. Babcock & J. Jenkins, 1986. New records and first observation on adult flight activity for *Deuterophlebia coloradensis* Pennack (Diptera: Deuterophlebiidae) in Idaho. Pan-Pacific Entomologist 62(2): 111–118.

Vaillant, F., 1951. Les larves d'*Hermione*. Travaux du Laboratoire d'Hydrobiologie et Pisciculture de Grenoble 1951: 23–38.

Vaillant, F., 1952. Les larves d'*Hermione* d'Algérie (Diptères Stratiomyiidae). Bulletin de la Société Naturelle de l'Afrique du Nord 43: 8–15.

Vaillant, F., 1977. Les Diptères Thaumaleidae d'Europe. Annales de la Société entomologique de France 13: 695–710.

Vaillant, F., 1971–1983. Psychodidae Psychodinae. In Lindner, E. (ed.), Die Fliegen der paläarktischen Region 9d, Lieferungen 287, 291, 292, 305, 310, 313, 320, 326, 328. E. Schweitzerbart'sche Verlagsbuchhandlung, Stuttgart: 1–358 + 120 pl.

Vaillant, F., 2002. Insect: Diptera: Lonchopteridae. In Schwoerbel, J. & P. Zwick (eds), Süßwasserfauna von Mitteleuropa, Vol. 21, Insecta: Diptera, 22, 23, Lonchopteridae, Sciomyzidae. Spektrum Akademischer Verlag, Gustav Fischer: 1–14.

Varley, G. C., 1937. Aquatic insect larvae which obtain oxygen from the roots of plants. Proceedings Royal Entomological Society of London (A) 12: 55–60.

Vockeroth, J. R., 1969. A revision of the genera of the Syrphini (Diptera: Syrphidae). Memoirs Entomological Society of Canada 62, 176 pp.

Wagner, R., C. Hoffeins & H. W. Hoffeins, 2000. A fossil nymphomyiid (Diptera) from the Baltic and Bitterfeld amber. Systematic Entomology 25: 115–120.

Wagner, R., 2002. Insecta: Diptera: Thaumaleidae. In Schwoerbel, J. & P. Zwick (eds), Süßwasserfauna von Mitteleuropa Vol. 21, Insecta: Diptera.: 10/11, Chaoboridae, Thaumaleidae. Spektrum Akademischer Verlag, Gustav Fischer: 41–110.

Wagner, R. & O. Gathmann, 1996. Long-term studies on aquatic dance flies (Diptera, Empididae) 1983–1993: Distribution and size patterns along the stream, abundance changes between years and the influence of environmental factors on the community. Archiv für Hydrobiologie 137: 385–410.

Wagner, R., F. Leese & A. R. Panesar, 2004. Aquatic dance flies from a small Himalayan mountain stream (Diptera: Empididae: Hemerodromiinae, Trichopezinae and Clinocerinae). Bonner Zoologische Beiträge 52: 3–32.

Werner, D. & A. C. Pont, 2003. Dipteran predators of Simuliid blackflies: a worldwide review. Medical and Veterinary Entomology 17: 115–132.

Werner, D. & A. C. Pont, 2005. The feeding and reproductive behaviour of the Limnophorini (Diptera: Muscidae). Studia Dipterologica, Supplement 14.

Whittington, A. E. & G. E. Rotheray, 1997. Afrotropical distribution and larval description of *Ornida obesa* (Fabricius, 1775) (Diptera; Syrphidae). Journal of African Zoology 111: 365–372.

Wiegmann, B. M., D. K. Yeates, J. L. Thorne & H. Kishino, 2003. Time flies, a new molecular time-scale for Brachyceran fly evolution without a clock. Systematic Biology 52: 745–756.

Wirth, W. W., 1965. Family Ceratopogonidae. In Stone, A., C. W. Sabrosky, W. W. Wirth, R. H. Foote & J. R. Coulson (eds), Catalog of Diptera of North America North of Mexico, Agriculture Handbook, Vol. 276. United States Department of Agriculture, Washington, D.C.: 121–142.

Wolf, B. & P. Zwick, 2001. Life Cycle, Production and Survival Rates of *Ptychoptera paludosa* (Diptera: Ptychopteridae). International Revue of Hydrobiology 86: 661–674.

Wood, H. G., 1934. Notes on the life history of *Dixa bicolor*, Wood (Dipt.). Stylops 3: 205–208.

Wood, D. & A. Borkent, 1989. Phylogeny and classification of the Nematocera. In McAlpine, J. F. (ed.), Manual of Nearctic Diptera, Vol. 3, Monograph 32. Research Branch, Agriculture Canada, Ottawa: 1333–1370.

Woodley, N. E., 1989. Phylogeny and classification of the "orthorrhaphous" Brachycera. In McAlpine, J. F. (ed.), Manual of Nearctic Diptera, Vol. 3, Monograph 32. Research Branch, Agriculture Canada, Ottawa: 1371–1395.

Woodley, N. E., 1995. The genera of Beridinae (Diptera: Stratiomyinae). Memoirs of the Entomological Society of Washington 18: 1–231.

Woodley, N. E., 2001. A World Catalog of the Stratiomyidae (Insecta: Diptera). Backhuys Publishers, Leiden, 473 pp.

Zatwarnicki, T., 1992. A new classification of Ephydridae based on phylogenetic reconstruction (Diptera Cyclorrhapha). Genus 3: 65–119.

Zwick, P., 1977. Australian Blephariceridae (Diptera). Australian Journal of Zoology, Supplementary Series 46: 1–121.

Zwick, P., 1998. The Australian Net-Winged Midges of the Tribe Apistomyiini (Diptera: Blephariceridae). Australian Journal of Entomology 37: 289–311.

Zwick, P., 2004. Fauna Europaea: Ptychopteridae. In de Jong, H. (ed.), Fauna Europaea: Diptera: Nematocera. Fauna Europaea version 1.1, http://www.faunaeur.org.

Zwick, P. & J. Starý, 2003. *Ptychoptera delmastroi* sp. n. (Diptera: Ptychopteridae) from Italy. Aquatic Insects 25: 241–246.

Hydrobiologia (2008) 595:521–528
DOI 10.1007/s10750-007-9038-9

FRESHWATER ANIMAL DIVERSITY ASSESSMENT

Global diversity of butterflies (Lepidotera) in freshwater

Wolfram Mey · Wolfgang Speidel

Abstract In Lepidoptera, the subfamily Acentropinae and Pyraustinae of Crambidae (Pyraloidea) and the family Arctiidae (Noctuoidea) contain species with true aquatic larvae, which live submerged during larval development. In Pyraustinae and Arctiidae only a few species exhibit an aquatic life-history. From the latter, aquatic larvae are known from the Neotropical genus *Paracles*. The number of aquatic *Paracles* species is unknown. The Acentropinae are predominantly aquatic. They are distributed worldwide, and reach the highest diversity in tropical regions of South East Asia/Malesia and in the Neotropical Region. At present, the Acentropinae include a total of 50 genera and 737 described species. All genera, assigned to the subfamily, are listed in a table, and the numbers of included species are indicated. The taxonomy and phylogeney of the genera are inadequately known. The species have a minor economic importance, however, they are very sensitive to degradation of water quality and habitat destruction.

Guest editors: E.V. Balian, C. Lévêque, H. Segers & K. Martens
Freshwater Animal Diversity Assessment

W. Mey (✉)
Museum für Naturkunde, Humboldt-Universität Berlin, Invalidenstr. 43, 10115 Berlin, Germany
e-mail: wolfram.mey@museum.hu-berlin.de

W. Speidel
Museum Witt, Tengstrasse 33, 80796 Munchen, Germany

Keywords Lepidoptera · Acentropinae · Pyraloidea · Arctiidae · Diversity · Distribution · Taxonomy · Genera · Species numbers

Introduction

For an outsider it is surprising to learn that there are indeed some Lepidoptera which are aquatic. Even for limnologists or aquatic entomologists aquatic moths do not belong to the well known and instantly recognised types of aquatic insects. There is a special reason for the poor recognition of aquatic Lepidoptera: the handling of the adult moths during sampling and processing differs completely from the methods used for other freshwater insects. The wing patterns provide features, which are diagnostic at the species level and, therefore, must be preserved intact, i.e. in a dry state. As a consequence, individuals have to be treated very carefully. Correct handling needs to use killing jars, pins, pinning boxes, setting boards, and insect drawers. These tools do not belong to the regular equipment of aquatic entomologists or limnologists. However, they are regularly used by lepidopterists, who are usually specialised on certain families. The study of aquatic Lepidoptera has always been a subject of lepidopterists alone, who on the other side, however, rarely have a limnological background.

Today, the accumulated knowledge on aquatic species has a focus on adults and is scattered in the vast Lepidoptera literature, which is not easily accessible for limnologists. Data on larvae is limited. The larvae are caterpillars with a small degree of intrageneric morphological differentiation, which makes the identification at the species level very difficult or impossible. The genera and species of the Holarctic Region are rather well known. But Holarctic species make up only a minority of the world fauna in contrast to the overwhelming diversity of taxa present in the tropics. Many species were only provisionally assigned to a genus and remained there up to now. Several aquatic species are exceptional cases within their systematic groups. However, there is a group of Lepidoptera, whose larvae are almost exclusively aquatic; these are the so-called China-Mark-Moths (Pyraloidea: Acentropinae). The larval stages of most tropical species are undescribed, but might be expected as aquatic. Information on the biology of immatures is very poor. At this stage, available data on the distribution, phylogeny and taxonomy of aquatic moths are much too incomplete to embark seriously into a deeper analysis of the group concerning historical biogeography, faunal changes, centres of endemicity, etc. However, a summary of the currently recognised taxa and their distribution on the planet can be provided. The present account concentrates on families, which have true aquatic species, i.e., with submerged living larvae. These are the families Crambidae (Pyraloidea) and Arctiidae. Some further families contain semi-aquatic species. They are not considered here.

Species and generic diversity

Family Arctiidae: Arctiinae

The subfamily includes about 3,600 species (Heppner, 1991). One species is well-known for its aquatic caterpillars: *Paracles laboulbeni* (Bar, 1873). It was formerly placed in the genus *Palustra*. The species is widely distributed in the Neotropical Region. The caterpillars feed on submerged plants in stagnant or slow flowing waters (Adis, 1983). They lack tracheal gills. Respiration is mediated by air-holding hairs forming a plastron on the dorsal side. The genus *Paracles* is very speciose. It is unclear, whether *P. laboulbeni* is the only aquatic species or whether further congeners have aquatic immatures.

Family Crambidae: Pyraustinae

The subfamily is a mega-diverse group with over 7,000 species word-wide (Heppner, 1991), also sometimes retained in a larger concept of Pyralidae. A few species are reported to have aquatic larvae (Buckingham, 1994):

Samea multiplicalis Guenée, 1854—Neotropical Region,

Niphograpta albiguttalis Warren, 1889—Neotropical Region.

Both species have been used in biological control programme to suppress mass developments of aquatic weeds in stagnant and running waters. In South–East Asia there are several, hitherto unidentified species, which were collected only close to running waters. They form aggregations of adults on the underside of boulders and overhanging rocks, a behaviour known for aquatic moths. The larvae of these species are supposed to be aquatic, but have not been traced yet.

Family Crambidae: Acentropinae

The group is well known under its synonymous name Nymphulinae. Heppner (1991) has compiled species totals for each of the over 300 subfamilies and families of the Lepidoptera of the world. They were tabulated according to major faunal regions whose delimitations are essentially based on Darlington (1957). According to his figures the Acentropinae comprise 716 described species. During recent years a number of new checklists, catalogues and taxonomic changes were published, which provide current figures for Acentropinae of nearly all faunal regions or continents: Henning (2003), Li et al. (2003), Munroe (1983, 1995), Shaffer et al. (1996), Speidel (2005), Speidel & Mey (1999), Yen (2004), You et al. (2002). In addition, a large number of new species have been described especially from the Oriental Region in the last decade (Li et al., 2003; Mey, 2006; Mey & Speidel, 2005; Speidel, 2003; Yoshiyasu, 1987; You & Li, 2005; You et al., 2003, etc.). Based on these publications and our own studies a new synthesis was undertaken at the generic level. We have used Heppner's scheme with slight changes: the northern Mexican border line is the border of the Neotropical Region and the Australian Region is divided into a Newguinean-Polynesian and proper

Australian–New Zealand Region. The new figures are summarised in Table 2. Up to date, the Acentropinae encompass 737 species in 50 genera. Interestingly, the counts are similar to Heppner's ones. In the past, many more genera were included, mainly because of superficial resemblance of wing patterns of the adults. They has been subsequently excluded, but this process is not complete. Exclusion of misplaced taxa and synonymies have not equalled the number of new species descriptions since 1991 (Figs. 1, 2).

The counts for the faunal regions are not by type localities. Thus, the problem of double counting of species which are distributed in several faunal regions has to be considered. The problem is notoriously evident in the East Palaearctic and Oriental Faunal Regions, which have a broad, common transition zone. Widespread species were indicated for all faunal regions in which they occur, but recorded only once in the "total" column of Table 2.

Phylogeny and historical processes

Lepidoptera are essentially terrestrial insects. Only a tiny fraction of the 165,000 described species can be termed aquatic, which means that they have immature stages that live under water. The adults are terrestrial as in the majority of water insects. The conquest of the aquatic environment by lepidopteran larvae is not that astonishing as it may appear at first glance. Larvae of Lepidoptera—the caterpillars—have an enormous adaptive potential (Scoble, 1992). Primitive species like Micropterigidae or ancestral Pyraloidea are detritivorous and moss feeders sometimes are associated

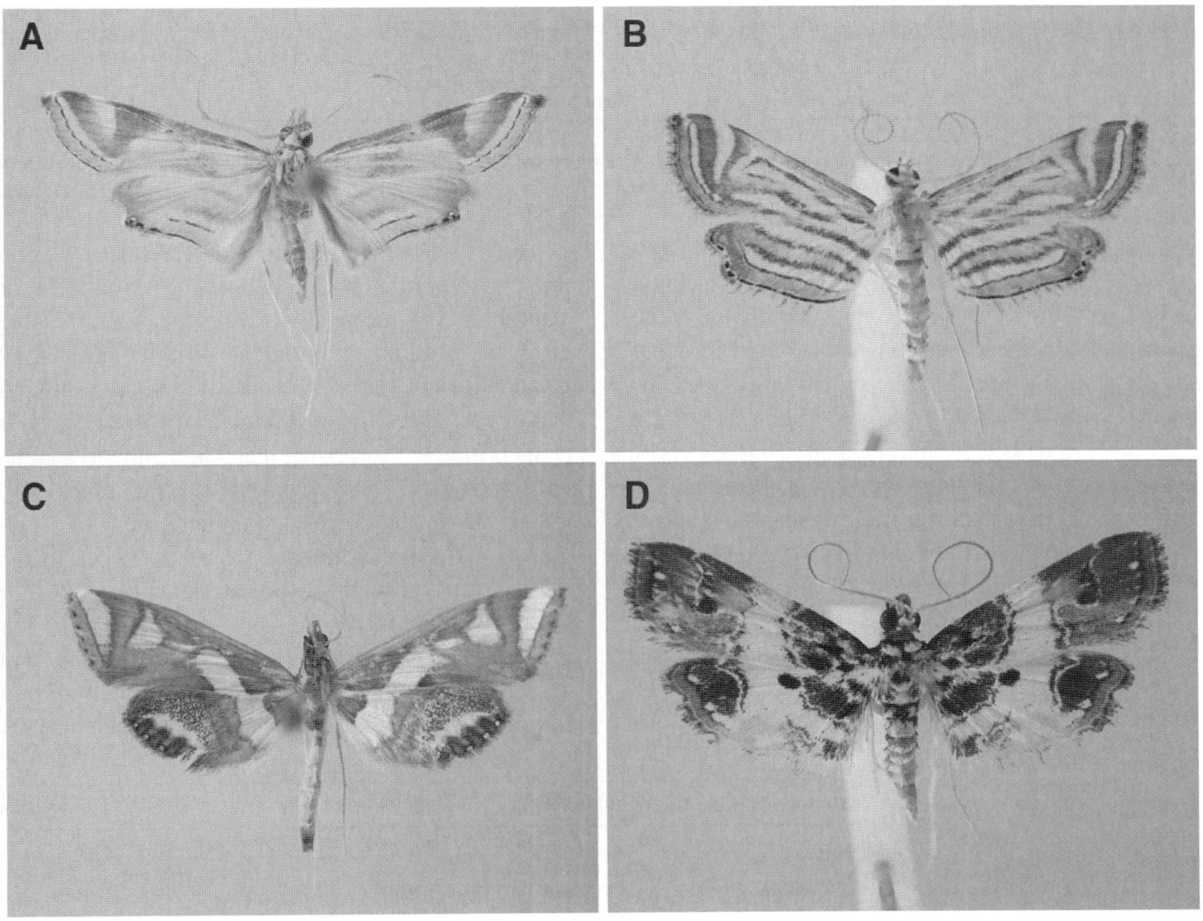

Fig. 1 (**A**) *Eoophyla boernickei* Mey, 2006, male (Borneo), (**B**) *Parapoynx leucographa* Speidel, 2003, male (Borneo) (**C**) *Margarosticha* spec., male (Sulawesi), (**D**) *Paracymoriza* spec., male (Borneo)

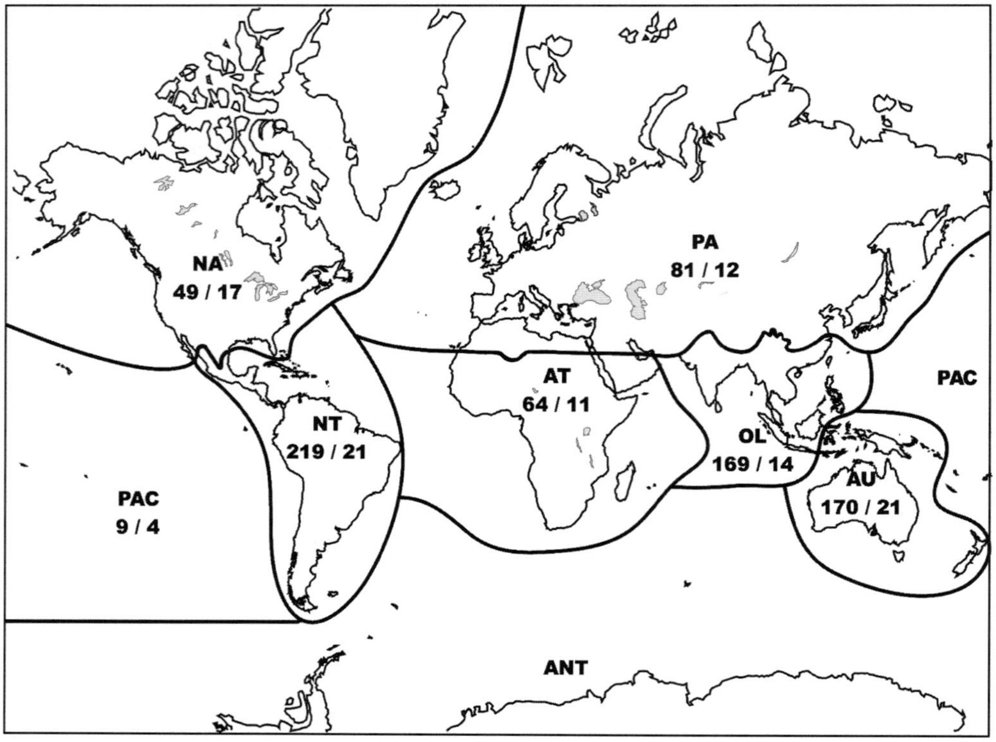

Fig. 2 Distribution of freshwater Lepidoptera species and genera by zoogeographical region (species number/genus number). PA—Palaearctic; NA—Nearctic; NT—Neotropical; AT—Afrotropical; OL—Oriental; AU—Australasian; PAC—Pacific Oceanic Islands

with fungi or animal matter (Munroe & Solis, 1998: 235). The larvae often live in a humid environment on the ground. With the development and radiation of the angiosperms in the Cretaceous a new food resource became widely available. Caterpillars of early evolutionary lineages started to use green-plant tissue as a food source and became phytophagous. In a process of co-evolution and radiation together with the angiosperms, the Lepidoptera were able to adapt to all plant groups and developed a wide array of life histories, forms and strategies which allowed them to exploit all parts of a plant: leaves, stems, roots, bark, twigs, flowers, seeds, etc. Aquatic or semiaquatic plants, being no exceptions, are hosts of a number of lepidopteran species, too.

The caterpillars usually feed on green leaves or on the periphyton in rocky habitats. The cuticule of caterpillars has morphological preadaptions to some kind of plastron respiration. With these larval preadaptions it was only a small step in the evolution from a wet terrestrial to a semiaquatic life history. It is, therefore, not surprising that a transition to aquatic or semiaquatic life occurred several times and independently in different, unrelated taxa of Lepidoptera. Aquatic larvae are known to occur in the mega-diverse families Crambidae and Arctiidae (Table 1). There are probably more families with aquatic representatives. The poor exploration and knowledge of tropical Lepidoptera, especially of rain forest areas with their huge variety of aquatic habitats

Table 1 Generic and species diversity of currently known Lepidoptera taxa with aquatic larvae

Superfamily/Family	Subfamily	Number of genera	Number of species
Pyraloidea			
Crambidae	Acentropinae	50	737
	Pyraustinae	2	2
Noctuidoidea			
Arctiidae	Arctiinae	1	1

Table 2 The genera of Acentropinae—their distribution and diversity in the major Biogeographical Regions of the world (PA = Palaearctic, NA = Nearctic, NT = Neotropical, AT = Afrotropical, OL = Oriental, AU = Australasian, PAC = Pacific Oceanic Islands (Micronesia-Polynesia))

Genus	PA	NA	NT	AT	OL	AU	PAc	Total
Acentria Stephens, 1829	1	1						1
Agassiziella Yoshiyasu, 1989				2	10			12
Ambia Walker, 1859*						7*		7
Anydraula Meyrick, 1885						2	3*	5
Araeomorpha Turner, 1908						2		2
Argyractis Hampson, 1897		1	12	8*				21
Argyractoides Lange, 1956			12					12
Argyrophorodes Marion, 1957*				7*				7
Aulacodes Guenée, 1854			35				3	38
Callilitha Munroe, 1959						2		2
Cataclysta Hübner, [1825]	1			9*		9*	2*	21
Chrysendeton Grote, 1881		3	16					19
Cryptocosma Lederer, 1863			1					1
Contiger Lange, 1956		1						1
Elophila Hübner, 1822	13	4	2	2	7*	6		31
Eoophyla Swinhoe, 1900	11			12*	54	73		142
Eoparargyractis Lange, 1956		3						3
Ephormotris Meyrick, 1933				2*	2			4
Eristena Warren, 1896	12				26	3*		41
Giorgia Clarke, 1965			1					1
Hemiloba Swinhoe, 1901					1			1
Hygraula Meyrick, 1885						2		2
Hylebatis Turner, 1908						1		1
Kasania Krulikowsky, 1910	1							1
Langessa Munroe 1972		1						1
Lathroteles Clarke, 1971*							1*	1
Margarosticha Lederer, 1863					2	13		15
Neargyractis Lange, 1956		1	7					8
Neocataclysta Lange, 1956		1						1
Neoschoenobia Hampson, 1900	1					1		2
Nyctiplanes Turner, 1937						1		1
Nymphicula Snellen, 1880	7			10*	22	7		46
Nymphula Schrank, 1802	4	1			2	2*		9
Nymphuliella Lange, 1956		1						1
Nymphulodes Hampson, 1919			1					1
Oligostigma Guenée, 1854			7	2*		2*		11
Oligostigmoides Lange, 1956		1	5					6
Oxyelophila Forbes, 1922		1	7					8
Paracataclysta Yoshiyasu, 1983				1*	1	1		3
Paracymoriza Warren, 1890	11*				22*	4*		37
Parapoynx Hübner, [1825]	17	8	13	9*	16	22		76
Petrophila Guilding, 1830		15	82					97
Potamomusa Yoshiyasu, 1985	2							2

Table 2 continued

Genus	PA	NA	NT	AT	OL	AU	PAc	Total
Pseudolithosia Hampson, 1907			1					1
Strepsinoma Meyrick, 1897					2	9		11
Synclita Lederer, 1863		4	3					7
Synclitodes Munroe, 1974			1					1
Tetrernia Meyrick, 1890						1		1
Thysanoidma Hampson, 1891					2			2
Usingeriessa Lange, 1956		2	10					12
Number of genera	12	17	18	11	14	21	4	50
Number of species	81	49	216	64	169	170	9	737

Numbers with * denote unclear generic or subfamily associations

suggests that we can expect further families with aquatic members.

The most advanced adaptations to an aquatic existence can be found in the pyraloid subfamily Acentropinae (=Nymphulinae). The majority of aquatic species belong to this subfamily. Hasenfuss (1991) published a convincing scenario for the evolution of the aquatic habit of Acentropinae. According to the plant species or food resource, the caterpillars developed a wide spectrum of adaptations towards a semiaquatic and finally towards a fully aquatic life: species with plastron respiration and open tracheal system feed on floating or partly submerged leaves, whereas species with tracheal gills and closed spiracles are able to live on fully submerged plants or are rock-dwellers in streams. A comparative, taxonomic analysis of the genera on a world-wide scale was not attempted so far, and thus, makes the validity of some taxa questionable. Phylogenetic relationships among genera were studied only on a restricted geographical scale (cf. Speidel, 1984, 2005; Yoshiyasu, 1985). A subdivision of the subfamily into two tribes was proposed by Lange (1956). He included in Nymphulini the case-making shredders and green-plant tissue feeders, and in Argyractini the free living scrapers and web spinning algae feeders. The division, however, was not found to be an adequate reflection of phylogenetic relationships (Munroe & Solis, 1998) and, thus, is not followed here. Consequently, the genera are listed alphabetically in Table 2.

The sister group relationship of the Acentropinae is unresolved (Solis & Maes, 2002). Traditionally, Acentropinae and Schoenobiinae have been considered closely related (Speidel, 1984; Munroe & Solis, 1998). They have semiaquatic larvae, which usually feed as internal borers in the stems of water monocots. Species of both subfamilies often occur together in a wide range of aquatic habitats.

Present distribution and main areas of endemicity

As mentioned in the introduction, our inadequate knowledge only allows rough conclusions from the numbers in Table 2. The data show that tropical regions have many more species than other faunal regions. With approximately 40% of the world total of described species, the peak diversity is in Malesia, including the Pacific islands from Micronesia to Polynesia. This is mainly due to the radiation of the genus *Eoophyla* in maritime South East Asia and, even more pronounced, in New Guinea. More than 140 Acentropinae species are endemic to this islands and the Wallacea, which had to be included into the Australian Region here in Table 2. *Eoophyla* is the largest group in Acentropinae. It is rather homogeneous but includes some isolated forms that may necessitate the establishment of separate genera. A comparably large genus in the Neotropical Region is *Petrophila*, which also gets into the Nearctic. It makes up nearly 50% of all Neotropical species. In contrast to *Eoophyla*, the radiation of *Petrophila* developed in a continental setting. The poor figures for tropical Africa are in accordance with figures from other Lepidoptera groups. However, Africa south of the Sahara, including Madagascar, is a poorly sampled continent and should contain many more species. There are two genera which occur in all faunal regions:

Elophila and *Parapoynx*. Their species are mainly inhabitants of stagnant waters.

Most genera have ranges restricted to one continent or to a single faunal region. The high number of generic taxa in the New World and Australia appears to demonstrate a higher diversity of evolutionary lines on these continents. This can be genuine or is merely an artifact due to the poor systematic investigation of the Afrotropical and Oriental faunas.

Human related issues

In general, aquatic moths are of little economic importance. A few species are known to damage rice in Asian countries: *Parapoynx fluctuosalis* (Guenée, 1854), *P. vittalis* (Bremer, 1864) and *P. stagnalis* (Zeller, 1852). Sometimes the injury can be serious, but usually remains localised. The species are clearly of less importance than rice stem borers of the subfamily Schoenobiinae. The species from rice paddies have a wide distribution. Some tropical species were incidentally introduced together with aquatic plants to other continents. In temperate countries they are unable to adapt to the low winter temperatures, and remain minor pests in greenhouses. In warmer regions, some species have managed to establish permanent populations in the field. The phytophageous caterpillars have been regarded as a promising means for the biological control of aquatic weeds. They have been successfully used in the suppression of *Hydrillia* and *Elodea* in Florida (Buckingham, 1994).

Acentropine species from running waters are very susceptible to water pollution and water channel regulations. They are among the first organisms which would disappear when facing a drop in water quality.

References

Adis, J., 1983. Eco-entomological observations from the Amazon IV. Occurrence and feeding habits of the aquatic caterpillar *Palustra laboulbeni* Bar, 1873 (Lepidoptera: Arctiidae) in the vicinity of Manaus, Brazil. Acta Amazonica 13: 31–36.

Buckingham, G. R., 1994. Biological control of aquatic weeds. In Rosen, D., Bennett, F. D. & Capinera, J. L. (eds), Pest Management in the Subtropics: Biological Control—a Florida Perspective. Intercept Ltd, Andover, 737.

Darlington, P. J. Jr., 1957. Zoogeography: The geographic distribution of animals. Wiley, New York, 675.

Hasenfuss, I., 1991. Evolution der aquatischen Larven und Puppen der Acentropinae (=Nymphulinae/Crambidae/Pyraloidea/Lep.). Verhandlungen der Deutschen Zoologischen Gesellschaft 84: 444.

Henning, S. F., 2003. Lepidoptera. In de Moor, I. J., J. A. Day & F. C. de Moor (eds), Guides to the freshwater invertebrates of Southern Africa, Vol. 8: Insecta III, 209.

Heppner, J. B., 1991. Faunal regions and the diversity of Lepidoptera. Tropical Lepidoptera 2 supplement, 1: 85.

Li, Hou-Hun, P. You & S.-X. Wang, 2003. A systematic study of the genus *Eoophyla* Swinhoe in China, with descriptions of two new species (Lepidoptera, Crambidae, Nymphulinae). Acta Zootaxonomica Sinica 28: 295–301.

Lange, H. W., 1956. A generic revision of the aquatic moths of North America (Lepidoptera: Pyralidae, Nymphulinae). The Wasmann Journal of Biology 14(1): 59–144.

Mey, W., 2006. Eine neue Art der Gattung *Eoophyla* Swinhoe, 1900 von Borneo (Lepidoptera: Crambidae, Acentropinae). Entomologische Zeitschrift (Stuttgart) 116(2): 91–93.

Mey, W. & W. Speidel, 2005. Two new species of *Eoophyla* Swinhoe, 1900 from continental South East Asia (Lepidoptera: Crambidae, Acentropinae). Bonner zoologische Beiträge 53(1/2), 115–119.

Munroe, E., 1983. Pyralidae. In Hodges, R. W. (ed), Check List of the Lepidoptera of America North of Mexico. E.W. Classey Ltd., London, xxiv + 284.

Munroe, E., 1995. Nymphulinae. In Heppner, J. B. (ed), Atlas of Neotropical Lepidoptera. Checklist: part 2, Gainesville, liv + 243.

Munroe, E. & A. M. Solis, 1998. The Pyraloidea. In Kristensen, N.P. (ed), Lepidoptera, Moths and Butterflie Handbuch der Zoologie, Bd. 4, 35, Vol. 1. Handbuch der Zoologie, Bd. 4, 35, de Gruyter, Berlin, 487.

Scoble, M. J., 1992. The Lepidoptera—form, function and diversity. The Natural History Museum, London.

Shaffer, M., E. S. Nielsen & M. Horak, 1996. Pyraloidea. In Nielsen, E. S., Edwards, E. D. & Rangsi, T. V. (eds), Checklist of the Lepidoptera of Australia.—Monographs on Australian Lepidoptera, Vol. 4. CSIRO, Canberra, xiv + 529.

Solis, A. M. & K. V. N. Maes, 2002. Preliminary phylogenetic analysis of the subfamilies of Crambidae (Pyraloidea, Lepidoptera). Belgian Journal of Entomology 4: 53–95.

Speidel, W., 1984. Revision der Acentropinae des palaearktischen Faunengebietes (Lepidoptera: Crambidae). Neue entomologische Nachrichten 12: 157 pp.

Speidel, W., 2003. New species of aquatic moths from the Philippines (Lepidoptera, Crambidae). Insecta Koreana 20: 7–49.

Speidel, W., 2005. Acentropinae. In Huemer, P. & Karsholt, O. (eds), Microlepidoptera of Europe, Vol. 4: 33–68.

Speidel, W. & W. Mey, 1999. Catalogue of the Oriental Acentropinae (Lepidoptera, Crambidae). Tijdschrift voor Entomologie 142: 125–142.

Yen, S.-H., 2004. Insecta: Lepidoptera, Crambidae, Acentropinae. In Yule, C. M. & Seng, Yong-Hoi (eds), Freshwater Invertebrates of the Malaysian Region. Academy of Sciences Malaysia, Kuala Lumpur, 861.

Yoshiyasu, Y., 1985. A systematic study of the Nymphulinae and the Musotiminae of Japan (Lepidoptera: Pyralidae). Scientific Report of the Kyoto Prefectural University 37: 162 pp.

Yoshiyasu, Y., 1987. The Nymphulinae (Lepidoptera: Pyralidae) from Thailand, with descriptions of a new genus and six new species. Microlepidoptera of Thailand 1: 133–184.

You, P., H.-H. Li & K.-L. Chen, 2002. Catalogue of the Nymphulinae and Musotiminae of China (Lepidoptera: Crambidae). In Zhang, Z.-Q. (ed), Fauna of China, Vol. 4: 37–64.

You, P., H.-H. Li & S.-X. Wang, 2003. A taxonomic study of the genus *Eristena* Warren, 1896 from China (Lepidoptera, Crambidae, Nymphulinae). Acta Zootaxonomica Sinica 28(2): 302–306.

You, P. & H.-H. Li, 2005. Two new species of Nymphulinae (Lepidoptera: Crambidae) from China. Entomotaxonomia 27(2): 131–134.

Hydrobiologia (2008) 595:529–534
DOI 10.1007/s10750-007-9036-y

FRESHWATER ANIMAL DIVERSITY ASSESSMENT

Global diversity of hymenopterans (Hymenoptera; Insecta) in freshwater

Andrew M. R. Bennett

Abstract A summary of the known species of aquatic Hymenoptera is presented. In total, 150 species from 11 families are recognized as aquatic (0.13% of the total described species). This number is likely an underestimate, because of the high percentage of undescribed species and a lack of knowledge of host range and behaviour for most species. All aquatic Hymenoptera are parasitoids. Many species have relatively dense pubescence to trap air and elongate, tarsal claws to grip the substrate, when underwater. Most species are known from the Holarctic and Oriental regions, but this is likely an artefact caused by lack of knowledge of other regions of the world. Aquatic behaviour has evolved independently at least 50 times within the order.

Keywords Global diversity · Aquatic Hymenoptera

Introduction

The order Hymenoptera includes such well-known insects as ants, bees and wasps. Hymenopterans undergo complete metamorphosis and most species may be recognized by possession of four membranous wings (Fig. 1) coupled together with a series of hook-like hairs known as hamuli (Goulet & Huber, 1993). The plesiomorphic biological condition is herbivory (most sawflies) with transitions to parasitoidism and predation in more derived groups. Most hymenopterans are parasitoids that develop in or on one host arthropod, killing the host in the process (Waage & Greathead, 1986). Some parasitoid species parasitize eggs or pupae, but most oviposit in or on host larvae.

Whereas almost all hymenopterans are terrestrial, development in water has evolved independently in at least 11 families (Table 1). All known aquatic Hymenoptera are parasitoids. A conservative approach has been adopted in this chapter such that only species for which unequivocal evidence exists that they spend at least some of their life in water are included. This includes species in which female adults enter the water to search for hosts, those with endoparasitoid larvae inside aquatic larval hosts (even if oviposition is terrestrial); and those in which freshly eclosed adults must travel to the water's surface following pupation (even if they develop inside stems of emergent vegetation). Only described species with known hosts are included, except for a few described species that lack host records, but belong to genera that are unquestionably monophyletic and have a clear host range that includes only aquatic hosts. Parasitoids that develop exclusively on

Guest editors: E. V. Balian, C. Lévêque, H. Segers & K. Martens
Freshwater Animal Diversity Assessment

A. M. R. Bennett (✉)
Canadian National Collection of Insects, Agriculture Canada, 960 Carling Ave., Ottawa, ON, Canada, K1A 0C6
e-mail: bennetta@agr.gc.ca

terrestrial stages of aquatic insects are excluded (e.g. in pupae of whirligig beetles or spongilla-flies).

Species diversity

The order Hymenoptera has over 115,000 described species (Gaston, 1993) and with estimates of one million (Grissell, 1999; Ulrich, 1999) to 2.5 million (Stork, 1996) total species, it may rival Coleoptera as the most species-rich order. Table 1 includes 150 aquatic species (0.13% of all described species). These figures have been compiled mainly from discussions with colleagues, from databases (Noyes, 2002; Yu et al., 2005) and from surveys of aquatic species for the Nearctic (Bennett, in press) and Palaearctic (Hedqvist, 1978) regions. Accordingly, most aquatic species in this chapter are Holarctic, but this is likely an artefact caused by poor sampling in other regions of the world and little or no knowledge of biology for the vast majority of species. This lack of knowledge makes it practically impossible to predict how many species of aquatic Hymenoptera might exist, especially considering that many aquatic species belong to genera that are predominantly terrestrial (i.e. knowledge of biology at the generic level does not always help predict existence of aquatic species). The use of water-filled yellow pan traps placed around water or on emergent rocks in water indicates that a great many more water-dependent species may exist, but until host associations and/or observations of aquatic behaviour are made, these species cannot be confirmed as truly aquatic. The number of confirmed aquatic species and genera in each biogeographic region are summarized in Tables 1 and 2 and Fig. 2.

Fig. 1 *Agriotypus gracilis* Waterston (Hymenoptera: Ichneumonidae) female habitus. Scale bar = 1 mm

Aquatic Hymenoptera are known from six superfamilies and 11 families (Table 1). The two largest families, Ichneumonidae and Braconidae, have the two greatest numbers of aquatic species: 39 and 26, respectively. Within the Ichneumonidae, species in at least five genera appear to be exclusively aquatic: *Agriotypus* Curtis (Agriotypinae) on Trichoptera and *Rhachioplex* Bischoff (Campopleginae), *Pseuderipternus* Viereck and *Tanychela* Townes (both Cremastinae) and *Apsilops* Förster (Cryptinae) on Lepidoptera. The braconid genus *Ademon* Haliday (Opiinae) is almost exclusively associated with *Hydrellia* spp. larvae in emergent vegetation. Next in richness of aquatic species are two families of egg parasitoids: Trichogrammatidae (22) and Scelionidae (17). *Centrobiopsis* Girault, *Hydrophylita* Ghesquière and *Prestwichia* Lubbock (Trichogrammatidae) have only aquatic species and *Pseudanteris* Fouts, *Thoron* Haliday, *Thoronella* Masner and *Tiphodytes* Bradley (Scelionidae) are all aquatic. The remaining families have 12 or fewer aquatic species recorded within each of them. *Caraphractus cinctus* Walker (Mymaridae), a parasitoid of dytiscid beetle eggs, has been studied extensively and is known to swim with its wings and mate inside submerged eggs (Jackson, 1966), although both of these habits are unusual for the order. Most aquatic female wasps walk along the substrate when searching for aquatic hosts and mate out of water. The four species of *Aspidogyrus* Yoshimoto (Figitidae) are endemic to Hawaii and parasitize fly larvae in streams (Beardsley, 1992). Finally, *Anoplius depressipes* Banks (Pompilidae) is the only wasp known to parasitize "non-insects" (water-dwelling pissaurid spiders of the genus *Dolomedes* Latreille) and the only one known to move its hosts to land prior to oviposition (Roble, 1985). Aquatic Hymenoptera parasitize at least 25 insect families in the orders Odonata, Hemiptera, Megaloptera, Coleoptera, Diptera, Lepidoptera and Trichoptera. No cases of aquatic hyperparasitism are yet known. Most aquatic Hymenoptera have been

Table 1 Number of species of aquatic Hymenoptera by biogeographic region

	PA	NA	NT	AT	OL	AU	PAC	ANT	World
Superfamily: Chalcidoidea	24	22	7	–	8	7	2	–	51
Family: Chalcididae	3	3	-	-	-	-	-	-	6
Family: Eulophidae	8	3	1	-	-	-	-	-	12
Family: Mymaridae	4	4	2	-	1	1	-	-	6
Family: Pteromalidae	4	4	-	-	-	-	1	-	5
Family: Trichogrammatidae	5	8	4	-	7	6	1	-	22
Superfamily: Cynipoidea	2	3	–	–	–	–	7	–	10
Family: Figitidae	2	3	–	–	–	–	7	–	10
Superfamily: Ichneumonoidea	29	20	2	1	16	1	–	–	65
Family: Braconidae	16	9	–	1	2	–	–	–	26
Subfamily: Alysiinae	12	4	–	–	–	–	–	–	16
Subfamily: Braconinae	–	1	–	–	–	–	–	–	1
Subfamily: Microgastrinae	1	–	–	–	1	–	–	–	2
Subfamily: Opiinae	3	4	–	1	1	–	–	–	7
Family: Ichneumonidae	13	11	2	–	14	1	–	–	39
Subfamily: Agriotypinae	6	–	–	–	10	–	–	–	16
Subfamily: Campopleginae	1	–	–	–	3	–	–	–	3
Subfamily: Cremastinae	–	6	1	–	–	–	–	–	7
Subfamily: Cryptinae	6	4	1	–	1	1	–	–	12
Subfamily: Metopiinae	–	1	–	–	–	–	–	–	1
Superfamily: Platygastroidea	2	5	7	–	4	–	–	–	17
Family: Scelionidae	2	5	7	–	4	–	–	–	17
Superfamily: Proctotrupoidea	2	3	1	–	–	–	–	–	6
Family: Diapriidae	2	3	1	–	–	–	–	–	6
Superfamily: Vespoidea	1	–	–	–	–	–	–	–	1
Family: Pompilidae	1	–	–	–	–	–	–	–	1
Total: Hymenoptera	60	53	17	1	28	8	9	–	150

PA: Palaearctic, NA: Nearctic, NT: Neotropical, AT: Afrotropical, OL: Oriental, AU: Australasian, PAC: Pacific and oceanic islands, ANT: Antartic

collected from lentic environments, but at least one-third of the species are associated with lotic habitats. For example, the 16 species of *Agriotypus* are only found in fast-running streams in Europe and Asia (Bennett, 2001) and species of the diapriid genus *Psychopria* Masner and García occur in running water in the new world (Masner & García, 2002).

Phylogeny and historical processes

The earliest fossil Hymenoptera are herbivorous sawflies from the late Triassic (230 mya) (Riek, 1955) and they were almost certainly terrestrial, as are all extant sawfly species. Aquatic development is, therefore, apomorphic within Hymenoptera and has evolved at least 50 times based on the number of unrelated genera of aquatic Hymenoptera (Table 2). Indeed, the only suprageneric aquatic radiation within the order may be within several putatively related genera of aquatic Scelionidae (Johnson & Masner, 2004).

How did aquatic parasitism evolve? It is likely that it did not evolve de novo from terrestrial parasitism, but rather through transition from wasps parasitizing semi-aquatic hosts around the water's surface. Females of these semi-aquatic species would benefit from characteristics that help movement on and

Table 2 Number of genera of Hymenoptera with at least one aquatic species by biogeographic region

	PA	NA	NT	AT	OL	AU	PAC	ANT	World
Superfamily: Chalcidoidea	11	15	5	–	5	5	2	–	18
Family: Chalcididae	1	1	–	–	–	–	–	–	1
Family Eulophidae	2	3	1	–	–	–	–	–	4
Family: Mymaridae	2	4	1	–	1	1	–	–	4
Family: Pteromalidae	3	2	–	–	–	–	1	–	3
Family: Trichogrammatidae	3	5	3	–	4	4	1	–	6
Superfamily: Cynipoidea	2	1	–	–	–	–	3	–	3
Family: Figitidae	2	1	–	–	–	–	3	–	3
Superfamily: Ichneumonoidea	12	12	2	1	6	1	–	–	21
Family: Braconidae	6	5	–	1	2	–	–	–	8
Subfamily: Alysiinae	3	2	–	–	–	–	–	–	3
Subfamily: Braconinae	–	1	–	–	–	–	–	–	1
Subfamily: Microgastrinae	1	–	–	–	1	–	–	–	2
Subfamily: Opiinae	2	2	–	1	1	–	–	–	2
Family: Ichneumonidae	6	7	2	–	4	1	–	–	13
Subfamily: Agriotypinae	1	–	–	–	1	–	–	–	1
Subfamily: Campopleginae	1	–	–	–	2	–	–	–	2
Subfamily: Cremastinae	–	4	1	–	–	–	–	–	4
Subfamily: Cryptinae	4	2	1	–	1	1	–	–	5
Subfamily: Metopiinae	–	1	–	–	–	–	–	–	1
Superfamily: Platygastroidea	1	4	2	–	2	–	–	–	5
Family: Scelionidae	1	4	2	–	2	–	–	–	5
Superfamily: Proctotrupoidea	2	1	1	–	–	–	–	–	3
Family: Diapriidae	2	1	1	–	–	–	–	–	3
Superfamily: Vespoidea	1	–	–	–	–	–	–	–	1
Family: Pompilidae	1	–	–	–	–	–	–	–	1
Order Hymenoptera	29	33	10	1	13	6	5	–	51

PA: Palaearctic, NA: Nearctic, NT: Neotropical, AT: Afrotropical, OL: Oriental, AU: Australasian, PAC: Pacific and oceanic islands, ANT: Antartic

around the water's surface. These features, in turn, would be pre-adaptations once these species moved into water. There are at least two characteristics of aquatic Hymenoptera that appear to be adaptive to life underwater: (1) short, dense pubescence on the body and wings that allows the wasp to be hydrophobic, as well as to maintain a plastron of air around the body; (2) elongated, strongly curved claws that allow gripping of the substrate, so that the female does not float to the surface or be swept in currents when searching for hosts. Not all aquatic wasps have either or both of these features, but many do. Specific aquatic adaptations of the egg and larva are not known, perhaps mainly because most aquatic Hymenoptera are endoparasitoids and, therefore, the habitat where the host develops has little or no effect on the parasitoid larva. The ectoparasitoid genus *Agriotypus* has evolved a unique pupal respiratory filament that is necessary because water flow in the host pupal Trichoptera case is compromised once the host is killed (Bennett, 2001).

Present distribution and main areas of endemicity

Hymenoptera have been collected in almost every region of the world from the high Arctic to small, subantarctic Islands. Aquatic Hymenoptera are known

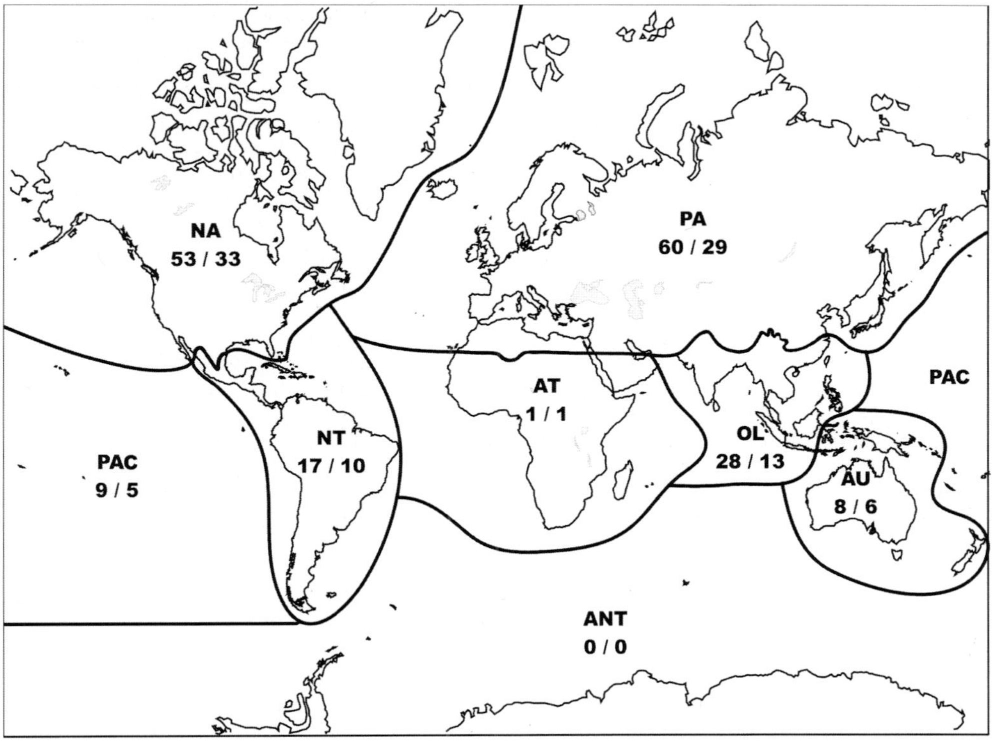

Fig. 2 Distribution of aquatic Hymenoptera species and genera by zoogeographical region (species number/genus number). PA—Palaearctic; NA—Nearctic; NT—Neotropical; AT—Afrotropical; OL—Oriental; AU—Australasian; PAC—Pacific Oceanic Islands, ANT—Antartic

from every major biogeographical region except Antarctica (Tables 1 and 2; Fig. 2). It is unclear whether the 150 known aquatic species are a representative sample of the total distribution of aquatic Hymenoptera, because so little is known about the behaviour, ecology and host range of the vast majority of species of Hymenoptera. The data in Table 2 are strongly biased by the amount of study of Hymenoptera in each region and extrapolations should not be made regarding distributional trends of aquatic Hymenoptera throughout the world. For example, only one described species of aquatic Hymenoptera is known from the Afrotropics (the braconid *Ademon angolanus* Fischer), but this is likely because of lack of sampling and biological information for subsaharan Africa and not because the Afrotropics are actually devoid of aquatic Hymenoptera. The most noteworthy area of aquatic wasp endemicity is Hawaii, with *Aspidogyrus* and two endemic species of the figitid genus *Kleidotoma* Westwood. Aquatic Hymenoptera, like that of Hymenoptera in general, require much greater study in terms of collections, rearings and phylogenetic analyses, which will allow greater understanding of this fascinating niche and the wasps that have managed to exploit it.

References

Beardsley, J. W., 1992. Review of the genus *Aspidogyrus* Yoshimoto, with descriptions of three new species (Hymenoptera: Cynipoidea: Eucoilidae). Proceedings, Hawaiian Entomological Society 31: 139–150.

Bennett, A. M. R., 2001. Phylogeny of Agriotypinae (Hymenoptera: Ichneumonidae) with comments on the subfamily relationships of the basal Ichneumonidae. Systematic Entomology 26: 329–356.

Bennett, A. M. R., (2007, in press). Aquatic Hymenoptera, Chapt. 21, In: Merritt, R. W., K. W. Cummins & M. B. Berg (eds), An Introduction to the Aquatic Insects of North America. Kendall/Hunt Publ. Co., Dubuque, IA.

Gaston, K. J., 1993. Spatial patterns in the description and richness of the Hymenoptera. In LaSalle, J. & I. Gauld (eds), Hymenoptera and Biodiversity. C.A.B. International, Wallingford, UK, 277–293.

Goulet, H. & J. T. Huber, 1993. Hymenoptera of the World: An Identification Guide to Families. Research Branch of Agriculture Canada Publication 1894/E, 668 pp.

Grissell, E. E., 1999. Hymenopteran biodiversity: some alien notions. American Entomologist 45: 235–244.

von Hedquist, K.-J., 1978. Hymenoptera. In Illies, J. (ed.), Limnofauna Europaea. Gustav Fischer Verlag, New York, USA, 286–289.

Jackson, D. J., 1966. Observations on the biology of *Caraphractus cinctus* Walker (Hymenoptera: Mymaridae), a parasitoid of the eggs of Dytiscidae (Coleoptera). Transactions of the Royal Entomological Society of London 118: 23–49.

Johnson, N. F. & L. Masner, 2004. The genus *Thoron* Haliday (Hymenoptera: Scelionidae), egg-parasitoids of waterscorpions (Hemiptera: Nepidae), with key to world species. American Museum Novitates 3452: 1–16.

Masner, L. & J. L. García, 2002. The genera of Diapriinae (Hymenoptera: Diapriidae) in the new world. Bulletin of the American Museum of Natural History 268: 1–138.

Noyes, J. S., 2002. Interactive catalogue of world Chalcidoidea 2001. CD-ROM. Second edition. Taxapad, Vancouver, Canada. www.taxapad.com.

Riek, E. F., 1955. Fossil insects from the Triassic beds of Mt. Crosby, Queensland. Australian Journal of Zoology 3: 654–691.

Roble, S., 1985. Submergent capture of *Dolomedes triton* (Araneae, Pisauridae) by *Anoplius depressipes* (Hymenoptera: Pompilidae). The Journal of Arachnology 13: 391–392.

Stork, N. E., 1996. Measuring global biodiversity and its decline. In Reaka-Kudla, M. L., D. E. Wilson & E. O. Wilson (eds), Biodiversity II. Joseph Henry Press, Washington, DC, 41–68.

Ulrich, W., 1999. The number of species of Hymenoptera in Europe and the assessment of the total number of Hymenoptera in the world. Polskie Pismo Enotomologiczne 68: 151–164.

Waage, J. & D. Greathead, (eds), 1986. Insect Parasitoids. Academic Press, London, 398 pp.

Yu, D. S., K. van Achterberg & K. Horstmann, 2005. World Ichneumonoidea 2004. Taxonomy, Biology, Morphology and Distribution. CD/DVD. Taxapad, Vancouver, Canada. www.taxapad.com.

Hydrobiologia (2008) 595:535–543
DOI 10.1007/s10750-007-9132-z

FRESHWATER ANIMAL DIVERSITY ASSESSMENT

Global diversity of true and pygmy grasshoppers (Acridomorpha, Orthoptera) in freshwater

Christiane Amédégnato · Hendrik Devriese

Abstract A small percentage of Orthoptera Acridomorpha is comprised species dependant on continental water ecosystems. However, as phytophagous insects, they are important at the basis of the trophic chain, mainly in regions with large permanent biota resulting from the pulses of the rivers. An assessment of the composition and origin of the populations from different biogeographical regions is attempted, and the state of our current knowledge is given in Tables 1, 2 and 3.

Keyword Acridoidea · Tetrigoidea · Adaptation · Distribution · Ecology

Guest editors: E. V. Balian, C. Lévêque, H. Segers & K. Martens
Freshwater Animal Diversity Assessment

Electronic supplementary material The online version of this article (doi:10.1007/s10750-007-9132-z) contains supplementary material, which is available to authorized users.

C. Amédégnato (✉)
Departement Systématique et Evolution, Museum National d'Histoire Naturelle, Entomologie, 45 rue de Buffon, Paris 75005, France
e-mail: amedeg@mnhn.fr

H. Devriese
Institut Royal des Sciences Naturelles, Entomologie, 29 rue Vautier, Bruxelles 1000, Belgium

Introduction

The Orthoptera are primitive hemimetabolous insects. They are usually not considered to be aquatic insects. However, some of their members are somehow linked to freshwater habitats mainly by a relation to an aquatic host plant. Species that cannot develop without freshwater, especially for egg laying and nymphal development are considered as being primary inhabitants of freshwater biota. Species living in or at the border of continental freshwater habitats, but without being strictly linked to a dependant plant will only be shortly mentioned.

The freshwater Orthoptera community is mainly represented by Acridomorpha Acridoidea. However, some Tetrigoidea are water dependant, at least at some stages of their life cycle. Among Ensifera, the predator Katydid genus *Phlugis* specialises on nymphs of aquatic grasshoppers in South America.

Tetrigoidea

Most species of Tetrigoidea can be considered as limno-terrestrial. However, some groups are clearly dependent on water: they dive under water, swim very well and feed on aquatic algae. Since the life cycle of only a few species (perhaps 1%) of Tetrigoidea is known—most of them from Europe—it is currently extremely difficult to assess the degree of water dependency of this group.

Acridoidea

South America and Asia have the most important large river systems with considerable variations in water levels (10 m) which generate immense surfaces of freshwater biota. Especially in tropical South America, rich aquatic radiations of Acridoidea have been developed that are not found elsewhere. These radiations include insects living on:

- floating plants, attached or not to the bottom (two families and four genera);
- grasses and related plants species adapted to flooded zones (several tribes);
- palm trees specific of these biota (inundated tropical forest: Varzea, Igapo) (one subfamily);
- margins of inundation zones and shores of running waters, but with no other special need except for a link to the vegetation (marginal species of different groups). The Oriental and Afrotropical regions are less studied, but the fauna appears to be quite similar in diversity to the American fauna. On the other hand, the sub-Saharan region seems to be poorer; this is also the case for the palearctic fauna, which comprises only few species inhabiting marshes, moreover often not strictly linked to these ecosystems.

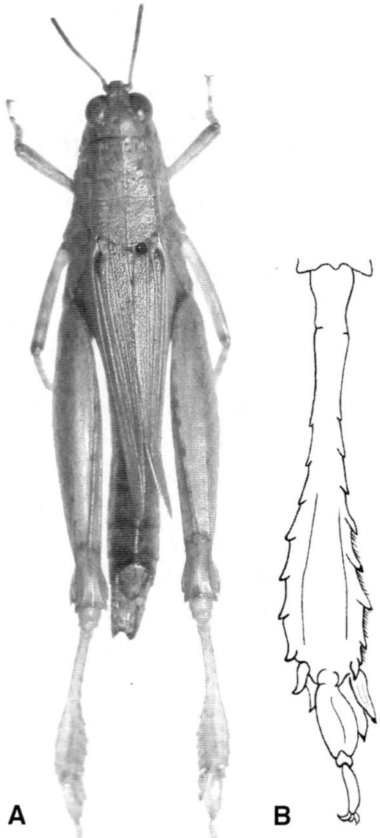

Fig. 1 (**A**) Habitus of *Marellia remipes* (Pauliniidae) (**B**) Specialisation of hind tibia and tarsus in *Marellia remipes*

Adaptations

The morphology of subaquatic species does not differ markedly from the general morphology of the group, with the exception of particular adaptations in some structures. These adaptations include a general morphology with a fusiform habitus for species living only in water and that are used to swim under the water (Fig. 1A). These species also developed a strong hind femora, expanded hind tibiae, including their spines and modified spurs (Fig. 1B). In some species, the presence of dense hairs on the margins of the tibiae also help to reinforce their efficiency to the aquatic habitat. This modification of the hind tibiae is more or less generalised within all the groups linked to water, but varies from weakly to strongly developed in the most aquatic species (Amédégnato, 1977; Carbonell, 1957; Roberts, 1978). Another adaptation is a morphological and chromatic mimicry to the plant habitus, often to Gramineae. Linked to physiology and ethology, occurs a modification of the ovipositor valves. Usually, grasshoppers' ovipositors are shaped to burrow in the soil. Water inhabiting grasshoppers have to oviposit within or at the surface of plants. Epiphylle egg lying is linked to a strong regression of the valves, while the endophytic egg lying, in the stems of the plants is linked to cutting edges, and depending on the behaviour of the species, it is also linked to different kinds of modifications of the upper and lower surfaces of the valves, modified to rasp the plant tissues.

Diversity of Orthoptera

The number of Orthoptera species is estimated to be over 25,000 (OSF2). They are distributed over all geographic zones, except Antarctica. The suborder Acridomorpha comprises Acrididea and Tetrigidea.

For the Tetrigidea 1,400 species have been described in 250 genera (OSF2).

Among Acrididea, only the superfamily Acridoidea is actually known to include families or subfamilies involved in the aquatic habitats. A recent estimate of the Acridoidea mentions 7,000 species in 1,500 genera (Otte, 1995a, b). With regard to this latest list, the groups concerned are the families Pauliniidae (neotropical) and Acrididae (world wide). It is within this last family that most of the genera with subaquatic adaptation have evolved, through several subfamilies, the most important being the neotropical Leptysminae and the Afrotropical and Oriental Oxyinae, Tropidopolinae and Hemiacridinae. Most Tetrigoidea groups—currently divided into nine subfamilies and one unassigned tribe (OSF2), but in strong need for a suprageneric revision—comprise limnoterrestrial species in all geographic zones, but only one, Scelimeninae, has true aquatic species, which occur in the Oriental region.

Phylogeny and historical process

Tetrigoidea

The oldest taxa that have been attributed to the Tetrigidae are *Archaeotetrix* and *Prototetrix*, from the lower Cretaceous of Transbaikalia (Sharov, 1971). These taxa have long elytra, a character that is not present in extant Tetrigidae. It is, therefore, not sure that these taxa belong to the same family. Apart from a poorly described taxon, *Succinotettix*, from the Baltic amber (Piton, 1938), no fossils of the family are known. Provisional analysis of the phylogeny indicates that the most primitive taxa have a leaflike pronotum with a split median carina (Devriese, unpublished). Such species occur on moist spots, but seldom near rivers. They have evolved into different lineages, of which the most aquatic are long-winged and have expanded hind tibia with reduced spines.

Acridoidea

Orthoptera of acridomorphoid type are known from the Carboniferous. However, extant groups are recent and do not antedate the Cenozoic (Amédégnato, 1993). As they are phytophagous insects, the coevolution with plants has led to a high degree of specialisation. Thus, in the whole group, ecology and phylogeny are narrowly linked, and groups associated to freshwater environments appear to be the result of such adaptive radiations. It is, especially, striking in the neotropics. Within Old world Acrididae, phylogenetic relations are not entirely clear.

The heterogeneity and bad definition of the subfamily Hemiacridinae (type genus very different from most of the other genera assigned to "Hemiacridinae"), is associated to the instability in the different sources of the lists provided for the Oxyinae, Tropidopolinae and "Hemiacridinae". As a result, the boundaries among these three subfamilies, in need of phylogenetic clarification, are imprecise. Thus, a complex OTH (Oxyinae, Tropidopolinae, Hemiacridinae) is used in the illustrations. However, Oxyinae, as well as some water dependant Hemiacridinae (inconstantly classified as Oxyinae), and some Tropidopolinae, could be closely related. In the Oriental and Afrotropical regions, water dependant grasshoppers could thus also result from a single radiation. However, phylogenetic studies and field work needs to be carried, in order to test this hypothesis.

At the present state of our knowledge, none of the groups mentioned are closely related to New world subfamilies. However, the morphological convergence is striking, as well for forest biota than for grassland biota.

Except for the Copiocerinae of palm trees, which could be distantly related to the wide-ranging Cyrtacantacridoid group, the bulk of Acrididae of humid biota belongs to only one subfamily, the Leptysminae, with two main radiations, linked to the main ecosystems: one mostly in forest biota, the Tetrataeniini, and the other in the grassland biota, the Leptysmini. It is clear that the two tribes originated in South America. The subfamily itself is part of a group of subfamilies of Acrididae also including the Ommatolampinae and the Rhytidochrotinae (Amédégnato, 1977), which have radiated in most environments.

Paulinia and *Marellia*, two monospecific genera, either considered as two subfamilies of Acrididae (Eades, 2000), partly *incertae sedis* (Carbonell, 2000), or gathered for practical reason in the same group Pauliniidae (Amédégnato, 1974; Amédégnato, 1977; Dirsh, 1961; Dirsh, 1975; Otte, 1995) are not closely related (Amédégnato, 1977; Carbonell, 2000). They only seem to be closer to the Acrididae than to another family of Acridomorpha (Amédégnato, 1977; Carbonell, 2000; Eades, 2000), but it is sure that they are not true actual Acrididae. Their high degree of divergence is probably linked to their narrow adaptation in a conservative biota, which allowed the survival of ancient phyla with no more relatives in the actual fauna, the reason why they are not related to any other subfamily.

A true continental fauna is characterised by its composition made of structured groups and subfamilies, all linked to each other in coherent entities, as the result of the biodiversity evolution. So, the insertion of heterogeneous elements (well-known paraphyletic Pauliniidae) into a coherent one (neotropical phyla of Acrididae) is only hampering the understanding of evolution. These two genera will continue here to be considered under the same name "Pauliniidae".

Present distribution and main areas of Endemicity (Fig. 2, 3)

Tetrigoidea (Tables 1, 2, 3)

Among the Tetrigoidea, only the "Scelimenae verae" (sensu Günther, 1938) seem, to be truly aquatic. They dive under water to hide and to feed on algae that grow on the underside of boulders (Bhalareo & Paranjape, 1986; Blackith & Blackith, 1987; Blackith, 1988; personal unpublished observations). However, an examination of the gut content of one species revealed few aquatic food (Reynolds et al., 1988). They consist of 79 species that occur in the Oriental and Australian regions, with a few also present on the Southeastern border of the Palearctic region.

Many other species are limno-terrestrial, and some have adaptations to swimming and are found at the margins of rivers and lakes. The Neotropical genera *Amorphopus*, *Eomorphopus* and *Crimisus*, the Oriental genera *Criotettix*, *Eucriotettix* and the Afrotropical *Morphopus* and *Afrocriotettix* all occur on very wet places adjacent to water bodies. In Europe, *Tetrix tüerki* is only found in wet-river valleys (Nadig, 1991; personal unpublished observations). But although all these species are encountered near rivers and lakes, it is not sure, whether they can be considered to be water dependent. In Western Africa, most species of the subfamily Tetriginae occur in seasonaly inundated grassland (Roy, 1982), but they are also found elsewhere.

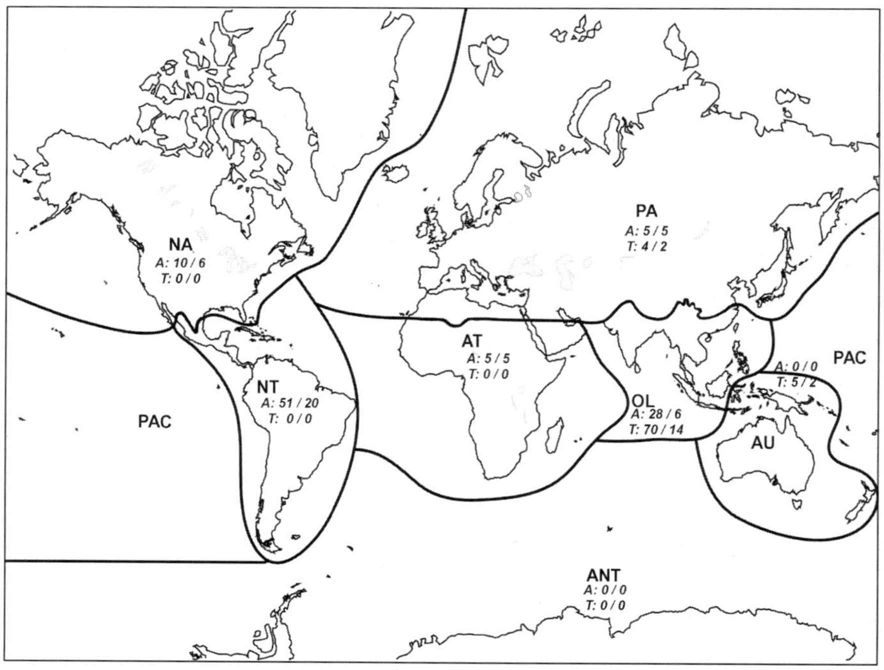

Fig. 2 Distribution of freshwater (aquatic + water dependant) Orthoptera Acridoidea (A) and Tetrigoidea (T) species and genera by zoogeographical region (species number/genus number). PA, Palaearctic Region; NA, Nearctic Region; NT, Neotropical Region; AT, Afrotropical Region; OL, Oriental region; AU, Australasian Region; PAC, Pacific Region Oceanic Islands

Table 1 Number of Freshwater Orthoptera species per zoogeographic region (Total FW, water dependent plus aquatic or subaquatic species; Wdpdt, Water dependent species). PA, Palaearctic Region; NA, Nearctic Region; AT, Afrotropical Region, NT, Neotropical Region; OL, Oriental Region; AU, Australasian Region; PAc Pacific Region and Oceanic Islands

Species number	PA		NA		NT		AT		OL		AU + PAC		World	
	Total FW	WDpt	Total FW	WDpt	Total FW	WDpt	Total FW	WDpt	Total FW	WDpt	Total FW	WDpt	Total FW	WDpt
ACRIDIDAE	5	5	10	10	52	48	14	14	28	27			107	102
Leptysminae			3	3	39	35							41	37
Copiocerinae					10	10							10	10
Oxyinae Hemiacridinae Tropidopolinae							5	5	28	27			32	31
Other subfamilies	5	5	7	7	3	3	9	9					24	24
PAULINIIDAE					2								2	
Acridoidea	5	5	10	10	54	48	14	14	28	27			109	102
Tetrigoidea	4	4							70		5	5	79	9
ORTHOPTERA	9	9	10	10	54	48	14	14	98	27	5	5	188	111

Acridoidea (Tables 1, 2, 3, Fig. 3)

In the Afrotropical and Oriental regions, where there are very few studies on aquatic Orthoptera, three subfamilies of Acrididae are involved in the colonisation of the marshy zones. They belong mainly to Oxyinae, Hemiacridinae and Tropidopolinae. In this region, Acridids of the main water surfaces are not known, with the exception, in India and Southwest Asia of *Gesonula punctifrons* (Oxyinae), which is described as the most aquatic species of Acridoidea, with egg pods deposited within Water Hyacinth, on which it also feeds (Kapur & Dutta, 1952; Sankaran, 1976; Sankaran et al., 1966). The inundation zones seem to be occupied by six other Oxyinae (Oxyini) and their related groups (members of the subfamilies Tropidopolinae and Hemiacridinae), but little is known about the life history of the species, and the species associations can hardly be recognised from the literature. For Oxyini, a total of five genera and 23 species are involved (*Gesonula punctifrons*, *Hygracris palustris; Oxya* and *Oxyina*: 18 species; *Quilta*: three species (Hollis, 1971, 1975).

The genus *Oxya*, the main component, extends over the whole paleotropical region but only one species occurs in Africa *(Oxya hyla)*. The same is true for the Hemiacridinae *Hieroglyphus*: 10 species mostly Asiatic with only one hygrophilous species *(H. daganensis)* in sub saharian Africa. On the contrary *Mesopsera filum* (Hemiacridinae) is only known from central Africa (Uvarov, 1977). Also mentioned for Africa are some members of the Tropidopolinae (*Tropidopola nigerica*) and another Hemiacridinae, *Spathosternum pygmaeum* in subsaharian Africa (Phipps, 1966, 1970). It is clear that the diversity is centred on India and South western Asia and that it is sharply decreasing towards Australia (one species) but also towards Africa. On the other hand, while the Asiatic fauna is represented by true hygrophilous Oxyiinae, such forms are poorly represented in Africa (five species, distributed in five genera and three subfamilies) (Phipps, 1970). Other species mentioned in the literature for Africa (COPR 1982; Fishpool & Popov, 1984; Uvarov, 1977) belong mostly to groups mainly adapted to other ecosystems: Eyprepocnemidinae with *Cataloipus*, *Phyllocercus* and *Eyprepocnemis*, Acridinae with *Duronia chloronata*, *Orthoctha grossa*, *Jasomenia sansibarica* (Phipps, 1966), *Calephorus compressicornis*, *Paracinema*

Table 2 Number of freshwater orthoptera genera (water dependant plus aquatic or subaquatic species). PA, Palaearctic Region; NA, Nearctic Region; AT, Afrotropical Region; NT, Neotropical Region; OL, Oriental Region; AU, Australasian Region; PAc Pacific Region and Oceanic Islands

GN: Genera Number	PA	NA	NT	AT	OL	AU + PAc	Total
ACRIDIDAE	5	6	18	5	6		34
Leptysminae		3	15				15
Copiocerinae			3				3
Oxyinae Hemiacridinae Tropidopolinae				5	6		9
Other subfamilies	5	3					**7**
PAULINIIDAE			2				**2**
Acridoidea	5	6	20	5	6		36
Tetrigoidea	2				14	2	14
ORTHOPTERA	7	6	20	5	20	2	50

Table 3 Taxonomic origin of water dependant Acridoidea and Tetrigoidea: a biogeographic perspective within the spectrum of the group diversity. Total = FW + Marine + terrestrial species, FW = Aquatic/subaquatic + water dependent. Species numbers are widely provisional; the source is Otte (1995a, b); For Acridoidea, the Ommexechidae, in the Neotropical region are not included, as well as some small groups in fact belonging to Pamphagoidea, in all continents

Species number	PA		NA		NT		AT		OL		AU + PAC		World	
	Total	FW	Total	FW	Total	FW	Total	FW	Total	FW	Total	FW	Total	Fw
Acrididae	596	5	797	10	1083	52	1459	14	1776	28	572		6283	107
Leptysminae			3	3	80	39							80	41
Copiocerinae					65	10							65	10
Oxyinae, Hemiacridinae Tropidopolinae	42						117	5	58	28	106		323	32
Other subfamilies	554	5	794	7	938	(3)	1342	(9)	1718	?	466		5812	24
Pauliniidae					2	2							2	2
Acridoidea	596	5	824	10	1539	54	1459	14	1776	28	572		6766	109
Tetrigoidea[a]	140	4	20	0	195	0	200	0	700	70	145	5	1400	79
ORTHOPTERA	736	9	844	10	1734	54	1659	14	2476	98	717	5	8166	188

[a] N.B. The number of species occurring on the edge of the Palaearctic and the Oriental region in China being substantial, the figures for both regions have been based on the number of species in resp. predominantly Palearctic (e.g., *Tetrix* and *Formosatettix*) and Oriental regions

tricolor and *Mesopsis abbreviatus*. The poverty of the fauna seems to be real, but it cannot be excluded that, in Africa, flood plains have not been sufficiently studied, notably those of the central basin of Congo. Also, the localisation of the others inundation zones in highlands or in desertic environments is probably determinant in hampering the ecological diversification.

The Palearctic and Nearctic zones have closely related and poor faunas. The Palearctic region, close to the Oriental one, suffered a drastic impoverishment, notably in Oxyiinae. The groups involved are of wide distribution, mainly euryapt and are only marginally adapted to freshwater habitats, by few species. It is the case for:

- the Melanoplinae (nearctic bush subfamily) with two genera and several species: *Gymnoscirtetes pusillus*, *Paroxya clavuliger*, *P. atlantica*, in the South east of the USA (Squitier & Capinera, 2002) and *Paroxya hoosieri*, in the great lakes region (Bright, 2006);
- the Gomphocerinae (a world wide ranging subfamily of grassland insects) with the genus *Stethophyma*, adapted to marshes and peat bogs

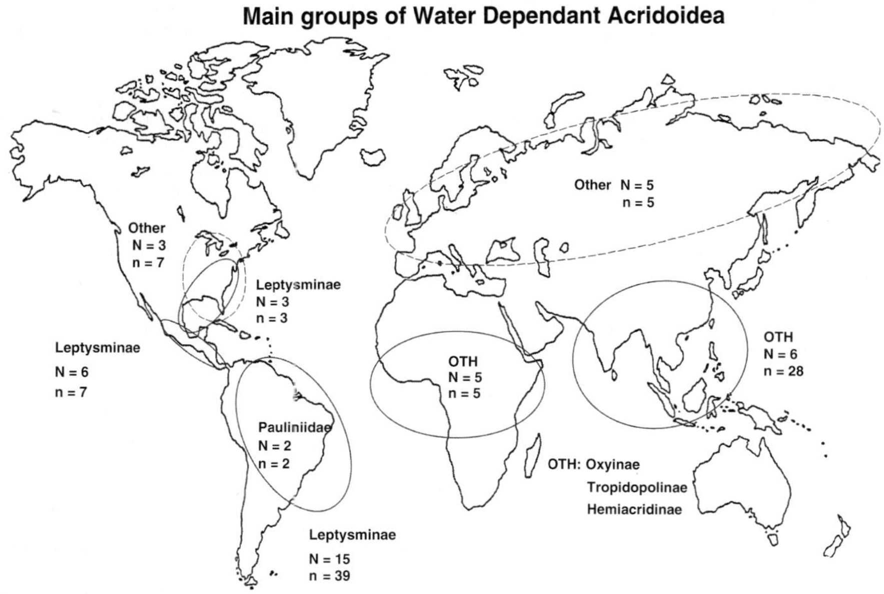

Fig. 3 Diversity and endemism of the main groups of water dependant grasshoppers. *N*: number of genera; *n*: number of species; OTH: subfamilies Oxyinae, Tropidopolinae, Hemiacridinae

with *Stethophyma grossum* (palearctic) and the neartic *S. gracilis* and *S. lineata* (Bright, 2006);
- the Acridinae, with *Epacromius tergestinus* and *Paracinema tricolor* (palearctic).

However, the Nearctic region has benefited from the colonisation of some members of the South American fauna that have reached the Southeast of North America (three genera, three species of Leptysmini, Leptysminae).

The Neotropical region is very rich, all the groups being completely endemic and none of them appearing to be directly related to the Old World *Taxa*. Typically, aquatic biota basically gather into the following groups: Pauliniidae, Acrididae, Leptysminae (Tetrataeniini and Leptysmini) and, in Amazonia, the group Copiocerae of Copiocerini (Acrididae, Copiocerinae). The other components are not constant. Inhabitants of water and floating plants are the Pauliniidae and the Tetrataeniini (*Cornops*); Water flooded plants around (Gramineae, Cyperaceae and Typhaceae) harbour Tetrataeniini (*Cornops* and *Haroldgrantia*) and Leptysmini: *Leptysma, Tucayaca, Stenacris, Leptysmina, Belosacris and Cylindrotettix*. They all have different ways to share this restricted habitat; for exemple *Typha* harbour egg lying for *Cornops paraguayense, Haroldgrantia lignosa*, and *Leptysmina gracilis* but these genera seem to have different habits for food resources (Turk & Aquino, 1995, 1996, 1998). At the same time, riverine palm trees support Copiocerae (Copiocerinae) in Amazonia. At the level of the continent, the group evolution in freshwater environments, besides being linked to plants, is markedly influenced by the paleogeography of South and Central America. Notably, the Andes was a barrier to most of the genera. Only the most mobile sub aquatic grasshoppers with a behaviour allowing a better active (flying) or passive (plant gramineous rafts) dispersal, reach Panama and the rest of central America. This is the case for several genera of Leptysmini which are present far to the north (seven species, sometimes only represented by subspecies, vicariant of their South American homologous) in comparison with the 27 present in South America. Among the Tetrataeniini, only *Cornops aquaticum* reaches Central America.

Human related issues

Afrotropical and Oriental, but also palearctic inundation zones, especially the seasonally flooded delta plains have to be mentioned here, as they are the outbreak zones of the important swarming locusts *Locusta migratoria*, in Asia, Africa, but also in the palearctic zone, and of *Nomadacris septemfasciata* in sub-Saharan East Africa; However, water dependant grasshoppers are concerned by human activities in different ways. The marsh species, which markedly specialised on Gramineae, and sometimes precisely on *Oriza*, have naturally become pests in rice's cultures. In Asia, most of the species belong to *Oxya*, *Gesonula*, *Hieroglyphus* and *Quilta*. In the neotropics, Leptysmini are also found in rice fields, but do not actually seem to be very important pests. In contrast, the neotropical species living on the floating plants with a significant feeding impact on their host (specially *Paulinia acuminata* and *Cornops aquaticum*) have been studied with the purpose of introduction as potential biological control agents. In fact, during the last century, their host plants have been introduced in the Nearctic and paleotropical regions and, without natural enemies they became invasive. *Paulinia acuminata* has been released in different lakes of Africa and also in India and Sri lanka to control *Salvinia*. *Cornops aquaticum* is still under study for water Hyacinth control as it is able to cause important damages followed by secondary pathogenetic infestations. It has recently been restrictedly released under control in South Africa (Hill & Oberholzer, 2000; Hill & Olckers, 2001; Oberholzer & Hill, 2001) Another side of human activity impact on water dependant grasshoppers is negative. Reclamation of marshes and riverine zones causes the disappearance of some species. In the palearctic region this is the case of *Epacromius tergestinus* but also of *Stetophyma grossum* and *Tetrix tuerki*. In the tropical regions, only the most common species are currently known, even if several species of *Oxya* were described only recently. A significant number of genera and species are known by very few specimens, if not only the type (they were not mentioned here); many other are still unknown, because of the lack of studies in these biota, which conceal particular microhabitats which completely escape to our knowledge. It is particularly, the case of mountain rivers banks.

References

Amédégnato, C., 1974. Les genres d'acridiens néotropicaux, leur classification par familles, sous familles et tribus. Acrida 3: 113–204.

Amédégnato, C., 1977. Etude des Acridoidea centre et sud américains (Catantopinae sensu lato), Anatomie des génitalia, Classification, répartition, phylogénie. Universite Pierre et Marie Curie, Paris: 385 pp.

Amédégnato, C., 1993. African-American relationships in the Acridians (Insecta, Orthoptera). In George, W. & R. Lavocat (eds), The Africa-South America Connection. Oxford Science publications, Clarendon Press, Oxford: 59–75.

Bennett, F. D., 1977. Insects as agents for biological control of aquatic weeds. Aquatic Botany 3: 165–174.

Bhalareo, A. M. & S. Y. Paranjape, 1986. Studies on the Bio-ecology of a grouse-locust Euscelimena harpago Serv. (Orthoptera: Tetrigidae). Geobios 13: 145–150.

Blackith, R. E., 1987. Primitive Orthoptera and Primitive Plants. In Baccetti, B. (ed.), Evolutionary Biology of Orthopteroid Insects. Ellis Horwood Ltd., Chichester: 124–126.

Blackith, R. E. & R. M. Blackith, 1987. Tridactylids and Tetrigids (Orthoptera) from Sulawesi, Indonesia. Tijdschrift voor Entomologie 130: 1–10.

Blackith, R. E., 1988. The Tetrigidae (Orthoptera) of Sri Lanka. Entomologica Scandinavica, Supplement 30: 91–107.

Bright, E., 2006. http://insects.ummz.lsa.umich.edu/~ethanbr/aim/sp/Orthoptera/sp_orthom.html.

Carbonell, C. S., 1957. The external anatomy of the South American semiaquatic grasshopper *Marellia remipes* Uvarov (Acridoidea, Pauliniidae). Smithsonian Miscellaneous Collections 137, 61–97, 61 pl.

Carbonell, C. S., 2000. Taxonomy and a study of the phallic complex, including its muscles, of *Paulinia acuminata* (Acrididae, Pauliniinae) and *Marellia remipes* (Acrididae, incertae sedis). Journal of Orthoptera Research 9: 161–180.

COPR, 1982. Centre for Overseas Pest Research: The Locust and Grasshopper Agricultural Manual, 690 pp.

Dirsh, V. M., 1961. A preliminary revision of the families and subfamilies of Acridoidea (Orthoptera, Insecta). Bulletin of the British Museum (N.H.) Entomology 10: 351–419.

Dirsh, V. M., 1975. Classification of the Acridomorphoid Insects. E.W.Classey, Faringdon: vii + 171 pp.

Eades, D. C., 2000. Evolutionary relationships of phallic structures of Acridomorpha (Orthoptera). Journal of Orthoptera Research 9: 181–210.

Fishpool, L. D. C. & G. B. Popov, 1984. The grasshopper faunas of the savannas of Mali, Niger, Benin and Togo. Bulletin de l' Institut Fondamental d' Afrique Noire, Série A 43: 275–410.

Günther, K., 1938. Revision der Acrydiinae, I. Mitteilungen aus dem Zoolog. Museum in Berlin 23(2): 299–437.

Hill, M. P. & I. G. Oberholzer, 2000. Host-specificity of the grasshopper *Cornops aquaticum*, a natural enemy of water hyacinth. In Spencer, N. R. (ed), Proceedings of the 10th International Symposium on Biological Control of Weeds: 349–356.

Hill, M. P. & T. Olckers, 2001. Biological control initiatives against water hyacinth in South Africa: constraining factors, success and new courses of action. In Julien, M. H., M. P. Hill, T. D. Center & J. Ding (eds), Proceedings of the Meeting of the Global Working Group for the Biological Control and Integrated Control of Water Hyacinth, Beijing, China, 9–12 October 2000. Australian Centre for International Agricultural Research, Canberra, Australia: 33–38.

Hollis, D., 1971. A preliminary revision of the genus *Oxya* Audinet-Serville (Orthoptera: Acridoidea). Bulletin of the

British Museum (Natural History), Entomology Series 26, London: 267–343.

Hollis, D., 1975. A review of the subfamily Oxyinae (Orthoptera: Acridoidea). Bulletin of the British Museum (Natural History), Entomology Series, London 31: 189–234.

Kapur, A. P. & D. K. Dutta, 1952. Studies on the bionomics of a semi aquatic grasshopper *Gesonula punctifrons* Stal Proceeding of the 39th Indian Sci. Congress Calcutta pt 3, Calcutta: 339.

Nadig, A., 1991. Die Verbreitung der Heuschrecken (Orthoptera: Saltatoria) auf einem Diagonalprofil durch die Alpen (Inntal-Maloja-Bregaglia-Lago di Como-Furche). Jahresbericht der Naturforschenden Gesellschaft Graubünden. NF 106, 2. Teil: 1–380.

Oberholzer, I. G. & M. P. Hill, 2001. How safe is the grasshopper *Cornops aquaticum* for release on water hyacinth in South Africa? In Julien, M. H., M. P. Hill, T. D. Center, & J. Ding (eds), Proceedings of the Meeting of the Global Working Group for the Biological Control and Integrated Control of Water Hyacinth, Beijing, China, 9–12 October 2000. Australian Centre for International Agricultural Research, Canberra, Australia: 82–88.

OSF2: Eades, D. C., D. Otte & P. Naskrecki. Orthoptera Species File Online. Version 2.3 [21.07.2006]. <http://osf2.orthoptera.org>.

Otte, D., 1995a. Orthoptera species file. 5. Grasshoppers [Acridomorpha]. D. Acridoidea, viii + 630 pp, and references there in.

Otte, D., 1995b. Orthoptera species file. 4. Grasshoppers [Acridomorpha]. C. Acridoidea, vii + 518 pp, and references there in.

Phipps, J., 1966. The habitat and seasonal distribution of some east African grasshoppers. Proceedings of the Royal Entomological Society of London (A) 41: 25–36.

Phipps, J., 1970. Notes on the biology of grasshoppers (Orthoptera: Acridoidea) in Sierra Leone. Journal of Zoology 161: 317–349.

Piton, L., 1938. *Succinotettix Chopardi* Piton, Orthoptère (Tetricinae) inédit de l'ambre de la Baltique. Bulletin de la Société entomologgique de France 43: 226–227.

Reynolds, J. D., R. E. Blackith & R. M. Blackith, 1988. Dietary observations on some tetrigids (Orthoptera: Caelifera) from Sulawesi (Indonesia). Journal of Tropical Ecology 4: 403–406.

Roberts, H. R., 1978. A revision of the tribe Leptysmini except the genus *Cylindrotettix* (Orthoptera: Acrididae: Leptysminae). Proceedings of the Academy of Natural Sciences of Philadelphia 129: 33–69.

Roy, R., 1982. Recherches scientifiques dans les parcs nationaux du Sénégal. VIII. Orthoptera Tetrigidae du parc national du Niokolo-Koba. Mémoires de l'IFAN, 92: 133–139.

Sankaran T., 1976. Possibilities of Biological Control of the Aquatic Weeds *Eichornia crassipes* and *Salvinia auriculata* in India. In Varshnay C. K. & J. Rzoska (eds), Aquatic Weeds in S.E. Asia. Dr. W. Junk B.V., The Hague: 323–329.

Sankaran, T., D. Srinath & K. Krishna, 1966. Studies on *Gesonula punctifrons* Stal (Orthoptera: Acrididae, Cyrtacanthacridinae) attacking water hyacinth in India. Entomophaga 11: 433–440.

Sharov, A. G., 1971. Phylogeny of the Orthopteroidea. Israel Program for Scientific translations, Jerusalem: 1–251.

Squitier, J. M. & J. L. Capinera, 2002. Habitat associations of Florida grasshoppers (Orthoptera: Acrididae). Florida Entomologist 85: 235–244.

Turk, S. Z. & A. L. Aquino, 1995. ACRIDOIDEOS DEL N.O.A. VII : Estudios bioecológicos en *Haroldgrantia lignosa* Carbonell, Ronderos y Mesa (Acrididae:Leptysminae: Tetrataeniini).Un nuevo caso de oviposición endofítica en el noroeste argentino. Acta Zoologica Lilloana 43: 99–103.

Turk, S. Z. & A. L. Aquino, 1996. ACRIDOIDEOS DEL N.O.A. VIII: Nuevo aporte a la bioecología y disribución del género *Cornops* Stal: *Cornops paraguayense* (Br.). (Acrididae: Leptysminae:Tetrataeniini). Acta Zoologica Lilloana 43: 427–432.

Turk, S. Z. & A. L. Aquino, 1998. ACRIDOIDEOS DEL N.O.A. IX: Contribución al conocimiento de los Leptysminae Neotropicales: Ciclo de vida de *Leptysmina gracilis* Bruner (Acrididae: Leptysminae: Leptysmini). Acta Zoologica Lilloana 44: 185–190.

Uvarov, B. P., 1977. Grasshoppers and locusts. A handbook of general acridology, Vol. 2., behaviour, ecology, biogeography, population dynamics, London ix + 613 pp.

Hydrobiologia (2008) 595:545–567
DOI 10.1007/s10750-007-9034-0

FRESHWATER ANIMAL DIVERSITY ASSESSMENT

Global diversity of fish (Pisces) in freshwater

C. Lévêque · T. Oberdorff · D. Paugy · M. L. J. Stiassny · P. A. Tedesco

Abstract The precise number of extant fish species remains to be determined. About 28,900 species were listed in FishBase in 2005, but some experts feel that the final total may be considerably higher. Freshwater fishes comprise until now almost 13,000 species (and 2,513 genera) (including only freshwater and strictly peripheral species), or about 15,000 if all species occurring from fresh to brackishwaters are included. Noteworthy is the fact that the estimated 13,000 strictly freshwater fish species live in lakes and rivers that cover only 1% of the earth's surface, while the remaining 16,000 species live in salt water covering a full 70%. While freshwater species belong to some 170 families (or 207 if peripheral species are also considered), the bulk of species occur in a relatively few groups: the Characiformes, Cypriniformes, Siluriformes, and Gymnotiformes, the Perciformes (noteably the family Cichlidae), and the Cyprinodontiformes. Biogeographically the distribution of strictly freshwater species and genera are, respectively 4,035 species (705 genera) in the Neotropical region, 2,938 (390 genera) in the Afrotropical, 2,345 (440 genera) in the Oriental, 1,844 (380 genera) in the Palaearctic, 1,411 (298 genera) in the Nearctic, and 261 (94 genera) in the Australian. For each continent, the main characteristics of the ichthyofauna are briefly outlined. At this continental scale, ichthyologists have also attempted to identify ichthyological ''provinces'' that are regions with a distinctive evolutionary history and hence more or less characteristic biota at the species level. Ichthyoregions are currently identified in each continent, except for Asia. An exceptionally high faunal diversity occurs in ancient lakes, where one of the most noteworthy features is the existence of radiations of species that apparently result from intralacustrine speciation. Numerous fish-species flocks have been identified in various ancient lakes that are exceptional natural sites for the study of speciation. The major threats to fish biodiversity are intense and

Guest editors: E. V. Balian, C. Lévêque, H. Segers & K. Martens
Freshwater Animal Diversity Assessment

Electronic supplementary material The online version of this article (doi:10.1007/s10750-007-9034-0) contains supplementary material, which is available to authorized users.

C. Lévêque (✉) · T. Oberdorff · D. Paugy
Institut de Recherches pour le Développement (UR-IRD 131), Muséum National d'Histoire Naturelle DMPA-USM 403, CP 26, 43 rue Cuvier, Paris 75005, France
e-mail: cleveque@mnhn.fr

M. L. J. Stiassny
Department of Ichthyology, American Museum of Natural History, 81st Street & Central Park West, New York, NY 10024, USA

P. A. Tedesco
UMR CNRS 5023, UR-IRD 131, Université Lyon 1, Bat. Forel, 43, Bd. du 11 Novembre 1918, 69622 Villeurbanne, Cedex, France

have been relatively well documented: overexploitation, flow modification, destruction of habitats, invasion by exotic species, pollution including the worldwide phenomena of eutrophication and sedimentation, all of which are interacting.

Keywords Fish · Freshwater · Distribution · Diversity · Ichthyoregions · Global scale

General issues

The word fish is applied to a heterogeneous grouping of aquatic chordates comprised of hagfishes and lampreys, sharks, rays and chimaeras, and the finned bony fishes. The latter is by far the most diverse group and is well represented in freshwaters, while the others are predominantly marine groups.

Each continent has a distinctive freshwater fish fauna and the observed patterns of fish distribution (see summary in Berra, 2001) are the result of physical barriers disrupting past fish dispersal, as well as to difference in temperature adaptations amongst the different groups. Most species occur in the tropical and subtropical regions and there is an overall reduction in diversity towards temperate and polar regions. Although some temperate regions, particularly those that were never glaciated are relatively rich in species, the continental areas that have experienced glaciations, such as northern North America, Europe and Asia, tend to have relatively depauperate fish faunas.

The freshwater fishes of the equatorial zone are extremely diverse and are not readily characterised by any particular clades. While the freshwaters of the northern temperate/cold regions are characterised by salt-tolerant salmonids, sturgeons, smelts, northern lampreys and several primary families including pikes, leuciscine cyprinids and perches. The southern temperate/cold regions have a low diversity of fishes including salt-tolerant galaxids and southern lampreys. Most oceanic islands are inhabited by species of predominately marine groups that have adapted to (or remained in) freshwaters.

Freshwater fishes, which tend to be more-or-less confined to drainage systems, provide a relatively conservative system for examining patterns of distribution that may reflect the imprint of past continental and climate changes.

Global species and taxonomic diversity

Fresh, brackish and saltwater fishes

Ichthyologists used to distinguish three major groups of freshwater fish according to their tolerance to saltwater and their hypothesised ability to disperse across marine barriers (Myers, 1949): the *primary division* fish being strictly intolerant of salt water; the *secondary division* able occasionally to cross narrow sea barriers; and the *peripheral division* including representatives of predominantly marine families that have colonised inland waters from the sea. This classification scheme has been challenged partly because of the subjectivity in distinguishing between divisions. However, the scheme is still widely used by many fish biogeographers and has the advantage of common usage. The popular internet site FishBase (http://www.fishbase.org) adheres to a slightly different classification with fresh and brackishwater fish species falling into three categories: (1) exclusively freshwater, (2) occurring in fresh and brackishwaters, (3) or in fresh, brackish and marine waters. The first category covers more or less the primary and secondary divisions of Myers, while categories 2 and 3 cover the peripheral division.

Global estimates of fish species diversity

The precise number of extant fish species remains to be determined. However, since Linneaus' listing of 478 species of teleost fish in 1758, our knowledge has increased considerably and some global estimates are available.

The Catalog of Fishes established by Eschmeyer (2005) provided an estimate of 27,300 valid fish species, with a prediction of about 31,500 species when all inventories are completed (Berra, 2001). In September 2005, 28,900 species were already listed in FishBase. Nelson (2006) suggested a total of almost 28,000 species (freshwater and marine), which is 51% of the 54,711 then recognised living vertebrate species. The eventual number of extant fish species may be projected to be close to, conservatively, 32,500 (Nelson, 2006). At the global scale, the fresh and brackishwater fish belong to 207 families and 2,513 genera.

Tables 1 and 2 provide an estimate of the number of fish species inhabiting inland waters, by continents

Table 1 Fresh and brackishwater fish species richness by continents or large sub-continental units (based on Fishbase, September 2005)

	Freshwater		Brackish/salt		Total	
	Families	Species	Families	Species	Families	Species
Africa	48	2,945	66	295	89	3,240
Asia	85	3,553	104	858	126	4,411
Europe	23	330	36	151	43	481
Russia	28	206	28	175	40	381
Oceania	41	260	74	317	85	577
North America	74	1,411	66	330	95	1,741
South America	74	4,035	54	196	91	4,231
Total		12,740		2,322		15,062

or large sub-continental units, recorded in FishBase. Nelson (2006) listed only 11,952 strictly freshwater species, and 12,457 using freshwater. FishBase uses Eschmeyer's classification and the difference with Nelson's estimate may result differing views on the definition of species. The unclear status "brackish species" may probably explain the differences in the total number of fish species using freshwaters (12,457 according to Nelson) or inhabiting fresh and brackishwaters (15,062 according to Fishbase) (Table 2).

Following Nelson (2006) and Eschmeyer (2005) about 40–43% of all fishes occur in, or almost always in, freshwaters. The current data from FishBase provide an even higher figure of 45%. Whatever the precise number, it is noteworthy that the estimated 13,000 freshwater species live in lakes and rivers that cover only 1% of the earth's surface, while the remaining 16,000 species live in marine habitats which cover a full 70%.

Taxonomic diversity

Table 2 provides an evaluation of the number of families and species, of inland water fishes in different taxonomic orders. Altogether fresh and brackishwater species are included in about 207 families (170 for strictly freshwater fish). The bulk of families and species occur in a few groups: the ostariophysan Characiformes, Cypriniformes, Siluriformes and Gymnotiformes, the Perciformes (including the family Cichlidae) and the Cyprinodontiformes. While supraspecific taxonomic ranks such as that of "family" are arbitrary, nonetheless family diversity is generally a reasonable indicator of taxonomic (i.e. species) diversity.

At the level of the biogeographic realms (Tables 3, 4) and taking into account only fully freshwater fish families (i.e. the primary and secondary divisions), the largest number of families by far (43) is found in the Neotropical region, with a high proportion of endemic families (33% or 77%) mainly belonging to the orders Characiformes and Siluriformes. Then, follows the Oriental region (33 families, 15 endemic) and the Afrotropical region (32 families, 17 endemic). The Nearctic and Palaearctic regions are relatively depauperate, as a result of Quaternary climatic events: 22 families in the Nearctic region (nine endemic) and 17 families in the Palaearctic (with a single endemic family). Figure 1 provides an approximation of the worldwide distribution of selected freshwater fish groups and illustrates the existence of groups distributed only in the North, and groups more-or-less widely distributed in the inter-tropical zone.

For strictly freshwater fishes, at the generic and species levels in the different biogeographic realms (Fig. 2A) the overall pattern is quite similar to that at the family level with 4,035 species (705 genera) in the Neotropical region, 2,938 (390 genera) in the Afrotropical, 2,345 (440 genera) in the Oriental, 1,844 (380 genera) in the Palaearctic, 1,411 (298 genera) in the Nearctic, and 261 (94 genera) in the Australian.

When taking into account the fresh and brackishwater fishes (Tables 3, 4; Fig. 2B), the figures are, respectively, 4,231 species (769 genera) in the Neotropical region, 3,272 (542 genera) in the Afrotropical, 2,948 (609 genera) in the Oriental, 2,381

Table 2 Number of families and species for fish orders with representatives in fresh and brackishwater. Data from FishBase (September 2005)

Class	Order	Fresh			Fresh–brackish		
		FishBase, 2005		Nelson, 2006	FishBase, 2005		Nelson, 2006
		Families	Species	Species	Families	Species	Species
Holocephali (chimaeras)	Chimaeriformes	1	1	0	1	1	0
Cephalaspidomorphi (lampreys)	Petromyzontiformes	1	33	29	2	57	38
Elasmobranchii (sharks and rays)	Carcharhiniformes			1	1	13	8
	Orectolobiformes			0	1	2	0
	Pristiformes			0	1	24	1
	Pristiophoriformes			0			1
	Rajiformes	2	24	0	3	35	2
	Myliobatiformes			23			28
Sarcopterygii (lobe-finned fishes)	Ceratodontiformes[a]	2	8	6	2	8	6
Actinopterygii (ray-finned fishes)	Acipenseriformes	2	8	14	2	56	27
	Albuliformes			0	1	5	0
	Amiiformes	1	1	1	1	1	1
	Anguilliformes	2	8	6	5	76	26
	Atheriniformes	7	181	210	7	224	240
	Batrachoidiformes	1	5	6	1	9	7
	Beloniformes	3	71	98	3	132	104
	Characiformes	17	1794	1674	17	1801	1674
	Clupeiformes	5	72	79	5	209	85
	Cypriniformes	7	3451	3268	7	3664	3268
	Cyprinodontiformes	9	964	996	9	1096	1008
	Elopiformes			0	2	12	7
	Esociformes	2	15	10	2	20	10
	Gadiformes			1	2	10	2
	Gasterosteiformes	2	13	21	2	30	43
	Gobiesociformes	1	9	0	1	9	0
	Gonorynchiformes	2	31	31	3	36	32
	Gymnotiformes	5	133	134	5	133	134
	Hiodontiformes			2			2
	Lepisosteiformes	1	4	6	1	7	7
	Lophiiformes			0	1	2	0
	Mugiliformes			1			7
	Ophidiiformes	1	4	5	1	6	6
	Osmeriformes	3	31	82	5	82	86
	Osteoglossiformes	7	219	218	7	221	218
	Perciformes	34	2402	2040	51	3368	2335
	Percopsiformes	3	9	9	3	9	9
	Pleuronectiformes	4	23	10	5	81	20
	Polyptériformes	1	16	16	1	17	16
	Salmoniformes	1	161	45	1	295	66
	Scorpaeniformes	4	75	60	6	105	62
	Siluriformes	34	2835	2740	34	2992	2750

Table 2 continued

Class	Order	Fresh			Fresh–brackish		
		FishBase, 2005		Nelson, 2006	FishBase, 2005		Nelson, 2006
		Families	Species	Species	Families	Species	Species
	Synbranchiformes	3	90	96	3	105	99
	Syngnathiformes	1	20	0	1	61	0
	Tetraodontiformes	1	29	14	1	48	22
Total		170	12740	11952	207	15062	12457

[a] Ceratodontiformes include Lepidosireniformes

Table 3 Number of fresh and brackishwater fish species per biogeographic realm (data from Fishbase, September 2005)

Order	PA	NA	NT	AT	OL	AU
Chimaeriformes	–	–	1	–	–	–
Petromyzontiformes	17	24	2	1	–	1
Carcharhiniformes	3	2	2	2	3	3
Orectolobiformes	1	–	–	1	1	1
Pristiformes	4	4	4	5	4	4
Pristiophoriformes[a]						
Rajiformes	1	1	18	4	9	3
Myliobatiformes[b]						
Ceratodontiformes[c]	–	–	1	7	–	1
Acipenseriformes	23	10	1	1	–	–
Albuliformes	1	1	1	1	1	1
Amiiformes	–	1	–	–	–	–
Anguilliformes	9	2	4	14	29	23
Atheriniformes	2	55	31	15	76	41
Batrachoidiformes	–	4	5	–	–	–
Beloniformes	9	12	14	4	83	14
Characiformes	1	88	1493	212	–	3
Clupeiformes	34	24	32	38	60	17
Cypriniformes	1394	392	17	539	1381	15
Cyprinodontiformes	30	377	346	309	10	12
Elopiformes	4	1	1	3	3	3
Esociformes	7	9	–	1	–	–
Gadiformes	3	4	–	–	–	–
Gasterosteiformes	11	7	–	1	3	–
Gobiesociformes	–	6	3	–	–	–
Gonorynchiformes	1	1	1	32	1	1
Gymnotiformes	–	7	126	–	–	–
Hiodontiformes[d]						
Lepisosteiformes	–	7	–	–	–	–
Lophiiformes	–	–	–	–	1	1
Mugiliformes[a]						

Table 3 continued

Order	PA	NA	NT	AT	OL	AU
Ophidiiformes	–	5	1	–	–	–
Osmeriformes	31	10	7	1	9	25
Osteoglossiformes	–	2	3	208	7	1
Perciformes	348	455	413	1274	604	327
Percopsiformes	–	9	–	–	–	–
Pleuronectiformes	14	11	21	9	24	8
Polypteriformes	–	–	–	17	–	–
Salmoniformes	161	47	11	6	1	7
Scorpaeniformes	61	31	–	–	4	4
Siluriformes	189	121	1666	496	535	45
Synbranchiformes	5	4	2	52	43	3
Syngnathiformes	10	6	2	12	24	12
Tetraodontiformes	7	1	2	7	32	4
Total	2381	1741	4230	3272	2948	580

[a] Mugiliformes & Pristiophoriformes are considered as stricly marine in Fishbase

[b] Myliobathiformes are included in Rajioformes in Fishbase

[c] Ceratodontiformes include Lepidosireniformes

[d] Hiodontiformes are included in Osteoglossiformes in Fishbase

(551 genera) in the Palaearctic, 1,741 (402 genera) in the Nearctic, and 580 (1,232 genera) in the Australian.

Some general characteristics of the fish fauna at the global scale

Many hypotheses have been proposed to explain spatial variability species richness at broad spatial scales (see for example Ricklefs, 2004 for a review). These may be grouped into three main hypotheses: the first, the "area" hypothesis, states that species richness increases as a function of surface area through size-dependent extinction/colonisation rates (MacArthur & Wilson, 1967) and/or habitat diversity (MacArthur, 1964). The second, the "productivity" hypothesis (Wright, 1983) predicts that species richness of a region will be positively correlated with the total energy available. The third, the "historical" hypothesis, which includes many variants (Ricklefs, 2004), explains diversity patterns by differential speciation or extinction rates, coupled with dispersal limitation, due to historical contingency. Concerning freshwater fishes it is clear that area per se explains a large portion of richness variability. In rivers, species richness increases with area (basin area and amount of discharge) as it generally does in terrestrial biomes (Hugueny, 1989; Oberdorff et al., 1995; Guégan et al., 1998) (Fig. 3). Nevertheless, when area is controlled for, energy availability and history also explain a significant part of the observed richness patterns, even if the contribution of the latter is generally weaker (Oberdorff et al., 1995, 1997; Tedesco et al. 2005).

Phylogeny and historical processes

The present distribution of freshwater fishes has been shaped by millions of years of changes in the global water cycle. In relation to climate change, the nature and dynamics of surface freshwater systems have evolved continuously, at various spatial and temporal scales. Many of the surface freshwater systems have, therefore, been transient; their fauna and flora usually disappeared when the systems disappeared, or were able to survive by developing adaptations to the changing circumstances. The dual processes of speciation and extinction have interacted with climatic and geological events that have both isolated fish populations and provided opportunities for migration and colonisation of new habitats.

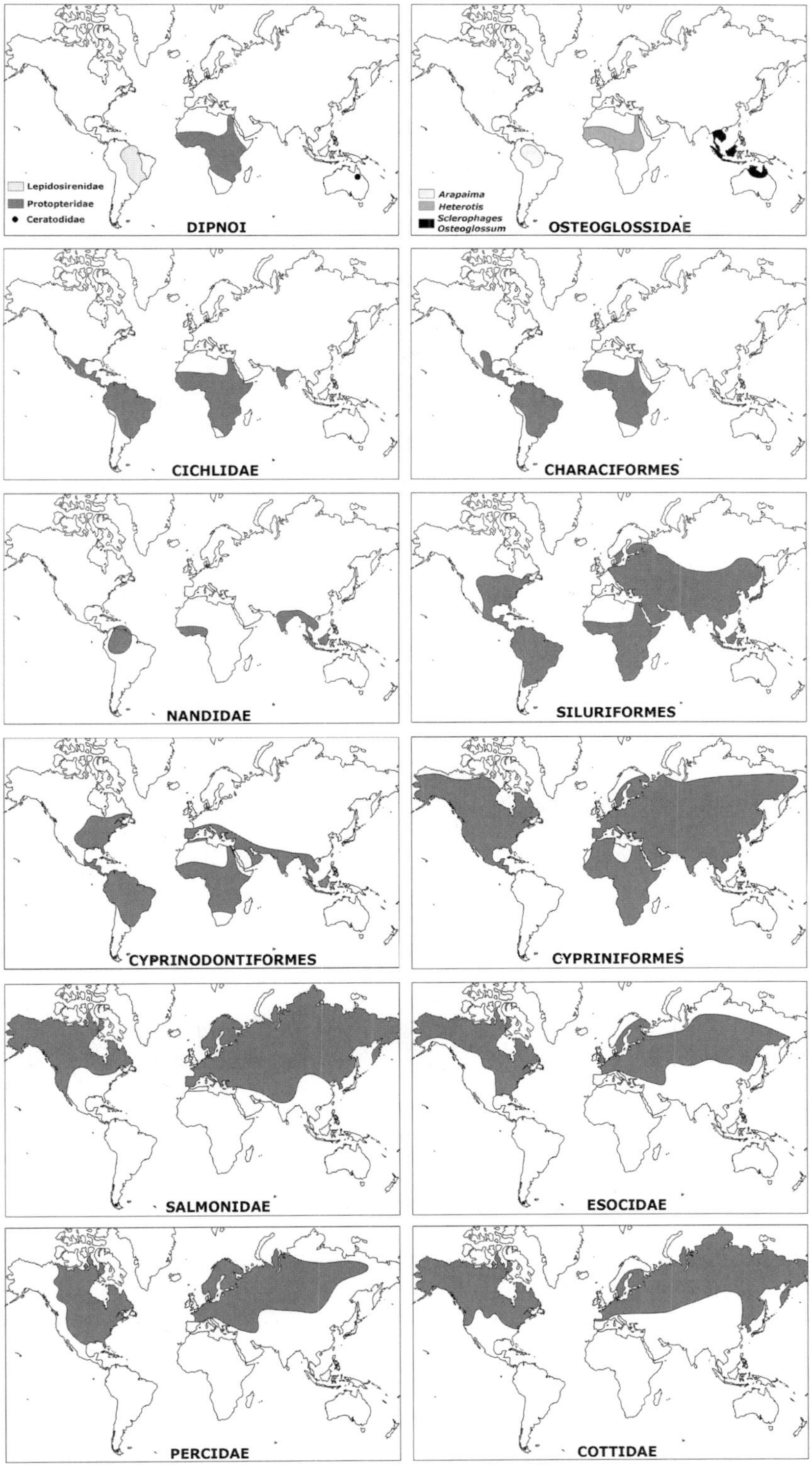

Fig. 1 Geographical distribution of selected fish groups (adapted from Berra, 2001). These maps represent only a simplified approach and need checking and refining

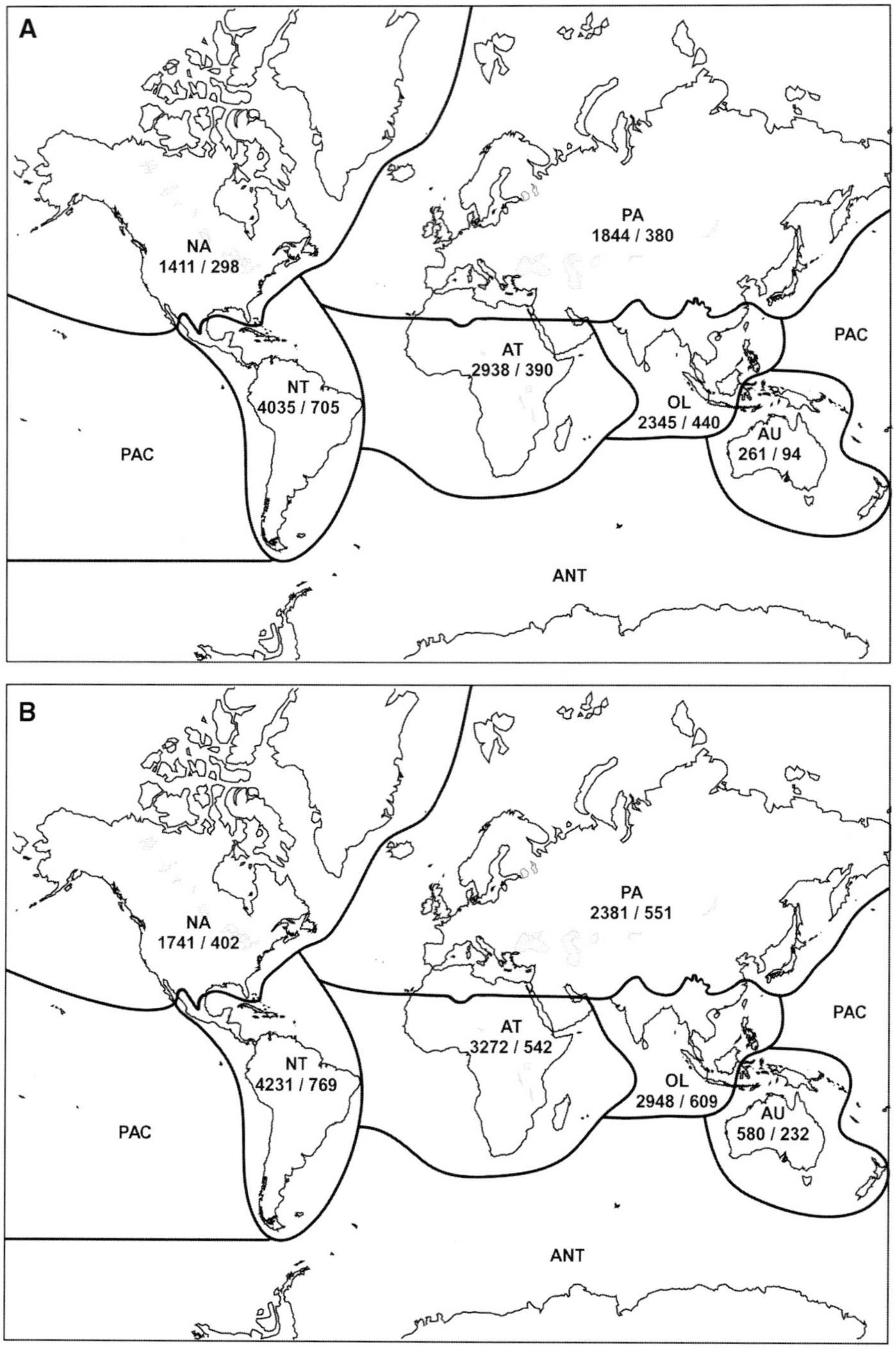

Fig. 2 Freshwater fish diversity: current number of species and genera (Sp/Gn) per zoogeographic region for strictly freshwater fishes (**A**) and for fresh and brackishwater fishes (**B**) (data from Fishbase, September 2005). PA—Palaearctic; NA—Nearctic; NT—Neotropical; AT—Afrotropical; OL—Oriental; AU—Australasian; PAC—Pacific Oceanic Islands, ANT—Antarctic

Some major groups of jawed fishes (chondrichthyes, dipnoi, some chondrostei) presently living were in existence by the middle of the Devonian, more than 350 Myrs ago and the relationship between continental drift and freshwater fish distribution has been widely discussed.

Several freshwater fish lineages appear to have a Gondwanan origin i.e. members of the lineage were present on Gondwana prior to its fragmentation. Consequently they have widespread distributions with living representatives and/or fossils present on different continents. Such is the case for the Dipnoi

Table 4 Number of fresh and brackishwater fish families and genera per biogeographic realm (some families and genera may be found in several realms) (data from Fishbase, September 2005)

Order	PA		NA		NT		AT		OL		AU	
	Families	Genera	Families	Genera	Families	Genera	Families	Genera	Families	Genera	Families	Genera
Chimaeriformes	–	–	–	–	1	1	–	–	–	–	–	–
Petromyzontiformes	1	5	1	5	1	2	1	1	–	–	1	1
Carcharhiniformes	1	3	1	2	1	2	1	2	1	3	1	2
Orectolobiformes	1	1	–	–	–	–	1	1	1	1	1	1
Pristiformes	1	2	1	1	1	1	1	2	1	2	1	2
Pristiophoriformes[a]												
Rajiformes	1	1	1	1	1	3	1	4	2	4	2	3
Myliobatiformes[b]												
Ceratodontiformes[c]	–	–	–	–	1	1	1	1	–	–	1	1
Acipenseriformes	2	5	2	3	1	1	1	1	–	–	–	–
Albuliformes	1	1	1	1	1	1	1	1	1	1	1	1
Amiiformes	–	–	1	1	–	–	–	–	–	–	–	–
Anguilliformes	5	5	1	1	2	2	4	6	5	10	4	7
Atheriniformes	2	2	2	8	2	5	2	4	5	15	3	7
Batrachoidiformes	–	–	1	1	1	4	–	–	–	–	–	–
Beloniformes	2	3	3	4	2	5	1	3	3	13	3	7
Characiformes	1	1	9	37	15	233	3	42	–	–	2	3
Clupeiformes	3	15	3	9	3	15	4	25	4	23	3	11
Cypriniformes	6	253	3	78	1	11	3	45	5	226	2	13
Cyprinodontiformes	4	6	7	56	4	47	3	28	3	3	3	7
Elopiformes	2	2	1	1	1	1	2	3	2	2	2	2
Esociformes	2	3	2	4	–	–	1	1	–	–	–	–
Gadiformes	2	2	2	3	–	–	–	–	–	–	–	–
Gasterosteiformes	1	3	1	4	–	–	1	1	1	1	–	–
Gobiesociformes	–	–	1	1	1	1	–	–	–	–	–	–
Gonorynchiformes	1	1	1	1	1	1	3	6	1	1	1	1
Gymnotiformes	–	–	4	4	5	30	–	–	–	–	–	–
Hiodontiformes[d]												
Lepisosteiformes	–	–	1	2	–	–	–	–	–	–	–	–
Lophiiformes	–	–	–	–	–	–	–	–	1	1	1	1
Mugiliformes[a]												
Ophidiiformes	–	–	1	2	1	1	–	–	–	–	–	–
Osmeriformes	3	11	1	5	1	3	1	1	2	5	2	5
Osteoglossiformes	–	–	1	1	2	2	5	24	2	3	1	1
Perciformes	38	138	20	91	19	89	29	252	39	172	36	122
Percopsiformes	–	–	3	7	–	–	–	–	–	–	–	–
Pleuronectiformes	3	5	3	6	2	9	4	7	3	7	3	6
Polypteriformes	–	–	–	–	–	–	1	2	–	–	–	–
Salmoniformes	1	11	1	7	1	4	1	4	1	1	1	3
Scorpaeniformes	5	19	1	3	–	–	–	–	2	3	2	3
Siluriformes	13	40	11	45	16	291	10	64	16	88	6	15
Synbranchiformes	2	4	1	3	1	1	2	4	3	12	1	2
Syngnathiformes	1	4	1	3	1	1	1	5	1	4	1	2

Table 4 continued

Order	PA		NA		NT		AT		OL		AU	
	Families	Genera	Families	Genera	Families	Genera	Families	Genera	Families	Genera	Families	Genera
Tetraodontiformes	1	5	1	1	1	1	1	2	1	8	1	3
Total	106	551	95	402	90	768	90	542	106	609	86	232

[a] Mugiliformes & Pristiophoriformes are considered as stricly marine in Fishbase

[b] Myliobathiformes are included in Rajioformes in Fishbase

[c] Ceratodontiformes include Lepidosireniformes

[d] Hiodontiformes are included in Osteoglossiformes in Fishbase

(lungfishes) a monophyletic group that first appeared in the Devonian, diversified in the Mesozoic, and is now represented by a few extant species in Africa, Australia and South America (Fig. 1) (Lundberg, 1998; Lundberg et al., 2000) and the Osteoglossomorpha, a distinctive clade of teleosts, at least late Jurassic in age, whose subgroups are scattered among tropical freshwater regions and North America (Fig. 1).

The Polypteridae (bichirs and reedfishes) also had a large distribution; however, today, living species occur only in Africa, while fossils have been identified in South America (Gayet & Meunier, 1991).

The Otophysi (superorder Ostariophysi) is a monophyletic group that includes very speciose families in freshwater: the Siluriformes (catfishes), Gymnotiformes (knifefishes), Cypriniformes (minnows) and Characiforms (characins) whose current distribution present some interesting contrasts (see Fig. 1). They do not occur in Australia, Madagascar or the West Indies. Recent molecular studies supported the hypothesis that Cypriniformes are the sister group to the remaining three orders (Saitoh et al., 2003). The divergence time of cypriniforms from the otophysan stock has been estimated at 250 mya (Kumazawa et al., 1999).

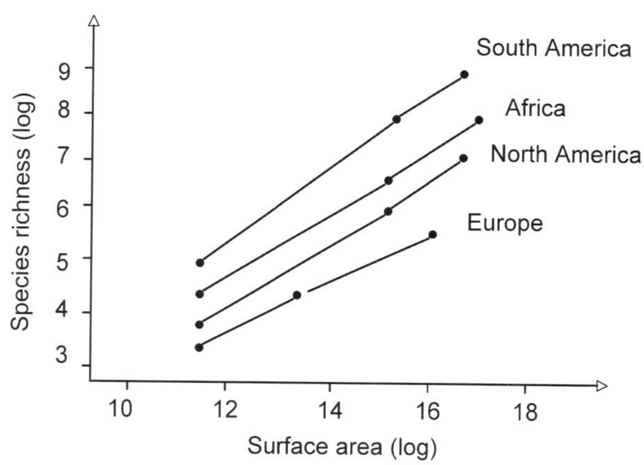

Fig. 3 Intercontinental comparison of species–area relationships. Each line links, respectively, the estimated species richness in a 10,000 km^2 drainage basin, the species richness of the largest continental river (i.e. Danube river for Europe, Mississippi river for North America, Congo river for Africa, and Amazone river for South America), and the continental species richness. Modified after Hugueny (2003) using data from Oberdorff et al. (1995)

The understanding of Otophysan biogeography has been a matter of debate, since the beginning of the 20th century. Asia, where the group is the most speciose, was long considered to be the centre of origin of the Cypriniformes. However, current evidence argues for a hypothesis that the cypriniforms, and probably also the siluriforms, originated in South America and migrated to Asia in the late Jurassic (ca 150–160 Myrs ago) along the northern shore of theThethys sea (Briggs, 2005). Diogo (2004) also supported the origin of catfishes in the South American region during the late Cretaceous period, at a time when there were still some remaining Pangean connections between Gondwana and Laurasia. Then catfishes would have dispersed to other areas with some subgroups migrating via predrift dispersion to Laurasian regions (Sullivan et al., 2006).

Recent palaeontological studies demonstrated that several archaic fish families disappeared from South America during the Cretaceous and the entire Cenozoic. Much diversification of modern Neotropical fishes occurred during the 70 Myrs period from the late Cretaceous through the Miocene (Lundberg, 1998). By the late Miocene (10 Myrs) the fish fauna was essentially modern. Fish diversification continued but at finer-taxonomic levels. Apparently geological events during the Pliocene and Pleistocene played little role in forging the great diversity of genera and families (Lundberg, 1998).

Present distribution: main characteristics of the ichthyofauna at the continental level

North America

Spanning the continent from Alaska and Canada in the north to the Transvolcanic Axis just south of the Mexican plateau, North America harbours some 1,050 freshwater fish species. Compared with other temperate zone regions this is an extremely species-rich area, and the continental US alone, with a little over 800 species, ranks seventh in World. As elsewhere, these fish species are not evenly distributed over the continent. Watershed boundaries, local geological and climatic forces shape the landscape and hydrology, and historical factors, particularly the imprint of past glaciations, have moulded ichthyofaunal distributions. Fish richness is greatest in the southeastern US where some 500 species reside, with a focus of endemism in the upland regions of southern Appalachia, where at least 350 species are concentrated. At the continental scale species richness tends to decline markedly to both the north and west of these southeastern foci. The extraordinary diversity of the southeastern US is probably the result of a combination of factors including a diverse physical geography, a favourably moist climate, and a long but dynamic history of zoogeographical interactions. Furthermore, and perhaps most critically, these southeastern regions escaped the repeated Pleistocene glacial advances that effectively denuded ichthyological landscapes in the northern third of the continent. In the southwest aridity and a harsh climatic regime, perhaps accounts for the relatively low levels of species richness, although locally high levels of endemism characterise much of the Pacific south western US. Diversity over the Mexican plateau is high with an estimated 250 species of which over 200 are considered endemic, and the majority of which have neotropical affinities.

Main characteristics

- Around 35 strictly freshwater families are represented in North American waters, 13 of which are endemic to the region. Numerically dominant families include the Cyprinidae, Percidae, Poeciliidae and Catostomidae, which together comprise over 50% of the total number of fish species.
- Radiations of freshwater species belonging to otherwise predominantly marine families (e.g. Cottidae, Atherinidae, Clupeidae and Moronidae) form an important component of the North American freshwater fauna.
- Anadromous trouts and salmons, particularly along the Pacific Coast, seasonally dominate freshwater ichthyofaunal communities mediating a significant nutrient passage between the ocean and inland areas. Similarly, at least historically, catadromous anguillid eels, *Alosa* and *Brevoortia* formed an important component of the coastal and inland waters of the Atlantic Seaboard.
- North America harbours a relatively rich representation of living basal actinopterygian fishes including sturgeons, paddlefish, gars and the enigmatic bowfin, *Amia calva*, as well as the only extant northern hemisphere osteoglossomorphs in the endemic Hiodontidae.
- The North American fauna includes a predominance of elements with Eurasian affinities, such as the Esocidae, Umbridae, Cottidae, Cyprinidae, Catostomidae, Percidae and Gasteosteidae, but Neotropical elements are also clearly represented, particularly in southern regions, for example the Cichlidae, Characidae, Pimelodidae, Rivulidae, Anablepidae and Synbranchidae.

Smith in Lundberg et al. (2000) provides an excellent summary review of fish diversity and notes that compared with tropical regions species diversity in North America is relatively well documented. While a few new species continue to be discovered and described each year, in general the potential for significant biodiscovery is probably limited to the Mexican regions in the south and overall the North American ichthyofauna can be considered to be well known. In addition to numerous regional works on the ichthyofauna the treatises of Mayden (1992) and Hocutt & Wiley (1986) provide excellent overviews.

Abell et al. (2000) provide a regional assessment of the conservation status of the fauna.

South and Central America

South and most of Central America corresponds to the Neotropical realm of bio-geographers. Its freshwater fish fauna is the most diversified in the world (around 3,600 freshwater fish species according to Reis et al. 2003, 4,164 according to FishBase). Despite the proximity of the Neotropical region to North America, there seems to be little relationship with the Nearctic fish fauna, and more with the Afrotropical region.

The aquatic fauna of Central America, from the isthmus of Tehuantepec to the border of Panama/ Columbia, consists of a mixture of North American and South American lineages, and includes some endemic groups of cyprinodontiform and many members of peripheral fish families. Only two North American families (Catostomidae and Ictularidae) occur in the Central America freshwater fish fauna.

Main characteristics

- The great majority of Neotropical fishes belong to five dominant groups: Characiforms (some 1,500 described, probably 2,000), Siluriforms (at least 1,400 known species, probably 2,000), Gymnotiforms (some 180 species), Cyprinodontiforms (some 400 species) and cichlids (some 450 species). There is a spectacular radiation of characoids and siluroids in South America.
- South America lacks many of the rather primitive fish families endemic to Africa, with the exception of Lepidosireniformes and Osteoglossiformes.
- There are no native Cypriniformes.
- The Gymnotiforms are electroreceptive fishes, showing a remarkable convergence with the unrelated mormyrids of Africa, and the notopterids of Africa and Asia.
- Characteristic of the Neotropical fish fauna is the abundance of very small species (size from 20 mm to 30 mm) among characiforms, siluriforms and cyprinodontiforms. At the other end of the scale, some very large fish occur here as well, such as the goliath catfish *Brachyplatystoma* in the Amazon (up to 3 m long and 140 kg), or the osteoglossid *Arapaima gigas*, (up to 4.5 m long and 200 kg).

Vari & Malabarba (1998) pointed out that some 800 new freshwater species have been described during the last two decades from South America, and they anticipate an increase in the rate of description. They forecasted a final total of some 8,000 Neotropical fish species. Lundberg et al. (2000) gave estimates of 5,000–8,000 species for the Neotropical ichthyofauna. Only 4,500 species are currently known (see Table 1) but many new species have been described recently, and many more await description.
At the moment no comprehensive review exists for the Neotropical fish fauna as a whole. Much of the available information was summarised and published in a symposium volume (Malabarba et al., 1998).

Europe

The European freshwater fish fauna is impoverished compared to other continents, as a result of recent glaciations. Biogeographic evidence indicates that the glaciated areas were recolonised (north and westward dispersal) during the interglacial and post-glacial periods mainly from the Ponto-Caspian region, and particularly from the middle and lower sections of the Danube basin, which served as a major refugia (Banarescu, 1992; Griffiths, 2006). Other refugia during the last ice age were located in the southern peninsulas of Iberia, Italy, the Balkans and Greece. However, post-glacial northern expansions of fishes from these regions, or recolonisation of these regions from the north, were prevented by mountain ranges (Pyrenees, Alps, for example). This geographical isolation explains the high level of endemism found in the southern regions (Banarescu, 1992; Durand et al., 2003; Reyjol et al., 2007).

Main characteristics

- Twenty families but the major groups are Cyprinidae—more than 50% of the species, (Revenga & Kura, 2003; Reyjol et al., 2007), Salmonidae, Coregonidae, Gobiidae and Cobitidae.
- Very few endemic species; the largest numbers are all located in areas which have served as major refugia in the last ice age (see above).

- The river Danube with about 90 species has the most diverse fauna of the continent.

In Europe, according to Maitland (2000) there are over 250 native freshwater fish species, while Kottelat (1997) recognised 358 species west of the former USSR. The species level taxonomy of salmonoids (Salmonidae and Coregonidae), which is far from resolved, may partly explain this difference.

Africa and Madagascar

Most of the African continent has remained above sea level since the Precambrian, more than 600 Myrs ago, though large areas such as the Sahara, Somalia and Ethiopia, have been at times inundated by the sea (Lévêque, 1997) Such a long period of exondation may explain why Africa has a diverse fish fauna and an unparalleled assemblage of archaic, mostly endemic, families.

Main characteristics

- There are about 48 families of freshwater fishes in tropical and southern Africa, 15 of which are endemic. The African ichthyofauna has fewer families and species than South America but it includes a higher number of basal and archaic families.
- The archaic groups include the Polypteridae, recorded since the Cretaceous only from Africa; the Denticipidae, considered as the sister group of Clupeiformes (Lavoué et al., 2005); and the Phractolemidae, Kneriidae, Cromeridae and Grasseichthydae.
- The African fauna also includes remnants of archaic elements of wider distribution, such as the Protopteridae, Notopteridae and Osteoglossidae. Three other families of the predominantly Gondwanan Osteoglossomorpha are endemic to Africa: the speciose family Mormyridae, and the monotypic Gymnarchidae and Pantodontidae.
- Two large lineages of secondary division freshwater fishes are present in Africa: the Cyprinodontiformes and the Cichlidae, both extremely diversified.
- Peripheral freshwater fish families are relatively poorly represented in African inland waters in comparison to other continents. Only a few families include exclusively freshwater genera or several freshwater resident species: Clupeidae, Ariidae, Synbranchidae, Latidae (ex Centropomidae), Gobiidae, Eleotridae, Mugilidae, Syngnathidae and Tetraodontidae.
- True diadromous species are rare in Africa. The genera *Anguilla* occurs in the Maghreb and five species are known from the east coast. However, the genus is completely absent from western and central Africa. This is also the case for many gobioids (e.g. *Sicydium*).

Madagascar's freshwater fish fauna contrasts with the continental Africa one (Sparks & Stiassny, 2003). Of a total of 135 native fish species, 84 are endemic to the island itself. Many new species have been described during the last decade and many more are waiting identification. The origins of the freshwater fish fauna remains unclear. Most of the species belong to widely distributed peripheral families. Many of the major groups of freshwater fish present in Africa, as well as those present in Africa and Asia (anabantids, bagrids, clariids, mastacembelids, notopterids, etc.), are absent in Madagascar (Stiassny & Raminosoa, 1994). Noteworthy is the absence of primary freshwater families such as the cyprinids, characins and mormyrids. Only three secondary freshwater families (Aplocheilidae, Cichlidae and Poeciliidae) have been recorded from the island. Many Malagasy species are phylogenetically basal within their respective families, and the ichthyofauna is apparently of relict nature.

Currently 3,255 species of fresh and brackishwater fish species have been described from Africa, belonging to 95 families. Numerous others are awaiting description.

Asia

Tropical Asia covers the Oriental region extending from the Indus basin to South China and Indonesia. In addition to continental areas the region includes many large islands such as Borneo, Sumatra and Java, as well as numerous smaller islands.

Main characteristics

- High number of fish families: 121 recorded from inland waters including 34 primary and secondary

division freshwater fishes (18 endemic to South-east tropical Asia) and 87 peripheral usually represented only by a few species (Lundberg et al., 2000).
- The dominant primary groups are Cypriniforms including Cyprinidae (about 1,000 spp.), Cobitidae (about 100 spp.) and Balitoridae (about 300 spp.), Siluriforms including Bagridae (about 100 spp.) and the Osphronemidae (85 spp.).
- On oceanic islands and in coastal river basins, peripheral families dominate the fish communities. In Sulawesi, the Moluccas and most Philippine islands, there are no primary freshwater fishes. The dominant peripheral group is the Gobiidae (about 300 spp.).
- As in South America, the discovery of miniature species is fairly common. *Boraras micros* from northern Thailand is adult at about 12 mm SL, and *Paedocypris progenetica*, recently described from peat swamp pools in Indonesia, is sexually mature at 7.9 mm SL. At the opposite extreme, the Mekong stingray *Himantura chaophraya*, has a disk width of 2 m, a total length of 4 m, and used to weigh up to 600 kg. *Pangasianodon (Pangasius) gigas* of the Mekong is also one of the largest freshwater fishes with historical records of up to 3 m and 300 kg.

An estimated total number of 3,000 species has been suggested (Lundberg et al., 2000), but incomplete surveys in many countries render this a probable under-estimate. FishBase records 3,553 freshwater species or some 4,400 species if peripheral species are included.
Knowledge of the fish fauna of tropical Asia is still in its exploratory phase particularly in China and India where survey work is incomplete. Many species have still to be described or to be discovered.

Oceania

Australia, Tasmania, New Guinea and the islands of the Australian continental shelf represent a well-delimited biogeographic entity. New Zealand belongs to this realm as well.

The terrestrial connections between Australia and other continents broke some 100 Myrs ago. The last 15 Myrs have seen increased drying, resulting in decreased surface water in drainages mostly established during the Palaeocene (Unmack, 2001). Most patterns of distribution were almost certainly established in the distant past, perhaps as early as Miocene. Pleistocene glaciations were geographically limited, with probably little effect on the aquatic biota (Unmak, 2001).

Main characteristics

- The freshwater fish fauna of Australia is depauperate and lacks all Otophysan primary freshwater families found elsewhere in the world.
- The majority of freshwater fishes are representatives of marine families with many catadromous species.
- The only primary freshwater fishes are the Australian lungfish, *Neoceratodus* and some osteoglossids of the genus *Scleropages*. Neoceratodus is restricted to Queensland where it survives in swamps and permanent rivers.
- High endemicity in several provinces: most provinces in southern, central, and western parts of the continent have a large proportion of endemics, whereas Northern and Eastern provinces have few. The pattern results in part from isolation, due to aridity and drainage divides.

New Zealand has no primary freshwater fishes and low species richness: only about 50 species, belonging to seven families (e.g. Galaxiidae, Eleotridae, Anguillidae) that probably colonised New Zealand by sea. More than 60% fish species are diadromous. The degree of endemism is high.

Continental ecoregions and main areas of endemicity

At continental scales, ichthyologists have attempted to identify biogeographic or ichthyological ''provinces'' based on their distinctive evolutionary history and more-or-less characteristic biota at the species levels (e.g. Abell et al., 2000; Thieme et al., 2005). The present pattern, however, may also be the imprint of a long evolutionary history that has resulted in differentiation at higher taxonomic levels.

Main ecoregions in North America

Based mainly upon distribution data for freshwater fishes, mussels and crayfish, Abell et al. (2000) recognise 10 regional complexes (herein designated as provinces) for North America within which 76 ecoregions are delineated. In North America fishes, mussels and crayfish tend to display rather similar distributional patterns and here we take the basic map of Abell et al. (2000), as a reasonable surrogate for a regional subdivision of North America into ichthyofaunal provinces (Fig. 4):

1. Pacific Coastal province (ca. 40 endemic fish species). Comprised mainly of temperate coastal rivers, lakes and springs. With xeric elements in the Sonoran and southern Pacific regions.
2. Great Basin province (23 endemic fish species). Comprised mainly of endoreic rivers, lakes and springs.
3. Colorado province (20 endemic species). A mix of the large temperate Colorado River and xeric region rivers, lakes and springs.

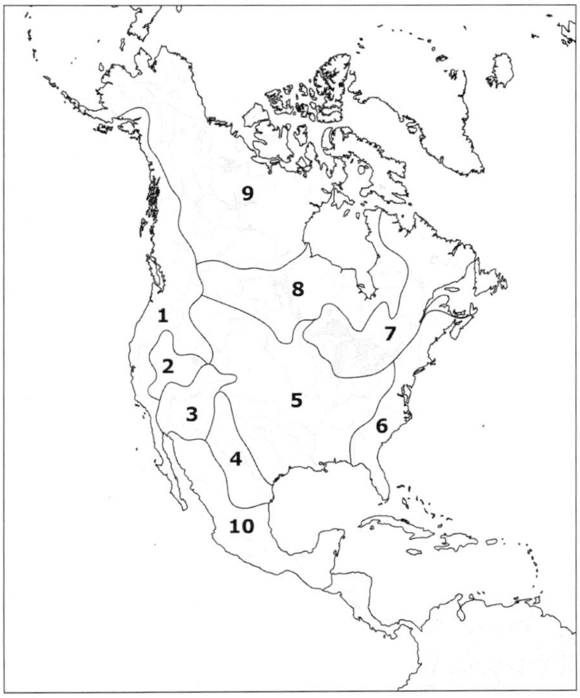

Fig. 4 Main ichthyological provinces in North and Central America (modified after Abell et al., 2000). See text for ecoregions

4. Rio Grande province (ca. 80 endemic species). Comprised mainly of the Upper and Lower Rio Grande River and a mosaic of xeric rivers, lakes and springs.
5. Mississippi province (ca. 130 endemic species). The largest river in the US, the Mississippi drains a basin of about one eighth the area of North America. Its mainstream, tributary systems, headwaters, embayments and karsts harbour at least 375 fish species. This basin has provided a refuge during times of glaciation and acted as a source for northern recolonisations.
6. Atlantic province (ca. 65 endemic species). A mosaic of subtropical coastal rivers, lakes and springs in the south, extending northward along the Atlantic seaboards temperate coastal rivers and lakes into maritime Nova Scotia.
7. St. Lawrence province (at least three endemic species). Temperate coastal rivers and lakes of northern Nova Scotia and the Gulf of St Lawrence, and the Great Lakes.
8. Hudson Bay province (no endemic species). Temperate headwaters and lakes of southern Canada, extending east to the Arctic rivers and lakes of the Hudson and James Bays.
9. Arctic province (four endemic species). Arctic rivers and lakes of the Yukon and Alaskan Arctic and Arctic islands.
10. Mexican Transition Bioregion (ca. 200 endemic species). A mosaic of xeric rivers, lakes and springs, and subtropical coastal rivers, volcanic crater lakes, sink holes and extensive wetlands.

Main ecoregions in Central America

More than 350 species are found in Central America, an area lying between the isthmus of Tehuantepec to the north and the Colombian border to the south. Based on the work of Bussing (1998) and supported by a preliminary analysis of freshwater fishes distribution at the basin scale (Tedesco et al., unpublished data), four ichthyofaunal provinces are recognised (Fig. 4):

1. The Usumacinta province comprises Atlantic rivers from Honduras, Guatemala, Belize and southern Mexico including the relatively large Usumacinta drainage.

2. The Pacific province includes coastal streams of the Pacific versant from Costa Rica to Mexico.
3. The San Juan province includes rivers of the Atlantic versant of Nicaragua and Costa Rica.
4. The Isthmian province sharing various species with the Magdalenean province from South America. Includes Pacific and Atlantic rivers from Panama and some Pacific rivers of Costa Rica.

Main ecoregions in South America

Based on a preliminary analysis of freshwater fishes distribution at the basin scale (Tedesco et al., unpublished data) and on the work of Gery (1969), 10 ichthyofaunal provinces are recognised for South America (Fig. 5):

1. South Patagonian province (12 species, 1 endemic). Southern Argentinean rivers extending from Tierra del Fuego to the Chubut river.
2. North Patagonian province (23 species, 5 endemics) includes the Colorado and Negro rivers from Argentina.
3. Trans-Andean (South) province (19 species, 13 endemics) includes all the small coastal rivers from Chile.
4. Lake Titicaca province (32 species, 30 endemics). Endorheic drainages from the Bolivian and Peruvian Andes where sympatric speciation of the genus *Orestias* seems to be ongoing.
5. Paranean province (847 species, 517 endemics). Mainly including coastal rivers from central Argentina and two large rivers, the La Plata drainage and the Sao Francisco drainage from Brazil.
6. South-East Brazilian province (194 species, 90 endemics) comprised southern Brazilian coastal streams and a large coastal lagoon system.
7. East Brazilian province (131 species, 50 endemics) includes coastal rivers from eastern Brazil. Existing information is scarce.
8. Amazonian province (2,416 species, 2,072 endemics). This great province includes the Amazon and the Orinoco drainages, and the coastal rivers from the Guyanas and northern Brazil. Many new discoveries are anticipated.
9. North Venezuelan province (61 species, 9 endemics).
10. Trans-Andean (North)/Magdalenean province (423 species, 326 endemics). Includes rivers from Ecuador, Colombia and the Maracaibo drainage from Venezuela.

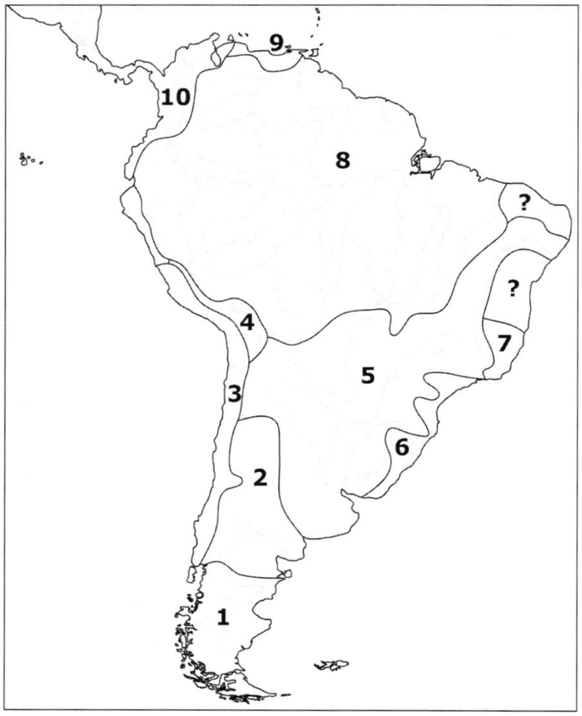

Fig. 5 Main ichtyofaunal provinces in South America. See text for ecoregions

Main ecoregions in Europe

The only study using Europe-wide data at the basin scale was performed by Reyjol et al. (2007). These authors used species list from 406 basins (233 species) fairly evenly distributed across Europe to define geographical regions (provinces) having homogenous fish fauna. Figure 6 illustrates these provinces:

1. Ponto-Caspian Europe province (98 species, 36.7% endemic),
2. Northern Europe province (42 species, 9.5% endemic),

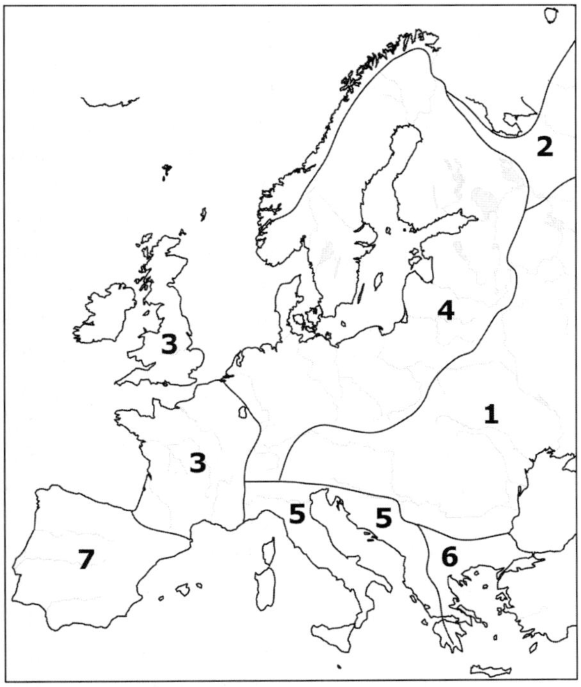

Fig. 6 Main ichthyological provinces in Western Europe. See text for ecoregions. Modified after Reyjol et al. (2007)

3. Western Europe province (47 species, 6.4% endemic) and
4. Central Europe province (57 species, 1.8% endemic).
5. Central peri-Mediterranean province (93 species, 64.5% endemic),
6. Eastern peri-Mediterranean province (64 species, 31.2% endemic),
7. Iberian Peninsula province (50 species, 60% endemic).

Main ecoregions in Africa

Several ichthyological provinces have been identified in Africa (Roberts, 1975; Levêque, 1997; Thieme et al., 2005) (Fig. 7):

1. The Maghreb has a very depauperate fauna with Paleartic affinities.
2. The Nilo Sudan province extends from the Atlantic coast to the Indian Ocean and includes the major drainage basins of the sahelian zone: Nile, Chad, Niger, Senegal, Volta. The fish fauna is relatively rich (Paugy et al., 2003).

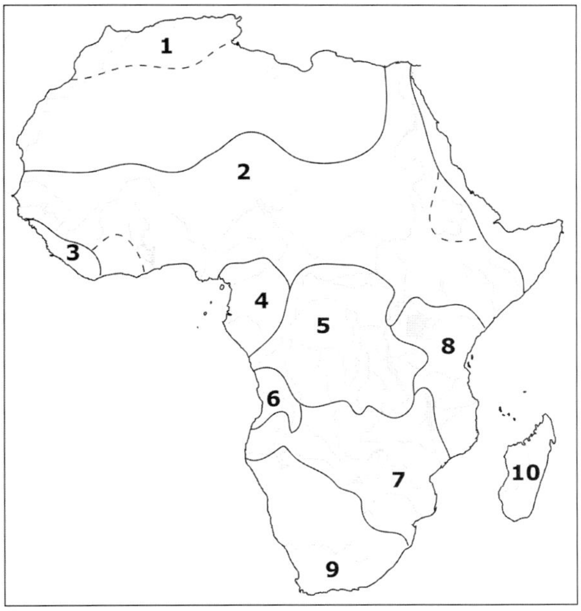

Fig. 7 Main ichtyological provinces in Africa (Lévêque, 1997, modified from Roberts, 1975). See text for ecoregions

3. The Upper Guinea province includes the coastal rivers from Guinea to Liberia and exhibits faunistic affinities with the Lower Guinea Province and the Congo. Fauna well diversified with many endemic taxa (Paugy et al., 2003).
4. The Lower Guinea covers coastal rivers from Cameroon to the mouth of the Congo river, with a well diversified fauna (Stiassny et al., in press).
5. The Congo province includes the entire Congo basin, which is the largest in Africa. The ichthyofauna is rich and diversified, but existing information needs to be synthetised and many new discoveries are anticipated.
6. The Quanza province which covers the Angolan coastal drainages is still extremely poorly known.
7. The Zambezi including the river Cunene, Okavango and Limpopo has a moderately rich fauna and is fairly well documented (Skelton, 2000)
8. The East Coast covers the coastal drainages from the Juba in the North to the Zambezi in the south. The fauna is moderately rich and a new synthesis is needed.
9. The Southern province includes the basins of the Orange-Vaal and all the coastal systems to the south. The fauna is moderately rich and well known (Skelton, 2000).

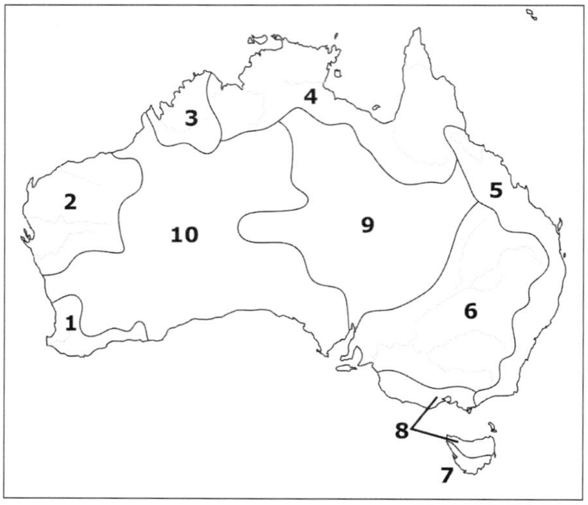

Fig. 8 Freshwater fish biogeographic provinces in Australia (from Unmack, 2001). See text for ecoregions

Ecoregions in Asia

We present no information about ecoregions in Asia, as work is in progress under the leadership of WWF-USA and will be published soon.

Main ecoregions in Australia

In a recent work, Unmack (2001) identified 10 biogeographical provinces (Fig. 8). Most patterns were almost certainly established in the distant past, perhaps as early as Miocene. Influences of Plio–Pleistocene events on broad patterns of freshwater fish distributions seem minimal (Unmack, 2001).

1. *South-western* province has few species, and no species in common with any other, suggesting long-term isolation. Faunal relationships, although distant, lie with south-eastern Australia;
2. *Pilbara* province is one of the hotest in Australia. Five of 12 recorded species are endemic, the remainder are widespread in Australia;
3. *Kimberley* province is characterised by high endemicity. Only 16 species, out of 29 are widespread;
4. *Northern* province has 38 endemic species (25 shared with New Guinea) out of 75 (50% endemic). A strong relationship exists between Fly River, New Guinea and the Northern Province of Australia, with 34 out of 75 freshwater fishes (45%) in common (Allen, 1991).
5. *Eastern* province is distinctive for its lack of faunal breaks, its boundary with the Northern Province being due to a sharp decline in richness. Fifteen out of 47 species (31%) are endemic.
6. *Bass* province has a depauperate fauna;
7. *Southern Tasmanian Province* has no shared species, all eight being endemic including three *Galaxias* and three *Paragalaxias*. All have restricted ranges, often one or a few lakes and/or streams.
8. *Murray-Darling* province appears to have experienced mixing of faunas from surrounding regions, while maintaining a high degree of endemism.
9. *Central Australian* province (30 species);
10. *Palaeo* province contains former connections to surrounding drainages now dried. Only one species recorded from this vast region.

Endemism and fish species flocks in ancient lakes

Exceptionally high faunal diversity occurs in certain ancient lakes ("long-lived" lakes that are more than 100,000 years old). One of the most noteworthy features is the existence of "species-flocks" that are aggregates of disproportionally high numbers of species, sharing a common ancestor and endemic to each lake (Table 5). The longevity of ancient lakes, compared to younger lakes, may explain the abundance of endemic evolutionary radiations they harbour. However, the processes accounting for these radiations are a matter of debate, but there is a growing body of evidence that suggests that in addition to intra-lacustrine allopatric speciation sympatric speciation may have also occurred (Schliewen et al., 2001). At present, several rich fish-species flocks have been identified in various ancient lakes that are exceptional natural sites for the study of speciation processes.

In Africa, the most striking feature of the Eastern African Great Lakes (Victoria, Tanganyika, Malawi) is that each has its own highly endemic lacustrine cichlid fauna that apparently evolved independently from riverine ancestors. Some 550 endemic haplochromine cichlids occur in Lake Victoria, probably more than 800 in Lake Malawi (Fryer, 1996), and 325

Table 5 Current fish species flocks in selected ancient lakes

	AFRICA					ASIA	SOUTH AMERICA			ASIA	EUROPE
	Matawi	Victoria	Tanganyika	Kivu	Tana	Lanao	Titicaca	Chapala	Chichancanab	Baïkal	Ohrid
Salmoniformes											
Salmonidae											■
Cypriniformes											
Cyprinidae					■	■					
Siluriformes											
Bagridae			■								
Mochokidae			■								
Atheriniformes											
Atherinidae								■			
Cyprinodontiformes											
Cyprinodontidae							■		■		
Synbranchiformes											
Mastacembelidae			■								
Scorpaeniformes											
Cottidae										■	
Comephoridae										■	
Abyssocottidae										■	
Perciformes											
Cichlidae	■	■	■	■							
Centropomidae			■								

in Lake Tanganyika ans some 1,000 species are still awaiting scientific description((Snoeks, 2000). Intra-lacustrine cichlid speciation has also occurred, to a lesser extent, in other smaller lakes of the Rift Valley (Kivu, Albert, Edward and Turkana), and in certain crater lakes of Cameroon (Stiassny et al., 1992; Schliewen et al., 1994). Rates of speciation in cichlids can be extremely fast, with some estimates at about 100,000 yrs (e.g. Verheyen et al. 2003).

The non-cichlid fauna of the East African Lakes is less speciose and has lower levels of endemism, but is still noteworthy. For instance, Lake Tanganyika harbours small species flocks within a few other families: 7 mastacembelid species, 6 species of the clarotid *Chrysichthys*, 7 species of the mochokid *Synodontis* and 4 species of the latid, *Lates* (De Vos & Snoeks, 1994).

A species flock of lacustrine cyprinids has recently been discovered in Lake Tana, Ethiopia, where Nagelkerke & Sibbing (1996, 2000) carried out detailed studies of the morphology, reproduction and feeding habits of the large hexaploid barbs belonging to the species complex, *Labeobarbus intermedius*. They identified at least 15 biologically distinct species differing in food niche and habitat preferences, as well as in spawning grounds.

In South America, the native fish fauna of Lake Titicaca includes 24 *Orestias* species (Cyprinodontidae) (Lauzanne, 1992). Lake Chapala, located in Mexico is a remnant of a series of Tertiary lakes where atherinids (silversides) of the genus *Chirostoma* are prominent with eight species (Echelle & Kornfiel 1984). In the Laguna Chichancanab in the Yucatan Peninsula (Mexico), a flock of six species of the genus *Cyprinodon* (pupfishes) has also been identified (Echelle et al., 2005).

In northern Asia, lakes existed in the Baikal Rift zone (East Siberia) for at least the last 60 million years. Since about 28 Myrs, one or more continuous large lakes have evolved into the extant lake (Mats, 1993). Today, Lake Baikal hosts a very diverse fauna, with some 2,500 described animal species (most of them endemic) including 56 species and subspecies of fish belonging to 14 families (Sideleva, 1994). Noteworthy is the presence of a sculpine (Cottotoidei) species flock, comprising 29 species (11 genera) of sculpins endemic to the lake. Through adaptive radiations, sculpins have colonised the most diverse habitats such as the abyssal and the pelagic zones of the lake. According to recent molecular studies sculpine fish comprise a fairly young species cluster, which have most likely diverged since the beginning of the Pleistocene (2 Myrs) when the climate in the region generally became much cooler (Yu Sherbakov, 1999).

In tropical Asia, Lake Lanao (Philippines) was formed 3.6–5.5 Myrs ago. The cyprinid flock is a widely acknowledged example of adaptive radiation while its age is a matter of debate (Rainboth, 1991). Unfortunately over-exploitation and exotic introductions have decimated the fauna, so that now only three or fewer endemic cyprinids are still surviving (Harrison & Stiassny, 1999). In central Sulawesi (Indonesia) a series of tectonic lakes host a fauna of 26 native species (22 endemic) (Kornfield & Carpenter, 1984).

The tectonic Lake Biwa is the largest and oldest lake in Japan. The Old-Biwa lake was established as a small, shallow lake about 5–6 million years ago (Kawanabe, 1996) with the present deep basin forming around 300,000 years ago. At present there are 71 species and subspecies of freshwater fishes in the lake and its tributaries (Yuma et al., 1998). More than half belong to the family Cyprinidae (37 species) and the rest to Cobitidae (6), Gobiidae (6), Salmonidae (5), Siluridae (3) among others.

Human related issues and conservation

The number of fish species able to use freshwaters, totals to about 13–15,000 species, which is 40–45% of the global fish diversity estimated at about 29–30,000 described species. Fish inhabiting freshwaters comprise therefore ca. 25% of living vertebrates (about 55,000 described species) and represent 13–15% of the 100,000 freshwater animal species currently known (Lévêque et al. 2005).

It is often claimed that freshwater ecosystems are the most endangered ecosystems in the world (Sala et al., 2000). The particular vulnerability of freshwater fish to global changes reflects the fact that both fish and freshwater are resources humans need and that have been heavily impacted by human usage and regulation. Asia supports over half of the global human population, with enormous consequent pressures on inland waters and freshwater fish biodiversity vities (Dudgeon et al., 2006). Conversely, in areas such as the Amazon and Congo basins with lower population densities, human impacts are relatively less marked although increasing nonetheless.

The major threats to fish biodiversity have been well identified: overexploitation, flow modification, destruction of habitats, invasion by exotic species, pollution including the worldwide phenomenon of eutrophication (Harrison & Stiassny, 1999; Dudgeon et al., 2006), all of which are interacting.

Freshwater fishes are important and valued resources for food, sport and ornament. Overexploitation occurs all over the world with the use of more and more sophisticated fishing gear, and the decrease of many fish stocks has been documented as a result of expanding fisheries (Allan et al., 2005). Illegal fishing using pesticides, electrofishing, dynamite, etc. are also major threats to fish diversity all over the world.

Other serious threats are flow modifications of running waters (diversion, extraction, storage) and water engineering such as impoundment by dams. The loss, or modification of aquatic habitats, are both responsible for extinction of native species (Harrison & Stiassny, 1999). The demand for reliable sources of fresh water and flood control has encouraged prolific dam-building (45,000 large dams and possibly 800,000 smaller ones—mainly since the last century) that has resulted in fragmentation and destruction of habitat, and loss of species (e.g., Vörostmary et al., 2006).

Over the past decades, excessive nutrient loading has emerged as an important direct driver of freshwater ecosystem change. World consumption of nitrogenous fertilisers grew nearly eightfold between 1960 and 2003, from 10.8 million tons to 85.1 million tons (MEA, 2005). Eutrophication is probably the most widespread problem affecting lake and reservoir waters. A direct impact of eutrophication is a change in the structure of fish species communities, or even elimination of fish populations (Seehausen et al., 1997).

Species have been introduced throughout the world for different purposes including stocking of fishes for aquaculture and fisheries, sport fishing, use of baitfish and their release after fishing, intentional or unintentional releases of aquarium species, environmental management for pest/weed control etc. In Europe, translocations are believed to date from Roman times, when carp, *Cyprinus carpio*, from the River Danube were reared in ponds in Italy and western and southern Greece (Balon, 1995). According to historical sources, it seems likely that during Roman times and later in the Middle Age, other freshwater fish species were moved from one system to another. Carp was probably not always introduced alone and other species were likely included in the carp transportations, as has been the case with the worldwide translocation of tilapias in the past century.

Biotic homogenisation, the process of gradual replacement of native biotas by non-indigenous and locally expanding non-native species is rapidly diminishing the regional distinctiveness of aquatic systems (Olden & LeRoy Poff, 2004). Such a process is well documented in United States where Rahel (2000) showed that states share, on average, 15.4 more species than before the European settlement. In documented cases, states that formerly had no species in common, now share an average of 25.2 species.

Inventories of freshwater biodiversity are incomplete in many parts of the world, especially the tropics, and rates of species loss may be higher than currently estimated. Today, hundreds of freshwater fish are close to extinction. Large tropical lakes like those in Eastern Africa, but also the Palaearctic Lake Baikal, have a very high heritage value. They are natural laboratories to study evolution and they should be given high priority for conservation. Maintenance of freshwater fish biodiversity is a critical test of whether water use and ecosystem modifications are sustainable. However, to be fully effective, protection of freshwater requires control over the upstream drainage network, the surrounding land, the riparian zone, and—in the case of migrating aquatic fauna—downstream reaches. Such prerequisites are hardly ever met!

Acknowledgements This work was partly supported by the French ANR "Freshwater fishes diversity" (ANR-06-BDIV-010).

References

Abell, R. A., et al., 2000. Freshwater Ecoregions of North America. A Conservation Assessment. Island Press, Washington, DC.

Allan, J. D., R. Abell, Z. Hogan, C. Revenga, B. W. Taylor, R. L. Welcomme & K. Winemiller, 2005. Overfishing of inland waters. Bioscience 55: 1041–1051.

Allen, G. R., 1989. Freshwater Fishes of Australia. TFH Publications, Neptune City, NJ, p. 240.

Allen, G. R., 1991. Field Guide to the Freshwater Fishes of New Guinea. Christensen Research Institute, Madang, p. 268.

Balon, E. K., 1995. Origin and domestication of the wild carp, Cyprinus carpio: From Roman gourmets to the swimming flowers. Aquaculture 129: 3–48.

Banarescu, P., 1992. Zoogeography of Fresh Waters. Vol. 2. Distribution and Dispersal of Freshwater Animals in North America and Eurasia. AULA-Verlag, Wiesbaden.

Berra, T. M., 2001. Freshwater Fish Distribution. Academic Press.

Briggs, J. C., 2005. The biogeography of otophysan fishes (Ostariophysi: Otophysi): A new appraisal. Journal of Biogeography 32(2): 287–294.

Bussing, W. A., 1998. Freshwater fishes of Costa Rica. International Journal of Tropical Biology 46(Suppl. 2): 1–468.

De Vos, L. & J. Snoeks, 1994. The non-cichlid fishes of the Lake Tanganyika basin. Archiv für Hydrobiologie Beiheft Ergebnisse der Limnologie 44: 391–405.

Diogo, R., 2004. Phylogeny, origin and biogeography of catfishes: Support for a Pangean origin of "modern teleosts" and re-examination of some Mesozoic pangean connections between the Gondwanan and Laurasian supercontinents. Animal Biology 54(4): 331–351.

Dudgeon, D., A. H. Arthington, M. O. Gessner, Z. I. Kawabata, D. J. Knowler, C. Lévêque, R. J. Naiman, A.-H. Prieur-Richard, D. Soto, M. L. J. Stiassny & C. A. Sullivan, 2006. Freshwater biodiversity: Importance, threats, status and conservation challenges. Biological Reviews 81: 163–182.

Durand, J.-D., P. G. Bianco, J. Laroche & A. Gilles, 2003. Insight into the origin of endemic Mediterranean ichthyofauna. Phylogeography of Chondrostoma genus (Teleostean, Cyprinidae). Journal of Heredity 94: 315–328.

Echelle, A. A., E. X. Caeson, A. F. Eckhelle, R. A. Van den Busche, T. E. Dowling & A. Meyer, 2005. Historical biogeography of the new-world pupfish genus *Cyprinodon* (Teleostei: Cyprinodontidae). Copeia 2005(2): 320–339.

Echelle, A. A. & I. L. Kornfield, 1984. Evolution of Fish Species Flocks. University of Maine at Orono Press, Orono.

Eschmeyer, W. N. (ed.), 2005. Catalog of fishes. Updated database version of May 2005. Catalog databases as made available to FishBase in May 2005.

Fryer, G., 1996. Endemism, speciation and adaptive radiation in great lakes. Environmental Biology of Fishes 45: 109–131.

Gayet, M. & F. J. Meunier, 1991. Firts discovery of Polypteridae Pisces, Cladistia, Polypteriformes, outside of Africa. Geobios 24(4): 463–466.

Gery, J., 1969. The freshwater fishes of South America. In Fittkau, E. J., J. Illies, H. Klinge, G. H. Schwabe & H. Sioli (eds), Biogeography and Ecology in South America. The Hague Publishers, 827–848.

Griffiths, D., 2006. Pattern and process in the ecological biogeography of European freshwater fishes. Journal of Animal Ecology 75: 734–751.

Guégan, J. F., S. Lek & T. Oberdorff, 1998. Energy availability and habitat heterogeneity predict global riverine fish diversity. Nature 391: 382–384.

Harrison, I. J. & M. L. J. Stiassny, 1999. The quiet crisis: A preliminary listing of freshwater fishes of the World that are either extinct or "missing in action". In MacPhee, R. D. E. (ed.), Extinctions in Near Time: Causes, Contexts, and Consequences. Plenum Press, New York and London, 271–331.

Hocutt, C. H. & E. O. Wiley (eds), 1986. The Zoogeography of North American Freshwater Fishes. John Wiley and Sons, Inc., New York, 866 pp.

Hugueny, B., 1989. West African rivers as biogeographic islands: Species richness of fish communities. Oecologia 79: 236–243.

Hugueny, B., 2003. (Macro) Ecologie des (méta) communautés: les poissons en rivière. Mémoire d'Habilitation à Diriger des Recherches (HDR). Université Lyon I, 92 pp.

Kawanabe, H., 1996. Asian Great Lakes, especially Lake Biwa. Environmental Biology of Fishes 47: 219–234.

Kornfield, I. L. & K. E. Carpenter, 1984. The cyprinidsn of lake Lanao, Philippines: Taxonomic validity, evolutionary rates and speciation scenarios. In Echelle, A. A. & I. L. Kornfield (eds), Evolution of Fish Species Flocks. University of Maine at Orono Press, Orono, 69–84.

Kottelat, M., 1997. European freshwater fishes. Biologia 52(5): 271.

Kumasawa, Y., M. Yamaguchi & M. Nishida, 1999. Mitochondrial molecular clocks and the origin of teleostean biodiversity: Familial radiation of perciforms may predated the Cretaceous/Tertiary boundary. In Goldblatt, P. (ed.), The Biology of Biodiversity. Yale University Press, New Haven, 159–199.

Lauzanne, L., 1992. Native species: The Orestias. In Dejoux, C. & A. Iltis (eds), Lake Titicaca, a Synthesis of Limnological Knowledge. Kluwer Academic Publishers, Boston, 405–419.

Lavoué, S., M. Miya, J. G. Inoue, K. Saitoh, N. B. Ishiguro & M. Nishida, 2005. Molecular systematics of the gonorynchiform fishes (Teleostei) on whole mitogenome sequences: Implications for higher-level relationships within the Otocephala. Molecular Phylogenetics and Evolution 37(1): 165–177.

Lévêque, C., 1997. Biodiversity and Conservation: The Freshwater Fish of Tropical Africa. Cambridge University Press, 432 pp.

Lévêque, C., E. V. Balian & K. Martens, 2005. An assessment of animal species diversity in continental waters. Hydrobiologia 542: 39–67.

Lundberg, J. C., 1998. The temporal context for the diversification of Neotropical fishes. In Malabarba, L. R., et al. (eds), Phylogeny and Classification of Neotropical Fishes. EDIPUCRS, Porto Alegre, 49–68.

Lundberg, J. G., M. Kottelat, G. R. Smith, M. Stiassny & A. C. Gill, 2000. So many fishes, so little time: An overview of recent ichthyological discovery in continental waters. Annals Missouri Botanical Garden 87: 26–62.

M. E. A. (Millenium Ecosystem Assessment), 2005. Ecosystems and Human Well-being Synthesis. Island Press, Washinton DC.

MacArthur, R. H., 1964. Environmental factors affecting bird species diversity. American Naturalist 98: 387–397.

MacArthur, R. H. & E. O. Wilson, 1967. The Theory of Island Biogeography. Princeton University Press, Princeton, New Jersey.

Maitland, P. S., 2000. Freshwater Fishes of Britain and Europe. Hamlyn, London.

Malabarba, L. R., R. R. Reis, R. P. Vari, Z. M. S. Lucena & C. A. S. Lucena, 1998. Phylogeny and Classification of Neotropical Fishes. EDIPUCRS, Porto Alegre.

Mats, V. D., 1993. The structure and development of the Baikal Rift depression. Earth Science Review 34: 81–118.

Mayden, R. L. (ed.), 1992. Systematics, Historical Ecology, and North American Freshwater Fishes. Stanford University Press, CA, 969 pp.

Myers, G. S., 1949. Salt-tolerance of fresh-water fish groups in relation to zoogeographical problems. Bijdragen tot de Dierkunde 28: 315–322.

Nagelkerke, L. A. J. & F. A. Sibbing, 1996. Reproductive segregation among *Barbus intermedius* complex of Lake Tana, Ethiopia. An example of intralacustrine speciation? Journal of Fish Biology 49: 1244–1266.

Nagelkerke, L. A. J. & F. A. Sibbing, 2000. The large barbs (*Barbus* spp., Cyprinidae, Teleostei) of Lake Tana (Ethiopia) with a description of a new species, *Barbus osseensis*. Netherlands Journal of Zoology 50(2): 179–214.

Nelson, J. S., 2006. Fishes of the World, 4th edn. John Wiley & Sons, Hoboken, New Jersey, 601 p.

Oberdorff, T., J. F. Guégan & B. Hugueny, 1995. Global scale patterns in freshwater fish species diversity. Ecography 18: 345–352.

Oberdorff, T., B. Hugueny & J. F. Guégan, 1997. Evidence of historical effects on freshwater fish species diversity: Comparison between Europe and North America. Journal of Biogeography 24: 461–467.

Olden, J. D. & N. LeRoy Poff, 2004. Ecological processes driving biotic homogenization: Testing a mechanistic model using fish faunas. Ecology 85(7): 1867–1875.

Paugy, D., C. Lévêque & G. G. Teugels, 2003. Poissons d'eaux douces et saumâtres de l'Afrique de l'Ouest. Tomes 1 et 2. IRD Editions.

Rahel, F. J., 2000. Homogenization of fish faunas across the United States. Science 288: 854–856.

Rainboth, W. J., 1991. Cyprinids of South East Asia. In Winfield, I. J. & J. S. Nelson (eds), Cyprinid Fishes: Systematics, Biology, and Exploitation. Chapman and Hall, London, 156–210. Fish and Fisheries Series 3, 667 pp.

Reis, R. O., S. O. Kullander & C. J. Ferraris, 2003. Check List of the Freshwater Fishes of South and Central America. EDIPUCRS, Porto Alegre.

Revenga, C & Y. Kura, 2003. Status and Trends of Biodiversity of Inland Water Ecosystems. Technical Series 11. Secretariat of the Convention on Biological Diversity, Montreal, Canada.

Reyjol, Y., B. Hugueny, D. Pont, P. G. Bianco, U. Beier, N. Caiola, F. Casals, I. Cowx, A. Economou, T. Ferreira, G. Haidvogl, R. Noble, A. Sostoa, T. Vigneron & T. Virbickas, 2007. Patterns in species richness and endemism of European freshwater fish. Global Ecology and Biogeography 16: 65–75.

Ricklefs, R. E., 2004. A comprehensive framework for global patterns in biodiversity. Ecology Letters 7: 1–15.

Roberts, T. R., 1975. Geographical distribution of African freshwater fishes. Zoological Journal of the Linnean Society of London 57(4): 249–319.

Saitoh, K., K. M. Miya, J. G. Inoue, N. B. Ishiguro & M. Nishida, 2003. Mitochondrial genomics of ostariophysan fishes: Perspectives on phylogeny and biogeography. Journal of Molecular Evolution 56: 464–472.

Sala, O. E., F. S. Chapin III, J. J. Armesto, R. Berlow, J. Bloomfield, R. Dirzo, E. Huber-Sanwald, L. F. Huenneke, R. B. Jackson, A. Kinzig, R. Leemans, D. Lodge, H. A. Mooney, M. Oesterheld, N. L. Poff, M. T. Sykes, B. H. Walker, M. Walker & D. H. Wall, 2000. Global biodiversity scenarios for the year 2100. Science 287: 1770–1774.

Schlieven, U. K., D. Tautz & S. Pääbo, 1994. Sympatric speciation suggested by monophyly of crater lake cichlids. Nature 368: 629–632.

Schliewen, U., K. Rassmann, M. Markmann, J. Markert, T. Kocher & D. Tautz, 2001. Genetic and ecological divergence of a monophyletic cichlid species pair under fully sympatric conditions in Lake Ejagham, Cameroon. Molecular Ecology 10(6): 1471–1488.

Seehausen, O., J. J. M. van Alphen & F. Witte, 1997. Cichlid fish diversity threatened by eutrophication that curbs sexual selection. Science 277: 1808–1811.

Sideleva, V. G., 1994. Speciation of endemic Cottoidei in Lake Baikal. Advances in Limnology 44: 441–450.

Skelton, P. H., 2000. A Complete Guide to the Freshwater Fishes of Southern Africa. Struik Publishers, Cape Town, 395 pp.

Snoeks, J., 2000. How well known is the ichthyiodiversity of the large East African lakes? Advances in Ecological Research 31: 17–38.

Sparks, J. S. & M. L. J. Stiassny, 2003. Introduction to Madagascar's freshwater fishes. In Goodman, S. M. & J. P. Benstead (eds), The Natural History of Madagascar. University of Chicago Press, Chicago, IL, USA, 849–863.

Stiassny, M. L. J. & N. Raminosoa, 1994. The fishes of the inland waters of Madagascar. Biological diversity in African fresh- and brackish water fishes. Geographical overviews. Annales Musée royal de l'Afrique centrale, Zoology 275: 133–149.

Stiassny, M. L. J., U. K. Schliewen & W. J. Dominey, 1992. A new species flock of cichlid fishes from Lake Bermin, Cameroon with a description of eight new species of *Tilapia* (Labroidei; Cichlidae). Ichthyological Exploration of Freshwaters 3(4): 311–346.

Stiassny, M. L. J., G. G. Teugels & C. Hopkins, in press. The Fresh and Brakish Water Fishes of Lower Guinea, West-Central Africa. IRD éditions, collection Faune et Flore tropicales.

Sullivan, J. P., J. G. Lundberg & M. Hardman, 2006. A phylogenetic analysis of the major groups of catfishes using rag1 and rag2 nuclear gene sequences. Molecular Phylogenetics and Evolution 41: 636–662.

Tedesco, P. A., T. Oberdorff, C. A. Lasso, M. Zapata & B. Hugueny, 2005. Area and history versus contemporary energy in explaining diversity patterns in tropical riverine fish. Journal of Biogeography 32: 1899–1907.

Thieme, M. L., R. Abell, M. L. J. Stiassny, P. Skelton, et al., 2005. Freshwater Ecoregions of Africa and Madagascar. A Conservation Assessment. Island Press.

Unmack, P. J., 2001. Biogeography of Australian freshwater fishes. Journal of Biogeography 28(9): 1053–1089.

Vari, R. P. & L. R. Malabarba, 1998. Neotropical ichthyology: An overview. In Malabarba, L. R., R. P. Reis, et al. (eds), Phylogeny and Classification of Neotropical Fishes. Edipucrs, Porto Alegre, 1–11.

Verheyen, E., W. Salzburger, J. Snoeks & A. Meyer, 2003. Origin of the superflock of cichlid fishes from lake Victoria, East Africa. Science 300: 325–329.

Vörösmarty, C. J., C. Lévêque, C. Revenga, et al., 2006. Chapter 7 – fresh water. In Scholes, H. R. & N. Ash (eds), Ecosystems and Human Well-being. Vol. 1. Current State and Trends. Millennium Ecosystem Assessment. Island Press, Washington, 165–207.

Wright, D. H., 1983. Species-energy theory: An extension of the species-area theory. Oikos 41: 496–506.

Yu Sherbakov, D., 1999. Molecular phylogenetic studies on the origin of biodiversity in Lake Baikal. TREE 14: 92–95.

Yuma, M., K. Hosoya & Y. Nagata, 1998. Distribution of the freshwater fishes of Japan: An historical overview. Environmental Biology of fishes 52: 97–124.

Hydrobiologia (2008) 595:569–580
DOI 10.1007/s10750-007-9032-2

FRESHWATER ANIMAL DIVERSITY ASSESSMENT

Global diversity of amphibians (Amphibia) in freshwater

Miguel Vences · Jörn Köhler

Abstract This article present a review of species numbers, biogeographic patterns and evolutionary trends of amphibians in freshwater. Although most amphibians live in freshwater in at least their larval phase, many species have evolved different degrees of independence from water including direct terrestrial development and viviparity. Of a total of 5,828 amphibian species considered here, 4,117 are aquatic in that they live in the water during at least one life-history stage, and a further 177 species are water-dependent. These numbers are tentative and provide a conservative estimate, because (1) the biology of many species is unknown, (2) more direct-developing species e.g. in the Brachycephalidae, probably depend directly on moisture near water bodies and (3) the accelerating rate of species discoveries and descriptions in amphibians indicates the existence of many more, yet undescribed species, most of which are likely to have aquatic larvae. Regional endemism in amphibians is very high, with only six out of 348 aquatic genera occurring in more than one of the major biogeographic divisions used herein. Global declines threatening amphibians are known to be triggered by an emerging infectious fungal disease and possibly by climate change, emphasizing the need of concerted conservation efforts, and of more research, focused on both their terrestrial and aquatic stages.

Guest editors: E. V. Balian, C. Lévêque, H. Segers & K. Martens
Freshwater Animal Diversity Assessment

M. Vences (✉)
Division of Evolutionary Biology, Zoological Institute, Technical University of Braunschweig, Spielmannstr. 8, Braunschweig 38106, Germany
e-mail: m.vences@tu-bs.de

J. Köhler
Department of Zoology, Hessisches Landesmuseum Darmstadt, Friedensplatz 1, Darmstadt 64283, Germany
e-mail: j.koehler@hlmd.de

Keywords Amphibia · Anura · Urodela · Gymnophiona · Species diversity · Evolutionary trends · Aquatic species · Biogeography · Threats

Introduction

Amphibians are a textbook example of organisms living at the interface between terrestrial and aquatic habitats. They fulfil this role both in an ecological context, with typically a strictly aquatic larval and largely terrestrial adult phase, and in an evolutionary context, representing the intermediate *bauplan* level between aquatic and fully terrestrial vertebrates ("fishes" vs. amniotes). Most amphibians are strictly dependent from water for their larval development, and water for this group of animals is synonym to freshwater. Although a few amphibians are able to tolerate high-salinity levels (Balinsky, 1981), there are no marine representatives of this class.

Although existence of an aquatic larval phase is probably the ancestral condition for recent amphibians, there are only few amphibian taxa with also fully aquatic adult phases. In contrast, multiple evolutionary trends towards more terrestrial reproduction have led to a plethora of reproductive modes (Duellman & Trueb, 1986) which make it difficult, in some instances, to decide if and to what degree a particular species is indeed strictly dependent on freshwater.

Recent amphibians are often named Lissamphibia. They are divided in three orders: frogs (Anura), salamanders (Urodela), and caecilians (Gymnophiona). Dubois (2004) recommended abandoning several other higher taxa names based on arguments of nomenclatural priority. Although these priority rules do not strictly apply to names above the family level, we here follow Dubois (2004) in not using the names Apoda (for caecilians), Caudata (for salamanders), Salientia (for Recent frogs), and Archaeobatrachia and Neobatrachia (for basal and modern frogs). However, we decided to here continue using Lissamphibia for the clade containing all three recent amphibian orders, and we use "Archaeobatrachia" and "Neobatrachia" in quotation marks since these established terms make discussion of anuran relationships easier.

Due to the large diversity of extinct Paleozoic amphibians, the phylogenetic relationships of lissamphibians relative to amniotes has in the past been questioned. Current evidence converges on their monophyly, based on morphological characters such as, for instance, their pedicellate teeth, special visual cells (green rods) in the retina, or the ear structure (Duellman & Trueb, 1986), and on molecular characters (e.g. Meyer & Zardoya, 2003; San Mauro et al., 2005). The paucity of fossils, especially from the Mesozoic, makes it difficult to trace the early evolution of lissamphibians, but they appear to be a very old group according to molecular clocks which date the separation among the three orders back into Paleozoic times before Pangaean break-up (San Mauro et al., 2005; Roelants & Bossuyt, 2005). Furthermore, deep divergences are also typical for amphibians at the species level. As already noted by Wilson et al. (1974), amphibian species have much larger molecular divergences (and, consequently, ages) than other vertebrates e.g. mammals and birds, and also the large divergences among populations considered to be conspecific are typical for amphibians (e.g. Vences et al., 2005a, b).

The (aquatic) larvae of caecilians and salamanders are morphologically largely similar to their adults, except for the presence of external gills which usually are reduced in the adults. In contrasts, the larval stage of frogs, the tadpole, is a larval phase radically different from the adults (Altig & McDiarmid, 1999). Especially the oral and digestive system of tadpoles is composed of numerous features which are not homologous to the corresponding structures in the adult, such as a horny beak, oral papillae and keratinous labial teeth, mainly due to the fact that typically tadpoles are omnivorous suspension feeders, ingesting a high degree of vegetable matter, while adult frogs are strictly carnivorous (with only a single known exception, *Xenohyla truncata*, a species that also eats fruits).

Several excellent resources on amphibians were available over the world wide web at the time of preparation of the present article. The Amphibian Species of the World database (Frost, 2004), hosted by the American Museum of Natural History, continues previous efforts (Frost, 1985; Duellman, 1993) to document from a taxonomic point of view all amphibian taxa. Amphibiaweb (2005, www.amphibiaweb.org), hosted by the University of California at Berkeley, provides a full species list of amphibians, too, but aims at providing also additional information such as distribution, photographs, and biological data. The Global Amphibian Assessment (www.globalamphibians.org) has compiled, during 2002–2004, the expertise of regional and taxonomic experts worldwide and provides an estimate of threat status (according to IUCN criteria) and distribution for all amphibian species.

In this article, we summarize species diversity and distribution, and zoogeography, of extant amphibians, based on a species list and distributional information compiled from these three online data sources as accessed in December 2005. Furthermore, we categorize all amphibian species according to their water dependence on a regional and taxonomical level. For taxonomy we generally follow Frost et al. (2006). In the following sections, we use the definitions of the freshwater diversity assessment project in defining (1) aquatic species as those with at least part of their life cycle in or on the water, (2) water dependent species as those which do not live directly in the water but closely depend on it e.g. for habitat or food, (3) water related species as aquatic plus water dependent species, and (4) nor water related species as those which are neither aquatic nor water dependent.

Species diversity

A striking aspect of amphibian taxonomy is the increasing rate of new species discoveries (Glaw & Köhler, 1998; Köhler et al., 2005). As of December 2005, a total of 5,828 amphibian species (aquatic+ non-aquatic) were known, but still at the end of 1992, there were only 4,533 species (Duellman, 1993). The absolute number of newly described amphibian species per decade (not only the cumulative number of valid and described species) has been steadily increasing since the decade of the 1960s, with especially steep increases since the 1990s (Glaw & Köhler, 1998; Köhler et al., 2005). The new species are in part known populations of described species that are found to be genetically or bioacoustically distinct, and hence recognized as different species. However, the largest proportion of new species are genuine new discoveries, as exemplified by the recent spectacular findings of a new frog lineage, genus and species in India, *Nasikabatrachus* (Biju & Bossuyt, 2003; considered to be part of the family Sooglossidae by Frost et al, 2006), and of a new genus and species of plethodontid salamander, *Karsenia*, in Korea, being the first Asian representative of this family (Min et al., 2005). This taxonomic progress has been made possible by the combination of increased field exploration in tropical regions, together with the application of molecular and bioacoustic techniques becoming routine. A case study in Madagascar indicated that newly discovered species since 1990 are genetically not less distinct from already described species than species discovered in the research periods before, and that the increase in new species is not a sign of taxonomic inflation due to exaggerated splitting approaches (Köhler et al., 2005).

Of the total of 5,828 amphibian species considered here, 4,117 are aquatic in that they live in the water during at least one life-history stage, and a further 177 species are water dependent. By the end of 2005, there were a total of 168 species of caecilians, 514 salamanders and 5,146 frogs. While it is relatively straightforward to decide which of these are, under the definitions used here, real aquatics (i.e. with at least part of the life cycle in or on the water), it is more difficult to decide which of the non-aquatic species may be water dependent, i.e. with close/ specific dependence on aquatic habitats (see Figs. 1 and 2). In our categorization, all amphibians characterized by direct development, viviparity with terrestrial birth, or tadpole development in terrestrial jelly or foam nests are considered to be non-aquatic, while species with tadpoles in water-filled leaf axils of plants or tree holes were included in the aquatic category (Figs. 2 and 3). To be able to categorize all species, we have here assumed that species share the reproductive mode of their closest relatives (usually congeneric taxa). Although certainly not fully precise, this estimate should be a relatively reliable approximation to the real situation. In our analysis, species were categorized as "unknown" with respect to their dependence from freshwater only when no data at all were available to us concerning the life history of the particular genera. This concerns a small portion of 67 amphibian species only (Fig. 2).

Several of the non-aquatic species are certainly water dependent. For instance, the South African pyxicephalids of the genera *Arthroleptella* and *Anhydrophryne* have direct development but usually live in dense, mossy vegetation around springs and cannot colonize other habitats. Many plethodontid salamanders have direct development and do not live in the water, but are predominantly found along streams because they rely on the humid substrate nourished by the water. These species were placed in category 'water dependent'. Nevertheless, there certainly are species which are fully independent from water, such as the desert-dwelling species of *Breviceps* which occur far from any water body and depend on air humidity only. Several direct-developing frogs of the genera *Craugastor*, *Euhyas* and *Eleutherodactylus* live in bromeliads, and probably depend on the moisture provided by water-filled phytotelmes. However, since the life habits of very few species in this species-rich genera were studied, we here found it premature to decide which and how many species are non-aquatic but water dependent. As we did in the genera *Craugastor*, *Euhyas* and *Eleutherodactylus*, we considered all non-aquatic species where natural history observations are sparse also as non-water dependent, which is probably true for the majority of these species. However, it is implicit in this procedure that the numbers provided in Tables 1 and 2 will be slight underestimates.

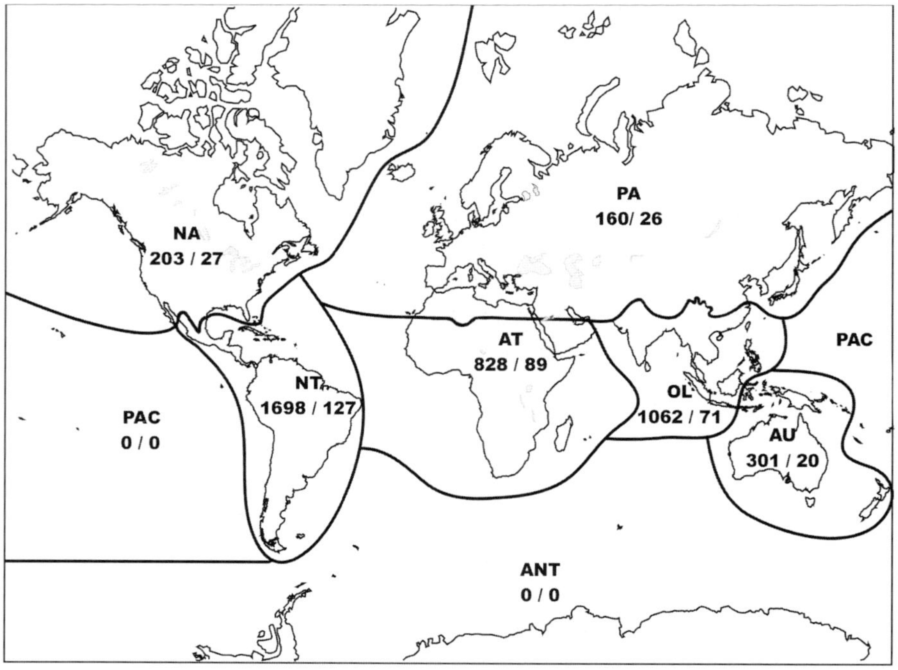

Fig. 1 Map showing species and genus diversity of water related amphibian species by major zoogeographic divisions. PA, Palearctic; NA, Nearctic; NT, Neotropical; AT, Afrotropical; OL, Oriental; AU, Australasian; PAC, Pacific Oceanic Islands; ANT, Antarctic. Numbers include aquatic amphibian species (with at least one aquatic life-history stage) plus those that are water dependent (e.g. some direct-developing species)

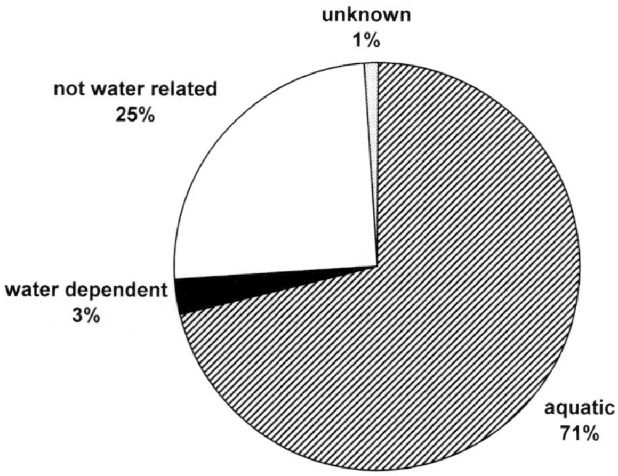

Fig. 2 Percentages (rounded) of amphibian species in categories aquatic, water dependent, not water related and unknown according to the definitions used herein. Together, the categories aquatic and water dependent are summed up as 'Total' in Tables 1 and 2

Historical processes: evolutionary trends away from and towards water

As outlined above, fully aquatic adult amphibians are rare. Among anurans, only representatives of the family Pipidae qualify as such plus very few species of other families, while among salamanders and caecilians, an aquatic life history is more common. In caecilians, species of the family Caeciliidae are reported to have aquatic adults with viviparous reproduction. However, ecological studies suggest that at least some species may not be strictly aquatic but actually display a semi-aquatic behaviour with resting periods out of water and foraging in aquatic habitat (Moodie, 1978). Among salamanders, fully aquatic families are the Cryptobranchidae, Amphiumidae, Sirenidae, and Proteidae, and aquatic adults also occur in the genus *Ambystoma*. In many of these species, for example in the well-known Axolotl, *Ambystoma mexicanum* (Ambystomatidae), the aquatic adults are paedomorphic (neotenic), and retain larval features such as external gills and a fully functional lateral-line system.

The most extreme modifications of reproductive modes are those that completely eliminate the free-living larval phase: direct development, and sometimes viviparity. Viviparous and ovoviviparous amphibians are relative rare, but caecilians are an exception where about 75% of the known species are considered to have a viviparous mode of reproduction (Himstedt 1996). Members of the Neotropical genus *Typhlonectes* and of the Afrotropical genus *Scolecomorphus* are exclusively viviparous, and viviparity also occurs in some other genera of the family Caeciliidae. In salamanders, viviparous species are

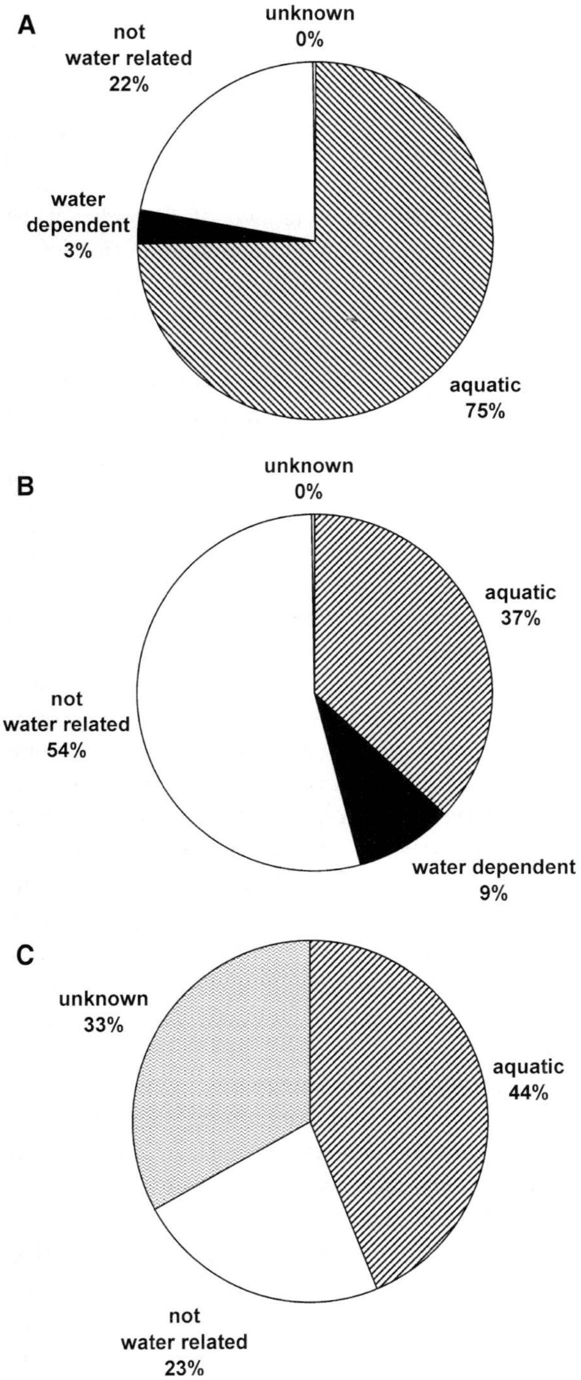

Fig. 3 Percentages (rounded) of frog (**A**), salamander (**B**) and caecilian (**C**) species in categories aquatic, water dependent, not water related and unknown. Together, the categories aquatic and water dependent are summed up as 'Total' in Tables 1 and 2

found exclusively in the Salamandridae (genera *Salamandra* and *Lyciasalamandra*). Among frogs, viviparity is known only in the two African bufonid genera *Nectophrynoides* and *Nimbaphrynoides*, and in the brachycephalid *Eleutherodactylus jasperi* (see Wake, 1977; 1989). All viviparous frogs bear fully metamorphosed young and therefore have no aquatic larval stage. Viviparous salamanders can bear metamorphosed young or (aquatic) larvae. Caecilians of the genus *Typhlonectes* are viviparous with aquatic larvae, whereas other viviparous caecilians have no larval stage. In pipid frogs of the genus *Pipa*, the eggs are embedded in the dorsum of the (aquatic) female and thus indirectly undergo development in an aquatic environment; in all other cases, direct-developing amphibians lay terrestrial eggs.

Direct development has evolved independently in many amphibian lineages: in salamanders once in the family Plethodontidae, which contains almost exclusively direct developers; in caecilians, in the family Caeciliidae; and among anurans in many of the major lineages: (1) in the basal Leiopelmatidae, genus *Leiopelma* from New Zealand; (2) in sooglossids (genus *Sooglossus*); (3) in some species of the genus *Pipa* (Pipidae); (4) in some genera of myobatrachids; (5) in brachycephalids; (6) in Hemiphractidae (genus *Hemiphractus*); (7) in Cryptobatrachidae (genera *Cryptobatrachus, Stefania*); (8) in Amphignathodontidae (genera *Flectonotus, Gastrotheca*); (9) in several genera of bufonids (e.g. *Oreophrynella, Osornophryne, Rhamphophryne*); (10) in the genus *Platymantis* (Ceratobatrachidae); (11) in Pyxicephalidae (*Arthroleptella* and related genera); (12) in at least one dicroglossid species (*Limnonectes hascheanus*); (13) in several species of one genus of mantellids (*Gephyromantis*; see Glaw & Vences, 2006); (14) in at least one genus of rhacophorids (*Philautus*); (15) in brevicipitids (*Breviceps* and related genera); (16) in Australasian microhylids (subfamily Asterophryinae); (17) in few Neotropical microhylids (*Myersiella, Synapturanus*); (18) in several arthroleptids (genera *Arthroleptis, Schoutedenella*). Hence, altogether there are at least 18 independent evolutionary events towards direct development in anurans, while this reproductive mode has evolved only once in salamanders and probably twice in caecilians. It needs to be emphasized that these are minimum estimates, and especially in frogs it is likely that more complete phylogenetic data will provide evidence that direct development evolved even more commonly in parallel.

Table 1 Numbers of water related amphibian species by order

Order	PA		NA		NT		AT		OL		AU		PAC		ANT		World	
	FW	WDpt	FW	WDpt	FW	WDpt	FW	WDpt	FW	WDpt	FW	WDpt	FW	WDpt	FW	WDpt	FW	WDpt
Anura	68	0	89	0	1,661	21	816	73	1,001	0	301	35	0	0	0	0	3,978	129
Urodela	92	0	114	48	14	0	0	0	17	0	0	0	0	0	0	0	237	48
Gymnophiona	0	0	0	0	23	0	12	0	44	0	0	0	0	0	0	0	79	0
Total	160	0	203	48	1,698	21	828	73	1,062	0	301	35	0	0	0	0	4,294	177

'FW' refers to aquatic plus water dependent species, WDpt to water dependent only (of a total of 5828 amphibian species considered; see text for definitions). PA : Palearctic; NA : Nearctic; NT : Neotropical; AT : Afrotropical ; OL : Oriental; AU : Australasian; PAC : Pacific Oceanic Islands, ANT: Antarctic

Interestingly, some of the direct-developing amphibian lineages are characterized by a very high species richness. The Brachycephalidae contain approximately 800 direct-developing species, and among salamanders, the largely direct-developing Plethodontidae encompass by far the largest number of species (349 out of a total of 514 salamander species). This may be seen as indication that water-independence is a particularly successful strategy for amphibians. It could also be a by-product of a putative higher fragmentation into isolated demes in direct developers, which may lead to an increased rate of species formation (Dubois, 2005). Studies that apply comparative methods to test against null models, and population genetic studies of direct developers are necessary to clarify this question.

Remarkably, recent phylogenetic evidence indicates that in a number of groups of predominantly direct development, some lineages have reversed their reproductive mode and re-acquired an aquatic larval stage. This appears to be the case in plethodontid salamanders, *Desmognathus*, as well as in some amphignathodontid frogs, *Gastrotheca*, and possibly in mantellid frogs as well (Duellman & Hillis, 1987; Vences & Glaw, 2001; Chippindale et al., 2004). These reversed trends emphasize the selective advantage, under at least some evolutionary conditions, of biphasic aquatic-terrestrial life cycles, and the importance of freshwaters for amphibian diversity.

Distribution and endemicity

Amphibians are in general considered to be poor dispersers, and the strong phylogeographic structure encountered in many amphibian species (e.g. Avise, 2000) appears to support this view. Due to their limited osmotic tolerance, overseas dispersal was long neglected as dispersal mechanism for amphibians, and their zoogeographic patterns explained largely by vicariance and dispersal over land connections (e.g. Duellman & Trueb, 1986). Evidence from molecular clocks and the discovery of endemic amphibians on oceanic islands, such as Mayotte on the Comoros, provide strong support that amphibians are able to colonize landmasses over the sea (e.g. Hedges et al., 1992; Vences et al., 2003; 2004). This concerns frogs, but may also apply to salamanders

Table 2 Numbers of amphibian genera including aquatic and water dependent species according to the definitions used herein by order

Order	PA	NA	NT	AT	OL	AU	PAC	ANT	World
Anura	8	9	119	85	64	20	0	0	294
Urodela	18	18	1	0	4	0	0	0	40
Gymnophiona	0	0	7	4	3	0	0	0	14
Total	26	27	127	89	71	20	0	0	348

For the subfamily Hylinae, the classicatory scheme of Faivovich et al. (2005) is followed. However, all other numbers in genera have not been updated and refer to a taxonomy prior to the publication of Frost et al. (2006). Therefore, some genera were supposed to occur in more than one biogeographic region and thus the sums of all regional numbers are higher than the total numbers. PA : Palearctic; NA : Nearctic; NT : Neotropical; AT : Afrotropical; OL : Oriental; AU : Australasian; PAC : Pacific Oceanic Islands, ANT: Antarctic

and even caecilians, as indicated by the presence of an endemic caecilian species, *Schistometopum thomense*, on the fully volcanic São Tomé island in the Gulf of Guinea (Measey et al., 2007). Nevertheless, it remains true that amphibian distributions have certainly largely been shaped by vicariance, as shown by relationships of relict forms such as the Seychellean sooglossid frogs and the Indian *Nasikabatrachus* (Biju & Bossuyt, 2003).

At the deep phylogenetic levels, there are clear distributional trends of salamanders and basal frogs having their centres of diversity in the Holarctis and caecilians and modern frogs in the tropics. Since some phylogenetic reconstructions placed caecilians as sister group of salamanders, and both basal and modern frogs ("archaeobatrachians" and "neobatrachians") as monophyletic groups, the distributional patterns of these lineages were interpreted by some phylogeneticists as indicative of vicariance during the break-up of Pangaea into the Laurasia and Gondwana supercontinents, with caecilians and "neobatrachians" evolving and diversifying on Gondwana, and "archaeobatrachians" and salamanders diversifying on Laurasia (Feller & Hedges, 1998).

Recent phylogenetic evidence, however, does not support this hypothesis. Evidence from complete mitochondrial sequences and nuclear genes indicates that frogs and salamanders, not salamanders and caecilians, are sister groups (Meyer & Zardoya, 2003; San Mauro et al., 2005; Frost et al., 2006). In addition, phylogenetic reconstructions based on different nuclear genes are concordant in establishing paraphyly of basal frogs ("archaeobatrachians") relative to the monophyletic "neobatrachians" (Hoegg et al., 2004; San Mauro et al., 2005; Roelants & Bossuyt, 2005), in accordance with morphological hypotheses (e.g. Duellman & Trueb, 1986). Furthermore, also the distributional patterns observed leave room for the assumption that not only causes of vicariance biogeography, but also of ecological requirements have shaped the distribution of the three amphibian orders.

Salamanders are almost exclusively distributed on previous Laurasian landmasses to which 9 of the 10 salamander families are restricted, if the presence of a few representatives of salamandrids (*Salamandra* and *Pleurodeles*) in northern Africa is disregarded. This pattern is obscured by the fact that one large radiation of one family, the Plethodontidae, has colonized the Neotropics and attains a high species diversity in Mexico and Central America. Indeed, 252 salamander species occur in the Neotropics as defined here, more than in any other biogeographic region. However, only few species of two genera, *Bolitoglossa* and *Oedipina*, have penetrated further into South America, leaving no doubts that northern America was the initial centre of diversification of this family. Almost all plethodontids are characterized by direct development, and are therefore less relevant in the present survey of freshwater diversity, most species not being included in Table 1. Salamanders have almost not penetrated into tropical areas of Asia, although there are salamandrids occurring as far to the south as Laos and Vietnam.

Caecilians have a distribution fully restricted to the tropics. They are found in the Neotropical, Afrotropical and Oriental regions. Interestingly, although endemic caecilians are present on the Seychelles, they are absent from Madagascar. Caecilians do not occur in southernmost South America, or in southern

Africa, indicating that the limiting factor for their distribution is indeed the presence of tropical-humid environments. Apart from climate which obviously triggers the current distribution of caecilians, it has been assumed that radiation in caecilians largely took place before Gondwana split into sub-continents. Some families like the South American Rhinatrematidae and the Asian Ichthyophiidae are supposed to represent relict distributions of formerly widespread Gondwanan ancestors (Duellman & Trueb, 1986), whereas some Neotropical members of the Caeciliidae (e.g. *Typhlonectes*) possibly are the product of subsequent radiations on the already isolated South American continent (Himstedt, 1996).

Basal frog lineages ("archaeobatrachians", defined as a paraphyletic group of all extant frogs not belonging to the modern frogs or "neobatrachians") are mainly distributed in the Holarctis, with four notable exceptions, however. (1) Pipids, the only frogs which are fully aquatic also in their adult stage, have a clearly Gondwanan distribution, with genera in Africa (*Xenopus*, *Silurana*, *Pseudhymenochirus* and *Hymenochirus*; 23 species) and in South America (*Pipa*; 7 species). (2) Leiopelmatidae: the genus *Leiopelma* (4 species) occurs in New Zealand, although its closest relative, *Ascaphus*, is restricted to the Nearctis. (3) The discoglossid genus *Barbourula* occurs on Borneo and the Philippines, whereas its closest relatives, the genus *Bombina*, has a Palearctic distribution. And (4) the Megophryidae, with 72 species by far the largest "archaeobatrachian" family, has radiated in the Oriental region and is common in tropical environments. Based on a robust molecular phylogeny and molecular clock dating, Roelants & Bossuyt (2005) found evidence for three major cladogenetic events between a Laurasia- and a Gondwana-associated lineage, represented by *Ascaphus* and *Leiopelma*, Rhinophrynidae and Pipidae and Pelobatoidea and "Neobatrachia", respectively, all these splits being very close to the onset of Pangaean break-up at 180 mya. Although this pattern substantiates a high-biogeographic relevance of "archaeobatrachians", they altogether make up only a negligible part of overall frog diversity, with a total of 191 of the total of 5,146 frog species (including non-water related taxa).

"Neobatrachians", with 4,955 species, do not only form the most speciose anuran subgroup, but also include by far more species than all other amphibian groups together. Monophyly of these "modern frogs" is well established by molecular and morphological characters (Duellman & Trueb, 1986; Hoegg et al., 2004; Roelants & Bossuyt, 2005; San Mauro et al., 2005). Their largest diversity belongs into two subgroups, the hyloids, with a centre of diversity in the Neotropics, and the ranoids, with a centre of diversity in Africa and Asia.

Ranoids, according to the scheme of Frost et al. (2006), include families restricted to Africa and/or Madagascar, such as the Arthroleptidae, Hemisotidae, Hyperoliidae, Mantellidae, Phrynobatrachidae, Ptychadenidae and Pyxicephalidae; two mainly Asian families with few African representatives, the Rhacophoridae and Dicroglossidae; two South Asian families, the Nyctibatrachidae and Microxalidae; one South-East Asian family, the Ceratobatrachidae; one family present in Africa and Asia, and that succeeded to colonize also North and South America, the Microhylidae; and the species-rich Ranidae that colonized Europe as well as North and South America. The Pedropeditae sensu Frost et al. (2006) contain African and South Asian species, although these relationships require further corroboration.

Hyloids include several families restricted to the Neotropics, such as the Amphignathodontidae, Brachycephalidae, Ceratophryidae, Centrolenidae, Cryptobatrachidae, Cycloramphidae, Dendrobatidae, Hemiphractidae, Leptodactylidae, Thoropidae; one family, the Bufonidae, common in the Neotropis with also many representatives in the Palearctic, Nearctic, Oriental, and Afrotropical regions, including genera endemic to the main biogeographical regions; and one family, the Hylidae, with many species in the Neotropis, which has representatives also in the Nearctic, Palearctic and Australian region.

Besides hyloids and ranoids, a number of further "neobatrachian" families exist. Into this assemblage belong the sooglossids from the Seychelles and southern India, as well as heleophrynids from South Africa, limnodynastids and myobatrachids from Australia and New Guinea and the Batrachophrynidae from South America.

While providing a general zoogeographic picture for amphibians or discussing the possible prevalence of vicariance vs. dispersal hypotheses, is clearly beyond the scope of this article, a number of general patterns can still be discerned from the distributions outlined above:

(1) At a very fundamental level, the influence of vicariance is very clearly visible in a number of amphibian distributions. Although salamanders and caecilians are certainly limited in their distribution by adaptations to temperate vs. tropical environments, their general patterns of geographic occurrence and the restriction of salamanders to temperate regions of the northern hemisphere make it likely that the basal diversification of salamanders occurred on Laurasia and that of caecilians on Gondwana (Feller & Hedges, 1998). "Archaeobatrachians" are separated in a number of lineages of alternatingly Laurasian or Gondwanian distribution (Roelants & Bossuyt, 2005). The distribution of basal neobatrachians in the southern hemisphere indicates that they initially had a Gondwanan distribution. And the diversity centres of hyloid vs. ranoid neobatrachian frogs in the New World vs. the Old World (and here, especially Africa) are likely to correspond to the separation of South America and Africa, which also roughly agrees with molecular clock calculations (e.g. San Mauro et al., 2005; Roelants & Bossuyt, 2005).

(2) The initial pattern originated by vicariance has been modified extensively by dispersal. The occurrence of some endemic amphibians on oceanic islands is a clear evidence for the possibility of overseas dispersal also in this group. The phylogenetic split of several lineages like the reed frogs, Hyperoliidae, are so young according to molecular clocks that their occurrence in Madagascar and on the Seychelles can only be seen in colonization by ancestors rafting on flotsam over the sea (Vences et al., 2003). The few genera occurring in more than one biogeographic region and continent provide further evidence for the possibility and potential speed of amphibian dispersal. The genus *Hoplobatrachus* has a number of species in Asia, and has one Afrotropical species that colonized, out of Asia, vast areas of the African savannas in short time spans, as to judge from the low molecular differentiation among Asian and African *Hoplobatrachus* species (Kosuch et al., 2001). Similar examples can be found in other genera and families as well. The colonization of South America by plethodontid salamanders and ranid frogs, and the colonization of the Palearctis by hylid frogs, almost certainly represent such instances. At the interface between dispersal and vicariance, the hylid subfamilies Pelodryadinae (Australian) and Phyllomedusinae (Neotropical) are sister groups (Hoegg et al., 2004; Frost et al., 2006), and probably are witnesses of a vaster distribution of these tree frogs while South America and Australia were connected over Antarctica in the Early Cenozoic. During this time, probably, the ancestor of these frogs dispersed from South America to Australia, and the two groups evolved in vicariance after the continental connections were severed.

(3) A third factor that should not be underestimated is (natural) extinction. There is impressive evidence in current amphibian distributions for formerly larger distribution areas of groups of today relictual occurrence. Among the examples are the relationships between *Ascaphus* and *Leiopelma* (Leiopelmatidae), the most basal of the extant anurans, and with two and four species restricted to the Western North America, and to New Zealand respectively. Plethodontid salamanders are today most diverse in the Nearctis and Neotropis, but one genus, *Speleomantes*, is known from Italy and France. The very recent discovery of the first Asian plethodontid *Karsenia* by Min et al. (2005) clearly demonstrates that this group had a wider distribution in Asia and Europe before, and probably went extinct over most of its Palearctic distribution area. Again, also the relictual distribution of the basal "neobatrachian" frogs in southern South America, Australia, South Africa, the Seychelles and India probably witnesses a previous, much wider Gondwanian distribution. Probably, and especially in frogs, successive waves of radiation of more modern groups have largely replaced the more basal groups which survived, if at all, as species-poor relicts in very restricted and fragmented distribution areas.

Due to their relatively limited dispersal ability, by far most amphibian genera, and almost all amphibian species, are endemic to single continents or biogeographic units. These units largely correspond to the biogeographic regions used here. Some additional subdivisions are obvious (of course not considering introductions): In the Afrotropical region, all species and all genera but one occurring in Madagascar are endemic to the island (with two genera of mantellid frogs also having one species each endemic to the Comoro island of Mayotte); and all genera and species of Seychellean caecilians and frogs are endemic to the archipelago. Sub-Saharan Africa shares no amphibian species with Asia or Europe, and the species-level of endemism of Australia is above 90%. Further islands with a degree of endemism of 100% at the species level are Jamaica, São Tomé and Principe, New Zealand, Fiji and Palau (percentages based on analyses including non-water related amphibian species; see http://www.globalamphibians.org/patterns.htm for a more detailed analysis based on the data from the Global Amphibian Assessment).

Most of the few genera with distributions extending over more than one of the main zoogeographic regions, all of them frogs, were classical "dump bin" genera which were recently split into various genera after comprehensive revisions (Faivovich et al., 2005; Frost et al., 2006). For example, the formerly 340 species of *Hyla* with representatives in the Neotropis, Orientalis, Nearctis and Palearctis have recently been taxonomically revised and were split into 15 genera, with the genus *Hyla* now being restricted to few species in the Nearctic and Palearctic regions (Faivovich et al. 2005). In addition, only the frog genera *Hoplobatrachus* (Orientalis with four and Afrotropis with one species) and *Ptychadena* (one species in Palaearctis as defined herein, all others in Afrotropis) have a distribution across biogeographic regions, and the salamander genus *Ambystoma* has 15 Nearctic and 14 Neotropical representatives; however, the Neotropical species are restricted to Mexico and the genus did not further disperse into Central or South America.

Human related issues: global amphibian declines

More new amphibian species are being discovered every year than ever, but at the same time, amphibians are paradoxically declining at a very fast rate (Hanken, 1999). Multi-causal declines have been recorded worldwide. The most obvious and immediate threat to most amphibians in a threatened IUCN category is habitat destruction, and for some species overexploitation (as pets or food) constitutes an imminent danger as well.

The Global Amphibian Assessment (Stuart et al., 2004) classified 1856 amphibian species (32.5%) into one of the IUCN threat categories (Vulnerable, Endangered or Critically Endangered), many more than in other groups such as mammals (23%) or birds (12%). About 43% of all species were recorded to experience some sort of population decline. A total of 32 amphibian species have become extinct, and 122 species were considered to be "possibly extinct", with no recent sightings.

Most alarmingly, so-called enigmatic declines have also been reported from unaltered and largely undisturbed habitats, especially in South America and Australia, but also in North America and Europe (Blaustein et al., 1994). Furthermore, it has been shown that the absence of aquatic larvae in declining anuran populations may significantly alter freshwater ecosystems (Ranvestel et al. 2004). Most likely, emerging infectious diseases, especially chytridiomycosis, play a key role in these declines which in many cases apparently have led to full extinctions already (Daszak et al., 2003). The chytrid fungus *Batrachochytrium dendrobatidis* especially affects species that are ecologically predisposed in that their natural history (high-altitude occurrence, stream breeding) coincides with the preferences of the pathogen, and if combined with low fecundity and habitat specialization, a species can quickly be driven to extinction (Daszak et al., 2003), and this process may be furthered by climatic change (Pounds et al., 2006). Interestingly, despite high degrees of chytrid infection in the wild, no African frogs have yet been reported to have enigmatically declined, which led Weldon et al. (2004) to hypothesize that the disease may have originated in Africa and spread to other continents by exported clawed frogs, *Xenopus*, as carriers.

The important role of chytridiomycosis in amphibian declines has been asserted, and obvious measures include the control of amphibian introductions into unaffected areas as well as the disinfection of fishing gear and similar equipment by limnologists working on different continents. However, the influence of other agents such as pesticides or increased UV radiation should not be disregarded, and multi-causal hypotheses may well be most powerful to explain declines in some cases. Certainly, the simultaneous dependence of many amphibian species from both aquatic and terrestrial environments make them especially vulnerable to a multitude of factors. The importance of the freshwater environment for the survival of amphibian populations is paramount, and more studies on the specific requirements of amphibians in their aquatic, mostly larval phase are necessary to develop integrated conservation strategies.

Acknowledgements We are grateful to Francisco Hita García for his help with updating our amphibian species database, and to Frank Glaw for numerous discussions and comments.

References

Altig, R. & R. W. McDiarmid, 1999. Body plan: development and morphology. In McDiarmid, R. W. & R. Altig (eds), Tadpole: The Biology of Anuran Larvae. University of Chicago Press, Chicago: 24–51.

AmphibiaWeb, 2005. Information on Amphibian Biology and Conservation. [web application]. Berkeley, California. AmphibiaWeb. Available: http://amphibiaweb.org/. (Accessed, 2005).

Avise, J. C., 2000. Phylogeography. The History and Formation of Species. Harvard University Press, Cambridge, Massachusetts.

Balinsky, J. B., 1981. Adaptation of nitrogen metabolism to hyperosmotic environment in Amphibia. Journal of Experimental Zoology 215: 335–350.

Biju, S. D. & F. Bossuyt, 2003. New frog family from India reveals an ancient biogeographical link with the Seychelles. Nature 425: 711–714.

Blaustein, A. R., D. B. Wake & W. P. Sousa, 1994. Amphibian declines: judging stability, persistence, and susceptibility of populations to local and global extinctions. Conservation Biology 8: 60–71.

Chippindale, P. T., A. S. Baldwin, R. M. Bonett & J. J. Wiens, 2004. Phylogenetic evidence for a major reversal of life history evolution in plethodontid salamanders. Evolution 58: 2809–2822.

Daszak, P., A. A. Cunningham & A. D. Hyatt, 2003. Infectious disease and amphibian population declines. Diversity and Distributions 9: 141–150.

Dubois, A., 2004. The higher nomenclature of recent amphibians. Alytes 22: 1–14.

Dubois, A., 2005. Developmental pathway, speciation and supraspecific taxonomy in amphibians 1. Why are there so many frog species in Sri Lanka. Alytes 22: 19–37.

Duellman, W. E., 1993. Amphibian Species of the World: Additions and Corrections, Vol. 21. University of Kansas, Museum of Natural History, Special Publication, pp. 1–372.

Duellman, W. E. & D. M. Hillis, 1987. Marsupial frogs (Anura: Hylidae: *Gastrotheca*) of the Ecuadorian Andes: resolution of taxonomic problems and phylogenetic relationships. Herpetologica 43: 141–173.

Duellman, W. E. & L. Trueb, 1986. Biology of Amphibians. McGraw-Hill, New York.

Faivovich, J., C. F. B. Haddad, P. C. A. Garcia, D. R. Frost, J. A. Campbell & W. C. Wheeler, 2005. Systematic review of the frog family Hylidae, with special reference to Hylinae: phylogenetic analysis and taxonomic revision. Bulletin of the American Museum of Natural History 294: 1–240.

Feller, A. E. & S. B. Hedges, 1998. Molecular evidence for the early history of living amphibians. Molecular Phylogenetics and Evolution 9: 509–516.

Frost, D. R. (ed.), 1985. Amphibian Species of the World. Association of Systematic Collections, Allen Press, Lawrence, Kansas.

Frost, D. R., 2004. Amphibian Species of the World: an Online Reference. Version 3.0 (22 August, 2004). Electronic Database accessible at http://research.amnh.org/herpetology/amphibia/index.html. American Museum of Natural History, New York, USA.

Frost, D. R., T. Grant, J. Faivovich, R. H. Bain, A. Haas, C. F. B. Hadad, R. De Sa, A. Channing, M. Wilkinson, S. C. Donnellan, C. J. Raxworthy, J. A. Campbell, B. L. Blotto, P. Moler, R. C. Drewes, R.A. Nussbaum, J. D. Lynch, D. M. Green & W. C. Wheeler, 2006. The amphibia tree of life. Bulletin of the American Museum of Natural History 297: 1–370.

Glaw, F. & J. Köhler, 1998. Amphibian species diversity exceeds that of mammals. Herpetological Review 29: 11–12.

Glaw, F. & M. Vences, 2006. Phylogeny and genus-level classification of mantellid frogs (Amphibia, Anura). Organisms Diversity & Evolution 6: 236–253.

Hanken, J., 1999. Why are there so many new amphibian species when amphibians are declining? Trends in Ecology and Evolution 14: 7–8.

Hedges, S. B., C. A. Hass & L. R. Maxson, 1992. Caribbean biogeography: molecular evidence for dispersal in west Indian terrestrial vertebrates. Proceedings of the National Academy of Sciences of the USA. 89: 1909–1913.

Himstedt, W., 1996. Die Blindwühlen. Neue Brehm-Bücherei, Vol. 630.

Hoegg, S, M. Vences, H. Brinkmann & A. Meyer, 2004. Phylogeny and comparative substitution rates of frogs inferred from sequences of three nuclear genes. Molecular Biology and Evolution 21: 1188–1200.

Kosuch, J., M. Vences, A. Dubois, A. Ohler & W. Böhme, 2001. Out of Asia: mitochondrial DNA evidence for an oriental origin of tiger frogs, genus *Hoplobatrachus*. Molecular Phylogenetics and Evolution 21: 398–407.

Köhler, J., D. R. Vietes, R. M. Bonett, F. Hita Garcia, F. Glaw, D. Steinke & M. Vences, 2005. New amphibians and global conservation: a boost in species discoveries in a highly endangered vertebrate group. BioScience 55: 693–696.

Measey, G. J., M. Vences, R. C. Drewes, Y. Chiari, M. Melo & B. Bourles, 2007. Freshwater paths into the ocean: molecular phylogeny of the frog *Ptychadena newtoni* gives insights into amphibian colonization of oceanic islands. Journal of Biogeography 34: 7–20.

Meyer, A. & R. Zardoya, 2003. Recent advances in the (molecular) phylogeny of vertebrates. Annual Reviews of Ecology, Evolution and Systematics 34: 311–338.

Min, M. S., S. Y. Yang, R. M. Bonett, D. R. Vieites, R. A. Brandon & D. B. Wake, 2005. Discovery of the first Asian plethodontid salamander. Nature 435: 87–90.

Moodie, G. E. E., 1978. Observations on the life history of the caecilian *Typhlonectes compressicaudus* (Duméril & Bibron) in the Amazon basin. Canadian Journal of Zoology 56: 1005–1008.

Pounds, J. A., M. R. Bustamante, L. A. Coloma, J. A. Consuegra, M. P. L. Fogden, P. N. Foster, E. LaMarca, K. L. Masters, A. Merino-Viteri, R. Puschendorf, S. R. Ron, G. A. Sánchez-Azofeifa, C. J. Still & B. E. Young, 2006. Widespread amphibian extinctions from epidemic disease driven by global warming. Nature 439: 161–167.

Ranvestel, A. W., K. R. Lips, C. M. Pringle, M. R. Whiles & R. J. Bixby, 2004. Neotropical tadpoles influence stream benthos: evidence for the ecological consequences of decline in amphibian populations. Freshwater Biology 49: 274–285.

Roelants, K. & F. Bossuyt, 2005. Archaeobatrachian paraphyly and Pangaean diversification of crown-group frogs. Systematic Biology 54: 111–126.

San Mauro, D., M. Vences, M. Alcobendas, R. Zardoya & A. Meyer, 2005. Initial diversification of living amphibians predated the breakup of Pangaea. The American Naturalist 165: 590–599.

Stuart, S. N., J. S. Chanson, N. A. Cox, B. E. Young, A. S. L. Rodriguez, D. L. Fishman & R. W. Waller, 2004. Status and trends of amphibian declines and extinctions worldwide. Science 306: 1783–1786.

Vences, M. & F. Glaw, 2001. Systematic review and molecular phylogenetic relationships of the direct developing Malagasy anurans of the *Mantidactylus asper* group (Amphibia, Mantellidae). Alytes 19: 107–139.

Vences, M., J. Kosuch, M. -O. Rödel, S. Lötters, A. Channing, F. Glaw & W. Böhme, 2004. Phylogeography of *Ptychadena mascareniensis* suggests transoceanic dispersal in a widespread African-Malagasy frog lineage. Journal of Biogeography 31: 593–601.

Vences, M., M. Thomas, A. van der Meijden, Y. Chiari & D. R. Vieites, 2005a. Comparative performance of the 16S rRNA gene in DNA barcoding of amphibians. Frontiers in Zoology 2: article 5.

Vences, M., M. Thomas, R. M. Bonett & D. R. Vieites, 2005b. Deciphering amphibian diversity through DNA barcoding: chances and challenges. Philosophical Transactions of the Royal Society London Series B 360: 1859–1868.

Vences M., D. R. Vieites, F. Glaw, H. Brinkmann, J. Kosuch, M. Veith & A. Meyer, 2003. Multiple overseas dispersal in amphibians. Proceedings of the Royal Society of London Series B 270: 2435–2442.

Wake, M. H., 1977. The reproductive biology of caecilians: an evolutionary perspective. In Taylor, D. H. & S. I. Guttman (eds), Reproductive Biology of Amphibians. Plemum Press, New York: 73–101.

Wake, M. H., 1989. Phylogenesis of direct development and viviparity in vertebrates. In Wake, D. B. & G. Roth (eds), Complex Organismal Functions: Integration and Evolution in Vertebrates. John Wiley & Sons Ltd.: 235–250.

Weldon, C., L. H. du Preez, A. D. Hyatt, R. Muller & R. Speare, 2004. The origin of the amphibian chytrid fungus. Emerging Infectious Diseases 10: 2100–2105.

Wilson, A. C., L. R. Maxson & V. M. Sarich, 1974. Two types of molecular evolution. Evidence from studies of interspecific hybridization. Proceedings of the National Academy of Sciences of the USA 71: 2843–2847.

Hydrobiologia (2008) 595:581–586
DOI 10.1007/s10750-007-9115-0

FRESHWATER ANIMAL DIVERSITY ASSESSMENT

Global diversity of lizards in freshwater (Reptilia: Lacertilia)

Aaron M. Bauer · Todd Jackman

Abstract No lizards are strictly aquatic, but at least 73 species in 11 different families can be considered to regularly utilize freshwater habitats. There are no aquatic lizards in the Nearctic or Palearctic regions, whereas the Neotropics, Southeast Asia, and the Indo-Australian Archipelago support the greatest diversity of freshwater forms, particularly in the families Gymnophthalmidae, Scincidae and Varanidae. A number of larger aquatic lizards are harvested for food and for the reptile skin trade and several are CITES listed.

Keywords Lizards · Distribution · Natural history · Scincidae · Varanidae · Gymnophthalmidae

Introduction

Approximately 5,000 species of lizards (including amphisbaenians) are distributed between the Arctic Circle and Tierra del Fuego, with most species occuring in the tropics and subtropics. Most are arboreal, saxicolous, terrestrial, or fossorial. Many tropical lizards prefer or require mesic environments and, as a consequence, some are largely or entirely restricted to habitats bordering streams, rivers, or pools (e.g., agamids of the genus *Gonocephalus*). A much smaller subset of lizards regularly utilize freshwater habitats. No lizards are known to be strictly aquatic in that they must remain submerged for extended periods or depend exclusively on aquatic prey. Those lizards here considered to be aquatic typically retreat to water when alarmed and regularly include aquatic prey in their diets. Many are also partly arboreal and perch in branches overhanging water bodies. The designation "aquatic" is necessarily subjective, however, as some individuals of many other lizard species may retreat to or feed in freshwater at some time.

Aquatic lizards vary significantly in size and body form and few generalizations can be made as to morphological features indicative of an aquatic lifestyle; even within the genus *Anolis*, the seven aquatic forms share few features in common (Leal et al. 2002). A laterally compressed tail, however, is common to most aquatic lizards (Bedford & Christian 1996) and rugose or keeled body scalation is exhibited by many (e.g., *Lanthanotus*, *Neusticurus*, *Tropidophorus*, and *Cophoscincopus*). A series of rectangular toe fringes characterizes two of the lineages (*Basiliscus*, *Hydrosaurus*) capable of running on the water surface (Luke 1986). Given the lack of reliable morphological markers, only natural

Guest editors: E. V. Balian, C. Lévêque, H. Segers & K. Martens
Freshwater Animal Diversity Assessment

A. M. Bauer (✉) · T. Jackman
Department of Biology, Villanova University,
800 Lancaster Avenue, Villanova, PA 19085, USA
e-mail: aaron.bauer@villanova.edu

T. Jackman
e-mail: todd.jackman@villanova.edu

history data can reveal if a lizard species is a freshwater-dweller. We present a summary of lizards which, on the basis of literature records and our own observations, appear to be partly dependent on freshwater habitats or resources for their existence.

Species diversity and distribution

Utilization of freshwater habitats and resources has evolved numerous times within lizards, occurring in 11 families (Tables 1, 2), chiefly in the tropics (Fig. 2). Higher order relationships of these families to one another are discussed by Townsend et al. (2004) and Lee (2005). Information on freshwater lizard systematics, distribution and natural history may be found in the works of Pianka and Vitt (2003) (World), Blanc (1967), Glaw & Vences (1994), Böhme et al. (2000) (Afrotropical), Beebe (1945), Howland et al. (1990), Avila-Pires (1995), Vitt & Avila-Pires (1998), Leal et al. (2002), Doan & Castoe (2005) (Neotropical), Sprackland (1972), Mägdefrau (1987), Darevsky et al. (2004), Honda et al. (2006) (Oriental), Shine (1986), Daniels (1987), and Greer (1989) (Australasian).

Palearctic region

No aquatic lizards occur in Europe, North Africa, or Palearctic Asia.

Nearctic region

Lizard diversity in the Nearctic is heavily biased toward arid and semi-arid adapted species. No aquatic lizards are known or are likely to occur in this region.

Afrotropical region

Five mainland African lizards, the Nile monitor, *Varanus niloticus*, and the ornate monitor, *V. ornatus*, as well as all three species of the west African skink genus *Cophoscincopus*, are semi-aquatic. Another four water-associated species: the gerrhosaurid *Zonosaurus maximus* and three skinks, *Amphiglossus astrolabi*, *A. waterloti*, and *A. reticulatus*, are endemic to Madagascar.

Neotropical region

The most striking Neotropical aquatic lizards are the members of the corytophanid genus *Basiliscus*. Although none are restricted to aquatic habitats, all possess fringed toes, which help to support the body when the animal runs bipedally across the water surface, as part of its spectacular escape behavior. A total of 7 out of 358 species of *Anolis* lizards (two in

Table 1 Global distribution of freshwater lizard species by biogeographic region

SP: Species number	PA	NA	NT	AT	OL	AU	PAC	ANT	World Aquatic Species	World species in family
Lizards										
Agamidae					4	2			5	395
Corytophanidae			4						4	9
Gerrhosauridae				1					1	32
Gymnophthalmidae			7						7	204
Lanthanotidae					1				1	1
Polychrotidae			7						7	395
Scincidae				6	21	4	1		32	1,320
Teiidae			3						3	123
Tropiduridae			1						1	317
Varanidae				2	1	8	1		11	62
Xenosauridae					1				1	7
Total	0	0	22	9	28	14	2	0	73	

Note that totals may be lower than the sum of all cells because some species are shared between regions

PA: Palaearctic; NA: Nearctic; NT: Neotropical; AT: Afrotropical; OL: Oriental; AU: Australasian; PAC: Pacific Oceanic Islands; ANT: Antarctic

Table 2 Global distribution of freshwater lizard genera by biogeographic region

GN: genus Number	PA	NA	NT	AT	OL	AU	PAC	ANT	World aquatic
Lizards									
Agamidae					2	2			2
Corytophanidae			1						1
Gerrhosauridae				1					1
Gymnophthalmidae			2						2
Lanthanotidae					1				1
Polychrotidae			1						1
Scincidae				2	2	1	1		6
Teiidae			2						2
Tropiduridae			1						1
Varanidae				1	1	1	1		1
Xenosauridae					1				1
Total	0	0	7	4	7	4	2	0	19

Note that totals may be lower than the sum of all cells because some genera are shared between regions

PA: Palaearctic; NA: Nearctic; NT: Neotropical; AT: Afrotropical; OL: Oriental; AU: Australasian; PAC: Pacific Oceanic Islands; ANT: Antarctic

the Greater Antilles and five in mainland Central America) are semi-aquatic, representing four independent derivations of this lifestyle. Several species are capable of running on the water surface, but lack specialized toe fringes. The tropidurid lizard *Uranoscodon superciliosus*, likewise, can run on water and feeds chiefly on invertebrates in the flotsam at the water's edge. Many species of gymnophthalmids of the genera *Neusticurus* and *Potamites* are also semi-aquatic and are usually associated with small streams. Several large teiid lizards of South America are also typically restricted to water courses. These include *Crocodilurus amazonicus* and two species of *Dracaena.* Members of the latter genus forage underwater, while walking on the bottom and are predators on snails. Other Neotropical lizards, including the iguanid *Iguana iguana* and the teiid *Tupinambis teguixin*, are often associated with freshwater habitats, but are not restricted to them.

Oriental region

This region contains the most species of semi-aquatic lizards. The majority of species are in the scincid genus *Tropidophorus*. Many of the species for which data are available live along streambeds and are the ecological equivalents of the Neotropical *Neusticurus* and *Potamites*. The Vietnamese skink *Sphenomorphus cryptotis* is also restricted to watercourses and has a laterally compressed tail. The agamid *Physignathus cocincinus,* as well as all species of *Hydrosaurus* are semi-aquatic and the latter possess toe fringes like those of *Basiliscus.*

Fig. 1 Endangered freshwater xenosaurid lizard, *Shinisaurus crocodilurus*, from Yen Tu Nature Reserve, Quang Ninh Province, Northeast Vietnam with body immersed in pool. Photo courtesy of Le Khac Quyet/ FFI Vietnam

One species, *H. amboinensis*, extends into the Australasian region as far as New Guinea. *Varanus salvator* is the largest aquatic lizard, reaching total lengths of more than 3.0 m. It uses a broad range of aquatic habitats and may be found in brackish or even salt water, as well as freshwater. The Oriental fauna also includes two highly distinctive and phylogenetically isolated taxa, *Shinisaurus* and *Lanthanotus*. *Shinisaurus crocodilurus* (Fig. 1) is distributed in China and Vietnam and is among the most aquatic of lizards. The biology of *Lanthanotus* is poorly known, but its diet and limited natural history observations indicate that it is both fossorial and semi-aquatic.

Australasian region

Australasian freshwater lizards occur in the Agamidae, Scincidae, and Varanidae. The agamid *Physignathus lesueurii* is at least semi-aquatic as are four members of the Australian lygosomine skink genus *Eulamprus* (*E. quoyii*, *E. leuraensis*, *E. kosciuskoi*, *E. tympanum*). Three Australian species of *Varanus* (*V. mertensi*, *V. mitchelli*, *V. semiremex*) are typically semi-aquatic as are some members of the Indo-Australian *V. indicus* complex (*V. caerulivirens*, *V. cerambonensis*, *V. jobiensis*, *V. melinus*; ecology unknown in some other recently described species) although the degree of reliance on freshwater varies significantly among these species. One primarily Oriental species, *V. salvator*, extends into Australasia in the region of Maluku, Indonesia.

Oceanic islands Pacific

Another member of the *Varanus indicus* complex, *V. juxtindicus* of Rennell Island in the Solomons, is also semi-aquatic and a single New Caledonian skink, *Lioscincus steindachneri*, is strongly associated with stream courses, particularly as juveniles.

Antarctica

No lizards of any kind occur in Antarctica.

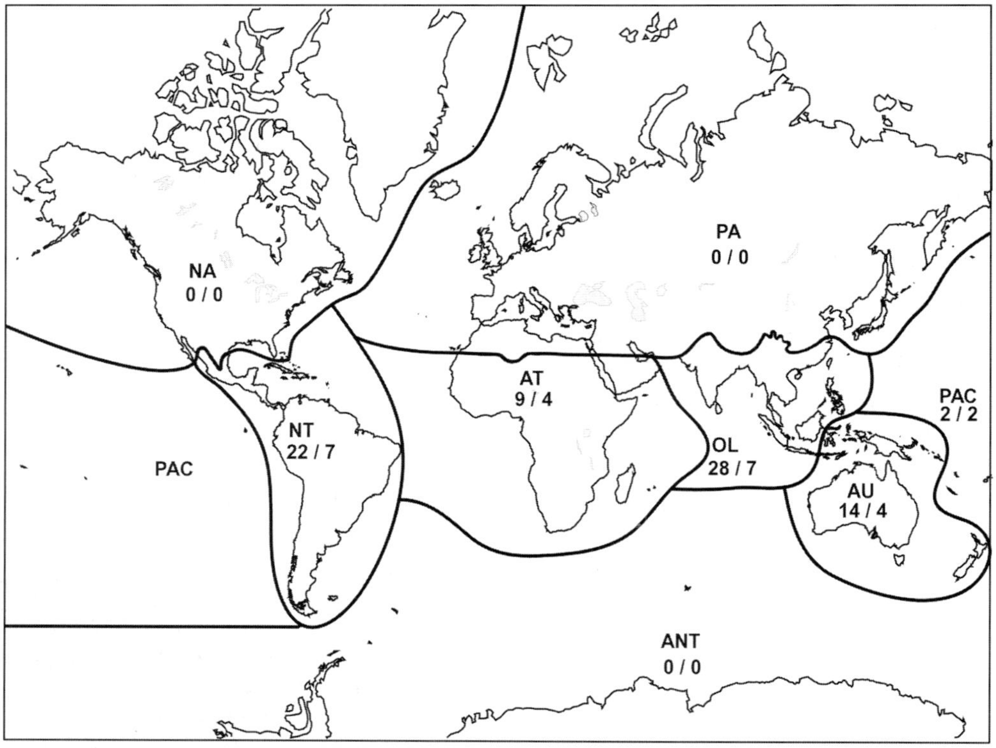

Fig. 2 Continental distribution of freshwater lizards. PA—Palaearctic; NA—Nearctic; NT—Neotropical; AT—Afrotropical ; OL—Oriental; AU—Australasian; PAC—Pacific Oceanic Islands; ANT—Antarctic

A total of 73 lizard species, or slightly less than 1.5% of all currently known taxa, is strongly associated with freshwater habitats (see Table 1). No lizard is known from the Antarctic Region and no freshwater lizards have been recorded from either the Nearctic or Palearctic regions. Aquatic lizards are most speciose in the tropics, particularly in association with humid forest regions of Central and South America, Southeast Asia, and the Indo-Australian Archipelago (Fig. 2, Tables 1, 2). Semi-aquatic habits have evolved in many unrelated groups of lizards and are typically not associated with highly specialized morphologies. However, most species that run on the water surface are members of the Iguania, a large clade of diurnal, visually oriented, ambush predators, whereas those lizards that regularly swim are members of a group of chemosensory active foragers—the Autarchoglossa. The fundamental differences between these two groups in activity, foraging mode, and dominant sensory modalities have undoubtedly contributed to their alternative aquatic adaptations, as well as many other ecological differences (Vitt et al. 2003).

Human related issues

Most lizards are too small to be consumed or otherwise used commercially by humans. However, larger lizards, including species of *Varanus*, *Dracaena* and *Crocodilurus* are harvested for their skins and may be regularly eaten by people. Harvesting for the skin trade is especially high (>500,000/annum) for the two semi-aquatic African *Varanus* species (de Buffrénil 1993) and for *V. salvator* (Luxmoore & Groombridge 1990). *Shinisaurus* has been negatively impacted by deforestation and is becoming rare within its range (Le Khac Quyet & Ziegler 2003). All aquatic species of *Varanus*, *Shinisaurus*, *Dracaena* and *Crocodilurus* are CITES Appendix II listed.

References

Avila-Pires, T. C. S., 1995. Lizards of Brazilian Amazonia (Reptilia: Squamata). Zoologische Verhandelingen 299: 1–706.

Bedford, G. S. & K. A. Christian, 1996. Tail morphology related to habitat of varanid lizards and some other reptiles. Amphibia–Reptilia 17: 131–140.

Beebe, W., 1945. Field notes on the lizards of Kartabo, British Guiana and Caripito, Venezuela. Part 3. Teiidae, Amphisbaenidae and Scincidae. Zoologica (New York) 30: 7–31.

Blanc, C. P., 1967. Notes sur les Gerrhosaurinae de Madagascar, I. - Observations sur *Zonosaurus maximus*, Boulenger, 1896. Annales de l'Université de Madagascar (Sciences) 5: 107–116.

Böhme, W., A. Schmitz & T. Ziegler, 2000. A review of the West African skink genus *Cophoscincopus* Mertens (Reptilia: Scincidae: Lygosominae): resurrection of *C. simulans* (Vaillant, 1884) and description of a new species. Revue suisse de Zoologie 107: 777–791.

Daniels, C. B., 1987. Aspects of the aquatic feeding ecology of the riparian skink *Sphenomorphus quoyii*. Australian Journal of Zoology 35: 253–258.

Darevsky, I. S., N. L. Orlov & TC Ho, 2004. Two new lygosomine skinks of the genus *Sphenomorphus* Fitzinger, 1843 (Sauria, Scincidae) from northern Vietnam. Russian Journal of Herpetology 11: 111–120.

de Buffrénil, V., 1993. Les Varans Africains (*Varanus niloticus* et *Varanus exanthematicus*). Données de Synthèse sur leur Biologie et leur Exploitation. Secrétariat CITES, Genève.

Doan, T. M. & T. A. Castoe, 2005. Phylogenetic taxonomy of the Cercosaurini (Squamata: Gymnophthalmidae), with new genera for species of *Neusticurus* and *Proctoporus*. Zoological Journal of the Linnean Society 143: 405–416.

Glaw, F. & M. Vences, 1994. A Fieldguide to the Amphibians and Reptiles of Madagascar, second edition. M. Vences & F. Glaw Verlags GbR, Köln.

Greer, A. E., 1989. The Biology & Evolution of Australian Lizards. Surrey Beatty & Sons Pty. Ltd., Chipping Norton.

Honda, M., H. Ota, R. W. Murphy & T. Hikida, 2006. Phylogeny and biogeography of water skinks of the genus *Tropidophorus* (Reptilia: Scincidae): A molecular approach. Zoologica Scripta 35: 85–95.

Howland, J. M., L. J. Vitt & P. T. Lopez, 1990. Life of the edge: The ecology and life history of the tropidurine iguanid lizard *Uranoscodon superciliosum*. Canadian Journal of Zoology 68: 1366–1373.

Le Khac Q. & T. Ziegler, 2003. First record of the Chinese crocodile lizard from outside of China: Report on a population of *Shinisaurus crocodilurus* Ahl, 1930 from north-eastern Vietnam. Hamadryad 27: 193–199.

Leal, M., A. K. Knox & J. B. Losos, 2002. Lack of convergence in aquatic *Anolis* lizards. Evolution 56: 785–791.

Lee, M. S., 2005. Squamate phylogeny, taxon sampling, and data congruence. Organisms, Diversity & Evolution 5: 25–45.

Luke, C., 1986. Convergent evolution of lizard toe fringes. Biological Journal of the Linnean Society 27: 1–16.

Luxmoore, R. & B. Groombridge, 1990. Asian Monitor Lizards: A Review of Distribution, Status, Exploitaion and Trade in Four Selected Species. Secrétariat CITES, Lausanne.

Mägdefrau, H., 1987. Zur Situation der Chinesischen Krokodilschwanz-Höckerechse, *Shinisaurus crocodilurus* Ahl, 1930. Herpetofauna 51: 6–11.

Pianka, E. R. & L. J. Vitt, 2003. Lizards, Windows to the Evolution of Diversity. University of California Press, Berkeley.

Shine, R., 1986. Diets and abundances of aquatic and semi-aquatic reptiles in the Alligator Rivers region. Supervising Scientist for the Alligator Rivers Region Technical Memorandum 16: 1–54.

Sprackland, G. B., 1972. A summary of observations of the earless monitor, *Lanthanotus borneensis*. Sarawak Museum Journal 20: 323–327.

Townsend, T. M., A. Larson, E. Louis & J. R. Macey, 2004. Molecular phylogenetics of Squamata: the position of snakes, amphisbaenians, and dibamids, and the root of the squamate tree. Systematic Biology 53: 735–757.

Vitt, L. J. & T. C. S. Avila-Pires, 1998. Ecology of two sympatric species of *Neusticurus* (Sauria: Gymnophthalmidae) in the western Amazon of Brazil. Copeia 1998: 570–582.

Vitt, L. J., E. R. Pianka, W. E. Cooper Jr. & K. Schwenk, 2003. History and global ecology of squamate reptiles. The American Naturalist 162: 44–60.

Hydrobiologia (2008) 595:587–591
DOI 10.1007/s10750-007-9030-4

FRESHWATER ANIMAL DIVERSITY ASSESSMENT

Global diversity of crocodiles (Crocodilia, Reptilia) in freshwater

Samuel Martin

Abstract Living crocodilians include the 24 species of alligators, caimans, crocodiles and gharials. These large semi-aquatic ambush predators are ubiquitous in freshwater ecosystems throughout the world's tropics and warm temperate regions. Extant crocodilian diversity is low, but the group has a rich fossil record in every continental deposit. Most populations suffered from over-hunting and habitat loss during the twentieth century and even though some species remain critically endangered others are real success stories in conservation biology and have become important economic resources.

Keywords Crocodile · Alligator · Gharial · Archosauria

Introduction

The living crocodilians belong to the order Crocodylia which is now represented by three families: the Crocodylidae, the Alligatoridae and the Gavialidae (Brochu, 2003). The 24 species of the group are all-amphibious and share morphological, anatomical, and physiological features, which make them more adapted to water than to land (Lang, 1976).

They all live in tropical and subtropical areas in various aquatic habitats (forest streams, rivers, marshes, swamps, elbow lakes, etc.) and can be considered as the largest fresh water dwellers. They can occasionally adapt to salty waters (mangroves or estuaries) (Dunson, 1982; Mazzotti & Dunson, 1984). They are nocturnal carnivorous opportunistic predators, whose diet depends on their developmental stage, species and potential prey diversity (Magnusson et al., 1987). All crocodilian species may be considered as totally water dependent since they can only mate in water. Crocodilians appear to be very important for freshwater ecosystems as they maintain, during the dry season waterholes that are used as reservoir for many arthropods, crustacean, fishes and amphibians (Gans, 1989; Kushlan, 1974).

Species/generic diversity

With only 24 living species, the order Crocodylia is the smallest taxonomic group of the class Reptilia. The three families, Crocodylidae, Alligatoridae and Gavialidae are quite homogeneous taxa as they contain between two and four genera.

The highest level of species diversity is to be found in the genus *Crocodylus* which gathers 13 species, whereas other genera only display one or two species.

Guest editors: E. V. Balian, C. Lévêque, H. Segers & K. Martens
Freshwater Animal Diversity Assessment

S. Martin (✉)
La Ferme aux Crocodiles, Pierrelatte 26700, France
e-mail: s.martin@lafermeauxcrocodiles.com

The taxonomic place of *Tomistoma schlegelii* is also subject to debate among specialists. It used to be placed with the subfamily Crocodylinae, whereas other created the subfamilily Tomistominae. Based on morphological features and on the latest DNA studies, we chose to place them together with *Gavialis* in the subfamily Gavialinae (Groombridge, 1987; Gatsby & Amato, 1992; Brochu, 2003). (Tables 1, 2)

Phylogeny and historical processes

Crocodilians belong to the great group of archosaurs which includes two extinct clades: the pterosaurs and the dinosaurs (Blake, 1982). The history of the crocodilians has been well reviewed by Buffetaud (1982), Taplin (1984), Taplin & Grigg (1989) and Brochu (2003). The very first crocodilians called Protosuchians are from the early Jurassic, whereas the Eusuchians (the modern crocodilians) appeared in the upper Triassic around 220 Million years ago under the form of terrestrial carnivores gathered in the group of Pristichampsines. The eight surviving genera of crocodilians are only a tiny rest of the past diversity of the group which has been revealed by at least 150 fossile genera (Brochu, 2003). The crocodilian diversity showed two peaks—one in the early Eocene and the other one in the early Miocene (Taplin, 1984; Markewick 1998). These fossils suggest that crocodiliomorphs were adapted to terrestrial, sub-aquatic, and even to marine environment (cf. Thalattosuchians). Until the end of the Tertiary, the geographical distribution of the crocodilians was much broader. The more restricted current distribution is due to the climatic deterioration, which narrowed the tropical and subtropical zones (Markewick, 1998).

Table 1 Freshwater crocodilian species in the zoogeographical regions

	PA	NA	AT	NT	OL	AU	PAC	ANT	World
Alligatoridae	0	1	0	5	1	0	0	0	7
Crocodylidae	2	1	3	4	5	4	0	0	14
Gavialidae	1	0	0	0	2	0	0	0	2
Total	3	2	3	9	8	4	0	0	

PA: Palaearctic, NA: Nearctic, NT: Neotropical, AT: Afrotropical, OL: Oriental, AU: Australasian, PAC: Pacific Oceanic Islands, ANT: Antarctic

Table 2 Freshwater crocodilian genera in the zoogeographical regions

	PA	NA	AT	NT	OL	AU	PAC	ANT	World
Alligatoridae	0	1	0	3	1	0	0	0	4
Crocodylidae	1	1	2	1	1	1	0	0	2
Gavialidae	1	0	0	0	2	0	0	0	2
Total	2	2	2	4	4	1	0	0	8

PA: Palaearctic, NA: Nearctic, NT: Neotropical, AT: Afrotropical, OL: Oriental, AU: Australasian, PAC: Pacific Oceanic Islands, ANT: Antarctic

Present distribution and main areas of endemicity (Groombridge, 1987; Ross 1998)

Except the two *Alligator* species which are to some extend more tolerant to colder temperatures, crocodilians are distributed in inter-tropical wetlands. As shown in Table 3, most crocodilians are endemic of a zoogeographical region. Only three species of the genus *Crocodylus* (*C. niloticus*, *C. porosus* and *C. siamensis*) and a gavialid (*Gavialis gangeticus*) are found in two adjacent regions. The range of distribution of crocodilians can be very variable in size. Some species, such as the Nile crocodile (*Crocodylus niloticus*) in Africa, the saltwater crocodile, (*Crocodylus porosus*) in the indopacific region or the spectacled caiman (*Caiman crocodilus crocodilus*) in South America are widely represented at a continental level, whereas most species are living in more restricted areas. This is one of the reasons that today half of the existing crocodilian species are considered either as being endangered or threatened according to the Red List criteria of the World Conservation Union IUCN. The Chinese alligator (*Alligator sinensis*), the Siamese crocodile (*Crocodylus siamensis*), the Orinoco crocodile (*Crocodylus intermedius*) and the Philippine crocodile (*Crocodylus mindorensis*) may be considered as the most endangered crocodilians species. The first one is only found in a few spots along the lower part of Yangtze River with a remaining stronghold in the province of Anhui in People's Republic of China, the second one is restricted to five Asian countries (Cambodia, Thailand, Vietnam, Indonesia and Laos) with

Table 3 Distribution of the 24 crocodilian species in the eight zoogeographical regions

Family	Genus	Species	Sub species	Distribution region	Common name, IUCN red list
Alligatoridae (4 genera, 7 species)	*Alligator*	*A. mississippiensis*		Nearctic	American alligator/LR
		A. sinensis		Oriental	Chinese alligator/CR
	Caiman	*C. crocodilus*	*C.c. apaporiensis, C. c. crocodilus, C. c. fuscus C. c. yacare*	Neotropical	Spectacled caïman/LR
		C. latirostris		Neotropical	Broad snouted caiman/LR
	Melanosuchus	*M. niger*		Neotropical	Black caiman/LR
	Palaeosuchus	*P. palpebrosus*		Neotropical	Cuvier's smooth fronted caiman/LR
		P. trigonatus		Neotropical	Schneider's smooth fronted caiman/LR
Crocodylidae (2 genera, 14 species)	*Crocodylus*	*C. acutus*		Nearctic; Neotropical	American crocodile/Vu
		C. cataphractus		Afrotropical	African slender snouted crocodile/DD
		C. intermedius		Neotropical	Orinoco crocodile/CR
		C. johnsoni		Australasia	Australian freshwater crocodile/LR
		C. mindorensis		Oriental	Philippines crocodile/CR
		C. moreletii		Neotropical	Morelet's crocodile/LR
		C. novaeguineae		Australasia	New guinea crocodile/LR
		C. niloticus		Palearctic, Afrotropical	Nile crocodile/LR
		C. palustris		Palearctic, Oriental	Marsh crocodile/LR
		C. porosus		Oriental, Australasia	Estuarine crocodile/LR
		C. raninus		Oriental	Borneo crocodile/DD
		C. rhombifer		Neotropical	Cuban crocodile/EN
		C. siamensis		Oriental, Australasia	Siamese crocodile/CR
	Osteolaemus	*O. tetraspis*	*O. t. tetraspis & O. t. osborni*	Afrotropical	African dwarf crocodile/ Vu
Galvialidae (2 genera, 2 species)	*Gavialis*	*G. gangeticus*		Palearctic, Oriental	Gharial/EN
	Tomistoma	*T. schlegelii*		Oriental	False gharial/EN

Four species CR (*Critically endangered*), three species E (*Endangered*), 14 species LR (*Low risks*), two species DD (*Data deficiency*)

scattered extremely small populations, the third one is restricted to the Orinoco water system of Venezuela and Colombia only and the fourth one is endemic to the archipelago of Philippines. The reasons for their being endangered are due to human pressure on habitat, but inversely linked to the adaptability to habitat variations. For instance the Nile crocodile is able to live in diverse aquatic environments such as streams, forest rivers, swamps, marshes, lagoons and even small desert water holes like a few known small populations lost in the Mauritanian Sahara (Pooley & Gans, 1976; Shine et al., 2001). This species like many other Crocodylidae and Alligatoridae are able to walk long distances on dry land. When necessary during long periods of drought they are able to migrate to find new water spots. Other species such as the mugger crocodile (*Crocodylus palustris*) will dig burrows during the dry season to protect themselves from the sun and wait in the shade the next raining season (Rao, 1994). Other crocodilian species adapt at burrowing are *Alligator sinensis* and *Osteolaemus tetraspis*. According to IUCN criteria, out of 24

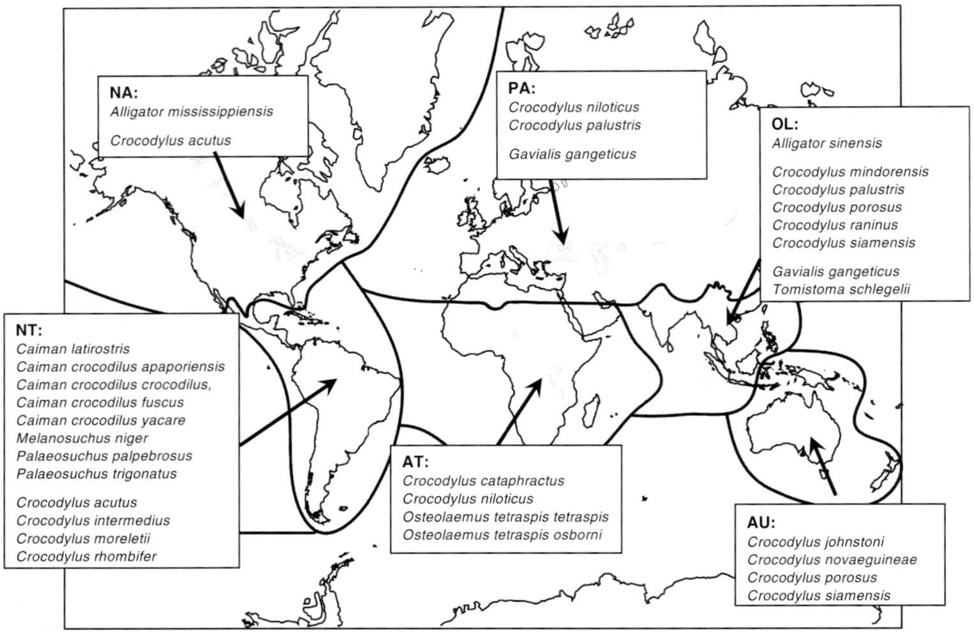

Fig. 1 Distribution of the 24 crocodilian species in the eight zoogeographical regions. PA: Palaearctic, NA: Nearctic, NT: Neotropical, AT: Afrotropical, OL: Oriental, AU: Australasian, PAC: Pacific Oceanic Islands, ANT: Antarctic

crocodilian species, four are critically endangered, three endangered and two species are considered as vulnerable (IUCN red list of threatened species, 2004) (Fig. 1).

Human related issues

Humans and crocodilians have been interacting since the dawn of civilization. Large crocodilians are potentially dangerous to man as they can prey on humans. Their populations have been depleted until the mid 60'ies because the high prices paid for their hides. In order to limit harvesting of wild populations, farming and ranching programs have been set up (Blake, 1982). Today several hundreds of farms around the world are breeding and raising crocodilians for leather and meat production (Brazaitis et al., 1998). Despite these efforts, some wild crocodilian populations are still depleting due to competition with humans for habitat and food. Dam construction on water streams has blocked seasonal migration of aquatic species when their prey was going down-stream during the rainy season and up-stream when the water level lowers (Gans, 1989). The draining of swamps for agricultural purposes has increased drastically habitat fragmentation and pollution. On a worldwide scale, the Crocodile Specialist Group of the I.U.C.N. Species Survival Commission coordinates crocodilian conservation programmes. The most successful ones are based on local community involvement combined with education (Ross, 1998).

Complete bibliography can be found on: http://utweb.ut.edu/faculty/mmeers/bcb/index.html

Reliable website: http://www.flmnh.ufl.edu/cnhc/csl.html

UNC/Species Survival Commission Crocodile Specialist Group News letter http://www.flmnh.ufl.edu/natsci/herpetology/CROCS/CSGnewsletter.htm

References

Blake, D. K., 1982. Crocodile ranching in Zimbabwe. The Zimbabwe Science News 26: 208–209.

Brazaitis, P., M. E. Watanabe & G. Amato, 1998. Trafic de Caïmans. Pour la Science 247: 84–90.

Brochu, C. A., 2003. Phylogenetic approaches toward crocodilian history. Annual Revue of Earth Planet Science 31: 357–397.

Buffetaud, E., 1982. The evolution of crocodilians. American Science 241:124–132.

Dunson, W. A., 1982. Salinity relations of crocodiles in Florida Bay. Copeia 2: 374–385.

Gans, C., 1989. Crocodilians in perspective! American Zoologist 29: 1051–1054.

Gatsby, J. & G. D. Amato, 1992. Sequence similarity of 12S ribosomal segment of mitochondrial DNAs of gharial and false gharial. Copeia 1992: 241–244.

Groombridge, B., 1987. The distribution and status of world crocodilians. In Webb, G. J. W., S. C. Manolis & P. J. Whitehead (eds), Wildlife Management: Crocodiles and Alligators. Chipping Norton, Australia, Surrey Betty and Sons Printing in association with the Conservation Commission of the Northern Territory, 9–21.

IUCN Red List of Threatened Specie, 2004. IUCN. Gland. Switzerland.

Kushlan, A. J., 1974. Observations on the role of the Americain alligator (*Alligator mississippiensis*) in the Southern Florida Wetlands. Copeia 4: 993–996.

Lang, W. J., 1976. Amphibious behavior of *Alligator mississippiensis*. Science 191: 575–577.

Magnusson, E. W., E. V. Da Silva & A. P. Lima, 1987. Diets of Amazonian crocodilians. Journal of Herpetology 21: 85–95.

Markewick, P. J., 1998. Crocodilian diversity in space and time: the role of climate in paleoecology and its implication for understanding K/T extinctions. Paleobiology 24: 470–497.

Mazzotti, J. & W. A. Dunson, 1984. Adaptations of *Crocodylus acutus* and Alligator for life in saline water. Comparative Biochemistry and Physiology 79: 641–646.

Pooley, A. C. & C. Gans, 1976. The Nile crocodile. Scientific American 234: 114–124.

Rao, R. J., 1994. Ecological studies of Indian crocodiles: an overview. CSG Proceedings, Pattaya, Thailand, 2–6 May 1994. Crocodile Specialist Group, Vol. 1, 259–273.

Ross, J. P. 1998. Crocodiles-Status, Survey and Conservation Action Plan. International Union for Nature Conservation Switzerland.

Shine, T., W. Böhme, H. Nickel, D. F. Thies & T. Wilms, 2001. Rediscovery of relict populations of the Nile crocodile *Crocodylus niloticus* in South-Eastern Mauritania, with observations on their natural history. Oryx 35: 260–262.

Taplin, L., 1984. Evolution and zoogeography of crocodilians: a new look at an ancient order. In Archer, M. & G. Clayton (eds), Vertebrate Evolution and Zoogeography in Australasia. Hesperian Press, Perth, 361–370.

Taplin, L. E. & G. C. Grigg, 1989. Historical zoogeography of the eusuchian crocodilians: a physiological perspective. American Zoology 29: 885–901.

Hydrobiologia (2008) 595:593–598
DOI 10.1007/s10750-007-9244-5

FRESHWATER ANIMAL DIVERSITY ASSESSMENT

Global diversity of turtles (Chelonii; Reptilia) in freshwater

Roger Bour

Abstract The turtles are an evolutionary ancient group of tetrapod vertebrates, and their present-day distribution and diversity reflects the long and complex history of the taxon. Today, about 250 of the total of 320 species recognized are freshwater turtles; most of these inhabit tropical and subtropical zones. Diversity hotspots occur in Southeast North America, regarding Emydidae, and in the Indo-Malayan region, mostly Geoemydidae and Trionychidae. Chelidae are predominantly Neotropical and Australasian, while Pelomedusidae are African. The majority of genus- and species-level taxa are regional or even local endemics. A majority of freshwater turtles are threatened in varying degrees, mostly by habitat modification and collection.

Keywords Biodiversity · Zoogeography · Chelonii · Review

Introduction

Turtles or Chelonians (order Chelonii, class Reptilia) are very ancient tetrapod vertebrates, their first members being known from Keuper (Triassic) deposits of ca. 230 M years old. The extant families have a relict distribution pattern that reflects their long evolutionary history. Morpho-functional analysis of their fossil remains, especially their limbs, reveals that the oldest known Chelonians most likely inhabited swamps or marshlands, and their present relatives are mostly freshwater species.

Soon after its emergence (Triassic–Jurassic), the order split into two groups diagnosed by several anatomical features, amongst which are the articulation of the cervical vertebrae and bending of the neck. These groups, namely Pleurodira (neck bending in a horizontal plane: "side neck") and Cryptodira (neck bending in a vertical plane, and neck more or less retractable inside the shell: "hidden neck"), are still present today, although they have different and unequal areas of occurrence and species richness. Cryptodira includes freshwater turtles in addition to marine turtles and terrestrial tortoises, while all Pleurodira species are more or less completely freshwater dependant (Fig. 1).

Reliance of turtles on freshwater is quite variable, depending on the species, and also on the age for some species. Roughly, the typical habitat may vary from large rivers and lakes, sometimes estuaries, to swamps, marshes, bogs, occasionally brackish waters, and some species are nearly as terrestrial as true land tortoises, which, themselves, include a few species that require a very damp environment. "Terrestrial" freshwater turtles are encountered in both large

Guest editors: E. V. Balian, C. Lévêque, H. Segers & K. Martens
Freshwater Animal Diversity Assessment

R. Bour (✉)
Reptiles et Amphibiens, Systématique et Evolution, Muséum national d'Histoire naturelle, 25 rue Cuvier, Paris 75231, France
e-mail: bour@mnhn.fr

Fig. 1 (**a**) *Mesoclemmys nasuta* (Pleurodira), (**b**) *Mauremys rivulata* (Cryptodira)

families, i.e. Emydidae and Geoemydidae. Their adaptation towards a more or less complete life outside of the water seems to be a rather fast process, because in one genus, one can recognize true freshwater species, amphibious or semi-aquatic species and terrestrial turtles. This condition is especially conspicuous in the Neartic genus *Terrapene* (box turtles), family Emydidae, the extreme being *T. coahuila*, which is fully aquatic, and *T. ornata*, which can live in very dry areas. Some other species that have adopted a rather terrestrial life, however often in forested and damp areas, belong to the family Geoemydidae and are encountered in South-east Asia (*Heosemys*, *Geoemyda*, *Pyxidea*, *Cuora*), and in Tropical America (*Rhinoclemmys*). Outside these families, turtles that regularly leave their water home and are comfortable on land belong to the genera *Pelomedusa* (Pelomedusidae), *Kinosternon*, *Chelonia*, *Glyptemys* and some others. These can wander long distances, especially by night, from one swamp to another. Most other species only leave the water either for basking (on logs, rocks, sandy shores: e.g. *Platysternon*) or for laying eggs. Table 1 gives data on the typical environment of the extant genera of turtles.

Many turtles are opportunistic regarding their diet, eating any available food, whether of animal or vegetal origin, but they usually probe potential food items by smelling before eating. Quite often, during growth, there is a shift from a rather carnivorous feeding towards a vegetarian one. Preys mainly consist of invertebrates (e.g. insects, mollusks), or carrion: many turtles are unable to capture adult vertebrates, except for some specialized taxa (e.g. Chelidae like *Chelus*, Trionychidae, *Macrochelys*) that are capable of ambushing fishes. Vegetarian turtles eat aquatic plants, grasses, leaves or fallen fruits.

Species and generic diversity of freshwater turtles

A little more than 320 species of Turtles are recognized today, depending on the taxonomy followed (Pritchard, 1979; Ernst & Barbour, 1989; Iverson, 1992; David, 1994; Spinks et al., 2004; Bonin et al., 2006). A single family includes the true terrestrial members, called Tortoises (18 genera, ca. 60 species), and two families include all the present marine Turtles or true Turtles (6 genera, 7–8 species). All others Chelonians are freshwater turtles and belong to one of the other 11 families (Table 2). Three families (Chelidae, Pelomedusidae and Podocnemididae) belong to suborder Pleurodira, which is considered to be the most primitive suborder notwithstanding that several specialized features are present. Pleurodira has a typical Gondwanan range, occurring mainly in the Southern hemisphere. The other families belong to Cryptodira; 3 families are monotypic (Carettochelyidae, Dermatemydidae and Platysternidae), with a limited distribution; 3 other families include numerous species and have a wide range, covering several biogeographical regions (Trionychidae, and the related Emydidae and Geoemydidae). Finally, 2 families (Chelydridae and Kinosternidae), with only 6 genera together, are limited to the New World, with a mainly Nearctic range (Vetter, 2005). Table 1 gives the list of the genera considered in this review, with their respective number of species.

Table 1 Summary of the diversity, distribution and ecology of the extant genera of freshwater turtles

Suborder, family	Genus	Number of species	Biogeographical region	Typical environment
Pleurodira				
Chelidae	*Chelodina*	8	Australasian	Rivers, swamps
	Elseya	10	Australasian	Rivers, swamps
	Elusor	1	Australasian	Rivers
	Emydura	7	Australasian	Rivers, swamps
	Macrochelodina	5	Australasian	Rivers
	Pseudemydura	1	Australasian	Swamps
	Rheodytes	1	Australasian	Rivers
	Acanthochelys	4	Neotropical	Small rivers, swamps
	Chelus	1	Neotropical	Swamps
	Hydromedusa	2	Neotropical	Small rivers
	Mesoclemmys	10	Neotropical	Rivers, swamps, ponds
	Phrynops	4	Neotropical	Rivers
	Platemys	1	Neotropical	Swamps, ponds,
	Rhinemys	1	Neotropical	Rivers (black waters)
Pelomedusidae	*Pelomedusa*	1	Afrotropical (incl. Madagascar)	Ponds, swamps
	Pelusios	18	Afrotropical (incl. Madagascar)	Lakes, quiet rivers
Podocnemididae	*Erymnochelys*	1	Madagascar (Afrotropical)	Lakes, quiet rivers
	Peltocephalus	1	Neotropical	Swamps, rivers
	Podocnemis	6	Neotropical	Rivers and tributaries
Cryptodira				
Carettochelyidae	*Carettochelys*	1	Australasian	Rivers and tributaries
Chelydridae	*Chelydra*	3	Nearctic, Neotropical	Swamps, rivers
	Macrochelys	1	Nearctic	Lakes, quiet rivers
Dermatemydidae	*Dermatemys*	1	Neotropical (North)	Lakes, quiet rivers
Emydidae	*Actinemys*	1	Nearctic	Swamps, ponds, rivers
	Chrysemys	1	Nearctic	Rivers, Lakes
	Clemmys	1	Nearctic	Swamps, ponds, rivers
	Deirochelys	1	Nearctic	Swamps, rivers
	Emydoidea	1	Nearctic	Swamps, rivers
	Emys	1	Palaearctic	Swamps, ponds, rivers
	Glyptemys	2	Nearctic	Swamps, bogs, rivers
	Graptemys	15	Nearctic	Rivers, lakes
	Malaclemys	1	Nearctic	Brackish waters only
	Pseudemys	3	Nearctic	Rivers, lakes, ponds
	Terrapene	6	Nearctic	Swamps or *terrestrial*
	Trachemys	15	Nearctic, Neotropical	Rivers, lakes, ponds
Geoemydidae	*Batagur*	1	Oriental	Rivers and tributaries, estuaries
	Callagur	1	Oriental	Rivers and tributaries, lakes
	Chinemys	2	Oriental, Palaearctic	Rivers, swamps, ponds
	Cuora	12	Oriental	Rivers, swamps or *terrestrial*
	Cyclemys	4	Oriental	Rivers, swamps
	Geoclemys	1	Oriental	Rivers and tributaries
	Geoemyda	3	Oriental	Swamps or *terrestrial*

Table 1 continued

Suborder, family	Genus	Number of species	Biogeographical region	Typical environment
	Hardella	2	Oriental	Rivers, lakes
	Heosemys	4	Oriental	Rivers, swamps or *terrestrial*
	Hieremys	1	Oriental	Rivers, swamps
	Kachuga	3	Oriental	Rivers and tributaries
	Leucocephalon	1	Oriental	Rivers, swamps
	Malayemys	2	Oriental	Rivers, swamps, ponds
	Mauremys	6	Oriental, Palaearctic	Rivers, swamps, ponds, wadis
	Melanochelys	2	Oriental	Rivers, swamps or *terrestrial*
	Morenia	2	Oriental	Rivers and tributaries, swamps
	Notochelys	1	Oriental	Rivers, swamps
	Ocadia	1	Oriental	Rivers, swamps, ponds
	Orlitia	1	Oriental	Rivers, lakes
	Pangshura	5	Oriental	Rivers, swamps
	Pyxidea	1	Oriental	Swamps or *terrestrial*
	Rhinoclemmys	9	Neotropical	Rivers, lakes, swamps or *terrestrial*
	Sacalia	2	Oriental	Rivers, swamps
	Siebenrockiella	1	Oriental	Rivers, swamps
Kinosternidae	*Claudius*	1	Neotropical (North)	Swamps, quiet rivers
	Kinosternon	19	Nearctic, Neotropical	Swamps, quiet rivers
	Staurotypus	2	Neotropical (North)	Lakes, quiet rivers
	Sternotherus	4	Nearctic	Swamps, quiet rivers
Platysternidae	*Platysternon*	1	Oriental	Small rivers in hilly areas
Trionychidae	*Amyda*	1	Oriental	Rivers and tributaries, lakes
	Apalone	3	Nearctic	Rivers and tributaries, lakes
	Aspideretes	4	Oriental	Rivers and tributaries, lakes
	Chitra	3	Oriental	Rivers and tributaries
	Cyclanorbis	2	Afrotropical	Rivers and tributaries, lakes
	Cycloderma	2	Afrotropical	Rivers and tributaries, lakes
	Dogania	1	Oriental	Rivers and tributaries, lakes
	Lissemys	2	Oriental	Rivers and tributaries, lakes
	Nilssonia	1	Oriental	Rivers and tributaries, lakes
	Oscaria	1	Oriental, Palaearctic	Rivers and tributaries, lakes
	Palea	1	Oriental	Rivers and tributaries, lakes
	Pelochelys	2	Oriental	Rivers and tributaries, estuaries
	Pelodiscus	1	Oriental, Palaearctic	Rivers and tributaries, swamps
	Rafetus	1	Palaearctic	Rivers and tributaries
	Trionyx	1	Afrotropical, Palaearctic	Rivers and tributaries, lakes
Total	N = 79	N = 257		

Present distribution and main areas of endemicity

Most freshwater turtles live in the tropical and subtropical zones, only few reach the Northern temperate zone. Even in warm areas, their distribution is quite unbalanced, and two hot spots, both of endemicity and diversity, can be discerned in the Northern hemisphere. One covers the Southeast of North America (Emydidae), the other one is Southeast Asia, more specifically the Indo-Malayan biogeographical

Table 2 Distribution of extant species and genera of freshwater turtles, per Biogeographical region. (number of genera in parentheses)

	PA	NA	NT	AT	OL	AU	PAC	ANT	World
Chelidae			23 (7)			33 (7)			56 (14)
Pelomedusidae				19 (2)					19 (2)
Podocnemididae			7 (2)	1 (1)					8 (3)
Carettochelyidae						1 (1)			1 (1)
Chelydridae		2 (2)	2 (1)						4 (2)
Dermatemydidae			1 (1)						1 (1)
Emydidae	1 (1)	42 (11)	6 (1)						48 (12)
Geoemydidae	3 (1)		9 (1)		57 (23)				68 (24)
Kinosternidae		12 (2)	17 (3)						26 (4)
Platysternidae					1 (1)				1 (1)
Trionychidae	4 (4)	3 (1)		5 (3)	17 (10)				26 (15)
Total	8 (6)	59 (16)	65 (16)	25 (6)	72 (34)	34 (8)	0 (0)	0 (0)	257 (79)

PA, Palaearctic region; NA, Nearctic region; NT, Neotropical region; AT, Afrotropical region; OL, Oriental region; AU, Australasian region; PAC, Pacific region and Oceanic Islands; ANT, Antarctic region

realm (Geoemydidae and Trionychidae) (Vetter & van Dijk, 2006). The relative paucity of turtles in the Palaearctic region, especially compared with the Nearctic, is most probably a consequence of the Pleistocene glaciations, in conjunction with the East-West orientation of geographical barriers (Mediterranean basin, Himalayas) preventing climate-induced range shifts and recolonization. In the Southern hemisphere, one family (Chelidae) predominates, and is almost equally represented in South America and in Australasia, while the Pelomedusidae is the dominant family of freshwater turtles in Africa. Endemicity at the regional, or even local, scale is the rule for freshwater turtles at the genus and, especially, species level; there are relatively few genera or species that have a range that spans more than one region. Figure 2 gives an arrangement of the taxa of Table 1 classified according to biogeographical regions.

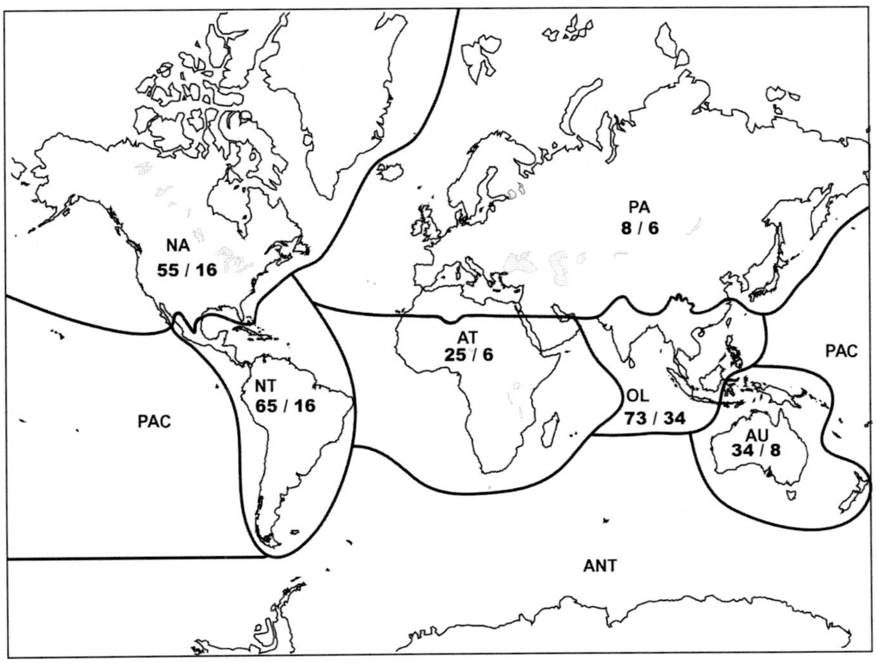

Fig. 2 Distribution of freshwater turtle species and genera per biogeographic region (number of species / number of genus). PA, Palaearctic region; NA, Nearctic region; NT, Neotropical region, AT, Afrotropical region; OL, Oriental region; AU, Australasian region; PAC, Pacific region and oceanic islands; ANT, Antarctic region

Human related issues

Turtles (and tortoises) are very sensitive to modifications of their environment, and are among the first vertebrates to disappear when a habitat is destroyed. Another threat is increasing collection, for food, for perceived medical purposes and for the pet trade. According to the last IUCN classification (IUCN, 2006), 159 species, or ca. 60%, of freshwater turtles are accounted as being more or less threatened. Data are deficient for 10 species including 5 probable hybrids, 1 species is at low-risk but conservation dependant (*Podocnemis expansa*, a large river turtle of South America), 11 are considered as of least concern, 40 as vulnerable, 38 as near threatened, 36 as endangered, 20 as critically endangered, 1 as extinct in the wild (*Aspideretes nigricans*, a large soft shell turtle restricted to a tank in Bangladesh) and 2 as extinct (*Pelusios seychellensis* and *Cuora yunnanensis*, although a specimen of the later has recently been observed in a market). About one third of these threatened species are South-east Asian, China being a major consumer of turtles. It is revealing that among the species listed as endangered or critically endangered there are at least five, resp. two species, which were quite abundant in the pet trade no more than twenty years ago: *Chinemys reevesii*, *Geoemyda spengleri*, *Heosemys spinosa*, *Pyxidea mouhotii*, *Platysternon megacephalum* and *Cuora galbinifrons* and *Cuora trifasciata*, respectively), all from South-east Asia.

Acknowledgements I thank Estelle Balian for her patience and support, and Peter Paul van Dijk for his fruitful comments as reviewer.

References

Bonin, F., B. Devaux & A. Dupré, 2006. Turtles of the World. Transl. Pritchard, The Johns Hopkins University Press, P. C. H. Baltimore.

David, P., 1994. Liste des reptiles actuels du monde I. Chelonii. Dumerilia 1: 7–127.

Ernst, C. H. & R. W. Barbour, 1989. Turtles of the world. Washington, Smithsonian Inst. press.

IUCN, 2006. 2006 IUCN Red List of Threatened Species. http://www.iucnredlist.org, downloaded on 6 November 2007.

Iverson, J. B., 1992. Revised Checklist with Distribution Maps of the Turtles of the World. Richmond, Indiana, privately printed.

Pritchard, P. C. H., 1979. Encyclopedia of Turtles. T.F.H. Publications Co., Neptune City, New Jersey.

Spinks, P. Q., H. B. Shaffer, J. B. Iverson & W. P. McCord, 2004. Phylogenetic hypothesis for the turtle family Geoemydidae. Molecular Phylogenetics and Evolution 32: 164–182.

Vetter, H., 2005. Terralog, Turtles of the World, Vol. 3. Central and South America. Frankfurt am Main and Rodgau, Chimaira and Aqualog.

Vetter, H. & P. P. van Dijk, 2006. Terralog, Turtles of the World, Vol. 4. East and South Asia. Frankfurt am Main and Rodgau, Chimaira and Aqualog.

Hydrobiologia (2008) 595:599–605
DOI 10.1007/s10750-007-9118-x

Global diversity of snakes (Serpentes; Reptilia) in freshwater

Olivier S. G. Pauwels · Van Wallach · Patrick David

Abstract A list of the snake species inhabiting freshwaters is provided. It includes 153 species, which represents about 5% of all known snakes. These freshwater snakes belong to 44 genera distributed among the families Acrochordidae, Boidae, Colubridae, Elapidae, Homalopsidae and Viperidae. The highest diversities in freshwater snakes are found in the Oriental (64 species) and Neotropical (39 species) Regions. Conservation actions are needed for several overcollected species with a limited distribution.

Keywords Snakes · Serpentes · Freshwater · Biodiversity

Guest editors: E. V. Balian, C. Lévêque, H. Segers, & K. Martens
Freshwater Animal Diversity Assessment

O. S. G. Pauwels
Monitoring and Assessment of Biodiversity Program, National Zoological Park, Smithsonian Institution, Gamba, Gabon

O. S. G. Pauwels (✉)
Institut Royal des Sciences Naturelles de Belgique, Rue Vautier 29, Brussels 1000, Belgium
e-mail: osgpauwels@yahoo.fr

V. Wallach
Museum of Comparative Zoology, Harvard University, Cambridge, MA 02138, USA
e-mail: vwallach@oeb.harvard.edu

P. David
Département Systématique et Evolution, USM 602 Taxonomie-collection—Reptiles & Amphibiens, Case postale 30, Muséum National d'Histoire Naturelle, 25 rue Cuvier, F-75231 Paris Cedex 05, France
e-mail: pdavid@mnhn.fr

Introduction

With about 3000 species known so far, snakes are a successful group of predatory vertebrates that occupy a wide range of environments in tropical and temperate areas, from deserts and mountain summits to oceans. Many snake species live close to water and often venture into it; probably all can swim, but only a limited number can be considered strictly aquatic. One can regard the truly aquatic snakes as those foraging in water and which are unable to survive without aquatic prey and frequent to constant submersion. Subtle external anatomical characters betray aquatic habits in freshwater snakes. These morphological specializations include position of nostrils on the snout top, allowing the snake to breath at the surface without being seen by birds and other predators; a banded pattern, useful for camouflage; valvular nostrils; dorsolaterally oriented eyes; keeled scales; etc. However none of these characters are common to all freshwater snakes, and some of these characters, like a banded pattern, are shared with many strictly terrestrial as well as with many exclusively marine species. Many freshwater species belong to genera also including marine or terrestrial species. Only the combination of morphology and natural history data can reveal if a species is a freshwater dweller, and much work remains to be done in these fields. Hereafter we briefly present the results of a thorough study of the literature and our experience in the field to provide a list as accurate as possible of the freshwater snakes of the World. The list must be regarded as provisional, since biological information which could confirm freshwater habits is lacking for many genera and species (Fig. 1).

Species diversity and distribution

Details on freshwater snake systematics, distribution and natural history are to be found in the works of, a.o., Anonymous (no date) (World), Boulenger (1913) (Palaearctic), Manthey & Grossmann (1997), Pauwels et al. (2001), Cogger et al. (1987), Murphy & Voris (2005), Vogel et al. (2004), Voris et al. (2002) (Australasian and Oriental), do Amaral (1978), Cei (1993) and Roze (1996) (Nearctic and Neotropical). Familial and subfamilial allocations of colubroid genera mostly follow Lawson et al. (2005) (Fig. 2).

Australasian Region

Among the most specialized aquatic snakes are the acrochordids, containing three species found in brackish and/or freshwater. Their morphological adaptations towards aquatic life are so radical (a.o. absence of enlarged ventral scales) that these snakes are unadapted for terrestrial locomotion. One freshwater *Acrochordus* species (*A. arafurae*) lives in the Australasian Region. *Laticauda crockeri*, endemic to Rennell Island (Solomon Archipelago), is the only known freshwater member of this marine elapid genus. The freshwaters of the region are also home to four homalopsid snakes: *Enhydris polylepis*, *Cantoria annulata*, *Cerberus rynchops* and *Heurnia ventromaculata*. The Australasian Region shares one freshwater snake species with the Oriental Region: *Cerberus rynchops*, which actually mainly lives in brackish and sea water. The Australasian freshwater snake diversity is as poor as that of the Palearctic Region, and about nine times less rich than that of the neighbouring Oriental Region.

Afrotropical Region

Relatively few aquatic snakes live in the Afrotropical Region's continental waters. They are distributed among the families Colubridae and Elapidae. Several genera are specialized towards aquatic life and contain exclusively freshwater species: *Afronatrix*, *Grayia* (*sensu lato*, including *Xenurophis*), *Helophis*, *Hydraethiops* (including two species, one, *H. laevis*, being known so far by only three specimens) and *Limnophis* (Colubridae), *Boulengerina* and *Lycodonomorphus* (including a.o. *L. bicolor*, endemic to Lake Tanganyika) (Elapidae). Madagascar houses only two freshwater species, both endemic to the island, *Liopholidophis lateralis* and *L. sexlineatus*. It is interesting to note the extreme colour and habitus

Fig. 1 *Enhydris jagori* (photo credit: Olivier S.G. Pauwels)

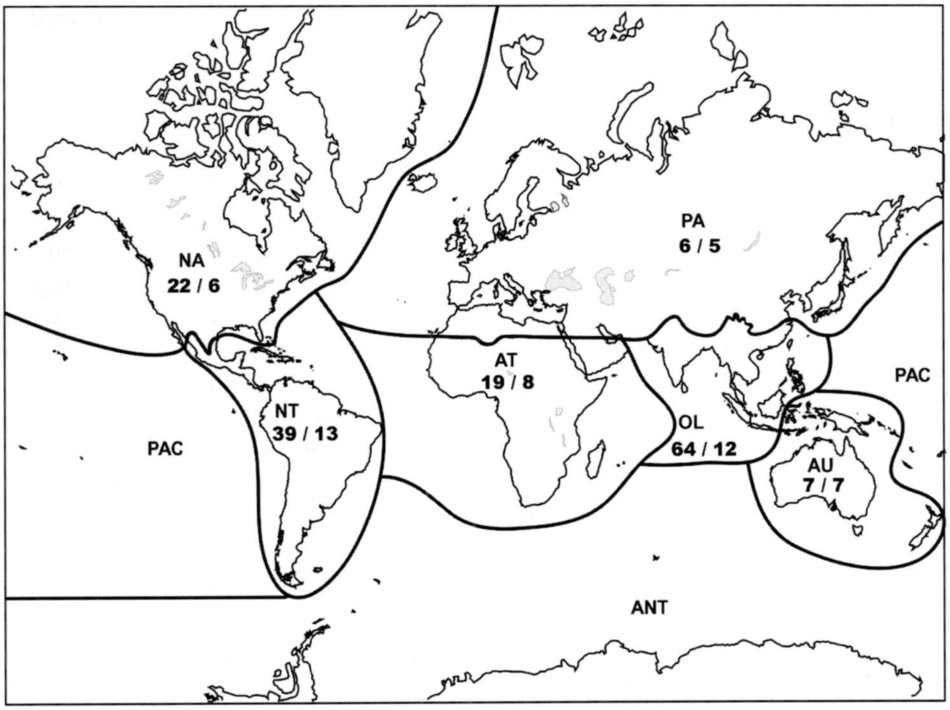

Fig. 2 Distribution of freshwater snake species and genera per biogeographic region (species number/genus number). PA: Palaearctic Region, NA: Nearctic Region, NT: Neotropical Region, AT: Afrotropical Region, OL: Oriental Region, AU: Australasian Region, PAC: Pacific Region and oceanic islands, ANT: Antartic Region

resemblance between the harmless *Grayia ornata* and the *Boulengerina* cobras, with which they share most of their distribution. In contrast with the Australasian Region, Afrotropical brackish waters do not house any specialized mangrove-dwelling species, although mangroves are widespread there. Besides the strictly freshwater Afrotropical taxa, a number of mostly terrestrial snakes use aquatic systems as hunting fields, at least occasionally, including: *Dromophis lineatus*, *Natriciteres* spp. (Colubridae), *Naja* spp., *Pseudohaje goldii* (Elapidae) and *Python* spp. (Pythonidae). None of the Afrotropical freshwater snake species is shared with another Region.

Nearctic Region

Nearctic freshwater snakes belong to the Colubridae, Elapidae and Viperidae. The colubrid genera *Farancia*, *Nerodia*, *Regina* and *Seminatrix* include only freshwater dwellers. The natricine genus *Thamnophis* includes 34 species, of which only four are freshwater dwellers. The Nearctic Region is home to a single venomous freshwater species, the pitviper *Agkistrodon piscivorus*. No Nearctic aquatic species is shared with the Neotropical Region. The diversity is relatively low and comparable to that of the Afrotropical Region.

Neotropical Region

The most remarkable Neotropical freshwater snakes are the anacondas (Boidae: *Eunectes* spp.), with *E. murinus* being one of the world's longest snakes, with official records of specimens above 8 m long. The venomous New World elapid genus *Micrurus* includes 70 species; a single species (*M. surinamensis*) being aquatic. All other Neotropical freshwater snakes belong to 11 xenodontine colubrid genera. Some of these genera exclusively contain aquatic species (e.g., *Helicops*); in others aquatic species are the exception (e.g., *Echinanthera*). None of the New World freshwater snake species is shared with the Old World. The Neotropical freshwater snake diversity is the world's second highest, although still much lower than the Oriental one.

Oriental Region

Among the two acrochordids inhabiting the Oriental Region, only *Acrochordus javanicus* penetrates freshwaters. The genus *Hydrablabes* is composed of two freshwater species, both endemic to Borneo. The genera *Opisthotropis* and *Parahelicops*, sometimes considered to be synonymous but here regarded as two distinct genera, comprise a total of 16 species from the Oriental Region, and one (*O. kikuzatoi*) from Japan. Many are restricted to unpolluted fast streams in undisturbed montane forest. All four *Sinonatrix* species, a genus endemic to the Oriental Region, are living in freshwater. The genus *Xenochrophis* includes 11 species, all freshwater dwellers and belonging to the Oriental Region, except one, *X. vittatus*, whose distribution extends east to Sulawesi, and is thus shared with the Australasian Region. Besides sea snakes, the most aquatic snakes are the Homalopsidae, which include 10 genera and 33 species, living in sea, mangroves or freshwater. They are mainly distributed in the Oriental Region, but some inhabit the Australasian Region. *Enhydris pakistanica* occurs at the limit between the Oriental and the Palaearctic Regions. Only one of the two known *Cerberus* inhabits freshwaters, and this species (*C. microlepis*) is endemic to Lake Buhi, Luzon, Philippine Islands. The most morphologically peculiar homalopsid is the genus *Erpeton*, characterized by two soft nasal appendices, the role of which is still unclear. The marine elapid snake *Hydrophis torquatus diadema* often penetrates into Tonle Sap Lake in Cambodia (Ineich, 1996). The “sea snake” *Hydrophis semperi* seems endemic to the freshwater lake Taal in the Philippines. *Hydrophis sibauensis* is known from Sibau River in Borneo. No information on the biology of the monotypic genera *Anoplohydrus*, *Fimbrios* and *Iguanognathus* (Colubridae; future morphological studies might attribute *Anoplohydrus* to homalopsids) are currently available, which would allow to list them with certainty among the freshwater snakes.

Palaearctic Region

In Europe, only two strictly aquatic snakes occur, both being harmless colubrids: *Natrix maura*, found in Europe and north Africa, and *N. tessellata*, which has a wide Eurasian distribution. Besides the latter species, Palaearctic Asia is home to four other natricine freshwater snakes. Two of them (*Sinonatrix annularis* and *Xenochrophis piscator* [including *X. sanctijohannis*]) are shared with the Oriental Region. The distributions of some of the species of *Opisthotropis* extend to the limits between the Oriental and the Palaearctic Regions. None of the Palaearctic freshwater species or genera are shared with the Afrotropical Region.

A total of 153 freshwater snake species, i.e., *circa* 5% of all currently known snakes, is found in the Australasian, Afrotropical, Nearctic, Neotropical, Oriental and Palaearctic Regions (see Table 1). No snakes are known from the Antarctic Region. The most specialized species towards aquatic life are the elapid sea snakes, of which two genera contain species living in freshwater, and Homalopsidae. All homalopsids show strong morphological adaptations to aquatic life (all notably have dorsolaterally oriented eyes and valvular nostrils) and contain marine as well as brackish and freshwater species. Most of the remaining freshwater snakes belong to the colubrid Natricinae and Xenodontinae, but these subfamilies contain terrestrial as well as freshwater species. Freshwater snakes are found in various groups in six families: Acrochordidae, Boidae, Colubridae, Elapidae, Homalopsidae and Viperidae. Except the Acrochordidae and Homalopsidae, no freshwater species belong to strictly aquatic snake families. Among the 44 genera including freshwater species, 13 (29.5%) include non-freshwater species, i.e. terrestrial or marine (compare total number of species in each genus *versus* number of freshwater species per Region in Table 1).

Human related issues

Few works on freshwater snake–human interactions are available, but field studies might reveal locally important links. For instance, some homalopsids are intensively collected for food or skin trade, as is happening in Tonle Sap Lake in Cambodia, where they are also used to feed crocodiles in neighbouring breeding farms; this overcollection notably puts at risk *Enhydris longicauda*, endemic to the lake (Stuart et al., 2000). In China, snakes like *Sinonatrix annularis*, some *Enhydris* spp. and *Homalopsis* are sold in large numbers in food markets and restaurants (Zhou & Jiang, 2005). Fuchs & Fuchs (2003) documented the

Table 1 Global distribution of freshwater snake species per Region

Taxa/Region	PA	NA	NT	AT	OL	AU	WORLD
ACROCHORDIDAE							
Acrochordus (3)					1	1	2
BOIDAE							
Boinae							
Eunectes (4)			4				4
COLUBRIDAE							
Colubrinae							
Grayia (4)				4			4
Natricinae							
Afronatrix (1)				1			1
Atretium (2)					2		2
Hydrablabes (2)					2		2
Hydraethiops (2)				2			2
Limnophis (1)				1			1
Natrix (4)	2						2
Nerodia (10)		10					10
Opisthotropis (15)	1				14		15
Parahelicops (2)					2		2
Regina (4)		4					4
Seminatrix (1)		1					1
Sinonatrix (4)	1				4		4
Thamnophis (34)		4					4
Xenochrophis (11)	1				11	1	11
Pseudoxyrhophiinae							
Liopholidophis (9)				2			2
Xenodontinae							
Coniophanes (12)			1				1
Echinanthera (6)			1				1
Farancia (2)		2					2
Gomesophis (1)			1				1
Helicops (17)			17				17
Helophis (1)				1			1
Hydrodynastes (2)			2				2
Hydromorphus (2)			2				2
Hydrops (3)			3				3
Liophis (44)			1				1
Pseudoeryx (1)			1				1
Sordellina (1)			1				1
Tetranorhinus (4)			4				4
Thermophis (1)	1						1
ELAPIDAE							
Boodontinae							
Lycodonomorphus (6)				6			6
Elapinae							
Boulengerina (2)				2			2

Table 1 continued

Taxa/Region	PA	NA	NT	AT	OL	AU	WORLD
Micrurus (70)			1				1
Hydrophiinae							
Hydrophis (29)					3		3
Laticauda (7)						1	1
HOMALOPSIDAE							
Cantoria (2)						1	1
Cerberus (2)					1	1	1
Enhydris (23)					22	1	23
Erpeton (1)					1		1
Heurnia (1)						1	1
Homalopsis (1)					1		1
VIPERIDAE							
Crotalinae							
Agkistrodon (4)		1					1
Total species	6	22	39	19	64	7	153
Total genera	5	6	13	8	12	7	44

Genera are followed between brackets by the total number of species they contain. Brackish and sea water species are considered if they also tolerate pure freshwater. PA: Palaearctic; NA: Nearctic; NT: Neotropical; AT: Afrotropical ; OL: Oriental; AU: Australasian; PAC: Pacific Oceanic Islands; ANT: Antarctic

use in leather trade of not less than ten freshwater snake species. In western central Africa, *Grayia ornata* plays an important role in local culture, food, traditional medicine and magic (Pauwels et al., 2002). Besides local overcollecting, the main threats to conservation might come from intensive logging and freshwater habitat degradation. Field studies on the most localized species (e.g., some *Enhydris* spp.) would be necessary to evaluate specific threats. Most freshwater snakes are harmless, but some are poisonous among the Elapidae, Homalopsidae and Viperidae. *Boulengerina* venoms have not been studied so far, and there thus exists no specific antivenom, although the venom might have a strong neurotoxic action (Spawls & Branch, 1995), like *Micrurus surinamensis*, *Hydrophis* spp. and *Laticauda* spp. Although venomous, Homalopsidae are rear-fanged and do not represent a medical problem; none is deadly. *Agkistrodon piscivorus*'s bite is potentially lethal to humans.

References

Anonymous, no date. The EMBL Reptile Database. http://www.embl-heidelberg.de/~uetz/LivingReptiles.html.

Boulenger, G. A., 1913. The Snakes of Europe. Methusen & Co. Ltd., London.

Cei, J. M., 1993. Reptiles del noroeste, nordeste y este de la Argentina. Herpetofauna de las selvas subtropicales, puna y pampas. Museo regionale di Scienze naturali, Torino, Monografie 14: 1–949, pl. 1–126.

Cogger, H. G., H. Heatwole, Y. Ishikawa, M. McCoy, N. Tamiya & T. Teruuchi, 1987. The status and natural history of the Rennell Island sea krait, *Laticauda crockeri* (Serpentes: Laticaudidae). Journal of Herpetology 21: 255–266.

do Amaral, A. P. B., 1978. Serpentes do Brasil. Iconografia colorida—Brazilian snakes: a color iconography. 2a ed. Ed. Melhoramentos & Editoria da Universidade de São Paulo, São Paulo.

Fuchs, K. & M. Fuchs, 2003. Die Reptilhaut. Ein wichtiger Merkmalsträger bei der Identifizierung von Echsen und Schlangen. The reptile skin. A key-feature in the identification of lizards and snakes. Edition Chimaira, Frankfurt am Main.

Ineich, I., 1996. *Hydrophis torquatus diadema*. Herpetological Review 27: 154.

Lawson, R., J. B. Slowinski, B. I. Crother & F. T. Burbrink, 2005. Phylogeny of the Colubroidea (Serpentes): new evidence from mitochondrial and nuclear genes. Molecular Phylogenetics and Evolution 37: 581–601.

Manthey, U. & W. Grossmann, 1997. Amphibien & Reptilien Südostasiens. Natur und Tier-Verlag, Münster.

Murphy, J. C. & H. K. Voris, 2005. A new Thai *Enhydris* (Serpentes: Colubridae: Homalopsinae). Raffles Bulletin of Zoology 53: 143–147.

Pauwels, O. S. G., P. David, W. Nutphand & C. Chimsunchart, 2001. First record of *Xenochrophis punctulatus* (Günther, 1858) (Serpentes: Colubridae: Natricinae) from Thailand. Hamadryad 26: 259–264.

Pauwels, O. S. G., A. Kamdem Toham & V. Mamonekene, 2002. Ethnozoology of the *dibomina* (Serpentes: Colubridae: *Grayia ornata*) in the Massif du Chaillu, Gabon. Hamadryad 27: 136–141.

Roze, J. A., 1996. Coral snakes of the Americas: biology, identification, and venoms. Krieger Publishing Co., Malabar, Florida.

Spawls, S. & B. Branch, 1995. The Dangerous Snakes of Africa. Natural History, Species Directory, Venoms and Snake Bite. Blandford, London.

Stuart, B. L., J. Smith, K. Davey, P. Din & S. G. Platt, 2000. Homalopsine watersnakes: the harvest and trade from Tonle Sap. Traffic Bulletin 18: 115–124.

Vogel, G., P. David, O. S. G. Pauwels & N. Brachtel, 2004. On the occurrence of the watersnake *Sinonatrix aequifasciata* (Barbour, 1908) (Serpentes, Colubridae, Natricinae) in Vietnam. Hamadryad 29: 110–114.

Voris, H. K., M. E. Alfaro, D. R. Karns, G. L. Starnes, E. Thompson & J. C. Murphy, 2002. Phylogenetic relationships of the Oriental-Australian rear-fanged water snakes (Colubridae: Homalopsinae) based on mitochondrial DNA sequences. Copeia 2002: 906–915.

Zhou, Z. & Z. Jiang, 2005. Identifying snake species threatened by economic exploitation and international trade in China. Biodiversity and Conservation 14: 3525–3536.

Hydrobiologia (2008) 595:607–617
DOI 10.1007/s10750-007-9122-1

FRESHWATER ANIMAL DIVERSITY ASSESSMENT

Global diversity of mammals (Mammalia) in freshwater

Geraldine Veron · Bruce D. Patterson ·
Randall Reeves

Abstract Species that are dependant on, or adapted to, freshwater environments are found in almost all mammalian orders, and two orders, the Cetacea and the Sirenia, are strictly aquatic and include some freshwater-dependant species. Overall, the aquatic and freshwater-dependant species represent around 70 of the more than 1,200 living or recent genera of mammals, and occur in all continents except Antarctica. They include some of the most endangered species of mammals, and several have gone extinct or become critically endangered in recent decades. One of the main threats is habitat loss or degradation. This chapter provides an overview of the freshwater species within each order of mammals, their evolutionary history, their relations to humans and their conservation status.

Keywords Mammalia · Freshwater · Diversity · Conservation

Guest editors: E. V. Balian, C. Lévêque, H. Segers & K. Martens
Freshwater Animal Diversity Assessment

Electronic supplementary material The online version of this article (doi:10.1007/s10750-007-9122-1) contains supplementary material, which is available to authorized users.

G. Veron (✉)
Département Systématique et Evolution, Muséum National d'Histoire Naturelle, CP 51, USM 601-UMR CNRS 5202, Origine, Structure et Evolution de la Biodiversité, 57 Rue Cuvier, 75231 Paris Cedex 05, France
e-mail: veron@mnhn.fr

B. D. Patterson
Department of Zoology, Field Museum of Natural History, 1400 S. Lake Shore Drive, Chicago, IL 60605-2496, USA

R. Reeves
Okapi Wildlife Associates, 27 Chandler Lane, Hudson, Quebec, Canada JOP 1HO

Introduction

The mammals are tetrapod vertebrates characterized by their jaw articulation (between the dentary and the squamosal), presence of three bones in the middle ear (malleus, incus, and stapes), mammary glands that produce milk, and presence of hair. Living forms are divided into three clades, the monotremes (Monotremata), the marsupials (Metatheria), and the placentals (Eutheria). Mammals have adapted to most kinds of habitat and have assumed various modes of existence—terrestrial, arboreal, aerial, fossorial, and aquatic, and inhabit all regions of the world. At least some forms in most of the orders depend on water for habitat and food, many also for protection from predators. Two orders, the Cetacea and the Sirenia, are strictly aquatic. For most partially or semi-aquatic mammals, at least giving birth or rearing of the young takes place outside the water.

Species diversity and geographical distribution

Freshwater mammals occur on all continents except Antarctica, some species widely distributed but many with very restricted ranges (Fig. 1). Many are threatened and these are listed in Table 1 with information on their distribution.

Order Monotremata (2 families; Groves, 2005)

Among the five species of living monotremes, the duck-billed platypus (*Ornithorhynchus anatinus*) lives in streams, lakes and lagoons in eastern Australia and Tasmania, but it has been extirpated from much of its range due to hunting and habitat degradation, river fragmentation from dams, and entrapment or entanglement in fishing gear (Nowak, 1999).

Metatheria (or Marsupialia) (7 Orders, 22 families; Wilson & Reeder, 2005)

Among the 331 species of metatherians, the only truly semi-aquatic species is the South American water opossum (*Chironectes minimus*). It possesses webbed feet and differs from other marsupials in having its rear-facing pouch equipped with a sphincter muscle to make it watertight for the attached young (Marshall, 1978). The thick-tailed opossum (*Lutreolina crassicaudata*) is an excellent swimmer that exploits wetlands for food and nesting habitat (Nowak, 1999).

Eutheria (placental mammals)

Order Chiroptera (1116 species in 18 families; Simmons, 2005): Many bats are associated with fresh water but one group, the fishing bats (*Noctilio albiventris* and *Noctilio leporinus*), belonging to the family Noctilionidae, feed on aquatic insects, and *N. leporinus* also eats fish, frogs, and crustaceans. Like other bats, they use echolocation to locate prey, detecting the ripples on the water surface (Schnitzler et al., 1994). They can swim and take flight from the water (Revenga & Kura, 2003). Several species of vesper bats (Vespertilionidae) share the enlarged hind legs and claws of noctilionids, including *Myotis (Pizonyx) vivesi* of Mexico, *Myotis adversus* of

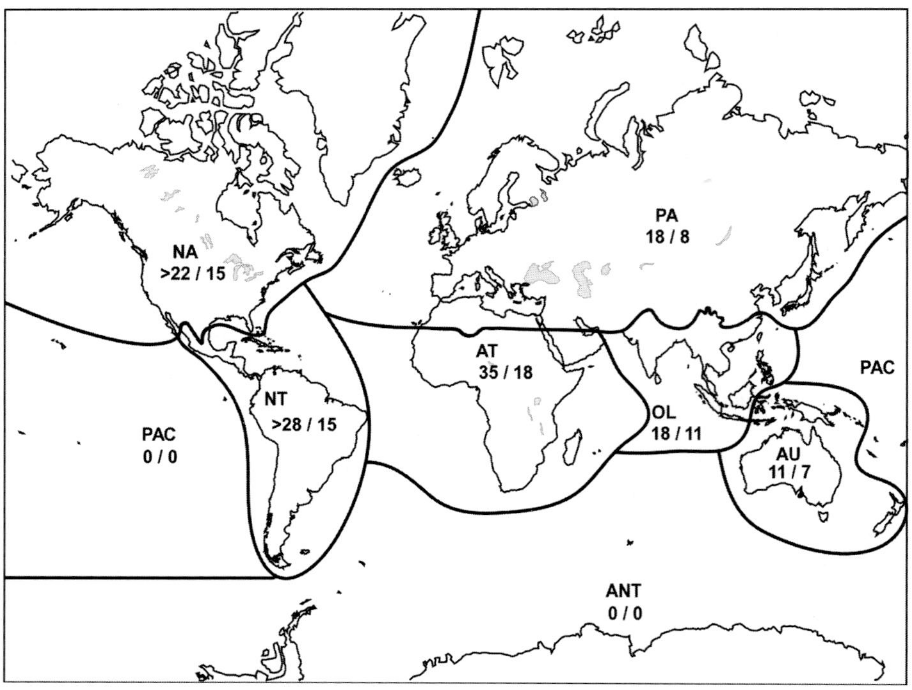

Fig. 1 Species and genus number of freshwater (aquatic and water dependent) mammals by zoogeographic region : species number/genus number. PA: Palearctic Region; NA: Nearctic Region; AT: Afrotropical Region; NT: Neotropical Region; OL: Oriental Region; AU: Australasian Region; PAC: Pacific Region and oceanic islands, ANT: Antarctic Region

Table 1 The most threatened freshwater mammals (2006 IUCN Red List, including species in categories higher than lower risk; subspecies and subpopulations are not included)

Taxon	IUCN Red List status	Distribution
Order Didelphimorphia		
Chironectes minimus	Near threatened	Central and South America
Order Soricomorpha		
Chimarrogale hantu	Critically endangered	South East Asia (tropical forest of Malay peninsula)
Chimarrogale phaeura	Endangered	South East Asia (tropical forest of Borneo)
Chimarrogale sumatrana	Critically endangered	South East Asia (tropical forest of Sumatra)
Desmana moschata	Vulnerable	Central Eurasia
Galemys pyrenaicus	Vulnerable	West Europe (Pyrenees and Northern mountains of Iberian Peninsula)
Order Afrosoricida		
Potamogale velox	Endangered	Tropical Africa
Micropotamogale lamottei	Endangered	West Africa (Mount Nimba in Guinea, Liberia, Ivory Coast)
Micropotamogale ruwenzorii	Endangered	Central Africa (Uganda, Zaire)
Limnogale mergulus	Endangered	East Madagascar
Order Lagomorpha		
Bunolagus monticularis	Critically endangered	South Africa (Cape Province)
Order Rodentia		
Castor fiber	Near threatened	Eurasia
Dasymys foxi	Vulnerable	Africa (endemic to South plateau of Nigeria)
Dasymys montanus	Vulnerable	Africa (endemic to Ruwenzori Mountains, Uganda)
Dasymys nudipes	Near threatened	Africa (S Angola, South West Zambia, North East Namibia, North Botswana)
Pelomys hopkinski	Vulnerable	Africa (Rwanda, Uganda, South West Kenya)
Pelomys isseli	Endangered	Africa (endemic to islands of Lake Victoria, Uganda)
Nectomys parvipes	Critically endangered	South America (known only from Comte River, French Guiana)
Ichthyomys pittieri	Vulnerable	South America (North Venezuela)
Neusticomys mussoi	Endangered	South America (known only from Paso Hondo, Rio Potosi, Venezuela)
Neusticomys oyapocki	Endangered	South America (known only from Trois sauts, French Guyana)
Neusticomys peruviensis	Endangered	South America (known only from Balta, Peru)
Nectomys parvipes	Critically endangered	South America (known only from Cacao, Comte River, French Guyana)
Mesocapromys angelcabrerai	Critically endangered	Cuba
Mesocapromys auritus	Critically endangered	Cuba
Mesocapromys nanus	Critically endangered	Cuba
Mesocapromys sanfelipensis	Critically endangered	Cuba
Order Carnivora		
Mustela lutreola	Endangered	Europe (in isolated small populations)
Aonyx congicus	Data deficient	Equatorial Africa (distribution scarce and not fully known)
Lutra lutra	Near threatened	Eurasia
Lontra longicaudis	Data deficient	Central and South America

Table 1 continued

Taxon	IUCN Red List status	Distribution
Lutra provocax	Endangered	South America (Patagonia)
Pteronura brasiliensis	Endangered	South America
Lutra sumatrana	Data deficient	South-East Asia
Lutrogale perspicillata	Vulnerable	Asia and Middle East
Aonyx cinerea	Near threatened	Asia
Herpestes palustris	Endangered	Asia (West Bengal, India)
Cynogale bennettii	Endangered	South-East Asia
Genetta piscivora	Data deficient	Africa (Zaire)
Prionailurus viverrinus	Vulnerable	Asia
Prionailurus planiceps	Vulnerable	South East Asia
Order Sirenia		
Trichechus manatus	Vulnerable	Southeastern North America, Central America and northern South America
Trichechus inunguis	Vulnerable	South America (Amazon basin)
Trichechus senegalensis	Vulnerable	West Africa
Artiodactyla		
Hexaprotodon liberiensis	Vulnerable	West Africa
Hippopotamus lemerlei	Extinct	Madagascar
Hippopotamus laloumena	Extinct	Madagascar
Hexaprotodon madagascariensis	Extinct	Madagascar
Bubalus bubalis	Endangered	Asia (formerly India to Indochina)
Elaphurus davidianus	Critically endangered (extinct in the wild)	Asia (formerly North East China)
Blastocerus dichotomus	Vulnerable	South America
Hyemoschus aquaticus	Data deficient	West and Central Africa
Cetacea		
Lipotes vexillifer	Critically endangered	Asia (Yangtze and Qiantang rivers, China)
Neophocaena phocaenoides	Data deficient	Asia
Platanista gangetica	Endangered	Asia (major rivers of South Asian subcontinent)
Orcaella brevirostris	Data deficient	South East Asia and Australasia
Inia geoffrensis	Vulnerable	Amazon, Madeira and Orinoco systems of central and northern South America
Sotalia fluviatilis	Data deficient	Central and northern South America

Australia, and *Myotis ricketti* of China, and are known to eat fish (Ma et al., 2003), often being included among the fishing bats.

Order Afrosoricida (51 species in 2 families; Bronner & Jenkins, 2005): The Tenrecidae includes some semi-aquatic forms. The otter shrews are represented by three African species (giant otter shrew, *Potamogale velox*, Nimba otter shrew *Micropotamogale lamottei*, and Ruwenzori otter shrew *Micropotamogale ruwenzorii*) with restricted distributions in the Congo Basin and Equatorial Africa. The remaining Tenrecidae are endemic to Madagascar. Some species live in marshy areas or burrow in banks of rice paddies, while the aquatic or web-footed tenrec (*Limnogale mergulus*) lives along large fast-flowing rivers and feeds mainly on aquatic invertebrates (Benstead & Olson, 2003, Fig. 2, Tables 2 and 3).

Order Soricomorpha (428 species in 4 families; Hutterer, 2005): The water shrews live in boreal North America (*Blarina brevicauda, Sorex palustris, Sorex bendirii*, and *Sorex alaskanus*), in Southeast Asia (*Nectogale elegans, Chimarrogale*, six species), and in the Palearctic (*Neomys*, three species). Water

Fig. 2 Web-footed tenrec *Limnogale mergulus* from Ranomafana National Park in Madagascar, Dec 1996. Photo by S. Zack & B. D. Patterson

shrews are found along streams and lakes and in wet forests; many other shrews also exploit mesic microhabitats yet are not water-dependent. The desmans are specialized aquatic insectivores living in Europe. The Russian desman, *Desmana moschata*, prefers quiet lakes and streams while the Pyrenean desman, *Galemys pyrenaicus*, requires fast-flowing streams. Among several American moles that frequent damp habitats, the star-nosed mole (*Condylura cristata*) is semi-aquatic, living in wet meadows and marshes and near streams. It is an accomplished diver and swimmer, eating aquatic insects, crustaceans, small fish, and earthworms (Nowak, 1999).

Order Lagomorpha (92 species in 3 families; Hoffmann & Smith, 2005): 2 of 61 species of Leporidae are closely associated with water, living in marshes, swamps, lake margins, and coastal waterways in North America (*Sylvilagus palustris* and *Sylvilagus aquaticus*). The riverine rabbit (*Bunolagus monticularis*) inhabits dense riverine scrub along seasonal rivers in South Africa but is entirely terrestrial.

Order Rodentia: This is the largest order of mammals, including 2277 species in 33 families (Wilson & Reeder, 2005). The Holarctic beavers, *Castor canadensis* and *Castor fiber*, are two of the best-known semi-aquatic mammals because of their former economic importance and their visible effects on the environment (Veron, 1992a,b). By building dams and lodges, they impound water, converting streams and rivers into wetlands. Beavers are the prototypical "landscape engineers."

Other water-dependent rodents include many of the common rats and mice (Muroidea; 1518 species, Musser & Carleton, 2005). Among the voles and lemmings (Arvicolinae), some are associated with water-dependent habitats (e.g., three species of Palearctic *Arvicola* and Nearctic *Microtus richardsoni*) and others are more strongly aquatic or semi-aquatic; the muskrats (*Ondatra* and *Neofiber*) live and forage in freshwater habitats. The American marsh rats (*Holochilus*), water rats (*Nectomys* and *Amphinectomys*), and some species of rice rats (*Oryzomys*) also inhabit marshy or swampy habitats, although they often forage in adjacent grasslands and forests. One tribe of Cricetidae [fish-eating mice or Ichythyomyini; *Anotomys* (1 sp) *Chibchanomys* (2 sp), *Ichthyomys* (4 sp), *Neusticomys* (5 sp), and *Rheomys* (4 sp)] lives mainly in streams and rivers of tropical America, foraging on aquatic invertebrates and small vertebrates (Voss, 1988), and the two species of Neotropical swamp rats (*Bibimys*) are accomplished swimmers well adapted to aquatic life. In Australasia, a parallel radiation has produced eight species of semi-aquatic murines: *Hydromys* (4 sp, among which 1 sp in Australia), *Parahydromys* (sp), *Crossomys* (1 sp), and *Baiyankamys* (2 sp), the New Guinea waterside rat (*Parahydromys asper*), and the earless water rat (*Crossomys moncktoni*). In Africa, the lone species of *Nilopegamys* is the only murid that seems to fill this swimming, pursuit-predator niche (Kerbis Peterhans & Patterson, 1995), although *Colomys* (1 sp), *Dasymys* (9 sp), *Malacomys* (3 sp), and *Pelomys* (5 sp) also live in close association with fresh water, some even foraging for aquatic organisms in shallow pools.

Members of several families of porcupine relatives have become aquatic or semi-aquatic. In Central and South America, one of the best-known aquatic rodents is the capybara (*Hydrochaeris hydrochaeris*), which grazes on vegetation near rivers, lakes, ponds, and wetlands. Another Neotropical rodent, the paca (*Cuniculus*), feeds on seeds and vegetation in riverine forest and is an accomplished swimmer, taking refuge from predators in water. The nutria or coypu (*Myocastor*), native to Patagonia but introduced worldwide to control weeds and provide food, is consummately adapted to life in fresh water. The five species of hutia (*Mesocapromys spp.*) inhabit the wetlands of Cuba. The African greater cane rat (*Thryonomys swinderianus*) lives near water in swamps, reed beds, and tall, dense grass and is a proficient swimmer.

Order Carnivora (286 species in 15 families, Wozencraft, 2005): All seals are carnivores highly

Table 2 Number of species of freshwater (aquatic and freshwater-dependent) mammals

	PA		NA		AT		NT		OL		AU		PAC		ANT		World FW Species number	Total mammal species number
	FW aq	FW depdt	FW aq	FW depdt	FW aq	FW depdt	FW aq	FW depdt	FW aq	FW depdt	FW aq	FW depdt	FW aq	FW depdt	FW aq	FW depdt	FW aq+depdt	
Monotremata											1						1	3
Metatheria				1?			1										2	331
Eutheria																		
Chiroptera				2													2	1,116
Insectivora	2	3		4		4				7							20	479
Insectivora-Soricomorpha (water shrews, desmans, moles)	2	3		4						7							16	428
Insectivora-Afrosoricida (otter shrews)						4											4	51
Lagomorpha				2		1											3	92
Rodentia	2	3	5	>3	2	18	>20				9						>60	2,277
Rodentia-castor	2		2														2	
Rodentia-Muroidea		3		>3	1	18	>20				9						>54	1,518
Rodentia-others			3		1												4	
Carnivora	6	2	2	2	7		3		5	2							25	286
Carnivora-seals	2		1														3	
Carnivora-otters	2	2	1	2	4		3		3								17	
Carnivora-mongooses					2				1								3	
Carnivora-Viverrids					1				1								2	
Carnivora-felids	2									2							2	
Sirenia						1	1	1									3	4
Artiodactyla						2											2	220
Cetacea							2		4		1						6	~80
Total	10	8	7	>15	9	26	>27	1	9	9	11	0	0	0	0	0	>124	~5,500

PA: Palearctic, NA: Nearctic, AT: Afrotropics, NT: Neotropics, OL: Orient, AU: Australasia, PAC: Oceanic Islands Pacific, ANT: Antarctic

Table 3 Genera of freshwater (aquatic and freshwater-dependent) mammals

	PA	NA	AT	NT	OL	AU	PAc	ANT	World FW genus number
Monotremata						1			1
Metatheria		1		1					2
Eutheria									
Chiroptera		1							1
Insectivora	3	2	3		2				10
Insectivora-Soricomorpha (water shrews, desmans, moles)	3	2			2				7
Insectivora-Afrosoricida (otter shrews)			3						3
Lagomorpha		1	1						2
Rodentia	2	7	6	8		4			26
Rodentia-castor	1	1							1
Rodentia-Muroidea	1	3	5	8		4			21
Rodentia-others		3	1						4
Carnivora	3	3	5	3	5	1			14
Carnivora-seals	1	1							2
Carnivora-otters	2	2	3	2	3				8
Carnivora-mongooses			2		1				2
Carnivora-Viverrids				1		1			2
Carnivora-felids					1				1
Sirenia			1	1					1
Artiodactyla			2						2
Cetacea				2	4	1			6
Total	8	15	18	15	11	7	0	0	65

PA: Palearctic, NA: Nearctic, AT: Afrotropics, NT: Neotropics, OL: Orient, AU: Australasia, PAC: Oceanic Islands Pacific, ANT: Antarctic

modified for aquatic life. Most are marine, but several live in fresh water. The Baikal seal (*Pusa sibirica*) is the only species of seal restricted to fresh water. The Caspian seal (*Pusa caspica*) is endemic to the Caspian Sea, a saline lake. Two subspecies of ringed seal (*Pusa hispida*), the Ladoga seal (*P. h. ladogensis*), and the Saimaa seal (*P. h. saimensis*), are endemic to freshwater lake systems in Russia and Finland, respectively, and the Ungava seal (*Phoca vitulina mellonae*), a subspecies of the widely distributed harbor seal (*Phoca vitulina*), inhabits a network of rivers and lakes in Quebec's Ungava Peninsula (Reijnders et al. 1993). Other harbor seals are primarily marine and estuarine, but sometimes wander far up rivers and visit lakes, occasionally remaining in fresh water year-round.

Many other carnivores fish in rivers and lakes, but spend more time on land. All otters (subfamily Lutrinae) other than *Enhydra lutris* and *Lontra felina*, which are principally marine (also some individuals of *Lontra canadensis* forage in near-shore marine waters), are completely dependent on fresh water; their morphology and behavior reflects their aquatic way of life. Otters live in Africa (*Aonyx capensis, Aonyx congicus, Lutra lutra, Lutra maculicollis*), Central and South America (*Lontra longicaudis, Lontra provocax, Pteronura brasiliensis*), Eurasia (*Lutra lutra, Lutra nippon*), southeast Asia (*Lutra sumatrana, Lutrogale perspicillata, Aonyx cinerea*) and North America (*Lontra canadensis*). Their terrestrial relatives, the American mink (*Neovison vison*) and the European mink (*Mustela lutreola*), are also closely associated with freshwater ecosystems, as is the polecat (*Mustela putorius*).

Other groups of carnivores show lesser degrees of dependence on fresh water. The raccoons (*Procyon spp.*) and many mongooses (Herpestidae) live in marshy and riverine habitats and feed on freshwater animals; the Marsh mongoose (*Atilax paludinosus*) and the long-snouted mongoose (*Herpestes naso*) are among the most water-dependent. Two viverrids are adapted to freshwater habitats and have morphological adaptations for semi-aquatic life. One lives in Congo tributaries (the aquatic genet, *Genetta piscivora*), whereas the other (the otter civet, *Cynogale bennettii*) lives in south-east Asian streams and swamps (Veron et al., 2006). Two species of Asian felids, the fishing cat (*Prionailurus viverrinus*) and the flat-headed cat (*Prionailurus planiceps*), are strongly associated with wetlands and hunt aquatic animals. The jaguar (*Panthera onca*) also feeds on aquatic prey like turtles and caimans, and the jungle cat (*Felis chaus*) can dive to catch fish.

Order Sirenia (2 families, Shoshani, 2005): Sirenians are fully aquatic herbivores that live in coastal marine and fresh waters of the tropics. There are four living species in the order, the dugong (marine) and three species of manatees (*Trichechus* spp). All three manatee species occur to some extent in fresh water and those manatees that live primarily in marine environments may depend on at least occasional access to fresh water for drinking. The Amazonian manatee (*T. inunguis*) is an obligate freshwater species confined to lakes and rivers of South America. The West African (*T. senegalensis*) and West Indian (*T. manatus*) manatees are riverine, estuarine, and marine, respectively occurring in central-west Africa and in northern South America, eastern Central America, south-eastern North America, and the Caribbean islands.

Order Artiodactyla (240 species in 10 families; Grubb, 2005): The best-known freshwater artiodactyls are the two living hippopotamuses (the common hippopotamus, *Hippopotamus amphibius* and the pygmy hippopotamus, *Hexaprotodon liberiensis*), both of which live in Africa (Fig. 3). Other ungulates forage in and around fresh water but cannot be considered semi-aquatic: the moose or elk (*Alces alces*) in the Holarctic region, the South American marsh deer (*Blastocerus dichotomus*), the Chinese water deer (*Hydropotes inermis*), the Père David's deer (*Elaphurus davidianus*), the Asian water buffalo (*Bubalus bubalis*) in tropical Asia, Africa's sitatunga (*Tragelaphus spekei*), the five species of the genus *Kobus* (waterbuck, lechwes, kob, puku), and the African water chevrotain (*Hyemoschus aquaticus*).

Fig. 3 A pod of hippos (*Hippopotamus amphibius*) rests on the banks of the Mara River in SW Kenya. Remaining in or close to water throughout the day, hippos emerge at night to graze on nearby vegetation. Photo by B. D Patterson

Order Cetacea (84 species in 11 families; Mead & Brownell, 2005): The cetaceans are strictly aquatic; most of the species are marine but a few live in fresh water. There are two obligate freshwater species in Asia: the Yangtze River dolphin or baiji (*Lipotes vexillifer*) and the blind South Asian species *Platanista gangetica* in the Indus, Ganges, Brahmaputra, Meghna, and Karnaphuli river systems. The Irrawaddy dolphin (*Orcaella brevirostris*) and finless porpoise (*Neophocaena phocaenoides*) are considered facultative freshwater cetaceans as they have populations living in major river systems (e.g., Irrawaddy, Mekong, Yangtze) as well as estuaries and coastal marine waters (Jefferson & Smith, 2002). South America has two species of river dolphin, the Amazon dolphin or boto (*Inia geoffrensis*) and the tucuxi (*Sotalia fluviatilis*). An estuarine and coastal marine form of tucuxi (*S. guianensis*) also exists, and it occurs far up some major rivers such as the Orinoco (Cunha et al., 2005). Another so-called "river dolphin," the La Plata dolphin or franciscana (*Pontoporia blainvillei*), actually lives in coastal waters and estuaries of eastern South America.

Phylogeny and historical process

The earliest representative of the Eutherian mammals is about 125 millions years old, but the ecomorphological diversification of placental mammals began in earnest with the demise of dinosaurs 65 millions years ago (Rose & Archibald, 2005).

The approximate ages of the two strictly aquatic orders of mammals are 52 million years for the cetaceans and 61 million years for the sirenians (Rose & Archibald, 2005). Within the Artiodactyla, the specialized freshwater family Hippopotamidae arose in the middle Miocene (ca 15 million years ago) (Rose & Archibald, 2005). Within the carnivores, the otters split from other Mustelidae between 15 and 8 million years ago (Marmi et al 2004). Within the Rodentia, the beavers were believed to be related to the Sciuridae, while recent studies now place them close to the Geomyoidea (pocket gophers and allies) with a possible Eocene origin in North America (Huchon et al., 2002). There are only two extant species, but the family was more diverse in the past, with species not adapted to aquatic life, including a species living underground (Miocene), and giant species during the Pleistocene (Müller-Schwarze & Sun, 2003). The Ichthyomyine forms a diversified group of freshwater rodents within the neotropical muroids; their ecomorphological adaptation to a semi-aquatic way of life and carnivory is peculiar among rodents (Voss, 1988).

In some cases, the morphological differentiation of an aquatic form has occurred in a short time, within a group of terrestrial forms. The aquatic genet is a specialized genet, and recent phylogenetic results (Gaubert et al., 2004) show that this species is included within the genet clade, suggesting its rapid adaptation to piscivory and freshwater habitat occurred via extreme morphological modification.

Human related issues

Many of the freshwater mammals are threatened (IUCN Red List species; IUCN, 2006), both by the destruction of their habitat and by direct or specific menaces. Also, very little is known about many of the species and there are insufficient data to determine the status of their populations. The status of threatened freshwater mammals is given in Table 1.

The major threats to the freshwater mammals are the modification or destruction of their habitat (deforestation, canalisation of rivers, removal of bankside vegetation, dams, draining of wetlands), pollution, trapping, and hunting.

Among the threatened freshwater mammals, the European mink suffers from competition with the American mink (escaped from fur farms) and from over-hunting and habitat loss. In France, an additional threat comes from unintentional poisoning and trapping. The American mink is not the only invasive species in Europe to cause serious threats to indigenous wetland species. The South American nutria, for instance, escaped from fur farms, and feral populations have become established in North America, Europe, and Asia, causing damage to river banks and feeding on wetland plants.

In Africa and Asia, otters are killed for skins and meat and they are considered responsible for poultry losses. Bush clearing and deforestation have destroyed or modified their habitat. Otters also suffer from canalisation of rivers, removal of bankside vegetation, dams, draining of wetlands, and other man-made habitat changes. Pollution is another major threat, as are trapping and hunting.

Some more specific threats affect the hippopotamus. The historical range of the common hippopotamus extended throughout Africa, but is now limited, with an estimated total population in 2004 of only 120,000–148,000 individuals (IUCN/SSC Hippo Specialist Group, 2006). The numbers in some countries are unknown. Common hippos have probably disappeared from Liberia and Mauritania. Major threats are habitat loss and illegal hunting for meat and ivory, which have been increasing in recent years. The trade ban on elephant ivory has increased illegal hunting of hippos for their teeth—annual exports of hippopotamus teeth increased by 530% within 2 years following the ivory ban (IUCN/SSC Hippo Specialist Group, 2006). The pygmy hippopotamus (*H. liberiensis*) also suffers from hunting and habitat loss and its total population was estimated in 2004 at no more than 2,000–3,000 individuals in the wild (IUCN/SSC Hippo Specialist Group, 2006). Three species of hippopotamus are recently extinct (*Hippopotamus lemerlei*, *Hippopotamus laloumena* and *Hexaprotodon madagascariensis*), but dating of their remains and times of extinction are debated.

Most of the water-dependent ungulates are threatened. Wild Asian buffalo suffer from interbreeding with feral and domestic buffalo and habitat loss and degradation. Transmission of diseases and parasites by domestic livestock and competition with them are also serious threats.

Certain of the river dolphins and porpoises are among the most threatened mammals of the world.

The Yangtze River dolphin has been regarded for many years as the world's most endangered cetacean (IWC, 2001; Reeves et al., 2003) and the results of a range-wide survey in late 2006 suggest that it already may be extinct. River cetaceans suffer from the loss or fragmentation of habitat (land 'reclamation', channelization, impoundment, diversion, etc.), pollution, depletion of food resources by fishing, incidental mortality from fishing operations (entanglement in nets, electrocution, snagging on hooks, etc.), boat collisions, and disturbance or displacement by intensive vessel traffic.

The beaver played a very important economic and historical role during the 16th to 18th century, particularly in North America. Competition for fur trade among the European powers resulted in conflict over territory and trade hegemony (Müller-Schwarze & Sun, 2003). As a consequence of trapping and hunting, by the end of the 19th century the populations of beavers in both North America and Eurasia had been reduced and even extirpated from large areas portions of their range (Veron, 1992a,b).

References

Benstead, J. P. & L. E. Olson, 2003. *Limnogale mergulus*, Wed-footed or aquatic tenrec. In Goodman, S. & J. P. Benstead (eds), The Natural History of Madagasca. University of Chicago Press, Chicago: 1267–1273.

Bronner, G. N. & P. D. Jenkins, 2005. Order Afrosoricida. In Wilson, D. E. & D. A. M. Reeder (eds), Mammal Species of the World: A Taxonomic and Geographic Reference, 3rd edn. The Johns Hopkins University Press, Baltimore: 1: 71–81.

Cunha, H. A., V. M. F. da Silva, J. Lailson-Brito Jr, M. C. O. Santos, P. A. C. Flores, A. R. Martin, A. F. Azevedo, A. B. L. Fragoso, R. C. Zanelatto & A. M. Solé-Cava, 2005. Riverine and marine ecotypes of *Sotalia* dolphins are different species. Marine Biology 148: 449–457.

Gaubert, P., M. Tranier, A. S. Delmas, M. Colyn & G. Veron, 2004. First molecular evidence for reassessing phylogenetic affinities between genets (*Genetta*) and the enigmatic genet-like taxa *Osbornictis*, *Poiana* and *Prionodon* (Carnivora, Viverridae). Zoologica Scripta 33: 117–129.

Groves, C., 2005. Order Monotrema. In Wilson, D. E. & D. A. M. Reeder (eds), Mammal Species of the World: A Taxonomic and Geographic Reference, 3rd edn. The Johns Hopkins University Press, Baltimore: 1: 1–2.

Grubb, P., 2005. Order Artiodactyla. In Wilson, D. E. & D. A. M. Reeder (eds), Mammal Species of the World: A Taxonomic and Geographic Reference, 3rd edn. The Johns Hopkins University Press, Baltimore: 1: 637–722.

Hoffmann, R. S. & A. T. Smith, 2005. Order Lagomorpha. In Wilson, D. E. & D. A. M. Reeder (eds), Mammal Species of the World: A Taxonomic and Geographic Reference, 3rd edn. The Johns Hopkins University Press, Baltimore: 1: 185–211.

Huchon, D., O. Madsen, M. J. J. B. Sibbakd, K. Ament, M. J. Stanhope, F. Catzeflis, W. W. de Jong & E. J. P. Douzery, 2002. Rodent phylogeny and a timescale for the evolution of glires: evidence from an extensive taxon sampling using three nuclear genes. Molecular Biology and Evolution 19: 1053–1065.

Hutterer, R., 2005. Order Soricomorpha. In Wilson, D. E. & D. A. M. Reeder (eds), Mammal Species of the World: A Taxonomic and Geographic Reference, 3rd edn. The Johns Hopkins University Press, Baltimore: 1: 220–311.

IUCN/SSC Hippo Specialist group, 2006. http://www.moray.ml.duke.edu/projects/hippos/ Downloaded on 28 April 2006.

IUCN 2006. 2006 IUCN Red List of Threatened Species, http://www.iucnredlist.org. Downloaded on 28 February 2006.

IWC. 2001. Report of the standing sub-committee on small cetaceans. Journal of Cetacean Research and Management 3 (Supplement): 263–291.

Jefferson, T. A. & B. D. Smith (eds.), 2002. Facultative freshwater cetaceans of Asia: their ecology and conservation. Raffles Bulletin of Zoology Suppl. 10.

Kerbis Peterhans, J. C. & B. D. Patterson, 1995. The Ethiopian water mouse *Nilopegamys* Osgood, with comments on the evolution of semi-aquatic adaptations in African Muridae. Zoological Journal of the Linnean Society 113: 329–349.

Ma, J., G. Jones, S. Zhang, J. Shen, W. Metzner, L. Zhang & B. Liang, 2003. Dietary analysis confirms that Rickett's big-footed bat (*Myotis ricketti*) is a piscivore. Journal of Zoology 261: 245–248.

Marmi J., J. F. Lopez-Giraldez, X. Domingo-Roura, 2004. Phylogeny, evolutionary history and taxonomy of the Mustelidae based on sequences of the cytochrome b gene and a complex repetitive flanking region. Zoologica Scripta 33: 481–499.

Marshall, L. G., 1978. *Lutreolina crassicaudata*. Mammalian Species 91: 1–4.

Mead, J.G. & R.L. Brownell Jr, 2005. Order Cetacea. In Wilson, D. E. & D. A. M. Reeder (eds), Mammal Species of the World: A Taxonomic and Geographic Reference, 3rd edn. The Johns Hopkins University Press, Baltimore 1: 723–743.

Müller-Schwarze, D. & L. Sun, 2003. The Beaver. Natural History of a Wetlands Engineer. Comstock Publishing Associates, Cornell University Press, Ithaca and London.

Musser, G. G. & M. D. Carleton, 2005. Superfamily Muroidea. In Wilson, D. E. & D. A. M. Reeder (eds), Mammal Species of the World: A Taxonomic and Geographic Reference, 3rd edn. The Johns Hopkins University Press, Baltimore: 2: 894–1531.

Nowak, R. M., 1999. Walker's mammals of the world, 6th ed. Johns Hopkins University Press, Baltimore, MD.

Reeves, R. R., B. D. Smith, E. A. Crespo & G. Notarbartolo di Sciara, 2003. Dolphins, Whales and Porpoises:

2002–2010 Conservation Action Plan for the World's Cetaceans. IUCN/SSC Cetacean Specialist Group. IUCN, Gland, Switzerland and Cambridge, UK. ix + p. 139.

Reijnders, P., S. Brasseur, J. van der Toorn, P. van der Wolf, I. Boyd, J. Harwood, D. Lavigne & L. Lowry, 1993. Seals, Fur seals, Sea Lions, and Walrus. Status Survey and Conservation Action Plan. IUCN/SSC Seal Specialist Group, IUCN, The World Conservation Union, Gland, Switzerland and Cambridge, UK, p. 88.

Revenga, C. & Y. Kura, 2003. Status and Trends of Biodiversity of Inland Water ecosystems. Secretariat of the Convention of Biological Diversity, Montreal, technical Series no. 11, p. 116.

Rose, K. D. & J. D. Archibald, 2005. The Rise of Placental Mammals. The John Hopkins University Press, Baltimore and London, p. 259.

Shoshani, J., 2005. Order Sirenia. In Wilson, D. E. & D. A. M. Reeder (eds), Mammal Species of the World: A Taxonomic and Geographic Reference, 3rd edn. The Johns Hopkins University Press, Baltimore: 1: 92–93.

Simmons, N. B., 2005. Order Chiroptera. In Wilson, D. E. & D. A. M. Reeder (eds), Mammal Species of the World: A Taxonomic and Geographic Reference, 3rd edn. The Johns Hopkins University Press, Baltimore: 1: 312–529.

Schnitzler, H. U., E. K. V. Kalko, I. Kaipf & A. D. Grinnell,1994. Fishing and echolocation behavior of the greater bulldog bat, Noctilio leporinus, in the field. Behavioral Ecology and Sociobiology 35: 327–345.

Veron, G. 1992a. Les castors. Atlas Visuels, Payot, Lausanne, 64 pp.

Veron, G., 1992b. Histoire biogéographique du Castor en Europe, *Castor fiber* (Rodentia, Mammalia). Mammalia 56: 87–108.

Veron, G., P. Gaubert, N. Franklin, A. Jennings & L. Grassman, 2006. A reassessment of the distribution and taxonomy of the endangered otter civet, *Cynogale bennettii* (Carnivora: Viverridae) of South-east Asia. Oryx 1: 42–49.

Voss, R. S., 1988. Systematics and ecology of ichthyomyine rodents (Muroidea): patterns of morphological evolution in a small adaptive radiation. Bulletin of the American Museum of Natural History 188: 259–493.

Wilson, D. E. & D. M. Reeder, 2005. Mammal Species of the World: A Taxonomic and Geographic Reference, 3rd edn. Johns Hopkins University Press, Baltimore, MD.

Wozencraft, W. C., 2005. Order Carnivora. In Wilson, D. E. & D. A. M. Reeder (eds), Mammal Species of the World: A Taxonomic and Geographic Reference, 3rd edn. The Johns Hopkins University Press, Baltimore: 1: 532–628.

Hydrobiologia (2008) 595:619–626
DOI 10.1007/s10750-007-9121-2

FRESHWATER ANIMAL DIVERSITY ASSESSMENT

Global diversity of freshwater birds (Aves)

Olivier Dehorter · Matthieu Guillemain

Abstract Among the 10,000 birds species living on earth, 5% (e.g., 560) need imperatively freshwater habitat in order to satisfy at least one of their life history traits. About 11 completed families could even disappear if their wetland habitat left. About 10% (58) of these can be considered as endemic. Africa contains the biggest number of endemic (20) and more precisely Madagascar. Among freshwater species, ducks and geese have a major importance in human activities in northern hemisphere related to food resources (hunting) or birding.

Keywords Birds · Aves · Wetland · Emdemism

Guest editors: E.V. Balian, C. Lévêque, H. Segers & K. Martens
Freshwater Animal Diversity Assessment

O. Dehorter (✉)
Centre de Recherche sur la Biologie des Populations d'Oiseaux, MNHN, Case Postale 51, 55 rue Buffon, Paris 75005, France
e-mail: Dehorter@mnhn.fr

M. Guillemain
Office National de la Chasse et de la Faune Sauvage, CNERA Avifaune Migratrice, La Tour du Valat, Le Sambuc, Arles 13200, France
e-mail: m.guillemain@oncfs.gouv.fr

Introduction

Birds have colonized all continents (from the Arctic to the Antarctic) and all habitats on earth (from the desert to the open sea) since their separation from the Archosauria at the cretacean (end of the secondary era) (Xu et al. 2003). Today, with some few exceptions (ostriches, *Struthionidae*; kiwi, *Apterygidae*; penguins, *Spheniscidae*;...), most bird species are able to fly, and for the general public birds are the only animals wearing feathers (an evolution from reptilian scales). As compared to other taxa, e.g., insects, bird species are not particularly numerous. Approximately 10,000 species have been described so far. Even if some new species continue to be discovered (see details below), the large interest for this group and the amount of natural history surveys for centuries allow us to believe that only a few more bird species will be discovered in the future.

According to traditional taxonomical classification, it is especially the oldest bird families that inhabit water habitats. Seawater and freshwater were therefore most likely the oldest habitats colonized by birds. Many bird taxonomical families use freshwater for all or a part of their cycle life. However, within this habitat type several geographical regions are often used by a given species, due to the migratory behaviour of most of them.

In this article, we considered as freshwater birds those species in which all individuals spend at least a part of their annual lifecycle in freshwater habitats (i.e., habitats where salinity is generally below 0.5 g/l) (Sibley & Monroe 1991; Birdlife international 2000), contrary to the birdlife definition which is more flexible. Thus, a species breeding in marshes as well as in mangroves is not included here, whereas a species wintering only in reedbeds and breeding in forests is considered. Typically aquatic species relying only on seawaters, e.g., penguins, are not considered either. In others words, the rule we followed was to select the species in which disappearance of a freshwater habitat would lead to disappearance of the species with little chance to adapt. Freshwater species typically show morphological adaptations, especially for locomotion (e.g., del Hoyo et al. 1992). Webs between toes, an efficient tool to swim or dive, are present in many groups like ducks (*Anatidae*), gulls (*Laridae*) or cormorants (*Phalacrocoracidae*). Long legs are quite common in order to walk in more or less deep water, e.g., in herons (*Ardeidae*) or shorebirds (*Charadriiformes*). Long fingers, to walk on emergent vegetation, can be found too in species like Jacanas (*Jacanidae*). However, some species that are restricted to wetlands or freshwater habitats do not have obvious aquatic adaptations, such as in passerine species. Finally, in addition to morphological adaptations some waterbirds have developed physiological traits like the salt gland, that is not only present in marine birds but is also functional in some freshwater species like ducks (Ash 1969).

Specific diversity

In 1990, Sibley and Alquist (1990) published the first bird taxonomical classification based on a genetic approach (DNA–DNA hybridization). This new classification drastically changed the vision of bird phylogeny, for instance in putting penguins in the same group as divers (*Gaviidae*) and frigatebirds (*Fregatidae*), while they were considered as a superorder before. In the same way, the new world vultures were now closer to storks (*Ciconiidae*) than to old world vultures. This new genetic classification contains 22 orders. New progresses in genetic techniques could again change bird classification in the future, i.e., DNA barcoding using the cytochrome c oxidase sequence (Hebert et al. 2004a, http://www.barcodingbirds.org/), which is used for other organism like plants (Kress et al. 2005), or insects (Hebert et al. 2004b), will be generalized for birds in close future. New taxonomic techniques can increase biodiversity by simply splitting already known species into several new ones. This was for example the case for four north-American bird species (*Tringa solitaria*, *Surnella magna*, *Vireo gilvus*, *Cistothorus palustris*), that could each be split into two (Hebert et al. 2004a). In addition, new bird species are still discovered each year in the world (55 new ones from 1980 to 2006), especially in the tropics (van Rootselaar 1999, 2002). A new species as large as an albatross was described in 1983 in Amsterdam island, even though the area was occupied by a permanent scientific base since the sixties (Roux et al. 1983). Of course, some species are also considered as having disappeared (over 100 species since the 16th century), among which some were freshwater birds, for instance belonging to the *Anatidae* (ducks) or *Rallidae* (rails and crakes) families (Birdlife international 2000).

Depending on the five most widely recognized taxonomic lists, there are about 8,700–11,000 bird species alive (Sibley & Monroe 1991; Devillers et al. 1993; American Ornithological Union 1998; Clements 2000; Howard & Moore 2003). The number of orders varies between 22 and 31. In this work, we used the list by Sibley and Monroe (1991), which contains 183 families for 30 orders.

Among the 30 orders in the Aves group, 14 have members using freshwater, and two are strictly restricted to this habitat (Gaviiforms and Podicipediforms). At the family level, 45 of the 183 listed families need freshwater habitat for one part of their lifecycle, and 11 are completely dependent upon this habitat (Table 1). Many of these families are monospecific, like the *Opisthocomidae* with the Hoatzin (*Opisthocomus hoazin*) or the *Balaenicipitidae* with the Shoebill (*Balaeniceps rex*), or include few species. The largest family here is the *Podicipedidae*, with 19 species.

Present distribution and main areas of endemicity

About 566 species, i.e., more than 5% of all bird species, need freshwater to survive (Table 1). Birdlife International (2000) recorded twice as many species dependent on water; however the criteria used was far less drastic than what we retained here.

Unsurprisingly, water-dependent species are in low numbers in biogeographic regions where freshwater

Table 1 Number of freshwater-dependent bird species per family (among families with at least one species dependent upon freshwater) and per zoogeographic region

Order	Family	PA	NA	NT	AT	OL	AU	PAC	ANT	World
Anseriformes	*Anatidae*	44 (50)	40 (47)	29 (37)	20 (21)	9 (11)	20 (26)	0 (1)	1 (1)	134 (164)
	Anhimidae			**3 (3)**						**3 (3)**
Caprimulgiformes	*Caprimulgidae*	0 (6)	0 (9)	1 (37)	0 (23)	0 (10)	0 (5)			1 (85)
Charadriiformes	*Charadriidae*	4 (12)	0 (8)	3 (10)	4 (17)	1 (8)	1 (12)			13 (67)
	Glareolidae	1 (4)			2 (7)	2 (5)	0 (1)			4 (17)
	Jacanidae		**1 (1)**	**2 (2)**	**3 (3)**	**2 (2)**	**1 (1)**			**8 (8)**
	Laridae	9 (16)	5 (35)	4 (21)	1 (16)	2 (13)	3 (19)	0 (10)	0 (2)	19 (97)
	Recurvirostridae	1 (2)	3 (3)	3 (4)	0 (1)	2 (3)	3 (4)	1 (1)		9 (12)
	Rostratulidae			**1 (1)**	**1 (1)**	**1 (1)**	**1 (1)**			**2 (2)**
	Scolopacidae	24 (45)	19 (42)	3 (7)	3 (3)	0 (1)	0 (4)	0 (1)		43 (87)
Ciconiiformes	*Ardeidae*	8 (15)	3 (14)	9 (20)	8 (19)	7 (16)	3 (9)			31 (67)
	Balaenicipitidae				**1 (1)**					**1 (1)**
	Ciconiidae	0 (2)	0 (1)	1 (3)	3 (7)	3 (7)	0 (1)			7 (19)
	Scopidae				**1 (1)**					**1 (1)**
	Threskiornithidae	3 (6)	2 (3)	8 (13)	4 (10)	3 (5)	3 (5)			17 (34)
Coraciiformes	*Alcedinidae*	1(1)	3 (3)	2 (2)	11 (25)	0 (27)	0 (39)	0 (10)		14 (97)
Cuculiformes	*Cuculidae*	0 (4)	0 (5)	0 (29)	1 (35)	0 (45)	0 (35)			1 (140)
	Opisthocomidae			**1 (1)**						**1 (1)**
Falconiformes	*Accipitridae*	3 (50)	3 (21)	1 (63)	4 (56)	4 (45)	1 (41)	0 (3)		10 (243)
Galliformes	*Phasianidae*	0 (56)	0 (13)	0 (1)	2 (43)	0 (62)	2 (13)	1 (1)		4 (179)
Gaviiformes	*Gaviidae*	**5 (5)**	**5 (5)**							**5 (5)**
Gruiformes	*Aramidae*		**1 (1)**	**1 (1)**						**1 (1)**
	Gruidae	5 (9)	1 (2)		3 (3)		2 (2)			10 (15)
	Heliornithidae			0 (1)	1 (1)	1 (1)				2 (3)
	Rallidae	8 (11)	10 (13)	29 (46)	14 (29)	13 (23)	14 (36)	4 (10)		65 (133)
Passeriformes	*Cinclidae*	**1 (1)**	**1 (1)**	**2 (2)**		**1 (1)**				**5 (5)**
	Emberizidae	2 (38)	6 (151)	8 (664)	0 (6)					15 (834)
	Estrildidae				1 (95)	0 (18)	2 (41)	0 (4)		3 (156)
	Furnariidae			5 (239)						5 (239)
	Hirundinidae	1 (8)	1 (10)	1 (26)	4 (42)	1 (7)	0 (5)	0 (1)		7 (93)
	Malaconotidae				1 (50)					1 (50)
	Motacillidae	2 (21)	0 (4)	0 (8)	1 (30)	1 (8)	0 (2)			3 (66)
	Muscicapidae	2 (107)	0 (13)	0 (56)	0 (150)	10 (109)	2 (40)	0 (5)		11 (459)
	Paridae	2 (24)	0 (10)		0 (23)	0 (10)				2 (67)
	Ploceidae				17 (114)	0 (4)				17 (119)
	Pycnonotidae	0 (6)			0 (75)	1 (58)	0 (1)			1 (140)
	Sylviidae	18 (105)	0 (1)		16 (198)	4 (63)	0 (28)	0 (11)		36 (398)
	Thamnophilidae			2 (199)						2 (199)
	Timaliidae	2 (61)	0 (1)		0 (42)	1 (204)	0 (1)			3 (287)
	Troglodytidae	0 (1)	1 (13)	4 (67)						4 (80)
	Tyrannidae		0 (33)	10 (398)						10 (421)
Pelecaniformes	*Anhingidae*		**1 (1)**	**1 (1)**	**1 (1)**	**1 (1)**	**1 (1)**			**4 (4)**
	Pelecanidae	1 (2)	0 (2)	0 (1)	2 (2)		0 (1)			2 (8)
	Phalacrocoracidae	1 (9)	2 (6)	1 (7)	2 (4)	2 (4)	0 (12)	0 (1)	0 (3)	7 (39)

Table 1 continued

Order	Family	PA	NA	NT	AT	OL	AU	PAC	ANT	World
Podicipediformes	*Podicipedidae*	**5 (5)**	**8 (8)**	**9 (9)**	**5 (5)**	**3 (3)**	**3 (3)**			**19 (19)**
Strigiformes	*Strigidae*	1 (26)	0 (18)	0 (46)	1 (29)	1 (40)	0 (23)	0 (1)		3 (171)
Total		154 (708)	116 (498)	145 (2027)	138 (1188)	76 (816)	62 (412)	6 (60)	1 (6)	566 (5335)

In parentheses: Total number of bird species in the family including FW, marine and terrestrial. (in bold, families where all species members are water-dependent). PA, Palaearctic; NA, Nearctic; NT, Neotropical; AT, Afrotropical ; OL, Oriental; AU, Australasia; PAC, Pacific Oceanic Islands; ANT, Antarctic region

habitats are rare, like in the Antarctic and in the Pacific islands 1 and 6 species, respectively (Table 1 and 2 and Fig. 1). It is also quite low in the Australian region where freshwater habitats are submitted to drastic climatic conditions, and in the Oriental region, which is mostly covered by forests. The other biogeographic regions each contain roughly the same numbers of water-dependent species, between 116 and 154 (Fig. 1). Among the 183 families with water-dependent species, only six are restricted to only one biogeographic region (*Anhimidae*, *Opisthocomidae*, *Furnariidae*, *Balaenicipitidae*, *Scopidae*, *Malaconotidae*); the first three are in the Neotropics, and the latter ones in the Afrotropics.

Table 2 Number of genera including freshwater-dependent bird species per family (among families with at least one species dependent upon freshwater) and per zoogeographic region

Order	Family	PA	NA	NT	AT	OL	AU	PAC	ANT	World
Anseriformes	*Anatidae*	16 (17)	16 (17)	16 (17)	12 (13)	3 (4)	12 (14)		1 (1)	40 (46)
	Anhimidae			**2 (2)**						**2 (2)**
Caprimulgiformes	*Caprimulgidae*			1 (12)						1 (15)
Charadriiformes	*Charadriidae*	3 (4)		4 (4)	2 (2)	1 (2)	1 (6)			5 (10)
	Glareolidae	1 (2)			1 (4)	1 (3)				1 (5)
	Jacanidae		**1 (1)**	**1 (1)**	**2 (2)**	**2 (2)**	**1 (1)**			**6 (6)**
	Laridae	3 (7)	3 (7)	3 (6)	1 (5)	1 (2)	2 (6)			4 (13)
	Recurvirostridae	1 (2)	2 (2)	2 (2)		1 (2)	2 (3)	1 (1)		2 (3)
	Rostratulidae			**1 (1)**	**1 (1)**	**1 (1)**	**1 (1)**			**1 (1)**
	Scolopacidae	8 (12)	10 (16)		2 (2)					12 (21)
Ciconiiformes	*Ardeidae*	6 (9)	3 (11)	8 (15)	5 (9)	4 (9)	3 (8)			10 (20)
	Balaenicipitidae				**1 (1)**					**1 (1)**
	Ciconiidae			1 (3)	3 (5)	3 (5)				4 (6)
	Scopidae				**1 (1)**					**1 (1)**
	Threskiornithidae	3 (5)	1 (2)	3 (7)	4 (6)	2 (3)	2 (3)			8 (14)
Coraciiformes	*Alcedinidae*	1 (1)		1 (1)	3 (7)					4 (18)
Cuculiformes	*Cuculidae*				1 (9)					1 (28)
	Opisthocomidae			**1 (1)**						**1 (1)**
Falconiformes	*Accipitridae*	3 (20)	3 (12)	1 (24)	3 (25)	3 (19)	1 (16)			5 (65)
Galliformes	*Phasianidae*				2 (6)		1 (7)	1 (1)		2 (45)
Gaviiformes	*Gaviidae*	**1 (1)**	**1 (1)**							**1 (1)**
Gruiformes	*Aramidae*		**1 (1)**	**1 (1)**						**1 (1)**
	Gruidae	1 (1)	1 (1)		2 (2)		1 (1)			2 (2)
	Heliornithidae				1 (1)	1 (1)				2 (3)
	Rallidae	5 (8)	6 (7)	11 (14)	8 (14)	7 (9)	7 (14)	2 (5)		17 (34)

Table 2 continued

Order	Family	PA	NA	NT	AT	OL	AU	PAC	ANT	World
Passeriformes	*Cinclidae*	**1 (1)**	**1 (1)**	**1 (1)**		**1 (1)**				**1 (1)**
	Emberizidae	1 (7)	4 (53)	6 (170)						10 (201)
	Estrildidae				1 (20)		1 (11)			2 (30)
	Furnariidae			3 (53)						3 (53)
	Hirundinidae	1 (3)	1 (5)	1 (8)	2 (5)	1 (3)				3 (14)
	Malaconotidae				1 (8)					1 (8)
	Motacillidae	1 (3)			1 (4)	1 (2)				1 (5)
	Muscicapidae	1 (20)				5 (23)	2 (16)			5 (69)
	Paridae	1 (3)								1 (7)
	Ploceidae				2 (17)					2 (17)
	Pycnonotidae					1 (9)				1 (21)
	Sylviidae	3 (15)			5 (44)	2 (14)				8 (61)
	Thamnophilidae			2 (45)						2 (45)
	Timaliidae	2 (14)				1 (36)				3 (53)
	Troglodytidae		1 (7)	2 (15)						2 (16)
	Tyrannidae			8 (102)						8 (102)
Pelecaniformes	*Anhingidae*		**1 (1)**	**1 (1)**	**1 (1)**	**1 (1)**	**1 (1)**			**1 (1)**
	Pelecanidae	1 (1)	1 (1)	1 (1)	1 (1)	1 (1)	1 (1)			1 (1)
	Phalacrocoracidae	1 (1)	1 (1)	1 (1)	1 (1)	1 (1)	1 (1)			1 (1)
Podicipediformes	*Podicipedidae*	**2 (2)**	**4 (4)**	**4 (4)**	**2 (2)**	**2 (2)**	**2 (2)**			**6 (6)**
Strigiformes	*Strigidae*	1 (11)			1 (6)	1 (8)				2 (23)
Total		68 (170)	62 (151)	87 (512)	73 (224)	48 (163)	42 (112)	4 (7)	1 (1)	198 (1097)

In parentheses: Total number of bird genera in the area. (in bold, families where all species members are water-dependent). PA, Palaearctic Region; NA, Nearctic Region; NT, Neotropical Region; AT, Afrotropical Region; OL, Oriental Region; AU, Australasian Region; PAC, Pacific region and Oceanic Islands, ANT, Antarctic Region

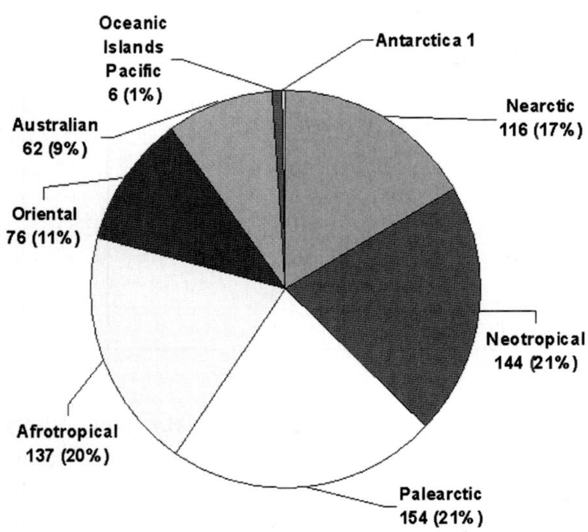

Fig. 1 Number of freshwater-dependent species according to biogeographic zones, with the percentage of waterbird species represented in brackets. Some species can occur in more than one biogeographic region

Approximately 10% of water-dependent bird species (i.e., 58 species) can be considered to be endemic. A batch of 12 and 10 endemic species belong to the *Anatidae* and *Rallidae* families, respectively, and eight families only have one endemic species (Table 3). Among water-dependent species, the proportion of endemic ones varies a lot between families, reaching more than 25% in *Podicipedidae* (grebes) and *Recurvirstridae* (Avocets and alias) (Fig. 2). Africa is especially rich in endemic birds (20 species) (Fig. 3A), most of them being found in Madagascar or on islands of the Gulf of Guinea. In fact, islands concentrate the majority of endemic species whatever the biogeographical region, except in the Neotropics where many endemic species are also found in the Andes, another region with relatively remote areas (Fig. 3B and 4).

Table 3 Number and percentage of endemic species according to taxonomic family

Order	Family	No of endemic sp	No of species	Percentage (%)
Anseriformes	*Anatidae*	12	164	7.3
Charadriiformes	*Jacanidae*	1	8	12.5
	Laridae	2	97	2.1
	Recurvirostridae	3	12	25
	Scolopacidae	1	87	1.1
Ciconiiformes	*Ardeidae*	4	67	6.0
	Threskiornithidae	3	34	8.8
Coraciiformes	*Alcedinidae*	2	97	2.1
Falconiformes	*Accipitridae*	1	243	0.4
Galliformes	*Phasianidae*	1	179	0.6
Gruiformes	*Gruidae*	1	15	6.7
	Rallidae	10	133	7.5
Passeriformes	*Emberizidae*	3	834	0.4
	Hirundinidae	1	93	1.1
	Ploceidae	1	119	0.8
	Sylviidae	3	398	0.8
	Timaliidae	1	287	0.3
	Troglodytidae	2	80	2.5
Podicipediformes	*Podicipedidae*	6	20	30

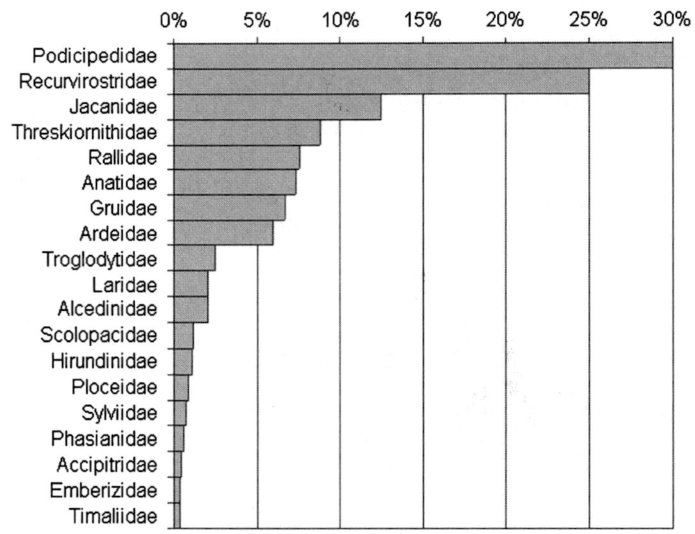

Fig. 2 Endemic rate in freshwater-dependent families

Human-related issues

Man has had close relationships with birds for a long time, probably before the Neolithic, be it for food, or for metaphysical aspects. Bird feathers were used for ornamental purposes by shamans, and birds were considered as gods in the Egyptian civilization (Horus is a falcon and Thôt is an Ibis), among others (e.g., south-American Indians). Following the example of the mammals, birds also were early domesticated. Chickens are an obvious case, but waterbirds like ducks (*Anas platyrhynchos*, *Cairina moschata*) and geese (*Anser anser*, *Anser cygnoides*) were also domesticated. Nowadays, the metaphysical

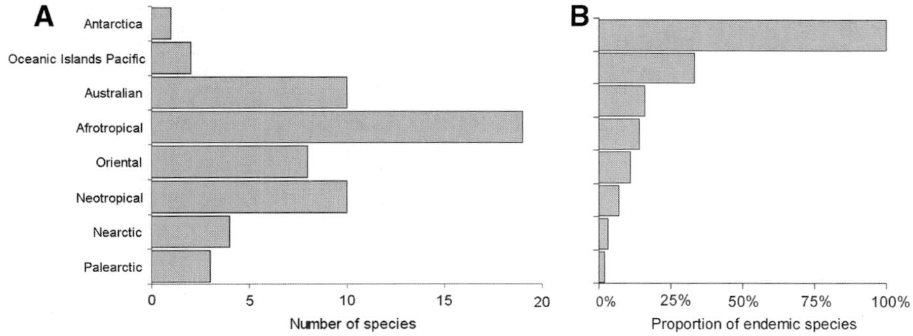

Fig. 3 Number (left, A) and proportion (right, B) of endemic freshwater-dependent bird species per biogeographic region

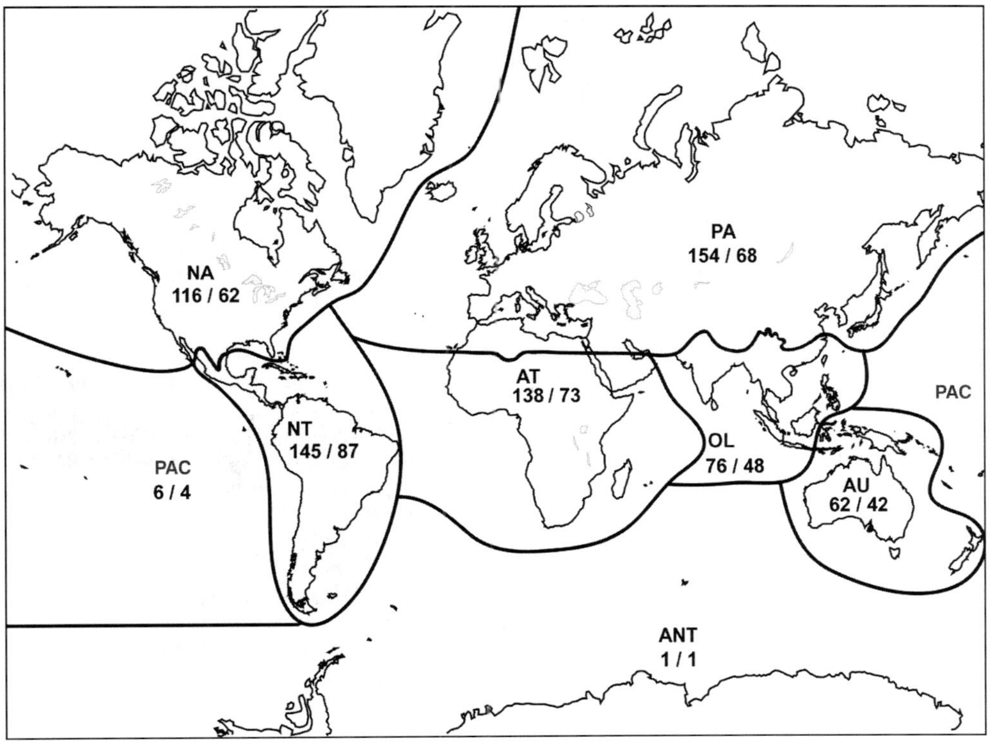

Fig. 4 Species and generic diversity of freshwater (aquatic and water dependent) birds by zoogeographic region: species number/genus number. PA, Palaearctic Region; NA, Nearctic Region; AT, Afrotropical Region; NT, Neotropical Region; OL, Oriental Region; AU, Australasian Region; PAC, Pacific Region and oceanic islands, ANT, Antartic Region

aspects have tended to disappear (except in first nations like in the Amazon where birds are still related with 'spirits'), and birds can only be considered as sources of food or pleasure. Wild birds are mostly considered as food resources in third world countries, whereas in northern countries these birds are essentially considered as sources of pleasures (leisure, hunting or bird watching).

Not surprisingly, birds are also affected by human activities, and many species are endangered or have already disappeared (the Dodo *Raphus cucullatus* for instance). According to Birdlife International (2000), 12% of total bird species are threatened, mainly by human activities, though these can also have an additive effect to intrinsic factors (small versus large populations size or range use). The same proportion is found among freshwater-dependent species (Birdlife international 2000). Habitat loss or degradation have major impacts on these species (Birdlife international 2000). Legal hunting is not considered as a major threat able to drive waterfowl species (ducks and geese) to extinction (Long et al. 2007), despite the generality of this practice on these species (e.g., Kalchreuter 1996; Mooij 2005). In compensation, birds have been the subject of numerous conservation policies for a long time (from the end of the 19th century onwards) in many countries (e.g., the Lacey act in 1900 in the USA). Waterbirds are especially concerned by several conventions and treaties at the international level. The RAMSAR convention, created in 1971 and ratified by 153 countries, focuses on "international cooperation for the conservation and wise use of wetlands and their resources". Similarly, AEWA (African-Eurasian Waterbird Agreement) was developed under the aegis of UN in the 1990s, under the framework of the Convention of Migratory Species (CMS). Analogous agreements were developed for the rest of the world (Asia, Americas). These agreements are in charge of protecting waterbirds and their habitats at the flyway level under the contracting states' responsibility.

References

American Ornithological Union, 1998. The American Ornithologists' Union Check-list of North American Birds, 7th edn. Washington, D.C.

Ash, R. W., 1969. Plsasma osmolality and salt gland secretion in the duck. Quarterly Journal of Experimental Physiology 54: 68–79.

Birdlife international, 2000. Threatened birds of the world. Barcelona and Cambridge. UK. Lynx Edicions and Birdlife International.

Clements, J. F., 2000. Birds of The World: A Checklist. 5th edn. Ibis Publishing, Vista, California.

Del Hoyo J., A. Elliott & J. Sargatal (eds), 1992. Handbook of the birds of the world, Vol. 1. Lynx Edicions, Barcelona.

Devillers, P., H. Ouellet, E. Benito-Espinal, R. Beudels, R. Cruon, N. David, C. Érard, M. Gosselin & G. Seutin, 1993. Noms français des oiseaux du Monde. MultiMonde et Chabaud Éd. MultiMondes Inc., Sainte-Foy, Québec & Éd. Chabaud, Bayonne, France.

Hebert, P. D. N., M. Y. Stoeckle, T. S. Zemlak & C. M. Francis, 2004a, Identification of Birds Through DNA Barcodes. PLoS Biology 2: 1657–1663.

Hebert, P. D. N., E. H. Penton, J. M. Burns, D. H. Janzen & W. Hallwachs, 2004b. Ten species in one: DNA barcoding reveals cryptic species in the .neotropical skipper butterfly Astraptes fulgerator. PNAS 101: 14812–14817.

Howard, R. & L. Moore, 2003. The Howard & Moore Complete Checklist of the Birds of the World. Princeton University Press.

Kalchreuter, H., 1996. Waterfowl harvest and population dynamics: an overview. Gibier Faune Sauvage—Game and wildlife 13: 991–1008.

Kress, W. J., K. J. Wurdack, E. A. Zimmer, L. A. Weigt & D. H. Janzen, 2005. Use of DNA barcodes to identify flowering plants. Proceedings of National Academy of Sciences 02: 8369–8374.

Long, P. R., T. Székely, M. Kershaw & M. O'Connell, 2007. Ecological factors and human threats both drive wildfowl populations decline. Animal conservation 10: 183–191.

Mooij, J. H., 2005. Protection and use of waterbirds in the European union. Beiträge zur Jagd und Wildforschung 30: 49–76.

Roux, F., P. Jouventin, J. F. Mougin, E. Stahl & H. Weimerskirch, 1983. Un nouvel albatros Diomedea amsterdamensis découvert sur l'ile Amsterdam (37°50′ S, 77°35′ E). Oiseau Rev Fr Orn 53: 1–11.

Sibley, C. G. & J. E. Ahlquist, 1990. Phylogeny and Classification of Birds. New Haven, Conn.: Yale University Press.

Sibley, C. G. & B. L. Monroe, 1991, Distribution and Taxonomy of Birds of the World. Yale University Press, New Haven and London.

van Rootselaar, O., 1999. New birds for the World: species discovered during 1980–1999 Birding World 12: 286–293.

van Rootselaar, O, 2002. New birds for the World: species described during 1999–2002 Birding World 15: 428–431.

Xu, X., Z. Zhou, X. Wang, X. Kuang, F. Zhang & X. Du 2003, Four-Winged Dinosaurs from China. Nature 421: 335–340.

Hydrobiologia (2008) 595:627–637
DOI 10.1007/s10750-007-9246-3

The Freshwater Animal Diversity Assessment: an overview of the results

E. V. Balian · H. Segers · C. Lévêque · K. Martens

Abstract We present a summary of the results included in the different treatments in this volume. The diversity and distribution of vertebrates, insects, crustaceans, molluscs and a suite of minor phyla is compared and commented upon. Whereas the available data on vertebrates and some emblematic invertebrate groups such as Odonata (dragonflies and damselflies) allow for a credible assessment, data are deficient for many other groups. This is owing to knowledge gaps, both in geographical coverage of available data and/or lack of taxonomic information. These gaps need to be addressed urgently, either by liberating date from inaccessible repositories or by fostering taxonomic research. A similar effort is required to compile environmental and ecological information in order to enable cross-linking and analysis of these complementary data sets. Only in this way will it be possible to analyse information on freshwater biodiversity for sustainable management and conservation of the world's freshwater resources.

Keywords Biodiversity · Continental aquatic ecosystems · Endemicity · Biogeography · Freshwater · Global · Assessment

Guest editors: E. V. Balian, C. Lévêque, H. Segers & K. Martens
Freshwater Animal Diversity Assessment

E. V. Balian · H. Segers
Belgian Biodiversity Platform, Brussels, Belgium

E. V. Balian (✉) · H. Segers · K. Martens
Freshwater Laboratory, Royal Belgian Institute of Natural Sciences, Vautierstraat 29 B-1000, Brussels, Belgium
e-mail: Estelle.Balian@naturalsciences.be

C. Lévêque
Antenne IRD, MNHN-DMPA, 43 rue Cuvier, Case Postale 26, Paris cedex 05 75231, France

K. Martens
Department of Biology, University of Ghent, K.L. Ledeganckstraat 35, Gent 9000, Belgium

Introduction

The fifty-eight chapters in this compilation aim to present a comprehensive and up-to-date review of animal (plus one chapter on macrophyte) diversity and endemism in the continental waters of the world. The treatises are diverse, and this is a consequence of the specific features of the different taxa they deal with. Nevertheless, owing to the standard approach all experts agreed to follow, it has, for the first time, become possible to compare patterns in the biodiversity of groups as diverse as nematodes, dragonflies and freshwater turtles. Clearly, one can imagine numerous approaches to study these data, and an in-depth analysis will be presented elsewhere. Here, we restrict ourselves to presenting a summary overview of the results.

The present overview focuses on species diversity and endemism. Data on the genus level are available and presented for all taxa except molluscs.

An overview of freshwater animal diversity

When we calculate the total number of described freshwater animal species, we obtain a total of 125,531 species (Tables 1, 2; plus one micrognathozoan) or approximately 126,000 species. This figure, obviously, represents present knowledge and significantly underestimates real diversity. Most authors, especially those dealing with less emblematic groups, point out that significant fractions of species remain to be discovered, and/or caution that cryptic diversity, the importance of which we can only speculate about, remains concealed because of the almost exclusive morphological approach to taxonomy. The record of 126,000 species represents 9.5% of the total number of animal species recognised globally (i.e., 1,324,000 species: UNEP, 2002). If it is taken into account that freshwaters (lakes, rivers, groundwater, etc.) take up only about 0.01% of the total surface of the globe, then it becomes evident that a disproportional large fraction of the world's total biodiversity resides in freshwater ecosystems.

The majority of the 126,000 freshwater animal species are insects (60.4%), 14.5% are vertebrates, 10% are crustaceans. Arachnids and molluscs represent 5 and 4% of the total, respectively. The remainder belong to Rotifera (1.6%), Annelida (1.4%) Nematoda (1.4%), Platyhelminthes (Turbellaria: 1%), and a suite of minor groups such as Collembola (the estimate of this taxon is based on a restricted subsample of species, see Deharveng et al., 2008, present volume) and some groups that are predominantly marine (e.g., Bryozoa, Porifera). On a regional scale, the Palaearctic appears to be the most speciose for most taxa, except for insects and vertebrates. The record for insects is fairly similar in the Palaearctic, the Oriental and the Neotropical regions, whereas vertebrates are most diverse in the Neotropical, followed by the Afrotropical, and Oriental regions.

Of freshwater macrophytes, there are 2,614 species distributed over 412 genera. This amounts to ca. 1% of the total number of vascular plants known to date (270,000: Chambers et al., 2008, present volume). This constitutes a considerable fraction, taking into account that macrophytes are primarily terrestrial. On the other hand, macrophytes play a key role in structuring freshwater ecosystems, as they provide habitat and food to many organisms. Macrophyte species diversity is highest (ca. 1,000 species) in the Neotropics, intermediate (ca. 600 species) in the Oriental, Afrotropical, and Nearctic, and relatively low (ca. 400–500 species) in the Australasian and the Palaearctic regions.

The present assessment of freshwater diversity is incomplete. Our focus is on animal taxa, and only vascular plants, of all other kingdoms, are also included. Micro-organisms such as bacteria (s.l.), viruses, Protozoa, Fungi, and algae are not treated although these groups clearly are as significant to freshwater ecology and diversity as the taxa here considered. Most of these groups, with the exception

Table 1 Total species diversity of the main groups of freshwater animals, by zoogeographic region

	PA	NA	AT	NT	OL	AU	PAC	ANT	World
Other phyla	3,675	1,672	1,188	1,337	1,205	950	181	113	6,109
Annelids	870	350	186	338	242	210	10	10	1,761
Molluscs	1,848	936	483	759	756	557	171	0	4,998
Crustaceans	4,499	1,755	1,536	1,925	1,968	1,225	125	33	11,990
Arachnids	1,703	1,069	801	1,330	569	708	5	2	6,149
Collembolans	338	49	6	28	34	6	3	1	414
Insects[a]	1,5190	9,410	8,594	14,428	13,912	7,510	577	14	75,874
Vertebrates[b]	2,193	1,831	3,995	6,041	3,674	694	8	1	18,235
Total	30,316	17,072	16,789	26,186	22,360	11,860	1,080	174	125,530

[a] The distribution of species by zoogeographic regions is incomplete for several families of Dipterans; as a result, the sum of the regional species numbers is lower than the number of genera known in the world (See chapter on Diptera families excluding Culicidae, Tipulidae, Chironomidae and Simulidae)

[b] Strictly freshwater fish species only are included (there are an additional ~2,300 brackish waters species)

Table 2 Total genus diversity of the main groups of freshwater animals, by zoogeographic region

	PA	NA	AT	NT	OL	AU	PAC	ANT	World
Other phyla	573	372	286	300	284	205	76	42	778
Annelids	190	121	78	109	90	77	4	11	354
Molluscs[a]	137	351	117	226	150	43	2	0	1,026
Crustaceans	634	294	288	424	381	325	76	25	1,533
Arachnids	152	148	171	120	102	139	5	2	456
Collembolans	71	22	5	15	10	3	2	1	78
Insects[b]	1,366	1,160	871	1,269	1,159	909	132	10	4,395
Vertebrates[c]	497	426	590	974	626	183	6	1	2,768
Total	3,620	2,894	2,406	3,437	2,802	1,884	303	92	11,388

[a] Gastropoda genera are not included

[b] The distribution of genera by zoogeographic regions is incomplete for several families of Dipterans; as a result, the sum of the regional genus numbers is lower than the number of genera known in the world (See chapter on Diptera families excluding Culicidae, Tipulidae, Chironomidae and Simulidae)

[c] Strictly freshwater fish genus number is estimated at around 2,000 (there are an additional ~500 brackish waters genera)

of algae and cyanobacteria, are dramatically understudied in aquatic biodiversity. As the key role of micro-organisms in ecosystem functioning and health is becoming more and more obvious, it is to be hoped that future assessments of micro-organismal diversity in freshwaters will complete the picture of freshwater biodiversity. Estimates on some groups are available, for example, there are 3,047 species on record for aquatic Fungi, 2,000 of which are probably restricted to freshwater (Shearer et al., 2007), and 2,392 species of freshwater protozoans (Finlay & Esteban, 1998).

Problems and knowledge gaps: state of the art

As noted above, the Palaearctic region has the highest number of species on record, for all taxa except vertebrates. For most groups, this remarkable result is very likely not factual, as indicated by many experts. The purported overwhelming biodiversity of the Palaearctic probably results from the fact that most taxonomic expertise and research efforts are centred in this region. Similarly, several authors highlight the lack of data from the Afrotropical and Oriental realms (e.g. Central Africa, parts of South America and Southeast Asia) The geographical gaps in knowledge are often linked to the extent (or limitation) of taxonomic expertise, which is greatly unequal from one group to another. On the other hand, there are several groups for which the current, Holarctic-centred distribution of species richness is suspected to be accurate: amphipods are typical of cool temperate climates and are notably rare in the tropics. Ephemeroptera or Plecoptera are predominantly Palaearctic and also this is congruent with the environmental preferences of these groups.

Similarly, a lack of knowledge on autecology of many species makes it difficult to decide whether a taxon is a true freshwater species or not, and hence whether they are to be included in the count. Such is the case for springtails, many water beetles and rotifers, amongst others. The current estimate for Collembola is based on the subset of species for which ecological information exists. It is likely that this number is an underestimate of the global number of freshwater-dependent springtails. In rotifers the problem is especially acute for bdelloids, often semi-terrestrial, many of which are known from single records only.

Diversity and distribution of vertebrates are clearly better documented than for other groups and even though it can be seen that new species of freshwater fish or even amphibians are still being described regularly, experts of all vertebrate groups are able to supply a fairly reliable estimate of the true number of extant species. Molluscs and crustaceans are generally also quite well documented, despite some geographical gaps in tropical areas. For insects, the situation is very different from one group to the next. The emblematic dragonflies are exemplary of an extensively studied group, and the current estimate of ca. 7,000 species can be considered reliable. Heteroptera and Culicidae (Diptera) also seem well documented. On the other hand, the knowledge and taxonomic expertise available for most of the numerous dipteran families vary a lot depending on the group, and it is clear that our current estimate of their diversity should be interpreted with care.

Amongst the least known groups are some phyla of primitive invertebrates such as Platyhelminthes/Turbellaria, Gastrotricha or Nematoda, to name a few, for which taxonomic knowledge and available data are critically limited. Problems relate to data mass, reliability and repeatability: unique, unvouchered or plainly dubious records are common in these little-studied groups. In addition, some of these taxa are often primarily marine or terrestrial and most of the available knowledge therefore concerns these habitats. Nematodes, for example, are likely to be the least known of all metazoan phyla. Experts currently estimate that the total diversity of extant nematodes stands at about one million species, 97% of which are undescribed (Hugot et al., 2001). As freshwater nematodes are relatively poorly studied when compared to marine or terrestrial ones, and as they represent only 7% (1,800 species) of the total number of described nematode species (27,000 species), the true diversity of freshwater nematodes is likely to be one or two orders of magnitude higher.

First results of the Freshwater Animal Diversity Assessment

In the following sections we summarise the information on species diversity and endemicity for five major groups above the level of the different chapters: vertebrates, insects, crustaceans, molluscs and a collection of several primitive phyla. Further, in-depth analyses on the FADA data will be presented elsewhere. All information and data have been extracted from the different contributions included in this special issue.

Vertebrates

The total number of freshwater vertebrate species, including water birds but excluding brackish fish species, is 18,235 species (Tables 3, 4). This represents 35% of all described vertebrates (52,000 species). Of these, a majority (69%) are fishes, followed by amphibians (24%). Considering that the total global number of fish species is presently estimated at ca. 29,000 species (Lévêque et al., 2008, present volume), this means that nearly 50% of all fish species inhabit fresh and brackish waters (15,062 species, 12,470 of which are strictly freshwater). Freshwater habitats support 73% of all amphibian species; other groups are less represented in freshwaters. Freshwater vertebrates are most diverse in the Neotropical region, followed by the Oriental and the Afrotropical regions, and this holds for both generic as well as species diversity (Fig. 1). The Palaearctic is more speciose than the Nearctic, but this holds for fishes and birds only; amphibians, reptiles and mammals are more diverse in the Nearctic. Australasia stands out by its relatively low vertebrate diversity, especially of fishes (Tables 3, 4).

The highest number of vertebrate endemics is found in the Neotropics, and, again, regards mostly fishes. Here, the Amazonian province is an endemicity hotspot for fishes: 2,072 of the 2,416 species recorded from the region are endemic. The Afrotropical ichthyofauna is notorious for the presence of several endemic species-flocks in a number of ancient lakes, complemented by high rates of endemicity in certain invertebrate groups. For birds, amphibians and reptiles, endemicity is highest in the Afrotropical region. The Oriental region is richest in endemic turtles, which also have an endemicity hotspot in the eastern Nearctic. Most species of mammals, amphibians and reptiles are endemic to a single continent or zoogeographical region; hence their diversity hotspots coincide with endemicity hotspots, which, for mammals, are the Neotropical and Afrotropical regions.

On a subregional scale, the island fauna's are notable as centres of endemicity for birds and amphibians. The Malagasy example is significant by its endemicity rates of 90–100% for fishes, amphibians and birds.

Insecta

Diptera, Coleoptera and Trichoptera are the major representatives of freshwater insects with 43, 18 and

Table 3 Species diversity of the main groups of freshwater vertebrates, by zoogeographic region

	PA	NA	AT	NT	OL	AU	PAC	ANT	World
Amphibia	160	203	828	1,698	1,062	301	0	0	4,294
Crocodilians	3	2	3	9	8	4	0	0	24
Lizards	0	0	9	22	28	14	2	0	73
Snakes	6	22	19	39	64	7			153
Turtle	8	55	25	65	73	34			260
Fish (FW only)	1,844	1,411	2,938	4,035	2,345	261			12,740
Mammals	18	22	35	28	18	11	0	0	124
Aves	154	116	138	145	76	62	6	1	567
Total	2,193	1,831	3,995	6,041	3,674	694	8	1	18,235

Table 4 Genus diversity of the main groups of freshwater vertebrates, by zoogeographic region

	PA	NA	AT	NT	OL	AU	PAC	ANT	World
Amphibia	26	27	89	127	71	20	0	0	348
Crocodilians	2	2	2	4	4	1	0	0	8
Lizards	0	0	4	7	7	4	2	0	19
Snakes	5	6	8	13	12	7			44
Turtle	6	16	6	16	34	8			86
Fish (FW only)	380	298	390	705	440	94			2,000
Mammals	10	15	18	15	10	7	0	0	65
Aves	68	62	73	87	48	42	4	1	198
Total	497	426	590	974	626	183	6	1	2,768

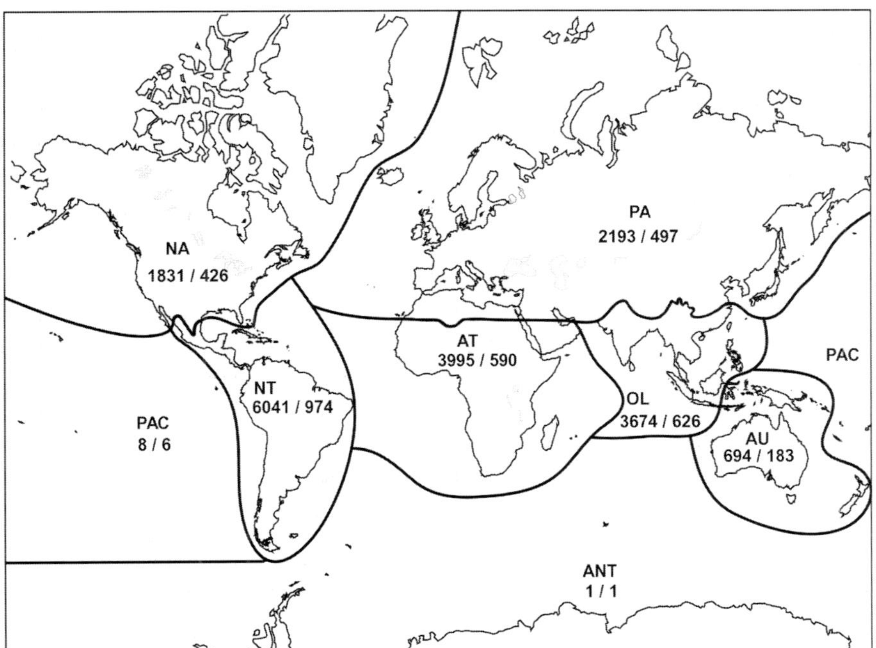

Fig. 1 Distribution of freshwater vertebrate species and genera, by zoogeographic regions (number of species/number of genera). Numbers include strictly freshwater fish (not brackish), amphibians, mammals, reptiles and water birds as defined in each specific contribution

15%, respectively, of the total of almost 76,000 freshwater insect species (Tables 5, 6). These numbers include some families of Diptera, such as Tabinidae, which are not addressed in specific chapters and whose diversity is estimated at around 5,000 species. Other important taxa are Heteroptera (6%), Plecoptera (5%), Odonata (7%) and Ephemeroptera (4%). In insects, there is a remarkable discrepancy between species- and genus-level diversity: Diptera account for 43% of total insect species-level diversity, against only 22% for genera. On the other hand, in Ephemeroptera, Odonata and Heteroptera, the genus-level diversity contributes about twice that of species-level diversity to total insect diversity.

The highest diversity of freshwater insects is recorded from the Palaearctic (20%), closely followed by the Neotropical (18.5%) and the Oriental realms (18.3%) (Fig. 2). The Afrotropical and Australasian regions represent 12 and 10%, respectively, of extant insect species diversity. As several experts did not treat the Pacific Oceanic Islands and Antarctic region separately, we here refrain from further commenting on the insect diversity of these regions.

The data on insect diversity should be interpreted with caution, as many experts report a strong sampling and study bias. Especially, the Holarctic insect fauna is notoriously better studied than that of the Neotropical, Afrotropical and Oriental regions, and this for most groups. This bias is less pronounced in two emblematic insect groups, namely butterflies and moths (Lepidoptera) and dragonflies (Odonata), and is reflected in the fact that for these groups, the Holarctic is not the most diverse region: Lepidoptera species diversity is highest in the Neotropical (30%), Australasian (23%) and Oriental (23%) realms, whereas for Odonata the Neotropical and Oriental regions have the most diverse fauna. In contrast, the fact that Hymenoptera are most diverse in the Holarctic region (Table 5) is most likely owing to a study bias. For insects, there are few species that occur in more than one region; hence hotspots of endemicity and diversity largely coincide.

Table 5 Species diversity of insect orders, by zoogeographic region

	PA	NA	AT	NT	OL	AU	PAC	ANT	World
Coleoptera	3,346	1,419	2,507	2,693	2,189	1,334			13,514
Diptera other families[a]	2,458	2,045	2,623	933	909	945	143	2	13,454
Diptera—Chironomidae	1,231	1,092	618	406	359	471	155	9	4,147
Diptera—Culicidae	492	178	1,069	795	1,061	764			3,492
Diptera—Simulidae	699	256	355	214	321	195	55	2	2,000
Diptera—Tabanidae[b]									5,000
Diptera—Tipulidae	1,280	573	805	339	925	385			4,188
Ephemeroptera	787	650	607	390	390	219			3,043
Heteroptera	496	424	1,289	799	1,103	654	37		4,801
Hymenoptera	57	53	17	1	28	8	9		147
Lepidoptera	81	49	219	64	169	170	9		737
Mecoptera			3			5			8
Megaloptera-Neuroptera	78	99	52	18	144	50	1	0	446
Odonata	560	451	1,636	889	1,665	870	168	1	5,680
Orthoptera	9	10	54	14	98	5			188
Plecoptera	1,156	650	474	95	828	295			3,497
Trichoptera	2,370	1,461	2,100	944	3,723	1,140			11,532
Total	15,190	9,410	14,428	8,594	13,912	7,510	577	14	75,874

[a] The distribution of species by zoogeographic regions is incomplete for several families of Dipterans; as a result, the sum of the regional species numbers is lower than the number of species known in the world (See chapter on Diptera families excluding Culicidae, Tipulidae, Chironomidae and Simulidae)

[b] Estimated

Table 6 Genus diversity of insect orders, by zoogeographic region

	PA	NA	AT	NT	OL	AU	PAC	ANT	World
Coleoptera	209	152	175	204	167	138			710
Diptera other families[a]	227	158	114	198	107	115	29	2	457
Diptera—Chironomidae	181	211	104	154	105	116	29	6	339
Diptera—Culicidae	19	13	15	24	25	22			42
Diptera—Simulidae	12	13	2	10	1	2	1	1	26
Diptera—Tipulidae	45	38	23	36	45	30			115
Ephemeroptera	77	94	93	84	78				405
Heteroptera	60	67	96	105	123	87	16		553
Hymenoptera	29	33	1	10	13	6	5		51
Lepidoptera	12	17	11	21	14	21	4		53
Mecoptera				1		2			2
Megaloptera-Neuroptera	14	10	5	11	16	10	1	0	45
Odonata	137	89	132	186	235	169	47	1	642
Orthoptera	7	6	5	20	20	2			50
Plecoptera	108	102	8	57	41	46			286
Trichoptera	229	157	87	148	169	143			619
Total	1,366	1,160	871	1,269	1,159	909	132	10	4,395

[a] The distribution of genera by zoogeographic regions was not complete for several families of Dipterans, (See chapter on Diptera families excluding Culicidae, Tipulidae, Chironomidae et Simulidae)

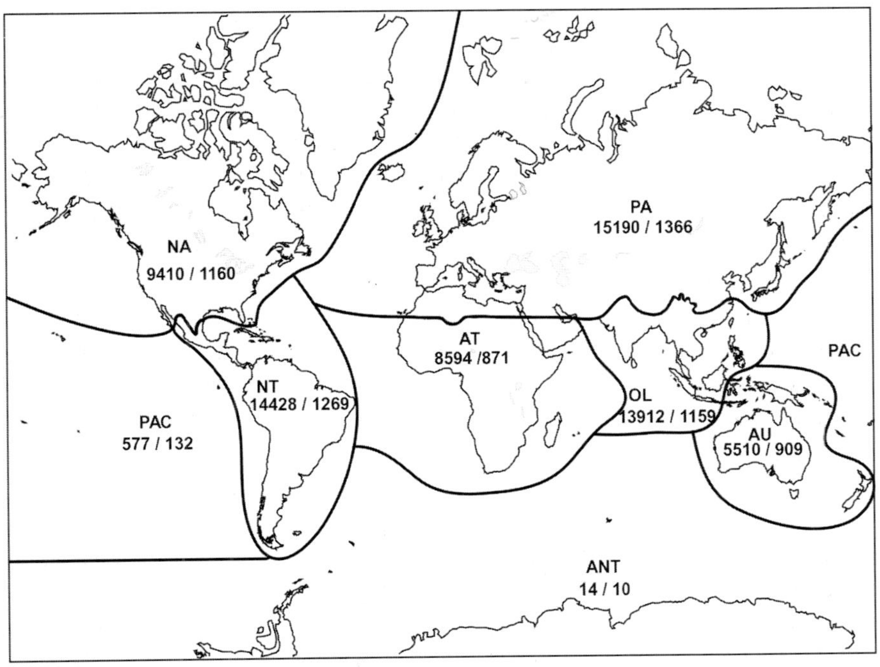

Fig. 2 Distribution of total insect species and genus diversity by zoogeographic regions (number of species/ number of genera). Numbers do not include some dipteran families (i.e. Tabanidae) that are not addressed in the specific contributions

Crustacea

The different chapters dealing with freshwater crustaceans report on a total of 11,990 described species, distributed over 1,533 genera (Tables 7, 8). This constitutes 30% of the total known diversity of crustaceans, which is estimated at about 40,000 species (Groombridge & Jenkins, 2002). Amongst

Table 8 Genus diversity of crustaceans, by zoogeographic region

	PA	NA	AT	NT	OL	AU	PAC	ANT	World
Amphipoda	185	23	17	35	10	34	9		293
Branchiopoda	28	20	14	18	14	12	2	1	43
Branchiura	1	1	3	3	1	2	1	0	4
Cladocera	60	52	46	50	44	52	21	7	95
Copepoda	134	87	60	104	79	50	15	14	257
Cumacea & Tanaidacea	10	2		2		2	1		14
Isopoda	45	18	8	42	11	50	4		194
Mysidacea	15	7	1	6	6	1	0	0	26
Ostracoda	87	57	73	55	46	57	4	3	189
Spelaeogriphacea	–	–	1	1	–	1	–		3
Syncarida	30	6	18	18	9	15	0	0	78
Thermosbaenacea	5	1	1	2	1	1	–		6
Aeglidae			1						*1*
Astacidea	6	11	1	6		9			33
Brachyura	14	4	27	65	139	24	13		238
Caridea	14	5	17	17	21	15	6	–	59
Decapoda	34	20	46	88	160	48	19	0	331
Total	634	294	288	424	381	325	76	25	1,533

Table 7 Species diversity of crustaceans, by zoogeographic region

	PA	NA	AT	NT	OL	AU	PAC	ANT	World
Amphipoda	1,315	236	56	127	17	107	10		1,866
Branchiopoda	175	93	81	61	47	75	2	1	508
Branchiura	8	18	40	33	16	3	1	0	113
Cladocera	245	189	134	186	107	158	33	12	620
Copepoda	1,204	347	405	561	381	205	29	17	2,814
Cumacea & Tanaidacea	20	2		2		3	1		25
Isopoda	475	130	22	109	31	134	5		942
Mysidacea	39	11	1	20	7	1	0	0	72
Ostracoda	702	298	455	275	199	176	5	3	1,936
Spelaeogriphacea	–	–	1	1	–	2	–		4
Syncarida	128	12	27	29	12	33	0	0	240
Thermosbaenacea	6	1	1	8	1	1	–		18
Aeglidae		63						63	
Astacidea	38	382	9	64		151			638
Brachyura	97	19	149	340	818	89	24		1,476
Caridea	47	17	92	109	349	87	25	–	655
Decapoda	182	418	313	513	1,167	327	49	0	2,832
Total	4,499	1,755	1,536	1,925	1,985	1,225	135	33	11,990

freshwater crustaceans, the most speciose taxa are the decapods (24%) and copepods (23%), closely followed by the ostracods and amphipods (both 16%). Branchiopods, Isopods and syncarids represent 9, 8 and 2%, respectively, of the total number of species. The remaining 2% is composed of representatives of smaller groups: mainly Branchiura and Mysidacea, with a few species of Cumacea, Tanaidacea, Spelaeogriphacea and Thermosbaenacea.

Again, the region with the highest number of species is the Palaearctic (37%). Second and third are the Oriental and Neotropical regions (both ca. 16%). This holds for most crustacean taxa, except for Brachyura and Caridea decapods, which are most diverse in the Oriental region, and Astacidea, which exhibit a diversity and endemicity hotspot in the Nearctic, and which are absent from the Oriental region. Aeglidae (Anomura) crabs form an endemic family in the Afrotropical region. All other crustacean taxa (Copepoda, Ostracoda, Branchiopoda, Isopoda, Amphipoda, Syncaridea) are most diverse in the Palaearctic. As for insects, sampling and study gaps most likely account for this.

Remarkable endemic crustacean faunas occur in the ponto-caspian basin and in Lake Baïkal. These are identified as hot spots of richness and endemicity for several crustacean taxa, including amphipods, ostracods, copepods and branchiopods. In amphipods, there is a large group of endemic taxa inhabiting subterranean habitats in the west Palaearctic, whereas crayfish exhibit a different pattern of endemicity, with a centre in the southeast of the Nearctic region, notably in the south of the Appalachean range.

Mollusca

The ca. 5,000 species of freshwater molluscs represent 4% of the total number of freshwater animal species, and account for only about 7% of the global total of described mollusc species, estimated at about 80,000 species (Groombridge & Jenkins, 2002). Eighty percent of the freshwater molluscs are gastropods, whereas 20% are bivalves. Gastropods and bivalves attain their highest diversity in the Palaearctic and Nearctic regions, respectively. However, the bivalve Unionidae family, of great economic importance, is most diverse in the Oriental region.

Freshwater gastropod faunas of underground systems, springs and small rivers are particularly rich, both in terms of species diversity and endemicity. Further noteworthy habitats are ancient oligotrophic lakes (e.g. Baikal, Ohrid, Tanganyika), which are key hotspots of gastropod diversity. The lower reaches of some river basins (e.g. Congo, Mekong, Mobile Bay) are also identified as areas of high species richness.

Minor invertebrate phyla

The most speciose amongst the "minor" invertebrate phyla are Rotifera (1,948 species), Nematoda (1,808 species), Annelidae (1,761 species) and Turbellaria (Platyhelminthes: 1,297 species). Gastrotricha, Nematomorpha and Porifera are less species rich in freshwater habitats (200–300 sp.), although they are very successful in marine environments. The same holds for Bryozoa and Tardigrada (60–80 species). The least diverse groups in freshwater are Nemertea (22 species) and Cnidaria (18 species). Rotifera, Nematomorpha and Annelida-Hirudinea are mainly freshwater, but there are also generally species-rich groups like Cnidaria (7,000+ species), or Annelida-Polychaeta (9,000+ species) that are, however, poorly represented in freshwater (Fig. 3).

All of these groups are generally ill-studied, and this was clearly emphasised by all experts. Nevertheless, Lake Baikal appears to have been studied more intensively for most of these groups and is identified as a hotspot of endemicity. Further generalisations are hard to make considering the lack of data, although the analysis of rotifer diversity and endemism reveals some intriguing patterns (Segers & De Smet, 2007; Segers, 2008, present volume).

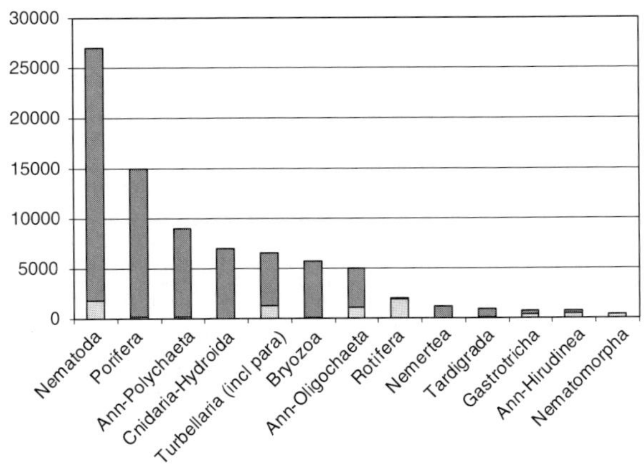

Fig. 3 species diversity in freshwater compared to total number of described species

Comparison with marine and terrestrial species diversity

As early evolution of all major animal phyla took place in the sea, it is not surprising that marine systems show higher diversity at the phylum and class level than terrestrial or freshwater systems. Of the total 33 metazoan phyla, 31 are found in the sea, with 11 being exclusively marine; whereas 17 phyla are present in freshwater and 12 on land (only 2 phyla, freshwater Micrognathozoa and terrestrial Onychophora have no marine species). At the species level, the diversity of terrestrial ecosystems, with more than 1.5 million species, largely exceeds the 280,000 species of marine organisms currently known. At habitat levels, the most diverse marine habitats—coral reefs—are far less diverse in terms of species number than the moist tropical forests that are often taken as their terrestrial counterparts.

Conclusion

A clear result of our survey is that increased sampling efforts are needed to address the obvious gaps, both geographical and taxonomical, the current assessment of freshwater biodiversity reveals. Especially in terms of richness and endemicity, hot spots are often located in less-studied areas of the Oriental, the Neotropical and the Afrotropical regions. The situation is especially critical for the least-known groups such as Nematoda. One possible cost-effective way to improve this situation is to make better use of the existing knowledge, shelved in museum collections, local laboratories or in scientists' drawers. This ongoing task is being carried out by several international initiatives including GBIF and the IUCN Freshwater Biodiversity Assessment Programme. However, additional surveys are also needed and will require a new generation of taxonomic experts and increased financial means.

This global assessment of freshwater species diversity and distribution is thus but a first step in the process of compiling and upgrading our knowledge on freshwater biodiversity. The regional or global-scale approach used here allows for the identification of knowledge gaps and is critical to come to a better understanding of evolutionary patterns in freshwater diversity and endemicity, in particular, for less-known invertebrate taxa.

In order to complement the present database on diversity and endemicity, a similar effort focussing on environmental information, from geographical to sociological, will be needed. It is clear that the results presented in this volume, apart of their inherent scientific value, should be interpreted in a broader ecological and evolutionary context, if they are to play a role in the development or improvement of sustainable management and conservation of freshwater resources. Indeed, the challenges society is confronted with in the face of global change and increased human utilisation of natural resources, are daunting and can only be dealt with successfully on the condition that sufficient and credible scientific knowledge is made available as a basis for action, in addition to the political will to implement the necessary measures (Dudgeon et al., 2006).

To facilitate usage and analysis of the data collected during the present Freshwater Animal Diversity Assessment (FADA) project, an on-line database is presently being developed. This resource, which can be consulted on http://FADA.biodiversity.be, will offer additional services including extraction of name lists, visualisation of geographical (GIS) records in an interactive environment and link to other datasets containing information of freshwater systems. All data will be made freely and universally accessible through the Internet. For this, FADA is developing links with global initiatives in the field, like the Global Biodiversity Information Facility (GBIF), Catalogue of Life (CoL), SpeciesBase and Encyclopedia of Life.

References

Chambers, P. A., P. Lacoul, K. J. Murphy & S. M. Thomaz, 2008. Global diversity of aquatic macrophytes in freshwaters. In Balian, E. V., C. Lévêque, H. Segers & K. Martens (eds), Freshwater Animal Diversity Assessment, Hydrobiologia, present volume. doi: 10.1007/s10750-007-9154-6.

Deharveng, L., C. A. D'Haese & A. Bedos, 2008. Global diversity of springtails (Collembola; Hexapoda) in freshwater. In Balian, E. V., C. Lévêque, H. Segers & K. Martens (eds), Freshwater Animal Diversity Assessment, Hydrobiologia, present volume. doi: 10.1007/s10750-007-9116-z.

Dudgeon, D., A. H. Arthington, M. O. Gessner, Z. -I. Kawabata, D. J. Knowler, C. Lévêque, R. J. Naiman, A.-H. Prieur-Richard, D. Soto, M. L. J. Stiassny & C. A. Sullivan, 2006. Freshwater biodiversity: importance, threats, status and conservation challenges. Biological Reviews 81: 163–182.

Finlay, B. J. & G. F. Esteban, 1998. Freshwater protozoa: biodiversity and ecological function. Biodiversity and Conservation 7: 1163–1186.

Groombridge, B. & M. Jenkins, 2002. World Atlas of Biodiversity: Earth's Living Resources in the 21st Century. University of California Press.

Hugot, J.-P., P. Baujard & S. Morand, 2001. Biodiversity in helminths and nematodes as a field of study: an overview. Nematology 3(3): 199–208.

Lévêque, C., T. Oberdorff, D. Paugy, M.L.J. Stiassny & P.A. Tedesco, 2008. Global diversity of fish (Pisces) in freshwater. In: Balian E. V., C. Lévêque, H. Segers & K. Martens (eds), Freshwater Animal Diversity Assessment, Hydrobiologia, present volume. doi:10.1007/s10750-007-9034-0.

Segers, H., 2008. Global diversity of rotifers (Phylum Rotifera) in freshwater. In Balian, E. V., C. Lévêque, H. Segers & K. Martens (eds), Freshwater Animal Diversity Assessment, Hydrobiologia, present volume. doi:10.1007/s10750-007-9003-7.

Segers, H. & W. H. De Smet, 2007. Diversity and Endemism in Rotifera: a review, and *Keratella* Bory de St Vincent. In W. Foissner (ed.), Protist diversity and geographic distribution. Biodiversity and Conservations. doi:10.1007/s10531-007-9262-7

Shearer, C. A., E. Descals, B. Kohlmeyer, J. Kohlmeyer, L. Marvanov, D. Padgett, D. Porter, H. A. Raja, J. P. Schmit, H. Thornton & H. Voglmayr, 2007. Fungal biodiversity in aquatic habitats. Biodiversity and Conservation 16, 49–67.

United Nations Environmental Programme, 2002. Global Environmental Outlook 3. Earthprint Ltd., Stevenage, Hertfordshire, England.